Microencapsulation
Second Edition

DRUGS AND THE PHARMACEUTICAL SCIENCES

DRUGS AND THE PHARMACEUTICAL SCIENCES
A Series of Textbooks and Monographs

1. Pharmacokinetics, *Milo Gibaldi and Donald Perrier*
2. Good Manufacturing Practices for Pharmaceuticals: A Plan for Total Quality Control, *Sidney H. Willig, Murray M. Tuckerman, and William S. Hitchings IV*
3. Microencapsulation, *edited by J. R. Nixon*
4. Drug Metabolism: Chemical and Biochemical Aspects, *Bernard Testa and Peter Jenner*
5. New Drugs: Discovery and Development, *edited by Alan A. Rubin*
6. Sustained and Controlled Release Drug Delivery Systems, *edited by Joseph R. Robinson*
7. Modern Pharmaceutics, *edited by Gilbert S. Banker and Christopher T. Rhodes*
8. Prescription Drugs in Short Supply: Case Histories, *Michael A. Schwartz*
9. Activated Charcoal: Antidotal and Other Medical Uses, *David O. Cooney*
10. Concepts in Drug Metabolism (in two parts), *edited by Peter Jenner and Bernard Testa*
11. Pharmaceutical Analysis: Modern Methods (in two parts), *edited by James W. Munson*
12. Techniques of Solubilization of Drugs, *edited by Samuel H. Yalkowsky*
13. Orphan Drugs, *edited by Fred E. Karch*
14. Novel Drug Delivery Systems: Fundamentals, Developmental Concepts, Biomedical Assessments, *Yie W. Chien*
15. Pharmacokinetics: Second Edition, Revised and Expanded, *Milo Gibaldi and Donald Perrier*
16. Good Manufacturing Practices for Pharmaceuticals: A Plan for Total Quality Control, Second Edition, Revised and Expanded, *Sidney H. Willig, Murray M. Tuckerman, and William S. Hitchings IV*
17. Formulation of Veterinary Dosage Forms, *edited by Jack Blodinger*
18. Dermatological Formulations: Percutaneous Absorption, *Brian W. Barry*
19. The Clinical Research Process in the Pharmaceutical Industry, *edited by Gary M. Matoren*
20. Microencapsulation and Related Drug Processes, *Patrick B. Deasy*
21. Drugs and Nutrients: The Interactive Effects, *edited by Daphne A. Roe and T. Colin Campbell*
22. Biotechnology of Industrial Antibiotics, *Erick J. Vandamme*
23. Pharmaceutical Process Validation, *edited by Bernard T. Loftus and Robert A. Nash*
24. Anticancer and Interferon Agents: Synthesis and Properties, *edited by Raphael M. Ottenbrite and George B. Butler*
25. Pharmaceutical Statistics: Practical and Clinical Applications, *Sanford Bolton*
26. Drug Dynamics for Analytical, Clinical, and Biological Chemists, *Benjamin J. Gudzinowicz, Burrows T. Younkin, Jr., and Michael J. Gudzinowicz*
27. Modern Analysis of Antibiotics, *edited by Adjoran Aszalos*
28. Solubility and Related Properties, *Kenneth C. James*
29. Controlled Drug Delivery: Fundamentals and Applications, Second Edition, Revised and Expanded, *edited by Joseph R. Robinson and Vincent H. Lee*
30. New Drug Approval Process: Clinical and Regulatory Management, *edited by Richard A. Guarino*
31. Transdermal Controlled Systemic Medications, *edited by Yie W. Chien*
32. Drug Delivery Devices: Fundamentals and Applications, *edited by Praveen Tyle*

33. Pharmacokinetics: Regulatory • Industrial • Academic Perspectives, *edited by Peter G. Welling and Francis L. S. Tse*
34. Clinical Drug Trials and Tribulations, *edited by Allen E. Cato*
35. Transdermal Drug Delivery: Developmental Issues and Research Initiatives, *edited by Jonathan Hadgraft and Richard H. Guy*
36. Aqueous Polymeric Coatings for Pharmaceutical Dosage Forms, *edited by James W. McGinity*
37. Pharmaceutical Pelletization Technology, *edited by Isaac Ghebre-Sellassie*
38. Good Laboratory Practice Regulations, *edited by Allen F. Hirsch*
39. Nasal Systemic Drug Delivery, *Yie W. Chien, Kenneth S. E. Su, and Shyi-Feu Chang*
40. Modern Pharmaceutics: Second Edition, Revised and Expanded, *edited by Gilbert S. Banker and Christopher T. Rhodes*
41. Specialized Drug Delivery Systems: Manufacturing and Production Technology, *edited by Praveen Tyle*
42. Topical Drug Delivery Formulations, *edited by David W. Osborne and Anton H. Amann*
43. Drug Stability: Principles and Practices, *Jens T. Carstensen*
44. Pharmaceutical Statistics: Practical and Clinical Applications, Second Edition, Revised and Expanded, *Sanford Bolton*
45. Biodegradable Polymers as Drug Delivery Systems, *edited by Mark Chasin and Robert Langer*
46. Preclinical Drug Disposition: A Laboratory Handbook, *Francis L. S. Tse and James J. Jaffe*
47. HPLC in the Pharmaceutical Industry, *edited by Godwin W. Fong and Stanley K. Lam*
48. Pharmaceutical Bioequivalence, *edited by Peter G. Welling, Francis L. S. Tse, and Shrikant V. Dinghe*
49. Pharmaceutical Dissolution Testing, *Umesh V. Banakar*
50. Novel Drug Delivery Systems: Second Edition, Revised and Expanded, *Yie W. Chien*
51. Managing the Clinical Drug Development Process, *David M. Cocchetto and Ronald V. Nardi*
52. Good Manufacturing Practices for Pharmaceuticals: A Plan for Total Quality Control, Third Edition, *edited by Sidney H. Willig and James R. Stoker*
53. Prodrugs: Topical and Ocular Drug Delivery, *edited by Kenneth B. Sloan*
54. Pharmaceutical Inhalation Aerosol Technology, *edited by Anthony J. Hickey*
55. Radiopharmaceuticals: Chemistry and Pharmacology, *edited by Adrian D. Nunn*
56. New Drug Approval Process: Second Edition, Revised and Expanded, *edited by Richard A. Guarino*
57. Pharmaceutical Process Validation: Second Edition, Revised and Expanded, *edited by Ira R. Berry and Robert A. Nash*
58. Ophthalmic Drug Delivery Systems, *edited by Ashim K. Mitra*
59. Pharmaceutical Skin Penetration Enhancement, *edited by Kenneth A. Walters and Jonathan Hadgraft*
60. Colonic Drug Absorption and Metabolism, *edited by Peter R. Bieck*
61. Pharmaceutical Particulate Carriers: Therapeutic Applications, *edited by Alain Rolland*
62. Drug Permeation Enhancement: Theory and Applications, *edited by Dean S. Hsieh*
63. Glycopeptide Antibiotics, *edited by Ramakrishnan Nagarajan*
64. Achieving Sterility in Medical and Pharmaceutical Products, *Nigel A. Halls*
65. Multiparticulate Oral Drug Delivery, *edited by Isaac Ghebre-Sellassie*

66. Colloidal Drug Delivery Systems, *edited by Jörg Kreuter*
67. Pharmacokinetics: Regulatory • Industrial • Academic Perspectives, Second Edition, *edited by Peter G. Welling and Francis L. S. Tse*
68. Drug Stability: Principles and Practices, Second Edition, Revised and Expanded, *Jens T. Carstensen*
69. Good Laboratory Practice Regulations: Second Edition, Revised and Expanded, *edited by Sandy Weinberg*
70. Physical Characterization of Pharmaceutical Solids, *edited by Harry G. Brittain*
71. Pharmaceutical Powder Compaction Technology, *edited by Göran Alderborn and Christer Nyström*
72. Modern Pharmaceutics: Third Edition, Revised and Expanded, *edited by Gilbert S. Banker and Christopher T. Rhodes*
73. Microencapsulation: Methods and Industrial Applications, *edited by Simon Benita*
74. Oral Mucosal Drug Delivery, *edited by Michael J. Rathbone*
75. Clinical Research in Pharmaceutical Development, *edited by Barry Bleidt and Michael Montagne*
76. The Drug Development Process: Increasing Efficiency and Cost Effectiveness, *edited by Peter G. Welling, Louis Lasagna, and Umesh V. Banakar*
77. Microparticulate Systems for the Delivery of Proteins and Vaccines, *edited by Smadar Cohen and Howard Bernstein*
78. Good Manufacturing Practices for Pharmaceuticals: A Plan for Total Quality Control, Fourth Edition, Revised and Expanded, *Sidney H. Willig and James R. Stoker*
79. Aqueous Polymeric Coatings for Pharmaceutical Dosage Forms: Second Edition, Revised and Expanded, *edited by James W. McGinity*
80. Pharmaceutical Statistics: Practical and Clinical Applications, Third Edition, *Sanford Bolton*
81. Handbook of Pharmaceutical Granulation Technology, *edited by Dilip M. Parikh*
82. Biotechnology of Antibiotics: Second Edition, Revised and Expanded, *edited by William R. Strohl*
83. Mechanisms of Transdermal Drug Delivery, *edited by Russell O. Potts and Richard H. Guy*
84. Pharmaceutical Enzymes, *edited by Albert Lauwers and Simon Scharpé*
85. Development of Biopharmaceutical Parenteral Dosage Forms, *edited by John A. Bontempo*
86. Pharmaceutical Project Management, *edited by Tony Kennedy*
87. Drug Products for Clinical Trials: An International Guide to Formulation • Production • Quality Control, *edited by Donald C. Monkhouse and Christopher T. Rhodes*
88. Development and Formulation of Veterinary Dosage Forms: Second Edition, Revised and Expanded, *edited by Gregory E. Hardee and J. Desmond Baggot*
89. Receptor-Based Drug Design, *edited by Paul Leff*
90. Automation and Validation of Information in Pharmaceutical Processing, *edited by Joseph F. deSpautz*
91. Dermal Absorption and Toxicity Assessment, *edited by Michael S. Roberts and Kenneth A. Walters*
92. Pharmaceutical Experimental Design, *Gareth A. Lewis, Didier Mathieu, and Roger Phan-Tan-Luu*
93. Preparing for FDA Pre-Approval Inspections, *edited by Martin D. Hynes III*
94. Pharmaceutical Excipients: Characterization by IR, Raman, and NMR Spectroscopy, *David E. Bugay and W. Paul Findlay*

95. Polymorphism in Pharmaceutical Solids, *edited by Harry G. Brittain*
96. Freeze-Drying/Lyophilization of Pharmaceutical and Biological Products, *edited by Louis Rey and Joan C. May*
97. Percutaneous Absorption: Drugs–Cosmetics–Mechanisms–Methodology, Third Edition, Revised and Expanded, *edited by Robert L. Bronaugh and Howard I. Maibach*
98. Bioadhesive Drug Delivery Systems: Fundamentals, Novel Approaches, and Development, *edited by Edith Mathiowitz, Donald E. Chickering III, and Claus-Michael Lehr*
99. Protein Formulation and Delivery, *edited by Eugene J. McNally*
100. New Drug Approval Process: Third Edition, The Global Challenge, *edited by Richard A. Guarino*
101. Peptide and Protein Drug Analysis, *edited by Ronald E. Reid*
102. Transport Processes in Pharmaceutical Systems, *edited by Gordon L. Amidon, Ping I. Lee, and Elizabeth M. Topp*
103. Excipient Toxicity and Safety, *edited by Myra L. Weiner and Lois A. Kotkoskie*
104. The Clinical Audit in Pharmaceutical Development, *edited by Michael R. Hamrell*
105. Pharmaceutical Emulsions and Suspensions, *edited by Francoise Nielloud and Gilberte Marti-Mestres*
106. Oral Drug Absorption: Prediction and Assessment, *edited by Jennifer B. Dressman and Hans Lennernäs*
107. Drug Stability: Principles and Practices, Third Edition, Revised and Expanded, *edited by Jens T. Carstensen and C. T. Rhodes*
108. Containment in the Pharmaceutical Industry, *edited by James P. Wood*
109. Good Manufacturing Practices for Pharmaceuticals: A Plan for Total Quality Control from Manufacturer to Consumer, Fifth Edition, Revised and Expanded, *Sidney H. Willig*
110. Advanced Pharmaceutical Solids, *Jens T. Carstensen*
111. Endotoxins: Pyrogens, LAL Testing, and Depyrogenation, Second Edition, Revised and Expanded, *Kevin L. Williams*
112. Pharmaceutical Process Engineering, *Anthony J. Hickey and David Ganderton*
113. Pharmacogenomics, *edited by Werner Kalow, Urs A. Meyer and Rachel F. Tyndale*
114. Handbook of Drug Screening, *edited by Ramakrishna Seethala and Prabhavathi B. Fernandes*
115. Drug Targeting Technology: Physical • Chemical • Biological Methods, *edited by Hans Schreier*
116. Drug–Drug Interactions, *edited by A. David Rodrigues*
117. Handbook of Pharmaceutical Analysis, *edited by Lena Ohannesian and Anthony J. Streeter*
118. Pharmaceutical Process Scale-Up, *edited by Michael Levin*
119. Dermatological and Transdermal Formulations, *edited by Kenneth A. Walters*
120. Clinical Drug Trials and Tribulations: Second Edition, Revised and Expanded, *edited by Allen Cato, Lynda Sutton, and Allen Cato III*
121. Modern Pharmaceutics: Fourth Edition, Revised and Expanded, *edited by Gilbert S. Banker and Christopher T. Rhodes*
122. Surfactants and Polymers in Drug Delivery, *Martin Malmsten*
123. Transdermal Drug Delivery: Second Edition, Revised and Expanded, *edited by Richard H. Guy and Jonathan Hadgraft*
124. Good Laboratory Practice Regulations: Second Edition, Revised and Expanded, *edited by Sandy Weinberg*

125. Parenteral Quality Control: Sterility, Pyrogen, Particulate, and Package Integrity Testing: Third Edition, Revised and Expanded, *Michael J. Akers, Daniel S. Larrimore, and Dana Morton Guazzo*

126. Modified-Release Drug Delivery Technology, *edited by Michael J. Rathbone, Jonathan Hadgraft, and Michael S. Roberts*

127. Simulation for Designing Clinical Trials: A Pharmacokinetic-Pharmacodynamic Modeling Perspective, *edited by Hui C. Kimko and Stephen B. Duffull*

128. Affinity Capillary Electrophoresis in Pharmaceutics and Biopharmaceutics, *edited by Reinhard H. H. Neubert and Hans-Hermann Rüttinger*

129. Pharmaceutical Process Validation: An International Third Edition, Revised and Expanded, *edited by Robert A. Nash and Alfred H. Wachter*

130. Ophthalmic Drug Delivery Systems: Second Edition, Revised and Expanded, *edited by Ashim K. Mitra*

131. Pharmaceutical Gene Delivery Systems, *edited by Alain Rolland and Sean M. Sullivan*

132. Biomarkers in Clinical Drug Development, *edited by John C. Bloom and Robert A. Dean*

133. Pharmaceutical Extrusion Technology, *edited by Isaac Ghebre-Sellassie and Charles Martin*

134. Pharmaceutical Inhalation Aerosol Technology: Second Edition, Revised and Expanded, *edited by Anthony J. Hickey*

135. Pharmaceutical Statistics: Practical and Clinical Applications, Fourth Edition, *Sanford Bolton and Charles Bon*

136. Compliance Handbook for Pharmaceuticals, Medical Devices, and Biologics, *edited by Carmen Medina*

137. Freeze-Drying/Lyophilization of Pharmaceutical and Biological Products: Second Edition, Revised and Expanded, *edited by Louis Rey and Joan C. May*

138. Supercritical Fluid Technology for Drug Product Development, *edited by Peter York, Uday B. Kompella, and Boris Y. Shekunov*

139. New Drug Approval Process: Fourth Edition, Accelerating Global Registrations, *edited by Richard A. Guarino*

140. Microbial Contamination Control in Parenteral Manufacturing, *edited by Kevin L. Williams*

141. New Drug Development: Regulatory Paradigms for Clinical Pharmacology and Biopharmaceutics, *edited by Chandrahas G. Sahajwalla*

142. Microbial Contamination Control in the Pharmaceutical Industry, *edited by Luis Jimenez*

143. Generic Drug Product Development: Solid Oral Dosage Forms, *edited by Leon Shargel and Izzy Kanfer*

144. Introduction to the Pharmaceutical Regulatory Process, *edited by Ira R. Berry*

145. Drug Delivery to the Oral Cavity: Molecules to Market, *edited by Tapash K. Ghosh and William R. Pfister*

146. Good Design Practices for GMP Pharmaceutical Facilities, *edited by Andrew Signore and Terry Jacobs*

147. Drug Products for Clinical Trials, Second Edition, *edited by Donald Monkhouse, Charles Carney, and Jim Clark*

148. Polymeric Drug Delivery Systems, *edited by Glen S. Kwon*

149. Injectable Dispersed Systems: Formulation, Processing, and Performance, *edited by Diane J. Burgess*

150. Laboratory Auditing for Quality and Regulatory Compliance, *Donald Singer, Raluca-Ioana Stefan, and Jacobus van Staden*

151. Active Pharmaceutical Ingredients: Development, Manufacturing, and Regulation, *edited by Stanley Nusim*

152. Preclinical Drug Development, *edited by Mark C. Rogge and David R. Taft*
153. Pharmaceutical Stress Testing: Predicting Drug Degradation, *edited by Steven W. Baertschi*
154. Handbook of Pharmaceutical Granulation Technology: Second Edition, *edited by Dilip M. Parikh*
155. Percutaneous Absorption: Drugs–Cosmetics–Mechanisms–Methodology, Fourth Edition, *edited by Robert L. Bronaugh and Howard I. Maibach*
156. Pharmacogenomics: Second Edition, *edited by Werner Kalow, Urs A. Meyer and Rachel F. Tyndale*
157. Pharmaceutical Process Scale-Up, Second Edition, *edited by Lawrence Block*
158. Microencapsulation: Methods and Industrial Applications, Second Edition, *edited by Simon Benita*

Microencapsulation

Methods and Industrial Applications

Second Edition

edited by
Simon Benita
Hebrew University of Jerusalem
Israel

Published in 2006 by
CRC Press
Taylor & Francis Group
6000 Broken Sound Parkway NW, Suite 300
Boca Raton, FL 33487-2742

Printed in the United States of America on acid-free paper
10 9 8 7 6 5 4 3 2 1

International Standard Book Number-10: 0-8247-2317-1 (Hardcover)
International Standard Book Number-13: 978-0-8247-2317-0 (Hardcover)
Library of Congress Card Number 2005052195

Library of Congress Cataloging-in-Publication Data

Microencapsulation : methods and industrial applications / edited by Simon Benita. -- 2nd ed.
p. cm. -- (Drugs and the pharmaceutical sciences ; v. 158)
Includes bibliographical references and index.
ISBN-13: 978-0-8247-2317-0 (alk. paper)
ISBN-10: 0-8247-2317-1 (alk. paper)
1. Microencapsulation. I. Benita, Simon, 1947- . II. Series.
[DNLM: 1. Drug Compounding QV 778 M6258 2005]

RS201.C3M27 2005
615'.19--dc22 2005052195

Visit the Taylor & Francis Web site at
http://www.taylorandfrancis.com

and the CRC Press Web site at
http://www.crcpress.com

Preface to the Second Edition

Prior to writing of the preface for the second edition, I read the preface of the first edition published a decade ago. The assumptions that research, development, and sales of drug delivery systems would intensify in the following years were fully verified and even exceeded expectations. As far as particulate delivery systems are concerned, the research and development has been moving from the micro- to the nano-size scale. There is no doubt that microparticulate controlled delivery systems mainly for topical and oral administration have been successfully exploited by the cosmetic and pharmaceutical industry respectively. However, the pharmaceutical industry is facing an uncertain future in which high clinical development costs coupled with declining drug discovery success rates are decreasing the flow of new products in the R&D pipeline. Experts are now recommending that pharmaceutical companies move from the blockbuster model to a more extensive product portfolio model that focuses on diseases with insufficient therapies mainly in specific populations such as the aging population. Furthermore, investigators are attempting to reformulate and add new indications to existing blockbuster drugs to maintain a reasonable scientific and economic growth rate. I believe that scientists have achieved remarkable successes in the field of oral delivery. There are practical solutions to improving the oral bioavailability of poorly absorbed lipophilic and hydrophobic drugs. Oral controlled microparticulate systems have succeeded in maintaining adequate and effective plasma levels over prolonged periods of time by controlling drug release following oral administration. However, there are still significant unmet medical needs in target diseases such as cancer, autoimmune disorders, macular degeneration and Alzheimer's disease. Most of the active ingredients used to treat these severe diseases can be administered only through the parenteral route. Indeed, molecular complexity associated with drugs and inaccessibility of most pharmacological targets are major constraints and the main reasons behind the increased interest and expanding research on nanodelivery systems, which can carry drugs directly to their site of action. Thus, drug targeting has evolved as the most desirable but elusive goal in the science of drug delivery. Drug targeting offers enormous advantages but is highly challenging and extremely complicated. A better understanding of the physiological barriers a drug needs to overcome should provide the pharmaceutical scientist with the information and tools needed to develop successful designs for drug targeting delivery systems. Optimal pharmacological responses require both spatial placement of the drug molecules and temporal control at the site of action. Many hurdles and drawbacks still need to be overcome through intensive efforts and concentrated interdisciplinary scientific collaboration to reach the desired goals.

The second edition of Microencapsulation, Methods and Industrial Applications comprises 11 expanded and revised chapters and 12 new chapters that reflect the evolution of this discipline in the past decade.

It is my hope that this multi-authored second edition of Microencapsulation, Methods and Industrial Applications will assist and enrich the readers in understanding the diverse types of particulate systems currently available or under development as well as highlight possible applications in the future.

I am deeply grateful to Ms. Madelyn Segev, the secretary of the Pharmaceutics Department of the School of Pharmacy of The Hebrew University of Jerusalem who spared no effort to help me in bringing this project to fruition.

To Einat, Yair and Maytal

Simon Benita

Preface to the First Edition

Research, development, and sales of drug-delivery systems are increasing at a rapid pace throughout the world. This worldwide trend will intensify in the next decade as cuts in public health expenses demand lower costs and higher efficacy. To meet this demand, many efficient drugs currently in use will be reformulated within delivery systems that can be value-added for optimal molecular activity. In addition to the health sector, the cosmetic, agricultural, chemical, and food industries operate in an open marketplace where free and aggressive competition demands novel coating techniques with enhanced effectiveness at the lowest possible cost. Currently, microencapsulation techniques are most widely used in the development and production of improved drug- and food-delivery systems. These techniques frequently result in products containing numerous variably coated particles. The exact number of particles needed to form a single administered dose varies as a function of the final particle size and can lie in either the micro- or nanometer size range for micro- and nanoparticulate delivery systems, respectively.

The microparticulate delivery systems include mainly pellets, microcapsules, microspheres, lipospheres, emulsions, and multiple emulsions. The nanoparticulate delivery systems include mainly lipid or polymeric nanoparticles (nanocapsules and nanospheres), microemulsions, liposomes, and nonionic surfactant vesicles (niosomes).

Generally, the microparticulate delivery systems are intended for oral and topical use. Different types of coated particles can be obtained depending on the coating process used. The particles can be embedded within a polymeric or proteinic matrix network in either a solid aggregated state or a molecular dispersion, resulting in the formulation of microspheres. Alternatively, the particles can be coated by a solidified polymeric or proteinic envelop, leading to the formation of microcapsules. The profile and kinetic pattern governing the release rate of the entrapped active substance from the dosage form depend on the nature and morphology of the coated particles, which need to be established irrespective of the manufacturing method used.

Microencapsulation techniques are normally used to enhance material stability, reduce adverse or toxic effects, or extend material release for different applications in various fields of manufacturing.

Until now, the use of some interesting and promising therapeutic substances has been limited clinically because of their restrictive physico-chemical properties, which have required frequent administration. It is possible that these substances may become more widely used in a clinical setting if appropriate microencapsulation techniques can be designed to overcome their intrinsic conveniences.

Investigators and pharmacologists have been trying to develop delivery systems that allow the fate of a drug to be controlled and the optimal drug dosage to arrive at the site of action in the body by means of novel microparticulate dosage forms. During the past two decades, researchers have succeeded in part in controlling the drug-absorption process to sustain adequate and effective plasma drug levels over a prolonged period of time by designing delayed- or controlled-release microparticulate-delivery systems intended for either oral or parenteral administration.

The ultimate objective of microparticulate-delivery systems is to control and extend the release of the active ingredient from the coated particle without attempting to modify the normal biofate of the active molecules in the body after administration and absorption. The organ distribution and elimination of these molecules will not be modified and will depend only on their physicochemical properties. On the other hand, nanoparticulate-delivery systems are usually intended for oral, parenteral, ocular, and topical use, with the ultimate objective being the alternation of the pharmacokinetic profile of the active molecule.

In the past decade, ongoing efforts have been made to develop systems or drug carrier capable of delivering the active molecules specifically to the intended target organ, while increasing the therapeutic efficacy. This approach involves modifying the pharmacokinetic profile of various therapeutic classes of drugs through their incorporation in colloidal nanoparticulate carriers in the submicron size range such as liposomes and nanoparticles. These site-specific delivery systems allow an effective drug concentration to be maintained for a longer interval in the target tissue and result in decreased side effects associated with lower plasma concentrations in the peripheral blood. Thus, the principle of drug targeting is to reduce the total amount of drug administered, while optimizing its activity. It should be mentioned that the scientific community was skeptical that such goals could be achieved, since huge investments of funds and promising research studies have in many cases resulted in disappointing and nonlucrative results and have also been slow in yielding successfully marketed therapeutic nanoparticulate dosage forms. With the recent approval by health authorities of a few effective nanoparticulate products containing antifungal or cytotoxic drugs, interest in colloidal drug carriers has been renewed.

A vast number of studies and review as well as several books have been devoted to the development, characterization, and potential applications of specific microparticulate- and nanoparticulate-delivery systems. No encapsulation process developed to date has been able to produce the full range of capsules desired by potential capsule users. Few attempts have been made to present and discuss in a single book the entire size range of particulate dosage forms covered in this book. The general theme and purpose here are to provide the reader with a current and general overview of the existing micro- and nanoparticulate-delivery systems and to emphasize the various methods of preparation, characterization, evaluation, and potential applications in various areas such as medicine, pharmacy, cosmetology, and agriculture. The systematic approach used in presenting the various particulate systems should facilitate the comprehension of this increasingly complex field and clarify the main considerations involved in designing, manufacturing, characterizing, and evaluating a specific particulate-delivery system for a given application or purpose. Thus, the chapters, which have been contributed by leading authorities in the field, are arranged logically according to the methods of preparation, characterization, and applications of the various particulate-delivery systems.

The first chapter is by C. Thies, a renowned scientist in the field of microencapsulation techniques. To provide an idea of which process is most appropriate for a specific application, the general principles of several microencapsulation processes are summarized and reviewed. This chapter focuses primarily on processes that have achieved significantly commercial use. S. Magdassi and Y. Vinetsky present an interesting technique of oil-in-water emulsion microencapsulation by proteins following adsorption of the protein molecules onto the oil–water interface. J. P. Benoit and Drs. H. Marchais, H. Rolland, and V. Vande Velde have contributed a chapter on advances in the production technology of biodegradable microspheres. This chapter deals mainly with the preparation and use of microspheres. The potential of the various technologies addressed is also discussed, with an emphasis on marketed products or those products currently under clinical evaluation. A. Markus demonstrates in his chapter the importance of applying microencapsulation techniques in the design of controlled-release pesticide formulations to meet the multifaceted demands of efficacy, suitability to mode of application, and minimal damage to the environment. The nanoparticulate-delivery systems are introduced by a chapter, authored by myself, B. Magenheim, and P. Wehrlé, that explains factorial design in the development of nanoparticulate systems. This chapter illustrates the application of the experimental design technique not only for optimization but also for elucidation of the mechanistic aspects of nanoparticle formation by spontaneous emulsification.

The second part of the book, which focuses on the evaluation and characterization of the various particulate-delivery systems, starts with an important chapter on microspheres morphology by J. P. Benoit and C. Thies. The chapter helps to clarify definitions and differences, which are very often confused. In addition, the chapter illustrates how morphology can be characterized by using different techniques. C. Washington provides his valuable expertise in the presentation of the various kinetic models used to characterize drug-release profiles from ensembles or population of microparticulate-delivery systems. It is worth noting that the release mechanism of a drug from multiparticulate systems such as microcapsules or microspheres cannot be identified by a study of global release profiles, since it has been shown that overall or cumulative release profiles form ensembles of microcapsules are entirely different from those of single microcapsules. The discrepancy arises from the heterogeneous distribution of the parameters determining release behavior in individual microcapsules, which is beyond the scope of the present chapter. The following chapter, by P. Couvreur, G. Couarraze, J. -P. Devissaguet, and F. Puisieux, presents a very detailed explanation of the preparation and characterization of nanoparticles. The authors first clearly define the morphology of nanocapsules and nanospheres, providing the background, information, and guidelines for choosing the appropriate methods for a given drug to be encapsulated.

K. Westesen and B. Siekmann have contributed an important chapter on biodegradable colloidal drug–carrier systems based on solid lipids. These new colloidal carriers different from the other well-known and widely investigated lipidic colloidal carriers, including liposomes, lipoproteins, and lipid or submicron oil-in-water emulsions by exhibiting a solid physical state as opposed to the liquid or liquid crystalline state of the above-mentioned and well-known lipidic colloidal carriers. The authors present different methods of preparation and point out the advantages of the novel dosage forms such as biodegradability, biocompatibility, ease of manufacture, lack of drug leakage, and sustained drug release. Despite three decades of intensive research on liposomes as drug-delivery systems, the number of systems that have undergone clinical trials and become products on the market is quite modest. Even though there have been few successes with liposomes, the need for drug-delivery systems is as acute as ever, and the potential that liposomes hold, although somewhat tarnished, has not been substantially diminished according to R. Margalit and N. Yerushalmi. An interesting and original approach is presented in their chapter on the pharmaceutical aspects of liposomes. Propositions are presented on how at least some of the hurdles in research and development can be overcome and in furthering the substantial strides that have been made in advancing liposomes from the laboratory to the clinic. An ingenious solution on how the drawbacks of liposomes in vivo can be overcome is presented by D. Lasic in the chapter on stealth liposomes. He explains how the stability of liposomes in liposomicidal environments of biological systems presented a great challenge, which was only recently solved by coupling polyethylene glycol to the lipid molecules. An example of the potential of niosomes (a colloidal vesicular system prepared from nonionic surfactants) for the topical application of estradiol is contributed by D. A. van Hal and J. A. Bouwstra, and H. E. Junginger. Niosomes have been shown to increase the penetration of a drug through human stratum corneum by a factor of 50 as compared with estradiol saturated in phosphate buffer solution, making this colloidal carrier promising for the transdermal delivery of drugs.

In the third part of the book, the potential applications of the various particulate-delivery systems are presented. The methods of preparation of microcapsules by interfacial polymerization and interfacial complexation and their applications are discussed by T. Whateley, an extremely knowledgeable scientist in this field. The fast-growing field of lipid microparticulate-delivery systems, particularly lipospheres, is explained and discussed by A. J. Domb, L. Bergelson, and S. Amselem. Lipospheres represent a new type of fat-based encapsulation technology developed for the parenteral delivery of drugs and vaccines and the topical administration of bioactive compounds. In their comprehensive and exhaustive chapter, N. Garti and A. Aserin underline the potential of pharmaceutical application of emulsions, multiple emulsions, and microemulsions, and emphasize the progress made in the last 15 years in understanding mechanism of stabilization of these promising liquid dispersed-delivery systems that open new therapeutic possibilities.

J.-C. Leroux and E. Doelker and R. Gurny in their chapter on the use of drug-loaded nanoparticles in cancer chemotherapy cover the developments and progress made in the delivery of anticancer drugs coupled to nanoparticles, and the interactions of the latter with neoplastic cells and tissues. This is probably the most promising and encouraging application of nanoparticles and by far the most advanced in the process of development into a viable commercial pharmaceutical product. G. Redziniak and P. Perrier have contributed a chapter on the cosmetic application of liposomes that have been successfully exploited over the last decade. To complete the whole range of applications of capsular products, a final chapter, by M. Seiller, M.-C. Martini, and myself, discusses cosmetic uses of vesicular particulate-delivery systems. Cosmetics are definitely the largest market, as manufacturers have demonstrated that marketed cosmetic products containing these vesicular carriers and tested by dermatologists improve cutaneous hydration and skin texture, increase skin glow, and decrease wrinkle depth. It is not taken for granted that liposomes and other vesicular carriers represent a major step in cosmetics formulation. However, this field requires numerous research studies coupled with strict controls.

It is my hope that the scientific information contained herein will modestly contribute to a better understanding of the various particulate systems of all sizes that are now available and to an improved comprehension of their current and potential applications.

Simon Benita

Contents

Contributors

Eric Allémann School of Pharmaceutical Sciences (EPGL), University of Geneva, Quai Ernest-Ansermet, Geneva, Switzerland

Abraham Aserin Casali Institute of Applied Chemistry, The Hebrew University of Jerusalem, Jerusalem, Israel

Simon Benita Department of Pharmaceutics, School of Pharmacy, Faculty of Medicine, The Hebrew University of Jerusalem, Jerusalem, Israel

Jean-Pierre Benoit INSERM U646, Ingénierie de la Vectorisation Particulaire, Université d'Angers, Angers, France

María José Blanco-Príeto Centro Galénico, Farmacia y Tecnología Farmacéutica, Universidad de Navarra, Pamplona, Spain

Amélie Bochot Université Paris-Sud, Faculté de Pharmacie, Châtenay-Malabry Cedex, France

Roland Bodmeier College of Pharmacy, Freie Universität Berlin, Kelchstr, Berlin, Germany

Leila Bossy-Nobs School of Pharmaceutical Sciences (EPGL), University of Geneva, Quai Ernest-Ansermet, Geneva, Switzerland

Frank Boury INSERM U646, Ingénierie de la Vectorisation Particulaire, Université d'Angers, Angers, France

J. A. Bouwstra Leiden/Amsterdam Center for Drug Research, Leiden University, Einsteinweg, RA Leiden, The Netherlands

Stéphanie Briançon LAGEP UMR CNRS 5007 and Laboratoire de Génie Pharmacotechnique et Biogalénique, Université Claude Bernard Lyon 1, Lyon, France

Franz Buchegger Service of Nuclear Medicine, University Hospital of Geneva, Rue Micheli-du-Crest, Geneva, Switzerland

Heike Bunjes Department of Pharmaceutical Technology, Institute of Pharmacy, Friedrich Schiller University Jena, Jena, Germany

Till Bussemer Sanofi-Aventis Deutschland GmbH, Pharmaceutical Sciences Department Industriepark Höchst, Frankfurt am Main, Germany

Florence Delie School of Pharmaceutical Sciences (EPGL), University of Geneva, Quai Ernest-Ansermet, Geneva, Switzerland

Eric Doelker School of Pharmaceutical Sciences (EPGL), University of Geneva, Quai Ernest-Ansermet, Geneva, Switzerland

Abraham J. Domb Department of Medicinal Chemistry and Natural Products, School of Pharmacy-Faculty of Medicine and the David R. Bloom Center for Pharmacy, Alex Grass Center for Drug Design and Synthesis, The Hebrew University of Jerusalem, Jerusalem, Israel

Dominique Duchêne Université Paris-Sud, Faculté de Pharmacie, Châtenay-Malabry Cedex, France

Nathalie Faisant INSERM U646, 'Ingénierie de la Vectorisation Particulaire', Université d'Angers, Immeuble IBT, Angers, France

Hatem Fessi LAGEP UMR CNRS 5007 and Laboratoire de Génie Pharmacotechnique et Biogalénique, Université Claude Bernard Lyon 1, Lyon, France

Jean-Sébastien Garrigue Novagali Pharma S.A., Batiment Genavenir IV, Evry, France

Nissim Garti Casali Institute of Applied Chemistry, The Hebrew University of Jerusalem, Jerusalem, Israel

Robert Gurny School of Pharmaceutical Sciences (EPGL), University of Geneva, Quai Ernest-Ansermet, Geneva, Switzerland

A. Atilla Hincal Hacettepe University, Faculty of Pharmacy, Department of Pharmaceutical Technology, Ankara, Turkey

P. L. Honeywell-Nguyen Leiden/Amsterdam Center for Drug Research, Leiden University, Einsteinweg, RA Leiden, The Netherlands

Thomas Kissel Department of Pharmaceutics and Biopharmacy, Philipps-University of Marburg, Ketzerbach, Marburg, Germany

Jörg Kreuter Institut für Pharmazeutische Technologie, Johann Wolfgang Goethe-Universität Frankfurt, Frankfurt/Main, Germany

Grégory Lambert Novagali Pharma S.A., Batiment Genavenir IV, Evry, France

Jean-Christophe Leroux University of Montreal, Centre ville, Montreal, Quebec, Canada

Charles Linder The Institutes for Applied Research, Ben-Gurion University of the Negev, Beer-Sheva, Israel

Philippe Maincent INSERM U734–EA 3452, Laboratoire de Pharmacie Galénique, Faculté de Pharmacie, Nancy, Cedex, France

Sascha Maretschek Department of Pharmaceutics and Biopharmacy, Philipps-University of Marburg, Ketzerbach, Marburg, Germany

Rimona Margalit Department of Biochemistry, The George S. Wise Faculty of Life Sciences, Tel Aviv University, Tel Aviv, Israel

Arie Markus The Institute of Chemisty and Chemical Technology, The Institutes for Applied Research, Ben-Gurion University of the Negev, Beer-Sheva, Israel

Marie-Claude Martini Institut des Sciences Pharmaceutiques et Biologiques, Lyon, France

Erem Memişoğlu-Bilensoy Hacettepe University, Faculty of Pharmacy, Department of Pharmaceutical Technology, Ankara, Turkey

Philippe Menei INSERM U646, 'Ingénierie de la Vectorisation Particulaire', Université d'Angers, Immeuble IBT and Department of Neurosurgery, CHU Angers, Angers, France

Anne-Marie Orecchioni Universite de Rouen, Rouen, France

Claudia Packhäuser Department of Pharmaceutics and Biopharmacy, Philipps-University of Marburg, Ketzerbach, Marburg, Germany

François Puel LAGEP UMR CNRS 5007, Université Claude Bernard Lyon 1, Lyon, France

Julia Schnieders Department of Pharmaceutics and Biopharmacy, Philipps-University of Marburg, Ketzerbach, Marburg, Germany

Nina Seidel Department of Pharmaceutics and Biopharmacy, Philipps-University of Marburg, Ketzerbach, Marburg, Germany

Monique Seiller Universite de Caen, Caen, France

Britta Siekmann Ferring Pharmaceuticals A/S, Ferring International Center, Kay Fiskers Plads, Copenhagen, Denmark

S. Tamilvanan Department of Pharmaceutics, School of Pharmacy, Addis Ababa University, Addis Ababa, Ethiopia and Department of Pharmaceutics, School of Pharmacy, Faculty of Medicine, The Hebrew University of Jerusalem, Jerusalem, Israel

Frédéric Tewes INSERM U646, Ingénierie de la Vectorisation Particulaire, Université d'Angers, Angers, France

Laury Trichard Université Paris-Sud, Faculté de Pharmacie, Châtenay-Malabry Cedex, France

Nathalie Ubrich INSERM U734–EA 3452, Laboratoire de Pharmacie Galénique, Faculté de Pharmacie, Nancy, Cedex, France

Angelica Vargas School of Pharmaceutical Sciences (EPGL), University of Geneva, Quai Ernest-Ansermet, Geneva, Switzerland

Clive Washington Pharmaceutical and Analytical Research and Development, AstraZeneca, Macclesfield Works, Hurdsfield Industrial Estate, Macclesfield, Cheshire, U.K.

Shicheng Yang KV Pharmaceutical Company, St. Louis, Missouri, U.S.A.

Noga Yerushalmi Department of Biochemistry, The George S. Wise Faculty of Life Sciences, Tel Aviv University, Tel Aviv, Israel

PART I: METHODS OF ENCAPSULATION AND ADVANCES IN PRODUCTION TECHNOLOGY

1

Biodegradable Microspheres: Advances in Production Technology

Frédéric Tewes, Frank Boury, and Jean-Pierre Benoit
INSERM U646, Ingénierie de la Vectorisation Particulaire,
Université d'Angers, Angers, France

INTRODUCTION

Research to find new or to improve microencapsulation techniques to process newly discovered active molecules is in constant progress because of the limitations of the current pharmacopeia. The new active molecules found with the help of advances in biotechnology and therapeutic science are more often peptides or proteins; they are very active in small doses, sensitive to unfolding by heat or organic solvents, available only in small quantities, and very expensive. Additionally, many new molecules that are synthesized have poor solubility in aqueous media, and some of them, when used in their typical therapeutic concentrations, such as paclitaxel, also have poor solubility in lipidic media.

Besides this, the regulatory authorities, such as the U.S. Food and Drug Administration (FDA), are restricting to greater degrees the amounts of additional components allowed such as organic solvents or tensioactive molecules.

For these reasons, in designing of new techniques one must take into account several new requirements: The stability and the biological activity of the drug should not be affected during the microencapsulation process, yield and drug encapsulation efficiency should be high, microsphere quality and the drug release profile should be reproducible within specified limits, microspheres should not exhibit aggregation or adherence, the process should be usable at an industrial scale, and the residual level of organic solvent should be lower than the limit value imposed by the European Pharmacopeia.

However, the commonly used techniques such as emulsification–solvent removal, polymer phase separation, spray drying, and milling methods are not always suitable in their original forms for these new requirements. Spray drying and milling can thermally denature some compounds. Milling produces a broad size distribution. Emulsion/solvent removal techniques can denature proteins at the interfaces. Polymer phase separation, spray drying, and emulsification processes often lead to amounts of residual solvent that are higher than the upper authorized values. Therefore, modifications to these processes or the development of new techniques is needed.

Near critical or supercritical fluid techniques are promising and fulfil some of the new requirements.

In this chapter, we present an overview of the existing techniques that usually make use of volatile solvents and we describe the improvements proposed to meet the new requirements. We then present techniques that do not use volatile organic solvents, with special attention to techniques using supercritical fluids.

TECHNIQUES USING ORGANIC SOLVENTS

Emulsification–Solvent Removal Processes

Initially, polymers are dissolved in a volatile organic solvent with low water miscibility, such as dichloromethane (DCM) or chloroform. The drugs are then dissolved or dispersed in the polymer solution. This mixture is then emulsified in a large volume of an aqueous phase containing tensioactive molecules such as poly(vinyl alcohol) (PVA), resulting in organic solvent droplets dispersed in a water phase—oil-in-water (O/W) emulsion. The emulsion is next subjected to solvent removal by either evaporation or an extraction process in order to generate microspheres. These particles are washed, collected by filtration, sieving, or centrifugation, and finally dried or lyophilized to provide free-flowing injectable microspheres.

The two processes of solvent removal influence microsphere size and morphology. When the solvent is evaporated, the emulsion is maintained at reduced or atmospheric pressure under low agitation. If the drug molecules are volatile and/or have a great affinity for the organic phase, they can be removed at the same time. During the extraction process, the emulsion is transferred into a large volume of water (with or without surfactant) or another quench medium, where the solvent diffuses out (1). The quenching medium must not make the polymer soluble and must have a great affinity for the aqueous phase. It should be noted that the solvent evaporation process is similar to the extraction method, in that the solvent must first diffuse out into the external aqueous dispersion medium before it can be removed from the system by evaporation (2). The rate of solvent removal influences the characteristics of the microspheres. Rapid solvent removal leads to the formation of porous structures on the microsphere surface and to a hardening of the polymers in the amorphous state. Solvent removal by the extraction method is faster (generally less than 30 min) than by the evaporation process and hence, the microspheres generated by the former method are more porous and more amorphous (2).

Hydrophilic Molecules

Anhydrous Emulsion. As described above, the O/W emulsification process is appropriate for encapsulating apolar drugs. It leads to poor encapsulation efficiency for polar water-soluble drugs (1). The polar drugs can partition between the two phases. Consequently, besides low drug entrapment, hydrophilic drugs are often deposited on the microsphere surface (3). This induces an initial rapid release of the drug, known as the burst effect (1).

Modifications of the conventional O/W solvent removal method have been suggested to avoid these problems. One of these is the use of totally anhydrous systems. They are constituted from an organic volatile phase (acetonitrile, acetone), containing drugs and polymers, emulsified in an immiscible oil (mineral or vegetable) or a nonvolatile organic solvent containing a surfactant with a low hydrophile–lipophile balance (HLB) [oil-in-oil emulsion (O/O emulsion)] (4,5). Microspheres are finally obtained by evaporation or extraction of the volatile solvent and washed in another solvent such as *n*-hexane to remove the oily dispersing media. This process essentially avoids the loss of drugs, being soluble in water, such as amino-alcohols, amino-acids and peptides, proteins, cytostatics like cisplatin, or anti-inflammatory drugs (4–9).

Multiple Emulsions. Another modification that allows the encapsulation of water-soluble molecules is the formation of a water-in-organic solvent-in-water (W/O/W) emulsion (10). An aqueous solution of the drug (sometimes containing a viscosity builder and/or protein acting as a tensioactive agent and carrier) is added to a volatile organic phase containing the polymers under intense shear (high pressure homogenizer, sonicator, etc.) to form a water-in-oil (W/O) emulsion. This emulsion is gently added under stirring into a large volume of water containing tensioactive molecules to form the W/O/W emulsion. The most commonly used tensioactive agents are PVA and poly(vinyl pyrrolidone) (11,12). The emulsion is finally subjected to solvent removal either by evaporation or extraction. The solid microspheres obtained by this process are washed, collected, and dried or lyophilized.

This technique has mainly been used for the encapsulation of peptides such as leuteinizing hormone–releasing hormone (LHRH) agonist (leuprolide acetate), somatostatin, proteins, and other hydrophilic molecules in poly(α-hydroxyacid)s such as poly(D,L-lactide) (DL-PLA), poly(D,L-lactide-co-glycolide) (PLGA), or poly(L-lactide) (L-PLA) microparticles (11–38).

Schugens et al. (36) have shown that an increase in the molecular weight M_w of L-PLA required the dilution of the polymer solution to prevent an exceedingly high viscosity, which led to formation of less stable primary emulsions and to more porous solid microspheres. They also concluded that the crystallinity of L-PLA affected the stability of the primary emulsion by exclusion of the internal aqueous droplets from the L-PLA matrix. This exclusion adversely impacted the encapsulation efficiency, and increased the microparticle porosity. Herrmann and Bodmeier (17) created microspheres composed of L-PLA, PLGA, or DL-PLA and found that an increase in the volume fraction of the internal aqueous phase in the primary W/O emulsion resulted in lower encapsulation efficiency when it is made with DCM as organic solvent. Crotts and Park (27) found that this increase induced the formation of pores in the shell layers of PLGA microspheres. Herrmann and Bodmeier (18) also showed that the addition of buffers or salts to the internal aqueous phase resulted in porous L-PLA microspheres and lowered somatostatin encapsulation efficiency. This is due to the increase of the difference of osmotic pressure between the two aqueous phases, and the promotion of an influx of water from the external phase toward the internal phase. The addition of salts to the external aqueous medium resulted in the formation of a dense and homogeneous polymer matrix (32).

This technique can induce an incomplete release of encapsulated protein as observed with PLGA microspheres by Park and Lu (28) and Pean et al. (32). Park and Lu determined that proteins were significantly and irreversibly denatured and aggregated during the emulsification step. Proteins are usually exposed to cavitation, heat, organic solvent, or shear during the microencapsulation process. In particular, the first emulsification step is considered as the main cause for protein denaturation and aggregation (24,39,40). Furthermore, a part of the protein strongly bound to the polymeric matrix was found by Boury et al. (41). Due to the protein denaturation at the water/organic solvent interface, alternative encapsulation procedures such as the anhydrous system in which the protein is dispersed (solid-in-oil-in-oil) have gained much attention (40,42). However, in these procedures, it is technically complex to ensure the homogeneous distribution of protein powder particles in the microspheres. Furthermore, in many protocols, protein powder particles have to be micronized prior to encapsulation, which can unfold the protein. Because many of these problems are avoided in W/O/W encapsulation procedures, developments of rational strategies to stabilize proteins upon W/O/W encapsulation are in progress. Several excipients such as dithiothreitol sodium dodecyl, polysaccharides such as dextran or heparin, trehalose, poloxamer 407, cyclodextrin, or bovine serum albumin (BSA) can increase protein stability (28,39,43–45). Polyethylene glycol (PEG) 400 added to the internal aqueous phase of the double emulsion prevented protein adsorption at the W/O interfaces, which avoided protein denaturation during the emulsification step and reduced protein anchorage in the PLGA layer during the microparticle preparation step (46).

Very high encapsulation yields for hydrophilic molecules such as proteins are obtained by using the W/O/W double-emulsion method (47). However, it requires many steps, and a strict control of the temperature and viscosity of the inner W/O emulsion, and does not allow large quantities of hydrophilic drugs to be encapsulated (48).

Another type of multiple emulsion (W/O/O/O) was developed by Iwata and McGinity (49) to produce multiphase microspheres. An internal aqueous phase containing a hydrophilic drug and a surfactant was emulsified in soybean oil. This first emulsion was then dispersed in acetonitrile containing the polymers to form a W/O/O emulsion. Finally, the W/O/O emulsion was dispersed into a mineral-oil solution, acting as a hardener and containing tensioactive molecules.

The oil in the primary emulsion prevents contact between the internalized hydrophilic drug (protein) and the polymer–solvent system. The isolation of the protein from the polymer–solvent mixture prevents a possible unfolding of the protein by the polymer or the solvent. Moreover, the possibility of polymer degradation due to reactive proteins or drug compounds is also limited.

O'Donnell et al. (50) prepared multiphase microspheres of PLGA by following the same process but using a potentiometric dispersion technique in the last step. This technique of dispersion produces narrower dispersion and a better loading efficiency than the classical agitation method. O'Donnell and McGinity (51) produced PLGA microspheres containing thioridazine HCl through four types of emulsions: O/W, O/O, W/O/W, and W/O/O/O. They found some degradation of PLGA microspheres when produced using the O/W emulsion due to a hydrolysis catalyzed by thioridazine HCl. Microspheres produced using the W/O/O/O type did not exhibit a burst effect as compared to the other types.

Residual Solvents

Solvents commonly used in microencapsulation via the emulsion solvent removal method, particularly when using chlorinated solvents such as DCM or chloroform, may be retained in the microspheres as a residual impurity, sometimes at values above those authorized by the Pharmacopeia. These solvents are highly toxic and volatile, which can induce problems in large-scale manipulation. Furthermore, it has also been reported that an antigen becomes less immunogenic after contact with an organic solvent (52). Therefore, routine usage of these solvents is becoming complicated and an alternative is needed.

One of the major advances in this direction is the replacement of more toxic, chlorinated solvents by less toxic solvents such as ethyl acetate, a mixture of ethyl acetate/acetone, or ethyl formate (44,53–61). Lagarce et al. (54) successfully prepared oxaliplatin-loaded PLGA microspheres with ethyl acetate by using the O/W emulsification–solvent extraction process. The encapsulation efficiency was around 90% and the release profile could be managed by changing the nature of the PLGA polymer.

The relatively high solubility of ethyl acetate in water allows a fast diffusion of ethyl acetate from droplets into the outer aqueous phase during the O/W emulsification step, which can lead to polymer precipitation rather than to the formation of microparticles (53). On the other hand, as proposed by Kim et al. (62), one can slow down the diffusion rate by saturating the outer aqueous phase with ethyl acetate prior to emulsification. A control of the ethyl acetate diffusion rate was also performed by Sah (53), by managing the volume of the outer aqueous phase and solidifying the native microparticles step by step, by adding a small amount of water in series, and by slowly extracting ethyl acetate. This allowed the production of either hollow or matrix-type microspheres, with different size distributions. Meng et al. (56) applied the same procedure on a W/O/W emulsion for encapsulating proteins. They obtained a high level of protein entrapment (always above 94%) and the full preservation of the entrapped lysozyme bioactivity. The release was slow, without any burst effect. Furthermore, compared to the more hydrophobic DCM, ethyl acetate can have a low unfolding effect on the entrapped proteins in microspheres produced by the W/O/W emulsification process (24,44).

However, most researchers in this field still choose DCM as the organic solvent because of its physical properties such as its ability to dissolve large amounts of biodegradable polymers, its low solubility in water (2.0%, w/v), and its low boiling point (39.8°C), which is compatible with the evaporation step (63–67). Moreover, replacement of the DCM with ethyl acetate can reduce encapsulation efficiency (17).

Organic Phase Separation (Coacervation)

The phase separation process has allowed the production of the first microspheres encapsulating a peptide (nafarelin acetate) for a parenteral application at an industrial scale (68).

This process consists in decreasing the solubility of the encapsulating polymers (PLGA, PLGA-PEG, and PLA) solubilized in an organic solvent (DCM, ethyl acetate, chloroform, toluene) by varying the temperature, or by adding a third component that interacts with the organic solvent but not with the polymer [coacervating agent (CA)] (69–76,48). Classes of CA can be distinguished into: (i) nonsolvents of the polymer, that induce coacervation by extracting the polymer solvent, and (ii) polymers that are incompatible (nonmiscible) with the wall polymers.

For a particular area of the (solvent–polymer–CA) ternary phase diagram (Fig. 1), i.e., "stability window," two liquid phases have been obtained (phase separation): a rich polymer phase called coacervate droplet and a phase depleted in polymers (69,70). The drug that is dispersed or sometimes dissolved in the polymer solution is coated or entrapped by the droplet of coacervate. Then, the coacervate droplets are solidified by using a hardening agent such as heptane to produce microparticles (microcapsules or microspheres), which are collected by washing, sieving, filtration, or centrifugation, and are finally dried (69,70,72,77). The hardening agent should be relatively volatile and should easily remove the viscous CA on washing.

This process is suitable to encapsulate both hydrophilic and hydrophobic drugs. Hydrophilic drugs such as peptides and proteins can be dissolved in water

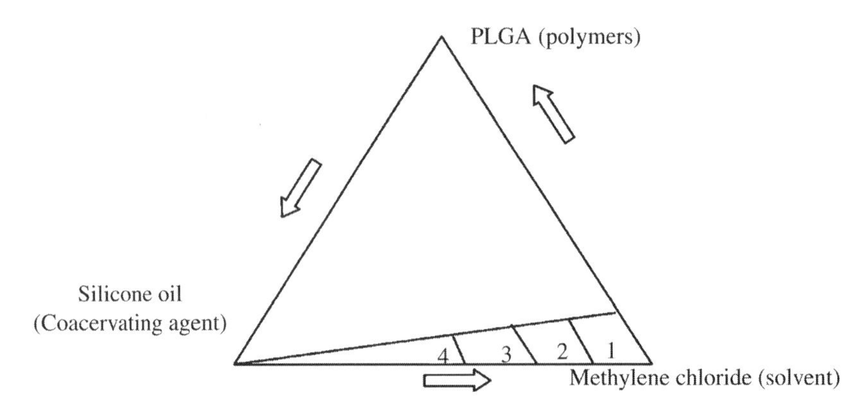

Figure 1 Ternary phase diagram of coacervation. (1) Emulsification of the CA–solvent phase in polymer–solvent phase, (2) beginning of a coacervation, but droplets of coacervate are unstable, (3) "stability window," (4) massive aggregation and precipitation of the coacervate droplets. *Source*: From Ref. 70. *Abbreviations*: CA, coacervating agent; PLGA, poly(D,L-lactide-co-glycolide).

and dispersed in the polymer solution to form a W/O emulsion, or directly dispersed in the organic phase (71,74–76). Hydrophobic drugs are usually dispersed or sometimes solubilized in the polymer solution.

The CA affects both phase separation and the solidification stages of the coacervation process. The CA should dissolve neither the polymer nor the drug and should be soluble in the solvent. It is added to the stirred polymer–drug–solvent system and gradually induces coacervation. The size of the resulting coacervate droplets is controlled by the stirring rate and by the rate of addition of the CA. The "stability window" determines the optimum volume of CA that is added to the polymer–drug–solvent system to obtain stable coacervate droplets. The process conditions, like viscosity of the CA and type of polymer coating or polymer concentration, that allow the obtaining of the "stability window" can be varied to obtain microparticles. For low CA volumes (windows 1 and 2 in Fig. 1), coacervation is not effective and stable coacervate droplets cannot be obtained. For high CA volumes (window 4 in Fig. 1), extensive aggregation and precipitation of coacervate droplets occur.

Typical CAs used with poly(α-hydroxyacid)s are low (M_w) silicone oil (poly (dimethylsiloxane)) (71,72,74,76,78), triglycerides, mineral oil, low (M_w) liquid polybutadiene, hexane and low (M_w) liquid methacrylic polymers. Hardening agents include aliphatic hydrocarbons such as hexane, heptane, octamethylcyclotetrasiloxane (OMCTS), and petroleum ether (71,72,74,76,79–82).

Temperature parameters, polymer concentrations, and (M_w) must be carefully adjusted to obtain physically stabilized droplets made of poly(α-hydroxy acid)s-rich phase surrounded by the poly(α-hydroxyacid)-poor phase. Thus, interfacial properties and viscosity will play key roles. Moreover, the entrapment of dissolved/dispersed drugs in the coacervate phase must be carried out with conditions that do not affect the activity of the compound.

In a sophisticated process, poly(lactide) (PLA) solution was injected into mineral oil, whereby particles of any desirable size were produced by varying the diameter of the injection equipment (80). Other modifications of the classical coacervation procedure were made for encapsulation of nafarelin acetate (79). These studies combined a phase-separation followed by a solvent-evaporation step to obtain freely flowing powders.

In the coacervation process, phase equilibrium is never reached. Therefore, the formulation and process variables significantly affect the kinetics of the entire process and ultimately the characteristics of the microspheres. Moreover, the coacervate droplets are extremely sticky and adhere to each other before completion of the phase separation operation or before hardening. Consequently, this technique tends to produce agglomerated particles. However, adjusting the stirring rate, temperature, or the addition of a stabilizer is known to rectify this problem (71).

Compared to the solvent evaporation–extraction method, the choice of solvents is less stringent because the solvent does not need to be immiscible with water and its boiling point can be higher than that of water. However, the method also requires large quantities of organic components (initial solvent and hardening agent), which are often difficult to remove from the final microspheres (77).

In an attempt to minimize the amount of residual solvents, a low ratio of solvent/CA [Polydimethylsiloxane (PDMS)] was claimed to be a key parameter in controlling the amount of the residual hardening agent heptane in histreline-loaded PLGA microspheres (83). The suitability of volatile siloxanes such as OMCTS or hexadimethylsiloxane, as hardening agents was shown in the encapsulation of the peptide drug triptorelin (84). The residual amount of siloxane in the final product was 2% to 5%. As shown by Thomasin et al. (85), the residual amounts of DCM and OMCTS depend greatly on the amount of PDMS used for coacervation.

Spray Drying

The principle of spray drying by nebulization is based on the atomization of a solution, containing drugs and carrier molecules, by using compressed air or compressed nitrogen through a desiccating chamber, and using a current of warm air for the drying process. This is performed in three steps: (i) formation of the aerosol, (ii) contact of the aerosol with the warm air and drying of the aerosol, and (iii) separation of the dried product and the air charged with the solvent.

This process has been applied to the creation of microparticles by spraying complex liquid mixtures containing an active principle that is dissolved/dispersed in an organic or aqueous polymer solution. The production of microspheres or microcapsules by spray drying depends on whether the initial formulation was whether in the form of a solution, suspension, or emulsion. It has been used successfully with several biodegradable polymers such as PLA, PLGA, poly(ε-caprolactone) (PCL), commercial Eudragit® (Degussa AG, Weiterstadt, Germany), gelatin, and polysaccharides or related biopolymers (86–98).

The first commercialized injectable microspheres, which encapsulated bromocriptine (Parlodel® LAR; Sandoz, Switzerland), were produced by spray drying. The formulation included PLGA branched to D-glucose (91). Other research described this process for the encapsulation of several hydrophilic or lipophilic drugs (86,99,100). In some cases, due to the incompatibility of the hydrophilic drug to the PLA, needle-shaped crystals of drug grew on the microsphere surface, whereas the lipophilic drug–PLA system provided smooth particles (86).

Spray drying usually leads to a broad distribution of particle size, with a Gaussian shape, centered on 10 μm. Flow rate, nozzle geometry, and solution viscosity are the most influencing parameters (1).

There may be a significant loss of the product during spray drying, due to adhesion of the microparticles to the inner wall of the spray-drier apparatus, or to agglomeration of the microparticles (101). To solve these problems, a novel technique has been developed; using two injection devices and mannitol as an antiadherent. A polymer–drug solution was sprayed through one nozzle and simultaneously an aqueous mannitol solution was sprayed through the second nozzle. The resulting microspheres exhibited surfaces coated with mannitol that decreased agglomeration (101). This method produced microspheres with higher yield and encapsulation ratio when compared to those prepared from a W/O/W emulsion–solvent removal method.

The use of a novel low-temperature spraying method has been reported by the company Alkermes® (ProLease® technology) for preparing PLA and DL-PLGA microspheres (102,103). A powder composed of protein (human growth hormone) and stabilizing excipients was suspended in a solution of polymer in acetone, ethyl acetate, or DCM. This suspension was then sprayed into a vessel containing liquid nitrogen. The liquid nitrogen was then evaporated and the organic solvent of the frozen droplets was extracted by liquid ethanol. Microspheres were then filtered and vacuum dried to eliminate residual solvents. The microspheres were 50 to 60 μm in size with drug encapsulation efficiency higher than 95% (102,104,105).

Contrary to coacervation and emulsification methods, the spray drying method is a one-step quick process, and is continuous, easy to scale-up, and inexpensive. It is less dependent on the solubility parameters of the drug and the polymer, and can be used without organic solvents. When used with organic solvents, the amount of residual solvent in particles is often lower than that reached with emulsification–solvent removal techniques (0.05–0.2%). They can, however, be higher than the limit value of the Pharmacopeia (0.06% for DCM). Nevertheless, a solvent less toxic than DCM, such as ethyl formate, has been widely used (88).

It should be mentioned that the formation of fibers instead of microspheres could occur when the sprayed solution is not sufficiently broken up (86). This occurs when the viscosity of the solution is too high (high polymer concentration, ramified polymers, etc.), and also when the geometry of the nozzle is not suitable, or when the flow rate is too low (1). In addition, the temperature necessary to dry particles is over 100°C when starting from aqueous solution, and can denature thermally labile drugs such as proteins.

TECHNIQUES WITHOUT ORGANIC SOLVENTS

Residual Solvent Considerations

Microparticles used as sustained-release dosage forms are mainly composed of biodegradable polymers. Unfortunately, the applied polyesters such as PLA, PLGA, and PCL, are only soluble in toxic organic solvents such as DCM, chloroform, or to a lesser degree, ethyl acetate or ethyl formate, that are commonly used to dissolve the coating polymer prior to microencapsulation.

Classical techniques such as emulsification–solvent removal, spray drying, and organic phase separation, involve an extensive use of organic solvents. This aspect leads to environmental problems of pollution, toxicity due to incomplete solvent removal, and solvent impurities that may cause chemical degradation of the bioactive substances within the polymer matrix. It also complicates the process and increases the cost of production.

Bitz and Doelker (106) compared the residual solvent traces (commonly known as organic volatile impurities or OVIs) in different formulations containing tetracosactide as a model drug. The residual amounts of solvent varied from 934 to 5998 ppm for chloroform, and from 281 to 705 ppm for DCM. Vacuum drying over three days decreased the concentration of DCM from 2 to 17 ppm. Spenlehauer et al. (107) found 30,000 ppm residuals of DCM in cisplatin-loaded microparticles manufactured by the emulsification/solvent evaporation technique. In a previous study, Spenlehauer et al. (108) showed dependence between the residual DCM content in microparticles and the drug content, the particle size, and the addition of a nonsolvent to the polymer solution.

Benoit et al. measured residual DCM content in progesterone-loaded microparticles prepared by the emulsification/solvent evaporation technique. They showed that vacuum drying led to a reduction of the solvent content from 18,000 to 360 ppm for PLA microparticles (109) and from 47,000 to 6800 ppm for polystyrene ones (110).

Owing to the toxicity of the generally employed solvents, the authorities [Pharmacopeia of the United States (USP), Europe (PhEur), and Japan (JP)] impose limitations in pharmaceuticals. Since 1997, the limited values imposed by the three Pharmacopeia are going to be harmonized by the International Conference on Harmonization (ICH), but only the PhEur and the JP have fully adopted the ICH guidelines (111).

The maximum limits for chloroform and DCM (which belong to Class 2 solvents according to the ICH because they are suspected of carcinogenicity as well as neurotoxicity and teratogenicity) imposed by the 2002 edition of the USP (112) and the guidelines of the ICH are 60 and 600 ppm, respectively. Most of the values of residual solvents mentioned before would not fulfill the guidelines of the Pharmacopeia, indicating that it is necessary to develop alternative production techniques.

Therefore, health problems can be caused by solvents such as DCM by environmental emissions and/or by the presence of final residues in the product. This led to some important research efforts, and "environmentally benign" processing techniques have been developed that either eliminate or significantly mitigate pollution at the source. Among the reported applications, the formation of drug particles using supercritical fluids (SF) and milling methods is very promising.

Techniques Using Supercritical Fluids (SF)

Recently, techniques using SF have emerged as promising methods to produce microparticles with environmental and processing advantages. In particular, these novel methods leave particles without or with very low amounts of residual organic solvent and provide a feasible and clean way to process thermolabile or unstable biological compounds. This therefore presents a promising application in the development of new drug-delivery systems (113–115).

Several processes using SF for the design of pure drug or composite particles (active molecules and carriers) have been investigated and several reviews have been published (116–118). Basic techniques can be distinguished into: (i) techniques using the SF as a solvent to solubilize active and/or carrier molecules, (ii) techniques using the SF as an antisolvent, where it is brought into contact with an organic solution to induce precipitation of the active molecules and/or the carrier, and (iii) techniques using the SF as a spray enhancer.

SF Properties

Generality. A substance is termed supercritical when its pressure and temperature are higher than its critical pressure (P_c) and critical temperature (T_c) (Fig. 2).

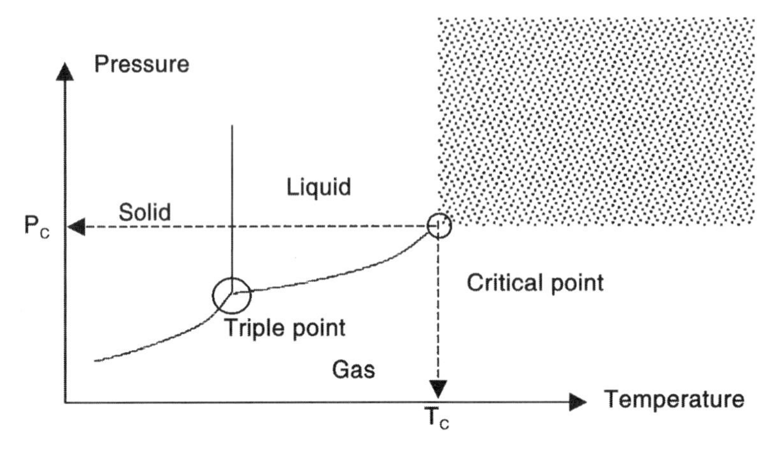

Figure 2 Phase diagram of a pure substance.

From a thermodynamic point of view, the critical point is defined as the point where the fluid becomes mechanically unstable and thus highly compressible (Table 1). This occurs near the critical temperature (between T_c and 1.2 T_c) and around the critical pressure (0.9–2.0 P_c), where the fluid density varies according to a continuum from gas-like to liquid-like state, but with a viscosity that remains close to that of a gas.

Therefore, the properties of the SF, such as viscosity and diffusivity (Table 1), and some other physical properties depending on the density (dielectric constant, solvent strength, and interfacial tension), can be finely and linearly controlled (120).

As a result, it is possible to obtain unique fluid properties to suit various processing needs. For example, the variation of the solvent power allows molecules to be solubilized when the density is close to liquid density and then to precipitate them by reaching a gas density. The ability to rapidly vary the solvent strength and thereby the rate of supersaturation and nucleation of dissolved compounds is a major aspect of supercritical technology for particle formation.

Control of the interfacial tension and the low viscosity of SF regulate the size of the liquid droplets sprayed in the supercritical fluid to be managed (121). Furthermore, the high diffusion coefficients in this medium (Table 1) lead to great mass transfer rates. Therefore, SFs are very appropriate for spray-based processes.

Another advantage compared to classical processes is the easy separation of the supercritical solvent from the particles. It is therefore possible to avoid large quantities of solvent by-products and to reduce the amount of residual solvents.

Table 1 Comparison of Some Physical Properties of Different Matter States

	Gas	SF	Liquid
Density ($kg\,m^{-3}$)	1–500	100–1000	600–1,600
Compressibility (MPa^{-1})	1–10	$\rightarrow\infty$ at the critical point	$1 \times 10^{-3} - 5 \times 10^{-3}$
Viscosity (Pa s)	$1 \times 10^{-5} - 4 \times 10^{-5}$	$1 \times 10^{-5} - 9 \times 10^{-5}$	$0.2 \times 10^{-3} - 3 \times 10^{-3}$
Diffusion coefficient of a solute ($m^2\,s^{-1}$)	$3 \times 10^{-5} - 1 \times 10^{-5}$	$7 \times 10^{-8} - 1 \times 10^{-8}$	$2 \times 10^{-9} - 0.2 \times 10^{-9}$
Thermal diffusivity ($m^2\,s^{-1}$)	5×10^{-4}	$\rightarrow 0$ at the critical point	$1 \times 10^{-7} - 2 \times 10^{-7}$

Source: From Ref. 119.

Carbon Dioxide (CO_2). CO_2 is the widely used SF for pharmaceutical applications. It is an ideal processing medium because it has low critical points ($T_c = 31.1°C$ and $P_c = 73.8$ bar). Furthermore, CO_2 is nontoxic, nonflammable, relatively inexpensive, recyclable, and "generally regarded as safe."

In the presence of supercritical CO_2 ($ScCO_2$), particle formation occurs in a nonoxidizing atmosphere at near ambient temperatures and without the need for the application of high shear forces—thanks to its low viscosity. This appears to be crucial for the microencapsulation of labile molecules.

CO_2 is nonpolar and exerts few van der Walls interactions. As such CO_2 is essentially a nonsolvent for many lipophilic and hydrophilic compounds (which covers most pharmaceutical compounds), but can solubilize low M_w lipophilic compounds and some polymers having low energy of cohesion (122). Consequently, $ScCO_2$ has been exploited both as a solvent and as an antisolvent in pharmaceutical applications.

$ScCO_2$ is known to swell and plasticize polymers, which lowers the glass transition (T_g) or melt (T_m) temperatures of polymers (123–125). The amount of swelling depends upon the chemical nature of the polymer and on its interaction with CO_2.

In the presence of an aqueous solution containing a pH-sensitive component, CO_2 is not always suitable because of its reaction with H_2O, forming carbonic acid and lowering the pH to as low as 3 (126). However, the use of a buffer can avoid this inconvenience (127). Furthermore, the inactivation of pathogenic bacteria and endotoxin by the $ScCO_2$ without denaturing the biomolecules is also an advantage (128). However, the inactivation mechanism is unclear and the sterilizant capacity of the processes using $ScCO_2$ is still in discussion.

Techniques Using SF as a Solvent

Rapid Expansion of Supercritical Solutions (RESS). *Description.* In this concept, molecules are dissolved in SF, which is then expanded adiabatically through a heated capillary nozzle in a low-pressure chamber. Due to SF properties, a sharp decrease of solvent density can be obtained by a relatively small change in pressure, leading to a large decrease in the solubility of the compounds initially dissolved.

In these conditions, very high supersaturation occurs, leading to a high nucleation rate. The nuclei grow gradually by the addition of single molecules on their surfaces by a mechanism of condensation. Particle growth can also occur via a coagulation mechanism, when particles collide and stick together.

Because expansion is a mechanical perturbation traveling at supersonic velocity (Fig. 3), a high rate of supersaturation and uniform conditions are ensured through the expansion device (129). The resulting particle mean sizes range typically from 0.5–20 μm, with a narrow distribution (117,130,131).

In Figure 4, the basic equipment consisting of two main units is shown: extraction and precipitation. In this configuration, pure CO_2 is pumped to the desired pressure and preheated to extraction temperature in order to reach a supercritical state. The SF is then passed through an extraction unit, where it is charged with the solute. After that, it is expanded in the low-pressure precipitation unit. Other configurations have been described for specific applications [coating, coprecipitation (113,131–137), etc.] and will be discussed hereafter.

Advantages and Disadvantages. The rapid expansion of supercritical solution (RESS) concept is close to that of the spray-drying process, but with the advantage of not using organic solvents and of applying only moderate processing temperatures

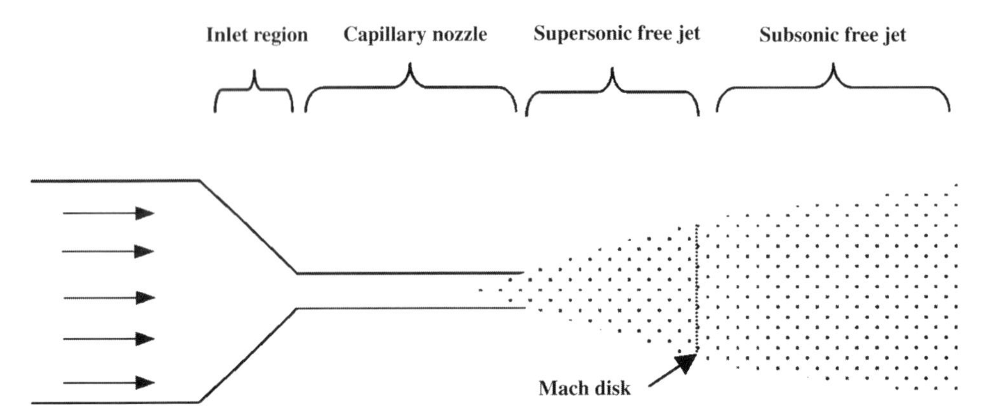

Figure 3 Scheme of an expansion device and of the different regions of the free jet. *Source*: From Ref. 129.

(113,136). This process, therefore, seems suitable to prevent substance denaturation and a loss of therapeutic activity. Dry and solvent-free particles can be obtained in a single step and the process can be undertaken with relatively simple equipment, although particle collection from the gaseous stream is not always easy (117). In addition, no surfactants or nucleating media are required to activate nucleation. Finally, the SF is removed by a simple mechanical separation.

Table 2 summarizes the main differences between the RESS and the "classical" spray-drying processes. It should be mentioned that RESS is sometimes performed at high temperatures (100°C), which can be problematic with sensitive molecules (144). Furthermore, due to low polymer solubility into the $ScCO_2$, most RESS processes are limited to polymers with high solubility in $ScCO_2$, i.e., with low cohesive energy, such as perfluoroethers and siloxanes, or to other couples of polymers/SF such as supercritical alkanes. Most of those polymers are not suitable for pharmaceutical applications, and due to their rapid expansion and their high solubility in SF, they can lead to sponge-like, nonuniform microparticles (137,143,145).

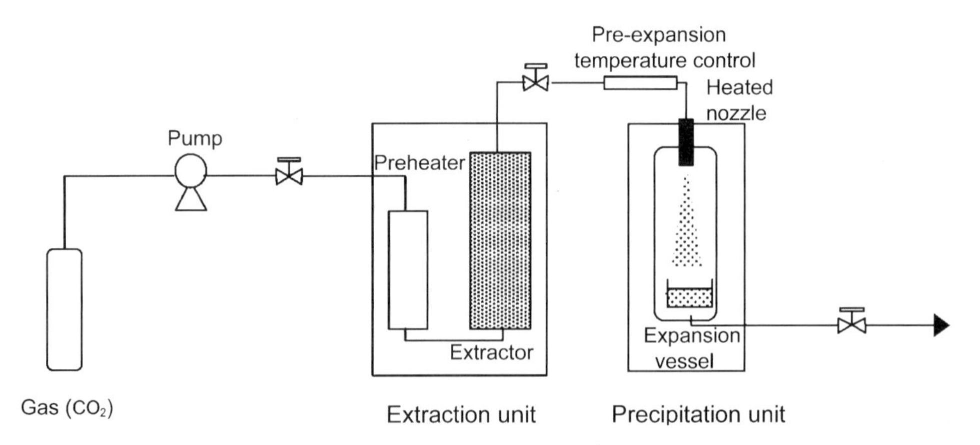

Figure 4 Scheme of a RESS apparatus. *Abbreviation*: RESS, rapid expansion of supercritical solution.

Table 2 Comparison of RESS and "Classical" Spray-Drying Processes on the Formulation of Low Polar Particles

Parameters	RESS	"Classical" spray drying
Propagation rate of the spray	Supersonic flow	Subsonic flow
Nucleation particle growth	Occurs during expansion characteristic time 10^{-7} sec	Occurs during solvent evaporation characteristic time 1 sec
Size distribution	RESS process produces smaller particles with narrower distribution	
Constraints related to the process	High shear stress during the expansion can induce denaturation of labile molecules	Use of organic solvent
	Low solubility of high M_w materials at moderate conditions (<60°C and 300 bar)	High temperature that can alter the properties of thermolabile molecules

Abbreviation: RESS, rapid expansion of supercritical solution.
Source: From Refs. 117, 119, 134, 136, 138–143.

Cosolvents such as methanol, DCM, butyl acetate, acetone, chlorodifluoromethane ($CHClF_2$), or ethanol may be used to improve polymer solubility, but their elimination from the resulting particles is not easy, it increases the cost of the process, and may cause environmental problems (134,136,146–149). This is because the current challenge is to obtain new CO_2-soluble polymers. For example, Sarbu, Styranec T, and Beckman et al. (150) have synthesized a new poly(ether-carbonate) with higher solubility in $ScCO_2$ than in fluoroether polymers.

Influencing Parameters. The morphology of the particles depends on the polymer and drug component characteristics (crystallinity, solubility, purity, polydispersity, etc.) and on the RESS parameters (temperature and pressure in the different units, geometry of the expansion device, etc.).

Components. In the RESS process, the dissolution of the polymer in $ScCO_2$ is a necessary condition. The commonly used polymers are polydisperse and the lowest M_w chains are preferentially extracted from the sample (136). Consequently, the M_w distribution of the polymer in the particles changes over time, and this leads to particles with different compositions. Consequently, a pre-extraction of low M_w polymer with $ScCO_2$ must be carried out in order to produce uniform particles.

High polymer concentrations or elevated temperatures can produce early phase separation, before reaching the precipitation unit, and cause the formation of undesirable structures such as fibers (151).

Temperature. Temperatures along the successive units of the RESS apparatus can be adapted to optimize the process (152). The temperature inside the precipitation unit controls the physical properties of the polymer and thus the particle shape. For example, at 50°C, the polymeric particles of L-PLA (M_w 5500 g/mol) are soft and thus coagulate easily, whereas at a lower temperature (15–37°C), the polymer is glassy and no adhesion occurs (136). The temperature of the expansion device must be sufficiently high to prevent phase changes (i.e., condensation and freezing) upon adiabatic expansion. In the extraction unit, the temperature must be adapted to obtain the desired solute concentration.

Expansion Orifice. Expansion devices are generally composed of a capillary nozzle. Its length varies from 2 to 40 mm and the inner diameter tubing is from 25 to 200 μm. The use of laser-drilled pinpoint orifices of 15 to 30 μm in diameter and larger orifices of 1600 μm has also been investigated (153). The nozzle geometry (length/diameter ratio) is an important factor that affects particle size and morphology (134). Kim et al. (135) found that a silica capillary (Ø 50 μm) with an upper length of 15 mm caused premature precipitation in the pipe and plugging problems even at high pressure (135). Furthermore, the distance between the exit of the nozzle and the particle collector must be sufficient; otherwise, particles pile up and do not flow out into the the collector.

Expansion induces a mechanical perturbation traveling into the device at a supersonic speed (129). The velocity of the resulting free jet remains supersonic at the outlet of the nozzle until a limit called mach disk (sonic flow); after that, the flow velocity becomes subsonic (Fig. 3).

In order to understand the mechanisms that control particle production under different conditions, many studies deal with the modelization of particle growth formed by RESS (129,139,152,154–156).

In 1993, Kwauk and Debenedetti (152) presented a mathematical model of nucleation and particle growth during the partial expansion of a dilute supercritical solution in a subsonic converging nozzle. The decrease in the solubility upon isobaric heating (which lead to a decrease of the CO_2 density) underlies the two most important tendencies predicted by the calculations: an increase in particle size upon increasing the preexpansion temperature, and decrease of the particle size upon increasing the extraction temperature and pressure.

By modeling particle formation and the flow field between the capillary inlet and the supersonic free jet, Helfgen et al. (154) and more recently Helfgen et al. (129) showed that particle nucleation mainly occurred in the supersonic free jet, with particles of a few nanometers in diameter at the mach disk (Fig. 3). Therefore, particle growth occurs mainly in the expansion chamber, and the conditions inside it are also important factors that influence particle characteristics.

Advances and Patents. The RESS technique was initially used for the micronization of drugs into fine particles. Then, a few studies revealed that that this process could also be used to obtain composite microparticles by playing on the solubility of wall and core materials. On the other hand, solubility can be modulated by adding an organic cosolvent to the SF, acting as a nonsolvent of the coating material when it is used alone (157). The resulting particles are then protected against agglomeration when suspended in this nonsolvent medium.

Thanks to the rapid elimination of CO_2 during the expansion, the modified techniques of RESS have been improved for optimizing coating operations or for the micronization of hydrophobic compounds. The main advances are presented in Table 3 and discussed thereafter.

Coprecipitation. When the RESS process is performed in the coprecipitation mode, the different components forming the particles can be extracted from a unique, or from several, extraction units (113). In the first case, particle composition is imposed by the equilibrium solubilities of the components during extraction. The second configuration allows for managing the relative composition of the particles, by separately controlling the solute concentration in the $ScCO_2$ prior to expansion. Furthermore, this configuration allows the application of a lower extraction temperature for drugs than for the encapsulating material, depending on their relative solubilities.

Table 3 Particle Production by the RESS-Based Process

Type of process	Objectives	Examples	References
Coprecipitation	Obtain composite microparticles with a narrow size distribution and a controlled drug content	Particles of various L-PLA composition and structure	(136)
		Microparticles of D,L-PLA loaded with lovastatin (heterogeneous product)	(113)
		Microparticles of L-PLA loaded with pyrene (homogeneous)	(134)
Coating	↑ Uniformity of the coating ↓ Particle agglomeration during the coating process	Microspheres of L-PLA (10–90 μm) loaded with naproxen (homogeneous)	(135)
		HYAFF® white microspheres (ø < 10 μm)	(131)
		Particles coated with paraffin	(132,133)
		SiO_2 particles surrounded by La_2O_2 (supercritical water)	(158)
RESS-N	↓ Particle agglomeration	Particles of Eudragit® E-100 and PEG 6000 loaded with 3-hydroxyflavone	(159,160)
		Particles of PEG, PMMA, L-PLA, DL-PLGA, PPG–PEG loaded with protein (lysozyme or lipase)	(149)
RESAS	Steric stabilization of hydrophobic compounds suspended in aqueous media	Fenofibrate, cyclosporine Lipoid® E-80, Tween® 80, phospholipids	(161,162)

Abbreviations: L-PLA, poly(L-lactide) microparticles; D,L-PLA, poly(D,L-lactide); PEG, polyethylene glycol; PMMA, polyacrylonitrile; PPG-PEG, polypropelene glycol-polyethylene glycol; RESAS, RESS into aqueous solution of supercritical solution; RESS-N, RESS-nonsolvent.

Biodegradable polymers such as esterified hyaluronic acid called HYAFF® or poly(α-hydroxy-acids) (134–136) were mostly studied. Tom and Debenedetti (136) were the first to study the RESS process by the coprecipitation of various poly(α-hydroxy-acids) (L-PLA, D,L-PLA, and PGA), and raised the problem of polymer polydispersity, which induces a variation of their solubility in SF during the process (131). They succeeded in producing only L-PLA particles with morphologies and compositions that varied during the process due to the L-PLA polydispersity.

In 1993, Debenedetti et al. (113) obtained composite microparticles of lovastatin and D,L-PLA. They used two experimental conditions: the first condition was made with one extraction unit charged with the two components. The second, with two extraction units, each overcharged with one of them. The former study showed a heterogeneous population of microparticles consisting of microspheres containing lovastatin needles, empty microspheres and needles without any polymer coating. The latter showed larger microspheres containing several needles, with an encapsulation yield of 27 to −36 wt% depending on the relative amounts of drugs and polymers initially dissolved in $ScCO_2$ solution.

The morphology of the final products obtained according to these two conditions indicated an independent and an unsatisfactory precipitation of the polymer and the drug. Ideal microspheres for controlled drug delivery systems would be microspheres containing many small, uniformly dispersed drug particles, rather than needles. This requires that the polymer precipitate before the drug particles are allowed to grow into large crystals.

In another study, Tom et al. (134) encapsulated a model solute (pyrene) into L-PLA by dissolving the polymer into a mixture of $ScCO_2$ and $CHClF_2$. This study yielded different morphologies and composition profiles compared to the previous study: confocal fluorescence microscopy showed a uniform distribution of pyrene within L-PLA particles indicating good mixing and a good coprecipitation. In this study, the two components were dissolved separately and the supercritical solutions were not mixed until they reached the precipitation unit. This configuration allows for the control of the relative composition of the particles.

A more recent study produced L-PLA microspheres (10–90 μm) loaded with naproxen from a sole mixture of naproxen/L-PLA dissolved in $ScCO_2$ (135). In this configuration, particle composition was set by the relative solubility of the polymers and of the drugs during extraction.

Coating. *Fluidized Bed Coating.* Conventional coating processes in fluidized beds are carried out by atomizing droplets of coating material solution through a nozzle into a hot gas fluidized particle bed. After the atomized droplets of coating material came into contact with the core particles, the solvent has to evaporate quickly; hence an excess of droplets in the bed would induce agglomeration. Therefore, suitable droplet atomizing and drying conditions are essential to avoid the unplanned agglomeration of solid particles. However, uncontrollable agglomeration often takes place especially for fine particles ($ø < 70$ μm) because of their strong cohesive force (163). Therefore, it is difficult to coat fine particles by using conventional fluidized-bed coating processes.

$ScCO_2$ has higher thermal diffusivity than liquids and the solvent strength can be modulated by changing the density by depressurization. The advantages of this process over conventional coating methods are that (i) a coating of cohesive particles is possible without the formation of agglomerates because of the absence of liquid droplets; (ii) expansion of the SF permits drying at relatively low temperatures;

(iii) uniform thin films of the solute can be deposited on the surface of core particles due to fine tuning of solubility.

Recently, such a RESS process was patented by Krause et al. (164) for the production of fine particles (of diameter less than 100 μm) with smooth and regular coatings. It was also employed by Tsutsumi et al. (132,133) who coated microparticles (catalyst particles ø 55 μm, SiO_2, ø 1 μm, and glass beads ø 130 μm) by the rapid expansion of $ScCO_2$ solutions of paraffin in a circulating fluidized bed with an internal nozzle at the center of the riser. Stable and uniform particle coating was produced, with no significant agglomeration during the process because the coating material was deposited directly on the surface of the core particles without the presence of liquid droplets, which act as binders for particles. Therefore, the fluidized-bed coating process by RESS allows a fine free solvent coating to be obtained, provided the coating materials must be soluble in SF.

Batch. The batch mode is not exactly a RESS process because expansion does not occur through a nozzle. Kitagawa K. (158) patented this new concept where he described the formation of SiO_2 particles surrounded by La_2O_2, using supercritical water. Primary microparticles are formed by the rapid decompression of a supercritical solution of the core material. These particles are mixed with a second supercritical solution of the coating agent and are then depressurized to form microparticles. Coating with multiple layers can be achieved using this process. Applications of this process are limited, as the core material and the coating agent have to be soluble in the SF throughout the whole process.

RESS-NonSolvent. In recent studies, polymer particles were produced by RESS in the presence of a cosolvent, which increases polymer solubility in the SF and becomes a nonsolvent after expansion (148,149,157,159,160). Therefore, the nonsolvent prevents the agglomeration of polymeric particles. Among the organic cosolvents studied, ethanol is acceptable environmentally, but implies the use of special equipment to prevent detonation.

This modified method was first patented by Mishima (157) for the formation of polymeric microcapsules by making a suspension of active molecules in a supercritical solution of polymer. This was then applied for the encapsulation of various compounds (Table 3). The thickness of the polymer coating around the core material, as well as the mean particle diameter and particle size distribution, was controlled by changing the feed composition.

RESAS. In order to overcome problems of hydrophobic particle agglomeration via van der Walls interactions, when RESS is performed in air, the rapid expansion of a supercritical solution containing the drug into an aqueous solution containing a surfactant such as Tween® 80 or phospholipids [RESS into Aqueous Solution (RESAS)] was patented in 1997 (165). Tension modifiers can sometimes be added to the supercritical solution. Steric stabilization by a surfactant impedes particle growth and agglomeration. This process therefore allows aqueous suspensions of water-insoluble drugs to be stabilized and leads to a narrow size distribution. This process has been applied with various coating and drug materials. The particles were of an order of magnitude smaller than those produced by RESS into air without the presence of a surfactant aqueous solution.

Other. In 1998, Bausch et al. (166) claimed to have developed a technique where a drug was dissolved under elevated pressure in $ScCO_2$, containing a surfactant, and then rapidly expanded. They produced particles of tetrahydrolipstatin surrounded by Brij®96 with an average size of 1.5 μm.

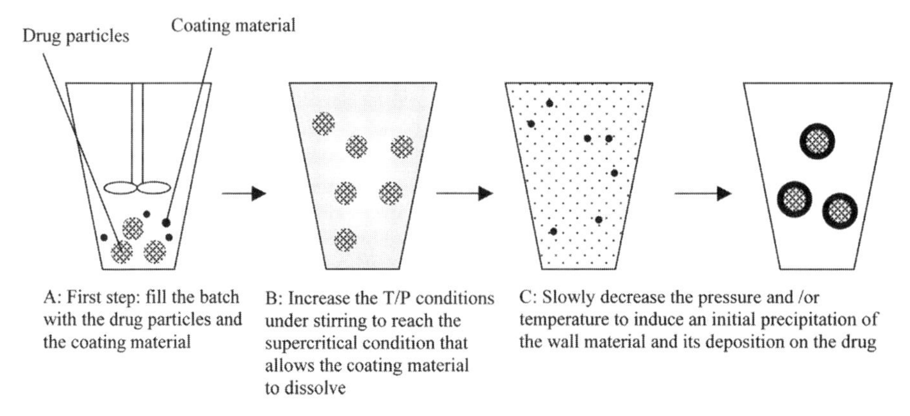

Figure 5 Scheme of the process of controlled coating. *Source*: From Refs. 167–169.

Controlled Coating. A new process, developed by Benoit et al. (167–169), consists of dissolving coating materials in $ScCO_2$ and then slowly adjusting the time/pressure (T/P) conditions in the autoclave to precipitate it onto the surface of suspended particles (Fig. 5).

This process has been applied to coat bovine serum albumin (BSA) and sugar granules, with trimyristin (Dynasan® 114) or Gelucire® 50–02, a mixture of glycerides and glyceride esters of PEG. Different coating morphologies were obtained depending upon the material used. A discontinuous, micro-needle coating resulted when trimyristin was used. An initial burst release of approximately 70% in 30 minutes was observed.

At the opposite end, a smooth regular coating was obtained with Gelucire 50–02. This material induced a prolonged release of BSA in a phosphate buffer solution at 37°C over a 24-hour period. This process allowed good control of the particle structure (composition and thickness). Furthermore, it was shown that BSA did not undergo any degradation after $ScCO_2$ treatment, indicating that this process could be useful for the encapsulation of proteins.

Impregnation. Substrates soluble in $ScCO_2$ can easily be impregnated inside porous systems to prepare controlled drug delivery systems as demonstrated by Domingo et al. (170). The impregnation process includes three major steps: (i) the polymeric particles are exposed to $ScCO_2$ in order to swell the polymer and increase the particle porosity; (ii) $ScCO_2$ containing solutes is introduced and it leads to the subsequent solute transfer from $ScCO_2$ to the polymer phase, and (iii) the CO_2 is released in a controlled way that results in the entrapment of the solute in the polymer. When exposed to $ScCO_2$, polymers will exhibit various extents of swelling and their chain mobility will be enhanced, which significantly facilitates and accelerates the transport of components. The swelling ability of the $ScCO_2$ controls the impregnation process; this depends upon the nature of the polymers. Therefore, highly crystalline polymers that do not swell in the presence of $ScCO_2$ are not suitable for the impregnation process.

In the work carried out by Domingo et al. (170), an amorphous inorganic mesopore and a polymeric matrix (amberlite) were impregnated with various thermolabile organic compounds (benzoic acid, salicylic acid, aspirin, triflusal, and ketoprofen) by diffusion from saturated $ScCO_2$.

Another impregnation microencapsulation technique has been developed by Liu and Yates (171) in which $ScCO_2$ is used to facilitate the transport of a dye into polymer particles. Aqueous latexes (polystyrene particles) were impregnated with a dye by emulsifying CO_2 and water in presence of Pluronic® F108. Partitioning of the dye between the aqueous polymeric phase and the medium surrounding the polymer particles is the driving force for the dye to enter the polymer phase. Consequently, the partition coefficient is the most important factor that determines the maximum dye loading at the thermodynamic equilibrium. However, increasing the solubility of the dye in the medium improves its loading kinetics but results in low entrapment levels. Dye loading is improved by increasing the interfacial area, either by increasing surfactant concentration or by lowering particle size.

The impregnation of biodegradable polymers such as DL-PLGA and their derivatives has been studied extensively. Because of their relatively low T_g temperature, it appears difficult to impregnate PLGA without obvious deformation or foaming. Particles containing 5-fluorouracil or β-estradiol encapsulated into PLGA appeared like foam with large pores or led to porous particles when depressurization conditions were controlled (172).

This technique of microencapsulation is promising because it is performed at room temperature, without organic solvents and with a precise control of the particle size owing to the separation of the process of particle creation and the process of encapsulation.

Techniques Using Supercritical Fluid Antisolvent (SAS)

Liquid antisolvent processes (coacervation) are largely used in industry. Supercritical fluid antisolvent (SAS) processes have been proposed as alternatives to these processes to limiting the residual organic solvent. In addition, it is suitable for processing nonsoluble components in $ScCO_2$ for which RESS is impossible, or for encapsulating materials sensitive to the high shear stresses caused by RESS.

These processes are used with components with a relatively low solubility in $ScCO_2$ but which are soluble in a conventional organic solvent or in a mixture of organic–aqueous phases. At the optimum in the (solvent–CO_2–solute) ternary phase diagram, the $ScCO_2$ is highly soluble in the solvent; therefore, evaporation and volume expansion occur when the two fluids come into contact, leading to a decrease of solvent density as well as a decrease in solvent capacity. Consequently, the components become nonsoluble in the mixture of the initial solvent and $ScCO_2$. This leads to high levels of supersaturation, solute nucleation, and particle formation.

These processes have some potential advantages when compared to coacervation and other conventional microencapsulation techniques. Firstly, the obtention of dry particles can be ideally carried out in one step by extracting the solvent with $ScCO_2$, and then by completely removing the $ScCO_2$ just by pressure reduction. Removing the liquid antisolvents in usual coacervation techniques requires more complex postprocessing treatments.

Secondly, $ScCO_2$ diffusivities can be up to two orders of magnitude higher than those of liquids. Therefore, considerable mass transfer between the organic solution and $ScCO_2$ occurs. This produces a fast supersaturation of the solute and its precipitation in micronized particles whose sizes are not possible to reach using conventional liquid antisolvent techniques. Moreover, because of its low viscosity, its high density relative to the air, and its low interfacial tension, $ScCO_2$ is able to generate smaller droplets than the traditional spray-drying process.

However, a constraint of the SAS processes is that the components must be soluble in one or more organic solvents, which are miscible with the SF at operating conditions, and must be insoluble (or very slightly soluble) in the mixture of both SF and the initial solvent(s). This condition is successfully met in most of the studies, due to the low solubility of the components in $ScCO_2$, but an extraction of the drug can occur, particularly for nonpolar drugs (173), or for polymers such as PLGA or PCL with low M_w, which can be dissolved in the organic solvent–$ScCO_2$ mixture (173–175). The solubility of the components in the mixture is controlled by $ScCO_2$ density and by the T/P conditions in the apparatus.

With the aim of good manufacturing practice (GMP) perspective, several additional attractive features can be recognized for SAS particle formation processes. The equipment for the single step, totally enclosed, free of moving parts for spray processes, is constructed from high-grade stainless steel with 'clean in place' facilities available for larger scale equipment. In addition to the low quantity of solvent required compared with conventional coacervation procedures, particle formation occurs in a light-free and oxygen-free environment and, if necessary, a moisture-free atmosphere. Although further engineering input is necessary to achieve true continuous collection and recovery of material at operating pressures, 'quasi-continuous' processing can be already feasible by switching to a device mounted in parallel with the particle-collection vessels (176,177).

Batch Operation. *Description.* These techniques are frequently called as gas antisolvent precipitation (GAS). Different configurations can be used with this work mode. In the classical mode, the precipitation vessel is loaded with a given quantity of the liquid solution and is then expanded several fold by adding the $ScCO_2$ at the desired final T/P conditions. In most T/P conditions, the $ScCO_2$ density is lower than the liquid solution. Thus, for better mixing, the $ScCO_2$ is added from the bottom of the precipitator (Fig. 6).

Because the CO_2-expanded solvent has a lower solvent strength than the pure solvent, the mixture becomes supersaturated, forcing the solute(s) to precipitate as microparticles. Once the solute(s) has been precipitated, the expanded solution is drained under isobaric conditions, and then additional $ScCO_2$ is added to remove the residual solvent and to produce dry particles upon depressurization (131,177–179).

It is also possible to charge the precipitation chamber with the antisolvent and then to perform a slow and discontinuous addition of the liquid solution (116). The difference between these two operational modes is that, in the first case, precipitation occurs in the liquid phase, whereas in the second case, it occurs in the supercritical phase. The results can vary significantly. However, in both cases, the operation is not

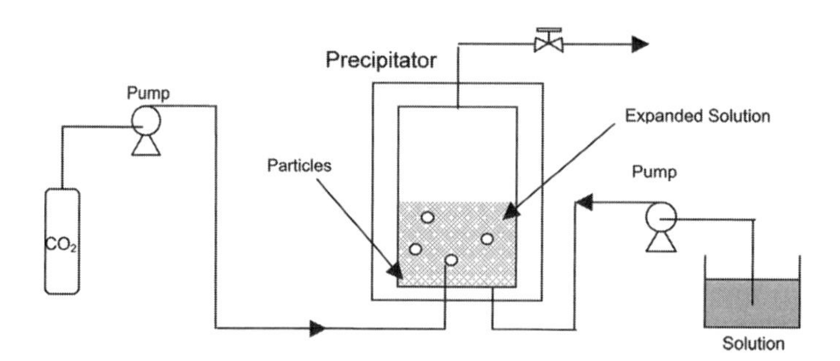

Figure 6 Diagram of the GAS process. *Abbreviation*: GAS, gas antisolvent precipitation.

performed at a steady state; it is therefore difficult to analyze the effect of the process parameters on the final characteristics of the powders.

Influencing Parameters. Particle formation from the solute–solvent–antisolvent system is governed by two main physical influencing parameters such as the volumetric expansion of the liquid solvent and two mass transfers: diffusion of the SF inside the solvent and evaporation of the solvent into the SF.

Volumetric Expansion: A prerequisite for success in the antisolvent process is the miscibility of the liquid solvent and $ScCO_2$. The increase of CO_2 solubility in the liquid as the pressure increases (at a fixed temperature) is a rule expressed in terms of the volumetric expansion of the liquid phase (113). Therefore, the measurement of the solvent volume expansion in function of the $ScCO_2$ pressure allows for determinating, experimentally, the best conditions to perform a SAS experiment. This relative volume expansion is classically defined by the following equation (180):

$$\frac{\Delta V}{V} = \frac{V(T, P) - V_0(T, P_o)}{V_0(T, P_0)} \tag{1}$$

where $V(T, P)$ is the total volume of the liquid phase at the T, P conditions and V_0 is the total volume of the pure solvent at the same temperature and at a reference pressure (normally atmospheric pressure).

This effect is illustrated here by Kordikowski et al. (180) who show that CO_2 produces a remarkable volumetric expansion of dimethylsulfoxide (DMSO) near the critical point of the DMSO–CO_2 mixture. Reverchon et al. (175) showed that the volumetric expansion of DMSO was similar for the ternary system DMSO–dextran–CO_2. Consequently, dextran does not modify the solubility behavior of the binary system, and dextran precipitation leads to a two-phase system composed of a solid solute phase and a solvent–antisolvent fluid phase. Therefore, from a thermodynamic point of view, this ternary system shows pseudobinary behavior: the complete miscibility of both the solvent and antisolvent is maintained.

Other phase behaviors like the partial solubilization of solute and the formation of two or more liquid and fluid phases in equilibrium could occur. For instance with inulin, a low M_w polymer (5000 Da), initially solubilized in DMSO, three phases were obtained with the appearance of two phases containing inulin (175). Indeed, the formulation of a complex vapor–liquid equilibrium is one of the reasons that the use of SAS process is made difficult (175).

In order to distinguish the differences of the liquid-phase expansivities between various solvents expanded by the same antisolvent, de la Fuente Badilla et al. (181) proposed a new definition for the relative volume expansion of a solvent–solute solution caused by an antisolvent [Eq. (2)]:

$$\frac{\Delta V}{V} = \frac{\tilde{V}(T, P) - \tilde{V}_0(T, P_o)}{\tilde{V}_0(T, P_0)} \tag{2}$$

This equation only considers molar volumes of the mixed liquid phase $\tilde{V}$ and of the pure solvent $\tilde{V}_0$. According to the volumetric expansion defined by Eq. (2), the liquid phase does not expand for all antisolvent pressures, and volume contraction often occurs at low pressure.

Furthermore, for some solvent–antisolvent systems, volume contraction can change into volume expansion when pressure increases. So, different behaviors of various solvents which are expanded by the same antisolvent can be obtained.

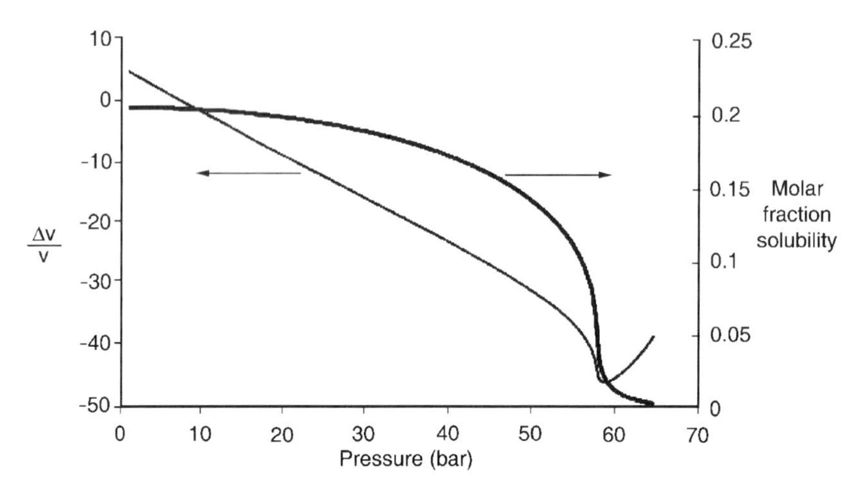

Figure 7 Representation of the volume expansion as defined by Eq. 2, and the simultaneous molar fraction solubility of a drug. *Source*: From Ref. 181.

From Figure 7, it appears that the optimal operational pressure for the GAS process has to be chosen in such a way that the relative expansion of the liquid phase occurs at a minimum value (181,182). In addition, for different solvents, the best solvent–antisolvent combination is achieved when the minimum volume expansion curve occurs at the comparatively low pressures. Studies carried out at various temperatures, indicated that the best temperatures were close to, but below, the critical temperature of the antisolvent (182).

On the other hand, Elvassore et al. (179) found that precipitation of a polymer, owing to expansion of the solvent, occurred at similar expanded solvent densities for different polymer–solvent systems. Moreover, the density values of the expanded liquid have been found to be independent of the temperature. This indicates that solvent density controls polymer solubility.

Mass Transfer During Gas Antisolvent. The phase behavior and the relevant mass transfer pathways for microparticle formation can be represented in a ternary phase diagram of the polymer, solvent, and CO_2 system at given T/P conditions (Fig. 8).

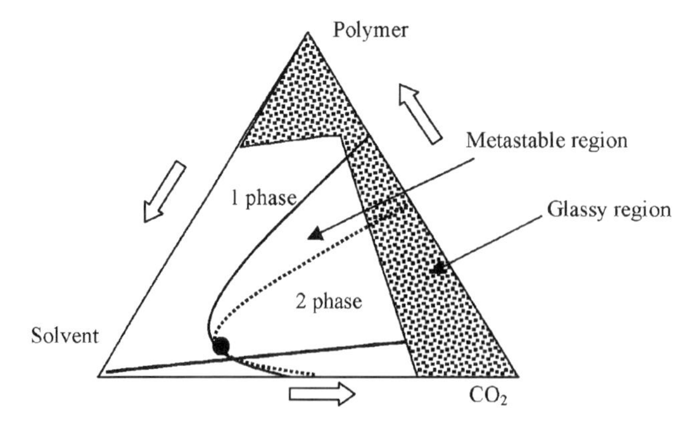

Figure 8 Schematic (polymers–solvent–CO_2) ternary phase diagram showing the mass transfer pathway for microparticle formation. (–) Binodal line, (...) spinodal line, (•) plait point.

The phase diagram is bounded by three regions (183,184); the one-phase region is located between the polymer–solvent axis and the binodal line, the metastable region is located between the binodal and spinodal lines, and the unstable or two-phase region is located between the spinodal line and the polymer–CO_2 axis. The intersection of the binodal and spinodal curves occurs at the plait or critical point, which corresponds to the polymer concentration at which the polymer chain coils begin to be packed and start to interact together (181). For mass transfer pathways starting from the one-phase region of the phase diagram, polymer-rich domains nucleate and grow in the metastable region. After the mass transfer pathway crosses the spinodal line, phase separation proceeds by spinodal decomposition. Below the plait point (dilute solution), there is little opportunity for polymer chains to overlap and spinodal decomposition results in a discontinuous polymer network. The polymer-rich domains solidify and give particles at the T_g or T_m temperature, which can be decreased by the presence of solvent and CO_2. Above the plait point (for high concentrations), the entanglement of the polymer chains prevents the formation of discrete particles. This type of precipitation pathway often results in the formation of a continuous network such as a porous membrane or fibres (123,179). We can therefore see that the solute concentration strongly influences the SAS process; it controls product morphology as well as the fluid pressure necessary to induce precipitation (123,175,179,185). Precipitation starts at lower pressures when the solute concentration is increased, because a lower content of CO_2 inside the liquid solution is required to reach saturation concentration.

On the other hand, the mass transfer of solvent molecules in $ScCO_2$ is governed by the diffusion coefficient (D), which decreases when the CO_2 density decreases (Table 1). In the GAS process, the size and the morphology of the particles are determined by the solution supersaturation rate and then by the volume expansion rate. The latter depends on two phenomena—the rate of CO_2 addition and the $CO_2 \leftrightarrow$ liquid mass-transfer rate (186). Particle size decreases when both the CO_2 addition rate and mass transfer increase. After the addition of $ScCO_2$, a mass transfer occurs as follows: the solvent evaporates from the bulk liquid phase at the same time as CO_2 is absorbed in the bulk liquid phase.

During a typical GAS experiment, the quantity of solute always remains significantly smaller than the combined amounts of solvent and antisolvent, so we can see that the solute has negligible effects on mass transfer (186). An influencing parameter is the stirring rate, which favors CO_2 transfer in the liquid phase when it increases. Above a given rate, this effect becomes nonsignificant. Vessel geometry and solvent physical properties, such as viscosity and polarity, must also be taken into consideration.

Advantages and Disadvantages. At the end of the precipitation process, particles must be washed with $ScCO_2$ to strip the liquid solvent. In batch mode, the particles are formed in the liquid phase, so their drying is performed by flushing them with pure $ScCO_2$ for a long period of time (at least two hours) (151). If this step is not well performed, the solvent is released from the supercritical solution during the depressurization step and may resolubilize the solute (117).

Furthermore, the batch process does not allow for the treatment of samples at specified pressure values, since they undergo all the pressures from atmospheric to the final one (179). For this reason, the batch mode is in many cases unsuitable for industrial production.

These aspects, coupled with problems associated with heat generation during the addition of $ScCO_2$ to organic solutions, led to process modifications to improve both the process and the product control and to achieve a semicontinuous or continuous operation, involving the spraying of an organic solution in $ScCO_2$. However, compared to continuous processes, the batch process ensured the total recovery of the polymer placed into the precipitator, with a simple experimental setup (131).

Advances and Patents. *Coating.* The feasibility of coating a solute dispersed in a solvent, using the GAS process has been demonstrated with several drugs (178,187,188). Benoit et al. (178) patented the production of microcapsules by means of a suspension of active substance dispersed in an organic solution of a slightly polar polymer. This suspension is put into contact with $ScCO_2$ that solubilizes the organic solvent in such a manner that there is coacervation of the polymer coating onto the solid particles.

Microparticles of insulin coated with DL-PLGA, with an average size of 3 μm and a drug loading greater than 80%, have also been patented by Richard et al. (188). Coprecipitation: Pallado et al. (189) patented in 1996, the preparation of HYAFF-11 particles. Three active molecules [calcitonin, insulin, and granulocyte macrophage colony stimulating factor (GM-CSF)] were embedded in HYAFF microspheres. Then in 1997, Benedetti et al. (131) compared the formation of HYAFF-11 particles by the SAS process in batch mode, with the RESS and continuous SAS process. SEM observation of the microspheres produced by the batch process showed the production of homogeneous particles with an average size of 0.4 μm. From the three types of process, the authors concluded that the batch mode produced the best particles, and that it had the easiest experimental preparation. They found that HYAFF-11 concentration markedly affected the nucleation step. A dilute solution (0.5 wt%) produced aggregated particles, whereas a solution with concentration of 1% led to microspheres with a narrow size distribution. Greater concentrations led to the formation of a polymer network because of the fact that a concentration above the plait concentration was reached (Fig. 8) (179). From this study, Elvassore et al. (179) used the GAS process to produce particles of different types of HYAFF polymer (HYAFF-11 with various proportions of benzyl esterification, and HYAFF-302), solubilized in DMSO at a concentration lower than the plait concentration value, i.e., 1% w/w (179). Particles obtained from the partially esterified HYAFF polymers were bigger (20–40 μm) than those obtained from complete esterified HYAFF 11 (0.4 μm). This effect was explained by the presence of free OH groups on the partially esterified polymer chain that could influence the intermolecular polymer interactions during the precipitation process. Furthermore, stirring would reduce particle agglomeration and size heterogeneity. This could be due to the enhancement of the CO_2 mass transfer in DMSO as has been explained with toluene by Lin et al. (186).

In 1997, Chou and Tomasko (190) studied GAS coprecipitation of naproxen with L-PLA. They produced very small spherical particles made of a naproxen core surrounded by a polymer shell.

When incorporating insulin into L-PLA, Elvassore et al. (191) reported that better results could be obtained by using a mixture of DCM and DMSO. Furthermore, the residual DCM content in particles was only 8 ± 2 ppm, a value significantly lower than the limit value imposed by the USP (112) and the ICH guidelines (600 ppm) (111). DMSO, which is difficult to remove during classical pharmaceutical processes because of its high boiling point, was also extensively eliminated in a washing step with $ScCO_2$ (residual content 300 ± 10 ppm). To modify the low biodegradability and high hydrophobicity of L-PLA, they incorporated various PEG with M_w ranging from 350 to 20,000 Da or one type of PEG (i.e., PEG 6000) in different proportions.

Although only PEG with low M_w could be efficiently entrapped, its role on the coprecipitation of polymers and peptides and on release behavior was demonstrated. Low PEG induced a slow and constant release over 62.5 hours, whereas a burst effect was obtained with higher M_w, while only low insulin amounts were released in the absence of PEG molecules. PEG has also been used to coprecipitate carbamazapine, but without obtaining any microparticles (192,193).

Continuous Operation (Spray Processes). In semicontinuous or continuous antisolvent techniques, a carrier and an active substance dissolved in an organic solvent are sprayed together or separately in $ScCO_2$ that acts as an antisolvent. The antisolvent expands and dissolves the solvent(s) that causes the microparticle formation. Continuous modes involve several processes that have few differences between them, and are named by diverse acronyms. The different processes are known as aerosol solvent extraction system (ASES), precipitation with a compressed fluid antisolvent (PCA), and solution enhanced dispersion by supercritical fluids (SEDS).

Although the differences between the various spraying processes seem trivial, there is growing evidence that they operate in different hydrodynamic regimes and thus provide different results.

After describing the principle of these processes, we will focus on the main influencing parameters and will present the relevant applications developed over the last few years.

Description. In the ASES technique, solids are dissolved in conventional solvent(s) and the solution(s) is sprayed continuously through an atomization device into a flowing or static supercritical fluid. The dispersion of the solution in the supercritical fluid leads to an expansion of the solvent droplets and, at the same time, extraction of the solvent into the supercritical fluid takes place. The solvent power of conventional solvents decreases dramatically and supersaturation leads to the formation of small and uniform particles.

The PCA and SEDS processes are similar to ASES in principle. For PCA, a liquid solution of a polymer is delivered via an injection device into the antisolvent that is in liquid subcritical or supercritical state. For SEDS, only the injection device which is claimed to be designed to enhance the spray distinguishes it from ASES.

A typical apparatus is shown in Figure 9. The SF is pumped into the precipitator through an isolated or coaxial injection device, as is indicated by the dotted line

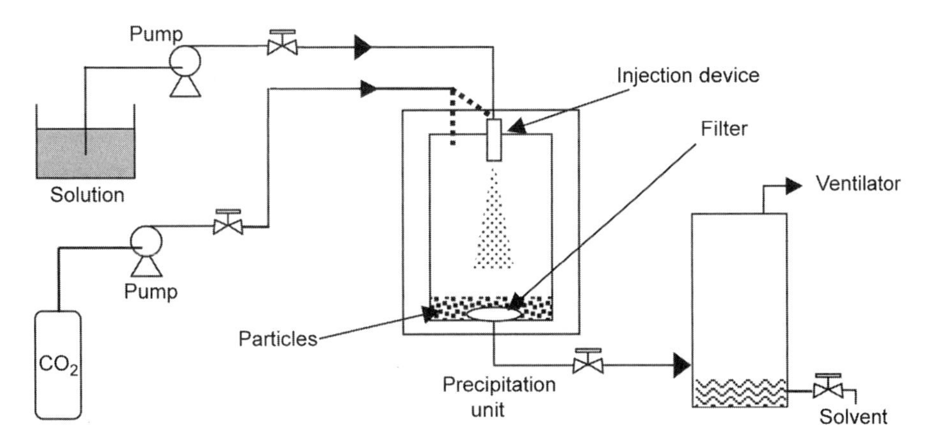

Figure 9 Continuous supercritical antisolvent apparatus.

in Figure 9. Thereafter, when the system reaches T/P steady state, the organic solution is introduced into the precipitation unit through the same coaxial device or via another isolated injection system. The carrier and active components can be introduced within the same solution or sprayed separately.

To produce small liquid droplets in the injection device, the liquid solution is typically pumped at a pressure that is 20 bar higher than the precipitator operating pressure. Particles are typically collected in a filter at the bottom of the vessel. The fluid mixture (SF and solvent) exits the vessel and flows to a depressurization tank where the T/P conditions lead to gas–liquid separation. After the particles are formed, the pumping of the liquid solution is stopped and pure supercritical fluid is allowed to flow continuously through the vessel for a long time so as to strip any residual solvent from the particles.

Advantages and Disadvantages. In comparison to batch mode techniques, these spraying processes lead to a maximum exposure of small amounts of organic solution to large quantities of $ScCO_2$, which increases the $CO_2 \leftrightarrow$ solvent mass transfer, leading to faster volume expansion and solvent dissolution; hence, these methods have a greater production rate and a lower drying time in comparison to the batch process. Furthermore, the enhanced mass transport and the hydrodynamic mixing provide additional variables that may be used to control particle size, size distribution, and morphology.

However, since these processes use organic solvents, the washing step with pure $ScCO_2$ at the end of the process remains crucial to remove the residual solvent for continuous operation. This washing step should also prevent condensation of the liquid phase that otherwise rains down on the particles, thereby modifying their characteristics.

Influencing Parameters. To make the spray processes commercially viable, a method must be found whereby there will be a continuous production of particles with the desired product characteristics and consistency. In particular, the continuous harvesting of particles at high yields remains a challenge, especially with submicron particles. The characteristics of the particles produced by the spray processes are not only influenced by mass transfer and thermodynamic factors as is the case for the batch mode, but are also influenced by hydrodynamic and jet break-up phenomena. Due to this, the process is extremely complex. Indeed, before successful particle formation can be achieved, it is necessary to understand the mechanisms that control particle size in the near-critical and supercritical regions.

Hydrodynamics. The physical properties of $ScCO_2$ including density, viscosity, and interfacial tension, as demonstrated by Tewes and Boury (120), are strongly dependent on the temperature and the pressure near the critical point. A small change in temperature and pressure near the critical point affects the velocity of CO_2 in the nozzle, the interfacial tension in the precipitation unit, and consequently, the droplet size; however, there is hardly any modification of the mass flow rate (diffusion coefficient) (121).

To characterize spray efficiency, hydrodynamic theory can be applied in the form of a Weber number (121,183). The Weber number (N_{We}) is the ratio of the deforming external forces on the reforming surface forces exerted by a liquid droplet. It is numerically defined by the following equation:

$$N_{We} = \frac{\rho_{CO_2} U^2 D}{\gamma} \tag{3}$$

where ρ_{CO2} is the antisolvent density (kg/m^3), U is the velocity difference ($U_{CO2} - U_{solvent}$ m/s), D the initial droplet diameter (m), and γ the interfacial tension (N/m). Therefore, when the external pressure forces are larger than the surface tension forces, N_{we} is high and jet break-up is enhanced. In the $ScCO_2$ process, N_{we} is significantly larger than in usual spraying processes owing to a low interfacial tension and a high density of CO_2 compared to that of air (183).

Several authors found a decrease in particle size when CO_2 density is increased. This shows that hydrodynamic factors, i.e., jet break-up, can control particle size, which is in accordance with the N_{we} theory [Eq. (3)] (137,183,194).

Mass Transfer. In the continuous SAS procedure, the mass transfer of CO_2 into polymer solutions and conversely solvent transfer into $ScCO_2$ are much greater than for conventional liquid antisolvents, so that the phase separation is extremely fast at less than 10^{-5}s. These diffusion rates are at least two orders of magnitude faster than is the case in conventional liquid antisolvents, and approach those of RESS (183,195). Consequently, SAS spraying processes may be used to produce particles exhibiting high surface area, which can be quenched in a glassy or crystalline state before the surface area relaxes (123).

In the model of ternary-phase behavior for the antisolvent–solvent–polymer system (Fig. 8), there are different possible mass transfer pathways. The mass transfer between the droplet and the $ScCO_2$ phases occurs between the two limit pathways: (i) solvent evaporation with little CO_2 penetration into the droplet phase, (ii) and swelling of the droplet phase due to CO_2 penetration with no solvent evaporation. Therefore, final product morphology created by the continuous SAS process can be controlled in order to obtain microspheres, interconnected bicontinuous networks, or fibers (123).

Several authors have found that according to the operating conditions in which the microspheres are obtained, there was no major influence of the ASES processing conditions upon L-PLA microparticle size. In the same way, Tu et al. (184) found that the change of nozzle diameter had no influence on particle size. Furthermore, increasing CO_2 density increased the size of particles produced by Randolph et al. (197). The latter observation is contrary to the N_{we} [Eq. (3)] and hydrodynamic rules, and could indicate that interphase mass transfer rates rather than initial droplet size, control particle size (174,175,196,197).

In order to investigate the effect of the initial sprayed droplet size on the final particle size, Rantakylä et al. (121) compared the size of L-PLA particles with a theoretical size, assuming that one microparticle was produced from one droplet. According to droplet formation models, they found that the initial droplet sizes do not have an effect on the final particle size.

Consequently, ASES processes using polymeric materials may be considered as a precipitation process whereby particle size mainly depends on the polymer properties, as reflected by several studies carried out with L-PLA and dextran polymers. Therefore, continuous spray processes in the $ScCO_2$ seem to be mainly controlled by mass transfer phenomena (174,175,196,197).

Considerations of Whether Hydrodynamic or Mass Transfer Phenomena Predominate in the Continuous SAS Processes. From the previous literature overview, it is difficult to know whether the continuous SAS processes are controlled only by mass transfer or are controlled by hydrodynamic factors. These contradictions probably result from the experimental conditions used (137,174,175,183,196–200).

Firstly, these apparent contradictions could be explained by a variation in temperature (183). Solvent density influences particle size because two competing effects—of changes in atomization (which increases as density increases) and in the solvent diffusion coefficients (which increase as density decreases). At subcritical temperatures, the changes in diffusion coefficients are relatively small and the effect of atomization is dominant. Therefore, when the CO_2 density increases, atomization becomes intense and particle size decreases. At supercritical temperatures, a contrasting behavior, or reduced effects of CO_2 density on particle size, has been observed since the diffusion effect or mass transfer is dominant.

The second point concerns the CO_2 densities at which the experiment is performed, as suggested by Thies and Müller (137). They found that the size of PLA particles remained constant for a CO_2 density higher than $580\,kg/m^3$. Below this CO_2 density, the particle size increases when the density decreases. Such effects indicate that particle formation is linked to a precipitation process influenced by $CO_2 \leftrightarrow$ solvent mass transfer at high CO_2 densities and to a spray process at low CO_2 densities.

Injection Device and Flow Rate. Under specific experimental conditions, a key role in continuous operation is played by the liquid solution injection device. The injector is designed to produce the break-up of the liquid jet and the formation of small micronic droplets that expand in the precipitator.

Various injection devices have been proposed in the literature. A simple nozzle was adopted by many authors (131,137,173,175,179). Other authors used small internal diameter capillaries or a vibrating nozzle (123,197,200–202). The latter apparatus produces a spray by superimposing a high frequency vibration on the liquid jet that exits from an orifice.

Coaxial devices in which two capillary tubes continuously deliver the liquid solution and the supercritical antisolvents have also been proposed (183,203). The formation of small droplets in this case depends upon the turbulent mixing of the two flows. Mawson et al. (183) used a coaxial two-way nozzle (Fig. 10) to produce L-PLA microparticles and compared the results obtained with a standard nozzle.

The velocities of the two coaxial flows can be controlled independently by allowing a convenient circulation of droplets inside the precipitator that prevents particle agglomeration.

For the standard nozzle, the resulting suspension does not fill the entire vessel and the particles do not circulate very rapidly; this can lead to poor mixing and agglomeration.

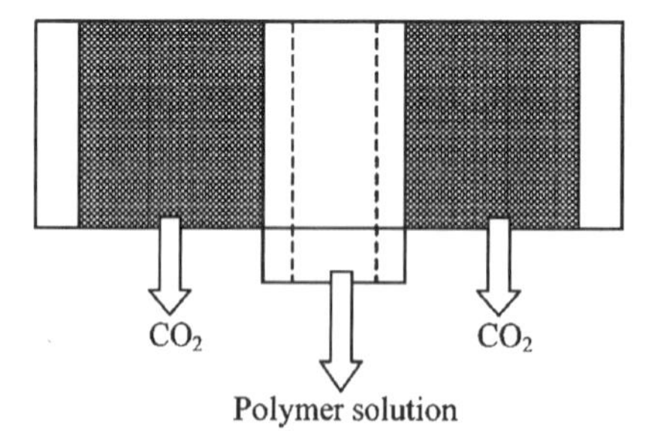

Figure 10 Diagram of a coaxial nozzle end. The polymer solutions are sprayed through a capillary tube, with CO_2 flowing cocurrently in an annular region around the capillary tube. *Source*: From Ref. 183.

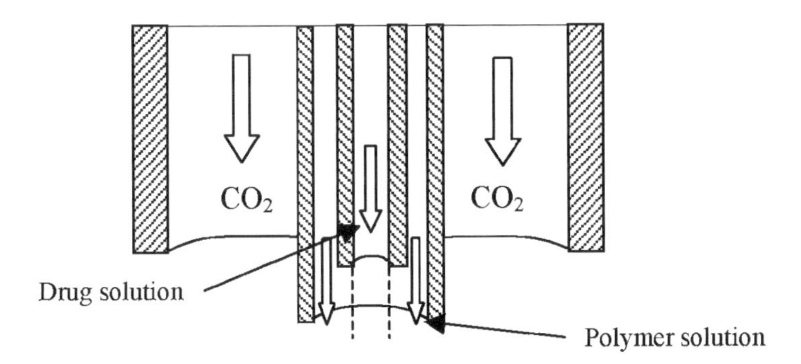

Figure 11 Scheme of a three way nozzle. *Source*: From Refs. 184, 203.

When $ScCO_2$ is not in static mode, i.e., coflowed in co- or countercurrent with the liquid solution using a one or a multiple injection device, the CO_2 flow rate permits to modulate the spray. The CO_2 flow acts on the break-up of the droplet solution. At slow CO_2 rates, atomization forces can be too low to break up the jet of solution, and fibers are obtained. At fast flow rates, the solution breaks up and microparticles are obtained more easily (179,200).

In order to perform encapsulation of the core material within the wall material, Tu et al. (184,203) have designed a coaxial three-way nozzle (Fig. 11) allowing the cointroduction of the drug solution, polymer solutions, and $ScCO_2$.

In all previous cases, SF is mainly used for its antisolvent properties, but some studies also used it for a mechanical effect. Subramaniam et al. (187,204,205) and Foster et al. (206) used a special continuous SAS process in which the nozzle is designed to produce high frequency sonic waves when the solution of coating polymers comes into contact with the SF (Fig. 12).

Simultaneous to spraying, an energizing gas is injected through a passageway, perpendicular to the direction of the solute-loaded solvent flow. This energizing gas stream generates high frequency waves that break up the liquid solvent and generate extremely small droplets, leading to small final particles ranging in size from 0.1 to 10 μm. The energizing gas can be the same as the antisolvent or can be selected from

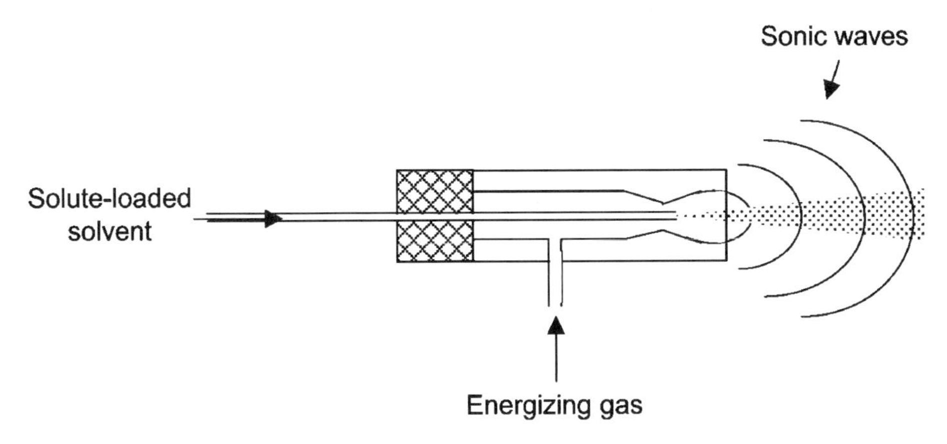

Figure 12 Nozzles producing sonic waves. *Source*: From Refs. 187, 204, 205.

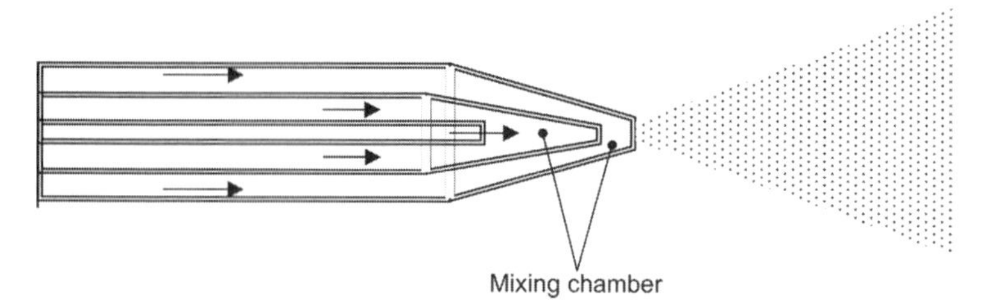

Figure 13 Example of an SEDS three-way nozzle. *Abbreviation*: SEDS, solution enhanced dispersion by supercritical solution.

air, or mixtures of gases oxygen, nitrogen, helium, carbon dioxide, propane, butane, isobutane, trifluoromethane, nitrous oxide, sulfur hexafluoride.

In the ASES and PCA techniques, particle agglomeration is influenced by the rate of solvent mass transfer into the SF from the droplet. This mass transfer is dependent on dispersal and mixing phenomena between the solution droplet and the SF (151). Thus, in order to minimize this agglomeration and to reduce or eliminate drying times, faster mass transfer rates are required. This has been successfully achieved in the SEDS process, developed by Hanna and York (207).

In this technique, the solution is sprayed continuously and cocurrently with the $ScCO_2$ through a coaxial nozzle, with two or three coaxial passages, containing a mixing chamber (Fig. 13). In this configuration, the organic solvent solution interacts and combines with the $ScCO_2$ in the mixing chamber of the nozzle prior to dispersal, and flows into a particle-formation vessel via a restricted orifice. Thus, high mass transfer rates are obtained with a high antisolvent–solvent ratio. The high velocities of the SF facilitate the breaking up of the solution into very small droplets. In this case, the SF is used as an antisolvent and as a "spray enhancer," providing a mechanical effect, similar to that in the process developed by Subramaniam et al. (187).

Spraying Pressure. Assuming that the polymer solution atomizes into droplets at the outlet of the nozzle and also that one microparticle originates from one droplet by "evaporation" of the solvent, particle size should be influenced by the spraying pressure. Thies and Muller (137) raised the pressure difference between the inlet of the nozzle and extraction column to 60 bar. The atomization of an injected solution without pulsation led to a decrease in particle size with increasing spraying pressure (137,184). As the spraying pressure was increased, an increasing amount of energy was brought into the system; this resulted in a decrease in particle size. However, when eddies were created in the outlet region of the nozzle, particle size increased due to the collision of the polymer-solvent droplets in this region.

Pre-Saturation of Spraying Solution. The predissolution of CO_2 in the polymer solution reduces both the viscosity of the liquid and the agglomeration tendency of the product. Particle formation from a CO_2 saturated solution might be faster because a lower amount of CO_2 diffuses into the droplets, leading to faster hardening of the surface and avoidance of agglomeration of the particles (137).

By spraying dilute polystyrene solutions in toluene into liquid CO_2, Dixon and Johnston (123) produced small particles (100 nm). They showed that the preaddition of CO_2 in polystyrene solution influenced particle morphology because of dilution and viscosity reduction (123).

Volumetric Expansion. In spraying mode, as in batch mode, the volumetric expansion of the solvent defines the range of exploitable pressures. This was shown by

Reverchon et al. (175) who carried out ASES experiments with dextran-DMSO and L-PLA-DCM solutions. As DCM is more soluble than DMSO in $ScCO_2$, its volumetric expansion curve with CO_2 at 40°C is located at lower pressures than for DMSO.

Advances and Patents. STUDIED WALL POLYMERS. Among the available pharmaceutical polymers, aliphatic polyesters were the most studied wall materials for particle formation by SAS continuous process (Table 4). Some of them, such as PCL, were found to be incapable of processing by ASES and PCA because of their plasticizing in the presence of $ScCO_2$ (175,200). The derivatives of lactide and glycolide lead to different results according to their M_w, crystallinity, or thermal behavior. For instance, D,L-PLA and PLGA polymers, with levels of M_w that are too low or that are too apolar, can be extracted by the SF–solvent mixture in the ASES process (174). L-PLA and PHB mainly led to the formation microparticles with a low polydispersity and a diameter from 1 to 10 μm (173,174,183,184,196,199–202,208–210).

Breitenbach et al. (211) synthesized a series of various polymers with a combined structure for ASES experiments; L-PLA, D,L-PLA, and PLGA chains were grafted onto polyol backbones (dextran sulfate sodium or poly vinyl alcohol). They found that the degree of crystallinity was the determining factor on particle morphology. The more crystalline the polymers are, the more spherical and more uniform the particles will be. Semicrystalline PLGA 3-block polymers yielded individual and round-shaped particles, whereas amorphous PLGA 50/50 yielded small soft and agglomerated particles (212,213). The crystallinity of L-PLA could explain its high efficiency to produce particles (Table 4).

An increase in the L-PLA concentration leads to the formation of fibers as in the study carried out by Tu et al. (184). This was explained by the increase of the solution viscosity and consequently the insufficient break-up of the solution into droplets. Furthermore, at the highest polymer concentrations, less CO_2 was required to precipitate the polymers, and so the precipitation occurred more rapidly than droplet formation. Saim et al. (214) observed similar results when spraying other classes of polymers such as hyaluronic acid ester (HYAFF-7)/DMSO solution into $ScCO_2$.

Surprisingly, the size of L-PLA microparticles produced by Mawson et al. (183) decreased by a factor of 2 to 5 when the polymer concentration was increased from 1.0 to 3.0 wt%. This was explained by the decrease in width of the metastable region [between the binodal and spinodal curves (Fig. 8)]. Less time was then available for the growing of polymer-rich domains.

By increasing the temperature for a given pressure, several authors reported that L-PLA particle size increased (121,184,197,200). This phenomenon can be explained by the increase of particle growth rate with an increase in the temperature, or by the plasticization of the polymers because of the reaching of T_g temperature and the consequent production of soft particles that aggregate more easily.

In contrast to ASES and PCA studies, PCL or PLGA microparticles have been obtained by Ghaderi et al. (210) who used SEDS and chose polymers exhibiting good inherent viscosity. The saturation of the polymer solution by adding small amounts of organic nonsolvent (isopropanol or hexane) increased the yield of particle formation from all tested polymers (174,175,200,210).

To improve the control of particle size, the same authors modified the SEDS process by using a combination of supercritical N_2 (ScN_2) and $ScCO_2$ for the production of PLGA, L-PLA, D,L-PLA, and PCL microparticles (209). The combination of ScN_2 and $ScCO_2$ led to a more efficient dispersion of the polymer solutions than $ScCO_2$ alone. Compared to the previous study, this resulted in a reduction of the

Table 4 Examples of Polyester Particle Formulation by Continuous SAS

Wall polymer	Type of process	Organic solvent	Injection device	Particle characteristics	References
	ASES	DMSO or DCM	One-way nozzle (60 μm ø int)	No particles	(175)
PCL	SEDS	DCM/acetone/isopropanol (1.1/6.2/2.7)	Two-way SEDS nozzle	Microspheres (30–210 μm)	(210)
			Three-way SEDS nozzle with $ScCO_2$ and ScN_2	aggregated at high CO_2 pressure	(209)
		DCM	One-way nozzle (CO_2 in static mode)	No particles	(174)
		DCM/trifluoroethanol Methanol for drug	One-way nozzle	Agglomerated particles	(212,213)
PLGA	ASES	DCM for PLA methanol for drug	Three-way nozzle (Fig. 11)	Isolated microspheres of drug coated with PLGA	(184,203)
		Ethyl acetate/DMSO	Energizing-gas nozzle (Figure 12)	Microspheres (mean ø 1 μm)	(187,204,205)
	SEDS	Acetone, ethyl acetate, isopropanol, DCM, methanol	Two-way SEDS nozzle	Microspheres (mean ø 130 μm)	(210)
			Three-way SEDS nozzle with $ScCO_2$ and ScN_2	Microspheres (mean ø 10 μm)	(209)
	ASES	DCM	One-way nozzle (CO_2 in static mode)	No particles	(174)
DL-PLA	SEDS	Acetone; ethyl acetate; hexane	Two-way SEDS nozzle	Agglomerated particles (150 μm)	(210)
			Three-way SEDS nozzle with $ScCO_2$ and ScN_2	Microspheres (mean ø 10 μm)	(209)
3-block-polymer (b-poly-L-lactide-co-D, L-lactide-co-glycolide)	ASES	DCM/trifluoroethanol methanol for drug	One-way nozzle	Individual microspheres	(212,213)
PHB	ASES	DCM	One-way nozzle (CO_2 in static mode)	Individual microspheres	(174)
Comb polymers (PVA- grafted- DL-PLA)	ASES	DCM	One-way nozzle	Size uniformity increases with crystallinity	(211)

(PVA- grafted- L-PLA) (PVA- grafted- PLGA)					
			Various one-way nozzle and a two-way nozzle	Microspheres (mean ø 5–60 μm by adjusting process parameters)	(137)
		DCM	One-way nozzle (CO_2 in static mode)	Individual microspheres (1–10 μm)	(174)
			One-way nozzle	Depends upon the conditions	(199)
	ASES		One-way nozzle (CO_2 in sprayed in cocurrent)	Drug-loaded microspheres	(173)
		DCM/methanol	One-way nozzle (CO_2 in sprayed in countercurrent)	Drug-loaded microspheres (ø 6 μm)	(208)
		DCM	One-way nozzle	Microspheres (5 μm)	(196)
L-PLA		DCM for PLA methanol for drug	Three-way nozzle	Fibrous drug network embedded into empty microspheres	(184)
		DCM	Quartz capillary	Microsphere (1–3 μm)	(197)
			One-way nozzle	Agglomerated particles	(183)
			Coaxial two-way nozzle	Individual microspheres	
	PCA	DCM	One-way silica capillary coated with polyimide (100 μm ø int)	Individual microspheres	(200)
			One-way ultrasonic nozzle, vibrating at 120 KHz; CO_2 and DCM are cocurrently injected into the precipitator	Drug-loaded microspheres (0.2–1 μm)	(201)
				Gentamycin-loaded microspheres	(202)
	SEDS	DCM/acetone/ isopropanol (3.3/6.5/0.2)	Two-way SEDS nozzle	Microspheres (0.5–5 μm)	(210)
			Three-way SEDS nozzle with $ScCO_2$ and ScN_2	Microspheres with mean ø of 10 μm	(209)

Abbreviations: SAS, $ScCO_2$ as an antisolvent; ASES, aerosol solvent extraction system; DCM, dichloromethane; DL-PLA, poly(D,L-lactide); DMSO, dimethylsulfoxide; PCA, precipitation with a compressed fluid antisolvent; PCL, poly(ε-caprolactone); PLA, poly lactide; PLGA, poly(D,L-lactide-co-glycolide); SEDS, solution enhanced dispersion by supercritical solution; PHB, poly-β-hydroxybutyrate.

microparticle size produced with all the amorphous polymers (210). Microspheres with a mean volumetric diameter of less than 10 μm were obtained from PLGA, D,L-PLA, and L-PLA.

It can be mentioned that other types of polymers (ethylcellulose, hydroxypropyl-methylcellulose (HPMC), polymethyl-methacrylate (PMMA), polyacrylonitrile, polystyrene) have been investigated. SEDS has also been developed to process water-soluble materials such as carbohydrates (123,151,183,194,215,216). The accurate and controlled coalimentation of an aqueous solution, ethanol, and $ScCO_2$ into the modified SEDS coaxial nozzle overcomes the problems associated with limited water solubility in $ScCO_2$.

Drug Encapsulation. Controlled release formulations have been investigated by several authors and they emphasize the effect of drug polarity on encapsulation efficiency (200). Bleich and Muller (173) coprecipitated four different drugs of various polarities (indomethacin, hyoscine butylbromide, piroxicam, and thymopentin) with L-PLA. They found that the greater the polarity of the drug, the higher the yield of encapsulation, with values up to 20% for polar indomethacin but only 5% for low polar thymopentin. Nonpolar drugs can be completely extracted with the SF-organic solvent mixture, depending on T/P conditions, which render their processing difficult when using ASES. However, polar drugs such as proteins can be easily incorporated into particles using the ASES process, due to their very low affinity for $ScCO_2$ at all T/P conditions (184,212). Foster et al. (206) produced lysozyme–lactose microparticles, coprecipitated from aqueous solution, using the ASES process with a gas-energizing nozzle for spraying the solution. The temperature used was only 20°C, which is suitable for the encapsulation of a protein. At a ratio of 92/8 lysozyme–lactose, the process produced spheres with sizes ranging from 0.1 to 0.5 μm.

L-PLA microspheres containing ionic pharmaceutical agents (gentamycin, naloxone, or naltrexone) were prepared by Falk et al. (201,202) using the PCA process. The ionic components were solubilized in DCM at concentrations up to 1 mg/mL, by the stoichiometric replacement of the polar counter ions by an anionic detergent. The release of the ion-paired drugs at 37°C into phosphate buffer displayed minimal burst effects and exhibited release kinetics that were approximately linear with the square root of time, indicating a drug release controlled by diffusion through the matrix. For gentamycin, linear release from the L-PLA microspheres was observed over seven weeks, even for drug loads reaching 25% (w/w). As it has already been shown in the previously cited studies, nonpolar drugs such as rifampin, which was not ion paired, was poorly encapsulated (199–201).

The SEDS process was successfully tested for the production of microparticles of indomethacin in combination with hydroxypropyl methylcellulose, ethylcellulose, or polyvinylpyrrolidone (215). Interestingly, the drug molecules were in an amorphous state and high encapsulation efficiency was obtained (over 60% w/w for indomethacin–PVP systems), together with a free flowing nature for the powder products. These two features are often absent in products prepared by traditional coprecipitation techniques, which have limited their commercial application.

Hydrocortisone was entrapped within PLGA microparticles with an entrapment efficiency of 22%, without any optimization (209,210). However, the particle yield was still low; 30% to 40% of the initial loaded polymers led to particles.

Recently, AstraZeneca R&D used the SEDS process for encapsulating a model drug in two different types of carriers, mannitol and Eudragit E100 (217). However, a true one-phase solid dispersion was impossible to obtain with either of the two types of carrier, because they precipitated separately.

The successful preparation of supercoiled, plasmid DNA–loaded particles from aqueous solutions by SEDS has been reported recently (127): in this article the use of SEDS to coformulate 6.9 kb pSVb plasmid with mannitol as a carrier was reported. After preliminary experiments showing a high degradation of the plasmid during powder formation, a systematic investigation of the process revealed that pH effects were crucial for the recovery of intact DNA. The use of buffer to counteract the acidic nature of $ScCO_2$ led to an increase in the recovered supercoiled proportion from 7% to 80%. Thus, by 'fine-tuning' the aqueous supply, the modified SEDS process can be used both to control particle formation and to protect labile biological materials during processing, thus providing a viable alternative to freeze- and spray-drying processes.

Residual Solvents and SAS Processes. The purity of the particles and the elimination of residual solvents is one of the major advantages of SAS over conventional precipitation in the pharmaceutical industry processes. However, if washing is ineffective, condensation of the residual solvent during system depressurization (i.e., during product collection) will lead to particle agglomeration. Therefore, the design of a process must involve consideration for vessel hydrodynamics and control of the washing process which often occur in a separate unit. Washing efficiency may be optimized at high CO_2 densities, which increases solvent miscibility, and at high temperatures, which increases the solvent vapor pressure (114). In most cases, the production processes lead to a compacted powder bed, so that, the residual solvent stripping is a separate operation to obtain a good circulation of the $ScCO_2$ between the particles.

In the GAS process, Thiering et al. (218) found that washing efficiency was improved by performing the operation in the two-phase region where gaseous and liquid CO_2 are present. The authors also noted the effect of increasing the quantity of CO_2 during the washing of residual DMSO from lysozyme precipitates (218). Washing with 70 g of CO_2 (35°C, 86 bar) yielded a residual DMSO concentration up to 120,000 ppm, whereas washing with 350 g reduced the concentration to 300 ppm.

Ruchatz et al. (196) found that the CO_2 pump rate influenced the encapsulating yield and the amount of residual solvent in the L-PLA microparticles produced by the ASES process. In all cases, microparticles had lower amounts of residual solvent than those specified by the USP (600 ppm) and this, after a very short period of production.

Products with low residual solvent quantities are easier to produce with continuous processes by maintaining a large excess of CO_2 with respect to the solution stream. As a result, the concentration of residual solvent in the precipitation vessel is reduced during the whole operation, as is particle agglomeration. Bleich and Muller (173) and Ruchatz et al. (196) made similar observations with PLA precipitated from DCM by ASES. In these studies, the residual solvent concentration was reduced to the range of 30 to 100 ppm. Despite the time spent washing these products, continuous and batch SAS remain fast techniques compared to conventional drying techniques that can take days.

In some cases, when the solvent is incorporated into a crystal lattice, the high solvent–crystal affinity requires an extension of washing time and the use of high temperatures and excessive volumes of $ScCO_2$ in order to remove the solvent from the product (202). The injection of a cosolvent into the CO_2 stream washing solution may also further reduce traces of poorly soluble or tightly bound solvents. For example, ethanol can be added to CO_2 in order to remove water that might be present in the system. Thiering et al. (218) used 5 wt% ethanol to remove traces of DMSO which was tightly bound within the lysozyme structure by hydrogen bonds. For the same washing volume the residual DMSO concentration was decreased four-fold.

Since the recycling of $ScCO_2$ is essential for economic reasons in continuous or batch processes, the presence of residual solvents in the stream needs to be controlled. If solvent condensation by depressurization or cooling occurs before recompression, problems can appear in the pumping system. This can be avoided by the adsorption of the recycled CO_2 stream on activated charcoal or a scrubber.

Techniques Using SF as Spray Enhancers

These processes consist of dissolving $ScCO_2$ in melted substance(s) or in a solution/suspension of substances and depressurizing this mixture through a nozzle, causing the formation of solid particles or liquid droplets. These processes allow particles to form from melt polymers that are not soluble in $ScCO_2$, but which absorb a large amount of CO_2 (1–30 wt%) (125).

During expansion, the increasing volume of CO_2 causes the liquid substance to disintegrate into tiny droplets through a turbulent flow. Additionally, owing to the Joule–Thomson effect (adiabatic decompression), the solution cools down below the solidification temperature of the solute, and fine particles are formed (117,219).

$ScCO_2$ allows viscous materials to be processed via a reduction in viscosity and interfacial tension (220,221). Furthermore, it allows the polymer T_g or T_m temperature to be reduced in comparison to those processes carried out in atmospheric conditions (222). This allows polymers to be liquefied at lower temperature (10–50°C), which is useful when encapsulating heat-sensitive components.

Several processes used this method based on SF dissolution for reasons already cited. They are known as particles from gas saturated solutions (PGSS), polymer liquefaction using supercritical solvation (PLUSS), supercritical-assisted atomization (SAA), Depressurization of an expanded organic solution (DELOS), CO_2-assisted nebulization and bubble drying (CAN-BD) (223–225).

Weidner (226) presented a pilot apparatus [Fig. 14; (I)], which can operate continuously to generate porous particles of PEG with sizes ranging from 3 to 500 µm. The PEG is melted in a stirred vessel, and is introduced in a static mixer where it is placed in contact with compressed and heated CO_2. The mixture is then expanded through a nozzle mounted immediately after the mixer. When using a PEG of 6,000 g/mol, particle size decreases when the CO_2/polymer mass ratio increases. The formation of spheres is favored at low pressure, but at high CO_2/polymer ratio, the influence of pressure on particle size is less pronounced than at low ratios. Moreover, the pre-expansion temperature strongly influences particle morphology. An increase of temperature enhanced the formation of spheres. This technique was used by Sencar-Bozic et al. (227) to coprecipitate nifedipine and PEG 4000.

As seen in Figure 14 (I), the pilot uses CO_2 like a spray enhancer for melt polymers. It can be modified to use the ability of CO_2 to reduce the polymer T_m or T_g temperature [Fig. 14 (II)]. In this last configuration, two liquids [melt polymers (A) and active molecules (B)] can be mixed with cold pressurized CO_2. The CO_2 induces T_m/T_g reduction and maintains the compounds in a liquid state, so that both constituents can be mixed intensively in a short time at temperatures below their T_m/T_g. After sufficient homogeneity of the mixture has been achieved, it is expanded in a nozzle and composite particles are obtained. By using this process, either solid–solid

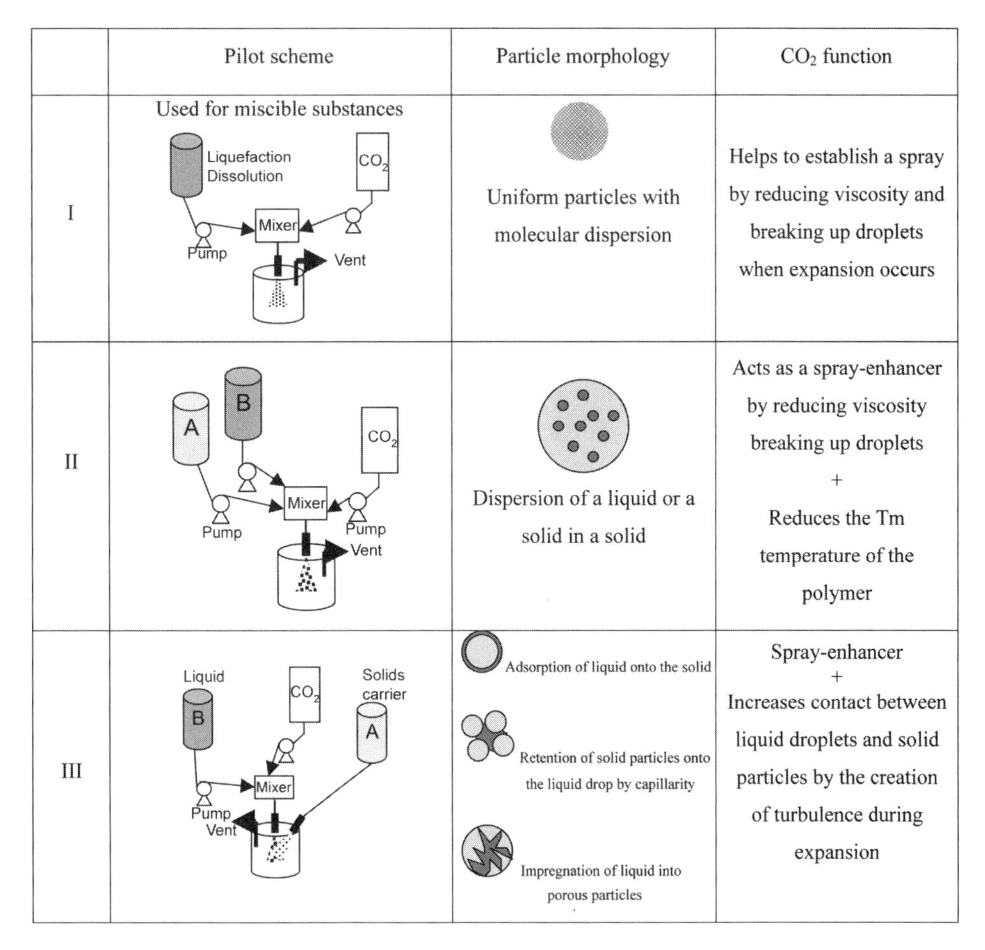

Figure 14 Scheme of different pilots using CO_2 as a spray enhancer.

blends or encapsulated liquids can be obtained. The suitability of such a process was demonstrated by the formation of microspheres (228). Shine and Gelb (224) patented this process in 1997 and called it PLUSS. They gave the example of infectious bursal disease virus (IBDV) vaccine encapsulated inside PCL (M_w 4000) microcapsules.

In another conformation [Fig. 14 (III)], the $ScCO_2$ can be solubilized in a solution, emulsion, or a suspension of active substance and just used as a spray-enhancer to produce very fine droplets when the liquid is sprayed. A micronized solid carrier is added concomitantly to the sprayed liquid, and free-flowing particles with a liquid content of up to 90 wt% are formed. This process is usually named concentrated powder form (CPF) (226). Different particle morphologies can be obtained according to the interaction between the solid carrier and the liquid droplets [Fig. 14 (III)]. The most interesting morphology is obtained by using porous particles. The very small droplets formed by CO_2 expansion may penetrate into the pore network. Typically, 0.5 to 5 kg of CO_2 is required to produce 1 kg of particles. Therefore the running costs are lower than for conventional processes like fluidized bed coating or spray drying.

Due to the high viscosity of the polymers used, the sorption of CO_2 into the molten polymers is always diffusion controlled in the absence of mixing. The good

Table 5 Summary of $ScCO_2$ Properties and Functionalities Involved in PGSS Processes

Properties of $ScCO_2$	Functionality
Reduction of T_m temperatures	Mixes and dissolves sensitive, reactive and/or immiscible substances in liquid.
Reduction of viscosity and surface tension	Mixes/dissolves liquid compounds even with high viscosities
	Generates sprays from gases containing liquids with a broad range of viscosities
Large volume increase during expansion	Generates small droplets in sprays
	Admixes and agglomerates additives at high turbulence in expanding sprays
Rapid and considerable temperature reduction by expansion and direct heat transfer	Rapid droplet solidification
	Establishes a high degree of supersaturation of solute dissolved in melt polymer
	Reduces losses of volatile compounds

Abbreviation: $ScCO_2$, supercritical CO_2.

mixing action provided by an extruder for highly viscous materials can be used to overcome the limited diffusion of CO_2 in the polymer (229). All the exploited CO_2 properties and resulting advantages are listed in Table 5.

Concluding Remarks

$ScCO_2$ allows developing new processes with promising application in the field of drug delivery systems based on biodegradable particles. Three main concepts have been developed taking into account the physical properties of $ScCO_2$ such as its tuning solvent properties and its high diffusibility. Depending on drug or polymer solubilities, $ScCO_2$ can be used as a solvent (RESS, impregnation), nonsolvent (SAS), or spray enhancer (PGSS, PLUSS, etc.). In comparison with classical processes, particles with lower residual solvent content can be designed and filled the criteria imposed by regulation authority. The particle features such as morphology and size distribution can be varied by adjusting the process parameters. On the other hand, playing on the pressure parameter at relatively low temperatures allows for processing of fragile molecules.

For each type of process, one can evidence some specific limits. In the case of RESS, the solubility in $ScCO_2$ of both polymer and drug must be high, but unfortunately very few biodegradable polymers have this property. The same remark is applicable to promising drugs that are proteins and peptides. It is why the design of new CO_2-philic biodegradable polymer is an important subject of research (122,150). For these reasons, RESS remains a technique studied essentially at the lab-scale.

Although the processes based on the antisolvent properties of $ScCO_2$ (SAS) were the most studied, they have the main disadvantage to use quantities of organic solvent which can remain as residual quantities in the final product that implies an additional extraction step. This complicates the design of equipment for industrial production under pharmaceutical conditions. Some attempts of scale-up, with a focus on pharmaceutical industry requirements, are reviewed and discussed by Deschamps and Richard (230). For instance, Thies and Muller (137) reported the use of an ASES large-scale process, a pilot apparatus of 50 dm^3, capable of producing up to 200 g of L-PLA empty microparticles. Polymer solutions were sprayed into a bulk of static $ScCO_2$ or a near-critical CO_2 gas phase through nozzles of different diameters (between 0.1 and 0.8 mm) and a coaxial nozzle. The L-PLA particle diameters increased from 6 to 50 μm by reducing the $ScCO_2$ density. With the same idea of scaling-up, the company Separex (231) built facilities for the ASES process with a 0.5, 4, and 50 dm^3 precipitation vessel, the CO_2 flow-rate being 5 to 500 kg/h, and the capacity of production ranging from 2 to 200 g/h product.

Recently, the SEDS process has been used for several drug formulation applications, with evidence of successful scaling-up. Hanna et al. (207) patented the preparation of particles of salmeterol xinafoate with a polymer matrix. Two distinct solutions of the active substance and of the polymer (hydroxypropylcellulose) in acetone were prepared and co-introduced with $ScCO_2$ in a precipitator using a three-passage nozzle. Analyses confirmed the inclusion of the product into the polymer matrix material.

The third type of process (PGSS) is very promising in the pharmaceutical field since it can be applied to a lot of active substances without using organic solvents. Moreover, its scalability has been evidenced and pilots are now available for producing large amounts of particles (230).

Dry Processes

In a constant search to answer the problems of residual solvent limit values in pharmaceutical formulations, other solvent-free techniques have been investigated. Among them, dry processes based on the milling or grinding of polymers that can be followed by a spheronization step have been developed. The milling technique was first introduced by Boswell and Scribner in 1973 (232). They melted PLA in which they dispersed a labile drug. They then congealed and milled the mixture into particles ranging from 1 to 200 μm in size.

As described, mechanical processes are performed in several steps. The preparation of the carrier and drug compounds is followed by mixing and shaping steps (melt extrusion, compression etc.). The resulting solid mass is then broken up and rounded into microparticles (milling, spheronization, etc.).

Pre-Milling

Mixing. The first step in mechanical processes consists of mixing the carrier and the drug. This can be carried out in the presence of a solid or molten carrier. With a carrier in the solid state, the mixing is more frequently performed in a granulator, in dry or aqueous conditions. Different types of granulators, such as a planetary mixer, sigma blade mixers, etc. have been tested (233).

In melt conditions, successive cycles of heating and cooling can be performed in order to obtain a homogeneous paste (234). In this condition, mixing is generally performed in an extruder (235).

Extrusion. Extrusion is a process that converts raw materials into a product of required shape and uniform density by forcing them through a die under controlled conditions (235). The extrusion process can be performed using four main classes of extruder: screw, sieve and basket, roll, and ram extruders (233). Extrusion may be classified as either molten systems, under temperature control, or as semisolid viscous systems. Semisolid systems are concentrated dispersions containing a high proportion of solid mixed with a liquid phase.

In molten extrusion, heat is applied to the material in order to enable it to flow through the die. The molten extrusion can be applied to disperse drugs in a melt polymer to form a true molecular solution of the active agent in the matrix.

Carrier systems including different classes of biodegradable polymers such as PLA, PGA, DL-PLGA, PEG, and cellulose-ethers have been tested (236). The basic prerequisite in melt extrusion is the thermoplasticity of the polymer. Polymers with a high T_g temperature require a high process temperature, and this can give rise to potential drawbacks.

Milling

Many milling techniques such as ball milling, jet milling, cryogenic pulverization, air milling, spin milling, etc. have been described. We will focus on the recent ones.

Jet milling is a new grinding technique used for the preparation of solvent-free microparticles (234). This technique is a combination of three consecutive steps: (i) Melting the polymer and drug disperser or solvation in the melt. The required temperature is adjusted so as to be slightly higher than the T_m or the T_g temperature of the chosen polymer. The melt is then congealed; several melting–congealing cycles are necessary to obtain a homogeneous solution or suspension. (ii) The congealed material undergoes a pregrinding step using a rotor speed mill. (iii) Several cycles of jet milling are performed to provide microparticle reduction and surface smoothening.

Different types of PLA and PLGA were used to produce particles with a mean size distribution of 4 to 6 μm (234). SEM imaging showed nonporous and smooth microparticles. The number of jet milling cycles strongly influences particle size distributions. Beyond three cycles, the particle size increases due to particle collision. This technique allows for a very high microparticle drug loading, exhibiting a zero order release. Carstensen and Mueller (237) used also jet milling to produce drug-loaded microparticles with a size below 25 μm.

At the industrial level, PLGA microparticles were developed containing the hydrophobic fatty acid salt of peptides comprising 3 to 45 amino acids (238). In this process, pulverulent forms of the polymer and drug are mixed in a ball mill at lower than room temperature, and are then progressively compressed and heated before being extruded at a temperature between 80°C and 100°C. The extrusion product is cooled and pulverized at a temperature between 0°C and −30°C to obtain microparticles around 15 μm in diameter (239). Microparticles of triptorelin pamoate Trelstar®/ Decapeptyl® produced by this technique delivered the triptorelin over a period of three months (240). Other peptides such as vapreotide (Octastatins®) are in phase II trials (241).

Many milling processes lead to particle surfaces that display irregularities; microparticles formed thereafter are not spherical and display a wide range of sizes. This implies the use of a complementary spheronization step.

Spheronization

A spheronization step can be carried out by a mechanical process, which provides large particles (up to 200 μm), or by heating the previously suspended particles.

Mechanical Spheronization. During the mechanical spheronization process, the extrudated cylinders are put on a striated spinning plate and broken up into

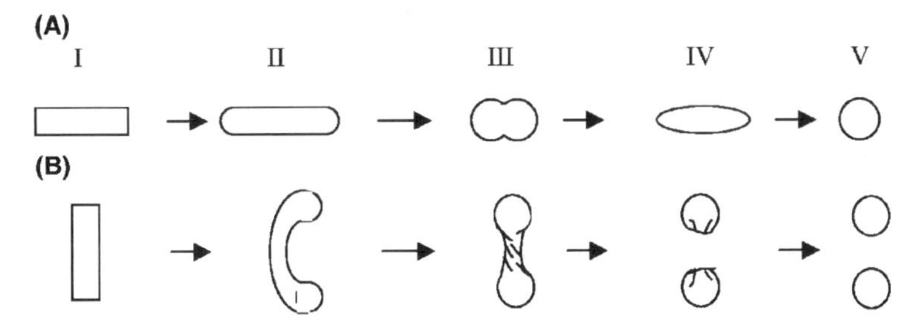

Figure 15 Spheronization mechanism according to: (**A**) I, cylinder; II, cylinder with rounded edges; III, dumb-bells; IV, elliptical particles; V, spheres. (**B**) I, cylinder; II, rope; III, dumb-bells; IV, spheres with a cavity outside; V, spheres. *Source*: From Refs. 242, 243.

smaller particles due to frictional forces. The size of particles obtained is generally quite high (up to 200 μm). During this process, different steps can be distinguished depending upon the shape of the particles. Starting from a cylinder, two main mechanisms are suggested (Fig. 15).

The parameters influencing the spheronization of a product are the rotational speed of the friction plate, usually between 200 and 400 rpm, the time of spheronization (2–10 min), and the spheronizer load (233). Grooves can exhibit two main geometries, i.e., cross-hatch geometry or radial geometry.

Heating Spheronization. Wichert and Rohdewald (244) performed a spheronization process on particles obtained from the milling of a cold melt composed of a mixture of PLA and vinpocetine. The particles were dispersed in a hot (180°C) Tween® buffered solution and were then spray-dried and rapidly cooled to obtain spherical microparticles.

Ruiz (245,246) used a similar concept to formulate microparticles loaded with LHRH and a somatostatin analogue. Cryogenic grinding of extruded polymer containing the dispersed peptide led to heterogeneous and porous particles. These microparticles were suspended in a gel under stirring and were softened by heating at a temperature above the polymer's T_g. The suspension was then rapidly cooled and filtered to obtain free microballs.

ACKNOWLEDGMENT

The authors wish to thank Dr. Frantz S. Deschamps for helpful discussions.

REFERENCES

1. Benoit JP, Marchais H, Rolland H, Van de Velde V. Biodegradable microspheres: advances in production technology. In: Benita S, ed. Microencapsulation: Methods and Indusstrial Applications: Marcel Dekker, Inc., 1996:35–72.
2. Arshady R. Preparation of biodegradable microspheres and microcapsules: 2. Polylactides and related polyesters. J Contr Rel 1991; 17:1–21.
3. Cavalier M, Benoit JP, Thies C. The formation and characterization of hydrocortisone-loaded poly((+/−)-lactide) microspheres. J Pharm Pharmacol 1986; 38:249–253.
4. Sturesson C, Carlfors J, Edsman K, Andersson M. Preparation of biodegradable poly (lactic-co-glycolic) acid microspheres and their in vitro release of timolol maleate. Int J Pharm 1993; 89:235–244.

5. Pradhan RS, Vasavada RC. Formulation and in vitro release study on poly(-lactide) microspheres containing hydrophilic compounds: glycine homopeptides. J Contr Rel 1994; 30:143–154.
6. Gardner D. Process of preparing microcapsules of lactides or lactides copolymers with glycolides and/or ε-caprolactones. U.S. Patent 4,637,905; 1987.
7. Wang HT, Schmitt E, Flanagan DR, Linhardt RJ. Influence of formulation methods on the in vitro controlled release of protein from poly (ester) microspheres. J Contr Rel 1991; 17:23–31.
8. Ike O, Shimizu Y, Wada R, Hyon S-H, Ikada Y. Controlled cisplatin delivery system using poly(-lactic acid). Biomaterials 1992; 13:230–234.
9. Sato T, Kanke M, Schroeder H, DeLuca P. Porous biodegradable microspheres for contolled drug delivery. I. Assessment of processing conditions and solvent removal techniques. Pharm Res 1988; 5:21–30.
10. Okada H, Ogawa Y, Yashiki T. Prolonged release microcapsule and its production. U.S. Patent 4,652,441–1987.
11. Singh M, Singh O, Talwar G. Biodegradable delivery system for a birth control vaccine: immunogenicity studies in rats and monkeys. Pharm Res 1995; 12:1796–1800.
12. Cohen S, Yoshioka T, Lucarelli M, Hwang L, Langer R. Controlled delivery systems for proteins based on poly(lactic/glycolic acid) microspheres. Pharm Res 1991; 8: 713–720.
13. Ogawa Y, Yamamoto M, Okada H, Yashiki T, Shimamoto T. A new technique to efficiently entrap leuprolide acetate into microcapsules of polylactic acid or copoly(lactic/glycolic) acid. Chem Pharm Bull 1988; 36:1095–1103.
14. Toguchi H. Formulation study of leuprorelin acetate to improve clinical performance. Clin Ther 1992; 14:121–130.
15. Okada H, Doken Y, Ogawa Y, Toguchi H. Preparation of a three-month depot injectable microspheres of leuproline acetate using biodegradable polymers. Pharm Res 1994; 11:1143–1147.
16. Kamijo A, Kamei S, Saikawa A, Igari Y, Ogawa Y. In vitro release test system of (D,L-lactic-glycolic) acid copolymer microcapsules for sustained release of LHRH agonist (leuprorelin). J Contr Rel 1996; 40:269–276.
17. Herrmann J, Bodmeier R. Somatostatin containing biodegradable microspheres prepared by a modified solvent evaporation method based on w/o/w-multiple emulsions. Int J Pharm 1995; 126:129–138.
18. Herrmann J, Bodmeier R. Effect of particle microstructure on the somatostatin release from poly(lactide) microspheres prepared by a W/O/W solvent evaporation method. J Contr Rel 1995; 36:63–71.
19. O'Hagen D, Jeffery H, Davis S. Long-term antibody responses in mice following subcutaneous immunization with ovalbumin entrapped in biodegradable microparticles. Vaccine 1993; 11:965–969.
20. O'Hagen D, McGee J, Lindblad M, Holmgren J. Cholera toxin B subunit (CTB) entrapped in microparticles shows comparable immunogenicity to CTB mixed with whole cholera toxin following oral immunization. Int J Pharm 1995; 119:251–255.
21. Uchida T, Martin S, Foster T, Wardley R, Grimm S. Dose and load studies for subcutaneous and oral delivery of poly(lactide-co-glycolide) microspheres containing ovalbumin. Pharm Res 1994; 11:1009–1015.
22. Sah H, Toddywala R, Chien Y. Biodegradable microcapsules prepared by a w/o/w technique: effects of shear force to make primary w/o emulsion on their morphology and protein release. J Microencapsul 1995; 12:59–69.
23. Blanco-Prieto M, Leo E, Delie F, Gulik A, Couvreur P, Fattal E. Study of the influence of several stabilizing agents on the entapment and in vitro release of pBC 264 from poly(lactide-co-glycolide) microspheres prepared by a w/o/w solvent evaporation method. Pharmac Res 1996; 13:1127–1129.

24. Cleland J, Jones A. Stable formulations of recombinant human growth hormone and interferon-γ for microencapsulation in biodegradable microspheres. Pharm Res 1996; 13:1464–1475.
25. Cleland J, Mac A, Boyd B, et al. The stability of recombinant human growth hormone in poly(lactic-co-glycolic acid) (PLGA) microspheres. Pharm Res 1997; 14:420–425.
26. Soriano I, Delgado A, Diaz RV, Evora C. Use of surfactants in polylactic acid protein microspheres. Drug Dev Ind Pharm 1995; 21:549–558.
27. Crotts G, Park TG. Preparation of porous and nonporous biodegradable polymeric hollow microspheres. J Contr Rel 1995; 35:91–105.
28. Lu W, Park TG. Protein release from poly(lactic-co-gly-polycolic acid) microspheres: protein stability problems. PDA J Pharm Sci Technol 1995; 49:13–19.
29. Jollivet C, Aubert-Pouessel A, Clavreul A, et al. Striatal implantation of GDNF releasing biodegradable microspheres promotes recovery of motor function in a partial model of Parkinson's disease. Biomaterials 2004; 25:933–942.
30. Jollivet C, Aubert-Pouessel A, Clavreul A, et al. Long-term effect of intra-striatal glial cell line-derived neurotrophic factor-releasing microspheres in a partial rat model of Parkinson's disease. Neurosci Lett; 2004; 19:207–210.
31. Gouhier C, Chalon S, Venier-Julienne MC, et al. Neuroprotection of nerve growth factor-loaded microspheres on the D2 dopaminergic receptor positive-striatal neurones in quinolinic acid-lesioned rats: a quantitative autoradiographic assessment with iodobenzamide. Neurosci Lett 2000; 288:71–75.
32. Pean J-M, Venier-Julienne M-C, Boury F, Menei P, Denizot B, Benoit J-P. NGF release from poly(-lactide-co-glycolide) microspheres. Effect of some formulation parameters on encapsulated NGF stability. J Contr Rel 1998; 56:175–187.
33. Pean JM, Menei P, Morel O, Montero-Menei CN, Benoit JP. Intraseptal implantation of NGF-releasing microspheres promote the survival of axotomized cholinergic neurons. Biomaterials 2000; 21:2097–2101.
34. Pean JM, Venier-Julienne MC, Filmon R, Sergent M, Phan-Tan-Luu R, Benoit JP. Optimization of HSA and NGF encapsulation yields in PLGA microparticles. Int J Pharm 1998; 166:105–115.
35. Perez C, De Jesus P, Griebenow K. Preservation of lysozyme structure and function upon encapsulation and release from poly(lactic-co-glycolic) acid microspheres prepared by the water-in-oil-in-water method. Intl J Pharm 2002; 248:193–206.
36. Schugens C, Laruelle N, Nihant N, Grandfils C, Jérôme R, Teyssié P. Effect of the emulsion stability on the morphology and porosity of semicrystalline poly l-lactide microparticles prepared by w/o/w double emulsion-evaporation. J Contr Rel 1994; 32:161–176.
37. Erden N, Celebi N. Factors influencing release of salbutamol sulphate from poly (lactide-co-glycolide) microspheres prepared by water-in-oil-in-water emulsion technique. Int J Pharm 1996; 137:57–66.
38. Ghaderi R, Sturesson C, Carlfors J. Effect of preparative parameters on the characteristics of poly(D,L-lactide-co-glycolide) microspheres made by the double emulsion method. Int J Pharm 1996; 141:205–216.
39. Morlock M, Koll H, Winter G, Kissel T. Microencapsulation of rh-erythropoietin, using biodegradable poly(lactide-co-glycolide): protein stability and the effects of stabilizing excipients. Eur J Pharm Biopharm 1997; 43:29–36.
40. Pérez C, Castellanos IJ, Constantino RC, Al-Azzam W, Griebenow K. Recent trends in stabilizing protein structure upon encapsulation and release from bioerodible polymers. J Pharm Pharmacol 2002; 54:301–303.
41. Boury F, Ivanova T, Panaiotov I, Proust JE. Dilatational properties of adsorbed Poly(D,L-lactide) and bovine serum albumin monolayers at the dichloromethane water Interface. Langmuir 1996; 11:1636–1644.
42. Carrasquillo KG, Stanley AM, Aponte Carro JC, et al. Non-aqueous encapsulation of excipient-stabilized spray freeze-dried BSA into poly(lactide-co-glycolide) microspheres results in release of native protein. J Contr Rel 2001; 76:99–208.

43. Sanchez A, Villamayor B, Guo Y, McIver J, Alonso MJ. Formulation strategies for the stabilization of tetanus toxoid in poly(lactide-co-glycolide) microspheres. Int J Pharm 1999; 185:255–266.
44. Sturesson C, Carlfors J. Incorporation of protein in PLG-microspheres with retention of bioactivity. J Contr Rel 2000; 67:171–178.
45. Sah H. Stabilization of proteins against methylene chloride/water interface-induced denaturation and aggregation. J Contr Rel 1999; 58:143–151.
46. Péan J-M, Boury F, Venier-Julienne M-C, Menei P, Proust J-E, Benoit J-P. Why Does PEG 400 Co-Encapsulation Improve NGF Stability and Release from PLGA Biodegradable Microspheres? Pharm Res 1999; 16:1294–1299.
47. Benoit JP, Faisant N, Venier-Julienne MC, Menei P. Development of microspheres for neurological disorders: From basics to clinical applications. J Contr Rel 2000; 65: 285–296.
48. Jain RA. The manufacturing techniques of various drug loaded biodegradablepoly (lactide-co-glycolide) (PLGA) devices. Biomaterials 2000; 21:2475–2490.
49. Iwata M, McGinity JW. Preparation of multi-phase microspheres of poly(D,L-lactic acid) and poly(D,L-lactic-co-glycolic acid) containing a W/O emulsion by a multiple emulsion solvent evaporation technique. J Microencapsul 1992; 9:201–214.
50. O'Donnell PB, Iwata M, McGinity JW. Properties of multiphase microspheres of poly(D,L-lactic-co-glycolic acid) prepared by a potentiometric dispersion technique. J Microencapsul 1995; 12:155–163.
51. O'Donnell PB, McGinity JW. Influence of processing on the stability and release properties of biodegradable microspheres containing thioridazine hydrochloride. Eur J Pharm Biopharm 1998; 45:83–94.
52. Alonso MJ, Gupta RK, Siber GR, Langer R. Biodegradable microspheres as controlled release tetanus toxoid delivery systems. Vaccine 1994; 12:299–306.
53. Sah H. Microencapsulation techniques using ethyl acetate as a dispersed solvent: effects of its extraction rate on the characteristics of PLGA microspheres. J Contr Rel 1997; 47:233–245.
54. Lagarce F, Cruaud O, Deuschel C, Bayssas M, Griffon-Etienne G, Benoit JP. Oxaliplatin loaded PLAGA microspheres: design of specific release profiles. Int J Pharm 2002; 242:243–246.
55. Lagarce F, Cruaud O, Benoit JP, Deuschel C. Method for preparing microspheres containing a water-soluble substance. France Patent WO0228386–2002.
56. Meng FT, Ma GH, Qiu W, Su ZG. W/O/W double emulsion technique using ethyl acetate as organic solvent: effects of its diffusion rate on the characteristics of microparticles. J Contr Rel 2003; 91:407–416.
57. Freytag T, Dashevsky A, Tillman L, Hardee GE, Bodmeier R. Improvement of the encapsulation efficiency of oligonucleotide-containing biodegradable microspheres. J Contr Rel 2000; 69:197–207.
58. Tinsley-Bown AM, Fretwell R, Dowsett AB, Davis SL, Farrar GH. Formulation of poly(-lactic-co-glycolic acid) microparticles for rapid plasmid DNA delivery. J Contr Rel 2000; 66:229–241.
59. Ruan G, Feng S-S, Li Q-T. Effects of material hydrophobicity on physical properties of polymeric microspheres formed by double emulsion process. J Contr Rel 2002; 84: 151–160.
60. Sah H. Ethyl formate–alternative dispersed solvent useful in preparing PLGA microspheres. Int J Pharm 2000; 195:103–113.
61. Walter F, Scholl I, Untersmayr E, et al. Functionalisation of allergen-loaded microspheres with wheat germ agglutinin for targeting enterocytes. Biochem Biophys Res Commun 2004; 315:281–287.
62. Kim HK, Park TG. Microencapsulation of dissociable human growth hormone aggregates within poly(lactic-co-glycolic acid) microparticles for sustained release. Int J Pharm 2001; 229:107–116.

63. Wang J, Wang BM, Schwendeman SP. Mechanistic evaluation of the glucose-induced reduction in initial burst release of octreotide acetate from poly(lactide-co-glycolide) microspheres. Biomaterials 2004; 25:1919–1927.
64. Ouchi T, Sasakawa M, Arimura H, Toyohara M, Ohya Y. Preparation of poly[-lactide-co-glycolide]-based microspheres containing protein by use of amphiphilic diblock copolymers of depsipeptide and lactide having ionic pendant groups as biodegradable surfactants by W/O/W emulsion method. Polymer 2004; 45:1583–1589.
65. Aubert-Pouessel A, Venier-Julienne MC, Clavreul A, et al. In vitro study of GDNF release from biodegradable PLGA microspheres. J Contr Rel 2004; 95:463–475.
66. Bouissou C, Potter U, Altroff H, Mardon H, van der Walle C. Controlled release of the fibronectin central cell binding domain from polymeric microspheres. J Contr Rel 2004; 95:557–566.
67. Martinez-Sancho C, Herrero-Vanrell R, Negro S. Optimisation of aciclovir poly(lactide-co-glycolide) microspheres for intravitreal administration using a factorial design study. Int J Pharm 2004; 273:45–56.
68. Kent JS, Sanders LM, Lewis DH, Tice TR. Microencapsulation of water-soluble polypetides. Eur Patent 052,510; 1982.
69. Ruiz J-M, Busnel J-P, Benoit J-P. Influence of average molecular weights of poly(D,L-lactic acid-co-glycolic acid) copolymers 50/50 on phase separation and in vitro drug release from microspheres. Pharm Res 1990; 7:928–934.
70. Ruiz JM, Tissier B, Benoit JP. Microencapsulation of peptide: a study of the phase separation of poly(-lactic acid-co-glycolic acid) copolymers 50/50 by silicone oil. Int J Pharm 1989; 49:69–77.
71. Nihant N, Grandfils C, Jerome R, Teyssie P. Microencapsulation by coacervation of poly(lactide-co-glycolide) IV. Effect of the processing parameters on coacervation and encapsulation. J Contr Rel 1995; 35:117–125.
72. Geze A, Venier-Julienne MC, Mathieu D, Filmon R, Phan-Tan-Luu R, Benoit JP. Development of 5-iodo-2′-deoxyuridine milling process to reduce initial burst release from PLGA microparticles. Int J Pharm 1999; 178:257–268.
73. Hausberger AG, Kenley RA, DeLuca PP. Gamma iIrradiation effects on molecular weight and in vitro degradation of poly(D,L-lactide-co-glycolide) microparticles. Pharm Res 1995; 12:851–856.
74. Thomasin C, Merkle HP, Gander BA. Physico-chemical parameters governing protein microencapsulation into biodegradable polyesters by coacervation. Int J Pharm 1997; 147:173–186.
75. McGee JP, Davis SS, O’Hagan DT. The immunogenicity of a model protein entrapped in poly(lactide-co-glycolide) microparticles prepared by a novel phase separation technique. J Contr Rel 1994; 31:55–60.
76. Mallarde D, Boutignon F, Moine F, et al. PLGA-PEG microspheres of teverelix: influence of polymer type on microsphere characteristics and on teverelix in vitro release. Int J Pharm 2003; 261:69–80.
77. Thomasin C, Ho NT, Merkle HP, Gander B. Drug microencapsulation by PLA/PLGA coacervation in the light of thermodynamics. 1. Overview and theoretical considerations. J Pharm Sci 1998; 87:259–268.
78. Thomasin C, Corradin G, Ying M, Merkle HP, Gander B. Tetanus toxoid and synthetic malaria antigen containing poly(lactide)/poly(lactide-co-glycolide) microspheres: importance of polymer degradation and antigen release for immune response. J Contr Rel 1996; 41:131–145.
79. Niwa T, Takeuchi H, Hino T, Nohara M, Kawashima Y. Biodegradable submicron carriers for peptide drugs: Preparation of -lactide/glycolide copolymer (PLGA) nanospheres with nafarelin acetate by a novel emulsion-phase separation method in an oil system. Int J Pharm 1995; 121:45–54.

80. Leelarasamee N, Howard SA, Malanga CJ, Ma JKH. A method for the preparation of polylactic acid microcapsules of controlled particle size and drug loading. J Microencapsul 1988; 5:147–157.
81. Johnson RE, Lanaski LA, Gupta V, et al. Stability of atriopeptin III in poly(-lactide-co-glycolide) microspheres. J Contr Rel 1991; 17:61–67.
82. Huang Y-Y, Chung T-W, Tzeng T-W. Drug release from PLA/PEG microparticulates. Int J Pharm 1997; 156:9–15.
83. Lewis DH, Sherman JD. Low residual solvent microspheres and microencapsulation process. Patent WO 89/03678;- 1989.
84. Lawter JR, Lanzilotti MG. Hardening agent for phase separation microencapsulation. Eur Patent No. 292,710; 1992.
85. Thomasin C, Merkle HP, Gander B. Drug microencapsulation by PLA/PLGA coacervation in the light of thermodynamics. 2. Parameters determining microsphere formation. J Pharm Sci 1998; 87:269–275.
86. Bodmeier R, Chen H. Preparation of biodegradable poly(+/-)lactide microparticles using a spray-drying technique. J Pharm Pharmacol 1988; 40:754–757.
87. Baras B, Benoit M-A, Gillard J. Parameters influencing the antigen release from spray-dried poly(-lactide) microparticles. Int J Pharm 2000; 200:133–145.
88. Prior S, Gamazo C, Irache JM, Merkle HP, Gander B. Gentamicin encapsulation in PLA/PLGA microspheres in view of treating Brucella infections. Int J Pharm 2000; 196:115–125.
89. Wagenaar BW, Muller BW. Piroxicam release from spray-dried biodegradable microspheres. Biomaterials 1994; 15:49–54.
90. Dickinson PA, Kellaway IW, Taylor G, Mohr D, Nagels K, Wolff H-M. In vitro and in vivo release of estradiol from an intra-muscular microsphere formulation. Int Jl Pharm 1997; 148:55–61.
91. Montini M, Pedroncelli A, Tengattini F, et al. Medical application of intramuscularly administered bromocriptine microspheres in pharmaceutical particulate carriers. In: Rolland A, ed. Therapeutic Applications. New York: Marcel Dekker, 1993:227–274.
92. Blanco-Prieto MJ, Besseghir K, Zerbe O, et al. In vitro and in vivo evaluation of a somatostatin analogue released from PLGA microspheres. J Contr Rel 2000; 67:19–28.
93. Blanco-Prieto MJ, Besseghir K, Orsolini P, et al. Importance of the test medium for the release kinetics of a somatostatin analogue from poly(-lactide-co-glycolide) microspheres. Int J Pharm 1999; 184:243–250.
94. Baras B, Benoit M, Gillard J. Influence of various technological parameters on the preparation of spray-dried poly(epsilon-caprolactone) microparticles containing a model antigen. J Microencapsul 2000; 17:485–498.
95. Amiet-Charpentier C, Gadille P, Benoit JP. Rhizobacteria microencapsulation: properties of microparticles obtained by spray-drying. J Microencapsul 1999; 16:215–229.
96. Bruschi ML, Cardoso MLC, Lucchesi MB, Gremiao MPD. Gelatin microparticles containing propolis obtained by spray-drying technique: preparation and characterization. Int J Pharm 2003; 264:45–55.
97. Billon A, Bataille B, Cassanas G, Jacob M. Development of spray-dried acetaminophen microparticles using experimental designs. Int J Pharm 2000; 203:159–168.
98. Takeuchi H, Yasuji T, Hino T, Yamamoto H, Kawashima Y. Spray-dried composite particles of lactose and sodium alginate for direct tabletting and controlled releasing. Int J Pharm 1998; 174:91–100.
99. Schmiedel R, Sandow J. Microencapsule production containing soluble protein or peptide using mixture of poly:hydroxy-butyric acid and poly:lactide-co-glycolide. Eur Patent 315875A; 1989.
100. Cohen G, Dubois J. Injectable microspheres containing antiestrogenic and antiprogestomimetic steroids. German Patent 4 036 425; 1991.

101. Takada S, Uda Y, Toguchi H, Ogawa Y. Application of a spray drying technique in the production of TRH-containing injectable sustained-release microparticles of biodegradable polymers. PDA J Pharm Sci Technol 1995; 49:180–184.
102. Johnson OL, Jaworowicz W, Cleland JL, et al. The stabilization and encapsulation of human growth hormone into biodegradable microspheres. Pharm Res 1997; 14:730–735.
103. Johnson OL, Ganmukhi MM, Bernstein H, Auer H, Khan MA. Composition for sustained release of human growth hormone. U.S. Patent 5,667,808; 1997.
104. Herbert P, Murphy K, Johnson O, et al. A large-scale process to produce microencapsulated proteins. Pharm Res 1998; 15:357–361.
105. Tracy MA. Development and scale-up of a microsphere protein delivery system. Biotechnol Prog 1998; 14:108–115.
106. Bitz C, Doelker E. Influence of the preparation method on residual solvents in biodegradable microspheres, 1st World Meeting APGI/APV, Budapest, 9–11 May, 1995.
107. Spenlehauer G, Vert M, Benoit JP, Boddaert A. In vitro and In vivo degradation of poly(D,L-lactide/glycolide) type microspheres made by solvent evaporation method. Biomaterials 1989; 10:557–563.
108. Spenlehauer G, Veillard M, Benoit JP. Formation and characterization of cisplatin-loaded poly(D,L-lactide) microspheres for chemoembolisation. J Pharm Sci 1986; 75: 750–755.
109. Benoit JP, Courteille F, Thies C. A physicochemical study of the morphology of progesterone-loaded poly(D,L-lactide) microspheres. Int J Pharm 1986; 29:95–102.
110. Courteille F, Benoit JP, Thies C. The morphology of progesterone-loaded polystyrene microspheres. J Contr Rel 1994; 30:17–26.
111. ICH Harmonized Tripartite Guideline for Residual Solvents, step 4, 1997.
112. Organic Volatile Impurities. United States Pharmacopeia 25-NF 20. Rockville, Maryland, USA, 2002.
113. Debenedetti PG, Tom JW, Sang-Do Yeo, Gio-Bin Lim. Application of supercritical fluids for the production of sustained delivery devices. J Contr Rel 1993; 24:27–44.
114. Thiering R, Dehghani F, Foster NR. Current issues relating to anti-solvent micronisation techniques and their extension to industrial scales. J Supercritical Fluids 2001; 21:159–177.
115. Subramaniam B, Rajewski RA, Snavely K. Pharmaceutical processing with supercritical carbon dioxide. J Pharm Sci 1997; 86:885–890.
116. Reverchon E. Supercritical antisolvent precipitation of micro- and nano-particles. J Supercritical Fluids, 1999; 15:1–21.
117. Jung J, Perrut M. Particle design using supercritical fluids: Literature and patent survey. J Supercritical Fluids 2001; 20:179–219.
118. Richard J, Deschamps F. Supercritical fluid processes for polymer particle engineering. In: Elaisari A, ed. Colloidal Biomolecules, Biomaterials, and Biomedical Applications. New York: Marcel Dekker, 2003:429–487.
119. Subra P, Jestin P. Powders elaboration in supercritical media: comparison with conventional routes. Powder Technol 1999; 103:2–9.
120. Tewes F, Boury F. Thermodynamic and dynamic interfacial properties of binary carbon dioxide-water systems. J Phys Chem B 2004; 108:2405–2412.
121. Rantakylä M, Jäntti M, Aaltonen O, Hurme M. The effect of initial drop size on particle size in the supercritical antisolvent precipitation (SAS) technique. J Supercritical Fluids 2002; 24:251–263.
122. Taylor DK, Corbonell R, DeSimone JM. Opportunities for pollution prevention and energy efficiency enabled by the carbon dioxide technology plateform. Annu Rev Energy Environ 2000; 25:115–146.
123. Dixon DJ, Johnston KP. Formation of microporous polymer fibers and oriented fibrils by precipitation with a compressed fluid antisolvent. J Applied Polymer Sci 1993; 50:1929–1942.

124. Condo PD, Paul DR, Johnston KP. Glass transitions of polymers with compressed fluid diluents: type II and III behavior. Macromolecules 1994; 27:365–371.
125. Tomasko DL, Li H, Liu D, et al. A review of CO_2 applications in the processing of polymers. Ind Eng Chem Res 2003; 42:6431–6456.
126. Holmes JD, Ziegler KJ, Audriani M, et al. Buffering the aqueous phase pH in water-in-CO2 microemulsions. J Phys Chem B. 1999; 103:5703–5711.
127. Tservistas M, Levy MS, Lo-Yim MYA, et al. The Formation of Plasmid DNA Loaded Pharmaceutical Powders Using Supercritical Fluid Technology. Biotechnology and Bioengineering 2001; 72:12–18.
128. Dillow AK, Langer RS, Foster NR, Hrkach JS. Supercritical fluid sterilization method. WO9966960 – 1999.
129. Helfgen B, Turk M, Schaber K. Hydrodynamic and aerosol modelling of the rapid expansion of supercritical solutions (RESS-process). J Supercritical Fluids 2003; 26:225–242.
130. Debenedetti PG, Kwauk X, Tom JW, Yeo SD. Rapid expansion of supercritical solutions (RESS): fundamentals and applications. Fluid Phase Equilibria 1993; 82: 303–310.
131. Benedetti L, Bertucco A, Pallado P. Production of micronic particles of biocompatible polymer using supercritical carbon dioxide. Biotechnol Bioeng 1997; 53:232–237.
132. Tsutsumi A, Nakamoto S, Mineo T, Yoshida K. A novel fluidized-bed coating of fine particles by rapid expansion of supercritical fluid solutions. Powder Technol 1995; 85:275–278.
133. Wang T-J, Tsutsumi A, Hasegawa H, Mineo T. Mechanism of particle coating granulation with RESS process in a fluidized bed. Powder Technol 2001; 118:229–235.
134. Debenedetti PG, Tom JW, Jerome R. Precipitation of poly(L-lactic acid) and composite poly(L-lactic acid)-pyrene particles by rapid expansion of supercritical solutions. J Supercritical Fluids 1994; 7:9–29.
135. Kim J-H, Paxton TE, Tomasko DL. Microencapsulation of naproxen using rapid expansion of supercritical solutions. Biotechnol Prog 1996; 12:650–661.
136. Tom JW, Debenedetti PG. Formation of bioerodible polymeric microspheres and microparticles by rapid expansion of supercritical solutions. Biotechnol Progress 1991; 7:403–411.
137. Thies J, Muller BW. Size controlled production of biodegradable microparticles with supercritical gases. Eur J Pharm Biopharm 1998; 45:67–74.
138. Tom JW, Debenedetti PG. Particle formation with supercritical fluids–a review. J Aerosol Sci 1991; 22:555–584.
139. Sun X-Y, Wang T-J, Wang Z-W, Jin Y. The characteristics of coherent structures in the rapid expansion flow of the supercritical carbon dioxide. J Supercritical Fluids 2002; 24:231–237.
140. Daneshvar M, Kim S, Gulari E. High-pressure phase equilibria of polyethylene glycol-carbon dioxide systems. J Physical Chem 1990; 94:2124–2128.
141. Mertdogan CA, Byun H-S, McHugh MA, Tuminello WH. Solubility of poly(tetrafluoroethylene-co-19 mol hexafluoropropylene) in supercritical CO_2 and halogenated supercritical solvents. Macromolecules 1996; 29:6548–6555.
142. Rindfleisch F, DiNoia TP, McHugh MA. Solubility of polymers and copolymers in supercritical CO_2. J Physical Chem 1996; 38:15581–15587.
143. Yan X, Kiran E. Miscibility, density and viscosity of poly(dimethylsiloxane) in supercritical carbon dioxide. Polymer 1995; 36:4817–4826.
144. Phillips EM, Stella VJ. Rapid expansion from supercritical solutions: application to pharmaceutical processes. Int J Pharm 1993; 94:1–10.
145. O'Neill ML, Cao Q, Fang M, et al. Solubility of homopolymers and copolymers in carbon dioxide. Ind Eng Chem Res 1998; 37:3067–3079.
146. Larson KA, King ML. Evaluation of supercritical fluid extraction in the pharmaceutical industry. Biotechnology Progress 1986; 2:73–82.

147. Tavana A, Chang J, Randolph AD, Rodriguez N. Scanning of cosolvents for supercritical fluid solubilization of organics. AIChE J 1989; 35:645–648.
148. Matsuyama K, Mishima K, Umemoto H, Yamaguchi S. Environmentally benign formation of polymeric microspheres by rapid expansion of supercritical carbon dioxide solution with a nonsolvent. Environ Sci Technol 2001; 35:4149–4155.
149. Mishima K, Matsuyama K, Tanabe D, Yamauchi S, Young TJ, Johnston KP. Microencapsulation of proteins by rapid expansion of supercritical solution with a nonsolvent. AIChE Journal 2000; 46:857–865.
150. Sarbu T, Styranec T, Beckman EJ. Non-fluorous polymers with very high solubility in supercritical CO2 down to low pressures. Nature 2000; 405:165–168.
151. Palakodaty S, York P. Phase behavioral effects on particle formation processes using supercritical fluids. Pharm Res 1999; 16:976–985.
152. Kwauk X, Debenedetti PG. Mathematical modeling of aerosol formation by rapid expansion of supercritical solutions in a converging nozzle. J Aerosol Sci 1993; 24:445–469.
153. Chang CJ, Randolph AD. Precipitation of microsize organic particles from supercritical fluids. AIChE J. 1989; 35:1876–1882.
154. Helfgen B, Hils P, Holzknecht C, Turk M, Schaber K. Simulation of particle formation during the rapid expansion of supercritical solutions. J Aerosol Sci 2001; 32: 295–319.
155. Weber M, Russell LM, Debenedetti PG. Mathematical modeling of nucleation and growth of particles formed by the rapid expansion of a supercritical solution under subsonic conditions. J Supercritical Fluids 2002; 23:65–80.
156. Turk M. Formation of small organic particles by RESS: experimental and theoretical investigations. J Supercritical Fluids 1999; 15:79–89.
157. Mishima K, Yamaguchi S, Umemoto H. Production of fine particles for coating. Japan; NIPPON PAINT CO LTD; Patent JP8104830–1996.
158. Kitagawa K. Production of coated fine particle. Japan; Kobe Steel Ltd; Patent JP5057166 – 1993.
159. Mishima K, Matsuyama K, Uchiyama H, Ide M. Microcoating of flavone and 3-hydroxyflavone with polymer using supercritical carbon dioxide. The 4th International Symposium on Supercritical Fluids, Sendai, Japan, May 1997:11–14.
160. Mishima K, Matsuyama K, Yamauchi S, Izumi H, Furudono D. Novel control of crystallinity and coating thickness of polymeric microcapsules of medicine by cosolvency of supercritical solution. 5th International Symposium on Supercritical Fluids, Atlanta, GA, USA, April 2000:8–12.
161. Pace GW, Vachon GM, Mishra KA, Henriksen IB, Krukonis V, Godinas A. Processes to generate submicron particles of water-insoluble compounds. U.S. Patent WO9965469–1999.
162. Young JT, Mawson S, Johnston KP, Henriksen IB, Pace GW, Mishra AK. Rapid expansion from supercritical to aqueous solution to produce submicron suspensions of water-insoluble drugs. Biotechnol Prog 2000; 16:402–407.
163. Iley WJ. Effect of particle size and porosity on particle film coatings. Powder Technol 1991; 65:441–445.
164. Krause H, Niehaus M, Teipel U. Method for microencapsulation of particules. Patent EP0865819–1998.
165. Pace GW, Mawson S, Henrikson IB, Johnston KP, Mishra AK. Insoluble Drug Delivery. U. S. WO9714407–1997.
166. Bausch A, Hidber P. Process for the manufacture of (sub)micron sized particles by dissolving in compressed gas and surfactants. Patent EP1071402 - WO9952504 - US6299906–1999.
167. Benoit JP, Rolland H, Thies C, Van de Velde V. Method of coating particles. France Patent EP0706821 - WO9611055 - US6087003–1996.

168. Ribeiro Dos Santos I, Richard J, Thies C, Pech B, Benoit JP. A supercritical fluid-based coating technology. 3: preparation and characterization of bovine serum albumin particles coated with lipids. J Microencapsul 2003; 20:110–128.
169. Thies C, Ribeiro Dos Santos I, Richard J, VandeVelde V, Rolland H, Benoit JP. A supercritical fluid-based coating technology 1: process considerations. J Microencapsul 2003; 20:87–96.
170. Domingo C, Garcia-Carmona J, Libre J, Rodriguez-Clemente R. Impregnation of porous supports by solute diffusion from SC-CO_2. A way of preparing controlled drug delivery systems. 6th Meeting on Supercritical Fluids, Chemistry and Materials, Nottingham, U. K., April 10–13, 1999.
171. Liu H, Yates MZ. Development of a carbon dioxide-based microencapsulation technique for aqueous and ethanol-based latexes. Langmuir 2002; 18:6066–6070.
172. Guney O, Akgerman A. Synthesis of controlled-release products in supercritical medium. AIChE J 2002; 48:856–866.
173. Bleich J, Müller BW. Production of drug loaded microparticles by the use of supercritical gases with the aerosol solvent extraction system (ASES) process. J Microencapsul 1996; 13:131–139.
174. Bleich J, Muller BW, Wassmus W. Aerosol solvent extraction system – a new microparticle production technique. Int J of Pharm 1993; 97:111–117.
175. Reverchon E, Della Porta G, De Rosa I, Subra P, Letourneur D. Supercritical antisolvent micronization of some biopolymers. J Supercritical Fluids 2000; 18:239–245.
176. Sieber R, Zehnder B. The 3rd International Symposium on High Pressure Chemical Engineering, Zurich, Switzerland, 1996.
177. Dehghani F, Foster NR. Dense gas anti-solvent processes for pharmaceutical formulation. Curr Opin in Solid State and Materials Sci 2003; 7:363–369.
178. Benoit JP, Richard J, Thies C. Method for preparing microcapsules comprising active materials coated with a polymer and novel microcapsules in particular obtained according to the method. France; Patent FR2753639 - WO9813136 - US6183783–1998.
179. Elvassore N, Baggio M, Pallado P, Bertucco A. Production of different morphologies of biocompatible polymeric materials by supercritical CO_2 antisolvent techniques. Biotechnol Bioeng 2001; 73:449–457.
180. Kordikowski A, Schenk AP, van Nielen RM, Peters CJ, de Swaan Arons J. Volume expansions and vapor-liquid equilibria of binary mixtures of a variety of polar solvents and certain near-critical solvents. J Supercritical Fluids 1995; 8:205–216.
181. De la Fuente Badilla JC, Peters CJ, de Swaan Arons J. Volume expansion in relation to the gas-antisolvent process. J Supercritical Fluids 2000; 17:13–23.
182. de la Fuente Badilla JC, Shariati A, Peters CJ. Guidelines for the design of an optimum gas process: measurements, modeling and thermodynamic interpretation. The 6th International Symposium on Supercritical Fluids, Versailles, France, France, April 28–30, 2003. Vol. 3.
183. Mawson S, Kanakia S, Johnston KP. Coaxial nozzle for control of particle morphology in precipitation with compressed fluid antisolvent. J Appl Polym Sci 1997; 64:2105–2118.
184. Tu LS, Dehghani F, Foster NR. Micronisation and microencapsulation of pharmaceuticals using a carbon dioxide antisolvent. Powder Technol 2002; 126:134–149.
185. Shekunov BY, Hanna M, York P. Crystallization process in turbulent supercritical flows. J Crystal Growth 1999; 198–199:1345–1351.
186. Lin C, Muhrer G, Mazzotti M, Subramaniam B. Vapor-liquid mass transfer during gas antisolvent recrystallization: Modeling and experiments. Ind Eng Chem Res 2003; 42:2171–2182.
187. Subramaniam B, Said S, Rajewski RA, Stella V. Methods and apparatus for particle precipitation and coating using near-critical and supercritical anti-solvents. WO9731691–1997.

188. Richard J, Dulieu C, Le Meurlay D, Benoit JP. Microparticles for pulmonary administration comprising at least a coating agent. WO 0112160–2001.
189. Callegaro L, Benedetti L, Pallado P. Nanospheres comprising a biocompatible polysaccharide. Italian Patent WO9629998 - EP0817620 - US6214384–1996.
190. Chou Y, Tomasko D. Gas crystallization of polymer-pharmaceutical composite particles. The 4th International Symposium on Supercritical Fluids, Sendai, Japan, May 11–14, 1997.
191. Elvassore N, Bertucco A, Caliceti P. Production of insulin-loaded poly(ethylene glycol)/poly(l-lactide) (PEG/PLA) nanoparticles by gas antisolvent techniques. J Pharm Sci 2001; 90:1628–1636.
192. Moneghini M, Kikic I, Voinovich D, Perissutti B, Filipovic-Grcic J. Processing of carbamazepine-PEG 4000 solid dispersions with supercritical carbon dioxide: preparation, characterisation, and in vitro dissolution. Int J Pharmaceutics 2001; 222:129–138.
193. Sethia S, Squillante E. Physicochemical characterization of solid dispersions of carbamazepine formulated by supercritical carbon dioxide and conventional solvent evaporation method. J Pharm Sci 2002; 91:1948–1957.
194. Dixon DJ, Johnston KP, Bodmeier R. Polymeric material formed by precipitation with a compressed fluid antisolvent. AIChE J 1993; 39:127–139.
195. Mawson S, Kanakia S, Johnston KP. Metastable polymer blends by precipitation with a compressed fluid antisolvent. Polymer 1997; 38:2957–2967.
196. Ruchatz F, Kleinebudde P, Müller BW. Residual solvents in biodegradable microparticles. Influence of process parameters on the residual solvent in microparticles produced by the aerosol solvent extraction system (ASES) process. J Pharm Sci 1997; 86:101–105.
197. Randolph TW, Randolph AD, Mebes M, Yeung S. Sub-micrometer-sized biodegradable particles of poly(l-lactic acid) via the gas antisolvent spray precipitation process. Biotechnol Prog 1993; 9:429–435.
198. Dixon DJ, Luna-Barcenas G, Johnston KP. Microcellular microspheres and microballoons by precipitation with a vapour-liquid compressed fluid antisolvent. Polymer 1994; 35:3998–4005.
199. Bleich J, Kleinebudde P, Muller BW. Influence of gas density and pressure on microparticles produced with the ASES process. Int J Pharm 1994; 106:77–84.
200. Bodmeier R, Wang H, Dixon DJ, Mawson S, Johnston KP. Polymeric microspheres prepared by spraying into compressed carbon dioxide. Pharm Res 1995; 12:1211–1217.
201. Falk R, Randolph TW, Meyer JD, Kelly RM, Manning MC. Controlled release of ionic compounds from poly (-lactide) microspheres produced by precipitation with a compressed antisolvent. J Contr Rel 1997; 44:77–85.
202. Falk RF, Randolph TW. Process variable implications for residual solvent removal and polymer morphology in the formation of gentamycin-loaded poly(L-lactide) microparticles. Pharm Res 1998; 15:1233–1236.
203. Tu LS, Dehghani F, Dillow AK, Foster NR. Applications of dense gases in pharmaceutical processing, 5th Meeting on Supercritical Fluids, Nice, France, March 23-25 1998.
204. Subramaniam B, Saim S, Rajewski RA, Stella V. Methods for a particle precipitation and coating using near-critical and supercritical antisolvents. US; U.S. Patent 5,83,3891; 1998.
205. Subramaniam B, Saim S, Rajewski RA, Stella V. Methods for particle micronization and nanonization by recrystallization from organic solutions sprayed into a compressed antisolvent. US; U.S. Patent 5874029–1999.
206. Foster N, Dehghani F, Bustami R, Chan H. Generation of lysozyme-lactose powders using the ASES process, the 6th International Symposium on Supercritical, Versailles, France, April 28–30, 2003.
207. Hanna M, York P. Method and apparatus for the formation of particles. U.K. Patent WO9501221–1994.

208. Witschi C, Doelker E. Influence of the microencapsulation method and peptide loading on poly(lactic acid) and poly(lactic-co-glycolic acid) degradation during in vitro testing. J Contr Rel 1998; 51:327–341.
209. Ghaderi R, Artursson P, Carlfors J. A new method for preparing biodegradable microparticles and entrapment of hydrocortisone in DL-PLGA microparticles using supercritical fluids. Eur J Pharm Sci 2000; 10:1–9.
210. Ghaderi R, Artursson P, Carlfors J. Preparation of biodegradable microparticles using solution-enhanced dispersion by supercritical fluids (SEDS). Pharm Res 1999; 16:676–681.
211. Breitenbach A, Mohr D, Kissel T. Biodegradable semi-crystalline comb polyesters influence the microsphere production by means of a supercritical fluid extraction technique (ASES). J Contr Rel 2000; 63:53–68.
212. Engwicht A, Girreser U, Muller BW. Critical properties of lactide-co-glycolide polymers for the use in microparticle preparation by the aerosol solvent extraction system. Intl J Pharm 1999; 185:61–72.
213. Engwicht A, Girreser U, Müller BW. Characterization of co-polymers of lactic and glycolic acid for supercritical fluid processing. Biomater 2000; 21:1587–1593.
214. Saim S, Subramaniam B, Rajewski RA, Stella VJ. Particle micronization with compressed gas antisolvents. Pharm Res 1996; 13:S-273.
215. York P. Strategies for particle design using supercritical fluid technologies. Pharm Sci Technol Today 1999; 2:430–440.
216. Luna-Barcenas G, Kanakia SK, Sanchez IC, Johnston KP. Semicrystalline microfibrils and hollow fibres by precipitation with a compressed-fluid antisolvent. Polymer 1995; 36:3173–3182.
217. Juppo AM, Boissier C, Khoo C. Evaluation of solid dispersion particles prepared with SEDS. Intl J Pharm 2003; 250:385–401.
218. Thiering R, Dehghani F, Dillow A, Foster NR. The influence of operating conditions on the dense gas precipitation of model proteins. J Chem Technol Biotechnol 2000; 75:29–41.
219. Knez Z, Weidner E. Particles formation and particle design using supercritical fluids. Curr Opin in Solid State and Materials Sci 2003; 7:353–361.
220. Li H, Lee LJ, Tomasko DL. Effect of Carbon Dioxide on the Interfacial Tension of Polymer Melts. Ind Eng Chem Res 2004; 43:509–514.
221. Li H, Han X, Wingert MJ, James Lee L, Koelling KW, Tomasko DL. The role of CO2 in enhancing polymer extrusion processes, The 6th International Symposium in Supercritical Fluids, Versailles, France, April 28–30, 2003. Vol. 3.
222. Kazarian SG. Polymer processing with supercritical fluids. Polymer Sci 2000; 42:78–101.
223. Weidner E, Knez Z, Novak Z. A process and equipment for production of fine particles from gas saturated solutions. Slovenian Patent 940079, WO 09521688 A1 - 1994.
224. Shine A, Gelb J. Microencapsulation process using supercritical fluids. Patent WO9815348 - EP0948396 - US5766637 - 1997.
225. Reverchon E. Supercritical-assisted atomization to produce micro- and/or nanoparticles of controlled size and distribution. Ind Eng Chem Res 2002; 41:2405–2411.
226. Weidner E. Powderous composites by high pressure spray processes, The 6th International Symposium in Supercritical Fluids, Versailles, France, April 28–30, 2003. Vol. 3.
227. Sencar-Bozic P, Srcic S, Knez Z, Kerc J. Improvement of nifedipine dissolution characteristics using supercritical CO_2. Int J Pharm 1997; 148:123–130.
228. Weidner E, Petermann M, Blatter K, Rekowski V. Manufacture of powder coatings by spraying of gas-enriched melts. Chem Eng Technol 2001; 24:529–533.
229. Nalawade SP, Janssen LPBM. Production of polymer particles using supercritical carbon dioxide as a processing solvent in an extruder. The 6th International Symposium on Supercritical Fluids, Versailles, France, April 28–30, 2003. Vol. 3.
230. Deschamps F, Richard J. Processing of pharmaceutical products using SCF: Industrial considerations and scaling-up. In: Bonnaudin N, Cansell F, Fousassier O, eds. Supercritical Fluids and Materials, 2003:331–367.

231. Jung J, Clavier J-Y, Perrut M. Gram to kilogram scale-up of supercritical anti-solvent process, The 6th International Symposium on Supercritical Fluids, Versailles, France, April 28–30, 2003. Vol. 3.
232. Boswell G, Scribner R. Polylactide-drug mixtures. U.S. Patent 3,773, 919; 1973.
233. Vervaet C, Baert L, Remon JP. Extrusion-spheronisation A literature review. Int J Pharm 1995; 116:131–146.
234. Nykamp G, Carstensen U, Muller BW. Jet milling – a new technique for microparticle preparation. Intl J Pharm 2002; 242:79–86.
235. Breitenbach J. Melt extrusion: from process to drug delivery technology. European Jl Pharm Biopharm 2002; 54:107–117.
236. Yano K, Kajiyama A, Hamada M, Yamamoto K. Constitution of colloidal particles formed from a solid dispersion system. Chem Pharm Bull 1997; 45:1339–1344.
237. Carstensen U, Mueller B. New process for the preparation of microparticles, useful e.g., for controlled drug release, comprises encapsulating active agent in biodegradable polymer under heating, cooling and milling in two stages to a fine powder. PHARMATECH GMBH DE10061932 – 2002.
238. Rothen-Weinhold A, Gurny R, Orsolini P, Heimgartner F. Pharmaceutical compositions for the sustained release of insoluble active principles. France; Debio Recherche Pharmaceutique S.A.; U.S. Patent 6,245,346 – 2001.
239. Orsolini P. Method for preparing a pharmaceutical composition in the form of microparticles. Helvetica; Debiopharm S.A.; U.S. Patent 5,134,122–1992.
240. Bouchot O, Soret JY, Jacqmin D, Lahlou N, Roger M, Blumberg J. Three-month sustained-release form of triptorelin in patients with advanced prostatic adenocarcinoma: results of an open pharmacodynamic and pharmacokinetic multicenter study. Horm Res 1998; 0:89–93.
241. Vapreotide: BMY 41606, RC 160, Sanvar. Drugs R D 2003; 4: 326–330.
242. Rowe RC. Spheronization: a novel pill-making process? Pharm Int 1985; 6:119–123.
243. Baert L, Remon JP. Influence of amount of granulation liquid on the drug release rate from pellets made by extrusion-spheronisation. Int J Pharm 1993; 95:135–141.
244. Wichert B, Rohdewald P. A new method for the preparation of drug containing polylactic acid microparticles without using organic solvent. J Contr Rel 1990; 14:269–283.
245. Ruiz JM. Sustained release particles preparation. U.K. Patent GB 2,246,514 A; 1992.
246. Ruiz JM. Preparation process of sustained release compositions and the compositions thus obtained. U.S. Patent 5 213 812 – 1993.

2

Advances in the Technology for Controlled-Release Pesticide Formulations

Arie Markus
The Institute of Chemisty and Chemical Technology, The Institutes for Applied Research, Ben-Gurion University of the Negev, Beer-Sheva, Israel

Charles Linder
The Institutes for Applied Research, Ben-Gurion University of the Negev, Beer-Sheva, Israel

INTRODUCTION

Categories of Pesticide Formulations

The science of pesticide formulation covers a very broad field, as it deals with the development, production, and storage of the formulations, as well as with the interaction of pesticides with the environment, including plants, insects, animals, soil, air, and water (1–5).

Pesticide formulations can be classified into the following types:

- Aqueous solutions
- Emulsifiable concentrates
- Dispersion concentrates (aqueous and nonaqueous flowables)
- Wettable powders
- Dry flowables (water-dispersible granules)
- Controlled-release formulations
- Others (dusts, aerosols, etc.)

The choice of formulation is influenced by the following factors: the physical properties of the pesticide (melting point, solubility, and volatility), the chemical properties of the pesticide (hydrolytic stability, thermal stability, and irradiation stability), the mode of application of the formulation (soil, foliar), the crop to be treated and the agricultural practices, the biological properties of the pesticide (crop selectivity, transport through soil or grass cover, and LD_{50} for mammals and nonmammalian species), and economic considerations.

This chapter covers methods of pesticide encapsulation and formulation, controlled-release techniques for pesticide application, and the current trends in the use of encapsulated natural products. Pesticides are conventionally applied to crops by periodic broadcasting or spraying. Very high, and possibly toxic, concentrations are applied initially, and these often decrease rapidly in the field to concentrations that fall below the minimum effective level. As a result, repeated applications are needed to maintain pest control (6). The formulation of a pesticide must thus be designed to meet the multifaceted demands of efficacy, suitability to mode of application, and minimization of damage to the environment. Controlled-release formulations meet these demands in that they enable smaller quantities of pesticide to be used more effectively over a given time interval and in that their design enables them to resist the severe environmental processes, i.e., leaching, evaporation, and photolytic, hydrolytic, and microbial degradation, that act to eliminate conventionally applied pesticides (6).

Release Patterns from Encapsulated Formulations

In most instances, the rate of removal of a conventionally formulated pesticide follows first-order kinetics (6–9). The time t_e that an effective level of pesticide is maintained after a single application is given by:

$$t_e = (1/k_r)\ln(M_\infty/M_e) \qquad (1)$$

where M_e is the minimum effective level, M_∞ the amount of agent applied initially, and k_r the removal rate constant (6).

It follows from Equation (1) that for an increase in the effective duration of action of a conventionally applied pesticide, an exponentially greater quantity of the pesticide would be required. If, however, the pesticide could be maintained at the minimum effective level by a continuous supply from a controlled-release system, then the optimum performance of the pesticide would be realized, and in that case, the duration t_0 of action of the pesticide would be given by:

$$t_0 = (M_\infty - M_e)(k_d M_e) \qquad (2)$$

where k_d is the rate constant for pesticide delivery from the controlled-release device (6).

The relationship between the concentration of application and the duration of action of a pesticide is shown in Figure 1 for a conventional formulation and for a controlled-release formulation (6). The area between the two curves represents the amount of pesticide that is wasted. It is apparent that for a short duration of effectiveness, e.g., one week or less, the efficiency of the conventional method is adequate. As the duration of effectiveness increases, the efficiency of the conventional system decreases exponentially (Fig. 1).

The different release rates, from an encapsulated pesticide formulation, are a function of the concentration of the remaining encapsulated pesticide. For the case of a pesticide core enclosed in a polymeric membrane, when the release rate is independent of the pesticide concentration (in effect, the thermodynamic activity remains constant), pesticide release is governed by zero-order kinetics; in effect, $dC/dt = k_d[C]^0 = k_d$, where C is the concentration of pesticide remaining in the microcapsule. When first-order kinetics are followed, the release rate is directly proportional to the remaining C; in effect, $dC/dt = k_d[C]^1$, indicating that the concentration of the remaining pesticide decreases exponentially with time. In the third type of release kinetics, the release rate is proportional to the square root of time, i.e., $t^{-1/2}$. A comparison of the kinetic release patterns (Fig. 2) shows that the release rates of zero-order and $t^{-1/2}$ formulations exceed that of the first-order formulations for equivalent pesticide quantities. The optimal rate of solute delivery from controlled-release devices is often a constant rate of zero-order release. Thus, a benefit of pesticide formulations comprising a core of pure pesticide encapsulated in a

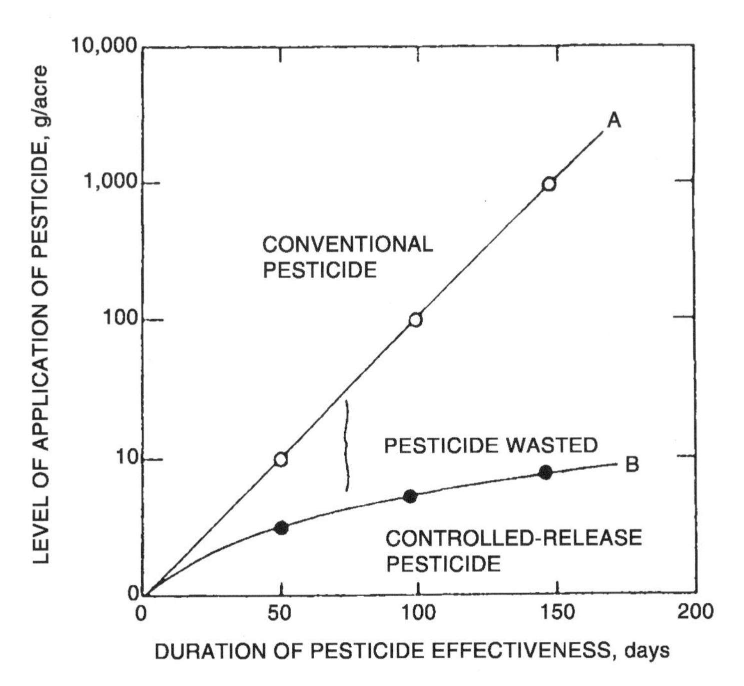

Figure 1 Relationship between the level of application and the duration of action for conventional and controlled-release formulations. The graphs are plotted using Equations (1) and (2) and assuming that the half-life of the pesticide is 15 days and the minimum effective level is 1 g/acre. *Source*: From Ref. 6.

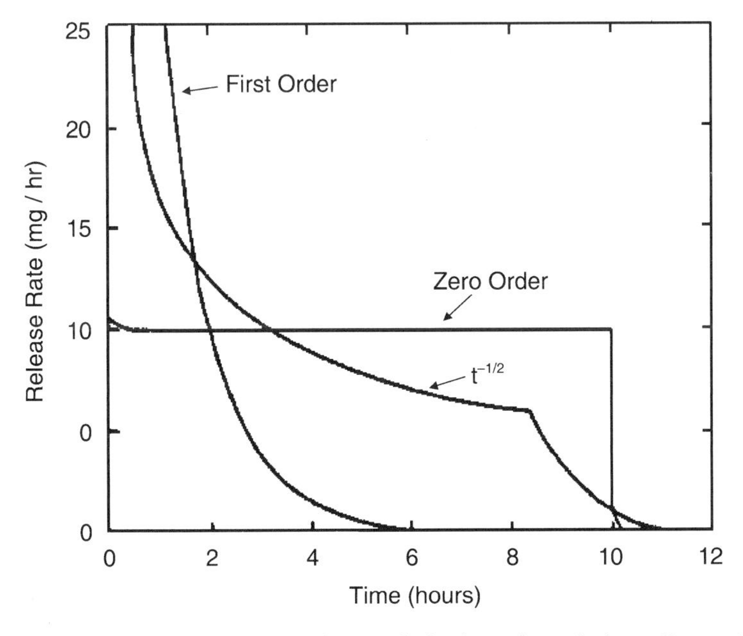

Figure 2 Release kinetics of controlled release formulations. *Source*: From Ref. 10.

polymeric membrane is related to zero-order kinetics. This implies that the release rate and duration of release are not related to each other and can be independently optimized. Release rates can be readily controlled by adjusting membrane thickness, microcapsule surface area, and membrane permeability, while duration of activity is controlled by the quantity of pesticide encapsulated.

Two frequently used material combinations in controlled-release systems are membrane-coated reservoirs and monolithic matrices, which are based on diffusion of the pesticide ingredient through polymer barriers. Other less common configurations are liquid membranes comprising a liquid within the pores of inert porous matrices, and laminated polymeric membranes whose release profiles are also based on diffusion through homogeneous barriers. Irrespective of the geometry of the matrix, i.e., flat wafer, hollow tube, microcapsule, or tablet) zero-order release (in effect ($dC/dt = k_d[C]^0 = k_d$) for encapsulated pesticides is expected under the following conditions: (i) when the thermodynamic activity of the pesticide solute within the reservoir remains constant and (ii) if pesticide transport to and from the membrane surfaces is rapid. Under these conditions, controlled release for the three most common geometries are given by the following equations (10):

$$\text{Slab: } dQ/dt = ADS/l$$

$$\text{Cylinder: } dQ/dt = 2\Pi hDS/\text{In}(r_o - r_i)$$

$$\text{Sphere: } dQ/dt = 4\Pi DS[r_o r_i/(r_o - r_i)]$$

where dQ/dt is the quantity of material released per unit time (g/sec), D the diffusivity of the pesticide through the polymer membrane (cm^2/sec), S the solute solubility within the polymer membrane (g/cm^3), A the surface area of the slab, l the membrane thickness, h the length of the cylinder, and r_o and r_i are the outer and inner radii, respectively, of a cylinder or sphere.

The minimum effective dose M_e, the kinetics k_d of pesticide release from the source, and the rate of elimination k_r of the pesticide from the site of application determine the quantity of pesticide to be applied for a given duration of activity. For example, for a pesticide formulation that is applied to a water surface, the rate of removal of free pesticide from the surface is expected to decrease exponentially; in effect, $dC_{pes}/dt = -k_r C_{pes}$ where dC_{pes}/dt is the rate of change of pesticide concentration, k_r the rate constant of removal, and C_{pes} the concentration of the free pesticide. Encapsulated formulations with a zero-order release rate from the microcapsules may be described kinetically by $dQ/dt = k_d$, where k_d is the constant input rate from the microcapsules into water and dQ/dt the quantity of pesticide released per unit time. Upon application of such a depot, a steady state will be reached when the rate of pesticide release to the surface equals the rate of removal, then $dQ/dt = k_r C_{pes} V = k_d$ and C_{pes} is proportional to k_d/k_r, where V is the apparent water surface volume containing the microcapsules. From this equation, it can be seen that C_{pes} is a function of two kinetic constants, k_r, which is determined by the solubility of the free pesticide in water, the evaporation rate, the sensitivity of the pesticide to ultraviolet (UV) radiation, and oxygen, the temperature, and the wind velocity, and k_d, which is determined by the physicochemical properties of the microcapsules. A steady state close to the minimum effective concentration can be maintained only if k_d is sufficiently large relative to k_r. However, this k_d must be achievable with a single dose of reasonable size.

Advantages of Controlled Release

In addition to the advantages described above, controlled-release pesticides have other important advantages over conventional formulations (6):

- reduction of mammalian toxicity for highly toxic substances,
- extension of duration of activity for an equal level of active agent,
- reduction of evaporative losses and of flammability of liquids,
- reduction of phytotoxicity,
- protection against environmental degradation,
- reduction of leaching into the earth and transportation into streams,
- reduction of contamination of the environment,
- increased convenience of use by conversion of liquid materials into solids and flowable powders,
- separation of reactive components,
- control of the release of active agents,
- decreased costs as less active material is needed,
- convenience of handling.

For the most part, microencapsulation of a pesticide is utilized when slow or controlled release of the active material is desired. It is also possible to produce microencapsulated formulations of pesticides for which the object is not necessarily controlled release, but rapid release, as is required in formulations for foliar applications. Such compositions contain the pesticide in small microcapsules with relatively thin encapsulating wall membranes, allowing for relatively quick release of the total contents of the microcapsules. Even though controlled release may not be the objective of such formulations, they nevertheless carry the advantages of reduced oral and dermal toxicity and minimization of dust formation in comparison with the non-microencapsulated material. In addition to the pesticide itself, the core of the microcapsules may also contain solvents, surfactants, and materials such as titanium dioxide or zinc oxide to minimize ultraviolet damage.

Although the advantages of controlled release are impressive, the disadvantages of the technology must not be forgotten. Each formulation has to be examined individually, and the positive and negative aspects weighed carefully (6). The following aspects of controlled-release formulations—some of which may be deleterious—thus require careful appraisal: (i) the costs of the materials and processing of the controlled-release preparation, which may be substantially higher than those of standard formulations, (ii) the fate of the polymer matrix (see below) and its effect on the environment, (iii) the fate of polymer additives such as plasticizers, stabilizers, antioxidants, and fillers, (iv) the environmental impact of the degradation products of the polymer matrix produced in response to heat, hydrolysis, oxidation, solar radiation, or biological agents, and finally (v) the cost, time, and probability of success in securing government registration of the product, if required (6).

STANDARD MICROENCAPSULATION PROCESSES FOR PESTICIDES

General Methods of Encapsulation

A variety of techniques have been proposed for the production of microcapsules (11–14). Indeed, one source suggests that hundreds of methods are to be found in the scientific and patent literature. In general, the methods for microencapsulation may be classified as follows:

- separation from an aqueous solution,
- formation by polymer–polymer incompatibility,
- interfacial polymerization,
- polymerization in situ,
- drying from a liquid state,
- solvent evaporation from an emulsion,
- gelation in the liquid state by cooling, and
- desolvation.

Physical techniques, such as spray drying or fluidized bed reactions, are not usually suitable for encapsulation of pesticides. Similarly, addition polymerization is not suitable as many pesticides contain P, N, or S atoms, which are known to be radical scavengers, or because the compounds may not be stable in acid or alkali media. Interfacial polymerization is the method of choice for encapsulation of highly toxic insecticides, as the active ingredient is completely enveloped by the polymer, and in most cases, release takes place via diffusion. This technique facilitates a dramatic decrease in toxicity. Both phase separation and interfacial-polymerization techniques may be used for nontoxic pesticides. The only technique suitable for biological pesticides is phase separation; with any other technique the active ingredient, which in this case is a biological microorganism, would not be able to cross the envelopes surrounding it.

Interfacial Polymerization

The technique for the microencapsulation of pesticides by interfacial polymerization comprises two stages. In the first stage, a liquid, molten, or dissolved pesticide containing a dissolved monomer—together comprising the organic phase—is agitated at a high speed in an aqueous solution, with emulsifiers and stabilizers. In the second, interfacial polymerization at the droplet surface is completed by the addition of a second water-soluble monomer to the continuous aqueous phase. The resulting aqueous slurry may be used as such or the pesticide-containing capsules may be recovered as a dry powder. When the pesticide is water miscible, it is also possible to carry out interfacial polymerization by emulsifying the liquid or molten pesticide in an organic phase, such as high-boiling petroleum ether.

The following factors influence the performance of the capsules:

1. *Composition of the polymeric capsule wall*: The monomers can be chosen to produce a variety of interfacial polycondensation products to be used as wall materials, e.g., polyamides, polyesters, polyureas, polyurethanes, or polycarbonates. Within each of these categories, a range of polymers may be produced, depending on the starting monomers. It is also possible to vary the wall composition by forming copolymers in a simple process that is based on a mixture of oil-soluble or water-soluble monomers.
2. *Degree of cross-linking*: For cases in which a higher degree of wall integrity is essential, multifunctional monomers are used.
3. *Capsule wall thickness*: The thickness of the capsule wall is a function of the concentration of the monomers. In the production of the capsule, the wall continues to thicken until all of the available monomer is consumed.

4. *Capsule size*: The size of capsules is determined by the degree of agitation and by the type of emulsifying agent used in the continuous phase. Capsule size can be varied from an average diameter of a few microns to a millimeter. If the formulation is to be sprayed, then the diameter of the capsule must be less than 100 μm.
5. *Physical form of the product*: The final product may take the form of an aqueous slurry or microcapsules that can be filtered off, washed, and dried to a free flowing powder. Aqueous slurries are indeed useful in the formulation of pesticides, as they need only to be diluted with water in order to be ready for field application as a sprayable product. In addition, they are cheaper, being more economical to manufacture.
6. *Additives*: A range of additives, such as UV light absorbers, antioxidants, and synergists, may be dissolved in the oil phase (pesticide) being encapsulated. It should be remembered that these additives must be soluble in the oil and must not react with the monomers.

Typically, effective pesticide formulations with zero-order release kinetics comprise microcapsules containing a pesticide core surrounded by a thin polyurea membrane formed by interfacial polymerization between a polyisocyanate and a polyamine (15,16). The latter polymers are generally preferred because they can form thin but strong and robust films around the pesticide core, and the conditions of formation are relatively mild. In a typical procedure for the microencapsulation of a pesticide, a hydrophobic oil phase containing an aromatic polyisocyanate as the membrane former is dissolved in a suitable nonwater miscible liquid containing the active component. The immiscible phase containing the isocyanate and the active ingredient is then emulsified (to approximately the desired size of the final microcapsules) in an aqueous solution containing stabilizing components such as surfactants or steric stabilizers, for example, polyvinyl alcohol (PVA). To the aqueous solution are added water-soluble polyfunctional reagents, such as amines, which rapidly react with the isocyanate groups at the interface between the immiscible droplet and the water phase to form a thin polyurea film. Sufficient time is allowed for all the isocyanate to react with the polyamine to form urea moieties. The resulting microcapsules have a thin, but strong, polyurea membrane surrounding the immiscible droplets. The membrane thickness is directly proportional to the mole fraction of the polyisocyanate (17). This interfacial-polymerization process is especially suitable for encapsulating water-dispersible formulations containing water-insoluble liquid pesticides, for example, insecticidal pyrethroids such as lambda-cyhalothrin, permethrin, cypermethrin, and many others, or relatively low-melting pesticides, such as the herbicides napropamide and fluazifopbutyl.

A recent example of the use of polyurea microcapsules for reducing toxicity and environmental impact is the encapsulation of the organophosphate insecticide, nematicide cadusafos (S,S-di-*sec*-butyl *O*-ethyl phosphorodithioate). The encapsulated product features significantly reduced human toxicity, less environmental impact, and enhanced stability but is as effective as the nonencapsulated formulation (18). Microcapsules of cadusafos are formed by microemulsion, followed by interfacial polymerization of a polyfunctional monomer such as polymethylene polyphenyl isocyanate with a polyfunctional amine to form an aqueous suspension of microcapsules with a polyurea shell surrounding a core of cadusafos. The components of the membrane are chosen such that the shell is sufficiently impenetrable to the cadusafos to reduce mammalian toxicity (skin, oral, and inhalation toxicity) of the formulation—in comparison with aqueous microemulsion cadusafos formulations of equivalent concentrations—and without loss of pesticidal activity or physical and chemical stability—in comparison with nonmicroencapsulated formulations.

Microencapsulation by interfacial polymerization may reduce the initial efficacy of pesticides exhibiting a low "knockdown effect." It is thus desirable to develop formulations with reduced toxicity that have good initial activity together with long-term effectiveness; for example, haloacetanilides [e.g., alpha-chloroacetanilides, such as alpha-chloro-6′-ethyl-*N*-(2-methoxy-1-methylethyl)-acetanilide] that have been encapsulated in polyurea shells, produced by interfacial methods using phase-transfer catalysts or using different polyamines in the aqueous phase, have reduced initial effectiveness (19,20). It is possible to perform interfacial polymerization, which results in a polyurea shell, without adding amines or phase-transfer catalysts to the water. In this case, the interfacial polymerization is carried out by using the reaction of the polyisocyanate (e.g., polymethylene polyphenylisocyanate and isomeric mixtures of toluene diisocyanate) with water to form an amine, which then reacts with other isocyanate groups to give a polyurea, without the addition of catalysts of other reactive components (21). With this approach, the encapsulation method can produce formulations with equivalent initial herbicidal activity while at the same time reducing both the number of applications required and the mammalian toxicity, compared with the nonencapsulated formulations.

The issue of large microcapsules with slow sustained release versus small approximately 5-μm capsules that provide both rapid initial release (due to the larger surface to volume area) and reduced mammalian toxicity becomes important when a rapid knockdown effect is required or when small capsules are needed to facilitate migration through soil or surface grass. Despite the demand for controlled-release formulations that exhibit an early knockdown effect followed by sustained release of the active material, little work has been done on the production of stable, very small, rapid-release pesticide microcapsules. The reason for this situation is that such small microcapsules (typically less than 5 μm), which have extremely thin polymer shell walls, are difficult to produce uniformly and reproducibly. In a recent patent assigned to Dow, U.S. small, microencapsulated pesticide (e.g., chlorpyrifos) formulations were produced by an interfacial polycondensation reaction between PVA and a polyamine (e.g., diethylene triamine) in the aqueous phase with a polyisocyanate (e.g., Voranate M220) (22). It was shown that it is indeed possible to produce such small microcapsules with improved storage stability, i.e., with no leaching of the active material from the microcapsules. The use of PVA is important in the production of such microcapsules. The pendant –OH groups of the PVA, present during the polycondensation reaction that forms the microcapsule walls, react with the isocyanate to form a polyurethane group. The mixture of polyurethane and polyurea (formed between the polyamine and the isocyanate) components in the encapsulating membrane changes and gives a membrane permeability that differs from that of a polyurea membrane. By using PVA as a reactant in the encapsulation reaction and varying the time before the addition of the polyamine, the polyurethane/polyurea ratio can be varied as a way of optimizing the characteristics of the encapsulating membrane; for example, as each polymer type has a different permeability, the polyurethane/polyurea ratio can thus control active ingredient release rates as well as the microcapsule diameter and the thickness of the encapsulating membrane wall. In addition, PVA forms—on the surface of the capsule—hydrophilic pendants, which act as a steric barrier to prevent aggregation during production and storage, to facilitate spray drying for dry microcapsule formulations, and to enhance subsequent redispersion in water just before use.

Microencapsulated pesticides are generally sold in the form of the aqueous suspensions in which they are produced. There is, however, a need for dry water-dispersible powders for certain applications. Dry formulations that can easily be mixed with water to produce a sprayable material have the following advantages: (i) higher pesticide loading, (ii) easy removal from containers, (iii) less environmental contamination, (iv) longer shelf life, and (v) smaller volumes providing for more economical transportation and storage. Dry formulations are usually prepared by spray drying an aqueous suspension of microcapsules of the water-insoluble pesticide, together with a spray-drying adjuvant; for example, the water-soluble polymer PVA or polyvinyl pyrrolidone (PVP), a salt such as ammonium sulfate or sodium, potassium, or calcium chloride, a surfactant, or some other water-soluble or water-dispersible material, such as gum, clay, or silica (23). While all the above mentioned spray drying adjuvants can be used for relatively large thick-walled microcapsules of pesticides, the list of adjuvants, suitable for producing water-dispersible formulations of microcapsules of relatively small particle size with relatively thin walls, is limited to synthetic and natural polymers such as PVP, PVA, polyethylene oxides, ethylene/maleic anhydride copolymer, methyl vinyl ether-maleic anhydride copolymer, water-soluble cellulose, water-soluble polyamides or polyesters, copolymers or homopolymers of acrylic acids, water-soluble starches and modified starches, natural gums such as alginates and dextrins, and proteins such as gelatins and caseins (24).

To enhance the efficiency of pesticide formulations, signal substances such as pheromones, kairomones, or other attractants may be added to insecticides in microencapsulated form or bound in water-soluble polymers (25). The release of the signal substances at the infested sites leads to migration of the pests to these sites, where they are exposed to the pesticidally active compounds and destroyed. This type of insect control is known as the "attract-and-kill" method. For the compositions to be effective over a sufficiently long period of time, the attractants must be formulated in such a way that they can be released in a controlled manner and are protected against environmental degradation by factors such as light, oxygen, moisture, and high temperatures.

To date, a wide range of insecticides, herbicides, and fungicides have been encapsulated by interfacial polymerization within a variety of polymeric shells, with polyurea and polyurethane being particularly popular. The company that introduced encapsulated pesticidal formulations to the market is the Pennwalt Corporation. Their first insecticidal product, Penncap-M (encapsulated methylparathion), was followed by KNOX OUT 2FM (encapsulated diazinon), Penncaptrin (encapsulated permethrin), and Pennphos (encapsulated chlorpyriphos), and later by the herbicide trifluralin (26). The Pennwalt polyurea envelope, synthesized by the reaction of polyisocyanate with amines and acid chlorides, and the method of encapsulation are described in the patents (26–29).

Another company active in the market is the Stauffer Chemical Company, whose interest lies mainly in encapsulated herbicides. The Stauffer process for the production of the polyurea envelope differs from that of Pennwalt in that the isocyanate also serves as a source of amine. In the reaction process, some of the isocyanate is converted into the amine, which then reacts with the remaining isocyanate (30–33).

A number of other companies are also involved in the production of encapsulated biocides and herbicides. Among these, Monsanto Chemical Company markets encapsulated alachlor under the trade name of Lasso. The envelope for the alachlor formulation is also based on polyurea and polyurethane, and the method and chemicals—isocyanates, amines, and emulsifiers—used by Monsanto are described in the patent literature (34–38). Lever Israel produces NO-ROACH (encapsulated diazinon), Chimgat 2000 produces De-Bugger® (encapsulated pyrethroids) and Effective Ultra (encapsulated propoxur), and Mahkteshim manufactures Master 25 (encapsulated chlorpyrifos) (39). In all the products mentioned above, the toxicity of the formulation is at least one-tenth that of the emulsifiable concentrate, and the duration of efficacy is longer without a reduction of insecticidal activity. Several other companies (40–44) have used polyurethane or polyurea shells as envelopes for pesticides; among the biggest of those are Hoechst, Bayer AG (45), and Ciba-Geigy (46).

The process for the production of encapsulated pesticides—in this case diazinon and pyrethroids—is presented in Scheme 1 (47).

So far, we have discussed the interfacial-polymerization technique in which polyureas, polyurethanes, and polyamides are used as the envelope for the pesticides. We have also made a passing reference to envelopes made from other polymers, including urea–formaldehyde and melamine–formaldehyde polymers, polyesters, epoxy, and polysilane. The formulations based on these polymers as envelopes have, so far, not been successfully commercialized.

Phase Separation

The second technique used in pesticide encapsulation is phase separation. The capsule obtained by this method is generally of the matrix type. This technique is suitable only for nontoxic and biological pesticides, as encapsulation into matrix-type capsules does not considerably reduce the toxicity of the product. An advantage of this mode of encapsulation is that, should the need arise, it is possible to grind the formulation, as in most cases the active ingredient is absorbed onto and within the polymer matrix.

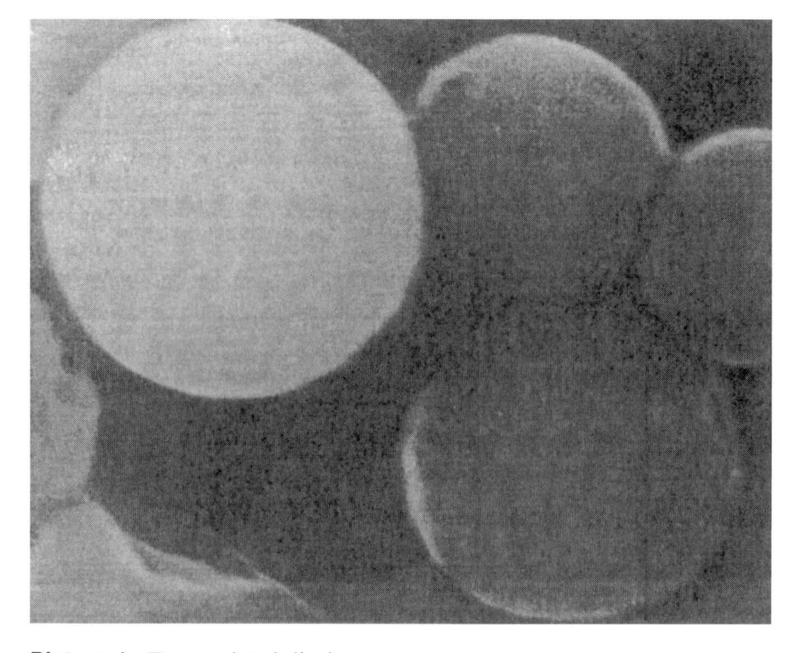

Picture 1 Encapsulated diazinon.

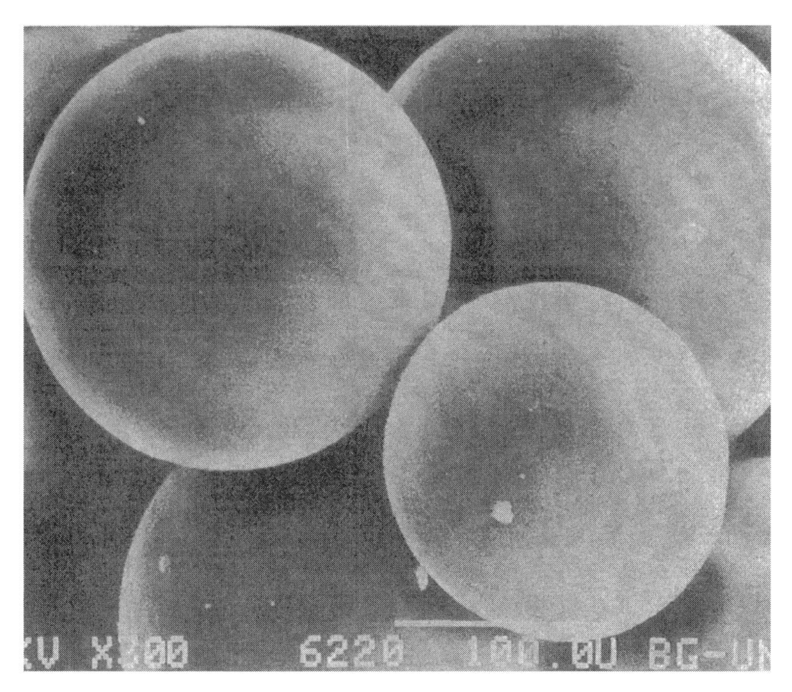

Picture 2 Encapsulated diazinon.

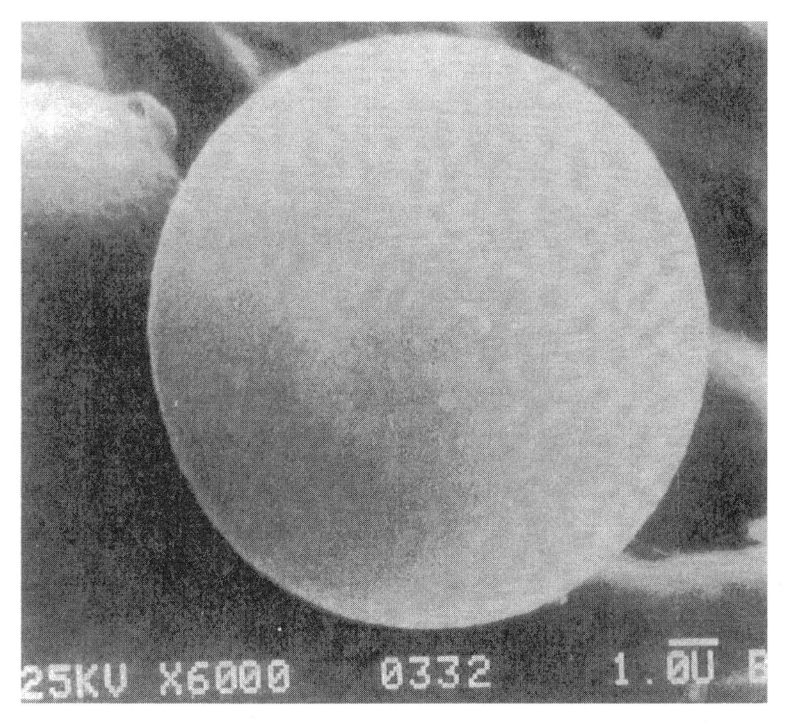

Picture 3 Encapsulated pyrethroids.

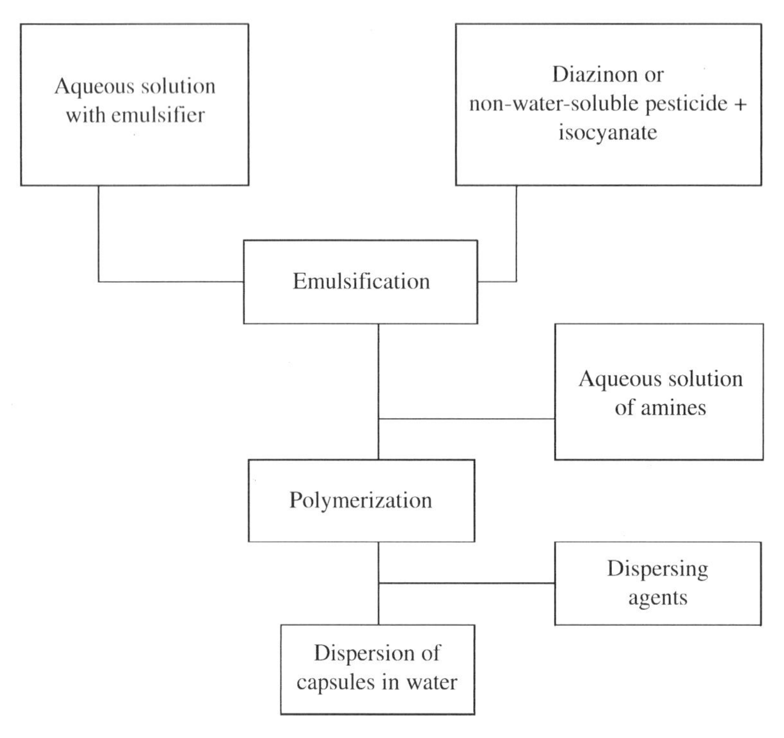

Scheme 1 Production of an encapsulated pesticide.

In the phase separation technique, the active ingredient is suspended in a solution of the wall material. The wall polymer is then induced to separate out as a viscous liquid phase by coacervation (i.e., by adding a nonsolvent), by lowering the temperature, by adding a second polymer, or by a combination of these methods. Coacervation, which may be simple, complex, or of the "salting out" type, is manifested as turbidity, droplet formation, or the separation of liquid layers as described below.

Simple coacervation occurs when a water-miscible nonsolvent (e.g., ethanol) is added to an aqueous polymer solution, causing the formation of a separate polymer-rich phase. A typical example of a simple coacervation system is water/gelatin dissolved in water; this system is, however, difficult to control, so it is little used in practice.

Complex coacervation takes place by the mutual neutralization of two oppositely charged colloids in aqueous solution. One of the most widely used methods of microencapsulation by this process is the use of a solution of positively charged gelatin ($pH < 8$), which forms a complex coacervate on being neutralized with negatively charged gum arabic. Other polymer systems may be employed, and other electrolytes may be used with gelatin. Complex coacervation is closely related to the precipitation of colloidal material from solution in that it immediately precedes precipitation. The process was originally developed in the 1950s—with gelatin and gum arabic as the two colloids—for the coating of carbonless copying paper.

In salt coacervation, the polymer is separated out from the aqueous solution by salting out, typically by adding an electrolyte to an aqueous polymer solution. The method may be used to encapsulate water-insoluble oils or dispersed solid particles, but it is difficult to control the microcapsule size and the agglomeration of particles. The system may be stabilized by altering the pH or temperature.

The polymers that can be used in the phase separation technique are summarized below (6):

Natural polymers	Synthetic elastomers	Synthetic polymers
Alginate	Butyl rubber	Chlorinated polyethylene
Bark	Chloroprene	Ethylenevinylacetate copolymer
Carboxymethylcellulose	Ethylene/propylene/diene terpolymer	Poly(*p*-xylylene)
Cellulose acetatephthalate	Hydrin rubber	Polyacrylate
Cellulose nitrate	Neoprene	Polyacrylonitrile
Ethylcellulose	Polybutadiene	Polyether
Galacturonic acid salt	Polyisoprene	Polyethylene
Gelatin	Polysiloxane	Polyhydroxyethyl methacrylate
Gum arabic	Silicone rubber	Polymethylmethacrylate
Kraft lignin	Styrene-butadiene rubber	Polypropylene
Methocel		Polystyrene
Methylcellulose		Polyvinyl chloride
Natural rubber		Polyvinylidene chloride
Propylhydroxycellulose		Polyvinyl acetate
Proteins		PVA
Shellac		PVP
Starch		
Waxes—paraffin		
Zein		

ADVANCES IN ENCAPSULATION TECHNOLOGIES

Biological Pesticides

In addition to currently used synthetic pesticides, there are strong environmental and regulatory incentives to use "natural pesticides." For example, biological pesticides, such as *bacillus thuringiensis israeliensis* (Bti) and *trichoderma harzianum*, have been encapsulated in matrix-type capsules in which envelopes are produced from two types of polymer—natural polymers such as alginates or synthetic polymers such as polyethylene (48–51). The methods of encapsulation are shown in Schemes 2 and 3. For example, the most common tactic used to control the larvae of the gypsy moth *Lymantria dispar* (L.) (Lepidoptera: Lymantridae), a destructive pest infesting cork oak forests in the eastern United States, Europe, and Asia, relies on the use of Bti-based insecticides (48).

Insect Growth Regulators (IGR)

In addition to currently used synthetic pesticides such as organophosphates, organochlorines, carbamates, and petroleum oils, use of encapsulated insect growth regulators (IGR), for example, methoprene or pyriproxyfen, is increasing. For example, ALTOSID®, produced by Wellmark International, and developed in its encapsulated slow-release form by Southwest Research Institute (SwRI), is a mosquito larvicide used in the United States to reduce mosquito infestations by preventing immature mosquito larvae from becoming disease-spreading adults. The active

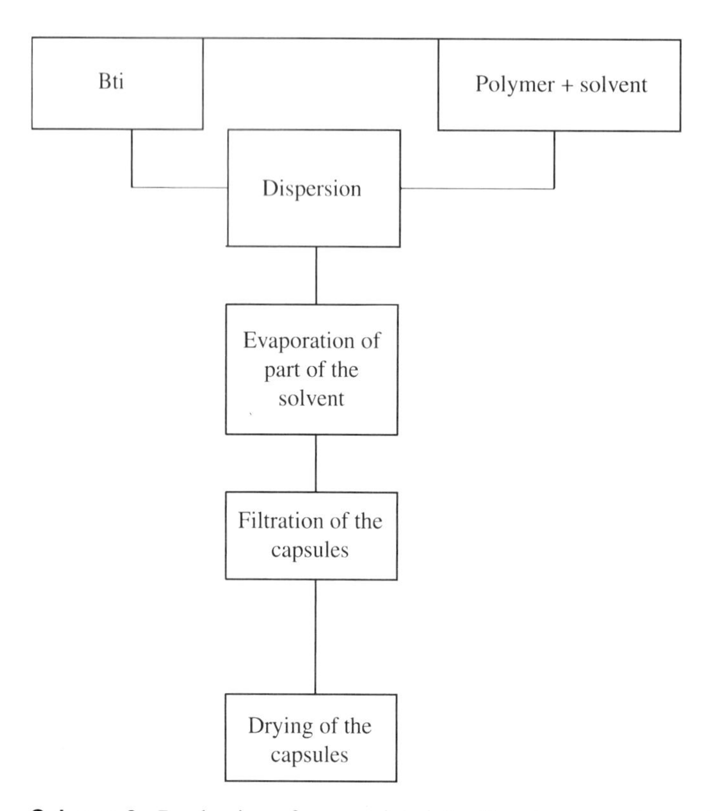

Scheme 2 Production of encapsulated *Bacillus thuringiensis israeliensis* (Bti).

ingredient, methoprene, is an insect growth regulator that interferes with normal mosquito development.

Essential Oils as Natural Pesticides

Alongside the use of natural pesticides like pyrethroids and bacterial larvicides, there has been a recent upsurge in the development of essential oils as "green pesticides" for such applications as larvicides, and moth and lice repellants (52–54). Formulations of essential oils are designed to compete with currently used pesticide products that are either highly toxic or very expensive or are harmful to the environment. In fact, before the development of the modern chemical and pharmaceutical industries, essential oils were used in many areas of daily life as antiseptic and disinfectant materials in pharmaceutical and cosmetic applications, e.g., as antimicrobial (antiviral, antibacterial, and antifungal) and larvicidal agents. With time, the toxic effects modern synthetic chemicals have on the environment have become apparent, and there is now an effort to replace them with the same essential oil agents that they replaced.

Many "green" materials, including essential oils, are less efficient and more expensive than the synthetic chemicals they seek to replace. There is thus a need to produce these green materials with a smaller effective dosage and increased effectiveness by enhancing the duration of activity per dose. This need is met by encapsulation of the oils, which are then released at a constant rate over a long period of time, thus increasing the duration of activity per dose and lowering the

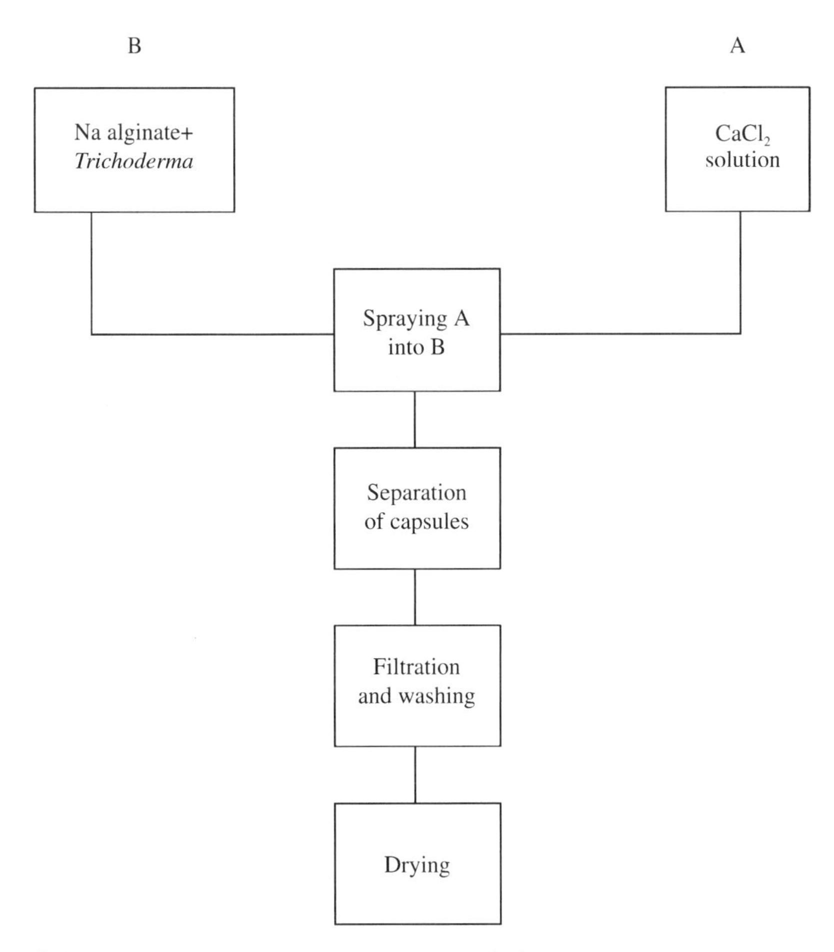

Scheme 3 Production of encapsulated *Trichoderma.*

quantity needed and hence reducing the cost of the product. Encapsulation also stabilizes the volatile essential oils with respect to evaporation, UV degradation, and oxidation, and facilitates the ease of application.

The patent literature on encapsulated essential oils can be divided into four categories as follows: (i) patents describing all methods of encapsulation and a wide range of polymer encapsulants but giving limited examples and claims, (ii) patents based on a solid core containing the essential oils, with and without subsequent coatings, (iii) patents describing the encapsulation of essential oil droplets or emulsions in a polymer shell by coacervation or adsorption of preformed polymers, and (iv) patents relating to the encapsulation in microorganisms.

A number of examples from the patent literature follow. Encapsulated formulations of essential oils containing terpenoids are good lice repellents; these are safe, efficacious, pleasant smelling, and relatively inexpensive and are not toxic to man and animals (54). The encapsulation facilitates both quick action and delayed release. A number of synthetic and natural agents, including biodegradable natural or synthetic polymeric microparticles (including microspheres and microcapsules), liposomes, cyclodextrins, various surfactants, and polymers that decrease the volatility of the terpenes, have been developed for delaying the release of the essential oils. For example, positively charged chitosan microcapsules have been used to facilitate the sustained release of essential oils that act as biodegradable lice-repelling agents (55). In the encapsulation process, which is based on coacervation, an oil-in-water emulsion is first produced by homogenizing the oil in an aqueous solution containing an anionic emulsifier. The emulsion so obtained is then added to an aqueous chitosan solution with continuous homogenization of the mixture to give a stable dispersion. Finally, the emulsion is converted into microcapsules by the initiation of coacervation by the addition of a suitable electrolyte and changing the pH.

Another example is the use of *Rosemarinus officinalis* and *Thymus herbabarona* oils as effective larvicides instead of Bti-based preparations against the larvae of the gypsy moth (56). For formulating these oils, encapsulation is carried out by means of the phase separation process of coacervation, followed by freeze-drying. The essential oil is dispersed in an aqueous solution of gelatin (10%, w/w) at 40°C and emulsified with a high-shear mixer. Na_2SO_4 is then added to start the coacervation. The resultant gelatin microcapsules are cross-linked with glutaraldehyde at a basic pH, filtered off, rinsed with cold water, and dehydrated by freeze-drying. The process gives aggregates containing spherical units of about 0.2 μm in diameter in a high encapsulation yield (over 98%) with both rosemary and thyme oils. In these particles, the core of essential oil is encapsulated in a cross-linked gelatin shell. It was found that the microcapsules exerted a larvicidal effect at a concentration similar to that usually employed for localized treatments with microgranular synthetic pesticides. As formulations of essential oils, prepared as described above, appear to be able to protect the core material against environmental degradation, they therefore could be considered for use as controlled-release systems.

Water-Soluble Pesticides

Most technologies for pesticide encapsulation have been developed for pesticides with low water solubility. There is, however, a class of water-soluble pesticides that would also benefit from microencapsulation, such as the acid or salt form of paraquat, diquat, glyphosate, dicamba, ioxynil, bromoxynil, bentazon, acifluorfen, and fomesafen. In one approach to encapsulation of such water-soluble ingredients, the active ingredient is embedded in a matrix of urea/formaldehyde polymer. The first step in this encapsulation procedure is the preparation of a water-in-oil emulsion from a water-immiscible liquid, a water-soluble urea–formaldehyde or melamine–formaldehyde prepolymer with methylol groups ($-CH_2OH$), the water-dispersible material to be encapsulated, an emulsifier, and water (57). The emulsion is then cured or treated to produce microcapsules by solidification of the urea–formaldehyde prepolymer resin to form a matrix encapsulating the droplets and permitting the separation of solid polymeric capsules containing the water-dispersible material. Curing is initiated with an amphiphatic catalyst, i.e., a catalyst that is soluble in both the water and oil phases of the emulsion. In a more recent approach to forming a shell surrounding the aqueous core—instead of a matrix—a surface-active proton-transfer catalyst that is soluble in the organic liquid, but only slightly soluble in the aqueous phase, is used to initiate an interfacial condensation reaction of the prepolymer at the water/oil interface (58). In this condensation reaction of the prepolymer, a polyurea–formaldehyde shell is formed around the water droplet instead of a matrix being formed within the water droplet. In a typical procedure, a urea–formaldehyde and/or melamine–formaldehyde prepolymer with methylol groups is dissolved in water, and the aqueous phase is emulsified in an organic liquid phase containing one or more surface-active agents. Self-condensation of the prepolymer in the aqueous phase of the discrete droplets adjacent to the interface is obtained by heating the emulsion to a temperature between 20°C and about 100°C with a proton-transfer catalyst.

QUALITY CONTROL

For any biocide formulation, the following parameters must be determined: the amount of active ingredient, particle-size distribution, shelf life, biological release rate, toxicity, and efficacy. As the first two parameters are well documented, they will not be described here. The tests for determining the other parameters are described in detail below.

Storage Stability

A sample of the microencapsulated pesticide formulation is kept in a closed vessel in an oven maintained at 54°C for two weeks or at 40°C for six weeks. The amount of the active ingredient in the product is analyzed before and after heating, and the formulation is considered stable if the amount of active ingredient does not change.

Release Rate of Pesticide

The apparatus used for this test is illustrated in Figure 3. One gram of encapsulated pesticide is placed on a plate, which is then sealed in a cylindrical glass vessel. The vessel is provided with a digital thermometer, and inlet and outlet tubes. The inlet

Controlled-Release Pesticide Formulations

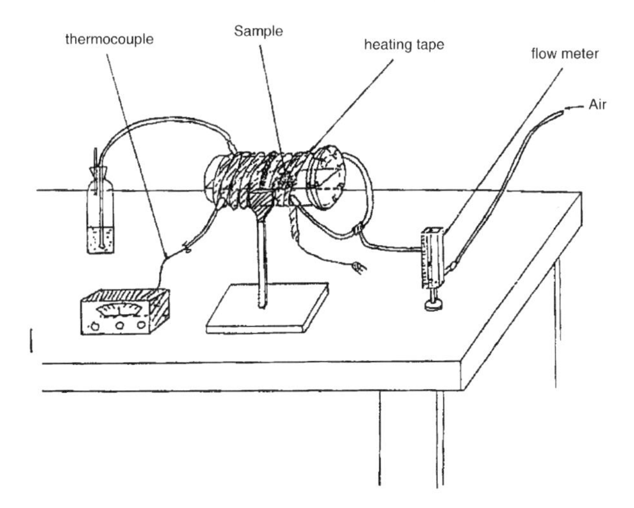

Figure 3 Apparatus for the determination of release rate of a pesticide from an encapsulated formulation.

tube is equipped with a flowmeter to control the volume of air introduced into the vessel. The temperature of the system is maintained at a predetermined value by an external heater. An air stream carries the released pesticide out of the vessel into a beaker containing ethanol. The amount of pesticide in the ethanol is determined, and the release rate calculated as a function of time.

Toxicity to Fish

A solution of the pesticide formulation is obtained by mixing the formulation in water in a high-shear mixer for five minutes. Ten fish are kept in a 16-L aquarium under the following optimal conditions: water quality as close as possible to pH 7.0 (neutral), water temperature 23–25°C, and good strong light for least 12 hours a day (more light makes them grow too fast). The test is performed in a room free of insecticidal contamination. Adult fish of either sex are supplied with adequate standardized food (Europet basic food) before and after the experiment. Food is withheld for two days before the experiment. After addition of the pesticide to the fish tank, mortality is checked after 3, 6, 24, 48, 72, and 96 hours. The time taken for 50% and 95% mortality (LT_{50} and LT_{95}) are then determined from the results.

Acute Oral Toxicity to Mice

Adult male mice (2–2.5 months of age) weighing 25 to 30 g are supplied with standardized mouse food for the period of the experiment. A solution of the pesticide formulation is obtained by placing a small amount of pesticide in water in a vortex mixer for five minutes. Pesticide solution, 1 mL of solution per 20 g body weight, is taken in a 2-mL syringe and introduced into the stomach of the mouse via the

Picture 4 Close-up of cockroach leg showing a diazinon capsule stuck to its leg.

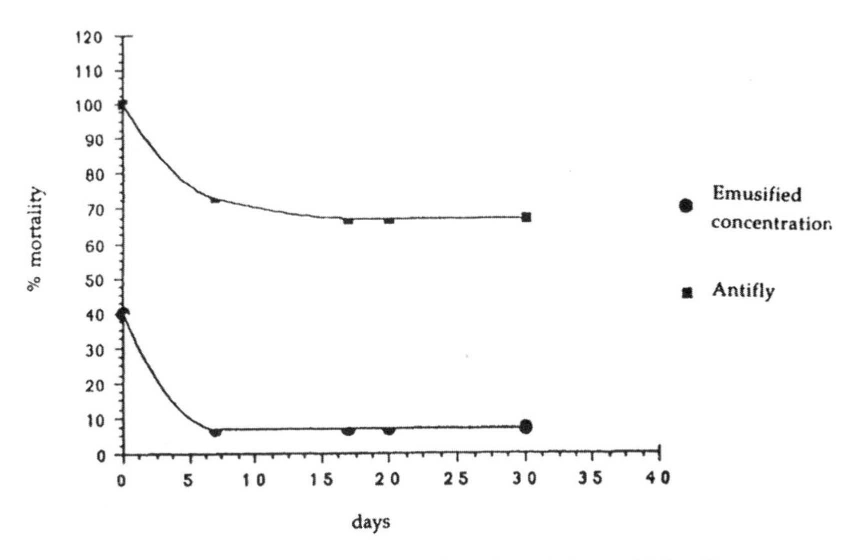

Figure 4 Percentage of mortality as a function of time of *Blattella germanica* exposed to pyrethroids.

mouth. Mortality is checked at 0, 5, 24, 48, 72, 96, 120, 144, and 168 hours, and LD_{50} is determined.

Efficacy

In this method, which measures the susceptibility of a population of cockroaches (*Blattella germanica*) to a given insecticide, the cockroaches are exposed to standard insecticide residues in a Petri dish, and mortality is determined. Solutions of the insecticide formulation are obtained by mixing it with water in a high-shear mixer for five minutes. Filter papers are put into this solution, removed, placed in a Petri dish, and dried in a hood. The cockroaches should be obtained, as far as possible, from the same area, and kept in a suitable container until required. It is preferable to use adult males for the test, but if it is not possible to obtain enough males, females may be used. The test should be carried out in a room free of insecticidal contamination. The cockroaches are given adequate standardized food before the

Table 1 Biological Efficacy After 24 hours of De-Bugger® Against the Flour Beetle (*Tribolium castaneum*)

Pyrethroid formulation	% Mortality at the concentration (ppm)			
	0	250	500	750
De-Bugger®	0	75	80	95
EC		50	68	74

Note: 0 (*zero*), control (without pesticide—water only) in the case of the capsules and solvent in the case of the EC product.
Abbreviation: EC, emulsifiable concentrate.

Table 2 Biological Efficacy After 24 hours of De-Bugger® Against the House Fly (*Musca domestica*)

Pyrethroid formulation	% Mortality at the concentration (ppm)						
	0	25	75	125	250	500	750
De-Bugger®	0	100	100	100	100	100	100
EC		60	60	60	68	80	95

Note: 0 (*zero*), control (without pesticide—water only) in the case of the capsules and solvent in the case of the EC product.
Abbreviation: EC, emulsifiable concentrate.

experiment. They are anesthetized with carbon dioxide before being put into the insecticide-containing Petri dish. The Petri dish is then placed in an experimental vessel, which is maintained at 25°C to 30°C and relative humidity greater than 25%.

Table 3 Biological Efficacy of De-Bugger® Against Fleas and Ticks[a]

Dogs		Parasite		Presence of parasites after days					
Sex	Type	Type	Level of infestation	1	7	14	21	28	60
First spraying									
M	Mongrel	Fleas + ticks	High	+	0	0	0	0	
F	Mongrel	Fleas + ticks	High	+	0	0	0	0	
F	Ridgeback	Fleas + ticks	High	+	0	0	0	0	
F	Mongrel	Fleas + ticks	High	0	0	0	0	0	
M	Mongrel	Fleas + ticks	High	0	0	0	0	0	
F	Mongrel	Fleas + ticks	High	+	0	0	0	0	
M	Pointer	Fleas + ticks	High	+	0	0	0	0	
M	German shepherd	Fleas + ticks	High	0	0	0	0	0	
M	German shepherd	Fleas + ticks	High	+	0	0	0	0	
F	Belgium shepherd	Fleas + ticks	High	+	0	0	0	0	
Second spraying									
M	Spitz	Fleas	2	0	0	0	0	0	0
F	Spitz	Fleas	2	0	0	0	0	0	0
F	Spitz	Fleas	1–2	0	0	0	0	0	0
F	Terrier	Fleas	2	0	0	0	0	0	1
F	Terrier	Fleas	2	0	0	0	0	0	0
M	German shepherd	Fleas	2	0	0	0	0	0	0
F	German shepherd	Fleas	2	0	0	0	0	0	0

[a]The experiment was performed in an animal shelter in Tel-Aviv, Israel under the supervision of the Chief Veterinarian. The spraying was done directly on the dogs. No untoward side effects were observed. Spraying on dogs heavily infested with ticks and fleas with De-Bugger® gave excellent results: after the first spraying the dogs remained without fleas and ticks for 28 days. After the second spraying, the dogs were free of fleas and ticks for 60 days. It is well known that the commercial product keeps the dogs free of ticks and fleas for a very short time (i.e., a few days).

From the results, the time taken for 50% and 95% knockdown (LT_{50} and LT_{95}) can be determined. (A cockroach is considered knocked down if it fails to move, on being returned to its normal posture.) Picture 4 shows the capsule attached to the leg of a cockroach.

CASE STUDY: DE-BUGGER®

De-Bugger® (encapsulated pyrethroids) is manufactured and distributed by Chimgat 2000 under license from Ben-Gurion University of the Negev, Beer-Sheva, Israel. The LD_{50} of De-Bugger® for mice, determined by the test described above, is 12,500 mg/kg, and the LD_{50} for golden orfe fish 2200 μg/L. The biological efficacy of this encapsulated formulation versus that of the emulsifiable concentrate against cockroaches, flour beetles, house flies, and fleas and ticks is shown in Figure 4, Tables 1, 2, and 3, respectively.

SUMMARY

This chapter covers methods of encapsulation, controlled-release techniques for pesticide application, and current trends in the use of encapsulated natural products. The benefits incurred from the microencapsulation of many pesticides such as improved shelf life, reduced toxicity and environmental damage, reduction in the number of required applications, and enhanced efficacy far outweigh the cost of the encapsulation process. In the case of "biological, natural, or green pesticides," encapsulation is necessary, if the materials are to compete with synthetic pesticides, such as the limited stability of the former under field conditions occurring because of radiation, oxidation, and evaporation, and because of the high cost, which hinders multiple applications. Physical techniques, such as spray drying or fluidized bed reactions, are not usually suitable for encapsulation of pesticides. For most commercial encapsulated pesticides, the process of choice is interfacial polymerization, followed by phase separation and coacervation. Interfacial polymerization is used for highly toxic insecticides, as the active ingredient is completely enveloped by the polymer, and in most cases, release takes place via diffusion. Both phase separation and interfacial-polymerization techniques may be used for nontoxic pesticides. The only technique suitable for biological pesticides such as Bti and *T. harzianum* is phase separation: with any other technique the active ingredients would not be able to cross the capsule envelope. In addition to currently used synthetic pesticides, such as organophosphates, organochlorines, carbamates, and petroleum oils, the use of encapsulated IGR such as methoprene or pyriproxyfen is continuously increasing.

Alongside the use of natural pesticides, there has been a recent upsurge in the development of essential oils as green pesticides for such applications as larvicides, and moth and lice repellants. Essential oils may be encapsulated by a variety of techniques including interfacial polymerization and phase separation. Ongoing R&D to improve the final formulations of encapsulated pesticides has produced dry products that may be readily reconstituted in water for spray application, the combination of microcapsules with long-term release properties with more rapidly releasing microcapsules for a quick knockdown effect, and the addition of slow-release attractants to encapsulated pesticides.

REFERENCES

1. Geissbuhler H. Advances in Pesticide Science, Part 3. U.K.: Pergamon Press, 1979: 717–826.
2. Miyamoto J, Keierney PC. Pesticide Chemistry–Human Welfare and the Environment. Vol. 4. U.K.: Pergamon Press, 1983:241–400.
3. Van Valkenburg W. Pesticide Formulations. New York: Marcel Dekker Inc., 1973.
4. Seymour KC. Pesticide Formulation and Application Systems, Second Conference. Philadelphia, PA: ASTM, 1983.
5. Scher HB. Advances in Pesticide Formulation Technology. Washington, D.C.: A.C.S., 1984.
6. Kydonieus AF. Controlled Release Technologies: Methods, Theory and Applications. Boca Raton: CRC Press Inc., 1980:1.
7. Fanger GO. General background and history of controlled release. In: Cardarelli NF, ed. Pesticide Symposium. Ohio: University of Akron, 1974.
8. Baker RW, Lonsdale HK. Principles of controlled release. In: Harris FW, ed. Proceedings of Controlled Release Pesticide Symposium. Dayton, Ohio: Wright State University, 1975.
9. Lewis DH, Cowsar DR. Principles of Controlled Release Pesticides. A.C.S. Symposium Series 53, A.C.S., Washington D.C., 1977.
10. Smith KL. Controlled Release: Theory and Practice. 3rd Annual Membrane Technology Planning Conference, Cambridge Mass., October 30–31, 1985.
11. Arshady R. Preparation of nano- and microspheres by polycondensation techniques. J Microencap 1989; 6:1–12.
12. Arshady R. Preparation of microspheres and microcapsules by interfacial polycondensation techniques. J Microencap 1989; 6:13–28.
13. Arshady R. Albumin microspheres and microcapsules: methodology of manufacturing techniques. J Contr Release 1991; 14:111–131.
14. Finch CA. Korsa DR, Stephenson RA, eds. Encapsulation and Controlled Release. Royal Society of Chemistry, 1993.
15. Scher HB. U.S. Patent 4,285,720, August 25, 1981.
16. Koestler RC. U.S. Patent 4,360,376, April 6, 1981.
17. Klug G, Weisser J. U.S. Patent 6,586,107, July 1, 2003.
18. Lee Fui-Tseng H, Nicholson P, Szamosi, J, Sommer WT. U.S. Patent 6,440,443, August 27, 2002.
19. Scher HB. U.S. Patent 4,140,516, February 20, 1979.
20. Beestman GB, Deming JM. U.S. Patent 4,280,833, July 28, 1981.
21. Scher HB, Rodson M. U.S. Patent 6,340,653, January 22, 2002.
22. Mulqueen PJ, Smith G, Lubetkin SD. U.S. Patent 5,925,464, July 20, 1999.
23. Deming JM, Surgant Sr JM. U.S. Patent 5,354,742, October 11, 1994.
24. Lo RJR, Chen JL, Scher HB, Van Koppenhagen JE, Shirley IM. U.S. Patent 6,555,122, April 29, 2003.
25. Quong, D. U.S. Patent 6,562,361, May 13, 2003.
26. Koestler RC. U.S. Patent 4,360,376, 1982.
27. Vandegaer JE. Wayne NJ. U.S. Patent 3,577,515, May 4, 1971.
28. Shimkin J. U.S. Patent 4,497,793, 1985.
29. Cantilli PA Ger. Patent 2,722,973, 1977.
30. Scher HB. Ger. Patent 2,312,059, 1973.
31. Scher HB. Ger. Patent 2,648,562, 1977.
32. Scher HB. Ger. Patent 2,706,329, 1977.
33. Scher HB. Belg. Patent 867,646, 1978.
34. Beestman GB. European Patent 17409, 1980.
35. Allert PJ. Ger. Patent 2726539, 1977.
36. Beestman GB. European Patent 148,169, 1985.

37. Becher ZB, Magin RW. U.S. Patent 4,563,212, 1986.
38. Beestman GB. U.S. Patent 4,640,709, 1987.
39. Markus A, Pelah Z. U.S. Patent 4,851,271, 1990.
40. Hitsu MN, Kabushiki K. U.K. Patent 1,091,077, 1967.
41. Crodts HP, Karloske JE. U.S. Patent 4,517,326, 1983.
42. Riecke K. U.S. Patent 4,622,267, 1986.
43. Garber G, Chatenet B, Pellenard P. U.S. Patent 4,309,213, 1982.
44. Nemeth CH. U.S. Patent 4,107,292, 1978.
45. Bayer. European Patent 50,264, 1981.
46. Ciba-Geigy-AG. European Patent 214,936, 1986.
47. Markus A, Pelah Z, Aharonson N. Israel Patent 84,219, 84,910, 1992.
48. Margalit J, Markus A, Pelah Z. Effect of encapsulation on the persistence of *Bacillus thuringiensis* var. *israelensis* serotype H-14. Appl Microbiol Biotechnol 1984; 19:1982.
49. Connick JR, Walker WR, Goynes Jr. WR. Sustained release of mycoherbicides from granular formulations. 10th Int. Symp. Controlled Release Bioactive Materials. San Francisco, 1983:283.
50. Walker HL, Connick Jr. Sodium alginate for production and formulation of mycoherbicides. Weed Science 1983; 31:333.
51. Farvel DR, Marois RD, Lumsden WJ, Connick J. Encapsulation of potential biocontrol agents in an alginate-clay matrix. Phytopathology 1985; 75:775.
52. Hsu HJ, Zhou J, Chang C-HL. U.S. Patent 6,231,865, May 15, 2001.
53. Magdassi S, Mumcuoglu K, Bach U. U.S. Patent 5,518,736, May 21, 1996.
54. Eini M, Tamarkin D. U.S. Patent 5,411,992, May 2, 1995.
55. Magdassi S, Mumcuoglu K, Bach U, Rosen Y. U.S. Patent 5,753,264, May 19, 1998.
56. Moretti MOL, Sanna-Passino G, Oemontis S, Bazzoni E. Essential oil formulations useful as a new tool for insect pest control. AAPS Pharm Sci Tech 2002; 3(2):article 13.
57. Golden R. U.S. Patent 4,157,983, June 12, 1979.
58. Rodson M, Scher HB. U.S. Patent 6,113,935, September 5, 2000.

3
Multiparticulate Pulsatile Drug Delivery Systems

Till Bussemer
Sanofi-Aventis Deutschland GmbH, Pharmaceutical Sciences Department Industriepark Höchst, Frankfurt am Main, Germany

Roland Bodmeier
College of Pharmacy, Freie Universität Berlin, Kelchstr, Berlin, Germany

INTRODUCTION

Most oral modified- or controlled-release drug delivery systems (DDS) are based on single unit or multiple unit reservoir- or matrix-type systems with constant- or variable drug-release rates. These dosage forms offer many advantages when compared to immediate-release delivery systems, such as

- Nearly constant drug levels at the site of action
- Minimization of peak–trough fluctuations of the drug concentration
- Avoidance of undesirable side effects
- Reduced dose
- Reduced frequency of administration, and
- Improved patient compliance

Many approaches to achieve controlled-release of drugs have been developed in recent years. However, in a number of cases, a continuous drug release and consequently a nearly constant drug concentration at the targeted site of action is not optimal for the therapy. Many efforts have been made during recent years to optimize drug delivery (1). In this context, an approach of fluctuating drug levels may mimic physiological conditions in a better way. The dependence of several diseases and body function on circadian rhythms is well-known. A number of hormones, such as renin, aldosterone, or cortisol, show distinct daily fluctuations (Fig. 1) (3). The onset and extent of symptoms often vary with circadian rhythms with respect to diseases such as the following:

- Angina pectoris
- Bronchial asthma
- Hypertension
- Myocardial infarction

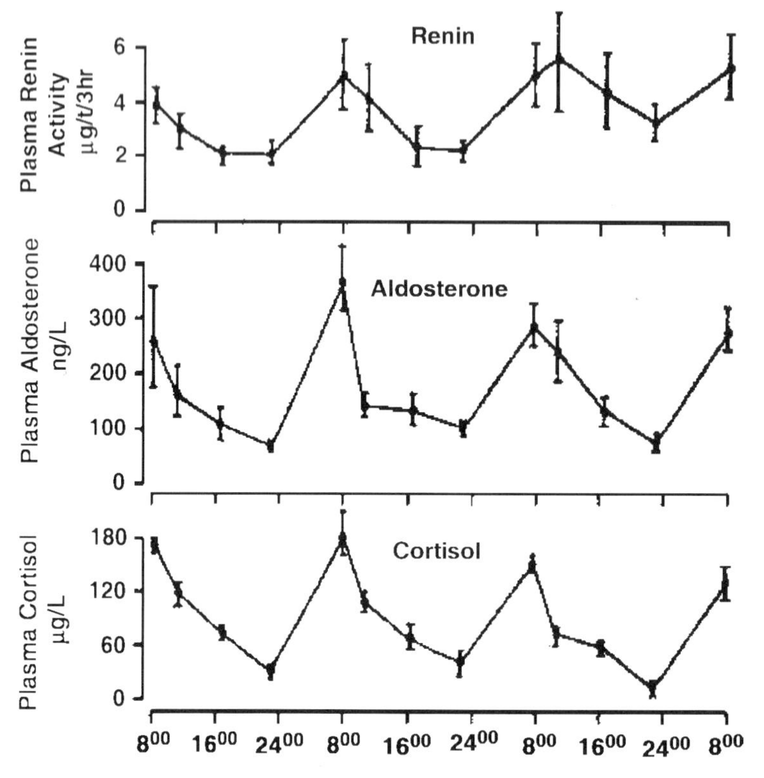

Figure 1 Circadian rhythm of physiological plasma levels of hormones. *Source*: From Ref. 2.

- Rheumatic disease
- Ulcer disease

Chronopharmacological aspects have to be considered also for the treatment of tumors (4,5). Blood flow to tumor tissue and cancer growth have a chronobiological behavior. Optimal circadian timing of 5-fluorouracil (5-FU)/leukovorin and oxyplatin can double the objective response frequency of this combination in patients with metastatic colo-rectal cancer (6). In addition, the toxicity of 5-FU can be reduced by chronomodulated therapy (7).

Dethlefsen and Repges (8) reported a sharp increase in the incidence of asthmatic attacks during the early morning hours, with a maximum occurring at 4 A.M. (Fig. 2). A treatment based on a theophylline–controlled–release dosage form resulting in a constant drug plasma level would not be optimal; a therapeutic scheme that gives consideration to diurnal variation should be more effective. Such a scheme could be realized by a pulsatile dosage form, taken at bedtime with a programmed start of drug release in the early morning hours. Circadian effects were also observed for the pH and acid secretion in the stomach (9). A continuous infusion of famotidine, an H2-antagonist, resulted in distinct diurnal fluctuation of the pH, whereas the famotidine plasma level was relatively constant (10). An effective elevation of the pH was achieved only during the first hours (to pH 7), but during the remaining day, the pH decreased again (to pH 2–3), especially after lunchtime, and increased again after midnight (up to pH 7). Therefore, a constant blood level, as obtained

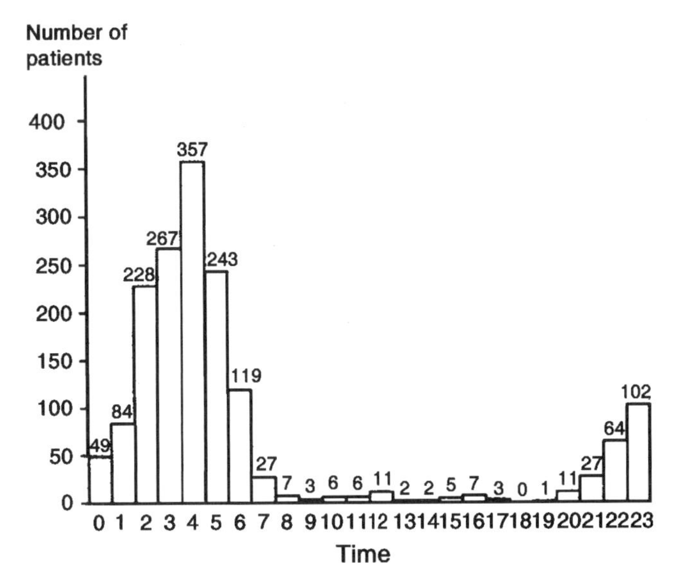

Figure 2 Incidence of asthmatic attacks in about 1600 patients during a 24-hour cycle. *Source*: From Ref. 8.

with a drug infusion and possibly achieved with conventional controlled DDS, does not always lead to a constant pharmacological effect. Similar results were published by Sanders et al. (11) for ranitidine. Especially, the progression of the midnight gastroesophageal reflux disease (GERD) depends on circadian rhythms of gastric acid secretion (12). Patients with GERD often develop complicated reflux diseases (ulcerative oesophagitis, esophageal strictures, and Barrett's syndrome) and erosive reflux esophagitis. Supine nocturnal esophageal acid reflux is considered to be critically involved in this phenomenon.

Another example for the need of diurnal adaptation is the treatment of pain in a study with patients suffering from postoperative pain. When the patients were allowed to control the administration of the drug with an IV infusion pump, Burns et al. (13) showed peaks for the need of morphine at 9 A.M. and 8 P.M.

Ritschel and Forusz (14) gave a broad overview on drugs that were studied with respect to chronopharmacological characteristics. Drug chronopharmacology can affect the therapy in two ways, either in daily variations in effects or in pharmacokinetics (15). All body functions involved in absorption, distribution, and elimination of drugs can be dependent on circadian rhythms (14). For instance, gastric emptying time of solids is faster in the morning than in the afternoon (16). Blood perfusion of the gastrointestinal tract was also found to be higher during early morning hours, which could affect drug absorption via passive diffusion. Especially for lipophilic drugs, the time to reach maximum plasma level (T_{max}) may decrease and the maximum drug concentration in plasma (C_{max}) may increase when applied in the morning hours (4).

A so-called master clock is located in the nucleus suprachiasmaticus (SCN) in the hypothalamus, which is involved in the rhythmic organization of the body (17). The cycle duration generated at the SCN is calibrated by the light–dark rhythm, and further transmitted via the paraventricular nucleus to the pineal body under melatonin secretion. The generated rhythm signals which influence many body systems are determined by a core set of circadian clock genes (18).

Chronopharmacological studies on different kinds of antihypertensive drugs also point to a genetically bound variation in the circadian pattern of blood pressure and its regulation (19). Differences in pharmacokinetics were described for ibuprofen administered by a time–controlled-release formulation at different times of the day (20–23). Maximum plasma level was found six hours after morning dosing compared to dosing after four hours when the drug was applied in the evening.

These findings lead to the requirement of a time-programmed therapeutic scheme in which the drug is at the site of action at the right time in the required amount. This can be realized with pulsatile DDS.

Another possible application of pulsatile systems is in the treatment of attention-deficit hyperactivity disorder with methylphenidate in school-aged children, where a second daily dose, which currently must be administered by a nurse during the school hours, could thus be avoided (24,25).

A pulsatile drug release is characterized by a lag phase with no drug released, followed by a period of drug delivery. Alternative terms to describe pulsatile release or sigmoidal release are delayed. Besides such one-pulse systems, multipulse systems release the drug in subsequent pulses. An ideal pulsatile drug delivery system should release the drug completely and rapidly after a lag time, as shown in an idealized release profile (Fig. 3A). In comparison, conventional controlled-release systems (Fig. 3B: matrix system; C: depot system) generally show a quick onset of drug release without lag time. One challenge of pulsatile-release systems is to achieve a rapid release after the lag time.

A pulsatile drug release pattern in which the drug is released completely, after a well-defined lag time without drug release, is advantageous for the following drugs or therapies (26,27).

- An extensive first-pass metabolism of drugs, e.g., beta-blockers, could be overcome by saturation of metabolic enzymes with pulses of high peak levels instead of constant drug flow below the saturation threshold,
- By avoiding the constant presence of the drug, to which patients develop biological tolerance, at the site of action, which diminishes the therapeutic effect,
- Drug targeting a specific site in the intestinal tract, e.g., to the large intestine or to the colon for the treatment of inflammatory diseases,
- Protection of the gastric or upper intestinal mucosa from irritating drugs,

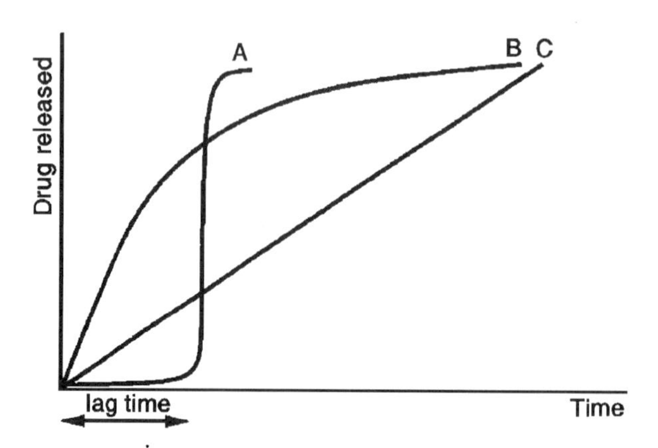

Figure 3 Drug release profiles: (**A**) pulsatile release; (**B** and **C**) controlled release.

- Protection of drug from enzymatic or hydrolytic degradation (e.g., peptide drugs),
- Adaptation of drug delivery to circadian rhythms of body functions or diseases.

PULSATILE SYSTEMS

A large variety of pulsatile systems with different mechanisms have been developed. The first system from the class of capsular-shaped systems, the Pulsincap® system (Scherer DDS Ltd., U.K.), consisted of a water-insoluble body (a hard gelatin capsule coated with polyvinyl chloride) filled with the drug formulation (28,29). The capsule half was closed at the open end with a swellable hydrogel plug. Upon contact with dissolution media or gastrointestinal fluids, the plug swells and pushes itself out of the capsule after a lag time; a rapid release of the capsule content follows. The lag time prior to the drug release can be controlled by the dimension and the position of the plug. To assure a rapid release of the drug content, effervescent agents or disintegrants can be included in the drug formulation, in particular, with water-insoluble drugs. Studies in animals and healthy volunteers proved the tolerability of the formulation (e.g., the absence of gastrointestinal irritation) (30). To overcome the potential problem of variable gastric residence time of a single unit dosage form, the Pulsincap® system was coated with an enteric layer that dissolves upon reaching the higher pH regions of the small intestine (29). This allowed a more precise control of the drug release after passage through the stomach, because the transit time in the intestinal tract is less variable (31,32). The major drawbacks of the Pulsincap® system, which led to the withdrawal of commercial activities connected with this system, are the complicated manufacturing process, reproducibility problems, and the use of a plug material, a cross-linked polyethylene glycol-based polymer, which has not been approved in pharmaceutical products.

A modification of the system was developed using compressed plugs of a water-soluble, swellable polymer, coated with an organic polymer solution (e.g., Eudragit® RS, RL, NE) (33). Only approved materials were used in forming this swellable, insoluble plug. As an alternative to swellable, cross-linked plugs, erodible plugs were investigated for Pulsincap-like systems. The plugs were prepared either by compression of a water-soluble, swellable polymer [hydroxypropyl methylcellulose (HPMC), polyvinyl alcohol, or polyethylene oxide] or by a congealing method in which the plugs were formed from melts of lipids.

Even more sophisticated capsules were developed for academic use or special investigations instead regular application in drug therapy. The high frequency capsule is a drug system that releases the drug in a pulsed fashion after a high-frequency signal is applied externally to the human body (34). Other systems were designed for site-specific application of particulate formulations (35). Drug release of all these systems is triggered by an external signal (magnetic or electromagnetic fields) applied after the capsule reaches the desired position within the gastrointestinal tract.

The capsular-shaped systems are single dose systems and were later improved by the development of multiparticulate systems. Multiparticulate systems comprise different dosage forms such as pellets (beads) or small tablets (minitablets), usually filled into an outer vehicle, e.g., hard gelatin capsules. Alternatively, pulsatile-release pellets can be compressed into tablets together with separating fillers to prevent damage of the individual release-controlling units.

Generally, multiparticulate systems have the advantage of minimizing the risk of dose dumping, which could be caused by burst release. High local drug concentrations which possibly could irritate the gastrointestinal mucosa are avoided because the units are spread over more uniformly along the intestine. They easily pass through pylorus; therefore drug release becomes less dependent because of the varying residence time in the stomach.

SITE-SPECIFIC SYSTEMS

Site-specific DDS release the drug at the desired site within the intestinal tract (36,37). The release is usually controlled by environmental factors, such as the presence of pH or enzymes in the intestinal tract. The simplest pulsatile drug delivery system is an enterically coated dosage form, which releases the drug rapidly after dissolution of the enteric coating. The lag time prior to the release depends primarily on the type and coating level of the enteric polymer and the residence time in the stomach. Longer lag times can be achieved with polymers, which dissolve at a higher pH, and at higher coating levels. The lag times under in vivo conditions are, however, quite variable, especially with single unit systems (e.g., tablets or capsules), because of the variable gastric residence time (31,38). Ewe et al. (39) measured gastric residence times between 40 minutes with an empty stomach, 250 minutes after a breakfast of 2200 kJ and up to more than nine hours after ingestion of breakfast, lunch, dinner, and additional snacks. An enteric-coated single unit dosage form, after application in the morning, may be delayed until late night. The colon is an interesting site for both local and systemic delivery of drugs. A local therapy for inflammatory diseases, such as Crohn's disease and ulcerative colitis, can be achieved with glucocorticoids (e.g., dexamethasone, budesonide, and prednisolone), mesalazine, and sulfasalazine (40). The colon may also be a potential absorption site for peptide drugs because of higher membrane permeability and a lower enzymatic activity. Colonic targeting has been investigated for peptide drugs such as insulin and vasopressin (41–43). Intact insulin was found to be absorbed from the colon as well as from the duodenum, but the colonic absorption led to higher and longer insulin blood levels (44). The transcytotic pathway was proved by gold-labeling of the protein. Insulin introduced in the lumen of the rat duodenum and colon appeared to be rapidly internalized by the epithelial cells and transferred through a transcytotic pathway.

The colon was described to possess "closed compartment conditions" (45). A slow propulsive movement not only results in longer exposure times of the drug to the proteolytic enzymes, but also in a prolonged residence time and a slow dilution of absorption enhancers or peptidase inhibitors. This is the explanation for the observed flat pharmacokinetic profiles.

pH-Controlled Systems

A site-specific drug targeting to the colon could be possible because of pH-differences between the acidic pH of the stomach and the pH of the ileum (37). Enteric polymers used for colon-targeted dosage forms include poly(methacrylic acid-methylmethacrylate) copolymer (1:1) (Eudragit® L) and poly(methacrylic acid-methylmethacrylate) copolymer (1:2) (Eudragit® S), which dissolve at pH 6 and 7, respectively (36). However, problems of failure of coated dosage because of a lack of disintegration have been reported (44). In addition, the drug release in the colon was not sufficiently reproducible (46,47). The drug release of Eudragit® S-coated tablets was different in different buffer media with identical pH, but different concentrations of buffer salts. In another dissolution test, mimicking the extremes of pH-values likely to be found during the transit of the dosage form through the gastrointestinal tract, drug release from the dosage form, originally designed to deliver to the colon, occurred as early as in the duodenum or not at all (46).

Another reason for the lack of in vivo reproducibility could be the drop in pH at the ileo-caecal junction because of the presence of short-chain fatty acids, e.g., butyric, acetic, and propionic acids, which are weak acids with pK_a values of about 4.8 produced by the microbiological degradation of nonabsorbed oligo- and polysaccharides (48). The pH dropped from 7.4 to 5.7 in the terminal ileum in the right colon in healthy volunteers. A decrease in colonic pH could be also related to inflammatory diseases. Fallingborg et al. (49) described pH values of 2.3 to 4.7 in the proximal parts of the colon, possibly caused by ulcerative colitis. If the coating of a dosage form is designed to dissolve at a higher pH, the system would not disintegrate at the desired target site of the intestine, if an unexpected drop in pH, as described above, occurs. To overcome this problem, for patients with inflammatory bowel diseases, Leopold and Eikeler (50) suggested the use of an acid-soluble polymer [poly(dimethylaminoethyl methacrylate-methylmethacrylate-butylmethacrylate) copolymer, Eudragit® E] for a drug-release system, which delivered the drug to the site of inflammation, but did not dissolve at higher pH values. This system would be able to resist the higher pH of the small intestine and the noninflamed regions of the colon. An additional outer enteric coating was needed for the protection against the acidic environment in the stomach.

Enzymatically Controlled Systems

The colonic microflora produces a number of enzymes, which are not found in the stomach or small intestine and hence are able to induce drug release in the colon. Various coating compositions, which degrade or increase permeability in the colon, including azo-polymers, polysaccharide-derived polymers with glycosidic bond, and also conventional acrylic or cellulosic polymers with incorporated colon-degradable pore formers have been investigated (36). The first work in this field was published in 1986 by Saffran et al. (41) and described the synthesis of polymers of polystyrene and hydroxypropyl methacrylate cross-linked with divinylazobenzene. In a subsequent study, it was reported that the reproducibility of colonic delivery in dogs was a problem with the azo-polymer-coated capsules (42). Depending on the batch of the azo-polymer, some capsules were bacterially degraded in the small intestine; others were not degraded and passed the gastrointestinal tract intact. The premature degradation was explained by an insufficient cross-linking of the polymer, while the resistance to bacterial degradation was caused by too much cross-linking. This was improved by the use of more hydrophilic, linear azo-polymers (51–53). The reduction of azo-compounds does not necessarily depend on the presence of the enzyme azoreductase. It was stated that the azoreduction could also be mediated through electron carriers such as NADPH (54). The safety of azo-polymers (formation of primary amines) has not been established yet. Pectin, a polysaccharide consisting of D-galacturonic acid and its methyl ester, can be used as calcium pectinate or as amidated pectin. Because of poor film-forming properties, pectin derivatives can only be used in press-coated dosage forms or matrix systems. The problem of gel formation, which hindered penetration of intestinal fluids, was overcome by the use of a pectin–chitosan complex (55). A pectin/chitosan/HPMC combination was used to achieve a biphasic release profile (56). Other polysaccharides such as dextran esters and galactomannanes were also evaluated as coating materials (57,58). Chitosan, a deacetylated chitin, is an acid-soluble polymer, hence a colon-delivery system with chitosan must possess an additional enteric layer (59). Chitosan capsules coated with enteric polymers have also been reported (60). The combination of insoluble polymers with colon-degradable materials formed good films. This was shown with Eudragit® RS plus β-cyclodextrin and with ethylcellulose plus glassy amylose (61–63). One common challenge to colonic systems is that the drug should not be released prior to reaching the colon, but upon arrival should then be released rapidly. Frequently, these systems release the drug during the gastrointestinal passage by diffusion through the coating prior to reaching the colon; if the drug is not released prior to the colon, the release in the colon is too slow.

TIME-CONTROLLED PULSATILE SYSTEMS

In time-controlled systems, the drug release is controlled primarily by the delivery system, and, ideally, not by the environment. The release should therefore be independent of gastrointestinal transit time or motility, pH, enzymes, and food intake. Time-controlled systems are classified into single unit systems (e.g., tablets or capsules) and multiple unit systems (also multiparticulate) (e.g., pellets, minitablets). For better acceptance and handling by patients and medicinal staff, multiple unit systems are often assembled into larger dosage forms by filling into hard capsules or compression into tablets. In contrast to single units, these systems disintegrate quickly after administration and release a multitude of individual units with pulsatile-release characteristics.

Systems with Eroding or Soluble Barrier Layers

Most pulsatile delivery systems are reservoir devices coated with a barrier layer. The barrier dissolves or erodes after a specified lag period, following which the drug is released rapidly from the reservoir core. The chronotopic system consists of a drug-containing core and an HPMC layer, optionally coated with an outer enteric coating (64,65). The lag time prior to drug release was controlled by the thickness and the viscosity grade of the HPMC layer. After erosion or dissolution of the rubbery HPMC layer, a distinct pulse was observed. To avoid retarding effects in the drug-release phase, the thickness as well as the viscosity grade of the HPMC layer should be limited. The system probably worked best for poorly water-soluble drugs, because highly water-soluble drugs could diffuse through the swollen HPMC layer prior to complete erosion. This system is not particularly well-suited for the application to multiparticulate systems, because relatively thick barrier layers were needed and the resulting drug loading of the system, often more critical in multidose systems, could be further decreased.

Systems with Rupturable Coatings

The other class of reservoir-type pulsatile systems is based on rupturable coatings in contrast to the swellable or erodible layers that characterized the systems discussed in the previous section. The drug is released from a core (tablet or capsule) after rupturing of a surrounding polymer layer; the rupture is caused by a pressure build-up within the system. The pressure necessary to rupture the coating can be achieved with gas-producing effervescent excipients, an increased inner osmotic pressure, or swelling agents such as cellulose ethers, polysaccharides, ion-exchange resins, or superdisintegrants

(67–71). Drug-containing tablets with an effervescent mixture of citric acid and sodium bicarbonate coated with ethylcellulose resulted in a pulsatile release after the coating was ruptured by the development of carbon dioxide after water penetrated the core (71).

A reservoir system with a semipermeable coating was developed especially for drugs with a high first-pass effect, to obtain in vivo drug patterns similar to those obtained by the administration of several immediate-release doses (72). The core contained the drug and a disintegrant, and was coated with cellulose derivatives such as ethylcellulose or cellulose acetate. The core protection was defined as the time until the coating ruptured and released the drug.

Another system was based on a swelling core and a surrounding coating that consisted of a combination of hydrophobic and hydrophilic polymers (73). The insoluble hydrophilic polymer, such as calcium pectinate or calcium alginate, was dispersed in the coating and served as a channel former to control water penetration. The core contained a swellable, water-insoluble polymer, a hardness enhancer (microcrystalline cellulose), and a disintegrant to achieve a fast disintegration after the membrane burst.

MULTIPARTICULATE SYSTEMS

The release mechanisms of multiparticulate pulsatile systems are in principle the same mechanisms as described for the drug release of single units. The general advantages of multiple unit systems were described above. Additionally, combinations of subunits with different release profiles are possible, e.g., blends of subunits of immediate, pulsatile release, and extended release characteristics allow a design of different overall release patterns. Zero-order release could be achieved by mixing units with successive individual pulses.

Pulsatile-release pellets can be manufactured by standard fluidized bed-coating methods, using aqueous or organic film coating. Minitablets can be prepared not only by compression (press coating), but also by coating in pans or fluidized bed equipments.

Eroding Systems

Eroding pulsatile systems contain a core with the drug usually covered by a polymer layer which erodes over a longer period of time serving as a barrier to prevent drug release during the lag phase (Fig. 4). This principle was developed, i.e., for pellets

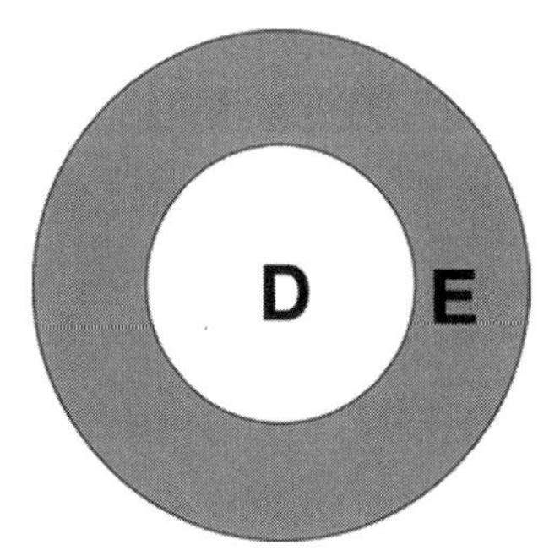

Figure 4 Principle of an eroding pulsatile drug delivery system. (D) Drug-containing core; (E) eroding barrier layer.

with a drug-containing core, an intermediate "prime" layer (HPMC), and an eroding layer, consisting of ethylcellulose (EC), hydroxypropyl methylcellulose phthalate (HPMCP), and the plasticizer diethyl phthalate (9). The ratio of EC to HPMCP was 1:1 resulting in pores forming due to HPMCP-leaching in the intestinal environment. A relatively high coating level (weight gain approximately 40%) was needed to achieve a lag time of 3.5 to 5 hours and probably to prevent premature drug release.

A clinical study on capsules containing pulsatile pellets with the drug nizatidine was performed on patients suffering from GERD. A significant nighttime relief of symptoms was noted in comparison with placebo capsules. To obtain extended release after the lag phase, the system could be modified by applying a controlled-release membrane such as ethylcellulose onto the core, followed by the eroding layer.

Omeprazole, which is a drug susceptible to acidic environment, was often formulated in combination with enteric coating (74). These enteric materials generally are polymers with acidic functional groups. Under acidic conditions, the acidic polymers are protonated and insoluble whereas at higher pH, they become ionized and soluble. Unfortunately, the acidic polymer interfered with the stability of the acid-labile drug omeprazole. Therefore, intermediate layers are necessary to separate the drug from the acidic enteric polymer. It was the idea from Robinson's patent to circumvent this phenomenon by the application of an insoluble eroding layer instead of an enteric plus separating layer (75). Pellets or minitablets were spray-coated with Eudragit® NE 30D with high amounts of talcum added. Lag times of 0.5 to 5 hours could be achieved.

In another approach, minitablets were prepared by press coating (76). The drug core was coated with high viscosity grades of HPMC and microcrystalline cellulose in different ratios to adjust the lag time. A relatively high amount of coating material in relation to the core was used (around 70% of the total tablet weight). Pulsatile release with a fast onset of drug release was observed, but the release rate gradually decreased over time, probably due to the release-sustaining effect of the coating polymer, which was not completely eroded. In a modification of the system, the drug was released in a near zero-order fashion after the lag time. The system was evaluated with a poorly soluble drug, nifedipine, but no proof for evaluation with water-soluble drugs was given.

In general, eroding systems require relatively thick eroding layers to achieve the desired lag times, especially when the individual units are small (pellets). Premature drug release of water-soluble drug substances through the swollen barrier was critical. On the other side, the drug release after the lag time often was sustained due to the remaining barrier materials.

An interesting system with oscillating delivery, every 50 minutes, of the drug dibucaine was proposed (77). The erodible matrix consisted of κ-carageenan, and the drug was dispersed in this matrix. After contact with water, the outer layer swelled, forming a hydrogel which served as a diffusion barrier. Oscillating swelling and deswelling of the hydrogel caused oscillating release rates.

Rupturing Systems

In contrast to eroding systems, wherein an outer release-delaying system is slowly dissolved or eroded as a function of time, other pulsatile dosage forms were developed with a burst mechanism. An outer membrane is ruptured after a programmed lag time due to a volume increase in the inner core. Typical designs of these systems are shown in Figure 5. In most cases, a water-permeable, insoluble membrane controls

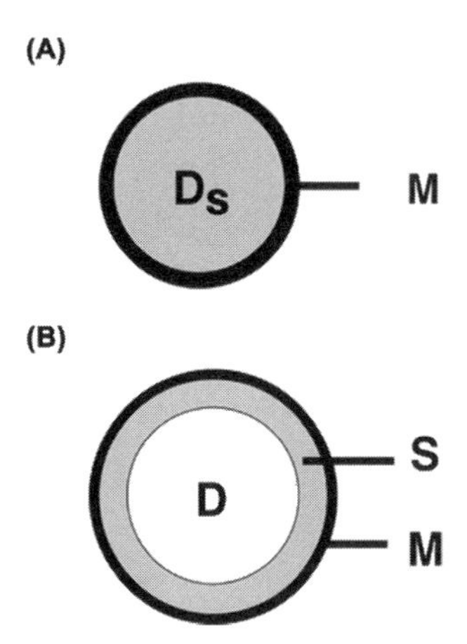

Figure 5 Typical designs of rupturing pulsatile DDS. (**A**) System with a core possessing sufficient swelling capacity; (**B**) system with a drug-containing core and an additional swelling layer. *Abbreviations*: D, drug-containing core; D_S, drug containing core with swelling capacity; S, swelling layer; M, water-permeable insoluble membrane. DDS, drug delivery systems.

water-penetration into the core after contact of the delivery system with body fluids, especially intestinal fluids. At the same time, this membrane provides integrity of the pulsatile system and consequently prevents drug release during the lag time. Premature drug release can be prevented when a membrane with minimum permeability for the drug is employed. Different mechanisms were investigated with multiparticulate systems: swelling of polymers and swelling due to an osmotic imbibe of water.

The time-controlled explosion system (TES) was a multiparticulate system, wherein the drug was layered on an inner core, followed by a swellable layer (e.g., hydroxypropyl cellulose) of optimal thickness (at least 180 μm) and an insoluble polymeric top layer (e.g., ethylcellulose) (78,79). Upon water ingress, the swellable layer expanded resulting in film rupturing with subsequent rapid drug release. The release was described to be independent of the environmental pH and drug solubility. The lag time increased with increasing membrane thickness and higher amounts of talc or lipophilic plasticizer in the coating; and the release rate increased with increasing concentrations of the osmotically active agents. In vivo studies of the TES containing the drug metoprolol, which had an in vitro lag time of three hours, showed first metoprolol blood levels after three hours and maximal blood levels after five hours (80). When the same system contained diclofenac, which has a markedly lower solubility than has metoprolol (greater than 100 mg/mL for metroprolol tartrate vs. 1.2 mg/mL for diclofenac sodium), the performance of the system became highly susceptible to alterations in the small intestinal transit time and the presence of water at the location of drug delivery within the gastrointestinal tract (81). Especially with a TES with an in vitro onset of drug release after six hours, the bioavailability declined compared to a lag time of three hours (concentration in vitro release and corresponding in vivo plasma are shown in time profiles Figs. 6 and 7).

The TES was manufactured using organic ethylcellulose solutions (solvents: ethanol, dichloromethane) in the final coating step. A further improvement of the system was achieved by the application of aqueous coating technology (82). Rupturing pellets with lag times between one and six hours were obtained from pellets

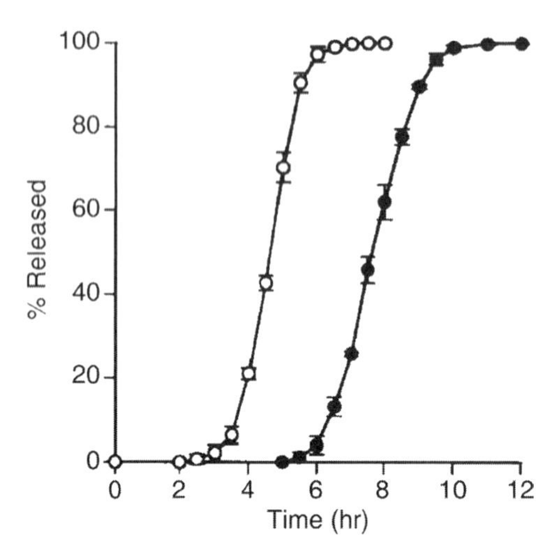

Figure 6 In vitro drug-release profiles of TES with a lag time of three hours (*open circles*) and six hours (*closed circles*) containing diclofenac sodium as a drug layer, a swelling layer with low substituted hydroxypropylcellulose at a constant thickness of 225 μm. The lag time was adjusted by different thickness of the outer ethylcellulose-talc coating (35 μm for three hours lag time, 53 μm for six hours). *Abbreviation*: TES, time-controlled explosion system. *Source*: From Ref. 81.

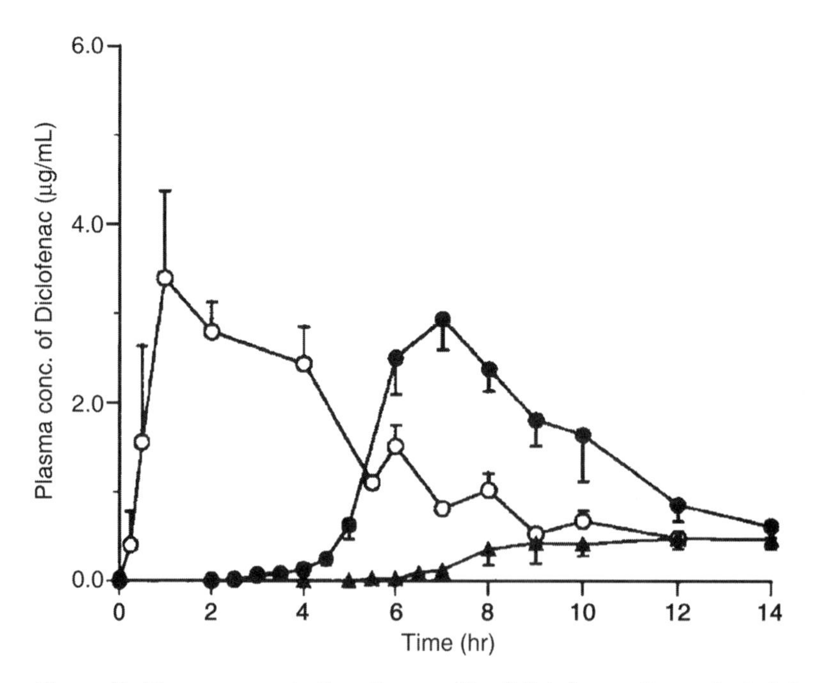

Figure 7 Plasma concentration–time profile of diclofenac after oral administration of (*open circles*) immediate-release tablet (*closed circles*) TES with three hours (*closed triangles*) TES with six hours lag time to dogs. *Abbreviation*: TES, time-controlled explosion system. *Source*: From Ref. 81.

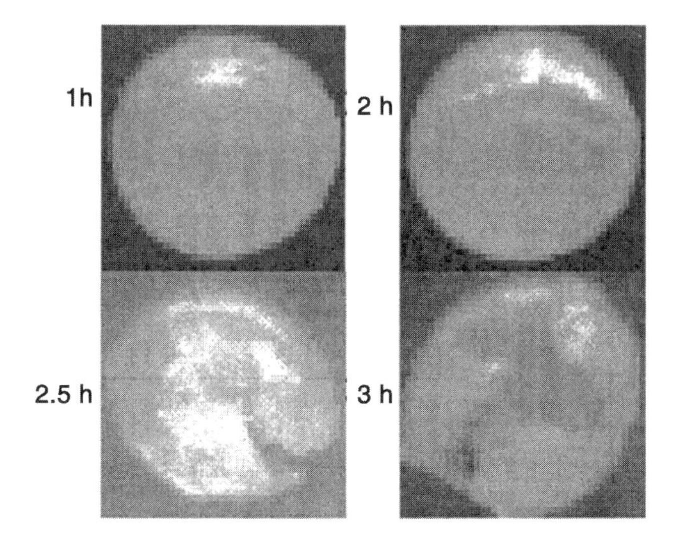

Figure 8 Rupturing of single pulsatile-release pellets after 1, 2, 2.5, and 3 hours of incubation in phosphate buffer at 37°C. *Source*: From Ref. 82.

containing drug (theophylline), a swelling layer (croscarmellose sodium), and an outer insoluble layer prepared from aqueous ethylcellulose dispersion. Typically, first, cracks occurred on the outer membrane when the pressure exerted by the swelling layer exceeds the mechanical resistance of the membrane at the end of the lag time. With progressing water flux into the swelling layer and the resulting expansion, the crack spread over the pellet surface and opened the membrane for fast and complete drug release (Fig. 8). The lag time increased with increasing thickness of the ethylcellulose membrane, which was expressed as percentage of coating level based on the pellet weight and was in the range between 10% and 30%, from 1 to 5.5 hours.

On one hand, the performance of the rupturing pulsatile DDS depends on the water permeability and the mechanical resistance of the insoluble membrane, and, on the other hand, on the pressure exerted by the swelling compartment inside the system. These parameters are described in detail in Table 1.

Traditionally, swelling agents were used as disintegrants in tablet formulations to ensure disintegration of the dosage form within the desired time. Highly efficient swelling materials such as croscarmellose, low-substituted hydroxypropylcellulose, and starch glycolate became well-known as "superdisintegrants" and were used as swelling excipients in rupturing pulsatile systems.

Sufficient swelling depends on several parameters (Table 2). For example, for the aqueous-coated rupturing system described above, a minimum quantity of 25% croscarmellose sodium was necessary to achieve complete membrane rupturing and fast drug release after the lag time (82). In some cases, a higher amount of superdisintegrant was necessary to compensate the decreasing hardness of the core caused by the ingress of water during the lag phase (92).

A combination of osmotic and swelling effects was used in the system developed by Amidon and Leesman (72). The permeability-controlled systems consisted of a core, containing an osmotically active substance, a swelling substance, and the drug. NaCl and sorbitol were used as osmotic substances, and Na carboxymethylcellulose as the swelling material. These cores were coated with an insoluble,

Table 1 Compartments of a Rupturable Pulsatile System and Their Properties That Control the Lag Time Before Drug Release

Compartment	Property	Impact on lag time	References
Core	Water uptake	Ideally, the lag time should be independent of the core, usually consisting of drug and excipients. However, the core may contribute to the overall water uptake rate of the DDS, e.g., by osmotic or swelling effects, thus reducing the lag time. In some cases also the mechanical rigidity of the core could be decreased and lag time could be prolonged	(83–85)
Swelling layer	Swelling pressure, swelling energy	Swelling pressure and energy were continuously increased upon progressing water influx. The time to achieve a critical threshold necessary to break the outer membrane depended on the nature of the membrane material (Fig. 9) and a number of environmental parameters (Table 2).	(71)
Outer membrane	Water permeability	A higher rate of water penetration through the outer membrane accelerated the achievement of the critical pressure to rupture this membrane which was developed by the swelling layer, thus shortening the lag time. Polymers with higher water penetration rate, thinner film layers, and the addition of pore formers (e.g., HPMC) decreased the lag time, whereas the addition of hydrophobic additives (e.g., magnesium stearate, talc) prolonged the lag time	(83,86,87)
	Mechanical properties	Polymer with higher mechanical strength, measured in a puncture test, required a higher critical pressure to burst, thus increasing the lag time. On the other hand, the strength decreased as a function of time upon contact with aqueous fluids due to the plasticizing effect of water and the leaching of coating additives (pore formers, plasticizers)	(66,88)

Abbreviations: DDS, drug delivery systems; HPMC, hydroxypropyl methylcellulose.

semipermeable polymer such as cellulose acetate. The time T_p, when the insoluble film ruptured, was described by the following equation:

$$T_p = \frac{V_e^* + V_d}{AaL_p\,\Delta\pi}$$

where T_p represented the time until film rupture (lag time), V_{*e} the volume to which the tablet has to be expanded for film break, V_d the displaceable volume inside the core, A the surface area of the film, a a constant, L_P the water permeability of the film, and $\Delta\pi$ the osmotic pressure difference across the film. The constant a is

$$a = \frac{\Delta r^*/r_0}{e^*}$$

Table 2 Parameters Influencing the Swelling Behavior of Superdisintegrants

Parameter	Effect	References
Amount of superdisintegrant	A minimum of superdisintegrant is necessary for the development of sufficient swelling pressure to the outer membrane	(71,82)
Additives, e.g., binders	Polymeric binders can reduce swelling pressure by spacial separation of superdisintegrant particles or competition for free water	(71,89,90)
Ionic strength of the medium	Competition of the ions for free water	(71)
pH value	Swelling can be influenced for superdisintegrants with ionizable groups (e.g., carboxylic groups in croscarmellose)	(71,91)

where Δr^* is the critical radius increase, r_0 the initial radius of the unit, and e^* the critical strain. This model was based on the following assumptions: the units (tablets, pellets) have a spherical shape, the thickness of the film and the osmotic pressure difference are constant, and the values of the critical stress and strain for film rupture do not change. In practice, the changing mechanical properties of the film, the changing permeability and weak points (defects) of the film would have to be considered.

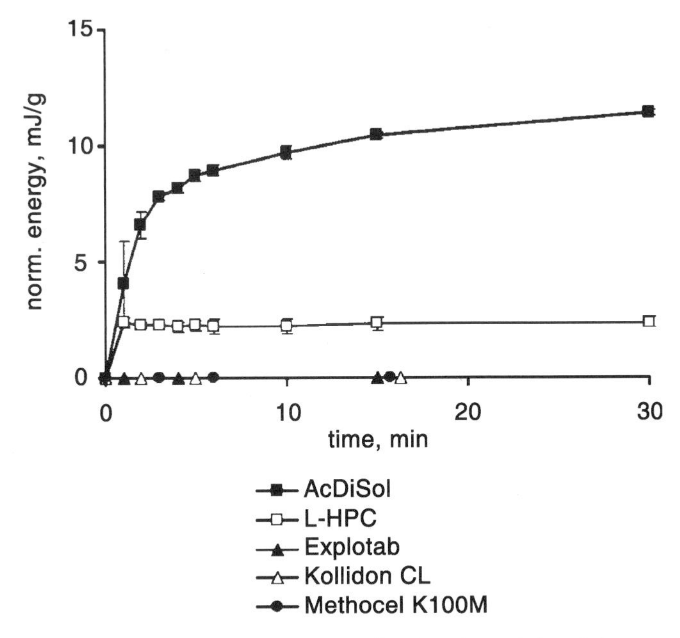

Figure 9 Swelling energy (normalized to the weight of the material) of different superdisintegrants as a function of time of incubation in water at 37°C, measured under load in a special device. AcDiSol (croscarmellose) showed the highest swelling energy, followed by L-HPC, Explotab® (starch glycolate), Kollidon CL (crosspovidone), and Methocel K100M (a high viscosity grade of HPMC). *Abbreviations*: HPMC, hydroxypropyl methylcellulose; L-HPC, low-substituted hydroxypropyl cellulose. *Source*: From Ref. 71.

Chen (92) has proposed a system with a core, containing an osmotically active substance (NaCl) and the drug, which was coated by an insoluble, permeable membrane. The coatings included different types of polyacrylate–polymethacrylate copolymers (Eudragit® NE30D, Eudragit® RS30D) and magnesium stearate, which reduced the water permeability of the membrane, thus allowing the use of thinner films for long lag times. Otherwise, thicker films had to be used, which were more difficult to rupture.

The lag time of a similar system was controlled by the addition of an enteric polymer to the surrounding insoluble membrane polymer (e.g., ethylcellulose). The enteric polymer, poly(methacrylic acid-methylmethacrylate) copolymer (Eudragit® S) or HPMCP, did not dissolve in the acidic pH of the stomach, but dissolved in the small intestine, thus weakening the membrane and resulting in rupturing after a predetermined time. Expanding of the core upon water ingress was achieved by the presence of Explotab® (starch glycolate), a swelling substance. Again, the inclusion of water-insoluble agent (magnesium stearate) in the outer membrane resulted in thinner, less permeable films with better rupturing properties (93).

An osmotically active, drug-containing pellet core has been coated with cellulose acetate, a semipermeable polymer, which is permeable for water, but impermeable for drugs (84,85). The lag time increased with increasing coating level and higher amounts of talc or lipophilic plasticizer in the coating; the release rate increased with increasing concentration of the osmotically active agents. The addition of osmotically active salts, such as NaCl, was necessary to achieve a pulsatile release. Otherwise, the release was extended after the lag time because of the lower degree of core swelling, which resulted in only small fissures and not in a complete rupturing of the coating.

OUTLOOK

The importance of chronopharmacological adaptation of drug treatment has received increasing attention from researchers involved in drug delivery. Pulsatile drug-release systems are not only a highly interesting approach for chronotherapy, but also for drug targeting to specific sites in the gastrointestinal tract. These innovative delivery systems could contribute to the success and patent life of drug products.

Benefits of chronotherapeutic drug regimens were described for several cases. The treatment with a verapamil delivery system (COER), which has an extended drug release after a programmed lag time, showed equivalence to a standard therapy with calcium-antagonists in the CONVINCE trial (94). However, the superiority of pulsatile DDS over standard therapy still has to be investigated in larger clinical trials.

REFERENCES

1. Langer R. Drug delivery. Drugs on target. Science 2001; 293(5527):58–59.
2. Williams GH, Cain JP, Dluhy RG, Underwood RH. Studies of the control of plasma aldosterone concentration in normal man. I. Response to posture, acute and chronic volume depletion, and sodium loading. J Clin Invest 1972; 51(7):1731–1742.
3. Breimer DD. The need for rate and time programming in future drug delivery. In: Pulsatile Drug Delivery Gurny R, Junginger HE, Peppas NA Stuttgart, Germany: Wissenschaftliche Verlagsgesellschaft mbh, 1993.
4. Lemmer B. Chronopharmacokinetics: implications for drug treatment. J Pharm Pharmacol 1999; 51(8):887–890.
5. Hrushesky WJ, Langevin T, Kim YJ, Wood PA. Circadian dynamics of tumor necrosis factor alpha (cachectin) lethality. J Exp Med 1994; 180(3):1059–1065.

6. Hrushesky WJ. Tumor chronobiology. J Contr Rel 2001; 74(1–3):27–30.
7. Milano G, Chamorey AL. Clinical pharmacokinetics of 5-fluorouracil with consideration of chronopharmacokinetics. Chronobiol Int 2002; 19(2):177–189.
8. Dethlefsen U, Repges R. Ein neues Therapieprinzip bei nächtlichem Asthma. Med Klin 1985; 80(2):44–47.
9. Percel PJ, Vyas NH, Vishnupad KS, Venkatesh GM. Pulsatile-release histamine H2 antagonist dosage form. US Pat Appl Publ 2,003,113,374, 2003.
10. Merki HS, Witzel L, Kaufman D, et al. Continuous intravenous infusions of famotidine maintain high intragastric pH in duodenal ulcer. Gut 1988; 29(4):453–457.
11. Sanders SW, Buchi KN, Moore JG, Bishop AL. Pharmacodynamics of intravenous ranitidine after bolus and continuous infusion in patients with healed duodenal ulcers. Clin Pharmacol Ther 1989; 46(5):545–551.
12. Frazzoni M, De Micheli E, Savarino V. Different patterns of oesophageal acid exposure distinguish complicated reflux disease from either erosive reflux oesophagitis or non-erosive reflux disease. Aliment Pharmacol Ther 2003; 18(11–12):1091–1098.
13. Burns JW, Hodsman NB, McLintock TT, Gillies GW, Kenny GN, McArdle CS. The influence of patient characteristics on the requirements for postoperative analgesia. A reassessment using patient-controlled analgesia. Anaesthesia 1989; 44(1):2–6.
14. Ritschel WA, Forusz H. Chronopharmacology: a review of drugs studied. Meth Find Exp Clin Pharmacol 1994; 16(1):57–75.
15. Lemmer B. Why are so many biological systems periodic? In: Gurny R, Junginger HE, Peppas NA, eds. Pulsatile Drug Delivery. Stuttgart, Germany: Wissenschaftliche Verlagsgesellschaft mbh, 1993; 11–24.
16. Goo RH, Moore JG, Greenberg E, Alazraki NP. Circadian variation in gastric emptying of meals in humans. Gastroenterology 1987; 93(3):515–518.
17. Eriguchi M, Levi F, Hisa T, Yanagie H, Nonaka Y, Takeda Y. Chronotherapy for cancer. Biomed Pharmacother 2003; 57(suppl 1):92s–95s.
18. Lowrey PL, Takahashi JS. Mammalian circadian biology: elucidating genome-wide levels of temporal organization. Annu Rev Genomics Hum Genet 2004; 5:407–441.
19. Lemmer B. Genetic aspects of chronobiologic rhythms in cardiovascular disease. In: Zehender M, Breithardt G, Just H, eds. From Molecule to Men. Darmstadt: Steinkopff Verlag, 2000:201–213.
20. Halsas M, Ervasti P, Veski P, Jurjenson H, Marvola M. Biopharmaceutical evaluation of time-controlled press-coated tablets containing polymers to adjust drug release. Eur J Drug Metab Pharmacokinet 1998; 23(2):190–196.
21. Halsas M, Hietala J, Veski P, Jurjenson H, Marvola M. Morning versus evening dosing of ibuprofen using conventional and time-controlled release formulations. Int J Pharm 1999; 189(2):179–185.
22. Halsas M, Penttinen T, Veski P, Jurjenson H, Marvola M. Time-controlled release pseudoephedrine tablets: bioavailability and in vitro/in vivo correlations. Pharmazie 2001; 56(9):718–723.
23. Halsas M, Simelius R, Kiviniemi A, Veski P, Jurjenson H, Marvola M. Effect of different combinations of hydroxypropyl methyl cellulose on bioavailability of ibuprofen from press-coated time-controlled tablets. STP Pharma 1998; 8(3):155–161.
24. Mehta AM, Zeitlin AL, Dariani MM. Delivery of multiple doses of methylphenidate. US Patent 5,837,284, 1998.
25. Crison JR, Vieira ML, Kim JS, Siersma C, Amidon GL. Pulse delivery of methylphenidate in dogs using an osmotic drug delivery system. 28th International Symposium on Controlled Release of Bioactive Materials and 4th Consumer and Diversified Products Conference, San Diego, CA, 2001.
26. Chang RK, Guo X, Burside BA, Couch RA, Rudnic EM. Formulation approaches for oral pulsatile drug delivery. Am Pharm Rev 1999; 2(1):6–13.
27. Bussemer T, Otto I, Bodmeier R. Pulsatile drug-delivery systems. Crit Rev Ther Drug Carrier Syst 2001; 18(5):433–458.

28. McNeill ME, Rashid A, Stevens HNE. Dispensing device. WO Patent 90/09168, 1990.
29. Wilding IR, Davis SS, Bakhshaee M, Stevens HN, Sparrow RA, Brennan J. Gastrointestinal transit and systemic absorption of captopril from a pulsed-release formulation. Pharm Res 1992; 9(5):654–657.
30. Binns J, Stevens HNE, McEwen J, et al. The tolerability of multiple oral doses of Pulsincap®TM capsules in healthy volunteers. J Contr Rel 1996; 38:151–158.
31. Khosla R, Davis SS. Gastric emptying and small and large bowel transit of non-disintegrating tablets in fasted subjects. Int J Pharm 1989; 52:1–10.
32. Davis SS. The design and evaluation of controlled release systems for the gastrointestinal tract. J Contr Rel 1985; 2:27–38.
33. Kroegel I, Bodmeier R. Pulsatile drug release from an insoluble capsule body controlled by an erodible plug. Pharm Res 1998; 15(3):474–481.
34. Graul EH, Loew D, Schuster O. Voraussetzung für die Entwicklung einer sinnvollen Retard- und Diuretikakombination. Therapiewoche 1985; 35:4277–4291.
35. Wilding IR, Prior DV. Remote controlled capsules in human drug absorption (HDA) studies. Crit Rev Ther Drug Carrier Syst 2003; 20(6):405–431.
36. Watts PJ, Illum I. Colon drug delivery. Drug Dev Ind Pharm 1997; 23(9):893–913.
37. Leopold CS. "Targeted delivery" in the gastrointestinaltract. Med Klin (Munich) 1999; 94(suppl 1):6–11.
38. Rouge N, Buri P, Doelker E. Drug absorption sites in the gastrointestinal tract and dosage forms for site-specific delivery. Int J Pharm 1996; 136:117–139.
39. Ewe K, Press AG, Bollen S, Schuhn I. Gastric emptying of indigestible tablets in relation to composition and time of ingestion of meals studied by metal detector. Dig Dis Sci 1991; 36(2):146–152.
40. Hanauer SB. Inflammatory bowel disease. N Engl J Med 1996; 334(13):841–848.
41. Saffran M, Kumar GS, Savariar C, Burnham JC, Williams F, Neckers DC. A new approach to the oral administration of insulin and other peptide drugs. Science 1986; 233(4768):1081–1084.
42. Saffran M, Pansky B, Budd GC, Williams FE. Insulin and the gastrointestinal tract. J Contr Rel 1997; 46:89–98.
43. Rao SS, Ritschel WA. Development and in vitro/in vivo evaluation of a colonic release capsule of vasopressin. Int J Pharm 1992; 86:35–41.
44. Bendayan M, Ziv E, Gingras D, Ben Sasson R, Bar-On H, Kidron M. Biochemical and morpho-cytochemical evidence for the intestinal absorption of insulin in control and diabetic rats. Comparison between the effectiveness of duodenal and colon mucosa. Diabetologia 1994; 37(2):119–126.
45. Rubinstein A, Tirosh B, Baluom M, et al. The rationale for peptide drug delivery to the colon and the potential of polymeric carriers as effective tools. J Contr Rel 1997; 46: 59–73.
46. Ashford M, Fell JT, Attwood D, Woodhead PJ. An in vitro investigation into the suitability of pH-dependent polymers for colonic targeting. Int J Pharm 1993; 91:241–245.
47. Ashford M, Fell JT, Attwood D, Sharma H, Woodhead PJ. An in vivo investigation into the suitability of pH-dependent polymers for colonic targeting. Int J Pharm 1993; 95: 193–199.
48. Fallingborg J, Christensen LA, Ingeman-Nielsen M, Jacobsen BA, Abildgaard K, Rasmussen HH. pH-profile regional transit times of the normal gut measured by a radiotelemetry device. Aliment Pharmacol Ther 1989; 3(6):605–613.
49. Fallingborg J, Christensen LA, Jacobsen BA, Rasmussen SN. Very low intraluminal colonic pH in patients with active ulcerative colitis. Dig Dis Sci 1993; 38(11):1989–1993.
50. Leopold CS, Eikeler D. Eudragit® E as coating material for the pH-controlled drug release in the topical treatment of inflammatory bowel disease (IBD). J Drug Target 1998; 6(2): 85–94.
51. Mooter GVD, Samyn C, Kinget R. Azo polymers for colon-specific drug delivery. Int J Pharm 1992; 87:37–46.

52. Mooter GVD, Samyn C, Kinget R. Use of azo polymers for colon-specific drug delivery 2: Influence of the type of azo polymer on the degradation by intestinal microflora. Int J Pharm 1993; 97:133–139.
53. Mooter GVD, Maris B, Samyn C, Augustijns P, Kinget R. Use of azo polymers for colon-specific drug delivery. J Pharm Sci 1997; 86(12):1321–1327.
54. Lloyd AW, Hodges NA, Martin GP, Soozandehfat SH. Azo reduction is mediated by enzymatically generated low molecular weight electron carriers. J Pharm Pharmacol 1993; 45(2):1107.
55. Fernandez-Hervas MJ, Fell JT. Pectin/chitosan mixtures as coatings for colon-specific drug delivery: an in vitro evaluation. Int J Pharm 1998; 169(1):115–119.
56. Ofori-Kwakye K, Fell JT. Biphasic drug release: the permeability of films containing pectin, chitosan and HPMC. Int J Pharm 2001; 226:139–145.
57. Kesselhut JF, Bauer KH. Development and characterization of water soluble dextran fatty acid esters as excipients for colon-targeting. Pharmazie 1995; 50(4):263–269.
58. Hirsch S, Binder V, Kolter K, Kesselhut JF, Bauer KH. lauroyldextran as a coating material for site-specific drug delivery to the colon: in vitro dissolution of coated tablets. Proc Int Symp Contr Rel Bioact 1997; 379–380.
59. Lorenzo-Lamosa ML, Remunan-Lopez C, Vila-Jato JL, Alonso MJ. Design of microencapsulated chitosan microspheres for colonic drug delivery. J Contr Rel 1998; 52(1–2): 109–118.
60. Tozaki H, Komoike J, Tada C, et al. Chitosan capsules for colon-specific drug delivery: improvement of insulin absorption from the rat colon. J Pharm Sci 1997; 86(9): 1016–1021.
61. Siefke V, Weckenmann HP, Bauer KH. β-Cyclodextrin matrix films for colon-specific drug delivery. Proc Int Symp Contr Rel Bioact. Deerfield, Illinois: Controlled Release Society, 1993.
62. Cummings J, Milojevic S, Harding M, et al. In vivo studies of amylose and ethylcellulose coated [13C] glucose microspheres as a model for drug delivery to the colon. J contr Rel 123–131.
63. Milojevic S, Newton JM, Cummings JH, et al. Amylose as a coating for drug delivery to the colon: preparation and in vitro evaluation using 5-aminosalicylic acid pellets. J Contr Rel 1996; 38:75–84.
64. Gazzaniga A, Sangalli ME, Giordano F. Oral chronotropic drug delivery systems: achievement of time and/or site specificity. Eur J Pharm Biopharm 1994; 40(4):246–250.
65. Poli S, Busetti C, Moro L. Oral pharmaceutical composition for specific colon delivery. EP Patent 0,572,942, 1993.
66. Kroegel I, Bodmeier R. Floating or pulsatile drug delivery systems based on coated effervescent cores. Int J Pharm 1999; 187:175–184.
67. Amsden B, Cheng YL. A generic protein delivery system based on osmotically rupturable monoliths. J Contr Rel 1995; 33(1):99–105.
68. Vazquez Lantes MC. Drug Delivery Systems with Controlled Drug Release. Ph D thesis, Freie Universität, Berlin, 2001.
69. Morita R, Honda R, Takahashi Y. Development of oral controlled release preparations, a PVA swelling controlled release system (SCRS). I. Design Of SCRS and its release controlling factor. J Contr Rel 2000; 63(3):297–304.
70. Bussemer T, Peppas NA, Bodmeier R. Evaluation of the swelling, hydration and rupturing properties of the swelling layer of a rupturable pulsatile drug delivery system. Eur J Pharm Biopharm 2003; 56(2):261–270.
71. Sungthongjeen S, Puttipipatkhachorn S, Paeratakul O, Dashevsky A, Bodmeier R. Development of pulsatile release tablets with swelling and rupturable layers. J Contr Rel 2004; 95(2):147–159.
72. Amidon GL, Leesman GD. Pulsatile Drug Delivery System. US Patent 5,229,131, 1993.
73. Lerner IE, Flashner M, Penhasi A. Delayed total release gastrointestinal drug delivery system. WO Patent 99/18938, 1999.

74. Pilbrant A, Cederberg C. Development of an oral formulation of omeprazole. Scand J Gastroenterol Suppl 1985; 108:113–120.
75. Robinson JR, McGinity JW. Delivery system for omeprazole and its salts. US Pat Appl Publ 2002160046, 2002.
76. Li Yh, Zhu Jb. Modulation of combined-release behaviors from a novel "tablets-in-capsule system". J Contr Rel 2004; 95(3):381–389.
77. Makino K, Idenuma R, Murakami T, Ohshima H. Design of a rate- and time-programming drug release device using a hydrogel: pulsatile drug release from kappa-carrageenan hydrogel device by surface erosion of the hydrogel. Colloids Surf B Biointerf 2001; 20(4): 355–359.
78. Ueda Y, Hata T, Yamaguchi H, Ueda S, Kotani M. Time controlled explosion system and process for preparation the same. US Patent 4,871,549, 1989.
79. Ueda S, Hata T, Yamaguchi H, Kotani M, Ueda Y. Development of a novel drug release system, time-controlled explosion system (TES). I. Concept and design. J Drug Target 1994; 2:35–44.
80. Hata T, Shimazaki Y, Kagayama A, Tamura S, Ueda S. Development of a novel drug delivery system (TES): V. Animal Pharmacodynamic Study and Human Bioavailability Study. Int J Pharm 1994; 110:1–7.
81. Murata S, Ueda S, Shimojo F, Tokunaga Y, Hata T, Ohnishi N. In vivo performance of time-controlled explosion system (TES) in GI physiology regulated dogs. Int J Pharm 1998; 161(2):161–168.
82. Mohamad A, Bussemer T, Dashevsky A, Bodmeier R. Development of multiparticulate pulsatile drug delivery system. AAPS PharmSci 2003; 5(4):W5140.
83. Bussemer T, Bodmeier R. Formulation parameters affecting the performance of coated gelatin capsules with pulsatile release profiles. Int J Pharm 2003; 267(1–2):59–68.
84. Schultz P, Kleinebudde P. A new multiparticulate delayed release system. Part I: dissolution properties and release mechanism. J Contr Rel 1997; 47(2):181–189.
85. Schultz P, Tho I, Kleinebudde P. A new multiparticulate delayed release system. Part II: coating formulation and properties of free films. J Contr Rel 1997; 47(2):191–199.
86. Ueda S, Ibuki R, Kimura S, Murata S, Takahashi T, Tokunaga Y. Development of a novel drug release system, time-controlled explosion system (TES). III. Relation between lag time and membrane thickness. Chem Pharm Bull 1994; 42(2):364–367.
87. Bussemer T, Dashevsky A, Bodmeier R. A pulsatile drug delivery system based on rupturable coated hard gelatin capsules. J contr Rel 2003; 93(3):331–339.
88. Bussemer T, Peppas NA, Bodmeier R. Time-dependent mechanical properties of polymeric coatings used in rupturable pulsatile release dosage forms. Drug Dev Ind Pharm 2003; 29(6):623–630.
89. Colombo P, Conte U, Caramella C, Geddo M, La Manna A. Disintegrating force as a new formulation parameter. J Pharm Sci 1984; 73(5):701–705.
90. Low BK, Mullen AB, Stevens HN. Inhibition of rupture of pellets containing disintegrant and PEG. AAPS PharmSci 2001; 3(3):W4166.
91. Khare AR, Peppas NA, Massimo G, Colombo P. Measurement of the swelling force in ionic polymeric networks. I. Effect of pH and ionic content. J contr Rel 1992; 22(3): 239–244.
92. Chen CM. Multiparticulate pulsatile drug delivery system. US Patent 5,260,068, 1993.
93. Chen CM. Pulsatile particles drug delivery system. US Patent 5,260,069, 1993.
94. Black HR, Elliott WJ, Grandits G, Grambsch P. Principal results of the Controlled Onset Verapamil Investigation of Cardiovascular End Points (CONVINCE) trial. JAMA 2003; 289(16):2128–2131.

4
Microencapsulation Techniques for Parenteral Depot Systems and Their Application in the Pharmaceutical Industry

Thomas Kissel, Sascha Maretschek, Claudia Packhäuser, Julia Schnieders, and Nina Seidel
Department of Pharmaceutics and Biopharmacy, Philipps-University of Marburg, Ketzerbach, Marburg, Germany

INTRODUCTION

Microencapsulation has been widely used in many industrial applications, including graphics and food and agriculture. Application of microparticulate delivery systems for pharmaceutical and medical problems has also been extensively studied. Microencapsulation technology allows protection of the drug from the environment, stabilization of sensitive drug substances, elimination of incompatibilities, or masking of unpleasant taste. Microparticles are particularly interesting for the development of controlled or prolonged release dosage forms. They play an important role as drug delivery systems aiming at improved bioavailability of conventional drugs and minimizing side effects.

Pharmaceutical applications of microspheres involve different routes of administration; among them oral, pulmonary, and parenteral delivery can be distinguished. Microparticles are consequently administered either as dry powders, e.g., by inhalation, or in the form of an aqueous suspension, e.g., by injection. Our considerations will be mainly limited to the parenteral route of administration.

The term "microencapsulation" is used to designate a category of technologies used to entrap solids, liquids, or gases inside a polymeric matrix or shell. In contrast to film coating techniques, particle formation occurs in a single step. Usually the drug substance is encapsulated in a biocompatible or biodegradable polymer forming particles with a diameter in the range of 1 to 1000 μm. For parenteral delivery system, the diameter of the microparticles should be less than 250 μm, ideally less than 125 μm, to allow injections with acceptable needle diameters.

Two general micromorphologies of microparticles can be distinguished—microcapsules and microspheres. The term "microcapsule" should be used to describe particles in which a drug-containing core is completely surrounded by a polymer shell. The core can be solid, liquid, or gas; the shell is a continuous, porous or nonporous, polymeric layer. On the other hand, "microspheres" are defined as microparticles in which the drug substance is either homogeneously dissolved or dispersed in a polymeric matrix. These monolithic delivery systems can also be described as solid solutions or solid dispersions of the drug in a microscale polymeric matrix. Microspheres show different release properties compared to true microcapsules, and an additional feature is that catastrophic drug burst due to rupture of the shell cannot occur.

With the recent progress in biotechnology and genetic engineering, a large number of protein drugs have found therapeutic applications. In most cases, these bioactive agents are ineffective after peroral administration, as they are rapidly degraded and deactivated by the proteolytic enzymes in the gastrointestinal tract. Furthermore, their molecular weights are usually too high for oral absorption. Due to their short biological half-lives, frequent injections or infusion therapy are needed to achieve therapeutic plasma levels for prolonged periods of time. Biodegradable parenteral delivery systems, such as implants or microparticles have already been proven to be suitable parenteral depot systems for the administration of proteins and polypeptides. DNA and oligonucleotides, are another challenging class of therapeutic agents which have also been successfully encapsulated.

In this chapter, we describe microencapsulation techniques which are relevant for the preparation of microparticles for parenteral depot delivery systems on an industrial scale. After a short summary of the most common biodegradable polymers which are used for this purpose, different techniques will be explained in more detail and examples will be given about pharmaceutical dosage forms prepared by these methods.

BIODEGRADABLE POLYMERS

Introduction

Many different polymers are used in pharmaceutical sciences and technology as excipients, drug delivery systems, bandage, suture, or packing materials. For parenteral depot systems, initially synthetic nondegradable polymers were used as drug carriers, diffusion barriers, or protective coatings. But, after the 1970s, new polymer materials, which degrade in a biological environment, emerged. For drug delivery, a biodegradable polymer and its degradation products have to be biocompatible and toxicologically harmless. Biodegradable polymers are advantageous for all cases, where the removal of the spent device is either inconvenient or even impossible.

Biodegradable polymers were first used as biomaterials for the manufacturing of absorbable sutures and orthopedic fixture materials (1–4). Later, they attracted a great deal of attention for drug delivery and tissue engineering (5,6). Since then, the search for new biodegradable polymers progressed to design polymers for specific applications (7).

Biodegradable polymers can be classified based upon the mechanism of erosions. It is essential to recognize that "degradation" refers to the chemical process of bond cleavage, whereas "erosion" is a physical phenomenon dependent on dissolution and diffusion processes. Two mechanisms of polymer erosion can be distinguished—surface and bulk erosion. Surface-eroding devices degrade by enzymatic or hydrolytic cleavage starting from the surface. In an ideal case, the interior of the device contains unchanged polymer in the dry state. Bulk-eroding devices degrade from the "inside," and hence water uptake precedes degradation of the polymer mostly by hydrolytic cleavage reactions. Consequently, the molecular weight of the polymer decreases until the cleavage products become water soluble. The interior of the devices retains drug and cleavage products in an aqueous milieu. In many cases, both mechanisms occur concurrently, but the relative extent of surface or bulk erosion varies depending on the structure and composition of the biodegradable polymer. Surface erosion is considered most desirable, because the kinetics of erosion, and hence the rate of drug release, is more predictable and the drug substance is not exposed to an acidic environment generated by hydrolytic cleavage products, e.g., from polyesters. The erosion mechanism depends, in particular, on the rate of water permeation in comparison to the rate of bond cleavage, and consequently on the diffusion of degradation products. In case the rate of water permeation is higher than that of erosion, bulk erosion can be expected. On the other hand, surface erosion occurs if the rate of erosion exceeds the rate of water permeation into the bulk of the polymer.

Controlled-release dosage forms improve the effectiveness of drug therapy by increasing the therapeutic activity while reducing the intensity of side effects and number of drug administrations required during treatment. A variety of biodegradable polymers have been synthesized for the controlled release of different drugs. The selection and design of a suitable biodegradable polymer is the first challenging step for the development of a parenteral drug delivery system. Several classes of synthetic polymers have been proposed, which include poly(ester)s, poly(anhydride)s, poly (carbonate)s, poly(amino acid)s, poly(amide)s, poly(urethane)s, poly(ortho-ester)s, poly(iminocarbonate)s, and poly(phosphazene)s. In the following sections, the most common classes of biodegradable polymers are briefly introduced; but a detailed discussion is beyond the scope of this chapter.

Poly(ester)s

The most widely investigated class of polymers with regard to toxicological and clinical data comprise aliphatic poly(ester)s consisting of lactic and glycolic acid. Their corresponding polymers, poly(lactic acid), poly(glycolic acid), and copolymer poly (lactic-co-glycolic acid) have found widespread commercial application as drug delivery devices. Some of these drug delivery devices will be discussed later in this chapter.

Because high molecular weight polymers of glycolic and lactic acid cannot be obtained by condensation of the related α-hydroxycarboxylic acids, the polymers are usually produced by ring-opening polymerization of their cyclic dimers in the presence of catalysts such as stannous octoate (Fig. 1).

Lactic acid contains an asymmetric α-carbon which is typically described as D or L form and thus three different polymers are conceivable: poly(L-lactic acid), poly(D-lactic acid), and poly(DL-lactic acid). To date, poly(L-lactide) and poly(DL-lactide) have received the most attention among these poly(lactide)s (PLAs). The stereochemistry of the polymers affects their mechanical, thermal, and biological

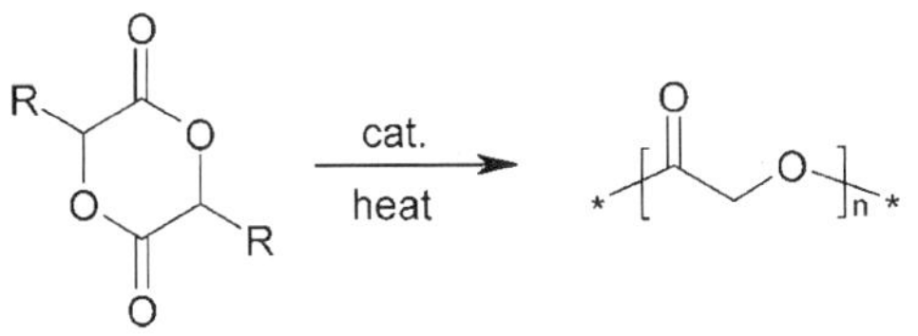

Figure 1 Ring-opening polymerization of PLA (R = H) or poly(glycolide) (R = CH_3). *Abbreviation*: PLA, poly(lactide).

properties. Poly(lactide-co-glycolide) (PLGA) represents the "gold standard" of biodegradable polymers. The properties of these copolymers can be modified by changing the ratio of PLA and poly(glycolide).

The mechanism of degradation in poly(ester)s is classified as bulk degradation with random hydrolytic scission of the ester bond linkages in the polymer backbone. In larger devices, the acidic degradation products can accumulate in the polymer matrix due to reduced diffusion. This leads to an autocatalytic effect on the degradation of the device.

Another polymer with excellent biocompatibility is poly(ethylene glycol). One reason for this high biocompatibility is its hydrophilic nature. This attribute leads to formation of hydrogen bonds between water and the polymer chains, which also inhibits protein adsorption. Thus, the presence of poly(ethylene glycol) chains at the surface of a parenteral device lead to increased blood circulation times by prolonging biological events such as endocytosis or phagocytosis. Furthermore, block copolymers of poly(ethylene glycol) and PLA or PLGA have been synthesized for the encapsulation of proteins (8). Proteins which are incorporated into devices made from these copolymers are less likely to adsorb on to the delivery system through hydrophobic interactions.

Poly(anhydride)s

Poly(anhydride)s are more hydrophobic than the poly(ester)s, even though it contains water-sensitive bonds. This reduces the water permeation into the polymer bulk, and hence poly(anhydride)s are believed to predominately undergo surface erosion by cleavage of the anhydride bonds at the surface of the device. The most widely studied poly(anhydride)s are based on sebacic acid, *p*-(carboxyphenoxy) propane, and *p*-(carboxyphenoxy) hexane (Fig. 2). Sipos et al. (9) were able to improve the interstitial administration of an antitumor agent using a loaded polymer disc composed of poly(carboxyphenoxypropane-sebacic acid).

For the design of parenteral drug delivery systems, proper selection of the biodegradable polymer is very important because solubility properties define the scope of microencapsulation technologies that can be used on one hand, and the drug release and degradation properties on the other hand. It should be noted that the biodegradable polymer is an integral part of the device and should not be regarded as an inert excipient. In the following sections, different microencapsulation techniques

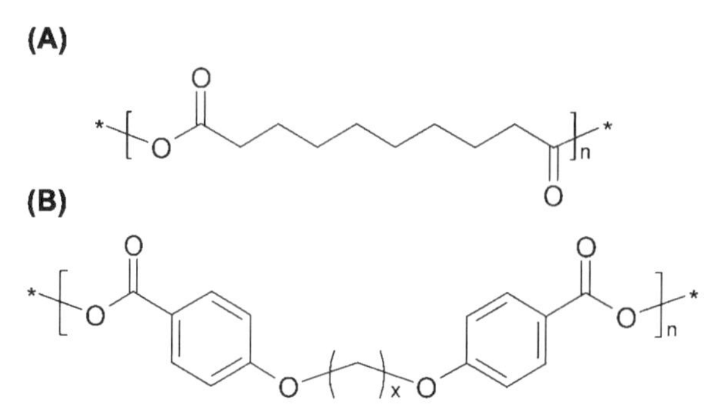

Figure 2 Structures of poly(anhydride)s based on monomers of sebacic acid (**A**), *p*-(carboxyphenoxy)propane (**B** for $x = 3$) and *p*-(carboxyphenoxy)hexane (**B** for $x = 6$).

will be discussed which were used to manufacture parenteral delivery systems on a commercial scale. These techniques can be classified into (water-in-oil)-in-water (W/O/W)-double emulsion method, phase separation techniques, and spray drying.

PHASE SEPARATION AND COACERVATION

Introduction

Carbonless copy paper was the first commercially successful product containing microcapsules. The original microencapsulation technique to produce these pressure-sensitive dye microcapsules was described by Green and Schleicher (10,11) in the 1950s and is based on macromolecular coacervation. The Dutch scientists Bungenberg de Jong and Kruyt (12) introduced the term "coacervation" in 1929. It is derived from the Latin word "coacervus," which means heap or pile. Coacervation is the macromolecular aggregation process brought about by partial desolvation of fully solvated macromolecules. A distinction is drawn between "simple coacervation" and "complex coacervation." In simple coacervation processes, phase separation is induced by addition of alcohol or salt, change in temperature, or change in pH. In complex coacervation, an oppositely charged polymer is added to the polymer solution leading to the formation of a coacervate phase via anion–cation interactions.

These phase separation processes can be used to encapsulate solid or liquid drug particles which are dispersed in the polymer solution. The encapsulation process involves three main stages (Fig. 3):

1. Phase separation of the primary wall polymer solution is induced by one of the above mentioned methods. This leads to a three-phase system consisting of a polymer-rich liquid phase (coacervate phase), a polymer-lean liquid phase, and a solid or liquid phase made up of the drug particles. This suspension or emulsion is usually maintained by continuous agitation.
2. The polymer-rich phase deposits as microdroplets on the interface of the dispersed drug particles. The microdroplets then start to spread leading to fusion into a membrane.
3. The polymer membrane is hardened through thermal, desolvation, or chemical methods.

A simple coacervation process can easily be demonstrated in an aqueous gelatin solution in which the solid or liquid drug is dispersed. Dropwise addition of a competing substance of greater hydrophilicity or concentration, such as alcohol, will lead to phase separation resulting in the two liquid phases: one phase in which the concentration of gelatin is high, and a second phase with a low gelatin concentration. This phase separation process takes place because the water–alcohol interactions in the gelatin solution are more favorable than the water–gelatin interactions. As a result, the macromolecules are partially desolvated. This, in turn, leads to increased interactions between the gelatin molecules themselves, hence to the formation of macromolecular aggregates or coacervates. The individual polymer coacervates form the polymer-rich coacervate phase which deposits on the interface of the dispersed solid or liquid drug particles.

The coacervation of gelatin and gum arabic (acacia) is an example for a complex coacervation process. The first step is to prepare a solution of gelatin and a solution of gum arabic in 45°C to 55°C water. The drug is dispersed or emulsified in the gelatin solution at, above 45°C. The resulting suspension or emulsion is then

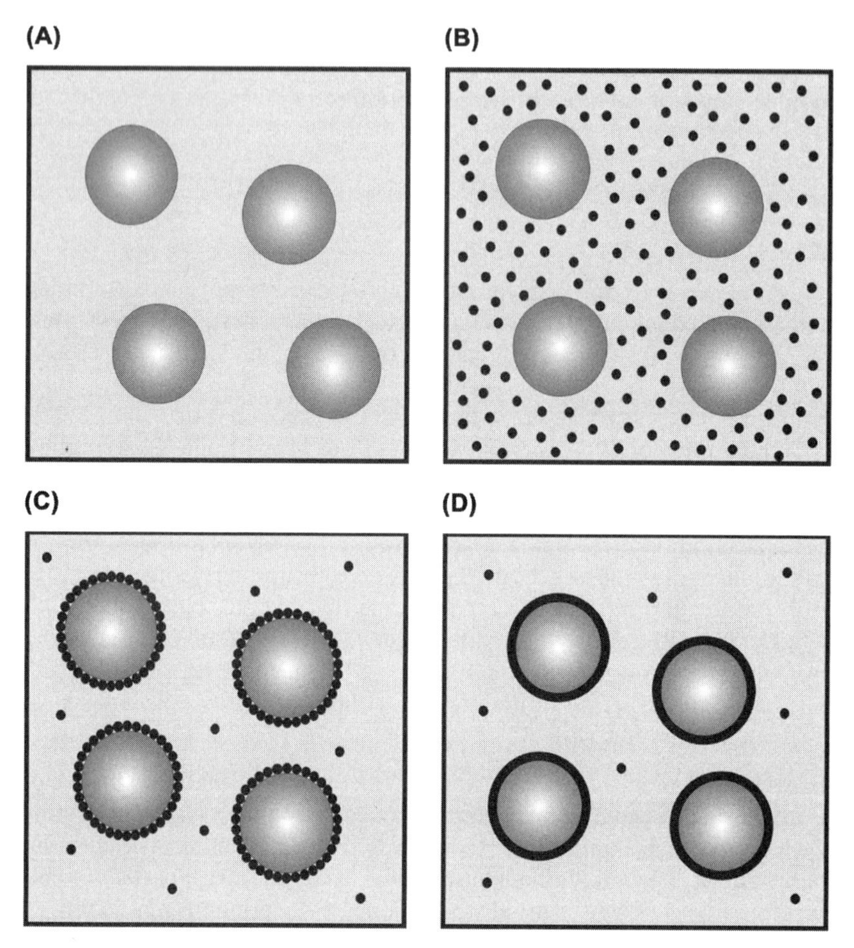

Figure 3 Schematic representation of microencapsulation through coacervation: (**A**) dispersed liquid or solid drug particles, (**B**) induction of phase separation, (**C**) deposition of microdroplets at the surface, and (**D**) fusion into a membrane.

diluted by the addition of water and gum arabic solution.Then, the pH is adjusted to below the isoelectric point of gelatin (3.8–4.4), so that its charge becomes positive, while that of the acacia is negative. At this point, the system is allowed to cool down to room temperature, while being stirred. During this cooling process, the liquid coacervates slowly deposits around the drug particles forming microparticles. As a last step, the mixture is cooled below 10°C, and glutaraldehyde is added, both to crosslink the polymers and harden the polymer shell.

Both gelatin and acacia are water-soluble, and hence the above mentioned classical methods are referred to as aqueous coacervation techniques. As the use of water-insoluble polymers (e.g., many biodegradable polymers) in pharmaceutical technologies have increased, nonaqueous coacervation techniques have been developed. These nonaqueous coacervation techniques are also referred to as organic phase separation. Desolvation is induced by temperature change, addition of a low molecular weight nonsolvent to the polymer, or, mostly, by polymer–polymer incompatibility.

If two chemically different polymers are dissolved in a common solvent, they usually separate into two phases because of incompatibilities. One phase is rich in one polymer, whereas the second phase is rich in the second polymer. For microencapsulation, the polymer that serves as the capsule shell is dissolved in an organic solvent. Then, the material to be encapsulated is introduced and a suspension or an emulsion is formed. To this mixture, the second polymer solution is added to induce phase separation by polymer–polymer incompatibility. The desired coating polymer has to be deposited preferentially on the surface of the drug for successful microencapsulation. Then, the mixture is usually cooled down, and the embryonic capsule walls are either chemically cross-linked or hardened by removal of the solvent.

Application in Pharmacy

The preparation of microcapsules by coacervation for pharmaceutical purposes started in the 1960s. Phares and Sperandio (13,14) were among the pioneers to investigate the phenomenon of coacervation as a new way to coat pharmaceuticals. They used a simple coacervation method to encapsulate five solids and two liquids by the addition of sodium sulfate to an aqueous gelatin solution. Among those solids, were drugs like acetylsalicylic acid and procaine penicillin G. Khalil et al. (15) and Nixon et al. (16,17) investigated the role of pH and the effect of electrolytes on coacervation in the gelatin–water–ethanol systems and gelatin–water–sodium sulfate systems. They, then presented an improved aqueous simple coacervation method for microcapsules containing sulphamerazine. Luzzi and Gerraughty (18,19) used the classical aqueous complex coacervation technique with gelatin and acacia to produce microcapsules containing pentobarbituric acid, and investigated the effect of additives and formulation techniques on controlled release of the drug from the microcapsules.

Other polymeric systems have also been described for the preparation of microcapsule products employing aqueous coacervation techniques. Merkle and Speiser (20) prepared phenacetine-loaded microcapsules by simple coacervation of a cellulose acetate phthalate solution through the addition of sodium sulfate. Mortada et al. (21,22) used Gantrez-AN, a hydrolyzed copolymer of maleic anhydride and methyl vinyl ether, and gelatin for a complex coacervation technique to produce nitrofurantoin-loaded microcapsules.

Today, biodegradable polymers like polylactic acid (PLA) and polylactic-co-glycolide acid (PLGA) are the most attractive polymers for microencapsulation of bioactive drugs like peptides, proteins, and hormones. As these polymers are water insoluble, many organic phase separation techniques have been developed. The first method to prepare polyester microspheres was described, probably for the first time, in 1978 (23). In the early 1980s, the luteinizing hormone–releasing hormone analogue, nafarelin was encapsulated into poly(D,L-lactic-co-glycolic) acid by phase separation process (24,25). These microspheres represented one of the earliest industrial developments of a controlled peptide release system for parenteral application. Since then, a great variety of peptide and protein compounds have been investigated.

Two commercially available pharmaceuticals containing microparticles which have been prepared by a phase separation technique are Decapeptyl® Depot (Ferring Pharmaceuticals, Suffern, New York, U.S.A.) and Sandostatin®-LAR® Depot (Novartis Pharma, Basel, Switzerland). Decapeptyl Depot, which is composed of triptorelin-loaded PLGA microspheres is used for the treatment of either prostate cancer of men or infertility of women. Sandostatin-LAR Depot consists of octreotide

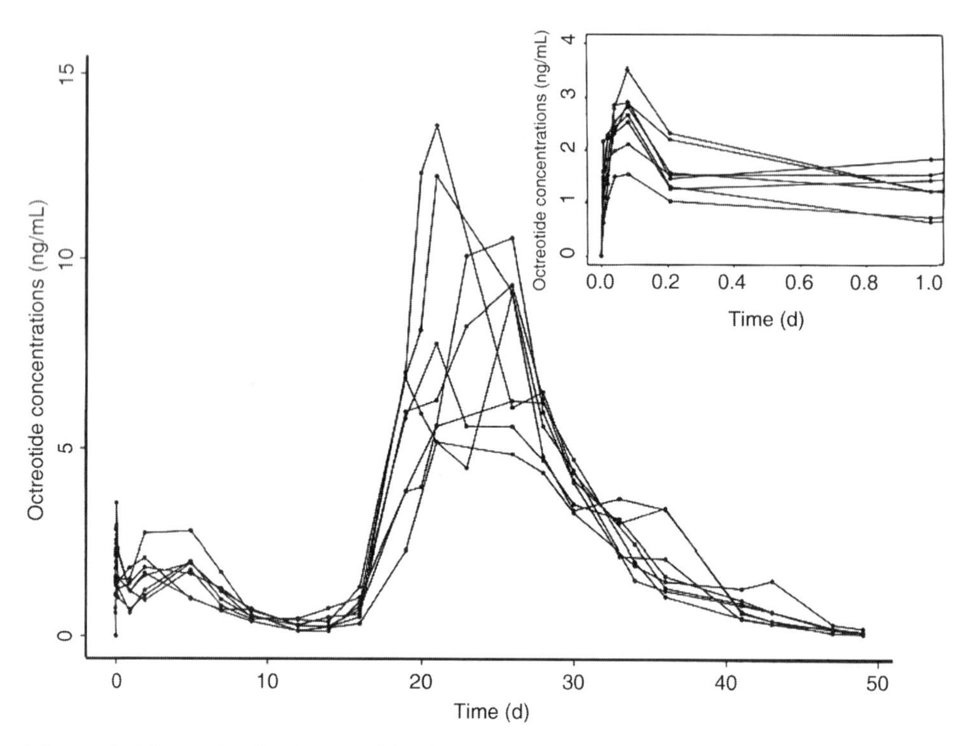

Figure 4 Plasma levels of octreotide after an IM bolus injection of 5 mg/kg of octreotide pamoate–loaded microspheres in eight rabbits. The inset shows the data collected during the first day of the experiment. *Abbreviation*: IM, intramuscular.

acetate–loaded PLGA microspheres which is used for the treatment of acromegaly— a disease resulting from a growth hormone–secreting pituitary adenoma (26–28).

Long-term treatment of acromegaly with octreotide would require three subcutaneous injections a day. Comets et al. (29) described the IM absorption profiles of octreotide pamoate–loaded microspheres (Fig. 4) in comparison to an intravenous (IV) application of octreotide solution.

For IV bolus injection, sufficient plasma levels of octreotide can be only achieved for a few hours. The release of octreotide from the microparticles leads to a triphasic absorption profile. During the initial phase, about 5% of the drug is released. After a lag-phase, the major part of the drug is released. A high interindividual variability, especially in the magnitude of the absorption, can be observed.

Octreotide release from microspheres has been studied under in vitro conditions. After 27 days, more than 90% of the drug had been released. The release profile depended on the ionic strength of the medium (Fig. 5). A drug release for more than 40 days has been observed in vivo. The data highlights the problem of "in vitro–in vivo correlation." A direct prediction of the in vivo release based on in vitro data is not possible in most cases (30).

W/O/W-DOUBLE EMULSION TECHNIQUE

The W/O/W-double emulsion technique, also designated as the in-water drying method, is a two-step microencapsulation process belonging to the solvent evaporation

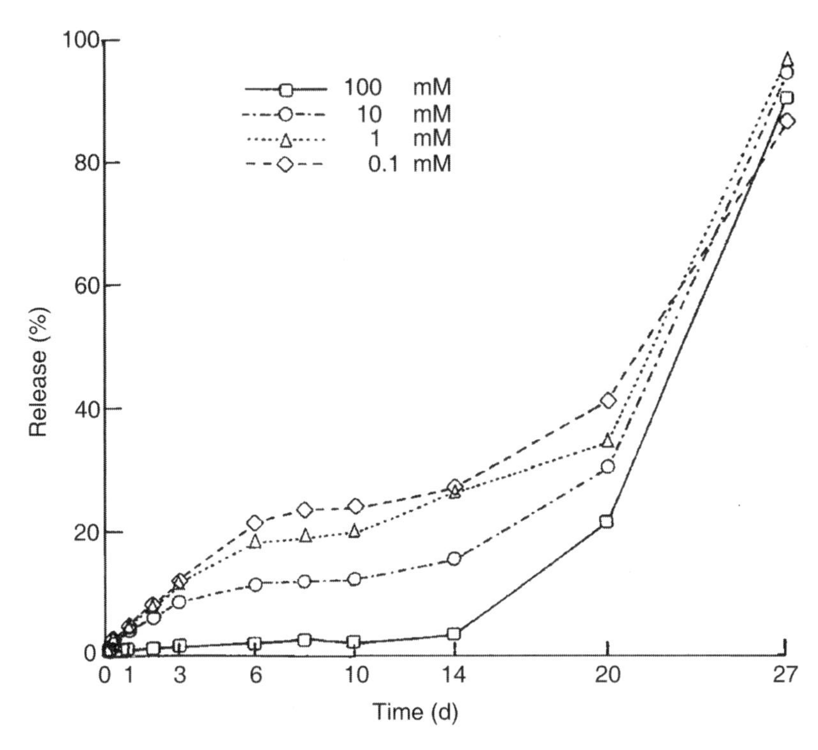

Figure 5 Cumulative in vitro release from octreotide-loaded microspheres in acetate buffer with different salt concentrations.

method. It is one of the most widely used methods of microencapsulation, because of its advantage in being able to handle drugs to be encapsulated in aqueous solution (31–34). Therefore, this method is especially suited for the microencapsulation of water-soluble and sensitive compounds such as proteins, peptides, vaccines, or other macromolecules. With oil-in-water (O/W) emulsion techniques, microspheres cannot be efficiently loaded, because drug substances are distributed into the external water phase (W) and not retained in the polymer phase.

A two-stage emulsification process is used to generate transient $W_1/O/W_2$ emulsions. The principle of encapsulation is based upon inducing phase separation of the polymer dissolved in an organic solvent (O phase) due to partial extraction of the solvent in a large volume of an external water phase (W_2 phase) and evaporation of the volatile solvent (Fig. 6). The polymer then forms a coacervate enclosing the internal aqueous phase (W_1) containing the active compound, and microparticles are hardened under removal of residual solvent. Solvents suitable for this technique are those that are good solvents for the polymer, volatile, and immiscible with water. A limited solubility of the solvent in the external water phase is beneficial. Dichloromethane was frequently employed because of its low boiling point, which facilitates the removal of residual solvent from the finished product. The other aspects include its partial solubility in water, high dissolving power for a wide range of polymers, and toxicity profile. The hydrophilic drug substance is dissolved either in water or buffer and represents the internal water phase (W_1). Stabilizing excipients can be added to protect the drug in the core of the microspheres. This W_1 phase is homogenized with the O phase yielding a W_1/O emulsion. Homogenization is carried out by using high-speed or high-pressure mixers, ultrasonication, or static mixers.

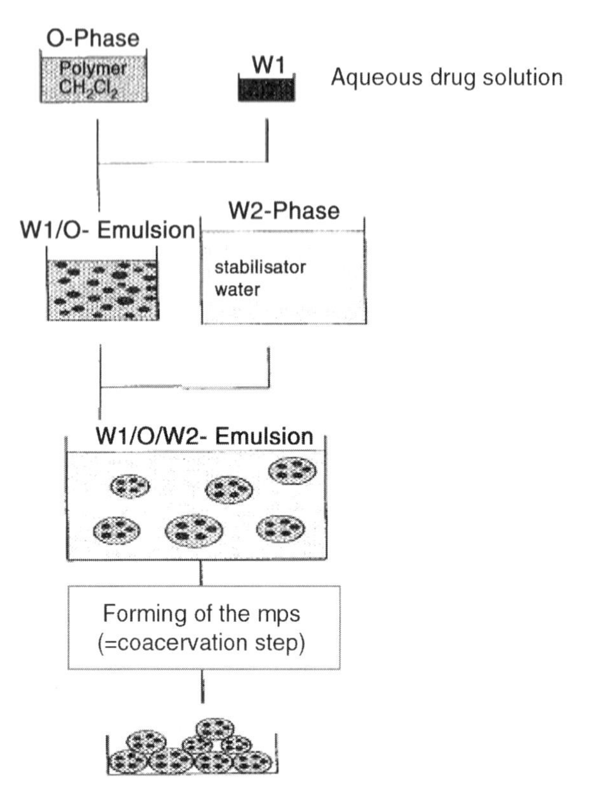

Figure 6 Preparation principle of microparticles according the W/O/W technique. *Abbreviation*: W/O/W, (water-in-oil)-in-water. *Source*: From Ref. 35.

This primary emulsion is then rapidly emulsified in a large volume of an external aqueous solution (W_2) that usually contains surfactants or stabilizers such as poly(vinyl alcohol) (PVA), leading to a transient double $W_1/O/W_2$ emulsion. The organic O phase acts as a barrier between the two aqueous compartments. Thus, the migration of the drug into the external W_2 phase is impeded.

The particle formation itself is based on coacervation. The solvent is extracted from the polymer containing O phase because of its initial diffusion into the continuous W_2 phase inducing phase separation of the polymer. The organic solvent is eliminated by two steps: firstly by extraction and secondly by evaporation (36).

For further removal of the organic solvent and particle hardening, the system is stirred at a constant rate for several hours, usually three to five hours. After complete solidification, the microparticles are isolated either by centrifugation or filtration and washed several times with water to remove PVA and nonencapsulated drug. Depending on the glass transition temperature of the polymer, the microspheres are lyophilized or dried in a vacuum at room temperature.

This first description of a W/O/W solvent evaporation method can be found in Japanese patents developed by the companies Upjohn and Fuji, in 1962 and 1967, and abstracted by Kondo (37). For pharmaceutical products, this microencapsulation technique was used to develop a parenteral depot formulation for leuprolide acetate (31). Takeda enterprises, secured several patents between the years 1984 to 1991 and obtained marketing approval for this product under the trade name, Enantone™ for the treatment of prostate cancer.

Since this first report by Ogawa et al. (38), the influence of various formulation and process parameters on microspheres formation has been investigated intensively which will be discussed below. Different peptides and proteins such as bovine serum albumin (BSA) or ovalbumin (OVA), and several antigens such as diphtheria toxoid, tetanus toxoid, ricin toxoid, influenza A hemagglutinin, and other therapeutically relevant proteins like erythropoietin, have been successfully encapsulated (Table 1) (53–56).

The main advantage of the W/O/W-double emulsion technique resides in the fact, that hydrophilic drug substances can be encapsulated under very gentle process conditions.

Furthermore, high yields and encapsulation efficiencies can be achieved also on small-scale laboratory equipment.

Although the W/O/W microencapsulation technique seems to be quite simple to carry out, the particle formation process is quite complicated and a host of process parameters influence on properties of microspheres. These parameters are summarized in Table 2.

Although some general trends, as outlined above, are known, the influence of these process variables on a system must be determined and optimized experimentally for each combination of drug and coating polymer. To characterize microspheres as parenteral depot systems, the following parameters routinely evaluated are yield, particle size and distribution, particle shape, drug loading, encapsulation efficiency, drug release profile, particle degradation, and sterility.

The microencapsulation of proteins still involves considerable difficulties. Proteins are very unstable and their complex three-dimensional structure, necessary for their biological activity, is extremely sensitive to changes in pH, temperature,

Table 1 Examples of Various Drugs Incorporated into Microparticles by W/O/W Method

Drug	Polymer	Reference (year)
Leuprorelin acetate	PLGA	(31) (1988)
Thyreotropic releasing hormone	PLGA	(38) (1987)
Octreotide	Branched PLGA	(39) (1991) (40) (2002) (41) (2004)
Diphtheria toxoid	PLA	(42) (1991)
D,L Sotalol	PLGA	(43) (1992)
Somatostatin acetate	PLGA	(44) (1993)
Ricin toxoid	PLGA	(45) (1993)
Tetanus toxoid	PLGA, PLA	(46) (1993)
Erythropoietin	PLGA	(47) (1997)
Insulin-like growth factor	PLGA	(48) (2001)
Insulin	PLA, PLGA, PLCA	(49) (2000) (50) (2001) (51) (2003)
Lysozyme, trypsin, heparinase, ovalbumin, albumin, immunglobulin	Polyanhydrides	(52) (1993)

Abbreviations: PLA, poly(lactide); PLCA, poly(L-lysine citramide); PLGA, poly(D,L-lactic-co-glycolide).
Source: From Ref. 35.

Table 2 Process Parameters of the W/O/W Technique and Their Influence on Particle Characteristics

Composition of the phases	Parameters	Common effects on particle characteristics
W_1 phase		
Properties of the drug (IEP, molecular weight)	Diffusivity, solubility, size, interactions with the polymer	Influence on drug release
Drug loading: amount of water, concentration, viscosity	Increased loading	Increased particle size Increased porosity
Additives (stabilizers, buffer salts, salts, solvent)	Increased osmotic pressure	Decreased entrapment efficiency
O phase		
Properties of the polymer (molecular weight, constitution, swellability)	Increased M_w Increased viscosity	Increased particle size Increased entrapment efficiency
Properties of the solvent (vapor pressure)	Decreased solidification time	Increased entrapment efficiency
	Too fast hardening aggregation before homogenization	Influence on porosity
Amount of the polymer, volume of the organic solvent, concentration, viscosity, and solubility in the continuous phase	Increased polymer concentration Increased viscosity	Increased particle size Increased entrapment efficiency Decreased burst effect
Additives (stabilizers, plasticizers)	Surfactants	Particle size reduction Decreased entrapment efficiency Higher burst effect
W_2 phase		
Properties of the stabilizer, type	Ability of stabilizing the dispersed phase	Influence on entrapment efficiency
Concentration of the stabilizer, viscosity, pH		
Additives (buffer salts)	Increased osmotic pressure	Increased entrapment efficiency
Volume ratios of the phases		
$W_1/O \leftrightarrow W_2$, $W_1 \leftrightarrow O$, $O \leftrightarrow W_2$	Decreased W_2	
	Increased W_1	Decreased entrapment efficiency Increased release rate
Test conditions		
Temperature		
Homogenization technique, tools and intensity (e.g., stirring rate, duration)	Increased intensity, time	Decreased particle size, decreased entrapment efficiency
Removal of the organic solvent		
Pressure ratio	Decreased solidification time	Increased entrapment efficiency Influence on porosity

(*Continued*)

Table 2 Process Parameters of the W/O/W Technique and Their Influence on Particle Characteristics (*Continued*)

Composition of the phases	Parameters	Common effects on particle characteristics
Scheduled quantity	Scaling up	
Vessel geometry		
Further processing of the microparticles		
Isolation (filtration, centrifugation)		
Washing		
Drying (vacuum, lyophilization)		

Abbreviations: IEP, isoelectric point; M_W, molecular weight.
Source: From Refs. 35, 57–59.

shear stress, etc. leading to deactivation and aggregation. During the encapsulation process, proteins are exposed to several interfaces that exist in systems such as, organic solvents, extreme shear forces or ultrasonication, and additives destabilizing proteins. Slight pH shifts in the W_1 phase and adsorption onto the polymer can lead to aggregation and denaturation of the proteins, requiring addition of buffer salts or other excipients for protein stabilization (60). This may lead to an increased osmotic pressure in the W_1 phase causing poor encapsulation efficiency. Increased volumes of the W_1 phase (to offset osmotic pressure) also lead to low drug entrapment (61). Also, the viscosity of the inner water phase plays an important role in the prevention of the above mentioned loss of the active agent.

As mentioned above, a parenteral depot delivery system for leuprorelin acetate has been marketed using the W/O/W-double emulsion technique. Leuprorelin is a member of a class of drugs called GnRH (gonadotropin-releasing hormone) agonists. GnRH agonists act by interfering with the production or activity of specific hormones in the body like luteinizing hormone (LH), follicle-stimulating hormone (FSH), and consequently oestradiol. Because of this effect, the medicament can be used to treat different conditions such as prostate cancer, endometriosis, fibroids, and central precocious puberty.

In the past, leuprorelin acetate had to be injected subcutaneously every day. The microsphere-depot form releases the drug at a constant rate either for one or three months following a single injection. This can be achieved using biodegradable polymers, which elicit different release profiles depending on the composition (62). The one-month release injection consists of a PLGA at a ratio of 3:1 with a molecular weight of 14,000 (63). Concerning the three-month depot formulation, the microcapsules were prepared using poly(DL-lactic acid) with a molecular weight of 15,000 (64). The products were available in Europe under the name of Enantone® (Takeda Pharma, Pharmaceutical Company Ltd., Osaka, Japan), Trenantone® (Takeda Pharma, Pharmaceutical Company Ltd.) and Enantone Gyn® (Takeda Pharma, Pharmaceutical Company Ltd.), accordingly in the U.S. Lupron Depot® (TAP Pharmaceutical Products Inc., Illinois, U.S.A.).

The in vivo release patterns of three different doses of leuprorelin acetate is demonstrated in Figure 7. PLGA microparticles for one-month release were injected subcutaneously into rats. Microcapsules remaining at the injection site were

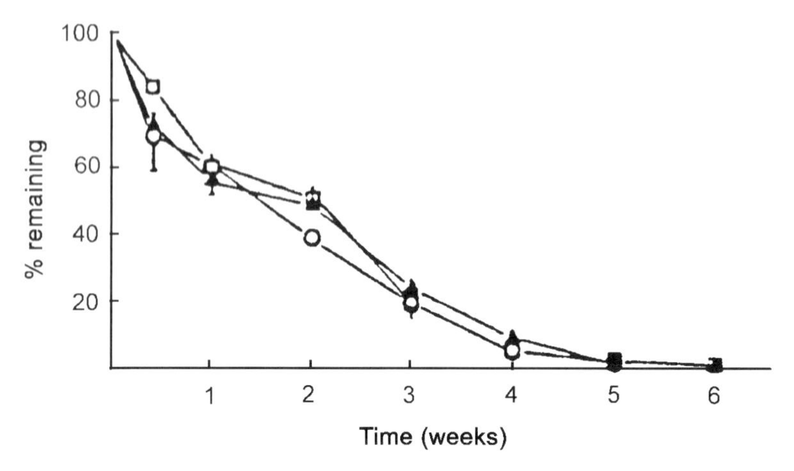

Figure 7 Leuprorelide acetate remaining in PLGA microspheres after subcutaneous injection at three different doses in five rats. *Abbreviation*: PLGA, poly(lactide-co-glycolide).

periodically excised to determine the percentage of the residual drug, which was quite similar for each dose. After an initial burst, the release profiles followed an approximately pseudo-zero-order kinetics. This release kinetics appeared to be rapid in comparison with that of the in vitro release kinetics (Fig. 8) (62,65).

The serum level of leuprorelin acetate is shown in Figure 9. The initial release of the drug produced a sharp increase in the serum level, but following this, the level was reasonably sustained. After three weeks, the level gradually declined till about six weeks (62).

The levels of LH, FSH, and oestradiol increased at an early stage because of the stimulatory effects of leuprorelin, but they decreased to below the normal level because of the paradoxical effect, and were maintained at the suppressed level for

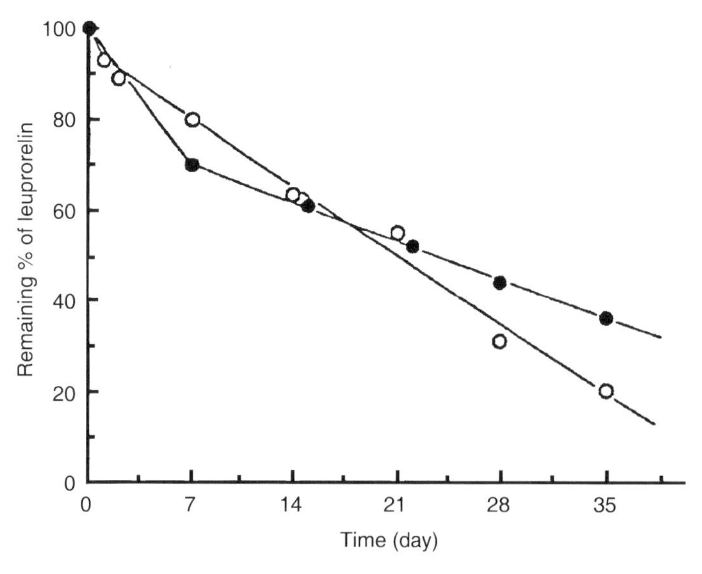

Figure 8 Leuprorelide acetate remaining in PLA (●) and PLGA (○) under in vitro release conditions. *Abbreviations*: PLA, poly(lactide); PLGA, poly(lactide-co-glycolide).

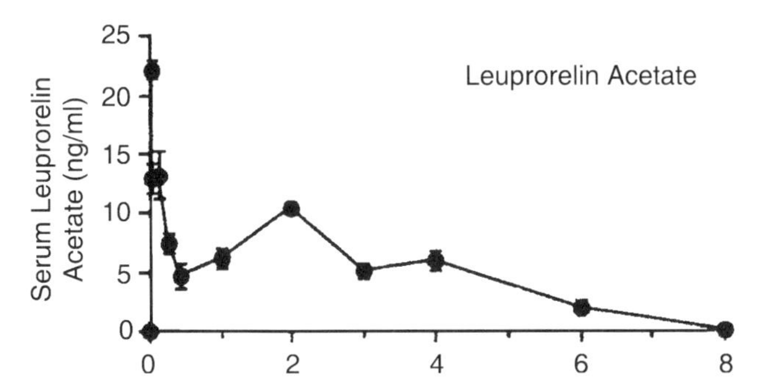

Figure 9 Serum level of leuprolin acetate after subcutaneous injection of microspheres in five rats.

more than six weeks (Fig. 10). No elevation was observed after the second and third administration (62).

It can be concluded that the aim of a continuous release kinetic can be achieved, but zero-order release kinetics cannot be expected.

SPRAY DRYING

Introduction

Spray drying is traditionally used in the chemical, pharmaceutical, biochemical, and food industries for the drying of substances. For example, it is commonly used to process milk, ceramics, and various chemicals into a dry powder form. Therefore, information from food processing and chemical engineering is particularly relevant to pharmaceutical problems. The spray drying technique can be used to encapsulate sensitive substances such as fragrances, essential oils, or vitamins. The internal structure of the microparticles depends on the solubility of the drug in the polymer leading to monolithic microspheres with dispersed or dissolved drug substance. True microcapsules can be obtained by spraying a suspension of a drug in an organic solution of the polymer. By contrast, microspheres of solid solution or solid dispersion morphology, are obtained by selection of appropriate process conditions and solvents. In the following sections, the different principles of the spray drying process, equipment (Fig. 11), and pharmaceutical applications will be discussed. For a detailed discussion, the reader is referred to the literature (66–68).

Principles of the Spray Drying Process

Spray drying is defined as the transformation of a feed from a fluid state (solution, dispersion, or paste) into a dried particulate form by spraying the feed into a hot gaseous drying medium. It is a continuous one-step processing operation in which four different phases can be distinguished, namely:

1. atomization of the feed,
2. mixing of spray and air,
3. solvent evaporation, and
4. product separation.

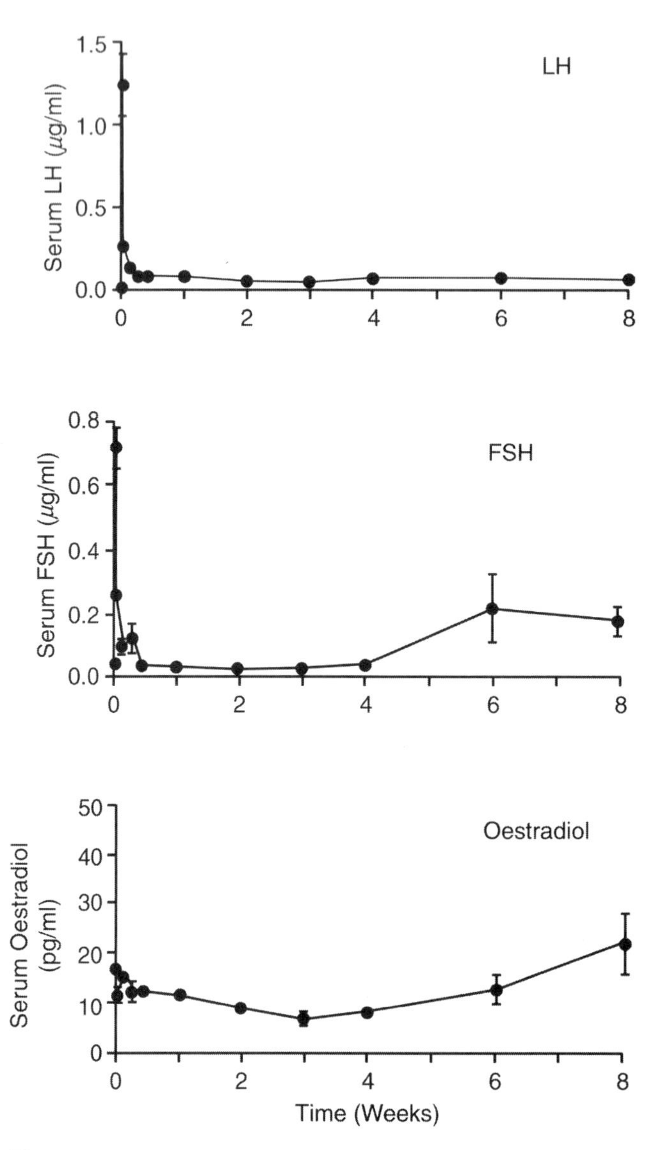

Figure 10 Serum level of LH, FSH, and oestradiol after subcutaneous injection of microspheres in five rats. *Abbreviations*: FSH, follicle-stimulating hormone; LH, luteinizing hormone.

A variety of atomization systems are available, which may be classified according to the nozzle design as rotary atomization, pressure atomization, and two-fluid atomization.

1. In rotary atomization, the feed fluid is sprayed into the drying chamber by means of a spinning disc or wheel which creates a spray of droplets.
2. Pressure atomization occurs when the feed is fed to the nozzle under pressure which causes the fluid to be disrupted into droplets.
3. In two-fluid nozzles, the feed fluid and atomizing air are passed separately to the nozzle, where they mix, and the air causes the feed to break up into a spray.

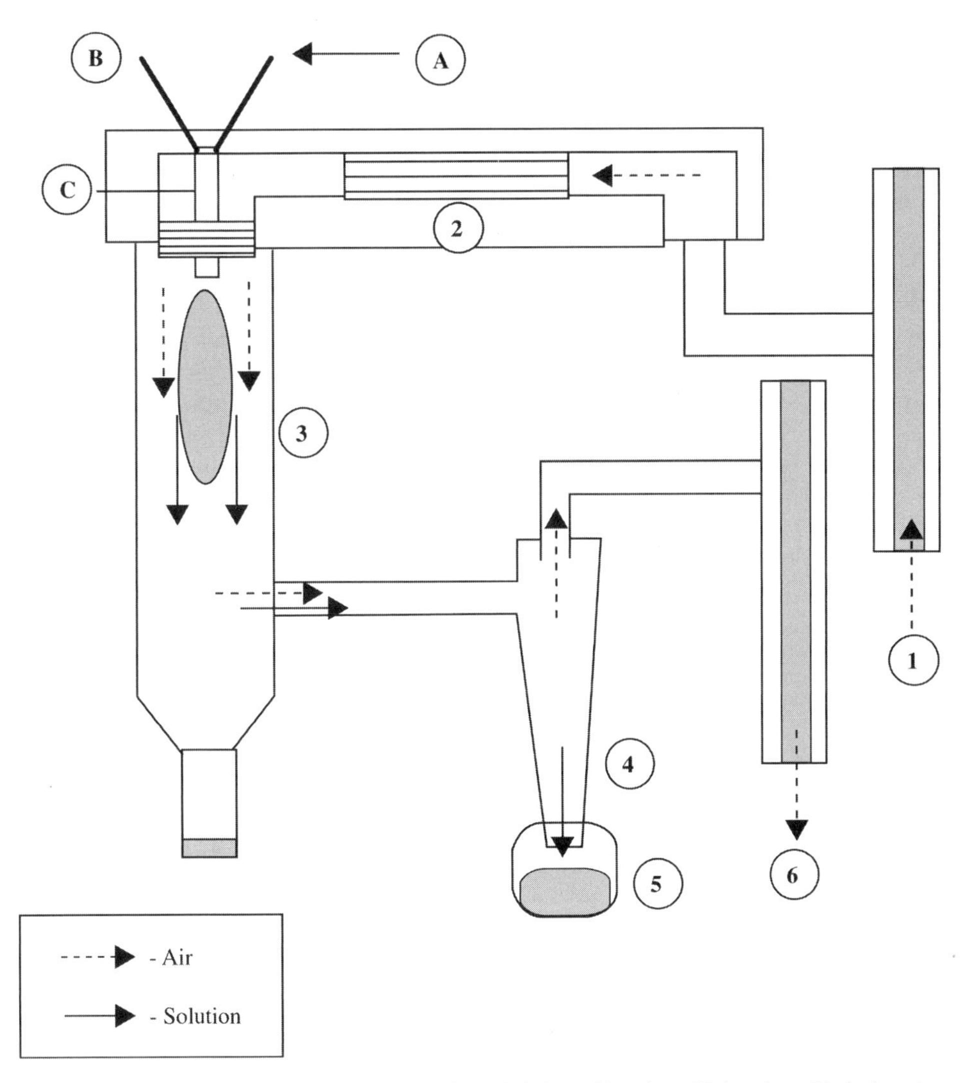

Figure 11 Laboratory spray dryer: (1) drying air inlet + filtration; (2) heating; (3) desiccation chamber; (4) cyclone; (5) collector for drying powder-microspheres; (6) filtration + air outlet; (**A**) solution, suspension, emulsion to spray; (**B**) compressed or nitrogen air; and (**C**) spray nozzle (e.g., pneumatic, ultrasonic).

The selection of the atomizer affects directly the desired particle size. Rotating disks can be used with very viscous feeds, and hence allow for the formation of microspheres up to 200 μm. Two-fluid nozzle atomizers with nozzles of a typical internal diameter between 0.5 and 1.0 μm, lead usually to small microparticles with a diameter less than 10 μm.

Two different spray dryer designs can be distinguished. If spray and drying air pass through the dryer in the same direction, they operate in a cocurrent manner. If spray and drying air enter the drying chamber at opposite ends, they operate in a counter-current fashion. The latter design is preferred when "sticky" products such as microspheres from polymers with low glass transition temperatures need to be processed. Because of the rapid evaporation of the solvent, the temperature of the droplets can be kept far below the drying air temperature, and is therefore applicable to both heat-resistant and heat-sensitive materials. For environmental protection, closed loop systems with recovery of the evaporated solvents are commercially available. Also, spray drying in an atmosphere of inert gas or under aseptic conditions is technically feasible (69).

Pharmaceutical Applications for Parenteral Depot Systems

Developing a parenteral depot formulation for drug substance, one must consider the desired kinetics profile for specific in vivo application. The release kinetics can be adjusted by the design of the drug delivery system by choosing different kinds of polymers, or varying the manufacturing process. One target for sustained drug delivery using microspheres is a zero-order drug release profile. By manipulating drug loading and polymer degradation, continuous release profiles can be attained, but again zero-order kinetics have been rather the exception than the rule. The correlation of in vitro release profiles to in vivo release kinetics is still poorly understood, and further work is necessary to solve this problem.

The spray drying method was used for the preparation of microparticles using biodegradable polymers as matrix on an industrial scale starting in the mid-1970s. The first product was bromocriptine, which was encapsulated in poly(L-lactic acid) using the above mentioned technology. Subsequently Parlodel LAR™ (Sandoz Pharma, now Novartis Pharma, Basel, Switzerland) (Figs. 12 and 13) was developed using a star-branched polyester poly(D,L-lactide-co-glycolide-D-glucose) which allowed repeated injections at monthly intervals (30,70,71).

Bodmeier and Chen (72) used poly(D,L-lactic acid) as polymer and encapsulated different substances, e.g., progesterone and theophylline. Microparticles containing captopril were prepared by spray drying using PLGA as polymer. In this study, the susceptibility of the drug substance to gamma-sterilization was affected by the micromorphology of the microspheres (73). The mild production conditions allow the encapsulation of proteins and peptides by spray drying (74). For example, the encapsulation via spray drying of recombinant human growth hormone (rh-GH) and tissue-type plasminogen activator was reported by Mumenthaler et al. (75). They demonstrated the feasibility to microencapsulate rh-GH and tissue plasminogen activator (t-PA) by spray drying using additives to prevent oxidation, aggregation, and

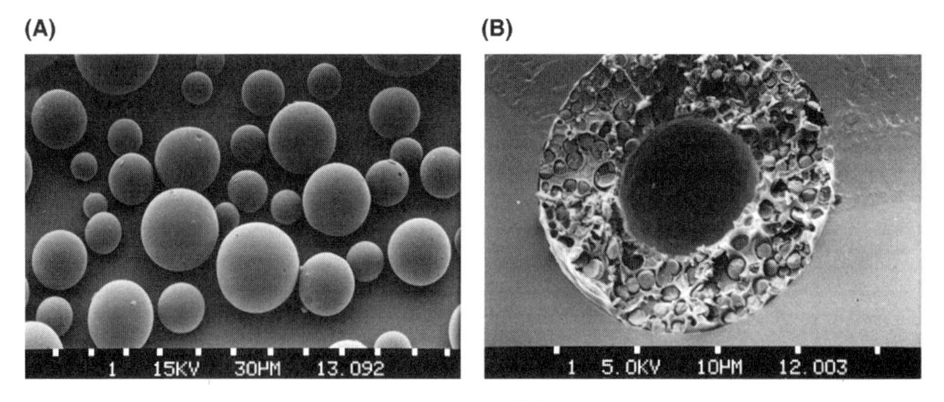

Figure 12 SEM photographs of Parlodel LAR™. (**A**) Morphology of microparticles and (**B**) cross-section of microparticle before drug release.

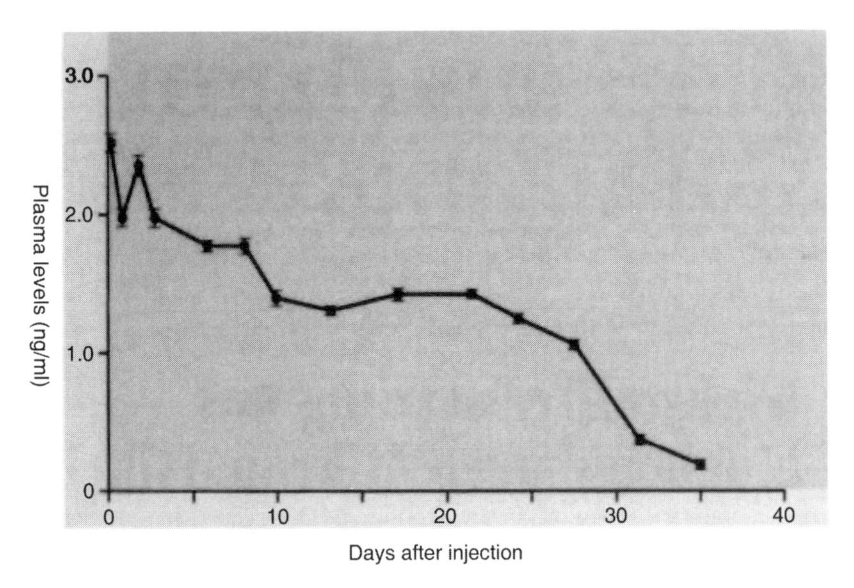

Figure 13 Plasma concentration of bromocriptine in rabbits after i.m. injection as a function of time ($n = 6$).

denaturation of the proteins. Also recombinant human erythropoietin (rhEPO) was successfully encapsulated using PLGA and spray drying technique providing a number of advantages compared to the W/O/W method (76). The microencapsulation of DNA into stearylamine-containing cationic microspheres for gene delivery was also reported (77). Compared to other microencapsulation techniques, spray drying offers some attractive features. This technique is reproducible, rapid, and easy to scale-up. Aseptic production is also possible. The main disadvantage of spray drying is its cost, both for its operation and the investment. Another difficulty frequently encountered in preparing the spray-dried microspheres was the formation of polymer fibers as a result of inadequate forces to disperse the filaments into droplets; the successful atomization into droplets is dependent on both the type of polymer used and the viscosity of the spray solution (66).

Pharmaceutical Products

In the year 1992, the pharmaceutical company Hoechst AG (Germany), now Sanofi-Aventis, patented "long-acting biodegradable peptide drug microparticles" with buserelin acetate as drug substance, which was spray dried with PLGA as a matrix-polymer and is currently available on the market as Profact® (Hoechst AG, now Sanofi-Aventis, Frankfurt, Germany) (78). Genentech microencapsulated recombinant human nerve growth factor (rh-NGF) in PLGA for promoting nerve cell growth, repair, survival, differentiation, maturation, or function (79). The microencapsulated formulation provided a continuous release of rh-NGF over 7 to 14 days. Yamanouchi Pharmaceutical Co. (Japan) spray dried a physiological active protein while sustaining its physiological activity and enclosing it in a biodegradable polymer substance at a high efficiency (80). Microencapsulation technology also provides the benefit for antigen vaccine delivery, e.g., recombinant hepatitis B surface antigen which was encapsulated using oligosaccharide ester derivatives by a spray drying method (81).

CONCLUSION

From this chapter, it has become apparent that a number of microencapsulation methods are available today for the preparation of microparticles on an industrial scale. In fact, parenteral drug delivery systems based upon biodegradable microspheres are a true success story for the concept of drug delivery. In view of the renaissance of recombinant proteins as biogenerics, parenteral delivery systems will receive increasing attention also from a commercial point of view.

Microencapsulation methods are based on different physicochemical principles, but are generally concerned with either liquid–liquid interfaces or with liquid–air interfaces. All methods seem to be suitable for a wide range of drugs or bioactive macromolecules. The selection of the technology, the polymer, and the drug substance can be based on a vast body of information available now in the literature.

Areas of further research are not only associated with advances in technology and a better understanding of the microencapsulation technologies, but also with their performance under in vitro and in vivo conditions. Especially, the prediction of drug release profiles from in vitro data is an area where increased efforts seem to be necessary. An in vitro release model mimicking the fate of microspheres applied by the parenteral route would be highly desirable.

Also, new strategies to stabilize proteins in PLGA microspheres during manufacture, shelf life, or in vivo could be of general interest.

Microencapsulation technologies have been transferred successfully to the industrial scale and further progress can be expected in the years to come.

REFERENCES

1. Schmitt EE, Rocco A. Polistina: surgical sutures. U.S. Patent No. 3,297,033, 1967.
2. Herrmann JB, Kelly RJ, Higgins GA. Polyglycolic acid sutures. Laboratory and clinical evaluation of a new absorbable suture material. Arch Surg 1970; 100:486–490.
3. Miller ND, Williams DF. The in vivo and in vitro degradation of poly(glycolic acid) suture material as a function of applied strain. Biomaterials 1984; 5:365–368.
4. Leenslag JW, Pennings AJ, Bos RR, Rozema FR, Boering G. Resorbable materials of poly(L-lactide). VI. Plates and screws for internal fracture fixation. Biomaterials 1987; 8:70–73.
5. Langer R. New methods of drug delivery. Science 1990; 249:1527–1533.
6. Langer R, Vacanti JP. Tissue engineering. Science 1993; 260:920–926.
7. Peppas NA, Langer R. New challenges in biomaterials. Science 1994; 263:1715–1720.
8. Youxin L, Volland C, Kissel T. In vitro degradation and bovine serum albumin release of the ABA triblock copolymers consisting of poly (L(+) lactic acid) or poly (L(+) lactic acid-co-glycolic acid) A-blocks attached to central polyoxyethylene B-blocks. J Contr Release 1994; 32:121–128.
9. Sipos EP, Tyler B, Piantadosi S, Burger PC, Brem H. Optimizing interstitial delivery of BCNU from controlled release polymers for the treatment of brain tumors. Cancer Chemother Pharmacol 1997; 39:383–389.
10. Green BK, Schleicher L. Manifold record material. U.S. Patent No. 2,730,456, 1956.
11. Green BK, Schleicher L. Pressure-responsive record materials. U.S. Patent No. 2,730,457, 1956.
12. Bungenberg de Jong HG, Kruyt HR. Proc Kungl Ned Acad Wetensch 1929; 32:849.
13. Phares RE Jr, Sperandio GJ. Coating pharmaceuticals by coacervation. J Pharm Sci 1964; 53:515–518.

14. Phares RE Jr, Sperandio GJ. Preparation of a phase diagram for coacervation. J Pharm Sci 1964; 53:518–521.
15. Khalil SA, Nixon JR, Carless JE. Role of pH in the coacervation of the systems: gelatin-water-ethanol and gelatin-water-sodium sulphate. J Pharm Pharmacol 1968; 20:215–225.
16. Nixon JR, Khalil SA, Carless JE. Effect of electrolytes on coacervation of the systems gelatin-water-ethanol and gelatin-water-sodium sulphate. J Pharm Pharmacol 1968; 20:348–354.
17. Nixon JR, Khalil SA, Carless JE. Gelatin coacervate microcapsules containing sulphamerazine: their preparation and the in vitro release of the drug. J Pharm Pharmacol 1968; 20:528–538.
18. Luzzi LA, Gerraughty RJ. Effects of selected variables on the microencapsulation of solids. J Pharm Sci 1967; 56:634–638.
19. Luzzi LA, Gerraughty RJ. Effect of additives and formulation techniques on controlled release of drugs from microcapsules. J Pharm Sci 1967; 56:1174–1177.
20. Merkle HP, Speiser P. Preparation and in vitro evaluation of cellulose acetate phthalate coacervate microcapsules. J Pharm Sci 1973; 62:1444–1448.
21. Mortada SA, Motawi AM, el Egaky AM, el-Khodery KA. Preparation of microcapsules from complex coacervation of Gantrez-gelatin. I. Development of the technique. J Microencapsul 1987; 4:11–21.
22. Mortada SA, el Egaky AM, Motawi AM, el Khodery KA. Preparation of microcapsules from complex coacervation of Gantrez-gelatin. II. In vitro dissolution of nitrofurantoin microcapsules. J Microencapsul 1987; 4:23–37.
23. Fong JW. Process for preparation of microspheres. U.S. Patent No. 4,166,800, 1978.
24. Sanders LM, Kent JS, McRae GI, Vickery BH, Tice TR, Lewis DH. Controlled release of a luteinizing hormone-releasing hormone analogue from poly(D,L-lactide-co-glycolide) microspheres. J Pharm Sci 1984; 73:1294–1297.
25. Sanders LM, McRae GI, Vitale KM, Kell BA. Controlled delivery of an LHRH analogue from biodegradable injectable microspheres. J Contr Release 1985; 2:187–195.
26. Bouchot O, Soret JY, Jacqmin D, Lahlou N, Roger M, Blumberg J. Three-month sustained-release form of triptorelin in patients with advanced prostatic adenocarcinoma: results of an open pharmacodynamic and pharmacokinetic multicenter study. Horm Res 1998; 50:89–93.
27. Davies PH, Stewart SE, Lancranjan L, Sheppard MC, Stewart PM. Long-term therapy with long-acting octreotide (Sandostatin-LAR) for the management of acromegaly. Clin Endocrinol (Oxf) 1998; 48:311–316.
28. Hunter SJ, Shaw JA, Lee KO, Wood PJ, Atkinson AB, Bevan JS. Comparison of monthly intramuscular injections of Sandostatin LAR with multiple subcutaneous injections of octreotide in the treatment of acromegaly: effects on growth hormone and other markers of growth hormone secretion. Clin Endocrinol (Oxf) 1999; 50:245–251.
29. Comets E, Mentre F, Nimmerfall F, et al. Nonparametric analysis of the absorption profile of octreotide in rabbits from long-acting release formulation OncoLAR. J Contr Release 1999; 59:197–205.
30. Bodmer D, Kissel T, Traechslin E. Factors influencing the release of peptides and proteins from biodegradable parenteral depot systems. J Contr Release 1992; 21:129–138.
31. Ogawa Y, Yamamoto M, Okada H, Yashiki T, Shimamoto T. A new technique to efficiently entrap leuprolide acetate into microcapsules of polylactic acid or copoly(lactic/glycolic) acid. Chem Pharm Bull 1988; 36:1095–1103.
32. Okada H, Toguchi H. Biodegradable microspheres. Crit Rev Ther Drug Carrier Syst 1995; 12:1–99.
33. Watts PJ, Davies MC, Melia CD. Microencapsulation using emulsification/solvent evaporation: an overview of techniques and applications. Crit Rev Ther Drug Carrier Syst 1990; 7:235–259.
34. O'Donell PB, McGinity JW. Preparation of microspheres by the solvent evaporation technique. Adv Drug Del Rev 1997; 28:25–42.

35. Görich SM. Untersuchungen zu verfahrenstechnischen Parametern des W/O/W-Tripelemulsionsverfahrens für die Mikroverkapselung hydrophiler Makromoleküle. Ph.D. dissertation, University of Marburg, Germany, 1994.
36. Thies C. Formation of degradable drug-loaded microparticles by in-liquid drying processes. In: Donbrow M, ed. Microcapsules and Nanoparticles in Medicine and Pharmacy. Boca Raton, Florida: CRS Press, 1991:47–71.
37. Kondo A. Microcapsule, Processing and Technology. New York and Basel: Marcel Dekker Inc., 1979.
38. Heya T, Okada H, Ogawa Y. Encapsulation of TRH or its analogues. Eur Pat No. 256726 Appl, 1988.
39. Bodmer D, Kissel T. Sustained release of the somatostatin analogue octreotide from microcapsule. Proc Int Symp Contr Rel Bioact Mater 1997; 18:597–598.
40. Wang J, Wang BM, Schwendeman SP. Characterization of the initial burst release of a model peptide from poly(D,L-lactide-co-glycolide) microspheres. J Contr Release 2002; 82:289–307.
41. Wang J, Wang BM, Schwendeman SP. Mechanistic evaluation of the glucose-induced reduction in initial burst release of octreotide acetate from poly(D,L-lactide-co-glycolide) microspheres. Biomaterials 1994; 25:1919–1927.
42. Singh M, Singh A, Talwar GP. Controlled delivery of diphtheria toxoid using biodegradable poly(D,L-Lactide) microcapsules. Pharm Res 1991; 8:958–961.
43. Labhasetwar V, Underwood T, Gallagher M, Langberg J, Levy J. DL-Sotalol polymeric controlled release preparations: in vitro characterisation, in vivo drug disposition and electrophysiologic effects. Proc Int Symp Contr Rel Bioact Mater 1992; 19:56–57.
44. Herrmann J, Bodmeier R. Peptide-containing biodegradable microspheres prepared by modified solvent evaporation methods. Proc Int Symp Contr Rel Bioact Mater 1993; 20:258–259.
45. Yan C, Hewetson J, Creasia D, et al. Enhancement of ricin toxoid efficacy by controlled rate-release from microcapsules. Proc Int Symp Contr Rel Bioact Mater 1993; 20:71–72.
46. Raghuvashi RS, Singh M, Talwar GP. Biodegradable delivery system for single step immunization with tetanus toxoid. Int J Pharm 1993; 93:1–5.
47. Morlock M, Koll H, Winter G, Kissel T. Microencapsulation of rh-erythropoietin, using biodegradable PLGA: protein stability and the effects of stabilizing exipients. Eur J Pharm Biopharm 1997; 43:29–36.
48. Meinel L, Illi OE, Zapf J, Malfanti M, Merkle HP, Gander B. Stabilizing insulin-like growth factor-I in poly(D,L-lactide-co-glycolide) microspheres. J Contr Release 2001; 70:193–202.
49. Mi L, Zhang Q, Li Y, Gu Z. Study on the preparation and pharmacodynamics of insulin-loaded polyester microparticles. Yaoxue Xuebao 2000; 35:850–853.
50. Trehan A, Ganga S. Effect of additives on internal aqueous phase of PDLLA microparticles loaded with insulin prepared by double emulsion method. Proceedings of the 28th International Symposium on Controlled Release of Bioactive Materials and 4th Consumer & Diversified Products Conference, San Diego, CA, June 23–27, 2001; 2:932–933.
51. Shenoy Dinesh B, D'Souza Reshma J, Sandip BT, Udupa N. Potential applications of polymeric microsphere suspension as subcutaneous depot for insulin. Drug Dev Ind Pharm 2003; 29:555–563.
52. Tabata Y, Gutta S, Langer R. Controlled delivery systems for proteins using polyanhydride microspheres. Pharm Res 1993; 10:487–496.
53. Freytag T, Dashevsky A, Tillman L, Hardee GE, Bodmeier R. Improvement of the encapsulation efficiency of oligonucleotide-containing biodegradable microspheres. J Contr Rel 2000; 69:197–207.
54. Cohen S, Yoshioka T, Lucarelli M, Hwang LH, Langer R. Controlled delivery systems for proteins based on poly(lactic/glycolic acid) microspheres. Pharm Res 1991; 8: 713–720.

55. Jeffrey H, Davis SS, O'Hagan DR. The preparation and characterization of poly(lactide-co-glycolide) microparticles. 2. The entrapment of a model protein using a (water-in-oil)-in-water emulsion solvent evaporation technique. Pharm Res 1993; 10:362–368.
56. Hilbert AK, Fritzsche U, Reers M, Kissel T. Release of influenza vaccine from biodegradable microspheres and microencapsulated liposomes. Proc Int Symp Contr Rel Bioact Mater 1996; 23:79–80.
57. Ghaderi R, Sturesson C, Carlfors J. Effect of preparative parameters on the characteristics of poly(D,L-lactide-co-glycolide) microspheres made by the double emulsion method. Int J Pharm 1996; 141:205–216.
58. Pistel KF, Kissel T. Effects of salt addition on the microencapsulation of proteins using W/O/W double emulsion technique. J Microencapsul 2000; 17:467–483.
59. Jeyanthi R, Mehta RC, Thanoo BC, DeLuca PP. Effect of processing parameters on the properties of peptide-containing PLGA microspheres. J Microencapsul 1997; 14:163–174.
60. Manning MC, Patel K, Borchardt RT. Stability of protein pharmaceuticals. Pharm Res 1998; 6:903–918.
61. Alex R, Bodmeier R. Encapsulation of water-soluble drugs by a modified solvent evaporation method. I. Effect of process and formulation variables on drug entrapment. J Microencapsul 1990; 7:347–355.
62. Ogawa Y. Monthly microcapsule-depot form of LHRH agonist, leuprorelin acetate (Enatone® Depot): formulation and pharmocokinetics in animals. Eur J Hosp Pharm 1992; 2:120–127.
63. Okada H, Yamamoto M, Heya T, et al. Drug delivery using biodegradable microspheres. J Contr Rel 1994; 28:121–129.
64. Okada H, Doken Y, Ogawa Y, Toguchi H. Preparation of three-month depot injectable microspheres of leuprorelin acetate using biodegradable polymers. Pharm Res 1994; 11:1143–1147.
65. Ogawa Y, Okada H, Heya T, Shimamoto T. Controlled-release of LHRH agonist, leuprorelide acetate, from microcapsules: serum drug level profiles and pharmacological effects in animals. J Pharm Pharmacol 1989; 41:439–444.
66. Rouan SKE, Broadhead J, Rhodes CT. The spray drying of pharmaceuticals. Drug Dev Ind Pharm 1992; 18:1169–1206.
67. Giunchedi P, Conte U. Spray-drying as a preparation method of microparticulate drug delivery systems: an overview; S.T.P. Pharma Sci 1995; 5:276–290.
68. Wendel SC, Celik M. An overview of spray-drying applications. Pharm Technol 1997; 21:124–156.
69. Masters K. Spray Drying Handbook. New York: John Wiley & Sons, 1991.
70. Kissel T, Brich Z, Bantle S, Lancranjan I, Nimmerfall F, Vit P. Parenteral depot-systems on the basis of biodegradable polyesters. J Contr Rel 1991; 16:27–42.
71. Kissel T, Rummelt A, Bier HP. Wirkstofffreisetzung aus bioabbaubaren Mikropartikeln. Deutsche Apotheker Zeitung 1993; 133:23.
72. Bodmeier R, Chen HG. Preparation of biodegradable poly(+/−)lactide microparticles using a spray-drying technique. J Pharm Pharmacol 1988; 40:754–757.
73. Volland C, Wolff M, Kissel T. The influence of terminal gamma-sterilization on captopril containing poly(D,L-lactide-co-glycolide) microspheres. J Contr Rel 1994; 31:293–305.
74. Lee G. Spray-drying of proteins. Pharm Biotechnol 2002; 13:135–158.
75. Mumenthaler M, Hsu CC, Pearlman R. Feasibility study on spray-drying protein pharmaceuticals: recombinant human growth hormone and tissue-type plasminogen activator. Pharm Res 1994; 11:12–20.
76. Bittner B, Morlock M, Koll H, Winter G, Kissel T. Recombinant human erythropoietin (rhEPO) loaded poly(lactide-co-glycolide) microspheres: influence of the encapsulation technique and polymer purity on microsphere characteristics. Eur J Pharm Biopharm 1998; 45:295–305.

77. Kusonwiriyawong C, Atuah K, Alpar OH, Merkle HP, Walter E. Cationic stearylamine-containing biodegradable microparticles for DNA delivery. J Microencapsul 2004; 21:25–36.
78. Lill N, Sandow J. Long-acting biodegradable peptide drug microparticles. Eur Pat No. 505966 Appl, 1992.
79. Cleland JL, Lam XM, Duenas ET. Controlled release microencapsulated formulations of nerve growth factor. US Pat Appl Publ, 2002.
80. Maeda A, Takaishi Y, Nakanishi K, Saito K, Yamashita N, Takagi A. Sustained-release compositions for injection and process for producing the same. PCT Int Appl, 2003.
81. Alcock R, Bibby DC, Gard TG. Encapsulation of recombinant hepatitis B surface antigen in oligosaccharide ester derivatives by spray drying. J Microencapsul 2003; 20:759–766.

5

Coupling Methods to Obtain Ligand-Targeted Liposomes and Nanoparticles

Leila Bossy-Nobs, Robert Gurny, and Eric Allémann
School of Pharmaceutical Sciences, (EPGL), University of Geneva, Quai Ernest-Ansermet, Geneva, Switzerland

Franz Buchegger
Service of Nuclear Medicine, University Hospital of Geneva, Rue Micheli-du-Crest, Geneva, Switzerland

INTRODUCTION

In recent years, an increasing number of studies have been devoted to the development of more and more sophisticated targeted delivery systems. First attempts in this direction were accomplished using monoclonal antibodies or antibody fragments coupled with active compounds, such as antitumor drugs (1). The antibodies were used to deliver the drug to the target cells expressing specific antigens on their surface. One of the major problems in the development and utilization of these drug–antibody conjugates was the preservation of both pharmacological activity and binding ability after production of the drug–antibody conjugates. In fact, to maintain the binding activity of the antibody, only a relatively low amount of a drug can be coupled to it, which might be insufficient to obtain the desired therapeutic effect. If a higher amount of a drug is coupled to the antibody, the risk of inactivating the recognition sites of the antibody is generally rather high. This fact promoted the investigation of other strategies and one of the most interesting and challenging strategies consists of the entrapment of drugs into liposomes or nanoparticles coated with specific ligands, such as antibodies.

These colloidal carriers have proved to be versatile for a wide variety of i.v. delivered active molecules (2). They present several advantages such as an increased drug versus antibody ratio, a high-loading efficiency of drugs, and protection of the drug from early degradation in the organism. Altogether, they offer a suitable means for delivering drugs combined with the potential of improving the therapeutic index while greatly reducing the side effects.

Initially, these carrier systems presented two major drawbacks: the rapid uptake within seconds or minutes by phagocytic cells of the reticuloendothelial system (RES) and the lack of target specificity (3,4). To avoid the rapid clearance, several authors have developed different strategies to obtain the so-called "stealth" or "sterically stabilized" liposomes or nanoparticles, which present greatly reduced reactivity towards serum proteins and reduced uptake by the RES. One of the major means to obtain stealth particles has been to coat them with amphipathic polyethyleneglycol (PEG), which provides a steric barrier on the surface of nanoparticles or liposomes because of the high mobility and flexibility of the PEG chains (5–9). The resulting particles have an increased lifetime in the blood compartment compared to plain particles. Furthermore, it has been demonstrated that "stealth" particles accumulate better in several solid tumors than plain carriers (10,11). In fact, the capillary permeability of the endothelia in tumor vessels is usually greater compared to healthy tissues, and thus, very small particles with prolonged lifetime in the blood stream preferentially extravasate and reach the target cells in tumors.

Although capable of enhanced accumulation in tumor tissue compared with conventional particles, tumor uptake of the sterically stabilized carriers and their interaction with tumor cells appears to be insufficient (12). Therefore, the need for exploring new strategies in the development of particles capable of delivering active compounds to specific sites is the primary challenge in this field. The active targeting of particles to specific sites other than liver and spleen (passive targeting) appears to be one of the most interesting and promising applications of colloidal carriers. Particles exhibiting specific targeting properties can be obtained by coupling ligands, such as antibodies, to their surface.

While in the field of liposomes many studies have been performed to enhance active targeting, only a few studies have been carried out in the field of nanoparticles. Therefore, the major part of this chapter is devoted to liposomes. Excellent review papers have been published regarding active targeting using liposomes, particularly for tumor targeting (12–16). The aim of this chapter is to illustrate and describe in detail coupling strategies used for the attachment of ligands to colloidal carriers.

LIPOSOMAL MODIFICATION TECHNIQUES

For active targeting, liposomes have been produced by the conjugation of antibodies, carbohydrates, or other ligands. Essentially two methods are used for attaching ligands to the liposomal surface—covalent and noncovalent coupling. Covalent linkage can be obtained by different chemical reactions, leading to a stable bond between ligand and liposomes. In the case of noncovalent associations, the ligand is bound to the liposome through electrostatic and hydrophobic interactions, and here also, several different approaches have been developed.

In this first section, the frequently used covalent reactions are described.

Covalent Binding of Ligand to the Liposomal Surface

Generally, ligands are bound to the surface of liposomes through hydrophobic anchors having functional groups. The anchors are used to convert normally soluble ligands into a form that can be incorporated into the liposomal membrane. One of the first attempts was accomplished by using a long-chain fatty acid, such as palmitic acid, which was covalently associated to an antibody. Several types of ligands have been coupled to the liposomes using palmitic acid, but the technique of artificially increasing the hydrophobicity is the same. Generally, the ligand is derivatized with palmitic acid using an activated ester of succinimydyl ester of the fatty acid, which reacts with primary amines of the ligand to form a stable amide bond (17–21). The coupling reaction can be controlled to obtain the desired degree of hydrophobicity. The modified ligand is either incorporated into the liposomal membrane during the formation of liposomes or added directly to the preformed liposomes (19,22). This approach has been successfully used with antibodies as ligand (17–20). Results indicated that from 50% up to 100% of the coupled antibodies were exposed on the outer surface of the liposomes (19,20). In vitro studies have shown that the incorporated antibody preserved specific binding properties to target cells and that the coupling of the antibody to the liposomes did not affect its functionality. Furthermore, in vitro and in vivo comparisons with uncoated liposomes or free antibodies demonstrated the efficacy of the antibody coupled to liposomes by this approach (17–19).

This method was also extended to lectins, and more specifically wheat germ agglutinin (WGA) (21). WGA was derivatized with palmitic acid to give a fatty acid–coupled WGA that was then incorporated into liposomes during their preparation. The activity of the lectin bound to the liposome was maintained and, globally, results demonstrated that lectins could be interesting moieties for target-specific drug delivery.

Phospholipids such as phosphatidylethanolamine (PE) and phosphatidylinositol (PI) have been also used as anchors (23,24). Here again, there are essentially two approaches to incorporate the ligand into the membrane of the liposomes. The first consists of binding the ligand to the anchor and mixing the resulting ligand with the other constituents of the liposome. In the second case, the anchor is already included in the liposomes bilayer and the coupling reaction occurs on the surface of preformed liposomes. Under these latter conditions, it is essential that the reagents chosen are nonaggressive so as to avoid irreversible alterations to the structure of the liposomes or damage to the active molecule carried by the liposomes.

Among the hydrophobic anchors, PE is more frequently used as it can be easily derivatized to offer functional groups. Each covalent reaction used to attach ligands to phospholipid anchors would be described seperately.

Ligands must also present functional groups enabling them to react with the anchor. Some ligands contain one or two diverse functional groups (primary amine and thiol groups) not needing any addition of such groups. Certain ligands lack desired specific functional groups, such as thiol, for which various methods have been developed that allow their introduction. The most common covalent reactions used to bind the ligand to the surface of the liposome with their advantages and drawbacks are presented here (Tables 1 and 2).

Coupling of Ligands to the Surface of Liposomes Using Thioether Bonds

The reaction between thiol functions and maleimide groups is a highly efficient reaction which gives a stable thioether bond (Fig. 1). Native thiol groups are present in some proteins, but in many cases, thiol functions are either absent or present in insufficient amounts. Thus they have to either be added via diverse heterobifunctional cross-linking agents or be obtained by reducing existing disulfide bonds (25,26). Disulfide bonds are present, in the case of the $F(ab')_2$ antibody fragments (27–29). Intact IgG do not contain free thiol groups and therefore, functional cross-linkers must be added to provide these functions. The coupling reaction can present a potential risk of damage to the binding site since the degree of reaction and the site of substitution are not easily controlled. Fab′ fragments can be used directly as each fragment has a thiol function available. Commonly, *N*-hydroxysuccinimidyl-3-(2-pyridyldithio)propionate (SPDP) and *N*-succinimidyl-*S*-acetylthioacetate (SATA)

Table 1 Common Cross-Linkers Used

Compounds	Structure	References
SATA	N—O—C(=O)—CH_2—S—C(=O)—CH_3	(25,26,30)
SPDP	—S—S—$(CH_2)_2$—C(=O)—O—N	(39,43,44)
SMPB	N—O—C(=O)—(CH_2)—	(36,39,41)
EDAC	$(CH_3)_2$N—$(CH_2)_3$—N=C=N—CH_2CH_3	(45–53)
Glutaraldehyde	H—C(=O)—CH_2—CH_2—CH_2—C(=O)—H	(59,60)
Suberimidate	HN=C(OCH_3)—CH_2—$(CH_2)_4$—CH_2—C(OCH_3)=NH	(60)

Abbreviations: EDAC, 1-ethyl-3-(3-dimethylaminopropyl)carbodiimide; SATA, succinimidyl-*S*-acetylthioacetate; SMPB, succinimidyl-4-(*p*-maleimidophenyl)butyrate; SPDP, *N*-hydroxysuccinimidyl-3-(2-pyridyldithio) propionate.

are used as cross-linkers (25,26,30). Both offer one primary amine reactive residue for the coupling with the ligand. In both cases, thiol functions are not directly available and a deprotection of these functions is necessary before the reaction with liposomes bearing maleimide groups. In the case of SPDP, dithiothreitol (DTT) or an alternative-reducing reagent is used to reduce the disulfide bond to thiol functions (Fig. 2). However, DTT presents a major inconvenience, as it can also react with maleimide functions and thus compete with the thiolated ligand. Therefore, DTT has to be removed thoroughly before the thiolated ligand reacts with maleimide-liposomes. This problem can be avoided by using SATA, which offers one protected thiol function that can be made available by deacetylation with hydroxylamine (Fig. 3). The deprotection is performed under comparatively milder conditions than with those of reducing reagent and without the inconvenience of an interference in further thiol reaction chemistries, as hydroxylamine does not react with maleimide groups.

Table 2 Covalent Coupling Techniques Used to Bind Ligands to Liposomes

Covalent linkage	Ligand	Derivatized anchor	References
Thioether	Fab′ fragments (–SH)	MPB-PE	(25–34)
	SATA-modified ligand	MP-PEG-DSPE	(35,38)
	SPDP-modified ligand	MCC-PEG-DSPE	(35,38)
		MPB-DOPE	(39)
		PDP-DOPE	(39)
		PDP-PEG-DSPE	(39,41)
		PDP-PEG-PE	(36)
Disulfide	PDP-modified ligand	PDP-PE	(40,44)
	Fab′ fragments (–SH)		
	SATA-modified ligand		
Amide	Ligand ($-NH_2$)	NHS ester of palmic acid	(17–22)
		DSPE-PEG-COOH	(45–47)
		NGPE	(23,48–53)
Hydrazone	Oxidized ligand	Hz-PEG-DSPE	(31,39,55,56)
Amine-amine cross-linking	Ligand ($-NH_2$)	PE ($-NH_2$)	(59,60)

Abbreviations: SATA, succinimidyl-*S*-acetylthioacetate; SPDP, *N*-hydroxy succinimidyl-3-(2-pyridyldithio)propionate; PDP, *N*-(3-(2-pyridyldithio)propionyl); MPB-PE, *N*-(4-(P-maleimidophenyl)butyrl)-phophatidylethanolamine; MP-PEG-DSPE, Maleimido-phenyl propinate-polyethyleneglycol-distearoyl phosphatidylethanolamine; MCC-PEG-DSPE, Maleimidomethyl-cyclohereance carboxylate-PEG-DSPE; MPB-DOPE, *N*-(4-(P-maleimidophenyl)butyryl)dioleoyl phosphoethanolamine; PDP-DOPE, *N*-(3-(2-pyridyldithio)propionyl)dioleoylphosphoethanolamine; PDP-PE, *N*-(3-(2-pridyldithio)propinyl)phosphatidylethanolamine; NHS, *N*-hydroxy succinimide; DSPE-PEG-COOH, Distearoyl-*N*-(3-carboxypropionoyl-poly(ethyleneglycol)succinyl)phosphatidylethanolamine; NGPE, *N*-glutaryl-phosphatidylethanolamine; HZ-PEG-DSPE, Hydrazide-PEG-DSPE.

Most frequently *N*-(4-(*p*-maleimidophenyl)butyryl)phophatidylethanolamine (MPB-PE) has been used as a functionalized anchor, which is merely mixed during the liposome formation with the other components of the bilayer membrane (25–28, 30,34). The extended spacer arm between the phospholipid head group and the maleimide moiety reduces the possibility of steric hindrance at the bilayer interface and thereby ensures favorable thiol reactivity. Once the liposomes with maleimide groups have been formed, the homing ligand, having thiol functions, can be easily coupled to the liposomes by a simple addition (26,28,32). Nassander et al. (27) used Fab′ fragments of OV-TL3 (monoclonal antiovarian carcinoma IgG) which have free thiol functions available as homing ligands for the coupling with maleimido-functionalized PE. The immunoliposomes formed were shown to bind specifically to OVCAR-4 target cells in vitro and in vivo, but no internalization of immunoliposomes by tumor target cells was observed. Absence of internalization was probably because of the chosen homing ligand, which does not sufficiently induce the internalization. Despite this negative result, it has been shown, in this study, that the coating of liposomes with Fab′ fragments using maleimido-functionalized PE is feasible and that the antibodies maintain their specific antigen-binding properties for the antigens. Park et al. (32) demonstrated that the internalization of immunoliposomes bearing another antibody, anti-p185^{HER2}, is possible and that the internalization is due to the specificity of the antibody coupled to the liposomes, as control liposomes lacking Fab′ showed no internalization by SK-BR-3 cells. Furthermore, immunoliposomes appeared to induce a high antiproliferative effect that was

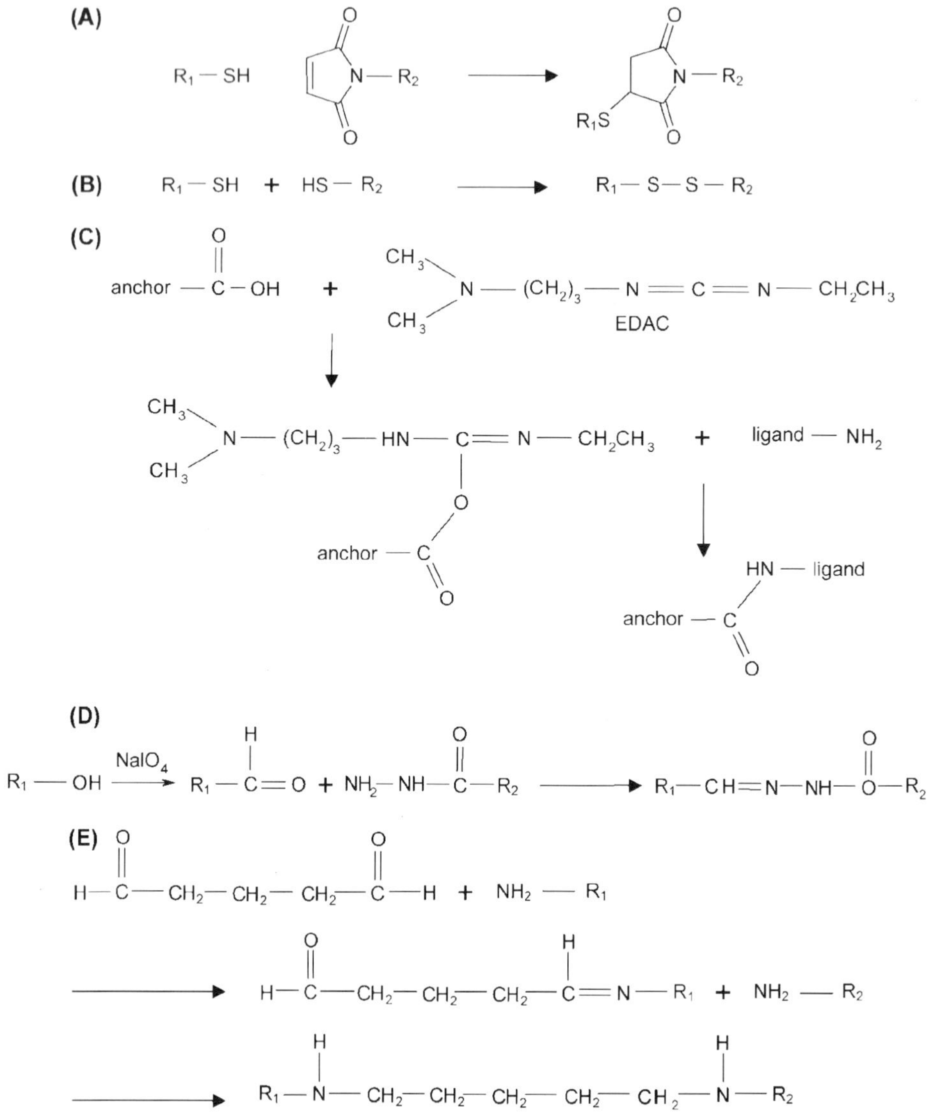

Figure 1 Schematic diagram of the different coupling methods used. Reaction between maleimide and thiol functions (**A**), formation of a disulfide bond (**B**), reaction between carboxylic acid and primary amine group (**C**), reaction between hydrazide and aldehyde function (**D**), and cross-linking between two primary amine (**E**).

superior to that of free monoclonal antibody, indicating that these kinds of liposomes are suitable carriers for therapeutic agents for the treatment of $p185^{HER2}$ overexpressing human cancers.

Another interesting study confirmed the efficiency of in vitro and in vivo coupling in antibodies with liposomes via a thioether bond (25). In this case, AR-3 monoclonal antibodies recognizing human colon carcinoma antigen were chosen. Compared with free drugs, liposomes bearing antibody were shown to be highly effective as antitumor agents and to induce less systemic toxicity.

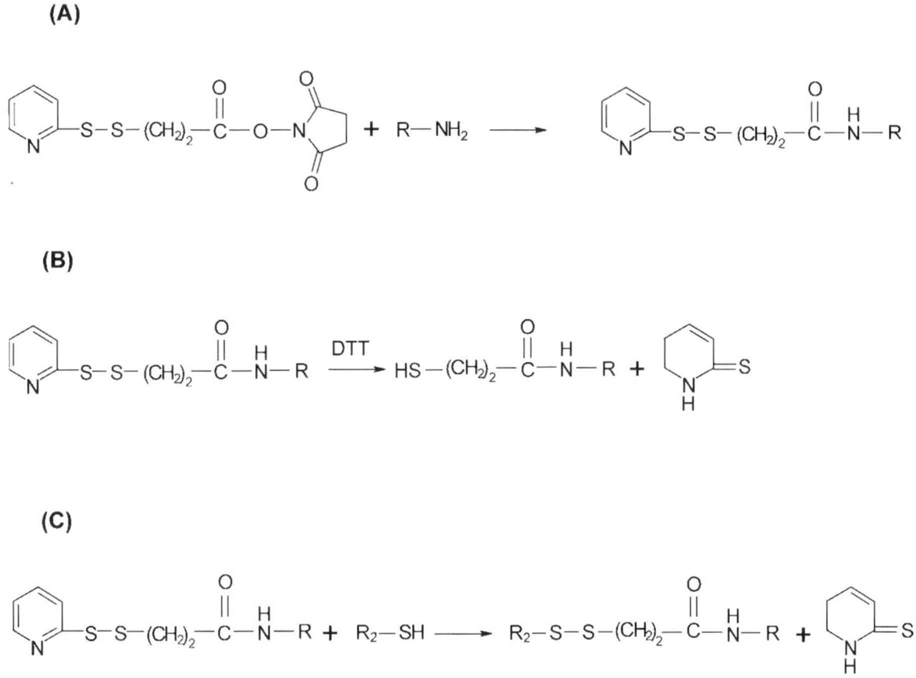

(A)

(B)

(C)

Figure 2 Reaction between SPDP and the ligand and/or anchor. Reaction between primary amine and succinimidyl ester of SPDP (**A**), reduction of the disulfide bonds by DTT to form free thiol function (**B**), and reaction with thiol function (**C**). *Abbreviation*: DTT, dithiothreitol; SPDP, *N*-hydroxysuccinimidyl-3-(2-pyridyldithio)propionate.

Immunoliposomes technology can be combined with sterically stabilized liposomes technology to obtain long circulating vesicles capable of selectively delivering active compounds to target cells. There are two ways of obtaining this kind of liposomes (i) by binding antibodies to the surface of liposome in parallel with PEG or (ii) by linking antibodies to the distal end of PEG chains (32,35,36). For both approaches, PEG is incorporated into the bilayer membrane via an anchor such as distearoylphosphatidylethanolamine (DSPE) during the liposomes preparation (31,35,37–39). When antibodies were coupled to the termini of PEG, the latter had to be derivatized to offer functional groups for the coupling reaction. Principally, two different maleimide-terminated PEG-PE have been investigated, maleimidomethylcyclohexanecarboxylate-PEG-DSPE (MCC-PEG-DSPE) and maleimido-phenylpropionate-PEG-DSPE (MP-PEG-DSPE) (35,38). Thiol groups of the Fab′ fragments or activated monoclonal antibody (entire IgG) were then attached to the distal end of the PEG chains via the maleimide groups (35,37). A great advantage, when Fab′ fragments are used, is that their attachment to the liposome results in a favorable orientation of each Fab′ fragments (40). Both MP and MCC end-groups proved to be equally effective for preparation of immunoliposomes (38). When antibodies were coupled directly at the liposomes surface, binding affinity for the target cells was reduced because of the steric hindrance caused by the high density of PEG coating that reduces dramatically the availability of antibodies for the target antigen (37,38). This problem was clearly circumvented by attaching antibodies to the distal end of PEG chains (37,38). In fact, in

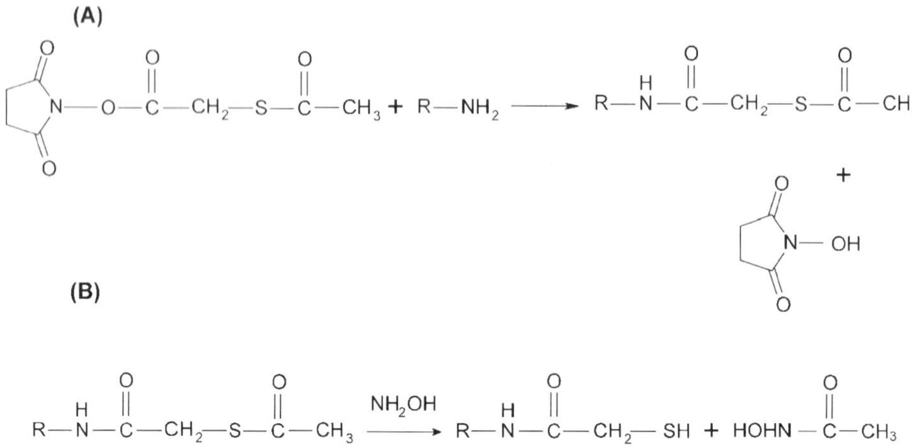

(A)

$$\text{(succinimidyl)N–O–C(=O)–CH_2–S–C(=O)–CH_3} + \text{R–NH_2} \longrightarrow \text{R–NH–C(=O)–CH_2–S–C(=O)–CH_3} + \text{(succinimidyl)N–OH}$$

(B)

$$\text{R–NH–C(=O)–CH_2–S–C(=O)–CH_3} \xrightarrow{NH_2OH} \text{R–NH–C(=O)–CH_2–SH} + \text{HOHN–C(=O)–CH_3}$$

Figure 3 Formation of thiol functions using SATA. Reaction between succinimidyl ester of SATA and the primary amine of the ligand or the anchor (**A**) and deprotection of the disulfide bond with hydroxylamine (**B**). *Abbreviation*: SATA, *N*-succinimidyl-*S*-acetylthioacetale.

this case a linker with a long spacer arm is used which provides attachment of the antibody that is distant from the liposome bilayer.

Attaching thiolated antibodies to maleimide-anchor is not the only coupling procedure via a thioether bond. In fact, it is also possible to link maleimido-antibodies to liposomes offering thiol functions (39,41). In the case of long-circulating liposomes, antibodies can be attached to the carrier either directly to the surface of the vesicles via *N*-(3-(2-pyridyldithio)propionyl)dioleoylphosphoethanolamine (PDP-DOPE) or at the PEG terminus via PDP-PEG-DSPE or PDP-PEG-PE (36,39,41). As seen before, thiol functions are not directly available when SPDP is used as cross-linker and an additional reaction is required to reduce the disulfide linkage to thiol functions. Maleimido-antibodies are obtained using a heterobifunctional cross-linker, succinimidyl 4-(*p*-maleimidophenyl)butyrate (SMPB), that is reactive with amino groups of antibodies and sulfhydryl groups. Hansen et al. (39) compared various coupling methods and results showed that the reactions involving the formation of a thioether linkage at the PEG end chain yielded a coupling efficiency of approximately 60% to 70% (36,39,41). In contrast, when antibodies were directly bound to the surface of liposome via *N*-maleimidophenylbutyrate dioleoylphosphatidylethanolamine (MPB-DOPE) in the presence of PEG-DSPE, the coupling efficiency was only 10%. This investigation confirmed that the presence of PEG, which reduces considerably the access of the functionalized antibody to the surface of the vesicles when the ligand is intended to be directly coupled to the surface of the liposome. Mercadal et al. (36) obtained coupling efficiencies of nearly 100% by attaching antibodies to the terminus of PEG, which confirms that this approach can greatly increase the accessibility of the functional groups and thus enhance the amount of ligand bound to the liposomes.

PEG can also be directly linked to cholesterol (Chol) instead of PE or DSPE (42). Pegylated Chol (PEG-Chol) derivatives are more easily obtained and with lower production costs. Despite these advantages, the amount of antibodies coupled to liposomes is inferior to that of PEG-PE vesicles (42). Most probably, this is because of the fact that the functional groups are less accessible owing to the reduced mobility of PEG-Chol, which is located deeper in the liposome membrane.

Attachment of Ligands to Liposomes via a Disulfide Linkage

One of the most rapid and easy coupling chemistries involves the conjugation of two thiol functions to form a disulfide bond (Fig. 1). However, it has to be pointed out that disulfide bonds are relatively unstable under the reductive conditions in serum (40). For this reason, the disulfide linkage has been progressively replaced by other more stable ones. Nevertheless, some studies that report the coupling of ligands to liposomes via this type of reaction are briefly discussed. Thiolated ligands can be generated either by reduction of disulfide bonds, if present, or by using heterobifunctional cross-linker such as SATA or SPDP which react with molecules containing primary amine groups to form an amide bound (39,40,43,44). The resulting conjugated protein–bound dithiopyridine protects the thiol functions in a disulfide linkage, which can be easily reduced by DTT just before the coupling reaction with liposomes. Under these conditions, intramolecular cross-linking between thiol functions is reduced to a minimum. The thiolated ligands can then either react with the maleimide-activated anchor, as seen previously, or with the pyridyldithio moiety of the anchor to form a disulfide linkage. Here again the most common used anchor is PE but coupled this time to SPDP to form the stable derivate PDP-PE which can easily be incorporated into the liposomal membranes during the formation of the vesicles (40,44). The coupling method resulted in an efficient binding of antibody to the liposomes without denaturation of the coupled ligand. Furthermore, in vitro studies demonstrated that liposomes bearing Fab′ fragments via disulfide bonds recognized and bound selectively to target cells (40,43,44).

Cross-Linking Between Carboxylic Acid Functions on the Surface of Liposomes and the Primary Amine of Ligands

Ligands can be attached to the surface of liposome by an amide bond (Fig. 1) using an anchor functionalized with carboxylic acid end groups. For this purpose, distearoyl-*N*-(3-carboxypropionoylpoly(ethyleneglycol)succinyl)phosphatidylethanolamine (DSPE-PEG-COOH), which offers carboxylic acid groups at the distant end of surface-grafted PEG chains is commonly used (45,46). The coupling reaction is carried out in the presence of 1-ethyl-3-(3-dimethylaminopropyl)carbodiimide (EDAC) (cross-linker) to form an acyl amino ester which would subsequently react with the primary amine of the ligand, yielding a stable amide bond (45–47). The presence of *N*-hydroxysuccinimide (NHS) during the reaction is important for the stabilization of the acyl amino ester intermediate. The major advantage of this method is that no prior ligand modification is required, thus reducing the risk of denaturation and loss of its specific activity. Maruyama et al. (46) used this coupling procedure for attaching the monoclonal antibody 273-32A (34A), specific for the pulmonary endothelial cells of a mouse, to the sterically stabilized liposome. Results in vivo indicated that the immunoliposomes recognized and bound specifically to target cells. Here again, a comparison between immunoliposomes bearing antibodies at the distal end of the PEG chains and immunoliposomes having antibody directly linked to the bilayer revealed that these latter were less efficient because of the steric hindrance of PEG chains. Antibodies were consequently less accessible for the recognition and binding to specific target cells. Ishida et al. (45) have investigated the possibility of using transferrin (TF) as a ligand for the active targeting to tumor cells. TF is a glycoprotein, which transports ferric ion in the body. It has been described that the concentration of TF receptors is higher in tumor cells than in normal cells. Studies in vitro on mouse colon carcinoma cells (colon 26) indicated that TF-liposomes bound to these cells, and were efficiently internalized by receptor-mediated endocytosis. In vivo investigation on colon 26 tumor–bearing mice confirmed these results and showed that liposomes can extravasate from the blood compartment into the solid tumor. The potential advantage of using TF as homing moiety instead of antibodies might be because of the presence of TF in the organism in high concentrations and thus the risk of an immunological response is rather low. Whether the abundance of TF receptors in human tumors is sufficient to provide efficient and specific therapy remains to be shown.

PE can also be functionalized to give *N*-glutaryl-phosphatidylethanolamine (NGPE) (23,48–52). Here again, the cross-linker used is EDAC and the ligand can be coupled to the vesicles via the amide bond either during the formation of the liposomes or on preformed vesicles (48,52,53). Various studies demonstrated that immunoliposomes, prepared with this method, accumulated efficiently in the target organ, whereas liposomes without antibody did not accumulate at all in the target cells. Particularly, Torchilin et al. (49) conjugated antimyosin antibody (AM) Fab′ to the hydrophobic anchor (NGPE) and the resulting modified antibody was then incorporated in the liposomal membranes during the formation of liposomes. The efficacy of the coupling reaction when measured was between 65% and 75% of the Fab′ fragments were bound to the liposomes. Several hundred Fab′ fragments were coupled per liposome. Immunoreactivity of the modified Fab′ coupled to the liposome surface was measured and results showed that unfortunately the immunoreactivity of the antibody decreased by 15 to 20 times compared to free antibody. This reduced immunoreactivity of a single Fab′ fragment was partially counterbalanced by the presence of a high number of antibodies per liposome. The efficacy of active targeting was shown in vivo, as liposomes bearing AM accumulated selectively in the infarcted myocardium, compared with native liposomes (without AM) that were significantly present only in noninfarcted myocardium.

Blume et al. (53) used this coupling method to bind ligand either directly on the surface of the liposomes or to the end of the PEG chains. Results showed that in the first case, liposomes had a long circulating time in the blood, but lost their specificity and binding capacity to target sites. In the second case, liposomes not only had a prolonged circulation time in the blood but also efficiently bound to the target sites.

Binding of Ligands to Liposomes via Hydrazone Bond

Another interesting technique for binding antibodies to the outer layer of liposomal membranes is by covalently binding antibodies, through their carbohydrate moieties, to hydrazide groups grafted onto the liposomal surface (Fig. 1). A mild oxidation of the carbohydrate groups on the constant region of the heavy chain of the immunoglobulin is required to produce aldehyde groups. The latter reacts with hydrazide groups of the anchor (31,54,55). The oxidation of the carbohydrate groups is done either by galactose oxidase or by sodium periodate (31,39,54–57). It is obvious that the oxidation reaction must be performed under mild conditions to avoid any loss of antibody activity. Once the oxidized antibodies are formed, they can be either directly coupled to the lipid bilayer containing a hydrazide–hydrophobic anchor such as lauric acid hydrazide, or attached to the distal end of the PEG chains of the sterically stabilized liposomes (31,39,55,54,56). In this latter case, a functionalized PEG-lipid hydrazide-PEG-DSPE (Hz-PEG-DSPE) is used (58). The reaction is simple and a stable hydrazone bond is formed. This coupling reaction theoretically offers the advantage that antibodies are correctly orientated once attached on the surface of the liposomes, as only Fc region is involved in the coupling reaction, leaving the antigen-binding site untouched. However, a comparative study between different coupling methods revealed that this technique is not very efficient; only a relatively small percentage of the antibodies (17%) used are attached to the liposomal membrane (39). In contrast, the thioether bond seemed promising by achieving approximately 61% of binding efficiency. This is reflected in the in vitro investigation, which showed that the approach allows only a small amount of liposomes to be attached to the cells because of the low antibody density in contrast with other coupling techniques (39). Nevertheless, positive results have been obtained in vitro with this coupling method by another research group. Koning et al. (31) have successfully used hydrazone linkage for the delivery of the anticancer prodrug, FUdR-dP, encapsulated in the bilayer of liposomes bearing monoclonal antibody against rat colon carcinoma CC531. It was observed in this study that the prodrug was selectively transferred to the surface of tumoral cells and then internalized into the lysosomal compartment where the prodrug was hydrolyzed to form the active drug.

Cross-Linking Between the Primary Amine on the Surface of Liposomes and the Primary Amine of Ligands

Among covalent coupling methods, direct amine–amine cross-linking has been also investigated to add moieties to the surface of liposomes (Fig. 1). Principally, two homobifunctional cross-linkers have been used, glutaraldehyde and suberimidate (59,60). The advantage of this coupling approach is that no prior modification is required to add functional groups to the ligand. In this coupling method, the primary amine of PE on the liposome surface via the cross-linker was first activated and subsequently the ligand was coupled. When suberimidate is chosen, primary amines react via imidoester to form a stable amide bond. The reaction with glutaraldehyde proceeds through the formation of Schiff's base with rearrangement to a stable product. Results have demonstrated that almost 60% of the antibody was coupled to the liposomes and that they retained the binding capacity for antigen (60). Despite the fact that the coupling reaction is efficient, the use of homobifunctional cross-linkers has been rarely exploited for the preparation of liposomes for active targeting owing to the uncontrollable homopolymerization of ligand or liposomes during the cross-linking reaction.

Multistep Attachment Using the Avidin–Biotin Interaction

Presently, one of the most interesting conjugation approaches that has been investigated is the avidin–biotin system, which is one of the strongest known noncovalent biological interactions (association constant 10^{15} M^{-1}). This stable interaction, comparable to the standard covalent reactions, allows various applications in immunology, affinity chromatography, histochemistry, and in other fields. Today, the avidin–biotin strategy has become an extremely useful and versatile tool for active targeting, especially owing to the possibility of a multistep strategy. Promising results have been achieved in which a two-step or a three-step approach was investigated (61–63). Commonly, in drug therapy, the two-step method consists in the administration of the biotinylated antibody or another targeting device followed by the injection of the avidin bearing the active molecule. Avidin conjugated to the antibody can also be used as a first step followed by the biotinylated molecule.

The three-step approach is based on the administration of a biotinylated antibody followed by excess of unlabelled avidin and, as a third step, biotin conjugated to an active compound is administrated. The main advantage of pretargeting approaches is that they allow antibodies to accumulate in target tissues and to minimize the amount of unlocated antibodies as well as active compounds (usually highly toxic) in nontarget tissues. The avidin–biotin technology has also been applied in the field of liposomes as an alternative to the covalent coupling methods, since the complexes formed are very stable. Commonly, biotin is covalently attached to the liposomal surface via an anchor, such as DOPE, and the biotinylated antibodies are obtained by covalent reaction (64–66). Here again, different reactions can be chosen. Frequently, primary amine groups, which react with NHS ester of biotin, are used (39). To avoid the risk of inactivation of the antibodies, the use of biotin derivates that react with thiol functions (after reduction of the immunoglobulin) can be considered.

Hansen et al. (39) compared the three-step targeting approach (Fig. 4) with covalent coupling methods. In this study, liposomes containing biotin–DOPE were incubated first with avidin and in a second step mixed with biotinylated antibodies. Results showed a low binding efficiency of antibodies to the liposomes ($9 \pm 3\%$) when compared with other coupling methods ($63 \pm 10\%$ with thioether linkage). The low amount of coupled antibodies is most probably because of the steric hindrance caused by the presence of two large proteins (avidin and immunoglobulin). Most likely, the use of longer linkers could have reduced the steric hindrance and increased the coupling efficacy.

Although the previous study was not very encouraging, other trials illustrated the interest in investigating this approach. Xiao et al. (67) demonstrated in vitro the efficacy of the three-step approach (Fig. 4) for targeting liposomes to human ovarian cancer cells (OVCAR-3). Biotinylated antibodies were first incubated with the cells; subsequently, streptavidin was added followed by biotinylated liposomes. Avidin can be replaced by streptavidin to avoid the nonspecific binding proteins of native avidin (68). The results showed that the liposomes specifically bound to the cell surface of OVCAR-3 cells, but not to cells that do not express the specific antigen. In another report, an in vitro study confirmed that liposomes bearing antibody were strongly attached to tumor cells using a two- or three-step strategy (68). Overall, the major inconvenience in using avidin–biotin technology is the presence of an additional large protein (avidin) that can greatly increase the size of the liposome.

It has to be kept in mind that the choice of the antibody or of any other targeting device is crucial for the multitargeting approach, as the device should remain available and not be rapidly internalized after binding to the target.

Noncovalent Methods Used to Bind Ligands to the Liposomal Surface

As stated previously, an alternative means to conjugate ligands to liposomes is the use of noncovalent techniques. These have the great advantage of being easy to carry out without the need of potential aggressive reagents. In this section, several noncovalent approaches will be described.

Direct Incorporation of Ligands into the Liposomal Bilayer

A simple method that does not require chemical reactions is to add the ligand to the mixture of phospholipids during the preparation of the carrier. IgG was used as a

(A) Direct approach

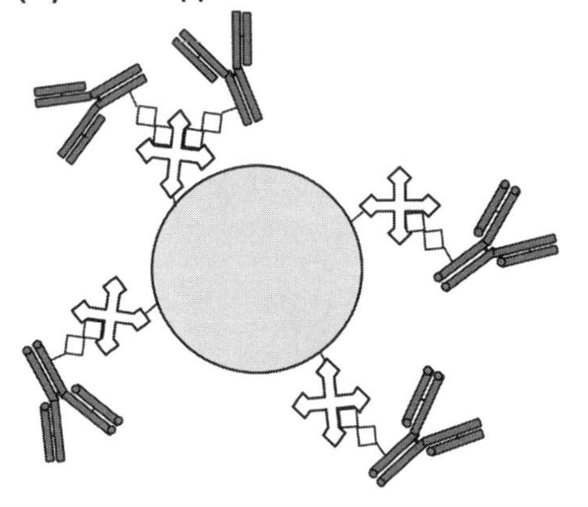

particle bearing antibodies

(B) Two-Step approach

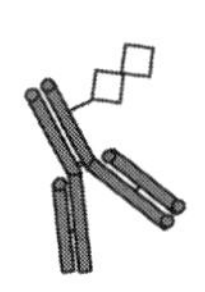

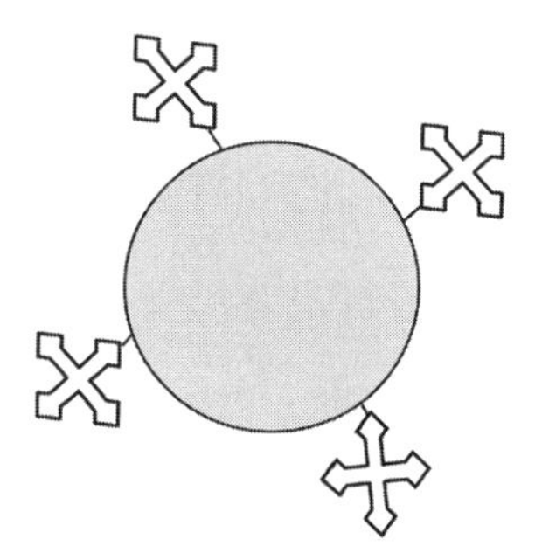

First step: biotinylated antibody

Second step: avidin-particle conjugate

(C) Third-Step approach

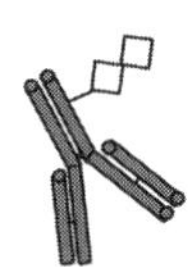

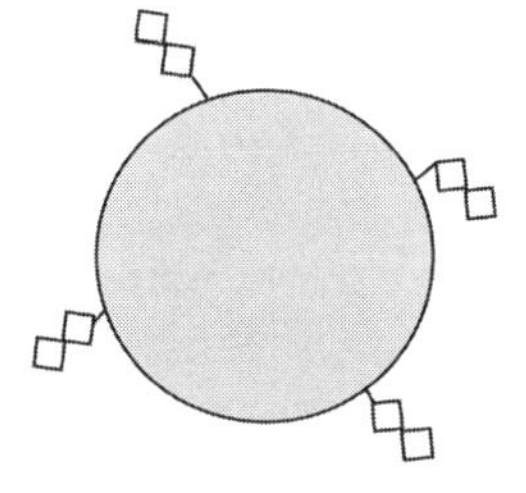

First step : biotinylated antibody

Second step : avidin

Third step : biotinylated particle

Figure 4 Attachment of antibodies to the carrier using the avidin–biotin interaction. Three different approaches are described. In the direct approach (**A**), biotinylated antibodies are attached to the particles by the avidin–biotin interaction and the particles are directly administrated. In the two-step approach (**B**), first the biotinylated antibodies are administrated to allow accumulation and binding to target the cells, and then, the avidin-conjugated particles are administered. In the three-step targeting (**C**), biotinylated antibodies are administrated first, followed by free avidin and then by biotinylated particles.

model and results showed that the antibody bound firmly to the vesicles and maintained the ability to recognize its antigen (69). However, the percentage of ligand attached to the carrier is relatively low (4–40%) and in addition, an aggregation of the liposomes was observed. Furthermore, the final amount of ligand linked to the liposome was not easily controllable and the correct orientation of the antibodies was not ensured. This might be the reason why this coupling method has not been extensively exploited.

Introduction of Lipid-Tagged Antibodies to the Liposomal Membrane

As an alternative to the chemical conjugation, the use of lipid-tagged antibodies obtained by genetic engineering is an interesting approach (70). The technique consists of the biosynthesis of lipoprotein single-chain antibodies using genetic engineering and followed by the incorporation of the lipid-tagged antibody into the liposomes membranes or by its adsorption on the preformed liposome (71). Lipoprotein single-chain antibodies are expressed in *Escherichia coli*. The lipid-tagged antibodies thus obtained contain a single glycerolipid moiety that provides the necessary hydrophobicity to the antibody allowing its insertion into the membranes of liposomes. The liposomes bearing lipid-tagged antibody displayed specific hapten-binding activity, but, to our knowledge, this approach has not yet been investigated in vivo (70,72). Thus the practical utility of using lipid-tagged antibody remains to be proven.

It has to be mentioned that this approach avoids chemical reactions and therefore the risk of decrease or loss of the activity of the antibody. However, this technique requires good knowledge in genetic engineering to convert antibodies into membrane protein maintaining their specific activity. The question remains also whether the small single chain antibody directly bound to the liposomal surface (without spacer arm) will be functional in stealth liposomes.

Attachment of Heat-Aggregated Antibodies to the Surface of Liposomes

The first approach to prepare liposomes as a device for the delivery of active drugs to specific target sites consisted of adsorbing nonspecifically heat-aggregated antibodies to the surface of liposomes. The association between aggregated immunoglobulins and liposomes is preferentially obtained through both electrostatic and hydrophobic interactions. An example of this technique is given by Weissmann et al. (73) who prepared heat-aggregated immunoglobulin by simply heating the dissolved native protein in a 62°C water bath for 10 minutes. This procedure induces a conformational modification in the Fc regions of the immunoglobulin allowing them to interact with liposomes, most probably by enhancing their hydrophobic associations with membrane phospholipids. In another study, liposomes were coated with heat-aggregated IgM and enzyme horseradish peroxidase was incorporated into liposomes (74). Results indicated that the uptake of the enzyme by lysosomes of peroxidase-deficient phagocytes is significantly enhanced when liposomes are coated with aggregated IgM in contrast to uncoated liposomes or with liposomes bearing native IgM. Although this procedure has given some interesting results, the interaction between the antibody and the liposomes is not very strong in contrast to covalently coupling reactions. Furthermore, the amount of antibody bound to the vesicles cannot be controlled.

Binding of Antibodies via Hapten

Another noncovalent approach for binding antibody to the outer surface of liposome is the use of haptens that act as a binding point for the antigen-binding sites of antibodies. Some interesting studies have been carried out using this approach that has the advantage over the previous noncovalent method, that the amount of antibody on the surface of liposome can be more easily controlled. Leserman et al. (75) prepared vesicles displaying on their surface the dinitrophenyl-hapten (DNP-hapten). In vitro trials on murine myeloma cells, MOPC 315, that express immunoglobulin, with affinity for the nitrophenyl hapten, indicated specific binding of liposomes to these cells. In a further study, the same group confirmed these results, on other tumor cells line P388D$_1$ (76). Here again, hapten was derivatized with PE to allow the incorporation of the hapten into the liposomal bilayer (77). The resulting vesicles revealed to bind to the cells, in the presence of IgG antidinitrophenyl. But even though the liposomes in general were shown to interact selectively with target cells, the enhancement of cell binding observed in these studies was limited; in fact, only a very small proportion of the entrapped active molecule was delivered into tumor cells (75,78).

Immunoliposomes and Active Targeting

As seen before, the most common ligand used has been the antibody. Immunoliposomes represent a potentially powerful strategy in the active targeting. In recent years, immunoliposomes have especially gained recognition as potential carriers for drugs for tumor targeting. In vitro studies confirmed the efficacy of immunoliposomes to recognize and specifically bind target antigens. Furthermore, in vivo investigations have shown the usefulness of liposomes combined with potent anticancer molecules incorporated in the vesicles. The systemic toxic effects were dramatically reduced not only by the encapsulation of the drug but also by the efficacy, which was similar and in some cases even higher than with free drug (35). Moreover, stable liposomal drug formulations (liposomal doxorubicin, Doxil®, and liposomal daunorubicin, Daunosome®) have been already approved for medical use in oncology.

It is obvious that beside the technique used for attaching antibodies to liposomes, other different parameters are crucial, including the choice of the antibody, antibody density on the liposome and the use of intact antibody or fragments. If an internalization of the carrier is desired, it is important to choose an antibody that induces internalization. The advantage of using antibodies, which are internalized and mediate the internalization of the carrier, is that the encapsulated drug is directly delivered intracellularly, enhancing thus the efficacy and reducing the side effects. It has to be kept in mind that the internalization might depend on the model cells chosen for in vitro trials. For instance, immunoliposomes targeted to the HER-2 antigen were reported to be efficiently internalized by human breast cancer cells, less efficiently by MKN-7 human gastric or SKOV-3 ovarian carcinoma cells, and not at all by N-87 human gastric carcinoma cells (38,56). The choice of the antibodies depends also on whether they are expected to possess, therapeutic properties. A good example is given by anti-HER2 monoclonal antibodies that have an antiproliferative effect that can enhance the therapeutic efficacy of the encapsulated drug. Once the antibody has been chosen, it is essential to determine the amount of antibody to be coupled to each liposome to ensure binding and cell response. There is no "magic" number, as it can vary from 10 to 40 up to 240 or even higher (27,36,38,55).

The number probably depends directly on the selected antibody–antigen system, coupling reaction chosen, and on the available functional groups in the liposomes.

Another crucial point to decide is whether the intact antibody, fragments [Fab′ or $(Fab)_2$], or genetically engineered Fv shall be used. Fragments may be preferred, because the absence of the Fc portion might allow such immunoliposomes to circulate longer since less are taken up by RES. $F(ab')_2$ fragments can also provide available thiol groups for the coupling to liposomes. The thiol functions in the hinge region offer a unique site for conjugation far away from the antigen binding site, ensuring also the correct orientation of the conjugated Fab′ and its immunoreactivity. Although functional groups can be introduced by means of heterobifunctional cross-linkers, the site and the degree of substitution are frequently arbitrary leading to a decrease of binding activity of the antibodies. Finally, Fab′ or Fv fragments are relatively small (50 or 25 kDa, respectively) compared to whole IgG (150 kDa) or to $F(ab')_2$ fragments (100 kDa) and this could be a great advantage when a fusion mechanism is considered between the cell and liposomal membranes. It is not yet well established if the fragmentation of the antibody into $F(ab')_2$ or Fab′ has a positive or negative impact on the binding activity of the antibody.

LABELING POLYMERIC NANOPARTICLES WITH LIGANDS

So far nanoparticles have not been extensively investigated for active targeting of therapeutic agents, despite certain advantages over liposomes, particularly an improved stability in vivo and upon storage, a wide choice of materials (synthetic polymers, such as polylactides or polycaprolactones, or natural polymers, such as proteins or lipids) and various techniques of preparation (Table 3). They are also

Table 3 Methods Used to Bind Ligands to the Surface of Polymeric Nanoparticles

Coupling method	Composition of the nanoparticles	Studies	References
Adsorption	Poly(cyanoacrylic)	In vitro	(83)
		–	(82)
		In vitro	(80)
		–	(81)
		In vivo	(79)
	Poly(styrene)	In vitro	(84)
Covalent			
Schiff's base formation	Bovine serum albumin	In vitro/in vivo	(85)
	Poly(cyanoacrylic)	In vitro	(86)
Periodate oxidation	Poly(cyanoacrylic)	In vitro	(87)
Carbodiimide reaction	Poly(styrene)	In vitro/in vivo	(88,89)
	Gliadin	In vitro	(90)
Polymer conjugate	Poly(lactic acid)	In vivo	(91)
	Poly(lactic acid)		(92)
	Poly(styrene)	–	(93)
Multistep approach	Poly(lactic acid)	–	(94)
Biotin–avidin interaction	Gelatin	–	(95)
	Human serum albumin	–	(96)
	Poly(lactic acid)/caprolactone	In vitro	(97)

able to carry a higher amount of active molecules. Nevertheless, there are still some improvements necessary to achieve a selective targeting of nanoparticles to specific tissues or organs. Here, some attempts to bind ligands to the surface of nanoparticles will be discussed.

Adsorption of Ligands to the Surface of Polymeric Nanoparticles

The most common approach described to design nanoparticles for active targeting, is the adsorption of monoclonal antibodies to their surface (79–83). This noncovalent technique was widely investigated on cynoacrylic nanoparticles, and interesting results were obtained by the different research groups. One of the first attempts was accomplished by Couvreur and Aubry (83) with 17-1-A antibody adsorbed onto the surface of preformed nanoparticles. High amounts of antibody were bound to the particles and the resulting nanoparticles were stable during storage. To optimize the coupling of the antibody to the nanoparticles, especially with respect to the orientation of the antibody, spacer molecules were used. Protein A (protein isolated from the cell wall of *Staphylococcus aureus*) was chosen as a spacer arm, which has the particularity of specifically interacting with Fc fragment of the IgG. After adsorption of Protein A on the surface of the nanoparticles, the antibody is simply added. Under these conditions, the IgG is correctly oriented with the Fab′ fragment accessible for the binding to the specific antigen. Results demonstrated that almost 95% of the monoclonal antibody was coupled to the nanoparticles, which then interacted with target cells (human colorectal carcinomas HT-29 and HRT-18). However, in the absence of in vitro experiments in the presence of serum proteins, the competitive displacement of the adsorbed antibody was not studied. This investigation was subsequently carried out but results were controversial (80–82). In fact, Manil et al. (82) observed no significant desorption of the antibody (anti-α-fetoprotein antibody) when the nanoparticles were incubated with human serum protein. In contrast, other in vitro and in vivo studies provided evidence of competitive displacement of the adsorbed antibody by serum proteins (79–81). Furthermore, the in vivo results were disappointing, as the nanoparticles accumulated mostly in the liver and spleen.

Another interesting work dealing with the adsorption of antibodies (humanized mAb HuEP5C7.g2) via Protein A suggested the potential of this binding method for targeting purpose (84). Here again, Protein A was used as a spacer arm between the surface of polystyrene nanoparticles and the antibodies, allowing the good orientation of the latter. Results indicated that these coated nanoparticles bound selectively to cellular expressed E- and P-selectin. These results suggest that this approach, using HuEP5C7.g2 could be potentially used to target nanoparticles to endothelium expressing E- and P-selectin. Here again, further investigations must be carried out to confirm the efficacy of this approach.

Covalent and Multistep Approaches Used to Bind Ligands to the Surface of Nanoparticles

To the best of our knowledge, very few studies dealing with the covalent coupling of ligands to the nanoparticles, and even fewer dealing with in vitro and in vivo experiments have been published. One of theses studies was carried out by Akasaka et al. (85) who attached rabbit antihuman antibodies to bovine serum albumin nanoparticles by formation of the Schiff's base between primary amine of the antibody and free aldehyde groups of glutaraldehyde (bifunctional cross-linker) present on the surface of the preformed albumin nanoparticles. Results indicated that antibodies maintained their binding activity after the coupling reaction. Furthermore, in vitro and in vivo experiments demonstrated the specific targeting ability of the antibody-coated nanoparticles, although with low affinity.

Another example of covalent binding of monoclonal antibodies to polymethacrylic nanoparticles was described by Rolland et al. (86). In this study, an antiperipheral human T lymphocyte antibody (CD3 monoclonal antibody) was linked to the surface of nanoparticles using the glutaraldehyde method. In vitro experiments showed that the covalent binding was stable and that the specificity of the monoclonal antibody was maintained.

A recent study on PEG-coated poly(cyanoacrylate) nanoparticles conjugated to TF revealed that this carrier system was useful for delivery of pDNA to target cells (87). TF was coupled to PEG-coated nanoparticles by a periodate oxidation technique and results revealed that 1% to 3% of the total PEG chains were linked to TF molecules.

Interesting work has been accomplished with respect to the oral delivery of nanoparticles to increase their gastrointestinal uptake (88–90). Although a discussion of techniques to enhance the bioadhesion of orally administrated nanoparticles is not the aim of this chapter, these studies described interesting coupling reactions that might be used for the coupling of ligands, such as antibodies, for active targeting. Various lectins have been successfully used, based on their primary amines available for reaction with carboxylic acid groups exposed on the surface of polystyrene or gliadin particles (88–90), and able to form an amide bond via a carbodiimide reaction.

Another interesting approach is the synthesis of a conjugate of a polymer and a ligand, which is then directly used for the preparation of nanoparticles. Gautier et al. (91) used this method for the preparation of biotinylated poly(lactic acid) (PLA) nanoparticles. A conjugate of biotin and poly(methyl methacrylate-*co*-methacrylic acid) was synthesized and coprecipitated with PLA into stable nanoparticles. For active targeting, such nanoparticles could be used for a multistep approach using the avidin–biotin interaction. Another possibility could be to replace biotin with a bifunctional cross-linker for coupling of the ligands.

A promising strategy for active targeting with nanoparticles was described by Maruyama et al. (92). Here, nanoparticles exposing surface carbohydrates, which could bind to target cells via the carbohydrate-binding proteins present on the surface of cells were prepared for the delivery of genetic material into cells. Various polysaccharides were investigated, and their binding to the nanoparticle was guaranteed by the presence of poly(L-lysine)-grafted–polysaccharide polymer in the PLA matrix of the nanoparticle.

An interesting work by Serizawa et al. (93) proved the potential of carbohydrate-conjugated nanoparticles for site-specific drug carriers. They prepared poly (vinylamine)-grafted polystyrene nanospheres by the free radical polymerization of styrene and poly(*N*-vinylacetamide). The lactose was bound to the particles by an amide linkage. Results suggested that the use of lactose-conjugate nanoparticles could be a useful approach for site delivery of a drug carrier, as these particles recognized and bound specifically to lectin RCA_{120}.

As a multistep targeting approach, our group is investigating the possibility of coating nanoparticles with homing moieties. The idea is to use the avidin–biotin interaction (Fig. 4) to attach specific antibodies on the surface of PLA nanoparticles. As a first step, functional thiol groups were added to the surface of nanoparticles with three different techniques having been investigated (94). The second step was to covalently attach NeutrAvidin® to the nanoparticles, which allows the subsequent binding of biotinylated antibodies to the carrier. A similar concept has already been used to prepare gelatine and human serum albumin nanoparticles as carriers for biotinylated compounds (95,96). More recently, Gref et al. (97) have prepared nanoparticles composed of PLA and biotin–PEG–poly(ε-caprolactone). Avidin was coupled to the surface of the nanoparticles followed by biotinylated WGA. In vitro studies revealed that the nanoparticles interacted specifically with target cells.

HOW TO CHOOSE THE COUPLING METHOD

Various coupling methods have been described in this chapter with their advantages and drawbacks. The crucial point is to choose the right method among the different coupling reactions.

Covalent reactions appear to be an effective way to irreversibly fix ligands to colloidal carriers, as the linkage formed is much more stable and reproducible when compared with noncovalent methods, such as adsorption. It has to be kept in mind that the coupling reaction must not affect the biological activity of the ligand. In the case of antibodies, all conjugations through primary amines will certainly involve a part of amine groups within the antigen-binding region, thus enhancing the probability of affecting the antibody activity, especially if highly modified antibodies are prepared. The use of antibody fragments having available thiol functions is a good alternative, as no modifications are required to add functional groups, thus preserving the antigen-binding activity. Furthermore, a good orientation of the fragments on the liposomes surface is guaranteed. No covalent coupling reaction is universal as the efficacy of the conjugation depends on various relevant parameters such as the type and the localization of the functional groups on the ligands and the number of available ligands attached to the carrier and the type of cross-linker. The length of the cross-linker is also crucial, as the accessibility of the ligand is directly related to the length of the cross-linker. The use of extended spacer arms such as PEG can greatly reduce the steric hindrance and thus improve the accessibility of the ligand.

It has to be kept in mind that covalent reactions require chemical reagents (e.g., DTT), which can interfere with subsequent reactions. In addition, the use of some cross-linkers, such as glutaraldehyde, can induce undesired aggregation of ligand and/or liposomes. Furthermore, often the reactions and the purification steps are carried out during a relatively longer period leading, in some cases, to an alteration of the liposome or the ligand.

The ligand can be chosen to be attached to the surface of the liposome either during its preparation, or late on preformed vesicles. The major drawback of the incorporation by mixing with other components of the bilayer membrane is the presence of immobilized ligands on the inner surface of the liposomes and consequently not available for the binding activity. Furthermore, the number of ligands coupled per liposomes can be quite heterogeneous and thus the binding capacity to target cells will not be the same (17,18).

On the other hand, if the coupling reaction is achieved on preformed liposomes, there is some risk of altering the structure of the liposomal membrane and in some cases, also the compound that is encapsulated. Therefore, it is important to choose nonaggressive reagents and to work under mild conditions.

Finally, in the authors' opinion, some basic requirements should be considered as essential for successful specific targeting. (i) The linkage between the carrier and the ligand needs to be stable; (ii) a sufficient amount of ligand must be bound to the particles; (iii) the specific binding properties of the ligand has to be preserved; (iv) the particle structure must not be affected by the coupling reaction; (v) the attachment of the ligand should not cause a significant increase in the size of the carrier.

CONCLUDING REMARKS

In recent years, significant improvement in the field of colloidal carriers has been achieved. Nanoparticles and liposomes have emerged as versatile tools for delivery of active compounds, such as antitumor drugs. The formulation of long circulating particles has given reliance to the usefulness of these carriers, and provides a solid basis for the development of particles for selective targeting. Many challenging questions arise when dealing with active targeting, such as those concerning the optimal size of the particles, the choice and the density of the ligand, and the optimal coupling reaction. In this chapter, the authors have mainly reviewed several coupling methods that have been explored for the modification of the surface of the carriers and for the attachment of ligands. In general, binding of the carriers, particularly of liposomes, bearing homing moieties to target cells has been widely demonstrated. Furthermore, in some cases, the efficacy of the encapsulated drug has been also shown, although results were not always optimal as expected. In fact, liposomes that bind to target cells are not necessarily internalized. In vitro, the internalization of particles is definitely favorable where each individual tumor cell can be targeted. The question remains open whether internalization is a prerequisite of efficiency in vivo where the large majority of tumor cells might not be accessible to the particles but where drugs could freely diffuse and reach all tumor cells. Thus, investigations still have to be carried out to improve the targeting efficiency of these colloidal carriers, especially regarding the delivery of anticancer drugs. Furthermore, the mechanisms involved in the delivery of drugs from the carrier into target cells are still not fully understood.

One has to keep in mind that the ability to achieve active targeting in vivo depends directly on whether or not the target is accessible from the vasculature. In the first case, particles bearing homing moieties and having a long circulating half-life have a high probability of reaching the target. When the target is extravascular, such as solid tumors, penetration of carriers to the bulk of the tumor cells might be more difficult. In this case, only repeated administration of particles might progressively open the way to the deep cell layers. Furthermore, the size of the carriers has also to be well established for their extravasation and accumulation in solid tumors considering that the pore sizes of discontinuous tumor microvasculature vary between 100 and 780 nm (98–100).

Although, these systems are far from being perfect, the authors conclude, as many authors have demonstrated, that the proof-of-concept for active-targeted drug delivery has been well established.

REFERENCES

1. Weiner LM. Monoclonal antibody therapy of cancer. Semin Oncol 2003; 26:43–51.
2. Martin FJ. Clinical pharmacology and antitumor efficacy of DOXIL (pegylated liposomal doxorubicin). In: Lasic DD, Papahadjopoulos D, eds. Medical Applications of Liposomes. New York: Elsevier Science BV, 1998:635–688.

3. Allémann E, Gurny R, Doelker E. Drug-loaded nanoparticles-preparation methods and drug targeting issues. Eur J Pharm Biopharm 1993; 39(5):173–191.
4. Senior JH. Fate and behavior of liposomes in vivo: a review of controlling factors. Crit Rev Ther Drug Carrier Syst 1987; 3(2):123–193.
5. Stolnik S, Dunn SE, Garnett MC, et al. Surface modification of poly(lactide-co-glycolide) nanospheres by biodegradable poly(lactide)-poly(ethylene glycol) copolymers. Pharm Res 1994; 11(12):1800–1808.
6. Gref R, Minamitake Y, Peracchia MT, Trubetskoy V, Torchilin V, Langer R. Biodegradable long-circulating polymeric nanospheres. Science 1994; 263:1600–1603.
7. Allen TM, Hansen C, Martin F, Redemann C, Yau-Young A. Liposomes containing synthetic lipid derivatives of poly(ethylene glycol) show prolonged circulation half-lives in vivo. Biochim Biophys Acta 1991; 1066(1):29–36.
8. Klibanov AL, Maruyama K, Torchilin VP, Huang L. Amphipathic polyethylenegly cols effectively prolong the circulation time of liposomes. FEBS Lett 1990; 268(1): 235–237.
9. Papahadjopoulos D, Allen TM, Gabizon A, et al. Sterically stabilized liposomes: improvements in pharmacokinetics and antitumor therapeutic efficacy. Proc Natl Acad Sci USA 1991; 88(24):11,460–11,464.
10. Vaage J, Barbera-Guillem E, Abra R, Huang A, Working P. Tissue distribution and therapeutic effect of intravenous free or encapsulated liposomal doxorubicin on human prostate carcinoma xenografts. Cancer 1994; 73(5):1478–1484.
11. Northfelt DW, Martin FJ, Working P, et al. Doxorubicin encapsulated in liposomes containing surface-bound polyethylene glycol: pharmacokinetics, tumor localization, and safety in patients with AIDS-related Kaposi's sarcoma. J Clin Pharmacol 1996; 36(1):55–63.
12. Harasym TO, Bally MB, Tardi P. Clearance properties of liposomes involving conjugated proteins for targeting. Adv Drug Deliv Rev 1998; 32(1–2):99–118.
13. Forssen E, Willis M. Ligand-targeted liposomes. Adv Drug Deliv Rev 1998; 29(3):249–271.
14. Maruyama K, Ishida O, Takizawa T, Moribe K. Possibility of active targeting to tumor tissues with liposomes. Adv Drug Deliv Rev 1999; 40(1–2):89–102.
15. Drummond DC, Meyer O, Hong K, Kirpotin DB, Papahadjopoulos D. Optimizing liposomes for delivery of chemotherapeutic agents to solid tumors. Pharmacol Rev 1999; 51(4):691–743.
16. Park JW, Hong K, Kirpotin DB, Papahadjopoulos D, Benz CC. Immunoliposomes for cancer treatment. Adv Pharmacol 1997; 40:399–435.
17. Hughes BJ, Kennel S, Lee R, Huang L. Monoclonal antibody targeting of liposomes to mouse lung in vivo. Cancer Res 1989; 49(22):6214–6220.
18. Huang A, Huang L, Kennel SJ. Monoclonal antibody covalently coupled with fatty acid. J Biol Chem 1980; 255(17):8015–8018.
19. Harsch M, Walther P, Weder HG. Targeting of monoclonal antibody-coated liposomes to sheep red blood cells. Biochem Biophys Res Commun 1981; 103(3):1069–1076.
20. Shen DF, Huang A, Huang L. An improved method for covalent attachment of antibody to liposomes. Biochim Biophys Acta 1982; 689(1):31–37.
21. Carpenter-Green S, Huang L. Incorporation of acylated wheat germ agglutinin into liposomes. Anal Biochem 1983; 135(1):151–155.
22. Torchilin VP, Omel'yanenko VG, Klibanov AL, Mikhailov AI, Gol'danskii VI, Smirnov VN. Incorporation of hydrophilic protein modified with hydrophobic agent into liposome membrane. Biochim Biophys Acta 1980; 602(3):511–521.
23. Weissig V, Lasch J, Klibanov AL, Torchilin VP. A new hydrophobic anchor for the attachment of proteins to liposomal membranes. FEBS Lett 1986; 202(1):86–90.
24. Torchilin VP, Klibanov AL, Smirnov VN. Phosphatidylinositol may serve as the hydrophobic anchor for immobilization of proteins on liposome surface. FEBS Lett 1982; 138(1):117–120.

25. Crosasso P, Brusa P, Dosio F, et al. Antitumoral activity of liposomes and immunoliposomes containing 5-fluorouridine prodrugs. J Pharm Sci 1997; 86(7):832–839.
26. Derksen JT, Scherphof GL. An improved method for the covalent coupling of proteins to liposomes. Biochim Biophys Acta 1985; 814:151–155.
27. Nassander UK, Steerenberg PA, De Jong WH, et al. Design of immunoliposomes directed against human ovarian carcinoma. Biochim Biophys Acta 1995; 1235(1): 126–139.
28. Martin FJ, Papahadjopoulos D. Irreversible coupling of immunoglobulin fragments to preformed vesicles. An improved method for liposome targeting. J Biol Chem 1982; 257(1):286–288.
29. Maruyama K, Takahashi N, Tagawa T, Nagaike K, Iwatsuru M. Immunoliposomes bearing polyethyleneglycol-coupled Fab' fragment show prolonged circulation time and high extravasation into targeted solid tumors in vivo. FEBS Lett 1997; 413(1): 177–180.
30. Derksen JT, Morselt HW, Scherphof GL. Uptake and processing of immunoglobulin-coated liposomes by subpopulations of rat liver macrophages. Biochim Biophys Acta 1988; 971(2):127–136.
31. Koning GA, Morselt HW, Velinova MJ, et al. Selective transfer of a lipophilic prodrug of 5-fluorodeoxyuridine from immunoliposomes to colon cancer cells. Biochim Biophys Acta 1999; 1420(1–2):153–167.
32. Park JW, Hong K, Carter P, et al. Development of anti-p185HER2 immunoliposomes for cancer therapy. Proc Natl Acad Sci USA 1995; 92(5):1327–1331.
33. Nassander UK, Steerenberg PA, Poppe H, et al. In vivo targeting of OV-TL 3 immunoliposomes to ascitic ovarian carcinoma cells (OVCAR-3) in athymic nude mice. Cancer Res 1992; 52(3):646–653.
34. Vingerhoeds MH, Steerenberg PA, Hendriks JJ, et al. Immunoliposome-mediated targeting of doxorubicin to human ovarian carcinoma in vitro and in vivo. Br J Cancer 1996; 74(7):1023–1029.
35. Park JW, Hong K, Kirpotin DB, Meyer O, Papahadjopoulos D, Benz CC. Anti-HER2 immunoliposomes for targeted therapy of human tumors. Cancer Lett 1997; 118(2): 153–160.
36. Mercadal M, Domingo JC, Petriz J, Garcia J, de Madariaga MA. A novel strategy affords high-yield coupling of antibody to extremities of liposomal surface-grafted PEG chains. Biochim Biophys Acta 1999; 1418(1):232–238.
37. Park JW, Hong K, Kirpotin DB, et al. Anti-HER2 immunoliposomes: enhanced efficacy attributable to targeted delivery. Clin Cancer Res 2002; 8(4):1172–1181.
38. Kirpotin D, Park JW, Hong K, et al. Sterically stabilized anti-HER2 immunoliposomes: design and targeting to human breast cancer cells in vitro. Biochemistry 1997; 36(1): 66–75.
39. Hansen CB, Kao GY, Moase EH, Zalipsky S, Allen TM. Attachment of antibodies to sterically stabilized liposomes: evaluation, comparison and optimization of coupling procedures. Biochim Biophys Acta 1995; 1239(2):133–144.
40. Martin FJ, Hubbell WL, Papahadjopoulos D. Immunospecific targeting of liposomes to cells: a novel and efficient method for covalent attachment of Fab' fragments via disulfide bonds. Biochemistry 1981; 20(14):4229–4238.
41. Allen TM, Brandeis E, Hansen CB, Kao GY, Zalipsky S. A new strategy for attachment of antibodies to sterically stabilized liposomes resulting in efficient targeting to cancer cells. Biochim Biophys Acta 1995; 127(2):99–108.
42. Carrion C, Domingo JC, de Madariaga MA. Preparation of long-circulating immunoliposomes using PEG-cholesterol conjugates: effect of the spacer arm between PEG and cholesterol on liposomal characteristics. Chem Phys Lipids 2001; 113(1–2):97–110.
43. Ivanov VO, Preobrazhensky SN, Tsibulsky VP, Babaev VR, Repin VS, Smirnov VN. Liposome uptake by cultured macrophages mediated by modified low-density lipoproteins. Biochim Biophys Acta 1985; 846(1):76–84.

44. Leserman LD, Barbet J, Kourilsky F, Weinstein JN. Targeting to cells of fluorescent liposomes covalently coupled with monoclonal antibody or protein A. Nature 1980; 288(5791):602–604.
45. Ishida O, Maruyama K, Tanahashi H, et al. Liposomes bearing polyethyleneglycol-coupled transferrin with intracellular targeting property to the solid tumors in vivo. Pharm Res 2001; 18(7):1042–1048.
46. Maruyama K, Takizawa T, Yuda T, Kennel SJ, Huang L, Iwatsuru M. Targetability of novel immunoliposomes modified with amphipathic poly(ethylene glycol)s conjugated at their distal terminals to monoclonal antibodies. Biochim Biophys Acta 1995; 1234(1):74–80.
47. Maruyama K, Holmberg E, Kennel SJ, Klibanov A, Torchilin VP, Huang L. Characterization of in vivo immunoliposome targeting to pulmonary endothelium. J Pharm Sci 1990; 79(11):978–984.
48. Torchilin VP, Klibanov AL, Huang L, O'Donnell S, Nossiff ND, Khaw BA. Targeted accumulation of polyethylene glycol-coated immunoliposomes in infarcted rabbit myocardium. FASEB J 1992; 6(9):2716–2719.
49. Torchilin VP, Narula J, Halpern E, Khaw BA. Poly(ethylene glycol)-coated anti-cardiac myosin immunoliposomes: factors influencing targeted accumulation in the infarcted myocardium. Biochim Biophys Acta 1996; 1279(1):75–83.
50. Holmberg E, Maruyama K, Litzinger DC, et al. Highly efficient immunoliposomes prepared with a method which is compatible with various lipid compositions. Biochem Biophys Res Commun 1989; 165(3):1272–1278.
51. Khaw BA, da Silva J, Vural I, Narula J, Torchilin VP. Intracytoplasmic gene delivery for in vitro transfection with cytoskeleton-specific immunoliposomes. J Contr Rel 2001; 75(1–2):199–210.
52. Torchilin VP, Levchenko TS, Lukyanov AN, et al. p-Nitrophenylcarbonyl-PEG-PE-liposomes: fast and simple attachment of specific ligands, including monoclonal antibodies, to distal ends of PEG chains via p-nitrophenylcarbonyl groups. Biochim Biophys Acta 2001; 1511(2):397–411.
53. Blume G, Cevc G, Crommelin MD, Bakker-Woudenberg IA, Kluft C, Storm G. Specific targeting with poly(ethylene glycol)-modified liposomes: coupling of homing devices to the ends of the polymeric chains combines effective target binding with long circulation times. Biochim Biophys Acta 1993; 1149(1):180–184.
54. Chua MM, Fan ST, Karush F. Attachment of immunoglobulin to liposomal membrane via protein carbohydrate. Biochim Biophys Acta 1984; 800(3):291–300.
55. Harding JA, Engbers CM, Newman MS, Goldstein NI, Zalipsky S. Immunogenicity and pharmacokinetic attributes of poly(ethylene glycol)-grafted immunoliposomes. Biochim Biophys Acta 1997; 1327(2):181–192.
56. Goren D, Horowitz AT, Zalipsky S, Woodle MC, Yarden Y, Gabizon A. Targeting of stealth liposomes to erbB-2 (Her/2) receptor: in vitro and in vivo studies. Br J Cancer 1996; 74(11):1749–1756.
57. Lopes de Menezes DE, Pilarski LM, Allen TM. In vitro and in vivo targeting of immunoliposomal doxorubicin to human B-cell lymphoma. Cancer Res 1998; 58(15):3320–3330.
58. Zalipsky S. Synthesis of an end-group functionalized polyethylene glycol-lipid conjugate for preparation of polymer-grafted liposomes. Bioconjug Chem 1993; 4(4):296–299.
59. Torchilin VP, Khaw BA, Smirnov VN, Haber E. Preservation of antimyosin antibody activity after covalent coupling to liposomes. Biochem Biophys Res Commun 1979; 89(4):1114–1119.
60. Torchilin VP, Goldmacher VS, Smirnov VN. Comparative studies on covalent and non-covalent immobilization of protein molecules on the surface of liposomes. Biochem Biophys Res Commun 1978; 85(3):983–990.
61. Casalini P, Luison E, Ménard S, Colnaghi MI, Paganelli G, Canevari S. Tumor pretargeting: role of avidin/streptavidin on monoclonal antibody internalization. J Nucl Med 1997; 38(9):1378–1381.

62. Magnani P, Paganelli G, Modorati G, et al. Quantitative comparison of direct antibody labeling and tumor pretargeting in uveal melanoma. J Nucl Med 1996; 37(6): 967–971.
63. Moro M, Pelagi M, Fulci G, et al. Tumor cell targeting with antibody-avidin complexes and biotinylated tumor necrosis factor1 α^1. Cancer Res 1997; 57(10):1922–1928.
64. Mao S-Y. Biotinylation of antibodies. Meth Mol Biol 1994; 34:49–52.
65. Portnoy J, Brothers D, Pacheco F, Landuyt J, Barnes C. Monoclonal antibody-based assay for Alt a1, a major Alternaria allergen. Ann Allergy Asthma Immunol 1998; 81(1):59–64.
66. Pieri I, Barritault D. Biotinylated basic fibroblast growth factor is biologically active. Anal Biochem 1991; 195:214–219.
67. Xiao Z, McQuarrie SA, Suresh MR, Mercer JR, Gupta S, Miller GG. A three-step strategy for targeting drug carriers to human ovarian carcinoma cells in vitro. J Biotechnol 2002; 94:171–184.
68. Loughrey H, Bally MB, Cullis PR. A non-covalent method of attaching antibodies to liposomes. Biochim Biophys Acta 1987; 901:157–160.
69. Huang L, Kennel SJ. Binding of immunoglobulin G to phospholipid vesicles by sonication. Biochemistry 1979; 18(9):1702–1707.
70. Laukkanen ML, Teeri TT, Keinanen K. Lipid-tagged antibodies: bacterial expression and characterization of a lipoprotein-single-chain antibody fusion protein. Protein Eng 1993; 6(4):449–454.
71. Laukkanen ML, Orellana A, Keinanen K. Use of genetically engineered lipid-tagged antibody to generate functional europium chelate-loaded liposomes. Application in fluoroimmunoassay. J Immunol Meth 1995; 185(1):95–102.
72. Laukkanen ML, Alfthan K, Keinanen K. Functional immunoliposomes harboring a biosynthetically lipid-tagged single-chain antibody. Biochemistry 1994; 33(38):11664–11670.
73. Weissmann G, Brand A, Franklin EC. Interaction of immunoglobulins with liposomes. J Clin Invest 1974; 53(2):536–543.
74. Weissmann G, Bloomgarden D, Kaplan R, et al. A general method for the introduction of enzymes, by means of immunoglobulin-coated liposomes, into lysosomes of deficient cells. Proc Natl Acad Sci USA 1975; 72(1):88–92.
75. Leserman LD, Weinstein JN, Blumenthal R, Sharrow SO, Terry WD. Binding of antigen-bearing fluorescent liposomes to the murine myeloma tumor MOPC 315. J Immunol 1979; 122(2):585–591.
76. Leserman LD, Weinstein JN, Blumenthal R, Terry WD. Receptor-mediated endocytosis of antibody-opsonized liposomes by tumor cells. Proc Natl Acad Sci USA 1980; 77(7): 4089–4093.
77. Weinstein JN, Blumenthal R, Sharrow SO, Henkart PA. Antibody-mediated targeting of liposomes. Binding to lymphocytes does not ensure incorporation of vesicle contents into the cells. Biochim Biophys Acta 1978; 509(2):272–288.
78. Leserman LD, Weinstein JN, Moore JJ, Terry WD. Specific interaction of myeloma tumor cells with hapten-bearing liposomes containing methotrexate and carboxyfluorescein. Cancer Res 1980; 40:4768–4774.
79. Illum L, Jones PD, Baldwin RW, Davis SS. Tissue distribution of poly(hexyl 2-cyanoacrylate) nanoparticles coated with monoclonal antibodies in mice bearing human tumor xenografts. J Pharmacol Exp Ther 1984; 230(3):733–736.
80. Illum L, Jones PDE, Kreuter J, Baldwin RW, Davis SS. Adsorption of monoclonal antibodies to polyhexylcyanoacrylate nanoparticles and subsequent immunospecific binding to tumour cells in vitro. Int J Pharm 1983; 17:65–76.
81. Kubiak C, Manil L, Couvreur P. Sorptive properties of antibodies onto cyanoacrylic nanoparticles. Int J Pharm 1988; 41:181–187.
82. Manil L, Roblot-Treupel L, Couvreur P. Isobutyl cyanoacrylate nanoparticles as a solid phase for an efficient immunoradiometric assay. Biomaterials 1986; 7:212–216.
83. Couvreur P, Aubry J. Monoclonal antibodies for the targeting of drugs: application to nanoparticles. Topics Pharmaceut Sci 1983:305–316.

84. Blackwell JE, Dagia NM, Dickerson JB, Berg EL, Goetz DJ. Ligand coated nanosphere adhesion to E- and P-selectin under static and flow conditions. Ann Biomed Eng 2001; 29(6):523–533.
85. Akasaka Y, Ueda H, Takayama K, Machida Y, Nagai T. Preparation and evaluation of bovine serum albumin nanospheres coated with monoclonal antibodies. Drug Des Deliv 1988; 3(1):85–97.
86. Rolland A, Bourel D, Genetet B, Le Verge R. Monoclonal antibodies covalently coupled to polymethacrylic nanoparticles: in vitro specific targeting to human T lymphocytes. Int J Pharm 1987; 39:173–180.
87. Li Y, Ogris M, Wagner E, Pelisek J, Rüffer M. Nanoparticles bearing polyethylenglycol-coupled transferrin as gene carriers: preparation and in vitro evaluation. Int J Pharm 2003; 259:93–101.
88. Hussain N, Jani PU, Florence AT. Enhanced oral uptake of tomato lectin-conjugated nanoparticles in the rat. Pharm Res 1997; 14(5):613–618.
89. Ezpeleta I, Arangoa MA, Irache JM, et al. Preparation of Ulex europaeus lectin–gliadin nanoparticle conjugates and their interaction with gastrointestinal mucus. Int J Pharm 1999; 191:25–32.
90. Irache JM, Durrer C, Duchêne D, Ponchel G. Preparation and characterization of lectin–latex conjugates for specific bioadhesion. Biomaterials 1994; 15(11):899–904.
91. Gautier S, Grudzielski N, Goffinet G, de Hassonville SH, Delattre L, Jérôme R. Preparation of poly(D,L-lactide) nanoparticles assisted by amphiphilic poly(methyl methacrylate-co-methacrylic acid) copolymers. J Biomater Sci Polym Ed 2001; 12(4):429–450.
92. Maruyama A, Ishihara T, Kim J-S, Kim SW, Akaike T. Nanoparticle DNA carrier with poly(L-lysine) grafted polysaccharide copolymer and poly(D,L-lactic acid). Bioconjug Chem 1997; 8(5):735–742.
93. Serizawa T, Uchida T, Akashi M. Synthesis of polystyrene nanospheres having lactose-conjugated hydrophilic polymers on their surfaces and carbohydrate recognition by proteins. J Biomater Sci Polym Ed 1999; 10(3):391–401.
94. Nobs L, Buchegger F, Gurny R, Allémann E. Surface modification of poly(lactic acid) nanoparticles by covalent attachment of thiol groups by means of three methods. Int J Pharm 2003; 250:327–337.
95. Langer K, Coester C, Weber C, Von Briesen H, Kreuter J. Preparation of avidin-labeled protein nanoparticles as carriers for biotinylated peptide nucleic acid. Eur J Pharm Biopharm 2000; 49(3):303–307.
96. Coester C, Kreuter J, Von Briesen H, Langer K. Preparation of avidin-labelled gelatin nanoparticles as carriers for biotinylated peptide nucleic acid (PNA). Int J Pharm 2000; 196:147–149.
97. Gref R, Couvreur P, Barratt G, Mysiakine E. Surface-engineered nanoparticles for multiple ligand coupling. Biomaterials 2003; 24:4529–4537.
98. Ishida O, Maruyama K, Sasaki K, Iwatsuru M. Size-dependent extravasation and interstitial localization of polyethyleneglycol liposomes in solid tumor-bearing mice. Int J Pharm 1999; 190(1):49–56.
99. Hobbs SK, Monsky WL, Yuan F, Roberts WG, Griffith L, Torchilin VP, Jain RK. Regulation of transport pathways in tumor vessels: role of tumor type and microenvironment. Proc Natl Acad Sci USA 1998; 95:4607–4612.
100. Yuan F, Dellian M, Fukumura D, et al. Vascular permeability in human tumor xenograft: molecular size dependence and cutoff size. Cancer Res 1995; 55(17):3752–3756.

6

Industrial Technologies and Scale-Up

François Puel
LAGEP UMR CNRS 5007, Université Claude Bernard Lyon 1, Lyon, France

Stéphanie Briançon and Hatem Fessi
LAGEP UMR CNRS 5007 and Laboratoire de Génie Pharmacotechnique et Biogalénique, Université Claude Bernard Lyon 1, Lyon, France

INTRODUCTION

The field of microencapsulation and nanoencapsulation is concerned with a large number of processes used to produce particles of defined size and structure. Encapsulation processes are generally classified under chemical, physicochemical, and mechanical processes, including many subcategories such as polymerization, polycondensation, spray drying, coacervation, solvent evaporation, etc. However, these terms do not completely describe the principle of particle formation and the mechanisms involved. These mechanisms include phase separation, precipitation, interfacial polymerization or polycondensation, crosslinking, and gelation. All these refer to the generation of particles from an initial phase, which can be a solution or more frequently a dispersion (emulsion or suspension). This initial stage is a fundamental step in the encapsulation process, especially when an emulsion or a suspension is concerned, as it directly influences the final properties and, most importantly, the size of the particles. Other classifications could be based on the mechanisms involved, on the nature of the continuous phase, or on the starting materials. A distinction is commonly made between the processes using polymer as encapsulating material and the others that involve monomers and are based on a chemical reaction (so called chemical processes). It can also be considered that a great difference exists between processes occurring in liquid phases or in gas phases. The former are the majority of chemical and physicochemical processes, generally based on the formation of an initial dispersion in an aqueous phase followed by a transformation of the droplets or fine particles in capsules or spheres. The latter belong to the mechanical processes category, based on the dispersion of the initial phase (solution or dispersion) in a gas stream. Whatever the classification adopted, an encapsulation process can be defined by the different basic steps of the particles formation, the mechanisms involved at each stage, and the technical means employed. The simultaneous consideration of technical aspects and mechanisms occurring is necessary because of the influence of the production techniques and conditions and the mechanisms (1).

The number of encapsulation processes cited in the literature has increased significantly since the 1950s, reflecting the growth of research and development in this field. This is a field in which simultaneous research works in several scientific disciplines are done. There are also numerous commercial applications involving the development of manufacturing processes. The first processes developed at an industrial scale were the mechanical ones, spray drying, and pan coating (2). The development of microcapsules for carbonless copy paper at the National Cash Register Company was the first recognized microencapsulation process. Today, a large number of physical, chemical, and physicochemical processes are identified, and their number increases continuously, as shown by the number of patents in this field. The number of documents in the Chemical Abstract Database on microencapsulation rose by 1500 in 2002, almost 1000 of these were patents (1). Some of these patents concern really original processes; others are improvements or adaptations of old ones. This large number of patented documents gives evidence to the secrets surrounding the commercial applications of microencapsulation. Concerning the manufacture of microencapsulated products, industry tends to guard information about scale-up and all the needed knowledge about equipment and manufacturing constraint as internal "know how."

The literature concerning microencapsulation processes generally focuses on the methodology of particles preparation, with the objectives of maximum loading and controlled release. Articles describe the influence of formulation parameters (physicochemical) such as the nature and concentration of constituents, viscosity of the phases, etc. Process parameters such as stirring rate, addition rate of the coacervation or polymerization agent, temperature, etc. are also studied. But a great majority of the published work leads to qualitative or semi-empirical relations between parameters and particles properties (3–10). The information obtained from this type of study is undoubtedly very important for the design of microencapsulation systems. The main problem is that it does not allow a real control of the particles properties and always leads to some "trial and error" research approach. Moreover, there is a question of whether the results obtained at laboratory scale can be extrapolated to a higher scale (pilot or plant). In cases where no specific methodology has been followed in the lab for producing microcapsules, there is a high risk of evolution of the main characteristics of the capsules (size, structure, and content) produced at larger scales. The only solution is then to adapt the operating process defined in the laboratory to the pilot or to the plant. It is undoubtedly time-consuming and significantly increases the cost of production.

Scale-up may be defined as the procedure of increasing the batch size. It is sometimes understood as applying the same process to obtain different output volumes. There are some differences between the two definitions: batch size–enlargement does not mean a systematic increase of the processing volume (11). To our knowledge, scaling up of microencapsulation processes is obtained by an increase of the volume of the equipment that is used in the laboratory such that it is suitable to be employed for batch production in a manufacturing plant.

A rational scale-up methodology has been used in chemical engineering for about 50 years and has gained a wide recognition in the chemical industry. It has been applied to chemical reactions taking part in monophasic medium (liquid or gas) or in multiphasic medium (liquid/solid, liquid/gas, gas/solid, and gas/liquid/solid). The objective was to obtain the same kind of progress in all the reactions.

As soon as it became necessary to predict the size distribution of bubbles, droplets, or crystals, the difficulty increased because several physical phenomena were competing with each other. For example, in the liquid–liquid contacting process, dispersing and coalescing droplets phenomena are progressively balanced.

The scale-up procedure can be summarized as having two stages: characterize the dominant phenomena occurring in the process at a given scale (laboratory scale most of the time) with a dimensional analysis procedure, and then apply process similarities between scales to keep the intensity of these main phenomena constant.

Dimensional analysis is a method for producing dimensionless numbers that completely characterize the process. The dimensionless numbers are made of parameters that are grouped together. The analysis can be applied even when the equations governing the process are not known: two processes may be considered completely similar if all the dimensionless numbers necessary to describe the process have the same value (11). Using dimensionless numbers simplifies the description of the process as it reduces the number of variables to be considered.

Process similarities are achieved between two processes when they accomplish the same process objectives by the same mechanisms and produce the same product, according to the required specifications (12). The two processes could be based on the same operating procedure, performed at different scales. In such process translation, processes may be similar only if different types of similarities have been followed. Some of them are absolutely necessary (e.g., geometric and dynamic similarities) for keeping fluid motion and internal forces in the medium. Others are more optional (e.g., thermal and chemical similarities) and depend on the requirements of the process.

The scale-up methodology is then, in practice, rather simple: express the process using a complete set of dimensionless numbers at lower scale, and match them at upper scales (11).

For microencapsulation processes, there is a lack of publications on process scale-up and manufacturing development in the literature (2). The more recent papers were found to be concerned with the coacervation process, emulsion–solvent evaporation or extraction, and nanoprecipitation for the production of nanoparticles (4,5,13–16). By using dimensional analysis approach, these studies aim to establish quantitative relationships between process parameters and physical properties of the particles, basically size distribution and morphology. The objective is to propose some general correlations between the system parameters (process and formulation) and the particles' characteristics. These correlations are sometimes used to investigate large-scale production systems, but the second step of the scale-up methodology is not really applied (4,15). Recently, this scale-up procedure was applied and evaluated to three different microencapsulation methods used for producing nanoparticles (17).

On looking into the bibliographic databases, it appears that the three manufacturing methods used most often are coacervation, spray drying, and interfacial polymerization, in that order (1). The coacervation and interfacial processes both include an emulsification step prior to encapsulation. This stage of emulsification generally determines the achievement of the right particle properties because with some reasonable assumptions, one can suppose that the microcapsule size is practically equal to the one of the droplet. The choice of the technologies for emulsification at a laboratory scale is a key issue, as these technologies must be employed at an industrial scale. Keeping the performance of the emulsification stage by selecting the right operating conditions during the scale-up is then crucial for the success of the process transfer.

In this chapter, we propose to describe the industrial technologies used in encapsulation processes including an emulsion phase, and then to apply scale-up concepts. The first part is dedicated to the emulsification process in a stirred reactor. We provide some insight into technology, we apply dimensional analysis for describing emulsification phenomena, and we discuss the prediction of droplet mean size. This part ends with the modeling of the influence of the stirring rate on the micro- and nanocapsules (NC) mean size. The scale-up methodology applied to emulsification processes is presented in a second part. Examples are given in the final two parts concerning microcapsules production by interfacial polycondensation and nanoparticles production by emulsion–diffusion (EMDF). Both the encapsulation processes have been first studied at a very small production scale. A pilot plant was designed to produce the particles at the lab-scale, i.e., a few liters with the equipment geometrically similar to those used for large-scale production. The final (two) parts give some example of scale-up results.

Our final objective is that at the end of this chapter, readers will be convinced that by following a few practical rules in the laboratory and during the scale transfer, the methodology of scaling up microencapsulation processes presented could be used routinely.

EMULSIFICATION PROCESSES

Generation of Droplets: Few Considerations

The formation of droplets under mechanical stirring could be described with hydrodynamic considerations, to link the phenomena occurring with the description of the equipment used for performing emulsification. It is a global description, which allows the formulation and process parameters to be taken into consideration. However, this description remains quite basic, and for more details the reader is recommended to refer elsewhere (18).

A drop suspended in a continuous phase undergoes two types of forces, which act in opposite ways. Under the influence of the fluid motion, the droplet is distorted by deformation forces. On the other hand, because of the interfacial tension σ and droplet viscosity μ_d, there are some internal droplet forces of cohesiveness that prevent droplet fragmentation. The deformation force is external, and the latter one is internal. If the local instantaneous stresses generated by turbulent motions in the continuous phase exceed the stabilizing forces, the droplets break up. On the contrary, as soon as the fluid motion goes through a quiescent zone, droplets grow by coalescence. When equilibrium between these forces is reached, the droplet size remains kinetically stable.

With the increase of the volume of emulsion, external forces are not exactly the same from one droplet to the other, because levels of stress are not uniform in the volume of the vessel. Consequently, droplet size distribution is established. Stable size distribution is obtained progressively. Its rate of evolution is determined by the equilibrium between drop breakage and the coalescence for each droplet.

Emulsification Process in Stirred Reactor

Hydrodynamics in Stirred Vessel

Emulsification stage requires the transport of unmixed fluids, partially nonmiscible via flow currents. The contact is achieved by mixing, which is promoted by a mechanical stirring system. The fluid motions of these incompressible phases can be classified by the ratio of the inertial forces to the viscous forces (19). This ratio is represented by a dimensionless number named Reynolds number, *Re*. In the case of a stirred reactor, the expression of the Reynolds number is

$$Re = \frac{\rho_C N D_A^2}{\mu_C} \tag{1}$$

where ρ_C is the density of the continuous phase, N the rotating velocity, D_A the diameter of the mechanical agitator, and μ_C the viscosity of the continuous liquid phase.

The fluid flow is considered to be laminar at low Reynolds number (<10). Viscous forces are the dominating inertial forces. The fluid flow could be represented by an elongational motion of the liquid phases. Other authors describe the mixing as a streamline flow, involving well-defined paths.

At higher Reynolds numbers ($>10^4$), the flow is considered to be turbulent. Turbulence is characterized by a fluctuating velocity, randomly with time, at each point in the flow field. It involves innumerable, different sizes of eddies or swirling flow motions. According to Kolmogorov's theory of local isotropic turbulence (1941), the energy spectrum of eddies depends on the rate of energy dissipation ε and the kinematic viscosity ν_C of the continuous liquid phase. The kinematic viscosity determines the rate at which the kinetic energy can be dissipated into heat. Large eddies decay to small eddies. The scale of large eddies, denoted macroscale by l_{macro}, can be approximated by the width of the fluid ejected by the agitator, i.e., the width of the stirrer blade (20). On the other hand, the scale of small eddies, denoted microscale by l_{micro}, is independent of agitator and tank size and is defined as:

$$l_{micro} = \left[\frac{\nu_C^3}{\varepsilon}\right]^{1/4} \tag{2}$$

where ε is the rate of dissipation of energy with time per unit mass of fluid, also denoted in the case of stirred reactor specific power input (W/kg); ν_C, the kinematic viscosity of the continuous liquid phase ($m^2\ s^{-1}$). l_{micro} is in the range of tens of μm.

Most of the highly turbulent mixing takes place in the stirrer region; fluid motion elsewhere serves primarily to bring fresh fluid into this region (12). Because of the rapid eddy motion in turbulent flow, shear stress is much more intense than that resulting from laminar flow. Thus, droplet breakup is much more efficient under turbulent flow conditions. For droplet sizes greater than the microscale of turbulence ($l_{micro} < d < l_{macro}$), the viscous forces can be neglected in comparison to inertial forces. The energy spectrum is independent of the kinematic viscosity and solely determined by the specific power input ε. On the contrary, for droplet sizes smaller than the microscale of turbulence ($d < l_{micro}$) the inertial forces are of the same magnitude as the viscous shear forces. Viscous effects may be important in the region of high shear stresses near the stirrer, where there is more probability for the occurrence of breakage events (18).

Technology of Tank and Stirrers

Although continuous flow mixing operations are employed to a limited extent in the pharmaceutical and cosmetic industries, the processing of emulsion often involves batch processing in a particular kind of tank or vessel, called the reactor. In the following parts, we are going to focus on batch operations in which mixing is performed by the use of dynamic mechanical mixers, such as turbines, propellers, or rotor/stator devices. Other equipments, such as static mixing devices, are discussed sparingly in the literature for cosmetic applications (5).

The power imparted by the stirrer to the emulsion develops an internal pumping flow and region of high shear. The pumping effect of the stirrer creates large flows that sweep the vessel from the top to bottom and from the center to outside, in an effort to homogenize the contents. When the flows go through regions of high shear, droplet breakups occur. So the stirrer must bring enough pumping effect because a multiple phase fluid has to be homogeneized, and sufficient shear stress is needed to obtain sufficient droplet breakup for the final size desired.

There are always discussions as to what particular type of stirrer is best for emulsification of nonviscous fluids. There are two considerations. Firstly, one must decide whether a radial or axial flow stirrer, and if some combination of stirrers is needed.

Radial flow stirrers have blades, which are parallel to the axis of the drive shaft. The first class of agitators is multiblades turbine (generally three, four, or six blades) with straight or curved flat blades (Fig. 1A and B). Some of them present a radial disk as part of their construction, such as the well-known Rushton turbine (Fig. 1C). Radial flow stirrers produce a high shear/high turbulence region. They do a good job producing a flow in the vicinity of the stirrer. The radial flow predominates over the axial flow, but this latter still exists along the stirrer shaft. There is a shortcoming in using the Rushton type or high shear rate stirrer (Fig. 1D), the reactor is divided into upper and lower sections. When a more intense droplet break up is necessary, the use of an ultrahigh shear device such as rotor–stator (Fig. 1E) is preferred. The pumping effect in the reactor is then almost negligible.

Axial flow stirrers include all stirrers in which the blades make an angle of less than 90° with the plane of rotation. One can find various classes of agitators: propellers (marine type or large thin blades propeller) (Fig. 2A and B), pitched blade

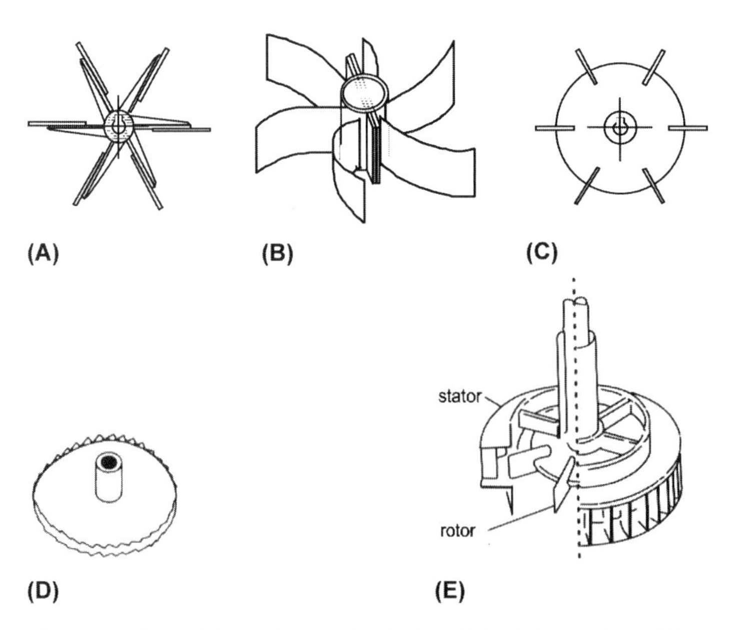

Figure 1 Radial flow stirrers: (**A**) straight flat blades turbine, (**B**) curved flat blades turbine, (**C**) Rushton turbine, (**D**) cutting, and (**E**) rotor-stator.

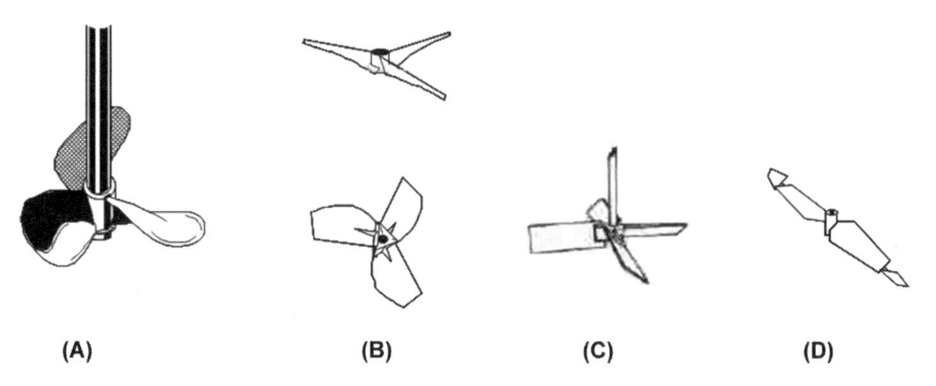

Figure 2 Axial flow stirrers: (**A**) marine propeller, (**B**) large thin blades propeller, (**C**) pitched blade turbine, and (**D**) double flux propeller.

turbine (four blades, mostly with an angle of 45°) (Fig. 2C), and double flux propeller (Fig. 2D). A large thin blades propeller allows for a constant profile of the velocity of the emulsion through the agitator with a minimum of shear. A four-bladed pitched blade turbine is very often used because it offers a good balance between shear and pumping flow. A double flux stirrer provides a good mixing when the level of viscosity reaches tens of Pa's. The opposed blade angles at the tip of the double flux stirrer provide an upward flow and good mixing, close to the wall. Axial flow stirrers produce a less radial flow and a more axial flow for similar power input. They, therefore, produced more vertical mixing when oriented in that way. They also do a better job mixing between the regions above and below the agitator.

Typical equipment for performing emulsification is based on a rounded bottomed vessel, fitted with a lid so that all air is excluded. The vessel is jacketed to promote heat transfer at the wall, for controlling temperature. The use of vertical baffles to provide a complete mixing is practically a requirement. In an unbaffled vessel with a stirrer rotating in the center, the centrifugal force acting on the fluid raises the fluid level at the wall and lowers the level at the shaft in the case of a nonviscous fluid. A vortex is created at the top surface of the liquid. Incorporation of gas in the liquid may occur. Moreover, a tangential flow dominates over an axial flow (Fig. 3A). In a baffled vessel, the presence of a vortex is avoided when baffles emerge from the liquid. Baffles convert a great amount of the tangential flow from the stirrer to an axial/radial flow (Fig. 3B). The absorption of gas in the liquid is then strongly minimized (19). The vessel is equipped with three to four vertical, equally spaced baffles, the number of baffles being different from the number of blades of the stirrer, to avoid mechanical resonance at some specific agitation speeds. Four baffles are almost always adequate (20). When a glass-lined reactor is used, only one baffle is present, which is often insufficient. A common baffle is one-tenth to one-twelfth of the tank diameter T. In a simple case, the level H of liquid is recommended to be made equal to T. To obtain uniform circulation and mixing, a single multipurpose stirrer such as a pitched blade turbine should be placed in the liquid at a height H_A from the vessel bottom, which is equal to one-third of the total height reached by the level of the liquid (Fig. 4). Few studies in the literature present the influence of the types of the stirrer on droplet mean size (21). The diameter of this stirrer D_A is recommended to be in the range of one-third ($T/3$) to one-half ($T/2$) of the vessel's diameter. The agitator is normally placed at a height H_A from the bottom, which is

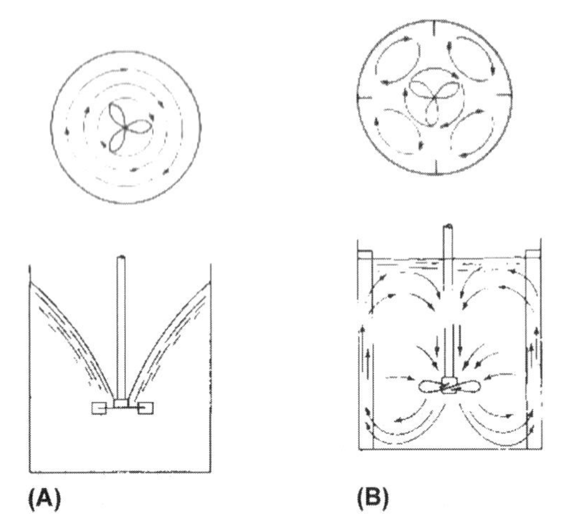

Figure 3 Hydrodynamics of (**A**) unbaffled tank [tangential flow (*bottom view*) vortex (*side view*)] and (**B**) baffled tank [radial and axial flow (*side and bottom view*) absence of vortex (*side view*)].

equal to one-third of the total height reached by the level of the emulsion in the reactor. When the performance of droplet breakup is not sufficient, a high shear rate stirrer or rotor–stator stirrer could be associated with a more axial flow stirrer (propeller or pitched blade turbine) (Fig. 4) (22). When the viscosity varies from few mPa's to several Pa's, some recommend the combining of pitch blade turbine with double flux stirrer in an unbaffled tank (23). Another solution proposed by suppliers of agitated

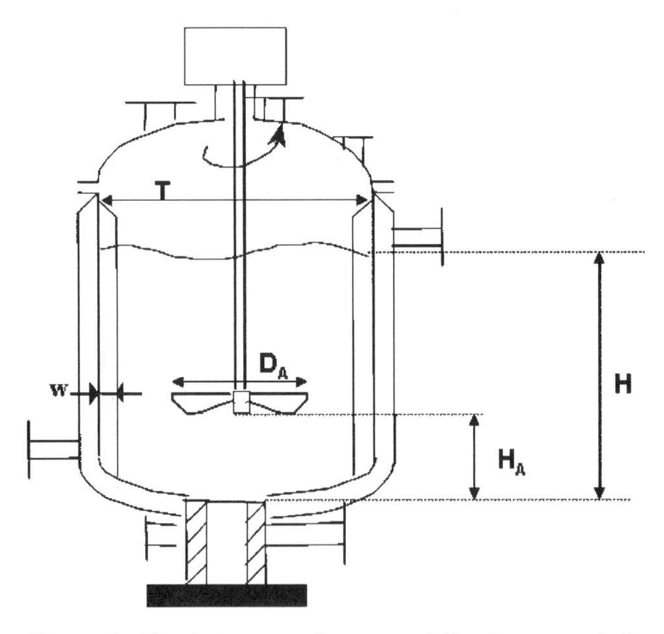

Figure 4 Classical reactor for an emulsification stage: jacketed vessel stirred with pitched blades turbine. *Abbreviations*: T, vessel diameter; w, width of baffles; D_A, agitator diameter; H_A, position of agitator; H, height of liquid.

reactors for emulsifying liquids of different viscosities is a combination on a same shaft of a pitched blade turbine rotating at high speed, creating a downward flow motion with an anchor rotating at a lower speed in an opposite direction creating an upward flow motion close to the wall.

Pumping Flow Rate, Power Consumption, and Velocity Head

Mechanical stirrers share a common functionality with pumps. The power imparted by the stirrer to the fluid is characterized in terms of the pumping flow rate and velocity head produced. The following expressions bring out the relation of the pumping flow rate, power consumption, and velocity heads imparted by the stirrer under turbulent flow conditions (19):

$$Q_P = N_Q N D_A^3 \tag{3}$$

$$P = N_P \rho_C N^3 D_A^5 \tag{4}$$

$$H = \frac{N_P}{N_Q} \frac{N^2 D_A^2}{g} = \frac{P}{\rho_C g Q_P} \tag{5}$$

where Q_P is the pumping flow rate ($m^3 s^{-1}$); N_Q the pumping number (discharge coefficient), dimensionless; P the power consumption (W); N_P the power number, dimensionless; H the velocity head (m); D_A the diameter of the agitator (m); N the agitator rotational velocity (s^{-1}); g the gravitational acceleration (ms^{-2}); ρ_C the density of the continuous liquid phase (kg m^{-3}). The pumping flow rate of a stirrer is the flow rate perpendicular to the stirrer discharge area. The velocity head of the pumping flow is the measure of the maximum force that the fluid can exert when its velocity is changed. With an increase of velocity head, shear rates and shear stresses are higher in the smallest eddies.

From Equation (5), we can assert that for a given power P, there is an inverse relationship between velocity head (or shear) and pumping flow rate. Depending on the objectives of mixing, one can balance between high shear and high pumping flow rate. For a same level of power P transmitted to the fluid, one can demonstrate by mixing Equations (3) to (6) such that $H \approx D_A^{-4/3}$ and $Q_P \approx D_A^{4/3}$.

If a higher level of shear (or velocity head H) is desired at the same power consumption P, a smaller stirrer diameter D must be used at a higher rotational speed N. Consequently, there is a decrease in the pumping flow rate. Such a situation is useful for: (i) reduction of concentration differences in the stirrer pumping flow (rapid mixing), (ii) production gas or liquid droplets presenting large interfacial area, (iii) solids desagglomeration, (iv) promotion of mass transfer between phases (19). Such a configuration has to be favored with the objective of emulsifying two immiscible liquid phases.

On the other hand, if a higher level of pumping flow rate is desired at the same power consumption P, a greater stirrer diameter D must be used at a lower rotational speed N. Shear rates are minimized at the same time. Such a situation is useful for: (i) promotion of heat transfer, (ii) reduction of concentration and temperature differences in all parts of the vessel, (iii) suspension of particles (crystals or droplets) (19). Such a configuration has to be favored with the objective of maintaing the droplet size distribution of emulsion, which has already been stabilized with surfactants.

The pumping flow rate Q_P has been measured for several types of stirrers, and their respective pumping numbers have been calculated. N_Q ranges from 0.4 to 0.5

for large thin blade propeller, and from 0.7 to 2.9 for turbines depending on the number of blades, blade-height-to-stirrer-diameter ratio, and stirrer-to-vessel-diameter ratio. The data for numerous stirrers have been reviewed (24–26).

For the calculation of the power consumption P, it is necessary to know the power number N_P. This number which is representative of a given stirrer/reactor configuration is calculated by inverting the expression (4):

$$N_P = \frac{P}{\rho_C N^3 D_A^5} \tag{6}$$

Although fluid density, viscosity, mixing vessel diameter, and stirrer rotational speed are often viewed by formulators as independent variables, their interdependency is quite evident when looking into the relationship between the Reynolds number and the power number N_P. This relationship is expressed in the log graph (Fig. 5).

We need to remember that flow motion cloud be classified using Reynolds numbers in three regimes: laminar, transitional, and turbulent.

- In the laminar regime ($Re < 10$), the power number N_P depends on Reynolds number in an inverse relationship: $N_P = K/Re$ (K = constant function of the combination stirrer/reactor). The power input can be expressed as $P = K\mu N^2 D^3$. As shear stress is proportional to rotational speed, shear stress can be increased at the same power consumption P, by increasing N in proportion to the decrease in $D_A^{-3/2}$ which is the stirrer diameter.
- In the transitional regimes, the power number N_P depends on Reynolds number in a nonlinear way: any general relationship could be discussed.
- In the turbulent regime ($Re > 10^4$), the power number N_P is constant. The comparison between radial flow and axial flow stirrers is straightforward. Turbines present higher power numbers [$1 < N_P < 10$] than propellers ($N_P < 1$). Enhancing turbulence and thus the droplet breakup require a higher power input P. That is why Rushton turbine, flat blade turbine, and pitched blade turbine are the most commonly used stirrers with nonviscous liquids for an emulsification stage.

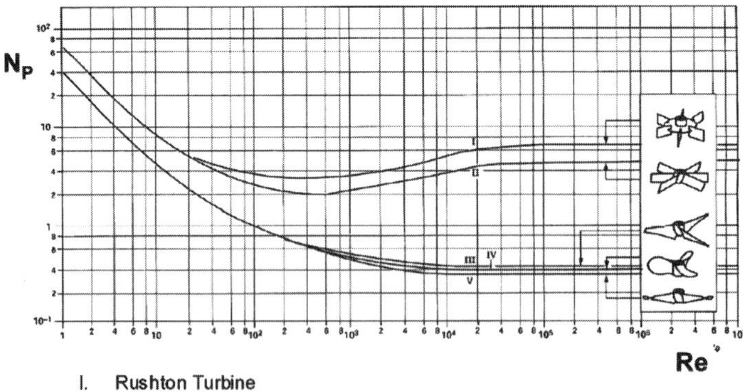

I. Rushton Turbine
II. Six blades turbine
III. Pitch blades tubine
IV. Marine Propeller
V. Double flux propeller

Figure 5 Relationships between power number, N_P, and Reynolds number, Re, for different stirrers. *Source*: From Ref. 26.

It is now possible to calculate the specific power input ε (W kg^{-1}) by dividing the power consumption P by the total volume of the fluids V_T (m^3)

$$\varepsilon = \frac{P}{V_T}$$

Two situations have to be considered depending on the flow regimes:

1. In a laminar flow regime,

$$\varepsilon = \frac{K\mu N^2 D_A^3}{V_T} \tag{7}$$

where K is a constant.

2. In a turbulent flow regime,

$$\varepsilon = \frac{N_P \rho N^3 D_A^5}{V_T} \tag{8}$$

where N_P is a constant.

For high values of dispersed phase volume fraction, the overall viscosity of the system increases, and therefore the specific power input decreases. This phenomenon has been explained as energy consumption from the droplets convection, interface oscillation, breakage, etc. (27). A correction is proposed as follows:

$$\frac{\varepsilon_E}{\varepsilon_C} = \left[\frac{\nu_C}{\nu_E}\right]^3 \tag{9}$$

where ε_C and ε_E are the specific power input for the pure continuous phase and the emulsion, respectively, ν_C and ν_E are their corresponding kinematic viscosity.

This specific power input is in fact a mean value. It has been widely recognized in the literature that the value is varying significantly according to the position within the agitated vessel (28). Locally, low levels traduce a quiescent zone, and high levels traduce a turbulence region. As a result, considerably different breakage and coalescence rates may be encountered in the stirrer and the circulation zones. Droplets breakup dominate in the high turbulence stirrer zone and droplets coalescence in the more quiescent recirculation region (27).

Droplet Size Distribution

In half of the past century, there have been numerous academic works in the field of chemical engineering, in studying the phenomena of dispersing and coalescing of droplets (29–32). These fundamental studies are still relevant, and for a good review of them we recommend Maggioris et al. (18).

The first difficulty in characterizing these phenomena arises from the mutual interaction between droplets breakup and coalescence. A second difficulty comes from the lack of experimental data in the field of emulsification in stirred vessel owing to crucial experimental problems (19). This is gradually being overcome by the use of in situ measurements (33).

Originally, the first approach developed was based on dimensional analysis. One of the objectives was to predict mean droplet size, frequently characterized by its Sauter mean diameter.

Later on, firstly, with the emergence of the population balance approach in the 1960s that was widely applied to systems containing a continuous phase and at least one dispersed phase and secondly with the continuous development of numerical methods for solving integropartial differential equations, a new methodology has become available. Combined with models of coalescence and breakup that are more and more precise, it is now possible to predict the droplet size distribution, and to extract the mean Sauter diameter. For growth phenomena characterized by the collision of discrete entities such as aggregation, agglomeration, and coalescence, and for size reduction phenomena such as breakage of solid and droplet breakup, many examples in the literature of modelling may be found (34). In the field of droplet coalescence and breakup, the work of Alvarez et al. (35), Spicer et al. (36), Semiao et al. (37), Serra et al. (38), White et al. (39), and Hounslow and Ni (40) are particularly relevant. This approach is now using a computational fluid dynamics tool to design the main hydrodynamical features of liquid flows with more precision (41,42). To our knowledge, nothing in the literature at the moment has been devoted to the modelling of microcapsules from a process including an emulsification step prior to the encapsulation. It is certain that studies on suspension polymerization processes would be interesting to examine, because the final particle size is the result of the initial drop size distribution of the monomer/water dispersion (18). Nevertheless, this second approach is much more complex to operate.

The approach based on dimensional analysis has given nice results in the field of emulsification. It is still used for the prediction of mean size of microcapsules (4,17). This practical approach is going to be presented in the following paragraph.

Dimensional Analysis Applied to Emulsion

Dimensional analysis is concerned with the nature of the relationship among various quantities involved in a physical problem. As an intermediate approach between formal mathematics and empiricism, it offers the pharmaceutical and chemical engineers an opportunity to generalize from experience and apply knowledge to a new situation (43,44). Dimensional analysis is based on the fact that if a theoretical equation exists among the variables affecting a physical process, such as emulsification, that equation must be dimensionally homogeneous. Thus, many factors can be grouped in an equation into groups with a smaller number of dimensionless variables (44).

The Weber number, We, is a dimensionless number that is widely used in the case of emulsification. It takes into account the balance between breakup and coalescence of the droplets. It represents the ratio of the driving force causing partial disruption to the resistance caused by the interfacial tension. Hinze (29) gave a general expression:

$$We = \frac{\tau d}{\sigma} \tag{10}$$

where τ is the stress at the surface of the droplet, d the diameter of the droplet, and σ the interfacial energy at the droplet interface.

Increased Weber number is associated with a greater tendency for droplet deformation to occur at higher shear.

Depending on the flow regime of the continuous liquid phase around the droplet, many expressions of the Weber number arise. In a stirred vessel, in a turbulent flow regime when the inertial forces are higher than the viscous effects ($Re > 10^4$), the Weber number is expressed as:

$$We = \frac{\rho_C N^2 D_A^3}{\sigma} \tag{11}$$

where ρ_C is the density of the continuous liquid phase, N and D_A are the rotational velocity and the diameter of the agitator, respectively, and σ the interfacial energy at the droplet interface.

Correlations Readily Used

Developing correlations of droplet mean size presents two stakes:

1. To predict a mean size by just considering the physicochemical data of the emulsion and characteristics of the stirring system.
2. From experimental data obtained in specific operating conditions, estimate the evolution of the mean droplet size according to the evolution of the operating conditions.

The second stake is very attractive, because it is possible to evaluate the evolution of the droplet mean size when a change of operating condition is engaged or when a scale-up of the equipment is performed.

Owing to the competition between breakup and coalescence, the correlations of droplet mean size could be classified according to the phenomenon that prevails.

Breakup Is the Dominating Phenomenon. The classical theory for the maximum stable droplet size d_{max} in a dilute turbulent dispersion, because of Kolmogorov's (45) and Hinze's (29) studies, are based on the theory of local isotropic turbulence (46). From this, d_{max} could be related to the maximum local energy dissipation rate ε_{max} in a stirred vessel by the relationship that follows (21,47):

$$d_{max} = K_1 \varepsilon_{max}^{-0.4} \sigma^{-0.6} \rho_C^{-0.6} \tag{12}$$

where K_1 is a constant. For a droplet diameter higher than d_{max}, the droplet does not resist to further breakup.

Equation (12) has been shown to apply to experimental data by many workers, such as Calabrese et al. (30). They assumed that ε_{max} was proportional to the mean specific power input ε. It was also assumed that at equilibrium, d_{max} is proportional to the Sauter mean diameter d_{32}. The equation (12) could be rearranged into the form of the stirrer Weber number, We.

$$\frac{d_{32}}{D_A} = K_2 We^{-0.6} \tag{13}$$

where K_1 is a constant; K_1 and K_2 are dimensionless constant, depending on the stirrer type, especially its power number N_P.

Two points have to be underlined:

- The exponent –0.6 on the Weber number in the droplet size correlation comes from the theory used for establishing the Equation (12). Within the last decade, experimental results in the literature have shown that the exponent can drift from –0.6 to –0.93 at long agitation times.

- In larger stirrer tanks, scale-up considering a constant power per unit volume ε (section "Emulsification Process in Stirred Reactor") produces faster breakup rates, leading to smaller droplets.

These deviations from the initial theory are now fully explained using an extended theory, which includes turbulent intermittency (fluctuations of ε at a point about its mean value at this point.) For more information, see Baldyga et al. (47,48).

Balance Between Coalescence and Breakup. In the previous development, the dispersed phase volume fraction was considered to be sufficiently small that the rate of coalescence was negligible compared to breakup, so that dispersed phase concentration was not included in the equation. With increasing dispersed phase concentration, coalescence occurs especially in the quiescent region of the vessel away from the stirrer.

The first modification appears with Calderbank (49), who proposed to empirically modify Equation (13), by considering the volume fraction of the dispersed phase and a ratio of viscosities of the two phases:

$$\frac{d_{32}}{D_A} = K_3 \phi We^{-0.6} \left[\frac{\mu_d}{\mu_c}\right]^{0.25} \tag{14}$$

with K_3 a constant, μ_d and μ_c are the viscosities of dispersed and continuous phases, respectively.

More recently, authors proposed another modification of Equation (13) [see reviews by Pacek et al. (21), Davies (50), and Peters (51)]:

$$\frac{d_{32}}{D_A} = K_2[1 + \phi K_4] We^{-0.6} \tag{15}$$

where K_4 is a constant that is a measure of the tendency to coalesce.

Coalescence is the Dominating Phenomenon. The theory of local isotropic turbulence is no more valid for the prediction of the droplets Sauter mean diameter, which can be estimated from empirical correlation such as:

$$\frac{d_{32}}{D_A} = K_5[\sigma D]^{-0.375} We^{-0.375} \tag{16}$$

where K_5 is a constant (45).

Costaz (52) presents a review of the main correlations found in the literature in the two first cases (domination of breakup or equilibrium breakup/coalescence). Most of the time, the stirrer used is a Rushton turbine, and the droplets sizes are situated in the micrometric size range (from a few μm to hundreds of μm). Only a few studies have investigated the prediction of nanometric droplets size with these correlations (53).

Specific Correlation for Rotor-Stator Stirrer (52). Already, for a decade, the use of a high shear device such as rotor–stator stirrer to attain a low value in the micrometric range (a few microns) or even the nanometric size range is widespread. So it was interesting to set up a correlation and to validate that a correlation based on Equation (13) could represent the experimental data.

Experimental diluted dispersions of organic phase in aqueous phase were performed in a thermostated vessel, with a capacity of half litre, that was agitated with a rotor–stator device. The ratio of the stirrer diameter to the vessel diameter was 0.5. Similar to the procedure mentioned in the literature, several organic liquids were selected to cover a range from 700 to 900 kg m^{-3} for the density, from 0.5×10^{-3}

to 0.5×10^{-2} Pa s for the viscosity and from 4.5×10^{-3} to 51×10^{-3} Nm^{-1} for the interfacial energy. The stirring rate of the rotor-stator varies from 3600 to 5300 rpm, leading to Reynolds number values between 1.4×10^4 and 2.0×10^4. The flow regime was turbulent. After one or two hours of agitation so as to be sure that the droplet size dispersion was stable, the droplet size distribution was measured by laser diffraction (circulation of the emulsion via an external loop through the measurement cell). Each experimental condition was performed twice. Measurements were accepted when the reproducibility on the mean Sauter diameter d_{32} did not exceed 5% relatively. Otherwise the experimental results were rejected. The mean Sauter diameter d_{32} varies from 2 to 10 μm.

The results were successfully expressed following the Equation (13) with a correction of viscosity:

$$\frac{d_{32}}{D_A} = 0.02\, We^{-0.6} \left[\frac{\mu_d}{\mu_c}\right]^{0.5} \tag{17}$$

Modeling the Effect of the Stirring Rate on the Mean Size of Micro- and Nanocapsules

By expressing the Weber number and reducing the Equation (12), it is possible to express the evolution of the droplet mean Sauter diameter with the stirring rate as follows

$$d_{\text{mean drop}} = K_3 N^{-1.2} \tag{18}$$

when K_3 is a constant including K_1, stirrer diameter, the density of the continuous liquid phase, and the interfacial energy at the droplet interface.

Assuming that the microcapsule size is proportional to the droplet size when the manufacture of the capsules requires an emulsification stage, this relationship was successfully applied by different authors (14,17). In Figure 6 was presented

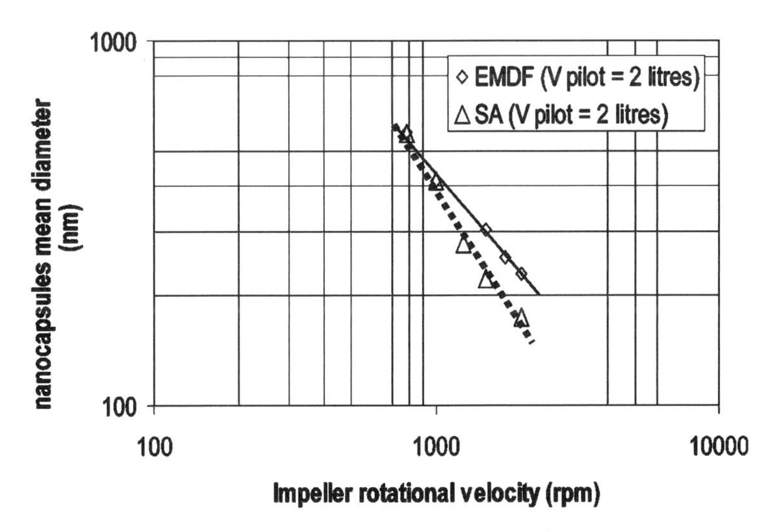

Figure 6 Evolution of the nanocapsules mean diameter with stirrer rotational velocity. Solid line represents regression line of EMDF process data; dotted line represents regression line of SA process data. *Abbreviations*: EMDF, emulsification/diffusion process; SA, salting out process. *Source*: From Ref. 17.

the evolution of mean diameter of NC with the stirring rate obtained by an EMDF process and salting out process (SA). By regression, the calculated slopes are equal to −0.93 (EMDF process) and −1.29 (SA process), respectively. These values are quite close to the theoretical −1.2, which indicates that this model can adequately describe the influence of the stirrer rotational velocity on the NC size. We underline that such a relationship is useful while breakup and coalescence are balanced. In the case when coalescence prevails over breakup, the exponent of the relationship is no longer valid (18).

SCALE-UP APPROACH: THE CASE OF EMULSIFICATION

In this part, the scale-up procedure used to increase the volume of the equipment from the laboratory size to the plant size is presented. We would like to underline that, for a decade now, an alternative approach of scale-up in some microencapsulation processes has been available. It consists of dividing the input volume into several smaller volumes to apply the process to each small volume. That is, for instance, the case in using an external flow through the homogenization process: thanks to a peristaltic pump, an emulsion located in a tank circulates in an external loop having a rotor-stator device performing droplet breakup (54). Every moment, a small volume of emulsion is dispersed, and the total volume is homogeneized gradually. The duration of external recirculation increases with the volume of emulsion. The problem of scale-up is then eliminated. This alternative is suitable if the droplet size distribution remains stable after the flow goes through the homogenizer.

Principles of Similarity

As soon as the emulsification processes have been characterized in a laboratory scale, with a dimensional analysis procedure, one can then apply the principles of similarities: Process similarities are achieved between two processes when they accomplish the same process objectives by the same mechanisms and produce the same product, to the required specifications (12). Originally, the importance of four similarities had been stressed. The four similarities being: (i) geometric, (ii) dynamic, (iii) thermal, and (iv) chemical. Each of these similarities presupposes the attainment of other similarities.

Two systems are geometrically similar when the ratio of the linear dimensions of the small-scale vessel and scale-up vessel is constant. In geometric similarity the fluid motion in a vessel is similar for both scales (23). The dynamic similarity is based on the conservation of the forces ratio defined by dimensional analysis performed on the process. In other words, the ratios of forces moving masses (e.g., pressure, gravitational, and centrifugal) between corresponding points in the two scales are equal. It is, however, clear that no more than one ratio can be kept constant. Consequently, all other criteria are violated. That is why it is important to recognize which criterion must be satisfied at each scale at the lower scale itself, before scaling up the procedure. Thermal similarity must be considered when heat transfer in the process is one of the main phenomena. Chemical similarity is concerned with the variation in chemical composition, from point to point, as a function of time.

It is too restricting to keep four similarities in the same time. In practice, for the case of emulsification, only geometric and dynamic similarities are taken into account.

Geometric Similarity

Geometric similarity involves proportional scale-up of geometric parameters of the vessel.

As already described in section "Emulsification Process in Stirred Reactor," the ratios of dimensions of vessels to be considered are:

- ratio of the height reached by liquid to the diameter of the vessel, i.e., H/T;
- ratio of agitator diameter to vessel diameter, i.e., D/T;
- position of the agitator to vessel diameter ratio, i.e., H_a/T;
- ratio of the width of the baffles to vessel diameter w/T.

These ratios are constant for both the small-scale equipment (T_1, H_1, D_{A_1}, H_{A_1}, and w_1) and the large-scale equipment (T_2, H_2, D_{A_2}, H_{A_2}, and w_2):

$$\frac{H_1}{T_1} = \frac{H_2}{T_2}; \quad \frac{D_{A_1}}{T_1} = \frac{D_{A_2}}{T_2}; \quad \frac{H_{A_1}}{T_1} = \frac{H_{A_2}}{T_2}; \quad \frac{w_1}{T_1} = \frac{w_2}{T_2} \tag{19}$$

The procedure is to express the total volume occupied by the liquids in the vessel having a sphere-shaped bottom as:

$$V_T = \prod \frac{T^2}{4} H \propto T^3 \tag{20}$$

When the volume of the larger scale V_{T_2} is defined, the dimensions (T_2, H_2, D_{A_2}, H_{A_2}, and w_2) are easily calculated from the Equations (19) and (20).

Knowing the volume V_{T_1} of the vessel at small scale and the volume V_{T_2} of the larger vessel, it is then possible to define the geometric scaling factor k as:

$$k = \frac{T_2}{T_1} = \left[\frac{V_{T_2}}{V_{T_1}}\right]^{(1/3)} \tag{21}$$

The authors would like to emphasize two points:

1. This first similarity involves an important constraint: it is not possible any more to use any vessel in the laboratory without considering the possibility of its application at a larger scale. For example, many formulators use beakers, having a capacity of hundreds of millilitres, for optimizing the emulsion used in the production of microcapsules. The risk involved here is that the flow motion at a larger scale would be too different from the lab-scale, thus it might lead to an unsuccessful scale-up. In such cases, optimization should be done again , which would involve a loss of time and possible loss of large amounts of materials. The easier way would be to use a representative vessel, having a capacity of half a liter, in the lab, scaled down from the larger one in the plant.
2. Many authors consider that enhancing the volume of the vessel to ten times that of the lab vessel is an important method of scale-up (4). In fact, in terms of geometric scaling factor ($k = 10^{1/3} = 2.15$) it is not such an important scale transfer. We can consider that the scale-up starts to be significant when k is equal to or above 10 (volume enhances by a factor of 1000).

Dynamic Similarity

The dynamic similarity involves conservation of forces ratio. For the emulsification process, following the dimensional analysis, one can expect to keep the The Reynolds number or Weber number for this stage. The Reynolds number is useful for distinguishing flow regimes and for characterizing mass transfer. The recommendation is to be in the same flow regime at both scales: laminar with viscous fluids and conversely turbulent with nonviscous liquids. If the flow regime changes from the smaller scale to the larger one, the application of the criteria becomes unsuitable. For predicting the mean droplets size, Weber number could be quite efficient. Nevertheless, it is not used very often in practice.

In the case of mixing in a stirred reactor, the most widely used scale-up criterion is the specific power input ε. It also denotes the rate of dissipation of energy per unit volume of fluid. In practice, applying this criterion has given accurate results for chemical reactions. The reason for this is that the specific power input plays a major role in the classical theory of turbulence, this latter being useful for the mixing of reactants in a vessel (47). However, it is also selected in the case of divided systems because ε influences the maximum stable drop size when dispersing immiscible liquids (see section "Correlations Readily Used"), as well as the coagulation and granulation of finely divided particles (23,47,50,55,56).

Let us consider the first criterion: the specific power input ε. According to the flow regimes, two situations have to be considered:

From the Equation (20), the total volume V_T is proportional to the diameter of the vessel T. Let us remember that the diameter of the stirrer D_A had been chosen in proportion to the diameter of the vessel. Consequently, V_T is also proportional to D_A ($V_T \propto D_A^3$) nature of the fluid is unchanged from one scale to the other. The Equations (7) and (8) can then be simplified as:

1. In a laminar flow regime,

$$\varepsilon \propto \frac{N^2 D_A^3}{D_A^3} \propto N^2 \tag{22}$$

2. In a turbulent flow regime,

$$\varepsilon \propto \frac{N^3 D_A^5}{D_A^3} \propto N^3 D_A^2 \tag{23}$$

The specific power input will be constant for both scales ($\varepsilon_1 = \varepsilon_2$), when the stirrer velocity N_2 at larger scale is adapted from the stirrer velocity N_1 used at the smaller scale in the following way:

1. In a laminar flow regime,

$$N_2 = N_1 \tag{24}$$

2. In a turbulent flow regime,

$$N_2 = N_1 \left[\frac{D_{A_1}}{D_{A_2}}\right]^{(2/3)} \tag{25}$$

One should note that the linear peripheral velocity V_A (m s^{-1}) of the stirrer is another criterion proposed for the dispersion of liquid without coalescence, which has been mentioned also in the literature. This velocity is directly linked with the shear forces responsible for the droplets breakup. Its expression is independent from the flow regime:

$$V_A = \pi N D_A$$

The simplification of this equation gives:

$$V_A \propto ND_A \tag{26}$$

The linear peripheral velocity will be constant for both scales ($V_{A_1} = V_{A_2}$) when the stirrer velocity N_2 at larger scale is equal to:

$$N_2 = N_1 \frac{D_{A_1}}{D_{A_2}} \tag{27}$$

The two criteria (ε and V_A) cannot be kept constant at the same time. In the case of emulsification, the choice between ε and V_A is linked to the type of stirrers chosen for producing the droplets dispersion. In a laminar flow regime, a high shear stirrer is useless, because the shear stresses created will be located only in the close vicinity of the blades. A specific axial flow stirrer for viscous fluid is used. The constant criterion to consider is the specific power input ε. In a turbulent regime with nonviscous liquids, two types of stirrers may be used: a classical radial flow stirrer (such as pitched blade turbine) and a high shear stirrer (such as a rotor–stator). The concept of a constant specific power input ε can be applied to the stirrer used alone. On the other hand, it is recommended to keep a constant linear peripheral velocity V_A when a high shear stirrer is employed. When a narrow size distribution in the nanometric size range is needed, a combination of the two types of stirrers is compulsory.

APPLICATIONS—EXAMPLES

Example 1: Microencapsulation by Interfacial Polycondensation

This method involves the formation of a polymeric shell around the phase to be encapsulated. The latter is generally an emulsion, the drug being dissolved or dispersed in the droplets. The drug can also be the only constituent of the droplets. It is a widely used method for production of various microcapsules applied to sundry commercial products (agriculture, adhesives, pharmaceuticals, and cells) (57,58). Currently, many commercially available encapsulated agrochemical formulations are prepared by interfacial polycondensation, and a large variety of biological materials can also be encapsulated by this process (59–62).

Interfacial polycondensation involves two monomers, particularly reactive when present together, reacting at the interface of two immiscible phases and leading to the formation of the polymeric membrane around the dispersed inner phase. The microcapsules are usually obtained from a water-in-oil or oil-in-water emulsion, where each phase contains one monomer, and the solubility and the partition coefficients of the two monomers control the localization of the reaction (at one side or the other of the interface).

The most common example of this method is polyamides microcapsules, formed by the reaction between a multifunctional amine and a multifunctional acid chloride. Polyureas and polyesters have also been developed (63–65).

The interfacial polycondensation is a two-step process. An emulsion is first prepared with the droplets containing one of the two monomers. The drug is dissolved or dispersed and if necessary a surfactant added. The external phase is a nonsolvent one, containing a surfactant to stabilize the droplets. After achieving the right droplet size distribution, the second monomer dissolved in a small quantity of the external phase is added. Owing to the high functionality of the monomers, the polycondensation reaction is very rapid and a first membrane is formed at the interface (66). Depending on the solubility of the oligomers in the droplets, the reaction can take place in the droplets in some cases, leading to the formation of microspheres instead of microcapsules. The reaction is also controlled by the partition coefficient of the monomers between the two phases, as their concentration level at the interface must be enough for the reaction to proceed. The mass transfer of one monomer to the reaction place should exceed the rate of reaction, as this reaction tends to remove the monomers. The mass transfer limitations increase when the reaction progresses, with the increase of the wall thickness.

The first emulsification step is a major part of this process, as it will determine the final size distribution of the particles. The physicochemical properties of the two phases and the choice of surfactant are obviously the key parameters in determining the equilibrium droplets size. But only the association between these properties and the process parameters can allow a real control of the size distribution. The process of emulsification can be described in terms of energy input in the system, including the way of mixing (mechanical stirring, pressure homogenisation, ultrasonic, etc.) and the duration of this step.

The process parameters involved in the second step, i.e., the polycondensation reaction, are the hydrodynamic conditions and the rate of monomer addition. The monomer added in the external phase must in fact be homogeneously distributed in all the external phases as quickly as possible. This is the only way to ensure that the reaction can take place at the interface of all the droplets simultaneously and in the same conditions, leading to similar properties for the particles. The rule of agitation is also to favor the diffusion of the monomer in the outer phase through the solution and its renewal at the interface to allow the reaction progression.

It has been found that stirring is a critical variable that can affect the molecular weight of various types of polymers (67). As stirring influences the mass transfer in the reactor, this mass transfer can be considered as a critical parameter that should be kept constant during scale-up.

Microcapsules Production

This part presents an example of scale-up study applied to the encapsulation of a chemical reagent in microcapsules, with a mean size around 30 μm. The reagent to be encapsulated is water soluble, so the process is based on an inverse emulsion, with an oily external phase. The objective was to produce microcapsules with a size less than 100 μm and a relatively narrow size distribution. These capsules will then be mixed in a complete formulation; so, the polymeric membrane must be mechanically rigid to resist to the mixing by kneading. The capsules must also be stable during storage in the final formulation.

The formulation optimisation was done previously at a lab-scale. The nature of the polymer, the concentrations of the monomers, and the ratio of one monomer to the other as well as the nature and concentrations of the surfactant were fixed during this study. The volume fractions of the emulsion (ratio dispersed/continuous phases) and the volume of external phase used to introduce the second monomer were also determined. After a first study, a mixed polymer polyamide–polyurea was chosen to optimise the stability and the resistance of the capsules. This means that a reaction between an amine, an acid chloride, and an isocyanate is involved. The aqueous phase used to encapsulate contains the reagent, the amine, and sodium carbonate to neutralize the chloride acid formed during the polycondensation reaction. It is first emulsified in a part of the oil containing a lipophilic surfactant. Then the second part of the oil in which the two lipophilic monomers have been previously dissolved is added to the emulsion to initiate the polycondensation reaction.

The scale-up study begins with the conception of a production system using classical equipment of chemical engineering, based on the standard geometrical methods, which can be extrapolated later by geometrical similitude. The system used for this part of the study is represented in Figure 7.

It is mainly constituted of a five-liter stirred vessel (reactor 1) in which the emulsion and the polycondensation reaction are done. The reactor, equipped with four baffles, is stirred by a propeller (Fig. 2B). The dimensions of the reactor and the stirrer are listed in Table 1. The other reactors that are used for the preparation of the aqueous phase and the dissolution of the lipophilic monomers in a part of the oil are also described. The physicochemical properties and the volumes of the two phases are given in Table 2 .

The process can be described as follows:

- Introduction of the first part of the oil in to the reactor. This quantity corresponds to about 2/3 of the total quantity of oil engaged.
- Dissolution of the surfactant under agitation at 300 rpm in 15 minutes
- Emulsification: duration 15 minutes, stirring rate variable between 300 and 500 rpm. The rate of addition of the aqueous phase is variable, but the total duration of the emulsification is maintained constant.
- Introduction of the second part of the oily phase containing the second monomer: the duration of this step is variable (5–17 minutes), the stirring rate is maintained at the same as the previous level.
- Polycondensation: The stirring rate is reduced to 300 or 400 rpm, the polycondensation reaction continues for six hours.

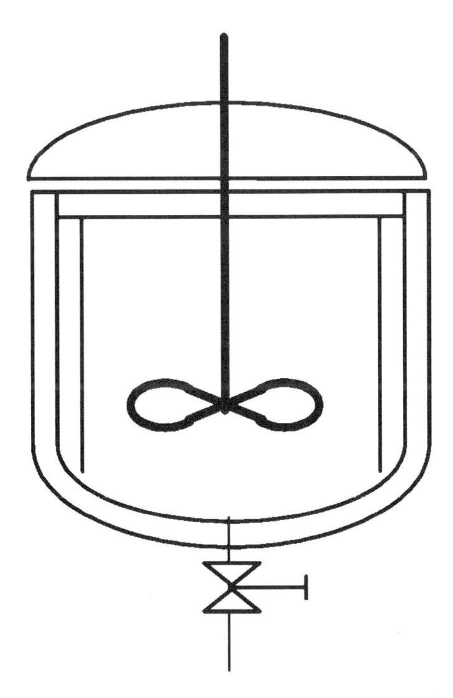

Figure 7 Reactor used for the production of microcapsules (5L).

Table 1 Main Characteristics of the Laboratory and Pilot Agitation Systems

Characteristics	Lab scale	Pilot scale
Reactor 1: emulsification and reaction		
Reactor capacity (L)	5	100
Reactor internal diameter (cm)	18	50
Volume of the emulsion (L)	2.9	80
Stirrer diameter (cm)	9	21
Number of baffles	4	4
Stirring rate during emulsification (rpm)	300–500	210–190
Stirring rate during polycondensation (rpm)	300–400	120–190
Reactor 2: preparation of the aqueous phase		
Reactor capacity (L)	0.2 (beaker)	5
Reactor internal diameter (cm)	–	18
Volume of the solution (L)	0.18	5
Stirrer diameter (cm)	Magnetic	9
Number of baffles	–	4
Stirring rate during dissolution (rpm)	–	300
Reactor 3: preparation of the lipophilic monomers solution		
Reactor capacity (L)	1	50
Reactor internal diameter (cm)	11	27
Volume of the solution (L)	0.9	25
Stirrer diameter (cm)	5	13
Number of baffles	4	4
Stirring rate during dissolution (rpm)	400	300

The size distribution is controlled by microscopic observations followed by images analysis during the emulsification and the addition of the second monomer. Then the progression of the polycondensation reaction is evaluated by sample microscopic observations. A test of mechanical resistance of the capsules has also been developed to control the evolution of the membrane solidity.

Results Obtained in the 5-L Reactor

Influence of the Stirring Rate. As expected, the mean size distribution of the emulsion decreases with the increase in the stirring rate. At the same time, the size distribution is narrowed. The results obtained with a stirring rate of 300, 350, 400, and 500 rpm are shown in Figure 8. It can also be observed that the size distribution presents two populations, one population of small droplets of around 10 μm and

Table 2 Properties of the Dispersion

	Aqueous phase	Organic phase	Dispersion
Specific gravity (kg m^{-3})	1000	860	880
Viscosity (Pa s) at 25°C	10^{-3}	140×10^{-3}	–
Mass fraction (%)	6.3	62 + 31.7[a]	100

[a]The first part of oil used to form the emulsion corresponds to 62% of the total mass; the second part (31.7%) corresponds to the quantity used to dissolve the lipophilic monomer, introduced after emulsification.

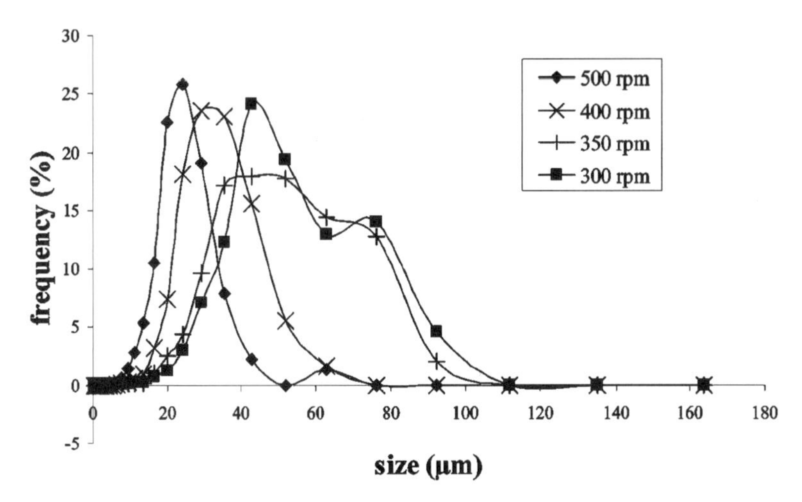

Figure 8 Emulsion size distributions in 5-L reactor.

bigger droplets that are about 50 µm, when the stirring rate is less than 400 rpm. However, after analysis of the others microcapsules properties, it appears that those obtained at low stirring rate present satisfying mechanical resistance. The objective for a process is obviously to minimize the energy dispensed to the system, so the minimum stirring rate will be used for the scale-up calculations.

Influence of the Addition Rate of Aqueous Phase. The addition rate has been found to have no influence on the emulsion size distribution, provided that the total duration of 15 minutes is respected for the emulsification step. If the aqueous phase is added very quickly, it has been observed that the emulsion is very coarse at first. After 15 minutes, the mean size decreased and the droplet size distribution narrowed. This result is in agreement with the representation of droplets generation by an equilibrium between coalescence and breakup forces, leading to a stable droplet size (see section "Emulsification Processes"). In the case of rapid addition of the aqueous phase, the mechanical stirring does not provide enough turbulence to achieve droplets breakup instantaneously. So, the emulsion is coarse and the distribution is large. Hence, at least 15 minutes of stirring is necessary to achieve the maximum break up of the droplets in all the reactor volume. The images in Figure 9 show the droplets after five minutes of stirring and at the end of the emulsification stage.

Influence of the Addition Rate of the Second Monomers. The addition rate of the second part of the oil containing the two lipophilic monomers has an influence neither on the size distribution nor on the capsules properties.

Scale-Up

The scale-up calculations are based on the conservation of the specific power input brought by the stirrer to the suspension, as presented in section "Scale-up Approach: The case of Emulsification." The calculation of the Reynolds number shows that the flow regime is intermediate between laminar and turbulent. We consider the two correlations (Eqs. 24 and 25) and we apply a mean value for the calculation of the stirring rate for the 100-L reactor:

Equation (24) leads to $N_2 = N_1 = 300$ rpm.

Equation (25) leads to $N_2 = N_1[D_1/D_2]^{2/3} = 170$ rpm.

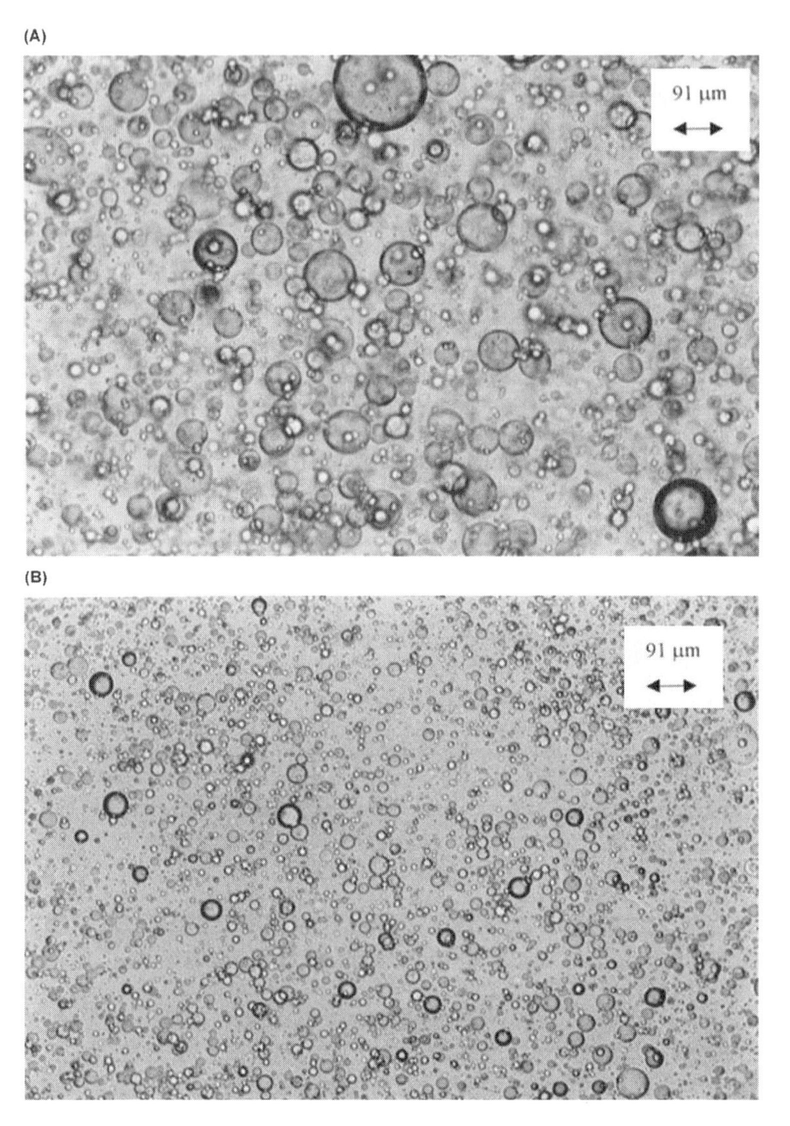

Figure 9 Images of the emulsion after (**A**) 5 minutes and (**B**) 15 minutes.

The reactor used for the validation of these assumptions is geometrically similar to the small one, with a volume of 80 L. The stirrer design (propeller) (Fig. 2B) is also similar and the ratio of the stirrer diameter to the reactor diameter is conserved. The calculations lead to a stirring rate of 220 rpm during the emulsification and 190 rpm during polymerization. The agitation was maintained for 30 minutes to achieve the equilibrium size distribution. The first results show that the particles size is smaller in the big reactor. This is a common result; the faster breakup rates are produced in larger tanks, if the specific power is constant, and lead to smaller droplets (see section "Correlations Readily Used"). Thus the stirring rate was decreased to 190 rpm during emulsification and 120 rpm during polycondensation. The size distributions of the produced microcapsules are represented in Figure 10, and are compared to the one obtained in the small reactor. The mean size of the microcapsules

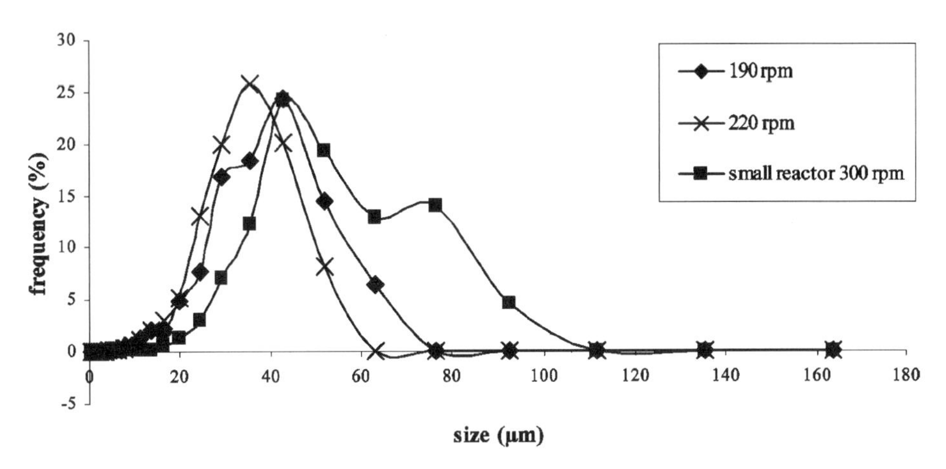

Figure 10 Emulsion size distributions in the 80l reactor.

produced in the 80-L reactor is slightly smaller and the distribution is narrower, particularly the second population of big particles are absent. Moreover, these particles present similar mechanical resistance and stability in the formulation than the ones prepared in the small reactor.

The experiments in the 80-L reactor also allowed the determination of a reaction time of 10 hours to ensure sufficient mechanical properties.

These results are the basis for further industrial process development. The main parameters have been fixed, especially the design and operating conditions of the reactor in which the capsules are produced. Another part of the work is to consider all the equipment around this major reactor. Smaller reactors, one for the preparation of the aqueous phase (reactor 2) and another for mixing the 1/3 of oil with the two lipophilic monomers (reactor 3) should also be designed. Then it is important to design a way that would enable the transfer of these phases into the big reactor (gravity, pumps, etc.). The 80-L experiments have shown that better results are obtained when the second monomers are added in the reactor near the stirrer instead of adding them at the upper liquid surface.

The separation of the particles from the oily external phase before their introduction in to the final formulation is another key point of the process. This was done by natural sedimentation at a small scale; an industrial decanter centrifuge can be used for the larger production.

Example 2: Nanoencapsulation by Emulsification Diffusion

Nanoparticles are defined as encapsulation particles whose size is less than one micrometer. The number of techniques referenced for the nanoparticles preparation is very important, and in constant progression (68). The methods used for the nanoparticles preparation are sometimes similar to those applied in microencapsulation, as for example emulsion solvent evaporation or polymerization. The challenge here is to achieve an emulsion size small enough to submicronic particles formation. The emulsion size is generally hundreds of nanometers; this means that a considerable quantity of energy is supplied to the systems (mechanical stirring, high pressure homogenization, ultrasonic energy, etc.) to reach high shearing forces and large amounts of emulsifiers have to be employed (section "Emulsification Process in Stirred Reactor").

There are also many methods used specifically for the production of nanoparticles. These methods are generally based on very rapid solvent transfer, leading to the formation of fine droplets or particles. The mechanisms involved are similar to those of spontaneous emulsification (69–71). This phenomenon is very interesting, and it leads to the formation of submicronic particles, without any energy input in the system. Nanometric monodispersed particles can be obtained without any emulsification step. The most important drawback of this method is that the size distribution of the particles cannot be controlled. It is dependent on the formulation parameters, and therefore on the physicochemical properties rather than the process parameters (72,73).

To overcome this drawback, some methods have been developed using an emulsification step to fix the size distribution, the rapid solvent diffusion occurring later to transform the droplets in particles. The emulsification–diffusion is one of these methods.

The emulsification–diffusion technique, patented by Quintanar-Guerrero et al. in 1999, is used to produce NC based on preformed polymers (74). It is a two-step process, based on the production of an emulsion that is followed by a dilution, leading to the deposition of the polymer around the droplets, and therefore the formation of NC. This method involves the use of a partially water-soluble solvent, which is previously saturated with water to ensure the initial thermodynamic equilibrium between the two liquids (water and solvent). The polymer, oil, and drug are dissolved into the saturated solvent producing the organic phase. This organic phase is then emulsified, under vigorous agitation, in an aqueous solution containing a stabilizing agent. The droplets of the emulsion should present a low mean size, around or below one micrometer, to ensure the formation of nanoparticles. The subsequent addition of water to the system, under moderate stirring, moves the solvent diffusion into the external phase, which induces the interfacial deposition of the polymer to form the NC (75). Depending on its boiling point, the solvent can be eliminated by distillation or ultrafiltration.

The emulsification–diffusion technique allows the production of nanoparticles in a large range of size, between 200 nm and several micrometers. The size distribution is controlled by the emulsification step. A direct correlation exists between the size distribution of nanoparticles and that of the emulsion (76). The agitation time, the stirrer type, and the rotational speed are the most influential parameters. During the dilution of the emulsion, the agitation is less important. It must be sufficient to homogenize the mixture; but because this condition is maintained, it has no influence on the final size distribution (22).

The objective of this work was to design a pilot plant to study the process of emulsification–diffusion. Starting from the laboratory scale, this pilot is designed with classical chemical engineering equipment, with the aim of finding the parameters necessary to scale-up the process. The properties of the nanoparticles produced at the laboratory scale and at the pilot scale are compared. This concerns essentially the size distribution, a mean size less than 0.5 μm, and quite a narrow distribution. The second parameter is the encapsulation rate, which must be maximized. The first part of the study was carried out with the small-scale equipment (production of 0.06 L) to optimize the formulation and the overall process conditions (77). The method of emulsification–diffusion can be described as follows:

1. Solvent and water mutual saturation.
2. Preparation of the aqueous phase: The stabilizing agent is dissolved in water saturated with the solvent.

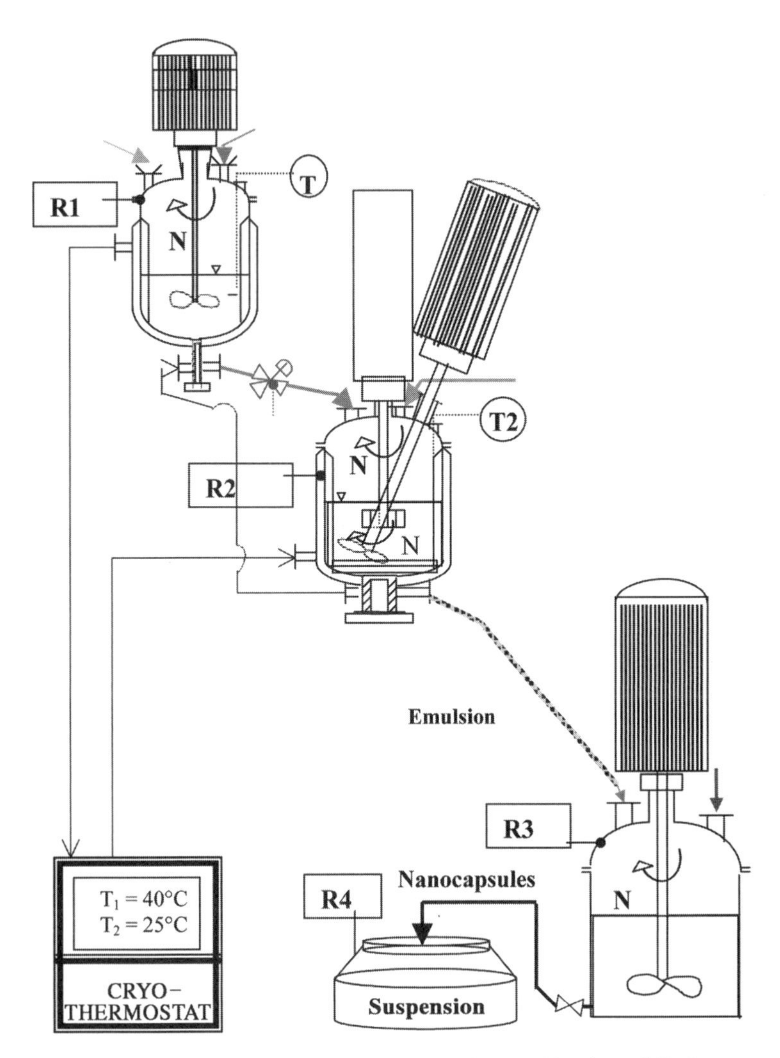

Figure 11 Pilot plant used for the study of emulsification–diffusion.

3. Preparation of the organic phase: The polymer, oil, and drug are dissolved in the solvent saturated with water.
4. Emulsification (oil-in-water) using a high speed drive unit (rotor–stator system).
5. Addition of the emulsion to the distilled water, under moderate stirring, with subsequent formation of the NC.
6. Elimination of the solvent and a part of water by evaporation under reduced pressure or by cross-flow filtration (ultrafiltration).

The pilot plant used is represented in Figure 11, and the characteristics of the equipment are listed in Table 3.

Results Obtained with the Pilot Plants

The first part of the study, at the lab-scale, has lead to the optimization of the formulation. The quantity of the two phases, the volume ratio of the emulsion, and the

Table 3 Main Characteristics of the Laboratory and Pilot Equipment

Characteristics	Lab scale	Pilot scale
Reactor 2: emulsification and preparation of the aqueous phase		
Reactor capacity (L)	0.15 (beaker)	2.5
Reactor internal diameter (cm)	–	15
Volume of the emulsion (L)	0.05	2
Stirrer and diameter (cm) Rotor-stator Ultra Turrax®	T25	T50
Turbine 4 paddles	–	7
Number of baffles	–	4
Stirring rate during emulsification (rpm)		
Rotor-stator	8000	5200–1000
Turbine	–	600–2000
Reactor 1: saturation of the phase and preparation of the organic phase[a]		
Reactor capacity (L)	0.30 (beaker)	2.5
Reactor internal diameter (cm)	–	15
Volume of the organic solution (L)	0.01	0.6
Stirrer diameter—turbine 4 paddles (cm)	5	7
Number of baffles	–	4
Stirring rate during dissolution (rpm)	300	300
Reactor 3: dilution of the emulsion		
Reactor capacity (L)	0.30 (beaker)	6
Reactor internal diameter (cm)	–	20
Volume of the solution, i.e., water + emulsion (L)	0.200	5 (4 + 1)[b]
Stirrer diameter (cm)	5	10
Number of baffles	–	4
Stirring rate during dilution (rpm)	300	0–800

[a]At the lab-scale, the saturation is done in a settling flask, the organic phase is then transferred to a beaker for the dissolution of polymer, oil, and drug. At the pilot scale, saturation and preparation of the organic phase are done in the same reactor R1.
[b]Only half of the emulsion is diluted, the other part in stored for further experiments.

concentrations of surfactant, polymers, and drugs are determined. Only the operating parameters have been studied, especially the stirring rate during emulsification and dilution.

The results show the necessity of coupling two stirrers in the emulsification reactor, a rotor–stator unit for the dispersion of the droplets, and a classical stirrer to promote the fluid homogenization. If the rotor–stator mixer is more efficient than a classical stirrer (turbine) for producing small emulsion droplets, it does not present enough pumping capacity. Its coupling with the homogenization turbine induces not only a size decay for the emulsion, but also a more uniform size distribution. Decrease in size and smaller polydispersity are also achieved by increasing the agitation time.

The optimum values are 10,000 rpm for the rotor–stator and 600 rpm for the turbine, and the agitation duration is fixed at 20 minutes (22).

During the dilution of the emulsion, no influence of the stirring rate on the particles size distribution is observed. However, an agitation must be performed to avoid particle agglomeration by maintaining the homogeneity of the suspension.

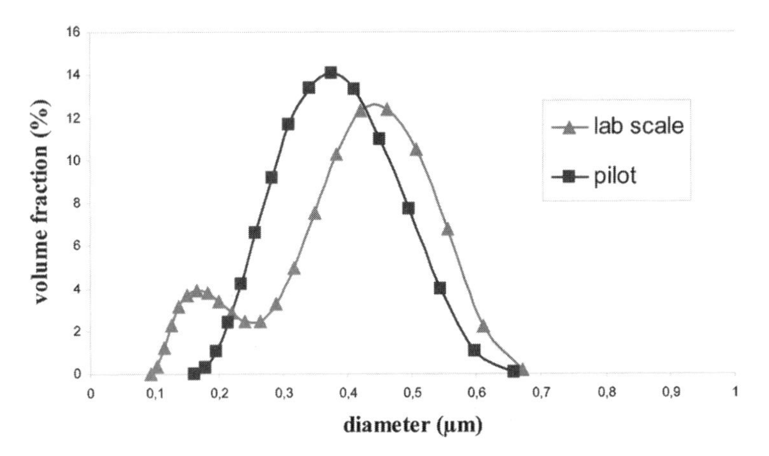

Figure 12 Size distributions of the nanoparticles produced at lab scale and with the pilot plant.

The size distributions of the nanoparticles obtained at the two scales are compared in Figure 12. There is a shift of the whole distribution to the smaller size.

Another study presented by Galindo-Rodriguez et al. (17) focuses also on scaling-up of such an EMDF process. Figure 13 presents the comparison between the nanoparticles produced at lab/pilot-scales according to the specific power input. The mean size tends to be smaller at the pilot-scale. As already mentioned in section "Correlations Readily Used" in a larger stirrer tank, a scale-up considering a constant power per unit volume ε produces faster breakup rates. The mean Sauter particles size tends to shift towards smaller values.

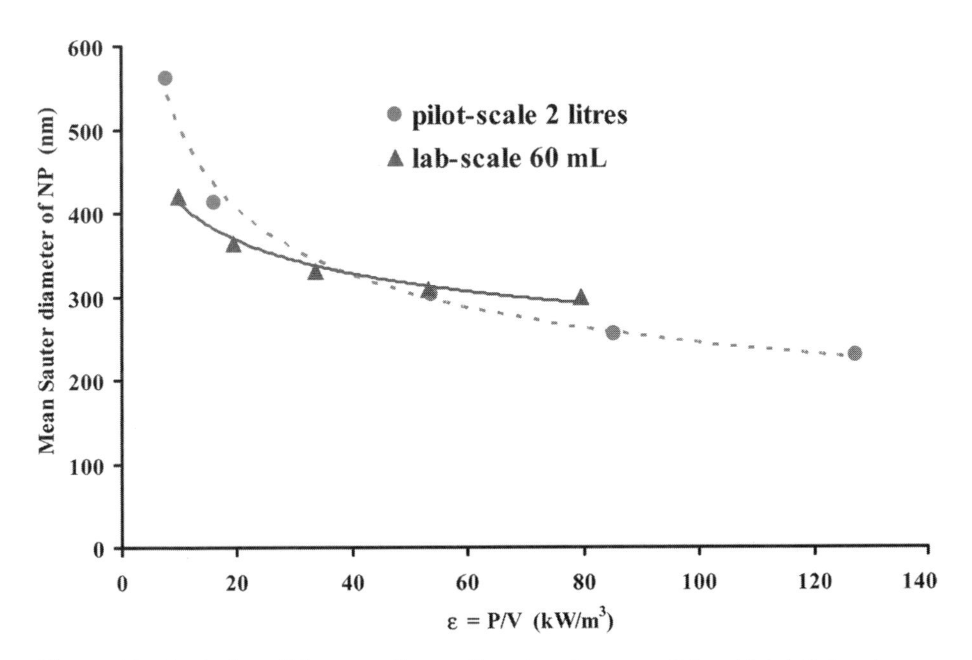

Figure 13 Evolution of the mean Sauter diameter of nanoparticles at lab/pilot scales versus the specific power input ε.

The encapsulation tests with a model drug (indomethacin) show that the efficacy of encapsulation is the same at the two scales: 80% (±10%) of the drug introduced in the formula is encapsulated or slightly less than 62% (17,22).

SUMMARY

The field of microencapsulation and nanoencapsulation is concerned with a large number of processes used to produce particles of defined size and structure. It appears that the three most often used manufacturing methods are spray drying, coacervation, and interfacial polymerization. Both of the last two processes include an emulsification step prior to encapsulation. The stage of emulsification generally determines the right particle properties because the microcapsule size is almost equal to the droplet size.

The emulsification stage requires mechanical mixing in the vessel, most of the time. The droplet size distribution results from a balance between breakup and coalescence. Break up is favored in the turbulent zone close to the agitator, while coalescence dominates in the quiescent recirculation regions. As the volume of the batch of the emulsion increases, the choice of the industrial mixing system is more and more crucial to adjust the balance between these two phenomena. Basically, the use of an axial flow turbine is recommended for low viscous dispersion, but other configurations with several mixers may be preferable according to the final droplet size distribution wanted or to the level of viscosity. The prediction of the droplet size distribution could be employed to predict the mean size of the microcapsules or at least to assess the effect of the modification of one parameter (a physicochemical one such as viscosity and interfacial energy, or a process one such as the stirring speed and size of the agitator). Maintaining the size distribution of the microcapsules for batch size enlargement by scaling up, requires adherence to the following two principal rules: (i) the use of representative lab equipment geometrically close to the pilot or industrial ones and (ii) the right selection of process parameters (especially stirring speed). The specific power input brought by the stirrer to the dispersion is the most relevant criterion to apply. This approach has given successful results in the case of microcapsules (1–10 μm) and nanocapsules (100–500 nm) production.

NOTATIONS

Re	Reynolds number	–
N	Rotational velocity	s^{-1}
D_A	Agitator diameter	M
P	Power consumption	W
N_P	Power number	–
N_Q	Pumping number	–
Q_P	Pumping flow rate	m^3s^{-1}
H	Velocity head	M
G	Gravitational acceleration	m^2s^{-1}
V_T	Total volume of the fluids	m^3
K	Linear scale-up factor	–
K, K_1, K_2, K_3, K_4, K_5	Constant	–

We	Weber number	–
d	Droplet diameter	m
d_{max}	Maximum stable droplet size	m
d_{32}	Sauter mean diameter	m
H	Height of liquid	m
T	Vessel diameter	m
H_A	Position of the agitator	m
w	Width of baffles	m
ρ	Density	$kg\,m^{-3}$
μ	Dynamic viscosity	Pa s
μ_C	Dynamic viscosity of the continuous phase	Pa s
μ_D	Dynamic viscosity of the dispersed phase	Pa s
ν	Kinematic viscosity	$m^2\,s^{-1}$
ε	Specific power input	$W\,kg^{-1}$
ε_E	Specific power input of the emulsion	$W\,kg^{-1}$
ε_c	Specific power input of the continuous phase	$W\,kg^{-1}$
η	Microscale	m
τ	Stress at the droplet surface	$N\,m^{-2}$
σ	Interfacial tension	$N\,m^{-1}$
Φ	Dispersed phase volume fraction	–
τ	Shear stress	$N\,m^{-2}$

REFERENCES

1. Arshady R, Boh B. Microcapsules patents and products. In: Arshady R, ed. MML Series. Vol. 6. London: Citus Books, 2003.
2. Crainich VA. Microencapsulation: scale up considerations and productions technology in specialized drug delivery systems. In: Tyle P, ed. Drugs and the Pharmaceuticals Sciences. Vol. 41. New York: Marcel Dekker, 1990.
3. Freitas S, Merkle HP, Gander B. Microencapsulation by solvent extraction/evaporation: reviewing the state of the art of microsphere preparation process technology. J Control Release 2005; 102:313–331.
4. Maa YF, Hsu C. Microencapsulation reactor scale-up by dimensional analysis. J Microencapsul 1996; 13(1):53–66.
5. Maa YF, Hsu C. (1996b) Liquid-liquid emulsification by static mixers for use in microencapsulation. J Microencapsul 1996; 13(4):419–433.
6. Embleton JK, Tighe BJ. Polymers for biodegradable medical devices. XI. Microencapsulation studies: characterization of hydrocortisone-loaded poly-hydroxybutyratehydroxyvalerate microspheres. J Microencapsul 2002; 19(6):737–752.
7. Benoit JP, Marchais H, Rolland H, van de Velde V. Biodegradable microspheres: Advances in production technology. In: Benita S, ed. Microencapsulation—Methods and Industrial Applications. New York: Marcel Dekker Inc., 1996:35–72.
8. Thies C. A survey of microencapsulation processes. In: Benita S, ed. Microencapsulation—Methods and Industrial Applications. New York: Marcel Dekker Inc., 1996: 133–154.
9. Thies C. Preparation and chemical applications. In: Arshady R, ed. MML Series. Vol. 1. London: Citus Books, 1999.
10. Nihant N, Grandfils C, Jérôme R, Teyssié P. Microencapsulation by coacervation of poly(lactide-co-glycolide). IV. Effect of the processing parameters on coacervation and encapsulation. J Control Release 1995; 35:117–125.
11. Levin M. Introduction in pharmaceutical process scale-up. In: Levin M, ed. Drugs and the Pharmaceutical Sciences. Vol. 118. New York: Marcel Dekker Inc., 2002.

12. Bloch LH, Levin, M. Chap. 3, in pharmaceutical process scale-up. In: Levin, M. ed. Drugs and the Pharmaceutical Sciences, Vol. 118. Newyork: Marcel Dekker Inc., 2002.
13. Dobetti L, Pantaleo V. Application of a hydrodynamic model to microencapsulation by coacervation. J Microencapsul 2002; 19(2):139–151.
14. Jégat C, Taverdet JL. Stirring speed influence study on the microencapsulation process and on the drug release from microcapsules. Polym Bulletin 2000; 44:435–351.
15. Brujes L, Legrand J, Carnelle G. Complete suspension of microcapsules in baffles and unbaffled stirred tanks. Chem Eng Technol 1998; 21(9):735–744.
16. De Labouret A, Thioune O, Fessi H, Devissaguet JP, Puisieux F. Application of an original process for obtaining colloidal dispersions of some coating polymers: preparation, characterization, industrial scale-up. Drug Dev Ind Pharm 1995; 21(2):229–241.
17. Galindo-Rodriguez S, Puel F, Briancon S, Allémann E, Doelker E, Fessi H. Comparative scale-up of three methods for producing ibuprofen-loaded nanoparticles. Eur J Pharm Sci 2005; 25(4–5):357–367.
18. Maggioris D, Goulas A, Alexopouos AH, Chatzi EG, Kiparissides C. Prediction of particle size distribution in suspension polymerisation reactors: effect of turbulence nonhomogeneity. Chem Eng Sci 2000; 55(20):4611–4627.
19. Perry RH, Green DW. Perry's Chemical Engineers Handbook. 7th ed. 1997.
20. Coulaloglou CA, Tavlarides LL. Drop size distributions and coalescence frequencies of liquid–liquid dispersions in flow vessels AIChE J. 1976; 22(2):289–297.
21. Pacek AW, Chamsart S, Nienow AW, Bakker A. The influence of impeller type on mean drop size and drop size distribution in an agitated vessel. Chem Eng Sci 1999; 54:4211–4222.
22. Colombo AP, Briancon S, Lieto J, Fessi H. Project, design, and use of a pilot plant for nanocapsule production. Drug Dev Ind Pharm 2001; 27(10):1063–1072.
23. Letellier B, Xuereb C, Swaels P, Hobbes P, Bertrand J. Scale-up in laminar and transient regimes of a multi stage stirrer, a CFD approach. Chem Eng Sci 2002; 57: 4617–4632.
24. Uhl V, Gray J. Mixing: theory and practice. San Diego: Academic Press, 1967.
25. Nagata S. Mixing: Principles and Applications. New York: John Wiley, 1975.
26. Roustan M. Agitation mélange: caractéristiques des mobiles d'agitation. Techniques de l'ingénieur 1997; J3802.
27. Tsouris C, Tavlarides LL. Breakage and coalescence models for drops in turbulent dispersions. AIChE J 1994; 40(3):395–406.
28. Kresta S. Turbulence in stirred tanks: anisotropic, approximate, and applied. Can J Chem Eng 1998; 76:563–576.
29. Hinze JO. Fundamentals of the hydrodynamic mechanism of splitting in dispersion processes. AIChE J 1955; 1(3):289–295.
30. Calabrese RV, Chang TPK, Dang PT. Drop break-up in turbulent stirred contactors, part I: effect of disperse phase viscosity. AIChE J 1986; 32(4):657–666.
31. Chen HT, Middleman S. Drop size distribution in agitated liquid–liquid systems. AIChE J 1967; 13:989–995.
32. Kumar S, Ganvir V, Satyanand C, Kumar R, Gandhi KS. Alternative mechanisms of drop break-up in stirred vessels. Chem Eng Sci 1998; 53(18):3269–3280.
33. O'Rourke AM, MacLouglhin PF. A comparison of measurement techniques in the analysis of evolving liquid–liquid dispersion. Chem Eng Process 2005; 44(8):885–894.
34. Ramkrishna D. Population Balances—Theory and Applications to Particulate Systems in Engineering. San Diego: Academic Press, 2000.
35. Alvarez J, Hernandez M. A population balance approach for the description of particle-size distribution in suspension polymerisation reactors. Chem Eng Sci 1994; 49(1): 99–113.
36. Spicer PT, Pratsinis SE, Trennepohl MD, Meesters GHM. Coagulation and fragmentation: the variation of shear rate and the time lag for attainment of steady state. Ind Eng Chem Res 1996; 35(9):3074–3080.

37. Semiao V, Andrade P, da Graca Carvalho M. Spray characterization: numerical prediction of sauter diameter and droplet size distribution. Fuel 1996; 75(15):1707–1714.
38. Serra T, Casamitjana X. Effect of the shear and volume fraction on the aggregation and break-up of particles. AIChE J 1998; 44(8):1724–1730.
39. White AJ, Hounslow MJ. Modelling droplet size distributions in polydispersed wet steam flows. Int J Heat Mass Transfer 2000; 43:1873–1884.
40. Hounslow MJ, Ni X. Population balance modelling of droplet coalescence and break-up in an oscillatory baffled reactor. Chem Eng Sci 2004; 59:819–828.
41. Alexopoulos AH, Maggioris D, Kiparissides C. CFD analysis of turbulence in non-homogeneity in mixing vessels A two compartment model. Chem Eng Sci 2002; 57: 1735–1752.
42. Vikhansky A, Kraft M. Modelling of a RDC combined CFD-population balance approach. Chem Eng Sci 2004; 59:2597–2606.
43. Taylor ES. Dimensional Analysis for Engineers. Oxford, UK: Clarendon Press, 1974.
44. McCabe WL, Smith JC, Harriott P. Unit Operations of Chemical Engineering. 5th ed. New York: MacGraw Hill, 1993:16–18.
45. Kolmogorov AN. Disintegration of drop in turbulent flow. Doklady Akademii Nauk SSSR 1949; 30:301–305.
46. Kolmogorov AN. The local structure of turbulence in incompressible viscous fluid for very large Reynolds numbers. Doklady Akademii Nauk SSSR 1941; 30:301–305.
47. Baldyga J, Bourne JR. Further applications: Drop break-up. Turbulent Mixing and Chemical Reactions. New York: J. Wiley Sons, 1999:841–860.
48. Baldyga J, Bourne JR, Pacek AW, Amanullah A, Nienow AW. Effects of agitation and scale-up on drop size in turbulent dispersions: allowance for intermittency. Chem Eng Sci 2001; 56(11):3377–3385.
49. Calderbank PH. Physical rate processes in industrial fermentation. Part I: the interfacial area in gas liquid contacting with mechanical agitation. Trans Int Chem Eng 1958; 36:443–463.
50. Davies GA. Mixing and coalescence phenomena in liquid–liquid systems. In: Thornton JD, ed. Science and Practice of Liquid–Liquid Extraction. Vol. 1. Oxford, UK: Clarendon Press,1992:245–342.
51. Peters DC. Dynamics of emulsification. In: Harnby N, Edwards MF, Nienow AW, eds. Mixing in the Process Industries. 2nd ed. Oxford, UK: Butterworth-Heinemann's, 1997:294–321.
52. Costaz H. PhD Thesis No. 041–96, Université Claude Bernard Lyon 1, 1996.
53. Rivautella L, Briançon S, Puel F, Barresi A. Formation of nanocapsules by emulsion-diffusion: prediction of the emulsion size. 4th European Congress of Chemical Engineering, Granada, Spain, September 20–25, 2003.
54. Maa YF, Hsu C. Liquid–liquid emulsification by rotor-sator homogeneization. J Control Release 1996; 38:219–228.
55. Kuster KA, Wijers JG, Thoenes D. Agregation kinetics of small particles in agitated vessels. Chem Eng Sci 1997; 52:107–121.
56. Leuenberger H. New trends in the production of pharmaceutical granules: the classical batch concept and the problem of scale-up. Eur J Pharm Biopharm 2001; 52:279–288.
57. Thies C. Microencapsulation, Kirk-Othmer Encyclopedia of Chemical Technology. Vol. 16. 4th ed. 1995:628–745.
58. Finch CA. Microencapsulation Ulmann's Encyclopedia of Industrial Chemistry. Vol. A21. 6th ed. Weinheim: Wiley–VCH Verlag GmbH&Co, 2003:733–748.
59. Scher HB. In: Miyamoto J, Kearny PC, eds. Pesticide Chemistry—Human Welfare and the Environment. Vol. 4. Oxford, UK: Pergamon Press 1983:295–300.
60. Sparks R. Microencapsulation. In: Kirch-Othmer, ed. Encyclopedia of Chemical Technology. Vol. 15. 3rd ed. 1991:470–493.

61. Chang TMS. Artificial cells for artificial kidney, artificial liver and artificial cells. Proceedings of the McGill Artificial Organs Research Unit International Symposium, McGill University, April 20, 1977, New York: Plenum Press, 1977:57–77.
62. Dubernet C, Benoit JP. La microencapsulation: ses techniques et ses applications en biologie. L'actualité chimique, décembre, 1986, 19–28.
63. Chang TMS. Semipermeable microcapsules. Sci 1964; 146:524–525.
64. Muramatsu N, Shiga K, Kondo T. Preparation of polyamide microcapsules having narrow size distributions. J Microencapsul 1994; 11(2):171–178.
65. Arshady R. General concepts and criteria. Microspheres, Microcapsules and Liposomes Preparation and Chemical Applications. The MML Series. Chapter 1. Vol. 1. London, UK: Citus Books, 1999.
66. Whateley T. Microcapsules: Preparation by interfacial polymerization and interfacial complexation and their applications. In: Benita S, ed. Microencapsulation—Methods and Industrial Applications. Drug and the Pharmaceutical Sciences Series. Vol. 73. New York: Marcel Dekker, Inc.1996:349–375.
67. Morgan PW. Condensation polymers: interfacial and solution methods. Polymers Review Series 10. New York: Interscience Publishers, John Wiley, 1965.
68. Montasser I, Briançon S, Fessi H, Lieto J. Méthodes d'obtention et mécanismes de formation de nanoparticules polymériques. J Pharmacie de Belgique 2000; 55(6):155–167.
69. Fessi H, Puisieux F, Devissaguet JP, Ammoury N, Benita S. Nanocapsule formation by interfacial polymer deposition following solvent displacement. Int J Pharm 1989; 55: R1–R4.
70. Fessi H, Devissaguet JP, Puisieux F, Thies C. Process for the preparation of dispersible colloidal systems of a substance in the form of nanoparticles. US Patent 5, 118, 528; 1992.
71. Sternling CV, Scriven LE. Interfacial turbulence: hydrodynamic instability and Marangoni effect. AIChE J 1959; 5:514–523.
72. Thioune O, Fessi H, Devissaguet JP, Puisieux F. Preparation of pseudolatex by nanoprecipitation: influence of the solvent nature on intrinsic viscosity and interaction constant. Int J Pharm 1997; 146:233–238.
73. Plasari E, Grisoni Ph, Villermaux J. Influence of process parameters on the precipitation of organic nanoparticles by drowning out. Trans I Chem E 1997; 75(Part A):237–244.
74. Quintanar-Guerrero D, Fessi H, Doelker E, Alleman E. Methods for preparing vesicular nanocapsules. PCT Patent No W09904766A, 1999.
75. Quintanar-Guerrero D, Alléman E, Doelker E, Fessi H. Preparation and characterization of nanocapsules from preformed polymers by a new process based on emulsification-diffusion technique. Pharm Res 1998; 15(7):1056–1062.
76. Galindo-Rodriguez S, Allémann E, Fessi H, Doelker E. Physicochemical parameters associated with nanoparticle formation in the salting-out, emulsification-diffusion and nanoprecipitation methods. Pharm Res 2004; 21:1428–1439.
77. Guinebretiere S, Briancon S, Lieto J, Mayer C, Fessi H. Study of the emulsion-diffusion of solvent: preparation and characterization of nanocapsules Drug Dev Res 2002; 57(1):18–33.

PART II: EVALUATION AND CHARACTERIZATION OF MICRO- AND NANOPARTICULATE DRUG DELIVERY SYSTEMS

7

Drug Release from Microparticulate Systems

Shicheng Yang
KV Pharmaceutical Company, St. Louis, Missouri, U.S.A.

Clive Washington
Pharmaceutical and Analytical Research and Development, AstraZeneca, Macclesfield Works, Hurdsfield Industrial Estate, Macclesfield, Cheshire, U.K.

INTRODUCTION

Microencapsulated and particulate systems for drug delivery have found wide application in pharmaceutics; initially for external use as creams and ointments, later for subcutaneous drug delivery, and in oral and intravenous administration (1). The release of drugs from microencapsulated systems, including micro- and nanocapsules, micro- and nanospheres, micro- and nanoparticles, and emulsion droplets, has been extensively reviewed (1–4). Release mechanisms are commonly inferred from kinetic measurement data of microencapsulated drug delivery systems by indirect methods based on the effects of solvent, buffer, agitation rate, or other variables. Drug-release data have a number of potential applications. At the simplest level, the data can be used for quality control purposes to ensure the constancy of behavior of a manufactured product. It can further be used to try to understand the physicochemical structure of the delivery system and the drug-release mechanism; however, this normally requires the applying of the data to that computed for a model of the drug delivery system (5–16). Finally, such data can be used in an attempt to predict the likely behavior of the system in vivo (17–29). Drug release or dissolution testing is an essential requirement for the development, establishment of in vitro dissolution and in vivo correlation (IVIVC), and registration and quality control of dosage forms. Under certain conditions, in vitro drug release can be used as a surrogate for the assessment of bioequivalence (30).

MEASUREMENT OF DRUG RELEASE

The measurement of drug release from microencapsulated systems poses many difficulties which are not encountered for larger particulate systems, which can be carried out using basket and paddle methods as described in the United States Pharmacopeia (USP) or other monographs (12,31–33). The main difficulty arises when an attempt is made to separate the released drug in the continuous phase from the suspended particulate carriers. This becomes increasingly difficult and slow as the particles become smaller; however, the release rates become faster for smaller particles, as one reaches a limiting lower size for which the separation cannot be carried out on a timescale that is faster than the release profile. Under these circumstances, any useful kinetic information in the release profile is lost.

The methods available to evaluate drug release from colloidal carriers fall into the following groups.

Dialysis Bag Diffusion Techniques

The dialysis bag diffusion technique is widely used to evaluate drug release from macro and nanosized carriers. A small volume of the concentrated drug–particle suspension is contained in a dialysis bag, which is immersed in a larger volume of continuous-phase acceptor fluid. Preferably, both compartments are stirred, and the drug then diffuses out of the particulate carrier into its local continuous phase, and then through the dialysis membrane into the acceptor phase, which is periodically sampled and assayed (34–36). This technique was criticized by Washington because the drug was not diluted in sink conditions in the dialysis bag (Fig. 1) (37). The measured rate of drug appearance in the external sink does not depend significantly on the particle release rate; instead, it depends largely on the partition coefficient of a drug between its carrier and the local acceptor phase (37).

Levy and Benita (38) confirmed the disadvantages of this method in studies of an emulsion containing diazepam in an intralipid-like base; they also demonstrated a reverse dialysis technique, in which the emulsion was diluted into a sink and small pre-equilibrated dialysis bags containing buffer were used to sample the continuous phase. This method has the advantage that a large sink can be used, but it was still found to be rather slow owing to the diffusion rate through the bags (approximately one hour).

Recently, Chidambaram and Burgess (39) claimed that sink conditions were maintained in a novel reverse dialysis bag technique because emulsions were diluted infinitely in the donor phase. In vitro clozapine release kinetics from solid lipid nanoparticles was measured in 0.1N HCl, double distilled water, and phosphate buffer, pH 7.4, using a modified Franz diffusion cell (40). Conventionally, the dialysis bag diffusion technique does not measure the true release rate but rather the partition of a drug between the various phases of a dispersed system. Although the reverse dialysis techniques do appear to be significantly faster than conventional

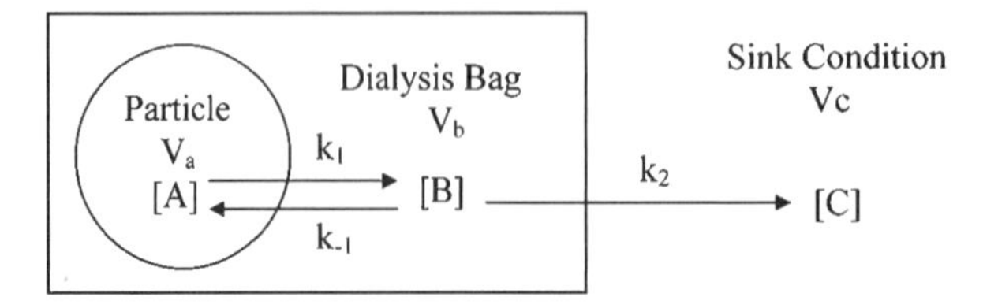

Figure 1 Kinetics of nonsink membrane dialysis technique.

dialysis, it should be realized that the diffusion of a typical drug molecule across the diameter of a small emulsion droplet would take only milliseconds. Although interfacial barriers may add one to two orders of magnitude to this, it is suspected that true release rates from emulsions will not be detectable until the time resolution of the experiment is shortened to the order of a second or less.

Sampling and Separation Methods

Examples of this method normally start by preparing a concentrated suspension of the particulate carrier and rapidly diluting it into a much larger bulk of acceptor phase. Samples of this diluted system are then removed at intervals, and the particles are separated so that the continuous phase may be assayed for the released drug. This technique is perfectly adequate for larger microparticles using filtration or centrifugation techniques, when the release times are measured in hours. Filtration through membrane filters is an attractive alternative; however, the possibility of released drug binding to the filters must be considered (18,29,34,41–44).

However, as the particle size decreases to nanosize ranges, the difficulty of separation increases and ultrafiltration and ultracentrifugation should be employed (45,46). Ultrafiltration membranes with low-binding properties are available, but these still adsorb small amounts of many drugs (47). Magenheim et al. (47) described a low-pressure ultrafiltration experiment to separate samples of the released drug from a relatively large volume of sink fluid. This apparatus had a time response of a few minutes, which made it suitable for studying microparticle formulations, but it was too slow for emulsion release studies. The ultrafiltration step can be accomplished rapidly by centrifugal filtration, and simple centrifugal ultrafilters are available which make this the most convenient method (48). The advantage of this method is that the particles in the sample are forced downward away from the ultrafilter, thus avoiding filter clogging.

An interesting final example of sampling and separation measurements was described by Widder et al. (49), who studied drug release from magnetite-albumin microspheres; these could rapidly be separated from the continuous phase using a magnet.

Continuous-Flow Methods

The continuous-flow method most commonly uses an ultrafiltration cell to recover samples of clean continuous phase; this is fed from a pressurized reservoir and flows continuously through a detector for a direct continuous measurement of drug release (50). This technique is conceptually simple but presents a number of sources of error, with the most significant being variations (usually a decrease) in the flow rate through the ultrafiltration membrane as the experiment proceeds, thus distorting the release profile.

The response time of such an experiment is approximately the time taken to replace the continuous phase in the ultrafiltration cell; consequently, a thin flat cell (i.e., one having a large ultrafiltration membrane area but a small total volume) is preferable. Typical flow rates of 1 mL min^{-1} in a 10 mL cell suggest a 10 mm time resolution; if the drug-release rate is comparable to this, the time resolution can be improved by deconvoluting the cell response function from the observed release function as described previously (51).

In Situ Methods

In situ methods depend on the use of analytical techniques, which are only sensitive to the drug in the solution and not to that which is in the solid particles or that is bound to their surface, so that the assay of the released drug may be performed without any separation step. Illum et al. (52) used a spectroscopic method to study the release of rose Bengal from cyanoacrylate microspheres. Scattering from the microspheres is a significant problem, and the method can only normally be performed if the drug is strongly colored; the carrier particles often absorb or scatter light too strongly for measurements to be made in the ultraviolet (UV) region of the spectrum. A novel in situ sampling method was described by Minagawa et al. (46), who attempted to measure the release of isocarbacyclin methyl ester from an emulsion formulation using silanized glass beads to adsorb the free hydrophobic drug. This technique resulted in a time resolution of approximately 15 to 30 seconds and an apparent release time of a few minutes from a typical triglyceride emulsion. As discussed below, techniques of this type are really too slow to study dynamic distribution processes in submicron emulsions, but they could prove to be very useful for solid microparticle studies.

Electrochemical techniques would appear to be valuable, but again have been little used. Polarography and ion-selective detection would have a response time of seconds, but very few drugs are amenable to detection using these methods. However, an attractive alternative is to use pH as an indicator of drug release. Many drugs are weak acids or bases, and their release can be monitored by measuring the neutralization of a suitable medium as the drug diffuses into it. Washington et al. used this method to study the release of model organic acids and the basic drug chlorpromazine from triglyceride emulsions (53). This experiment had a time resolution of approximately one second, and indicated that the triglyceride emulsions released their drug over periods of 10 seconds, and were dominated by interfacial transport barriers. The rapidity of drug release in these emulsion systems casts doubt on many measurements made by alternative techniques.

MECHANISMS OF DRUG RELEASE

In microencapsulated matrix systems, a drug is incorporated into a polymer matrix by either particulate or molecular dispersion. The diffusional release of drug from carrier has been assumed to be the controlling step, and usually can be described by Fick's first and second laws (54,55). Release is influenced by a range of factors, such as diffusion and partition coefficients, drug loadings and solubilities, boundary surface areas, layer thickness, and shape factors (3). Several of the drug properties, such as solubility, pK value, and partition coefficient, are closely dependent upon the properties of sink medium into which the drug diffuses (Fig. 2).

As a measure of the important Fickian parameter surface area, the size distribution will undoubtedly be relevant to the release profile provided it is measured under realistic conditions, as there may be time dependency involving differential swelling, degradation, erosion, or splitting (3). The particles can depart from a monosized distribution of spheres. They usually have a distribution of sizes, with the occasional large particle having a significant fraction of the total drug mass. Without going into details of release mechanisms, this will imply a range of release rates, with the smaller particles showing rapid release and the larger ones a slower component (56). In theory, it is straightforward to integrate the release rate expression over the

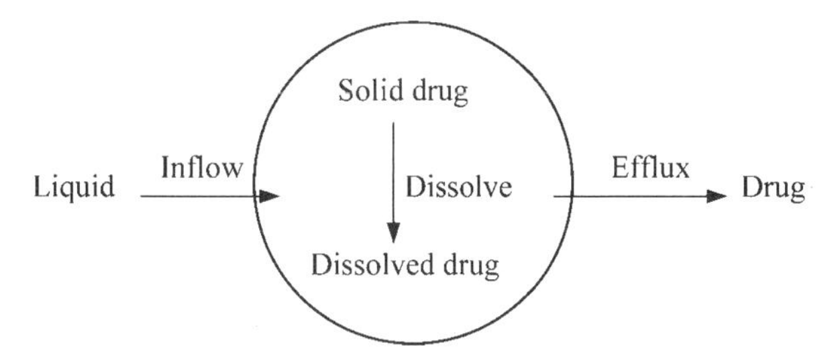

Figure 2 Schematic illustration of drug release from microparticulated systems.

particle size distribution if this is known (1,3). However, this relationship may not be straightforward, as the structure of the particle may vary with size. For example, Siepmann et al. (57) reported that the relative and the absolute release rates of 5-fluorouracil from poly(lactic-co-glycolic acid) (PLGA)-based microparticles prepared with a solid-in-oil-in-water solvent extraction technique increased with increasing microparticle radius, despite the increasing length of diffusion pathways. This was because of the fact that the initial drug loading significantly increased with increasing radius of the drug delivery system; thus, large microparticles became more porous during drug release than small microparticles, leading to higher apparent diffusivities and drug transport rates. This effect overcompensated the effect of the increasing diffusion pathway lengths with increasing microparticle radius, resulting in increased drug-release rates with increasing device dimension.

Shape variation is a further complicating factor. The particles may be nonspherical; many papers show electron micrographs of "spherical" particles which show minor or gross distortions; these can be induced or augmented by freeze-drying, so that the dried formulation shows very different release profiles to the suspension formulation. Also, the structure of the particles may not be uniform; they could display core–shell behavior, with the drug being nonuniformly distributed across the particle section, or multiple phases of drug and carrier may be present.

Wall–core ratios also vary in populations. Increasing the core versus wall ratio will increase the drug content of particles and consequently the release rate. Decrease of the core versus wall ratio leads, as expected, to a reduction in release rate. Porosity and cross-linking density have a direct relationship with drug release (58). In macroporous systems, where the polymer is hydrophilic, the drug release may occur in two phases: first by desorption from the polymer to pores, then by emptying of the drug from pores to the medium. The drug diffusion coefficient is slower through the beads with a higher degree of cross-linking, thus, cross-linking treatment on a porous matrix causes a modification of the pore tortuosity and the drug-release rate (58).

Drug location in microencapsulated systems also influences the drug-release rate and/or pattern. If the drug is surface active, it will adsorb to the outside of the particles during preparation, and may provide a burst effect on dilution. The release profiles depend on the number of washing steps carried out prior to the actual release measurement, because of the removal of the loosely deposited drug near the surface. The mechanism of adriamycin release from albumin microspheres depended on the location of the drug (i.e., adsorption on to the microsphere surface or inclusion in the matrix) and on the properties of the matrix itself (59). To ascertain the position of this equilibrium, it is usually helpful to isolate the continuous phase by ultracentrifugation and assay for unentrapped drug. Benita and Levy (60) also described how to determine drug locations in submicron emulsions.

Many variables that have been studied in nonparticulate controlled release devices, and in the field of polymer science, are also relevant to release from microencapsulated systems. Important factors include drug diffusivity and solubility in polymers and solutions, molecular weights of polymer and diffusant, diffusant size, degree of polymer crystallinity or cross-linking, presence of diluents, plasticizers and fillers, geometry and dimensions of polymer matrix and/or membrane thickness, degree of polymer swellability, polymer degradation, polymer erosion, thickness of aqueous boundary layer, porosity and tortuosity, partition coefficient of a drug between polymer and aqueous medium, drug loading, pH value of medium and pK value of drug, drug polymorphism, and the interplay among these variables (4,61).

DRUG RELEASE KINETIC MODELS

Fick's Law of Diffusion

In the development of microencapsulated drug delivery systems, mathematical modeling of the drug release process plays a clarifying role as it suggests the mechanism(s) of drug release and indicates which parameters may be critical in particle design. In most systems, some sort of diffusion process will be the rate-controlling step, and this can be described by Fick's first and second laws. Application of the structural features and boundary conditions then allows specific models to be developed (54,55). Written in its differential form in one dimension, Fick's first law can be expressed as

$$J = -D\frac{\partial c}{\partial z} \tag{1}$$

where J is the molar flux of drug (in mol/cm^2 sec), D the mutual diffusional coefficient of the drug (in cm^2/sec), c the concentration of drug (in mol/cm^3), and z the position in the device (in cm).

In the simplest case, where the molar flux of the drug and the mutual diffusion coefficient are constant, Equation (1) can be integrated to give the integrated form of Fick's first law

$$J = DK\frac{\Delta c}{\delta} \tag{2}$$

where K is the thermodynamic partition coefficient and δ the device thickness. The partition coefficient K is the ratio between the drug concentration at the interface and the drug concentration in the bulk.

In general, the matrix may be anisotropic and D will depend on position within the matrix and on time, because of the possible spatial and temporal variations, respectively, in porosity, solubility, etc. In some instances, the diffusion coefficient D may also be concentration-dependent. The distribution of drug concentration $c(x,t)$ within the system at any time t and position x is given by Fick's second law (62)

$$\frac{\partial c}{\partial t} = \frac{\partial}{\partial x}\left(D_x\frac{\partial c}{\partial x}\right) + \frac{\partial}{\partial y}\left(D_y\frac{\partial c}{\partial y}\right) + \frac{\partial}{\partial z}\left(D_z\frac{\partial c}{\partial z}\right) \tag{3}$$

In the simplest homogeneous isotropic matrix (i.e., having a constant scalar diffusion coefficient) and constant boundaries, the three-dimensional unsteady diffusion Equation (3) is

$$\frac{\partial c}{\partial t} = D\frac{\partial^2 c}{\partial x^2} \tag{4}$$

Release from Particulates

Although Equation (3) captures the essential physics of drug release, like any differential equation the difficulties lie in setting up the boundary conditions and finding solutions. This may be possible for simple systems such as flat plates and spheres, but becomes increasingly problematic for real systems, particularly inhomogeneous ones. Real colloidal systems may have drug dispersed molecularly or be phase-separated from the carrier; the distribution of drug may vary across the section of the particle; the particle may be porous. Under these conditions, analytic modeling of drug release becomes almost impossible and numerical simulation techniques can be applied with advantage. Fick's diffusional release is not the only mechanism by which a drug is released from the matrix; the erosion of the matrix is particularly important, and in fact is the dominant process for many biodegradable systems such as polyesters. Modeling of this process is much less well established.

Diffusion in a Homogenous Sphere

This is the simplest model of a drug-releasing particle matrix that does not degrade during the experiment, or present an interfacial barrier. Considering the spherical geometry of the investigated system (radius, r), and assuming: (i) constant drug diffusion coefficients, (ii) perfect sink conditions, (iii) a uniform initial drug concentration that is smaller than the solubility of the drug within the system, the following analytical solution of Fick's second law of diffusion can be used to describe the resulting drug-release rate:

$$\frac{M_t}{M_0} = 1 - \frac{6}{\pi^2}\sum_{n=1}^{\alpha}\frac{1}{n^2}\exp\left(-\frac{n^2\pi^2 Dt}{r^2}\right) \tag{5}$$

where M_t and M_0 denote the cumulative absolute amount of drug released at time t and at infinite time, respectively (62–64). Because the existence of "nanopores" cannot be excluded, D represents an apparent diffusivity, taking into account drug transfer through the matrix itself, as well as possible drug transfer through water-filled "nanopores."

Two further simplified forms are: For short times, valid for $M_t/M_0 < 0.4$

$$\frac{M_t}{M_0} = 6\sqrt{\frac{Dt}{r^2\pi}} - 3\frac{Dt}{r^2} \tag{6}$$

For long times, valid for $M_t/M_0 > 0.6$

$$\frac{M_t}{M_0} = 1 - \frac{6}{\pi^2}\exp\left(\frac{-\pi^2 Dt}{r^2}\right) \tag{7}$$

This expression can be rearranged into the linear form

$$\ln\left(1 - \frac{M_t}{M_0}\right) = \ln\left(\frac{6}{\pi^2}\right) - \frac{\pi^2 Dt}{r^2} \tag{8}$$

Thus, a graph of $\ln(1 - M_t/M_0)$ against time will have a limiting slope of $(-\pi^2 D)/r^2$ at long times, enabling the diffusion coefficient of the drug in the particle to be calculated (65).

Higuchi (66) developed a square root of time equation to describe drug release from spherical monolithic inert polymeric devices containing a drug

$$M_t = \sqrt{Dc_m(2c_t - c_s)t} \tag{9}$$

In this expression, c_m is the solubility of the drug in the matrix, c_t the initial drug concentration, and c_s the drug solubility in the sink phase.

An approximate relation derived from the Higuchi equation (66) describes the release of a dispersed solute from a rigid matrix of general shape where there is no swelling or erosion (67–69).

$$M_t = 4r^2\pi\left[\sqrt{2(c_0 - c_s)Dc_s t} - \frac{4c_s D}{9r}\left(\frac{c_s}{2c_0 - c_s} - 3\right)t\right] \tag{10}$$

or

$$M_t = B_1 t^{1/2} - B_2 t \tag{11}$$

Here, M_t is the cumulative absolute amount of drug released at time t; r represents the radius of the spherical device; c_0 and c_s are the initial drug concentration and the solubility of the drug within the system, respectively; D the constant diffusion coefficient of the drug. It follows that, regardless of the device shape, for short times (small values of t) the mass released is proportional to the product of $t^{1/2}$ and the area of the matrix. For large values of t the contribution of the second term becomes significant, and consequently the particle shape becomes important.

Heterogeneous Particle Mechanisms

The majority of microparticle systems are heterogeneous, and drug diffusion through them is more complex than that would be described by the simple Fickian processes. Early papers use the term "anomalous diffusion" to indicate that the lumped behavior does not follow Fickian kinetics. It is important to realize that this is simply because of the microscopic complexity of the system rather than any fundamental breakdown of the physics of homogeneous diffusion. Terms like this have gradually disappeared as our ability to model diffusion in complex media has developed.

Release of drugs from heterogeneous systems has often been treated using expressions similar to the Higuchi equation (66). Because this was one of the first models available to account for drug-release data it has often been inappropriately applied to much more complex systems.

Dispersed and Phase-Separated Systems. A potentially common system is that in which the drug has phase separated from the particle matrix into pure domains, and is then released by dissolution and diffusion through the matrix. Baker and

Lonsdale (65) studied a system of this type, and they found that the release rate was given by

$$\frac{3}{2}\left[1-\left(1-\frac{M_t}{M_0}\right)^{2/3}\right]-\frac{M_t}{M_0}=\frac{3C_s}{r^2}\frac{Dt}{A} \tag{12}$$

where C_s is the drug solubility in the dissolution medium and A the drug loading per unit volume. The drug particles are assumed to be smaller when compared with the particle radius, and the matrix is assumed to remain intact (unlike a swelling or eroding polymer system).

The drug release from a porous or granular matrix can be described by the Higuchi equation

$$M_t=\left[\frac{D\varepsilon}{\tau}(2c_0-\varepsilon c_s)c_s t\right]^{1/2} \tag{13}$$

where M_t is the amount drug released at time t, c_0 the initial drug concentration in matrix, c_s the drug solubility in the medium, D the diffusion coefficient of drug in the medium ε the polymer porosity, and τ the polymer tortuosity (66). The drug is able to pass out of the matrix through fluid-filled channels and does not pass through the polymer directly; again, this is a highly simplified version of the behavior of real polymer particles.

As the loading of solid drug in the matrix is increased, ultimately a point will be reached at which the drug particles are in contact with each other. This normally occurs at loadings of approximately 10% to 20%. In this case, as the drug diffuses out of the matrix, solvent-filled channels are left, which act as pathways through which the remaining drug is preferentially released. The system is said to have reached a "percolation threshold" and as a result, the drug release rate increases rapidly. The full expression for release from a spherical matrix of this type is rather complex and involves terms for the porosity and tortuosity of the matrix (66). These terms can only be accessed with difficulty, and it would appear that such treatments are straining the limits of the pure analytical approach to drug-release modeling. Ultimately, numerical simulation may prove to be more useful for these systems.

Examples of pure multiphase systems are difficult to find, because many fall into the category of erodable or degradable systems rather than diffusive systems. However, one important class of multiphase systems is multiple emulsions, in which the drug is carried in an inner (normally aqueous) phase within an oil droplet. These systems have been extensively studied, although they have not been widely exploited, primarily, owing to the stability problems of the inner aqueous droplets within the oil droplets. Magdassi and Garti (70) suggested that the Baker and Lonsdale (65) expression should be applicable to release from multiple emulsions, and they found a reasonable agreement with experiment, as did Raynal et al. (71), who studied the release of sodium lactate from a liquid paraffin/isopropyl myristate multiple emulsion system.

Erodible and Biodegradable Systems. These represent an increasingly important class of particles owing to the current interest in biodegradable particles for drug delivery, and particularly for the delivery of macromolecules such as polypeptides and DNA fractions. Erodible and biodegradable polymers can degrade in vitro and in vivo on a timescale of hours to weeks, resulting in the formation and growth of micropores and mesopores; thus increasing the drug release rate (14,72–74). The chemical reactions and mass transfer processes controlling drug release from bioerodible delivery systems strongly depend on the specific device characteristics, such as the type of polymer, particle size, polymer blend, solid-state drug–polymer solubility, type of drug, crystallinity of the polymer and composition (72,75–81). Drug-loaded microparticles prepared by salting-out and spontaneous emulsification methods showed a much diminished burst release both in vitro and in vivo (82,83).

Hopfenberg (84) derived expressions for drug release from erodible slabs, cylinders, and spheres. The main argument used is that the erosion rate is proportional to the continuously changing area of the device. The drug release from a surface eroding particle, in contrast to a bulk-eroding particle, is given by

$$\frac{M_t}{M_0} = 1 - \left(1 - \frac{K_e t}{c_0 r}\right)^3 \tag{14}$$

where K_e is an erosion rate constant, r is the initial radius of the sphere, and c_0 the concentration of drug in the sphere.

The mathematical modeling of drug release from bioerodible delivery systems is rather complex (72,85). In addition to physical mass transport phenomena, chemical reactions decreasing the average polymer molecular weight have to be considered. The drug diffusion coefficient depends on the decreasing average polymer molecular weight (M_w) in bioerodible drug delivery systems. It is possible to take an empirical approach to a problem of this type, for example (63,69).

$$D(M_w) = D_0 + \frac{k}{M_w} \tag{15}$$

where D_0 is the diffusion coefficient of the drug in the nondegraded polymer ($t = 0$), and k is a constant.

Considering pseudo first-order polymer degradation kinetics upon contact with the release medium, the decrease in polymer molecular weight of PLGA can be empirically described as follows

$$M_{(t)} = 78.4 \exp(-k_{degr} t) \tag{16}$$

where k_{degr} is the pseudo first-order degradation rate constant of the polymer (69).

For reasons of simplicity, polymer degradation and the subsequent diffusion of the degradation products out of the device are simulated as one event (polymer erosion). As soon as a polymer chain is exposed to water, its "lifetime" starts to decrease. After the latter has expired, the chain is then assumed to erode instantaneously. The "lifetime," $t_{lifetime}$, of a chain is calculated using randomizing functions controlled by a variable ε (integer between 0 and 99)

$$t_{lifetime} = t_{average} + \frac{(-1)^{\varepsilon}}{\lambda} \ln\left(1 - \frac{\varepsilon}{100}\right) \tag{17}$$

where $t_{average}$ is the average "lifetime" of the original polymer chains upon contact with water, and λ is a constant for a specific type and physical state of the polymer (69).

A more fundamental model to describe drug release from bioerodible microparticles was developed based on Fick's second law for spherical geometry, a Higuchi-like pseudo-steady-state approach, and the dependence of the drug diffusivity on the polymer molecular weight (63,69,73).

$$\frac{\partial c}{\partial t} = \frac{1}{r}\left[\frac{\partial}{\partial r}\left(rD\frac{\partial c}{\partial r}\right) + \frac{\partial}{\partial \theta}\left(\frac{D}{r}\frac{\partial c}{\partial \theta}\right) + \frac{\partial}{\partial z}\left(rD\frac{\partial c}{\partial z}\right)\right] \quad (18)$$

Here, c and D are the concentration and diffusion coefficient of the drug, r denotes the radial coordinate, z and θ are polar coordinates, and t represents time (73).

One of the most complete theoretical models of erosion and macromolecular drug release from biodegradable microspheres is that of Langer and coworkers (14). As the microsphere degrades, drug is released by desorption and diffusion. Desorption is assumed to originate with drug initially contained on the sphere surface, and in mesopores connected to the external surface of the microsphere, whereas drug diffusion is delayed by an induction time sufficient to allow for micropores to coalesce and permit the passage of the macromolecular drug out from the occlusions through the microporated matrix.

The initial burst of macromolecular release of the adsorbed drug from a microsphere is given by

$$C_{\mathrm{d}}^{\mathrm{s}}(t) = C_{\mathrm{d}}^{\mathrm{s}}(0)\mathrm{e}^{-k_{\mathrm{d}}t} \quad (19)$$

where $C_{\mathrm{d}}^{s}(0)$ and $C_{\mathrm{d}}^{s}(t)$ are the surface drug concentration at 0 and t time; k_{d} is the drug desorption rate constant (14). The mass of drug in the microsphere then decays exponentially during desorption

$$\frac{M_{\mathrm{d}}(t)}{M_{\mathrm{d}}(0)} = 1 - \phi_{\mathrm{d}}^{\mathrm{burst}}(1 - \mathrm{e}^{-k_{\mathrm{d}}t}) \quad (20)$$

Here, M_{d}(t) and M_{d}(0) are the mass of drug present in the microsphere at time 0 and t, respectively; $\phi_{\mathrm{d}}^{\mathrm{burst}}$ is the mass fraction of drug involved in the burst (adsorbed to the surface of mesopores and outer microsphere surface) relative to the drug initially present.

During the hydration and degradation of the bulk-eroding polymeric system, the mean pore radius leading from the macropores to the external bath grows until the radius is equal to the Stokes–Einstein radius of the macromolecular drug. The corresponding time is called the induction time (t^{d}). Fickian diffusion of the drug through the porous microsphere then allows the mass of drug remaining in the microparticle to be determined as follows:

$$\frac{M_{\mathrm{d}}(t)}{M_{\mathrm{d}}(0)} = 1 - \phi_{\mathrm{d}}^{\mathrm{burst}}(1 - \mathrm{e}^{-k_{\mathrm{d}}t}) - (1 - \phi_{\mathrm{d}}^{\mathrm{burst}})\left(1 - \frac{6}{\pi^2}\sum_{j=1}^{\infty}\frac{\mathrm{e}^{-j^2\pi^2 D(t-t^{\mathrm{d}})/r_0^2}}{j^2}\right) \quad (21)$$

where D is an effective drug diffusivity. This equation can apply for all the times because $D = 0$ for the times $t \leq t^{\mathrm{d}}$ (14).

Swellable Polymeric Particles. Many systems display swelling behavior, i.e., an absorption of the bath fluid, generally into a glassy matrix. Typical examples are amorphous polymers (such as some polyesters), and polysaccharides, both of which are widely used for drug delivery. The kinetics of the drug release process can be considerably influenced by this swelling process. It can change the drug distribution inside the polymeric particles, and the drug dissolution properties (for example, via the pH or the ionic strength). In addition, it changes the distribution of the polymer in the matrix, as the solvent gradually "dilutes" the system.

To simplify the analysis of data from swelling polymeric devices of varying geometry, an empirical power law expression was used to relate the fractional release of drug to the release time

$$\frac{M_t}{M_\infty} = kt^n \tag{22}$$

where M_t/M_∞ is the fractional solute release, t is the release time, k is a constant, and n is the exponent characteristic of the release mechanism, and is usually called the "diffusional exponent" (86,87). With a suitable value of n, this equation often adequately fits the data until around 60% of the total amount of drug is released from slabs, spheres, cylinders, and discs from both swellable and nonswellable matrices (Table 1).

More models that are recent have sought to take into account the polymer flow during swelling. One of the more successful approaches is that of Camera-Roda and Sarti, who described the polymer flow as a sum of Fickian diffusion and relaxation processes. The controlling factor in this model is a dimensionless group called the Deborah number (*De*). This was named after the Biblical prophet Deborah, who allegedly said that "the mountains flow on the Lord's time scale, not that of humans," and describes the rate at which the polymer is being deformed compared to its ability to relax. A Deborah number of less than one corresponds to Newtonian viscoelastic behavior, while higher values represent increasingly nonNewtonian systems.

Grassi et al. (88) developed a model of simultaneous dissolution and diffusion controlled drug release from a swellable and nonerodible viscoelastic polymer. The model assumes that the drug is loaded into the dry matrix as a solid phase. Two mass balances were required for both the drug and the incoming penetrant. The Camera-Roda and Sarti equation was used to describe the penetrant uptake and polymer behavior, while Fick's equation accounted for the concentration gradient developing in the matrix, and was used to calculate the drug diffusion. This model indicated that the features of the drug release kinetics were affected by the viscoelastic properties of the swollen matrixes and by the drug dissolution and phase transition rates.

Brazel and Peppas (89) modeled the water swelling and drug diffusion from initially hydrophilic and glassy polymer matrices using concentration-dependent

Table 1 Diffusional Exponent and Mechanism of Diffusional Release from Spherical Nonswellable and Swellable Controlled Release Systems

Controlled release system (Ref.)	Diffusional exponent (n)	Drug release mechanism
	<0.5	Release from porous material
Nonswellable (86)	0.5	Fickian diffusion
	0.5–1.0	Anomalous (non Fickian) transport
	1	Zero-order release
	0.43	Fickian diffusion
Swellable (87)	0.43–0.85	Anomalous (non Fickian) transport
	0.85–1	Case-II transport
	>1	Super-Case II transport

diffusion equations for water and drug. The transport equation for water incorporated a relaxation-dependent mechanism. These equations were solved with suitable boundary conditions and a relaxation-dependent Deborah number. Experimental results from drug release from PVA and poly(2-hydroxyethyl methacrylate) samples were used to determine the validity of the model. Grassi et al. (15) also described a model for drug release from drug delivery systems composed of an ensemble of drug loaded cross-linked polymer particles. The model accounted for the main factors affecting the drug release, such as particle size distribution, the physical state, and the concentration profile of the drug inside the polymeric particles, the viscoelastic properties of the polymer–penetrant system and the dissolution–diffusion properties of the loaded drug.

Many authors have reported experimental data for drug-loaded swelling systems, and space only permits a selection to be mentioned here. Giammona et al. (90) incorporated the anti-inflammatory suprofen in hydrogel biopolymer networks such as alpha, beta-polyasparthydrazide (PAHy) and alpha,beta-poly(*N*-hydroxyethyl)-D,L-aspartamide (PHEA) cross-linked by glutaraldehyde or gamma-rays, respectively. Swelling tests carried out in aqueous media showed that the pH affected the swelling degree of the hydrogels. Experimental data indicated that suprofen was released in a sustained way both from PAHy and PHEA microparticles.

The release of proxyphylline from water-swollen spheres and cylinders of a urethane cross-linked poly(ethylene oxide) showed "anomalous" release profiles for which the exponent of release into water from the dry xerogels was 0.8 compared with 1.0 for Fickian diffusion (91).

The dynamic swelling of gelatin or poly(acrylic acid) microspheres obeyed the "square root of time" law, and a shift from the diffusional to the relaxational process was observed, dependent on the content of poly(acrylic acid) in gelatin microspheres. Drug release was influenced by the poly(acrylic acid) content, the particle size, and by the pH of the medium. The mechanism of release was analyzed by applying an empirical exponential equation and by the calculation of the approximate contribution of the diffusional and relaxational mechanisms to the anomalous release process by fitting the data to the coupled Fickian/Case II equation (92).

Indomethacin-loaded biodegradable polymer hydrogel networks were based on a hydrophilic dextran derivative of allyl isocyanate and hydrophobic poly(D,L)-lactide diacrylate macromer (PDLLAM). As the PDLLAM content increased, the indomethacin diffusion coefficient and release half-life time decreased, while the release increased. The controlled release mechanism was determined by the combination of three factors: the rate and degree of formation of swelling-induced 3D porous structure in the hydrogel, the hydrolytic degradation of PDLLAM components, and the hydrophobic interaction between PDLLAM and indomethacin (93).

Microspheres of gelatin and hyaluronic acid normally swell prior to being digested, and so one must assume that some external enzyme would be transported into the core of the particle during this swelling phase (94,95). Magee et al. (96) demonstrated that the microspheres swelled to several times their initial diameter over 20 to 30 minutes before being degraded over a further 30 to 60 minutes in a trypsin-containing medium.

Jayakrishnan et al. (97) demonstrated that the release of methotrexate from casein microspheres was influenced by the degree of cross-linking as a result of varying the amount of glutaraldehyde cross-linker used in the particle preparation; similar results were found by Rubino et al. (98) and by Dilova and Shishkova (99) for chemically and thermally denatured albumin microspheres. Hollow polyelectrolyte microcapsules made of poly(allylamine hydrochloride) and sodium poly(styrene sulfonate), templated on various cores, manganese and calcium carbonate particles or polystyrene latexes, responded to a change of pH, leading to a swelling of the capsules in basic conditions and a shrinking when the pH was reduced (100).

Ion-Exchange Microspheres. Drug release profiles from ion-exchange microspheres could be modeled by combining ion-exchange kinetics and diffusion-controlled drug release. Cremers et al. (101) modeled this process by assuming that the ion transport into the particle was rapid and in equilibrium with the rapid solvent swelling of the particle. Under these conditions, the diffusion of the drug in the matrix was rate limiting, and so the release kinetics were similar to that of the simple diffusion model. However, there is one important difference; the activity of free drug (i.e., not bound to the polyelectrolyte) in the microsphere is controlled by the ionic strength of the release medium. Thus, the fraction of drug extracted increased as the ionic strength of the medium was increased, although the release rate is constant and controlled by the diffusion coefficient of the drug in the microsphere. Ion-exchange microspheres were designed as a drug delivery system for embolization, coupling the ability to occlude blood vessels with chemotherapy. They were used to evaluate a manufacturing process allowing the control of drug-release rate through reduction of diffusion rate of the drug within the particle by impregnation of calcium alginate inside the porous microspheres (102).

Release from Microcapsules

On immersion of a sample of microencapsulated carriers in an aqueous medium, three steps lead to drug release into the medium: (i) solvent penetration into the core, (ii) dissolution of drug in the core, and (iii) drug efflux caused by diffusion across the coating layer into the bulk aqueous phase (Fig. 2) (4,12). The cumulative fraction of released drug is determined by three rate constants, one for each process mentioned above, together with two dimensionless parameters. These parameters are related to the porosity of the core and the solubility of the drug in the dissolution medium (12).

The diffusion process under steady-state conditions may be characterized by means of Fick's first law. For spherical microcapsules, the permeation rate is given by

$$\frac{\mathrm{d}M_t}{\mathrm{d}t} = \frac{DKA}{h}(c_0 - c) = \frac{DKA\Delta c}{h} \tag{23}$$

where M_t is the mass of drug diffused, $\mathrm{d}M/\mathrm{d}t$ is the steady-state diffusion rate at time t, and c_0 and c are the drug concentrations inside and outside the microcapsules (4). A, the total surface area, is defined as the product of the number of capsules in a sample, N, and the surface area of an individual capsule, i.e., $A = 4N\pi r_{\mathrm{mc}} r_{\mathrm{c}}$ for microcapsules with a thin wall, and $A = N2\pi(r_{\mathrm{mc}}^2 + r_{\mathrm{c}}^2)$ for capsules with a thick wall, where r_{mc} and r_{c} are the microcapsule and core radii, respectively.

At the beginning of drug release c_0 is equal to the drug solubility (the saturation concentration), c_{s} for a certain period. Under sink conditions $c_0 >> c$, the concentration gradient is constant and a pseudo steady state is achieved, with zero-order release:

$$M_t = \frac{DKA \cdot c_{\mathrm{s}}}{h} t \tag{24}$$

It is possible to calculate the percentage of drug release up to which saturation is found in the internal microcapsule volume.

Guy et al. (64) treated the microcapsule problem as being a simple interfacial barrier, and stated that no complete solution existed, and presented approximate solutions for short and long times.

At short times, the release is zero-order and is given by

$$\frac{M_t}{M_0} = 3k\frac{Dt}{r^2} \tag{25}$$

where k is a reduced interfacial rate constant. It is difficult to verify this regimen for real systems, since the presence of a burst release component often overlaps on this timescale. In addition there may be an initial delay phase, as solvent may be needed to diffuse into the core prior to release.

At long times, the release is given by a single exponential decay

$$\frac{M_t}{M_0} = 1 - \exp\left(\frac{-3k_1 t}{r^2}\right) \tag{26}$$

This is a useful method for obtaining k_1, because it is straightforward to linearize:

$$\ln\left(1 - \frac{M_t}{M_0}\right) = \frac{-3k_1 t}{r^2} \tag{27}$$

Thus, a graph of $\ln(1 - M_t/M_0)$ against time will have a limiting slope at long times of $-(3k_1)/r^2$ enabling the interfacial transport rate constant of the drug between the particle and the release medium to be found.

A mathematical model was developed to describe the physical phenomena involved in drug-release from a system of polydisperse microencapsulated particles, based on the hypothesis of a progressive dissolution of the internal solid drug core (because of the solvent penetration through the coating) that produces a liquid solution in the region between the coating and the dissolving solid core (103). The existence of a concentration gradient between the inner solution and the outer release environment determines the drug diffusion through the coating. The model was a good fit for the theophylline release from solid cores coated by an insoluble polymeric layer of ethylcellulose (103).

A full analysis of diffusive release from a coated particle is complex, but it has been fully detailed (16,104). The full solution can be simplified by taking a number of specific cases; the most useful is that for release into an infinite sink. Solutions are also available for the cases where the diffusion coefficient in the core or the inner matrix is large compared to that in the wall. Pekarek et al. (105) described a method of making two-layer microspheres by the control of surface wetting and interfacial energetics. Double-walled microspheres with biodegradable poly(L-lactic acid) (PLLA) shells and PLGA cores were fabricated with highly water-soluble etanidazole entrapped within the core as solid crystals. Release profiles for normal double-walled samples had about 80% of drug released over 10 days after the initial time lag, while for irradiated double-walled samples, the sustained release lasted for more than three weeks (106). The release of albumin from pectin beads was retarded by coating with chitosan, and was dependent on the pH of coating solution and release medium, which affected the degree of swelling of pectin beads (107).

Regulated Drug Release

The most sophisticated drug-delivery systems now attempt to control or regulate the release of drug based on some localized chemical signal or biological need in the tissue. Such systems are called regulated-release devices, and they evidently pose significant difficulties for in vitro characterization, because it is necessary to confirm that the release rate responds to the external stimulus. Although the study of such devices is in its infancy, a number of systems have been developed; these have largely been macroscopic rather than particulate systems. It seems inevitable, however, that this technology will be applied to particulates in the near future. For example, Yui et al. (108) demonstrated that cross-linked hyaluronic acid was degraded by hydroxyl radicals, and that this could be used to trigger degradation of the matrix at inflammation sites. More recently, Napoli et al. (109) has developed microspheres made from oxidation-responsive polymers.

Release can also be triggered by an external stimulus. The best known of these systems are magnetic microspheres, in which the application of an external magnetic field causes both local accumulation and modifies the drug-release rate. Saslawski et al. (110) demonstrated that this technology could be used to deliver insulin from ferrite-containing alginate microspheres, with an oscillating magnetic field causing a 50-fold increase in the insulin-release rate. Edelman et al. (111) reported that the application of an oscillating magnetic field increased the rate of drug release from polymer-drug matrices containing an embedded magnet in vitro and in vivo. Responsive drug release from hydrogels results from the electro-induced changes in the gels, which may deswell, swell, or erode in response to an electric field (112).

Pulsatile Release Systems

There is much interest in producing systems that release their payload in one or more pulses over a controlled period. Such a system would be useful for the delivery of multiple-challenge antigens and peptide hormones. Pulsed systems are generally produced by engineering a time delay into the particle-degradation mechanism; for example, Kibat et al. (113) demonstrated that liposomes coencapsulated with phospholipase would release their drug payload after a delay, corresponding roughly to the time taken for the phospholipase to destroy the integrity of the liposomes. Delayed release can also be achieved with some of the larger PLGA microsphere systems, because these normally show a lag phase prior to release; this corresponds to the time taken for solvent swelling and partial internal polymer degradation (114). Other pulsed-release systems under study include polyiminocarbonates and poly (ortho esters) (115,116). A novel controlled release microcapsule with a thermosensitive coat composed of an ethylcellulose matrix containing nanosized thermosensitive poly(*N*-isopropylacrylamide) hydrogels, which could reversibly change the shell thickness in water with response to an environmental temperature change, made it possible to obtain an "on–off" pulsatile release, which could alter the release rate in the order of a minute, in response to stepwise temperature changes between 30°C and 50°C (117). Kikuchi and Okano (118) described drug delivery devices using hydrogels to achieve pulsed delivery of drugs to mimic the function of the living systems, while minimizing undesired side effects.

EMPIRICAL MODELS AND COMPARISON OF DRUG RELEASE PROFILES

The Food and Drug Administration (FDA) has placed emphasis on the meaningful comparison of drug dissolution profiles or release data in a number of guidance documents. For example, the FDA scale-up and postapproval changes for modified release guidance indicates that similar dissolution profiles for approved and modified formulations may be acceptable justification for certain levels of change, without prior FDA approval or the need to perform bioequivalence studies (119). As a result, interest has focused on methodology for the comparison of dissolution profile data. Comparison of dissolution profiles has extensive application throughout the product development process for various modified release dosage forms, such as in developing in vitro–in vivo correlations, establishing final specifications, and in establishing the similarity of pharmaceutical dosage forms (119). A number of methods to compare drug release profiles have been proposed (119–126). Those methods were classified into several categories.

Statistical Methods Based in the Analysis of Variance or Student's *t*-Tests

Methods based in the analysis of variance can be divided into one-way analysis of variance (ANOVA) and multivariate analysis of variance (MANOVA). The statistical methods assess the difference between the means of two drug release data sets in single time point dissolution (ANOVA or Student's *t*-tests) or in multiple time point dissolution (MANOVA).

Model-Independent Methods

Dissolution Time ($t_{x\%}$) and Assay Time (t_{xmin})

The $t_{x\%}$ parameter is the time necessary to release $x\%$ of drug (e.g., $t_{20\%}$, $t_{50\%}$, $t_{80\%}$). Pharmacopeias very frequently use this parameter as an acceptance limit of the dissolution test (e.g., $t_{45\ \text{min}} \geq 80\%$), but its shortcomings are obvious for devices or systems which are intended to have nonlinear or other "sophisticated" responses (122,123).

Statistical Moments

Statistical moment theory is commonly used as a noncompartmental approach for biopharmaceutical evaluations, e.g., bioavailability and bioequivalence studies, and in vitro–in vivo correlation studies. The fraction of drug released can be regarded as a cumulative frequency function and thus different statistical moments can be calculated. In vitro mean dissolution time of drug, variance of dissolution time (VR), and an associated statistical parameter, the relative dispersion of the concentration–time profile (RD) (127).

The first moment of the dissolution rate–time curve (MDT) is defined as follows (122,123,127):

$$\text{MDT} = \frac{\sum_{j=1}^{n} T_j \Delta M_j}{\sum_{j=1}^{n} \Delta M_j} \tag{28}$$

where j is the sample number, n is the number of dissolution sample times, T is the time at midpoint between t_j and t_{j-1} and ΔM_i is the additional amount of drug dissolved between t_i and t_{i-1}.

The second moment of the concentration–time curve can be used to obtain the VR (122,123,127).

$$\mathrm{VR} = \frac{\sum_{j=1}^{n} (T - \mathrm{MDT})^2 \Delta M_j}{\sum_{j=1}^{n} \Delta M_j} \tag{29}$$

The relative dispersion of dissolution times (RD) is given by (123),

$$\mathrm{RD} = \frac{\mathrm{VR}}{\mathrm{MDT}^2} \tag{30}$$

Difference Factor (f_1) and Similarity Factor (f_2)

Difference factor (f_1) and similarity factor (f_2) are methods that compare the data on a pairwise basis (120–123). The difference factor (f_1) measures the error between two curves over all time points:

$$f_1 = \frac{\sum_{j=1}^{n} |R_j - T_j|}{\sum_{j=1}^{n} R_j} \times 100 \tag{31}$$

where n is the number of points, and R_j and T_j are the percent dissolved of the two formulations at each time point j. The percent error is zero when the test and drug reference profiles are identical, and increase with the dissimilarity between the dissolution profiles. A significant problem with pairwise procedures of this type is that both formulations must be tested at the same set of time points. This is not a problem when designing new procedures, but may prevent its application to older legacy data, or that obtained by different investigators.

f_2 is a logarithmic transformation of the sum-squared error of differences between the test T_j and reference products R_j over all time points

$$f_2 = 50 \times \log\left\{\left[1 + (1/n)\sum_{j=1}^{n} |R_j - T_j|^2\right]^{-0.5} \times 100\right\} \tag{32}$$

To consider two dissolution profiles as "similar," the f_1 values should be close to 0 and values f_2 should be close to 100. Two profiles are never identical so there is the problem of deciding appropriate limits. Current FDA guidelines suggest that two profiles are similar if f_2 is between 50 and 100 (120,121).

f_2 was used to compare the measured and modeled dissolution profiles of matrix-controlled release theophylline pellets containing different ratios of microcrystalline cellulose and glyceryl monostearate, using a model based on artificial neural networks (128). The f_2 results indicated that the predicted dissolution profiles were closely similar to those obtained from the experiments for different matrix ratios. The dissolution profiles of ganciclovir from PLGA microspheres before and after the gamma radiation sterilization process had similar drug release behavior, with f_2 values in the range 51–55 (129).

Rescigno Index

Rescigno proposed a bioequivalence index to measure the similarity between a reference formulation and a test product, based on plasma concentration–time data. This Rescigno index (ξ_1) can also be used to compare drug dissolution concentrations (122)

$$\xi_i = \left(\frac{\int_0^\infty |d_R(t) - d_T(t)|^i \mathrm{d}t}{\int_0^\infty |d_R(t) + d_T(t)|^i \mathrm{d}t} \right)^{1/i} \tag{33}$$

where $d_R(t)$ is the amount of reference product dissolved, and $d_T(t)$ is the amount of test product dissolved, at each sample time point, and i is 1 or 2 for the first and second Rescigno index. This dimensionless index is 0 when the two release profiles are identical, and 1 when the drug from either the test or the reference formulation is not released at all.

Dissolution Efficiency

The dissolution efficiency (DE) of a dosage form is given by

$$\mathrm{DE} = \frac{\int_0^t y \times dt}{y_{100} \times t} \times 100 \tag{34}$$

where y is the drug percent dissolved at time t. It corresponds to the area under the dissolution-time curve, up to a certain time, t, as a percentage of the area of the rectangle described by 100% dissolution in the same time (122,123).

Chow and Ki's Method

Chow and Ki (130) describe a method for the comparison of dissolution profile data that could be regarded as similar to that used in an assessment of the average bioequivalence of two drug formulations. A test and reference formulation are bioequivalent if the ratio of the mean bioavailability parameters (AUC, C_{max}, etc.) lie within the 80% to 125% bioequivalence limits with 90% confidence. If the ratio of the mean dissolution rates for the test and reference formulations at a particular time point is within the equivalence limits, with a certain level of confidence, the dissolution data for the two formulations are declared to be "locally similar." The dissolution profiles for the two formulations are "globally similar" if they are similar at each dissolution time point.

Model-Dependent Methods

Empirical Mathematical Models

Some of the wide range of models of the drug release process have been previously discussed, many of which lead to useful mathematical descriptions of drug release, and which in turn allow the release process to reveal details of the underlying chemistry of the system. However, many systems are too complex in structure or behavior for an ab initio treatment, or are insufficiently well defined at a microscopic level to allow this type of modeling. In those cases, empirical descriptions can prove useful. Some of the most relevant and more commonly used empirical mathematical models describing the drug release profiles are shown in Table 2 (122).

Table 2 Empirical Mathematical Models Used to Describe the Drug-Release Profiles

Name	Equation	Parameters	References
Baker–Lonsdale	$(3/2)\{1-[-1(M_t/M)_\infty]^{2/3}\}-(M_t/M_\infty)=kt$	k	(65,122)
Biexponential	$M_t/M_0=1-(A\times e^{-at}+B\times e^{-bt})$	a and b	(1)
First-order	$\ln M_t=\ln M_0+kt$	k	(122,123)
Gopertz	$M_t=Ae^{-e-k(t-y)}$	k	(122)
Higuchi equation	$M_t=kt^{1/2}$	k	(66,122,123)
Hixon–Crowell	$M_0{}^{1/3}-M_t{}^{1/3}=k_st$	k	(1,122,123,131)
Hopfenberg	$M_t/M_0=1-[1-k_0t/c_0a_0]^n$	n	(122)
Peppas	$M_t/M_0=kt^n$	k and n	(122,123,132)
Logistic	$M_t=A/[1+e^{-k(t-y)}]$	k	(122)
Noyes–Whitney	$M_t=A(1-e^{-kt})$	k	(133)
Quadratic	$M_t=100(k_1t^2+k_2t)$	$k_1$and k_2	(122)
Weibull	$\log\{-\ln[1-(M_t/M_\infty)]\}=b\times\log t-\log a$	a and b	(122,123)
Zero-order	$M_t=M_0+kt$	k	(122,123)

Fitting Equations

Correlation Coefficients. Correlation coefficients are frequently used as measures of similarity, particularly between a dataset and an equation (123). These measures can also be applied to the comparison of dissolution curves. The correlation coefficient (R) quantifies the relationship between two sets of variables X and Y:

$$R=\frac{\sum(x-\bar{x})(y-\bar{y})}{\sqrt{\sum(x-\bar{x})^2\sum(y-\bar{y})^2}} \tag{35}$$

where x and y are the data points and $\bar{x}$ and $\bar{y}$ are the corresponding means. The correlation coefficient ranges from -1 to $+1$. For $R=\pm1$ all points (x,y) exist on a straight line. If $r=0$ then X and Y are completely uncorrelated, and intermediate values indicate a partial correlation, the closer $|R|$ is to 1. For the same number of parameters, the coefficient (R) can be used to determine the most appropriate of a collection of model equations (Table 2). For example, camptothecin-release data from solid–lipid nanoparticles were well fitted to a Weibull distribution. The correlation of the equation was 0.9936 and the half-life value was 23.1 hours (134). However, when comparing models with different numbers of parameters, the adjusted correlation coefficient (R^2_{adjusted}) should be used (122):

$$R^2_{\text{adjusted}}=1-\frac{n-1}{n-p}(1-R^2) \tag{36}$$

where n is the number of dissolution data points, and p is the number of parameters in the model. It is well-known that R will always increase as more parameters are added, whereas R^2_{adjusted} may decrease (indicating over-fit).

Akaike Information Criterion. The Akaike Information Criterion (AIC) is a measure of goodness-of-fit. When comparing several models to a given set of data, the model with the smallest value of AIC is regarded as giving the best fit out of that set of models. The AIC is only appropriate when comparing models using the same weighting scheme.

$$\mathrm{AIC} = n \times \ln(\mathrm{WSSR}) + 2 \times p \tag{37}$$

where n is the number of dissolution data points, p is the number of the parameters of the model, and WSSR is the weighed sum of square of residues, given by

$$\mathrm{WSSR} = \sum_{j=1}^{n} w_i (y_i - \bar{y}_i)^2 \tag{38}$$

where w_i is an optional weighting factor and $\bar{y}_i$ denotes the predicted value of y_i. The AIC has become a standard tool in model fitting, and is available in many statistical programs.

Frenning (135) modeled drug release from spherical matrix systems using the Noyes–Whitney and diffusion equations. An approximate analytical formula for calculating the amount of released drug was valid during the early stages of the release process. The analytical approximation provided a good description of the major part of the release profile, irrespective of the dissolution rate. Lin (136) tried 10 separate simple functions as fitting equations for the release of theophylline microcapsules.

Analyzing dissolution results with linear regression is a common practice. The data can be transformed into linear form, although it is preferable to use nonlinear fitting on the original data. Computer programs are available that allow a number of empirical fits to dissolution data to be rapidly compared (125,126,137).

IN VITRO–IN VIVO CORRELATION

In vitro/in vivo correlations (IVIVC) are relationships between in vitro dissolution and in vivo input rate (it is noted that this is *not* similar to the plasma concentration–time curve from which it must be obtained by deconvolution). In cases where a meaningful IVIVC could be developed, this can be used as a surrogate for bioequivalence and minimize the number of the necessary bioequivalence studies. Four categories (Level A, B, C, and Multiple level C) of IVIVC have been described in the FDA guidance (30).

For many controlled release systems, the in vivo release behavior is often not particularly well correlated to the in vitro release (18). For example, in vivo release from polymer microspheres depends on the amount and composition of the surrounding intracellular fluid, and its exchange or replenishment rate, all of which are difficult to reproduce in vitro. However, as our understanding of the behavior of disperse delivery systems has improved, more studies are designing and reporting experiments in which acceptable or good correlations are found. For example, the in vitro release in phosphate-buffered saline and the in vivo release in rats showed an excellent agreement independent of the release rate of 14C-methylated lysozyme from poly(ethylene glycol) terephthalate/poly(butylene terephthalate) microspheres. The IVIVC coefficients obtained from point-to-point analysis (Level A) were greater than 0.96 for all microsphere formulations (29). A linear IVIVC was established for each of the investigated topical emulsion formulations and application times (138). Heya et al. (139) evaluated the in vitro and in vivo release of subcutaneous thyrotrophin releasing hormone from PLGA microspheres, and found a correlation between the in vivo release rate and the in vitro dissolution rate in phosphate buffer (pH 7). A linear correlation between in vitro and in vivo release was found for a methadone implant (140). Tamura et al. (141) reported that the in vivo release rate of cisplatin from PLLA or PLGA microspheres at various blend ratios was in accord with the release rate in vitro. The results indicated that the in vitro dissolution test of cisplatin microspheres was a reasonable estimate of cisplatin release in vivo.

Chen et al. (142) developed an in vitro drug release model to predict the total drug release fraction in the gastrointestinal tract as arising from four consecutive compartments, i.e., stomach, duodenum, jejunum, and ileum. This model was well in accord with the existing in vivo dissolution data of controlled-release pellets of isosorbide-5-nitrate independently obtained through plasma analysis.

SUMMARY

In vitro drug release or dissolution testing is an essential requirement for the new drug development, establishment of in vitro dissolution and in vivo correlation, registration, and quality control of dosage forms. The understanding of microencapsulated systems has developed considerably over the past 10 years. Well-characterized systems are now available, with details of internal morphology, accurate size distributions, and surface properties; and expertise has been gained in the reproducible preparation of high-quality dispersed systems. Publications are increasingly exploring the drug-release mechanism, and models are incorporating increasingly sophisticated physics, including polymer swelling and multiphase compositions. The drug release from microencapsulated systems can be influenced by many factors, such as the structure, materials, morphology, particle size and distribution, porosity, property of encapsulated drug, release medium, etc. Mathematical modeling of the drug release process plays a pivotal role as it supports the identification of the mechanism(s) of drug release and provides guidelines in designing delivery systems. The diffusional process of drug from carrier can be described by Fick's first and second laws, and various model equations. For many polymeric microencapsulated systems, such as biodegradable, matrix erosion, and swelling systems, Fickian diffusion is not the only important factor in controlling release; the chemical reaction, polymer relaxation, and a range of surface chemical processes also contribute to the overall drug release in the polymeric matrix. Some mathematical models and simulating computer programs have been developed, which are helpful in designing the desired drug release formulation and understanding the drug release mechanism. There are few publications on the comparison of the drug release data for a given sample by more than one method. The drug release comparison would answer whether differences in drug release data obtained by various workers were functions of production diversity or different release methodologies.

Under certain conditions, in vitro drug release can be used as a surrogate for the assessment of bioequivalence. In most cases, the in vivo drug release profile shape and/or time course is significantly different from the in vitro data. However, a good IVIVC has been observed for a number of systems. A full understanding of the behavior in vitro and in vivo of drug release from a complex system such as a microparticle suspension is probably beyond the possibility of a rigorous analytic solution. Systems may ultimately be evaluated using numerical simulation of dissolution and diffusion three-dimensionally, with full details of their heterogeneous in vivo environment. Under these circumstances, it finally would become possible to design a microparticle system rationally by computational methods, based on the same philosophy that new chemical entities are developed currently.

Despite these achievements, the two central problems remain in the development and research. First, how do we design a microencapsulated system that will have specific desired drug release properties in vitro and in vivo? Second, even if we know exactly what is required, how do we manufacture it? There is an ever increasing number of articles published and patents applied for relating to these two problems, and we foresee that there will be more products with engineered drug release profiles and efficacy being launched into the market in the near future.

REFERENCES

1. Washington C. Drug release from microparticulate systems. In: Benita S, ed. Microencapsulation: Methods and Industrial Applications. New York: Marcel Dekker, 1996:155–181.
2. Benita S. Microparticulate drug delivery systems: release kinetic models. In: Arshady R, ed. Microspheres, Microcapsules & Liposomes. Vol. 2. Medical & Biotechnology Applications, 1999:155–181.
3. Donbrow M. The relation of release profiles from ensembles to those of individual microcapsules and the influence of types of batch heterogeneity on release kinetics. In: Donbrow M, ed. Microcapsules and Nanoparticles in Medicine and Pharmacy. Boca Raton: CRC Press, 1992:219–237.
4. Jalsenjak I. In vitro release from microcapsules and microspheres. In: Donbrow M, ed. Microcapsules and Nanoparticles in Medicine and Pharmacy. Boca Raton: CRC Press, 1992:193–218.
5. Varde NK, Pack DW. Microspheres for controlled release drug delivery. Expert Opin Biol Ther 2004; 4:35–51.
6. Lee WK, Park JY, Yang EH, et al. Investigation of the factors influencing the release rates of cyclosporin A-loaded micro- and nanoparticles prepared by high-pressure homogenizer. J Contr Rel 2002; 84:115–123.
7. Akbuga J, Bergisadi N. Effect of formulation variables on cis-platin loaded chitosan microsphere properties. J Microencapsul 1999; 16:697–703.
8. Heya T, Okada H, Ogawa Y, Toguchi H. Factors influencing the profiles of TRH release from copoly(dl-lactic/glycolic acid) microspheres. Int J Pharm 1991; 72:199–205.
9. Teixeira H, Dubernet C, Rosilio V, et al. Factors influencing the oligonucleotides release from O-W submicron cationic emulsions. J Contr Rel 2001; 70:243–255.
10. Zuleger S, Fassihi R, Lippold BC. Polymer particle erosion controlling drug release. II. Swelling investigations to clarify the release mechanism. Int J Pharm 2002; 247:23–37.
11. Lavasanifar A, Ghalandari R, Ataei Z, Zolfaghari ME, Mortazavi SA. Microencapsulation of theophylline using ethylcellulose: in vitro drug release and kinetic modelling. J Microencapsul 1997; 14:91–100.
12. Frenning G, Tunon A, Alderborn G. Modelling of drug release from coated granular pellets. J Contr Rel 2003; 92:113–123.
13. Polakovic M, Gorner T, Gref R, Dellacherie E. Lidocaine loaded biodegradable nanospheres. II. Modelling of drug release. J Contr Rel 1999; 60:169–177.
14. Batycky RP, Hanes J, Langer R, Edwards DA. A theoretical model of erosion and macromolecular drug release from biodegrading microspheres. J Pharm Sci 1997; 86:1464–1477.
15. Grassi M, Colombo I, Lapasin R. Drug release from an ensemble of swellable crosslinked polymer particles. J Contr Rel 2000; 68:97–113.
16. Manca D, Rovaglio M. Modeling the controlled release of microencapsulated drugs: theory and experimental validation. Chem Eng Sci 2003; 58:1337–1351.

17. Yenice I, Calis S, Atilla B, et al. In vitro/in vivo evaluation of the efficiency of teicoplanin-loaded biodegradable microparticles formulated for implantation to infected bone defects. J Microencapsul 2003; 20:705–717.
18. Jiang G, Qiu W, De Luca PP. Preparation and in vitro/in vivo evaluation of insulin-loaded poly(acryloyl-hydroxyethyl starch)-PLGA composite microspheres. Pharm Res 2003; 20:452–459.
19. Tuncel T, Bergsadi N, Akin L, Otuk G, Kuscu I. In vitro and in vivo studies on microcapsules and tabletted microcapsules of cephradine. Pharmazie 1996; 51:168–171.
20. Abazinge M, Jackson T, Yang Q, Owusu-Ababio G. In vitro and in vivo characterization of biodegradable enoxacin microspheres. Eur J Pharm Biopharm 2000; 49:191–194.
21. Abu-Izza K, Tambrallo L, Lu DR. In vivo evaluation of zidovudine (AZT)-loaded ethyl cellulose microspheres after oral administration in beagle dogs. J Pharm Sci 1997; 86:554–559.
22. Diaz RV, Llabres M, Evora C. One-month sustained-release microspheres of 125I-bovine calcitonin In vitro–in vivo studies. J Contr Rel 1999; 59:55–62.
23. Dickinson PA, Kellaway IW, Taylor G, Mohr D, Nagels K, Wolff H-M. In vitro and in vivo release of estradiol from an intra-muscular microsphere formulation. Int J Pharm 1997; 148:55–61.
24. Hasan M, Najib N, Suleiman M, El-Sayed Y, Abdel-Hamid M. In vitro and in vivo evaluation of sustained-release and enteric-coated microcapsules of diclofenac sodium. Drug Dev Ind Pharm 1992; 18:1981–1988.
25. Liu Y, Schwartz JB, Schnaare RL, Sugita ET. A multi-mechanistic drug release approach in a bead dosage form and in vivo predictions. Pharm Dev Technol 2003; 8:419–430.
26. Lin S, Chao PY, Chien YW, et al. In vitro and in vivo evaluations of biodegradable implants for hormone replacement therapy: effect of system design and PK-PD relationship. AAPS Pharm Sci Tech 2001; 2:E16.
27. Schliecker G, Schmidt C, Fuchs S, Ehinger A, Sandow J, Kissel T. In vitro and in vivo correlation of buserelin release from biodegradable implants using statistical moment analysis. J Contr Rel 2004; 94:25–37.
28. Ebel JP, Jay M, Beihn RM. An in vitro/in vivo correlation for the disintegration and onset of drug release from enteric-coated pellets. Pharm Res 1993; 10:233–238.
29. van Dijkhuizen-Radersma R, Wright SJ, Taylor LM, John BA, de Groot K, Bezemer JM. In vitro/in vivo correlation for 14C-methylated lysozyme release from poly(ether-ester) microspheres. Pharm Res 2004; 21:484–491.
30. Uppoor VR. Regulatory perspectives on in vitro (dissolution)/in vivo (bioavailability) correlations. J Control Release 2001; 72:127–132.
31. Sadeghi F, Ford JL, Rajabi-Siahboomi A. The influence of drug type on the release profiles from Surelease-coated pellets. Int J Pharm 2003; 254:123–135.
32. Cheu SJ, Chen RL, Chen JP, Lin WJ. In vitro modified release of acyclovir from ethyl cellulose microspheres. J Microencapsul 2001; 18:559–565.
33. Lecomte F, Siepmann J, Walther M, MacRae RJ, Bodmeier R. Polymer blends used for the coating of multiparticulates: comparison of aqueous and organic coating techniques. Pharm Res 2004; 21:882–890.
34. Kostanski JW, De Luca PP. A novel in vitro release technique for peptide containing biodegradable microspheres. AAPS Pharm Sci Tech 2000; 1:E4.
35. Yang S, Ge HX, Hu Y, Jiang XQ, Yang C. Doxorubicin-loaded poly(butylcyanoacrylate) nanoparticles produced by emulsifier-free emulsion polymerization. J Appl Polym Sci 2000; 78:517–526.
36. Yang S, Zhu JB. Preparation and characterization of camptothecin solid lipid nanoparticles. Drug Dev Ind Pharm 2002; 28:265–274.
37. Washington C. Evaluations of non-sink dialysis methods for the measurement of drug release from colloids: effects of drug partition. Int J Pharm 1989; 56:71–74.

38. Levy MY, Benita S. Drug release from submicronized o/w emulsion: a new in vitro kinetic evaluation model. Int J Pharm 1990; 66:29–37.
39. Chidambaram N, Burgess DJ. A novel in vitro release method for submicron sized dispersed systems. AAPS Pharm Sci 1999; 1:1–9.
40. Venkateswarlu V, Manjunath K. Preparation, characterization and in vitro release kinetics of clozapine solid lipid nanoparticles. J Control Release 2004; 95:627–638.
41. Jain RA, Rhodes CT, Railkar AM, Malick AW, Shah NH. Controlled release of drugs from injectable in situ formed biodegradable PLGA microspheres: effect of various formulation variables. Eur J Pharm Biopharm 2000; 50:257–262.
42. Jiao YY, Ubrich N, Hoffart V, et al. Preparation and characterization of heparin-loaded polymeric microparticles. Drug Dev Ind Pharm 2002; 28:1033–1041.
43. Woo BH, Na KH, Dani BA, Jiang G, Thanoo BC, De Luca PP. In vitro characterization and in vivo testosterone suppression of 6-month release poly(D,L-lactide) leuprolide microspheres. Pharm Res 2002; 19:546–550.
44. Pean JM, Venier-Julienne MC, Boury F, Menei P, Denizot B, Benoit JP. NGF release from poly(D,L-lactide-co-glycolide) microspheres. Effect of some formulation parameters on encapsulated NGF stability. J Control Release 1998; 56:175–187.
45. Magalhaes NS, Fessi H, Puisieux F, Benita S, Seiller M. An in vitro release kinetic examination and comparative evaluation between submicron emulsion and polylactic acid nanocapsules of clofibride. J Microencapsul 1995; 12:195–205.
46. Minagawa T, Kohno Y, Suwa T, Tsuji A. Entrapping efficiency and drug release profile of an oil-in-water (o/w) emulsion formulation using a polydimethylsiloxane-coated glass bead assay. Pharm Res 1994; 11:503–507.
47. Magenheim B, Levy MY, Benita S. A new in vitro technique for the evaluation of drug release from colloidal carriers-Ultrafiltration technique at low pressure. Int J Pharm 1993; 94:115–123.
48. Ammoury N, Fessi H, Devissaguet JP, Puisieux F, Benita S. In vitro release kinetic pattern of indomethacin from poly(D,L-lactide) nanocapsules. J Pharm Sci 1990; 79:763–767.
49. Widder KJ, Flouret G, Senyei AE. Magnetic microspheres: Synthesis of a novel parenteral drug carrier. J Pharm Sci 1979; 68:79–81.
50. Koosha F, Muller RH, Davis SS. A continuous flow system for in vitro evaluation of drug-loaded biodegradable colloidal carriers. J Pharm Pharmacol 1988; 40:131P.
51. Washington C, Koosha F. Drug release from microparticulates; deconvolution of measurement errors. Int J Pharm 1990; 59:79–82.
52. Illum L, Khan MA, Mak E, Davis SS. Evaluation of carrier capacity and release characteristics for poly(butyl-2-cyanoacrylate) nanoparticles. Int J Pharm 1986; 30:17–28.
53. Washington C, Evans K. Release rate measurements of model hydrophobic solutes from submicron triglyceride emulsions. J Control Release 1995; 33:383–390.
54. Narasimhan B. Accurate models in controlled drug delivery systems. In: Wise DL, Brannon-Peppas L, Klibanow AM, Mikos AG, Peppas NA, Trantolo DJ, Wnek GE, Yaszemski MJ, eds. Handbook of Pharmaceutical Controlled Release Technology. New York: Marcel Dekker, 2000:155–181.
55. Kim C. Controlled Release Dosage Forms Design. Lancaster: Technomic Publishing, 2000.
56. Borgquist P, Nevstenb P, Nilsson B, Wallenberg LR, Axelsson A. Simulation of the release from a multiparticulate system validated by single pellet and dose release experiments. J Control Release 2004; 97:453–465.
57. Siepmann J, Faisant N, Akiki J, Richard J, Benoit JP. Effect of the size of biodegradable microparticles on drug release: experiment and theory. J Control Release 2004; 96:123–134.
58. Bulgarelli E, Forni F, Bernabei MT. Can kinetic analysis be a tool for evaluating pore characteristics? J Microencapsul 2000; 17:701–710.

59. Gupta PK, Hung CT, Perrier DG. Albumin microspheres. I. Release characterization of adriamycin. Int J Pharm 1986; 33:137–146.
60. Benita S, Levy MY. Submicron emulsions as colloidal drug carriers for intravenous administration: comprehensive physicochemical characterization. J Pharm Sci 1993; 82:1069–1079.
61. Freiberg S, Zhu XX. Polymer microspheres for controlled drug release. Int J Pharm 2004; 282:1–18.
62. Crank J. The Mathematics of Diffusion. Oxford: Clarendon Press, 1975.
63. Faisant N, Siepmann J, Benoit JP. PLGA-based microparticles: elucidation of mechanisms and a new, simple mathematical model quantifying drug release. Eur J Pharm Sci 2002; 15:355–366.
64. Guy RH, Hadgraft J, Kellaway IW, Taylor MJ. Calculations of drug release rates from spherical particles. Int J Pharm 1982; 11:199–207.
65. Baker RW, Lonsdale HS. Controlled release: mechanisms and rates. In: Tanquary AC, Lacey RE, eds. Controlled Release of Biologically Active Agents. New York: Plenum Press, 1974:15–71.
66. Higuchi T. Mechanism of sustained-action medication: theoretical analysis of rate of release of solid drugs dispersed in solid matrices. J Pharm Sci 1963; 52:1145–1149.
67. Brophy M, Deasy PB. Application of the Higuchi model for drug release from dispersed matrices to particles of general shape. Int J Pharm 1987; 37:41.
68. Koizumi T, Panomsuk SP. Release of medicaments from spherical matrixes containing drug in suspension: Theoretical aspects. Int J Pharm 1995; 116:45–49.
69. Faisant N, Siepmann J, Richard J, Benoit JP. Mathematical modeling of drug release from bioerodible microparticles: effect of gamma-irradiation. Eur J Pharm Biopharm 2003; 56:271–279.
70. Magdassi S, Garti N. A kinetic model for release of electrolytes from W/O/W multiple emulsions. J Control Release 1986; 3:273–277.
71. Raynal S, Grossiord JL, Seiller M, Clausse D. A topical W/O/W multiple emulsion containing several active substances: formulation, characterization, and study of release. J Control Release 1993; 26:129–140.
72. Siepmann J, Gopferich A. Mathematical modeling of bioerodible, polymeric drug delivery systems. Adv Drug Deliv Rev 2001; 48:229–247.
73. Siepmann J, Faisant N, Benoit J-P. A new mathematical model quantifying drug release from bioerodible microparticles using monte carlo simulations. Pharm Res 2002; 19:1885–1893.
74. Lemaire V, Belair J, Hildgen P. Structural modeling of drug release from biodegradable porous matrixes based on a combined diffusion/erosion process. Int J Pharm 2003; 258:95–107.
75. Berkland C, Kim K, Pack DW. PLG microsphere size controls drug release rate through several competing factors. Pharm Res 2003; 20:1055–1062.
76. Blanco-Prieto MJ, Campanero MA, Besseghir K, Heimgatner F, Gander B. Importance of single or blended polymer types for controlled in vitro release and plasma levels of a somatostatin analogue entrapped in PLA/PLGA microspheres. J Control Release 2004; 96:437–448.
77. Cho KY, Choi SH, Kim CH, Nam YS, Park TG, Park JK. Protein release microparticles based on the blend of poly(D,L-lactic-co-glycolic acid) and oligo-ethylene glycol grafted poly(L-lactide). J Control Release 2001; 76:275–284.
78. Panyam J, Williams D, Dash A, Leslie-Pelecky D, Labhasetwar V. Solid-state solubility influences encapsulation and release of hydrophobic drugs from PLGA/PLA nanoparticles. J Pharm Sci 2004; 93:1804–1814.
79. Sendil D, Gursel I, Wise DL, Hasirci V. Antibiotic release from biodegradable PHBV microparticles. J Control Release 1999; 59:207–217.
80. Mallapragada SK, Peppas NA. Crystal dissolution-controlled release systems: I. Physical characteristics and modeling analysis. J Control Release 1997; 45:87–94.

81. Capan Y, Woo BH, Gebrekidan S, Ahmed S, De Luca PP. Influence of formulation parameters on the characteristics of poly(D,L-lactide-co-glycolide) microspheres containing poly(L-lysine) complexed plasmid DNA. J Control Release 1999; 60:279–286.
82. Allemann E, Leroux JC, Gurny R, Doelker E. In vitro extended-release properties of drug-loaded poly(D,L-lactic acid) nanoparticles produced by a salting-out procedure. Pharm Res 1993; 10:1732–1737.
83. Fu K, Harrell R, Zinski K, et al. A potential approach for decreasing the burst effect of protein from PLGA microspheres. J Pharm Sci 2003; 92:1582–1591.
84. Hopfenberg HB. Controlled release from erodible slabs, cylinders, and spheres. In: Paul DR, Harris FW, eds. Controlled Release Polymeric Formulations. ACS Symposium Series. Vol. 33. Washington, DC: America Chemical Society, 1976:182–194.
85. Lee JW, Gardella JA, Hicks W, Hard R, Bright FV. Analysis of the initial burst of drug release coupled with polymer surface degradation. Pharm Res 2003; 20:149–152.
86. Ritger P, Peppas N. A simple equation for description of solute release I. Fickian and non-Fickian release from non-swellable devices in the form of slabs, spheres cylinders or discs. J Control Release 1987; 5:23–36.
87. Ritger P, Peppas N. A simple equation for description of solute release. II. Fickian and anomalous release from swellable devices. J Control Release 1987; 5:37–42.
88. Grassi M, Lapasin R, Pricl S. The effect of drug dissolution on drug release from swelling polymeric matrices: mathematical modeling. Chem Eng Commun 1999; 173:147–173.
89. Brazel CS, Peppas NA. Modeling of drug release from swellable polymers. Eur J Pharma Biopharm 2000; 49:47–58.
90. Giammona G, Pitarresi G, Tomarchio V, De Guidi G, Giuffrida S. Swellable microparticles containing Suprofen: evaluation of in vitro release and photochemical behaviour. J Control Release 1998; 51:249–257.
91. McNeill ME, Graham NB. Properties controlling the diffusion and release of water-soluble solutes from poly(ethylene oxide) hydrogels: 3. Device geometry. J Biomater Sci Polym Ed 1996; 7:937–951.
92. Preda M, Leucuta SE. Oxprenolol-loaded bioadhesive microspheres: preparation and in vitro/in vivo characterization. J Microencapsul 2003; 20:777–789.
93. Zhang Y, Chu CC. Biodegradable dextran-polylactide hydrogel networks: their swelling, morphology and the controlled release of indomethacin. J Biomed Mater Res 2002; 59:318–328.
94. Narayani R, Rao KP. Controlled release of anticancer drug methotrexate from biodegradable gelatin microspheres. J Microencapsul 1994; 11:69–77.
95. Illum L, Farraj NF, Fisher AN, Gill I, Miglietta M, Benedetti LM. Hyaluronic ester microspheres as a nasal delivery system for insulin. J Control Release 1994; 29:133–141.
96. Magee GA, Willmott N, Halbert GW. Development of a reproduceable in vitro method for assessing the biodegradation of protein microspheres. J Control Release 1993; 25:241–248.
97. Jayakrishnan A, Knepp WA, Goldberg EP. Casein microspheres: preparation and evaluation as a carrier for controlled drug delivery. Int J Pharm 1994; 106:221–228.
98. Rubino OP, Kowalsky R, Swarbrick J. Albumin microspheres as a drug delivery system: relation among turbidity ratio, degree of crosslinking, and drug release. Pharm Res 1993; 10:1059–1065.
99. Dilova V, Shishkova V. Albumin microspheres as a drug delivery system for dexamethasone, pharmaceutical and pharmacokinetic aspects. J Pharm Pharmacol 1993; 45:987–989.
100. Dejugnat C, Sukhorukov GB. pH-responsive properties of hollow polyelectrolyte microcapsules templated on various cores. Langmuir 2004; 20:7265–7269.
101. Cremers HFM, Verrijk R, Noteborn HPJM, Bae YH, Kim SW, Feijen J. Adriamycin loading and release characteristics of albumin-heparin conjugate micro-spheres. J Control Release 1994; 29:143–155.

102. Chretien C, Boudy V, Allain P, Chaumeil JC. Indomethacin release from ion-exchange microspheres: impregnation with alginate reduces release rate. J Control Release 2004; 96:369–378.
103. Sirotti C, Colombo I, Grassi M. Modeling of drug-release from poly-disperse microencapsulated spherical particles. J Microencapsul 2002; 19:603–614.
104. Lu SM, Chen SR. Mathematical analysis of drug release from a coated particle. J Control Release 1993; 23:105–121.
105. Pekarek KJ, Jacob JS, Mathiowitz E. Double-walled polymer microspheres for controlled drug release. Nature 1994; 367:258–260.
106. Lee TH, Wang J, Wang CH. Double-walled microspheres for the sustained release of a highly water soluble drug: characterization and irradiation studies. J Control Release 2002; 83:437–452.
107. Kim TH, Park YH, Kim KJ, Cho CS. Release of albumin from chitosan-coated pectin beads in vitro. Int J Pharm 2003; 250:371–383.
108. Yui N, Nihira J, Okano T, Sakurai Y. Regulated release of drug microspheres from inflammation-responsive degradable matrixes of crosslinked hyaluronic acid. J Control Release 1993; 25:133–143.
109. Napoli A, Valentini M, Tirelli N, Muller M, Hubbell JA. Oxidation-responsive polymeric vesicles. Nat Mater 2004; 3:183–189.
110. Saslawski O, Weingarten C, Benoit JP, Couvreur P. Magnetically responsive microspheres for the pulsed delivery of insulin. Life Sci 1988; 42:1521–1528.
111. Edelman ER, Brown L, Taylor J, Langer R. In vitro and in vivo kinetics of regulated drug release from polymer matrices by oscillating magnetic fields. J Biomed Mater Res 1987; 21:339–353.
112. Murdan S. Electro-responsive drug delivery from hydrogels. J Contr Rel 2003; 92:1–17.
113. Kibat PG, Igari Y, Wheatley MA, Eisen HN, Langer R. Enzymatically activated microencapsulated liposomes can provide pulsatile drug release. FASEB J 1990; 4:2533–2539.
114. Eldridge JH, Staas JK, Meulbroek JA, McGhee JR, Tice TR, Gilley RM. Biodegradable microspheres as a vaccine delivery system. Mol Immunol 1991; 28:287–294.
115. Pulapura S, Li C, Kohn J. Structure–property relationships for the design of polyiminocarbonates. Biomaterials 1990; 11:666–678.
116. Wuthrich P, Ng SY, Roskos KV, Heller J. Pulsatile and delayed release of lysozyme from ointment-like poly-(ortho esters). J Control Release 1992; 21:191–200.
117. Ichikawa H, Fukumori Y. A novel positively thermosensitive controlled-release microcapsule with membrane of nano-sized poly(N-isopropylacrylamide) gel dispersed in ethylcellulose matrix. J Contr Rel 2000; 63:107–119.
118. Kikuchi A, Okano T. Pulsatile drug release control using hydrogels. Adv Drug Deliv Rev 2002; 54:53–77.
119. O'Hara T, Dunne A, Butler J, Devane J. A review of methods used to compare dissolution profile data. Pharm Sci & Technol Today 1998; 1:214–223.
120. Liu JP, Ma MC, Chow SC. Statistical evaluation of Similarity factor f2 as a criterion for assessment of similarity between dissolution profiles. Drug Inf J 1997; 31:1255–1271.
121. Shah VP, Tsong Y, Sathe P, Liu JP. In vitro dissolution profile comparison—statistics and analysis of the similarity factor, f2. Pharm Res 1998; 15:889–896.
122. Costa P, Sousa LJM. Modeling and comparison of dissolution profiles. Eur J Pharm Sci 2001; 13:123–133.
123. Costa FO, Sousa JJ, Pais AA, Formosinho SJ. Comparison of dissolution profiles of Ibuprofen pellets. J Control Release 2003; 89:199–212.
124. Fassihi R, Pillay V. Evaluation and comparison of dissolution data derived from different modified release dosage forms: an alternative method. J Control Release 1998; 55:45–55.
125. Adams E, Coomans D, Smeyers-Verbeke J, Massart DL. Application of linear mixed effects models to the evaluation of dissolution profiles. Int J Pharm 2001; 226:107–125.

126. Adams E, Coomans D, Smeyers-Verbeke J, Massart DL. Non-linear mixed effects models for the evaluation of dissolution profiles. Int J Pharm 2002; 240:37–53.
127. Passerini N, Perissutti B, Albertini B, Voinovich D, Moneghini M, Rodriguez L. Controlled release of verapamil hydrochloride from waxy microparticles prepared by spray congealing. J Control Release 2003; 88:263–275.
128. Peh KK, Lim CP, Quek SS, Khoh KH. Use of artificial neural networks to predict drug dissolution profiles and evaluation of network performance using similarity factor. Pharm Res 2000; 17:1384–1388.
129. Herrero-Vanrell R, Ramirez L, Fernandez-Carballido A, Refojo MF. Biodegradable PLGA microspheres loaded with ganciclovir for intraocular administration. Encapsulation technique, in vitro release profiles, and sterilization process. Pharm Res 2000; 17:1323–1328.
130. Chow SC, Ki FY. Statistical comparison between dissolution profiles of drug products. J Biopharm Stat 1997; 7:241–258.
131. Hixon AW, Crowell JH. Dependence of reaction velocity upon surface and agitation: I-Theoretical consideration. Ind Eng Chem 1931; 23:923–931.
132. Peppas NA. Analysis of Fickian and non-Fickian drug release from polymers. Pharm Acta Helv 1985; 60:110–111.
133. Macheras P, Dokoumetzidis A. On the heterogeneity of drug dissolution and release. Pharm Res 2000; 17:108–112.
134. Yang SC, Lu LF, Cai Y, Zhu JB, Liang BW, Yang CZ. Body distribution in mice of intravenously injected camptothecin solid lipid nanoparticles and targeting effect on brain. J Control Release 1999; 59:299–307.
135. Frenning G. Theoretical analysis of the release of slowly dissolving drugs from spherical matrix systems. J Control Release 2004; 95:109–117.
136. Lin SY. In vitro release behaviour of theophylline from PIB-induced ethylcellulose microcapsules interpreted by simple mathematical functions. J Microencapsul 1987; 4:213–216.
137. Lu DR, Abu-Izza K, Chen W. Optima: a Windows-based program for computer-aided optimization of controlled-release dosage forms. Pharm Dev Technol 1996; 1:405–414.
138. Welin-Berger K, Neelissen JA, Emanuelsson BM, Bjornsson MA, Gjellan K. In vitro-in vivo correlation in man of a topically applied local anesthetic agent using numerical convolution and deconvolution. J Pharm Sci 2003; 92:398–406.
139. Heya T, Okada H, Ogawa Y, Toguchi H. In vitro and in vivo evaluation of thyrotrophin releasing hormone release from copoly(D,L-lactic/glycolic acid) microspheres. J Pharm Sci 1994; 83:636–640.
140. Negrin CM, Delgado A, Llabres M, Evora C. In vivo–in vitro study of biodegradable methadone delivery systems. Biomaterials 2001; 22:563–570.
141. Tamura T, Imai J, Tanimoto M, et al. Relation between dissolution profiles and toxicity of cisplatin-loaded microspheres. Eur J Pharm Biopharm 2002; 53:241–247.
142. Chen X, Chen WY, Hikal AH, Shen BC, Fan LT. Stochastic modeling of controlled-drug release. Biochem Eng J 1998; 2:161–177.

8

Manufacture, Characterization, and Applications of Solid Lipid Nanoparticles as Drug Delivery Systems

Heike Bunjes
Department of Pharmaceutical Technology, Institute of Pharmacy, Friedrich Schiller University Jena, Jena, Germany

Britta Siekmann
Ferring Pharmaceuticals A/S, Ferring International Center, Kay Fiskers Plads, Copenhagen, Denmark

Dedicated to the Memory of Professor Kirsten Westesen.

INTRODUCTION: THE RATIONALE OF USING BIODEGRADABLE, NANOPARTICULATE SOLID LIPIDS IN DRUG DELIVERY

A large number of drug substances are characterized by poor solubility in aqueous media, which may cause formulation problems. Encapsulation in colloidal carrier systems is an alternative way to render poorly water-soluble drugs applicable by parenteral and nonparenteral routes. The particulate carrier may also protect the drug from degradation in biological fluids. Furthermore, incorporation of drugs in particulate carriers provides a possibility to manipulate drug release and to alter the biodistribution of drugs. In this context, colloidal carrier systems have attracted growing interest concerning drug delivery to site-specific targets, especially in cancer chemotherapy (1–3). During the last decades, several approaches have been used to develop submicron-size drug delivery systems. Based on the carrier material, the conventional vehicles can generally be divided into two groups: polymeric and lipidic systems.

Polymeric nanoparticles consist of nonbiodegradable synthetic polymers or, preferably, biodegradable macromolecular materials of synthetic, semisynthetic, or natural origin. The methods for the preparation of polymeric nanoparticles, such as emulsion polymerization and solvent evaporation techniques, often involve the use of toxicologically harmful excipients and additives, e.g., organic solvents, cancerogenic monomers, and reactive crosslinking agents, the complete removal of which from the product is very difficult (4). Moreover, the carrier material in itself can be a potential toxicological risk. Apart from polymer accumulation upon repeated administration due to slow biodegradation, toxic metabolites may be formed during the biotransformation of polymeric carriers, e.g., formaldehyde as a metabolite of polycyanoacrylates (5). Despite these shortcomings some polymeric microparticle products have reached the market, such as the injectable depot formulations Lupron® (leuprolide), Decapeptyl® (triptorelin), Nutropin® (recombinant human growth hormone), Sandostatin® (octreotide), and Parlodel® (bromocriptine). These products are based on biodegradable (co)polymers of polylactic acid and polyglycolic acid.

To avoid potential toxicological problems associated with polymeric nanoparticles, a great deal of interest is currently being focused on lipid-based carrier systems, inter alia liposomes, and lipid oil-in-water emulsions. These vehicles are composed of physiological lipids, such as phospholipids, cholesterol, and triglycerides, and, due to the biological origin of the carrier material, the toxicological risk is much lower than that of polymeric particles. There are, however, a number of drawbacks inherent in conventional lipid carriers, which are basically related to physicochemical instabilities; e.g., the storage stability of liposomes is limited. In particular, small unilamellar vesicles are in a thermodynamically unfavorable state due to the high curvature of the phospholipid bilayer (6). The incorporation of drugs into the bilayer may further decrease stability. Large-scale production and sterilization of these carriers is complicated, and the stability problems often require lyophilization to ensure adequate shelf life. In spite of these technological problems, some liposomal products have reached the market in recent years, e.g., AmBisome® (amphotericin B), Doxil®/Caelyx® (doxorubicin), and DaunoXome® (daunorubicin).

Submicron-size vegetable oil-in-water (o/w) emulsions have been used as a calorie source in parenteral nutrition for decades (7,8). These systems are manufactured in large scale and display an acceptable long-term stability. Lipid emulsions have therefore been extensively investigated as drug carrier systems (9–11). Despite these efforts and the available production technology, there are considerably few drug-containing colloidal lipid emulsions in the market till now [e.g., Diazemuls®/Diazepam-Lipuro® (diazepam), Liple® (alprostadil), Diprivan® (propofol), Limethason® (dexamethasone palmitate), Lipo-NSAID®/Ropion® (flurbiprofen axetil), Etomidat-Lipuro® (etomidate)], which points to formulation problems caused by the susceptibility of the carrier toward incorporation of drug. This can, for example be attributed to perturbations of the stabilizing emulsifier film caused by the diffusing drug molecules that have a high mobility in the liquid oil phase. These perturbations may induce instabilities of either mechanical (reduction of film elasticity, film rupture) or electrochemical nature (influence on the zeta potential), causing coalescence and particle growth (12). The high mobility of incorporated drug molecules allows them to equilibrate quickly into the aqueous phase ("drug leakage" from the droplets) and may thus cause a fast release of drug from the carrier in biological fluids preventing sustained release from the emulsion formulation (13,14).

Many drawbacks associated with conventional lipid drug carrier systems can be attributed to a fluid-like state of the dispersed lipid. It was therefore obvious to combine the superiorities of colloidal lipid carriers, such as the biodegradability and biocompatibility of the carrier material as well as the ease of manufacture of lipid emulsions, with the advantages of the solid-like state of polymeric nanoparticles with respect to stability and sustained drug release. One approach was to prepare aqueous suspensions of lipid nanoparticles by using solid lipids, thus circumventing the toxicological problems associated with polymer particles. Potential advantages of the solid physical state of the dispersed lipid phase are summarized in Table 1.

Table 1 Possible Advantages of the Solid Physical State of Lipid Carrier Systems

Solid particle core
Enhanced physical stability
Enhanced chemical stability (of core material and incorporated substances)
Reduced mobility of incorporated drug molecules
Reduction of drug leakage
Circumvention of instabilities due to interactions between diffusing drug molecules and emulsifier film
Sustained drug release potential
Static interface solid/liquid
Facilitated surface modification

The solid particle core is expected to provide better physical and, in particular, chemical stability than that of liquid or liquid crystalline carriers. Because the mobility of incorporated drug molecules is drastically reduced in a solid phase, leakage of the drug from the carrier and drug migration into the emulsifier film are counteracted. Provided the drug is distributed in the matrix and the matrix does not melt at body temperature, drug release from the carrier is controlled by degradation rather than by diffusion and can thus be controlled to a certain extent by the choice of matrix constituents. Moreover, the presence of a static interface may facilitate surface modification of the carrier particles, e.g., by adsorption of nonionic surfactants. This is of relevance to reduce carrier uptake by the reticuloendothelial system (RES), which is related to surface properties. Surface modification has also been used as an approach to drug targeting with colloidal carriers (15,16).

The production of solid lipid particles in the micrometer size range was reported in the late 1950s and the beginning of the 1960s (17,18). Different preparation techniques have been described, such as milling of drug-containing lipid phase, melt dispersion, solvent evaporation and extraction, and spray drying and congealing (17–28). Lipid microspheres are predominantly used as sustained release formulations for oral and parenteral administration (see chap. 10).

The early concepts to develop lipid-based colloidal suspensions attempted to transfer the production principles of submicron-size phospholipid-stabilized o/w emulsions to the preparation of colloidal lipid suspensions by substituting the dispersed liquid oil phase by solid triglycerides. However, the attempts to formulate long-chain triglycerides into phospholipid-stabilized aqueous suspensions failed due to stability problems such as the formation of semisolid gels (29,30). The first report on nanoparticulate solid lipid carriers is a patent application on the production of oral lipid nanopellets for persorption filed by Speiser in 1985 (31). Product stability is, however, not taken into consideration in the application. The development of colloidal lipid suspensions with satisfactory long-term stability had not been reported in pharmaceutical literature until the beginning of the 1990s. At that time, a small number of research groups focused their activities on the development of solid lipid-based colloidal carrier systems. The activities resulted in different methodologies for the preparation of stable solid lipid nanoparticles (29,32–34). In recent years, the concept of solid lipid nanoparticles has been adopted by an increasing number of research groups all over the world, which has led to refined manufacturing and characterization methodology as well as to numerous application-oriented studies. The increasing interest in solid lipid–based drug delivery systems is also reflected in the increasing amount of published information on these systems (Fig. 1).

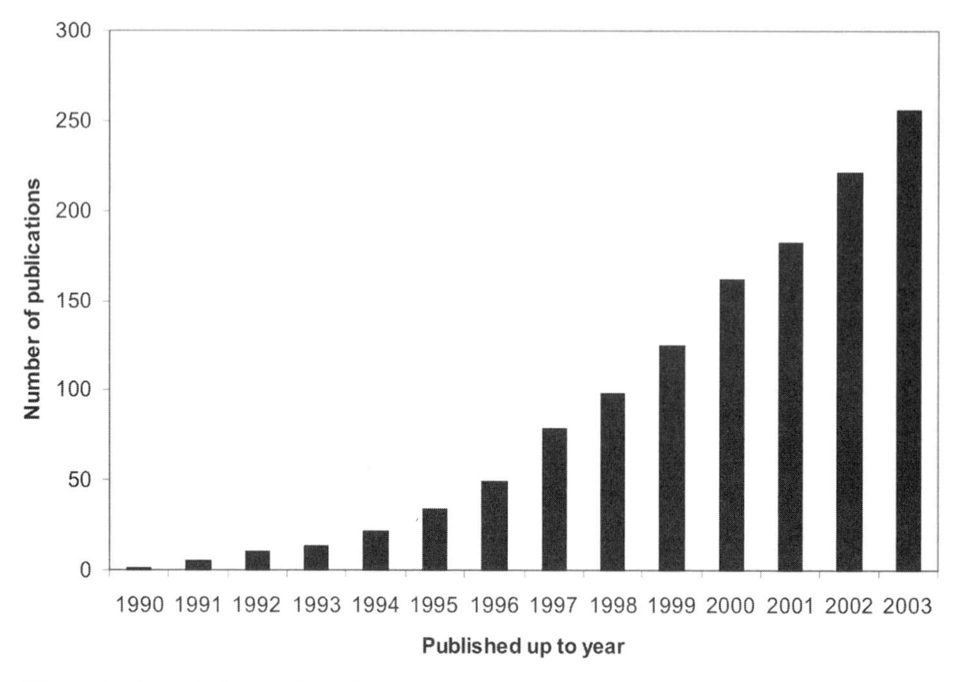

Figure 1 Cumulative number of publications on colloidal solid lipid–based delivery systems from 1990 to 2003 (according to a search performed in International Pharmaceutical Abstracts and Derwent Drug File for Nonsubscribers using the search strategy "solid lipid" OR "lipospheres" OR "lipid nanoparticles"; the search was restricted to abstract/summary).

MANUFACTURING METHODS FOR LIPID NANOPARTICLE SUSPENSIONS

A variety of preparation techniques for lipid nanoparticle suspensions has been developed since the onset of studies on these drug carrier systems around 1990. These techniques can roughly be divided into two groups: those involving high energy dispersion of the lipid phase (melt and cold homogenization, precipitation from solvent-in-water emulsions) and those based on precipitation from homogenous systems (warm microemulsions, solutions in water miscible organic solvents). The choice of the appropriate technique may depend on various parameters such as physicochemical properties and stability of drugs to be incorporated, desired particle size, concentration and stability of the colloidal formulation, and available equipment (Table 2).

Melt Homogenization

Colloidal dispersions of solid lipids can be obtained by dispersing the melted lipid in an aqueous phase with the aid of emulsifiers. As a first approach in this direction, lipid nanopellets as an oral drug delivery system with a particle size (predominantly 80–800 nm) small enough for persorption through the intestinal cell layer were described by Speiser (31). These carriers consisting of water insoluble solid lipids and surface-active agents are obtained by dispersing the lipid melt in the warm aqueous phase by high speed stirring and sonication. Nanopellet suspensions being

Table 2 Comparison of Different Preparation Methods

Method	Advantages	Disadvantages	Particularly suitable for the processing of:
Hot homogenization	Stable dispersions	Use of heat and shear forces	Highly lipophilic, thermally stable drugs
	Good reproducibility	Special manufacturing equipment needed	
	High concentration of lipid phase possible		
	No organic solvents required		
	Established technology		
Cold homogenization	Thermal stress limited	Use of shear forces	Heat sensitive, lipophilic (and hydrophilic) drugs
	Limited partitioning of drug during homogenization	High energy input required	
		Yields comparatively coarse dispersions	
	No organic solvents required	Special manufacturing equipment needed	
Precipitation from solvent-in-water emulsions	Can generate very small particles	Use of organic solvents and shear forces	Heat sensitive, lipophilic drugs
	No heat required	Low lipid concentration (i.e., low drug load)	
		Difficult to scale-up	
		Special manufacturing equipment needed	
Precipitation from warm microemulsions	Simple process, conventional equipment	Often very low initial lipid concentration (i.e., low drug load)	Shear sensitive, lipophilic drugs (highly potent drugs in laboratory environment)
	No shear forces involved	Use of heat	
		Particle growth upon storage	
		Usually requires further processing	
Precipitation from organic solution	Simple process, conventional equipment	Use of organic solvents	Heat and shear sensitive lipophilic drugs
	No heat or shear forces involved	Very low lipid concentration (i.e., low drug load)	

stable upon storage could, however, only be obtained at low lipid concentration; e.g., suspensions with 0.2% to 1% lipid content were stable on storage for three months, whereas higher concentrated suspensions displayed a considerable particle growth and tended to form semisolid gels (35).

In the beginning of the 1990s, melt-emulsification by high-pressure homogenization was introduced as preparation method for solid lipid nanoparticles by the groups of Westesen (29) and Müller (34) (Fig. 2). Dispersions based on a broad variety of solid lipid materials such as different kinds of glycerides, waxes, paraffin, and fatty acids have since then been under investigation (29,30,34,36–119). The solid matrix lipids are melted, and after predispersion (usually by Ultra Turrax vortexing or ultrasonication) the melt is dispersed in an aqueous phase by high-pressure homogenization in the heat with the aid of emulsifying agents. Subsequently, the droplets of the resulting hot colloidal emulsion have to be crystallized. This may occur by simply cooling the dispersion to room temperature but, depending on the composition, can also require specific thermal treatment such as cooling to refrigerator or even to subzero temperatures (e.g., in the case of C14-, C12-, or C10-monoacid triglycerides or many suppository masses) (41,42,48,60,118). Solidification of the matrix lipid should be confirmed experimentally for the composition and preparation conditions chosen. Some studies simply assume that the particles under investigation are in a solid state (e.g., 120–123) even though the matrix lipids used exhibit highly retarded crystallization in the colloidal state and most likely do not crystallize under the preparation conditions applied (40–42,124). The resulting supercooled lipid particles may have higher drug incorporation capacity and may be easier to stabilize than solid lipid nanoparticles but do not have the potential advantages expected for nanoparticles in the solid state (42). For some matrix materials used for the preparation of solid lipid nanoparticle dispersions (e.g., tricaprin, many hard fat suppository bases), the solid state is not stable at body temperature and such particles are going to melt upon administration. Although the advantages of the solid state are lost upon administration of such systems, they may still offer possibilities, e.g., to increase stability upon storage.

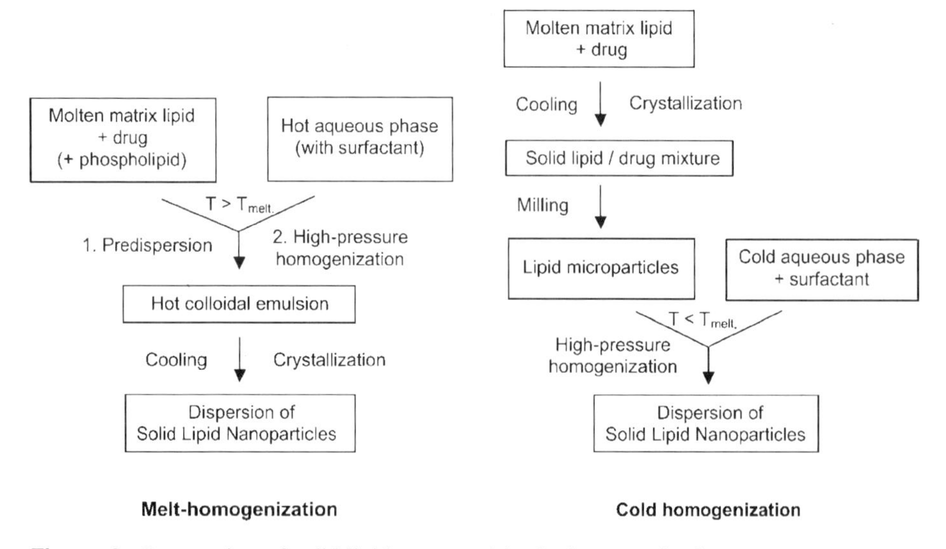

Figure 2 Preparation of solid lipid nanoparticles by homogenization.

The mean particle size of high-pressure melt-emulsified lipid suspensions is typically in the range of 50 to 400 nm. It depends on the composition of the suspensions, e.g., on the matrix constituents, the volume fraction of the dispersed phase, the type and amount of emulsifying agents as well as on homogenization parameters. The mean particle size usually decreases with increasing emulsifier/matrix lipid ratio and with energy input during homogenization; e.g., with application of increased homogenization pressure or prolonged homogenization times or cycle number, the mean particle size generally decreases until it reaches a minimum and then levels off, as does the amount of large particles (34,38,74,114,117). Melt viscosity may also play a role for homogenization efficiency because an increasing particle size was observed with increasing melting point of the matrix constituent or with decreasing homogenization temperature (29,41,94). In summary, the effects of composition and homogenization parameters are basically similar to those found for oil-in-water emulsions, and thus reflect the emulsion-like state of the dispersed molten lipids during emulsification (125,126).

While high-pressure homogenization is currently the most frequently employed technique for the final step in the preparation of solid lipid nanoparticles via the melt, other techniques such as prolonged probe ultrasonication and high-speed mixing are sometimes also used; e.g., a preparation procedure using a high-speed rotor-stator equipment has been described, leading to mean photon correlation spectroscopy (PCS) particle sizes between 100 and 200 nm with polydispersity indices mainly between 0.2 and 0.4 (127–135). Much higher polydispersity values (up to 0.6) have, however, also been observed in some cases, pointing to a considerable heterogeneity of the dispersions (130,132,135). Although no systematic studies on the content of larger particles in these dispersions have been published yet, the authors report on the absence of microparticles as evaluated by laser diffraction in some of their studies (131,133).

The effect of the type and amount of emulsifying agents is very pronounced with respect to both the particle size distribution and the storage stability of melt-emulsified solid lipid nanoparticles. Dispersions stabilized by phospholipids only tend to form semisolid, ointment-like gels upon crystallization of the matrix lipid, in particular when highly phosphatidylcholine-enriched phospholipids are employed. The formation of viscous systems with gelation tendency, particularly upon application of shear stress, has also been observed (29,30,42). This gel formation phenomenon has been attributed to a relative lack of highly mobile surface-active agents that would be required to stabilize the newly formed particle surfaces during crystallization, as the exchange kinetics of excess phospholipids (preferably localized in vesicles) are believed to be too slow to be effective in this process (30). For phospholipid-containing dispersions, gel formation can usually be prevented by the addition of highly mobile coemulsifying agents. A variety of different ionic and nonionic surfactants, such as bile and fatty acid salts, poloxamers, polysorbates, and tyloxapol, have been used for this purpose; e.g., the anionic bile salt glycocholate or the nonionic surfactant tyloxapol proved to be efficient coemulsifiers with regard to long-term storage stability of the liquid dispersions, indicating that electrostatic or steric barriers can counteract gel formation (29,38,39). Coemulsifiers also decrease the mean size of the nanoparticles, and, when properly composed and processed, lead to homogenous dispersions with mean sizes well below that of commercial lipid emulsions. The use of steric stabilizers such as tyloxapol or poloxamers requires relatively higher amounts (on a w/w basis) of surfactants for effective stabilization than does costabilization with ionic surfactants (38,136). The stabilization of triglyceride dispersions with a combination of lecithin with certain nonionic surfactants (e.g., Cremophor® EL, tyloxapol, and polysorbates) may lead to time- and temperature-dependent gel formation, which is, however, often reversible upon storage (46,47). Ionic as well as nonionic stabilizers are also often used without being combined with phospholipids. It has, however, been reported that the trend to form coarse particles, which would render the dispersion unsuitable for IV use, may be higher in dispersions stabilized without the aid of phospholipids (47). For glyceride dispersions stabilized solely with poloxamer, a tendency to form highly viscous or semisolid systems upon shear stress or storage, in particular under stress conditions like light exposure or increased temperature has been observed (36,62,65). The presence of electrolytes may also cause gelation in poloxamer-stabilized glycerolbehenate dispersions (64). On the other hand, storage stability over several years has been observed for optimized systems of this lipid–stabilizer combination under adequate storage conditions (62,65). Particle size stability upon storage over many months and even several years is not unusual for melt-homogenized solid lipid nanoparticle dispersions of optimized composition (29,34,39,45).

Surface-active agents can also be used to modify the surface properties of the nanoparticles; e.g., the particle surface can be covered by surfactants bearing polyoxyethylene chains (such as poloxamers, poloxamines, or PEG-stearate conjugates) for achieving longer circulation times or by surfactants which promote the adsorption of proteins that are assumed to direct the particles to specific body sites (such as Tween® 80) (104). In most cases, these surface-modifying agents are introduced already prior to homogenization and act as (co)emulsifiers during this process. Adsorption of poloxamer and poloxamine on preformed, phospholipid-stabilized particles being still in the emulsion state has, however, also been demonstrated (30). A proper choice of the lipid–surfactant combination can also stabilize the particles against the influence of gastrointestinal media (77). Solid lipid nanoparticles that are surface modified with cationic surfactants and lipids have been prepared by melt-homogenization for the potential use as DNA delivery systems (69,107). Small particles (in the range of 100–150 nm in most cases) with distinctly positive zeta potential were obtained in this way. Systematic studies on the effect of subsequent DNA loading on the particle size do not seem to be available for these systems till now. Characterization of the particle–DNA complexes by atomic force microscopy (AFM), however, indicates that the particle size may increase considerably upon loading with DNA (69). Surface-modified cationic paraffin and glycerolbehenate nanoparticles were prepared to serve as adjuvants for immunization (70). Charge-modifying agents have also been used to balance a negative effect of drug incorporation on the zeta potential (128).

Comparatively high concentrations of lipid phase (typically around 10%) can be processed into stable liquid dispersions of solid lipid nanoparticles by high-pressure melt homogenization. The matrix concentration of liquid dispersions is limited by an increase in viscosity with increasing concentration, which finally leads to gel formation, even in otherwise adequately stabilized dispersions (51,84,85). Other than on the lipid concentration, the flow properties of solid lipid nanoparticle dispersions depend on the type of matrix lipid and the stabilizer composition (probably via an influence on particle shape) and can be influenced by the presence of salt in the dispersion medium (51). In gels formed by cetylpalmitate nanoparticles, the particle size was found to be retained in spite of the high lipid concentration (85). Both the colloidal particle size (resulting in a high specific surface area) and the solid nature of the matrix lipid (leading to anisometric particle shape) are necessary to obtain the desired semisolid consistency of the dispersions (with up to 35% cetylpalmitate) because they increase the potential for particle–particle interactions (86).

Such highly concentrated semisolid nanoparticle formulations may offer potential for the preparation of topical dosage forms which do not require incorporation of the lipid nanoparticles into an additional semisolid system.

Different kinds of active agents such as vitamins and vitamin-like substances (36–38,41,42,45,75,88–95,103,135), sunscreens (74,97,99–101), corticoids (34,52,54, 82,83,106), benzodiazepines (42,111,128), anesthetics (34,53,60,61), and cytostatics (115–118) have been incorporated into melt-homogenized dispersions. In most cases, the active substances are simply dissolved in the lipid melt but mixing by coprecipitation with the lipids prior to melting (118,131–135) or by heating the drug–lipid mixture in ethanol above the melting-temperature under evaporation of solvent (115–117) has also been reported. Drug loads are usually up to 10%; high concentrations, leading to precipitation of drug crystals, the appearance of droplets of drug, or colloidal instability of the dispersions (41,42,61,93,117) drug loads lead to precipitation of drug crystals, the appearance of droplets of drug, or colloidal instability of the dispersions (41,42,61,93,117). Interestingly, higher drug loads have mainly been achieved with liquid lipophilic substances or with substances such as tocopherol (acetate) and ubidecarenone that do not readily recrystallize after melt-dispersion, indicating that the liquid state of incorporated substances may facilitate drug loading (42,45,73,96,103,129). For ubidecarenone, it was demonstrated that higher drug loads were not incorporated homogenously but formed a separate, liquid phase within the single particles (45). Drug loading is also facilitated when liquid triglycerides are incorporated into the composition of the particle matrix due to the higher drug solubility in liquid oils (59,89,90). In these oil-loaded solid lipid nanoparticles (also called nanocompartment carriers or nanostructured lipid carriers), the drug is presumably mainly localized in the liquid fraction. It has been shown that, for a solid–liquid lipid combination (glyceryl behenate–medium chain triglycerides) typically used for the preparation of such nanoparticles, at least a considerable fraction of the liquid lipid is not incorporated into the solid particles but attached to the surface of the solid matrix as a liquid film or a liquid droplet (112,113).

After preparation, melt-homogenized particles can be processed into other than liquid forms. Spray drying and lyophilization have been suggested to circumvent storage problems of liquid systems and to prepare dry formulations, e.g., for the use in oral administration (34,39,60,76,118). Freeze-drying requires the addition of a cryoprotectant to enable redispersion and to suppress particle growth. Even then, redispersion may not be possible by manual treatment and may require additional energy input such as ultrasonication, which would be unfavorable in clinical practice (34,60). In a freeze-drying study using carbohydrates and different polymers as cryoprotectants, trehalose proved to be most efficient for both glycerolbehenate and trilaurin nanoparticles, but in both cases the PCS mean particle size increased considerably during the freeze-drying process as did the number of particles in the micrometer range (60). An optimized formulation was, however, found still suitable for IV administration after redispersion. Freeze-drying of drug-loaded dispersions may even be more problematic, which was attributed to the presence of free drug in the dispersion medium (60). Also for tricaprin nanoparticles—drug-free as well as drug-loaded—a considerable increase in particle size was observed after freeze-drying, in this case without much effect of drug-loading (118). For melt-homogenized stearic acid nanoparticles, a combination of mannitol and glucose performed particularly well as cryoprotectants. The resulting particles could quickly be reconstituted by manual shaking without major changes in particle size as concluded from turbidity measurements (117). In the case of a highly instable drug-loaded dispersion, a complex freezing program helped to preserve the dispersed state of the formulation upon freeze-drying (76). Lyophilization may not necessarily be a suitable technique for improvement of long-term stability because it has been observed that lyophilized powders of melt-homogenized triglyceride nanoparticles displayed an impaired redispersibility and exhibited a pronounced particle growth after storage for 12 months, which was tentatively attributed to sintering processes during storage (39,60). When the dispersed lipid crystallizes only during the freeze-drying process additional complications may be expected, but this phenomenon seems not to have been investigated in detail till now (60,137).

Spray drying of solid lipid nanoparticle dispersions was mainly investigated with respect to incorporation into solid dosage forms (63). Dispersions of glycerol-behenate, cetylpalmitate, and Synchrowax HRSC stabilized with poloxamer 188 were spray dried in the presence of carbohydrates. A high melting point of the matrix lipid was found beneficial to withstand the thermal stress in the apparatus, as the particles should not melt during the process. A higher carbohydrate-to-lipid ratio also improved the quality of the product, with trehalose showing the best protecting properties. The fraction of larger particles could be reduced by the addition of alcohols to the dispersions prior to spraying. Optimized formulations had mostly retained their particle size characteristics after redispersion; but redispersion still required sonication to break down aggregates that had been formed upon drying. For use in peroral drug delivery, the incorporation of melt-homogenized solid lipid nanoparticles into pellet formulations by using the aqueous dispersion as granulation liquid has also been proposed (67).

Although lipid nanosuspensions are often prepared with respect to parenteral administration, which requires sterility, there is only little information on the effect of sterilization on the properties of melt-homogenized solid lipid nanoparticles till now. Müller et al. (34) report briefly that the stability upon autoclaving depends on the nature of the lipid/surfactant combination and that optimized systems can be autoclaved at 121°C. Results of the studies on trilaurin nanoparticles (120,137) that, however, remain in the liquid state after melt-homogenization (40,41,60) are only of limited transferability to solid particles undergoing a crystallization step after heat sterilization. Concerning the stability during heat treatment, information obtained on lipid emulsion systems can presumably be transferred at least to a certain extent to dispersions based on solid lipids. From such studies, it can be deduced that stabilization with certain ethoxylated surfactants such as, e.g., Tween® 80, Cremophor® EL, or Solutol® HS15 may be detrimental for dispersion stability upon autoclaving (138,139). A drastic increase in mean particle size was indeed observed upon autoclaving poloxamer 188–stabilized solid lipid nanoparticles as a result of the incompatibility of this steric stabilizer with the heat treatment employed (128).

Because a major application focus of melt-homogenized lipid nanoparticles is on their dermal administration, the incorporation into semisolid dosage forms such as creams and hydrogels has quite frequently been investigated (73,88,91,95,101, 102,119). In hydrogels, interactions with some gelling agents leading to aggregation of the solid lipid nanoparticles have been observed (88).

High-pressure melt-homogenization has developed into one of the most frequently used methods to prepare solid lipid nanoparticle dispersions. This technology is well established from its use in dairy processing as well as the preparation of parenteral emulsions and, depending on the available equipment, it can be used for the preparation of virtually any batch size from small to large-scale production. Several studies with liquid as well as semisolid nanoparticle dispersions demonstrated a comparatively noncritical scale-up process for these systems from the milliliter to the 20 to 50 kg range (74,86,94). When the appropriate composition and preparation conditions are chosen, the resulting nanoparticles are usually physically stable on long-term storage even in aqueous dispersion. The properties of the nanoparticles, e.g., size, surface characteristics, and melting temperature, can be varied over a wide range via the dispersion composition.

Cold Homogenization

Solid lipid nanoparticles can also be prepared by passing the predispersed matrix lipid through a high-pressure homogenizer at a temperature below its melting point (Fig. 2) (34,52–54,87,140–143). This allows processing also of lipid matrix materials with a melting temperature distinctly above 100°C, e.g., cholesterol (52–54). In most cases, however, lipids with lower melting points (e.g., glyceryl behenate, hard fats) have been processed. Active agents are incorporated into the matrix by dissolving or dispersing them in the melted lipid which is subsequently solidified and ground into a fine powder at low temperature (e.g., under liquid nitrogen or dry ice cooling). This powder is processed into a submicron suspension by high-pressure homogenization in an aqueous surfactant solution (e.g., of poloxamers, sodium cholate, or Tween 80). Because the dispersion of solid lipids requires a higher energy input than that of a liquid melt, harsher homogenization conditions with respect to homogenization pressure and number of homogenization cycles are often applied. The dispersions obtained are still typically of larger mean particle size and of broader size distribution than that resulting from processing of melted lipids, often with particle sizes in the upper nanometer or even in the micrometer range (52,87,140,141). The preparation of dispersions fulfilling the requirements for intravenous (IV) use will thus probably be difficult with this method. As cold homogenization reduces heat exposure to the time required for dissolution of active agents in the lipid melt it may be an alternative to melt homogenization for incorporation of thermally labile drugs, e.g., peptides (141). Moreover, a solid state of the particles prevents partitioning of incorporated drugs into the aqueous phase during homogenization. For prednisolone-loaded glyceryl behenate particles a distinctly higher entrapment efficiency was observed after cold homogenization (85%) than after melt homogenization (50–56%). The authors also observed a reduction of the burst effect in drug release for the cold-homogenized particles (54). The reduced potential for drug leakage into the aqueous phase may also be advantageous for the processing of hydrophilic drugs; e.g., lysozyme or iotrolan were incorporated by solubilizing the substances in the lipid melt with the aid of poloxamers (34,141). In the case of lysozyme, the presence of amphiphilic components (e.g., cetyl alcohol) facilitated solublization compared to pure hard fats. Even upon cold homogenization, the loading of both substances into the lipid phase decreased, however, to about half the initial amount. Friedrich and Müller-Goymann (142) employed cold homogenization to avoid loss of lecithin and solubilized drug into the aqueous phase during processing of solidified reversed micellar solutions (1:1 mixture of lecithin and hard fat) into lipid nanosuspensions. Cold homogenization was also used to incorporate magnetide particles (approximately 90 nm) into solid Compritol® particles (34,143).

Homogenization on cold eliminates the crystallization step for the dispersed lipid and can thus help to circumvent problems with supercooling and polymorphism. It does, however, not necessarily guarantee that the lipid particles indeed remain solid during processing. High-pressure homogenization usually leads to an increase in product temperature, in particular when the device is not actively cooled. If the lipid is processed at a temperature very close to its melting point, softening or (partial) melting during the homogenization step cannot be excluded; e.g., using hard fats with melting temperatures around 40°C in a mixture with phospholipids as matrix lipids, a distinct decrease in mean particle size and polydispersity index was observed when the solidified reversed micellar solutions were processed at room temperature compared to dispersions obtained under ice cooling of the high-pressure homogenizer (142). The measured product temperatures after homogenization at room temperature were in the range of the melting point of the matrix lipids. The authors concluded that a flexibilization or partial melting of the lipids leads to the observed increase in homogenization efficiency.

Precipitation from Solvent-in-Water Emulsions

A method to prepare submicron-size particles of cholesteryl acetate by precipitation in lecithin-stabilized solvent-in-water emulsions was presented by Sjöström and Bergenståhl in 1992 (33). Cholesteryl acetate and lecithin were dissolved in an organic solvent, and the organic solution was emulsified in an aqueous phase containing a cosurfactant by high-pressure homogenization to yield a submicron-size o/w emulsion. Upon removal of the organic solvent by evaporation under reduced pressure, cholesteryl acetate nanoparticles precipitated in the emulsion droplets. With a blend of phosphatidylcholine and sodium glycocholate as emulsifier, particles as small as 25 nm could be obtained. The particle size stability over 30 days decreased with increasing lecithin/bile salt ratio. It was not possible to prepare stable emulsions of the cholesteryl acetate-containing organic solvent in water using pure phosphatidylcholine without the addition of a cosurfactant.

Although the cholesteryl acetate particles were prepared as a model of drug nanoparticles, it should, in principle, also be possible to use this kind of colloidal dispersions as drug carrier systems by incorporating lipophilic drugs into the cholesteryl acetate core (144–146). Moreover, the method was successfully transferred to the manufacture of triglyceride nanoparticles. Tripalmitin suspensions with mean particle sizes ranging from 25 to 120 nm could be obtained; the particle size depended on the lecithin/cosurfactant blend, with sodium glycocholate being the most effective coemulsifier (147). The storage stability of precipitated tripalmitin nanoparticles is, however, poorer than that of melt-emulsified systems and precipitated suspensions tend to grow in particle size within several weeks or a few months. A major concern with this method is the use of organic solvents which may lead to toxicological problems. Residual amounts of solvent could be detected in tripalmitin dispersions by ^{1}H nuclear magnetic resonance (NMR) spectroscopy and differential scanning calorimetry (DSC). Another limitation of this precipitation method is the low lipid concentration in the suspension after evaporation of the solvent. In the case of tripalmitin precipitated in chloroform emulsions, the maximum lipid concentration achievable was not more than 2.5% (w/w) due to the limited solubility of the triglyceride in the organic solvent. These drawbacks may be the reason why the solvent evaporation method has not received much attention in the field of solid lipid nanoparticles, although it allows processing of high-melting lipids which cannot be used in melt-emulsification and although it may yield extremely small particles that could be advantageous with respect to drug targeting purposes. The absence of heat during the entire preparation process makes this preparation method also interesting for the formulation of heat sensitive drugs. Recently, a modified process using a w/o/w double emulsion approach has been proposed for the processing of hydrophilic macromolecules such as peptides (148). The initial lecithin-stabilized w/o emulsion containing a solution of tripalmitin in dichloromethane (and optionally insulin as a model peptide in the aqueous phase) was dispersed in a second water phase with the aid of poloxamer 188, polyethylene glycol (PEG)-stearate, or sodium cholate. Both emulsification steps were performed with ultrasonication, and the organic solvent was later evaporated under stirring or in a rotary evaporator. The resulting particles, which were mainly prepared with respect to peroral administration, were in most cases of comparatively large size (200–400 nm) at a final solid lipid concentration of approximately 1%.

Another modified procedure has been described by Trotta et al. (149) who used partially water-soluble solvents of comparatively low toxicity (benzyl alcohol, benzyl lactate) as the organic phase of their emulsions. Glycerol monostearate as matrix lipid was dissolved in the organic phase which had previously been equilibrated with the aqueous phase at about 47°C. An emulsion was prepared at the same temperature by Ultra-Turrax stirring using different surfactants and surfactant mixtures as stabilizers. Subsequently, the organic phase was extracted from the emulsion droplets into the continuous phase by dilution of the system with a large amount of water to precipitate solid lipid particles. Washing of the dispersion by diaultrafiltration removed most of the organic solvent (99.8% in the case of benzyl alcohol). The particle sizes obtained by this method varied considerably (from the lower nanometer into the micrometer range) with composition. The best results with respect to small particle sizes were obtained with combinations of lecithin and taurodeoxycholate or cholate as stabilizers, leading to particle sizes between about 150 and 350 nm for lipid concentrations of 2.5% to 5% in the primary emulsion.

Precipitation from O/W Microemulsions

The preparation of solid lipid nanoparticles by precipitation from a warm microemulsion (Fig. 3) was introduced by Gasco and Morel in 1990 (32) and has since then been employed in numerous studies (150–185). The solid lipid particle matrix typically consists of a fatty acid (usually stearic acid) which is formulated into an

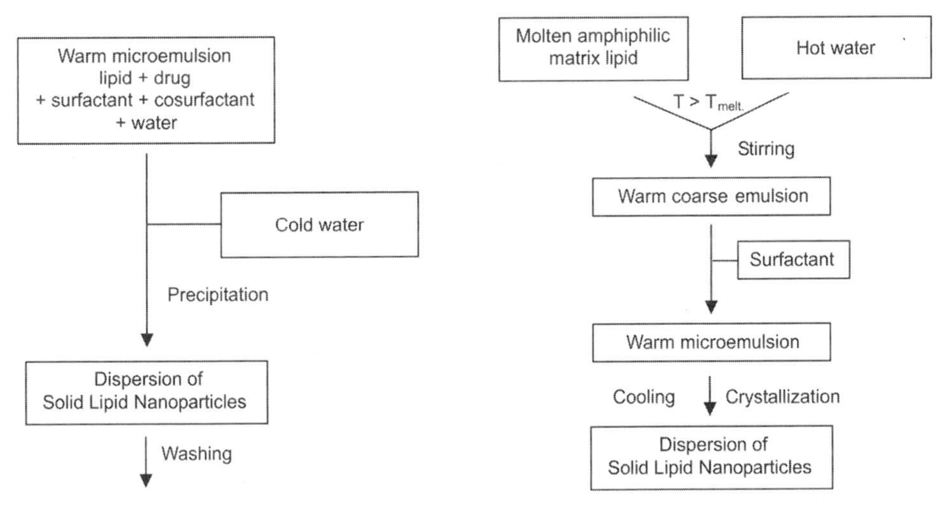

Figure 3 Preparation of solid lipid nanoparticles from warm microemulsions: *Left panel*: According to Gasco et al.; *Right panel*: According to Mumper et al.

aqueous microemulsion system in the heat (e.g., around 70°C) with the aid of a surfactant/cosurfactant system. Other matrix ingredients such as glycerides and a cholesterol ester have occasionally also been processed (162,165,181,182,186). In most cases, egg or soybean phosphatidylcholine, sometimes also Tween 20, is used as surfactant, while bile salts such as taurodeoxycholate, taurocholate, and glycocholate, optionally with the addition of butanol are used as cosurfactants. To precipitate the nanoparticles, the hot microemulsion is dispersed into a cold (2–3°C) aqueous phase under mechanical stirring. Typical microemulsion:water dispersion ratios are 1:10 or 1:5 but other ratios ranging from 1:1 and even 3:1 to 1:100 have also been used (150,158,171,172,184). Dispersing the microemulsion into a large volume of water of low temperature and pH favors the formation of small particles, at least in Tween/butanol-containing systems (150,158,161). The advantageous effect of a large volume of dispersion medium was confirmed for a phospholipid/bile salt system (171).

After precipitation, the nanoparticulate system is washed several times with water by ultrafiltration to remove a major part of the water-soluble compounds, in particular the cosurfactants used for microemulsion formation. Ultrafiltration can also be used to concentrate the systems (170,175,176,179,180,184). Purification of the nanoparticle dispersion by dialysis or centrifugation has also been described (183,185). The washed dispersions may be used directly or after subsequent treatment such as autoclaving, in some cases also freeze-drying (162,169,174,182,186). Freeze-drying is often performed to facilitate analytical investigations, such as determination of drug content. Considerable particle growth with respect to the original formulation has been observed for dispersions that had been freeze-dried without or with only little cryoprotectant (150,162). In spite of these results, there seem to be no intensive attempts to optimize the freeze-drying process with the aim to obtain a product for administration purposes. There are indications that high amounts of cryoprotectants or the presence of surfactants with polyoxyethylene chains may enhance the redispersability (162,166). Recently, an evaporative drying process has been proposed as an alternative approach to stabilize the nanoparticles, which eliminates freezing as a potential source of agglomeration (168,172).

While the first formulations prepared by the microemulsion method were based on Tween 20, taurodeoxycholate, and butanol as surfactant/cosurfactant system, those in more recent studies, particularly for biological investigations, often employ a simple phospholipid/bile salt system and can thus exclusively be based on physiological compounds (151,152,154,167,169,170,172–174,176,178,181). The particle sizes obtained with the phospholipid/bile salt system are much smaller (usually below 100 nm) than those observed for the Tween-based systems (>150 nm, often >200 nm) (32,150,160,161,164). The type of bile salt influences the particle size obtained: Taurocholate derivatives are more efficient in dispersing the matrix lipid into small particles, probably due to their higher degree of dissociation at the pH of preparation as they have lower p*K*a values (175). Moreover, Cavalli et al. (166) propose that undissociated glycocholate might be incorporated into the nanoparticles and thereby increase the particle size. The amount of bile salt needed for the formation of the microemulsion system is considerably high, but more than 70% of the initial amount of bile salt is typically removed during the washing steps (169,173,180).

Drugs can be incorporated into the systems by adding them to the hot microemulsion. A variety of active agents including steroids (150,165), anticancer drugs (154,176–179,181), antibiotics (167,169,170,174), ophthalmics (151,152,159) as well as hydrophilic and lipophilic peptides (155,156,182) but also different contrast agents (157,184) have been incorporated into nanoparticles this way. Labeling of the particles with fluorescent dyes and ^{131}I-17 iodoheptanoic acid as a radioactive marker has also been described (169,173,175,180,181). The concentration of incorporated drug is typically less than 10% as determined from the washed and dried systems. Many of the drugs under investigation, e.g., doxorubicin, idarubicin, tobramycin, or thymopentin are comparatively hydrophilic regarding their use in a lipophilic carrier system. Therefore, alkylphosphates are often added as counterions to increase drug lipophilicity and favor the partitioning into the lipid matrix (151,152,154,156,159,167,169, 170,174,176–179,181). The apparent partition coefficient increases with the lipophilicity of the alkylphosphate but a correlation of the drug loading capacity with the alkylphosphate chain length was not observed (152,159). The use of alkylphosphates also reduces the diameter of the nanoparticles (152). A "water-in-oil-in-water-microemulsion" approach has also been described for the incorporation of hydrophilic substances, e.g., hydrophilic peptides or contrast agents (155–157,184). Using this approach, about 90% of the hydrophilic peptide (D-Trp-6) LHRH could be recovered in a solid nanoparticle dispersion in spite of the washing steps (155). With respect to the overall loading efficiency, this procedure was, however, found less favorable than using the counterion approach because only a very small amount of internal, drug-containing aqueous phase can be processed (156).

The washing step after nanoparticle precipitation is expected to reduce the drug content in the aqueous phase and on the surface of the nanoparticles. Sometimes, considerable amounts of incorporated drug are removed during this process, but in other cases only small amounts of incorporated drug were detected in the washing waters (150,152,173,179).

The surface characteristics of nanoparticles prepared by this microemulsion method can be modified by stealth agents containing polyoxyethylene chains (166, 175,178–180). Preferably, stearic acid–PEG 2000 has been employed for this type of nanoparticles, but the use of dipalmitoylphosphatidylethanolamine–PEG 2000 has also been reported. The incorporation of stealth agents is usually reflected in a concentration-dependent increase in particle size and decrease of the zeta potential as compared to that of the blank nanoparticles.

Although a complex process of particle formation upon precipitation is to be expected, this mechanism has not been studied in much detail till now. Usually, a droplet structure of the microemulsion and crystallization of the droplets upon dispersion into water are assumed (162,164,171,187). A potential participation of phospholipid/bile salt mixed micelles in the solubilization of the matrix lipid in the hot system has not been studied so far. Boltri et al. (158) found a biphasic formation process of nanoparticles (prepared without washing process). The particle size increased rapidly over the first 20 hours followed by a period with slower particle growth until the equilibrium state was reached (158). They discussed deposition of solubilized stearic acid and Ostwald ripening as potential mechanisms for particle growth. Even though these results point to a highly dynamic system, particularly directly after preparation, the time point of particle size measurement is usually not mentioned in the studies employing this method of nanoparticle preparation. Moreover, and quite surprisingly with respect to their frequent use and comparatively long "history" in the field, there is almost no data on the long-term stability of the systems prepared by this method. As a rare exception, Cavalli et al. (163) report on the storage stability of aqueous systems. Their results, which show a dramatic increase in PCS particle size after one year of storage for most systems under investigation, indicate that the problem of particle size stability remains to be solved for this type of system.

Recent results on similar dispersions indicate a correlation between surfactant remaining after washing and storage stability of particle size (185). These authors conclude that ultrafiltration over a 100 kDa membrane was not sufficient to remove excess surfactant from their systems in contrast to dialysis (300 kDa) or ultracentrifugation.

Mumper et al. (188–190) have introduced a simplified approach to prepare lipid nanoparticles from warm microemulsions (Fig. 3). These nanoparticles are based on nonionic emulsifying wax (a mixture of cetyl alcohol and polsorbate 60 in a molar ratio of about 20:1) or Brij® 72 (polyoxyethylene 2 stearyl ether) as matrix material (188,189,191–203). The matrix lipid is melted at moderate temperature (e.g., 55°C) and emulsified into hot water by stirring. Addition of a sufficient concentration of surfactant [e.g., Brij® 78, Tween 80, cetyl trimethyl ammonium bromide (CTAB) or sodium dodecylsulfate (SDS)] transforms the systems into clear microemulsions. The formation of nanoparticles is induced by cooling the microemulsions in the preparation vessel to room temperature. Other procedures, such as dilution into cold water or cooling in a fridge or freezer, have also been tested, but did not show distinct advantages (188,198). Freshly prepared dispersions with optimized composition typically have mean particle sizes around or even distinctly below 100 nm. Plasmid DNA and a cationized protein have been adsorbed to the surface of cationic (CTAB-stabilized) or anionic (SDS-stabilized) nanoparticles, respectively, with the aim to use the systems for immunization (189,191–193,195,197). Also for these nanoparticles, a moderate to considerable increase in particle size was observed upon loading with DNA (189,193,195,197). Hydrophobic DNA complexes with dimethyl dioctadecyl ammonium bromide (DDAB) or dioleyl trimethyl ammonium propane (DOTAP) were incorporated into systems stabilized by nonionic surfactants by adding them to the matrix lipid prior to microemulsion formation (194). Poorly water-soluble gadolinium-containing compounds were also processed with this method (188,198,199). Till now, coenzyme Q_{10} is the only more traditional type of active substance which has been incorporated into this kind of nanoparticles (200). For use in genetic immunization, cell uptake, and brain delivery studies, the particle properties have been modified by incorporation of dioleylphosphatidylethanolamine or cholesterol in the microemulsion formulation and/or by coupling different cell-specific ligands to the particle surface (189,193–196,198,199,201). This was performed either by adsorbing the ligands via a lipophilic anchor to the particle surface after cooling the system to room temperature or by adding them to the microemulsion.

A drawback of these systems, in particular with respect to the use with less potent drugs, is the extremely low concentration of matrix lipid (typically 0.2%). The possibilities to directly prepare systems with higher concentration (e.g., 1% and more) seem to be limited (190). Moreover, in spite of their low concentration, the particles are not stable in aqueous dispersion, even upon short-term storage (189,191,192,196,200,202,203). Ostwald ripening has been considered as a potential cause for the increase in particle size observed (202,203). Freeze-drying in the presence of disaccharides such as lactose, sucrose, and trehalose improved the stability of the particle size values considerably (196,200). Processing of the particles without cryoprotectant led, however, to instabilities already upon freezing. Also in the process introduced by Mumper et al., comparatively high concentrations of surfactant are required for microemulsion formation. As there are indications that excess surfactant may impair particle size stability and because it could also interfere with the subsequent adsorption of DNA or protein, the particles are often purified by gel permeation chromatography (188,189,191,192,194–196). This may also separate the particles from free, nonassociated drug (200).

In conclusion, precipitating solid lipid nanoparticles from warm microemulsions is a comparatively simple process that does not require sophisticated preparation equipment. This simplicity may be particularly advantageous when highly active drugs, such as cytostatics or radioactive markers, are to be incorporated in a laboratory environment. Moreover, in contrast to the preparation by high-pressure homogenization, there is no risk of damaging shear sensitive ingredients such as DNA or proteins. On the other hand, the formulation of the initial microemulsion, which determines the properties of the resulting lipid dispersion, may be a challenging task. In some cases, the microemulsion system is stable only over a narrow temperature range and its composition may have to be adapted according to the desired system (161). Alterations of the microemulsion composition and the presence of incorporated (model) drugs can influence the size of the particles resulting from the dilution method (154,164,165,170). The addition of drugs may even completely inhibit microemulsion formation (164). As the formation of a clear microemulsion system requires temperatures close to the melting point of the matrix lipid, thermal stress is inevitable during the preparation process (162). Concerning larger-scale production, scaling up and control of the precipitation process—at least when using the dilution method—may be difficult even though some suggestions for the design of preparation equipment have been made (161,171). Mumper et al. (190) report that they did not see alterations upon scaling their process from 1 mL to 1 L. In any case, the formation of the initial microemulsion requires very high amounts of surface-active agents (which are cost intensive in the case of bile salts) and sometimes other compounds such as butanol and butyric acid. The excess of these components has later to be removed by time-consuming purification procedures. At least the initial particle concentration of the precipitated systems is in most cases very low compared with that achievable by high-pressure homogenization even though the preparation of more concentrated systems has also been described (172). Moreover, particle size stability seems to be a considerable problem in these dispersions that may demand further processing, e.g., by freeze-drying to achieve pharmaceutical products with acceptable shelf life.

Precipitation from Organic Solution

The precipitation of nanoparticulate solid material from water miscible organic solvents upon mixing with water, a technique commonly applied to prepare polymeric nanoparticles (204), has only recently been transferred to the preparation of solid lipid nanoparticles. Chen et al. (205) precipitated sterically stabilized, paclitaxel-loaded stearic acid nanoparticles from a solution in acetone. In the first step, stearic acid emulsion droplets were formed by injecting the organic solution of stearic acid, phospholipid, and drug into the same amount of a hot aqueous phase (70°C) which contained either Brij 78 (polyoxyethylene 20 stearyl ether) or a mixture of poloxamer 188 and polyethylene glycol-derivatized distearoylphosphotidylethanolamine (PEG-DSPE) to stabilize the emulsion. After concentration of the water phase to 25% of its original volume by evaporation, the emulsion droplets were transformed into solid particles by dispersing the emulsion in twice its volume of cold (0–2°C) water. The particle sizes were in the lower nanometer-region and the fraction of incorporated drug found to be associated with the nanoparticles was about 75% (stabilization with poloxamer 188/PEG-DSPE) or 47% (stabilization with Brij 78), respectively.

As an alternative to the microemulsion process and using the same composition (except for the solvent) Bondì et al. (183) precipitated palmitic acid nanoparticles from a warm (70°C) ethanolic solution with phospholipid and cloricromene by dispersing the solution 1:50 in taurocholate-containing water at 2°C to 3°C. The PCS mean particle sizes obtained this way (between about 100 and 150 nm) were comparable to those obtained by the microemulsion method.

Hu et al. (206,207) used the solvent diffusion method to prepare monostearin nanoparticles loaded with clobetasol propionate or gonadorelin as a model peptide. After dissolving the matrix lipid and drug in a mixture of acetone and ethanol at 50°C, the organic solution was poured into a 10-fold volume of aqueous phase under stirring. The precipitated nanoparticles were stabilized by polyvinylalcohol present in the water phase. The particles were agglomerated under acidic conditions to be separated by centrifugation and later redispersed in water at a final concentration of 1% with the help of ultrasonication. After purification by redispersion, mean PCS particle sizes were found to be typically between 400 and 500 nm (the original precipitates having smaller particle sizes) and were found to be stable over two weeks of storage. Incorporation of 1% of drug related to the matrix lipid led to a recovery of drug associated with the final particles close to 100% for clobetasol propionate. For gonadorelin, the particle associated fraction increased from about 50% to 70% upon decreasing the temperature of the dispersion medium from 25°C to 0°C.

A systematic investigation of the formation of drug-free lipid nanoparticles upon injection of organic lipid solutions into water at room temperature was reported by Schubert and Müller-Goymann (208). They screened different low melting triglyceride mixtures as well as cetylpalmitate as matrix materials. Lipid nanoparticles could be obtained from water miscible solvents like acetone, ethanol, isopropanol, and methanol with all solid lipids under investigation. The introduction of surface-active agents like lecithin, Tween 80, and poloxamer 188 decreased the particle size which varied between 80 and 300 nm depending on the preparation conditions such as lipid concentration, amount injected, emulsifier concentration, and solvent. Stability of the dispersions, which were of rather low lipid concentration (usually much below 1%) and which were not purified after preparation, was not an issue in the study described.

PHYSICOCHEMICAL CHARACTERIZATION OF COLLOIDAL LIPID SUSPENSIONS AND NANOPARTICLES

After preparation, it has to be ensured that the particles obtained have the desired properties and are thus suitable for the intended type of administration. The most obvious parameters to be investigated are the (colloidal) particle size and the (solid) state of the particle matrix. Other important features include the surface characteristics, the particle shape, and, in particular, the interaction with incorporated drugs. Detailed knowledge on the structural properties of the dispersions is not only a matter of quality control but also inevitable for a further development of optimized systems. Due to the complexity of the systems, a combination of different characterization techniques is the most promising approach to obtain a realistic image of the sample properties.

Particle Size and Size Distribution

Particle size measurements are routinely performed in studies on solid lipid nanoparticle dispersions to confirm the desired colloidal size range and to monitor the colloidal stability of the formulations during storage or further processing (e.g., freeze-drying, sterilization, or adsorption of macromolecules). Almost all particle size determinations on solid lipid nanoparticle dispersions are performed by light scattering methods, mainly with PCS. This method analyzes the Brownian motion of the particles in the dispersion medium via intensity fluctuations of scattered laser light (209–212). The resulting diffusion coefficient is subsequently transformed into particle size information via the Stokes–Einstein equation. The assumption of spherical particle shape usually made for this transformation may not be justified for solid lipid nanoparticles which frequently crystallize in a platelet-like shape (29,39,43,44,50,51,142,147,213). For such anisometric particles, a larger hydrodynamic diameter is observed in PCS compared to corresponding emulsion systems (in spite of the volume contraction upon crystallization) because the diffusion coefficient of anisometric particles is larger than that of a sphere of the same volume (213–215). In almost all cases, particle size data for solid lipid nanoparticle dispersions are reported as the results of the cumulants method of data evaluation, the so-called *z*-average diameter (*z*-ave, sometimes also referred to as the effective diameter) and the polydispersity index (PI) as an indication for the width of the particle size distribution (216). Although being a useful, simple and robust characterization tool in particle size analysis, *z*-ave and PI do not have much in common with the parameters normally used for the description of particle size distributions such as volume or number diameter. The transformation of the raw data into either monomodal or more complex size distributions weighted by volume or number has much less often been performed for the dispersions under discussion here, possibly due to the much higher demands of such type of analysis on knowledge of sample properties, quality of measured data, necessary assumptions, and high dependency of the results on the type of mathematical analysis performed, making comparison of results with those from other groups difficult.

In particular, for broader size distributions and dispersions that contain a considerable amount of particles in the upper nanometer and/or micrometer range, laser diffraction (LD) data obtained on an instrument equipped with adequate submicron instrumentation, e.g., polarization intensity differential scattering may give more detailed insight into the particle size distribution (217). With LD, the angular distribution of the light scattered from the particles upon irradiation with laser light is determined using an array of detectors (218). This angular intensity distribution is transformed into a particle size distribution and its characteristic particle size values (e.g., mean, mode, median diameter, diameter at 90% or 99% of the distribution), usually on the basis of Mie theory (requiring knowledge on the optical parameters—refractive index and absorbance—of the particles) for submicron particles. Lipid nanosuspensions have frequently been analyzed with such type of equipment (60,62–65,88,89,112,113). Uncertainties in the results may originate from not exactly known optical parameters, anisometric particle shape and inhomogeneous sample composition as well as a combination of physically different methods to investigate the particle size range of interest. As in PCS, the particle sizing results should thus be regarded as approximations rather than as absolute values. Combining the information obtained from different particle sizing techniques can help to get a better impression on the "real" particle size range.

The presence of particles larger than about 5 μm may cause serious problems if the dispersions are going to be administered intravenously as they may block small blood vessels. Even with LD, it may, however, be difficult to detect very small amounts of such microparticles in colloidal lipid suspension. So far, only very few studies have been addressing this problem for solid lipid nanoparticle dispersions using electrical zone sensing ("Coulter counter" method) to determine the absolute number of particles in the micrometer range (34,60,62). The measurement of alterations in electrical resistance when the particles pass a narrow glass pinhole requires dilution of the sample in a comparatively concentrated salt solution (e.g., 0.9% sodium chloride) which may cause problems for electrostatically stabilized colloidal particles due to the interference with the electric double layer and subsequent destabilization. The use of single particle optical sensing by light obscuration, which has frequently been applied for the measurement of lipid emulsions with respect to microparticulate contamination may be an interesting alternative (219). Concerning the presence of large aggregates, visual inspection and/or light microscopy may also be very helpful tools to obtain a realistic impression.

Information on particle size may also come from microscopic techniques sensitive in the colloidal range, such as electron microscopy and atomic force microscopy. Although the determination of particle size distributions from electron microscopic investigations has been reported (114–117), these techniques are not routinely used for this purpose, since they are tedious to perform with the required statistical accuracy and are also not free from methodological problems that can affect sizing results (See Section "Morphology and Microstructure; Particle Morphology and Ultrastructure of the Dispersions").

Morphology and Microstructure

Crystallinity and Polymorphism

The expected advantages of solid lipid particles, e.g., modified release properties, essentially rely on the solid state of the particles. After processing in the melted state (e.g., in melt-homogenization), some matrix materials do, however, not crystallize readily in colloidal dispersions. Retarded or suppressed crystallization has been observed for shorter chain monoacid triglycerides like tricaprin, trilaurin or trimyristin as well as for more complex glyceride mixtures, e.g., some hard fats (37,40–42,60,118,124,220). Without special thermal treatment, dispersions of such matrix lipids may remain in the emulsion state for months or even years instead of forming the desired solid particles. The admixture of longer chain triglycerides or the use of emulsifiers with long, saturated alkyl chains may reduce this effect (41,46). On the other hand, the presence of liquid drugs or oil can further decrease the crystallization tendency (45,89). Monitoring of the crystalline status is thus a very important point in the characterization of solid lipid nanoparticle dispersions, particularly when novel compositions or preparation procedures are introduced.

After crystallization, the particles may undergo polymorphic transitions, a phenomenon typical for lipidic materials (221–223). This process and other crystal aging phenomena may proceed over several days or weeks after solidification (37,65,213,220). The matrix material determines the types of polymorphs that can be formed in the suspensions. For fatty acid nanoparticles investigated in the dried state mainly polymorph C has been found along with the presence of polymorph B in some cases (153,160,162,164). For glycerides, the presence of the main polymorphs α, β' and β as well as an intermediate form β_i has been confirmed in nanoparticles prepared by melt-homogenization (36,41,42,89,112). In some cases, polymorphic forms not observable in the corresponding raw materials were detected in triglyceride nanosuspensions (42,44). The rate of polymorphic transitions was found to be accelerated in the nanoparticles compared to the bulk material (36,213,220). It depends on the type of matrix lipid, the stabilizer composition and also on the particle size; e.g., shorter chain saturated monoacid triglycerides transform more quickly from the least stable α- into the stable β-modification than triglycerides with longer chains and smaller particles have a higher transformation rate than larger ones (37,41,44,46,47). During polymorphic transitions, the particles undergo a rearrangement of the matrix molecules into a more dense packing and may also change their shape (47,213,220). Polymorphism can thus affect pharmaceutically relevant properties, in particular drug incorporation (42,88). Stability related issues have also been attributed to polymorphism and increase in crystallinity (65).

Differential scanning calorimetry (DSC) and X-ray diffraction (XRD) are the techniques most widely used for the characterization of crystallinity and polymorphism of solid lipid particles. Preferably, the dispersions should be investigated in their native state to avoid artifacts arising from sample preparation. If such preparation is inevitable, e.g., in cases of very diluted suspensions, great care has to be taken not to change the properties of the particles during the sample preparation procedure. In particular, the application of temperatures that may lead to phase transitions must be avoided. Moreover, as the properties of the nanoparticles are size dependent, the particle size should not change upon sample preparation; e.g., freeze or air drying of samples may lead to alterations in transition temperatures, crystallinity, and polymorphism (44,53,60,61). In some special cases such procedures may be inevitable, e.g., drying of samples to be checked for the presence of high melting drug crystals by DSC (163,165) but the potential alterations of the sample due to the preparation technique have to be considered upon data interpretation.

The presence of solid lipid particles can conveniently be confirmed by DSC via the detection of a melting transition upon heating and the amount of crystalline material quantified via determination of the melting enthalpy. This method has thus been frequently used to characterize the original liquid dispersions as well as processed formulations such as freeze-dried powders or semisolid formulations. Cooling curves give indications whether the dispersed material is likely to pose recrystallization problems and what kind of thermal treatment may be used to induce solidification (37,41,42,45). DSC is also well suited to monitor and quantify physical changes, e.g., due to polymorphism or increase in crystallinity, upon storage. The degree of crystallinity of solid lipid nanoparticles has been related to different application relevant parameters such as gelation tendency, enzymatic degradation or occlusive properties (65,71,98). The exact assignment of the transitions observed in the DSC scans may, however, not be an easy task for polymorphic materials in small colloidal particles because their thermal behavior may differ from that of the bulk material. A decreased melting temperature (and also melting enthalpy) has been observed for solid lipid nanoparticles, particularly for those in the lower nm size range (37,44,150). Moreover, the transitions of the dispersions are usually broader than those of the bulk material. For small (e.g., <150–200 nm) monoacid triglyceride nanoparticles a very peculiar, size-dependent melting behavior has been observed leading to multiple transitions which are not due to polymorphism (Fig. 4) (43,44,213). Eutectic behavior in mixtures of solid matrix lipids or with incorporated compounds such as oils or drugs may also lead to a shift in the peak positions (41,45,89,112).

Investigation with complementary methods, in particular XRD, are thus very helpful and often crucial for the detailed interpretation of thermoanalytical data, especially for the assignment of polymorphic forms. The matrix lipids of crystalline lipid nanoparticles give rise to small as well as wide angle X-ray reflections the wide angle reflections being particularly suitable to identify the polymorphic form of the

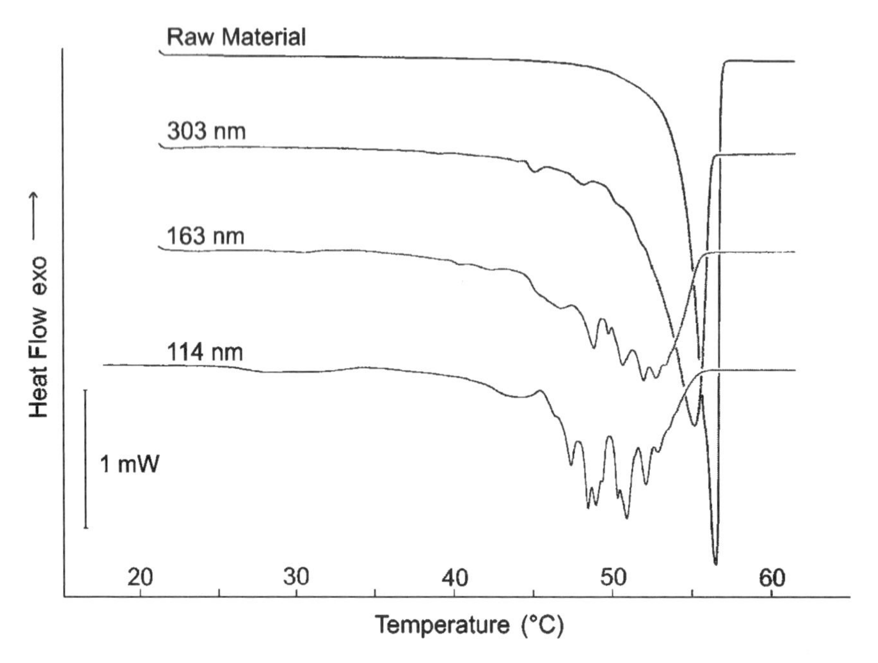

Figure 4 Differential Scanning Calorimetry heating curves of trimyristin dispersions (stabilized with different concentrations of a fixed soybean lecithin/sodium glycocholate combination) with different mean particle sizes. The raw material was dispersed in a bile salt solution. *Source*: From Ref. 213.

particles (36,41–47,50,89,92,93,109,112,153,160,162,164,213). The X-ray reflections of solid lipid nanoparticles are usually much broader than that of the bulk material as a result of the small particle size and potentially also of a decrease in crystalline order. This particle size-dependent line-broadening effect has been utilized to assign the single transitions in complex DSC heating curves of small triglyceride nanoparticles to particle fractions of different, distinct thickness via a sophisticated analysis of the line shape of the corresponding X-ray reflection (43,48).

Particle Morphology and Ultrastructure of the Dispersions

Knowledge on the morphology of single particles and the organization of the colloidal structures within the dispersion is an important support for the interpretation of the behavior of the dispersions, e.g., with respect to stability issues, drug incorporation and release. Spherical nanoparticles would offer the highest potential for protection of incorporated substances and for controlled release as they provide minimal contact with the aqueous environment and longest diffusion pathways. They would also require the smallest amount of surface-active agent for stabilization as a result of their comparatively small specific surface area. On the other hand, anisometric particles would be advantageous for the association with active agents, e.g., biological macromolecules such as proteins or DNA, by adsorption onto the particle surface.

The lipid nanoparticles may not be the only type of colloidal particles in the dispersions. Most emulsifiers (often used in considerable amounts for dispersion preparation) are able to form additional colloidal structures by self-assembly, such as vesicles and/or micelles. The lipophilic domains of these structures may represent alternative sites of drug localization within the dispersions. As a consequence, the presence of additional colloidal structures may affect drug loading capacity and release behavior.

Imaging of the nanoparticles can be performed by electron microscopy yielding an impression on the size, shape and some cases also internal structures of single particles as well as on the potential presence of different types of particles within the dispersions. While scanning electron microscopy has been employed only in a few cases (145,168) transmission electron microscopy (TEM) is much more common for the investigation of solid lipid nanoparticle dispersions. Specimen prepared by staining with uranyl acetate or phosphotungstic acid as contrast agents have often been used to demonstrate the presence of small particles in dispersions of solid lipids and to draw conclusions on their size and shape, in particular for nanoparticles prepared by the microemulsion method from fatty acids or emulsifying wax (115,149,152,163, 168,182,186,189,191,192,198,199,201,205,206). A spherical particle shape has often been concluded for such nanoparticles from the corresponding images but has not yet been confirmed by additional investigations. A great disadvantage of the usual staining methods is the major alteration of the original sample state during preparation by adsorbing the particles on a flat surface (which may lead to selective adsorption and preferred orientation) and removal of the aqueous phase by drying (which may cause the collapse of soft structures like phospholipid vesicles). Techniques such as freeze-fracturing (FF) or cryoelectron microscopy (cryo-TEM) aim at preserving the liquid-like state of the sample and the organization of the dispersed structures during preparation by freezing the sample so quickly that all liquid structures including the dispersion medium solidify in an amorphous state. In freeze-fracturing, the sample is subsequently fractured and a replica of the fracture plane is prepared after shadowing with carbon/platinum. In TEM, the replica gives a relief-like image of the fracture plane providing information on the size range and shape of the colloidal structures in the sample as well as on their internal structure. Solid triglyceride nanoparticles in the stable β-modification usually appear as sharply edged anisometrical structures displaying distinct internal layers corresponding to the layering of the triglyceride molecules in the crystal lattice (Fig. 5) (29,30,39,43,44,49–51,142,147,213,224). In phospholipid-containing dispersions, the presence of phospholipid vesicles, resulting from an excess of stabilizer, can also be detected by this technique (213). In cryo-TEM, frozen-hydrated samples are visualized directly, without additional preparation. Cryo-TEM is very well suited to study the coexistence of different colloidal structures (such as lipid nanoparticles and vesicles) and to obtain information on the particle shape because it shows the contours of the particles in two-dimensional projection (30,213,225). Depending on their position toward the electron beam, platelet-like crystalline triglyceride particles can be seen as circular to elongated, edged particles with low contrast or as dark, "needle" like structures. Unilamellar phospholipid vesicles can easily be recognized by the ring-like appearance of the phospholipid bilayer (30,213). For glyceride nanoparticles loaded with liquid drug or oil, the formation of two phases within single particles (caused by the expulsion of the liquid substance from the crystal lattice to the surface of the crystalline carrier particles) was demonstrated with cryo-TEM (45,113). Unfortunately, cryoelectron micrographs tend to be biased toward smaller particles so that it is problematic to draw definitive conclusions on the particle size distribution. Moreover, structures with only one or two large dimensions—such as platelets—may by accommodated stretched out parallel to the surface of the thin specimen film created upon sample preparation and thus cannot be viewed any more in a random orientation.

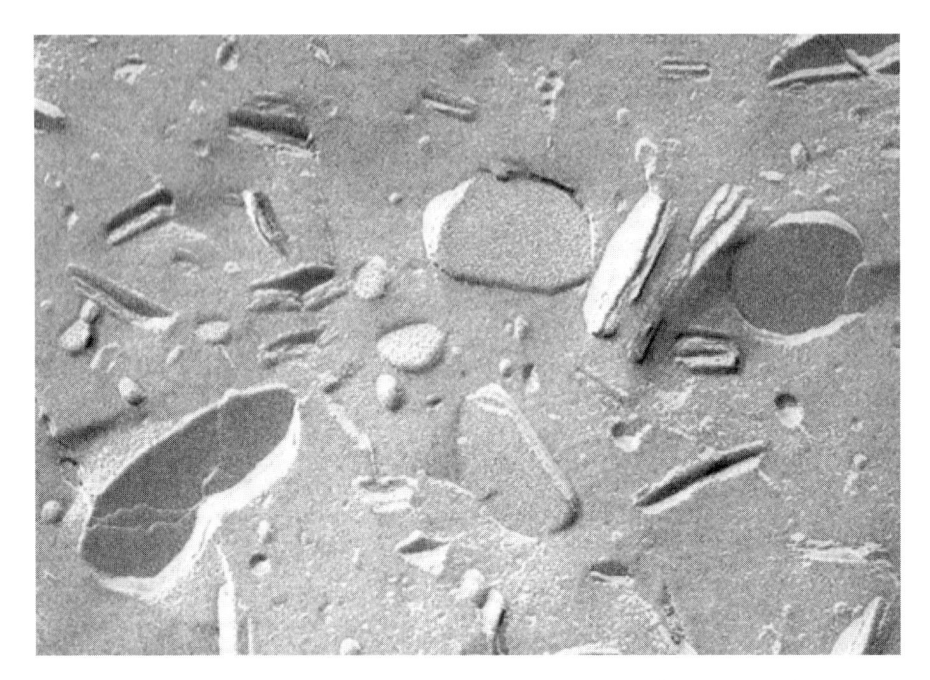

Figure 5 Freeze fracture electron micrograph of a trimyristin suspension prepared by melt homogenization (10% trimyristin stabilized with 2.4% soybean lecithin/0.6% sodium glycocholate).

The size and shape of lipid nanoparticles can also be studied by AFM which has, however, not been used much for the characterization of lipid nanosuspensions till now (52,54,69,73). In contrast to conventional electron microscopy, AFM provides information on the three-dimensional extension of single solid particles and does also allow conclusions on the mechanical properties at the particle surface. The investigation of cholesterol- and glyceryl behenate nanosuspensions by AFM, e.g., indicated a disk like shape of these nanoparticles (52).

Colloidal solid lipid particles may form superstructures, e.g., gels upon increasing lipid concentration or due to insufficient stabilization, which have been investigated rheologically as well as with electron microscopy (30,84–86). In triglyceride nanosuspensions, the platelet-like nanoparticles have been found to organize in stacks above a critical triglyceride concentration (49,50). Stack formation, which is reversible upon dilution, could not only be observed by electron microscopy but is also reflected in the appearance of additional small angle X-ray and neutron reflections with very large repeating unit. Analysis of these reflections allows to conclude on the distance of the particles in the stacks as well as to quantify the extent of stack formation.

Surface Properties

The surface characteristics of colloidal drug carriers play an important role for their in vivo behavior as well as for their stability against aggregation. Stabilization of the nanoparticles in aqueous dispersion can be achieved by electrostatic or steric repulsion or a combination of both principles. For solid lipid nanoparticles, the repulsion forces are imposed by the surfactant molecules on the particle surface and their interaction with the surrounding medium. Electrostatic repulsion arises from the presence of ionized groups on the particle surface, mainly as a result of the use of ionic surfactants such as, e.g., bile salts. The zeta potential usually serves as an easily accessible characteristic parameter for the nanoparticle charge (209,226). In most cases, solid lipid nanoparticles bear a negative surface charge with zeta potential values depending on the type of stabilizer system but cationic particles have also been prepared to be used as DNA carriers. For the development of carrier particles with high adsorption potential for oppositely charged (macro)molecules such as nucleic acids or peptides, zeta potential investigations are an important tool for the optimization of the surface charge and the investigation of the interaction of the particles with the adsorbing molecules (69,107,189,191–193,197). In most cases, zeta potential determinations have, however, been employed to draw conclusions about the physical stability or instability, respectively, of the formulations during storage or upon interaction with electrolytes or (simulated) biological fluids. They were, e.g., used in the investigation of gel formation and to evaluate different compositions with respect to electrolyte and pH stability (29,62,64,77,88). The effect of different preparation parameters and a potential influence of drug loading or processing of liquid dispersions such as freeze-drying or sterilization on the zeta potential have also been studied (60,67,118,128,149,163). Investigation of the impact of different surface-active agents on the zeta potential can help to study interaction with surfactants (30,118,166,175,178). For steric stabilization, which is of major importance when nonionic and macromolecular agents are employed as stabilizers, there is no such convenient experimental value as the zeta potential that could be used to characterize its extent. For this reason, investigation of the stability upon storage of the particles remains an important tool to assess the stabilization potential. The "hydrophilic cloud" around sterically stabilized particles, in particular those containing PEGylated compounds, is assumed to be of great importance to escape clearance from the circulation by phagocytosis. Reduced phagocytic uptake and longer circulation times have been demonstrated for sterically stabilized solid lipid nanoparticles in cell culture and in-vivo studies, respectively (175,178–180).

The differences in the biodistribution of nanoparticles stabilized with different surfactants is ascribed to variations in the pattern of plasma proteins adsorbing to the surface of the nanoparticles after intravenous administration (227). First investigations on the protein adsorption pattern of solid lipid nanoparticles incubated with human plasma by two-dimensional polyacrylamide gel electrophoresis revealed differences for poloxamer 188 and Tween 80–stabilized cetyl palmitate particles which may be relevant with respect to drug targeting: Nanoparticles stabilized with poloxamer 188 displayed an adsorption pattern typical of particles with prolonged circulation time (104,105). For Tween 80-stabilized particles the adsorption of apolipoprotein E (and other apolipoproteins) was detected. Apolipoprotein E is under discussion as potential targeting moiety for brain delivery as it may facilitate binding and uptake of nanoparticles via the LDL receptor in the brain (228).

Drug Incorporation and Release

Drug Incorporation in Solid Lipid Nanoparticle Dispersions

While the drug concentration in lipid nanodispersions is usually quite well known there is still comparatively little knowledge about the interaction of drugs with these complex and variable systems, e.g., with respect to the state and localization of the drug within the dispersions. In general, the highly ordered and tightly packed crystalline particle matrix should be expected to represent a rather unfavorable localization for the incorporation of at least larger amounts of drugs because the drug will disturb the order of the crystal lattice. The incorporation capacity will depend on the physicochemical properties of the drug, but also on the type of matrix material (e.g., pure triglycerides are assumed to have a lower incorporation capacity than complex lipids) and the matrix state (in particular the degree of crystallinity and polymorphic form). Drug molecules that cannot be accommodated within the crystalline matrix may adsorb to the nanoparticle surface or separate from the particles. This may lead to the formation of drug crystals or droplets (41,42,93) or to distribution into the aqueous phase or additional colloidal structures present in the dispersion (e.g., micelles or phospholipid vesicles). Investigations on drug loading of solid lipid nanoparticles are often performed on the whole dispersion, often after drying (165,169,170,174,177–179,181,182,206). As a result, the overall concentration of drug in the dispersion is obtained but the amount of drug actually incorporated into or adsorbed to the nanoparticles cannot be distinguished from drug that has phase separated or is localized in additional colloidal structures or in the aqueous phase. Investigations of the drug distribution between nanoparticles and aqueous phase have confirmed that the drugs are not always completely distributed toward the nanoparticles and may sometimes even be present in the aqueous phase in considerable amount (34,53,60,83,111,114–118,128,129,200,205,207). Depending on the separation technique applied in the distribution studies, additional colloidal structures may or may not contribute to the amount of drug determined in the aqueous phase. Phase separation of drug in the form of large crystals may be easily detectable by light microscopy or even visually (41,42,53,54,93) but detection of phase-separated drug becomes difficult at low concentrations, when it forms nanoparticles or does not crystallize at all. Conclusions on drug distribution have also been drawn via stability studies of an incorporated drug (retinol) being chemically unstable in the aqueous phase of the dispersion (93).

Experimental confirmation of drug association with the lipid nanoparticles may not be easy to obtain and is often rather indirect. It can be reflected in alterations of the thermal properties of the matrix lipid, such as the melting and recrystallization temperature or the kinetics of polymorphic transitions (45,53,93,163). Evidence for the presence of crystalline drug in the dispersions as reflected by the occurrence of a melting transition has only scarcely been reported (164,206). The absence of thermal transitions related to drug melting is usually attributed to a preferably amorphous or molecularly dispersed state of the incorporated substance (45,163,165). Also a decrease in molecular mobility of the drug upon crystallization of the matrix lipid as reflected by line broadening in high resolution NMR spectroscopy has been used as an indication for drug–particle association (42). Even then it remains to decide whether the drug is indeed incorporated into or only adsorbed to the surface of the nanoparticles. Additional information into this direction may be derived from drug release studies as they reflect the accessibility of the drug to the aqueous phase (53,54) but it can be difficult to separate the effects arising from the nanoparticles of interest from those of other colloidal structures in the dispersions such as liposomes and/or micelles.

Spectroscopic investigations are a very promising approach to study the characteristics of lipid nanoparticles on the molecular level, including their interaction with drugs. High resolution ^{1}H NMR investigations have been used to investigate the properties of the matrix lipids as well as the behavior of incorporated drugs

(42,90,112). As highly immobile molecules do not lead to an analyzable signal in high resolution NMR spectroscopy investigation by magic angle or off-magic angle spinning methods may be very valuable for a more detailed investigation of solid nanoparticles (110). Interesting information on these systems has also been obtained by a related technique, electron spin resonance (ESR). In contrast to NMR, ESR studies require labeling of the systems with paramagnetic probe molecules which usually carry a nitroxide radical group as spin label. Investigations with spin probes of different structure and physicochemical properties can provide insights into possible types of interaction of lipid nanoparticle dispersions with foreign substances, e.g., with regard to their distribution within the systems and local environment (112,131, 133). The chemical accessibility of a spin labeled model drug in solid lipid nanoparticle dispersions to the aqueous phase can be studied with an ascorbic acid reduction assay (112,131,133,134). As a drawback of ESR, it can only be used in model, spin probe containing systems (usually carrying only very small amounts of the spin probe) or would require spin labeling of "real" drugs to be incorporated.

Drug Release from Solid Lipid Nanoparticles

A major driving force behind the development of solid lipid nanoparticles was the need for a colloidal lipid-based sustained release system for lipophilic drugs. Slow drug release—at least compared to other, more fluid types of colloidal lipid drug carrier systems—would thus be expected to be a central property of lipid nanosuspensions and drug release has been studied for a number of systems. The assessment of drug release from colloidal systems is, however, not an easy task, in particular for lipophilic drugs (Chap. 7) (229). As a complication compared to simple dissolution testing, distribution of the drug between the carrier particles and the release medium as determined by the partition coefficient plays a major role in release studies from colloidal lipid particles. Assessment of a drug release profile that is solely governed by drug–carrier interactions requires dilution of the system to sink conditions which is difficult to achieve for lipophilic drugs. In some studies on solid lipid nanoparticle dispersions solubility enhancers such as surfactants or organic solvents had to be added to the release medium to achieve sufficient drug solubility (200,205,206). As a further complication, the release medium can usually not be assessed for released drug in the presence of the nanoparticles because they disturb the analytical procedure. As a consequence, the particles have to be separated from the release medium which becomes increasingly difficult with decreasing particle size. In particular for rapid release processes, e.g., to assess initial burst release, separation of the particles from the release medium must also be rapid to not distort the release profile.

To circumvent the problem of particle separation many release studies on solid lipid nanoparticle dispersions have been performed with dialysis setups using small or large volume symmetric dialysis cells, a dialysis bag filled with nanoparticle suspension suspended in a large volume of release medium or a Franz cell equipped with a dialysis membrane (32,115–117,128,152,154–156,159,165,170,176,177,179,181,182, 200,203,205). The "slow release" usually obtained as a result in such studies does, however, not reflect a true release profile because confinement of the suspension within the donor compartment does not allow a release under sink conditions and the "release" profile is further convoluted by slow diffusion of drug across the dialysis membrane (see also Chap. 7) (229). These studies, sometimes more adequately referred to as "in vitro diffusion" studies (32,170,182) may still be of some value to compare diffusion kinetics from different formulations but are, unfortunately, often presented for a single formulation only.

The so-called reverse dialysis bag technique may allow a more realistic view of the release situation as it is based on the dilution of the dispersion into a large volume of release medium. Release is assessed by the assay of release medium contained in small dialysis bags which are immersed in the large volume of release medium. Even though the release profile determined via the reverse dialysis bag technique is still influenced by the diffusion-driven equilibration process between large release volume and dialysis bags it was considered appropriate for systems that exhibit a not too fast drug release (230). The technique was used by Wang et al. (129) to study release of 5-fluoro-2′-deoxyuridine from lipid nanoparticles containing the prodrug 3′,5′-dioctanoyl-5-fluoro-2′-deoxyuridine in a trypsogen-containing release medium. After a considerable initial burst the drug continued to appear only slowly in the dialysis bags in this study.

Other release studies on solid lipid nanoparticles utilizing dilution into large volumes of release medium approached separation of the nanoparticles from the release medium in a rapid manner, e.g., by filtration through membrane filters, ultrafiltration with the aid of centrifugal filtration units or centrifugation (34,53,54, 140,148,183,203,207). Separation becomes increasingly difficult with decreasing particle size, leading, e.g., to prolonged centrifugation times which may lead to loss in time resolution as the release process continues during the separation process. Filtration usually provides good time resolution but a potential adsorption of free drug to the filter materials has to be checked carefully (54).

The first results with this sampling and separation approach (dilution into buffer or water as release medium and separation by filtration through membrane filters) were reported by Müller et al. in 1995 (34) and described in more detail in several subsequent publications (53,54). For etomidate and tetracaine, controlled release could not be observed. These drugs, which were released within minutes for particles prepared by hot or cold homogenization, were assumed to be localized at or near the surface of the nanoparticles (34,53). For prednisolone, sustained release over several weeks was reported for particles prepared by both methods. In contrast to cold homogenized nanoparticles of cholesterol or glyceryl behenate, glyceryl behenate dispersions prepared by melt-homogenization showed pronounced initial burst release of prednisolone the extent of which depended on the preparation conditions, in particular surfactant concentration and homogenization temperature. The biphasic release pattern of hot homogenized particles was attributed to drug distribution processes during the preparation procedure leading to different ratios of drug at the surface and in the matrix of the nanoparticles (34,53,54).

Bondi et al. (183) observed distinct differences between different release media for the release of cloricromene from palmitic acid nanoparticles of about 100 nm size. In buffer of pH 7.4 complete release was observed within one hour whereas in plasma only 70% of the drug was released quickly. At pH 1.1 only about 50% of the drug was released rapidly, the rest being released after changing the pH to 7.4. Hu et al. (206) investigated release of clobetasol propionate from monostearin nanoparticles redispersed into a surfactant/polyethyleneglycol containing release medium. After an initial burst release of about 45% during the first three hours the drug was released in a sustained manner over the monitored period of four days. The particles, which had previously been purified from the preparation medium by centrifugation following coagulation, were of comparatively large size with PCS mean diameters distinctly above 400 nm. For gonadorelin in similarly prepared monostearin nanoparticles with sizes about 400 nm a linear release over about two weeks was observed after an initial burst under pH conditions mimicking peroral or subcutaneous administration (207).

For nanoparticle dispersions developed with the aim of dermal administration, release is usually studied in Franz cells using excised skin or lipophilized filter membranes as barriers between formulation and release medium (88,127). The in vitro permeation of triptolide through excised rat skin was higher from tristearin or stearic acid nanoparticle formulations than from a water/propylene glycol solution and increased with decreasing particle size (127). In comparison to corresponding nanoemulsion formulations, Jenning et al. observed a lower flux of retinol from semisolid formulations containing glyceryl behenate nanoparticles into a receptor medium supplemented with bovine serum albumin whereas the liquid nanosuspension displayed a more complex permeation profile. Comparing the release of retinol from glyceryl behenate nanoparticles loaded with different amounts of triglyceride oil, sustained release in comparison to a nanoemulsion was mostly found for nanoparticles loaded with no or only little oil whereas particles with higher oil load approached or even exceeded the release from the nanoemulsion (89).

For dermal lipid nanoparticle formulations a so-called "test tube assay" has also been used to study the transfer of lipophilic drugs to a layer of medium chain triglycerides placed on top of the nanodispersion (88,100). To increase drug solubility in the aqueous phase and thus the transfer rate to the lipophilic receptor medium Tween 80 was added to the dispersions. Using this assay differences were observed for the release of retinol from glyceryl behenate nanoparticles in dependence on whether a buffer solution or water was used as the dispersion medium. The transfer from particles in buffer (which facilitated the polymorphic transformation of the nanoparticles) was distinctly faster than that from the particles in water. The transfer from emulsion nanoparticles was intermediate and independent on the dispersion medium (water or buffer). For retinol palmitate the transfer from the solid nanoparticles in water as well as in buffer was slower than from the nanoemulsion but again transfer was distinctly faster for the nanoparticles in buffer, approaching the release rate of the emulsion at longer release times (88). Using the test tube assay to monitor the behavior of the sunscreen oxybenzone, slightly slower transfer was observed for 5% oxybenzone incorporated into cetylpalmitate nanoparticles than from a nanoemulsion; at 10% oxybenzone a smaller fraction of the drug was released and almost no difference between solid nanoparticle and emulsion formulation could be detected (100). In contrast, release from the emulsion droplets loaded with 10% oxybenzone was distinctly higher than that from the nanoparticles when the release was studied into a medium chain triglyceride/ethanol mixture (1:9) separated from the dispersion by a membrane filter in a Franz cell setup (100).

The available data on drug release from solid lipid nanoparticles thus illustrates that the subject is rather complex. The presently available data is much too diverse to allow systematic conclusions on this topic. Till now, there is no standardized procedure (e.g., concerning general experimental setup, composition of, as well as maximum drug concentration in release medium), and the systems under investigation vary considerably in composition, preparation procedure and physicochemical properties. Results obtained in different studies are thus barely comparable. It can, however, be concluded that replacing the liquid oil in emulsion droplets with a solid lipid does not generally lead to a higher sustained release potential. This is not surprising when considering the physical processes in particle formation such as the formation of a tightly packed crystalline lattice, dynamic processes involved in polymorphic transitions or increase in particle surface when anisometric particles are formed. On the other hand, also more promising results have been reported that do point to a certain potential for sustained release. In conclusion, the controlled release potential of solid lipid nanoparticle formulations is expected to highly depend on the specific drug–particle interaction of a given formulation. The generation of drug release data obtained under more standardized conditions and on well characterized systems, preferably including drug localization studies, would be highly desirable to contribute to a realistic view on the potential of different types of solid lipid nanoparticles in controlled release applications.

APPLICATIONS IN DRUG DELIVERY

During recent years lipid nanoparticles have been extensively investigated for their use in drug delivery for both local and systemic administration. Toxicity aspects, interactions with biological fluids and cells, uptake and transport in the body as well as metabolism have been studied in vitro and in vivo. Pharmacokinetics, biodistribution, and pharmacological effects have been investigated in animal studies exploring different administration routes including parenteral, peroral, pulmonary, and topical administration. Table 3 provides examples of the increasing number of in vivo studies conducted using colloidal solid lipid-based drug delivery systems.

Safety and Tolerability Aspects

Solid lipid-based drug delivery systems can principally be designed using excipients that are generally regarded as safe (GRAS), and they are therefore supposed to be well tolerated. Like any other new delivery system, however, solid lipid nanoparticles need to undergo a careful toxicological safety evaluation before they can be used in humans. So far, a limited number of in vitro toxicity and in vivo tolerability studies have been performed. As will be summarized below, these studies generally demonstrated good tolerability of solid lipid-based delivery systems.

The safety requirements for new delivery systems depend on the route of administration. Requirements for topical and oral administration are lower than for parenteral administration. For topical use, the excipients should be nonirritating and nonallergenic. Solid lipid particles prepared from excipients which are currently used in pharmaceutical and cosmetic formulations are thus expected to be well tolerated when used on the skin.

For oral use, the excipients should be nontoxic and biodegradable when absorbed. Basically, all lipids and surfactants used in commercial oral dosage forms should be suitable for the preparation of oral solid lipid delivery systems from a toxicological point of view. In principle, this also applies to food lipids and surfactants used in food products. In the gastrointestinal tract (GIT) lipids are degraded by lipases or esterases. Pancreatic lipase acts together with colipase on triglyceride surfaces (239). Müller et al. (55) investigated the in vitro degradation of solid lipid nanoparticles in solutions of pancreatic lipase/colipase and found that the degradation rate depended on the nature of the lipid matrix with trimyristin being faster degraded than glycerol behenate, and cetyl palmitate being intermediate. Degradation was slowed down by using poloxamer 188 as a stabilizer compared to sodium cholate. An effect of surfactants on the degradation rate was also observed by Olbrich et al. (71,72). Trimyristin nanoparticles stabilized with sodium cholate or lecithin degraded rapidly whereas trimyristin nanoparticles prepared using steric stabilizers

Table 3 In Vivo Studies Investigating Solid Lipid–Based Delivery Systems

Drug	Carrier lipids	Route	Animal	Test	Reference
–	Cetyl palmitate	Topical	Human	Skin hydration	(102)
–	Glyceryl behenate	Pulmonary	Rat	Local tolerance	(241)
$^{99m}T_c$-HMPAO	Glyceryl behenate	Pulmonary	Rat	Biodistribution	(236)
Calcitonin	Tripalmitin	Oral	Rat	Efficacy	(237)
Camptothecin	Stearic acid	Oral	Mouse	PK, biodistribution	(115)
Cyclosporin A	Imwitor 900	Oral	Pig	PK, BA	(243)
Diclofenac	Cholesteryl palmitate/ palmitic acid	IV	Rat	PK, efficacy	(246)
Doxorubicin	Stearic acid	IV	Rat	PK, biodistribution	(176)
Doxorubicin	Stearic acid	IV	Rat	PK, biodistribution	(179)
Doxorubicin	Stearic acid	IV	Rabbit	PK, tissue distribution	(178)
Gadolinium	Emulsifying wax	IV	Mouse	Biodistribution	(247)
Idarubicin	Stearic acid	Duodenal IV	Rat	PK, biodistribution	(177)
Idarubicin	Stearic acid	Duodenal IV	Rat	PK, bioavailability	(245)
Mycoplasma bovis antigen	Solid paraffin	Injection	Sheep	Immune response	(232)
Paclitaxel	Stearic acid	IV	Mouse	PK	(205)
Piribedil	Glyceryl behenate	Oral	Rabbit	Bioavailability	(140)
Plasmid DNA	Emulsifying wax	SC	Mouse	Immune response	(189)
Plasmid DNA	Emulsifying wax	IV	Mouse	Biodistribution	(194)
Plasmid DNA	Emulsifying wax	Topical	Mouse	Immune response	(193)
Tobramycin	Stearic acid	Duodenal IV	Rat	Biodistribution	(174)
Tobramycin	Stearic acid	Ocular	Rabbit	Bioavailability	(169)
Triptolide	Tristearin; stearic acid	Topical	Rat	Efficacy	(127)
β-galactosidase	Emulsifying wax	SC	Mouse	Immune response	(191)

Abbreviation: IV, intravenous; SC, subcutaneous; PK, pharmacokinetics; BA, bioavailability.

such as poloxamer 407 or polysorbate 80 displayed a slower degradation rate (72). The authors explained the results by the steric hindrance of the attachment or interfacial activation of the lipase/colipase complex by the steric stabilizers. For surfactants that delay the degradation of the nanoparticles, the authors observed also an effect of particle size on the degradation rate with small particles being degraded faster than larger ones (72). The same authors also found an effect of the crystallinity of the lipid matrix on the degradation rate for lipid nanoparticles stabilized with surfactants that disturb the crystallization process of the lipid (71). These degrade faster and show a decrease in degradation rate with increasing storage time due to an increase in crystallinity.

The requirements for parenteral administration include—beside tolerability and biodegradability of the carrier components—also injectability and a particle size distribution that is suitable for the intended route of administration, in particular with respect to intravenous injection due to the risk of capillary blockage and embolism. Solid lipid–based delivery systems for intravenous administration should have a size distribution in the submicron range, and shearing-induced gelation that has been observed with certain solid lipid nanoparticle dispersions must be avoided (30,36,65).

The cytotoxicity of solid lipid nanoparticles and their interaction with phagocytizing cells has been investigated in vitro in different cell culture models. When comparing the phagocytic uptake of poloxamer 188-stabilized glyceryl behenate nanoparticles with that of polystyrene nanoparticles incubated with human granulocytes, the uptake of the solid lipid nanoparticles was significantly lower (58). In the same study the cytotoxicity of glyceryl behenate and cetyl palmitate nanoparticles stabilized with up to 5% poloxamer 188 was about 10-fold lower than that of poloxamer 188 modified polylactide/glycolide (PLA/GA) nanoparticles of similar size. Glyceryl behenate nanoparticles stabilized with poloxamine 908 or poloxamer 407 were found to be less phagocytized by human granulocytes than hydrophobic polystyrene nanoparticles surface-modified with the same stabilizers (56). Viability tests showed that the glyceryl behenate nanoparticles were 10-fold less cytotoxic than polylactide nanoparticles and 100-fold less cytotoxic than butylcyanoacrylate nanoparticles. The nature of the lipid had no effect on the viability of HL60 cells and human granulocytes after incubation with the nanoparticles whereas there were distinct differences observed for different stabilizers (57). Poloxamer 407 was well tolerated. A concentration-dependent decrease in viability was observed for other poloxamers, polysorbate 80 and sodium dodecylsulfate. Compared to the surfactants free in solution, the cytotoxicity was reduced by a factor of approximately 1000 when the surfactants were bound to the nanoparticle surface. The solid lipid nanosuspensions again showed a significantly lower cytotoxicity compared to polyalkylcyanoacrylate and PLA/GA nanoparticles. From these in vitro results it can thus be expected that solid lipid nanoparticles have a better in vivo tolerance than polymeric nanoparticles.

Schöler et al. (78) investigated possible cytokine-mediated immunomodulatory effects of glyceryl behenate and cetyl palmitate nanoparticles stabilized by poloxamer 188 on murine peritoneal macrophages. The nanoparticles did not result in stimulation of proinflammatory cytokine responses. However, at high nanoparticle concentrations a significant decrease in cytokine production caused by cytotoxic effects on peritoneal macrophages was observed. It has to be borne in mind that different cell lines may exert different sensitivity to nanoparticle formulations.

Heydenreich et al. (185) studied the cell toxicity of polysorbate 80-stabilized cationic solid lipid nanoparticles consisting of stearylamine and different triglycerides in three different cell lines. The formulation containing stearylamine only as matrix lipid showed the highest cell toxicity. However, partly replacing stearylamine with triglycerides led to higher acceptable doses of the formulations in total, indicating that the other lipids exert a toxicity reducing effect. Moreover, the removal of excessive polysorbate 80 by dialysis decreased the cytotoxicity of the formulations by a factor of 10 in agreement with earlier results indicating a higher cytotoxicity of free compared to surface-bound surfactant (57).

In a study on the effect of emulsifying wax and Brij 72 nanoparticles on blood–brain barrier (BBB) integrity using the in situ rat brain perfusion technique and bovine brain microvessel endothelial cells, no significant alterations in cerebral perfusion were observed even at high nanoparticle concentrations (202). In situ sucrose permeability was unchanged indicating that the integrity of the BBB was not impaired. Neither was thiourea permeation affected in vitro. There was no apparent change in protein expression of brain endothelial cells in vitro. The authors concluded that it was unlikely that tight junction integrity was altered in the presence of the lipid nanoparticles, and that these nanoparticles appeared to have minimal effect on primary BBB parameters making them suitable as brain delivery systems.

The in vivo toxicity of parenteral solid lipid nanoparticles has so far only been studied in mice (240). The animals were repeatedly injected (six times, every third day) with high doses of glyceryl behenate or, respectively, cetyl palmitate nanoparticles by IV bolus. The administered dose corresponded to a human dose of approximately 100 g lipid. During and immediately after injection no signs of acute neurotoxicity were observed. It was found that glyceryl behenate accumulated in liver and spleen concomitant with histomorphological changes of these organs. These effects were dose dependent and reversible. For cetyl palmitate nanoparticles no signs of acute inflammation or lipid accumulation in liver and spleen were found. Cetyl palmitate obviously degrades faster in vivo than glyceryl behenate in agreement with in vitro degradation studies in human plasma and in lipase solution (55,240).

The toxicity data available so far indicate that solid lipid-based carrier systems thus appear to have an in vivo tolerability which is at least comparable to that of approved polymeric systems. This conclusion is supported by data obtained upon subcutaneous, intramuscular or pulmonary administration of microparticulate glyceride systems (22,231,241).

Systemic Administration

The specific properties of solid lipid-based carrier systems make them promising candidates for systemic drug delivery where the following objectives want to be achieved:

1. Controlled drug release and prolonged drug action;
2. Administration of poorly soluble or poorly permeable drugs;
3. Effects on drug distribution in the body/drug targeting.

Uses in Controlled Release

Solid lipid nanoparticles have been suggested for controlled drug release (242), and they can be characterized as matrix type delivery systems releasing the incorporated drug as a function of drug diffusion and degradation of the carrier matrix. Drug that is enriched on the surface or in the outer particle shell will, however, lead to initial burst release, as has been observed in vitro with tetracaine- and etomidate-loaded solid lipid nanoparticles prepared by melt homogenization (53). On the other hand, sustained in vitro release has been found for other systems, e.g., for nanoparticles loaded with prednisolone or gonadorelin (sect. "Drug Incorporation and Release") (34,53,54,207). In vivo, drug release is also determined by degradation of the carrier particles by lipases and other esterases. Because the degradation rate depends on the composition of the solid lipid particles, it should therefore in theory be possible to fine-tune the in vivo release rate of drugs incorporated into the particle matrix by an optimal choice of lipids and surfactants (55,71,72). For example, Penkler et al. (243) found in a pig study that cyclosporin A incorporated into solid lipid nanoparticles composed of Imwitor® 900, Tagat® S and sodium cholate showed prolonged plasma levels of the drug upon oral administration compared to the commercial formulation Sandimmun Neoral®, in which cyclosporin is incorporated into liquid microemulsion droplets. The solid lipid-based formulation avoids the high peak blood levels of the commercial product which may lead to cyclosporin induced nephrotoxicity. The avoidance of high cyclosporin plasma levels as obtained with Sandimmun Neoral could also be demonstrated in humans when cyclosporin-loaded solid lipid nanoparticles were orally administered to healthy volunteers (243).

Evidence of the sustained release potential of orally and intravenously administered solid lipid nanoparticles can also be deduced from a number of in vivo results presented below (see succeeding two sections).

Administration of Poorly Soluble or Poorly Permeable Drugs

Solid lipid–based carrier systems offer the possibility to solubilize poorly water-soluble drugs in the lipid carrier matrix. Solid lipid nanoparticles thus present an alternative intravenous carrier for drugs which poor solubility in aqueous media prevents their administration by the intravenous route or which have to be delivered in cosolvent vehicles such as Cremophor EL, an excipient that may cause anaphylactic reactions. Paclitaxel is an anticancer drug that today is on the market as a Cremophor EL containing formulation due to its very low solubility. Cavalli et al. (234) demonstrated that paclitaxel can be incorporated into tripalmitin nanospheres prepared by the warm microemulsion technique. In an in vitro study these paclitaxel containing nanoparticles were tested for their cytotoxic effect in human malignant glioma cells and found to be more effective than paclitaxel solution on an equimolar basis (244). Human promyelocytic leukaemia cells (HL60) were less sensitive to paclitaxel in lipid nanoparticles than to free paclitaxel at low drug concentrations, whereas cytotoxicity was enhanced when the cells were exposed to higher concentrations of paclitaxel in the nanoparticles (181). On human breast carcinoma cells (MCF-7) paclitaxel incorporated into solid lipid nanoparticles was more cytotoxic than free paclitaxel even at very low concentrations (181).

For oral administration, solid lipid particles may present an alternative carrier for drugs that are poorly absorbed due to their low solubility in the GIT. These drugs will be solubilized in the lipid carrier matrix during transport through the GIT and prevented from precipitation until they are released and can be absorbed. A prerequisite for taking advantage of these specific properties of solid lipid particles for oral administration is, of course, their stability in the GIT. The gastric environment with high ionic strength and low pH may potentially destabilize the lipid particles and lead to aggregation. The electrolyte and pH-stabilities of solid lipid nanoparticles consisting of different lipid matrices and different surfactants have been investigated in vitro using artificial gastrointestinal media (77). It was found that some compositions already showed aggregation and particle growth in the presence of electrolytes at neutral pH whereas others tolerated electrolytes but were pH-sensitive. It was possible to prepare lipid nanoparticles that were resistant to both electrolytes and low pH using optimized compositions. A prerequisite for GIT stability seems to be a zeta potential of at least 8–9 mV in combination with a steric stabilization (77).

Solid lipid particles have been shown to increase the oral bioavailability of drugs with solubility/dissolution-limited absorption. Camptothecin™, an antitumor agent with poor aqueous solubility, has been incorporated into stearic acid nanoparticles stabilized with poloxamer 188 and administered orally to mice (115).

The plasma and organ levels of camptothecin were found to be higher for nanoparticles compared to a solution of camptothecin. There were two peaks in the plasma concentration versus time curve for camptothecin in nanoparticles. The first peak corresponded to the peak of the solution and was attributed to free camptothecin. The second peak appeared after approximately three hours and was attributed to drug released from the nanoparticles upon their degradation in the GIT and/or to drug from nanoparticles that were taken up intact across the intestinal mucosa. Bargoni et al. (245) found that idarubicin, an oral anticancer agent with moderate (20–30%) and highly variable bioavailability, produced significantly higher plasma levels compared to a solution when administered incorporated into stearic acid nanoparticles prepared from a warm o/w microemulsion. This was observed both after duodenal and after intravenous administration to rats. The AUC of the lipid nanospheres compared to the solution was three times higher after IV administration and 15 times higher after duodenal administration. Compared to idarubicin solution, the clearance was significantly reduced for drug incorporated into the nanoparticles. Interestingly, the AUC was higher upon duodenal than after intravenous administration of the nanoparticles.

Solid lipid nanoparticles have also been shown to increase the oral bioavailability of poorly absorbed drugs and peptides. Tobramycin is an antibiotic that needs to be administered parenterally because it is minimally absorbed from the GIT. However, when tobramycin was incorporated into solid lipid nanoparticles, high plasma concentrations after duodenal administration to rats were obtained and were still present after 24 hours (167). Also in this case the AUC upon duodenal administration was found to be higher than after intravenous application. The sustained plasma levels of drug incorporated into solid lipid nanoparticles found in the examples above also illustrates the controlled release potential of the particles.

Prego et al. (237) reported on the use of chitosan-coated tripalmitin nanoparticles for the oral delivery of the peptide drug calcitonin. Calcitonin has a negligible oral bioavailability. However, when calcitonin incorporated into chitosan-coated lipid nanoparticles was given orally to rats, a high degree of the serum calcium levels at one hour post-administration could be measured. The hypocalcemic effect was maintained for the observation period of 24 hours. In contrast, calcitonin solution or calcitonin incorporated into PEG-coated lipid nanoparticles did not lead to significantly reduced calcium serum levels, suggesting an absorption enhancing effect of the chitosan. The marked hypocalcemic effect was, however, more pronounced with the lipid nanoparticles than that reported for other chitosan-coated polymer nanoparticles or liposomes.

Effects on Drug Distribution

Colloidal particles are generally taken up and cleared by cells of the RES upon intravenous injection. Phagocytic uptake is mediated by the adsorption of certain plasma proteins on the particle surface, a process called opsonization (233). Clearance by the RES results in predominant accumulation of the particles in liver and spleen, partly also in lungs and bone marrow (15). Through modification of the particle surface properties, e.g., by using steric stabilizers, it is possible to escape clearance by the RES resulting in long circulating carriers (15). The biodistribution of a drug may hence be different when administered in particles compared to solution. This is of interest if the natural distribution pattern of colloidal particles can be exploited to target drugs to organs of the RES, e.g., in case the drug exerts its biological activity in these organs, or to avoid drug distribution to other organs which may be, e.g., a site of adverse drug action. Long-circulating particles may be targeted to, e.g., tumor sites by using targeting moieties attached to the colloidal drug carrier. Moreover, sustained release from long-circulating carriers as an intravenous depot formulation may be desired. Solid lipid nanoparticles may therefore represent an alternative parenteral delivery system to influence pharmacokinetics and drug distribution in the body. Modified pharmacokinetics compared to drug in solution can, however, only be expected when the drug remains associated with the carrier particles over a sufficient period after administration. Alterations in biodistribution as observed in many of the examples outlined in the following are thus an indirect indication for a sustained release from the nanoparticles.

Podio et al. (180) have investigated the biodistribution of radiolabeled stealth, i.e., sterically stabilized, and nonstealth stearic acid nanoparticles upon intravenous administration to rats. A polyethylene glycol 2000 conjugated stearic acid was used as a stealth agent. The radioactivity in the liver and lung 60 min after injection was much lower for stealth solid lipid particles compared to nonstealth particles, indicating that the stealth particles more efficiently escaped recognition and uptake by the RES. The spleen uptake was low with both types of nanoparticles, and radioactivity detected in heart, kidneys, bladder, and testicles was negligible in all animals. Brain radioactivity was also low with both types of lipid nanoparticles. However, the increase of the cerebrospinal fluid (CSF)-to-blood radioactivity was marked, and appeared to be higher with stealth-nanoparticles compared to nonstealth particles.

Pharmacokinetic results consistent with drug delivery to the brain were obtained on doxorubicin-loaded solid lipid nanoparticles (176,178,179). Zara et al. (176) compared the pharmacokinetics of doxorubicin incorporated as ion pair in stearic acid nanoparticles with that of a commercial solution of the drug after intravenous injection to rats. It was found that the doxorubicin blood level was significantly higher at each time point with the nanoparticles than with the solution. The plasma concentration versus time curve was biexponential, and doxorubicin-nanoparticles showed a more than 17-fold increase in plasma AUC (0–180 min), a lower rate of clearance and a smaller volume of distribution than the commercial drug solution. The drug concentration was also higher in the lung, spleen, and brain with lipid nanoparticle-incorporated doxorubicin, whereas drug levels were lower in liver, heart, and kidney. At 180 minutes, doxorubicin concentration in the heart, the target organ for doxorubicin-related cardiomyopathy, was 34% lower in nanoparticle-treated animals. Contrary to treatment with solid lipid nanoparticles, doxorubicin from the commercial solution was not detected in the brain. Fundaro et al. (179) investigated the pharmacokinetics and tissue distribution of doxorubicin incorporated into stealth and nonstealth stearic acid nanoparticles (80–90 nm) over 24 hours after intravenous administration to rats using stearic acid-PEG 2000 as stealth agent. Following IV injection, doxorubicin was still present in plasma when given as nanoparticles whereas with doxorubicin solution the drug was cleared from plasma after 180 min. Doxorubicin stealth nanoparticles gave generally higher drug concentrations in plasma than the nonstealth nanoparticles. Stealth-nanoparticles showed a lower clearance, a higher distribution volume and a significantly higher AUC than nonstealth nanoparticles. In all tissues, except the brain, less doxorubicin was present in rats injected with the nanoparticle dispersion than in those injected with the solution, and the nanoparticles prevented the accumulation of doxorubicin in the heart. A significant amount of doxorubicin was detectable in the brain and CSF of nanoparticle injected rats. The highest brain concentration was obtained with stealth-nanoparticles. Zara et al. (178) studied the pharmacokinetics and tissue distribution of doxorubicin in nonstealth and stealth solid lipid nanoparticles upon intravenous administration to rabbits. Doxorubicin bioavailability increased with increasing amounts of stealth agent present in the nanoparticles. All nanoparticle formulations significantly decreased accumulation of doxorubicin in heart and liver and demonstrated drug delivery to the brain, which was affected by stealth agent concentration. Whereas doxorubicin solution was not taken up by the brain, doxorubicin from the lipid nanoparticles was present in the brain 30 minutes and two hours after i.v administration. Six hours after injection, doxorubicin was detectable in the brain only with the stealth nanoparticles at the highest concentration of stealth agent.

The uptake of solid lipid nanoparticles into the brain could be demonstrated by Koziara et al. (203) using an in situ rat brain perfusion model. Two different nanoparticle compositions were tested consisting of emulsifying wax or Brij 72. Their particle size was below 100 nm. Uptake of tritium-labeled nanoparticles into the brain was evaluated from 0 to 60s. The integrity of the BBB was verified using [^{14}C]sucrose as a marker. Both nanoparticle formulations could be transported across the BBB. Brij 72 nanoparticles that contained polysorbate 80 resulted in a significantly higher brain transport over emulsifying wax nanoparticles. The authors suggested that the nanoparticles are transported across the BBB by endocytosis/transcytosis or passive diffusion of the nanoparticles. Lockman et al. (201) incorporated thiamine as a surface ligand on the nanoparticles to specifically target them to the brain via the BBB thiamine transporter. It was found that the thiamine-coated nanoparticles associated with the thiamine transporter and increased brain uptake rate.

The potential of solid lipid nanoparticles for drug delivery to the brain was also demonstrated by Yang et al. (116) who found that drug delivery to the brain from camptothecin-loaded stearic acid nanoparticles was increased about five-fold compared to camptothecin solution after intravenous administration to mice. The mean residence time in plasma of the nanoparticles was 18 times higher than the solution, indicating a substantial sustained release effect of the nanoparticles. Accumulation of camptothecin in the kidneys was reduced for the nanoparticles suggesting lower nephrotoxicity. Wang et al. (129) investigated brain uptake of 3′,5′-dioctanoyl-5-fluoro-2′-deoxyuridine (DO-FUdR), a lipophilic prodrug of 5-fluoro-2′-deoxyuridine (FUdR), when administered incorporated into solid lipid nanoparticles. Upon intravenous administration to mice, the brain AUC of DO-FUdR from nanoparticles was about 11-fold higher than that of FUdR solution, and $t_{1/2}$ in the brain was prolonged from 8.54 to 21.58 hours. Brain AUC of nanoparticle incorporated DO-FUdR was twice as high as free DO-FUdR.

The distribution of solid lipid nanoparticles to the brain can also be exploited for use in imaging. Peira et al. (184) have studied superparamagnetic iron oxide–loaded nanoparticles as contrast agent for magnetic resonance imaging (MRI). Upon intravenous administration to rats superparamagnetic nanoparticles showed slower blood clearance than Endorem®, a colloidal iron oxide suspension. MRI data are consistent with nanoparticle uptake into the central nervous system confirming the ability of solid lipid nanoparticles to cross the BBB.

Khopade and Jain (246) reported on the targeting of diclofenac diethylammonium incorporated into lipid nanoparticles to inflammatory sites. The nanoparticle composition was aimed at mimicking low-density lipoproteins (LDL), but it did not contain a protein component. Addition of sphingomyelin to the composition proved able to prolong the circulation time of the nanoparticles and to improve the anti-inflammatory efficacy of diclofenac diethylammonium. PEG-coated LDL-like lipospheres containing the anticancer agent 6-mercaptopurine (6-MP) were observed to circulate for prolonged times in plasma after IV injection into rats (235). Uncoated and stealth lipospheres generated 2.3 and 15.8 times higher AUC than free 6-MP. High drug concentrations could be maintained in plasma for the observation period of 24 hours with the nanoparticles whereas free 6-MP was cleared from the plasma within six hours. For the stealth lipospheres, low tissue-to-serum AUC ratio was observed for RES organs indicating avoidance of RES uptake. The uncoated lipospheres showed a high spleen-to-blood ratio indicating primary clearance by this organ. The potential use of solid lipid nanoparticles in anticancer therapy has also been demonstrated by Chen et al. (205) who investigated the pharmacokinetics of two types of paclitaxel-loaded stearic acid nanoparticles upon intravenous injection in mice compared to paclitaxel in Cremophor EL formulation. Encapsulation of paclitaxel in solid lipid nanoparticles significantly changed the pharmacokinetics of the drug with $t_{1/2}\beta$ being 3.6 and 7.4 times prolonged compared to paclitaxel solubilized with Cremophor EL.

Mumper et al. (190) described the use of folate-targeted lipid nanoparticles to deliver Gd complexes to tumors for potential applications in neutron capture therapy. The folic acid receptor is overexpressed in a number of human tumors. Folic acid was chemically linked to distearoyl phosphatidyl ethanolamine (DSPE) via a PEG-spacer, and the folate ligand was anchored into or onto lipid nanoparticles. In in vitro tumor cell uptake studies, uptake of the folate-coated nanoparticles was about 10-fold higher than uncoated nanoparticles after 30 minutes incubation in KB cells. In an in vivo study in tumor-bearing mice both folate- and PEG-coated gadolinium–loaded nanoparticles had comparable tumor accumulation (247). However, the cell uptake and tumor retention of folate-coated nanoparticles was significantly enhanced over PEG-coated nanoparticles suggesting a benefit of the folate ligand coating to facilitate tumor cell internalization and retention of gadolinium nanoparticles in the tumor tissue. These results demonstrate the potential of solid lipid nanoparticles as intravenous carriers for passive and active (i.e., via attachment of targeting moieties) drug targeting.

Interestingly, solid lipid nanoparticles have also been found to influence drug distribution upon duodenal administration. Zara et al. (177) reported that the pharmacokinetics and the tissue distribution of idarubicin-loaded stearic acid nanoparticles were different from those of idarubicin solution after duodenal administration to rats. The AUC was approximately 21 times higher, and elimination half-life was 30 times higher after nanoparticle administration compared to the solution. The AUC after duodenal administration of idarubicin nanoparticles was higher than after intravenous administration of the same formulation. Idarubicin concentrations were lower in heart, lung, spleen, and kidney when administered duodenally in the nanoparticles. Idarubicin was detected in the brain only after administration in nanoparticles, which is remarkable because idarubicin is generally regarded as unable to cross the blood–brain barrier because of its relatively low lipophilicity. The results indicate that nanoparticle-bound idarubicin is taken up from the GIT and passes the BBB as intact nanoparticles. Brain delivery of drug incorporated into solid lipid nanoparticles after oral or duodenal administration was also observed for camptothecin and tobramycin (115,174). Incorporation of tobramycin in solid lipid nanoparticles was found to modify the pharmacokinetics and tissue distribution when administered duodenally compared to tobramycin in solution or in lipid nanoparticles administered intravenously. In particular, tobramycin concentrations in the kidney, an organ of drug-induced toxicity, were lower after administration in nanoparticles duodenally or intravenously, than after IV administration of drug solution. These results indicate that the organ toxicity of anticancer and antibiotic agents can be reduced by incorporation of the drug into solid lipid nanoparticles owing to alterations of the pharmacokinetics and tissue distribution, both upon intravenous and oral administration, and that delivery to the brain can be achieved for drugs that are usually not transported across the blood–brain barrier.

Bargoni et al. (173) from their results suggest that solid lipid nanoparticles may be taken up as intact particles into blood and lymph after duodenal administration to fed rats. The authors reported that they found evidence for the presence of nanoparticles in rat lymph and plasma by transmission electron microscopy and laser light scattering. Using radiolabeled nanoparticles it was shown that peak radioactivity concentrations were obtained in the lymph about 130 to 150 minutes after duodenal administration (monitored period 180 minutes), and that peak levels were dose dependent. A low radioactivity was detected in the blood indicating that direct transport of nanoparticles into the blood is low. Cavalli et al. (170) reported the presence of tobramycin-loaded solid lipid nanoparticles in the lymph and in mesenteric lymph nodes upon duodenal administration to rats using transmission electron microscopy. Apart from indicating the way of uptake of poorly absorbable drugs from the GIT when loaded to lipid nanoparticles these results suggest the potential use of solid lipid nanoparticles for targeting to the lymph, e.g., in the treatment of lymphatic cancers or for lymphatic imaging. In this context, it can be noted that Videira et al. (236) observed ^{99m}Tc-labeled glyceryl behenate nanoparticles in regional lymph nodes after pulmonary administration. Upon inhalation of the aerosolized lipid nanodispersion to rats, radiolabeled nanoparticles were deposited in the deep regional small airways from where they began to translocate to regional lymph nodes within a few minutes and remained trapped there. The biodistribution of the lipophilic ^{99m}Tc marker was different in the nanoparticles compared to the free marker which accumulated mainly in liver and spleen upon clearance from the lungs. It is noteworthy that four hours after inhalation of the radiolabeled nanoparticles 25% of radioactivity was still present in the lungs, indicating that solid lipid nanoparticles might be a promising delivery system for pulmonary administration.

Immunization and DNA Delivery

Solid lipid nanoparticles have been suggested as carriers or adjuvants for protein antigens and DNA for use in immunization. Olbrich et al. (232) found adjuvant activity of solid lipid nanoparticles with *mycoplasma bovis* antigen in sheep. Adjuvant activity in antibody titer was moderate compared to Freund's incomplete adjuvant (FIA) but could be improved by addition of EQ1 (*N*,*N*-di-(β-stearoylethyl)-*N*, *N*-dimethyl-ammonium chloride). In contrast to FIA, the solid lipid nanoparticle–based antigen formulations were well tolerated. Cui and Mumper (191) coated a cationized model protein antigen on anionic lipid nanoparticles prepared from microemulsion templates. Compared to cationized antigen alone or noncationized antigen administered together with alum adjuvant, the cationized protein–coated nanoparticles produced the strongest and most reproducible antibody titer upon SC injection to mice. Moreover, in contrast to the reference formulations, it was surprisingly found that the nanoparticles elicited both enhanced T-helper type 1 and type 2 immune responses.

In recent years genetic immunization using plasmid DNA has emerged as an alternative to conventional vaccination using protein antigens, live attenuated viruses, or killed bacteria. An advantage of DNA vaccines is the ability to elicit both humoral, i.e., production of antibodies, and cellular immune responses, i.e., generation of cytotoxic T lymphocytes. However, "naked" plasmid DNA usually results in low transfection efficiency, and nanoparticulate systems such as liposomes, polymeric nanoparticles, and solid lipid nanoparticles have been proposed as nonviral vehicles for plasmid DNA to enhance transfection efficiency and immune response. Generally, positively charged particles are used to form complexes with the negatively charged DNA, and cationic lipids and surfactants are used to impose a positive charge on the particles. Olbrich et al. (69) have investigated cationic solid lipid nanoparticles as a transfection agent for plasmid DNA using EQ1 or cetylpyridinium chloride as cationic stabilizers. Typical DNA–nanoparticle complexes were in the size range of 300 to 800 nm. The DNA–nanoparticle complexes were shown to transfect Cos-1 cells in vitro whereas incubation with uncomplexed DNA did not lead to expression of the reporter gene. However, compared to known transfection agents, such as poly-L-lysine, the transfection efficiency of the investigated solid lipid nanoparticles was only moderate. Tabatt et al. (107) attempted to enhance in vitro transfection efficiency by optimizing cationic lipid and matrix lipid composition of solid lipid nanoparticles. Cytotoxicity was found to depend strongly on the cationic lipid used with one-tailed cationic lipids being highly cytotoxic whereas two-tailed cationic lipids were well tolerated. All cationic nanoparticle formulations increased transfection compared to naked DNA. The combination of cetylpalmitate and the cationic lipid DOTAP was found to exhibit significantly higher transfection efficiency than all other tested compositions. In the presence of chloroquine, the transfection efficiency was almost comparable to polyethyleneimine (PEI), one of the most effective cationic polymers for gene therapy currently available.

Cui and Mumper (189) were able to significantly enhance in vitro transfection efficiency of cationic solid lipid nanoparticles by incorporation of dioleyl phosphatidylethanolamine (DOPE), an endosomolytic lipid, and surface coating of a specific cell-targeting ligand, pullulan. The pullulan-coated DNA–nanoparticle complexes showed transfection levels that were comparable to that of Lipofectin®, a cationic lipid known as transfection agent. In a mice immunization study, DNA–nanoparticle complexes containing DOPE and a surface-deposited mannan-ligand, which is targeted toward dendritic cells, generated antibody titers that were 16-fold enhanced compared to naked DNA upon SC administration (189). Immunization with DNA solid-lipid nanoparticle complexes also resulted in strong Th1-type cytokine release.

Cui and Mumper (194) have developed a process to entrap hydrophobized plasmid DNA inside solid lipid nanoparticles prepared from microemulsion precursors in an attempt to protect the DNA from degradation by serum nucleases. Uniform plasmid DNA (pDNA)-entrapped nanoparticles of 100 to 160 nm in size could be obtained. The particles were negatively charged, and DNA entrapment efficiency was about 25% to 30%. Whereas naked DNA was completely digested upon incubation with mouse serum, nanoparticle-entrapped DNA remained intact. In Hep G2 cells plasmid DNA alone or DNA entrapped in nanoparticles did not result in measurable transfection. Incorporation of DOPE into the nanoparticles enhanced the transfection by fourfold compared to nanoparticles without DOPE. Deposition of pullulan on the surface of the DOPE-free nanoparticles significantly enhanced transfection, however, the overall magnitude of expression compared to Lipofectin® was considerably lower. The biodistribution of radiolabelled DNA-entrapped nanoparticles in mice after IV injection revealed that the nanoparticles circulated longer in the blood than naked DNA. The absence of a high level of radioactivity in the lung indicated that the nanoparticles did not aggregate in blood, a phenomenon that is common for cationic lipid complexes with plasmid DNA.

Mannan-coated lipid nanoparticle/DNA complexes have been developed for targeting to the very potent antigen presenting dendritic cells in skin epidermis for noninvasive topical immunization (193). In vitro studies confirmed the transfection efficiency of the complexes. Upon topical administration to mice, net negatively charged DNA-coated nanoparticles generated a sixfold-enhanced antigen-specific immunoglobin G (IgG) titer compared to naked DNA. Surface deposition of the mannan ligand on the nanoparticles resulted in an increase of the antibody titer by a factor of 13 over naked DNA. DNA-loaded nanoparticles with both DOPE and the mannan ligand enhanced the antibody titer 16-fold. All DNA–nanoparticle formulations induced significantly enhanced proliferation of splenocytes isolated from immunized mice compared to naked DNA. The DNA–nanoparticles with both DOPE and mannan significantly enhanced splenocyte proliferation over those with and without mannan. The DNA–nanoparticles thus demonstrated both enhanced humoral and cellular immune responses over naked DNA.

Local Administration

Solid lipid nanoparticles have been proposed for dermal administration of drugs and for use in cosmetics (238). Features that make solid lipid nanoparticles attractive for topical applications because of the solid nature of the matrix include protection of chemically labile compounds, possibility to modify drug release and skin penetration, and their film forming properties. For better applicability on the skin the nanoparticles can be incorporated into an ointment or gel.

As a result of film formation on the skin upon administration, solid lipid nanoparticles exert an occlusive effect that can be quantified in vitro as occlusion factor (96,119). It has been found that the occlusion factor is proportional to the degree of crystallinity of the lipid matrix whereas noncrystalline emulsion systems do not have occlusive properties (96). The effect of solid lipid nanoparticles on skin hydration and viscoelasticity of the skin has been tested in vivo in 25 healthy volunteers (102). A conventional o/w cream and the same cream enriched with cetyl palmitate nanoparticles were applied twice daily on the forearm for 28 days. Skin hydration was measured using a corneometer evaluating capacitance changes in the stratum corneum. The viscoelastic properties were investigated with a cutometer. After 28 days, the nanoparticle-enriched cream was found to be significantly more hydrative than the conventional cream, increasing skin hydration by 31% versus 24%. The viscoelastic parameters remained, however, unchanged by both formulations. The authors attributed the latter finding to the young age of the test subjects.

Owing to their film forming ability, solid lipid nanoparticles have been proposed for use in chronic atopic eczema. Prednicarbate, a topical corticosteroid, was incorporated into solid lipid nanoparticles to selectively target the drug to eczematous viable epidermis, and skin penetration was tested using excised human skin and reconstructed epidermis mounted in Franz flow-through diffusion cells (82). Prednicarbate penetration into human skin increased by 30% when using the nanoparticles compared to cream. Drug concentration reaching the dermis was almost twice as high with the nanoparticle formulation. With reconstructed epidermis prednicarbate penetration was increased about three-fold.

Jenning et al. (91) found a drug localizing effect of vitamin A (retinol and retinyl palmitate) in the upper layers of porcine skin when using glyceryl behenate nanoparticles. Following drug penetration from solid lipid nanoparticle dispersion for six hours, high vitamin A concentrations could be maintained in the stratum corneum and upper epidermis. A submicron-size emulsion formulation delivered significantly lower drug amounts to these locations. After 24 hours treatment the vitamin A concentration in the upper skin layers declined considerably for the solid nanoparticles while drug amount in the deeper layers increased. Retinol showed a better penetrability than retinyl palmitate. With the nanoemulsion the drug distribution pattern did not change. The results were attributed to the film formation of solid lipid nanoparticles upon water evaporation, suggesting that reduced transepithelial water loss and enhanced occlusion due to film formation might enable retinol to penetrate into the deeper layers and through the skin. After six hours a polymorphic transition occurs in the nanoparticles leading to drug expulsion and loss of sustained release properties of the carrier. When incorporating the nanoparticles into a hydrogel, this polymorphic transition is delayed. Use of an o/w cream vehicle delays the polymorphic transition even further so that the drug localizing effect of the nanoparticles is prolonged over a 24 hour period. This also applies to sustained drug release. When vitamin A–loaded solid lipid nanoparticles were incorporated into a hydrogel or cream vehicle, drug release was found to be sustained over 12 to 18 hours (88). Sustained release correlated to the presence of the metastable β' polymorph. The β' to β_i transition leads to drug expulsion from the carrier particles and increased release. The transition is controlled by the presence of surfactants, thickening agents, and humectants. Mei et al. (127) found a penetration enhancing effect of solid lipid nanoparticles for triptolide in vitro using excised rat skin. In the carageenan induced rat paw edema model, the anti-inflammatory activity of the nanoparticle dispersion was higher than that of a microemulsion formulation upon topical administration. In contrast, the microemulsion was found to be more efficient in complete Freund's adjuvant induced paw edema.

Recently, it was discovered that highly crystalline lipid nanoparticles act as physical sunscreens by scattering or reflecting UV radiation. The scattering properties depend strongly on the degree of crystallinity of the particles. Improved photoprotection compared to a placebo emulsion with the same lipid content was shown (99). It was also found that incorporation of chemical sunscreens, which absorb UV radiation such as benzophenone-3 or tocopherol acetate, further improved protection from UV light in a synergistic manner (95,99). It is therefore possible to reduce the amount of chemical sunscreens which may induce photoallergic or phototoxic reactions on the skin by incorporation in solid lipid nanoparticles. It is also possible to decrease the amount of physical sunscreens, such as titanium dioxide, by incorporation into the nanoparticle dispersions while maintaining the UV protection level (248).

When incorporating solid lipid nanoparticles into hydrogels or creams to facilitate their applicability on the skin in semisolid vehicles, a major disadvantage is dilution of the nanoparticle dispersion and hence reduction of the drug load in the carrier. This problem has recently been overcome by producing semisolid solid lipid nanoparticle dispersions by high-pressure homogenization using high lipid content up to 35%. Despite the semisolid consistency the lipid dispersions preserved their colloidal particle size (86). Another possibility to prepare semisolid lipid nanoparticle formulations without the need to use additional vehicles may be offered by the gelation phenomena observed in comparatively low concentrated lipid nanosuspensions stabilized solely with phospholipids or poloxamer (30,36,62,65).

Colloidal solid lipid particles have also been suggested for ophthalmic use. Cavalli et al. (159,169) proposed the use of stearic acid nanoparticles as a sustained ocular delivery system for pilocarpine or tobramycin loaded into the particles as ion pair. Tobramycin-loaded nanoparticles produced a significantly higher bioavailability in the aqueous humor upon topical administration to rabbits compared to administration as standard commercial eyedrops. The AUC was increased by a factor of four. Drug-free fluorescent nanoparticles administered in the rabbit eye were retained for longer times on the cornea and in the conjunctival sac compared to a fluorescent solution, suggesting that the increased bioavailability can be attributed to the prolonged preocular retention time.

CONCLUSIONS

Solid lipid nanoparticles are receiving growing interest as delivery systems in the pharmaceutical field. Various manufacturing methodologies are available and have been refined during recent years (Figs. 2 and 3) (Table 2). The choice of manufacturing method depends inter alia on the physicochemical properties of the drug to be incorporated, intended use and route of administration, as well as batch sizes. Scale-up experiments indicate that solid lipid particles can be produced in large scale (75,86,94,171). It has been demonstrated that melt-homogenized dispersions exhibit considerable long-term storage stability (29,34,39,45,62,65) and that problems associated with the crystallization process can be circumvented by a proper choice of matrix constituents, emulsifying agents, and production parameters (41,42,46,118). A general limitation of crystalline solid lipid nanoparticles remains, however, their limited inclusion capacity for many drugs. Developments to increase drug load include the preparation of nanoparticles consisting of lipophilic prodrugs or drug–lipid conjugates, oil-loaded solid lipid nanoparticles, or liquid-crystalline lipid nanoparticles (89,90,112,113,186,249–253,254). Colloidal dispersions based on biodegradable solid lipids have been found to be potential drug carrier systems of complex physicochemical properties as expressed, e.g., by their recrystallization behavior, polymorphism, particle morphology, and drug incorporation and release properties, emphasizing the importance of physicochemical characterization for product development. Despite their complexity solid lipid nanoparticles represent a feasible alternative to conventional drug carrier systems such as polymer nanoparticles, liposomes, and lipid emulsions. Low toxicity and good tolerability have been found both in vitro and in vivo (sect. "Safety and Tolerability Aspects").

Hence, it is not surprising that solid lipid-based carrier systems have been extensively investigated for various applications in drug delivery during recent years, both for lipophilic and hydrophilic drugs.It has been demonstrated in vivo that solid lipid nanoparticles can be used as an alternative to potentially harmful cosolvent formulations for the administration of lipophilic drugs such as paclitaxel by the parenteral route (181,234,244). For oral administration, solid lipid nanoparticles have been shown to increase the oral bioavailability of poorly soluble drugs, e.g., camptothecin, as well as of poorly permeable drugs such as tobramycin and the peptide calcitonin (115,167,237).

One of the possible advantages of solid lipid nanoparticles over lipid emulsions or liposomes is the reduced mobility of incorporated drug molecules in the carrier preventing drug leakage and allowing sustained drug release. The sustained release potential of solid lipid-based systems has been demonstrated in vivo upon oral/duodenal, parenteral, topical, and ophthalmic administration (88,159,170,174, 176–179,243). Prolonged retention of the drug in the carrier is of particular importance for the targeting of drugs using long-circulating particles or targeting moieties attached to the carrier particles.

As with liposomes and lipid emulsions, it has been found possible to change the intravenous pharmacokinetics of incorporated drugs by using solid lipid nanoparticles. The use of solid lipid nanoparticles was observed to prolong circulation of the incorporated drugs, reduce their clearance rate, increase bioavailability, and change biodistribution upon intravenous administration. These effects are usually more pronounced with stealth lipid particles that have polyethylene oxide chains on their surface, probably because of reduced interaction with components of the RES. The altered organ distribution may increase the therapeutic index of drugs which can be directed away from organs of drug-induced toxicity. This has been demonstrated, e.g., for doxorubicin (176). Moreover, it has been shown that solid lipid nanoparticles can be actively targeted to, e.g., tumor sites by the attachment of targeting moieties such as folic acid (190,247).

Interestingly, solid lipid nanoparticles have also been found to influence drug distribution upon oral administration, pointing to their uptake as intact particles from the gastrointestinal tract. Intact particles could be detected in blood and lymph upon duodenal administration to rats (173). Lymphatic uptake may enable circumvention of first pass metabolism.

A particularly interesting feature of solid lipid nanoparticles is their ability to deliver drugs to the brain both upon intravenous and oral administration, making solid lipid nanoparticles interesting as carriers for brain delivery and imaging. The exact mechanism of brain delivery needs, however, further clarification (177,178).

The extremely large surface area of anisometric lipid nanoparticles makes them suitable carriers for active substances that can be transported in adsorbed form, e.g., macromolecules like DNA and protein antigens, and recent results suggest a potential of solid lipid particles as adjuvants and in genetic immunization (107,189,193). The transfection efficiency of DNA bound to solid lipid nanoparticles is, however, relatively modest till now.

Solid lipid nanoparticles have also been investigated for local administration, in particular by the dermal route. Solid lipid nanoparticles form films when applied on the skin and thus exert an occlusive effect, which may improve drug penetration (91). Highly crystalline lipid nanoparticles have been found to scatter or reflect UV radiation, which makes them interesting as physical sunscreens (99).

In conclusion, solid lipid nanoparticles have recently found a wide range of applications and demonstrated large potential in systemic and local drug delivery including drug targeting. The increasing interest in solid lipid–based drug delivery systems is also reflected by attempts of the drug delivery industry to commercialize the drug carriers. For example, Eurand offers the use of lipid nanospheres prepared from microemulsions to enhance the oral absorption of poorly permeable drugs. Skyepharma develops solid lipid nanoparticles as a platform technology for topical drug delivery. So far, however, development has not yet reached the clinical phase and appears far from market authorization. Considering that it took the liposomes 30 years to reach the pharmaceutical market, it will probably take at least another decade until solid lipid nanoparticles make it there. By analogy with liposomes, it can also be expected that cosmetic products based on solid lipid nanoparticles may be commercially available before pharmaceuticals do.

REFERENCES

1. Gupta PK. Drug targeting in cancer chemotherapy: a clinical perspective. J Pharm Sci 1990; 79:949–962.
2. Barrat G. Colloidal drug carriers: achievements and perspectives. Cell Mol Life Sci 2003; 60:21–37.
3. Allen TA, Cullis P. Drug delivery systems: entering the mainstream. Science 2004; 3003:1818–1822.
4. Allémann E, Gurny R, Doelker E. Drug-loaded nanoparticles—preparation methods and drug targeting issues. Eur J Pharm Biopharm 39; 1993:173–191.
5. Kante B, Couvreur P, Dubois-Krack D, et al. Toxicity of polyalkylcyanoacrylate nanoparticles. I. Free nanoparticles. J Pharm Sci 1982; 71:786–790.
6. Lentz BR, Carpenter TJ, Alford DR. Spontaneous fusion of phosphatidylcholine small unilamellar vesicles in the fluid phase. Biochemistry 1987; 26:5389–5397.
7. Schuberth O, Wretlind A. Intravenous infusions of fat emulsions, phosphatides and emulsifying agents. Acta Chir Scand 1961; 278(suppl.):1–21.
8. Wretlind A. Development of fat emulsions. J Parent Enteral Nutr 1981; 5:230–235.
9. Collins-Gold LC, Lyons RT, Bartholow LC. Parenteral emulsions for drug delivery. Adv Drug Deliv Rev 1990; 5:189–208.
10. Prankerd RJ, Stella VJ. The use of oil-in-water emulsions as a vehicle for parenteral drug administration. J Parent Sci Technol 1990; 44:139–149.
11. Klang S, Benita S. Design and evaluation of submicron emulsions as colloidal drug carriers for intravenous administration. In: Benita S, ed. Submicron Emulsions in Drug Targeting and Delivery. Amsterdam: Harwood Academic Publishers, 1998:119–152.
12. Washington C. Stability of lipid emulsions for drug delivery. Adv Drug Del Rev 1996; 20:131–145.
13. Magenheim B, Levy MY, Benita S. A new in vitro technique for evaluation of drug release profile from colloidal carriers—ultrafiltration technique at low pressure. Int J Pharm 1993; 94:115–125.
14. Washington C, Evans K. Release rate measurements of model hydrophobic solutes from submicron triglyceride emulsions. J Control Release 1995; 33:383–390.
15. Müller RH. Colloidal Carriers for Controlled Drug Delivery and Targeting—Modification, Characterization and In vivo Distribution. Boca Raton: CRC Press, 1991.
16. Scholes PD, Coombes AGA, Davies MC, Illum, L, Davis SS. Particle engineering of biodegradable colloids for site-specific drug delivery. In: Park K, ed. Controlled Drug Delivery. Washington: American Chemical Society, 1997:73–106.
17. Robinson MJ, Bondi A, Swintosky J. Sulfamethylthiadiazole: human blood concentration and urinary excretion data following oral doses. J Am Pharm Assoc, Sci Ed 1958; 17:874–878.
18. Kowarski CR, Volberger B, Versanno J, Kowarski A. A method of preparing sustained release sulfamethazine in small size batches. Am J Hosp Pharm 1964; 21:409.
19. Del Curto MD, Chicco D, D'Antonio M, et al. Lipid microspheres as sustained release system for a GnRH antagonist (Antide). J Control Release 2003; 89:297–310.
20. Kawashima Y, Ohno H, Takenaka H. Preparation of spherical matrixes of prolonged-release drugs from liquid suspension. J Pharm Sci 1981; 70:913–916.
21. Domb AJ. Long acting injectable oxytetracycline-liposphere formulations. Int J Pharm 1995; 124:271–278.
22. Reithmeier H, Herrmann J, Göpferich A. Lipid microparticles as a parenteral controlled release device for peptides. J Control Release 2001; 73:339–350.
23. Cortesi R, Esposito E, Luca G, Nastruzzi C. Production of lipospheres as carriers for bioactive compounds. Biomaterials 2002; 23:2283–2294.
24. Sugiura S, Nakajima M, Tong J, Nabetani H, Seki M. Preparation of monodispersed solid lipid microspheres using a microchannel emulsification technique. J Colloid Interf Sci 2000; 227:95–103.

25. Erni C, Suard C, Freitas S, Dreher D, Merkle HP, Walter E. Evaluation of cationic solid lipid microparticles as synthetic carriers for the targeted delivery of macromolecules to phagocytic antigen-presenting cells. Biomaterials 2002; 23:4667–4676.
26. Eldem T, Speiser P, Hincal A. Optimization of spray-dried and -congealed lipid micropellets and characterization of their surface morphology by scanning electron microscopy. Pharm Res 1991; 8:47–54.
27. Akiyama Y, Yoshioka M, Horibe H, Hirai S, Kitamori N, Toguchi H. Noval oral controlled-release microspheres using polyglycerolesters of fatty acids. J Control Release 1993; 26:1–10.
28. Savolainen M, Khoo C, Glad H, Dahlqvist C, Juppo AM. Evaluation of controlled-release polar lipid microparticles. Int J Pharm 2002; 244:151–161.
29. Siekmann B, Westesen K. Submicron-sized parenteral carrier systems based on solid lipids. Pharm Pharmacol Lett 1992; 1:123–126.
30. Westesen K, Siekmann B. Investigation of the gel formation of phospholipid stabilized solid lipid nanoparticles. Int J Pharm 1997; 151:35–45.
31. Speiser P. Lipidnanopellets als Trägersystem für Arzneimittel zur peroralen Anwendung. European Patent Application EP 0,167,825 (1986).
32. Gasco MR, Morel S. Lipospheres from microemulsions. Farmaco 1990; 45:1127–1128.
33. Sjöström B, Bergenståhl B. Preparation of submicron drug particles in lecithin-stabilized o/w emulsions. I. Model studies of the precipitation of cholesteryl acetate. Int J Pharm 1992; 88:53–62.
34. Müller RH, Mehnert W, Lucks JS, et al. Solid lipid nanoparticles (SLN)—an alternative colloidal carrier system for controlled drug delivery. Eur J Pharm Biopharm 1995; 41:62–69.
35. Kecht-Wyrsch P. Hochdisperse Glycerid-Mikropartikel als Perorales Arzneiträgersystem. Ph.D. Thesis, Eidgenössische Technische Hochschule Zürich; 1987.
36. Westesen K, Siekmann B, Koch MHJ. Investigations on the physical state of lipid nanoparticles by synchrotron radiation X-ray diffraction. Int J Pharm 1993; 93: 189–199.
37. Siekmann B, Westesen K. Thermoanalysis of the recrystallization process of melt-homogenized glyceride nanoparticles. Colloids Surf B: Biointerfaces 1994; 3:159–175.
38. Siekmann B, Westesen K. Melt-homogenized solid lipid nanoparticles stabilized by the nonionic surfactant tyloxapol I. Preparation and particle size determination. Pharm Pharmacol Lett 1994; 3:194–197.
39. Siekmann B, Westesen K. Melt-homogenized solid lipid nanoparticles stabilized by the nonionic surfactant tyloxapol. II. Physicochemical characterization and lyophilisation. Pharm Pharmacol Lett 1994; 3:225–228.
40. Westesen K, Bunjes H. Do nanoparticles prepared from lipids solid at room temperature always possess a solid lipid matrix? Int J Pharm 1995; 115:129–131.
41. Bunjes H, Westesen K, Koch MH J. Crystallization tendency and polymorphic transitions in triglyceride nanoparticles. Int J Pharm 1996; 129:159–173.
42. Westesen K, Bunjes H, Koch MHJ. Physicochemical characterization of lipid nanoparticles and evaluation of their drug loading capacity and sustained release potential. J Control Release 1997; 48:223–236.
43. Unruh T, Bunjes H, Westesen K, Koch MHJ. Observation of size-dependent melting in lipid nanoparticles. J Phys Chem B 1999; 103:10373—10377.
44. Bunjes H, Koch MHJ, Westesen K. Effect of particle size on colloidal solid triglycerides. Langmuir 2000; 16:5234–5241.
45. Bunjes H, Drechsler M, Koch MHJ, Westesen K. Incorporation of the model drug ubidecarenone into solid lipid nanoparticles. Pharm Res 2001; 18:287–293.
46. Bunjes H, Koch MHJ, Westesen K. Effects of surfactants on the crystallization and polymorphism of lipid nanoparticles. Progr Colloid Polym Sci 2002; 121:7–10.
47. Bunjes H, Koch MHJ, Westesen K. Influence of emulsifiers on the crystallization of solid lipid nanoparticles. J Pharm Sci 2003; 92:1509–1520.

48. Unruh T, Bunjes H, Westesen K, Koch MHJ. Investigations on the melting behaviour of triglyceride nanoparticles. Colloid Polym Sci 2001; 279:398–403.
49. Unruh T, Westesen K, Bösecke P, Lindner P, Koch MHJ. Self-assembly of triglyceride nanocrystals in suspension. Langmuir 2002; 18:1796–1800.
50. Illing A, Unruh T, Koch MHJ. Investigation on particle self-assembly in solid lipid based colloidal drug carrier systems. Pharm Res 2004; 21:592–597.
51. Illing A, Unruh T. Investigations on the flow behavior of dispersions of solid triglyceride nanoparticles. Int J Pharm 2004; 284:123–131.
52. Zur Muehlen A, zur Mühlen E, Niehus H, Mehnert W. Atomic force microscopy studies of solid lipid nanoparticles. Pharm Res 1996; 13:1411–1416.
53. Zur Mühlen A, Schwarz C, Mehnert W. Solid lipid nanoparticles (SLN) for controlled drug delivery—drug release and release mechanism. Eur J Pharm Biopharm 1998; 45:149–155.
54. Zur Muehlen A, Mehnert W. Drug release and release mechanism of prednisolone loaded solid lipid nanoparticles. Pharmazie 1998; 53:552–555.
55. Müller RH, Rühl D, Runge SA. Biodegradation of solid lipid nanoparticles as a function of lipase incubation time. Int J Pharm 1996; 144:115–121.
56. Müller RH, Maassen S, Weyhers H, Mehnert W. Phagocytic uptake and cytotoxicity of solid lipid nanoparticles (SLN) sterically stabilized with poloxamine 908 and poloxamer 407. J Drug Target 1996; 4:161–170.
57. Müller RH, Rühl D, Runge S, Schultze-Forster K, Mehnert W. Cytotoxicity of solid lipid nanoparticles as a function of the lipid matrix and the surfactant. Pharm Res 1997; 14:458–462.
58. Müller RH, Maassen S, Schwarz C, Mehnert W. Solid lipid nanoparticles (SLN) as potential carrier for human use: interaction with human granulocytes. J Control Release 1997; 47:261–269.
59. Müller RH, Dobrucki R, Radomska A. Solid lipid nanoparticles as a new formulation with retinol. Act Pharm Polon 1999; 56:117–120.
60. Schwarz C, Mehnert W. Freeze-drying of drug-free and drug-loaded solid lipid nanoparticles (SLN). Int. J. Pharm. 1997; 157:171–179.
61. Schwarz C, Mehnert W. Solid lipid nanoparticles (SLN) for controlled drug delivery. II. Drug incorporation and physicochemical characterization. J Microencaps. 1999; 16: 205–213.
62. Freitas C, Müller RH. Effect of light and temperature on zeta potential and physical stability in solid lipid nanoparticle (SLN) dispersions. Int J Pharm 1998; 168:221–229.
63. Freitas C, Müller RH. Spray-drying of solid lipid nanoparticles (SLN). Eur J Pharm Biopharm 1998; 46:145–151.
64. Freitas C, Müller RH. Stability determination of solid lipid nanoparticles (SLN) in aqueous dispersion after addition of electrolyte. J Microencaps 1999; 16:59–71.
65. Freitas C, Müller RH. Correlation between long-term stability of solid lipid nanoparticles (SLN) and crystallinity of the lipid phase. Eur J Pharm Biopharm 1999; 47:125–132.
66. Radomska A, Dobrucki R, Muller RH. Chemical stability of the lipid matrices of solid lipid nanoparticles (SLN)-development of an analytical method and determination of long-term stability. Pharmazie 1999; 54:903–909.
67. Pinto JF, Müller RH. Pellets as carriers of solid lipid nanoparticles (SLN) for oral administration of drugs. Pharmazie 1999; 54:506–509.
68. Olbrich C, Müller RH. Enzymatic degradation of SLN—effect of surfactant and surfactant mixtures. Int J Pharm 1999; 180:31–39.
69. Olbrich C, Bakowsky U, Lehr CM, Müller RH, Kneuer C. Cationic solid-lipid nanoparticles can efficiently bind and transfect plasmid DNA. J Control Release 2001; 77:345–355.
70. Olbrich C, Müller RH, Tabatt K, Kayser O, Schulze C, Schade D. Stable biocompatible adjuvants—a new type of adjuvant based on solid lipid nanoparticles: a study on cytotoxicity, compatibility and efficacy in chicken. ATLA 2002; 30:443–458.

71. Olbrich C, Kayser O, Müller RH. Lipase degradation of Dynasan 114 and 116 solid lipid nanoparticles (SLN)—effect of surfactants, storage time and crystallinity. Int J Pharm 2002; 237:119–128.
72. Olbrich C, Kayser O, Müller RH. Enzymatic degradation of Dynasan 114 SLN—effect of surfactants and particle size. J Nanopart Res 2002; 4:121–129.
73. Dingler A, Blum RP, Niehus H, Muller RH, Gohla S. Solid lipid nanoparticles (SLN/ Lipopearls). A pharmaceutical and cosmetic carrier for the application of vitamin E in dermal products. J Microencaps. 1999; 16:751–767.
74. Dingler A, Gohla S. Production of solid lipid nanoparticles (SLN): scaling up feasibilities. J Microencaps 2002; 19:11–16.
75. Gohla SH, Dingler A. Scaling up feasibility of the production of solid lipid nanoparticles (SLN). Pharmazie 2001; 56:61–63.
76. Zimmermann E, Müller RH, Mäder K. Influence of different parameters on reconstitution of lyophilized SLN. Int J Pharm 2000; 196:211–213.
77. Zimmermann E, Müller RH. Electrolyte- and pH-stabilities of aqueous solid lipid nanoparticle (SLN) dispersions in artificial gastrointestinal media. Eur J Pharm Biopharm 2001; 52:203–210.
78. Schöler N, Zimmermann E, Katzfey U, Hahn H, Müller RH, Liesenfeld O. Effect of solid lipid nanoparticles (SLN) on cytokine production and the viability of murine peritoneal macrophages. J Microencaps 2000; 17:639–650.
79. Schöler N, Zimmermann E, Katzfey U, Hahn H, Müller RH, Liesenfeld O. Preserved solid lipid nanoparticles (SLN) at low concentrations do cause neither direct nor indirect cytotoxic effects in peritoneal macrophages. Int J Pharm 2000; 196:235–239.
80. Schöler N, Olbrich C, Tabatt K, Müller RH, Hahn H, Liesenfeld O. Surfactant, but not the size of solid lipid nanoparticles (SLN) influences viability and cytokine production of macrophages. Int J Pharm 2001; 221:57–67.
81. Schöler N, Hahn H, Müller RH, Liesenfeld O. Effect of lipid matrix and size of solid lipid nanoparticles (SLN) on the viability and cytokine production of macrophages. Int J Pharm 2002; 231:167–176.
82. Santos Maia C, Mehnert W, Schäfer-Korting M. Solid lipid nanoparticles as drug carriers for topical glucocorticoids. Int J Pharm 2000; 196:165–167.
83. Santos Maia C, Mehnert W, Schaller M, et al. Drug targeting by solid lipid nanoparticles for dermal use. J Drug Target 2002; 10:489–495.
84. Lippacher A, Müller RH, Mäder K. Investigation on the viscoelastic properties of lipid based colloidal drug carriers. Int J Pharm 2000; 196:227–230.
85. Lippacher A, Müller RH, Mäder K. Preparation of semisolid drug carriers for topical application based on solid lipid nanoparticles. Int J Pharm 2001; 214:9–12.
86. Lippacher A, Müller RH, Mäder K. Semisolid SLN dispersions for topical application: influence of the formulation and production parameters on viscoelastic properties. Eur J Pharm Biopharm 2002; 53:155–160.
87. Liedtke S, Wissing S, Müller RH, Mäder K. Influence of high pressure homogenisation equipment on nanodispersions characteristics. Int J Pharm 2000; 196:183–185.
88. Jenning V, Schäfer-Korting M, Gohla SH. Vitamin A loaded solid lipid nanoparticles for topical use: drug release properties. J Control Release 2000; 66:115–126.
89. Jenning V, Thünemann AF, Gohla SH. Characterisation of a novel solid lipid nanoparticle carrier system based on binary mixtures of liquid and solid lipids. Int J Pharm 2000; 199:167–177.
90. Jenning V, Mäder K, Gohla SH. Solid lipid nanoparticles (SLN) based on binary mixtures of liquid and solid lipids: a ^{1}H-NMR study. Int J Pharm 2000; 205:15–21.
91. Jenning V, Gyser A, Schäfer-Korting M, Gohla SH. Vitamin A loaded solid lipid nanoparticles for topical use: occlusive properties and drug targeting to the upper skin. Eur J Pharm Biopharm 2000; 49:211–218.
92. Jenning V, Gohla S. Comparison of wax and glyceride solid lipid nanoparticles (SLN). Int J Pharm 2000; 196:219–222.

93. Jenning V, Gohla SH. Encapsulation of retinoids in solid lipid nanoparticles (SLN). J Microencaps. 2001; 18:149–158.
94. Jenning V, Lippacher A, Gohla SH. Medium scale production of solid lipid nanoparticles (SLN) by high pressure homogenization. J Microencaps 2002; 19:1–10.
95. Wissing SA, Müller RH. A novel sunscreen system based on tocopherol acetate incorporated into solid lipid nanoparticles. Int J Cosmet Sci 2001; 23:233–243.
96. Wissing SA, Müller RH. Investigations on the occlusive properties of solid lipid nanoparticles (SLN). J Cosmet Sci 2001; 52:313–324.
97. Wissing SA, Muller RH. Solid lipid nanoparticles (SLN)—a novel carrier for UV blockers. Pharmazie 2001; 56:783–786.
98. Wissing SA, Müller RH. The influence of the crystallinity of lipid nanoparticles on their occlusive properties. Int J Pharm 2002; 242:377–379.
99. Wissing SA, Müller RH. The development of an improved carrier system for sunscreen formulations based on crystalline lipid nanoparticles. Int J Pharm 2002; 242:373–375.
100. Wissing SA, Müller RH. Solid lipid nanoparticles as carrier for sunscreens: in vitro release and in vivo skin penetration. J Contr Release 2002; 81:225–233.
101. Wissing SA, Müller RH. Cosmetic applications for solid lipid nanoparticles (SLN). Int J Pharm 2003; 254:65–68.
102. Wissing SA, Müller RH. The influence of solid lipid nanoparticles on skin hydration and viscoelasticity—in vivo study. Eur J Pharm Biopharm 2003; 56:67–72.
103. Wissing SA, Müller RH, Manthei L, Mayer C. Structural characterization of Q10-loaded solid lipid nanoparticles by NMR spectroscopy. Pharm Res 2004; 21:400–405.
104. Göppert TM, Müller RH. Plasma protein adsorption of Tween 80 and poloxamer 188-stabilized solid lipid nanoparticles. J Drug Target 2003; 11:225–231.
105. Göppert TM, Müller RH. Alternative sample preparation prior to two-dimensional electrophoresis protein analysis on solid lipid nanoparticles. Electrophoresis 2004; 25:134–140.
106. Haberland A, Santos Maia C, Jores K, et al. Albumin effects on drug absorption and metabolism in reconstructed epidermis and excised pig skin. ALTEX 2003; 20:3–9.
107. Tabatt K, Sameti M, Olbrich C, Müller RH, Lehr CM. Effect of cationic lipid and matrix lipid composition on solid lipid nanoparticle-mediated gene transfer. Eur J Pharm Biopharm 2004; 57:155–162.
108. Lukowski G, Pflegel P. Electron Diffraction of solid lipid nanoparticles loaded with aqciclovir. Pharmazie 1997; 52:642–643.
109. Lukowski G, Kasbohm J, Pflegel P, Illing A, Wulff H. Crystallographic investigation of cetyl palmitate solid lipid nanoparticles. Int J Pharm 2000; 196:201–205.
110. Mayer C, Lukowski G. Solid state NMR investigations on nanosized carrier systems. Pharm Res 2000; 17:486–489.
111. Sznitowska M, Gajewska M, Janicki S, Radwanska A, Lukowski G. Bioavailability of diazepam from aqueous-organic solution, submicron emulsion and solid lipid nanoparticles after rectal administration in rabbits. Eur J Pharm Biopharm 2001; 52: 159–163.
112. Jores K, Mehnert W, Mäder K. Physicochemical investigations on solid lipid nanoparticles and on oil-loaded solid lipid nanoparticles: a nuclear magnetic resonance and electron spin resonance study. Pharm Res 2003; 20:1274–1283.
113. Jores K, Mehnert W, Drechsler M, Bunjes H, Johann C, Mäder K. Investigations on the structure of solid lipid nanoparticles (SLN) and oil-loaded solid lipid nanoparticles by photon correlation spectroscopy, field-flow fractionation and transmission electron microscopy. J Contr Release 2004; 95:217–227.
114. Patravale VB, Ambarkhane AV. Study of solid lipid nanoparticles with respect to particle size distribution and drug loading. Pharmazie 2003; 58:392–395.
115. Yang S, Zhu J, Lu Y, Liang B, Yang C. Body distribution of camptothecin solid lipid nanoparticles after oral administration. Pharm Res 1999; 16:751–757.

116. Yang SC, Lu LF, Cai Y, Zhu JB, Liang BW, Yang CZ. Body distribution in mice of intravenously injected camptothecin solid lipid nanoparticles and targeting effect on brain. J Contr Release 1999; 59:299–307.
117. Yang SC, Zhu JB. Preparation and characterization of camptothecin solid lipid nanoparticles. Drug Devel Ind Pharm 2002; 28:265–274.
118. Lim S, Kim CK. Formulation parameters determining the physicochemical characteristics of solid lipid nanoparticles loaded with all-trans retinoic acid. Int J Pharm 2002; 243:135–146.
119. de Vringer T, de Ronde HAG. Preparation and structure of a water-in-oil cream containing lipid nanoparticles. J Pharm Sci 1995; 84:466–472.
120. Schwarz C, Mehnert W, Lucks JS, Müller RH. Solid lipid nanoparticles (SLN) for controlled drug delivery. I. Production, characterization and sterilization. J Contr Release 1994; 30:83–96.
121. Heiati H, Phillips NC, Tawashi R. Evidence for phospholipid bilayer formation in solid lipid nanoparticles formulated with phospholipid and triglyceride. Pharm Res 1996; 13:1406–1410.
122. Heiati H, Tawashi R, Shivers RR, Phillips NC. Solid lipid nanoparticles as drug carriers I. Incorporation and retention of the lipophilic prodrug 3′-azido-3′-deoxythymidine palmitate. Int J Pharm 1997; 146:123–131.
123. Heiati H, Tawashi R, Phillips NC. Solid lipid nanoparticles as drug carriers. II. Plasma stability and biodistribution of solid lipid nanoparticles containing the lipophilic prodrug 3%-azido-3%-deoxythymidine palmitate in mice. Int J Pharm 1998; 174:71–80.
124. Bunjes H, Siekmann B, Westesen K. Emulsions of supercooled melts—a novel drug delivery system. In: Benita S, ed. Submicron Emulsions in Drug Targeting and Delivery. Amsterdam: Harwood Academic Publishers, 1998:175–218.
125. Walstra P. Formation of emulsions. In: Becher P, ed. Encyclopedia of Emulsion Technology. Vol. 1. New York: Marcel Dekker, 1983:57–127.
126. Phipps LW. The High Pressure Dairy Homogenizer. Reading: College of Estate Management. 1985:46–78.
127. Mei Z, Chen H, Weng T, Yang Y, Yang X. Solid lipid nanoparticles and microemulsion for topical delivery of triptolide. Eur J Pharm Biopharm 2003; 56:189–196.
128. Venkateswarlu V, Manjunath K. Preparation, characterization and in vitro release kinetics of clozapine solid lipid nanoparticles. J Control Release 2004; 95:627–638.
129. Wang JX, Sun X, Zhang ZR. Enhanced brain targeting by synthesis of 3′,5′-dioctanoyl-5-fluoro-2′-deoxyuridine incorporation into solid lipid nanoparticles. Eur J Pharm Biopharm 2002; 54:285–290.
130. Ahlin P, Kristl J, Šmid-Korvar J. Optimization of procedure parameters and physical stability of solid lipid nanoparticles in dispersions. Acta Pharm 1998; 48:259–267.
131. Ahlin P, Kristl J, Šentjurc M, Štrancar J, Pečar S. Influence of spin probe structure on its distribution in SLN dispersions. Int J Pharm 2000; 196:241–244.
132. Ahlin P, Šentjurc M, Štrancar J, Kristl J. Location of lipophilic substances and ageing of solid lipid nanoparticles studied by EPR. STP Pharma Sci 2000; 10:125–132.
133. Ahlin P, Kristl J, Pečar S, Štrancar J, Šentjurc M. The Effect of lipophilicity of spin-labeled compounds on their distribution in solid lipid nanoparticle dispersions studied by electron paramagnetic resonance. J Pharm Sci 2003; 92:58–66.
134. Kristl J, Volk B, Ahlin P, Gombač K, Šentjurc M. Interactions of solid lipid nanoparticles with model membranes and leukocytes studied by EPR. Int J Pharm 2003; 256:133–140.
135. Kristl J, Volk B, Gašperlin M, Šentjurc M, Jurkovič P. Effect of colloidal carriers on ascorbyl palmitate stability. Eur J Pharm Sci 2003; 19:181–189.
136. Siekmann B. Untersuchungen zur Herstellung und zum Rekristallisationsverhalten schmelzemulgierter intravenös applizierbarer Glyceridnanopartikel. Ph.D. Thesis, Technical University of Braunschweig; 1994.

137. Heiati H, Tawashi R, Phillips NC. Drug retention and stability of solid lipid nanoparticles containing azidothymidine palmitate after autoclaving, storage and lyophilization. J Microencapsul 1998; 15:173–184.
138. Yamaguchi T, Nishizaki K, Itai S, Hayashi H, Ohshima H. Physicochemical characterization of parenteral lipid emulsions: Influence of cosurfactants on flocculation and coalescence. Pharm Res 1995; 12:1273–1278.
139. Jumaa M, Müller BW. The stabilization of parenteral fat emulsion using non-ionic ABA copolymer surfactant. Int J Pharm 1998; 174:29–37.
140. Demirel M, Yazan Y, Müller RH, Kilic F, Bozan B. Formulation and in vitro-in vivo evaluation of piribedil solid lipid micro- and nanoparticles. J. Microencaps 2001; 18:359–371.
141. Almeida AJ, Runge S, Müller RH. Peptide-loaded solid lipid nanoparticles (SLN): influence of production parameters. Int J Pharm 1997; 149:255–265.
142. Friedrich I, Müller-Goymann CC. Characterization of solidified reverse micellar solutions (SRMS) and production development of SRMS-based nanosuspensions. Eur J Pharm Biopharm 2003; 56:111–119.
143. Müller RH, Maaßen S, Weyhers H, Specht F, Lucks JS. Cytotoxicity of magnetide-loaded polylactide, polylactide/glycolide particles and solid lipid nanoparticles. Int J Pharm 1996; 138:85–94.
144. Sjöström B, Westesen K, Bergenståhl B. Preparation of submicron drug particles in lecithin-stabilized o/w emulsions. II. Characterization of cholesteryl acetate particles. Int J Pharm 1993; 94:89–101.
145. Sjöström B, Kronberg B, Carlfors J. A method for the preparation of submicron particles of sparingly water-soluble drugs by precipitation in oil-in-water emulsions. I. Influence of emulsification and surfactant concentration. J Pharm Sci 1993; 82:579–583.
146. Sjöström B, Bergenståhl B, Kronberg B. A method for the preparation of submicron particles of sparingly water-soluble drugs by precipitation in oil-in-water emulsions. II: Influence of the emulsifier, the solvent, and the drug substance. J Pharm Sci 1993; 82:584–589.
147. Siekmann B, Westesen K. Investigations on solid lipid nanoparticles prepared by precipitation in o/w emulsions. Eur J Pharm Biopharm 1996; 42:104–109.
148. Garcia-Fuentes M, Torres D, Alonso MJ. Design of lipid nanoparticles for the oral delivery of hydrophilic macromolecules. Colloids Surf B 2002; 27:159–168.
149. Trotta M, Debernardi F, Caputo O. Preparation of solid lipid nanoparticles by a solvent emulsification–diffusion technique. Int J Pharm 2003; 257:153–160.
150. Gasco MR, Morel S, Carpignano R. Optimization of the incorporation of deoxycorticosterone acetate in lipospheres. Eur J Pharm Biopharm 1992; 38:7–10.
151. Gasco MR, Cavalli R, Carlotti ME. Timolol in lipospheres. Pharmazie 1992; 47:119–121.
152. Cavalli R, Gasco MR, Morel S. Behaviour of timolol incorporated in lipospheres in the presence of a series of phosphate esters. STP Pharma Sci 1992; 6:514–518.
153. Aquilano D, Cavalli R, Gasco MR. Solid lipospheres from hot emulsions in the presence of different concentrations of surfactant: the crystallization of stearic acid polymorphs. Thermochim Acta 1993; 230:29–37.
154. Cavalli R, Caputo O, Gasco MR. Solid lipospheres of doxorubicin and idarubicin. Int J Pharm 1993; 89:R9–R12.
155. Morel S, Gasco MR, Cavalli R. Incorporation in lipospheres of (D-Trp-6)LHRH. Int J Pharm 1994; 105:R1–R3.
156. Morel S, Ugazio E, Cavalli R, Gasco MR. Thymopentin in solid lipid nanoparticles. Int J Pharm 1996; 132:259–261.
157. Morel S, Terreno E, Ugazio E, Aime S, Gasco MR. NMR relaxometric investigations of solid lipid nanoparticles (SLN) containing gadolinium(III) complexes. Eur J Pharm Biopharm 1998; 45:57–163.

158. Boltri L, Canal T, Esposito PA, Carli F. Relevant factors affecting the formation and growth of lipid nanospheres suspensions. Eur J Pharm Biopharm 1995; 41:70–75.
159. Cavalli R, Morel S, Gasco MR, Chetoni P, Saettone MF. Preparation and evaluation in vitro of colloidal lipospheres containing pilocarpine as ion pair. Int J Pharm 1995; 117:243–246.
160. Cavalli R, Aquilano D, Carlotti ME, Gasco MR. Study by X-ray powder diffraction and differential scanning calorimetry of two model drugs, phenothiazine and nifedipine, incorporated into lipid nanoparticles. Eur J Pharm Biopharm 1995; 41: 329–333.
161. Cavalli R, Marengo E, Rodriguez L, Gasco MR. Effects of some experimental factors on the production process of solid lipid nanoparticles. Eur J Pharm Biopharm 1996; 43:110–115.
162. Cavalli R, Caputo O, Ugazio E, Gasco MR. The effect of butanol and pentanol isomers on the crystallization of stearic acid polymorphs in solid lipid nanoparticles. Act Technol Leg Medicam 1996; 7:101–111.
163. Cavalli R, Caputo O, Carlotti ME, Trotta M, Scarnecchia C, Gasco MR. Sterilization and freeze-drying of drug-free and drug-loaded solid lipid nanoparticles. Int J Pharm 1997; 148:47–54.
164. Cavalli R, Caputo O, Marengo E, Pattarino F, Gasco MR. The effect of the components of microemulsions on both size and crystalline structure of solid lipid nanoparticles (SLN) containing a series of model molecules. Pharmazie 1998; 53:392–396.
165. Cavalli R, Peira E, Caputo O, Gasco MR. Solid lipid nanoparticles as carriers of hydrocortisone and progesterone complexes with β-cyclodextrins. Int J Pharm 1999; 182:59–69.
166. Cavalli R, Bocca C, Miglietta A, Caputo O, Gasco MR. Albumin adsorption on stealth and non-stealth solid lipid nanoparticles. STP Pharma Sci 1999; 9:183–189.
167. Cavalli R, Zara GP, Caputo O, Bargoni A, Fundarò A, Gasco MR. Transmucosal transport of tobramycin incorporated in SLN after duodenal administration to rats Part I. A pharmacokinetic study. Pharmacol Res 2000; 42:541–545.
168. Cavalli R, Gasco MR, Barresi AA, Rovero G. Evaporative drying of aqueous dispersions of solid lipid nanoparticles. Drug Dev Ind Pharm 2001; 27:919–924.
169. Cavalli R, Gaco RM, Chetoni P, Burgalassi S, Saettone MF. Solid lipid nanoparticles (SLN) as ocular delivery system for tobramycin. Int J Pharm 2002; 238:241–245.
170. Cavalli R, Bargoni A, Podio V, Muntoni E, Zara GP, Gasco MR. Duodenal administration of solid lipid nanoparticles loaded with different percentages of tobramycin. J Pharm Sci 2003; 92:1085–1094.
171. Marengo E, Cavalli R, Caputo O, Rodriguez L, Gasco MR. Scale-up of the preparation process of solid lipid nanoparticles. Int J Pharm 2000; 205:3–13.
172. Marengo E, Cavalli R, Rovero G, Gasco MR. Scale-up and optimization of an evaporative drying process applied to aqueous dispersions of solid lipid nanoparticles. Pharm Dev Technol 2003; 8:299–309.
173. Bargoni A, Cavalli R, Caputo O, Fundaro A, Gasco MR, Zara GP. Solid lipid nanoparticles in lymph and plasma after duodenal administration to rats. Pharm Res 1998; 15:745–750.
174. Bargoni A, Cavalli R, Zara GP, Fundaro A, Caputo O, Gasco MR. Transmucosal transport of tobramycin incorporated in solid lipid nanoparticles (SLN) after duodenal administration to rats. Part II. Tissue distribution. Pharm Res 2001; 43:497–502.
175. Bocca C, Caputo O, Cavalli R, Gabriel L, Miglietta A, Gasco MR. Phagocytic uptake of fluorescent stealth and non-stealth solid lipid nanoparticles. Int J Pharm 1998; 175:185–193.
176. Zara GP, Cavalli R, Fundaro A, Bargoni A, Caputo O, Gasco MR. Pharmacokinetics of doxorubicin incorporated in solid lipid nanospheres (SLN). Pharmacol Res 1999; 40:281–286.

177. Zara GP, Bargoni A, Cavalli R, Fundaro A, Vighetto D, Gasco MR. Pharmacokinetics and tissue distribution of idarubicin-loaded solid lipid nanoparticles after duodenal administration to rats. J Pharm Sci 2002; 91:1324–1333.
178. Zara GP, Cavalli R, Bargoni A, Fundaro A, Vighetto D, Gasco RM. Intravenous administration to rabbits of non-stealth and stealth doxorubicin-loaded solid lipid nanoparticles at increasing concentrations of stealth agent: pharmacokinetics and distribution of doxorubicin in brain and other tissues. J Drug Target 2002; 10:327–335.
179. Fundaro A, Cavalli R, Bargoni A, Vighetto D, Zara GP, Gasco MR. Non-stealth and stealth solid lipid nanoparticles (SLN) carrying doxorubicin: pharmacokinetics and tissue distribution after i.v. administration to rats. Pharmacol Res 2000; 42:337–343.
180. Podio V, Zara GP, Carazzone M, Cavalli R, Gasco RM. Biodistribution of stealth and non-stealth solid lipid nanospheres after intravenous administration to rats. J Pharm Pharmacol 2000; 52:1057–1063.
181. Miglietta A, Cavalli R, Bocca C, Gabriel L, Gasco MR. Cellular uptake and cytotoxicity of solid lipid nanospheres (SLN) incorporation doxorubicin or paclitaxel. Int J Pharm 2000; 210:61–67.
182. Ugazio E, Cavalli R, Gasco MR. Incorporation of cyclosporin A in solid lipid nanoparticles (SLN). Int J Pharm 2002; 241:341–344.
183. Bondì ML, Fontana G, Carlisi B, Giammona G. Preparation and characterization of solid lipid nanoparticles containing cloricromene. Drug Delivery 2003; 10:245–250.
184. Peira E, Marzola P, Podio V, Aime S, Sbarbati A, Gasco MR. In vitro and in vivo study of solid lipid nanoparticles loaded with superparamagnetic iron oxide. J Drug Target 2003; 11:19–24.
185. Heydenreich AV, Westmeier R, Pedersen N, Poulsen HS, Kristensen HG. Preparation and purification of cationic solid lipid nanospheres—effects on particle size, physical stability and cell toxicity. Int J Pharm 2003; 254:83–87.
186. Pellizzaro C, Cordini D, Morel S, Ugazio E, Gasco MR, Dandone MG. Cholesteryl butyrate as an alternative approach for butyric acid delivery. Anticanc Res 1999; 19:3921–3926.
187. Gasco MR. Solid lipid nanospheres from warm micro-emulsions. Pharm Technol Europe 1997; 9:52–59.
188. Oyewumi MO, Mumper RJ. Gadolinium-loaded nanoparticles engineered from microemulsion templates. Drug Dev Ind Pharm 2002; 28:317–328.
189. Cui ZR, Mumper RJ. Genetic immunization using nanoparticles engineered from microemulsion precursors. Pharm Res 2002; 19:939–946.
190. Mumper RJ, Cui ZR, Oyewumi MO. Nanotemplate engineering of cell specific nanoparticles. J Disp Sci Technol 2003; 24:569–588.
191. Cui ZR, Mumper RJ. Coating of cationized protein on engineered nanoparticles results in enhanced immune responses. Int J Pharm 2002; 238:229–239.
192. Cui ZR, Mumper RJ. Intranasal administration of plasmid DNA-coated nanoparticles results in enhanced immune responses. J Pharm Pharmacol 2002; 54:1195–1203.
193. Cui ZR, Mumper RJ. Topical immunization using nanoengineered genetic vaccines. J Control Release 2002; 81:173–184.
194. Cui ZR, Mumper RJ. Plasmid DNA-entrapped nanoparticles engineered from microemulsion precursors: In vitro and in vivo evaluation. Bioconj Chem 2002; 13:1319–1327.
195. Cui ZR, Baizer L, Mumper RJ. Intradermal immunization with novel plasmid DNA-coated nanoparticles via a needle-free injection device. J Biotechnol 2003; 102:105–115.
196. Cui ZR, Hsu CH, Mumper RJ. Physical characterization and macrophage cell uptake of mannan-coated nanoparticles. Drug Dev Ind Pharm 2003; 29:689–700.
197. Cui ZR, Mumper RJ. The effect of co-administration of adjuvants with a nanoparticle-based genetic vaccine delivery system on the resulting immune responses. Eur J Pharm Biopharm 2003; 55:11–18.

198. Oyewumi MO, Mumper RJ. Influence of formulation parameters on gadolinium entrapment and tumor cell uptake using folate-coated nanoparticles. Int J Pharm 2003; 251:85–97.
199. Oyewumi MO, Liu S, Moscow JA, Mumper RJ. Specific association of thiamine-coated gadolinium nanoparticles with human breast cancer cells expressing thiamine transporters. Bioconj Chem 2003; 14:404–411.
200. Hsu CH, Cui ZR, Mumper RJ, Jay M. Preparation and characterization of novel coenzyme Q10 nanoparticles engineered from microemulsion precursors. AAPS Pharm Sci Tech 2003; 4:Article 32.
201. Lockman PR, Oyewumi MO, Koziera JM, Roder KE, Mumper RJ, Allen DD. Brain uptake of thiamine-coated nanoparticles. J Control Release 2003; 93:271–282.
202. Lockman PR, Koziera J, Roder KE, et al. In vivo and in vitro assessment of baseline blood–brain barrier parameters in the presence of novel nanoparticles. Pharm Res 2003; 20:705–713.
203. Koziera JM, Lockman PR, Allen DD, Mumper RJ. In situ blood–brain barrier transport of nanoparticles. Pharm Res 2003; 20:1772–1778.
204. Quintanar-Guerrero D, Allemann E, Fessi H, Doelker E. Preparation techniques and mechanisms of formation of biodegradable nanoparticles from preformed polymers. Drug Dev Ind Pharm 1998; 24:1113–1128.
205. Chen DB, Yang TZ, Lu WL, Zhang Q. In vitro and in vivo study of two types of long-circulating solid lipid nanoparticles containing paclitaxel. Chem Pharm Bull 2001; 49:1444–1447.
206. Hu FQ, Yuan H, Zhang HH, Fang M. Preparation of solid lipid nanoparticles with clobetasol propionate by a novel solvent diffusion method in aqueous system and physicochemical characterization. Int J Pharm 2002; 239:121–128.
207. Hu FQ, Hong Y, Yuan H. Preparation and characterization of solid lipid nanoparticles containing peptide. Int J Pharm 2004; 273:29–35.
208. Schubert MA, Müller-Goymann CC. Solvent injection as a new approach for manufacturing lipid nanoparticles—evaluation of the method and process parameters. Eur J Pharm Biopharm 2003; 55:125–131.
209. Washington C. Photon correlation spectroscopy. In: Washington C, ed. Particle Size Analysis in Pharmaceutics and Other Industries. New York: Ellis Horwood, 1992: 135–167.
210. Finsy R. Particle sizing by quasi-elastic light scattering. Adv Colloid Interf Sci 1994; 52:79–143.
211. Allen T. Photon correlation spectroscopy. In: Allen T, ed. Particle size measurement Vol. 1. London: Chapman & Hall, 1997:426.
212. Tscharnuter W. Photon correlation in particle sizing. In: Meyers RA, ed. Encyclopedia of Analytical Chemistry. Chichester: John Wiley & Sons, 2000:5469–5485.
213. Westesen K, Drechsler M, Bunjes H. Colloidal dispersions based on solid lipids. In: Dickinson E, Miller R, eds. Food Colloids: Fundamentals of Formulation. Cambridge: Royal Society of Chemistry, 2001:103–115.
214. Cantor CR, Schimmel PR. Biophysical Chemistry. Part II. Techniques for the Study of Biological Structure and Function. New York: Freeman, 1980:539–590.
215. Perrin F. Mouvement Brownien d'un ellipsoide (II) Rotation libre et dépolarisation des fluorescences. Translation et diffusion de molécule ellipsoidales. J Phys Radium 1936; 7:1–11.
216. Koppel DE. Analysis of macromolecular polydispersity in intensity correlation spectroscopy: the method of cumulants. J Chem Phys 1972; 57:4814–4820.
217. Xu R. Improvements in particle size analysis using light scattering. In: Müller RH, Mehnert W, Hildebrandt GE, eds. Particle and Surface Characterizing Methods. Stuttgart: Scientific Publishers, 1997:27–56.

218. Washington C. The angular distribution of scattered light. In: Washington C, ed. Particle Size Analysis in Pharmaceutics and Other Industries. New York: Ellis Horwood, 1992:109–133.
219. Driscoll DF. The significance of particle/globule sizing measurements in the safe use of intravenous lipid emulsions. J Disp Sci Technol 2002; 23:679–687.
220. Bunjes H, Westesen K. Influences of colloidal state on physical properties of solid fats. In: Garti N, Sato K, eds. Crystallization Processes in Fats and Lipid Systems. New York: Marcel Dekker, 2001:457–483.
221. Hagemann J. Thermal behavior and polymorphism of acylglycerides. In: Garti N, Sato K, eds. Crystallization and Polymorphism of Fats and Fatty Acids. New York: Marcel Dekker, 1988:9–95.
222. Hernqvist L. Crystal structures of fats and fatty acids. In: Garti N, Sato K, eds. Crystallization and Polymorphism of Fats and Fatty Acids. New York: Marcel Dekker, 1988: 97–137.
223. Sato K, Ueno S. Molecular interactions and phase behavior of polymorphic fats. In: Garti N, Sato K, eds. Crystallization Processes in Fats and Lipid Systems. New York: Marcel Dekker, 2001:177–209.
224. Bunjes H. Characterization of solid lipid nano- and microparticles. In: Nastruzzi C, Esposito E, eds. Lipospheres in Drug Targets and Delivery. Boca Raton: CRC Press, 2005:41–66.
225. Sjöström B, Kaplun A, Talmon Y, Cabane B. Structures of nanoparticles prepared from oil-in-water emulsions. Pharm Res 1995; 12:39–48.
226. Hunter RJ. Electrokinetics of particles. In: Hubbard AT, ed. Encyclopedia of Surface and Colloid Science. New York: Marcel Dekker, 2002:1907–1919.
227. Malmsten M. Protein adsorption in intravenous drug delivery. In: Malmsten M, ed. Biopolymers at Interfaces. New York: Marcel Dekker, 1998:561–596.
228. Kreuter J. Nanoparticulate systems for brain delivery of drugs. Adv Drug Del Rev 2001; 47:65–81.
229. Washington C. Drug release from microdisperse systems: a critical review. Int J Pharm 1990; 58:1–12.
230. Levy MY, Benita S. Drug release from submicron o/w emulsion: a new in vitro kinetic evaluation model. Int J Pharm 1990; 66:29–37.
231. Domb AJ. Liposphere parenteral delivery system. Proc Int Symp Control Release Bioact Mater 1993; 20:121–122.
232. Olbrich C, Kayser O, Müller RH, Grubhofer N. Solid lipid nanoparticles (SLN) as vaccine adjuvant—study in sheep with mycoplasma bovis antigen and stability testing. Proc Int Symp Control Release Bioact Mater 2000; 27:6110.
233. Juliano RL. Factors affecting the clearance kinetics and tissue distribution of liposomes, microspheres and emulsions. Adv Drug Deliv Rev 1988; 2:31–53.
234. Cavalli R, Caputo O, Gasco MR. Preparation and characterization of solid lipid nanospheres containing paclitaxel. Eur J Pharm Sci 2000; 10:305–309.
235. Khopade AJ, Shelly C, Pandit NK, Banakar UV. Liposphere based lipoprotein-mimetic delivery system for 6-mercaptopurine. J Biomater Appl 2000; 14:389–398.
236. Videira MA, Botelho MF, Santos AC, et al. Lymphatic uptake of pulmonary radiolabelled solid lipid nanoparticles. J Drug Target 2002; 10:607–613.
237. Prego C, Garcia-Fuentes, Alonso AJ, Torrres D. Chitosan-coated lipid nanoparticles enhance the oral absorption of calcitonin. 2003 Controlled Release Society 30th Annual Meeting Proceedings, 308.
238. Müller RH, Radtke M, Wissing SA. Solid lipid nanoparticles (SLN) and nanostructured lipid carriers (NLC) is cosmetic and dermatological preparations. Adv Drug Del Rev 2002; 54(suppl. 1):S131–S155.
239. Borgström B. On the interaction between pancreatic lipase and colipase and the substrate, and the importance of bile salts. J Lipid Res 1975; 16:411–417.

240. Weyers H, Ehlers S, Mehnert W, Hahn H, Müller RH. Solid lipid nanoparticles—determination of in vivo toxicity. Proceedings of the First World Meeting APGI/APV, Budapest, 9–11 May 1995, 489–490.
241. Sanna V, Kirschvink N, Gustin P, et al. Solid lipid nanoparticles for drug pulmonary administration. 2003 Controlled Release Society 30th Annual Meeting Proceedings, 727.
242. Müller RH, Mäder K, Gohla S. Solid lipid nanoparticles (SLN) for controlled drug delivery—a review of the state of the art. Eur J Pharm Biopharm 2000; 50:161–177.
243. Penkler LJ, Müller RH, Runge SA, Ravelli V. Pharmaceutical cyclosporin formulation with improved biopharmaceutical properties, improved physical quality and greater stability, and method for preparing said formulation. US Patent 6,551,619 (2003).
244. Mauro A, Miglietta A, Cavalli R, et al. Enhanced cytotoxicity of paclitaxel incorporated in solid lipid nanoparticles against human glioma cells. Proc Int Symp Control Release Bioact Mater 2000; 27:#6152.
245. Bargoni A, Cavalli R, Fundaro A, Gasco MR, Vighetto D, Zara GP. Gastrointestinal absorption of idarubicin-loaded solid lipid nanoparticles. Proc Int Symp Control Release Bioact Mater 2000; 27:8409.
246. Khopade AJ, Jain NK. Long-circulating lipospheres targeted to inflamed tissue. Pharmazie 1997; 52:165–166.
247. Oyewumi MO, Yokel RA, Jay M, Coakley T, Mumper RO. Comparison of cell uptake, biodistribution and tumor retention of folate-coated and PEG-coated gadolinium nanoparticles in tumor-bearing mice. J Control Release 2004; 95:613–626.
248. Cengiz E, Wissing SA, Yazan Y, Müller RH. TiO_2-loaded solid lipid nanoparticles (SLN): UV-blocking efficiency of topical formulations. 2003 Controlled Release Society 30th Annual Meeting Proceedings, 224.
249. Ugazio E, Marengo E, Pellizzaro C, et al. The effect of formulation and concentration of cholesteryl butyrate solid lipid nanospheres (SLN) on NIH-H460 cell proliferation. Eur J Pharm Biopharm 2001; 52:197–202.
250. Salomone B, Ponti R, Gasco MR, et al. In vitro effects of cholesteryl butyrate solid lipid nanospheres as a butyric acid pro-drug on melanoma cells: Evaluation of antiproliferative activity and apoptosis induction. Clin Experim Metastas 2001; 18:663–673.
251. Yu BT, Sun S, Zhang ZR. Enhanced liver-targeting by synthesis of N_1-Steary-5-Fu and incorporation into solid lipid nanoparticles. Arch Pharm Res 2003; 26:1096–1101.
252. Gessner A, Olbrich C, Schröder W, Kayser O, Müller RH. The role of plasma proteins in brain targeting: species dependent protein adsorption patterns on brain-specific lipid drug conjugate (LDC) nanoparticles. Int J Pharm 2001; 214:87–91.
253. Olbrich C, Gessner A, Kayser O, Müller RH. Lipid-drug-conjugate (LDC) nanoparticles as novel carrier system for the hydrophilic antitrypanosomal drug diminazenediaceturate. J Drug Target 2002; 10:387–396.
254. Kuntsche J, Westesen K, Drechsler M, Koch MHJ., Bunjes H. Supercooled smectic nanoparticles—a potential novel carrier system for poorly water soluble drugs. Pharm Res 2004; 21:1836–1845.

9

Amphiphilic Cyclodextrins and Microencapsulation

Erem Memişoğlu-Bilensoy and A. Atilla Hincal
Hacettepe University, Faculty of Pharmacy, Department of Pharmaceutical Technology, Ankara, Turkey

Amélie Bochot, Laury Trichard, and Dominique Duchêne
Université Paris-Sud, Faculté de Pharmacie, Châtenay-Malabry Cedex, France

INTRODUCTION

Cyclodextrins are molecules of great interest to scientists because of their ring structure and capacity to include guest molecules in their internal cavity, which confers on them new physicochemical properties. Such an inclusion can be considered as a molecular encapsulation, resulting in better stability to air and light, higher apparent water solubility, possible increase in bioavailability, and decrease in side effects.

However, natural cyclodextrins, which have a rather high external hydrophilicity, can be considered either as not water soluble enough, for their appearance on the market of water-soluble derivatives, or too hydrophilic to be incorporated in biological membranes, leading to the development of amphiphilic derivatives.

A number of amphiphilic derivatives have been synthesized in the last 15 years. These new derivatives are still capable of including active guests and can carry them close to biological membranes where they can be easily absorbed.

Among amphiphilic cyclodextrins, those substituted on the primary or secondary face, have very interesting interfacial characteristics conferring on them the remarkable property of self-aggregating in the form of nanoparticles (nanocapsules and nanospheres). For these reasons, amphiphilic cyclodextrins can be used as a new raw material for drug encapsulation.

In this chapter, general information will be provided on natural cyclodextrins and their major derivatives, amphiphilic cyclodextrins will be described by pointing out their interfacial properties, ability to form nanocapsules and nanospheres, and factors influencing the physicochemical properties of these particles.

CYCLODEXTRINS AND DERIVATIVES

Natural cyclodextrins are obtained by the enzymatic degradation of starch as well as their water-soluble derivatives. They have a wide range of applications in the pharmaceutical industry mainly for their solubilizing and stabilizing properties. On the other hand, amphiphilic cyclodextrins, have been derived from these natural cyclodextrins, mostly β- and γ-cyclodextrin, to decrease the hydrophilic properties of cyclodextrins and confer on them better affinity for biological membranes.

Classical Cyclodextrins

Cyclodextrins are a family of oligosaccharides obtained by the enzymatic degradation of starch by cyclodextrin glucosyl transferase. Depending on the number of α-1, 4-D(+)-glucosidic linked glucopyranose units, several natural cyclodextrins have been described and are represented in Figure 1: α-cyclodextrin (six units), β-cyclodextrin (seven units), and γ-cyclodextrin (eight units) (1–3). Cyclodextrins with higher glucopyranose units, such as δ-cyclodextrins with nine units, have been described.

Physicochemical properties of the three major natural cyclodextrins are summarized in Table 1.

They are crystalline substances and their crystal structure analyses have demonstrated that all glucose residues in the torus-like ring possess the thermodynamically favored chair conformation because all substituents are in the equatorial position (1–3). Cyclodextrins behave more or less like rigid compounds with two degrees of freedom—rotation at the glucosidic links C4–O4 and C1–O4 and rotations at the O6 primary hydroxyl groups at the C5–C6 band. As a consequence of this chair conformation, all secondary hydroxyl groups on C2 and C3 are located at the broader side of the CD torus in the equatorial position. Hydroxyl groups on C2 point towards the cavity and hydroxyl groups on C3 point outwards. The primary hydroxyl groups at the C6 position are located at the narrower side of the torus. These hydroxyl groups

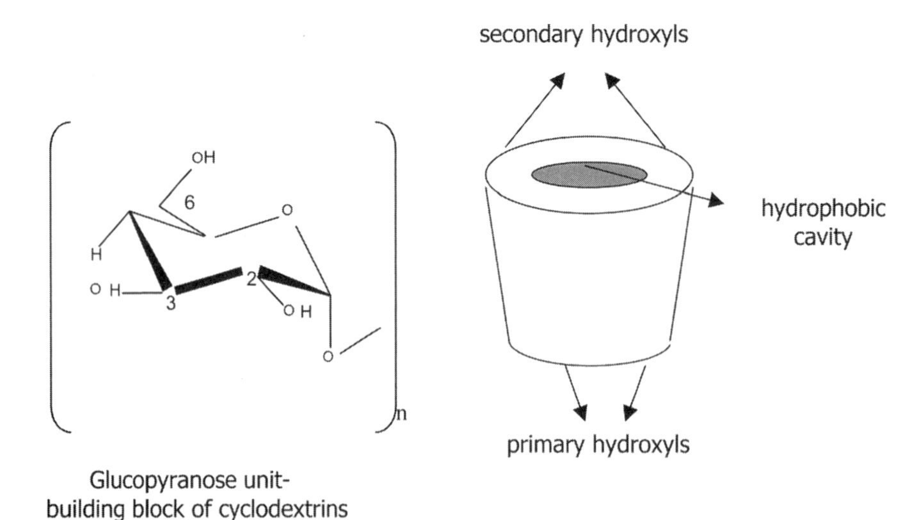

Figure 1 Schematic representation of natural cyclodextrin and glucopyranose unit that forms the cyclodextrin molecule ($n = 6$, 7, or 8 for α-, β-, and γ-cyclodextrins, respectively).

Table 1 Physicochemical Properties of Natural Cyclodextrins

	α-CD	β-CD	γ-CD
Number of glucose units	6	7	8
Molecular weight (g/mol)	972	1135	1297
Internal diameter (Å)	4.7–5.2	6.0–6.4	7.5–8.3
External diameter (Å)	14.2–15.0	15.0–15.8	17.1–17.9
Depth (Å)	7.9–8.0	7.9–8.0	7.9–8.0
Solubility in water (25°C, g/L)	145	18.5	232
Crystal water (wt.%)	10.2	13.2–14.5	8.13–17.7
Number of H_2O molecules in the cavity	6	11	17
Approximative cavity volume in 1 mol CD ($Å^3$)	174	262	472
Melting point (°C)	250–260	255–265	240–245
Surface tension (mN/m) (0.1 mM)	71	71	71
Half-life in 1 M HCl at 60°C (hr)	6.2	5.4	3.0
Diffusion constant at 40°C	3.443	3.224	3.000
Crystal forms (from water)	Hexagonal plate	Monoclinic parallelogram	Quadratic prism
p*K* (by potentiometry) at 25°C	12.332	12.202	12.081
Adiabatic compressibility in aqueous solutions (mL/mol × 10^4)	7.2	0.4	–5.0
Partial molar volumes in solution (mL/mol)	611.4	703.8	801.2
Hydrolysis by α-amylase	Negligible	Low	Rapid

Source: From Ref. 4.

ensure good water solubility for the natural cyclodextrins. The cavity of the torus is lined with a ring of C–H groups in C3, a ring of glucosidic oxygen atoms and another ring of C–H groups in C5. Thus, the cavity of cyclodextrins exhibits an apolar character. This is accompanied by a high electron density and Lewis base property. Due to their physicochemical characteristics, cyclodextrins are able to entrap within their cavity hydrophobic drugs leading to the formation of an inclusion complex. This inclusion complex exhibits new physicochemical characteristics such as increase in aqueous solubility, dissolution rate, and stability, and decrease in side effects and unpleasant odors and tastes (4,5).

$$\text{Drug} + \text{CD} \leftrightarrows \text{Drug}-\text{CD inclusion complex}$$

Owing to the equilibrium in aqueous medium between free components (cyclodextrin and drug) and the inclusion complex, poorly water-soluble drugs can be delivered in molecular form to biological membrane.

Safety is a major issue in the use of pharmaceutical excipient; two of the natural cyclodextrins (α and β) are reported to be parenterally unsafe because of hemolytic and nephrotoxic effects (6). The etiology of β-cyclodextrin nephrotoxicity is explained by the reprecipitation, as microcrystals in the kidneys, of cholesterol–β-cyclodextrin inclusion previously formed in the bloodstream (7). The ability of α- and β-cyclodextrins to cause red blood cell hemolysis and membrane irritation seems to depend on their ability to interact with lipid membrane components—cholesterol and phospholipids (7–9).

Modification of the natural cyclodextrins has been focused on improving safety while maintaining the ability to form inclusion complexes with various substrates. Some cyclodextrin derivatives produced and applied in the pharmaceutical field are briefly as follows:

Methyl cyclodextrins are among the first water-soluble cyclodextrin derivatives described. Methylation of cyclodextrins concerns either all C2 secondary and C6 primary hydroxyl groups (*dimethyl cyclodextrins*) or all the hydroxyl groups C2, C3, and C6 (*trimethyl cyclodextrins*), or results in the formation of a randomized methylated β-cyclodextrin (RAMEB). All methyl β-cyclodextrins have a water solubility of over 300 g/L, which unfortunately decreases with increasing temperature, thereby limiting their use in aqueous injectable solutions to be heat sterilized (10). It was found that partial methylation of the hydroxyls at the C2, C3, or C6 positions of β-cyclodextrin generally leads to stronger drug binding and also to better hemolytic properties (10).

Hydroxypropylated cyclodextrins are always a mixture of products with various degrees of substitution. The existence of so many types of hydroxypropyl cyclodextrins in the same reaction medium makes crystallization impossible, and, as a result, amorphous compounds are obtained. HP-β-CD are highly water-soluble (greater than 500 g/L) because of their amorphous properties as well as their chemical structure. Their dissolution is endothermic; so there is no decrease in solubility with increasing temperature (11,12). It was demonstrated that the higher the degree of hydroxypropyl substitution, the lower the drug binding, because of steric hindrance effects (13,14). They are usually designed to be used in parenteral formulations because of their low toxicity and high solubilizing effects. They are also commercially available as tablets, eye drops, and excipients under the trademarks of Encapsin® (Jansen Biotech, Belgium) and Molecusol® (Pharmatec Inc., U.S.A.) (15).

Sulfobutylether-β-cyclodextrins (SBE-β-CD) are also water-soluble and parenterally safe but unlike in HP-β-CD, higher sulfobutyl group substitution often results in higher drug complexing properties (16). Inability of the SBE-β-CD, especially $(SBE)_7$-β-CD Captisol® (CyDex Inc., U.S.A.), to form strong 1:2 complexes with cholesterol and other membrane lipids, because of their polyanionic nature, gives rise to coulombic repulsions and results in a little or no membrane disruption (7,12). This behavior provides a sound basis for the greater safety of $(SBE)_7$-β-CD compared to some other cyclodextrin derivatives (17). Sulfobutylether-β-cyclodextrins are proposed in formulations for parenteral (18) and ocular systems as well as osmotic tablets (19–21). Captisol is also used as a freeze-drying excipient (22).

Branched cyclodextrins such as mono- or di-glucosyl, maltosyl, and glucopyranosyl α- and β-cyclodextrins have also been described. Branched cyclodextrins are more resistant to the action of α-amylase than the natural cyclodextrins, probably due to steric reasons. The hemolytic activity of branched cyclodextrins is weaker than that of the natural cyclodextrins (23).

Cyclodextrin polymers with low molecular weight (3000–6000) are soluble in water while those with molecular weights above 10,000 can only swell in water and form insoluble gels (24,25). Solutions of 400 to 600 g/L are still moderately viscous. Insoluble cyclodextrin polymers, with a molecular weight of over 20,000, are often used as column material due to their fast swelling properties in organic solvents. Insoluble cross-linked bead polymers seem to be applicable as wound healing agents in the treatment of wounds like burns or ulcers. For controlled delivery of antiseptic agents, iodine has been complexed with the cyclodextrin polymer as a wound-healing agent (26).

Cyclodextrin derivatives, because of their various substitutions, behave like xenobiotics; they are not recognized as natural sugars and are not metabolized in the gastrointestinal tract. Long-term administration of cyclodextrin derivatives, can affect weight gain and lipid metabolism because they are able to form complexes with normal molecules of the gastrointestinal tract (27,28).

Amphiphilic Cyclodextrins

Classification of Amphiphilic Cyclodextrins

Until the last decade, natural cyclodextrins and their derivatives had been recommended for use in pharmaceutical formulations to increase drug solubility and to enhance bioavailability because of their external hydrophilic property. However, this external hydrophilicity may also be a disadvantage for absorption through biological membranes. The supramolecule inclusion displays a lack of affinity for the biological membranes. This has led to the investigation of molecules with a relative external hydrophobicity.

Amphiphilic cyclodextrins have been synthesized by grafting hydrocarbon chains on the hydroxyl groups of either the primary or the secondary face of α-, β-, γ-cyclodextrin. Chemical bonds used in these modifications were selected to provide the molecule a potential biodegradability in the organism. Ester bonds, for example, that could be detached by esterases have been preferred in most derivatives. On the other hand, due to the functionalization by relatively smaller apolar groups, the new cyclodextrins possess surface-active properties, which prove their amphiphilic character. Lollipop cyclodextrins are obtained by grafting an aliphatic chain to a 6-amino-β-cyclodextrin (29). However, to prevent self-inclusion of the pendant group, cup-and-ball cyclodextrins were synthesized by the introduction of a voluminous group such as the tertiary butyl group, which is linked to the end of the aliphatic chain (30,31). Medusa-like cyclodextrins are obtained by the grafting of hydrophobic groups—aliphatic chains with length between C4 to C16—to all the primary hydroxyls of the cyclodextrin molecule (32–34).

Skirt-shaped cyclodextrins consist of β- and γ-cyclodextrins grafted with aliphatic esters or amides (C2 to C14) on the secondary face (35–38). Bouquet-shaped cyclodextrins result from the grafting of 14 polymethylene chains to 3-monomethylated β-cyclodextrin implying the existence of seven chains on each side of the cyclodextrin ring molecule (39). Per (2,6-di-*O*-alkyl) cyclodextrins, where the alkyl chain may be propyl, butyl, pentyl, 3-methylbutyl, or dodecyl, are also part of the bouquet family (40). Some amphiphilic cyclodextrins are presented in Figure 2A–E.

Amphiphilic cyclodextrins can be characterized not only by the nature of substitution and position but also by the length and number of grafted chains and finally by the type of bond used in the modification (ester, ether, amide, thio). Figure 3 depicts some medusa-like and skirt-shaped amphiphilic β-cyclodextrins that we have more specifically studied (41); substituted amphiphilic cyclodextrins grafted with linear or branched acyl groups C_6 linked with an ester bond on the primary or secondary face (Fig. 3B and 3C) and derivatives with amide bonds (Fig. 3A, 3D and 3E).

More complicated neutral derivatives including cholesteryl cyclodextrins were recently synthesized (42–44). They have been designed with assuming that the cyclodextrin is the hydrophilic part and the cholesterol is the hydrophobic part. The three cholesteryl derivatives described until now are as follows: 6-(cholest-5-en-3β-yloxycarbonyl)amino-6-deoxycyclomaltoheptaose, 6-(cholest-5-en-3α-ylamido)

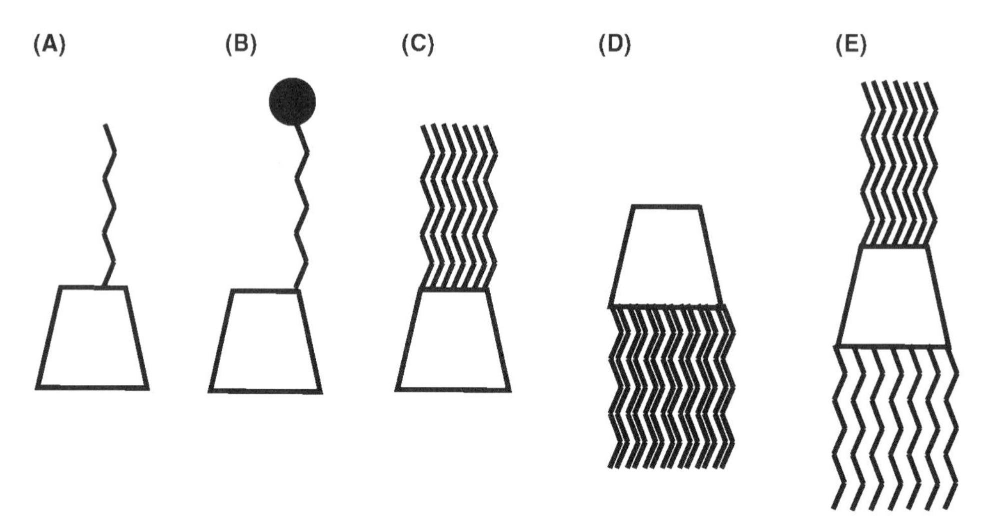

Figure 2 Schematic representation of some amphiphilic cyclodextrins. (**A**) Lollipop cyclodextrins, (**B**) cup-and-ball cyclodextrins, (**C**) medusa-like cyclodextrins, (**D**) skirt-shaped cyclodextrins, and (**E**) bouquet-shaped cyclodextrins.

succinyl-amido-6-deoxycyclomaltoheptaose, and (2,3-di-*O*-acetyl)-hexakis-(6-*O*-acetyl)-6^I-(cholest-5-en-3β-yloxycarbonyl)amino-6^I-deoxy cyclomaltoheptaose. They are differentiated by the length of the spacer arm between the cyclodextrin and the cholesterol and thus differentiating the space between the polar part and the hydrophobic part.

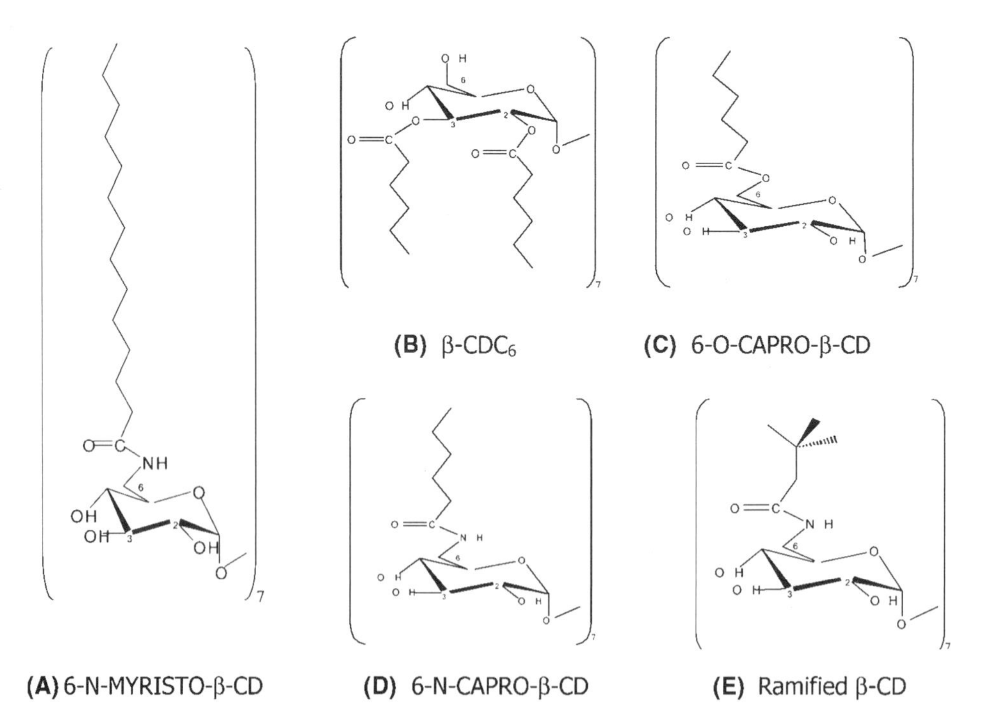

Figure 3 Amphiphilic β-cyclodextrins selectively modified on the secondary and primary face. *Source*: From Ref. 41.

Recently, ionic (cationic and anionic) amphiphilic derivatives, potentially more soluble in water, have been synthesized (45–47). The cationic derivatives carry an amine group as an ionic group. The anionic derivatives, on the other hand, have a sulphate group. Their chemical structure–property relation has not yet been well-elucidated. Particularly, the complexation activity of the ionic amphiphilic cyclodextrins has not been demonstrated and can be ignored because grafted chains or voluminous groups sterically hinder the entrance of the hydrophobic cavity.

Inclusion Capability of Amphiphilic β-Cyclodextrins

The capability of inclusion of a hydrophobic molecule after the functionalization of the parent cyclodextrins has been studied by few research groups (34,48–51).

In fact, access to the cavity is rendered more difficult by the hindrance caused by aliphatic chains grafted to the primary and/or secondary faces. Contrary to the substitution on the secondary face, substitution on primary face may facilitate the formation of inclusion complexes because it leaves the wider side of the cyclodextrin cavity open for the entry of molecules (50,51). However, drugs can also interact with hydrophobic acyl chains.

Interfacial Properties of Amphiphilic Cyclodextrins

Interfacial properties of amphiphilic cyclodextrins have been reported. Studies carried out on β-cyclodextrin diesters with hydrocarbon chains varying from 2 to 14 carbons showed that these compounds have real surfactant characteristics, and when their chloroform solution is spread at the water surface, they form monolayers (52,53). Surprisingly, the surface pressure of the products does not correlate directly with hydrocarbon chain length, but the maximum surface pressure was displayed by the derivative substituted with C_6 chain.

Products having hydrocarbon chains of C_6, C_{12}, and C_{14} that are water-insoluble, when dissolved in tetrahydrofuran present a solubility that increases slowly with temperature up to 50°C, at which they have almost the same solubility (500 g/L) (52). Above this value, a sharp increase is observed. At low temperatures, the solubility is inversely related to chain length and at high temperatures the contrary is observed. With an increase in concentration, the surface tension of the solutions decreases to a small extent with an inflection point resembling that observed for surfactants in aqueous solutions at critical micellar concentration. This suggests the formation of vesicles, such as those in the case of phospholipids (52).

The temperature dependence of solubility in tetrahydrofuran and ^{1}H NMR studies suggests a geometry of aggregates with the polar head groups interacting and the hydrophobic chains oriented towards the solvent. In pyridine, the proposed structure is a bilayer with polar head groups oriented towards the solvent (52).

However, there are contradictory studies about the influence of alkyl chain length on the formation of monolayers at the air–water interface. Kawabata et al. (54) studied the surface pressure–area isotherms of 12 amphiphilic sulfinyl cyclodextrins with different chain lengths (C_4, C_8, C_{12}). It was demonstrated that cyclodextrins with alkyl chains longer than C_8 formed stable condensed monolayers at the air–water interface. The observed limiting areas and the polarized IR spectra of the Langmuir–Blodgett films indicate that the amphiphilic cyclodextrins aligned themselves with the molecular orientation of cyclodextrin ring parallel and alkyl chain perpendicular to the film as shown in Figure 4.

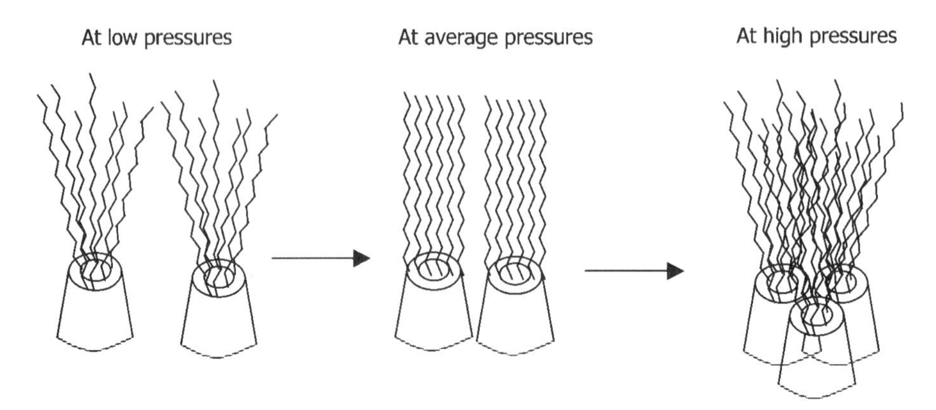

Figure 4 Schematic representation of amphiphilic cyclodextrin molecules at the air–water interface.

Kasselouri et al. (48) have shown that per-6-azido-cyclodextrins form organized assemblies at the air–water interface; the structural properties of these systems are highly dependent on the number of glucopyranose units in the macrocycle suggesting that β-cyclodextrin derivatives form relatively more stable monolayers. Per-6-azido α-, β-, and γ-cyclodextrin monolayers were studied using scanning force microscopy by Alexandre et al. (55). At the micrometer scale, the crystalline nature of β-cyclodextrin derivative was clearly observed whereas α- and γ-derivatives were more continuous. At the molecular scale, hexagonal arrangement of β-cyclodextrin derivative films could easily be imaged but it was not so for α- and γ-derivatives.

Cholesteryl cyclodextrins modified on the primary face are another amphiphilic cyclodextrin type extensively studied for its interfacial structure. It was reported that the grafting of a cholesterol group to a methylated cyclodextrin through a spacer arm produces an amphiphilic compound exhibiting high solubility in water (56). These molecules were characterized thoroughly, with regard to purity, by techniques including high resolution nuclear magnetic resonance and mass spectrometry. It was proved by surface tension measurements, small-angle X ray, and neutron scattering techniques that they self-assembled into monodisperse spherical micelles with an average aggregation number of 24. The micelles could be described as two-shell objects with the cyclodextrin moieties being exposed to the aqueous medium. The authors suggested that this property makes the molecules prone to inclusion of guest molecule in its cavity, and can be a useful delivery system for biologically active material. The same research group also investigated the incorporation of these amphiphilic cyclodextrins into phospholipid bilayers using small-angle X-ray scattering, differential scanning calorimetry, and ^{31}P NMR (42). The mode of incorporation depended on the spacer length between cholesterol and cyclodextrin.

Interfacial properties of amphiphilic cyclodextrins depicted in Figure 3 have been studied in air–water and oil–water systems (57). To characterize the interfacial behavior of the studied amphiphilic cyclodextrins, their monolayers spread at the air–water interface were compressed in dynamic conditions. The surface pressure (π)–molecular area ($Å^2$) isotherms seen in Figure 5 reveal that β-CDC_6 displayed a characteristic profile of amphiphilic β-cyclodextrins modified on the secondary face, for example, a high limiting molecular area and a large plateau (36,38). All cyclodextrins modified on the primary face exhibited a steeper rise in π relative

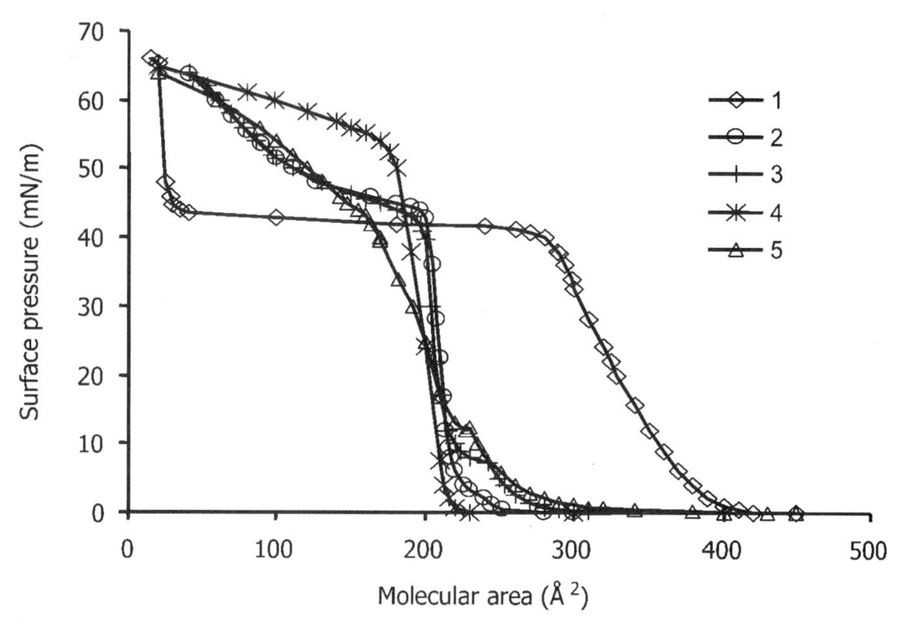

Figure 5 Surface pressure (π)–molecular area (Å^2) isotherms of amphiphilic cyclodextrins deposited at the air–water interface. (**A**) β-CDC_6, (**B**) 6-*O*-CAPRO-β-CD, (**C**) 6-*N*-CAPRO-β-CD, (**D**) 6-*N*-MYRISTO-β-CD, and (**E**) ramified β-CD. *Abbreviation*: CD, cyclodextrin. *Source*: From Ref. 57.

to that of β-CDC_6. In addition, their molecular areas were lower compared with that of β-CDC_6. This provided evidence for their inverted orientation at the interface with regard to cyclodextrins modified on the secondary face. For cyclodextrins modified on the primary face, it was the smaller face of the ring of the cyclodextrin molecule substituted with aliphatic chains that was oriented outward from the air phase.

From the data seen in Figure 6, it was immediately apparent that an important decay in the interfacial tension of the Miglyol® oil–water system occurred in amphiphilic β-CDs substituted at the primary face, resulting in the lowering of the interfacial tension to values between 6 and 10 mN/m. A relatively small decrease in the initial oil–water interfacial tension (to about 16 mN/m) was observed for β-CDC_6 substituted at the secondary face. It should also be noted that while at low cyclodextrin coverages ranging from 10^{13} to 3×10^{13} molecules/cm^2, the decay of the interfacial tension for all studied cyclodextrins was almost the same; at higher surface densities, substantial differences in the behavior of these molecules were observed. This strongly indicated that the chemical nature of cyclodextrins had a significant effect on their spontaneous arrangement at the interface, and consequently on the extent of the reduction in the interfacial tension of the system.

From the comparison of the interfacial tension–surface density relationships, it seems obvious that the modification of β-cyclodextrins on their primary face, with 14 OH groups on the secondary face lying on the aqueous phase, resulted in a pronounced decrease in the oil–water interfacial tension. On the contrary, substituting hydroxyl groups on the secondary face, which leaves only seven OH groups of the primary face in contact with the water phase, appeared to be insufficient to substantially reduce the interfacial tension of the system.

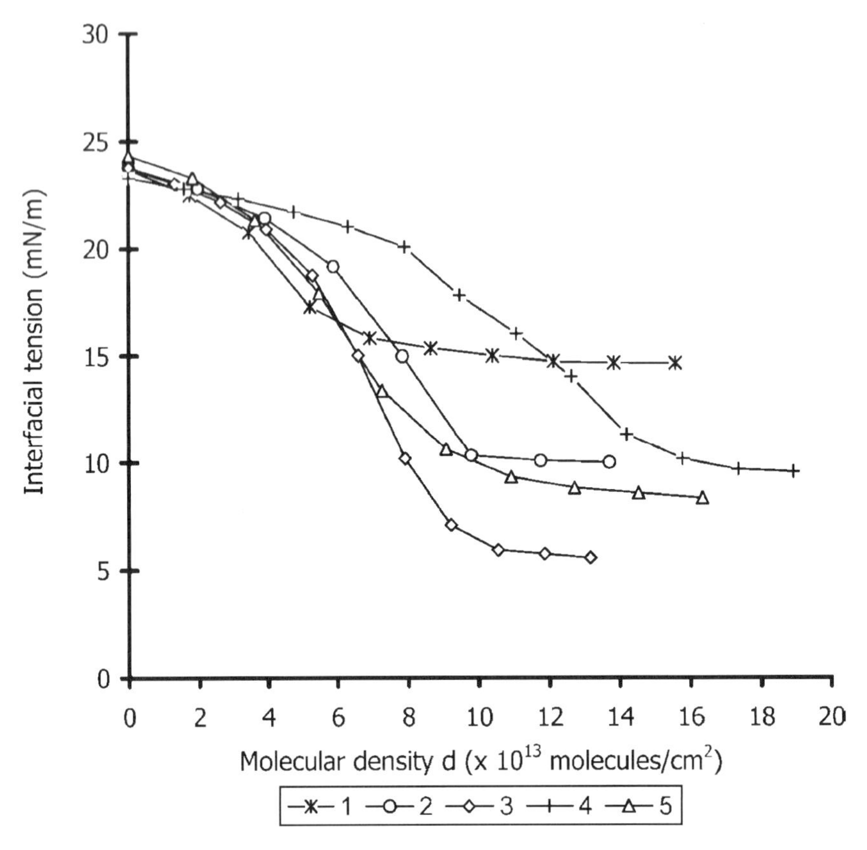

Figure 6 Interfacial tension of Miglyol oil–water as a function of molecular density of amphiphilic cyclodextrins: (**A**) β-CD-C_6, (**B**) 6-*O*-CAPRO-β-CD, (**C**) 6-*N*-CAPRO-β-CD, (**D**) 6-*N*-MYRISTO-β-CD, and (**E**) ramified-β-CD. *Abbreviation*: CD, cyclodextrin. *Source*: From Ref. 57.

AMPHIPHILIC CYCLODEXTRIN NANOPARTICLES

Various amphiphilic cyclodextrins were found to present self-association properties causing the formation of particulate systems (Table 2). β-Cyclodextrin or γ-cyclodextrins grafted on primary and/or secondary faces with C_2 to C_{20} chains can form nanoparticles.

In this chapter, preparation process factors influencing the feasibility of nanoparticles and the nature of cyclodextrin will be discussed, followed by a discussion on its capacity to encapsulate and release various active ingredients. Hemolytic properties of amphiphilic cyclodextrins have also been reviewed in this context.

Nanoparticle Preparation Methods

Nanoprecipitation Technique

Nanospheres are prepared according to the nanoprecipitation method introduced by Uekama et al. (10). An organic phase consisting of amphiphilic β-cyclodextrin

Table 2 Amphiphilic Cyclodextrins Forming Nanoparticulate Systems

Name	Family	Parent CDs	Grafted face	Grafted chains	Bond
2,3-Di-*O*-acyl-CD neutral or sulphated on C_6	Skirt	β and γ	Secondary	Alkyl chains with 6 to 14 carbons	Ester
6-*O*-acyl-CD	Medusa	β	Primary	Alkyl chains of 2 to 22 carbons	Ester
6-Deoxy-6-*N*-acyl-CD	Medusa	β	Primary	Alkyl chains of 2 to 22 carbons	Amide
2,3-Acyl-6-deoxy-6-*N*-cholesteryl-CD	Bouquet	β	Secondary and primary	Alkyl chains of 2 carbons and 1 cholesterol	Ester and Amide
2,3,6-Tri-*O*-perfluoroalkyl-CD	Bouquet	β	Secondary and primary	Fluoroalkyl chains	Ester

dissolved in water-miscible solvent (acetone or ethanol depending on cyclodextrin solubility) is added, with constant stirring, to an aqueous phase consisting only of deionised water. After stirring for one hour at room temperature, the organic solvent is evaporated under vacuum and the nanosphere dispersion is concentrated to the desired volume. Presence of a surfactant (e.g., poloxamer 188) can be necessary for preparing loaded nanospheres (58,59).

Using nanoprecipitation, nanospheres can be loaded with drug using different techniques:

Conventionally loaded nanospheres: The drug is added to the organic phase during the preparation.

Pre-loaded nanospheres: Preformed drug–cyclodextrin complex is added to the organic phase during the preparation.

Highly loaded nanospheres: Preloaded nanospheres are overloaded during preparation by dissolving an additional amount of drug in the organic phase.

To obtain nanocapsules, oil (Miglyol or benzyl benzoate with or without drug) is introduced along with the amphiphilic cyclodextrin in the organic phase (ethanol or acetone) (59).

Emulsion/Solvent Evaporation Technique

Amphiphilic cyclodextrin is dissolved in a water-immiscible organic solvent (e.g., methylene chloride) with or without an active ingredient. The organic phase is poured into water with constant stirring and finally, the organic solvent is evaporated under vacuum (60). This technique allows the formation of nanospheres.

Detergent Removal Technique

It is possible to reconstitute nanospheres from mixed micelles based on the detergent and amphiphilic cyclodextrin, by dilution or dialysis of *n*-octylglucoside, a nonionic detergent. This technique involves mild conditions and avoids the use of organic solvents and exposure to high temperatures (61).

Parameters Influencing Amphiphilic Cyclodextrin Nanoparticle Characteristics

Not all amphiphilic cyclodextrins are capable of forming stable nanoparticles (41,45,51,62). This property is affected mostly by the nature of the grafted chain. A summary of different publications pertaining to the obtaining of particles, spheres, or capsules is presented in Table 3. Most of the studies are focused on the use of amphiphilic cyclodextrin with C_6 chains grafted by either ester or amide bond. Among the factors affecting the capacity of amphiphilic cyclodextrins to form nanoparticles, we can cite the following:

Substituted Face

The first nanoparticles based on amphiphilic cyclodextrins were obtained by Alexandre et al. (55) using cyclodextrins grafted by alkyl chains on the secondary face. Following this, substitution was performed on the primary face so as to keep the larger side of the cavity, free. These amphiphilic cyclodextrins can also form nanoparticles (41,50,51,57).

Substitution on the primary or secondary face influences the characteristics of cyclodextrin nanoparticles. Cyclodextrins substituted on the primary face present a much lower surface tension as a function of their density compared with those grafted on the secondary face (57). Thus, cyclodextrins modified on the primary face stabilize the oil–water interface better, allowing the obtention of nanoparticles and especially stable nanocapsules, with lower cyclodextrin concentrations than in the case of cyclodextrins substituted on the secondary face (41).

Degree of Substitution

This parameter is measurable by mass spectrometry, which allows the determination of molecular weight of the products after ionization. Szejtli et al. (8) have shown that entirely acylated (peracylated) amphiphilic β-cyclodextrins (14 alkyl chains) do not allow the formation of nanoparticles even in the presence of Pluronic F68 that is known to stabilize this type of particles. On the other hand, the mixture of amphiphilic cyclodextrins carrying 11 or 13 alkyl chains (underacylated) with peracylated cyclodextrins form nanoparticles of 200-nm diameters even without stabilizers. This suspension maintained the stability of the particle diameters for seven months at 6°C (62). Similar results were obtained by Alexandre et al. (55) and Auzely-Velty et al. (56) who used a cyclodextrin mixture containing underacylated derivatives resulting from an imperfectly controlled synthesis.

However, selectively substituted derivatives grafted on secondary or primary faces, with linear or ramified alkyl chains, using ester or amide bonds with chain lengths varying from C_6 to C_{14}, have been proven to give stable nanocapsules for 12 months (41).

Length of Grafted Chains

The formation of nanoparticles from cyclodextrins grafted with alkyl chains of different lengths was first studied by Alexandre et al. (55) and Auzely-Velty et al. (56). They worked on cyclodextrins grafted on the secondary face and succeeded in obtaining nanoparticles without surfactants, by nanoprecipitation technique, with diameters around 100 nm for derivatives with alkyl chains of C_6, C_{12}, and C_{14} (65).

Table 3 Capability of Amphiphilic Cyclodextrins to Form Nanoparticles

	Face	Bond	Substitution	Nanoparticles	References
2,3-Di-*O*-C_6-β-CD (β-CDC_6)	Secondary	Ester	C_6 saturated linear	NC NS	41,57–59,62
2,3-Di-*O*-C_6-γ-CD	Secondary	Ester	C_6 saturated linear	NS	60,72
2,3-Di-*O*-C_{12}-β-CD	Secondary	Ester	C_{12} saturated linear	NC NS	58,59
2,3-Di-*O*-C_{14}-β-CD	Secondary	Ester	C_{14} saturated linear	NC NS	58,59
2,3-Di-*O*-$C_5F_4C_8$-β-CD	Secondary	Ester	C_5 fluorinated linear and C_8 saturated linear	NC NS	63
2,3-Di-*O*-C_6-6-$(OSO_3^-)_7$-β-CD	Secondary	Ester	C_6 sulphated linear	None	45,46
2,3-Di-*O*-C_6-6-$(OSO_3^-)_4$-β-CD	Secondary	Ester	C_6 sulfated linear	None	45,46
2,3-Di-*O*-$(C_6)_{21}$-6-$(OSO_3^-)_7$-β-CD	Secondary	Ester	C_6 saturated linear and/or $(2 \times C_6)$ = branched and/or $(3 \times C_6)$ = branched	None	45,46
2,3-Di-*O*-$(C_6)_{21}$-6-$(OSO_3^-)_4$-β-CD	Secondary	Ester	C_6 saturated linear and/or $(2 \times C_6)$ = branched and/or $(3 \times C_6)$ = branched	NS	45,46
6-*O*-C_6-β-CD	Primary	Ester	C_6 saturated linear	NC NS	41,50,51,57,64
6-*O*-C_{12}-β-CD	Primary	Ester	C_{12} saturated linear	NC NS	63,64
6-*O*-$C_{(4,8,10,14,16,18 \text{ or } 20)}$-β-CD	Primary	Ester	$C_{(4,8,10,14,16,18, \text{ or } 20)}$ saturated linear	NS	64
6-N-C_6-β-CD	Primary	Amide	C_6 saturated linear	NS NC	41,50,51,57
6-N-C_{14}-β-CD	Primary	Amide	C_{14} saturated linear	NS NC	41,57
6-N-C_6R-β-CD (ramified β-CD)	Primary	Amide	C_6 saturated ramified	NS NC	41,57
2,3-acyl-6-deoxy-6-N-cholesteryl-CD	Primary and secondary	Ester and amide	C_2 and cholesterol	NS NC	44

Memişoğlu et al. (41) also underlined the difficulty of nanocapsule preparation using derivatives grafted with long alkyl chains (C_{14} chain) on the primary face.

Nature of the Grafted Chain

Memişoğlu et al. (41,66) have tested the feasibility of preparation of nanocapsules using two derivatives with C_6 chains—linear and ramified. When the chain is linear, nanoparticles are obtained very easily using alcohol as solvent. On the other hand, cyclodextrins carrying a ramified chain give nanoparticles only at very low concentrations. Authors explain this phenomenon on the basis of the structure of the chain, which does not allow the correct alignment of cyclodextrin at the oil–water interface (57).

Nature of Chemical Bonds Used for Modification

Effect of the nature of chemical bond between grafted chain and cyclodextrin glucopyranose unit on the interfacial properties and consequently the formation of nanoparticles have also been studied by Memişoğlu et al. (41,66) and Ringard-Lefebvre et al. (57). 6-*N*-CAPRO-β-CD possesses a hydrophilic amide bond and presents a diminution of interfacial tension as a function of molecular area much lower than that of 6-O-CAPRO-β-CD, which possesses a hydrophobic ester bond (Fig. 6) (57). The smaller nanocapsule diameter and lower polydispersity index especially at higher concentrations of the cyclodextrin, for example, 6-N-CAPRO-β-CD can be correlated to this variation. In fact, as spontaneous emulsification occurs when the interfacial tension is low, it seems obvious that the higher the reduction in interfacial tension of the oil–water system, the smaller the radii of droplet curvature and easier the nanocapsule formation.

Charge of the Cyclodextrin

A systematic study has been conducted (45) with various anionic amphiphilic cyclodextrins to form nanoparticles. Solid lipid nanoparticles of sizes varying from 100 to 350 nm have been observed for two of the molecules in the presence of salts such as NaCl, KCl, $MgCl_2$, and $CaCl_2$.

Nanocapsules

Characteristics of Nanocapsules

Nanocapsules were prepared at a molar cyclodextrin concentration ranging from 0.1 to 1 mM in different organic solvents: acetone and absolute alcohol. The particle size depends both on the nature of amphiphilic β-cyclodextrins and their concentrations (41). Size of nanocapsules did not change significantly after 12-month storage except for nanocapsules prepared from more lipophilic derivatives 6-N-MYRISTO-β-CD and β-CD C_6 for which a size increase is observed (66).

Zeta potential of nanocapsules varied between –18 and –31 mV and did not display changes upon storage as aqueous dispersion for 12 months (41). The negative charge of different amphiphilic β-cyclodextrins was believed to result from the fact that, while enveloping the oil droplet, cyclodextrin aligns itself so that the aliphatic chains point towards the oil and the hydroxyl groups face the aqueous phase. It is possible that adsorption and desorption of protons may also occur and render the suspension a negative charge (67). Freeze-fracture photomicrograph of 6-*O*-CAPRO-β-CD nanocapsules is presented in Figure 7.

In Vivo Applications

Skiba et al. (68) reported the in vivo application of indomethacin-loaded nanocapsules prepared from β-CDC_6 containing benzyl benzoate as oil, in a rat model. In this study, the rat gastric ulcerative effect of encapsulated indomethacin was compared with that of a commercial suspension (Indocid®, MSD, U.S.A.). Independent of the administered dose, ulcerative effect was significantly decreased maintaining the bioavailability of indomethacin from the product. Gastrointestinal ulcer protection by encapsulation of indomethacin in β-CDC_6 nanocapsules was 82% for 5 mg/kg and 53% for 10 mg/kg of administered dose.

Nanospheres

Characteristics of Nanospheres

Particle size is the most frequently assessed characteristic of colloidal drug carrier systems. Amphiphilic cyclodextrin nanoparticles are mainly designated for injectable use. Thus, their particle size distribution and polydispersity index should be within certain limits (less than 1 μm).

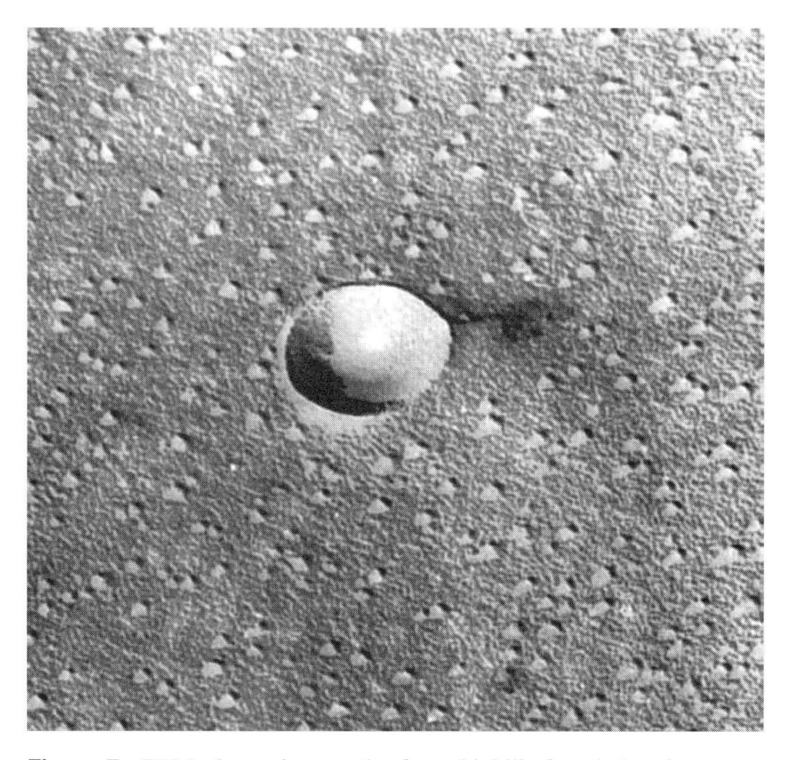

Figure 7 TEM photomicrograph of amphiphilic β-cyclodextrin nanocapsules after freeze-fracture (×31,500). *Abbreviation*: TEM, transmission electron microscopy. *Source*: From Ref. 41.

The effect of amphiphilic β-cyclodextrin molar concentration on nanosphere particle size and polydispersity index has been investigated by Memişoğlu et al. (51). Each amphiphilic β-cyclodextrin was evaluated in the most appropriate organic solvent (acetone or alcohol) according to its solubility as seen in Figure 8. Particle sizes are constant for amphiphilic β-cyclodextrin substitued on the primary face (below 400 nm) even at high concentrations (1 mM) for most amphiphilic β-cyclodextrins. In contrast, with β-CDC_6 (substituted on the secondary face) nanosphere size and polydispersity index dramatically increase over 0.6 mM cyclodextrin concentration.

Nanospheres prepared from β-CDC_6 were imaged by freeze-fracture electron microscopy (Fig. 9) and by photon scanning tunnelling microscopy (PSTM), scanning force microscopy (SFM) and no-contact mode scanning force microscopy (NC-SFM) to ensure spherical geometry and high degree of monodispersity (51,69).

Gulik et al. (70) reported the structural properties of α-CDC_{14}, β-CDC_6, β-CDC_8, β-CDC_{10}, β-CDC_{12}, β-CDC_{14}, and γ-CDC_{14} by freeze-fracture electron microscopy and X-ray scattering experiments. They have shown that amphiphilic cyclodextrins may associate in either hexagonal or cubic forms. The hexagonal structure consists of polar columns comprising the cyclodextrin moieties and separated by aliphatic chains. The cubic structure is body-centred and proposes two models comprising either a polar or an aliphatic aggregate. Electron microscopy reveals that the

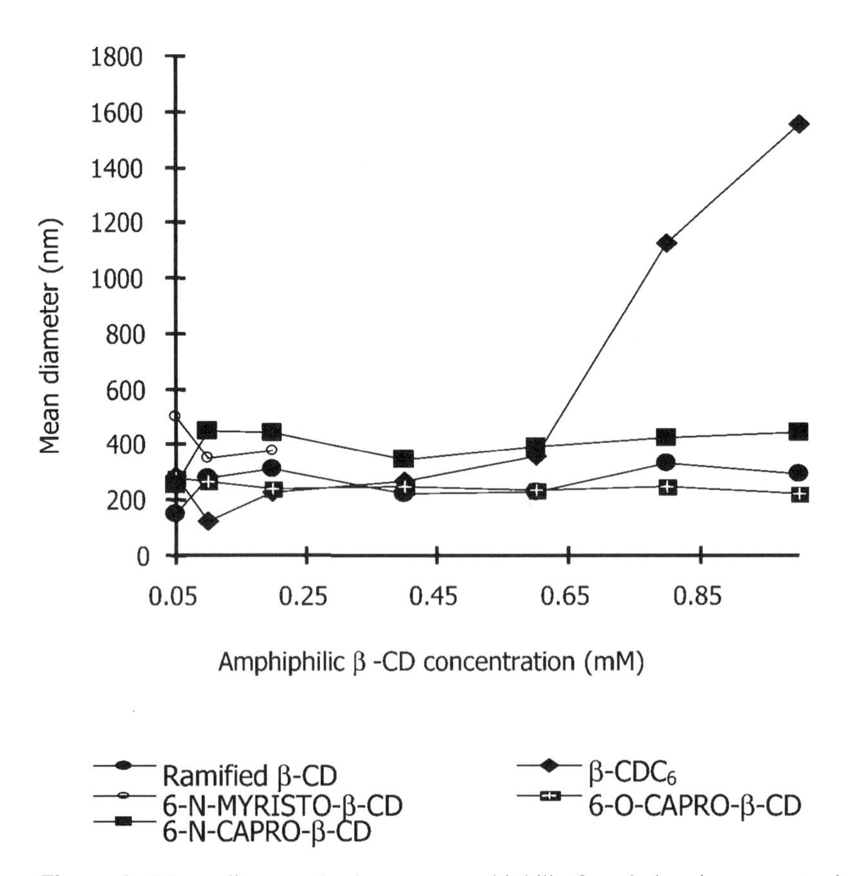

Figure 8 Mean diameter (nm) versus amphiphilic β-cyclodextrin concentration in nanospheres ($n = 3$). *Source*: From Ref. 51.

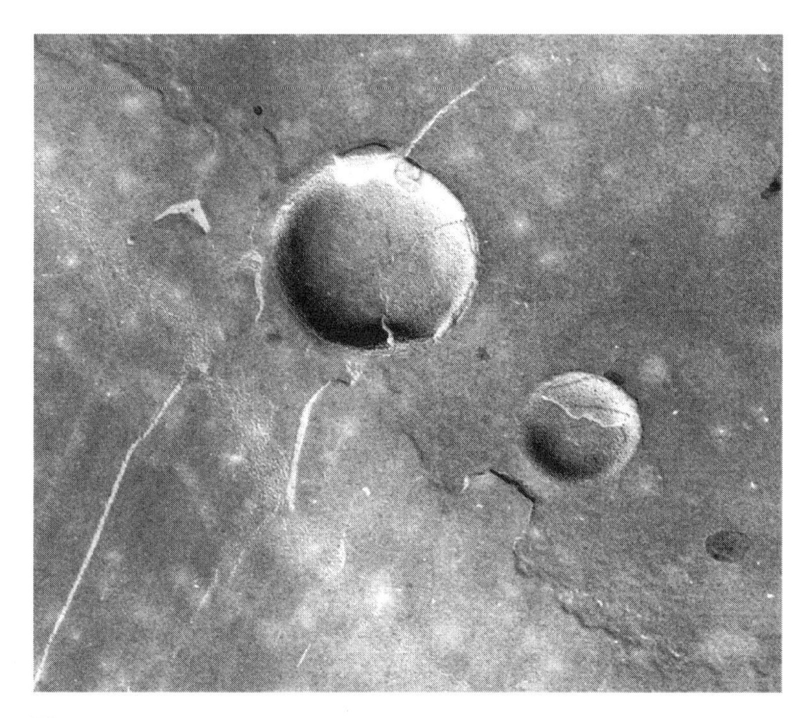

Figure 9 TEM photomicrograph of amphiphilic β-cyclodextrin nanospheres after freeze-fracture (×31,500). *Abbreviation*: TEM, transmission electron microscopy. *Source*: From Ref. 50.

second model is more likely to occur in nature. This type of association is unusual for cyclodextrins and results from the bulkiness of the aliphatic chain requiring a much greater area than the cyclodextrin unit. The small amount of water that can be associated with amphiphilic cyclodextrin can explain the easy formation of nanospheres by the nanoprecipitation technique. Generally, the internal structure of polymeric nanospheres does not present a crystalline structure. However, amphiphilic cyclodextrin nanospheres display a crystalline internal organization that might have interesting properties of drug loading and delivery.

Drug Encapsulation

The conditions necessary to obtain nanoparticles were previously described. After the formation and characterization of the particles, the studies have focused on their potential for encapsulation and controlled release of active ingredient.

Hydrophilic as well as hydrophobic active ingredients have been encapsulated in nanospheres. However, it is likely that the mechanism of drug association to the nanospheres is not the same. In the case of hydrophilic doxorubicin, interface adsorption through hydrophobic interactions can occur between the nanosphere surface and the lipophilic part of the drug. In addition, because doxorubicin can be protonated, strong electrostatic interactions may be established with the anionic part of the surfactant employed (sodium dodecyl sulphate) to prepare the nanospheres (71,72). For lipophilic drugs, nanosphere preparation and loading methods have to be considered.

Emulsion/solvent evaporation. The study performed by Lemos-Senna et al. (60) concerning the encapsulation of progesterone in the nanospheres of γ-CDC_6 using the emulsion/solvent evaporation technique reveals that the following factors influence drug encapsulation:

The initial quantity of progesterone present in the organic phase: Encapsulation increases when the initial quantity of progesterone is between 0 and 1 mg, and after 1 mg, 50% encapsulation of the initial quantity is expected.

The quantity of surfactant: An increase in the loading of progesterone is observed with a decrease in surfactant concentration. The authors explain this phenomenon by the modification of the partition of progesterone between oily phase and aqueous phase containing the mixed micelles of cyclodextrin and surfactant. Lower surfactant concentration induces lower diffusion of progesterone towards the aqueous phase thus increasing the drug encapsulation.

The quantity of cyclodextrin initially introduced: High concentration of initial γ-CDC_6 concentration favors retention of progesterone in the oily phase and thus in the spheres (59).

The influence of the drug association constant to parent cyclodextrin and of its water solubility has been studied by Lemos-Senna using steroids (progesterone, testosterone, and hydrocortisone) (61). Association constant with γ-CD is linearly correlated with loading while water-solubility inversely affects loading capacity of nanosphere prepared with γ-CDC_6.

In the case of progesterone, DSC analysis showed the drug to be amorphously or molecularly associated to the particle (51,60,73). Up to 90% of this drug is very rapidly released within 30 minutes suggesting its adsorption on the particle surface (60).

Nanoprecipitation. Factors influencing drug entrapment are the loading technique itself and the chemical structure of amphiphilic cyclodextrins.

Loading technique. The most frequently used loading technique is the simple dissolution of active ingredient in one of the two phases during preparation of nanospheres by nanoprecipitation (conventionally loaded nanospheres).

A novel technique was introduced by Memişoğlu (66). It is based on the fact that amphiphilic cyclodextrins still preserve the capacity of inclusion of active ingredients either in their cavity or entangled within the alkyl chains. Thus the technique consists of the preformation of inclusion complex between the drug and the cyclodextrin before the preparation of nanoparticles (preloaded nanospheres). However, this technique does not significantly increase entrapment efficiencies or associated drug percentages of progesterone, bifonazole, and clotrimazole, compared with conventional loading (51,50). A variation of this procedure combined the conventional loading method with the preloading methods (high loaded nanospheres). In this case, the amount of drug encapsulated within the nanospheres (progesterone, clotrimazole, or bifonazole) increases when compared with conventionally loaded nanospheres as seen in loading values listed in Table 4 (51,50).

Conventionally loaded nanospheres were reported to associate the drug largely by adsorption on the particle surface; so inclusion in cyclodextrin cavity is a very low possibility. In techniques based on the use of preformed inclusion complexes, drug is already included in the cyclodextrin cavity. A second way of drug entrapment is the interaction and possible drug loading in the aliphatic chains of the amphiphilic β-cyclodextrin. High loading increases entrapment of drug per cyclodextrin unit (μg drug/mg cyclodextrin) about three-fold when compared with conventional loading.

Amphiphilic β-cyclodextrin chemical structure. The type of amphiphilic β-cyclodextrin influences drug encapsulation efficiency. Entrapped drug quantity

Table 4 Loading Characteristics of Preloaded, Conventionally Loaded and Highly Loaded Amphiphilic β-Cyclodextrin Nanoparticles with Bifonazole and Clotrimazole ($n = 6$)

	Bifonazole			Clotrimazole		
Formulation	Entrapped drug quantity (μg ± SD)	Entrapment efficiency (μmol drug/μmol CD)	Associated drug (%)	Entrapped drug quantity (g ± SD)	Entrapment efficiency (μmol drug/ μmol CD)	Associated drug (%)
Preloaded						
1:1 β-CDC$_6$	35 ± 5	32	32	30 ± 2	25	25
1:1 6-N-CAPRO-β-CD	49 ± 5	32	34	36 ± 6	23	22
Conventionally loaded						
β-CDC$_6$	68 ± 3	55	34	74 ± 4	54	37
6-N-CAPRO-β-CD	81 ± 2	47	41	83 ± 3	44	42
Highly loaded						
1:1 β-CDC$_6$	112 ± 11	102	36	99 ± 1	83	31
1:1 6-N-CAPRO-β-CD	173 ± 8	117	50	124 ± 6	78	35

and associated drug percentage data suggest that 6-N-CAPRO-β-CD (cyclodextrin substituted on the primary face) has a higher loading capacity than β-CDC_6 (cyclodextrin substituted on the secondary face) (Table 4). Drug association is significantly higher ($p < 0.05$) for 6-N-CAPRO-β-cyclodextrin. This difference is even more pronounced in the case of bifonazole because it has a higher affinity to the cyclodextrin cavity. These results confirm that drug inclusion may be facilitated by leaving the secondary face unmodified and thus reducing the steric hindrance.

Drug Release

Only a few studies have evaluated the release kinetics of nanoparticles prepared from amphiphilic cyclodextrins (50,51,66,73). Parameters involved in in vitro release are nanosphere loading technique, drug:cyclodextrin association constant, and effect of substitution side.

Release profile of a model drug, bifonazole, was assessed in water:PEG400 (60:40) (Fig. 10). Conventionally loaded nanospheres liberated bifonazole immediately within the first 15 minutes due to burst effect resulting from drug adsorption onto particle surface (60,73). On the other hand, preloaded nanospheres prolonged the release up to 1 hour because of drug entrapment in the amphiphilic cyclodextrins. Consequently, intermediate release profiles observed with highly loaded nanospheres resulted in the addition of these two phenomena (drug adsorption and drug entrapment). Similar result was obtained with progesterone (51).

The affinity of mother β-cyclodextrin to model drugs (clotrimazole and bifonazole) plays a major role in drug release from amphiphilic β-cyclodextrin nanospheres (50). It

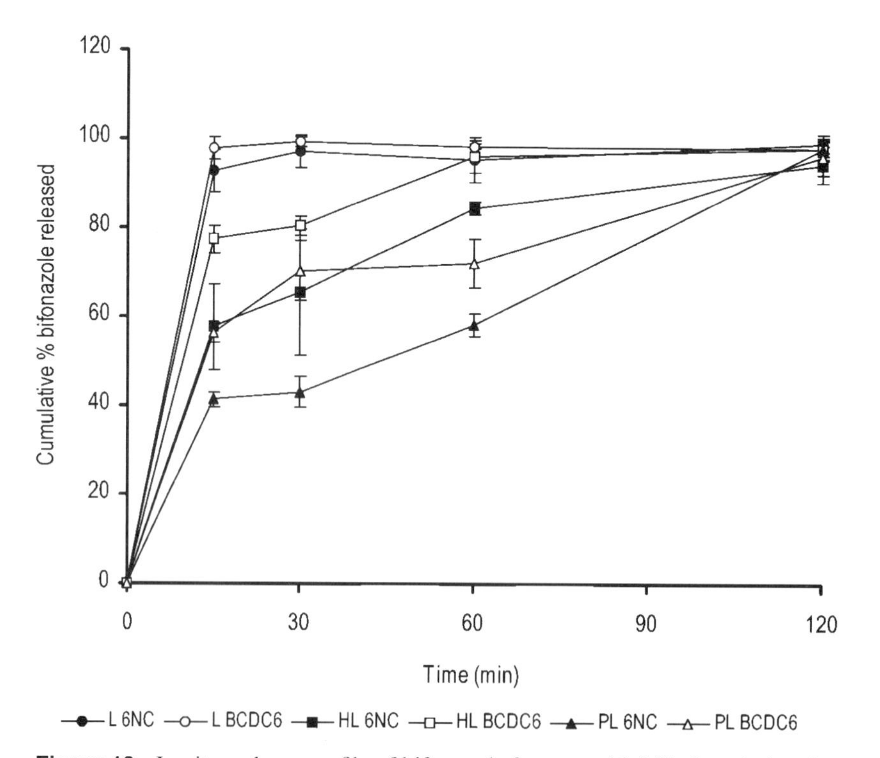

Figure 10 In vitro release profile of bifonazole from amphiphilic β-cyclodextrin nanospheres in water:PEG 400 ($n = 6$, SE). *Source*: From Ref. 50.

can be observed in Figures 10 and 11 that bifonazole, which has a very high association constant with β-cyclodextrin ($K_{1:1} = 11{,}000\ M^{-1}$), is released significantly more slowly than clotrimazole which has a lowest association constant ($K_{1:1} = 500\ M^{-1}$). This latter result explains the weak effect of the loading technique on the clotrimazole release.

Comparing bifonazole releases from 6-N-CAPRO-β-CD amphiphilic β-cyclodextrin (substitution on the primary face) or β-CDC_6 (substitution on the secondary face), it appears that 6-N-CAPRO-β-CD displayed slower release profiles than β-CDC_6. This difference is observed for preloaded and highly loaded nanospheres, both techniques using preformed inclusion complexes. This result is explained by the easiest access to the cyclodextrin when the widest secondary face remains free of substitution (50).

In Vitro Evaluation of Amphiphilic Cyclodextrin Nanosphere Biological Effects

A major drawback in the parenteral use of β-cyclodextrin for is its hemolytic effects owing to its interaction with blood cells. The hemolytic properties of amphiphilic β-cyclodextrin nanospheres, prepared with either 6-N-CAPRO-β-CD (substituted on primary face) or β-CDC_6, (substituted on secondary face) have been evaluated on whole blood and erythrocyte suspension samples (50,74).

As expected, amphiphilic β-cyclodextrin nanospheres were less hemolytic than the natural β-cyclodextrin because of their hydrophobic substituents in all concentrations covered by this study (Fig. 12). Besides hydrophobicity, self-assembly

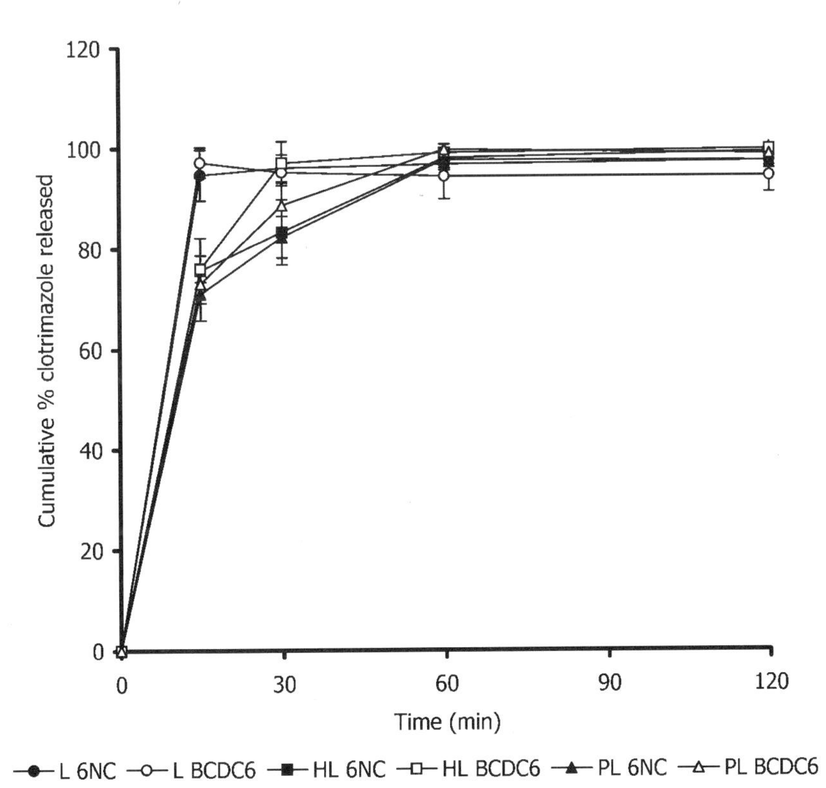

Figure 11 In vitro release profile of clotrimazole from amphiphilic β-cyclodextrin nanospheres in water:PEG 400 ($n = 6$, SE). *Source*: From Ref. 50.

of amphiphilic β-cyclodextrins in the form of nanospheres was also believed to reduce interaction and direct contact of cyclodextrin with red blood cells. Hemolytic effect was much more pronounced on red blood cell suspension, which is also expected because the presence of other blood components and proteins aids the disruption of erythrocyte cell membrane (8). When amphiphilic β-cyclodextrins were compared with regard to hemolysis, it was found that β-CDC_6 caused significantly lower hemolysis in red blood cells and total blood than 6-N-CAPRO-β-CD.

Memişoğlu et al. (50) reported the antimicrobial activity of model drugs (bifonazole and clotrimazole), amphiphilic β-cyclodextrins (β-CDC_6 and 6-N-CAPRO-β-CD), and nanospheres loaded with these model drugs. The activity was assessed by the determination of their minimum inhibitory concentration (MIC) values on *C. albicans* ATCC 90028. Clotrimazole has a lower MIC value than bifonazole probably because of higher water solubility which is in accordance with previous data (75). Interestingly, both amphiphilic β-cyclodextrin displayed considerable antimycotic activity as seen in Table 5. When drug-loaded amphiphilic β-cyclodextrin nanospheres were incubated with *C. albicans*, synergistic effect was very pronounced. Indeed, MIC values were lowered by 2- and 10-fold for bifonazole and clotrimazole, respectively.

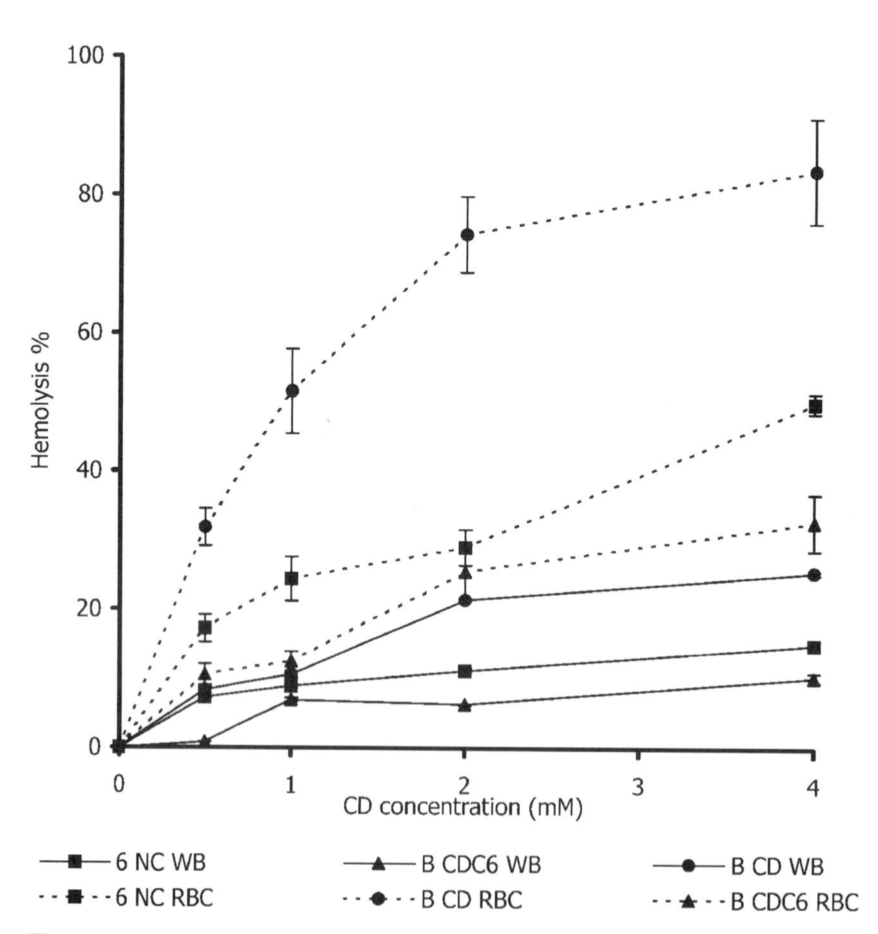

Figure 12 Hemolytic activity of amphiphilic β-cyclodextrin nanospheres in WB and RBC ($n = 3$, SE). *Abbreviations*: RBC, red blood cells; WB, whole blood. *Source*: From Ref. 50.

Table 5 MIC Values of Amphiphilic β-Cyclodextrins, Antimycotic Model Drugs and Drug-Loaded Amphiphilic β-Cyclodextrin Nanospheres Against *C. albicans* ($n = 3$, SD).

System component	MIC value (μg/mL ± SD)
Bifonazole ethanolic solution	0.48 ± 0.13
Clotrimazole ethanolic solution	0.12 ± 0.07
6-N-CAPRO-β-CD ethanolic solution	12.5 ± 3.6
β-CDC_6 ethanolic solution	25.0 ± 7.2
Bifonazole:6-N-CAPRO-β-CD nanosphere	0.19 ± 0.11
Bifonazole:β-CDC_6 nanosphere	0.19 ± 0.05
Clotrimazole:6-N-CAPRO-β-CD nanosphere	0.012 ± 0.006
Clotrimazole:β-CDC_6 nanosphere	0.012 ± 0.006
β-CD:bifonazole complex aqueous solution	125 ± 36
β-CD:clotrimazole complex aqueous solution	7.8 ± 4.5

CONCLUSION

Amphiphilic cyclodextrins synthesized by the grafting of various substituents to different faces of natural cyclodextrins, β- and γ-cyclodextrin in particular, have proved to be promising excipients for the formulation of various active ingredients in nanoparticulate systems (nanocapsules and nanospheres). Similar to parent cyclodextrins, amphiphilic cyclodextrins offer the interesting property of including drug molecules in their cavity and preserving this ability during the formation of nanoparticles.

Compared with polymers traditionally used for microencapsulation, amphiphilic cyclodextrins are self-assembling small molecules. However, they can be employed as polymers in the preparation of nanoparticles with the same techniques (nanoprecipitation and emulsion-solvent evaporation). Based on our current knowledge, the fast drug release observed with amphiphilic cyclodextrin nanoparticles can be attributed to their different organization compared to polymer nanoparticles. Presently, amphiphilic cyclodextrin nanoparticles cannot be considered as deposit systems but rather as modified release systems for various administration routes.

Until now, a great number of this new class of cyclodextrin derivatives has been widely studied for Physicochemical aspects but only a few biological studies have been undertaken. Biodegradability, pharmacokinetics, and toxicity have still to be thoroughly investigated and are the keys for the development of cyclodextrins in the pharmaceutical field.

REFERENCES

1. Wenz G. Cyclodextrins as building blocks for supramolecular structures and functional units. Angew Chem Int Ed Eng 1994; 33:803–822.
2. Albers E, Müller BW. Cyclodextrin derivatives in pharmaceuticals. CRC Crit Rev Ther Drug Carrier Syst 1995; 12(4):311–337.
3. Loftsson T, Brewster ME. Pharmaceutical applications of cyclodextrins. I. Drug solubilization and stabilization. J Pharm Sci 1996; 85(10):1017–1025.

4. Duchêne D, Glomot F, Vaution C. Pharmaceutical applications of cyclodextrins. In: Duchêne D, ed. Cyclodextrins and Their Industrial Uses. Paris: Editions de Santé, 1987:211–258.
5. Hirayama F, Uekama K. Methods of investigating and preparing inclusion compounds. In: Duchêne D, ed. Cyclodextrins and Their Industrial Uses. Paris: Editions de Santé, 1987:131–172.
6. Irie T, Uekama K. Pharmaceutical applications of cyclodextrins. III. Toxicological issues and safety evaluation. J Pharm Sci 1997; 86:147–162.
7. Thompson DO. Cyclodextrins as enabling excipients: their present and future use in pharmaceuticals. CRC Crit Rev Ther Drug Carrier Syst 1997; 14(1):1–104.
8. Szejtli J, Cserhati T, Szögyi M. Interactions between cyclodextrins and cell-membrane phospholipids. Carbohyd Polym 1986; 6:35–49.
9. Debouzy JC, Fauvelle F, Crouzy S, et al. Mechanism of α-cyclodextrin induced hemolysis 2. A study of the factors controlling the association with serine-, ethanolamine- and choline-phospholipids. J Pharm Sci 1997; 87(1):59–66.
10. Uekama K, Hirayama F, Irie T. Modification of drug release by cyclodextrin derivatives. In: Duchêne D, ed. New Trends in Cyclodextrins and Derivatives. Paris: Editions de Santé, 1991:409–446.
11. Uekama K, Hirayama F, Irie T. Cyclodextrin drug carrier systems. Chem Rev 1998; 98:2045–2076.
12. Stella VJ, Rajewski RA. Cyclodextrins: their future in drug formulation and delivery. Pharm Res 1997; 14(5):556–567.
13. Müller BW, Brauns U. Hydroxypropyl-β-cyclodextrin derivatives: influence of average degree of substitution on complexing ability and surface activity. J Pharm Sci 1986; 75:571–572.
14. Yoshida A, Yamamoto M, Irie T, Hirayama F, Uekama K. Some pharmaceutical properties of 3-hydroxypropyl/ and 2,3-dihydroxypropyl β-cyclodextrins and their solubilizing and stabilizing ability. Chem Pharm Bull 1989; 37:1059–1063.
15. Müller BW. Hydroxypropyl β-cyclodextrin in drug formulation. Proceedings of CRS Workshop What's New in Cyclodextrin Delivery?, Paris, July 7–8, 2000.
16. Zia V, Rajewski R, Bornancini ER, Luna EA, Stella VJ. Effect of alkyl chain length and substitution degree on the complexation of sulfoalkyl ether β-cyclodextrins with steroids. J Pharm Sci 1997; 86:220–224.
17. Stella VJ. SBE_{7M}-β-CD or Captisol®-possible utilisations. Proceedings of CRS Workshop What's New in Cyclodextrin Delivery?, Paris, July 7–8, 2000.
18. Xiang TX, Anderson BD. Stable supersaturated aqueous solutions of silatecan 7-t-butyldimethylsilyl-10-hydroxycamptothecin via chemical conversion in the presence of a chemically modified beta-cyclodextrin. Pharm Res 2002; 19:1215–1222.
19. Jarvinen T, Jarvinen K, Urtti A, Thompson D, Stella VJ. Sulfobutyl ether beta-cyclodextrin (SBE-beta-CD) in eyedrops improves the tolerability of a topically applied pilocarpine prodrug in rabbits. J Ocul Pharmacol Ther 1995; 11:95–106.
20. Okimoto K, Ohike A, Ibuki R, et al. Design and evaluation of an osmotic pump tablet (OPT) for chlorpromazine using (SBE)7m-beta-CD. Pharm Res 1999; 16:549–554.
21. Okimoto K, Rajewski RA, Stella VJ. Release of testosterone from an osmotic pump tablet utilizing (SBE)7m-beta-cyclodextrin as both a solubilizing and an osmotic pump agent. J Control Release 1999; 58:29–38.
22. Ma DQ, Rajewski RA, Stella VJ. Thermal properties and processing of (sulfobutylether)-7M-β-cyclodextrin as a freeze-drying excipient in pharmaceutical formulations. STP Pharma Sci 1999; 9:261–266.
23. Yamamoto M, Yoshida A, Hirayama F, Uekama K. Some Physicochemical properties of branched β-cyclodextrins and their inclusion characteristics. Int J Pharm 1989; 49: 163–171.
24. Friedman R. Cyclodextrin-containing polymers. In: Duchêne D, ed. New trends in Cyclodextrins and Polymers. Paris: Editions de Santé, 1991:157–178.

25. Sébille B. Cyclodextrin derivatives. In: Duchêne D, ed. Cyclodextrins and Their Industrial Uses. Paris: Editions de Santé, 1987:351–393.
26. Fenyvesi E, Szejtli J, et al. Complex formation of cyclodextrin polymers with iodine. In: Duchêne D, ed. Proceedings of the 5th International Symposium on Cyclodextrins. Paris: Editions de Santé, 1990:276–279.
27. Irie T, Uekama K. Pharmaceutical applications of cyclodextrins III. Toxicological issues and safety evaluation. J Pharm Sci 1997; 86:147–162.
28. Favier ML, Remesy C, Moundras C, Demigne C. Effect of cyclodextrin on plasma lipids and cholesterol metabolism in the rat. Metabolism 1995; 44:200–206.
29. Bellanger N, Perly B. NMR investigations of the conformation of new cyclodextrin-based amphiphilic transporters for hydrophobic drugs: molecular lollipops. J Mol Struct 1992; 15:215–226.
30. Dodziuk H, Chmurski K, Jurczak J, et al. A dynamic NMR study of self-inclusion of a pendant group in amphiphilic 6-thiophenyl-6-deoxycyclodextrins. J Mol Struct 2000; 519:33–36.
31. Lin J. Synthèse des cyclodextrines amphiphiles et étude de leur incorporation dans des phases phospholipidiques. Ph.D. thesis, University of Paris VI, Paris, 1995.
32. Kawabata Y, Matsumoto M, Tanaka M, et al. Formation and deposition of monolayers of amphiphilic cyclodextrin derivatives. Chem Lett 1983:1933–1934.
33. Djedaini F, Coleman AW, Perly B. New cyclodextrins based media for vectorization of hydrophilic drug, mixed vesicles composed of phospholipids and lipophilic cyclodextrins. In: Duchêne D, ed. Minutes of the 5th International Symposium on Cyclodextrins. Paris: Editions de Santé, 1990:328–331.
34. Liu FY, Kildsig DO, Mitra AK. Complexation of 6-acyl-O-β-cyclodextrin derivatives with steroids, effects of chain length and substitution degree. Drug Dev Ind Pharm 1992; 18:1599–1612.
35. Zhang P, Ling CC, Coleman AW, Parrot-Lopez H, Galons H. Formation of amphiphilic cyclodextrins via hydrophobic esterification at the secondary hydroxyl face. Tetrahedron Lett 1991; 32:2769–2770.
36. Zhang P, Parrot-Lopez H. Self-organizing systems based on amphiphilic cyclodextrins diesters. J Phys Org Chem 1992:518–528.
37. Memişoğlu E, Charon D, Duchêne D, Hincal AA. Synthesis of per(2,3-di-O-hexanoyl)-β-cyclodextrin and characterization amphiphilic cyclodextrins nanoparticles. In: Torres-Labandeira J, Vila-Jato J, eds. Proceedings of the 9th International Symposium on Cyclodextrins. Dordrecht: Kluwer Academic Publishers, 1999:622–624.
38. Lesieur S, Charon D, Lesieur P, et al. Phase behaviour of fully-hydrated DMPC-amphiphilic cyclodextrin systems. Chem Phys Lip 2000; 106:127–144.
39. Canceill J, Jullien L, Lacombe L, Lehn JM. Channel type molecular structures, Part 2, synthesis of bouquet-shaped molecules based on a β-cyclodextrin core. Helv Chim Acta 1992; 75:791–812.
40. Wenz G. Synthesis and characterization of some lipophilic per (2,6-di-O-alkyl) cyclomaltooligosaccharides. Carbohydr Res 1991; 214:257–265.
41. Memişoğlu E, Bochot A, Şen M, Charon D, Duchêne D, Hincal AA. Amphiphilic β-cyclodextrins modified on the primary face: synthesis, characterization and evaluation of their potential as novel excipients in the preparation of nanocapsules. J Pharm Sci 2002; 95:1214–1224.
42. Auzely-Velty R, Perly B, Tache O, et al. Cholesteryl-cyclodextrins: synthesis and insertion into phospholipid monolayers. Carbohyd Res 1999; 318:82–90.
43. Roux M, Auzely-Velty R, Djedaini-Pilard F, Perly B. Cyclodextrin-induced lipid lateral separation in DMPC membranes (2) H nuclear magnetic resonance study. Biophys J 2002; 82:813–822.
44. Weisse S. Complexes cyclodextrines/esters de Vitamine A: stabilisation, solubilisation et promotion de l'absorption cutanée. Ph.D. thesis, Université Paris-Sud, Paris, 2002.

45. Dubes A, Parrot-Lopez H, Shahgaldian P, Coleman AW. Interfacial interactions between amphiphilic cyclodextrins and physiologically relevant cations. J Coll Interf Sci 2003; 259:103–111.
46. Dubes A, Parrot-Lopez H, Abdelwahed W, et al. Scanning electron microscopy and atomic force microscopy imaging of solid lipid nanoparticles derived from amphiphilic cyclodextrins. Eur J Pharm Biopharm 2003; 55:279–282.
47. Sukegawa T, Furuike T, Niikura K, Yamagishi A, Monde K, Nishimura S. Erythrocyte-like liposomes by means of amphiphilic cyclodextrin sulfates. Chem Commun (Camb.) 2002; 7:430–431.
48. Kasselouri A, Munoz M, Parrot-Lopez H, Coleman AW. Synthesis and self-organisational properties of the per-9-azido-6-deoxy-cyclodextrins. Polish J Chem 1993; 67:1981–1985.
49. Takahashi H, Irinatsu Y, Kazuka S, Tagaki W. Host-guest complexation of a lipophilic heptakis(6-dodecyl amino-6-deoxy)-β-cyclodextrin with nitrophenol in chloroform. Mem Fac Eng Osaka City Univ 1985; 26:93–99.
50. Memişoğlu E, Bochot A, Özalp M, Şen M, Duchêne D, Hincal AA. Direct formation of nanospheres from amphiphilic β-cyclodextrin inclusion complexes. Pharm Res 2003; 20:117–125.
51. Memişoğlu E, Bochot A, Şen M, Duchêne D, Hincal A. Non-surfactant nanospheres of progesterone inclusion complexes with amphiphilic β-cyclodextrins. Int J Pharm 2003; 251:143–153.
52. Tchoreloff P, Boissonnade MM, Coleman AW, Baszkin A. Amphiphilic monolayers of insoluble cyclodextrins at the water/air interface. Surface pressure and surface potential studies. Langmuir 1995; 11:191–196.
53. Munoz M, Deschenaux R, Coleman AW. Observation of microscopic patterning at the air/water interface by mixtures of amphiphilic cyclodextrins-a comparison isotherm and Brewster angle microscopy study. J Phys Org Chem 1999; 12:364–369.
54. Kawabata Y, Matsumoto M, Nakamura T, et al. Langmuir Blodgett films of amphiphilic cyclodextrins. Thin Solid Films 1998; 159:353–358.
55. Alexandre S, Coleman AW, Kasselouri A, Valleton JM. Scanning Force Microscopy investigation of amphiphilic cyclodextrin Langmuir-Blodgett films. Thin Solid Films 1996; 284–285, 765–768.
56. Auzely-Velty R, Djedaini-Pilard F, Desert S, Perly B, Zemb T. Micellization of hydrophobically modified cyclodextrins 1. Micellar structures. Langmuir 2000; 16:3727–3734.
57. Ringard-Lefebvre C, Bochot A, Memişoğlu E, Charon D, Duchêne D, Baszkin A. Effect of spread amphiphilic β-cyclodextrins on interfacial properties of the oil/water system. Coll Surf B Biointerf 2002; 25:109–117.
58. Skiba M, Wouessidjewe D, Fessi H, Devissaguet JP, Duchêne D, Puisieux F. Préparation et utilisations des nouveaux systèmes colloidaux dispersibles à base de cyclodextrines, sous forme de nanosphères, French Patent, 9,207,287, 1992.
59. Skiba M, Wouessidjewe D, Fessi H, Devissaguet JP, Duchêne D, Puisieux F. Préparation et utilisations des nouveaux systèmes colloidaux dispersibles à base de cyclodextrines sous forme de nanocapsules, French Patent, 9,207,285, 1992.
60. Lemos-Senna E, Wouessidjewe D, Lesieur S, Duchêne D. Preparation of amphiphilic cyclodextrin nanospheres using the emulsion solvent evaporation method, influence of the surfactant on preparation and hydrophobic drug loading. Int J Pharm 1998; 170:119–128.
61. Lemos-Senna E. Contribution à l'étude pharmacotechnique et physicochimique de nanosphères de cyclodextrines amphiphiles comme transporteurs de principes actifs. Ph.D. thesis, Université Paris-Sud, Paris, 1998.
62. Gèze A, Aous S, Baussanne I, Putaux JL, Defaye J, Wouessidjewe D. Influence of chemical structure of amphiphilic β-cyclodextrins on their ability to form stable nanoparticles. Int J Pharm 2002; 242:301–305.

63. Skiba M, Skiba M, Duclos R, Combret JC, Arnaud P. Cyclodextrines mono-substituées à persubstituées par des groupements fluoroalkyles, leur préparation et leur utilisation. French Patent, 9,909,863, 1999.
64. Terry N, Rival D, Coleman A. Perreir E. Cyclodextrines substituées préférentiellement sur leur face primaire par des fonctions acide ou amine, French Patent, 0,006,102, 2000.
65. Skiba M, Duchêne D, Puisieux F, Wouessidjewe D. Development of a new colloidal drug carrier from chemically-modified cyclodextrin: nanospheres and influence of Physico-chemical and technological factors on particle size. Int J Pharm 1996; 129:113–121.
66. Memişoğlu E. Evaluation of amphiphilic β-cyclodextrins as novel excipients in the preparation of nanoparticulate drug delivery systems. Co-tutelle Ph.D. thesis, Université Paris-Sud, France, Hacettepe University, Ankara, 2002.
67. Barratt G. Characterization of colloidal drug carrier systems with zeta potential measurements. Pharm Technol (Europe) 1999; 1:26–32.
68. Skiba M, Morvan C, Duchêne D, Puisieux F, Wouessidjewe D. Evaluation of gastrointestinal behaviour in the rat of amphiphilic β-cyclodextrins nanocapsules loaded with indomethacin. Int J Pharm 1995; 126:275–279.
69. Skiba M, Puisieux F, Duchêne D, Wouessidjewe D. Direct imaging of modified β-cyclodextrin nanospheres by photon scanning tunneling and scanning force microscopy. Int J Pharm 1995; 120:1–11.
70. Gulik A, Delacroix H, Wouessidjewe D, Skiba M. Structural properties of several amphiphilic cyclodextrins and some related nanospheres. A X-ray scattering and freeze-fracture electron microscopy study. Langmuir 1998; 14:1050–1057.
71. Leroy-Lechat F, Wouessidjewe D, Duchêne D, Puisieux F. Stability studies of the doxorubicin association to new colloidal carriers made of amphiphilic cyclodextrins. In: Proceedings of the 1st World Meeting on Pharmaceutics, Biopharmaceutics and Pharmaceutical Technology, Budapest, 1995; 499.
72. Wouessidjewe D, Skiba M, Leroy-Lechat F, Lemos-Senna E, Puisieux F, Duchêne D. A new concept in drug delivery based on "skirt-shaped cyclodextrin aggregates". Present state and future prospects. STP Pharma Sci 1996; 6:21–26.
73. Lemos-Senna E, Wouessidjewe D, Lesieur S, Puisieux F, Couarrazze G, Duchêne D. Evaluation of the hydrophobic drug loading characteristics in nanoprecipitated amphiphilic cyclodextrins nanospheres. Pharm Dev Technol 1998; 3:1–10.
74. Yamaguchi H, Hiratani T, Plempel M. In vitro studies of a new imidazole antimycotic, bifonazole in comparison to clotrimazole and miconazole. Arzneim Forsch/Drug Res 1983; 33:546–551.
75. Memişoğlu E, Bochot A, Duchêne D, Hincal AA. Haemolytic evaluation of nanospheres prepared from amphiphilic β-cyclodextrins modified on the primary or secondary face. Eur J Pharm Sci 2002; 17(suppl 1):S81.

10

Liposphere for Controlled Delivery of Substances

Abraham J. Domb
Department of Medicinal Chemistry and Natural Products, School of Pharmacy-Faculty of Medicine and the David R. Bloom Center for Pharmacy, Alex Grass Center for Drug Design and Synthesis, The Hebrew University of Jerusalem, Jerusalem, Israel

INTRODUCTION

Lipospheres are fat-based encapsulation system, developed for parenteral and topical drug delivery of bioactive compounds (1–12). Lipospheres consist of water dispersible solid microparticles, which have their diameter between 0.1 and 100 μm. These are composed of a solid hydrophobic fat core (triglycerides) stabilized by a layer of phospholipid molecules embedded in their surface. The internal core contains the bioactive compound, dissolved or dispersed in the solid fat matrix. The liposphere system has been used for the controlled delivery of various types of drugs including anti-inflammatory compounds, local anesthetics, antibiotics, and anticancer agents. They have also been successfully used as carriers of vaccines and adjuvants (2–5). Recently, lipospheres have been used for the delivery of peptides and for oral drug–delivery (10,11). Similar systems based on solid fats and phospholipids have been described (13,14). Solid lipid nanospheres, which are essentially nanosize lipospheres when phospholipid is used, have also been recently reviewed (14).

Lipospheres have several advantages over other delivery systems such as emulsions, liposomes, and microspheres. Some of the advantages in using the lipospheres include better physical stability, low cost of ingredients, ease of preparation and scale-up, high dispersability in an aqueous medium, high entrapment of hydrophobic drugs, controlled particle size, and extended release of entrapped drug after a single injection, which can last from a few hours to several days.

Updated information on the preparation and physicochemical properties of lipospheres and their use for the parenteral delivery of various drugs and vaccines and topical and oral administration of active agents have been described in this chapter.

PREPARATION OF LIPOSPHERES

In contrast to certain oil emulsions, the liposphere approach utilizes naturally occurring biodegradable lipid constituents. The internal hydrophobic core of lipospheres is composed of fats, mainly solid triglycerides, while the surface activity of liposphere particles is provided by the surrounding lecithin layer that is composed of phospholipid molecules.

The neutral fats used in the preparation of the hydrophobic core of the liposphere formulations described here included: tricaprin, trilaurin, and tristearin, stearic acid, ethyl stearate, and hydrogenated vegetable oil. The polymers used in the preparation of polymeric biodegradable lipospheres include low molecular weight poly(lactic acid) and poly(caprolactone).

The phospholipids used to form the surrounding layer of lipospheres include pure egg phosphatidylcholine (PCE), soybean phosphatidylcholine (PCS), dimyristoyl phosphatidylglycerol (DMPG), and phosphatidylethanolamine (PE). Food grade lecithin (96% acetone insoluble) was used in the preparation of lipospheres for topical and veterinary applications.

Liposphere formulations are prepared by solvent or melt processes. In the melt method, the active agent is dissolved or dispersed in the melted solid carrier, i.e., tristearin or polycaprolactone (PCL), and a hot buffer solution is added at once, along with the phospholipid powder. The hot mixture is homogenized for about two to five minutes using a homogenizer or ultrasound probe, after which a uniform emulsion is obtained. The milky formulation is then rapidly cooled down to about 20°C by immersing the formulation flask in an acetone–dry ice bath, while homogenization is continued to yield a uniform dispersion of lipospheres.

Alternatively, lipospheres might be prepared by a solvent technique. In this case, the active agent, the solid carrier, and phospholipid are dissolved in an organic solvent such as acetone, ethyl acetate, ethanol, or dichloromethane. The solvent is then evaporated and the resulting solid is mixed with a warm buffer solution, and the mixing is continued until a homogeneous dispersion of lipospheres is obtained.

In a typical preparation, dexamethasone (200 mg), tristearin (400 mg), and propylparaben (5 mg) are added to a 50 mL round bottom glass flask. The flask is heated at 75°C to melt the tristearin–dexamethasone mixture, and a hot 0.1 M phosphate buffer solution of pH 7.4 (75°C, 9.3 g) is added along with PCE (200 mg). The mixture is homogenized for two minutes until a uniform milk-like formulation is obtained. The hot formulation is rapidly cooled to less than 20°C by immersing the flask in a dry ice–acetone bath with continued mixing, to yield a white thin dispersion. If needed, the pH of the formulation is adjusted to 7.4 with a 1N HCl solution. The formulation may contain antioxidants such as tocopherol and preservatives such as parabens. Submicron size lipospheres are prepared by passing (four times) the liposphere formulation, by extrusion, through a submicron series of filters at a temperature that is 5°C above the melting point of the liposphere core composition. Particle size may be reduced to about 200 nm.

Polymeric biodegradable lipospheres can also be prepared by solvent or melt processes. The difference between polymeric lipospheres and the standard liposphere formulations is the composition of the internal core of the particles. Standard lipospheres, as those previously described, consist of a solid hydrophobic fat core that is composed of neutral fats like tristearin, while in the polymeric lipospheres, biodegradable polymers such as polylactide (PLD) or PCL substitute the tryglycerides. Both types of lipospheres are thought to be stabilized by one layer of phospholipid molecules embedded in their surface (see section "Physical Characterization of Lipospheres").

Sterile liposphere formulations are prepared by sterile filtration of the dispersion in the hot stage during preparation using a 0.2-μm filter, at a temperature that is 5°C above the melting point of the liposphere core composition. Heat sterilization using a standard autoclave cycle decomposed the formulation. g-Sterilization of liposphere formulations did not affect their physical properties. Lipospheres formulations containing bupivacaine, tristearin, and phospholipid in the ratio of 1:4:2 and 2:4:2 (w/w% ratio) were irradiated with a dose of 2.33 Mrads and these samples were analyzed for particle size, bupivacaine content, in vitro release characteristics, and in vivo activity. The irradiated formulations had the particle size, bupivacaine content and release rate, and anesthetic effectiveness similar to that seen with bupivacaine HCl solution (Marcaine®), in the rat paw analgesia model. However, a more careful analysis of the formulation ingredients should be performed, as phospholipids may degrade during irradiation.

Nanosize lipospheres have been prepared by homogenization through serial filters of reduced pore size, as described above. An alternative method for in situ preparation of nanosize lipospheres, which have particle size below 100 nm, was recently developed using dispersible concentrate oil system (12). In this system, the drug, triglyceride, phospholipid, and other additives are dissolved in a mixture of common surfactants such as Tween® and Span® and an organic solvent that is miscible with all components. Such organic solvents include ethanol, propanol, propylene glycol, low molecular weight–polyethylene glycol (PEG), *N*-methyl pyrrolidone (NMP), ethoxilated castor oil (Cremophor®), propylene–ethylene glycol copolymers (Poloxamer®), and PEG conjugated α-tocopherol. This clear unhydrous solution spontaneously forms nanoparticles when gently mixed in aqueous solution. The particle size is controlled mainly by the formulation compositions. Cationic or anionic nanolipospheres can be obtained when adding a cationic or anionic lipid, such as stearyl amine, phosphatidilethanol amine, stearic acid, or phosphadilic acid, to the solution. This concept has been applied for improving the bioavaiability of cyclosporin (12).

PHYSICAL CHARACTERIZATION OF LIPOSPHERES

Structure

To determine the lamillarity of lipospheres obtained at different phospholipid to neutral lipids ratios, Nuclear Magnetic Resonance (NMR) spectroscopy in the presence of paramagnetic ions was used. The phospholipid content on the surface of lipospheres was determined by ^{31}P NMR, before and after manganese (Mn^{+2}) or preseodimium (Pr^{+3}) complexation, and by trinitrobenzenesulfonic acid (TNBS) labelling using liposphere formulations that contain PE (15,16). In both methods, the agents, Mn^{+2} or TNBS, interact specifically with the exposed phosphate or amino groups of the phospholipid or PE. Determination of the surface phospholipid using the TNBS method showed that 70% to 90% of the phospholipid polar heads are in the surface of the liposphere particles prepared from triglyceride and phospholipid at a 1:0.5 to 1:0.25 w/w ratio. Increasing the phospholipid content decreases the percentage of surface phospholipid polar heads, which indicates the formation of other phospholipid structures such as liposomes in the formulation. Similar results were obtained by ^{31}P NMR analysis using Mn^{+2} and Pr^{+3} ions for interaction with the phosphate polar heads. The decrease in the peak area after the addition of 5 mM Mn^{+2} represent the relative amount of phospholipid in the outer monolayer that is available for interaction with Mn^{+2}. Lipospheres composed of tricaprin and phospholipid in a weight ratio of 2:1 and 5:1 showed 75% and 90% of surface phospholipid polar heads, respectively. For comparison, liposomes of the same composition and size but without trilcaprin had only 40% of its polar heads in the surface (15).

This data suggests that the proposed structure of a liposphere is a spherical particle with a monolayer of phospholipid molecules surrounding the internal solid fat core, where the hydrophobic chains of the phospholipids are embedded onto the surface of the internal triglyceride core containing the active agent.

Particle Size

Analysis of the particle size distribution of lipospheres was performed using a LS 100 Coulter Counter Particle Size Analyzer (Beckman Coulter Corp., Hialeah, FL, U.S.). This instrument can measure particles ranging from 0.4 to 800 μm by particle size-dependent light diffraction patterns. The particle size of submicron lipospheres was estimated from the transmission electron microscope pictures and by using a particle size analyzer for the particle having a diameter in the range of 0.01 to 3 μm.

Blank and drug-loaded lipospheres prepared by the melt method, without further treatment, had a unimodal shape and the average particle size between 5 and 15 μm; less than two percentage of particles were greater than 100 μm. This particle size formulation is useful for subcutaneous (SC) or intramuscular injection or for topical applications.

In an effort to reduce the particle size, lipospheres with average diameter of 16.9-μm, prepared by the melt method, were passed through a laboratory Microfluidizer® (Microfluidics, Newton, MA, U.S.). Thereby, the average size was reduced to 10.8 μm after eight passes and to 9.4 μm after 40 cycles. Similar results were obtained when using a homogenizer.

Lipospheres containing the R32NS1 malaria antigen and differing in their fat composition were prepared by the melt procedure (2–5). Three groups of neutral fats were used: (i) solid fats such as tristearin and stearic acid with melting points in the range of 65°C to 70°C, (ii) semisolid fats such as tricaprin and ethylstearate with melting points in the range of 30°C to 35°C, and (iii) liquid fats such as olive oil and corn oil. Two populations of liposphere particles usually coexisted, one population having their diameter in the range of 1 to 10 μm (population A), and a second population with their diameter between 10 and 100 μm (population B). No correlation was found between fat physical state (solid, semisolid, and liquid) and particle size distribution of the lipospheres formed. Lipospheres made of tristearin were the most homogeneous formulations, with 100% of the particles having an average diameter of about 7 ± 3 μm.

Biodegradable polymeric lipospheres made of PLD and lecithin showed a very broad particle size distribution,ranging from 2 to 100 μm in diameter, with the average size of 30.6 ± 25.9 μm, and the median (% of particles >50 μm) equal to 21.6 (3). PCL lipospheres showed a similar range of particle size distribution, but with 1.5-fold mean average size (45.6 ± 29.5 μm) and double the median value (43.6 μm) compared to PLD lipospheres. Inclusion of lipid A in the composition of the polymeric lipospheres reduced their average particle size by a factor of 0.25, regardless of the polymer type (3).

All the liposphere formulations, with a polymeric core prepared, remained stable during the three months period of the study, and no phase separation or appearance of aggregates were observed.

Viscosity and Drug Loading

For drugs such as oxytetracycline (OTC), itraconazole, and dexamethasone, up to 20% drug loading into lipospheres was obtained, such that the formulations remained fluid enough to be injected. For other agents like bupivacaine, lidocaine, and chloramphenicol, drug loading above 10% produced a viscous lotion. The viscosity of the lipospheres formulation is dependent on the drug properties, the ionic strength and pH of the continuous aqueous solution, and the ratio and amount of phospholipids and triglycerides. Increase in the content of the insoluble ingredients (drug and lipids) and the salt concentration of the aqueous medium, increases the viscosity of the formulation. For topical applications, lotion and paste consistencies are desired, which are achieved by adding NaCl to the water phase and increasing the fat and phospholipid content of the liposphere lipid phase (10).

Drug Distribution

The drug distribution in a liposphere formulation was determined by isolating the particles by centrifugation and determining the drug content in the isolated cake after extraction of the free drug with, for example, acidic buffer solution for basic drugs. The amount of unencapsulated free drug can be estimated by microscopic analysis (the free drug appears usually as crystals while liposphere are seen in round shapes). A study was conducted to determine bupivacaine distribution in the 2% bupivacaine formulation (17). It was found that the nonencapsulated drug was observed as needle-shaped crystals, dispersed among the liposphere particles. These liposphere and drug crystals were isolated by centrifugation, the bupivacaine crystals were extracted from the mixture with acidic buffer solution, and the drug-loaded liposphere were analyzed for bupivacaine content after dissolution–extraction of the liposphere in Triton® solution. The drug loading was in the range of 70 to 85 wt% in the core and about 20% of the drug was extracted by the acidic solution appearing as a non-incorporated drug. About 4% of the drug was soluble in the aqueous solution, which is the solubility of bupivacaine in the buffer solution. In an attempt to determine the form of the non-incorporated drug, bupivacaine and phospholipid were dispersed in aqueous medium in the absence of the tristearin component. A uniform and stable submicron dispersion was obtained. Microscopic examination of this fat free preparation showed that the dispersed drug microparticles are nonspherical but are in the form of long needles. It should be noted that bupivacaine free base is not dispersible in buffer solution without the surfactants such as phospholipids. Thus, it is apparent that the unincorporated bupivacaine in the tristearin liposphere formulation is in a form of dispersible microparticles, which are composed of the solid drug and phospholipids.

The compatibility of the encapsulated drug with the solid core material is a key issue for maintaining the drug in the liposphere particles. When an incompatible solid core is used, the drug may migrate out of the liposphere and crystallize in the solution. This is demonstrated with a bupivacaine free base, incorporated in ethyl stearate. Bupivacaine migrates out of the particles and forms needle-like crystals, as shown in Figure 1. The migration process is a result of gradual dissolution of the drug by the aqueous medium to saturation, where the drug molecules start to precipitate from the solution to form crystals. To avoid this migration, which occurs in the presence of water, the liposphere dispersion was lyophilized and kept dry until reconstitution shortly before use.

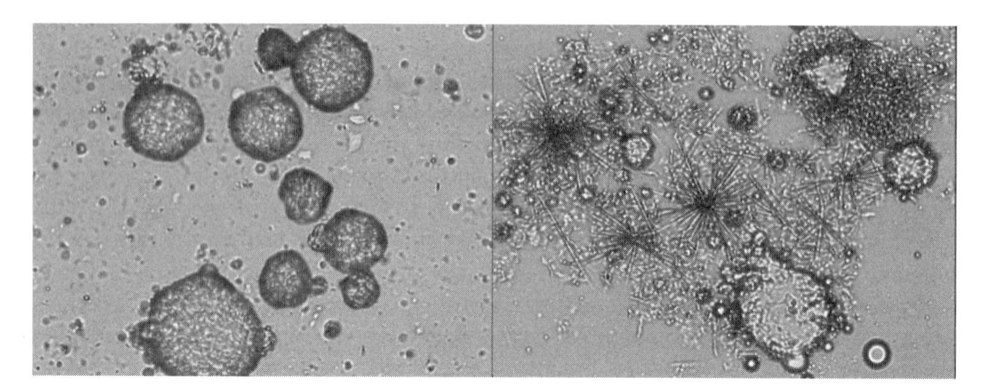

Figure 1 Lipospheres loaded with bupivacaine, 0.5 hour after preparation (*left*) and five hours after preparation (*right*).

In Vitro Drug Release

In vitro release studies were conducted, using a large pore dialysis tubing of 300,000 molecular weight cut off (MWCO), to minimize the effect of the tubing on the release rate from the formulation (a regular pore size tubing of 12,000 MWCO did affect the drug release rate). The control solutions of drugs such as bupivacaine (Marcaine, 0.75% in solution for injection) were released through the large pore tubing within a few hours. Both formulations released the drug for 48 hours, following a first-order kinetics ($r^2 = 0.97$). These release profiles are expected for formulations that contain 4% of the free drug soluble in the aqueous vehicle, which are released immediately through the dialysis tubing.

The release of etoposide, a water insoluble anticancer agent, from 2%-loaded lipospheres placed in a dialysis tubing (300,000 MWCO), was studied. Over 90% of the drug was constantly released during a period of 80 hours.

Lipospheres containing ^{14}C-diazepam were prepared by the melt technique. The lipospheres showed a very uniform particle size distribution, with an average particle size of 8 μm. The lipospheres were evaluated in vitro, by placing 0.5 mL of the formulation in dialysis tubing with 300,000 MWCO. The dialysis tubing was placed in a 0.1 M phosphate buffer, and the cumulative release of diazepam was determined by measuring the radioactivity in the release medium. The release medium was changed periodically to provide sink conditions. Sustained release of diazepam was obtained over a period of three days.

The in vitro release was also determined by mixing a sample of the formulation in excess buffer solution, taking specimens of the mixture every few hours, and determining the drug released into the solution after removal of the lipospheres by ultracentrifugation. The release rate by this method was faster than by dialysis tubing.

APPLICATIONS OF LIPOSPHERES

This section describes the use of lipospheres for parenteral administration of long acting local anesthetics and OTC, oral administration of cyclosporin, topical application of *N*,*N*-diethyl-*m*-toluamide (DEET) insect repellent, and delivery of luteinizing hormone releasing hormone (LHRH) from polymer based lipospheres.

Parenteral Delivery of Local Anesthetics

Local anesthetics are preferred over general anesthetics because of the serious complications that can occur during general anesthesia. However, even local anesthetics, which are usually injected as an aqueous solution, are eventually absorbed from the site of application into the circulation. Frequent administration of local anesthetics may result in the development of systemic toxicity. A long acting formulation that provides extended regional blockade may be useful for pain management following surgery or for chronic pain relief.

Lipospheres were used for the delivery of the common local anesthetics such as bupivacaine and lidocaine, for the purpose of extending their effectiveness to a few days after a single injection. Liposphere formulations containing local anesthetics were prepared by the melt and solvent methods, as described above. Sterile formulations were prepared by dissolving bupivacaine (100 g) and egg phospholipid (100 g) in ethanol, followed by sterile filtration through a 0.22-μm filter. Tristearin (200 g) was dissolved in hot ethanol and filtered into the same flask and the ethanol was evaporated to dryness. To the remaining semisolid, 5 L of hot (65°C), sterile 0.1 M phosphate buffer solution containing 0.05% methyl paraben and 0.1% propyl paraben as preservatives was added and the mixture was homogenized for five minutes at high speed. The uniform milk-like preparation was rapidly cooled down to below 20°C by immersing the flask in a dry ice–acetone bath while homogenization continued. The formulation was filled into 10-mL vials and stored under aseptic conditions at 4°C until used.

A variety of models have been used for the evaluation of peripheral analgesic agents. However, only three have been extensively used to evaluate pharmaceutical compositions. These are the Randall–Selitto test, the abdominal construction writhing response to intraperitoneal injection of an irritant, and the pain response of mice after formalin injection (18,19). The local effects as a function of time of liposphere formulations were tested using the Randall–Selitto experimental animal model.

To induce hyperalgesia, animals were first anesthetized, then a yeast or carrageenan solution was injected through the foot pad nearest the first digit. The foot withdrawal score was then measured at the indicated times after the injection, on a scale from 0 to 25 (0 = could not stand any pressure; 25 = could stand extensive pressure) using the Randall–Selitto instrument. Liposphere formulations to be tested were coadministered with the yeast solution. Control animals received an identical volume of water. For animals being tested for more than 24 hours, the liposphere formulation was injected at time zero and yeast was injected 24 hours prior to liposphere injection, as yeast-induced hyperalgesia is often not measurable when the time exceeds 24 hours postadministration. The hyperalgesic response was developed within the first hour after injection and maintained for 48 or 72 hours.

The effectiveness of several bupivacaine formulations was investigated. Liposphere formulations containing tristearin as the fat and PCE or PCS and loaded with 1% or 2% bupivacaine produced long lasting analgesia for at least 48 hours, and almost 72 hours (20). The blank liposphere formulation did not produce any analgesia; the Marcaine control solution (0.75% in saline) produced a strong analgesic effect, which lasted for less than three hours.

A rat sciatic nerve preparation was developed to study prolonged local anesthetic blockade from bupivacaine–lipospheres (6,7,9). Blockade and/or section of the rat sciatic nerve are time-honored systems to study regional anesthesia. The nerve is large and easily seen, and the effects of motor and sympathetic blockade in the foot can be detected. Sensory blockade can also be measured, provided that there is no spillover in the saphenous nerve distribution. After anesthesia was induced, bilateral posterolateral incisions were made in the upper thighs, and the sciatic nerves were visualized. Sham vehicle was injected around the nerve on one side, and vehicle containing either 5% or 10% bupivacaine (0.5 mL dose) were injected on the other side. The facia was then closed over the deep compartment, to partially restrict egress of drug formulation. Motor blockade was scored on a four-point scale based on the following visual observations: (i) normal appearance, (ii) impaired ability to splay toes when elevated by the tail, (iii) toes and foot remained plantar flexed with no splaying ability, and (iv) loss of dorsiflexion, flexion of toes, and impairment of gait.

Both 5% and 10% bupivacaine–liposphere formulations showed significant levels of motor blockade through day three, and in some cases, day four. Motor function in all animals returned to normal by day six. Sympathetic blockade was determined indirectly by foot pad–temperature measurements. The foot receives sympathetic innervation, largely from the sciatic nerve, though there may be a contribution from the saphenous nerve as well. Skin temperature measurements of the blockade side were also monitored as an indication of vascular tone. During the period when motor block was apparent, the blockade side was essentially always warmer than the side without the blockade. Temperature differences seemed to dissipate within one day after motor block resolved. As the sensory blockade measurements in this experiment are complicated by several factors, it was important to find a method for detecting sensory blockade that is relatively independent of motor responses. Vocalizations in response to defined transcutaneous electrical stimulation of points on the feet were used as a criterion for the measurement of sensory blockade. The thresholds to hind paw–electric shock and hind paw pad–temperature measures of sympathetic block were both increased for three to four days. No impairments were observed on the contralateral control side. One-week postliposphere administration, the sciatic nerves were removed and histologically evaluated. No evidence of nerve damage and very little inflammation of foreign body–response were observed.

A study was conducted to evaluate the efficacy of 2% bupivacaine–liposphere formulation to produce analgesia in the rat formalin model (7,9). The model was designed to assess the antinociceptive ability against chronic pain caused by a test compound. Another method of assessment, namely, the foot flick thermal method was also performed on the same animals, in addition to the formalin study. The formalin study compared the effects of administration of 2% loaded bupivacaine–lipospheres with blank lipospheres, standard bupivacaine solution (Marcaine, 0.5% with 1:200,000 epinephrine), and physiologic saline on nociception in the rat. Rats were pretreated at various times with test or control formulations by infiltration injection into the right popliteal fossa, and then injected with 5% formalin into the dorsal surface of the right hind paw. Nociception was then measured in the form of paw flinches in a five-minute period, during both the acute and tonic phases. Both 2% bupivacaine–liposphere and Marcaine had onset of action times within 10 minutes. However, the liposphere formulation was able to maintain significant antinociception for at least nine hours in both acute and tonic phases as compared to less than three hours for Marcaine. Similar results were obtained in the foot flick thermal stimulus model. Sensory blockade was measured by the time required for each rat to withdraw its hind paw from a 56°C plate. Latency to withdraw each hind paw from the hot plate was recorded by alternating paws. If no withdrawal occurred from the hot plate within 15 seconds, the trial was terminated.

Neither Marcaine nor the liposphere formulation caused a change in paw volume, and there was no significant influence of these formulations on the development of formalin-induced edema.

The long acting effect of bupivacaine–liposphere formulation as compared to Marcaine solution was further confirmed by the rat-tail-flick model. The study compared administration of the 2% bupivacaine–liposphere, blank liposphere, standard Marcaine, and saline. After administration of the formulations, tail flick latencies were determined. The liposphere formulation exceeded the anesthetic duration of Marcaine by 12-fold. Treatment with blank lipospheres was not significantly different from treatment with the saline control.

A pilot pharmacokinetic study was performed in rabbits to compare the C_{max} and T_{max} values, obtained after intramuscular injection of Marcaine HCl solution and a liposphere formulation. A total amount equivalent to 20 mg of bupivacaine was injected to rabbits ($n = 3$) and blood was collected for 72 hours. Bupivacaine blood concentrations were determined by HPLC following a USP method. Lidocaine was used as an internal standard, and the linear calibration curve for bupivacaine was between 50 and 1000 ng/mL. The maximal blood concentrations were 681 ± 246 and 200 ± 55 ng/mL for bupivacaine in solution and in lipospheres, respectively.

The toxicity of bupivacaine–lipospheres in rats was evaluated. As the liposphere formulation consists of natural inert components, phospholipids, and triglycerides, they are expected to be biocompatible in vivo (21). The incidence of microscopic observations after intramuscular injection of liposphere formulations was studied. Blank lipospheres, 1% and 5% bupivacaine in lipospheres, 5% dextrose solution, and bupivacaine solution (0.1 mL) were injected in rats followed by histological examinations of the sites of injection at day 3, 7, and 14. The degree of inflammation, necrosis, and fibrosis was scaled from zero to four, where zero means absent and four means marked. The degree of inflammation, necrosis, and fibrosis was similar to all formulations. At day three, some irritation and inflammation was observed, which was reduced after 14 days. In a second study, the local toxicity of a 2% bupivacaine formulation, after daily injections for two weeks in dogs, was estimated. Minimal local irritation was observed in these studies as determined by histology examination. However local inflammation was observed, presumably caused by an accumulation of lipospheres in the lung.

Parenteral Delivery of Antibiotics

Several antibiotics including ofloxacin, norfloxacin, chloramphenicol palmitate, and OTC and antigungal agents such as nystatin and amphotericin B have been incorporated into lipospheres in high encapsulation yield. The use of lipospheres for antibiotic delivery was demonstrated by the development of liposphere–OTC formulations for veterinary use (6).

Parenteral OTC therapy in farm animals requires daily administration of the drug over several days, usually three to five days, to provide prolonged therapeutic blood levels. Serum OTC concentrations of potential clinical and therapeutic values in the treatment of OTC sensitive organisms are estimated in the range of 0.15 to 1.5 mg/mL. The minimal inhibitory concentration (MIC) in mg/mL for certain farm animals pathogens are 0.15 for *Pasteurella multocida*, 0.3 for *Staphylococcus* and *Pasteurella anatipestifer*, 0.4 for *Hemophyllis paragallinarum*, 0.8 for *Mycoplasma gallisepticum*, and 1.5 for *Escherichia coli*. Blood levels of above 0.5 mg/mL are required for treatment of most bacterial infections.

Several long-acting OTC formulations have been reported (20,22,23). These formulations have been tested in various farm animals and showed adequate blood levels for 72 hours following a single injection, at a dose of 20 mg/kg.

OTC was encapsulated in good yields in solid triglyceride liposphere formulations. The microdispersion containing up to 15 wt% OTC was injectable through a 20G needle and was stable for at least one year when stored under refrigerated conditions. OTC was released in vitro from dialysis tubing for five days, while OTC was released from solution through the tubing in less than six hours. Blood levels in turkeys or rabbits were maintained for up to four to five days, for the 8% loaded formulation. Increase in OTC concentration required a decrease in the content of triglyceride and phospholipid in the formulation, resulting in a decrease in the duration of drug release. The formulation degraded in tissue but remnants of the formulations remained for periods longer than four weeks.

An animal study was conducted to evaluate the controlled release effect of the liposphere formulations by following the OTC blood levels and the elimination of the administered dose from the injection site. In the first study, four OTC loaded liposphere formulations based on tristearin and trilaurin and phospholipid, differing in their compositions, were compared to an OTC solution (10% OTC in acidic solution) used as reference and a blank liposphere formulation as control. The formulations were injected intramuscularly to a group of six turkeys, and the OTC blood levels were determined. The injection sites were observed for residuals 7, 11, and 28 days postinjection. All four liposphere formulations showed OTC levels above MIC for at least three days. Although tristearin showed good results, it is not preferred because it is less susceptible for elimination from the injection site, as discussed below.

The residual amounts evaluated at 7 and 11 days were maximal (about 90% of the original dose) for tristearin-based formulation, about 50% for the trilaurin based formulations, and about 20% for the blank formulation based on tristearin. The deposits contained less than 10% OTC of the original dose. After 28 days, only little amounts of deposits at the injection site were found for the trilaurin formulations. No OTC was detected in any of the deposits retrieved from the animals after 28 days (17).

All animals in these studies were healthy and gained weight like the the non-treated animals, with no pathological signs. In all the injection sites, there were no signs of damage, swelling, or inflammation. None of the injection sites showed any necrosis or encapsulation even when precipitates were observed.

Lipospheres have been used for the encapsulation of amphotericin B, a common antifungal and antileishmanial agent, for intraveneous (IV) and SC administration for treating systemic infections. Lipospheres with particle size in the range of 200 nm were prepared from solid triglycerides and soybean phospholipid, by using the dispersible concentrate formulation method. The formulation was stable as dry powder for reconstitution for one year, and showed to be effective and safe when injected intravenously. SC administration to dogs suffering from leishmanial infection resulted in some effectiveness, but with the formation of lesions caused by the corrosive effect of amphotericin B at the injection sites.

Parenteral Delivery of Vaccines and Adjuvants

Several reports describing the improvement of immune response achieved by the association of antigens with lipid carriers such as liposomes or microparticles like polymeric biodegradable microcapsules have been published (24,25). The ability of these delivery systems to enhance immunogenicity was related to the physicochemical characteristics of the particles.

The physicochemical properties and immunogenic activity of different liposphere–vaccine formulations containing a recombinant R32NS1 malaria antigen, derived from the circumsporozoite protein of *Plasmodium falciparum* as the model antigen, are described. Manufacture of lipospheres was accomplished by gently melting the neutral fat in the presence of phospholipid, and dispersing the mixture in an aqueous solution of the antigen by vigorous shaking, which results upon cooling in the formation of a phospholipid-stabilized solid hydrophobic fat core containing the antigen (4).

The effect of the type of fat used in the preparation of liposphere on their immune response to encapsulated antigen was tested. Mice were immunized twice, at weeks zero and four, with lipospheres containing R32NS1 malaria antigen. Although for all liposphere formulations, the first immunization at week zero caused a very small immune response. However, after the boost injection, a very marked increase of mean IgG antibody levels was observed for most of the liposphere–vaccine formulation tested, the immune response obtained remaining at very high levels of IgG antibody titers, even after the 12 weeks period of the experiment (2). The most immunogenic liposphere formulation was the one made of ethylstearate, while lipospheres made of stearic acid showed the lowest IgG ELISA titers. The complete order of immunogenic activity (based on fat composition) of the liposphere formulations tested was (2):

Ethylstearate > olive oil > tristearin > tricaprin > corn oil > stearic acid

No correlation between liposphere particle size or fat chemical characteristics and immunogenicity was found. It is worth noting that the IgG antibody ELISA titers obtained on immunizing rabbits with liposphere–R32NS1 were superior to those obtained following similar immunizations with the free antigen absorbed to alum, which showed no antibody activity at the same antigen concentrations. It was previously shown that this antigen was poorly immunogenic even in humans, when injected alone as an aqueous solution, or when adsorbed on alum (26).

Incorporation of a negatively charged phospholipid, DMPG, in the liposphere lipid phase caused a significant increase in the antibody response to the encapsulated R32NS1 antigen (2–4). Enhancement of immunogenicity by inclusion of charged lipids have also been observed with certain antigens in liposomes. Negatively charged liposomes produced a better immune response to diphtheria toxoid than positively charged liposomes (27). However, when liposomes were prepared with other antigens, positively charged liposomes worked well, on par with those bearing negative charge (27). Further studies are needed to determine whether negative charges in lipospheres have general abilities to enhance immunogenicity or whether, as with liposomes, charge effects are dependent on individual antigen composition.

An interesting correlation was observed between the liposphere fat to phospholipid (F/PL) molar ratio, particle size, and immunogenicity. Low F/PL ratios ($\leq$0.75) were found to induce the formation of lipospheres of small particle size (70% less than 10 μm in diameter), and this apparently resulted in increased antibody titers (2–4). Among the ratios tested, a maximal level of IgG antibody production was obtained at a F/PL ratio of 0.75, while at larger ratios, decreased antibody production was observed. Although the reason for this phenomenon is unknown, a possible explanation may be the occurrence of better antigen orientation and epitope exposures in the small lipospheres because of higher surface curvature.

Two populations of particles usually coexist in vaccine loaded liposphere formulations, one in the size range of 1 to 10 μm diameter (population A), and a second population with a diameter between 10 and 80 μm (population B). As mentioned previously, the particle size distribution of lipospheres depends on the F/PL molar ratio, and the immune response to liposphere-encapsulated R32NS1 was also dependent on the F/PL ratio. The average size of the particles increases with the increasing F/PL molar ratio. Under conditions where the F/PL ratio is high (≥2.5), the large particle population is predominant (approximately 80% of the particles had an average size of 73 μm); while for F/PL ratios of ≤0.75 most of the lipospheres have a diameter of less than 10 μm (2–4).

To examine the influence of different routes of administration of lipospheres on the immunogenicity of the lipospheres, rabbits were immunized orally or parenterally (by SC, intraperitoneal, intramuscular, and IV routes) with lipospheres made of tristearin and lecithin (1:1 molar ratio) and containing the malaria antigen (3,4). The immune response obtained was followed with time for a period of 12 weeks postimmunization. No antibody activity was found after oral immunization in any of the individual rabbits immunized with liposphere–R32NS1 vaccine formulation. However, rabbit immunization by all parenteral routes tested resulted in enhanced immunogenicity, with increased antibody IgG levels over the entire postimmunization period. The individual rabbit immune response shows that immunization by SC injection was the most effective vaccination route among all the parenteral routes of administration tested (3,4).

Incorporation of lipid A, the terminal portion of gram-negative bacterial lipopolysaccharide, in lipospheres significantly increased the immune response to R32NS1 malaria antigen, resulting in double IgG levels, when compared to the effect of R32NS1 lipospheres lacking the lipid A. The adjuvant effect of lipid A incorporated in lipospheres was observed even after 1600-fold dilution of the rabbit sera. The adjuvant effect of different doses of lipid A in lipospheres was also examined by immunizing rabbits with lipospheres containing R32NS1 and prepared at different final concentrations of lipid A. A gradual increase in IgG antibody titer with increasing lipid A dose was observed. The strongest antibody activity was obtained with lipospheres containing 150 μg of lipid A/rabbit. At higher lipid A dose (200 μg/rabbit), a decrease in ELISA units was observed (2.4).

The preparation and use of polymeric biodegradable lipospheres as a potential vehicle for the controlled release of vaccines was also studied. The immunogenicity of polymeric lipospheres composed of PLD or PCL and containing the recombinant R32NS1 malaria antigen was tested in rabbits after intramuscular injection of the formulations. High levels of specific IgG antibodies were observed in the sera of the immunized rabbits, up to 12 weeks after primary immunization, using a solid phase ELISA assay. PCL lipospheres containing the malaria antigen were able to induce sustained antibody activity after one single injection in the absence of immunomodulators. PCL lipospheres showed superior immunogenicity compared to PLD lipospheres, the difference being attributed to the different biodegradation rates of the polymers.

PLA Lipospheres for the Delivery of LHRH Analogs

The aim of this work was to develop alternative peptide-loaded microspheres using liposphere formulation—a lipid based microdispersion system—to improve the entrapment efficiency and release profile of triptorelin and leuprolide (LHRH analogues) in vitro. Peptides (2%, w/w) were loaded into lipospheres containing polylactic acid (PLA) or poly(lactic-*co*-glycolic acid) (PLGA) with several types of phospholipids. The effects of polymer and phospholipid type and concentration, method of preparation and solvents on the liposphere characteristics, particle size, surface and bulk structure, drug diffusion rate, and erosion rate of the polymeric matrix were studied (11). The use of L-PLA (M_w 2000) and hydrogenated soybean phosphatidylcholine (HSPC) with phospholipid and polymer in the ratio of 1:6 (w/w), was the most efficient composition that formed lipospheres of particle size in the range of 10 μm, with most of the phospholipid embedded on the particles surface. In a typical procedure, peptides were dissolved in NMP and dispersed in a solution of polymer and phospholipids, in a mixture of NMP and chloroform with the use of 0.1% poly (vinyl alcohol) (PVA) as the emulsified aqueous medium. Uniform microspheres were prepared after solution was mixed at 2000 rpm at room temperature for 30 minutes. Using this formulation, the entrapment efficiency of LHRH analogues in lipospheres was up to 80%, and the peptides were released for more than 30 days.

LHRH belongs to hypothalamic hormones that regulate the trophic function of the pitutary gland. The use of LHRH agonists has wide clinical use in the treatment of sexsteroid-dependent diseases, such as breast and prostate cancer, precocious puberty, and endometriosis. Chronic continuous treatment with large doses of LHRH or its agonists is the common therapy in these types of diseases. The development of sustained delivery systems for LHRH agonists, consisting of analogues in biodegradable polymers administered intramuscularly once a month or of implants injected subcutaneously, has greatly facilitated their therapeutic use in humans. The preparation was based on o/w solvent evaporation technique. In a typical preparation, L-PLA (200 mg) and HSPC (35 mg) were dissolved in chloroform (1 mL). A solution of drug (4 mg) in NMP (500 μL) was added to the organic solution to obtain a clear solution of three components. Aqueous solution of 0.25% (w/v) PVA (1 mL) was then poured into the organic phase, with vigorous stirring using a vortex. To the resulting solution, additional aqueous solution of 0.1% (w/v) PVA (5 mL) was added, with continuous stirring. A stirring motor mixed the resulting emulsion at 2000 rpm for 30 minutes. Mannitol (200 mg) was added following by lyophilization into a powder, under reduced pressure (10 mmHg) at –50°C for 24 hours.

Triptorelin release profile from lipospheres prepared from L-PLA, PLGA 50:50, and PLGA 75:25 with HSPC (phospholipids and polymer in the ratio of 1:6) was examined. Only with L-PLA formulation, the drug was trapped at about 80% and released for over 30 days in a fairly constant rate. These lipospheres were fine spherical particles with an average diameter size of 10 μm. Triptorelin release was examined also by using L-PLA with three EPC ratios (1:3, 1:6, and 1:10). The 1:10 ratio formulation had the slower release rate. However, all three formulations released about 60% to 70% of the peptide within three days, which is shorter than aimed for. From these results, it is evident that triptorelin release from L-PLA with HSPC 1:6 ratio is favorable, as it releases the drug for more than 30 days. When EPC was used as the phospholipid component, where EPC and L-PLA were in the ratios of 1:3, 1:6, or 1:10, no difference was found and 80% of the loaded drug was released within the first 48 hours. Figure 2 shows the effect of phospholipid type and concentrations on the cumulative release of leuprolide. When no HSPC was used (X, microspheres), there was a burst release of 80% within the first 24 hours. The addition of HSPC improved the entrapment efficiency up to 50% (1:6 ratio, ●), with a release for over 30 days. When comparing this result with the EPC and L-PLA (1:6, +) formulation, one can see that the HSPC formulation is superior over the EPC formulation in entrapment efficiency and drug release.

Formulations prepared from L-PLA as the polymer component and HSPC as the phospholipid component with 1:6 ratio were the most effective formulations

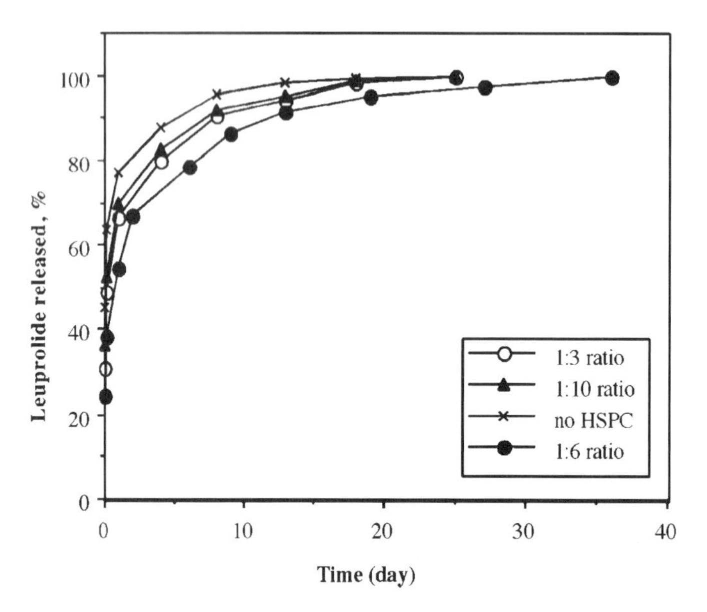

Figure 2 Effect of HSPC:polymer ratio on the cummulative release of leuprolide from L-PLA lipospheres. Release experiment was performed in pH 7.4 phosphate buffer, at 37°C, and analyzed by HPLC. *Source*: From Ref. 10.

resulting in a constant release of triptorelin and leuprolide for 30 days, as shown in Figure 3.

Topical Delivery of Insect Repellent

A common mean for repelling insects consists of applying the compound DEET to the skin. The commercially available liquid DEET formulations contain between 15% and 100% DEET and they are not recommended for use on children. The toxicity of DEET has been extensively reported and related to its high skin absorption after topical administration (28,29). Previous studies have demonstrated that as much as 50% of a topical dose of DEET is systemically absorbed (28,30). In an effort to develop a new topical formulation for DEET that possesses reduced skin absorption as well as an increase in the duration of repellency, we have encapsulated DEET into lipospheres and studied its skin absorption dynamics and duration of action (10). We hypothesized that encapsulation of DEET will reduce its contact surface area with skin and reduce its evaporation rate from the skin surface, resulting in reduced dermal uptake and extended repellent activity.

The liposphere microdispersions containing DEET incorporated in solid triglyceride particles were prepared by the melt method, using common natural ingredients in one step and without the use of solvents. The formulation was preserved by parabens, propylparaben in the oil phase, and methylparaben in the aqueous phase. The average particle size was in the range of 15 μm, and microscopic examination showed spherical particles. The formulations were stable for at least one year when stored at 4°C and 25°C in a closed glass container, and the DEET content, particle size, and viscosity remained almost constant. The residual efficacy of liposphere formulations was evaluated on volunteers, by applying the formulations to the skin and exposing the subjects to mosquitoes (10). The

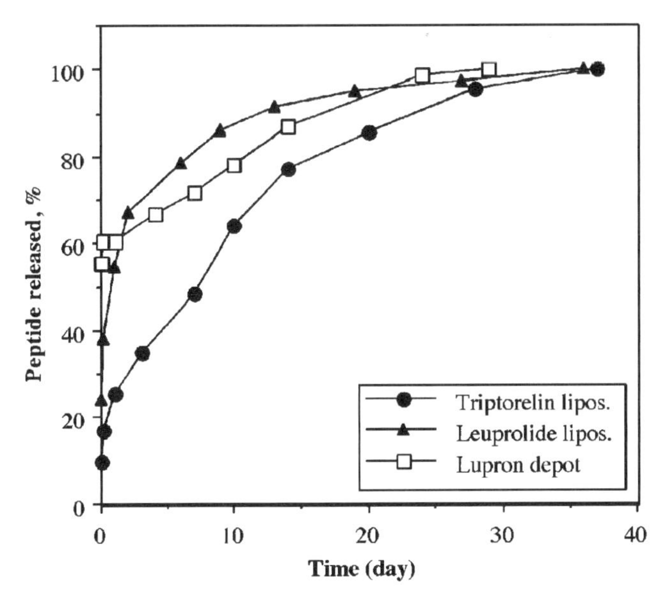

Figure 3 In vitro release of LHRH analogs from lipospheres. *Source*: From Ref. 10.

time of 100% repellency (zero biting) was the index for determining the effectiveness of a formulation.

The formulations were applied on four locations on the arm of volunteers with a dose of 2.5 mg/cm^2 on a total area of 12 cm^2 skin surface. Mosquitoes were placed in a screen bottomed (18 mesh netting, 10 cm^2 exposure area) cylindrical cup. The cup contained fifteen 5- to 15-day-old female mosquitoes that displayed host-seeking behavior, and had access to the skin through the netting. The forearm was placed on the mosquito netting for 10 minutes every 30 minutes and the number of biting mosquitoes (evident by a blood meal) were recorded. Prior to any efficacy experiment, the mosquitoes were tested on untreated skin to confirm their host-seeking behavior. Two mosquito species were tested: *Aedes aegypti* and *Anopheles stephensi*, both aggressive biters. The results of this experiment are given in Table 3. The formulations were repellent for a minimum of 2.5, 3.5, and 6.3 hours for the 6.5, 10, and 20% DEET-containing lipospheres, respectively. The DEET-free formulation (control) and the untreated groups did not show any activity against *A. aegypti*, and the 10% DEET solution in alcohol was repellent for about 1.5 hours.

The bioavailibility of DEET from a 10% ethanol solution was 45%, while the bioavailibility from DEET–lipospheres was only 16%, a three-fold reduction in the amount of DEET absorbed. About 74% of the IV administered dose was collected in the urine, and 39% and 19% of the topically administered doses were collected for the alcoholic and liposphere formulations, respectively. Assuming that the error in urine collection is similar in all experiments, the difference in radioactivity contained in the urine after topical administration of the liposphere dosage form is about 50% of that of the alcoholic dose, which corresponds to the blood bioavailability calculations. The total amount of DEET recovered from skin (washing of residual dose and extraction from skin) was similar for both formulations, indicating that both formulations were similarly exposed to the skin and thus the results are comparable.

A related application was the development of a liposphere-based moisturizer. Plain liposphere preparation made of tristearin and soybean phospholipid without any active agent were mixed with hydroxypropyl cellulose gel and tested as moisturizer. It was found in a human test that the moisturizing effect lasted for at least eight hours. Addition of titanium oxide to this formulation showed useful sunscreen effect, with ease of application and retention on skin. Lipospheres of particle size below 100 nm loaded with active agents have been shown to penetrate deep into the skin.

Nanolipospheres for Cell Targeting of Anticancer Drugs

The potential use of lipospheres loaded with paclitaxel, to overcome tumor cells acquired resistance to the drug, was investigated (31). It was assumed that, if absorbed by the cells, the encapsulation of paclitaxel will prevent the rapid expulsion of the drug outside the cell and a sufficient cytotoxic level of drug concentration will be maintained in cell plasma. Two liposphere formulations, one based on tricaprin and the other on PCL, were compared with liposomes of the same composition but without the core component (tricaprin or PCL). The formulations were prepared by the solvent method and extruded through a series of submicron filters to yield nanoparticles, each with a size of 200 nm. The rate and amount of uptake of particles by cells were determined using lipospheres or liposomes containing phycoerythrin (PE) fluorescein. The rate of uptake was followed by fluorescence microscopy visualization of the amount of fluorescence accumulated in cells, or by measuring the amount of fluorescence in each cell using the FACS system. The formulations were incubated with wild-type F-98 glioma cell-line or with F-98 cell-lines, which were resistant to 1×10^{-8}, 7.5×10^{-8}, and 8.0×10^{-8} mM taxol®. The results indicate that: (i) it takes about 24 hours of incubation of particles with cells, to reach saturation of particle uptake, (ii) cells accumulate higher concentrations of liposomes than lipospheres, and (iii) cells which are more resistant to taxol, accumulate higher taxol concentrations upon incubation with taxol–liposomes and taxol–lipospheres than on incubation with the free drug.

The cytotoxicity of liposphere or liposome encapsulated taxol on the % cell survival of two cell-lines (wild-type and F-98/1×10^{-8} resistant cells) after six hours of treatment with drug, and further incubation of cells up to 72 hours, was studied. F-98 cells were incubated with a range of drug concentrations and different preparations for six hours, and then washed with a fresh medium and incubated for an additional 66 hours. The number of cells in each plate was counted daily. The data indicate that taxol encapsulated in liposomes or lipospheres had a higher cytotoxic effect than free taxol. The results also demonstrate that while there is no significant difference between the cytotoxic effect of free taxol or the taxol encapsulated in liposomes on the wild type–cells, there is a significant difference in the effects on resistant cell-lines. Taxol encapsulated in liposomes is about 30% and 50% more cytotoxic than free taxol in cells resistant to 10^{-8} and 10^{-7} mM taxol, respectively. These preliminary results indicate that both lipospheres and liposomes were effective to overcome drug resistance, the liposomes being more effective.

The blood circulation time of nano-lipospheres was compared with that of liposomes (32). Lipospheres and liposomes of particle size between 100 and 200 nm containing radioactive cholesteryl hexadecyl ether were injected to mice and the tissue distribution was determined by following the radioactivity content in blood, liver, and spleen at 1, 3, 8, and 24 hours. The cholesteryl hexadecyl ether in organs was extracted with chloroform and the radioactivity in the extract was determined. The liposphere formulation was excreted faster than liposomes and was concentrated in the liver rather than in the spleen for liposomes.

To increase the retention time of lipospheres in the blood circulation, lipospheres were coated with PEG having a molecular weight of 5000; this has been shown to increase blood circulation (33). Liposphere formulation made from phospholipid containing 10% (w/w) PE were reacted with aldehyde terminated methoxy PEG to form PEG coated lipospheres. The imine-bound PEG was reduced to the amine linkage, which is more stable. Preliminary in vitro experiments indicate that PEG coated lipospheres are stable in serum (32).

In Situ Formation of Lipospheres for Oral Delivery of Cyclosporin

Cyclosporin was incorporated into lipospheres with varying particle size and nanoparticles, and the bioavailability of these formulations was tested in humans (12). Dispersible concentrated oil solution formulations of cyclosporin that, upon mixing in water, spontaneously form a nanodispersion of lipospheres were developed. The concentrated oils are clear solutions, composed of the drug, a solid triglyceride, a water miscible organic solvent, and a mixture of surfactants and emulsifiers.

Cyclosporin A (CyA) is a first-line immunosuppressive drug used to prevent transplant rejection and to threat autoimmune diseases. CyA is a highly lipophilic molecule, with poor absorption from the gastrointestistinal tract. Because of its limited water solubility, cyclosporin has been given in an oil solution, microemulsion, and complexes with cyclodextrin. CyA is available for clinical use as an oil solution, encapsulated in soft gelatin capsule that forms a microemulsion in the stomach. This medication (Neoral®) shows 25% oral bioavailability. Lipospheres have been investigated as an oral delivery system for CyA with improved gastrointestinal absorption. An oily solution of cyclosphorine was prepared in a mixture of fats, surfactants, and water-miscible organic solvent. This solution was loaded into gelatin capsules and administered orally. When gelatin capsules containing the concentrated solution are swallowed, their content is released to the stomach and the gastric juices spontaneously form a nanodispersion. In this method, the active agent is dissolved in a water miscible organic solvent, which is appropriate for oral use. Examples of such solvents are: NMP, PEG, and propylene glycol. Phospholipids are dissolved in this mixture. Other ingredients (surfactants, emulsifiers, and stabilizers), are added into the mixture and dissolved to form an oily, transparent, and uniform solution. A typical composition and method of preparation is as follows: 300 mg phospholipid is dissolved in 600 μL *N*-methylpyrrolidone, then 150 mg cyclosporin, 300 μL Tween® 80, and 300 mg tricaprin were dissolved. The particle size of the concentrated oil solution was determined by adding three drops of the solution, mixed in 5 mL distilled water at 37°C, and then the particle size of the obtained suspension was determined at 37°C for 200 seconds by Coulter N4 MD Submicron Particle Size Analyzer. TEM examination of a typical liposphere formulation showed spherical particles. To obtain a dried powder of the cyclosporin formulation suitable for reconstitution, the oily, dispersible concentrate was suspended in mannitol solution at different concentrations,> vortexed, and dried by lyophilization. The obtained powder was resuspended again with distilled water, at 37°C, resulting in a uniform nanodispersion of a similar particle size. The bioavailability in humans was conducted on healthy volunteers who received similar diets and were under similar conditions from the evening before

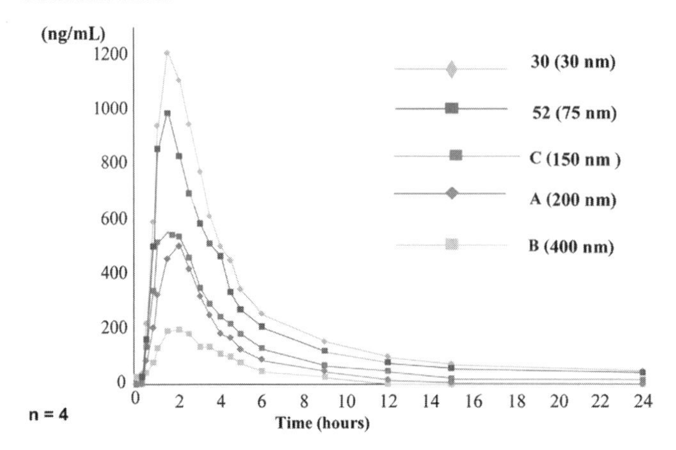

Figure 4 Cyclosporin absorption from developed formulations as a function of particle size.

and to the end of the experiment. At the start of the study, a single dose of 200 mg of cyclosphorin in various formulations was administered to fasting volunteers and blood samples were thereafter withdrawn during 24 hours postdose intake. The mean result of six healthy volunteers is shown in Figure 4. A concentration–time curve was constructed for each volunteer for each experiment. The observed maximal concentration was recorded as C_{max}. The area under the curve (AUC) was computed for each volunteer. Pharmacokinetic profile of preparations was determined and compared with that of a commercial product: Sandimmune Neoral® 100 mg soft gelatin capsules (Sandoz).

The decrease in particle size had a significant effect on the bioavailability. The highest blood levels and the AUC were obtained for the 25 nm, forming dispensible concentrate with a linear decreasing AUC and C_{max} with the increase in particle size. The blood level of the lowest particle size formulation was even slightly higher than that of the reference Neoral.

SUMMARY

Lipospheres are solid, water insoluble nano- and microparticles composed of a solid hydrophobic core having a layer of a phospholipid embedded on the surface of the core. The hydrophobic core is made of solid triglycerides, fatty acid esters, or bioerodible polymers containing the active agent. Liposphere formulations were effective in delivering various drugs and biological agents including: local anesthetics, antibiotics, vaccines, and anticancer agents with a prolonged activity of up to four to five days.

The feasibility of polymeric biodegradable lipospheres as carriers for the controlled release of a recombinant malaria antigen and LHRH was also demonstrated. Polymeric lipospheres containing R32NS1 malaria antigen were able to induce very high levels of antibody activity after one single injection, in the absence of immunomodulators. LHRH was constantly released for more than 30 days. New liposphere formulations that form in situ nanoparticles of particle size below 100 nm were found effective in oral delivery cyclosporin in humans.

REFERENCES

1. Domb AJ. Lipospheres for controlled delivery of substances. U.S. Patents 5,188,837; 5,227,165; 5,221,535; 5,340,588.
2. Amselem S, Domb AJ, Alving CR. Lipospheres as a vaccine carrier system: effect of size, charge, and phospholipid composition. Vaccine Res 1992; 1:383–395.
3. Amselem S, Alving CJ, Domb AJ. Biodegradable polymeric lipospheres as vehicles for controlled release of antigens. Polym Adv Technol 1993; 3:351–357.
4. Amselem S, Alving C, Domb A. Lipospheres for the delivery of vaccines. In: Bernstein H, Cohen S, eds. Microparticulate Systems for Drug Delivery. New York, NY: Marcel Dekker Inc., 1993:399–434.
5. Amselem S, Domb AJ. Liposphere delivery systems for vaccines. In: Bernstein H, Cohen S, eds. Microparticulate Systems for Drug Delivery. New York, NY: Marcel Dekker Inc., 1996:149–168.
6. Domb AJ. Long acting injectable oxytetracycline-liposphere formulations. Int J Pharm 1995; 124:271–278.
7. Hersh EV, Maniar M, Green M. Anesthetic activity of the lipospheres bupivaccine delivery system in the rat. Anesth Prog 1992; 39:197–200.
8. Masters D, Berde C. Drug delivery to peripheral nerves. In: Domb AJ, ed. Polymer Site-Specific Pharmacotherapy. Chichester, U.K.: Wiley, 1994:443–455.
9. Masters DB, Domb AK. Liposphere local anesthetic timed-release for perineural site application. Pharm Res 1998; 15:1038–1045.
10. Domb AJ, Marlinsky A, Maniar M, Teomim L. Insect repellent formulations of N,N-diethyl-m-touamide (deet) in liposphere system. J Am Mosq Control Assoc 1995; 124:271–278.
11. Rasiel A, Sheskin T, Bergelson L, Domb AJ. Phospholipid coated poly(lactic acid) microspheres for the delivery of LHRH analogues. Polym Adv Technol 2002; 13(2):127–136.
12. Bekerman T, Golenser J, Domb A. Cyclosporin nanoparticulate lipospheres for oral administration. J Pharm Sci 2004; 93(5):1264–1270.
13. Amselem A, Yogev A, Zawoznik E, Friedman D. Emulsions, a novel drug delivery technology. Proc Int Symp Control Rel Bioact Mater 1994; 21:668–669.
14. Muller RH, Mader K, Gohla S. Solid lipid nanoparticles (SLN) for controlled drug delivery—a review of the state of the art. Eur J Pharm Biopharm 2000; 50:161–177.
15. Fenske DB. Structural and motional properties of vesicles as revealed by nuclear magnetic resonance. Chem Phys Lipids 1993; 64:143–162.
16. Barenholz Y, Amselem S. Quality control assays in the development and clinical use of liposome-based formulations. In: Gregoriadis G, ed. Liposome Technology. Vol. 1. 2nd ed. Boca Raton, FL: CRC Press, 1993:527–616.
17. Maniar M, Burch R, Domb AJ. In vitro and in vivo evaluation of a sustained release local anesthetic formulation. AAPS Meeting, Washington, D.C., November, 1991.
18. Dubuisson D, Dannis SG. The formalin test: a quantitative study of the analgesic effects of morphine, mepetidine, and brain stem stimulation in rats and cats. Pain 1977; 4: 161–174.
19. Haunskaar S, Fasmer OB, Hole K. Formalin test in mice, a useful technique for evaluating mild analgeics. J Neurosci Meth 1985; 14:69–76.
20. Oukessou M, Uccelli-Thomas V, Toutain PL. Pharmacokinetics and local tolerance of a long-acting oxytetracycline formulation in camels. Am J Vet Res 1992; 53: 1658–1662.
21. Palham MJ. Liposome phospholipid. Toxicological and environmental advantages. In: Brown O, Korting HC, Maibach HY, eds. Liposome Dermatics. Berlin, Heidelberg: Springer Verlag, 1992:57–68.
22. Landoni MF, Errecalde JO. Tissue concentrations of a long-acting oxytetracycline formulation after intramuscular administration in cattle. Rev Sci Tech 1992; 11:909–915.

23. Adawa DA, Hassan AZ, Abdullah SU, Ogunkoya AB, Adeyanju JB, Okoro JE. Clinical trial of long-acting oxytetracycline and peroxicam in the treatment of canine ehrlichiosis. Vet Q 1992; 14:118–120.
24. Alving CR. Liposomes as carriers of vaccines. In: Ostro MJ, ed. Liposomes: From Biophysics to Therapeutics. New York: Marcel Dekker Inc., 1987:195–218.
25. Eldridge JH, Staas JK, Meulbroek JA, McGhee JR, Tice TR, Gilley R. Biodegradable microspheres as a vaccine delivery system. Mol Immunol 1991; 28:287–294.
26. Rickman LS, Gordon DM, Wistar R Jr, Krzych U, Gross M, Hollingdale M, Egan JE, Chulay JD, Hoffman SL. Use of adjuvant containing mycobacterial cell-wall skeleton, monophosphoryl lipid A, and squalene in malaria circumsporozoite protein vaccine. Lancet 1991; 337:998–1001.
27. Allison AC, Gregoriadis G. Liposomes as immunological adjuvants. Nature 1974; 252:252.
28. Clem JR, Havemann DF, Raebel MA. Insect repellent (N,N-diethyl-m-toluamide) cardiovascular toxicity in an adult. Ann Pharmacother 1993; 27:289–293.
29. Lipscomb JW, Kramer JE, Leikin JB. Seizure following brief exposure to the insect repellent N,N-diethyl-m-toluamide. Ann Emerg Med 1992; 21:315–317.
30. Snodgrass HL, Nelson DC, Weeks MH. Dermal penetration and potential for placental transfer of the insect repellent N,N-diethyl-m-toluamide. Am Ind Hygiene J 1982; 43:747–753.
31. Gur A. Taxol incorporated in nanoliposphere formulations against taxol resistant cells. M.Sc. thesis, The Hebrew University of Jerusalem, Jerusalem, Israel, 1994.
32. Lichtman-Teomim L. Injectable systems for the delivery of insoluble anticancer agents. M.Sc. thesis, The Hebrew University of Jerusalem, Israel, 1994.
33. Woodle MC, Newman MS, Martin FJ. Liposome leakage and blood circulation: comparison of absorbed block copolymers with covalent attachment of PEG. Int J Pharm 1992; 88:327–334.

11

Pharmaceutical Aspects of Liposomes: Academic and Industrial Research and Development

Rimona Margalit and Noga Yerushalmi
Department of Biochemistry, The George S. Wise Faculty of Life Sciences, Tel Aviv University, Tel Aviv, Israel

INTRODUCTION

In the last few decades liposomes have been developed for pharmaceutical applications in very different directions, from drug-delivery systems (DDS) to diagnostic tools, vaccine adjuvants, artificial blood, gene therapy vehicles, etc. However, the approval and use of a few drugs formulated in liposomes were introduced in the last decade. The scientific literature is rich with comprehensive reviews of liposomes as DDS (1–16). Each review presents a unique and specific point of view, which can range from the strictly physicochemical, through various biological levels, to the clinical. Furthermore, the prospect of liposomes as pharmaceutical products is inherent in these reviews even when not specifically identified and addressed. Given this background, it was deemed essential to first address a question that might be raised by the reader: "Why yet another manuscript on liposomes as pharmaceutical products?"

Liposomes as DDS are among the research topics that are being vigorously investigated in both academic and industrial laboratories, with different outlooks but common goals and end products. Addressing fundamental questions while being continuously innovative are perceived to be among the responsibilities of the academicians in this field. Addressing developmental issues such as scale-up, shelf-life, dosage forms, long-term stability, and cost-effective production are perceived to be among the responsibilities of the industrial scientists. Yet the R&D of liposomes as DDS encompass industry-oriented issues that should have been taken into account at the basic academic stage, as well as questions that arise at the industrial stage that have distinct "basic science" components.

Stemming from this premise, the objectives of this chapter are to address those issues and questions and attempt to provide an integrated view that would (i) be of use and contribute to the mutual understanding of the topics most critical to investigators in both industry and universities, and (ii) serve as a guideline for newcomers in the field. It is hoped that, through this, this chapter will become a modest addition to, rather than a repetition of, the existing literature.

To lay out a common ground for the discussion of liposomes as DDS, the first part of this chapter provides a brief definition of liposomes and discusses the unmet therapeutic needs that justify their pursuit as DDS and their advantages and drawbacks. The subsequent parts of this chapter attempt to provide the proposed integrated approach, addressing the following issues: (i) types of liposomes, with respect to scale-up, production, and cost, (ii) preparation and production of sterile and pyrogen-free liposomes, (iii) shelf life, stability, and dosage forms, (iv) targeted (mostly surface-modified) liposomes as exciting research tools and as pharmaceutical products, and (v) characterization of liposome–drug systems (such as encapsulation efficiencies, drug-release kinetics, and biological activity) for academic studies and for industrial quality assurance. Conclusions in the final part are made from the academician's point of view, but will hopefully be relevant to all liposome investigators; a short summary of today's approved drug formulations is also included.

The authors are well aware that the line of division between academic and industrial liposomal R&D is quite often indistinct. Investigators in academic research could be engaged in questions on liposomes as pharmaceutical products, whereas, more so, investigators in industry could be engaged in basic liposome research. To avoid any confusion or resultant grievances (however unintentional), it is stressed here that throughout this chapter the terms *academic* and *industrial research* will be used to indicate the nature and tendency of the research (i.e., basic or applied) and not the physical definition of the environment or community where it is performed.

LIPOSOMES: DEFINITION, NEEDS FOR, AND OUTLINE OF THEIR ADVANTAGES AND DRAWBACKS

Liposomes: A Family of Structurally Related Microparticles

Liposomes, frontline microparticulate carriers investigated and developed for drug delivery, are artificial microscopic and submicroscopic particles made of lipids and water alone that were developed 30 years ago (1). A variety of liposome species with respect to shape, type, size, and composition have been developed in the course of the last three decades, extending this type of particle into a family of structurally related delivery systems. For the novice in the field, it is worth noting that in the early days of liposomes, there was an ongoing debate between the terms *vesicle* and *liposome*, the legacy of which is present in the names, abbreviations, and acronyms used for various liposome types.

Multilamellar vesicles (MLV) are the oldest liposome species, and they have been described frequently enough that only a brief summary is merited here (1,2,5–7,17, and references therein). They are composed of concentric shells of lipid bilayers, with water between shells and an inner aqueous core. Liposomes of this type form spontaneously on the proper interaction of lipids with water, provided the right choice of lipids and technical conditions have been met. A typical MLV will have 8 to 15 concentric shells, and will run from (roughly) 0.5 to several micrometers in diameter. Consequently, a typical MLV preparation will be quite heterogeneous in terms of liposome sizes.

Unilamellar vesicles (ULV) are composed, as their name indicates, of a single lipid bilayer and an inner aqueous core (1,2,6–8,10,11,18,19 and references therein). Depending on the method of preparation, ULV can be made in various size ranges, from as small as 20 nm to several micrometers in diameter, with a significantly smaller size distribution within a preparation compared with MLV.

Both MLV and ULV can be made from a wide (although not infinite) range of lipid and lipid mixtures and can accommodate hydrophilic and hydrophobic drugs in their aqueous and lipid compartments, respectively (1–11,17–20). Furthermore, with careful selection of liposome type, encapsulation of matter that is as small as the lithium ion up to macromolecules and as large as genetic material (of several hundred thousand Daltons) can be achieved.

By scanning the four decades of liposome literature it is quite clear that besides the abbreviations MLV and ULV, there are many more names and types of liposomes, the distinction among and the classification of which are frequently based on methods and/or devices of preparation with/without the contribution of the resultant size-related properties. In fact, this "horn of plenty" can be a source of confusion to the novice in the field and a burden on both veterans and newcomers, as exemplified by the two cases outlined below.

The first case concerns liposomes that have been classified as large unilamellar vesicles (LUV). One of the major methods for producing such liposomes is the reverse-phase evaporation technique; such liposomes have been abbreviated REV and are usually classified as LUV (18). A REV preparation can have a relatively wide size distribution up to 600 nm in diameter. Subjected, as frequently recommended, to postpreparation fractionation steps, the size distribution of a REV preparation can be quite narrow together with a reduction in the dominant size. Yet, regardless of whether such fractionation has or has not been subjected to postpreparation steps, these systems will all be referred to as REV and considered to be LUV. Other approaches for producing unilamellar liposomes are based on extrusion devices, such as the LIPEX extruder (22). Liposomes produced with this device have been named LUVET and are also referred to as LUV; typical preparations of this type have a relatively narrow size distribution. Sizewise, they can run from 30 to several hundred nanometers in diameter, depending on the pore sizes of the filters used.

The second case concerns small unilamellar vesicles (SUV), which is the acronym for small unilamellar vesicles as well as for sonicated unilamellar vesicles. SUV with a size range of 20-nm diameters are usually obtained through the use of probe sonicators and are one of the oldest types of liposomes. One also finds the acronym SUV in use for unilamellar liposomes that are in size ranges of 40, 60, and even up to 100 nm that have been made by a variety of methods other than probe sonicators. Yet another device for the production of unilamellar liposomes is the Microfluidize, which yields (depending on operating conditions) liposome preparations with a narrow size distribution that (depending on operating conditions) can range in diameter from under 100 to several hundred nanometers (23). Microemulsified liposomes (MEL), such liposomes can qualify as either SUV or LUV.

Besides the multiplicity in the field with respect to liposome names and classifications that these few examples illustrate, there is also an inherent weakness in using methods and devices for liposome preparation as a major criterion for classification. Clearly some methods and devices can become obsolete, and others have not yet been invented.

In an attempt to ease the current liposome name/type situation and at the same time offer an informative classification, a simplified and straightforward two-tier approach that eliminates references to preparation methods and/or devices is hereby proposed. This two-tier approach is used throughout this chapter with the hope that it will find favor with and be of use to both veterans and novices in the field.

The first tier, which is the primary classification, is based on a major structural difference between liposome species which is simply the number of lamellae. The line of division is set between one and more than one lamella. Based on this premise, there are only two major types of liposomes, for which there is no need to use names other than the traditional terms MLV and ULV. Liposome preparations, whether MLV or ULV, might also contain a (usually small) share of liposomes denoted oligolamellar that have few, such as two to three, lamellae. These are considered to be a minor type of liposomes, as to date there have been neither reports on applications for which such liposomes are specifically sought, nor procedures for their preparation as the major dominant particle.

The second tier, which will be illustrated below by several examples, is a sub-classification according to the dominant size range and/or size distribution, whichever is the more relevant. For example, in a reported study, the investigated liposome type would be simply classified as ULV 80 ± 30 nm. These details would allow the reader to infer that this preparation is quite homogeneous and truly unilamellar and that the majority of these liposomes are sufficiently small so that, if endowed by proper receptor affinity, they can also be endocytosed by nonphagocytic cells through the coated pit mechanism (1,2,7). Identifying an investigated liposome preparation as ULV 650 ± 200 nm would be a clear indication to the reader that this preparation probably contains not only unilamellar but also oligolamellar liposomes (i.e., liposomes with more than one lamella but less than the number designated to define MLV). A preparation identified as MLV 0.5 to 2.5 μm would clearly indicate that this is a rather heterogeneous liposome preparation that has probably not been subjected to any fractionation procedures and contains particles that can differ fivefold in diameter.

Besides offering simplification and being informative, it is argued that this classification will leave descriptive terms such as small or large to the personal inclination of the investigators and the readers, and it will help clarify differences between liposomes and other lipid-based particles that are under development as delivery systems. The latter are, for the most, proprietary technologies such as plurilamellar liposomes that are reported to contain well over 100 lamellae and to range in size up to 100 μm; multivesicular liposomes (DepoFoam), also referred to as MLV, each particle of which consists of several wall-sharing large unilamellar liposomes, with a size that can reach 100 μm; solvent dilution microcarriers; and transferosomes developed specifically for the transdermal administration to intact skin (24–29).

Deficiencies in Treatment with Free Drugs that Justify the Pursuit of DDS and Consideration of Liposomes for the Task

Whether viewed from the clinical, the physiological, the financial, or any other of the many aspects involved, treatment with drug-loaded delivery systems is substantially more complex than with free drug. Therefore, the existence of therapeutic needs that are truly unmet with the free drug and the ability of the liposomes to provide significant improvements in clinical outcomes that can be developed into established treatment modalities justify the investment in research and development of liposomes.[a]

[a] Although most of the issues discussed with respect to the needs for liposomes apply to other drug-delivery system also, the discussion will refer to liposomes alone, as they are the topic of this chapter.

By tracing the fate of a drug administered in its free form, dissolved or dispersed in an appropriate vehicle, one can get a glimpse of those unmet needs that justify the pursuit of liposomes as DDS. Examples will be given below for two cases, from which the general picture can be inferred, not only for systemic but also for nonsystemic (such as topical and regional) administrations.

The first example is taken from tumor treatment by chemotherapy, where the deficiencies of treatment with free drug come from the effects of the drug on the biological environment and vice versa. It is well known that conventional chemotherapeutic drugs, such as doxorubicin, Vinblastine, Vincristine, 5-flourouracil (5-FU), and many others are virulent substances, a property which is an asset with respect to drug effects on malignant cells but is disastrous when the drug interacts with any other (healthy) cells (30). Most frequently, such drugs are administered in their free form systemically. Lack of targeting at the tumor sites results in the indiscriminate distribution of the drug over the whole body, reducing the efficacy of the treatment and at the same time leading to undesirable side effects and toxicity. Free-form administration of drugs exposes them to the elements of the biological environment, and they are often prematurely cleared, which is especially critical for drugs that act at a specific stage of the cell cycle. The drugs (in free form) are also susceptible to inactivation and/or degradation and to scavenging by endogenous carriers such as serum albumin and serum lipoproteins, which can aid as well as abet drug access to the target zones.

The second example concerns wound healing, where growth factors, such as epidermal growth factor (EGF), fibroblast growth factors α and β, platelet-derived growth, transforming growth factors α and β3, and many others, are under development as topical therapeutic agents for the acceleration and stimulation of the self-healing process in wounds and burns (31–39). Current dosage forms used in both basic and developmental growth factor studies that are appropriate for the future treatment of patients center on free growth factor in vehicles such as solutions of saline or other physiological buffers (poured onto the wound or soaked into a gauze dressing), cellulose gels, and collagen sponges (31–42). The use of such vehicles and procedures corresponds to the immediate (or almost immediate) exposure of the total growth factor dose to the wound. As in the previous example, the mutual effects of the drug and of the biological environment undermine successful therapy with the free drug.

Growth factors are especially vulnerable to the biological environment, which is enzymatically hostile to polypeptides. In addition to the enzyme-catalyzed degradation, growth factors that are free in the wound are subject to continuous clearance from the wound area and can be scavenged through binding to the various particulate and soluble wound fluid components. Owing to their susceptibility to proteolysis, the growth factors are also prevented (for the most part) from exerting any self-targeting within the wound area toward their sites of action (i.e., their specific receptors). The small share of the dose that does reach the target form is often too low to affect a significant difference, especially as growth factors are agents that act at a specific stage of the cell cycle (43–45). Hence, effective therapy also requires a continuous supply of active growth factor near enough to its sites of action for a sufficient duration (33,35,44). The other side of the coin, namely, the detrimental effects of growth factors on the biological system, is seen on attempts to increase the bioavailability at the target by dose escalation. It is well known, and not only for growth factors but also for hormones in general, that substantially high (local) concentrations, even if transient, can result in adverse effects and toxicity (35,43).

To wrest some benefit from the application of growth factors despite the deficiencies described above, current treatment regimens use multiple dosing at frequent intervals; for example, twice daily for 70 to 100 days (45). Such regimens prevent the implementation of state-of-the-art approaches to good wound management that advocate (i) minimal interference with the wound, (ii) maintenance of a moist wound environment, which was found to be most conducive to the self-healing, and (iii) the use of occlusive dressings that are changed once every few days (47–54). Obviously, treatment regimens that interfere with the healing will detract from any benefit that a medication for healing can provide.

Although the deficiencies of free-drug administration, defined above, clearly call out for ways and means to overcome them, and in principle a delivery system is among the most viable solutions, such systems need to meet particular attributes to qualify for the task. These attributes can be divided into three categories. The first two categories that have been amply discussed are considered to be within the realm of basic science, and comprise qualities that the carrier needs to function (denoted category I) and to avoid replacing one set of problems by another (denoted category II). The third (denoted category III) comprises qualities required to develop the drug–carrier systems into pharmaceutical products that can then be implemented into established treatment modalities.

Properties included in category I are (i) the ability to provide mutual protection of the carried drug and the biological environment, (ii) effective drug targeting, including the ability to reach and access the target zone, (iii) stability on route from the site of administration to the site(s) of drug action, and (iv) the ability to perform as a sustained-release drug depot. To meet the attributes that make up category II, the carrier should be biocompatible, biodegradable, nonimmunogenic, and nontoxic. A major share of the qualities that is dominant in category III has to do with the production of the drug–carrier systems. The carrier should be of the type that its preparation procedures (including drug loading) can be scaled up, yield a sterile and pyrogen-free product, and meet the requirements of quality assurance. Another major part of category III properties concerns postproduction requirements, such as dosage forms that can provide long-term shelf life and high stability. Needless to say, to ensure that as wide as possible patient population could have access to this new therapy, all cost-effective issues should be considered within category III, such as the cost of production itself, sources and cost of raw materials, shelf life length, expenses of storage, and others.

Among DDS that can address a major share of the unmet needs encountered when treatments are with free drug, liposomes are frontline candidates. Their advantages for the task have been well documented and include the following: Liposomes can meet that required mutual drug/biological environment protection. Liposomes are biocompatible and biodegradable and, to a significant extent, are also nonimmunogenic and nontoxic. Because of their versatility in terms of type, size, and lipid compositions the encapsulation and delivery of a wide range of drug species become possible, and liposomes can act as sustained-release depots. Despite all these qualities, significant problems in the implementation of therapies with liposomes still exist, particularly with respect to targeting, the responses of the biological system to liposomes in the blood stream, the stability of both particle and drug in vivo, the in vivo rates of drug release, shifts in rather than prevention of drug toxicity, and long-range effects of chronic liposome administration. The advantages of liposomes as well as these unresolved issues are addressed in the following sections, taking into consideration not only these category I and II aspects but also those of category III that are critical for the evolution of liposomes from entities under research and development into entities that are pharmaceutical products.

SELECTION OF THE LIPOSOME TYPE/SPECIES: VIEWS AND CRITERIA FROM ACADEMIC AND INDUSTRIAL RESEARCH AND THEIR PROPOSED INTEGRATION

Liposome Type, Size, and Production Issues

One of the initial decisions that needs to be made on venturing into the development of a liposomal drug-delivery system for a specific therapeutic case centers on the type or types of liposomes that will be explored. Whether that selection will be made from the abundance of existing liposome species, or the invention of new liposome species, is contemplated, the criteria and considerations that are used to make that decision constitute one of the major points of diversion between investigators in academic and industrial research.

The selection of liposome types in industry will be made following, in general, the guidelines for pharmaceutical products, and would be relatively free from the limitations of accessibility to devices for liposome production. The main concerns would be production and product issues. As discussed under category III (see section "Deficiencies of treatment with free drugs that justify the pursuit of DDs and considerations of liposomes for the task"), a critical criterion would be a liposome production method/procedure that could be scaled up and executed under GMP rules and regulations. Other critical criteria would stem from the requirements for a product that would have batch-to-batch reproducibility and be stable, sterile, and pyrogen free. As also discussed above, the liposome production method had to be cost effective, also taking into consideration sources, supplies, and cost of quality raw materials. While considering cost and environmental factors the issue of waste disposal, in particular, for those methods of liposome production that involve the use of organic solvents would also be taken into account. Needless to say, while selecting the liposome type the need for acceptable shelf life and dosage forms would also be taken into consideration.

The academician is perceived to have a considerably higher level of freedom, being able to choose, in principle, from all types of liposomes for which procedures of preparation are in existence. Physiological and clinical aspects are the drive for selection criteria. These include the specific therapy in question, drug properties, the route of liposome administration with all its inherent issues, and the mutual effects of the drug and the biological system. However, as exemplified by the following case, other criteria often become the deciding factors.

Liposomes have become such a common and prevalent delivery system that an investigator with expertise and a primary interest in using a particular drug—e.g., a 6000-Da polypeptide—seeks it. Screening the "horn of plenty" with respect to liposome types and species, certain dogmas in the field might drive that investigator not to consider MLV, although these liposomes might well fit his therapeutic objectives and the other requirements. Concerns of stability and retention of the native active conformation could lead that investigator to also exclude ULV made by probe sonicator. This selection can pose severe limitations on the efficiency of encapsulation and on other properties of the system that can result in low efficacy of treatment that would discourage the investigator from any further pursuit of the liposomal approach. Even if those small ULV would prove to be satisfactory with respect to physicochemical properties and successfully functional in vitro and/or in vivo, the question of whether they could be developed into a pharmaceutical product would still be faced. The same situation would have to be faced with numerous other liposome types that are currently used in academic research, such as those made by reverse-phase evaporation, ethanol injection, detergent dialysis, French press, detergent removal through gel exclusion chromatography, and others (2,3,5–7,17–19,22,23).

It is stressed that for objectives other than drug delivery, especially those where liposomes serve as models of cell membranes or as a solubilization media for lipophilic matter, most of the limitations discussed in this section are of no consequence. However, cases in which efforts and resources have been invested in developing liposomal drug systems that prove to be successful in animal model studies and even in clinical trials, with liposome types that cannot be produced as pharmaceutical products, are not far fetched. For such cases, the progress of that system into an established treatment modality would be (at best) significantly delayed until similar positive results would be obtained with another, product-suitable liposome type. Needless to say, a major share of the already completed studies would have to be repeated.

Hence the conclusion drawn is that, in the selection of liposome type, the burden of change in current considerations is mostly on the shoulder of the academician. It is proposed that the academician integrate the industrial criteria into his or her initial selection process, even if the requirements of a planned study could be accomplished with the use of small-scale, nonsterile, preparations that would be discarded at the end of a day's work. A few examples of implementing the integrated view are offered below. A comprehensive list of selected liposome types is deemed to be beyond the scope of this chapter. This is not only due to the abundance of liposome production methods but also to the realization (already introduced in the previous section) that this is a dynamic field where new species and/or production methodologies are expected to be continuously invented.

The first example concerns MLV, which have fallen out of favor with many investigators. Granted, there definitely are treatments and applications for which MLV would not do. For treatments of tumors and infectious diseases in which the targets are outside the reticuloendothelial system (RES) and where intravenous administration with long-term retention of the liposomes in circulation is required, MLV would not do. On the other hand, MLV can be eminently suitable for some therapies that require nonsystemic administration as well as for therapies that require intravenous administration and fast uptake of the liposomes by phagocytic cells in circulation can be easily made in the laboratory and can also meet the criteria for pharmaceutical products. The latter is not meant to imply that all industrial issues with respect to MLV production have been resolved, but that their resolution is feasible. A pertinent example of a current problem that needs resolution is the already mentioned issue of organic solvents, such as methanol, chloroform, dichloromethane, Freon, and similar ones, that are routinely used in MLV production. The use of such organic solvents is currently critical for proper MLV production. They are applied for the dissolution and optimal mixing of the different lipids that make up the formulation and of those drugs that are introduced through the lipid phase. They also make a major contribution to the attainment of the form and dispersion of dry lipids for optimal contact with the swelling solution. On the other hand, the specific solvents used present a problem on their own, both in the industrial and the academic laboratory level, as regulatory authorities phase more and more possible solvents out of use. Owing to the environmental and economic issues involved, waste disposal of these solvents, which could accumulate to substantial quantities in industrial-scale MLV production, is another cause for concern. Clearly, if MLV or any of the many types of ULV, for which MLV are the source material, are to be used, other means would be required to still produce the MLV, but with an altogether elimination of the need for organic solvents or their replacement by less hazardous materials that could be used with much lower quantities. Based on the issues discussed above, we propose that MLV should be revisited.

The second example is the case in which, based on the therapeutic objectives, MLV are ruled out, and it has been determined that the optimal ULV should have a dominant size of 200 nm in diameter. The many ways by which such liposomes are produced can be divided into two categories: The first is device oriented, using MLV, as source material, that are subject to extrusion, homogenization, or microemulsification devices to transform them into the desired ULV. The second category includes methods such as reverse-phase evaporation, ethanol injection, detergent dialysis, detergent removal through gel exclusion chromatography, and others (2,3,5–7, 17–19,22,23). In this category, the start materials are usually multicomponent systems that include lipids, drug, water, and an organic solvent or detergent. Despite the production of liposomes by the methods of the second category being relatively easy and of lower cost, even assuming the contact between a drug and an organic solvent or a detergent is not a limiting factor, it is argued here that the first category is the better choice. The ability to scale-up procedures of the second category is questionable and, at best, might need investment for its development while scaled-up models of the first category devices are already in existence. The issue of organic solvents and waste has already been discussed above for MLV, but unlike the case of MLV, it is questionable if the critical roles of the organic solvents/detergents in methods of the second category can be eliminated or even reduced in quantity (especially relative to the other components of the system). The recommendation here for the selection of ULV made by the device-oriented methods is not meant to ignore the cost limitations. Rather it is argued that ULV of a specified size range that are produced by either type device should be sufficiently similar, especially as all use the same source material (i.e., MLV). With the right choice, this should make it possible to perform the basic stages of a study with one device and proceed to the industrial R&D levels with another without significant changes in the liposomal properties. For example, it should be possible for the academician to use small, relatively inexpensive extrusion devices (as well as design and build one) and use for the industrial stage a microemulsifier operated under conditions that would give the same size liposomes under similar temperature, buffer, and other experimental parameters.

Lipid Composition

The source and cost of lipids have been concerns for liposomologists in both academic and industrial research. For the major run-of-the-mill lipids used in most liposome formulations, such as Phosphatidylcholine, the increase in available sources has somewhat eased those concerns. This trend can be expected to continue and grow as more liposomes successfully complete clinical trials and move into the arena of products, making the supply of high-purity lipids a viable business venture. The situation is different when it comes to the more rare lipids. The motivation for incorporating relatively rare lipids into a liposome formulation comes from the need to endow the liposomes with unique properties that will take into account the nature of the drug in question and address specific demands of the intended therapy. The perceived freedom of the academician in selecting liposome types might also be extended into selection of lipid composition; but in reality, the paucity of source materials for uncommon lipids together with their high cost might negate their use at the pharmaceutical product stage. Thus, even at the basic stage, a balance would have to be struck between the liposome formulation and the properties to be endowed by specific lipids. In general, the integrated approach would dictate giving up that singular lipid and finding an alternative means to endow the liposomes with the same desired attribute. On the other hand, it should be borne in mind and taken into consideration that basic studies might offer unanticipated solutions to particularly difficult pathological situations that would justify even exceptionally high costs.

The evolution within the Stealth® (Liposome Technology, Inc., Menlo Park, California, U.S.) liposomes is an instructive example. The unique feature of these liposomes is their ability to delay/evade uptake by the Reticuloendothelial system (RES) for longer periods than regular liposomes of similar size range. This allows for prolonged retention in circulation, and through that, higher shares of the dose can reach the non-RES targets. The first generation of Stealth liposomes required a specific ganglioside in their formulation (55). Subsequent generations of Stealth liposomes have seen the move from that lipid to the relatively less regular and hydrogenated phosphatidyl-inositol and specific hydrogenated Phosphatidylcholine cholesterol mixtures, ending up in formulations that do not require rare and/or modified lipids at all (56). In the current version of these liposomes, the "stealth" property is achieved through surface modification by polyethylene glycols (57). The extent to which elimination of the dependence on rare lipids has been a driving force in this evolution has not been specified by its developers. Yet, even if considerations of lipid raw materials were not a major driving force for this particular evolution, this case shows that such considerations can be accommodated without giving up on desirable formulation-dependent liposomal features.

TARGETED/MODIFIED LIPOSOMES: AN INTERESTING AND EXCITING SCIENTIFIC TOOL, BUT CAN THEY BE MADE INTO PRODUCTS (ESPECIALLY IMMUNOLIPOSOMES)?

The ability to target the sites of drug action, and to them alone, is among the critical properties required of liposomes. It is well known that despite enormous investigative efforts, the achievement of targeting that would be highly effective in vivo and applicable to a wide range of therapeutic objectives is still an elusive goal. As will be discussed below, complete targeting has not been achieved even in some of those cases defined as "passive targeting," where the targets are within the RES (58). Despite the frequent appearance of the term *targeting* in liposome research and literature, its very prolific use has somewhat blurred it to mean different things to different people. Owing to this current state of affairs, it was deemed advantageous to offer here a brief discussion and definition of targeting prior to a discussion of the means by which liposome targeting is attempted, and its evaluation from both the academic and industrial points of view.

Targeting: Definition, Stages, and Status

It is offered here that liposome (or any other drug-delivery system) targeting can be viewed as a two-tier process, the first of which takes place on the introduction of liposomes into a living system, and thus is termed here "targeting at the organ level."

The objectives and efforts within this tier are focused on getting the drug-loaded liposomes from the site(s) of administration to the organ/tissue/vicinity where the state(s) of disease resides. The second tier, termed "targeting at the cellular level," is a form of "fine tuning." It centers on pinpointing the drug-loaded liposomes, on their arrival at that location, as close as possible to the actual sites of drug action. It should be emphasized that to avoid interference with the therapeutic activity, the liposomes might need to be placed close to, but not at, the sites of drug action. Thus, targeting at the cellular level can involve two distinct types of sites: sites at which the liposomes reside, which will be termed "carrier sites," and those at which the drug binds to initiate its therapeutic effect, which will be termed "drug sites."

The targeting capabilities demanded of liposomes for therapies that require systemic administration depend to a large extent on the locations of the sites of drug action. For ultimate targeting at locations outside the RES, the liposomes need capabilities to address the needs of both targeting tiers. For targeting at locations within the RES, the liposomes need to provide the second tier alone, as the first one is provided by the biological system itself. For therapies that do not require this or are preferably achieved by nonsystemic drug administration, the situation is somewhat similar to RES targets, that is, the liposomes have to provide only for the second tier of targeting. In contrast to the previous case, in the nonsystemic therapies, the first tier is achieved (or one could say eliminated) by the mere selection of the route of administration. Included in such therapies would be topical drug administration in the treatment of wounds, burns, and ocular conditions and regional drug administrations such as intraperitoneal infusions and aerosol inhalations for treatment of disorders in the peritoneal cavity and in the lungs, respectively (59–62).

Liposomal specifications for the achievement of the first tier of targeting in systemic applications for non-RES targets have been comprehensively and extensively reviewed together with the inherent physiological and anatomical obstacles (2–11,17,55–58, and references therein). In an effort to refrain from redundancy, the reader is referred to these referenced sources and the discussion below is focused on the less-addressed specifications for the achievement of the second tier. The basic premise is that once the liposome arrives at, or is introduced into, the organ/tissue/location where the state(s) of disease resides, for effective targeting at the cellular level the adherence of liposome with high affinity to its designated carrier sites, despite cellular and fluid dynamics, is required for a sufficient duration to release an effective drug dose for the time span dictated by the specifics of drug and disease. Ideally, the liposome should also be instrumental in aiding drug access to sites of drug action. Tailoring liposomes to this end should be done with care and is rational for those cases where detailed knowledge of the cellular or tissue location of the drug site is available. Obviously, for intracellular sites of drug action, liposome internalization would be beneficial, but it is not an absolute necessity. For sites of drug action that are extracellular, such as certain membrane-embedded receptors or matrix-residing bacterial colonies, liposome internalization would defeat the purpose.

The Means Through which Liposome Targeting has been Attempted and the Feasibility of Developing Liposomes as Pharmaceutical Products

For endowing liposomes with targeting abilities, the main efforts have been focused on liposome features. Less attention has been paid to other aspects regarding treatment regimens such as liposome dose size and frequency of administration or pretreatment with "blank" liposomes. With respect to liposomal features, the two main avenues explored were liposome specifications such as size, type, and lipid composition and liposome surface modification.

The attempts to confer a targeting ability on liposomes through manipulation of their specifications, provided they do not stray from the integrated criteria for the selection of liposome types discussed in the previous section, will continue to result in liposomes that can be developed into pharmaceutical products. Some success has been found in this avenue of exploration, as shown by preferential accumulation of small daunomycin-encapsulating liposomes in solid tumors, although the operating mechanism has not yet been elucidated (63). It has been shown that small stealth liposomes accumulate, apparently by entrapment, in the intracellular spaces within solid tumors and are retained within those spaces for a sufficient duration to allow drug to diffuse from them into the tumor cells (13–16,64,65).

The attempts to confer targeting ability through liposome surface modification are an entirely different matter. Antibodies are, by far, the most extensively investigated and attempted class of targeting agents that have been attached to the liposomal surface for both tiers of targeting (2,3,7,8,17,66–69). A comprehensive discussion of these liposomes, for which the term *immunoliposomes* has been universally accepted, is beyond the scope of this chapter, and the reader is referred elsewhere and to those cited therein (1–11,17,55–59). The discussion here will touch briefly on the state of the field with respect to targeting with immunoliposomes and their feasibility as pharmaceutical products.

A fair share of the studies pursuing targeting with immunoliposomes has been conducted in cell cultures. Taken as studies aimed at the first tier of targeting, cell cultures are not the best choice for a model system, especially for systemic applications. Taken as studies aimed at the second tier of targeting, cell cultures are quite suitable models even if the goal of the investigators was to test the first tier rather than the second. Animal model studies have shown, to date, that despite being such an elegant approach, successful in vivo targeting with systematically administered immunoliposomes is quite limited and still far from being the overall solution to liposomal targeting. Moreover, the question of whether it might ever be a wide-range solution is still open. On the other hand, in vitro studies with immunoliposomes, which have shown specificity in binding and/or improved a biological response that was restricted to cells carrying the appropriate antigen, clearly demonstrate that it should be possible to achieve the second tier of targeting with this approach. Obviously, this speaks for a realistic potential of immunoliposomes in nonsystemic applications where, as already discussed, the first tier of targeting is achieved (or the need for it eliminated) by the selection of the administration route. Theoretically, targeting of intravenously administered immunoliposomes could also become possible for a wide range of applications if the means could be found to endow the immunoliposomes with the ability to provide the first tier, without compromising their ability to provide the second tier of targeting.

Because they have a future with at least some targeting applications, the current situation with immunoliposomes makes the question of their feasibility as a pharmaceutical product relevant and valid. This is a rather complex question comprising issues, to be discussed below, that concern the antibodies themselves as well as the antibody–liposome systems. It is stressed that the discussion is focused only on the issues that are related to the question at hand, namely, the feasibility of immunoliposomes as pharmaceutical products. Physiological and clinical advantages and drawbacks of immunoliposomes are not addressed here.

Obviously, to be a pharmaceutical product, immunoliposomes are required to meet all criteria that bear on antibody production, stability, retention of biological activity (for targeting, at least with respect to the antigen–antibody interaction), and appropriate shelf life. Another major factor in the making of an antibody into a pharmaceutical product (or part of it) is the combination of specificity and antigen abundance. The antibody has to be specific enough for its designated use and the antigen should be general and abundant enough so that the product would be applicable to a wide enough range of the patient population. The surface-modification process itself, such as the most frequently reported multistep process employing SPDP or similar agents, needs extensive streamlining to become an industrial process (69,70). Considerations of raw materials, waste disposal, stability, retention of encapsulated drug, and drug stability would all add to the difficulty, as would the need for an overall cost-effective process. On this basis, it seems clear that although immunoliposomes will continue to be a useful research tool, their development into pharmaceutical products, even for cases where they can provide targeting, will have to be decided case by case, taking into account not only the obstacles listed above, but also those that could arise from physiological and clinical aspects.

The realization that targeting of systemically administered liposomes requires both long circulation and target specificity has led to efforts to modify the liposomal surface with two agents—either bound individually or linked to one another—one responsible for long-term circulation and the other for high affinity to tumor sites. One example is binding antibodies to the edge of PEGylated liposomes (71,72). The prospects of this approach depend on whether the risks of mutual interferences can be satisfactorily resolved.

Other types of liposome surface modifications for the targeting purpose that are not based on the immune system might be more feasible as future pharmaceutical products. The case of polyethylene glycolated stealth liposomes for systemic administrations has already been discussed. Other potential opportunities are bioadhesive liposomes, which were originally developed for nonsystemic topical and regional applications (60–62,73–75). These are based on the use of ubiquitous surface-bound agents such as collagen, hyaluronan (hyaluronic acid) and growth factors, and on surface-modification procedures that are less complex than those discussed above for antibodies.

With respect to tumor targeting (via systemic administration) two types of surface-modified liposomes offer prospects. One type makes use of hyaluronan as a targeting agent positioned (by covalent binding) on the surface of unilamellar liposomes. Within this type are two versions, one is the bioadhesive liposomes discussed above where the targeting agent is the naturally occurring high M_r hyaluronan that has ~3000 repeats of the basic disaccharide unit made of *N*-acetyl glucosamine and glucuronic acid (60–62,73–77). In the other version, the targeting agent is a short fragment that has 6 to 12 repeats of that basic disaccharide (78). Hyaluronan candidacy as a tumor-targeting agent stems from its recently found ability to confer upon small liposomes the long circulating ability on a par with that of PEG and from the location and nature of hyaluronan receptors (76,77). These receptors, in active conformation for hyaluronan binding, are overexpressed in many types of tumors, whereas normal cells usually have poor expression of these receptors that are, furthermore, in nonactive conformations for hyaluronan binding (76,77 and citations within). In vivo active tumor targeting has recently been shown for the liposomes that have the full hyaluronan as their targeting agent (76,77). Tumor diversity carries the implication that there is place and need for several versions of targeted liposomes based on hyaluronan receptors, and future reports on in vivo performance of the liposomes that have the short hyaluronan fragment as their targeting agent will show if these liposomes can join the arsenal of such liposomes.

The overexpression of folate receptors in tumor cells opens the door to the other type of tumor-targeted liposomes that utilized folate as a targeting agent (79). Utilizing the approach of two agents, discussed above for PEG and antibodies, successful in vitro and in vivo data spell a potential for tumor targeting of small PEGylated liposomes that have folate bound to the free edges of PEG residues (79–81).

With respect to targeting, the first concern of an investigator developing a drug–liposome system for a specific therapeutic objective is to find the means to achieve targeting. Together with this concern, it is proposed that the investigator should weigh the options along the lines discussed above. If the therapy in question requires addressing both tiers, the best present options are hyaluronan-liposomes and the two-agent surface-modified liposomes of the type discussed above, or hyaluronan. If the therapy requires the second tier alone, targeting opportunities are offered by surface-modified liposomes, where a single agent of the types discussed above (i.e., antibodies, folate, hyaluronan, and other ligands for which the target has overexpressed receptors) suffices; also there is no need to provide the liposomes with an agent responsible for long circulation. In all cases, regardless of whether targeting is pursued with current or new surface-modifying agents, the feasibility for future pharmaceutical products, as exemplified above for immunoliposomes, should be taken into account.

Finally, despite "the heat of the battle," it is imperative not to lose sight of the critical objective of targeting: It is the drug which should be targeted, with the liposome a means to that end, even though it is often experimentally wiser to first pursue targeting with drug-free liposomes. Even if successful targeting of a given liposome system has been achieved, and the resultant liposomes can meet the criteria of becoming pharmaceutical products, the work is not yet done. Completion requires experimental verification whether those liposomes can reach the target while still carrying a drug load that is sufficient for effective therapy.

LIPOSOMES AS A STERILE, PYROGEN-FREE SYSTEM WITH PHARMACEUTICALLY ACCEPTABLE SHELF-LIFE, STABILITY, AND DOSAGE FORMS

Sterile, Pyrogen-Free Liposomes

In basic research, an investigator can forego the need for sterile, pyrogen-free liposomes provided the experiments conducted are such that the liposomes do not come into contact with living matter at all or that such contacts, in vitro or in vivo, are of short duration. For a pharmaceutical product, such qualities are critical. The steps that should be taken to obtain pyrogen-free liposomes are not essentially different from those implemented for other pharmaceutical products (5,83). Securing the sterility of liposomal products is an entirely different matter and has the makings of a major obstacle (83). The major approaches in use, such as heat sterilization and Y-irradiation, are not suitable for the end product (5,83). The two feasible approaches are sterilization by filtration and an aseptic production process. Of the two, the latter can potentially be applied to all species and types of liposomes, regular as well as surface modified. The sterile filtration is restricted, obviously, to liposomes small enough to pass the selected pore of the filter, usually of size 0.2 μm, provided obstructions such as filter clogging and filter adsorption (especially for surface-modified liposomes) do not turn into insurmountable complications or lead to significant losses in matter, Consequently, the issues surrounding the production of sterile liposomes, by being of mutual interest and need to liposome investigators in both academic and industrial research, constitute an area where cooperation will be needed to arrive at acceptable solutions.

Shelf-Life, Stability, and Dosage Forms

It is well known that the formation of liposomes from dry lipids suspended in an aqueous media is driven by thermodynamic stability (1,84). Hence, to survive and function as liposomes, these particles need to be surrounded by water, particularly at the start and end of the road, namely, at the end of a liposome production process and when called to action within the biological milieu. During the time interval between production and in vivo performance, the liposomes have to be stored for an acceptable period that, for pharmaceutical products, usually extends beyond one year. For refrigerated (but not frozen) aqueous suspensions and lyophilized powders, defining the major risks of long-term storage and reviewing, from the point of view of the academician, the advantages and drawbacks of each form of storage and for administration are the topics of this section.

Shelf-Life and Long-Term Stability

Breakdown in sterility, chemical destabilization, formation of undesirable degradants, and loss of therapeutic activity are among the major risks in long-term storage of any pharmaceutical entity. When it comes to liposomes, guarding against and prevention of such risks are inherently more difficult. More than one chemical entity is involved: lipids, drugs, and for modified liposomes, the surface-anchored agents. It also involves the preservation of particle integrity and the retention of the encapsulated drug within the liposome.

The major advantages of liposomal storage in the form of aqueous suspensions are (i) the retention of the original type and specifications of the liposomes selected and produced for the designated therapeutic task, and (ii) their "ready-to-use" mode. The former is of particular importance when unilamellar liposomes are concerned, especially when substantial efforts have been invested in their production and where the effective therapy is dependent on a specific size and narrow size distribution of the liposomal preparation. A critical component of the ready-to-use mode involves the distribution between encapsulated and unencapsulated drug. Obviously, effective therapy in which the liposomes will make a substantial difference compared with free drug requires an encapsulation of a sufficiently high share of the drug in the administered dose. This is particularly important for toxic drugs (such as chemotherapeutic agents) where protection of the biological environment from the drug is a major motivation for liposomal delivery. In such cases, even a relatively small share of unencapsulated drug could defeat that objective. The considerations outlined above favor the storage of liposome preparations that have been cleaned from unencapsulated drugs together with the retention of this state throughout the storage. Removal of unencapsulated drug, which constitutes a displacement of the equilibrium distribution of the drug between the liposomes and the external aqueous phase, will inevitably create an electrochemical gradient, driving the drug out of the liposomes (60–62,85). As will be discussed in greater detail in a following section, the rate of this diffusion will depend on several factors, with drug properties being foremost among them.

To evaluate the release of kinetics of the drug in question and of the effects of already recognized tools for implementation of satisfactory storage conditions, as well as the development of new tools for the task, it is required to start at the basic research level. The risks of loss of encapsulated drug on long-term storage of liposomes that have undergone separation from unencapsulated drug and are therefore in a nonequilibrium state are particularly high for small molecular weight drugs stored under those nonequilibrium conditions. For such cases, storage at high lipid concentrations, the introduction of additional intraliposomal barriers such as the ammonium sulfate gel, and utilization of remote-loading procedures could be used to reduce the risk of that loss (83,85,86). For other cases, even though it might take away from the ready-to-use advantage, the only recourse would be to store the preparation under equilibrium conditions (i.e., without the removal of the unencapsulated share) and use it as is or institute a separation procedure immediately prior to administration. The drawbacks of storage in suspension form have been resolved for at least one approved product, Doxil® which is a doxorubicin HCl liposome injection, marketed by Alza corporation. Besides being economically more attractive, storage of liposome–drug systems in dry form, as freeze-dried powders, seems the better choice, because it leads to significant reduction in the risks discussed above. On the other hand, except for selected situations (to be discussed in the next section), the dry powder is not the ready-to-use form, and liposomal reconstitution through rehydration would be required prior to use. Needless to say, reconstitution has to be performed under sterile and pyrogen-free conditions. The main concerns and drawbacks with this dosage form center on the nature of the liposomal species and on the situation of drug encapsulation that are obtained on reconstitution. Studies aimed at these issues have shown that unless specific steps—mainly the introduction of monosaccharides such as lactose, glucose, trehalose serving as cryoprotectants into the liposomal formulation—are taken prior to lyophilization, the reconstituted systems will revert to MLV, independent of the original liposome species (87–89). The feasibility of this approach has been proven, for example, in Myocet as an approved doxorubicin-encapsulating nonmodified small liposome, marketed by Elan and stored as a lactose-containing powder.

Recently, a novel option was found for unilamellar bioadhesive liposomes surface modified by hyaluronan, discussed earlier for their targeting ability (76,77). In these liposomes hyaluronan acts as an intrinsic cryoprotectant (75). In this role, hyaluronan apparently acts similar to the monosaccharides cryoprotectants, stabilizing the dried liposome powder with hydrogen bonds to replace those of the removed water molecules. Independent of liposome species, less is known about the fate of drug distribution in the reconstituted system between the liposome and the (new) aqueous medium. If the share of unencapsulated drug is beyond the level of acceptance for the system in question, this will require correction through actions such as revisions in conditions of lyophilization and/or reconstitution and post-reconstitution purification steps. For those liposomes that are surface modified, the distribution of that agent should also be considered. If big enough, steric hindrance might favor its relocation on reconstitution to the liposomal surface, but the situation could arise where the share of reconstituted liposomes having that agent inside is not negligible. This could result in problems that range from the mere loss in the fraction of targeted liposomes to the hazards of liposomes carrying toxic drugs circulating indiscriminately in the body.

The characteristics of the reconstituted liposome, including preservation of original structures through the use of cryoprotectants, the possible adverse effects due to those additives, and the distributions of the drug and the surface-bound agent in the reconstituted systems have been explored (87–89). However, it is argued here that whether such investigations have been conducted with drug models or with specific drugs, they need to be done anew for each specific liposome–drug system investigated and developed for a particular therapeutic objective. Undoubtedly, such questions have strong components that put them within the domain of basic research.

Dosage Forms for Administration

Suspended in an aqueous suspension form, regardless of the original storage form, the liposomes are ready for application utilizing not only the intravenous route but other routes of administration as well. They could be applied topically to the eyes, to wounds, and to burns. They could be injected subcutaneously or intramuscularly or introduced into the peritoneal cavity through injection or infusion. Other routes of administration require further processing of the aqueous liposome suspension, which touch on basic and industrial issues that have not been addressed yet in this chapter. A most prominent case is processing into aerosols of drug-encapsulating liposomes (regular and surface modified) that would be useful for pulmonary applications.

The making of liposomal aerosols for drug delivery opens new questions such as the effects of the aerosolization process (where relevant), aerosolization components on (i) all aspects of stability (drug, lipids, surface-bound agents), (ii) the retention of encapsulated drug, (iii) the retention of (liposome) particle integrity and size, and (iv) the retention of therapeutic activities, without which success in the retention of the other properties is immaterial. Liposome aerosols have been prepared and characterized with respect to some of these properties, and the results attest to both the feasibility of this dosage form and to the strong dependence of its success on the drug and the liposome species explored (90–98). Consequently, the data that have accumulated using drug models will not save an investigator developing a drug–liposome system aerosol form from investing in basic research of the stability and retention issues listed above. The physical features of the aerosol and their optimization with respect to the designated therapeutic goal would also be required, utilizing cascade impingers and similar devices, as well as a means for aerosol collection and analysis (90,93,94).

LIPOSOME CHARACTERISTICS (PERCENTAGE OF ENCAPSULATION, KINETICS OF RELEASE, BIOLOGICAL ACTIVITY)—IN BASIC RESEARCH AND IN QUALITY ASSURANCE

In this chapter, much has already been said about basic liposomal traits such as the efficiency of encapsulation, kinetics of drug release, and retention of biological activity. The investigation of these traits is in the realm of basic research, and among other objectives is the intention to serve in system optimization and in the development of in vitro models for pursuit of the liposomal in vivo fate. Once these encapsulation, release, and activity specifications are set for a given liposome–drug system, they can be used within the industrial environment in the evaluation of the retention of the liposome properties in the scaled-up production and as part of the battery of criteria for quality assurance. In consideration of the common ground these properties constitute for investigators from both academia and industry, several comments and proposals are offered below.

Efficiency of Drug Encapsulation

Throughout the four decades of liposome research, several definitions have been introduced as measures of this property. For example, the fraction of encapsulated drug from the total in the system, the ratio of drug to lipid, and the volume (per given lipid quantity) of the internal aqueous phase. The most traditional method of drug encapsulation is in the course of liposome preparation, which carries the risks of compromising drug integrity in the course of the preparation of regular and more-so surface-modified liposomes. Another drawback of this traditional mode is drug loss during purification and wash procedures that undermine cost-effective production. To counter these risks, methods of remote loading—loading the drug into preformed liposomes—have been developed. The more veteran approach relies on generating by the use of specific agents, an electrochemical gradient across the liposomal membrane that draws the drug inside (86). In a more recent approach, the drug-free liposomes are lyophilized (and can be stored as such), and drug encapsulation is performed in the course of liposome reconstitution by rehydration—using an aqueous solution of the drug as the rehydration solution (74,75).

For molecules serving as models of encapsulated matter, such as carboxyfluorescein (CF) and similar derivatives, elegant procedures have been devised to determine the efficiency of encapsulation (2,6,17). One of their most favorable features is the elimination of the need to separate the drug-encapsulating liposomes from the media containing the unencapsulated drug. For most real drugs, irrespective of the efficiency parameter pursued, the experimental process for determination of encapsulated matter will require such a separation step. Although simple in concept, in continuation of the discussion in "Shelf-life and long-term stability" above with respect to cleansing from unencapsulated drug, most methodologies contain steps that can lead to overestimation or underestimation of the quantity of liposome-encapsulated drug, especially for small molecular weight drugs.

Separations that make use of semi-permeable matrices such as exhaustive dialysis, various filtration devices, or gel-exclusion column chromatography all run the risk of diluting the external medium in the course of separation. This creates a driving force for efflux of the encapsulated matter, which would be most pronounced for small molecular weight drugs. The encapsulation level determined at the end of such procedures might be an underestimation of the initial level, with the deviation depending not only on drug properties but also on the specific experimental conditions used for the separation. This could become a problem not only with respect to accurate evaluation of encapsulation efficiencies, but also to batch-to-batch reproducibility. A quality assurance criterion would also be adversely affected, unless extensive care is taken in duplicating the separation process from batch to batch. Separation by centrifugation is supposedly free from the risks of external medium dilution. However, if a single cycle is used, there could be overestimation of the encapsulated fraction due to adsorption of unencapsulated drug to the liposomal pellet. Additional cycles of washing in drug-free buffer could reduce that error, but they need to be used with caution to avoid the risk of dilution, which can lead (as discussed above) to underestimation.

How then can the level of encapsulation be accurately assessed without the interferences that occur as a result of the separation steps? One approach exemplified by the aforementioned CF is to develop assays that can distinguish (quantitatively) between encapsulated and unencapsulated matter in the same system. Such assays would be strongly dependent on drug properties and probably would be useful for a limited number of cases. Another possibility, to be discussed in the following section, is to determine the encapsulation efficiencies through the studies of drug release.

Kinetics of Drug Release

Irrespective of the level of organization at which they are performed, studies of drug release/diffusion from liposomal systems are directed toward issues that are relevant to the in vivo as well as to the non–in vivo arenas. For liposomes in the in vivo arena, the drug-release studies are expected to yield data and understanding that will lead to (i) minimizing the loss of encapsulated drug on route from the site of administration to the site of drug action, and (ii) the ability to match the rate of release (once the liposomes arrive at the target) to the requirements of the therapy. The objectives of drug-release studies that concern the non–in vivo arena are (i) physicochemical characterization of the systems, including liposomes processed into aerosols or reconstituted from freeze-dried powders, (ii) various aspects of system optimization such as the selection of liposome type, lipid composition, and parameters of shelf life, and (iii) criteria for quality assurance. To derive relevant data from such studies, the experimental conditions should be set to fit the specific objectives, especially with respect to the extent of liposomes and drug (each, separately) dilutions that the system is anticipated to undergo. Detailed discussion of the in vivo objectives is not within the scope of this chapter, and discussion of the impact of such dilutions has been discussed elsewhere (59). The non–in vivo objectives are within the subject of the present discussion and will be addressed below.

Drug release from intact liposomes into the media within which the liposomes are placed (or suspended) is essentially a process of diffusion in a heterogeneous system, the latter made of several phases of water and lipid bilayers, the number of which will depend on the liposome type and on drug solubility properties. At the very least, two phases are involved: a lipophilic drug embedded within the lipid bilayer of a unilamellar liposome. In principle, initiation of drug diffusion simply requires setting the driving force (i.e., an electrochemical gradient of the drug from its liposomal location to the external medium) after which the process will proceed until equilibrium is established (or re-established). When the starting material is a drug–liposome system at equilibrium with respect to drug distribution, reduction in the drug concentration at the external medium suffices to trigger the release process. Several means can be used to generate that reduction: precipitation, enzyme-catalyzed degradation, and dilution of the liposomal system into the medium of choice such as buffer or body fluid (real or simulated). The latter is, by far, the easiest and most general means to generate such a drug concentration reduction that is suitable for a wide range of the objectives listed above. If, as discussed above with respect to encapsulation efficiencies, the liposomal system has been previously subjected to "cleaning procedures" to rid it of unencapsulated drug, the electrochemical gradient is already in existence and the diffusion process already operative prior to the specific kinetic experiment.

Independent of the state of the system on initiation of a particular experiment (i.e., equilibrium or nonequilibrium of drug distribution), it is proposed that sustaining a unidirectional flux of the drug from the liposomal system into the bulk medium during the entire experiment is a key element. Data from experiments done under such conditions represent the worst-case scenario for drug loss from the liposomes. This knowledge can then be used to design shelf-life conditions under which drug loss is minimized (60–62,85). Data of this type can also be used to set up, in the process of liposome surface modification, the conditions that will prevent depletion of the encapsulated drug.

Once the diffusion has been initiated (or is already in progress), the experiment itself consists of quantitative evaluation of drug accumulation in the medium and/or drug loss from the liposomes at designated periods. Critical to the successful execution of such studies is the quantitative separation of liposomes and the external medium in aliquots withdrawn from the reaction mixture at designated time points. Classic means for the quantitative separations of particles from their suspension medium, such as centrifugation and filtration, are often compromised by drug loss to the filtration matrix or by centrifugation procedures that are too slow with respect to the time that has elapsed between samplings. Additional experimental constraints might be encountered in drug and liposome assays. The dilution which is essential for the onset and maintenance of the gradient can result in lipid and drug levels in the separated fractions of the withdrawn aliquots that fall below the limits of detection and/or quantification.

Elimination altogether of the need for separation together with significant reduction of the risks of falling below detection limits is offered by the use of appropriate dialysis set-ups (60–62,85). In this approach, a liposome preparation is enclosed within a dialysis sac that is immersed in a drug-free medium. The wide variety among available semi-permeable dialysis membranes makes it possible to select, according to the specifications of the investigated system, a membrane that will not adsorb the drug and will not be a barrier to the drug but at the same time be a complete barrier to the liposomes.

Analysis of the raw data, rather than remaining at the phenomenological level, can not only give insights into the mechanism(s) but can also aid in identifying factors that are instrumental in gaining control of the release. Among the theoretical frameworks available for the analysis of diffusion data, two properties make the theoretical approach developed by Eyring particularly suitable for drug diffusion from liposomes. It allows drug release from homogeneous (unencapsulated drug) and heterogeneous (liposome-associated and liposome-encapsulated drug) systems to be dealt with simultaneously. It yields parameters that allow the direct determination of the fraction of encapsulated drug and of the half-life of drug release. In the long run, such parameters are especially useful for systems that are destined to serve as therapeutic entities, in defining optimization criteria, and for designing dose ranges and treatment regimens.

Studying the kinetics of drug release from liposomal (regular and surface-modified) systems, using the experimental and data processing approaches discussed above, the authors were able to sort several underlying principles that are general to these release processes (60–62,85). It was found that the release kinetics can be described as a series of parallel first-order processes, each representing a drug pool that exists in the system at time = 0. One pool represents the drug that is unencapsulated, and all others represent drug that is liposome associated. The mathematical expression for this type of mechanism is given by (85)

$$f = \sum_{j=l}^{n} f_j(1 - \mathrm{e}^{-kt}{}_j) \qquad (1)$$

where t represents time, the experimental independent parameter, and f represents the dependent experimental parameter, which is the cumulative release, normalized to the total drug in the system, at time = 0. The total number of independent drug pools given by n, f_j is the fraction of the total drug occupying the jth pool at time = 0 and k_j is the rate constant of the diffusion of the drug from the jth pool. For data derived from liposome–drug systems in which the drug was at equilibrium distribution at the onset of the experiment, this form of data analysis will yield the percentage of encapsulation through the magnitude of ~ for the encapsulated pool. It is proposed that, although extracted from kinetic data, this is the most accurate evaluation of this thermodynamic parameter, because it is free of the limitations imposed by separation procedures (see section "Efficiency of drug encapsulation").

Being diffusion processes, it was anticipated and verified experimentally that the properties of the drug are the primary parameters dictating the specifics of drug release (60–62,85). For a given drug, liposomal properties such as liposome type, lipid composition, and liposome concentration constitute means for some modulation of drug release. Among those liposomal parameters, liposome concentration was found to be the most useful tool for the task (60–62,85). It has been shown that the rate constant of the release of the encapsulated drug generally decreases with the increase in liposome concentration and that the phenomena and trend are independent of drug and of liposome species, whereas the actual magnitudes are system specific (85).

Besides providing an understanding of the liposomal system of interest, how then can such studies be used for the attainment of those non–in vivo objectives that are among the concerns of the industrial scientist?

With respect to shelf life, if the selected dosage form is liposome suspensions, storage should obviously be at high liposome concentrations. Furthermore, the kinetic parameters determined for the system of interest become product specifications (or "fingerprints") that will constitute critical input into the data base on which the decision of whether to store with or without unencapsulated drug will rest. If the selected dosage form is a freeze-dried powder of drug-encapsulating liposomes, these fingerprints can be useful in defining the optimal conditions for reconstitution and in verifying whether the reconstituted system has retained its original properties. Regardless of the dosage form selected for storage, retention of the same magnitudes of the kinetic parameters can be included within the battery of quality assurance tests.

When it comes to surface-modified liposomes, the processes of drug release add some concerns that are of interest to the liposome investigators in both academic and industrial research. A particular concern is the risk of drug (encapsulated) loss that can occur in the course of the modification itself, as well as in the subsequent procedures of separation and purification. Kinetic studies of the type discussed in this section can be used to determine whether such losses are significant at all and to evaluate their extent. For the systems at risk, inclusion of drug in the wash buffers could eliminate the problem. Whether the modification is done on preformed drug-encapsulating liposomes or on a single lipid component prior to liposome formation, such studies can also address the extent to which (if at all) the modification interferes with drug release and the optimal conditions for minimizing that interference.

In conclusion, studies of drug release from liposomes that can be categorized as basic research within the domain of the academician are also an essential part of many aspects and needs of product development and maintenance that are delegated to the domain of the industrial liposomologist. Such studies are needed anew for each drug liposome system, conducted with the specific drug of interest rather than with models.

Biological Activities of Liposome-Encapsulated Drugs

The need to retain the therapeutic activity of the liposome-encapsulated drug is so clear and unambiguous that it does not necessitate, besides this statement, any further discussion. The context in which this issue is discussed here centers on basic and product-oriented pre–in vivo studies that are deemed critical for the evaluation and application of liposome–drug systems.

Monitoring and ensuring the biological activity of the liposome-encapsulated drug have the highest priority whether the studies are conducted at the basic research level, at the stage of development, or in the course of routine production of a pharmaceutical product. The span of systems studied varies among these objectives, and are the most extensive at the basic stage, where it is attempted to best understand the system, optimize it, and elucidate the operating mechanisms. Even for products where the final decision might favor the administration of a liposomal system that contains both encapsulated and unencapsulated drug, the basic studies should explore both the complete system and a preparation free of unencapsulated drug. This is to verify that the biological activity is indeed that of the encapsulated drug, because if not, it defeats the objective of using liposomes. In light of the discussions in the previous sections, the time span between separation and initiation of the activity experiment should be selected according to the properties of the system at hand, with the goal of minimizing the loss of the encapsulated drug to the new drug-free medium. In addition to the obvious controls of free drug and vehicle, for further verification that it is the encapsulated drug which is responsible for all (or at the least most) of the therapeutic activity, drug-free liposomes suspended in the vehicle and in free drug should also be tested. Where modified liposomes are concerned, at least part of those studies should be duplicated for the (control) unmodified liposomes also.

For those cases in which the basic studies have ensured that the designated product (i.e., the encapsulated drug in the specified liposomes) has retained a satisfactory level of biological activity, the product development stage can forego controls of drug-free liposomes suspended in vehicle and in free drug. Similarly, if the product is a modified liposome, nonmodified liposomes can be abandoned. Yet the weight of separately evaluating the contributions of the system at equilibrium (i.e., containing both encapsulated and unencapsulated drug) and at nonequilibrium (i.e., containing encapsulated drug alone) states cannot be ignored. Such data should be especially sought for the final storage and administration dosage forms, namely, aqueous suspensions originally, freeze-dried powders that are rehydrated, or aerosols. Evidently, once the liposomal system reaches the stage of an established pharmaceutical product, the biological studies can be limited to continuous monitoring of that system alone.

It is indisputable that the ultimate preclinical level of organization at which the therapeutic activity of the liposome-encapsulated drug should be studied is that of the whole animal. Nevertheless, for the objectives listed above, it is well worth the effort to proceed with in vitro studies. If done properly in relevant systems, such studies can significantly reduce the investments (such as time, efforts, and resources) needed at the in vivo stage without compromising the quality and significance of the results and their conclusions. Furthermore, for quality assurance of established products, the in vitro studies might suffice.

As illustrated by several examples below, selecting and implementing in vitro systems for the evaluation of the in vivo designated therapeutic activity is an attainable task. In vitro testing of liposome-encapsulating chemotherapeutic drugs for tumor treatment has been extensively studied and reported, and a newcomer to the field can scan the liposome literature to select the systems suitable to the objectives. A similar situation exists with respect to liposomes encapsulating antibiotics for the treatment of intracellular infections of phagocytic cells. Two additional examples that have not been as extensively reported, and therefore are addressed here in more detail, concern wound healing and the treatment of extracellular bacterial infections.

The wound healing effects of several growth factors are exerted mainly (if not solely) through their stimulation of cell proliferation. This opens the door to in vitro testing of the biological activity of such liposome-encapsulated polypeptides through the selection of suitable cell lines. Basically, two types of cell lines can be used as the test system: those that have been specifically developed to have absolute dependence on a particular growth factor and those that can be made to have such dependence through the use of specific experimental conditions. For EGF, which is in the first line of growth factors in development for wound healing, the first type is represented by the cell line BALB/MK, which is the classic for EGF bioassay (99). The well-known NIH3T3 line, grown in very low serum (0.125%), is a representative of the second type (Yerushalmi and Margalit, in preparation). For both cases, the first task is to determine the dose–response curves to free EGF, to determine the range in which stimulation of proliferation is measurable yet still free from adverse effects (99; Okon et al., in preparation; Yerushalmi and Margalit, in preparation). Next, the liposomal systems including the various partial and control groups discussed above can be tested. Implementing such studies with both lines, it was found that EGF does indeed retain its biological activity when encapsulated in regular as well as in surface-modified liposomes.

Cultures of the relevant test organism can serve as the in vitro systems for testing liposomes that are designated for infection treatment of extracellular organisms, irrespective of the route of administration or anatomical locations. Adapting the paper disk version of the growth inhibition assay of free to liposome-encapsulated antibiotics, this approach was satisfactorily implemented for the following cases: liposome-encapsulated ampicillin and liposome-encapsulated cefazolin (Schumacher and Margalit, unpublished data) using *Micrococcus luteus* and *Staphylococcus aureus* as the respective test organisms (100,101). Furthermore, the adaptation to liposomal systems is general enough that it can be easily extended to other antibiotic–liposome systems and additional test organisms.

In all cases in which the in vivo designated therapeutic activity is tested in vitro, it is imperative to distinguish quantitatively between the total drug dose in the liposomal preparation and the actual dose that has become available to the cells during the course of the experiment. Treatment with equal total doses of free and of liposome-encapsulated drug, where the latter system yields lower response, need not be interpreted as liposome-associated loss in drug activity. Nor should similar levels of response to free and to liposome-encapsulated drug be taken to indicate that the liposomal system has no potential to improve clinical outcomes. In both cases, it might simply be that in actuality the comparison was between unequal doses, with the liposomal one being the lower. This situation could arise when not all of the encapsulated drug is released and thus made available to the cells during the experiment. Two approaches can be taken to distinguish between cases in which the equal or lower response of the liposomal treatment is an artifact of comparisons at unequal doses and when it is truly due to the loss of activity and/or inadequacy of liposomes for that specific treatment. Both approaches require previous data and analysis of the drug-release profile, especially at the level of dilution used in the bioassay. For the question of whether the liposome-encapsulated drug has retained its activity, such data can then be used to design bioassay conditions under which all of the encapsulated drug is made available. For the question of whether the liposomal systems have the potential to improve clinical outcomes, such data can be used to estimate the actual dose to which the cells have been exposed. Obviously, caution should be exercised in making quantitative use of drug-release profiles derived under conditions devoid of living matter. Determination of the kinetics of drug release in the cellular system tested under the conditions of bioassay, when possible, would augment the available data and increase the accuracy with which the dose–response curve of the liposomal system can be shifted to its true place. Finally, it is noted that a similar awareness of differences between total and actual liposomal doses is needed for dose–response comparisons in vivo, especially when the objective is to evaluate if liposomes can improve the clinical outcomes.

SUMMARY AND PROSPECTS

It is the opinion of the authors that any communication on liposomes as delivery systems cannot be concluded without a reminder of the current reality. Awareness of the substantial hurdles to the implementation of liposomes as DDS in established treatment modalities cannot be ignored. Nor can the current state of affairs where, despite four decades of research, the number of systems that have become products on the market is quite modest, although the number of systems that matures into clinical trials grows continuously. The need for DDS is still as acute as ever, and the potential that liposomes hold, although somewhat tarnished, has not been substantially diminished.

With this background, it is proposed that at least some of those hurdles can be overcome and substantial strides can be made in advancing liposomes from the laboratory to the clinic through attention and mutual awareness of the issues faced by liposome researchers in the academic and industrial environments and through increased collaborations and knowledge gained from the cases that have made it into the clinic. It is hoped that this chapter will make a contribution, however modest, in these directions.

REFERENCES

1. Bangham AD. Liposomes: the Babraham connection. Chem Phys Lipids 1993; 64: 275–285.
2. Gregoriadis G. Liposomes as Drug Carriers, Recent Trends and Progress. New York: Wiley, 1988.
3. Sato T, Sunamoto J. Recent aspects in the use of liposomes in biotechnology and medicine. Prog Lipid Res 1992; 31:345–372.
4. Litzinger DC, Huang L. Phosphatidylethanolamine liposomes: drug delivery, gene transfer and immunodiagnostic applications. Biochim Biophys Acta 1992; 1113:201–227.
5. Woodle MC, Lasic DD. Sterically stabilized liposomes. Biochim Biophys Acta 1992; 1113:171–199.
6. Senior JH. Fate and behavior of liposomes in vivo: a review of controlling factors. CRC Crit Rev Therap Drug Carrier Sys 1987; 3:123–193.

7. Machy P, Lesserman L. Liposomes in Cell Biology and Pharmacology. London: Libbey, 1987.
8. Gregoriadis G. Overview of liposomes. J Antimicrob Chem 1991; 28:39–48.
9. Karlowsky JA, Zhanel GG. Concepts on the use of liposomal antimicrobial agents: applications for aminoglycosides. Clin Infect Dis 1992; 15:654–667.
10. Gregoriadis G, Florence AT. Liposomes in drug delivery. Drugs 1993; 45:15–28.
11. Gabizon A. Tailoring liposomes for cancer drug delivery: from the bench to the clinic. Ann Biol Clin 1993; 51:811–813.
12. Hope MJ, Kitson CN. Liposomes. A perspective for dermatologists. Dermatol Clin 1993; 11:143–154.
13. Mayhew E, Lazo R, Vail WJ, King J, Green AM. Characterization of liposomes prepared using a microemulsifier. Biochim Biophys Acta 1984; 775:169–174.
14. Gruner SM, Lenk RP, Janoff AS, Ostro MJ. Novel multilayered lipid vesicles: comparison of physical characteristics of multilamellar liposomes and stable plurilamellar vesicles. Biochim 1985; 24:2833–2842.
15. Kim S, Martin GM. Preparation of cell-size unilamellar liposomes with high captured volume and defined size distribution. Biochemistry Biophys Acta 1981; 646:1–9.
16. Price CI, Horton JW, Baxter CR. Topical liposomal delivery of antibiotics in soft tissue infection. J Surg Res 1990; 49:174–178.
17. Price CI, Horton JW, Baxter CR. Liposome delivery of aminoglycosides in burn wounds. Surg Gynecol Obstet 1992; 174:414–418.
18. Planas ME, Gonzalez P, Rodriguez L, Sanchez S, Cevc G. Noninvasive percutaneous induction of topical analgesia by a new type of drug carrier, and prolongation of local pain insensitivity by anesthetic liposomes. Anesth Analg 1992; 75:615–621.
19. Cevc G, Blume G. Lipid vesicles penetrate into intact skin owing to the transdermal osmotic gradients and hydration force. Biochim Biophys Acta 1992; 1104:226–232.
20. Physician's Desk Reference. 48th ed. Montvale, NJ: Medical Economics Data Production Co., 1994.
21. Brown GL, Nanney LB, Griffen J, et al. Enhancement of wound healing by topical treatment with epidermal growth. N Engl J Med 1989; 321:76–79.
22. Davidson JM, Klagsbrun M, Hill KE, et al. Accelerated wound repair, cell proliferation, and collagen accumulation are produced by a cartilage-derived growth factor. J Cell Biol 1985; 100:1219–1227.
23. Hunt TK. Basic principles of wound healing. J Trauma 1990; 30:S122–S128.
24. Kingsnorth AN, Slavin J. Peptide growth factors and wound healing. Br J Surg 1991; 78:1286–1290.
25. Knighton DR, Fiegel VD. Regulation of cutaneous wound healing by growth factors and the microenvironment. Invest Radiol 1991; 26:604–611.
26. Schultz G, Rotatori DS, Clark W. EGF and TGF-alpha in wound healing and repair. J Cell Biochem 1991; 45:346–352.
27. Yates RA, Nanney LB, Gates RE, King LE Jr. Epidermal growth factor and related growth factors. Int J Dermatol 1991; 30:687–694.
28. Graves DT, Cochran DL. Mesenchymal cell growth factors. Crit Rev Oral Biol Med 1990; 1:17–36.
29. Feldman ST. The effect of epidermal growth factor on corneal wound healing: practical considerations for therapeutic use. Refract Corneal Surg 1991; 7:232–239.
30. Knighton DR, Ciresi K, Fiegel VD, Schumerth S, Butler E, Cerra F. Stimulation of repair in chronic, nonhealing, cutaneous ulcers using platelet-derived wound healing factors. Surg Gynecol Obstet 1990; 170:56–60.
31. Herndon DN, Hayward PG, Rutan RL, Barrow RE. Growth hormones and factors in surgical patients. Adv Surg 1992; 25:65–97.
32. Pastor JC, Calonge M. Epidermal growth factor and corneal wound healing. A multicenter study. Cornea 1992; 11:311–314.

33. Servold SA. Growth factor impact on wound healing. Clin Podiatr Med Surg 1991; 8:937–953.
34. Gartner MH, Shearer ID, Bereiter DF, Mills CD, Caldwell MD. Wound fluid amino acid concentrations regulate the effect of epidermal growth factor on fibroblast replication. Surgery 1991; 110:448–455.
35. Falcone PA, Caldwell MD. Wound metabolism. Clin Plast Surg 1990; 17:443–456.
36. Steed DL, Goslen JB, Holloway GA, Malone JM, Bunt TJ, Webster MW. Randomized prospective double-blind trial in healing chronic diabetic foot ulcers. CT-102 activated platelet supernatant, topical versus placebo. Diabetes Care 1992; 15:1598–1604.
37. Alvarez OM, Mertz PM, Eaglestein WH. The effects of occlusive dressings on collagen synthesis and re-epithelization in superficial wounds. J Surg Res 1983; 35:142–148.
38. Mertz PM, Eaglstein WH. The effect of a semiocclusive dressing on the microbial population in superficial wounds. Arch Surg 1984; 119:287–289.
39. Eaglstein WH, Mertz PM, Falanga V. Occlusive dressings. Am Fam Physician 1987; 35:211–216.
40. Swain SF. Bandages and topical agents. Vet Clin North Am Small Anim Pract 1990; 20:47–65.
41. Xakellis GC, Chrischilles EA. Hydrocolloid versus saline-gauze dressings in treating pressure ulcers: a cost-effectiveness analysis. Arch Phys Med Rehabil 1992; 73:463–469.
42. Hulten L. Dressings for surgical wounds. Am J Surg 1994; 176:42S–44S.
43. Field FK, Kerstein MD. Overview of wound healing in a moist environment. Am J Surg 1994; 167:2S–6S.
44. Burton CS. Venous ulcers. Am J Surg 1994; 167:37S–40S.
45. Allen TM, Chonn A. Large unilamellar liposomes with low uptake into the reticuloendothelial system. FEBS Lett 1987; 223:42–46.
46. Gabizon A, Papahadjopoulos D. Liposome formulations with prolonged circulation time in blood and enhanced uptake by tumors. Proc Natl Acad Sci USA 1988; 85:6949–6953.
47. Allen TM, Hansen C, Martin F, Redemann C, Yau-Young A. Liposomes containing synthetic lipid derivatives of poly(ethylene glycol) show prolonged circulation half-lives in vivo. Biochim Biophys Acta 1991; 1066:29–36.
48. Poste G. Liposome targeting in vivo: problems and opportunities. Biol Cell 1983; 47: 19–38.
49. Margalit R. Vesicles as topical drug delivery systems. In: Rossof M, ed. Vesicles, Surfactant Science Series. New York: Marcel Dekker, 1994:527–560.
50. Margalit R, Okon M, Yerushalmi N, Avidor E. Bioadhesive liposomes for topical drug delivery: molecular and cellular studies. J Control Release 1992; 19:275–287.
51. Yerushalmi N, Margalit R. Bioadhesive, collagen-modified liposomes: molecular and cellular level studies on the kinetics of drug release and on binding to cell monolayers. Biochim Biophys Acta 1994; 1189:13–20.
52. Yerushalmi N, Arad A, Margalit R. Molecular and cellular studies of hyaluronic-acid modified liposomes as bioadhesive carriers of growth factors for topical delivery in wound healing. Arch Biochem Biophys 1994; 313:267–273.
53. Forssen EA, Coulter DM, Proffitt RT. Selective in vivo localization of daunorubicin small unilamellar vesicles in solid tumors. Cancer Res 1992; 52:3255–3261.
54. Gabizon AA. Selective tumor localization and improved therapeutic index of anthracyclines encapsulated in long-circulating liposomes. Cancer Res 1992; 52:891–896.
55. Senior J, Gregoriadis G. Dehydration–rehydration vesicle methodology facilitates a novel approach to antibody binding to liposomes. Biochim Biophys Acta 1989; 1003:58–62.
56. Mori A, Klibanov AL, Torchilin VP, Huang L. Influence of the steric barrier activity of amphipatic poly(ethyleneglycol) and ganglioside GM1 on the circulation time of liposomes and on the target binding of immunoliposomes in vivo. FEBS Lett 1991; 284:263–266.

57. Sato T, Sunamoto J. Recent aspects in the use of liposomes in biotechnology and medicine. Prog Lipid Res 1992; 31:345–372.
58. Lesserman L, Langlet C, Schmitt-Verhulst A-M, Machi P. In: Tartakoff AM, ed. Vesicular Transport, Part B. San Diego, California: Academic Press, 1989:447–471.
59. Zhou X, Huang L. Targeted delivery of DNA by liposomes and polymers. J Control Release 1992; 19:269–274.
60. Amselem S, Cohen R, Barenholz Y. In vitro tests to predict in vivo performance of liposomal dosage forms. Chem Phys Lipids 1993; 64:219–237.
61. Jain MK, Wagner RC. Introduction to Biological Membranes. New York: Wiley, 1980.
62. Margalit R, Alon R, Linenberg M, Rubin I, Roseman TJ, Wood RW. Liposomal drug delivery: thermodynamic and chemical kinetic considerations. J Control Release 1991; 17:285–296.
63. Mayer LD, Bally MB, Hope MJ, Cullis PR. Uptake of antineoplastic agents into large unilamellar vesicles in response to a membrane potential. Biochim Biophys Acta 1985; 816:294–302.
64. Friede M, van Regenmortel MH, Schuber F. Lyophilized liposomes as shelf items for the preparation of immunogenic liposomes–peptide conjugates. Anal Biochem 1993; 211:117–119.
65. Carpenter C, Zendegui J. A biological assay for epidermal growth factor/urogastrone and related polypeptides. Anal Biochem 1986; 153:279–282.
66. Akimoto Y, Mochizuki Y, Uda A, et al. Concentrations of ampicillin in human serum and mixed saliva following a single oral administration of lenampicillin, and relationship between serum and mixed saliva concentrations. J Nihon Univ Sch Dent 1990; 32:14–18.
67. Schumacher I, Margalit R. Liposome-encapsulated ampicillin: physicochemical and antibacterial properties. J Pharm Sci 1997; 86:635–641.
68. Clark AP. Liposomes as drug delivery systems. Cancer Practice 1998; 6:251–253.
69. Allen C, Dos Santos N, Gallagher R, et al. Controlling the physical behavior and biological performance of liposome formulations through use of surface grafted poly(ethylene glycol). Biosci Rep 2002; 22:225–250.
70. Bakker-Woudenberg IA. Long-circulating sterically stabilized liposomes as carriers of agents for treatment of infection or for imaging infectious foci. Int J Antimicrob Agents 2002; 19:299–311.
71. Lasic DD. Liposomes in Gene Delivery. and NY: CRC Press, 1997.
72. Gabizon A, Shmeeda H, Barenholz Y. Pharmacokinetics of pegylated liposomal Doxorubicin: review of animal and human studies. Clin Pharmacokinet 2003; 42: 419–436.
73. Carrion C, Domingo JC, de Madariaga MA. Preparation of long-circulating immunoliposomes using PEG–cholesterol conjugates: effect of the spacer arm between PEG and cholesterol on liposomal characteristics. Chem Phys Lipids 2001; 113:97–110.
74. Hosokawa S, Tagawa T, Niki H, Hirakawa Y, Nohga K, Nagaike K. Efficacy of immunoliposomes on cancer models in a cell-surface-antigen-density-dependent manner. Br J Cancer 2003; 89:1545–1551.
75. Brignolè C, Marimpietri D, Gambini C, Allen TN, Ponzoni M, Pastorino F. Development of Fab fragments of anti-GD(2) immunoliposomes entrapping doxorubicin for experimental therapy of human neuroblastoma. Cancer Lett 2003; 197:199–204.
76. Yerushalmi N, Margalit R. Hyaluronic-acid modified bioadhesive liposomes as local drug depots: effects of cellular and fluid dynamics on liposome retention at target sites. Arch Biochem Biophys 1998; 349:21–26.
77. Peer D, Margalit R. Physicochemical evaluation of a stability-driven approach to drug entrapment in regular and in surface-modified liposomes. Arch Biochem Biophys 2000; 383:185–190.
78. Peer D, Florentin A, Margalit R. Hyaluronan is a key component in cryoprotection and formulation of targeted unilamellar liposomes. Biochim Biophys Acta 2003; 1612:76–82.

79. Peer D, Margalit R. Loading Mitomycin C inside long circulating hyaluronan targeted nano-liposomes increases its antitumor activity in three mice tumor models. Int J Cancer 2004; 108:780–789.
80. Peer D, Margalit R. Tumor-targeted hyaluronan nanoliposomes increase the antitumor activity of liposomal Doxorubicin in syngeneic and human xenograft mouse tumor models. Neoplasia 2004; 6:343–353.
81. Eliaz RE, Szoka FC Jr. Liposome-encapsulated doxorubicin targeted to CD44: a strategy to kill CD44-overexpressing tumor cells. Cancer Res 2001; 61:2592–2601.
82. Gosselin MA, Lee RJ. Folate receptor-targeted liposomes as vectors for therapeutic agents. Biotechnol Annu Rev 2002; 8:103–131.
83. Gabizon A, Horowitz AT, Goren D, et al. Targeting folate receptor with folate linked to extremities of poly(ethylene glycol)-grafted liposomes: in vitro studies. Bioconj Chem 1999; 10:289–209.
84. Gabizon A, Horowitz AT, Goren D, Tzemach D, Shmedaa H, Zalipsky S. In vivo fate of folate-targeted polyethylene-glycol liposomes in tumor-bearing mice. Clin Cancer Res 2003; 9:6551–6559.
85. Gregoriadi G. Liposome Technology. Vols. I-III. Boca Raton, FL: CRC Press, 1984.
86. Szoka F Jr, Papahadjopoulos D. Comparative properties and methods of preparation of lipid vesicles (liposomes). Ann. Rev Biophys Bioeng 1980; 9:467–508.
87. Hope MJ, Bally MB, Webb G, Cullis PR. Production of large unilamellar vesicles by a rapid extrusion procedure. Characterization of size, trapped volume and ability to maintain a membrane potential. 1985; 812:55–65.
88. Carnie S, Israelachvili JN, Pailthorpe BA. Lipid packing and transbilayer asymmetries of mixed lipid vesicles. Biochim Biophys Acta 1979; 554:340–357.
89. Kumar VV. Complementary molecular shapes and additivity of the packing parameters of lipids. Proc Natl Acad Sci USA 1991; 88:444–448.
90. Bogdanov AA, Gordeeva LV, Torchilin VP, Margolis LB. Lectin-bearing liposomes: differential binding to normal and to transformed mouse fibroblasts. Exp Cell Res 1989; 181:362–374.
91. Crowe LM, Womersley C, Crowe JH, Appel L, Rudolph A. Prevention of fusion and leakage in freeze-dried liposomes by carbohydrates. Biochim Biophys Acta 1986; 861:131–140.
92. Crowe LM, Crowe JH, Rudolph A, Womersley C, Appel A. Preservation of freeze-dried liposomes by trehalose. Arch Biochem Biophys 1985; 242:240–247.
93. Kellaway IW, Farr SJ. Liposomes as drug delivery systems to the lungs. Adv Drug Del Rev 1990; 5:149–161.
94. Gupta PK, Hickey AJ. Contemporary approaches in aerosolized drug delivery to the lung. J Control Release 1991; 17:127–147.
95. Schreier H, Gonzalez-Rothi RJ, Stecenko AA. Pulmonary delivery of liposomes. J Control Release 1992; 24:209–223.
96. Farr SJ, Kellaway IW, Carman-Meakin B. Comparison of solute partitioning and efflux in liposomes formed by a conventional and aerosolized method. Int J Pharmaceut 1989; 51:39–46.
97. Taylor KMG, Taylor G, Kellaway IW, Stevens J. The stability of liposomes to nebulization. Int J Pharmaceut 1990; 58:57–61.
98. Niven RW, Scherier H. Nebulization of liposomes. I. Effects of lipid composition. Pharm Res 1990; 7:1127–1133.
99. Niven RW, Speer M, Scherier H. Nebulization of liposomes. II. The effects of size and modeling of solute release profiles. Pharm Res 1991; 8:217–221.
100. Niven RW, Carvajal TM, Scherier H. Nebulization of liposomes. III. The effects of operating conditions and local environment. Pharm Res 1992; 9:515–520.
101. Debs SB, Straubinger RM, Brunette EN, et al. Selective enhancement of pentamidine uptake in the lungs by aerosolization and delivery in liposomes. Am Rev Respir Dis 1987; 135:731–737.

12

Microemulsions for Solubilization and Delivery of Nutraceuticals and Drugs

Nissim Garti and Abraham Aserin
Casali Institute of Applied Chemistry, The Hebrew University of Jerusalem, Jerusalem, Israel

THE RATIONALE

There were many reasons for writing this review article. Some relate to a continuous search for better and more efficient delivery vehicles. Others are of a more personal concern, derived from the need to design improved formulations for hundreds of new natural active molecules that have not yet been explored as drugs. Yet other reasons derive from the development of new emerging categories of food supplements that will soon become nutraceuticals and will need formulations and delivery vehicles. Liquid formulations based on nanoparticles or nanodroplets are, to the best of our understanding, the key to improved bioavailability of drugs. They provide better active molecules for systemic environmental protection of active compounds and are excellent vehicles for improved pharmacokinetics.

Recently, I was impressed by a very interesting article accusing pharmaceutical companies of ignoring "traditional medicine" and for not doing enough for the poor people in medication. This article alerted scientists dealing with delivery of drugs to consider doing more research on naturally occurring compounds with health benefits to improve the health of the poor in our world (1). I have taken the liberty to quote some of their concerns in the introduction to this chapter. Many estimates have shown that up to 40% of the new chemical entities discovered by the pharmaceutical industry today are poorly soluble or lipophilic compounds. The solubility issue, complicating the delivery of these new drugs, also affects the delivery of many existing drugs. Methods to deliver poorly soluble drugs will grow in significance in the coming years. Similarly, generic drug manufacturers will need to employ economically efficient methods of delivery, as more low solubility drugs go off patent, to maintain a competitive edge and to be able to compete as profit margins shrink.

The biopharmaceutical classification system groups poorly soluble compounds, class III and IV drugs, which feature poor solubility and high permeability and poor solubility and poor permeability, respectively. Highly soluble (class I and II) compounds are those where the largest dose is soluble in <250 mL water at a pH range of 1.0 to 7.5, and highly permeable compounds are those that demonstrate >90%

absorption of the administered dose. In contrast, compounds with solubility below 0.1 mg/mL face significant solubilization obstacles and often present difficulties related to solubilization during formulation. The most common approach to improve the solubility of drugs with a net negative charge is to form salts (e.g., hydrochlorides, sulfates, nitrates, maleates, citrates, and tartrates) of the basic drugs. Yet, for other drugs this method of improving solubility is not possible, as they do not form such salts. Another route is to reduce drug particle sizes by new technologies or by applying new crystallization processes to improve dissolution kinetics.

Modern pharmaceutical drugs have alleviated human suffering tremendously. Yet, the drug discovery processes utilized by most Western pharmaceutical companies are not designed to develop inexpensive drugs or natural compounds with pharmaceutical properties or food supplements with disease-preventive properties. Modern medicines are not going to effectively control most of the diseases affecting the world's poor. Traditional medicines, such as extracts from medicinal plants, have been used for thousands of years to treat a wide variety of infectious and somatic diseases. They are increasingly seen as viable alternatives to synthetic medicines. Unfortunately, there have been serious limitations to the use of plant extracts as therapeutic agents. Dried herbs and powders are rarely compositionally consistent and may, in fact, be toxic if consumed inappropriately. In the last 15 years, there has been a renewed interest in identifying compounds from plants, fungi, and bacteria that can be used to improve human health and to treat diseases. Thousands of studies have been conducted on plant and microbial extracts in the hope of finding suitable compounds for treating many of the diseases ravaging the world's population. Many of these studies have identified very promising therapeutic compounds. However, the identification of promising natural medicinal compounds is only the first step toward the use of those compounds by humans. Effective formulation is an essential additional step that must be considered.

Traditional medicine practitioners in China, Japan, Korea, India, Tibet, and many other countries have relied for centuries on herbs and medicinal plants as treatments for disease. Although it has been acknowledged in Western medicine that some medicinal plants contain potentially valuable medicines, Western scientists and pharmaceutical companies, in particular, have largely ignored "traditional" medicines, as many herbal extracts have had minimal effects on the diseases they are purported to treat. And, other extracts are toxic and should not be used in large doses. Modern medicine demands proof, not tradition, in evaluating the effectiveness of specific medicines. Most herbal medicines are processed into dried plant leaves, ground powders, and pastes. This is understandable, because fresh, nondried, plant materials will eventually degrade or oxidize into products that render the active ingredients useless. Many medicinal molecules are trapped in the dried cellulose mass and cannot be extracted by stomach acid when eaten. Humans do not have the ability to degrade cellulose enzymatically. Many medicinal molecules are actually nonpolar lipids. Other molecules, regardless of their solubility, will never be absorbed by intestinal cells because they do not have the correct molecular structure. Many molecules that do get absorbed are immediately modified by sulfatation or sulfonation or other processes in the intestinal cells and liver and are rapidly returned to the lumen of the intestines for destruction by intestinal bacteria. A synthetic or natural drug that shows activity in a culture dish may do nothing in the body. Pharmaceutical companies know these problems all too well.

In the conventional pharmaceutical industry and in modern medicine, it is well known to all dealing with formulations and delivery of drugs that many promising synthetic drugs never make it to the market because the stomach or intestines cannot absorb them. Others are destroyed by stomach acid or are cleared from the blood too rapidly to be effective. Plant extracts are by definition a composition of many different compounds. Often, the therapeutic utility of a natural herb or plant extract cannot be attributed to a single compound. It is the interaction of different compounds that often makes a plant extract therapeutically useful. The most desirable synthetic drugs are those that are highly specific in their mode of action.

Most plant extracts contain both oil- and water-soluble therapeutic compounds. Many essential oils, the volatile compounds that constitute the fragrance of plants, are very well known to have potent antimicrobial properties. Tea tree oil, pomegranate oil, and grape seed oil are composed of a complex blend of volatile compounds that have powerful antimicrobial properties. While many modern antibiotics are highly specific in their properties, antimicrobial essential oils often kill gram-negative and gram-positive bacteria, fungi, and protozoa with equal ease. Now there is a renewed interest in the use of antimicrobial oils because antibiotic-resistant bacteria are increasingly becoming a health problem. While dried plant material is easy to transport and store, it no longer contains the essential oils, as they are lost during the drying process. The technology described herein was designed to capture both water- and oil-soluble components from plant extracts without losing them to evaporation or heat denaturation.

We are slowly learning to fortify our food and drinks with important natural extracted ingredients known as "food supplements." In many countries, water-soluble additives essential to human health, such as vitamins B and C, simple phenols, and polyphenols, have been added to the foods to fortify individuals' health and boost their immune system. Many of those are sold in drugstores or supermarkets in the form of tablets, capsules, soft-gels, and so on. However, oil-soluble or water-insoluble vitamins are still very difficult to incorporate into water, soft drinks, and many of our foods and are lacking in the diet of those with imbalanced diet or with poor nutrition.

In recent years, there has been a tremendous change in individual and public attitudes toward those supplements. Many new products are seen in the market, and the varieties are growing very fast. One can find hundreds of new extracts and blends of naturally occurring ingredients overflowing the stores. Some of the food supplements show health benefits, whereas others are questionable. However, some of the reasons for the questionable benefit might be poor formulations and poor bioavailability.

Many new and reliable studies are showing encouraging results on new extracts that can contribute to human health. Recently, a new category of food additives has emerged. The term "nutraceuticals" is frequently used to describe extracts, mostly from plants, that have health benefits. Moreover, the more established population is "tempted" to consume large uncontrolled quantities of nutraceuticals because, according to studies, the health effects are detected only if the intake is high. Most capsules of vitamin E that are sold in the market are of 100 to 400 i.u., while the health regulations recommend a daily intake (mostly from food) of vitamin E of approximately 50 i.u. Similarly, the amounts of recommended consumption by the nutraceutical companies for other supplements are, in many cases, thousands of percent above the daily intakes recommended by the health authorities. What are the reasons for such discrepancies—poor formulations or new human needs?

In the last 10 years, an additional important trend has been identified—nutraceuticals not as health strengtheners but as remedies. Individuals consume "fast acting nutraceuticals" which are characterized by a prompt or almost immediately measurable effect on human health. Humans suffering from chronic diseases would like to see measurable relief and the beneficial effect of the nutraceuticals within short periods of time. Even healthy individuals expect to observe improvement in their "good health" within a short period of time after nutraceutical intake. As a result, a dramatic increase in the consumption of nutraceuticals is seen. We termed this shift in human attitude "the third paradigm in nutraceutical science." The slogan in the stores has changed from "healthier foods for a healthier tomorrow" to "healthier foods of health improvers for a healthier life today." In other words, use nutraceuticals that will allow you to "be healthier today and not tomorrow." Some of the nutraceuticals that are advertised in this category are glucosamines for healthy cartilage, fenugreek for management of glucose levels, lutein to reduce cataract risks, phytosterols as an anticholesterol agent.

It now seems clear that nutraceuticals may soon become preventive pharmaceuticals (or good health maintenance drugs) or in some cases even replace present drugs. However, as has been seen, some nutraceuticals have very poor bioavailability to our systems and they need treatment, purification, and characterization, but mostly delivery vehicles. Unlike many drugs, their bioavailability is restricted because of poor solubility and presystemic decomposition. Most of the new carotenoids (lycopene, lutein, zeaxanthin, astaxanthin, omega fatty acids) have shown strong antioxidation activity which is beneficial in retarding cardiovascular diseases and reducing the risk factors for aging, but they can easily be oxidized on the shelves and become noneffective or even pro-oxidants.

Scientists are searching for new and more efficient vehicles to carry active molecules into the blood stream. These new vehicles should solubilize or entrap nutraceuticals and modern drugs that are sensitive to environmental conditions, nonsoluble, poorly absorbed, and decompose in the digestive tract.

Some of the most promising vehicles are microemulsions. Oil-soluble components are dissolved into oils that substantially improve their half-lives in the body. As the liver or kidneys do not rapidly clear microdroplets of oil, the potency of the dissolved compounds is substantially increased. Water-soluble components can similarly be packaged into nanometer-size structures, nanovehicles, to increase their half-life and biological potency. Using the microemulsion technique, water- and oil-soluble components from different plant extracts can be combined to attain a specific therapeutic goal. Microemulsions can be introduced into the body orally, topically to the skin, nasally as an aerosol for direct entry into the lungs, or via an intravenous "drip."

MICROEMULSIONS AS NANOVEHICLES

In the market, nutraceuticals are seen in many forms: some in a powdered form (tablets) and some encapsulated within protecting capsules or softgels (powders, pastes, or liquids). Others are dissolved and sold as liquid solutions (syrups, injectable solutions, or liquids). However, drugs with a high propensity for decomposition, and drugs for which efficiency and effectiveness are essential parameters in their physiological activity, require more sophisticated delivery formulations. Microcapsules, microspheres and nanospheres, nanopowders, nanocrystals, and nanodispersions are only some of the options. Other options are to deliver the nutraceuticals or drugs in liquid vehicles such as emulsions, double emulsions, microemulsions, and micellar solutions. The most sophisticated systems are liposomes, cubosomes, hexosomes, and so on. These latter two vehicles are still in an experimental stage.

Emulsions are suspensions of oil and water in the presence of surfactants. Most emulsions are unstable and will eventually separate into oil and water phases. The microemulsions described herein are emulsions of nanodroplets that are extremely stable and will not separate into different phases. Also, and most importantly, the oil droplets are extremely small, of < 0.1 μm in diameter. This allows the emulsion to be filtered, sterilized, and injected intravenously into the body. Ordinarily, oils must be sterilized by steam autoclave at 120°C for 20 minutes before they can be injected into the body. The vast majority of therapeutically active ingredients purified from medicinal plants cannot withstand the autoclave sterilization procedure. Nonsterile plant extracts can be orally ingested, but the body often does not absorb the desired therapeutic components, thus they may require sterilization. Oils can only be sterilized by heat or filtration in the form of microemulsions. These oil droplets are smaller than bacteria and can freely pass through large pharmaceutical filters.

Chronic diseases are complex processes that can rarely be controlled by a single chemotherapeutic agent. Although oil- and water-based components could be introduced separately into the body, it would be more effective to introduce them together in the form of a stable microemulsion. The oil droplets in the microemulsions are small enough to traverse the entire capillary network of the body without generating the risk of blocking blood flow to the tissues. These nanodroplets can penetrate virtually every organ of the body, including lymph nodes and the deep vasculature of solid tumors. They are too large to be removed by the kidneys and too small to be fixed by complement or be readily removed by the liver. They are an ideal delivery vehicle that allows Nature's medicines to be optimally and economically used in the treatment of infectious and somatic diseases.

This chapter focuses on the use of microemulsions as drug and nutraceutical delivery vehicles. It is the authors' attempt to review some recent progress on solving some of the problems related to microemulsions and, at the same time, to draw the reader's attention to some new, modified, and more sophisticated microemulsions as delivery systems that have been developed in the last five years.

This chapter is divided into two major parts. In the first part, we summarize some of the major concepts related to the formation of microemulsions in general, and, in particular, we bring data related to the formation of a "new U-type microemulsion that is fully dilutable and of high solubilization capacity." The second part is devoted to applications of microemulsions as delivery systems for drugs and nutraceuticals. This part also stresses recent progress in the U-type microemulsions as delivery systems of poorly water-soluble drugs and nutraceuticals.

PART I—MICROEMULSION PREPARATION AND MICROSTRUCTURES

Drug Administration Routes

Pharmaceutical scientists are mostly concerned, initially, with how a drug is to be delivered orally into the gastrointestinal (GI) tract and its absorption into the blood stream. Oral administration is patient-friendly, and mass production of oral dosage forms, such as tablets or capsules, is relatively simple. However, the oral route has its limitations. Drugs can be degraded by the low pH found in the stomach or by digestive enzymes. Those problems can be overcome in some cases by coating the active compound. A drug may also be poorly absorbed across the GI lumen (this particularly applies to hydrophilic and high-molecular weight molecules). The immunosuppressant cyclosporin, for example, has a combination of poor water-solubility and poor absorption characteristics that result in a low and variable bioavailability unless solubilizing additives are present in the formulation.

Even if the drug is adequately absorbed through the GI tract, it may be rapidly inactivated by the body on its first pass through the GI tract wall and the liver. Retaining such a drug in the body would, therefore, require a very short interval between doses, and ultimately a highly patient-unfriendly dosing regimen. Taking a single full-day's dose in conventional tablet form is often impractical owing to the risk of overdose. By sustained delivery technology, the drug is gradually released over periods of up to one day. Such release patterns allow expansion of the dosing interval without inducing super-high initial blood levels.

To solve some of the problems, drugs can now be solubilized by surfactants and formulated in microemulsions. Thus, the bioavailability is expected to be much higher, and varies less within and between individuals, than the original formulation.

Definitions and Concepts

Almost any review on microemulsion-based media as a drug delivery system starts with acknowledging Hoar and Schulman, who had already noticed in 1943 that by titrating a milky emulsion (a mixture of water, oil, and surfactant) with hexanol, a clear single-phase and stable solution was formed (2a). Schulman and associates also coined the term microemulsion that holds up to this day (2b). The definition has changed slightly through the years, but the general concepts are still intact.

Most scientists use a general and broad definition for microemulsion as "a system of water, oil, and surfactant (or amphiphile) which is a single optically isotropic and thermodynamically stable solution." Such a definition does not require the existence of any microstructure within the system, although it is clear that without the existence of an interface separating oil from water by surfactant, the mixture will not have any solubilization and delivery capabilities. It is well established today that microemulsions can appear in at least three major microstructures: water-in-oil (w/o), bicontinuous structure, and oil-in-water (o/w) (Fig. 1).

For a microemulsion to exhibit delivery properties, the existence of microstructures in the mixture must be clearly demonstrated. Moreover, the transitions from

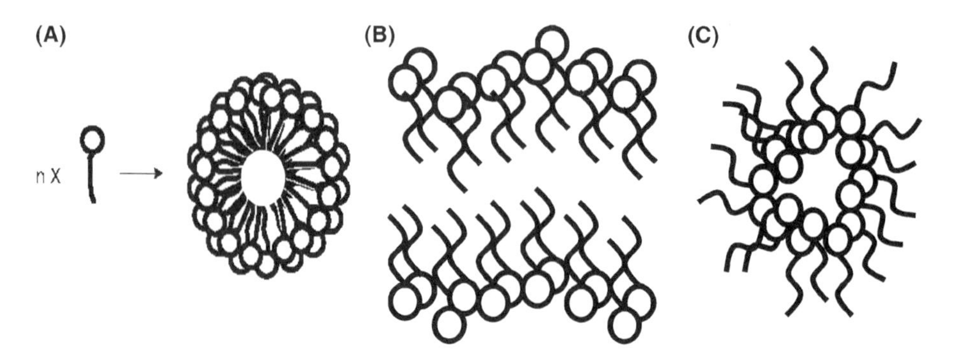

Figure 1 Crude and schematic representation of the three most commonly encountered microemulsion microstructures: (**A**) oil-in-water, (**B**) bicontinuous, and (**C**) water-in-oil.

one microstructure (e.g., w/o) to another (e.g., o/w) in empty systems and in systems consisting of droplets with solubilized drug molecules are very important. These key considerations must be addressed by researchers. However, it is clear from most pharmaceutical reports that these issues are not always specifically studied.

The main difference between emulsions and microemulsions is in the size and shape of the droplets that are dispersed in the continuous phase, reflecting the differences in the thermodynamic stability of the two systems (Table 1). Emulsions are kinetically stable but thermodynamically unstable, and after storage or aging, droplets will coalesce and the two phases separate. In contrast, microemulsions are thermodynamically stable and will not separate into the corresponding phases. It should be stressed that the term "mini-emulsions" was coined by some authors to describe emulsion droplets of submicron size with improved stabilities, other scientists may call those emulsions "nanoemulsions." Needless to say, these emulsions have been frequently adopted by formulators. While nanoemulsions do not have a long shelf life, they frequently are freshly prepared and used. It should also be stressed that in some studies, the authors neglect to test stability and consider mini- or nanoemulsions to be true microemulsions. Therefore, it is strongly recommended that the published works and mainly the experimental part of those vehicles be read with great attention.

The stability of microemulsions has been explained by many authors. The detailed and explicit theories explaining microemulsion stability will not be discussed in this chapter. We will only stress that the free energy gain of microemulsions is significant to keep the droplets intact and stable. The free energy gain is considered to be derived from: (i) the extent to which surfactant lowers the interfacial tension between the two phases and, (ii) the change in entropy of the system. The free energy, G_f, is composed of an enthalpy term $\gamma_i A$. The interfacial tension, γ_i, in the presence of excess surfactant and cosurfactant, is very small and close to zero. The parameter A reflects the surface area of the small droplets and is very large. The term $\gamma_i A$ contributes to the destabilization effect of the microemulsion, but it becomes very small with the

Table 1 Summary of the Main Differences Between Microemulsions, Nanoemulsions (Here, Termed Miniemulsion), and Emulsions.

Properties	Emulsions		Microemulsions
	Miniemulsion	Macroemulsion	NSSL
Visual aspect			
Typical characteristic size	20–200 nm	> 1 μm	10–100 nm
Stability	Kinetic		Thermodynamic
Formation	Energy input		Spontaneous
Surfactant concentration	Low		High

Source: Courtesy of Prof. C. Solans, taken from a conference presentation.

decrease in the interfacial tension. In contrast, the term $T\Delta S$ reflects the change in the entropy of the microemulsion, the effective dispersion entropy, as a result of the formation of very small droplets and is very significant. $T\Delta S$ is also significantly large owing to the mixing of one phase into the other and the formation of an enormous number of droplets. In addition, the surfactant diffusion in the interfacial layer and the monomer–micelle surfactant exchange also contribute to the entropy gain. Therefore, the term $\Delta G_f = \gamma_i A - T\Delta S$ in microemulsions is always negative.

Formulators have to carefully select the surfactant and the nature of the two phases so that the interfacial tension will always be close to zero, the oil is solubilized with the surfactant tails, and the aqueous phase will properly hydrate the head groups of the surfactant. If such selection is made, the destabilization effect because of the gain in surface area will be minimal, and the gain in entropy will be maximal. Such microemulsions will be formed spontaneously (self-associate or self-aggregate) and will be thermodynamically stable.

Another important consideration in the formation of microemulsions is related to the packing parameter, which is important for structures with high curvatures (Fig. 2). Surfactants must have the proper molecular volume dimensions and proportions to effectively pack into a micellar structure. Oil phases with high molecular volume fractions (such as triglycerides) will pack less efficiently and will have difficulties in entering between the surfactant tails. This is also reflected in the isotropic regions of a phase diagram. It should be stressed that the o/w microemulsion droplets generally have a larger effective interaction volume than w/o droplets. Also, whereas emulsions consist of roughly spherical droplets of one phase dispersed into the other, microemulsions constantly evolve into various structures ranging from "droplet-like" swollen micelles to bicontinuous structures, frequently making the usual o/w and w/o distinctions irrelevant.

Because the size of the particles is much smaller than the wavelength of visible light, microemulsions are transparent and their structure cannot be observed through an optical microscope.

Microemulsions have very high surface areas and, therefore, it is obvious that they can incorporate, in their core or at the interface, large quantities of molecules that are usually insoluble in the continuous phase. The molecules that are incorporated at the interface are solubilized rather than dissolved, and, therefore, microemulsion efficiency is measured based on the solubilization capacity of guest molecules.

As already stressed, to form microemulsions, the interfacial tension of the oil/water interface must be reduced to zero or almost zero, and the interfacial layer must

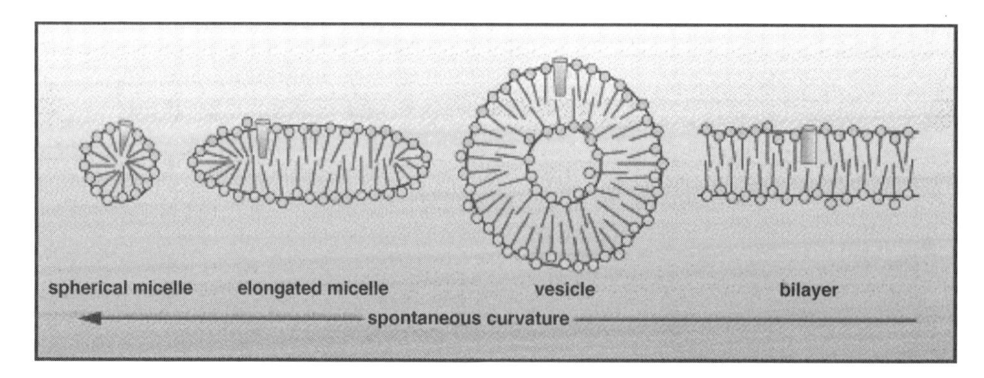

Figure 2 Schematic representation of the formation of the most commonly encountered microemulsion microstructures of oil-in-water and the packing parameter.

be kept flexible and fluid. The cosurfactant helps to achieve those two prerequisites. The cosurfactant, which is incorporated into the interface, keeps the film flexible, fluid, and tightly packed. Extensive studies have been done on microemulsions using cosurfactants, usually short or medium chain alcohols that are not edible, nonpharmaceutical grade, and an irritant to the skin. Only in a few "true microemulsions" were the alcohols replaced by a blend of high and low hydrophilic lipophilic balance (HLB) surfactants instead of cosurfactants.

Many review articles have been written in the last 10 years on microemulsions and their structural and physical properties in general, and their use for pharmaceutical applications (3–10).

Lawrence and Rees wrote two excellent reviews (4,5). The first, in 1994, on "microemulsions and vesicles as vehicles of drug delivery" stressed all the potential applications of these systems as well as the physical aspects related to the formation of microemulsions (4). We examined the examples of microemulsions that are listed in the review, and in the published literature thereafter, and the phase diagrams that were used by scientists in the 1990s as the basis of most formulations for solubilization and delivery of drugs. We found that, in most of the cases, the basic concepts of formation of microemulsions were known to the scientists and the formulators in the late 1990s, but the examples of types and compositions of microemulsions with relation to the nature of the solubilizates, solubilization capacities, selection of surfactants, oils, cosolvents, cosurfactants, and others were very limited and based on simple nonsophisticated compositions drawn from the basic research papers done by colloid chemists and physicists. Detailed examination of the initiatives that were taken in those years reveals that most authors searched for improved kinetics of release of existing drugs and existing formulations and compositions, and not for new compositions that could provide improved solubilization capacities of poorly soluble drugs.

It can also be clearly seen that the understanding of the relationship between the nature of the solubilizate, its actual location at the interface, and its effects on phase transitions and microstructures that are formed was very limited and practically nonexistent.

Lawrence and Rees wrote a review that stressed the progress made in the year 2000 and strongly emphasized the need for more basic science to be utilized by those making drug microemulsions (5). The review seems to have influenced some scientists, and recently some researchers have changed the scientific focus of their work and explored more the "relation between microstructure and the delivery properties of the microemulsion." Lawrence and Rees gave a new momentum to much of the work presently being done (5). It is worth mentioning that the authors stressed the differences between "self-microemulsifying drug delivery systems (SMEDDS)" that are actually not microemulsions (although they may be considered to be closely related systems) and true microemulsions. A SMEDD typically is comprised of a mixture of surfactant, oil, and drug (known also as concentrate) which, when incorporated into the body, is rapidly dispersed to form droplets of approximately the same size range as those in microemulsion systems. Once dispersed, such systems would be expected to behave in vivo much the same as o/w microemulsions. In most studies, the results were not tested by those criteria. The definition does not specify any structural request from the formulation. The presence of co-solvent in the concentrated system might facilitate or induce the formation within the digestive tract of a "true solution" and thus lose its internal droplet structure and no longer have a pool of water or oil and an interface between the inner reservoir and the continuous phase. Also, as the publications do not provide any simulation data, it is not very clear what happens to "the consumed concentrate" upon dilution. We know today, and it will be stressed at a later stage in this chapter that upon aqueous dilution, the reverse micellar (or o/w microemulsions) will turn into a bicontinuous microemulsion and upon further dilution might invert into an o/w ME or it might phase separately and the drug leaches out of the reservoir in a noncontrolled and premature stage. Such information is mostly lacking in many of the early studies.

It should be stressed that most of the pharmaceutical research publications do not include any evidence related to the microstructure of the systems and the reader cannot clearly correlate the solubilization or delivery results to the existence of surfactant interface, to the presence of boundaries between the oil and water phases, or to the existence of an inner core or reservoir within the nanosized droplets. Moreover, it is very clear that once concentrates are formed, the amphiphile, be it hydrophilic or hydrophobic, will self-aggregate to form reverse micelles and the drugs will be solubilized at both the interface of the reverse micelles and continuous oil phase. However, once dilution takes place in the digestive tract or in any other aqueous formulation, there are significant structural changes that occur in the self-assembled structure to sometimes form liquid crystalline structures and, in other cases, bicontinuous structures that then upon further dilution are inverted into an o/w microemulsion. The amphiphilic interface changes dramatically, and the solubilization capacity is expected to change. The hydrophobic solubilizates will gradually be released from the interface, and the hydrophilic drugs may be better solubilized and strongly attracted by the surfactant to the interface and be more difficult to release.

Microemulsions Composition and Ingredients

Several review articles have been written in the last few years on the methodology of creating microemulsions and on their microstructures (3–8,11–14). A very comprehensive review entitled "Influence of microemulsion phases on the preparation of fine-disperse emulsions" describes in great detail the physical meaning of phase behavior of ternary oil/surfactant/water systems, the Kahlweit-fish diagram and Winsor-type microemulsions with relation to macroscopic phase behavior, and an isotropicity of a ternary surfactant/oil/water system (6). The review discusses the concepts related to the microemulsion formation and the influence of additives on those microstructures. The authors explain in great detail the effects of minimal interfacial tension and the geometric aspects required for proper formation of microemulsions. In another very comprehensive review, Bagwe et al. described specifically how to achieve "improved drug delivery using microemulsions" (7). The review focuses on defining the current technology of microemulsions in relation to pharmaceutical applications. The authors devoted a full section to microemulsion formulations in the pharmaceutical industry, beginning with the introduction to various methods of drug delivery and continuing with the important factors that influence each method. Specifically, they discuss how to make microemulsions out of nonionic, cationic, and anionic surfactants.

It is well known that the most common way to form a microemulsion is to construct a phase diagram of at least three main ingredients: the oil phase, the aqueous phase, and a surfactant. Compositional variables can be studied more easily this way. A simple ternary phase diagram will be composed of oil, water, and amphiphile. Each corner in the phase diagram represents 100% of that particular component.

Each of the ingredients occupies one of the triangle corners. The surfactant is mixed with one of the phases at various ratios, and the mixture is titrated with the other immiscible phase. The isotropic clear regions are identified by optical observation after reaching equilibrium and a phase diagram is constructed. The isotropic regions can represent formulations where the inner core phase is water and the microemulsions are denoted as w/o ME; if the entrapped inner phase is oil, the microemulsions will be called o/w.

A hypothetical phase diagram of the three components where the liquid crystalline phase is located in the surfactant-rich corner and the two-phase region is located between the oil and water corners is demonstrated in Figure 3. The large area of the phase 1 region is contributed by the surfactant to the formation of w/o, bicontinuous, and o/w structures. It is demonstrated in Figure 4 a more realistic, or typical, three-component phase diagram from which one can clearly see the formation of all four categories of the isotropic regions separated by nonisotropic two phase regions.

Many of the ternary phase diagrams that have been constructed and studied have very narrow isotropic regions of either w/o (Fig. 5) or o/w (Fig. 6) microemulsions, even in the presence of a cosolvent. It is quite clear that the isotropic region will become larger if the two phases are more alike in their miscibility. Thus, when a hydrophilic surfactant, such as sucrose monostearate, is mixed with a blend of hydrocarbons and butanol, the one phase region will extend almost to the far corner of the water phase, and the structures will be of w/o and bicontinuous (Fig. 5). However, if the surfactant is hydrophilic and no cosolvent or cosurfactant is added, the o/w microstructure will be formed, but the isotropic region will be very limited in its area (Fig. 6).

Therefore, more commonly, in almost any pharmaceutical microemulsion, the microemulsion will contain additional components such as cosurfactant, cosolvent, and drug. The cosurfactant does not form microstructures by itself, but it helps to form a more tightly packed interface and to reduce the interfacial tension to zero. In cases where four or more components are incorporated, pseudoternary phase diagrams are used where one of the corners will typically represent a binary mixture of two components such as oil/cosurfactant, surfactant/cosurfactant, water/cosolvent, oil/drug, or water/drug.

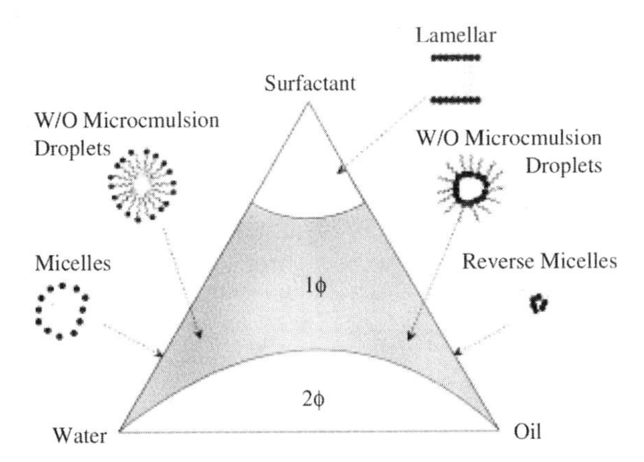

Figure 3 A hypothetical pseudoternary phase diagram of an oil/surfactant/water system with emphasis on microemulsion and emulsion phases. Within the phase diagram, existence regions are shown where micelles, reverse micelles, water-in-oil microemulsions, and oil-in-water microemulsions are formed, along with the bicontinuous phase. *Source*: From Ref. 5.

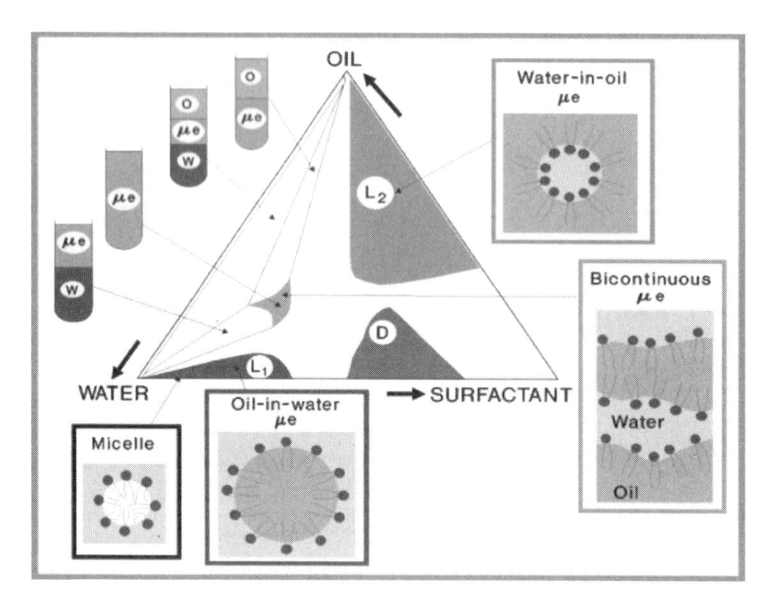

Figure 4 Classical phase diagram based on oil phase, AOT (Aerosol OT 100 [bis(2-ethylhexyl) sodium sufosuccinate]) as surfactant, and water. The different microstructures are denoted in the phase diagram.

Preparation

Microemulsions are formed virtually instantaneously, as the interfacial tension is close to zero and light mixing is sufficient to make the surfactants self-assemble. However, some equilibration time might be needed, mainly when the system is approaching the phase separation or phase transition boundaries. It is therefore recommended to allow time for such equilibration. In some cases, this might take hours (mainly if macromolecules are used).

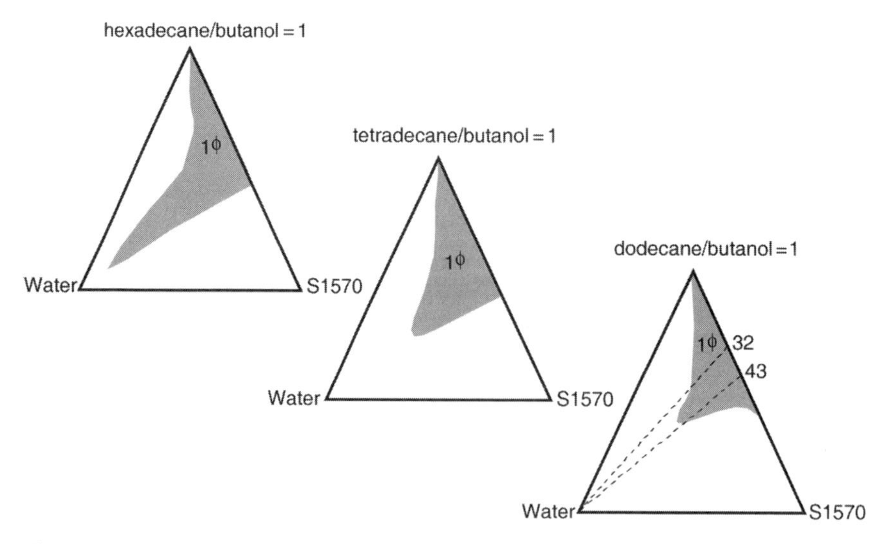

Figure 5 Example of phase diagrams composed of sugar ester, paraffin oils, butanol and water. The dark colored regions are the isotropic regions.

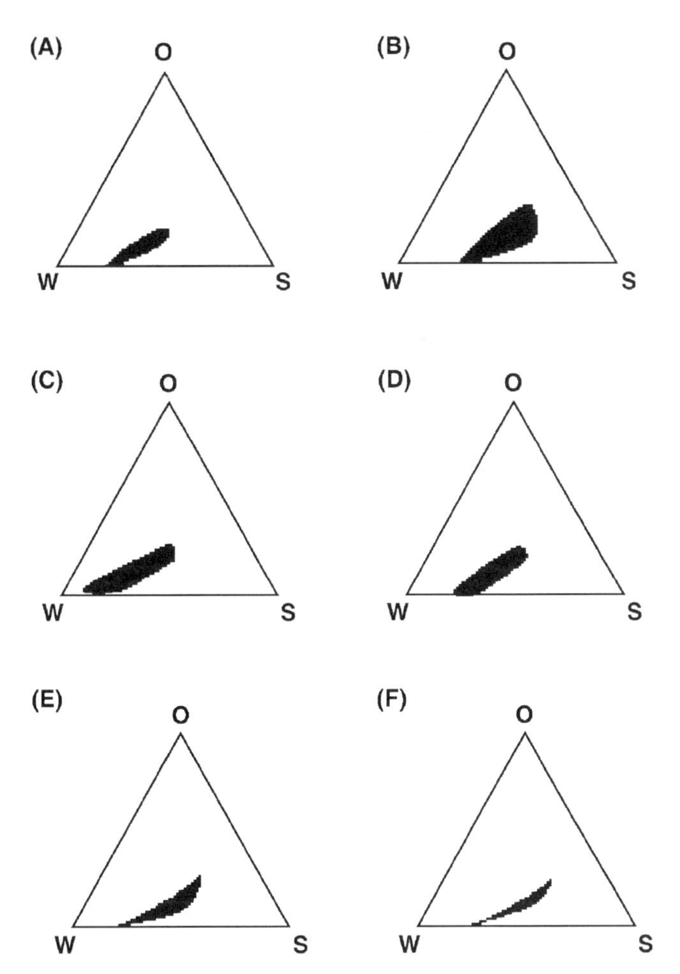

Figure 6 Partial phase diagrams of the isopropyl myristate (O), polysorbate 40, sorbitol (S) and water (W) systems, showing areas of existence of oil-in-water microemulsions at 37°C for polysorbate/sorbitol mass ratios of (**A**) 1/1; (**B**) 1/1.5; (**C**) 1/2; (**D**) 1/2.5; (**E**) 1/3; (**F**) 1/3.5. *Source*: From Ref. 5.

Constructing a full phase diagram is time-consuming, particularly when the aim is to accurately delineate a phase boundary, as it is necessary to prepare a large number of samples. It should be clear that true microemulsions form spontaneously and the surfactants self-assemble; therefore, there is no need for sophisticated machines that are commonly used in the formation of emulsions. However, some preparations might require some light stirring. Some authors have used sonication for forming the microemulsions. It should be stressed that the use of sonication or increased temperature is needed when solid drugs or other solid components are to be dispersed into the microemulsion (ME).

The solubilization of an active ingredient (drug or other additive) is sometimes a tricky process, as in many cases the pre-dissolution of the drug is essential, and in some cases the drug or the nutraceutical is totally insoluble in any of the individual ingredients. The solubilizate becomes soluble or solubilized only in the presence of the surfactant or the cosurfactant (or cosolvent); therefore, a proper formulation will require some knowledge (or pre-solubilization work) of the solubility of the drug in individual or multiple ingredients.

Examples of nonsoluble nutraceutical solubilization of lycopene, lutein, and phytosterols in o/w microemulsions are shown in Figure 7. Lycopene, for example, has practically no solubility (approximately 100 ppm) in water, has very limited solubility in any food-grade oil or solvent, and is solid at room temperature. To study and quantitatively determine the solubilization capacity of the microemulsions with lycopene, it is essential to add the lycopene in its molten form (heat over 120°C) to the oil/surfactant/cosurfactant mixture and keep the concentrate at high temperatures until full dissolution (solubilization) is reached and then titrate the concentrate with water.

One should bear in mind that usually the drug is solubilized into the microemulsions to its maximum solubilization as determined by optical observation (clear solutions). However, in many cases, prolonged storage might cause the drug to precipitate from the microemulsion; seed crystals start to appear and might grow to large crystalline materials that will precipitate out at the bottom of the vessel. Crystallization takes place mostly when the drug is a crystalline material of low solubility in all of the pure ingredients and the ingredient blends. Some reports that have shown solubilization data were found to be misleading, because on prolonged storage (days or months) the microemulsions turned cloudy or crystalline material was detected. The effect is more severe if, during storage, fluctuations of temperature occur. In other cases if, at the preparation stage, the solubilized material was not fully dissolved within the concentrate or in the final microemulsion, the remaining crystalline material can serve as a nucleation agent, and fast nucleation and growth followed by precipitation will be observed. In other cases, if the microemulsion is further diluted with water or aqueous systems, the solubilizate reaches its maximum solubilization values, and precipitation is seen.

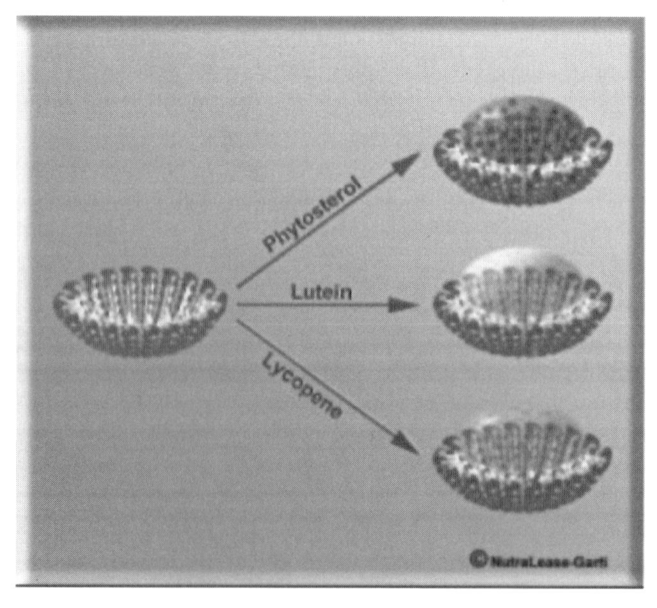

Figure 7 Schematic illustration of the solubilization of nutraceuticals onto microemulsion vehicles.

One should remember that in many of the drug delivery systems the amounts of solubilized drug are very small. With such small-solubilized drug concentrations it might be difficult, by the naked eye, to detect turbidity that is derived from colloidal precipitation of the drug, and the authors might consider it as minidroplet formation. However, such formulations, with time, will leach out or release and precipitate the drug, and its effectivity and bioavailability will be dramatically reduced. It is therefore advisable to use some advanced techniques to distinguish between colloidal precipitation and miniemulsion droplet formation.

Scientists have used hydrophilic drugs for the formation of o/w microemulsions and have taken advantage of the interactions that the surfactant might have with the drug. It is again essential to establish what the nature of such interactions is, as our work shows that, in many cases, the surfactant serves as a templating agent and the drug will precipitate from the microemulsion in different polymorphic structures. In many cases, the drug forms hydrates or amorphous solids and will have very different characteristics than the original one (dissolution kinetics, bioavailability, pre-systemic metabolism, environmental stability, etc.).

From the recently published papers, we realized that many authors do not report much about the formulation preparation stages (as microemulsions are considered self-associated structures) or the mode of solubilization of the drug, and in many cases there is no structural work to describe phase transitions, structural changes, and so on, that might occur upon aging or dilution.

Care must be taken to ensure that the observations are made on stable, and not metastable, systems. Clearly, even if microemulsions are thermodynamically stable when empty, time constraints impose a physical limit on the length of time required before stating that "the formulations are clear stable microemulsions" and moreover before claiming that the system can be stored indefinitely.

Some authors conveniently use centrifugation to speed up separation and to reach equilibrium. Such rapid screening procedures appear in the literature but in our opinion, should not be practiced (15).

Outside the microemulsion region, particularly for compositions close to the oil–water binary axis, there is insufficient surfactant to facilitate the formation of a single microemulsion phase. In this case, multiple phases may exist, the complexity of which increases with the number of components in the mixture. Within this region, and indeed other multiphase regions of the ternary phase diagram, microemulsions can exist in equilibrium with excess water or oil. These multiphase systems can be conveniently described using the Winsor classification; we will use only the Winsor IV systems, which are practically one-phase systems.

Temperature of Preparation

Temperature is an important parameter when making microemulsions because of two major concerns:

1. If microemulsions are made from nonionic surfactants, based on ethoxylated head groups, the HLB of the surfactant varies significantly with temperature. Increase in the system temperature will cause surfactant solubility in the aqueous solution to decrease, which might result in interfacial structural changes that might also lead to phase separation. Phase diagrams should, therefore, be constructed as a function of temperature. Once solubilizate is added, the nature of the interface becomes even more temperature-sensitive and certain drugs will be removed from the interface

while others will be better solubilized. Nonionic surfactants based on hydroxylated head groups (sugar esters or polyglycerol esters) are less sensitive to temperature and therefore, the total isotropic region in the phase diagram is less affected. However, as those surfactants are in part solid in nature, the Krafft points of those systems are very important parameters that must be considered. In others, the cloud points should also be taken into consideration when temperature variations are expected during the formulation and storage or at application.

2. Microemulsions might be very sensitive to temperature mainly if the phase transition temperature (PIT) is quite low and the operation conditions are close to the PIT. Anionic surfactant might lower the PIT of the nonionic surfactants and so do salts.

Isotropic Regions and Phase Transitions

Transition from w/o to o/w microemulsions may occur via a number of different structural states including bicontinuous or lamellar mesophase, and also via multiple phase systems (formation of emulsions). Nonionic-based microemulsions are temperature-sensitive, which might result in a temperature increase changing the PIT. These effects are also dependent on other components such as the nature of the oil and its content. When the amount of solubilized water increases, the w/o nanodroplets are no longer spherical. The structures have the characteristics of both the oil and the water phases as the continuous phases. These structures are denoted as bicontinuous structures. The isotropic regions in the phase diagram can be small, meaning that only small amounts of water are entrapped in w/o ME. Or, alternatively, the isotropic o/w regions can be in very small areas of the phase diagram indicating that only small amounts of the oil are solubilized within the surfactant layer. The total isotropic area in the phase diagram is termed A_T and the amount of solubilized water or oil along any dilution line is termed W_S. As one can see from the phase diagram (Fig. 8), the blend of AOT (Aerosol OT 100 [bis(2-ethylhexyl) sodium sufosuccinate)], decane, and water leads to small isotropic regions, a small bicontinuous region, and a small additional mesostructure called lyotropic liquid crystalline (LLC) structure. LLC structures are organized structures with interesting properties, but they will not be discussed in this chapter. To enlarge the isotropic regions in the phase diagram, the introduction of additional cosurfactants and, in some cases, cosolvents has been suggested. The cosurfactant will help to further reduce interfacial tension and will be incorporated within the surfactant layer to achieve better interfacial packing. Some of the most common cosurfactants are short-chain and medium-chain alcohols and, in some cases, those are surfactants with lower or higher hydrophilicity in comparison with the main surfactant. In many of the cases, the cosurfactant and the cosolvents are the same molecule. Transitions from one microstructure to another is very dilution-dependent, but might be dictated also by the presence of other additives such as electrolytes.

Phase Diagrams—U-Type Phase Diagrams

Making w/o microemulsions with up to 40% to 50% solubilized water in the core of the surfactant aggregates is a relatively easy task. Also, formulating an o/w microemulsion composed of 10 to 40 wt.% oil in the core of the direct swollen micelles is feasible. Yet, the isotropic regions for each of those preparations are small and discontinuous, that is, in most cases one cannot transform one microstructure into another by simple water or oil dilution. In other words, the dilution lines at any surfactant/oil ratio are discontinuous, and upon dilution the system goes through an area of emulsion or phase separation. Practically, this means that the oil-based concentrates are not freely dilutable. The compositions into which the isotropic region can be extended are of great importance to those who are trying to maximize the amount of solubilized dispersed phase in the droplet core ingredients and thereafter, the amount of the solubilized drug molecules. The larger and more extended the isotropic region, the better are the chances that the drug will be accommodated in the core of the microemulsion or at its interface. For practical reasons, one needs the highest levels of the dispersed phase and lower levels of the surfactant phase with maximum solubilization capacities. The size of the isotropic area, by percentage of the total area of a phase diagram, is termed A_T (total isotropic area within the phase diagram). Constructing phase diagrams with large isotropic regions (large A_T) has been always a challenge.

In addition, in many cases, it is important to form an "oil-based concentrate" that will contain the oil, surfactant, cosurfactant, and solubilized active matter. The selected formulation of the microemulsion oil–based components and the drug are the SMEDDS. The water dilution of such concentrate is termed "the water-dilution line," and the maximum amount of solubilized water is termed "the solubilization capacity of water," (W_m). As one can see from the phase diagram in Figure 8, the maximum amount of the solubilized water at SMEDDS 70/30 (70 wt.% surfactant and 30 wt.% oil phase) is 50 wt.% water. In practice, the drug will be delivered as a concentrate or diluted with some amount of water along the dilution line; prior to its intake, one should be able to dilute it with aqueous phase to the required dilution levels. Such flexibility of use is not obvious. Most compositions that have been offered by scientists making microemulsions are not dilutable with the aqueous phase. Therefore, in most cases, the microemulsions are of either w/o or of o/w droplets and, therefore, are very restricted in their use and difficult to use as "general purpose use" vehicles. Also, if the microemulsions are not fully dilutable, chances are that, upon aqueous dilution in the digestive tract, the microemulsion will disintegrate and the drug will precipitate or crystallize out. From Figure 8, one can also see that the term S_m is calculated reflecting "the maximum amount of solubilized inner phase (in this case water) that can be solubilized per amount of surfactant' (the maximum surfactant solubilization capacity).

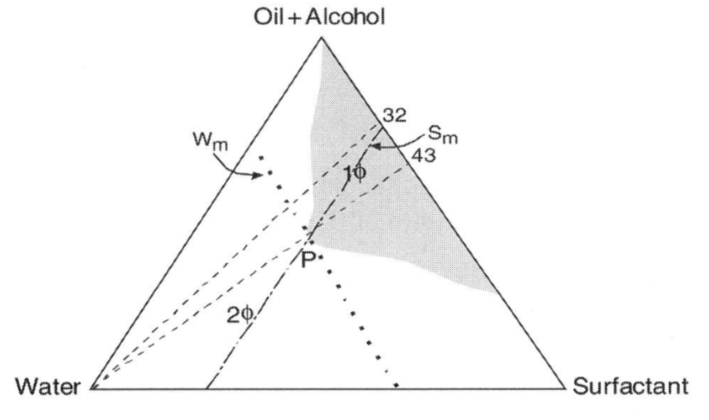

Figure 8 Typical phase diagram of water-in-oil prepared with sugar esters as surfactant, vegetable oil as oil phase, and butanol as cosurfactant. The dilution lines and the terms S_m and W_m are also illustrated in the diagram (see text).

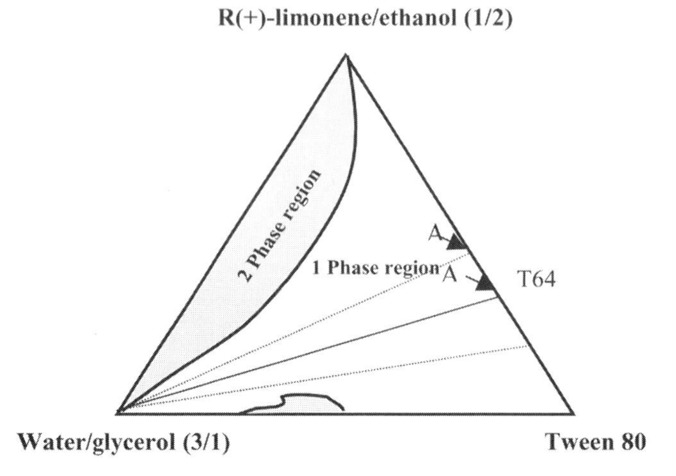

Figure 9 Typical U-type phase diagram of Winsor IV microemulsion based on five-component system showing full aqueous phase dilution along dilution lines 6:4, 7:3, and 8:2 (oil phase to surfactant weight ratios).

In our recent studies, we have found the criteria and proper blends of surfactants, cosolvents, and cosurfactants to make the so-called U-type phase diagram where the oil phase concentrate is fully dilutable with the aqueous phase to any desired levels up to the far water corner without any phase separation. Those unique microemulsions have been patented and are of importance to any practical application (Fig. 9) (16).

The Surfactant

The amount of surfactant in emulsions is very small, 0.1% to 1.0% of the total emulsion weight. The amount of surfactant in a microemulsion is a minimum of 10% of the total ME weight; such large surfactant levels are essential because of the large increase in interface area between the aqueous and oil phase.

Selection of a proper surfactant is the key to the formation of any microemulsion (17). The use of the HLB concept, similar to emulsion formation, is not relevant when microemulsions are formed. Hydrophobic surfactants will be suitable for the formation of w/o) microemulsions (ME), and the hydrophilic surfactants will form o/w ME. However, there are many examples in which hydrophilic surfactants will form w/o ME in the presence of a cosolvent; similarly, very hydrophilic surfactants, such as SLS, will require a cosurfactant to form an o/w ME. Some investigators, however, still use such an approach, whereas others prefer the measure of the critical packing parameter (CPP). The CPP is a measure of the preferred geometry adopted by the surfactant and, as a consequence, is predictive of the type of aggregate that is likely to form.

CPP is defined by the equation,

$$\mathrm{CPP} = V_H/al_H$$

where V_{H} is the partial molar volume of the hydrophobic portion of the surfactant, a the optimal head group area, and l_{H} the length of the surfactant tail. This measure has been elaborated by many scientists and it was claimed that for o/w ME, the CPP value should be of 1/3 to 1/2, while for w/o microemulsions the values are >1.

However, this measure has very little value for multicomponent MEs, as the geometry of the surfactant will change as a result of the presence of cosolvent or cosurfactant in its environment. Small oil molecules that solubilize the surfactant tail increase the effective surfactant hydrophobe volume, whereas large molecular volume oils would not affect the CPP of the surfactant. Similarly, an increase in ionic strength will decrease the effective head group area of an ionic surfactant, as the double layer will shrink. The presence of hydrophilic molecules that we sometimes term aqueous phase cosolvents, such as glycerol and sorbitol, will also influence optimal head group area by altering the solubility of the head group in the aqueous phase. For practical use, the concept of CPP has very little importance and should serve as a guideline only for first screening of formation of microemulsions of the two types. However, if large isotropic regions are needed, the use of cosolvents and cosurfactants is essential, and the packing parameter is no longer valid (Fig. 10).

In industrial applications, it is common to use inexpensive ionic surfactants but in food, pharmaceutical, and cosmetic applications, the ionic surfactant toxicity limits their use. Two examples are documented in the food industry, sodium stearoyl lactylate (SSL) and diacetyl tartaric acid esters of monoacylglycerols (DATEM). The most common anionic surfactants are the sodium di-isooctyl sulfosuccinate (AOT) and the sodium dodecylsulfate (SDS).

Nonionic surfactants are very often used in pharmaceutical microemulsion formation. Tweens (ethoxylated sorbitan esters) are well known and widely used. They are water-soluble and have high HLB values and, therefore, are used mainly for making o/wo/w microemulsions. Ethoxylated (with up to 40 EO units castor oil, ECO-40) and hydrogenated ethoxylated castor oil (HECO) with 8 to 40 EO groups

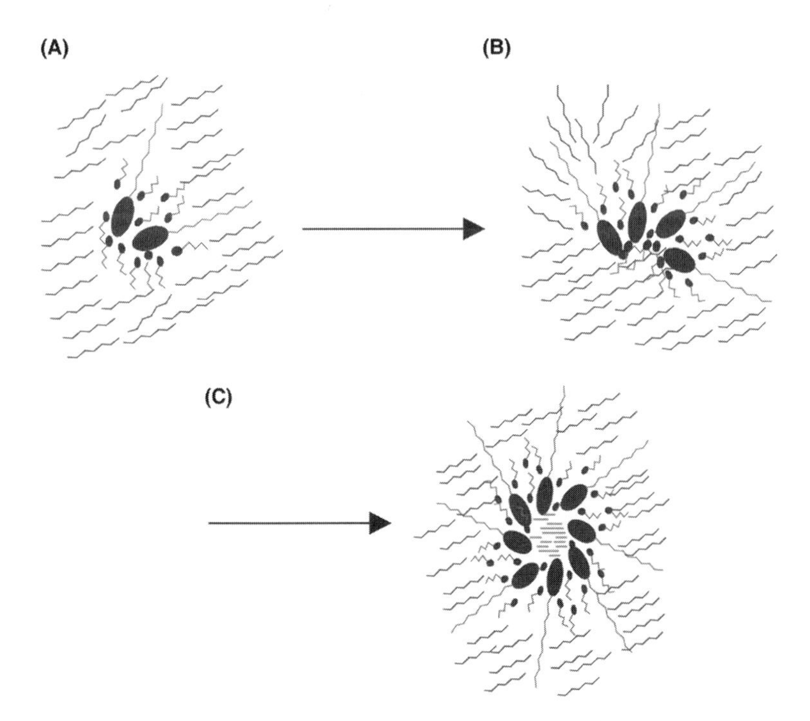

Figure 10 Formation of reverse micelles of hydrophilic surfactant oil phase, cosurfactant, and water. (**A**) reverse hemimicelles, (**B**) swollen micelles, (**C**) water-in-oil microemulsions.

attached to the hydroxyl group on the side chain of the triglyceride are regarded as very efficient surfactants. In ethoxylated fatty acids and fatty alcohols (Myrj and Brij, respectively) and also in polyethylene glycols (PEG of up to 200 EO units) or Poloxamers, polyoxyethylene glycol blocks appear in some of the formulations (18). In some more rare cases, Tyloxapol was used.

All the ethoxylated surfactants are temperature-sensitive and become more hydrophobic with increasing temperature. The two categories that are less temperature-sensitive are sugar esters and polyglycerol esters.

Sugar esters are nonirritant surfactants well known in Japan, but not widely studied in most of the Western countries to make microemulsions. Most scientists have used polyglycerol esters of fatty acids (PGEs) as surfactants because they felt they might have an advantage over other surfactants, as there were some indications in *early* studies that the PGE enhances permeation (19).

The PGEs are also less sensitive to temperature fluctuations. Ho et al. used Captex 300 (light mineral paraffinic oil), PGEs of HLB 8–13, and cosurfactants like ethanol, propanol, or butanol, and attempted to correlate the type of the PGE to the ME phase properties (19). They found that surfactants with very low or very high HLB do not form MEs, even in the presence of cosurfactants. Typical phase diagrams that were formulated with the PGEs are shown in Figure 11. The authors

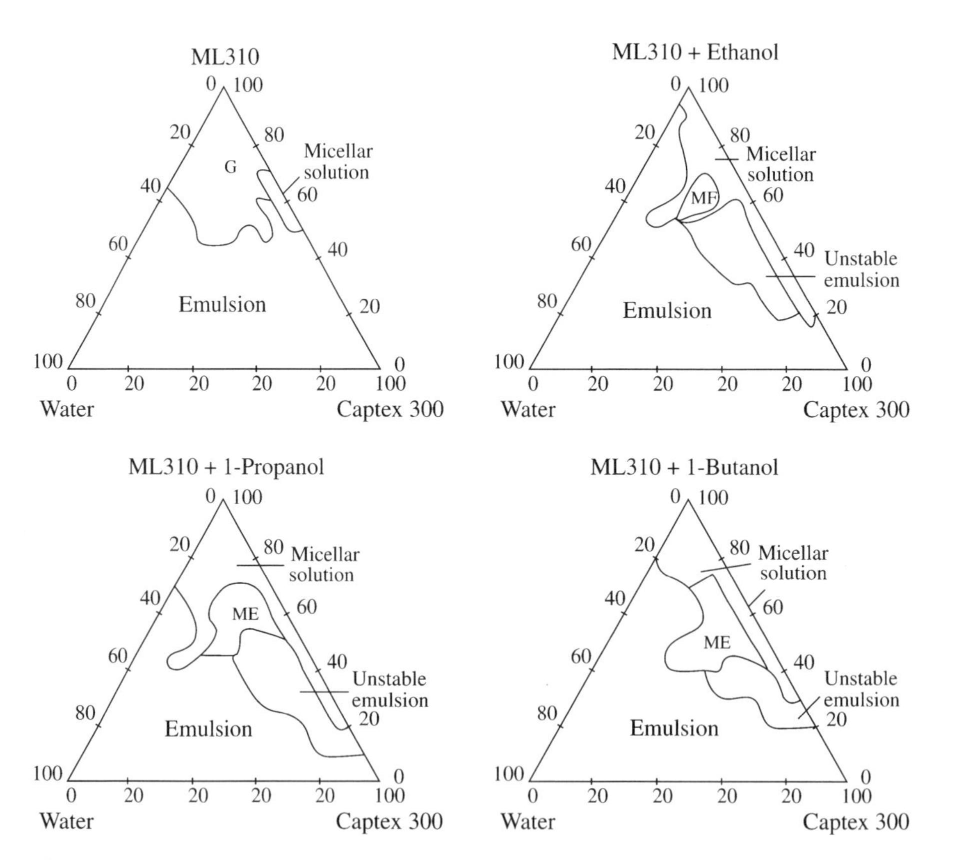

Figure 11 Phase diagrams of ML310 (tetraglycerol monolaurate)/Captex 300 (highly liquid paraffin)/water/alcohol system. *Source*: From Ref. 19.

dissolved insulin in the acidic aqueous solution of the ME, then added it to the oil/surfactant phase, and showed that the ME remained stable even if stored at 4°C. However, the formulation compositions clearly show that only w/o MEs have been formed by this emulsifier and only up to 49 wt.% water could be solubilized in those formulations. The authors did not provide any additional information on the behavior of those formulations upon further aqueous dilutions and there are no indications whether the insulin survives if the ME is further diluted (as in our digestive systems). The authors also fall short of studying the microstructure of the ME. Therefore, at first glance it seemed that the authors found a way to form pseudoternary phase diagrams with PGEs, which are very promising molecules for cosmetic and pharmaceutical applications, but much more work is needed to be able to formulate dilutable w/o MEs that can survive dilution effects.

Amphoteric surfactants like lecithin are of low toxicity and are considered as natural ingredients. There are many commercial lecithin products in the market with various degrees of purity and with a large variation in internal phospholipid compositions.

Combinations of ionic and nonionic surfactants may be effective in extending the isotropic regions in the phase diagram.

Quaternary ammonium alkyl salts are characterized by having a small head group in relation to their tails and, therefore, the CPP is a very suitable measure of for the formation of tightly packed spherical structures of w/o. Two main surfactants, cetyltrimethyl ammonium bromide (CTAB) and dodecyldimethyl ammonium bromide (DDAB), that form good microemulsions have been widely used (20–30). DDAB is an excellent surfactant that can form microemulsions without the need for alcohols, and has a very high solubilization capacity. Microemulsions based on DDAB were used to solubilize pilocarpine and chloramphenicol (26).

The Cosurfactant

Cosurfactants play a very important role in microemulsions. They help the surfactant reduce the interfacial tension to very low values to achieve thermodynamic stability. They both modify the curvature of the interface by incorporating additional apolar groups and provide more fluidity to the film, preventing crystallization of the tails of the surfactants, which could result in the formation of lyotropic liquid crystals. Cosurfactants are considered to be liquid crystal structure breakers. They are of less help to surfactants with unsaturated (double) bonds in their tails. However, they are especially essential when the surfactant has a saturated tail.

In most cases, the cosurfactants are short (ethanol) and medium-chain (propanol to octanol) alcohols. In some pharmaceutical applications (transdermal, ocular), we use mostly those that do not present irritation. Other molecules, such as amino acids and short organic acids, have also been utilized.

There is an interesting work by Fubini et al. on the evaluation of enthalpy in ME in the presence of alcohol (31). It shows the importance of the alcohol as a cosurfactant. Different amounts of butanol were added to two types of ME, differing only in the surfactant type and molar ratio of the components in the presence of butanol. Addition of different amounts of butanol yielded constant molar enthalpy changes, up to the critical microemulsion concentration, and the differences in the nature of the surfactant did not markedly influence the $-\Delta H$ values. A linear relationship between $-\Delta H$ and molar fraction of added cosurfactant was found at molar fractions below the one required for ME formation.

The Oil or the Organic Phase

For drugs that cannot be formulated as an aqueous solution, emulsions and microemulsions have typically been cost-effective and provided for ease of administration.

Hydrocarbons pack well within the surfactant tails and, therefore, they are the most recommended compounds for making large w/o and o/w MEs for pharmaceutical applications. Too long tails are not good oily phases because of lack of solubility. Other common oil phases include esters of fatty acids or fatty alcohol, depending on the nature of the application and the regulations.

The oil that typically comes to mind for pharmaceutical applications consists of digestible oils from the family of triglycerides, including soybean oil, sesame seed oil, cotton seed oil, and safflower oil. These oils are inexpensive and compatible with most surfactants, but are not stable for high temperature treatment or filtration sterilization because of either heat-induced destabilization of the ME or hydrolysis of the triglycerides. Triglycerides seem to be excellent oils for making MEs, but it was shown that making an ME of o/w from vegetable oils, such as soy, Canola, corn, and many others, is a very difficult task. These oils, although they look somewhat polar from their chemical structure, are actually too bulky, with very high molecular volume fractions, and form curved interfaces with difficulty, and their solubility (solvation) around the surfactant tails is limited. MEs made of vegetable oils are limited in their isotropic regions.

In a paper by Bagwe and Shah diglycerol monooleate (DGMO), glycerol monooleate (GMO), and diacetyl tartaric acid ester of monoglycerides (DATEM) were used and were found to solubilize Canola oil in a Canola oil/Tween 80/water (2% NaCl) system (32). It was found that use of these additives increased the oil solubilization by 16% to 20%. Addition of polyols, such as glycerol, erythritol, xylitol, dulcitol, and fructose, increased the solubilization by an additional 15–20% (32). However, all those microemulsions were of o/w and their behavior in the presence of solubilized drugs or upon dilution is not shown by the authors. It should be also stressed that such diluted formulations have very limited practical value.

However, medium-chain fatty acid esters of glycerol (MCT) and short-chain triglycerides (e.g. triacetin) are much better candidates to serve as the oil phase. MCT (Miglyol 812) is composed of a mixture of fatty acids of C_8 to C_{10} (capric and caprylic) carbon atoms, stable to oxidation, and relatively polar and capable of solubilizing the surfactant tail. Other very common oils are isopropyl myristate (IPM) and monoesters of fatty acids and alcohols such as Jojoba oil or methyl or ethyl oleates. Oleic acid esters of sucrose (sucrose mono-, di-, and tri-fatty acid esters known also as sucrose esters) are sometimes used both as the oil phase and as the surfactant.

A very interesting approach was taken by Constantinides et al. who used tocols as the oil phase in MEs for drug solubilization in parenteral delivery (33). Tocols are a family of tocopherols, tocotrienols, and their derivatives and are fundamentally derived from the simplest tocopherol. The most common tocol is d-α-tocopherol, also known as vitamin E. Tocols can be excellent solvents for water-insoluble drugs and are compatible with other cosolvents, oils, and surfactants. The tocols offer an appealing alternative for the parenteral administration of poorly soluble drugs including major chemotherapeutics (Paclitaxel). Tocols are novel, biocompatible solvents, cost-effective, nonirritating, and can be thermally sterilized. They can also provide acceptable shelf life under controlled storage conditions and avoidance of volatile solvents such as ethanol. The authors also suggest dissolving the drugs in

Table 2 Solubility of Drugs in Organic Solvents and Vitamin E, and SVE-Solubility in Vitamin E (see text)

	Solubility (mg mL^{-1})			Parameter		
Drug	Water	Methanol	Chloroform	Vitamin E	SVE	Solubility
Itraconazole	Insoluble	Insoluble	500	60	>1000	10.6
Paclitaxel	Insoluble	0.03	6	11	200	11.9
Cyclosporine	Slightly soluble	0.71	363	100	520	10.7
Ergosterol	Insoluble	1.5	32	50	25	9.6
Cholesterol	Insoluble	5	200	150	40	9.6
Prednisolone	0.22	3.3	5.0	Insoluble	0.02	13.6
Amphotericin	Insoluble	Soluble	Insoluble	Insoluble	<1	14.4

the tocols in the presence of a cosolvent, such as ethanol, that is removed at a later stage by simple evaporation. A comparison of the solubility of some drugs in tocopherols versus other solvents is given in Table 2. Constantinides et al. (33) and Illum et al. (34) proposed a solubility-in-vitamin E parameter (SVE) that is calculated as the solubility in chloroform divided by the solubility in methanol, expressed in mg/mL for both solvents. An SVE of at least 10, preferably 100, would indicate solubility in vitamin E. The results on shrinkage of tumor growth of mice implanted with B16 murine melanoma using those tocol-based microemulsions were very impressive (Fig. 12).

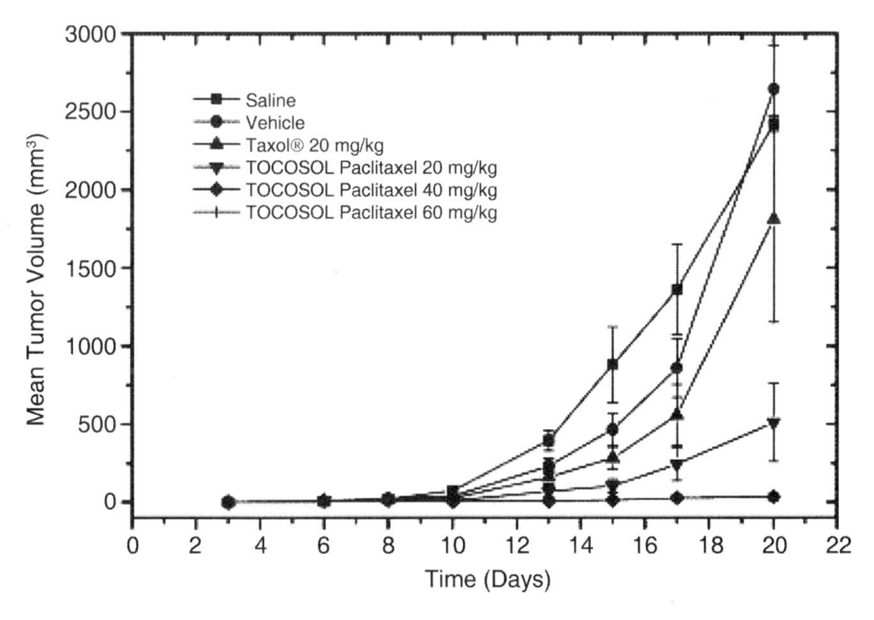

Figure 12 Comparative tumor growth of mice implanted with B16 murine melanoma using tocol-based microemulsions. Mice were administered saline, vehicle, tocol emulsions, and Taxol on a schedule of 3qdx5. *Source*: From Ref. 33.

The Aqueous Phase

The aqueous phase, in practice, is almost never just water because most formulations dictate the use of several additives such as buffers, isotonic agents, antibacterial compounds, and many others that might in some cases affect the ME. Salinity might strongly affect MEs made of anionic surfactants. Low pH might also damage the ME, mainly if esters of fatty acids are used as surfactants and/or oils, and, therefore, it is always recommended to keep the microemulsion as close as possible to neutral pH.

The effects of electrolytes as additives on microemulsion stability were studied in pharmacological reports mainly in relation to a specific system without considering the general trends of the effect of electrolytes on the ionic surfactant double layer or the salting out effects on other surfactants.

Solubilization Capacities

Enhancing the solubilization capacity of active matter within microemulsions is an important issue (35–45). If the active molecule (the drug or the nutraceutical) is lipophilic, it will be best solubilized at the w/o interface and within the oil continuous phase. However, upon dilution with water (in the digestive tract or while formulating, prior to drug administration) the amounts of the oil phase are reduced (more water is added) and eventually a bicontinuous structure may be formed. The curvature of the microdroplets changes, the packing parameter is altered, and the dilution becomes very significant. The amount of solubilized drug is expected to drop significantly. Upon further dilution, o/w microemulsions are formed together with direct micelles. As the water content increases, the droplets become poorer in oil, more micelles are formed, and the solubilization values decrease very significantly.

It is rare to find studies that evaluate the solubilization capacities of any drug upon dilution. Yet, from our experience, it is very obvious that upon dilution, the solubilized drug tends to become supersaturated in the microenvironment; thus it is solubilized in and tends to phase separately, precipitate, and, in most cases, grow into large crystals that have no bioavailability.

Two examples of the solubilization patterns of phytosterols and lycopene in U-type MEs are shown in Figures 13 and 14.

Regardless of the decrease in solubilization capacity upon dilution of the microemulsion, one can clearly see that the solubilization efficiency is very large, many-fold over the solubility of the drug in any of the individual phases. Moreover, the reduction in solubilization capacity is less than the dilution factor. By plotting

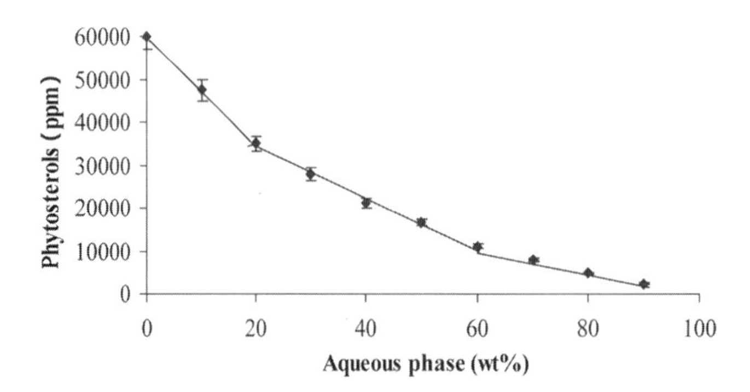

Figure 13 Solubilization capacity of phytosterols in microemulsions.

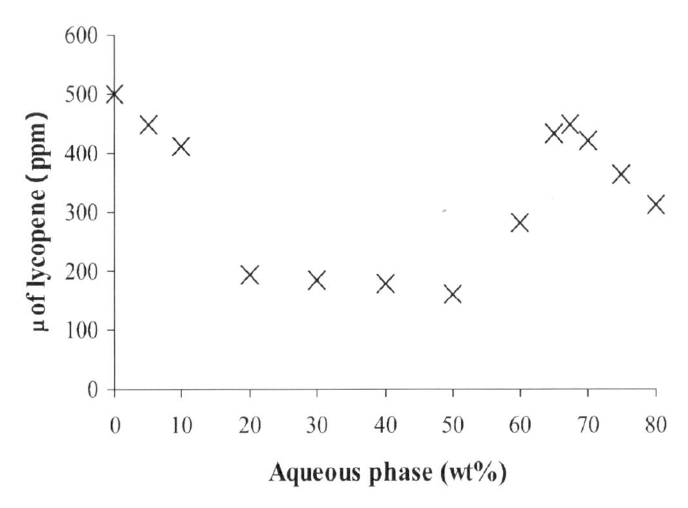

Figure 14 Solubilization capacity of lycopene in microemulsions.

the solubilization efficiency of lycopene, which is calculated on the basis of normalizing the amount of solubilized lycopene to the oil content (α) at each dilution point, one can see that the solubilization capacity is many-fold greater than the solubility of the solubilizate within any of the phases (Fig. 15). Similarly, we can represent the results as a function of the surfactant capabilities to solubilize the required active compound (γ) (Fig. 16), and it is very clear that the solubilization capacity is surfactant-dependent.

Modeling Compositions of Ingredients to Form ME

Agatonovic-Kustrin et al. (46) in their paper try to help formulators, who would like to incorporate insoluble drugs into microemulsions, to select their components (oil/surfactant/cosurfactant) on a sounder basis, rather than carrying out hundreds

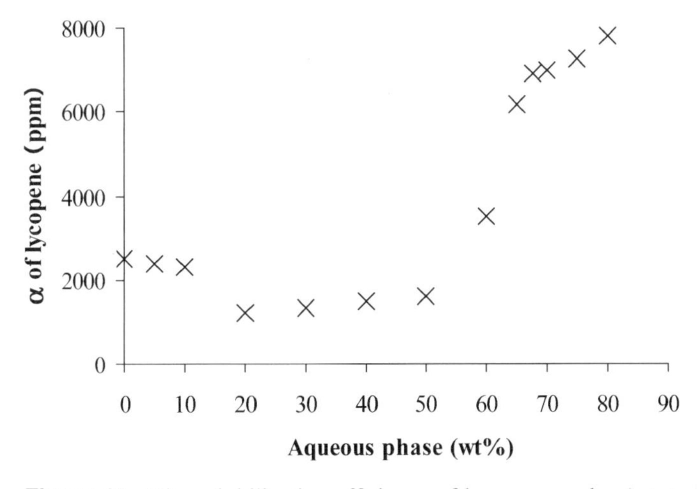

Figure 15 The solubilization efficiency of lycopene and α (normalized to the amount of oil phase at each dilution point tested). The parameters were plotted against the aqueous phase content along dilution line 64, at 25°C.

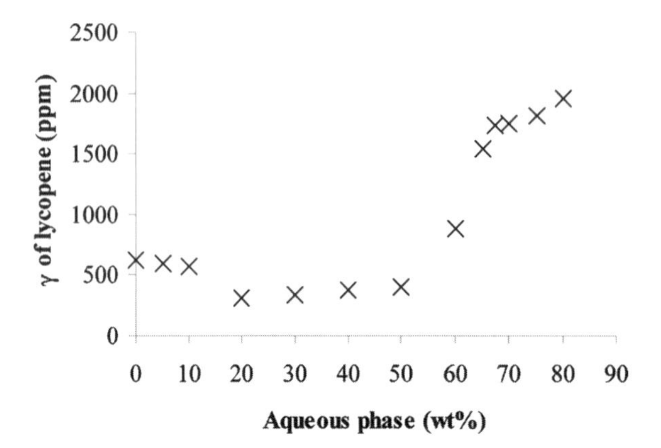

Figure 16 The maximum oil/surfactant efficiency (γ-parameter, in ppm) (amount of lycopene solubilized, normalized to the total amount of oil and surfactant). The parameters were plotted against the aqueous phase content along dilution line 64, at 25°C.

of experiments basically by trial and error. This work was done in view of attempts to solubilize both RIF and INH (antimalaria drugs with very different solubility patterns). The authors used a computer simulation for data modeling that is based on an artificial neural network. Data from 20 pseudoternary phase triangles containing MCT as the oil component and a mixture of surfactants or surfactant/cosurfactants were used to validate the model.

Agatonovic-Kustrin et al. (46) claimed that they successfully predicted the microemulsion regions, but failed to predict the multiphase liquid crystalline phases. It should be stressed that the results presented in this paper can be applied to the two model groups, but are by no means general or instructive for other cases, and there is no real theory beyond the study except the attempt to fit experimental results to computer simulations. Close examination of Figure 17 reveals that the regions of ME are, in most cases, very small and limited to very high concentrations of surfactant. All the surfactants used in the study are complex commercial blends, and in several of the blends, the results are nonreproducible. The surfactants in use were monoglyceride stearate (Inwitor), mono- and diglyceride caprylate and caprate (Inwitor 742), $C_{18:1}E_{10}$ (Brij 97) and Tween 20 (Crillet 3), and hydrogenated castor oil-40-ethoxylated (Cremophor RH). Most formulations were made with 2 to 5 wt.% water but some were diluted with additional water. A typical final formulation will have approximately 20% water, 23% MCT (Miglyol 812), 27% Inwitor 308, and 27% Crillet 3. The authors state that they made an o/w ME in which 150 mg/mL of rifampicin (RIF) and 100 mg/mL of Isoniazid (INH) were incorporated in the oil phase, the lipophilic chains of the surfactant/cosurfactant film, and the water, respectively. The RIF was not stable and separated to some extent from the INH. The authors suggest that the MEs can be diluted with water but may revert to mixed micellar solutions. They did not study the nature of the formulation at approximately 50% water and at highest water content and did not state whether the diluted system remains clear and homogeneous.

Model of Drug Release

Similarly, Sirotti et al. (47) attempted to model drug release from an ME (based on Tween 80, benzyl alcohol, IPM, and water) using a membrane diffusion technique

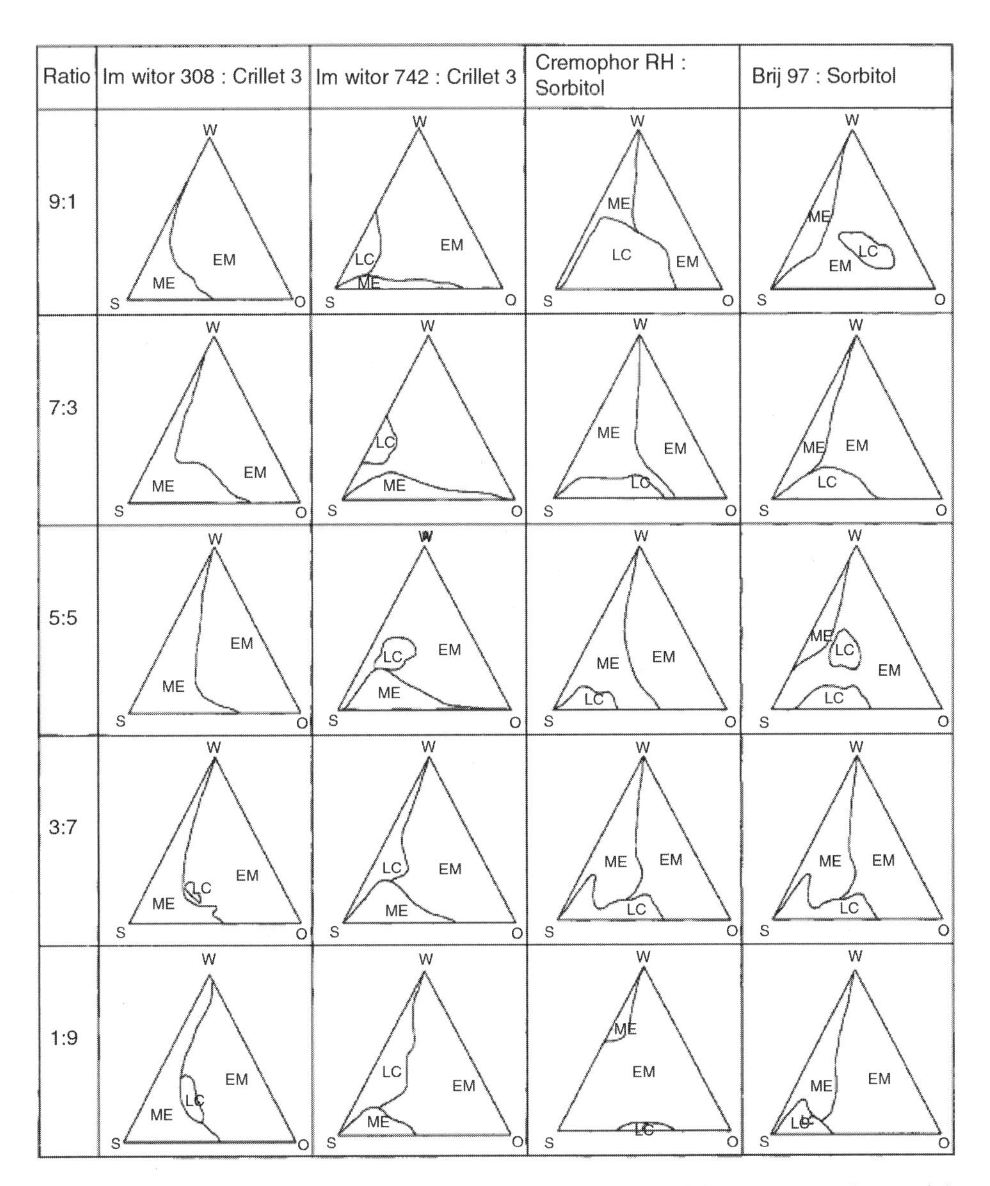

Figure 17 Pseudoternary phase of 20 different compositions of four sets at various weight ratios of surfactants. The isotropic regions were estimated by the model proposed. *Source*: Form Ref. 46.

and formulated a specific release model for a particular microemulsion (Fig. 18). They studied the drug release kinetics by permeation experiments through a synthetic membrane. They have shown that the drug transfer from the micellar-phase may be combined with the disruption of some micelles and the consequent solubilization of surfactant molecules in the microemulsion aqueous phase, which is responsible for an increase of drug solubility. They came to the conclusion that permeation was not linear as expected, because the drug (Nimesulide—an anti-inflammatory drug) interacted with the surfactant micelles. They supported this assumption by experimental evidence and a mathematical model. The model shows the donor

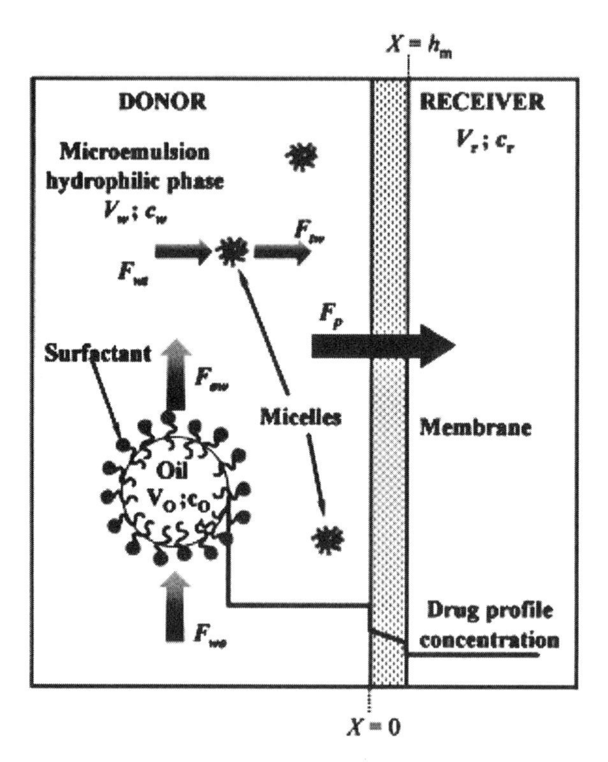

Figure 18 Schematic view of the phenomenon to be analyzed. The drug moves from the oil and micellar phases to the surrounding hydrophilic phase of the donor compartment and then crosses the interposed membrane to reach the receiver compartment.

and the receiver compartments (aqueous phase), exhibiting nonlinear profile characteristics of release through thick membrane. The surfactant–drug interactions increase drug solubility in the microemulsion aqueous phase. Although some of the parameters that were incorporated into the model might need corrections because they are difficult to measure, the work stresses the idea that if water-insoluble drugs are incorporated into the microemulsion, its solubility in the aqueous continuous phase and the permeation patterns will be enhanced.

Microemulsion Characterizations

Structural characterization of microemulsions is a difficult task because complementary studies using a combination of techniques are usually required to obtain a comprehensive insight into the physicochemical properties of these nanostructured fluids and their structure. This is because of their complexity, namely the variety of structures and components involved in these systems, as well as the limitations associated with each technique. Characterization of microemulsions requires macroanalytical tools, such as viscosity, conductivity, dielectric, and differential scanning calorimetry (DSC), together with scattering advanced techniques (DLS, SAXS, SANS) and spectroscopic advanced methods (HR-NMR, PSEG-NMR) and electronic microscopy (TEM and cryo-TEM) (48–60). Other sophisticated methods such as time-corrected single-photon counting methods (photophysics of a fluorescent drug) have also been utilized (61).

The commonly used techniques for structural information are small-angle neutron scattering (SANS), small-angle x-ray scattering (SAXS), and dynamic light scattering (DLS) (48–50,61–73).

Small-Angle Scattering

The techniques of small-angle scattering (SAS), in particular, small-angle X-ray scattering (SAXS) and small-angle neutron scattering (SANS), probe the pertinent colloidal length scales of 1 to 100 nm and therefore, are used to obtain information on the size, shape, and internal structure of colloidal particles. In addition, because the two kinds of scattering radiation are sensitive to different physical properties of the scattering, these techniques can be used in parallel to obtain a rich variety of information. The scattering intensity depends on the different scattering length densities of the particles and the solvent. The scattering length is a complex function of the atomic number for SANS and must be determined experimentally. For SAXS, the scattering length density is proportional to the electron density, which is a linear function of the number of electrons. So, SAXS is particularly useful for investigating the interfacial region of aggregates composed of surfactant molecules (48–50,74–80).

The scattering curves of highly diluted samples can be interpreted in real space in terms of their pair distance distribution function (PDDF) (48–50,81). The PDDF, $p(r)$, is the Fourier transformation of the angle-dependent scattering intensity $I(q)$, where q is the length of the scattering vector and is defined as $q = (4\pi/\lambda) \sin \theta/2$, where λ is the wavelength and θ the scattering angle (48–50,81). A quantitative numerical analysis of properly pretreated data involving model independent inverse Fourier transformation provides information not available elsewhere (7,13,48–50,81). The PDDF represents a histogram of distances inside the particle, weighted with the scattering length density differences and goes to 0 at the maximum particle dimension. So, one can directly read out the diameter from this function. For globular micelles with constant contrast, the PDDF is a bell-shaped function with its maximum at about half the diameter. The area under the PDDF is proportional to the forward-scattering intensity and thus to the squared scattering length density difference (contrast) between the particles and the solvent, to the concentration of the dispersed phase, and, at constant concentration and contrast, to the aggregation number. In highly concentrated samples, where the concentration of the sample in the dispersed and continuous phases are comparable, there is not a simple relation of concentration with the dispersed phase, but it follows the formula $\varphi^*(1-\varphi)$, where φ is the volume fraction. To allow comparison of results from different concentrations or volume fractions, φ, the scattering data must be normalized to φ. Under these conditions, the PDDFs are identical if the only change is an increase in the number of micelles with φ and no change in aggregation number or shape.

The interaction of the particles in solution must be taken into account because of the high volume fractions involved. In the case of interacting globular particles, the scattering intensity $I(q)$ can be written as the product of the particle from factor $P(q)$ and the structure factor $S(q)$. $S(q)$ describes the inter-particle interference. For ideally diluted solutions, $S(q)$ is a constant and equals one. The scattering curves show the development of a structure factor with a pronounced maximum for increasing concentrations, superposed on the particle form factor. The evaluation of such concentrated solutions is now possible with the recently developed "generalized indirect fourier transformation" (GIFT). With this technique, it is possible to simultaneously determine the form factor and the structure factor. This method is model-free for the form factor and is based on a hard sphere model for the structure factor; strictly speaking, it is only valid for spherical, monodispersed, and uncharged particles. However, the effects of polydispersity can be approximated, and recent applications have shown that the GIFT method can be used far beyond this theoretical limit, so the technique can be used for nonhomogeneous, nonglobular, and polydispersed systems. It is, for example, possible to follow the sphere to rod transition of binary and ternary systems. The typical parameters for the structure factor model are mean interaction radius, effective volume fraction of the dispersed phase, and polydispersity. The most important parameter that is determined in these calculations is the mean hard sphere interaction radius $R_{HS,}$ defined as half the center-to-center distance between two micelles at their closest contact. In the hard sphere model used (S_{ave}), the polydispersity is not completely correctly taken into account, as its impact on the structure factor is simply calculated by averaging structure factors of a size distribution (different R_{HS}) at constant volume fraction. It has no impact on the forward scattering of the structure factor, but only on the broadness of the peak.

Yaghmur et al. (81) in their work discuss two examples of the data that can be obtained from SAXS and SANS measurements of microemulsions based on Tween and Brij nonionic surfactants. Scattering curves of Brij 96 and Tween 60 microemulsions along two dilution lines (D64 and T64) are presented in Figure 19. These dilution lines are characterized by single continuous microemulsion regions transition from oil- to aqueous phase-rich microemulsions without the occurrence of phase separation. In this unique Winsor IV isotropic region, the system transforms from a w/o microemulsion via a bicontinuous phase to an o/w microemulsion (82,83). The figures show that with increasing aqueous phase content, the position of the maximum shifts to lower angles, corresponding to growing distances. The PDDFs for microemulsions in the aqueous phase–rich regime along dilution line D64 are shown in Figure 20. In principle, there is a minor increase in the overall size of all microemulsion droplets with increasing aqueous phase content. All the swollen micelles have approximately the same shape. The parameters of the structure factors for o/w microemulsions based on Brij 96v and Tween 60 are presented in Figure 21. The interaction radius (R_{HS}) increases, whereas the effective volume fraction decreases with increasing content of the aqueous phase. It should be noted from Figure 21(A) that there is a significant difference in the effective interaction radii (R_{HS}) of Brij 96v–based systems and those of Tween 60–based systems, while there is no significant difference in the effective volume fractions for either system. The values of R_{HS} are much smaller for the Brij 96v samples. With increasing surfactant concentration (moving from dilution line D64 to D73 and from T64 to T73), the effective volume fractions stay approximately constant, irrespective of the mixing ratio. Meanwhile, R_{HS} is strongly decreased with increasing surfactant content as expected, as the later generates more available interfacial area. Again, all microemulsion droplets based on Tween 60 are larger than those based on Brij 96v.

Small Angle Neutron Scattering

In recent years, small angle neutron scattering (SANS) has been considered a good tool to characterize and extract information from the structuring processes of microemulsions and liquid crystals. In these systems, the structural evolution as a function of the aqueous phase composition (water plus PG) was studied (starting from pure D_2O and subsequently, increasingly replacing water by PG up to a mixing

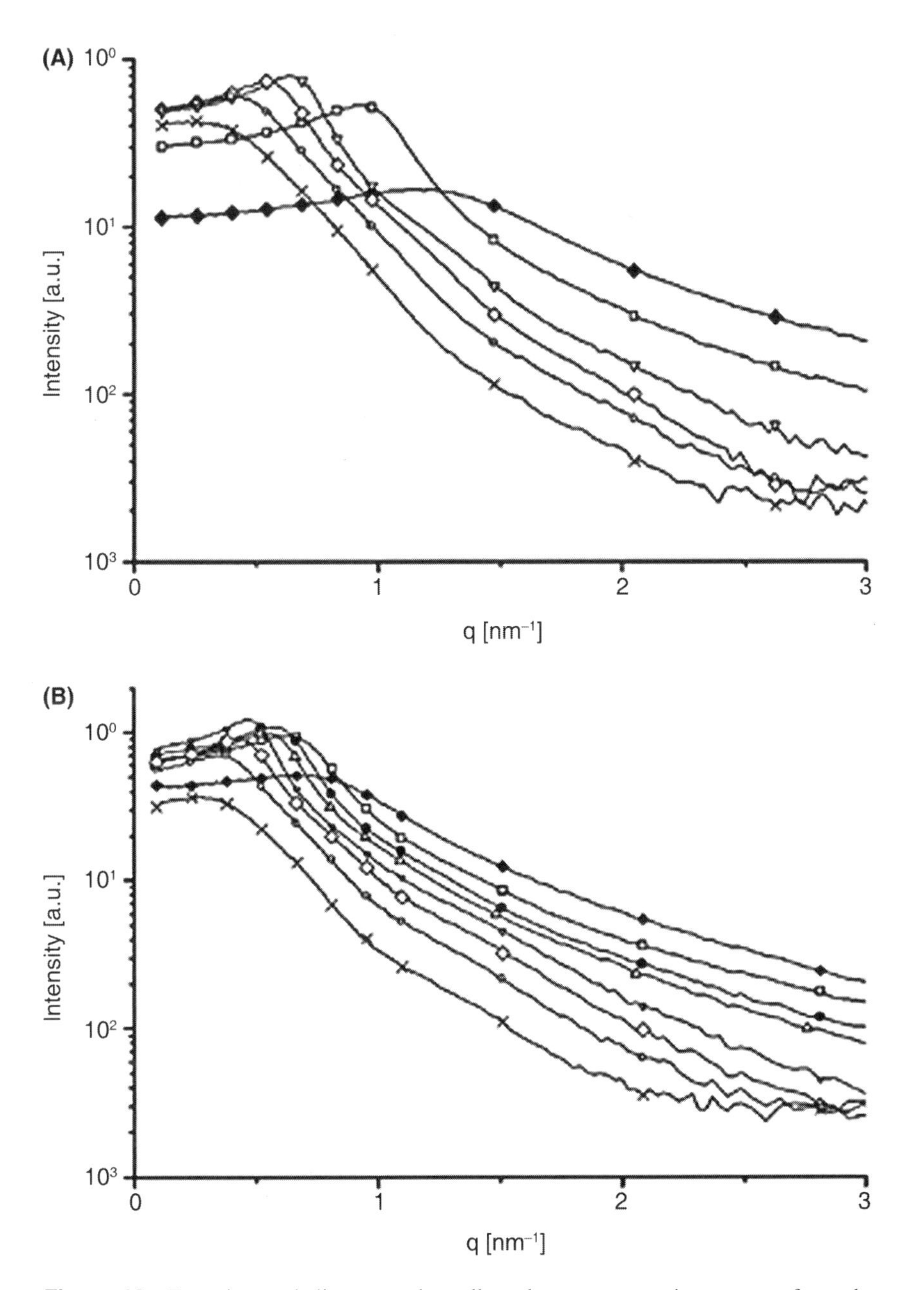

Figure 19 Experimental slit smeared small-angle x-ray scattering curves after subtraction of the background and solvent scattering for the samples along dilution lines T64 (**A**), and D64 (**B**), respectively. The aqueous phase (water plus PG at a constant weight ratio of 1/1) content of the investigated samples: (◆) 10 wt.%, (□) 30 wt.%, (●) 40 wt.%, (Δ) 50 wt.%, (▽) 60 wt.%, (◇) 70 wt.%, (O) 80 wt.%, and (×) 90 wt.%.

ratio of 1/1). In Figure 22 is shown the normalized PDDFs of these pseudobinary systems. It can be seen that increasing the PG content leads to shrinkage of micelle size. In addition, the micelles transform from a prolate to a quite globular shape with increasing PG content in the aqueous phase (ratio of the position of the maximum in

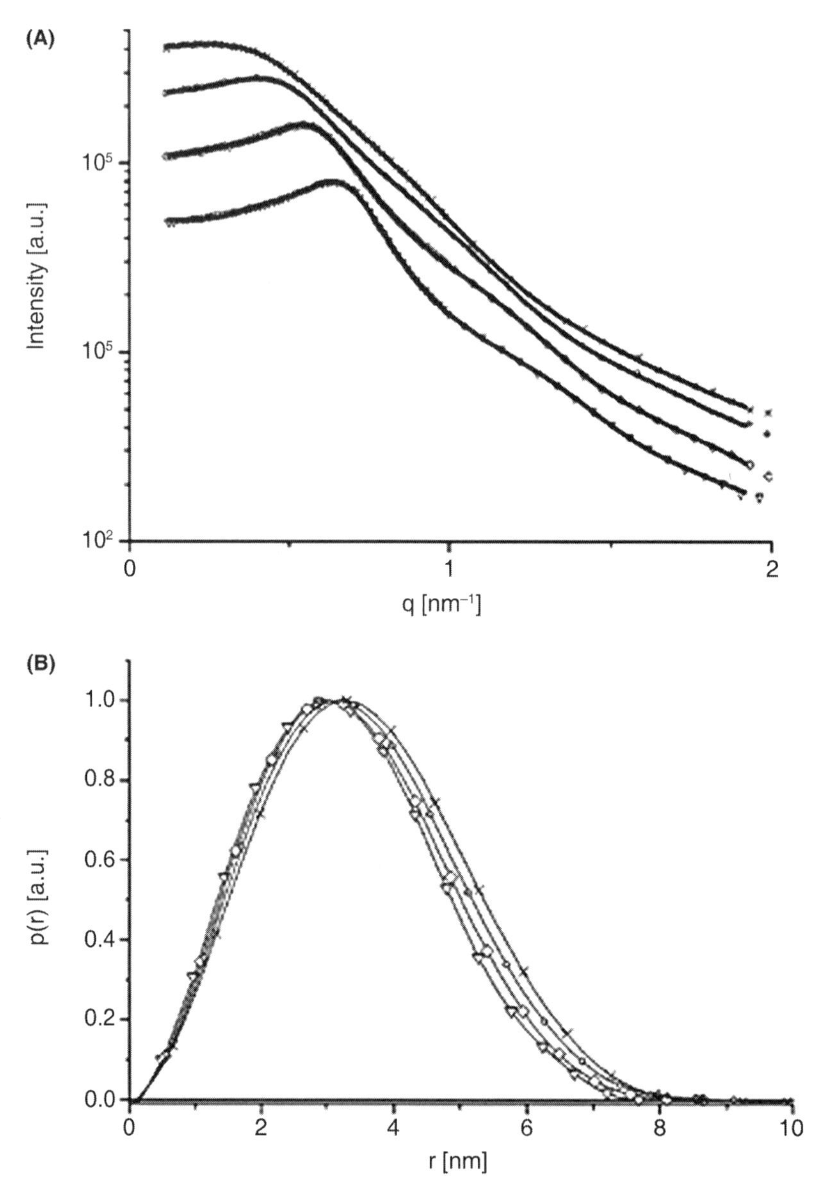

Figure 20 (**A**) Experimental slit smeared scattering curves (symbols) for microemulsion systems in the aqueous-rich region along the dilution line D64, together with the fit (*solid line*). The aqueous phase (water plus PG at a constant weight ratio of 1/1) content of the investigated samples: (▽) 60 wt.%, (◇) 70 wt.%, (○) 80 wt.%, and (×) 90 wt.%. For better visibility, the scattering curves were multiplied by 10^0, $10^{1/3}$, $10^{2/3}$, and 10, respectively, (**B**) PDDFs for these samples (normalized to the same height). *Abbreviation*: PDDFs, pair distance distribution functions.

the PDDF to D_{max} is increased). The interaction radius (R_{HS}) decreases from 4.8 to 4.2 nm, and, simultaneously, the volume fraction of the dispersed phase decreases to a lower extent (Fig. 22). By combining these two parameters, it is possible to calculate the particle number density (N_p/mm^3), which increases by about 30%. So, the

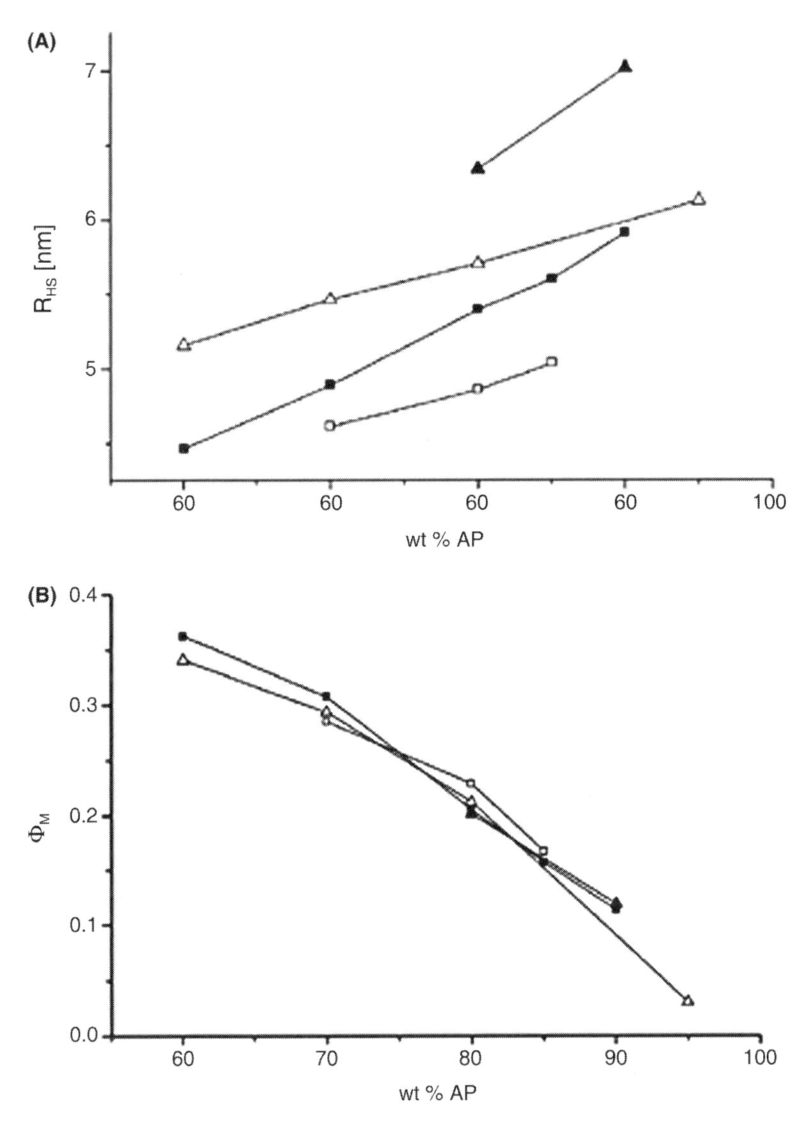

Figure 21 (**A**) Effect of aqueous phase content along the Brij 96v-based dilution lines D64 (■) and D73 (□), and the Tween 60-based dilution lines T64 (▲) and T73 (△) in the aqueous-rich region on the structure. Factor parameters as calculated from the evaluation of SAXS data by GIFT: (**A**) the interaction radius (R_{HS}) and (**B**) the effective volume fraction (Φ_{eff}). *Abbreviations*: GIFT, generalized indirect fourier transformation; SAXS, small-angle x-ray scattering.

presence of PG increases the number of micelles, and decreases their size and total volume. The larger number of micelles of smaller size and, consequently, lower surfactant aggregation number can be explained, as a considerable part of PG is incorporated into the interfacial film.

From the SAXS measurements it was concluded that: (1) upon increasing the aqueous phase content of microemulsions along the various dilution lines there is an increase in the overall size of all microemulsion droplets (which can be attributed to

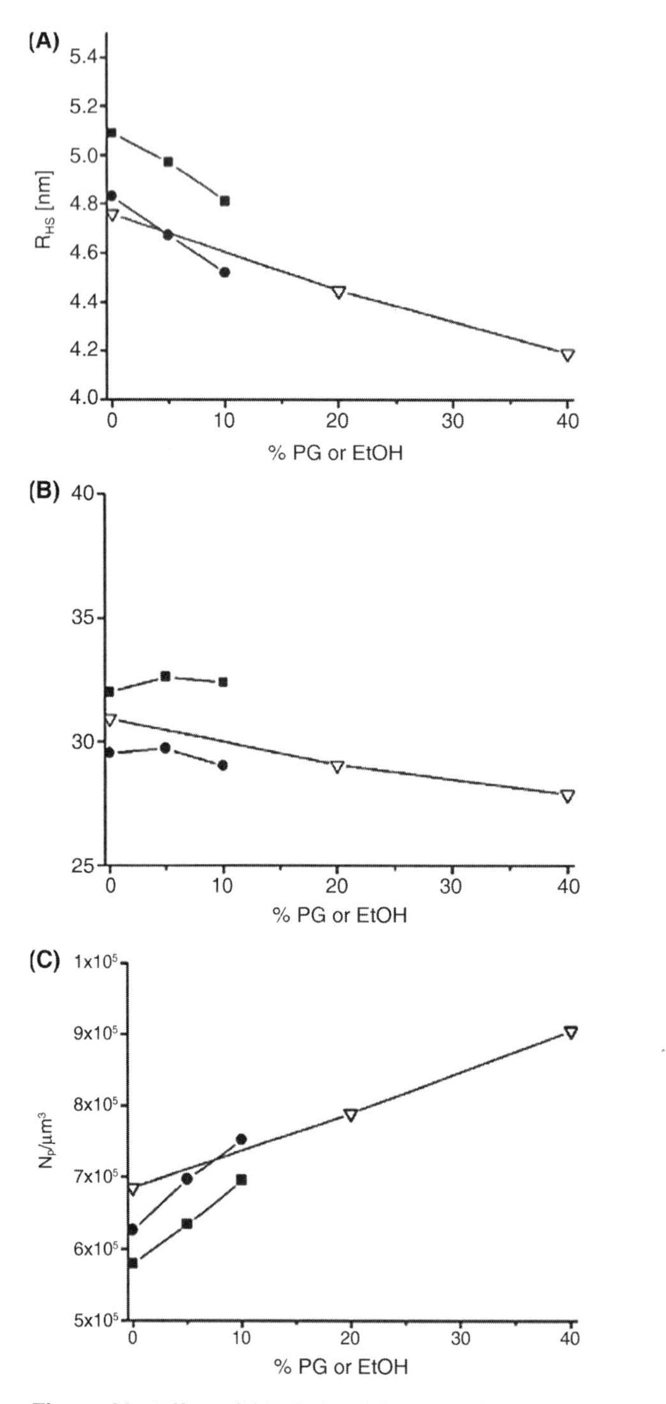

Figure 22 Effect of d-EtOH and d-PG on the structure factor parameters of three different structured systems based on Tween 80, as calculated from SANS data by GIFT: (▽) impact of PG on pseudobinary systems, (●) impact of EtOH on PG-based systems, and (■) impact of EtOH on PG-free systems. (**A**) The variation of the interaction radius, R_{HS}, (**B**) the effective volume fraction Φ_{eff}, and (**C**) the particle number density (N_p, μm^{-3}) with increasing d-EtOH or d-PG content. *Abbreviation*: GIFT, generalized indirect Fourier transformation.

a redistribution of ethanol); (2) the microemulsion droplet size based on Tween 60 is larger than that of Brij 96v, at equal weight fractions; (3) surfactant concentration plays an important role: upon increasing the surfactant concentration (mixing ratio of surfactant to oleic phase at constant aqueous phase content), the effective volume fractions of the dispersed phase stay constant, while the interaction radii decrease, leading to an increase in the number particle density. Therefore, the interfacial area increases as expected; (4) ethanol is highly soluble in the oil, but even more so in the aqueous phase, and we could prove that a considerable part of the alcohol also redistributes into the interface, that is, ethanol can act as a co-surfactant, depending on the ratio of the other components.

From the SANS measurements, it was possible to conclude the following: (1) upon increasing the PG content in the aqueous phase, there is an increase in micelle number, a transformation from prolate to quite globular droplets, and also shrinkage in the micellar size, leading to the conclusion that some of the PG is also incorporated into the interface; (2) there is a strong similarity between EtOH and PG: while most of these two components remains in the aqueous phase, both are partially incorporated into the interface, rendering the droplets smaller in size and their shape more globular. The only observed difference between these two water-soluble alcohols was that replacing the aqueous phase by EtOH does not lead to a significant change in the effective volume fraction, while PG decreases it because PG lowers the hydration; (3) EtOH and PG influenced the microstructure of our investigated samples independently.

Investigators of ME structures will always use combinations of some of the above techniques to characterize microemulsions.

Cryo-Transmission Electron Microscopy

Cryo-transmission electron microscopy (cryo-TEM) has emerged as a unique and powerful method for structural analysis of a wide range of biological macromolecules and complexes such as the ribosome, viruses, membrane proteins, and molecular chaperones. This method is equally attractive for studying molecular assemblies like micelles, microemulsions, vesicles, and liquid crystalline phases formed by surfactants, lipids, and amphiphilic polymers (84–88). Numerous reasons make cryo-TEM an excellent technique for studying these dynamic assemblies and structures. First, this is the only experimental method providing *direct* structural details on shape, approximate size, and internal ordering at a high resolution of approximately 1 nm that is comparable with the dimensions of the assemblies (a surfactant micelle and a lipid bilayer are each about 5 nm in thickness). Importantly, this fine structural information relates to the native state of the labile assemblies, as cryo-TEM involves ultra-rapid thermal (physical) fixation that leads to vitrification, thereby preserving the original microstructure of the aggregates (88). Moreover, the information obtained is detailed and aggregate specific and is not averaged over the whole volume as is the case with indirect methods. The unique advantages that cryo-TEM offers in the investigations of cubic phases and cubosomes are demonstrated in several studies as well as in our preliminary data (89).

Sample preparation and imaging of the vitrified specimens are rather complex and require experience. However, our advancement of digital-imaging cryo-TEM has overcome many of the difficulties encountered previously in imaging, making the method now much more practical, efficient, and reliable for morphological investigations of the kind we performed in this project (90). Some specific examples

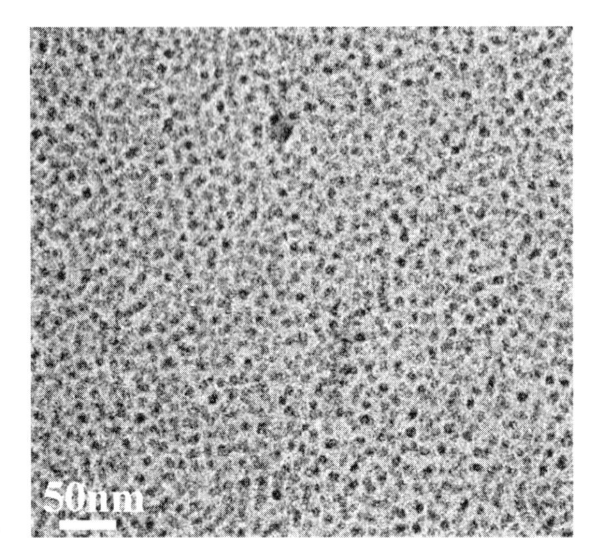

Figure 23 Cryo-TEM image of microemulsion of food-grade surfactants at 90% water dilution. *Abbreviation*: TEM, transmission electron microscopy.

of microemulsions of o/w at high water dilutions are shown in Figure 23 (of empty microemulsions) and in Figure 24 (in the presence of solubilized lycopene). It is clearly seen that the droplets are monodisperse, small in size, and do not exceed 15 nm. The lycopene has a very minor effect on the structure of the microdroplets.

Self-Diffusion Coefficients from PGSE-NMR

NMR diffusion measurements are potentially a valuable tool for probing the conformation and state of molecules within their environment. Self-diffusion measurements are sensitive to the shape and size of the diffusing molecule, the viscosity of the

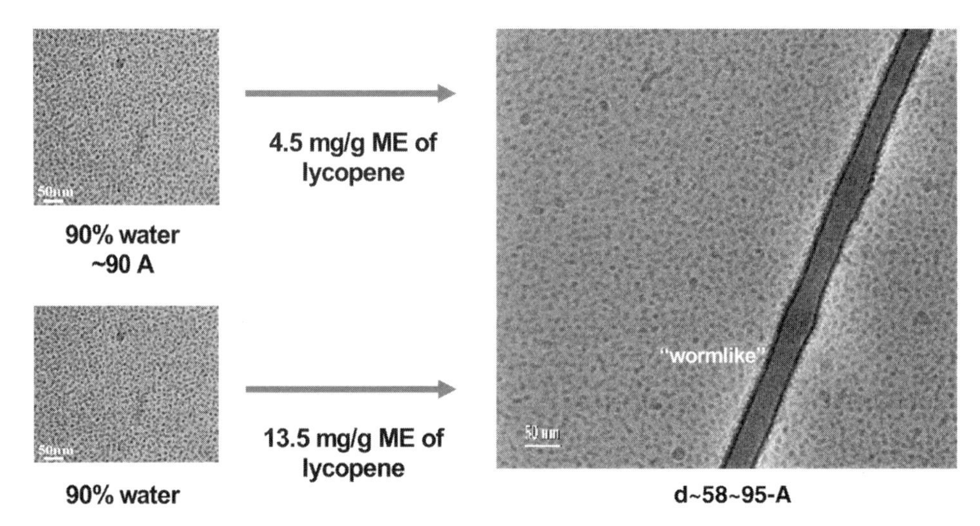

Figure 24 Microemulsions of lycopene at 90% and 95% dilutions.

medium through which it is diffusing, and the size and geometry of any barriers to its random motion. The restriction of motion of the molecule that is solubilized within a droplet of a micelle can be estimated from such measurements (91–99).

Diffusion is measured by NMR with pulse sequences based upon either spin or stimulated echoes in which the magnetization is first dephased and subsequently rephrased by a linear magnetic field gradient. When the magnetic gradient is applied along the z-axis, at coordinate z, the magnetization will process during the application of the gradient at a frequency ω. If the gradient is applied during the dephasing period for a time δ, each spin will accumulate a phase $\omega\delta$, which will be dependent upon its z-coordinate. Applying a second gradient pulse with the same area as the first during the rephrasing period of the experiment will result in formation of an echo. The attenuation of that signal will be solely because of relaxation. However, any random motion along the z-axis, such as that because of diffusion, will lead to incomplete rephrasing of the signal and hence to further attenuation of the observed echo. If the time between the start of the two gradient pulses is Δ, the signal attenuation $A(t)$ will be given by:

$$A(t) = A(0)\exp\left(-R(t) - \gamma^2 G^2 D \delta^2 (\Delta - \delta/3)\right)$$

The value of the diffusion coefficient, D, can be obtained from the equation.

When a guest molecule, or any ingredient of the microemulsion, is in the continuous phase, it will be free to move and the diffusion coefficient will be very high. However, if the molecules are restricted and confined within the droplet of the microemulsion, they will be constrained from movement, and the diffusion coefficient will be that of the other ingredients that move together. Once the water is confined in the inner phase, the diffusion coefficient will be small. Upon dilution, the water will move into a bicontinuous structure and its mobility will be enhanced; once it becomes the continuous phase, its diffusion coefficient will be similar to that of free water. The oil will behave just the opposite. Once oil is structured within the w/o droplet, it will have a high diffusion coefficient and upon inversion, where o/w droplets are formed, its movement will be restricted, and its SD coefficient will decrease dramatically. The rate of surfactant movement will remain practically unchanged throughout the total dilution process. At the bicontinuous region, the oil and the water will have similar diffusion coefficients. The mobility of the cosolvent or the cosurfactant also varies as a function of the water dilution; one can tell more about their locus within the droplets if they are displaced, and to what extent, upon dilution.

Kreilgaard et al. (99) investigated the influence of structure and composition of microemulsions on the transdermal delivery potential of lidocaine (representing a lipophilic drug) by using the self-diffusion coefficient determined from pulsed-gradient spin echo NMR spectroscopy and relaxation times. There was a direct correlation between the diffusion coefficients of the drugs and the transdermal flux (Fig. 25).

In our more recent work, we examined the SD coefficients U-type, dilutable, and microemulsions (oil, water surfactant and cosurfactant) in an empty microemulsion along full dilution lines (as the amount of the water content increases) of each of the components and compared those values to the SD coefficient of the same components when nutraceuticals (lycopene, phytosterols, or lutein) were present. The SD coefficients of an empty ME for three main water contents are given in Table 3.

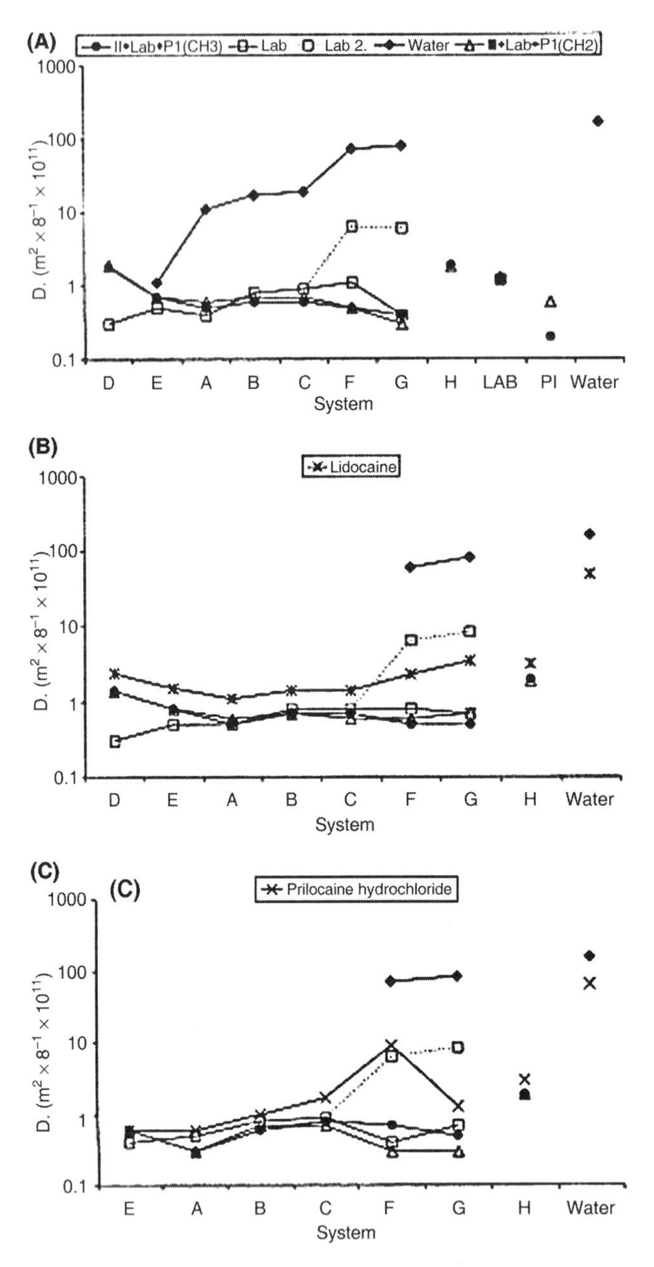

Figure 25 Self-diffusion coefficients (*D*, ms 310) of (**A**) unloaded, (**B**) lidocaine (4.8%), or (**C**) prilocaine hydrochloride (2.4%) loaded microemulsion systems, and single constituents [Labrasol (Lab), Plurol Isostearique (PI), water, and isostearylic isostearate (II)], determined by PGSE NMR at 258°C. The systems are arranged according to increasing water content. The self-diffusion coefficients for the two common signals for Lab, PI, and II were determined from the resonance of CH_3 and CH_2, respectively. The self-diffusion coefficients for the unique signal of Lab were determined from the resonance of the OCH_2 CH_2 group of the polyethylene glycol moiety. Symbols with dashed lines indicate biphasic behavior. The self-diffusion coefficient of prilocaine in neat isostearylic isostearate was determined with the free base form of the drug and that of lidocaine in neat water with 0.4% lidocaine. *Abbreviations*: II, isostearylic isostearate; Lab, labrasol; PI, plurol isostearique.

Table 3 Self-Diffusion Coefficient of Water (D^w) and Oil (D^o) (m^2/s) and Relative Diffusion Coefficients of Water and Oil (D_w/D_0^w and D^o/D_0^o) in Empty and Loaded Microemulsions at Two Different Levels of Solubilization [Water-Poor Corner (10 wt% Water) and Oil-Poor Corner (60–80 wt% Water)]

	Water diffusion coefficients				Oil diffusion coefficients			
	Empty ME		ME with nutraceutical		Empty ME		ME with nutraceutical	
Diffusion coefficients	$D^w 10^{-12} m^2/s^{-1}$	D^w/D_0^w	$D^w 10^{-12} m^2/s^{-1}$	D^w/D_0^w	$D^o 10^{-12} m^2/s^{-1}$	D^o/D_0^o	$D^o 10^{-12} m^2/s^{-1}$	D^o/D_0^o
Phytosterols (at 10% water)	2.0	0.003	4.2	0.01	206.0	0.53	168.5	0.44
Phytosterols (at 80% water)	431.0	0.77	443.4	0.79	1.6	0.004	1.84	0.0048
Lutein (at 10% water)	150.0	0.27	80.5	0.14	220.9	0.58	156.2	0.41
Lutein (at 60% water)	512.3	0.92	536.1	0.97	2.47	0.0065	1.22	0.003
Lycopene (at 10% water)	2.0	0.003	7.2	0.013	206.0	0.53	225.0	0.53
Lycopene (at 80% water)	431.0	0.77	392.0	0.70	1.6	0.004	3.5	0.01

It can be clearly seen that the water diffusion coefficient increases as the transition from the w/o ME into bicontinuous and to o/w ME takes place. The oil phase, in contrast, is free in the concentrate but is slowly entrapped within the droplets and its SD coefficient increases. The inversion points are easy to identify in most cases. The oil phase is part of the continuous phase in the w/o ME and is entrapped once the inversion occurs. From Figure 26, one can see the effect of the solubilizate on the SD coefficients for systems containing lycopene. One can clearly see the effect of the solubilizate on the transitional microstructure. The lycopene, although solubilized in very small amounts, affects the SD coefficients and the formation of the o/w ME is deferred until higher levels of those nanosized droplets are reached.

Other Methods

DSC and fluorescence measurements have also been used to study the nature of the water core in w/o MEs. Microemulsions are intriguing to scientists mainly because the core of the swollen micelle is generally composed of water or hydrophilic components dissolved in water. The water pool in the microemulsion is not necessarily free water, as the hydrophilic head groups of the surfactant, cosolvent, and cosurfactant all tend to be hydrated by water. The water is therefore not free, and its activity is very restricted. Many investigators attempted to take advantage of the water pool and to solubilize drugs, nutraceuticals, proteins, enzymes, and so forth into that water reservoir. However, it should always be realized that there are at least three types of water in the core of the ME: free water (with activity of 1) with high freedom of mobility, water bound to the surfactant or cosurfactant, and nonfreezable water. The differences between these types of water are expressed in the water activity and reactivity to interactions in the swollen micelle core. Some scientists distinguish between interfacial and bound water as well. Various methods have been applied to quantify these types of water at every dilution point and to correlate the bound water to the type of surfactant, surfactant/oil, or surfactant/water ratios, type and amounts of cosolvent and cosurfactant, and so forth. We used mainly DSC techniques to study the binding capacity of the microemulsion ingredients to the inner water and have demonstrated that all those parameters influence the mobility and reactivity of the ME ingredients (Figs. 27 and 28). One can clearly see that in w/o microemulsions made from sugar esters, there is no free water at dilution line 7/3 up to 14 wt.% of total water, and only at this level of total water does the water in the core start to have some mobility (100). For every dilution line, that is, for every surfactant/oil ratio, there will be a certain amount of bound water in the core of the microemulsion. As the amount of the surfactant increases, the surfactant/oil ratio increases and more bound water will exist in the core of the microemulsion. This phenomenon is very important, as various additives that are incorporated in the microemulsion core will have restricted mobility up to a certain amount of total water, and only upon exceeding these values will the ingredients gain mobility. It could be demonstrated that solubilized enzymes will not react until free water emerges in the microemulsion core, triggering the release, or the reactivity, of components that are solubilized in the core of the microemulsion.

It should be noted that simple analytical techniques, such as viscosity and conductivity, could also provide vital structural information on microemulsions that are empty versus those into which drugs have been solubilized. The techniques allow us to clearly see when the solubilizate is postponing transitions from one phase to the other, or when it is advancing the occurrence of such transformation at lower water dilution contents.

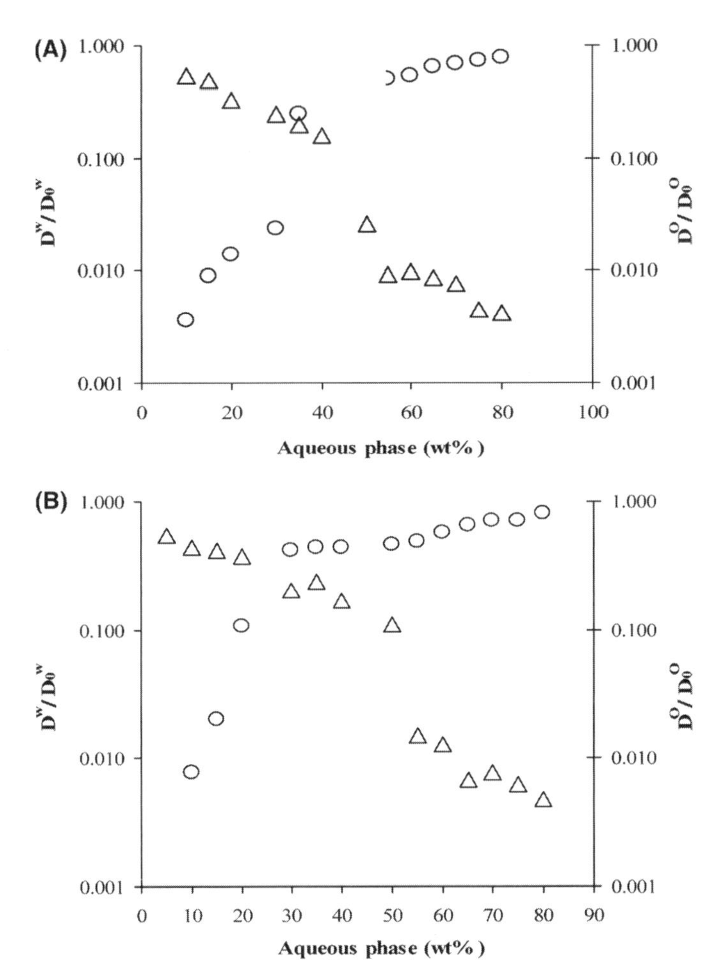

Figure 26 SD-NMR spectra at 25°C of (**A**) empty microemulsion and (**B**) microemulsion loaded with solubilized phytosterols.

Andrade et al. (61) examined the structural changes in w/o Triton X-100/cyclohexane/hexane/water ME probed by fluorescent drug Piroxicam using a transient (picosecond) technique. They have detected three distinct spectroscopic species with pH-dependent contributions. The species were detected only above $W_0 = 8$, confirming the existence of bound water at low concentrations and free water at higher water contents. The transition from w/o to o/w microemulsions with increasing amounts of water indicate structural changes in the ME.

PART II—POTENTIAL APPLICATIONS

Microemulsions have attracted considerable interest as potential drug delivery vehicles because of their ease of formation, solubilization capacities, low viscosity, transparency, and thermodynamic stability.

The objective of this part of the chapter is to highlight recent work on the state-of-the-art lipid microemulsions with focus on their use for nutraceutical delivery and

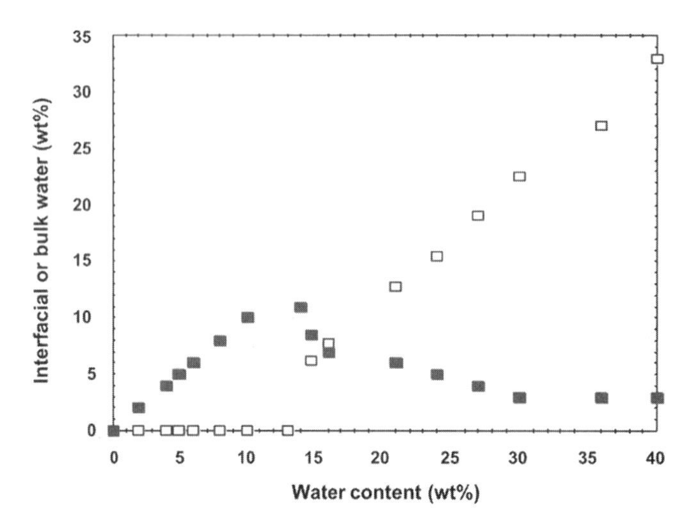

Figure 27 Variation of free and interfacial water content with total concentration of water. Concentrations are calculated in wt.%, relative to the weight of the microemulsion samples. (□) Bulk (free) water and (■) interfacial (bound) water.

as vehicles for oral, transdermal, and parenteral delivery of both lipophilic and water-soluble drugs. Many other potential applications of microemulsions have been studied such as pulmonary, intravaginal, or intrarectal administration delivery vehicles for lipophilic drugs, such as microcides, steroids, and hormones, and intramuscular formulations of peptide or cell targeting formulations; other drugs have been evaluated also (101–103,19). The space limitations that are imposed on this review, and the emphasis on new methods to make these vehicles prevent us from further elaborating on those applications.

Scientists explored the use of ME in many types of formulations, but the major applications that were mentioned are the transdermal delivery of hydrophilic drugs, using w/o microemulsions, and the lipophilic drugs, using o/w microemulsions. Oral delivery applications have also been extensively studied but are considered to be less easily adopted by pharmaceutical companies, because some of the problems related to oral intake of drugs, such as pre-systemic decomposition, environmental stability, and solubilization capacities in much-diluted systems, were not satisfactory resolved by this technology. Microemulsions are not solid particles and therefore cannot be considered "fool-proof protracting systems." However, recently more oral delivery microemulsion examples have emerged with better characteristics and improved environmental protection for sensitive drugs.

Surfactants of low toxicity and low irritation are required for most pharmaceutical applications, and little work has been done using pharmaceutically acceptable components for each of the potential applications. Lecithin is used as the primary nontoxic and nonirritant food and pharmaceutical grade emulsifier, but as the w/o microemulsions that it forms are not dilutable, and as it does not form o/w microemulsions, its use is restricted to transdermal and topical applications only (10,101,104–135). Nonionic surfactants are of two major types: ethoxylated and polyglycerol ethers (glucosides) or esters. The ethoxylated have a tendency to foam and to irritate skin and other sensitive tissues (vaginal, ocular, and muscular), and to suffer from hemolytic activity. Water-soluble ethoxylated triglycerides (e.g., castor

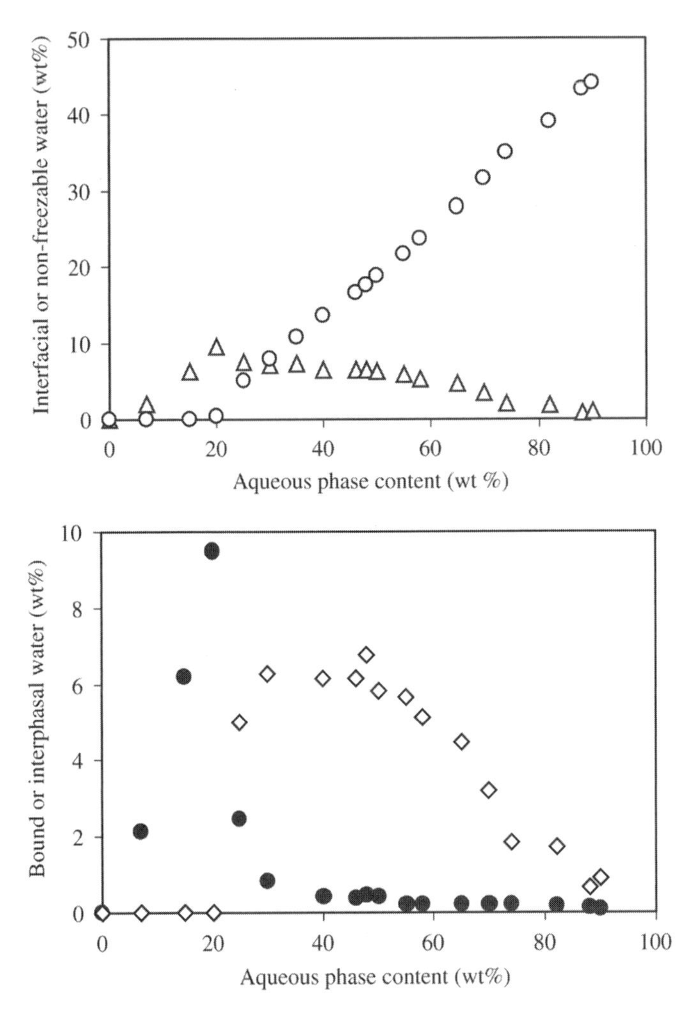

Figure 28 (*Upper*) Variation of "nonfreezable" (O) and "interfacial" ("interphasal" plus "bound") water (Δ) as a function of the aqueous phase content along the dilution line D64. (*Lower*) Variation of "bound" and interphasal water along the dilution line D64. (◇) Bound water and (●) interphasal water. Concentrations are calculated in mass%, relative to the mass of the microemulsion samples.

oil), monoglycerides, sorbitan esters (termed Tweens), alcohols (termed also CiEj or Brijs), or acids are widely used by scientists but found only in some oral delivery systems (17,137–145). Sugar esters are friendlier and biodegradable, but suffer from hydrolysis if used in acidic conditions and under certain haemolytic conditions (146–151). Alkyl glucosides are ethers with more hydrolysis resistance but smaller emulsification and aggregation capabilities (152–158). N-oxide surfactants (e.g., dimethyldodecylamine N-oxide) have been mentioned as the biodegradable version of surfactants (159). Some anionic surfactants (AOT) or cationic surfactants (DDAB) have been used also (27–30,160–167).

One should read some of these reports with great caution and must be aware of the disadvantages associated with each surfactant.

Oral Applications

Oral drug delivery systems can be classified into two categories: controlled-release delivery systems and bioavailability-enhanced delivery systems. Bioavailability-enhanced delivery systems have attracted increasing attention in the pharmaceutical field because high-throughput drug screening processes often identify water-insoluble drug candidates. Most of these hydrophobic drugs are not easily absorbed in the GI tract because of limitations in dissolution and solubility. By dissolving these types of drugs in a liquid carrier, preferably a self-emulsifying formulation, the drugs can be delivered in a presolubilized and more easily absorbed form. These systems are usually soft or hard gelatin capsules, which encapsulate the self-emulsifying liquid or semisolid drug formulation.

Microemulsions for oral applications of drugs were considered to be an excellent solution for many of the available and new drugs and drew the attention of many scientists around the world (168,169). Several excellent reviews have been written. As early as 1977, Prince wrote in his book that microemulsions have potential as delivery tools (170). Ziegenmeyer and Führer (171) in their early studies mentioned microemulsion formulations as future delivery vehicles. In 1989, Trotta et al. (169) noticed that AOT-based ME can solubilize drugs and could be used as oral delivery vehicles. They measured the release properties through a hydrophilic membrane. The initial mass transfer contents were linearly related to the portioning coefficients of the drugs in the oil–cosurfactant–water mixtures of five drugs with different lipophilic characteristics in o/w microemulsions of AOT, 1-butanol, IPM, and water. The authors concluded that the initial mass transfer constants of different drugs, calculated on the basis of these concentrations, were very similar to the mass transfer constants of the drugs obtained by permeation studies through the same membrane using aqueous phase, but had some minor advantages. Also in 1989, o/w microemulsions were used by Ritschel et al. (172) as carriers for the oral administration of cyclosporin, showing some very promising results. In a later work, Trotta et al. (162) used nonaqueous microemulsions constituted of lecithin, taurodeoxycholic acid, ethyl oleate, and six different glycols, as surfactant, cosurfactant, dispersion phase, and continuous phases, respectively (162). The results showed that the permeation membrane could be related to the qualitative and quantitative compositions of the ME, whose internal phase behaves as a reservoir for lipophilic drugs. In 1993, Pattarino et al. (118) explored in more detail the same four-component ME of lecithin as surfactant, taurodeoxycholic acid as cosurfactant, ethyl oleate as dispersing phase, and a cosolvent of 1,2-propylene glycol, in relation to the release of drugs. Drug permeability was modeled as a function of the mixture composition. The findings confirm the reservoir effect of the internal phase of the microemulsion. It has been demonstrated that micellization of the drug has an important influence on its release behavior. Also, at constant TDC levels, the permeability coefficient of the drug decreases significantly as the percentage of the oil rises, and when the oil is held at constant proportions, the permeation rate decreased with an increasing percentage of the cosurfactant (Fig. 29). The models predicted the permeability of retinol from the ME, which can be correlated to the proper mixture composition to achieve a desired drug release performance. The oral route is the preferred route for chronic drug therapy (5,166,173a,174).

One of the most persistent problems faced by formulation scientists has been to find methods of improving the bioavailability of poorly water-soluble drugs. As a drug must almost invariably be in solution within the GI tract before it can cross

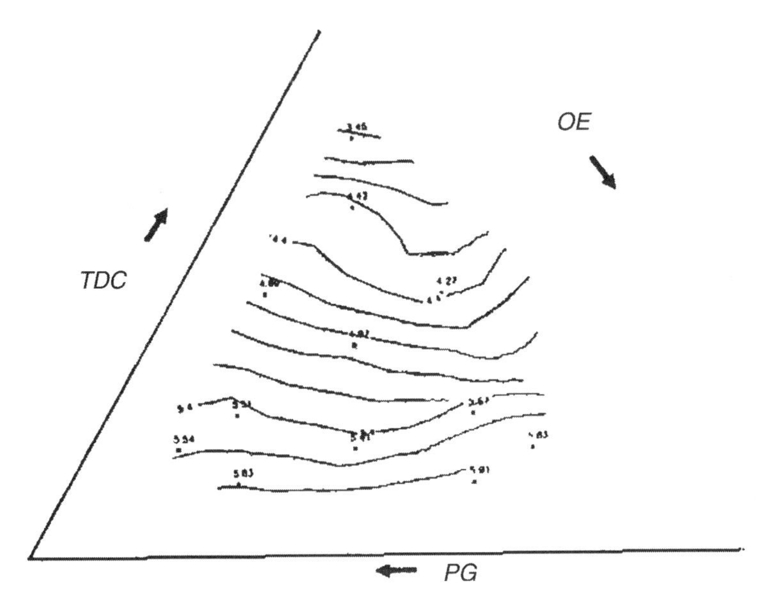

Figure 29 Contour-plot of iso-permeability curves obtained from the final PLS model with set B variables. *Abbreviation*: PLS, partial least squares.

the GI mucosa, poor water solubility may lead to incomplete and erratic absorption. Several methods are used to overcome these problems, like incorporation of the lipophilic active component into inner lipid vehicles such as oils, surfactant dispersions, self-emulsifying formulations, and liposomes. Among these, an important method is to incorporate the drug into self-emulsifying liquids, thereby preparing an o/w concentrate that, upon dilution with water, will spontaneously transform into o/w emulsions. Formulation of drugs in this manner represents a possible alternative to tablets and capsules, as the system may be expected to self-emulsify in the contents of the stomach. So, the drug may be rapidly emptied from the stomach and distributed in fine droplet form throughout the GI tract. In the past, these droplets were usually made of AOT or similar anionic surfactants and were negatively charged. This would theoretically confer the advantage of a large surface area from which dissolution could occur as well as a reduction in the irritation caused by prolonged direct contact between drugs in conventional dosage forms and the stomach wall. It was also shown that the formulation imparts an advantage as the absorptive cells, as well as the other cells in the body, are negatively charged with respect to the mucosal solution in the lumen.

Positively charged emulsion droplets that were made or formed by appropriate dilution, undergo electrostatic interaction with the Caco-2-monolayer and the mucosal surface of the everted rat intestine. Positively charged formulations differ from negatively charged formulations in their interaction with biological components in the GI environment. Positively charged droplets should be attracted to the negatively charged physiological compounds naturally occurring in the lumen. Larger droplets (a few micrometers in size range) are less neutralized by mucin solutions of different concentrations than smaller, nanosized droplets. A similar behavior was observed in different concentrates which, upon aqueous dilution, resulted in positively charged coarse or fine emulsions or microemulsions.

Nonionic surfactants were used to form dilutable w/o emulsions that invert and transform into o/w droplets that eventually will release the drug in the GI tract.

Pseudoternary phase diagrams have been constructed to evaluate the phase behavior of systems containing charged microemulsions, such as water/lecithin/polysorbate 80/IPM, at different polysorbate 80/lecithin weight ratios (175). O/W microemulsion regions were accurately determined and the influence of the effect of the nonionic/anionic ratio on the area of existence of such disperse systems was also examined. Viscosity studies and particle size analysis by dynamic light scattering were carried out in o/w microemulsions, and the influence of the oil phase content, the total amount of surfactants, and the surfactant ratios on the rheological behavior, viscosity, and droplet size of such dispersed systems were evaluated. All systems studied showed a water-rich isotropic region (o/w microemulsion area) that was seen to be highly dependent upon the surfactant/cosurfactant weight ratio.

Scientists have formulated a very large number of surfactant mixtures into microemulsions to improve the bioavailability of poorly soluble drugs (78,123,176). Most surfactants were propylene glycol monoalkyl esters and glycerol monoalkyl esters. In most cases, the solubilization of oil within the surfactant aqueous mixture was enhanced by mixing the two surfactants at a 1/1 ratio with hydrogenated ethoxylated castor oil (HECO-40, HECO-60), Tween® 80, ethoxylated (EO_9) lauryl alcohols ($C_{12}E_9$, BL-9EX), and block copolymers of ethylene and propylene glycols (Pluronic P84). The use of additional surfactants, such as SDS or sodium deoxycholate, significantly improved the solubilization capacity of the oils. The oils were long-chain triglycerides (LCT), medium-chain triglycerides (MCT), and monocapryl glycerol (MCG). The phase diagrams reveal certain areas of o/w and some areas where w/o microemulsions are formed, but the isotropic regions are not connected, exhibiting discontinuity and phase separation upon dilution (Fig. 4). The authors incorporated Nitrendipine, a poorly soluble model drug, into the microemulsions and showed that its administration was significantly enhanced compared with a suspension or an oil solution. The absorption behavior varied with the type of surfactant. The absorption from Tween 80–based formulation was very rapid, while HECO-60–based formulations showed a prolonged plasma concentration profile. The absorption from BL-EX–based formulation was hardly observed. HECO-60 and Tween 80–based formulations were mild to the organs and did not rupture membranes, while BL-EX formulation caused serious physiological damage (176).

Of special interest are the new lipophilic drugs or cyclic peptides (somatostatin and vasopressin) that are generally known to be unsuitable for oral administration because of low bioavailability as a result of enzymatic attachment by exo- and endopeptidases that digest the drugs, and physical barriers. According to Lee (177) the oral bioavailability of most peptides is less than 10%, with the exception of cyclosporin and thyrotropin-releasing hormone which are resistant to hydrolysis (168).

Cyclosporin has dramatically improved long-term graft survival in renal transplant patients and opened the possibility of successful heart and liver transplantations. However, the drug has also caused some difficulties to the users, of which the most important is "chronic graft failure" (rejection). Cyclosporin is very lipophilic drug which causes considerable variations in the pharmacokinetics of the drug among individuals—drug absorption is influenced by bile flow, food intake, and GI mobility, and low dosage causes insufficient immunosuppressive control, while overdosage might risk nephrotoxicity. These concerns have resulted in debates on dosage. Europeans give smaller doses than the Americans, who are more concerned about the efficacy than the side effects. Bristol Meyers formulated cyclosporin in a microemulsion and claimed it has significant advantages, almost like having a new drug. The drug was found to have a faster and more consistent rate of absorption with less pharmacokinetic variability. It showed improvement in dose linearity and had more advantageous efficacy and toxicity profile, was less affected by the physiological state of the GI tract and was less dependent on food intake and bile secretion, leading to more continuous absorption. Parameters like AUC increased, which might increase the risk of neurotoxicity, but parameters such as C_{max}, t_{max}, and C_{min} were reduced and the terminal variance after 24 hours is less variable in the ME formulation. The ME increased cyclosporin bioavailability by 30%. Those interesting practical results are the driving force for most of the recent work.

Many other studies have been done on oral microemulsion formulations that have demonstrated repeatedly that microemulsions are excellent vehicles for oral drug delivery, but in practice only a few found their way to the marketplace. It is still not clear why we do not see more such delivery systems.

IV Microemulsions

Few papers describe microemulsion formulations for intravenous (IV) drug delivery. Some show significant promise and should be further studied. However, one should bear in mind that, as in many cases that have been reported in the literature, the so-called microemulsions are not necessarily true microemulsions as explained in Part I of this chapter. In some cases, reports of mini-emulsions or highly sheared emulsions were made. High shear homogenization or filtered macroemulsions were formed to reach small droplet size (0.2 μm). By no means are those emulsions thermodynamically stable, and preparations must be used soon after the preparation stage otherwise they will coalesce and separate into two phases.

Neoplastic disease may have pronounced effects on the regulation of plasma lipid metabolism that leads to changes in the plasma concentrations of lipids and apolipoproteins. It has increasingly become recognized that these changes may bear important clues for the understanding of the pathophysiology, and even for the prognosis and treatment, of neoplastic diseases. Patients with acute myeloid leukemia develop lower plasma levels of LDL, the lipoprotein that carries most of the cholesterol present in the plasma. It has been shown that cholesterol-rich microemulsions (LDE) that bind to LDL receptors can concentrate in acute myeloid leukemia cells and in ovarian and breast carcinomas (178–184). Thus, LDE may be used as a vehicle for drugs directed against neoplastic cells. The authors used 131 volunteer patients with diagnosed multiple myeloma classified as clinical stage IIIA. Carmustine in ME was given over four weeks. It was found that LDL cholesterol was lower in all patients than in healthy controls. Moreover, entry of LDE into multiple myeloma cells was shown to be mediated by the LDL receptors. This means that LDE might target multiple myeloma.

Figure 30 shows the mean decay curves of the emulsion H3-cholesteryl esters obtained from 14 multiple myeloma patients and 14 controls. During the first six hours, the mean plasma decay of the emulsion was slightly slower in the patients than in controls, but from six hours onwards, the decay become faster until the final 24 hours radioactivity measurement, and at this time the patients retained much less radioactivity from emulsion in their plasma than the controls.

Figures 31 and 32 show changes in gamma globulin and cholesterol concentration peaks in the plasma protein electrophoretic profile under the formulated carmustine treatment.

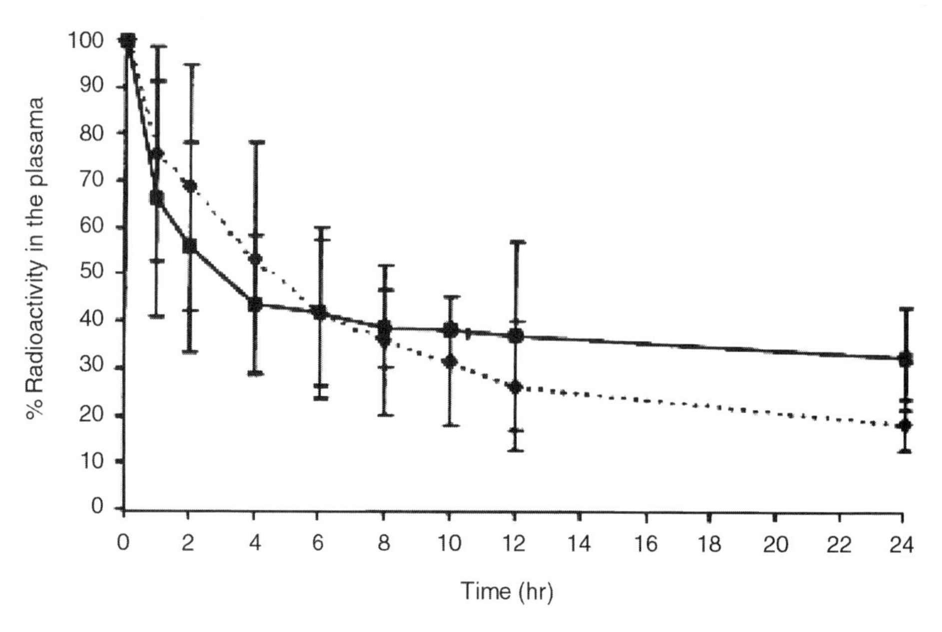

Figure 30 Decay curve of an H3-cholesteryl ester emulsion obtained from 14 patients with multiple myeloma (dotted line) and 14 controls (continuous line). The labeled emulsion was intravenously injected as a bolus and plasma samples were taken over 24 hours for radioactive counting in scintillation vials. *Source*: From Ref. 179.

The gamma globulin decreased by 10% to 70% within three treatment cycles and by 25% to 70% after five or six treatment cycles. Side effects of the treatment were negligible. Cholesterol levels increased, indicating partial destruction of neoplastic cells with receptor upregulation. The nanodroplet formulations (termed LDE) were prepared from a lipid mixture composed of cholesteryl oleate (40 mg), egg PC (20 mg), triolein (1 mg), and cholesterol (10.5 mg). The system is not a true ME system, as it requires prolonged ultrasonic irradiation in aqueous medium followed by a two-step ultracentrifugation of the crude emulsion with density adjustment by the addition of KBr. The LDE were dialyzed against saline solution and passed though a 0.22 μm filter for injection into patients.

Injectable forms of vitamin E as active matter and solvent were formed by several authors, but in most cases they formed emulsions with extremely small droplets of approximately 200 nm (33). In some cases, micellar solutions were also prepared. Kato et al. (185) prepared both micellar solutions and emulsions of vitamin E. The micellar solutions were prepared by dissolving a tocopherol in Tween, Brij 58, or HECO-60 with isopropyl alcohol or methanol. The cosolvents were removed by evaporation to form α-tocopherol/surfactant micelles. In a similar way, others have solubilized cyclosporin, Paclitaxel, certain steroids, and antibiotics, but in most cases emulsions were made from the micellar systems by dissolving α-tocopherols in soy PC and soybean oil, with IPA as cosolvent (removed at a later stage). The first commercially available product for parenteral application containing Paclitaxel was termed Taxol and was approved by the FDA in 1992. It is formulated in a vehicle containing 1:1 ECO and ethanol. ECO is a relatively toxic molecule and caused various side effects in patients. Constantinides et al. (33) concluded that, although several generic products have recently been approved for clinical use, a substantial need

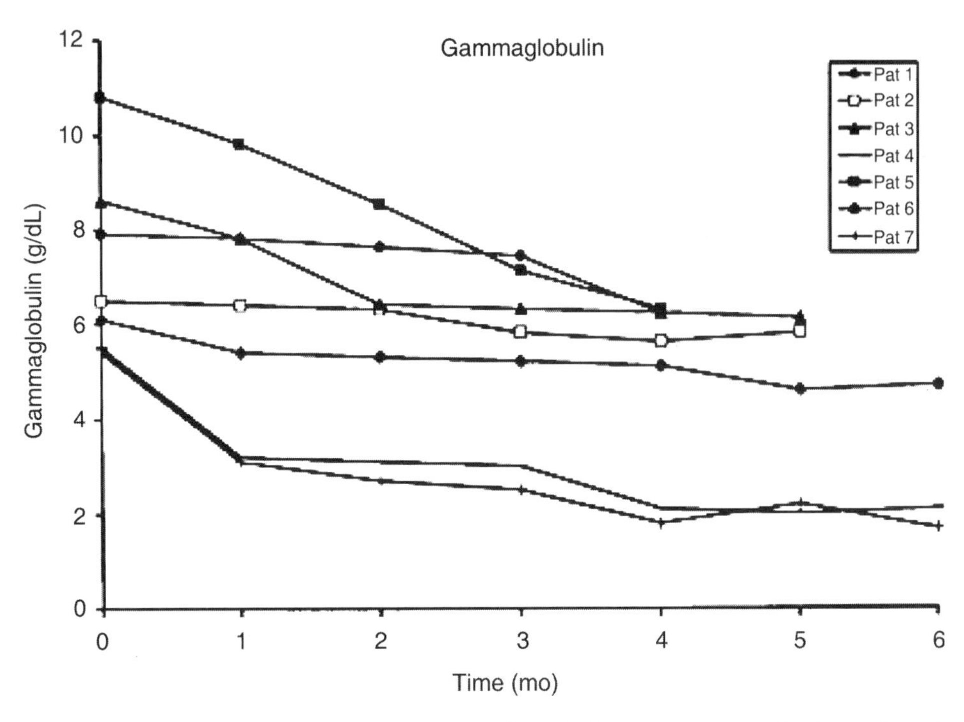

Figure 31 Monthly evolution of gamma globulin peak in the plasma protein electrophoresis profile under LDE-carmustine treatment. Determinations were performed a few days before the monthly administration of the LDE-carmustine dose. *Source*: From Ref. 179.

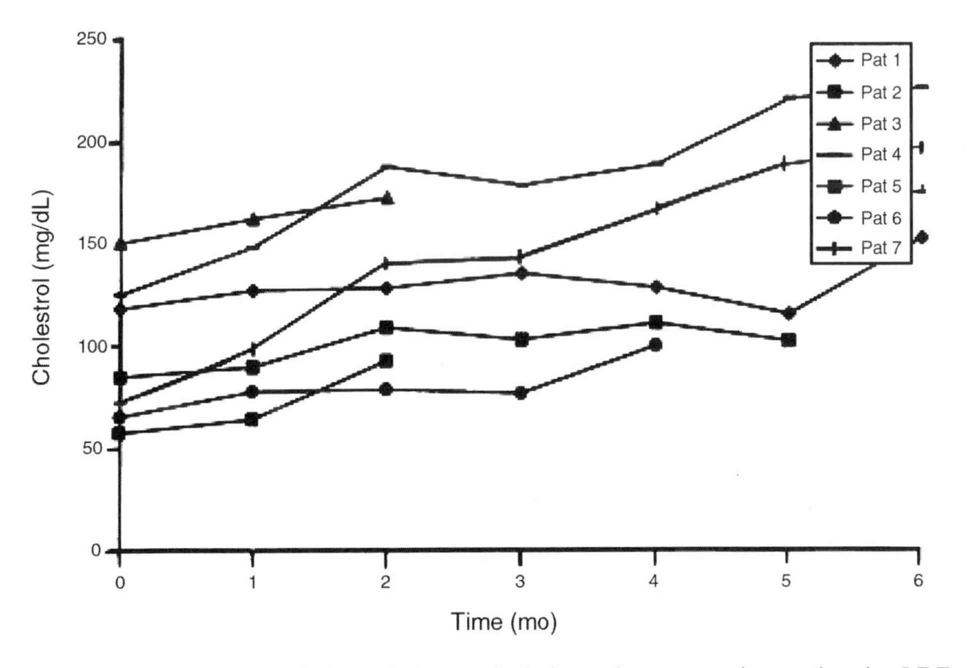

Figure 32 Monthly evolution of the total cholesterol concentration under the LDE-carmustine treatment. Determinations were performed a few days before the monthly administration of the LDE-carmustine dose. *Source*: From Ref. 179.

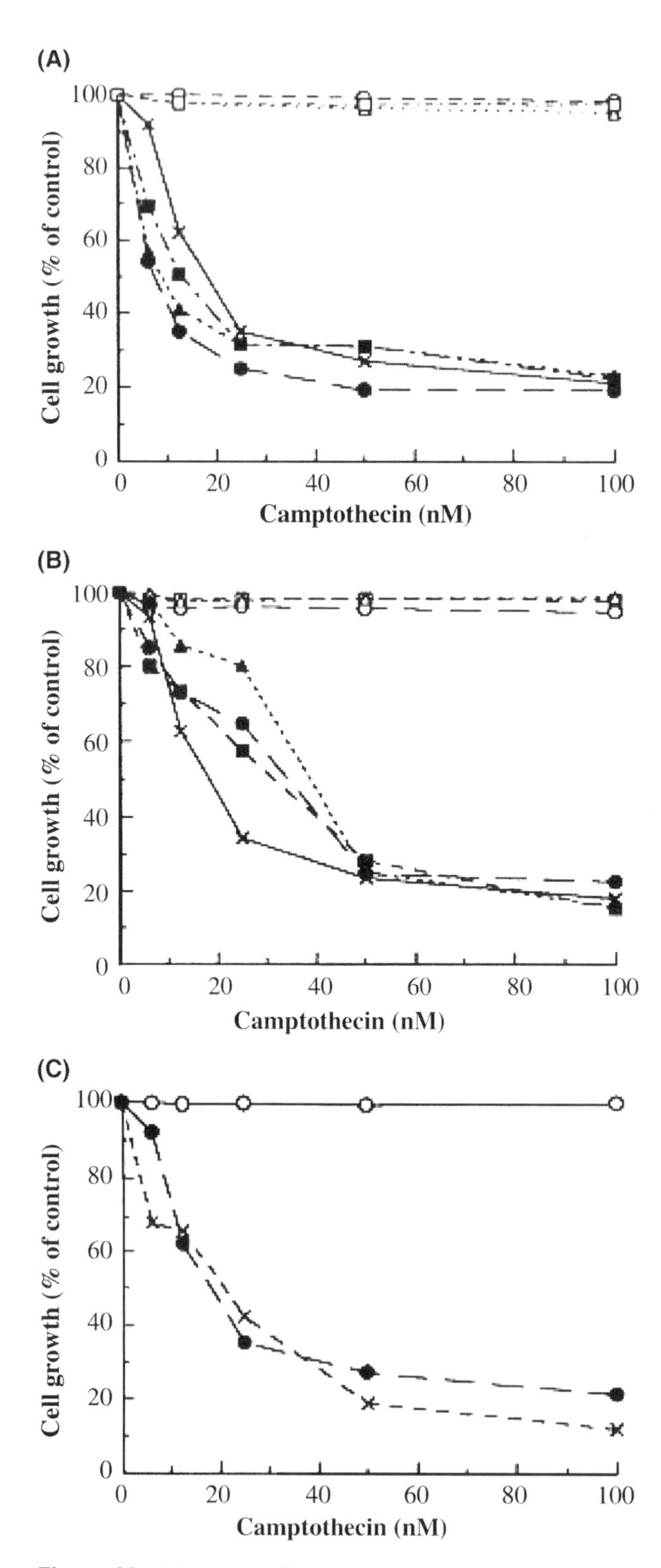

Figure 33 (*Caption on facing page*)

still exists for new formulations of Paclitaxel that will retain efficacy, improve tolerability, and provide advantages for preparation and administration. "There is still a great need for an ME formulation based on components that will cause no side effects and will transport the drug in the most efficient way possible by oral administration." Müller et al. (186) suggest a novel technology for the formulation of IV emulsions with poorly soluble drugs. They claim to have developed a solvent-free production method for lecithin-based emulsions called SolEmuls® for parenteral Lipofundin, Amphotericin B, and carbamazepine formulation. However, close examination of the technology reveals that the method requires high-pressure homogenization (Ultra-Turax and Micron Lab 40 homogenizer at 45°C, 1500 bar, and 1 to 20 homogenization cycles). After 20 homogenization cycles the droplets are approximately 190 to 200 nm in size, stable for over three months (for Amphotericin B), but are not true microemulsions. The solubilization capacities in nondiluted emulsions of Amphotericin B are of approximately 2 mg/mL. The high streaming velocities lead to accelerated drug dissolution and partitioning into the interfacial layer (the so-called solubilization by emulsification).

Although this method is interesting and gives good solubilization and stability results, such formulations will not be considered thermodynamically stable and will not be totally clear solutions.

Cortesi and Nastruzzi describe the production and characterization of specialized delivery systems for Camptothecin in liposomes, micellar solutions, and microemulsions (173b). All the formulations were designed to increase the solubility of the drug in an aqueous environment and to reduce toxicity problems. It was found that all three formulations showed similar cell growth results (Fig. 33) with some improved prospects on the toxicity of the drug.

Skin Applications

Microemulsions for application on human skin have been made both for cosmetic applications as well as for transdermal drug delivery systems. It is not always obvious how deep the additive or the drug penetrate, and there should be selectivity in the formulations that one selects for one application over the other. We will discuss some major applications.

Transdermal Delivery

As early as 1990, Friberg (187) and later Atwood and associates published works on micelles, microemulsions, liquid crystals, and the structure of stratum corneum lipids

Figure 33 (*Figure on facing page*) Comparative analysis of the effect of free CPT (×) and CPT formulated in specialized delivery systems on proliferation of K562 cells. The empty and CPT-containing formulations were added at the same indicated concentrations. Determinations were performed after six days of cell culture. Data represent the average of three independent experiments. (**A**) Liposome:CPT:Egg-PC (●), empty Egg-PC (○), CPT:Egg-PC:CH:DDAB18 [▲], empty Egg-PC:CH:DDAB18 (Δ), CPT:Egg-PC:CH:DPC [■], empty Egg-PC:CH:DPC (□). (**B**) Micellar solutions:CPT:Poloxamer 407 (●), empty Poloxamer 407 (○), CPT:Tween 80 [▲], empty Tween 80 (□), CPT:Tween 80:Tween 85 (1:1, w:w)[Δ], and empty Tween 80:Tween 85 (1:1, w:w) (Δ). (**C**) Microemulsion containing CPT (●) and empty microemulsion (○). *Source*: From Ref. 173b.

and on colloidal drug delivery systems for skin applications (105,113,137,188). The two groups predicted that microemulsions would serve as delivery systems for skin applications. These self-assembled aggregates have been shown to serve as reservoirs for hydrophobic and hydrophilic compounds and as potential vehicles for the release of drugs. Gasco and associates have also shown similar effects (118,189). Since that time, many additional studies have shown that ME can help to increase the permeability of drugs and facilitate transdermic transport (39,40,42,99,122,190–205). Others have shown improved bioavailability of the drugs (206–208). Several authors have discussed the therapeutic effects of drugs incorporated in MEs especially when formulated with balanced concentrations of surfactant/alcohol and oils with pharmaceutical use (31,116,209–212).

Transdermal drug delivery has drawn much attention in the last 15 years. Many papers were published with various suggested formulations. Kreilgaard in his very comprehensive review summarized the main studies published before 2002 in a table we include here (Table 4) (40,99,122,195–205,213–225).

Skin is very sensitive to surfactants. As a result, most scientists prefer to minimize the amount of surfactants and to avoid the use of cosurfactants. Valenta and Schultz (219) used three different microemulsion formulations as models for skin penetration and added carrageenan as stabilizer. They found there was a significant increase in skin penetration when microemulsions were used (Figs. 34 and 35). Formulation A is Brij 97 ($C_{18:1}E_{10}$) (2.7 grams), tributyrin (0.34 grams), and water (5.17 grams); B is Brij 97, water, and MCT; C is Brij 97, water, and soybean oil; A′, B′, C′ are the same formulae with carrageenan (1.5%) (219).

To reduce irritation, special types of microemulsions were suggested by Sintov and Shapiro (226). Microemulsions based on glycerol oleate and polyoxyl [40 EO] fatty acid derivatives/tetraglycol (cosurfactant)/isopropyl palmitate/water were used to construct a pseudoternary phase diagram at fixed cosurfactant/surfactant ratios (Fig. 36). Lidocaine was solubilized and penetration studies using rat skin in vitro showed that the transdermal flux of lidocaine was significantly improved by microemulsions composed of glycerol oleate–PEG 400 stearate over glycerol oleate–PEG 400 hydroxylated castor oil (Fig. 37).

Microemulsions for transdermal drug delivery were, in most cases, o/w formulations and the flux enhancement from these formulations was primarily because of an increase in drug concentration (220). The drug transport occurs only from the continuous outer phase (169,199). By this account, hydrophobic drugs show great flux from w/o emulsions, whereas o/w systems provide controlled release of the drug that is dependent on the partitioning of drug into the outer phase. This pathway was demonstrated using hydrophilic molecules, like glucose, which parallels the diffusion of water from the bulk phase (45). Lee et al. (220) composed water/IPM microemulsions with Tween 80 as the nonionic surfactant that previously had been used as a transdermal surfactant (221,222). Transdermal enhancers, like oleyl alcohol and *n*-methyl pyrrolidone (NMP), are believed to reduce the barrier properties of the skin by disrupting lipid bilayers within the stratum corneum. Lee et al. (220) have demonstrated that NMP is capable of significantly enhancing drug transport from both the organic and the aqueous phases. The NMP bound to the lidocaine enhanced the flux to approximately 10 $mg/cm^2/hr$. The study was carried out with lidocaine free-base, lidocaine hydrochloride, estradiol, and diltiazem hydrochloride. The water phase was found to be a crucial component for flux of hydrophobic (lidocaine free-base) as well as hydrophilic drugs. Furthermore, the simultaneous delivery of both hydrophilic and hydrophobic drugs from an ME system is indistinguishable from either of

Table 4 Overview of Cutaneous Drug Delivery Studies with Microemulsions In Vitro

Drug	Oil phase	Microemulsion			Ref.
		Surfactants	Aqueous phase	Membrane/skin	
[^{3}H]H_2O	Octanol	Dioctyl sodium sulphosuccinate	Water	Human	45, 203
8-Methoxsalen	Isopropyl myristate	Tween 80, Span 80, 1,2-octanediol	Water	Pig	213
Apomorphine hydrochloride	Isopropyl-myristate, Decanol	Epikuron 200, 1,2-propanediol benzyl alcohol	Water, Aerosil 200	Mouse	214
Diclofenac	Isopropyl-palmitate	Lecithin	Water	Human	195
Diclofenac	Diclofenac diethylamine	Lecithin	Water	Human	122
Diphenhydramine hydrochloride	Isopropyl-myristate	Tween 80, Span 20	Water	Human	200
Felodipine	Isopropyl-myristate, benzyl alcohol	Tween 20, tauro-deoxycholate	Water, Transcutol, Carbopol	Mouse	197
Glucose	Octanol	Dioctyl sodium sulphosuccinate	Water	Human	045
Hematoporphyrin	Isopropyl-myristate, decanol, hexadecanol, oleic acid, monoolein	Lecithin, sodium monohexyl-phosphate, benzyl alcohol	Water, PG	Mouse	198
Indomethacin	Isopropyl-palmitate	Lecithin	Water	Human	195
Ketoprofen	Triacetin, myvacet, oleic acid	Labrasol, Cremophor RH	Water	Rat	215
Lidocaine	Isostearylic isostearate	Labrasol, Plurol Isostearique	Water	Rat	099

(*Continued*)

Table 4 Overview of Cutaneous Drug Delivery Studies with Microemulsions In Vitro (*Continued*)

Drug	Oil phase	Microemulsion			Ref.
		Surfactants	Aqueous phase	Membrane/skin	
Methotrexate	Decanol	Lecithin, benzyl alcohol, Labrasol/Plurol	Water, PG	Mouse	199
Methotrexate	Ethyl oleate	Isostearique	Aq. 145 mM NaCl (pH 7.4)	Pig	209
Methotrexate	Isopropyl-myristate	Tween 80, Span 80, 1,2-octanediol	Water	Pig	209
Nifedipine	Benzyl alcohol	Tween 20, tauro-deoxycholate	Water, Transcutol, PG	Mouse	196
Prilocaine hydrochloride	Isostearylic isostearate	Labrasol, Plurol Isostearique	Water	Rat	099
Propanolol	Isopropyl-myristate	Polysorbate 80	Water	Artificial	216
Prostaglandin E_1	Oleic acid	Labrasol, Plurol Oleique	Water[a]	Mouse	201, 203
Prostaglandin E_1	Gelucire	Labrafac, Lauroglycol	Transcutol, water	Mouse	202
Sodium salicylate	Isopropyl-myristate	Tween 21/81/85, bis-2-(ethylhexyl) sulphosuccinate	Water, gelatin	Pig	217
Sucrose	Ethyl oleate	Labrasol, Plurol Isostearique	Aq. 154 mM NaCl	Mouse	195

[a]Seven different co-solvents were tested with each basic vehicle.
Source: From Ref. 9.

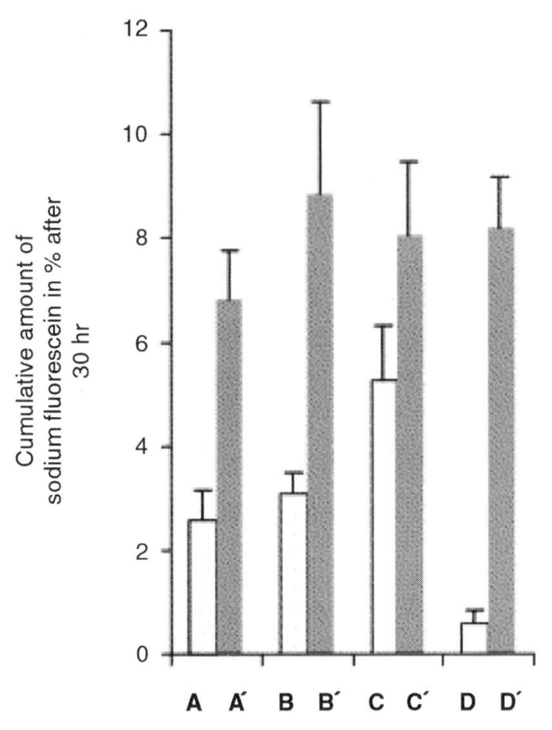

Figure 34 Comparison of the permeated cumulative amount of sodium fluorescein from different formulations. A–D (pure microemulsions, white bars); A′–D′ (with 1.5% carrageenan, gray bars), after 30 hours through porcine skin. Indicated values are mean (±SD) of at least three experiments. *Source*: From Ref. 219.

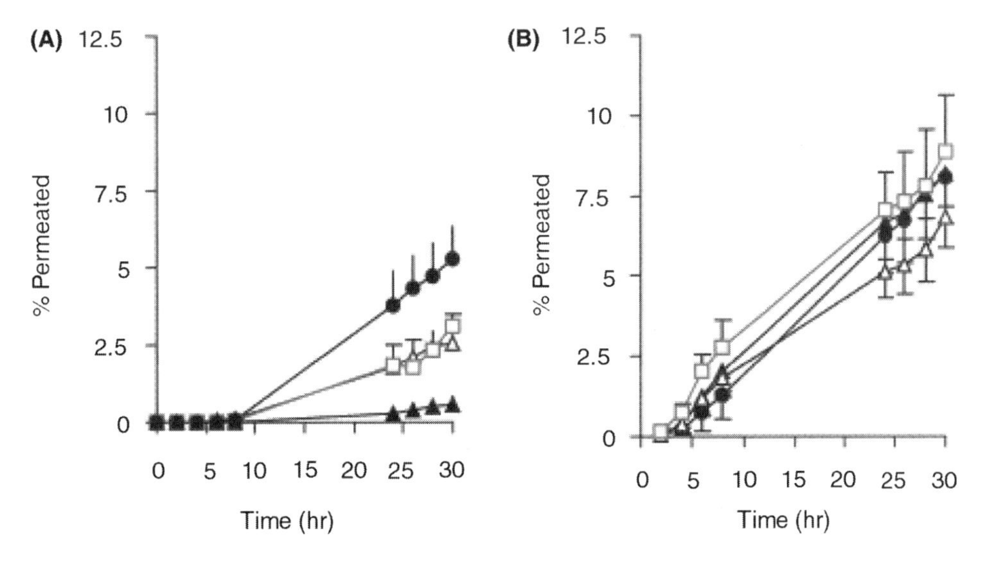

Figure 35 (**A**) Permeation profile of sodium fluorescein through porcine skin from pure microemulsions (-□-) A; (-□-) B; (-●-) C; (-▲-) D. (**B**) Permeation profile of sodium fluorescein through porcine skin from microemulsions and carrageenan (1.5%) (-□-) A′; (-□-) B′; (-●-) C′; (-▲-) D′. Indicated values are mean (±SD) of at least three experiments. *Source*: From Ref. 219.

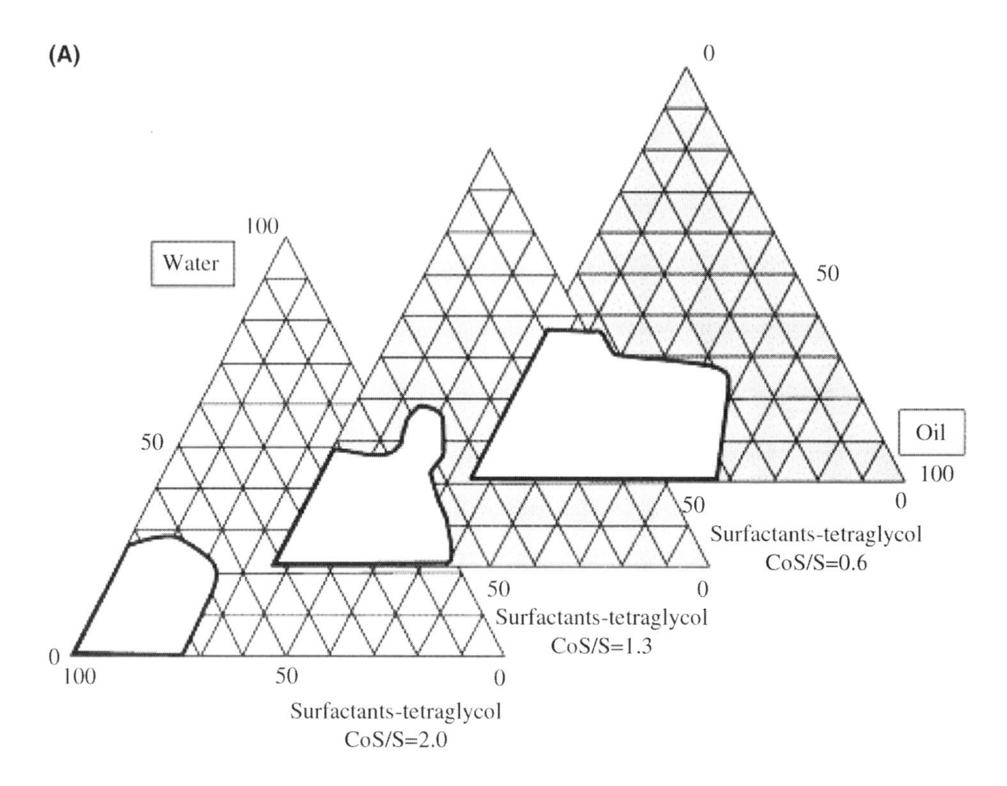

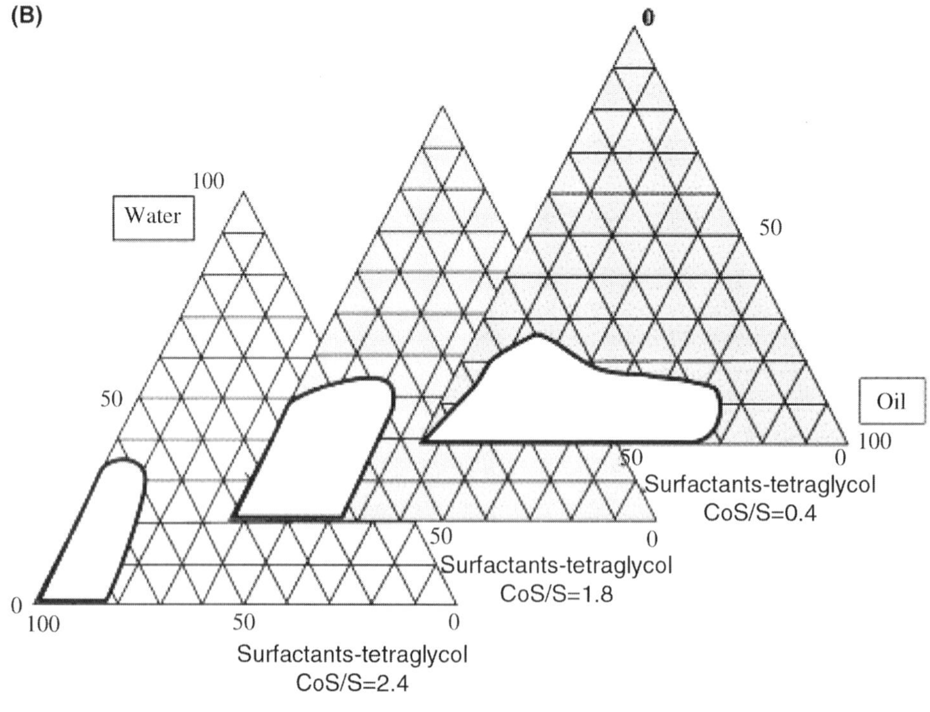

Figure 36 Pseudoternary phase diagrams of (**A**) microemulsion system a made of isopropyl palmitate/water/tetraglycol/glyceryl oleate + PEG-40 hydrogenated castor oil at three different CoS/S ratios and (**B**) microemulsion system b made of isopropyl palmitate/water/tetraglycol/glyceryl oleate + PEG-40 stearate at three different CoS/S ratios. *Source*: From Ref. 226.

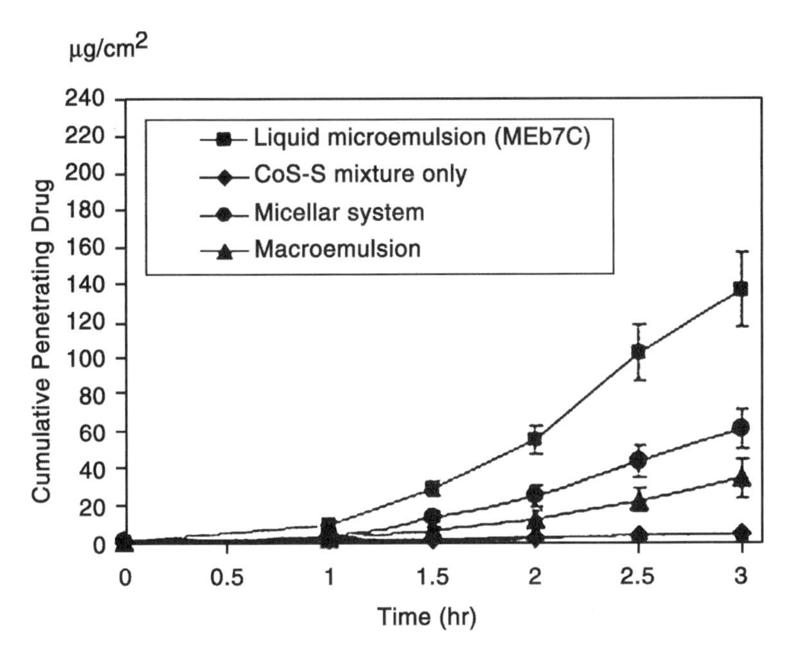

Figure 37 Percutaneous penetration of lidocaine from a microemulsion liquid vehicle of a specific formulation (MEb7C) through rat skin in vitro ($n = 5$) during three hours. The transport kinetics (■) is compared to the penetration profiles of lidocaine from a surfactants mixture in tetraglycol (◆), an oil-free micellar system (●), and a macroemulsion (▲). *Source*: From Ref. 226.

the drugs alone. Enhancement of drug permeability in the NMP system compared with the o/w ME system was 17-fold for lidocaine free-base, 30-fold for lidocaine hydrochloride, 58-fold for estradiol (Fig. 38), and 520-fold (Fig. 39) for diltiazem HCl. The general conclusion was that ME potentially offers many potential beneficial characteristics for transdermal drug delivery (Table 5).

Kreilgaard et al. (99) investigated the influence of the structure and composition of MEs on their transdermal delivery potential of a lipophilic (lidocaine) and a hydrophilic model drug (prilocaine hydrochloride) and compared the drug delivery potential of microemulsions to conventional vehicles. They used self-diffusion coefficients as determined by the pulsed-gradient spin-echo NMR (PGSE-NMR) to study the transdermal flux using Franz-type diffusion cells. The authors prepared phased diagrams where w/o and o/w MEs were formed (Fig. 40) (although they were not fully dilutable) and demonstrated clearly that MEs increased the transdermal flux of lidocaine up to four times compared to a conventional o/w emulsion and the prilocaine hydrochloride almost 10-times compared to hydrogels. The flux seems to be dependent on drug solubility in the ME and on drug mobility within the individual vehicle (Table 6).

In a very interesting paper, Schmalfuss et al. (200) studied the penetration of a hydrophilic substance, diphenhydramine hydrochloride, from a w/o microemulsion into human skin under ex vivo conditions. The focus of the study was to determine the amount of a model substance in the different skin layers, rather than to measure the transdermal rate. They studied the mechanism of drug penetration into the skin from ME systems, particularly into the stratum corneum and the influence of the constituents of the ME on this penetration process. The w/o ME consisted of

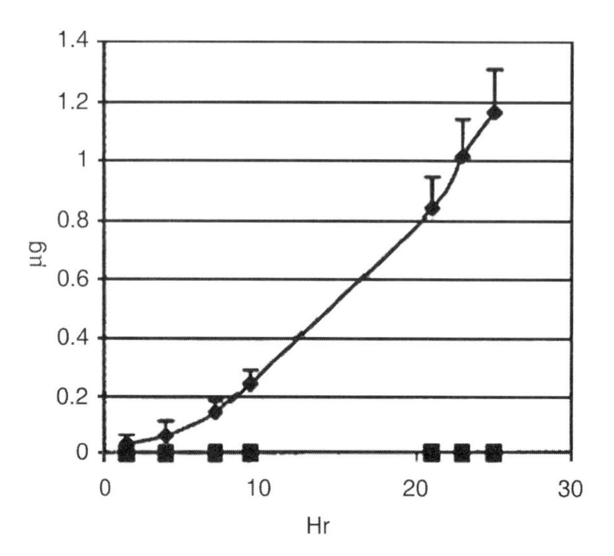

Figure 38 Estradiol transport across stripped human skin in o/w formulation. In vitro transport of the hydrophobic drug estradiol through stripped human skin (5 mm diameter) from the oil-in-water microemulsion (0.4% drug, w/w, ◆) and IPM (saturated drug, ■); $n = 3$. *Abbreviation*: IPM, isopropylmyristate. *Source*: From Ref. 220.

5 wt.% water and 15–20 Tween 80/Span 20 (2/3 ratio) and approximately 70 IPM as the oil phase in the presence of 2% to 5 % oleic acid or cholesterol and 1% of the DPH as the drug model. Figure 41 shows the penetration of DPH from IPM (200). A standard microemulsion showed an accumulation of penetrated drug in the dermis, indicating a potentially high absorption rate. Incorporation of cholesterol into the system leads to an even higher penetration rate and shifting of the concentration profile further toward the epidermis. Addition of oleic acid had no effect.

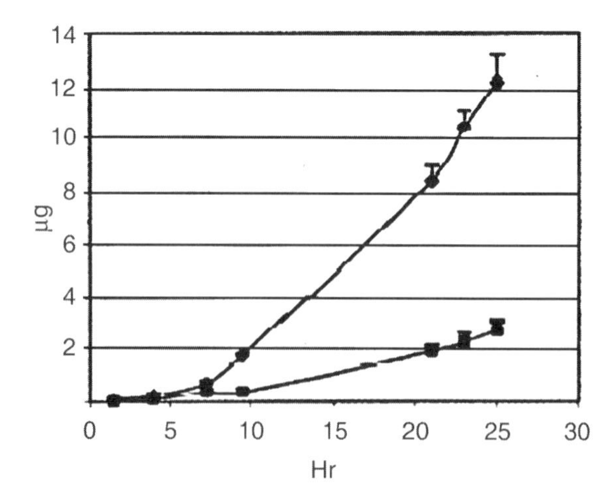

Figure 39 Diltiazem HCl in ME formulation transport across stripped human skin. Transport of the hydrophilic drug diltiazem HCl through stripped human skin (5 mm diameter) from oil-in-water microemulsion (2% drug, w/w, ◆) and water (2% drug, w/w, ■); $n = 3$. *Abbreviations*: HCl, hydrogen chloride; ME, microemulsion. *Source*: From Ref. 220.

Table 5 Estradiol and Diltiazem HCl Transport from ME Systems

Formulation	Estradiol Flux^{SS} (μg cm^{-2} hr^{-1})	Estradiol Permeability (cm hr^{-1} 10^5)	Diltiazem HCl Flux^{SS} (μg cm^{-2} hr^{-1})	Diltiazem HCl Permeability (cm hr^{-1} 10^5)
H_2O	0.015 ± 0.006	460 ± 183	0.05 ± 0.01	0.015 ± 0.004
W/O ME	0.053 ± 0.029	1.1 ± 0.6	0.25 ± 0.13	1.2 ± 0.6
W/O ME both drugs	0.12 ± 0.06	2.4 ± 1.2	0.24 ± 0.08	1.2 ± 0.4
W/O ME	0.27 ± 0.07	5.8 ± 1.5	1.6 ± 0.3	7.8 ± 1.3
W/O ME both drugs	0.23 ± 0.05	5.0 ± 1.2	1.6 ± 0.4	7.8 ± 1.9
Water phase	6.5 ± 1.7			6.1 ± 3.7

Abbreviation: SS, steady state; HCl, hydrogen chloride; ME, microemulsion; w/o; water-in-oil.
Source: From Ref. 220.

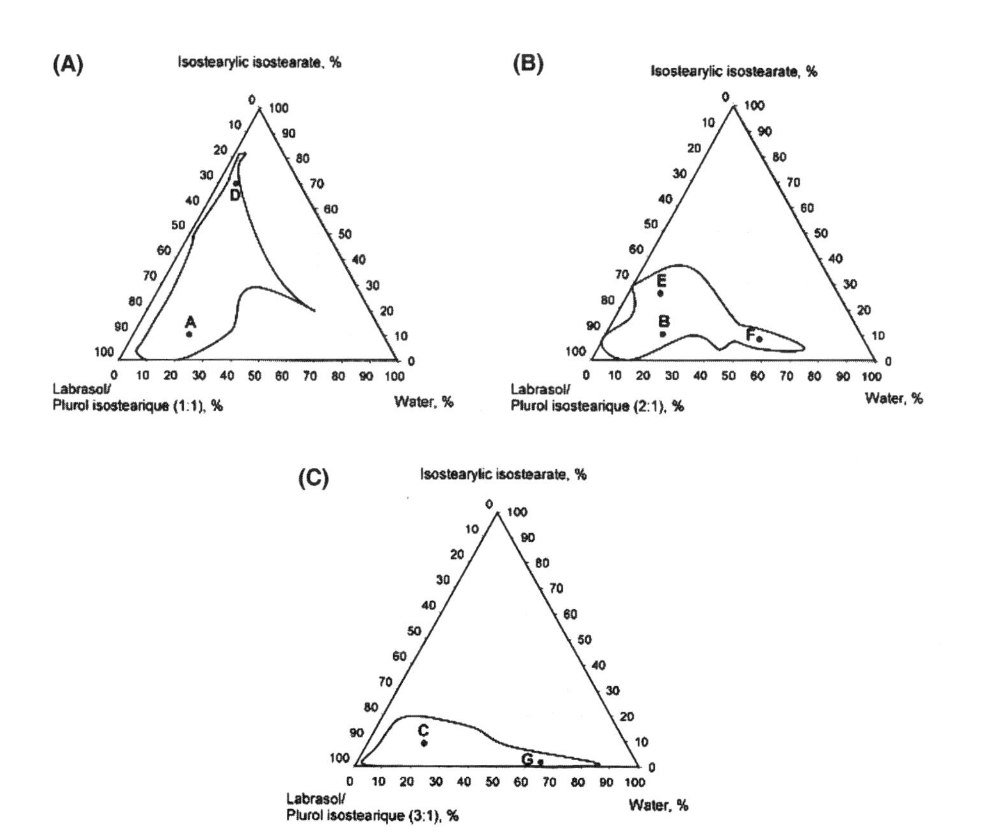

Figure 40 Pseudoternary phase diagrams of microemulsion regions of existence (within the connected lines) with different surfactant/co-surfactant ratios [(**A**) 1:1; (**B**) 2:1; (**C**) 3:1] composed of labrasol, plurol isostearique, isostearyle isostearate, and water (%, w/w). Composition of microemulsion systems A–G is indicated in the diagrams and listed in Table 6. *Source*: From Ref. 99.

Table 6 Steady-State Flux (J; 20–28 hr) and Permeation Coefficient (k) of Lidocaine and Prilocaine HCl Through Rat Skin from p-Microemulsion Systems A–G with near Maximum Drug Load and Comparison with EMLA 5%, Xylocaine 5%, o/w-Cream and Xylocaine 2% Hydrogel

	Lipophilic model drug			Hydrophilic model drug		
Formulation	Lidocaine (%, w/w)	J (μg hr^{-1} cm^{-2})	k_P (μg hr^{-1} cm^{-2})	Prilocaine HCl (%, w/w)	J (μg hr^{-1} cm^{-2})	k_P (μg hr^{-1} cm^{-2})
A	23	36.2 ± 4.2	1.6 ± 0.2	5	8.1 ± 2.6	1.6 ± 0.5
B	27	41.1 ± 8.9	1.5 ± 0.3	5	6.3 ± 1.3	1.3 ± 0.3
C	24	44.8 ± 4.3	1.9 ± 0.2	5	6.4 ± 2.1	1.3 ± 0.4
D	17	56.5 ± 4.6	3.3 ± 0.3	–	–	–
E	25	50.8 ± 8.6	2.0 ± 0.3	2.4[b]	6.2 ± 0.7	2.6 ± 0.3
F	12	44.6 ± 4.3	3.7 ± 0.4	13	24.7 ± 4.3	1.9 ± 0.3
G	9.1	78.3 ± 3.9	8.7 ± 0.4	14	29.7 ± 9.9	2.1 ± 0.7
EMLA[a]	2.5	52.3 ± 19	20.9 ± 7.6	–	–	–
EMLA[a,c]	2.5[a,c]	57.8 ± 18.8[c]	23.1 ± 7.5[c]	–	–	–
Xylocaine 5%	5	22.2 ± 6	4.4 ± 1.2	–	–	–
Xylocaine 2%	–	–	–	2[d]	3.1 ± 1.0	1.6 ± 0.5

[a] $n = 5$.
[b] Near maximum concentration equals 2.4%.
[c] Prilocaine-free base.
[d] Lidocaine hydrochloride.
Abbreviation: HCl, hydrochloride.
Source: From Ref. 99.

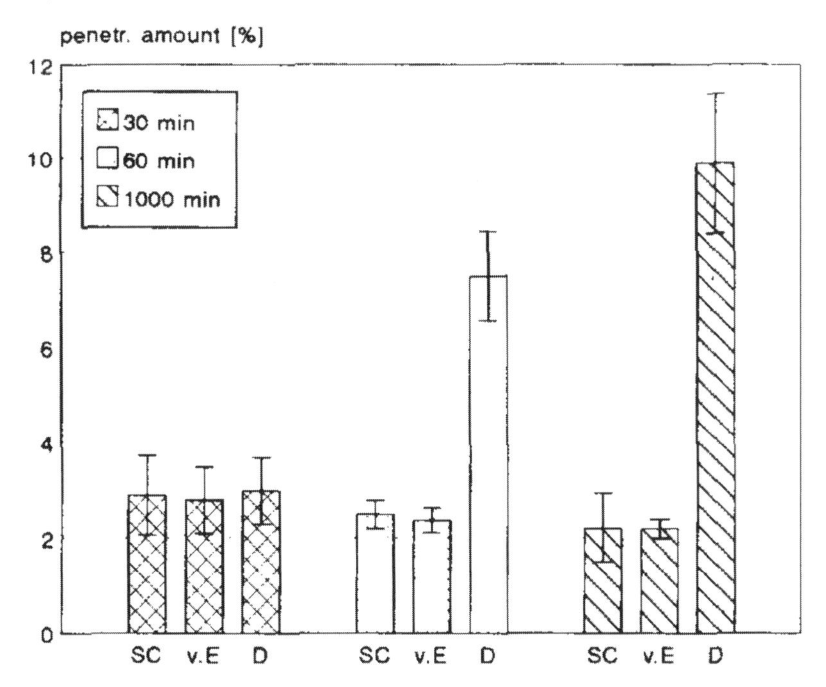

Figure 41 Penetration of DPH from ME 1 into human skin under ex vivo conditions. *Abbreviation*: ME, microemulsion. *Source*: From Ref. 200.

The hydrophilic drug follows hydrophilic structures into the stratum corneum. Therefore, an alteration of barrier properties can obviously be achieved only by influencing this hydrophilic pathway.

Piroxicam is a nonsteroidal anti-inflammatory compound with analgesic and antipyretic effects that has been used for rheumatoid arthritis, osteoarthritis, and traumatic contusion treatment. However, because piroxicam is associated with GI side effects, it is important to develop therapeutic systems that will minimize these effects (223).

Canto et al. (224) and Dalmora and Oliveira have shown that interaction of piroxicam with β-cyclodextrin (β-CD), cationic ME, and ME in the presence of β-CD are important factors for the delivery of the drug (223,225). A typical ME formulation included HTAB (*N*-hexadecyl-*N*,*N*,*N*-trimethyl ammonium bromide) as the cationic surfactant (30.05%), ethyl alcohol (30.05%), IPM (6.6%), and phosphate aqueous buffer (pH 5.5; 33.3 %). The piroxicam was solubilized at levels of 0.5% ion as the basis of the ME and the release patterns were determined through cellulose acetate membrane. The release profiles are shown in Figure 42 .

The effect of daily application of piroxicam formulations on six rats (F-5, F89 4 mg/kg) for six days on the formation of granulomatous tissues is shown in Figure 43, with control, with phosphate buffer (F-5), complexed with β-CD (F6), in ME (F7), and complexed with ME/β-CD, (F8).

Topical formulations of piroxicam were evaluated by determination of their in vitro release and in vivo anti-inflammatory effect. ME provided a reservoir effect for piroxicam release, but the incorporation of the drug into carboxyvinylic gels provoked a greater reduction in the release of piroxicam than the ME system alone. Subcutaneous administration of the drug formulations showed a significant effect on

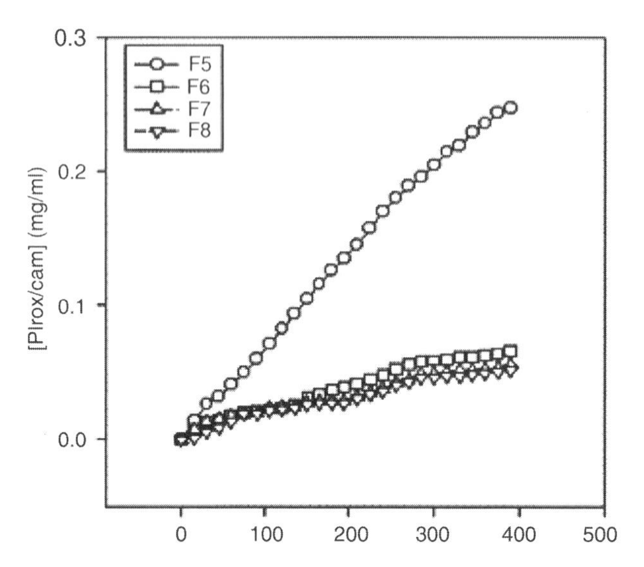

Figure 42 In vitro release profiles of piroxicam in the presence of Carbopol 940® through the cellulose acetate membrane at 37°C, in phosphate buffer pH 5.5 (F5), complexed with β-CD in phosphate buffer (F6), in a ME (F7), and complexed with β-CD in a ME (F8). *Abbreviation*: ME, microemulsion. *Source*: From Ref. 223.

the inhibition of inflammation, 68% and 70.5% when the piroxicam was incorporated in ME and in a combined system of β-CD (β-CD)/ME, respectively, relative to the buffered piroxicam (42.2%). These results demonstrated that the ME induced prolonged effects, providing inhibition of the inflammation for nine days after administration of a single dose.

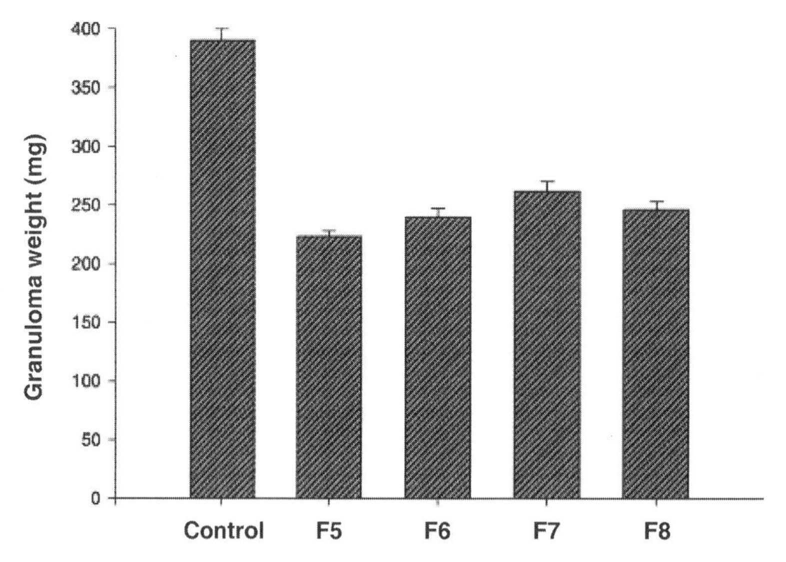

Figure 43 Effect of daily topical application of piroxicam formulations F5–F8 (4 mg kg^{-1}), for six days, on the formation of granulomatous tissues. Control carboxyvinilic polymer in phosphate buffer pH 5.5. Data represent the mean ± SE of six rats. *Source*: From Ref. 223.

Triptolide possesses immunosuppressive, antifertility, and anticancer activities. Because of its severe toxicity, microemulsions with controlled, sustained, and prolonged delivery of triptolide via a transdermal route are expected to reduce its adverse side effects. The purpose of these studies was to investigate microemulsions for transdermal delivery of triptolide (227,228). Pseudoternary phase diagrams were developed and various microemulsion formulations were prepared using oleic acid as an oil, Tween 80 as a surfactant, and propylene glycol as a cosurfactant. The droplet size of microemulsions was characterized by photocorrelation spectroscopy. The transdermal ability of triptolide from microemulsions was evaluated in vitro using Franz diffusion cells fitted with mouse skins, and triptolide was analyzed by high-performance liquid chromatography. The effect of menthol as a permeation enhancer and the loading dose of triptolide in microemulsions on the permeation rate were also evaluated. The triptolide-loaded microemulsions showed an enhanced in vitro permeation through mouse skin (227,228).

Topical administration of timolol as an ion-pair with octanoate was achieved via an o/w microemulsion containing lecithin as a surfactant (43). The microemulsion, a solution of the ion-pair, and a solution of timolol alone were inserted into the conjunctival sac of rabbits. A rapid method for the separation and determination of timolol by HPTLC was used. The bioavailability of timolol from the microemulsion and the ion-pair solution was higher than that obtained from timolol alone. The areas under the curve for timolol after administration of the microemulsion and the ion-pair solution were 3.5 and 4.2 times higher, respectively, than that observed after the administration of timolol alone.

Gelatin-stabilized microemulsion-based organogels (MBGs) (Fig. 44) were made for iontrophoretic transdermal drug delivery; the surfactants were Tween 85 and IPM. Unlike most organogels, the MBG are electrically conductive and have been successfully employed in the study of iontophoretic delivery of a model drug through excised pigskin. A high release rate of sodium salicylate compared to passive diffusion was obtained (Fig. 45) (217).

Water Transport and Hydrophilic Drugs

As early as in 1988, microemulsions showed some potential as topical drug delivery vehicles (229,230). Scientists attempted to make microemulsions to transport water across the skin. The authors used three microemulsions with a fixed weight ratio of surfactant to cosurfactant, and varied the water concentration from 15 to 35 and 68%. They found that most of the water in the 15 wt.% water ME was bound to the surfactant head group and was not transported across the skin. For MEs with higher water content, an approximately sixfold enhancement in water transport occurred, which resulted from a synergistic effect between the surfactant, dioctyl sodium sulphosuccinate (DSS), and the cosurfactant (octanol). In a later paper, Osborne et al. (45) studied the in vitro transdermal transport of hydrophilic compounds in the same ME compositions. They tested the ability of the ME to transport glucose across human cadaver skin and found that both the 35% and 68% water MEs produced enhanced (30-fold) transport of the glucose. No transport was detected for the 15% water ME (Fig. 46). The authors concluded that a hydrophilic drug would not be available for percutaneous transport from an ME unless the water in the ME was freely transported percutaneously. This means that such drug delivery requires sufficient mobility of water within the ME prior to being transported.

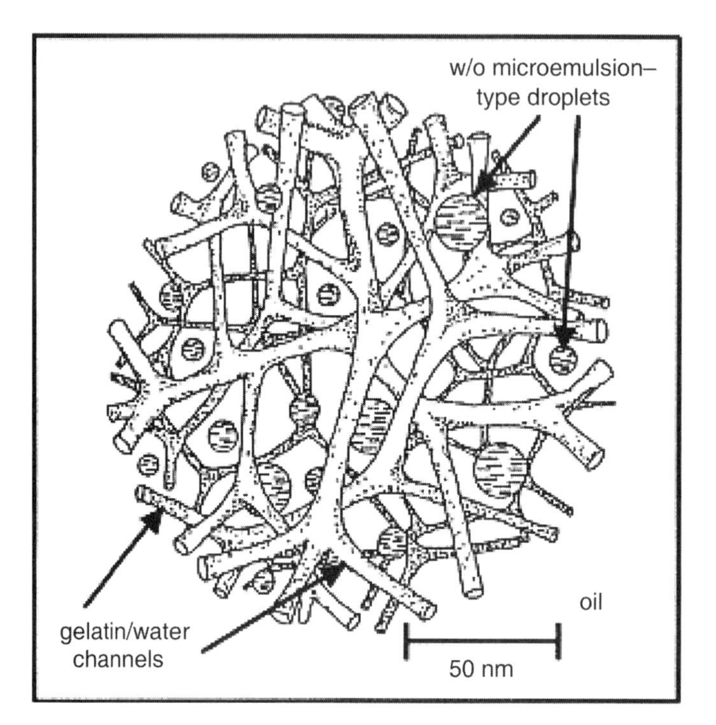

Figure 44 Proposed MBG structure based on small angle neutron scattering. *Abbreviation*: MBG, microemulsion-based organogel. *Source*: From Ref. 217.

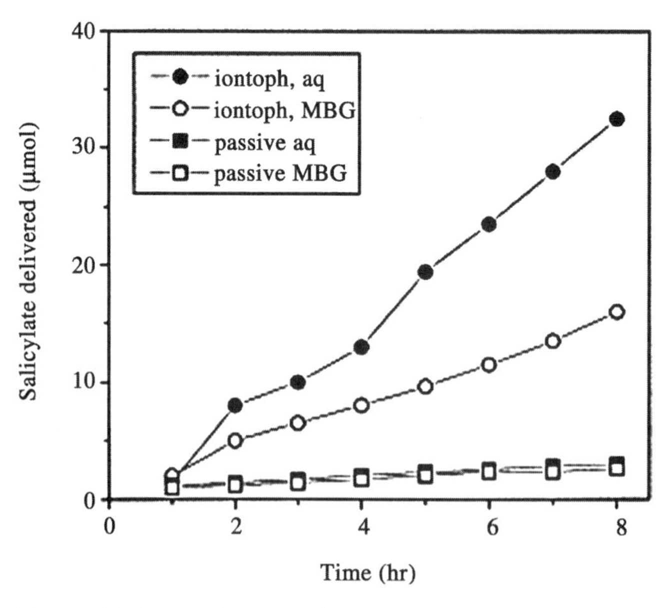

Figure 45 Drug release profile from an aqueous solution containing 2.14% (w/w) sodium salicylate and from an AOT/IPM organogel containing 10% (w/w) drug in the aqueous phase (2.14%, w/w overall) under passive and iontophoretic delivery conditions. Incubation temperature of 25°C. *Source*: From Ref. 217.

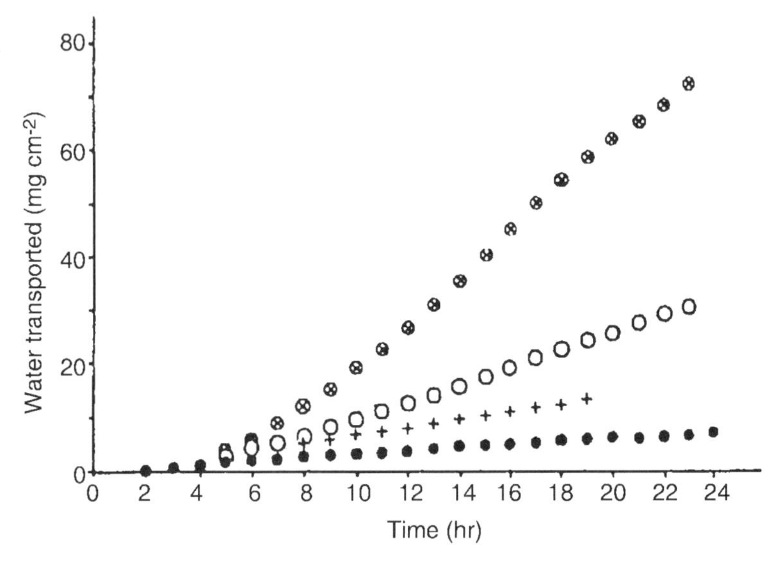

Figure 46 In vitro transdermal data for (+) pure water and the (•) 15, (○) 35, and (⊗) 68% water microemulsions formed from a 58:42 weight ratio of DSS/octanol. Cadaver skin was from the abdomen of a 57-year-old Caucasian female, using five replicates for each of the microemulsions and four replicates for pure water. *Abbreviation*: DSS, dioctyl sodiumsulphosuccinate.

Sonneville-Aubrun et al. (231) used a high shear device to form nanosized emulsions that are not true microemulsions, but are almost transparent emulsions with the slightest sign of destabilization in their visual appearance (derived from Ostwald ripening and induced flocculation typical of macroemulsions). Some of the formulations which are intended for skin care and not transdermal applications can be treated with various amounts of added polymers [Carbomer-crosslinked polyacrylic acid polymers or HEUR (hydrophobically modified urethane ethoxylate) types of associative polymer] to reduce destabilization, but for transdermal applications it is not a good solution.

Transdermal Skin Patches

Transdermal skin patches can allow a drug to be released over periods of up to one week, avoiding the first-pass metabolism effect and allowing the use of lower doses than in oral formulations. Not every drug is a suitable candidate for transdermal delivery. The skin, and in particular the stratum corneum, is a formidable barrier. Critical to success is the lipophilicity (lipid-loving nature) of the drug. In general, potent lipophilic drugs, within a certain size range, can pass through the stratum corneum well enough to achieve therapeutically effective blood levels. Basic compounds such as alkaloids (including morphine and nicotine), fentanyl, and buprenorphine also penetrate the intact skin at a sufficiently high rate. A limiting factor is the dose size—only drugs given in low doses (less than 1–2 mg/day) are candidates. The successful transdermal patches already on the market include estradiol as hormone replacement therapy and nicotine for smoking cessation. Estradiol provides a clear example of the benefits of transdermal delivery: oral administration results in low bioavailability as the drug undergoes extensive first-pass metabolism in the liver.

The half-life of the drug is only about one hour, so frequent oral dosing would be necessary. However, transdermal delivery completely avoids first-pass metabolism through the GI tract and liver, allowing therapeutic levels of estradiol to be reached at a dosage much lower than that from oral dosage. Patches for the local delivery of a drug have also been introduced—an example is the cytostatic-5-fluorouracil for the treatment of actinic keratoses.

A method called iontrophoresis has been under investigation as a way of delivering hydrophilic drugs transdermally. It involves an electric current being applied between two electrodes placed on the skin. This current helps drug molecules pass through the skin at much higher rates than are possible in the absence of this current. Low-molecular weight drugs, and high-molecular weight proteins such as insulin, might be administered successfully with iontrophoresis. The first prototype systems are currently being tested in the clinic both for systemic delivery of drugs and for local delivery of lidocaine for topical anesthesia. Iontrophoresis is clearly in an investigational stage, and its future position in our arsenal of delivery strategies is as yet unknown.

The use of a patch was also studied in a most recent work using a pseudoternary microemulsion for solubilizing lipophilic lidocaine. The results show that patches with the ME are very effective delivery mechanisms and that lidocaine accumulated in the epidermis layer after 60 minutes at higher concentrations than the liquid microemulsion and had much better effect than the EMLA (EMLA cream is a commercial formulation—Astra, Sweden). The lidocaine accumulation in the dermis layer was somewhat less pronounced than with the microemulsion, but still much more than from the corresponding cream.

Cosmetic Applications

Microemulsions were mentioned and examined for several cosmetic applications for topical use. The objectives of those studies were to design and develop topical microemulsions for poorly water-soluble compounds using antifungal agents such as miconazole, ketoconazole, and itraconazole as model drugs. Various sets of surfactants and cosurfactants have been tried and were evaluated individually. Either olive oil or mineral oil was used as the oil phase. Pseudoternary phase diagrams were constructed for each microemulsion system. Additionally, a gel formulation of ketoconazole (about 1% w/w) containing 90% v/w of alcohol was also developed. The release profiles of ketoconazole from the microemulsions (containing about 10% v/v alcohol) and the gel formulation were compared using Franz diffusion apparatus, and the release rates were calculated (232). Water-in-oil microemulsions were obtained using oil and surfactant concentrations that ranged from 8.3% to 33.3% v/v and 16.7% to 75.8% v/v, respectively. The release rates of ketoconazole from microemulsion and gel formulations were 766.8 and 677.6 μg/hour, respectively. No significant difference was seen between the release rates of ketoconazole from both formulations despite their differences in alcohol content. As a result, microemulsions of poorly water-soluble antifungal agents were successfully developed with in vitro release rates comparable to that of the gel formulation.

Ophthalmic Drugs

Approximately 90% of all ophthalmic drug formulations are now applied as eye-drops. While eye-drops are convenient and well accepted by patients, about 95% of the drug contained in the drops is lost because of absorption through the conjunctiva or through tear drainage. A major fraction of the drug eventually enters the blood stream and may cause side effects. The drug loss and the side effects can be minimized by using disposable soft contact lenses for ophthalmic drug delivery. The essential idea is to encapsulate the ophthalmic drug formulations in nanoparticles, and to disperse these drug-laden particles in the lens material. Upon insertion into the eye, the lens will slowly release the drug into the pre-lens (the film between the air and the lens) and the postlens (the film between the cornea and the lens) tear films, and thus provide drug delivery for extended periods of time.

A very extensive review article on "Microemulsions as ocular drug delivery systems" by Vandamme in 2002 (8), covers most of the most important work that has been done on ocular applications. This review covers various aspects of choice for various ingredients for such formulations, preparation processes, and the significance of microemulsions in ocular drug delivery devices. Vandamme also discusses some important physicochemical issues for improved solubilization of various drugs, and also in vitro results using diffusion cells and some in vivo trials and results on humans. Some of the most important studies that were mentioned in the review include works by Gasco et al., Ruth, Benita and Muchtar, Siebenbrodt and Keipert, and Jarvinen et al. (43, 233–236).

There are no significant new findings in the area of ocular delivery, therefore we have selected some examples of the more interesting and relevant studies to demonstrate potential ocular applications using microemulsions.

In Figure 47 is shown one example of eye drop formulation. Tables 7–9 are of typical eye drop concentrations and list the various ingredients (8,43,236–238). The authors describe in great detail the advantages of MEs in various formulations and compare the solubility vs. solubilization of various drugs in different media. Some of the most impressive examples are those of chloramphenicol, diclofenac, piroxicam, and indomethacin (Fig. 48).

Siebenbrodt and Keipert (236) compared the kinetic release of chloramphenicol, indomethacine, and sodium diclofenac which were obtained from aqueous solutions and ME, and found that 6.2, 4.2, and 5.4 times less indomethacine, chloramphenicol, and diclophenac were released from MEs after, 2, 4, and 6 hours, respectively. This paper focuses on dispersing stabilized microemulsion drops in poly-2-hydroxyethyl methacrylate (p-HEMA) hydrogels and modeling the release of the drug from the microemulsion-laden gels. The results of this study show that the p-HEMA gels loaded with a microemulsion that is stabilized with a silica shell are transparent and that these gels release drugs for a period of over eight days. There is an initial period of rapid drug release and the subsequent release is controlled by diffusion of the drug across the microemulsion surface. Contact lenses made of particle-laden gels are expected to deliver drugs at therapeutic levels for a few days. Delivery rates can be tailored by controlling the particle and drug loading. It may also be possible to use this system for both therapeutic drug delivery to eyes, and the provision of lubricants that might alleviate eye problems prevalent in extended lens wear.

Thermosetting Emulsions

Thermosetting emulsions, microemulsions, and mixed micellar systems were investigated in relation to drug delivery of lidocaine and prilocaine stabilized by copolymers such as Lutrol F127 and Lutrol 68 (238–240). The systems were made of dilute micelles or microemulsions. Effects of temperature-induced gelation, drug

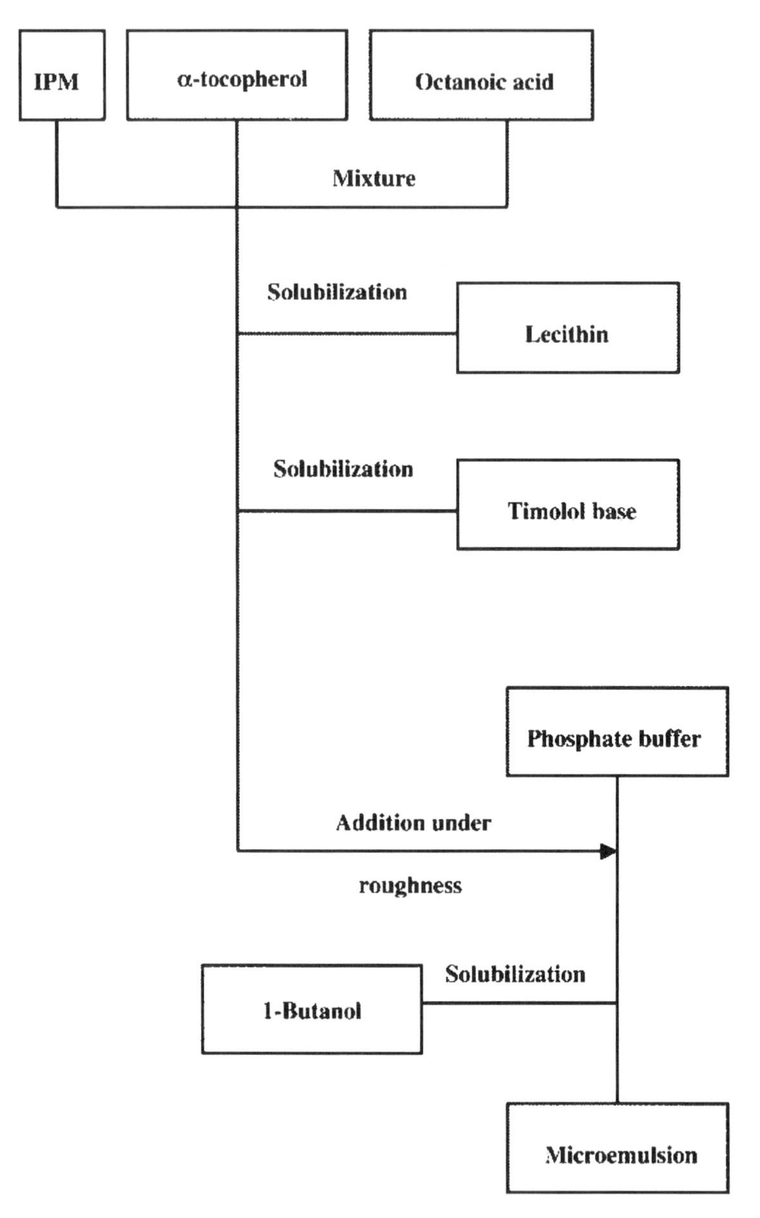

Figure 47 Diagram for the preparation of eye drop microemulsion containing timolol base. *Source*: From Ref. 8.

solubilization, and release were studied. The main disadvantage of those formulations is that they are very diluted systems when the microemulsions are formed and nonstable once emulsions a re formed (238–240).

Delivery of Nutraceuticals

The literature is saturated with examples of microemulsions as delivery technology for drugs in almost any possible mode of administration. The results of hundreds

Table 7 Example of Eye Drop Microemulsion

Component	Amount (%, w/w)	Function
Timolol base	2.60	Drug
Egg lecithin	28.70	Surfactant
IPM	11.70	Internal phase
Octanoic acid	4.70	Ion pairing agent
1-Butanol	14.90	Co-surfactant
Tocopherol	0.01	Antioxidant
Isotonic phosphate buffer (pH = 7.4)	QSP 100	External phase

Abbreviation: IPM, isopropylmyristate.
Source: From Ref. 8.

of studies and dozens of different formulations are promising but not free of problems, which are related to the fact that the formulations are rich in surfactant, alcohols, and other ingredients.

Recently, we have seen a new and very promising area of activity related to preventive medicine and reducing risk factors of chronic and other diseases. The emerging nutraceutical compounds—molecules with some health benefits derived mostly from plant extracts—require much attention to formulation. Some of those molecules are poorly water-soluble and some suffer from enzymatic or chemical degradation, and in many of the cases the bioavailability was shown to be very limited or close to zero. Some of the most well-known nutraceutical molecules are phytosterols, lycopene, vitamin E (tocopherols), lutein, CoQ10 and others.

There are practically no significant scientific studies that correlate the formulations of the nutraceuticals with their bioavailability or delivery advantages.

We have recently demonstrated that a new U-type phase diagram can be constructed into which concentrates of oil-based formulations can be placed that are fully dilutable with any amount of water or aqueous phase (238–244). Such formulations have significant advantages if the nutraceuticals have to be incorporated into food, especially clear beverages. It also permits the design of new soft gels with clear concentrates of nutraceuticals that are formulated in oil-based solutions, but totally dissolve or disperse in water to form clear microemulsions with improved solubilization capacities.

The most common way to construct such pseudoternary phase diagrams is by preparing an oil phase of the surfactant and the cosurfactant phase. The mixture is titrated with water until it turns turbid. The volume of water used is then recorded.

Table 8 Example of Eye Drop Microemulsion

Component	Amount (%, w/w)	Function
Chloramphenicol	0.5	Drug
Triacetin	20.0	Organic phase (internal phase)
Symperonic® L64	15.0	Surfactant
Propylene glycol	40.0	Co-surfactant
Water	24.5	Aqueous phase (external phase)

Source: From Ref. 237.

Table 9 Example of Eye Drop Microemulsion

Component	Amount (%, w/w)	Function
Indomethacin	0.125	Drug
MCT	10.0	Oily phase
Lipoid E-80®	0.75	Surfactant
Tocopherol	0.02	Antioxidant
Miranol MHT®	0.5	Co-surfactant
Glycerol	2.25	Osmotic agent
Distilled water	QSP 100	Aqueous phase

Source: From Ref. 237.

The pseudoternary phase diagram is constructed by plotting the amounts of water phase, oil phase, and surfactant/cosurfactant phase used in the experiment.

Several of these microemulsions have been titrated with nutraceutical compounds to the point where the microemulsion again turns turbid. The maximum amount of solubilized nutraceutical is recorded and plotted against the water content in each emulsion along a given dilution line. Figure 49 is an example of the maximum solubilization of lutein in U-type microemulsions.

FINAL REMARKS

One major advantage of a microemulsion over traditional emulsions is the ease of preparation, especially with regard to large batch manufacturing. Another advantage

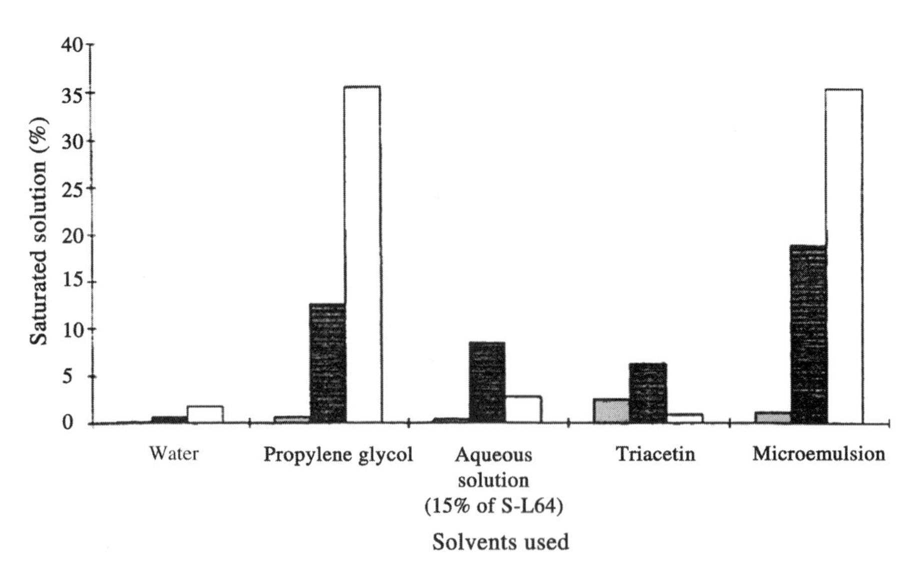

Figure 48 Histogram of the maximum percentage of indomethacin (*grey*), chloramphenicol (*black*), and sodium diclofenac (*white*) in each component of the microemulsion. *Source*: From Ref. 8.

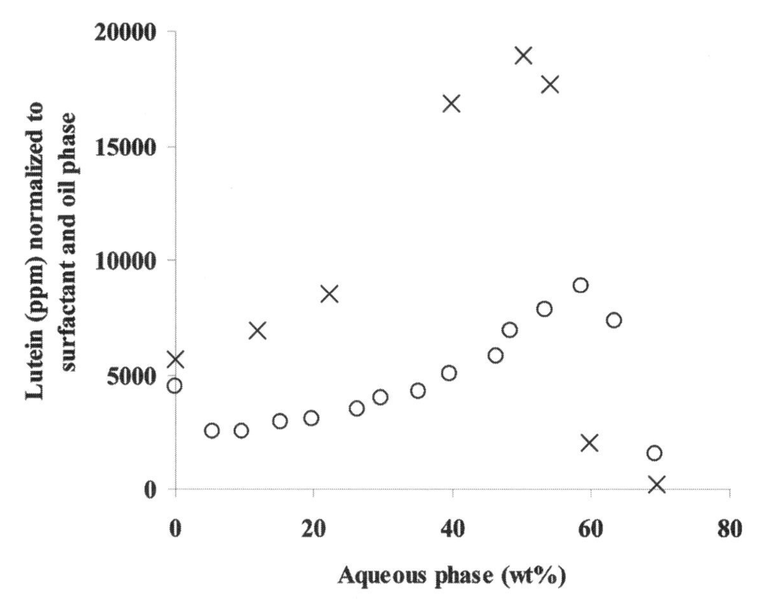

Figure 49 Total solubilization capacity. Amount of solubilized free lutein (×) and lutein ester (O) normalized to the total amount of oil, alcohol, and surfactant plotted against the aqueous phase content, along the dilution line 64, at 25°C.

is the physical stability of the formulation. In a traditional emulsion system, the larger droplet size favors a decrease in the surface area, which in turn favors a decrease in the free energy of the system. However, a microemulsion system has lower interfacial tension between water and oil because of the presence of surfactant and cosurfactant; therefore, the surface area of the dispersed droplets is very large. The lower interfacial tension compensates for the dispersion entropy; therefore, the microemulsion system becomes thermodynamically stable because of the low free energy of the system.

As mentioned earlier, a microemulsion normally contains droplets with diameters ranging from 1 to 100 μm. A microemulsion has a more elegant appearance than an emulsion, which may increase patient compliance. In addition, the presence of surfactant and cosurfactant in the system enables high drug loading, especially for lipophilic compounds. Regarding bioavailability, a lipophilic drug in a microemulsion system has better penetration of the physiological membrane because of several factors. These include the enhancing effect of surfactants, the proper balance between hydrophilicity and lipophilicity of the formulation, smaller particle size, the in vivo partition coefficient of the drug between the two immiscible phases, presence of the drug in an emulsified form, site of absorption, metabolism of the oil in the microemulsion, effect of lipid vehicle on gastric emptying, and drug solubility in the microemulsion excipients.

Microemulsions are potential drug delivery systems for several applications especially oral, nasal, topical, and transdermal. They represent an easy-to-manufacture, thermodynamically stable system with improved bioavailability, less alcohol content, and a transparent and elegant appearance. However, the toxicity of the surfactant and cosurfactant has to be investigated thoroughly for oral and nasal applications, especially for drugs requiring chronic use.

One must note that although a substantial number of reports exist in the literature for microemulsions as delivery vehicles, the studies are not very systematic and do not cover many relevant aspects that are required to make these systems into a commercially viable technology. Furthermore, structural studies are scarce and the investigations have mainly been confined to in vitro studies with only a few in vivo studies (9,40,204,205).

Much progress has been made in the last 15 years in exploring new types of formulations of microemulsions for drug delivery and there is a much better understanding of the relationship between the structure of the microemulsion droplets and the nature of the solubilized drug. As a result, much more attention is being paid to the quantitative questions of the maximum solubilization of a substance in the concentrate (with water added) and on its behavior upon dilution. Attempts were made to quantify the amount of solubilized matter to its locus on the interface or within the core of the emulsions.

As advanced analytical methods are becoming more available to scientists from other disciplines, pharmacologists and pharma scientists can better quantify the correlation between the nature of the microdroplets and their bioavailability. Scientists are attempting to understand the role of the bicontinuous phases in the solubilization activity of the microemulsion. Phase transitions are known to occur in the microemulsion microstructure as a result of changes in pH, temperature, dilution, etc., but mostly because of the presence of the guest molecules, solubilized drugs or nutraceuticals particularly (its type and its quantity). Phase transitions were minimally studied in past reports, but their effect on drug behavior in the microemulsion was not elucidated. Phase equilibria in the presence of drugs in microemulsions are key questions in oral or transdermal deliveries. Drugs tend to precipitate and crystallize as large crystals upon storage or dilution. Those aspects were not considered in the past, but are relevant issues in relation to sustained or slow deliveries.

It should also be stressed that nutraceuticals, a new emerging category of food supplements with health benefits, need formulations, and microemulsions are excellent vehicles for such poorly water-soluble molecules. Much more work is required in this area.

REFERENCES

1. Hite M, Turner S, Federici C. Part 1: oral delivery of poorly soluble drugs. PMPS Summer 2003; 3:38–40.

2a. Hoar TP, Schulman JH. Transparent water-in-oil dispersions: the oleopathic hydromicelle. Nature 1943; 152:102–104.

2b. Schulman JH, Stoeckenius W, Prince LM. Mechanism of formation and structure of microemulsions by electron microscopy. J Phys Chem 1959; 63:1677–1680.

3. Shukla A, Janich M, Jahn K, Neubert RHH. Microemulsions for dermal drug delivery studied by dynamic light scattering: effect of interparticle interactions in oil-in-water microemulsions. J Pharm Sci 2003; 92:730–738.

4. Lawrence MJ. Surfactant systems-microemulsions and vesicles as vehicles for drug-delivery. Eur J Drug Metab Pharmacokinet 1994; 3:257–269.

5. Lawrence MJ, Rees GD. Microemulsion-based media as novel drug delivery systems. Adv Drug Deliv Rev 2000; 45:89–121.

6. Förster T, Von Rybinski W, Wadle A. Influence of microemulsion phases on the preparation of fine-disperse emulsions. Adv Colloid Int Sci 1995; 58:119–149.

7. Bagwe RP, Kanicky JR, Palla BJ, Patanjali PK, Shah DO. Improved drug delivery using microemulsions: rationale, recent progress, and new horizons. Crit Rev Ther Drug Carrier Systems 2001; 18:77–140.
8. Vandamme ThF. Microemulsions as ocular drug delivery systems: recent developments and future challenges. Prog Retin Eye Res 2002; 21:15–34.
9. Kreilgaard M. Influence of microemulsions on cutaneous drug delivery. Adv Drug Delivery Rev 2002; 54:S77–S98.
10. Constantinides PP. Lipid microemulsions for improving drug dissolution and oral absorption: physical and biopharmaceutical aspects. Pharm Res 1995; 12:1561–1572.
11. Tadros ThF. Microemulsions–An overview. In: Mittal KL, Lindman B, eds. Symp. on Surfactants in Solution. 4th. New York: Plenum, 1984:1501–1532.
12. Ezrahi S, Aserin A, Fanun M, Garti N. Subzero temperature behavior of water in microemulsions. In: Garti N, ed. Thermal Behavior of Dispersed Systems. Surfactant Science Series. New York: Marcel Dekker, Inc., 2001:59–120.
13. Tenjarla S. Microemulsions: an overview and pharmaceutical applications. Crit Rev Ther Drug Carrier Systems 1999; 16:461–521.
14. Kumar P, Mittal KL, eds. Handbook of Microemulsion Science and Technology. New York: Marcel Dekker, 1999.
15. Rosano HL, Cavallo JL, Chang DL, Whittam JH. Microemulsions–a commentary on their preparation. J Soc Cosmet Chem 1988; 39:201–209.
16. Garti N, Aserin A, Spernath A, Amar I. Nano-sized self-assembled structured liquids. PCT/IL03/00498, 2002.
17. Garti N, Aserin A, Ezrahi S, Wachtel E. Water solubilization and chain-length compatibility in nonionic microemulsions. J Colloid Interface Sci 1995; 169:428–436.
18. Hofland HEJ, Van der Geest R, Bodde HE, Junginger HE, Bouwstra JA. Estradiol permeation from nonionic surfactant vesicles through human stratum corneum in vitro. Pharm Res 1994; 11:659–664.
19. Ho HO, Hsiao CC, Sheu MT. Preparation of microemulsions using polyglycerol fatty acid esters as surfactant for the delivery of protein drugs. J Pharm Sci 1996; 85:138–143.
20. Tan EL, Liu JC, Chien YW. Effect of cationic surfactants on the transdermal permeation of ionized indomethacin. Drug Dev Ind Pharm 1993; 19:685–699.
21. Rees GD, Robinson BH. Esterification reactions catalyzed by chromobacterium-viscosum lipase in CTAB-based microemulsion systems. Biotechnol Bioeng 1995; 45: 344–355.
22. Giustini M, Palazzo G, Colafemmina G, Della Monica M, Giomini M, Ceglie A. Microstructure and dynamics of the water-in-oil CTAB/*n*-pentanol/*n*-hexane/water microemulsion: a spectroscopic and conductivity study. J Phys Chem 1996; 100: 3190–3198.
23. Mehta SK, Kawaljit. Isentropic compressibility and transport properties of CTAB-alkanol-hydrocarbon-water microemulsion systems. Colloids Surf A: Physicochem Eng Aspects 1998; 136:35–41.
24. Palazzo G, Lopez F, Giustini M, Colafemmina G, Ceglie A. Role of the surfactant in the CTAB/water/*n*-pentanol/*n*-hexane water-in-oil microemulsion. 1. Pentanol effect on the microstructure. J Phys Chem B 2003; 107:1924–1931.
25. Hazra P, Chakrabarty D, Chakraborty A, Sarkar N. Solvation dynamics of coumarin 480 in neutral (TX-100), anionic (SDS), and cationic (CTAB) water-in-oil microemulsions. Chem Phys Lett 2003; 382:71–80.
26. Skodvin T, Sjoblom J, Saeten JO, Gestblom B. Solubilization of drugs in microemulsions as studied by dielectric spectroscopy. J Colloid Interface Sci 1993; 155:392–401.
27. Olla M, Monduzzi M, Ambrosone L. Microemulsions and emulsions in DDAB/W/oil systems. Colloids Surf A: Physicochem Eng Aspects 1999; 160:23–36.
28. Olla M, Monduzzi M. DDAB microemulsions: influence of an aromatic oil on microstructure. Langmuir 2000; 16:6141–6147.

29. Olla M, Semmler A, Monduzzi M, Hyde ST. From monolayers to bilayers: mesostructural evolution in DDAB tetradecane microemulsions. J Phys Chem B 2004; 108: 12,833–12,841.
30. Murgia S, Monduzzi M, Ninham BW. Hofmeister effects in cationic microemulsions. Curr Opin Colloid Interface Sci 2004; 9:102–106.
31. Fubini B, Gasco MR, Gallarate M. Microcalorimetric study of microemulsions as potential drug delivery systems. I. Evaluation of enthalpy in the absence of any drug. Int J Pharm 1988; 42:19–26.
32. Bagwe R, Shah D. Effect of various additives on solubilization, droplet size and viscosity of canola oil in oil-in-water food grade microemulsions. Abstract Papers Am Chem Soc 2002; 223:465.
33. Constantinides PP, Tustian A, Kessler DR. Tocol emulsions for drug solubilization and parenteral delivery. Adv Drug Del Rev 2004; 56:1243–1255.
34. Illum L, Washington C, Lawrence S, Watts P. Lipid vehicle drug delivery composition containing vitamin E. WO 97/03651 (1997).
35. Grossklaus S, Pfestorf R, Quitzsch K, Möhle L, Hauthal HG. Charakterisierung eines mikroemulsionsgebietes. Tenside Surf Det 1989; 26:387–389.
36. Constantinides PP, Scalart JP, Lancaster C, Marcello J, Marks G, Ellens H, Smith PL. Formulation and intestinal absorption enhancement evaluation of water-in-oil microemulsions incorporating medium-chain glycerides. Pharm Res 1994; 11:1385–1390.
37. Schomäcker R. Microemulsions as medium for chemical reactions. Nachr Chem Tech Lab 1992; 40:1344–1352.
38. Das ML, Bhattacharya PK, Moulik SP. Model biological microemulsions 1. Phase-behavior and physicochemical properties of cholesteryl benzoate and sodium deoxycholate contained microemulsions. Indian J Biochem Biophys 1989; 26:24–29.
39. Kemken J, Ziegler A, Mueller BW. Pharmacodynamic effects of transdermal Bupranolol and timolol invivo–comparison of microemulsions and matrix patches as vehicle. Meth Find Exp Clin Pharmacol 1991; 13:361–365.
40. Kemken J, Ziegler A, Mueller BW. Influence of supersaturation on the pharmacodynamic effect of Bupranolol after dermal administration using microemulsions as vehicle. Pharm Res 1992; 9:554–558.
41. Drewe J, Meier R, Vonderscher J, Kiss D, Posanski U, Kissel T, Gyr K. Enhancement of the oral absorption of cyclosporine in man. Br J Clin Pharmacol 1992; 34:60–64.
42. Kemken J, Ziegler A, Müller BW. Investigations into the pharmacodynamics effects of dermally administered microemulsions containing beta-blockers. J Pharm Pharmacol 1991; 43:679–684.
43. Gasco MR, Gallarate M, Trotta M, Bauchiero L, Gremmo E, Chiappero O. Microemulsions as topical delivery vehicles: ocular administration of timolol. J Pharm Biomed Anal 1989; 7:433–439.
44. Linn EEC, Pohland RC, Byrd TK. Microemulsion for intradermal delivery of cetyl alcohol and octyl dimethyl PABA. Drug Dev Ind Pharm 1990; 16:899–920.
45. Osborne DW, Ward AJ, O'Neill KJ. Microemulsions as topical drug delivery vehicles: in-vitro transdermal studies of a model hydrophilic drug. J Pharm Pharmacol 1991; 43:451–454.
46. Agatonovic-Kustrin S, Glass BD, Wisch MH, Alany RG. Prediction of a stable microemulsion formulation for the oral delivery of a combination of antitubercular drugs using ANN methodology. Pharm Res 2003; 20:1760–1765.
47. Sirotti C, Coceani N, Colombo I, Lapasin R, Grassi M. Modeling of drug release from microemulsions: a peculiar case. J Membrane Sci 2002; 204:401–412.
48. Bagger-Jörgensen H, Olsson U, Mortensen K. Microstructure in a ternary microemulsion studied by small angle neutron scattering. Langmuir 1997; 13:1413–1421.
49. Glatter O, Orthaber D, Stradner A, Scherf G, Fanun M, Garti N, Clement V, Leser ME. Sugar-ester nonionic microemulsion: structural characterization. J Colloid Interface Sci 2001; 241:215–225.

50. Brunner-Popela J, Mittelbach R, Strey R, Schubert KV, Kaler EW, Glatter O. Small-angle scattering of interacting particles. III. D_2O-$C_{12}E_5$ mixtures and microemulsions with n-octane. J Chem Phys 1999; 110:10623–10632.
51. Hellweg T, Brulet A, Sottmann T. Dynamics in an oil-continuous droplet microemulsion as seen by quasielastic scattering techniques. Phys Chem Chem Phys 2000; 2:5168–5174.
52. Langevin D, Rouch J. Light scattering studies of microemulsion systems. In: Kumar P, Mittal KL, eds. Handbook of Microemulsion Science and Technology. New York: Marcel Dekker, 1999:387–410.
53. Feldman Y, Kozlovich N, Nir I, Garti N. Dielectric relaxation in sodium bis(2-ethylhexyl)sulfosuccinate–water–decane microemulsions near the percolation temperature threshold. Phys Rev E 1995; 51:478–491.
54. Xu J, Li GZ, Zhang ZQ, Zhou GW, Ji KJ. A study of the microstructure of CTAB/1-butanol/octane/water system by PGSE-NMR, conductivity and cryo-TEM. Colloids Surf 2001; 191:269–278.
55. Bellare JR, Haridas MM, Li XJ. Characterization of microemulsions using fast freeze-fracture and cryo-Tem microscopy. In: Kumar P, Mittal KL, eds. Handbook of Microemulsion Science and Technology. New York: Marcel Dekker, 1999:411–436.
56. Cerichelli G, Mancini G. NMR techniques applied to micellar systems. Curr Opin Colloid Interface Sci 1997; 2:641–648.
57. Wennerström H, Söderman O, Olsson U, Lindman B. Macroemulsions versus microemulsions. Colloids Surf A: Physicochem Eng Aspects 1997; 123:13–26.
58. Bourrel M, Schechter RS. The R-ratio. In: Bourrel M, Schechter RS, eds. Microemulsions and Related Systems. New York: Marcel Dekker, 1999:457–482.
59. Kahlweit M, Busse G, Faulhaber B. Preparing microemulsions with alkyl monoglucosides and the role of *n*-alkanols. Langmuir 1995; 11:3382–3387.
60. Schubert KV, Kaler EW. Nonionic microemulsions. Ber Bunsen Phys Chem 1996; 100:190–205.
61. Andrade SM, Costa SMB, Pansu R. Structural changes in W/O Triton X-100/cyclohexane–hexanol/water microemulsions probed by a fluorescent drug Piroxicam. J Colloid Interface Sci 2000; 226:260–268.
62. Strey R. Microemulsion microstructure and interfacial curvature. Colloid Polym Sci 1994; 272:1005–1019.
63. Zana R. Microemulsions. Heterogen Chem Rev 1994; 1:145–157.
64. Salager JL, Antón RE. Ionic microemulsions. In: Kumar P, Mittal KL, eds. Handbook of Microemulsion Science and Technology. New York: Marcel Dekker, 1999:247–280.
65. Salager JL. Guidelines for the formulation, composition and stirring to attain desired emulsion properties (type, droplet size, viscosity and stability). In: Chattopadhyay AK, Mittal KL, eds. Surfactants in Solution. New York: Marcel Dekker, 1999:185–246.
66. Lindman B, Olsson U. Structure of microemulsions studied by NMR. Ber Bunsen Phys Chem 1996; 100:344–363.
67. Garti N, Clement V, Fanun M, Leser ME. Some characteristics of sugar ester nonionic microemulsions in view of possible food applications. J Agric Food Chem 2000; 48:3945–3956.
68. Acharya A, Moulik SP, Sanyal SK, Mishra BK, Puri PM. Physicochemical investigations of microemulsification of coconut oil and water using polyoxyethylene 2-cetyl ether (Brij 52) and isopropanol or ethanol. J Colloid Interface Sci 2002; 245:163–170.
69. Acharya A, Sanyal SK, Moulik SP. Formation and characterization of a useful biological microemulsion system using mixed oil (ricebran and isopropyl myristate), polyoxyethylene(2)oleyl ether (Brij 92), isopropyl alcohol, and water. J Dispersion Sci Technol 2001; 22:551–561.
70. Acharya A, Sanyal SK, Moulik SP. Physicochemical investigations on microemulsification of eucalyptol and water in presence of polyoxyethylene (4) lauryl ether (Brij-30) and ethanol. Int J Pharm 2001; 229:213–226.

71. Sjöblom J, Lindberg R, Friberg SE. Microemulsions-phase equilibria characterization, structures, applications and chemical reactions. Adv Colloid Interface Sci 1996; 65: 125–287.
72. Radomska A, Dobrucki R. The use of some ingredients for microemulsion preparation containing retinol and its esters. Int J Pharm 2000; 196:131–134.
73. Knackstedt MA, Ninham BW. Model disordered media provided by ternary microemulsions. Phys Rev E 1994; 50:2839–2843.
74. Lopez-Montilla JC, Herrera-Morales PE, Pandey S, Shah DO. Spontaneous emulsification: mechanisms, physicochemical aspects, modeling, and applications. J Dispersion Sci Technol 2002; 23:219–268.
75. Klier J, Tucker CJ, Kalantar TH, Green DP. Properties and applications of microemulsions. Adv Mater 2000; 12:1751–1757.
76. Dungan SR. Microemulsions in foods: properties and applications. In: Solans C, Kunieda H, eds. Industrial Applications of Microemulsions. New York: Marcel Dekker, 1997:148–170.
77. Engström S, Larsson K. Microemulsions in foods. In: Kumar P, Mittal KL, eds. Handbook of Microemulsion Science and Technology. New York: Marcel Dekker, 1999: 789–796.
78. Kawakami K, Yoshikawa T, Moroto Y, Kanaoka E, Takahashi K, Nishihara Y, Masuda K. Microemulsion formulation for enhanced absorption of poorly soluble drugs–I. Prescription design. J Control Release 2002; 81:65–74.
79. Gasco MR. Microemulsions in the pharmaceutical field: perspectives and applications. In: Solans C, Kunieda H, eds. Industrial Applications of Microemulsions. New York: Marcel Dekker, 1997:98–120.
80. Lawrence MJ. Microemulsions as drug delivery vehicles. Curr Opin Colloid Interface Sci 1996; 1:826–832.
81. Yaghmur A, De Campo L, Glatter O, Leser ME, Garti N. Structural characterization of five-component food grade oil-in-water nonionic microemulsions. Phys Chem Chem Phys 2004; 6:1524–1533.
82. Yaghmur A, Aserin A, Garti N. Phase behavior of microemulsions based on food-grade nonionic surfactants: effect of polyols and short-chain alcohols. Colloids Surf A: Physicochem Eng Aspects 2002; 209:71–81.
83. Spernath A, Yaghmur A, Aserin A, Hoffman RE, Garti N. Food-grade microemulsions based on nonionic emulsifiers: media to enhance lycopene solubilization. J Agric Food Chem 2002; 50:6917–6922.
84. Danino D, Kaplun A, Talmon Y, Zana R. Cryo-transmission electron-microscopy investigations of unusual amphiphilic systems in relation to their rheological properties. Structure and Flow in Surfactant Solutions. ACS Symp Series 1994; 578:105.
85. Danino D, Gupta R, Satyavolu J, Talmon Y. Direct cryogenic-temperature transmission electron microscopy imaging of phospholipid aggregates in soybean oil. J Colloid Interface Sci 2002; 249:180–186.
86. Konikoff FM, Danino D, Weihs D, Rubin M, Talmon Y. Microstructural evolution of lipid aggregates in nucleating model and human biles visualized by cryogenic transmission electron microscopy. Hepatology 2000; 31:261–268.
87. Ickenstein LM, Arfvidsson MC, Needham D, Mayer LD, Edwards K. Disc formation in cholesterol-free liposomes during phase transition. Biochim Biophys Acta 2003; 1614:135–138.
88. Danino D, Talmon Y, Zana R. Cryo-TEM of thread-like micelles: on-the-grid microstructural transformations induced during specimen preparation. Colloids Surf A: Physicochem Eng Aspects 2000; 169:67–73.
89. Spicer PT, Hayden KL, Lynch ML, Ofori-Boateng A, Burns JL. Novel process for producing cubic liquid crystalline nanoparticles (cubosomes). Langmuir 2001; 17:5748–5756.

90. Danino D, Bernheim-Groswasser A, Talmon Y. Digital cryogenic transmission electron microscopy: an advanced tool for direct imaging of complex fluids. Colloids Surf A: Physicochem Eng Aspects 2001; 183:113–122.
91. Tillett ML, Lian LY, Norwood TJ. Practical aspects of the measurement of the diffusion of proteins in aqueous solution. J Magnetic Reson 1998; 133:379–384.
92. Garti N, Amar-Yuli I, Spernath A, Hoffman RE. Transitions and loci of solubilization of nutraceuticals in U-type nonionic microemulsions studied by self-diffusion NMR. Phys Chem Chem Phys 2004; 6:2968–2976.
93. Stubenrauch C, Findenegg GH. Microemulsions supported by octyl monoglucoside and geraniol. 2. An NMR self-diffusion study of the microstructure. Langmuir 1998; 14:6005–6012.
94. Datema KP, Boltwesterhoff JA, Jaspers A, Daane JGR, Rupert LAM. Demonstration of bicontinuous structures in microemulsions using automatic-mode NMR (self-)diffusion measurements. Magn Reson Chem 1992; 30:760–767.
95. Lindman B, Shinoda K, Jonstromer M, Shinohara A. Change of organized solution (microemulsion) structure with small change in surfactant composition as revealed by NMR self-diffusion studies. J Phys Chem 1988; 92:4702–4706.
96. Das KP, Ceglie A, Lindman B. Microstructure of formamide microemulsions from NMR self-diffusion measurements. J Phys Chem 1987; 91:2938–2946.
97. Olsson U, Shinoda K, Lindman B. Change of the structure of microemulsions with the hydrophile–lipophile balance of nonionic surfactant as revealed by NMR self-diffusion studies. J Phys Chem 1986; 90:4083–4088.
98. Phillippi MA, Ward AJI, Hsieh JCL, Wiersema RJ. Deuterium NMR-study of hydrotropic action in lyotropic lamellar phases of a nonionic surfactant. J Am Oil Chemists Soc 1986; 63:460–460.
99. Kreilgaard M, Pedersen EJ, Jaroszewski JW. NMR characterization and transdermal drug delivery potential of microemulsion systems. J Control Release 2000; 69:421–433.
100. Garti N, Aserin A, Tiunova I, Fanun M. A DSC study of water behavior in water-in-oil microemulsions stabilized by sucrose esters and butanol. Colloids Surf A: Physicochem Eng Aspects 2000; 170:1–18.
101. Von Corswant C, Thoren P, Engström S. Triglyceride-based microemulsion for intravenous administration of sparingly soluble substances. J Pharm Sci 1998; 87:200–208.
102. D'Cruz OJ, Uckun FM. Gel-microemulsions as vaginal spermicides and intravaginal drug delivery vehicles. Contraception 2001; 64:113–123.
103. D'Cruz OJ, Yiv SH, Waurzyniak B, Uckun FM. Contraceptive efficacy and safety studies of a novel microemulsion-based lipophilic vaginal spermicide. Fertil Steril 2001; 75:115–124.
104. Evans RM, Farr SJ. The development of novel, pressurised aerosols formulated as solutions. J Biopharm Sci 1992; 3:33–40.
105. Attwood D, Mallon C, Taylor CJ. Phase studies of oil-in-water phospholipid microemulsions. Int J Pharm 1992; 84:R5–R8.
106. Shinoda K, Araki M, Sadaghiani A, Khan A, Lindman B. Lecithin-based microemulsions: phase behavior and microstructure. J Phys Chem 1991; 95:989–993.
107. Aboofazeli R, Lawrence MJ. Investigations into the formation and characterization of phospholipid microemulsions: I pseudo-ternary phase diagrams of systems containing water–lecithin–alcohol–isopropyl myristate. Int J Pharm 1993; 93:161–175.
108. Schurtenberger P, Peng Q, Leser ME, Luisi PL. Structure and phase behavior of lecithin-based microemulsions: a study of chain length dependence. J Colloid Interface Sci 1993; 156:43–51.
109. Angelico R, Palazzo G, Colafemmina G, Cirkel PA, Giustini M, Ceglie A. Water diffusion and head group mobility in polymer-like reverse micelles: evidence of a sphere-to-rod-to-sphere transition. J Phys Chem B 1998; 102:2883–2889.
110. Von Corswant C, Engström S, Söderman O. Phase behavior and microstructure. Langmuir 1997; 13:5061–5070.

111. Von Corswant C, Söderman O. Effect of adding isopropyl myristate to microemulsions based on soybean phosphatidylcholine and triglycerides. Langmuir 1998; 14:3506–3511.
112. Von Corswant C, Olsson C, Söderman O. Microemulsions based on soybean phosphatidylcholine and isopropylmyristate-effect of addition of hydrophilic surfactants. Langmuir 1998; 14:6864–6870.
113. Saint Ruth H, Attwood D, Ktistis G, Taylor CJ. Phase studies and particle size analysis of oil-in-water phospholipid microemulsions. Int J Pharm 1995; 116:253–261.
114. Aboofazeli R, Barlow DJ, Lawrence MJ. Particle size analysis of concentrated phospholipid microemulsions: I. Total intensity light scattering. AAPS Pharm Sci 2000; 2: 449–470.
115. Cho YW, Flynn M. Oral delivery of insulin. Lancet 1989; 2:1518–1519.
116. Fubini B, Gasco MR, Gallarate M. Microcalorimetric study of microemulsions as potential drug delivery systems. II. Evaluation of enthalpy in the presence of drugs. Int J Pharm 1989; 50:213–217.
117. Gallarate M, Gasco MR, Trotta M, Chetoni P, Saettone MF. Preparation and evaluation in vitro of solutions and o/w microemulsions containing levobunolol as ion-pair. Int J Pharm 1993; 100:219–225.
118. Pattarino F, Marengo E, Gasco MR, Carpignano R. Experimental design and partial least squares in the study of complex mixtures: microemulsions as drug carriers. Int J Pharm 1993; 91:157–165.
119. Murtha JL, Ando HY. Synthesis of the cholesteryl ester prodrugs cholesteryl ibuprofen and cholesteryl flufenamate and their formulation into phospholipid microemulsions. J Pharm Sci 1994; 83:1222–1228.
120. Schurtenberger P, Cavaco C. Polymer-like lecithin reverse micelles. 1. A light scattering study. Langmuir 1994; 10:100–108.
121. Constantinides PP, Yiv SH. Particle size determination of phase-inverted water-in-oil microemulsions under different dilution and storage conditions. Int J Pharm 1995; 115:225–234.
122. Kriwet K, Muller-Goymann CC. Diclofenac release from phospholipid drug systems and permeation through excised human stratum corneum. Int J Pharm 1995; 125: 231–242.
123. Lee MJ, Lee MH, Shim CK. Inverse targeting of drugs to reticuloendothelial system-rich organs by lipid microemulsion emulsified with Poloxamer 338. Int J Pharm 1995; 113:175–187.
124. Satra C, Thomas M, Lawrence MJ. The solubility of testosterone in oil-in-water microemulsions. J Pharm Pharmacol 1995; 47:1126.
125. Trotta M, Ugazio E, Gasco MR. Pseudo-ternary phase-diagrams of lecithin-based microemulsions–influence of monoalkylphosphates. J Pharm Pharmacol 1995; 47: 451–454.
126. Leser ME, Van Evert WC, Agterof WGM. Phase behavior of lecithin-water-alcohol-triacylglycerol mixtures. Colloids Surf A: Physicochem Eng Aspects 1996; 116:293–308.
127. Trotta M, Cavalli R, Ugazio E, Gasco MR. Phase behavior of microemulsion systems containing lecithin and lysolecithin as surfactants. Int J Pharm 1996; 143:67–73.
128. Hasse A, Keipert S. Development and characterization of microemulsions for ocular application. Eur J Pharm Biopharm 1997; 43:179–183.
129. Kahlweit M, Busse G, Faulhaber B. Preparing nontoxic microemulsions. 2. Langmuir 1997; 13:5249–5251.
130. Trotta M, Pattarino F, Grosa G. Formation of lecithin-based microemulsions containing n-alkanol phosphocholines. Int J Pharm 1998; 174:253–259.
131. Choi SY, Oh SG, Bae SY, Moon SK. Effect of short-chain alcohols as co-surfactants on pseudo-ternary phase diagrams containing lecithin. Kor J Chem Eng 1999; 16:377–381.
132. Park KM, Lee MK, Hwang KJ, Kim CK. Phospholipid-based microemulsions of flurbiprofen by the spontaneous emulsification process. Int J Pharm 1999; 183:145–154.

133. Trotta M. Influence of phase transformation on indomethacin release from microemulsions. J Control Release 1999; 60:399–405.
134. Trotta M, Gallarate M, Pattarino F, Carlotti ME. Investigation of the phase behavior of systems containing lecithin and 2-acyl lysolecithin derivatives. Int J Pharm 1999; 190:83–89.
135. Von Corswant C, Thoren PEG. Solubilization of sparingly soluble active compounds in lecithin-based microemulsions: influence on phase behavior and microstructure. Langmuir 1999; 15:3710–3717.
136. Kibbe AH., ed. Handbook of Pharmaceutical Excipients. 3rd ed. London: Pharmaceutical Press, 2000:407–423.
137. Attwood D, Mallon C, Ktistis G, Taylor CJ. A study on factors influencing the droplet size in nonionic oil-in-water microemulsions. Int J Pharm 1992; 88:417–422.
138. Malcolmson C, Satra C, Kantaria S, Sidhu A, Lawrence MJ. Effect of oil on the level of solubilization of testosterone propionate into nonionic oil-in-water microemulsions. J Pharm Sci 1998; 87:109–116.
139. Malcolmson C, Lawrence MJ. A comparison of the incorporation of model steroids into nonionic micellar and microemulsion systems. J Pharm Pharmacol 1993; 45: 141–143.
140. Malcolmson C, Lawrence MJ. Three-component non-ionic oil-in-water microemulsions using polyoxyethylene ether surfactants. Colloids Surf B: Biointerfaces 1995; 4:97–109.
141. Alander J, Warnheim T. Model microemulsions containing vegetable oils 1. nonionic surfactant systems. J Am Oil Chemists' Soc 1989; 66:1656–1660.
142. Kale NJ, Allen LV Jr. Studies on microemulsions using Brij 96 as surfactant and glycerin, ethylene glycol and propylene glycol as cosurfactants. Int J Pharm 1989; 57:87–93.
143. Kunieda H, Ushio N, Nakano A, Miura M. Three-phase behavior in a mixed sucrose alkanoate and polyethyleneglycol alkyl ether system. J Colloid Interface Sci 1993; 159:37–44.
144. Warisnoicharoen W, Lansley AH, Lawrence MJ. The cytotoxicity of nonionic oil-in-water microemulsions to a human cell line. In: Drug Delivery to the Lungs. Vol. VI The Aerosol Society, 1995:40–43.
145. Oh KH, Baran JR, Wade WH, Weerasooriya V. Temperature insensitive microemulsion phase-behavior with nonionic surfactants. J Dispersion Sci Technol 1995; 16:165–188.
146. Bolzinger MA, Thevenin MA, Grossiord JL, Poelman MC. Characterization of a sucrose ester microemulsion by freeze fracture electron micrograph and small angle neutron scattering experiments. Langmuir 1999; 15:2307–2315.
147. Bolzinger MA, Carduner TC, Poelman MC. Bicontinuous sucrose ester microemulsion: a new vehicle for topical delivery of niflumic acid. Int J Pharm 1998; 176:39–45.
148. Keipert S, Schulz G. Microemulsions with sucrose fatty ester surfactants 1 in-vitro characterization. Pharmazie 1994; 49:195–197.
149. Thevenin MA, Grossiord JL, Poelman MC. Sucrose esters/cosurfactant microemulsion systems for transdermal delivery: assessment of bicontinuous structures. Int J Pharm 1996; 137:177–186.
150. Garti N, Clement V, Leser M, Aserin A, Fanun M. Sucrose ester microemulsions. J Mol Liq 1999; 80:253–296.
151. Garti N, Aserin A, Fanun M. Non-ionic sucrose esters microemulsions for food applications. Part 1. Water solubilization. Colloids Surf A: Physicochem Eng Aspects 2000; 164:27–38.
152. Parker WO Jr, Genova C, Carignano G. Study of micellar solutions and microemulsions of an alkyl oligoglucoside via NMR spectroscopy. Colloids Surf A: Physicochem Eng Aspects 1993; 72:275–284.
153. Desai NB. Esters of sucrose and glucose as cosmetic materials. Cosmet Toilet 1990; 105:99–107.
154. Kahlweit M, Busse G, Faulhaber B. Preparing nontoxic microemulsions with alkyl monoglucosides and the role of alkanediols as cosolvents. Langmuir 1996; 12:861–862.

155. Ryan LD, Kaler EW. Role of oxygenated oils in n-alkyl β-d-monoglucoside microemulsion phase behavior. Langmuir 1997; 13:5222–5228.
156. Ryan LD, Schubert KV, Kaler EW. Phase behavior of microemulsions made with *n*-alkyl monoglucosides and *n*-alkyl polyglycol ethers. Langmuir 1997; 13:1510–1518.
157. Ryan LD, Kaler EW. Effect of alkyl sulfates on the phase behavior and microstructure of alkyl polyglucoside microemulsions. J Phys Chem B 1998; 102:7549–7556.
158. Stubenrauch C, Paeplow B, Findenegg GH. Microemulsions supported by octyl monoglucoside and geraniol. 1. The role of the alcohol in the interfacial layer. Langmuir 1997; 13:3652–3658.
159. Tolle S, Zuberi T, Lawrence MJ. Physicochemical and solubilization properties of *N,N*-dimethyl-*N*-(3-dodecylcarbonyloxypropyl)amineoxide: a biodegradable nonionic surfactant. J Pharm Sci 2000; 89:798–806.
160. Varshney M, Khanna T, Changez M. Effects of AOT micellar systems on the transdermal permeation of glyceryl trinitrate. Colloids Surf B: Biointerf 1999; 13:1–11.
161. Bhargava HN, Narurkar A, Lieb LM. Using microemulsions for drug delivery. Pharm Tech 1987; 11:46–52.
162. Trotta M, Gasco MR, Pattarino F. Diffusion of steroid-hormones from O/W microemulsions–influence of the cosurfactant. Acta Pharm Technol 1990; 36:226–231.
163. Eastoe J, Fragneto G, Robinson BH, Towey TF, Heenan RK, Lena FJ. Variation of surfactant counterion and its effect on the structure and properties of aerosol-OT-based water-in-oil microemulsions. J Chem Soc Faraday Trans 1992; 88:461–471.
164. Andrade SM, Costa SMB. Fluorescence quenching of Acridine Orange in microemulsions induced by the non-steroidal anti-inflammatory drug Piroxicam. Photochem Photobiol Sci 2003; 2:605–610.
165. Dickinson PA, Howells SW, Kellaway IW. Novel nanoparticles for pulmonary drug administration. J Drug Targeting 2001; 9:295–302.
166. Changez M, Varshney M. Aerosol-OT microemulsions as transdermal carriers of tetracaine hydrochloride. Drug Dev Industrial Pharm 2000; 26:507–512.
167. Andrade SM, Costa SMB, Pansu R. The influence of water on the photophysical and photochemical properties of Piroxicam in AOT/iso-octane/water reversed micelles. Photochem Photobiol Sci 2000; 71:405–412.
168. Sarciaux JM, Acer L, Sado PA. Using microemulsion formulations for oral drug delivery of therapeutic peptides. Int J Pharm 1995; 120:127–136.
169. Trotta M, Gasco MR, Morel S. Release of drugs from oil–water microemulsions. J Control Release 1989; 10:237–243.
170. Prince LM. Microemulsions, Theory and Practice. New York: Academic Press, 1977:91.
171. Ziegenmeyer J, Führer C. Mikroemulsionen als topische Arzneiform. Acta Pharm Technol 1980; 26:273–275.
172. Ritschel WA, Ritschel GB, Sabouni A, Wolochuk D, Schroeder T. Study on the peroral absorption of the endecapeptide cyclosporine-A. Meth Find Exp Clin Pharmacol 1989; 11:281–287.
173a. Cortesi R, Nastruzzi C. Liposomes, micelles and microemulsions as new delivery systems for cytotoxic alkaloids. Pharm Sci Technol Today 1999; 2:288–298.
173b. Cortesi R, Esposito E, Maietti A, Menegatti E, Nastruzzi C. Formulation study for the antitumor drug camptothecin: liposomes, micellar solutions and a microemulsion. Int J Pharm 1997; 159:95–103.
174. Eccleston J. Microemulsions. In: Swarbrick J, Boylan JC, eds. Encyclopedia of Pharmaceutical Technology. Vol. 9. New York: Marcel Dekker, 1994:375–421.
175. Moreno MA. Lecithin-based oil-in-water microemulsions for parenteral use: pseudoternary phase diagrams, characterization and toxicity studies. J Pharm Sci 2003; 92:1428–1437.
176. Kawakami K, Yoshikawa T, Takahashi K, Nishihara Y, Masuda K. Microemulsion formulation for enhanced absorption of poorly soluble drugs: II. In vivo study. J Control Release 2002; 81:75–82.

177. Lee VHL. Enzymatic barriers to peptide and protein-absorption. Crit Rev Drug Ther Drug Carrier Syst 1988; 5:69–97.
178. Constantinides PP. Lipid microemulsions in drug solubilization and delivery. Abstr Papers Am Chem Soc 2000; 220:28.
179. Hungria VTM, Latrilha MC, Rodrigues DG, Bydlowski SP, Chiattone CS, Maranhão RC. Metabolism of a cholesterol-rich microemulsion (LDE) in patients with multiple myeloma and a preliminary clinical study of LDE as a drug vehicle for the treatment of the disease. Cancer Chemother Pharmacol 2004; 53:51–60.
180. Hirata RDC, Hirata MH, Mesquita CH, Cesar TB, Maranhão RC. Effects of apolipoprotein B-100 on the metabolism of a lipid microemulsion model in rats. Biochim Biophys Acta 1999; 1437:53–62.
181. Maranhão RC, Cesar TB, Pedroso-Mariani S, Hirata MH, Mesquita CH. Metabolic behavior in rats of a nonprotein microemulsion resembling low-density-lipoprotein. Lipids 1993; 28:691–696.
182. Maranhão RC, Garicochea B, Silva EL, Llacer PD, Cadena SMS, Coelho IJC, Meneghetti JC, Pileggi FJC, Chamone DA. Plasma kinetics and biodistribution of a lipid emulsion resembling low-density lipoprotein in patients with acute leukemia. Cancer Res 1994; 54:4660–4666.
183. Maranhão RC, Graziani SR, Yamaguchi N, Melo RF, Latrilha MC, Rodrigues DG, Couto RD, Schreier S, Buzaid AC. Association of carmustine with a lipid emulsion: in vitro, in vivo and preliminary studies in cancer patients. Cancer Chemother Pharmacol 2002; 49:487–498.
184. Pinto LB, Wajgarten M, Silva EL, Vinagre CGC, Maranhao RC. Plasma kinetics of a cholesterol-rich emulsion in young, middle-aged, and elderly subjects. Lipids 2001; 36:1307–1311.
185. Kato Y, Watanabe K, Nakakura M, Hosokawa T, Hayakawa E, Ito K. Blood clearance and tissue distribution of various formulations of alpha-tocopherol injection after intravenous administration. Chem Pharm Bull 1993; 41:599–604.
186. Müller RH, Schmidt S, Buttle I, Akkar A, Schmitt J, Brömer S. SolEmuls®–novel technology for the formulation of i.v. emulsions with poorly soluble drugs. Int J Pharm 2004; 269:293–302.
187. Friberg SE. Micelles, microemulsions, liquid crystals and the structure of stratum corneum lipids. J Soc Cosmet Chem 1990; 41:155–171.
188. Attwood D. Microemulsions. In: Kreuter J, ed. Colloidal Drug Delivery Systems. New York: Marcel Dekker, 1994:31–71.
189. Gasco MR, Morel S. Lipospheres from microemulsions. Il Farmaco 1990; 45: 1129–1136.
190. Osborne DW, Amann AH. Topical Delivery Formulations. New York: Marcel Dekker, 1990.
191. Walters KA, Hadgraft J. Pharmaceutical Skin Penetration Enhancement. New York: Marcel Dekker, 1990.
192. Tadros ThF. Future developments in cosmetic formulations. Int J Cosmet Sci 1992; 14:93–111.
193. Bonina FP, Montenegro L, Scrofani N, Esposito E, Cortesi R, Menegatti E, Nastruzzi C. Effects of phospholipid based formulations on in vitro and in vivo percutaneous absorption of methyl nicotinate. J Control Release 1995; 34:53–63.
194. Delgado-Charro MB, Iglesias-Vilas G, Blanco-Mendez J, Lopez-Quintela MA, Guy RH. Delivery of a hydrophilic solute through the skin from novel microemulsion systems. Eur J Pharm Biopharm 1997; 43:37–42.
195. Dreher F, Walde P, Walther P, Wehrli E. Interaction of a lecithin microemulsion gel with human stratum corneum and its effect on transdermal transport. J Control Release 1997; 45:131–140.
196. Boltri L, Morel S, Trotta M, Gasco MR. In vitro transdermal permeation of nifedipine from thickened microemulsions. J Pharm Belg 1994; 49:315–320.

197. Trotta M, Morel S, Gasco MR. Effect of oil phase composition on the skin permeation of felodipine from o/w microemulsions. Pharmazie 1997; 52:50–53.
198. Trotta M, Gasco MR, Caputo O, Sancin P. Transcutaneous diffusion of hematoporphyrin in photodynamic therapy: in vitro release from microemulsions. STP Pharma Sci 1994; 4:150–154.
199. Trotta M, Pattarino F, Gasco MR. Influence of counter ions on the skin permeation of methotrexate from water–oil microemulsions. Pharm Acta Helv 1996; 71:135–140.
200. Schmalfuss U, Neubert R, Wohlrab W. Modification of drug penetration into human skin using microemulsions. J Control Release 1997; 46:279–285.
201. Ho HO, Chen LC, Chiang HS, Spur BP, Wong PYK, Sheu MT. The percutaneous delivery of prostaglandin E-1 carried by microemulsion system. Proc Int Symp Controlled Release Bioact Mater Soc 1998; 25:579–580.
202. Ho HO, Huang MC, Chen LC, Hsia A, Chen KT, Chiang HS, Spur BW, Wong PYK, Sheu MT. The percutaneous delivery of prostaglandin E-1 and its alkyl esters by microemulsions. Chin Pharm J Taiwan 1998; 50:257–266.
203. Osborne DW, Ward AJ, O'Neill KJ. Microemulsions as topical drug delivery vehicles. Part 1. Characterization of a model system. Drug Dev Ind Pharm 1988; 14: 1202–1219.
204. Kreilgaard M, Kemme MJ, Burggraaf J, Schoemaker RC, Cohen AF. Influence of a microemulsion vehicle on cutaneous bioequivalence of a lipophilic model drug assessed by microdialysis and pharmacodynamics. Pharm Res 2001; 18:593–599.
205. Kreilgaard M. Dermal pharmacokinetics of microemulsion formulations determined by in vivo microdialysis. Pharm Res 2001; 18:367–373.
206. Trotta M, Gallarate M, Carlotti ME, Morel S. Preparation of grisefulvin nanoparticles from water-dilutable microemulsions. Int J Pharm 2003; 254:235–242.
207. Priano L, Albani G, Brioschi A, Calderoni S, Lopiano L, Rizzone M, Cavalli R, Gasco MR, Scaglioni F, Fraschini F, Bergamasco B, Mauro A. Transdermal apomorphine permeation from microemulsions: a new treatment in Parkinson's disease. Movement Disorders 2004; 19:937—942.
208. Priano L, Albani G, Brioschi A, Guastamacchia G, Calderoni S, Lopiano L, Rizzone M, Cavalli R, Gasco MR, Fraschini F, Bergamasco B, Mauro A. Nocturnal anomalous movement reduction and sleep microstructure analysis in parkinsonian patient during 1-night transdermal apomorphine treatment. Neurol Sci 2003; 24:207–208.
209. Alvarez-Figueroa MJ, Blanco-Mendez J. Transdermal delivery of methotrexate: iontophoretic delivery from hydrogels and passive delivery from microemulsions. Int J Pharm 2001; 215:57–65.
210. Priano L, Albani G, Calderoni S, Baudo S, Lopiano L, Rizzone M, Astolfi V, Cavalli R, Gasco MR, F, Fraschini F, Bergamasco B, Mauro A. Controlled-release transdermal apomorphine treatment for motor fluctuations in Parkinson's disease. Neurol Sci 2002; 23(suppl 2):S99–S100.
211. Cavalli R, Caputo O, Marengo E, Pattarino F, Gasco MR. The effect of the components of microemulsions on both size and crystalline structure of solid lipid nanoparticles (SLN) containing a series of model molecules. Pharmazie 1998; 53:392–396.
212. Trotta M, Fubini B, Gallarate M, Gasco MR. Calorimetric study on the solubilization of butanol by alkylphosphate and alkylphosphate-lecithin systems. J Pharm Pharma 1993; 45:993–995.
213. Baroli B, Lopez-Quintela MA, Delgado-Charro MB, Fadda AM, Blanco-Mendez J. Microemulsions for topical delivery of 8-methoxsalen. J Control Release 2000; 69:209–218.
214. Peira E, Scolari P, Gasco MR. Transdermal permeation of apomorphine through hairless mouse skin from microemulsions. Int J Pharm 2001; 226:47–51.
215. Rhee YS, Choi JG, Park ES, Chi SC. Transdermal delivery of ketoprofen using microemulsions. Int J Pharm 2001; 228:161–170.

216. Ktistis G, Niopas I. A study on the in-vitro percutaneous absorption of propranolol from disperse systems. J Pharm Pharmacol 1998; 50:413–418.
217. Kantaria S, Rees GD, Lawrence MJ. Gelatin-stabilized microemulsion-based organogels: rheology and application in iontophoretic transdermal drug delivery. J Control Release 1999; 60:355–365.
218. Warnheim T, Henriksson U, Sjoblom E, Stilbs P. Phase diagrams and self-diffusion behavior in ionic microemulsion systems containing different cosurfactants. J Phys Chem 1984; 88:5420–5425.
219. Valenta C, Schultz K. Influence of carrageenan on the rheology and skin permeation of microemulsion formulations. J Control Release 2004; 95:257–265.
220. Lee PJ, Langer R, Shastri VP. Novel microemulsion enhancer formulation for simultaneous transdermal delivery of hydrophilic and hydrophobic drugs. Pharm Res 2003; 20:264–269.
221. Walters KA, Dugard PH, Florence AT. Non-ionic surfactants and gastric mucosal transport of paraquat. J Pharm Pharmacol 1981; 33:207–213.
222. Sarpotdar PP, Zatz JL. Evaluation of penetration enhancement of lidocaine by nonionic surfactants through hairless mouse skin in vitro. J Pharm Sci 1986; 75:176–181.
223. Dalmora ME, Dalmora SL, Oliveira AG. Inclusion complex of piroxicam with β-cyclodextrin and incorporation in cationic microemulsion. In vitro drug release and in vivo topical anti-inflammatory effect. Int J Pharm 2001; 222:45–55.
224. Canto GS, Dalmora SL, Oliveira AG. Piroxicam encapsulated in liposomes: characterization and in vivo evaluation of topical anti-inflammatory effect. Drug Dev Ind Pharm 1999; 25:1235–1239.
225. Dalmora ME, Oliveira AG. Inclusion complex of piroxicam with β-cyclodextrin and incorporation in hexadecyltrimethylammonium bromide based microemulsion. Int J Pharm 1999; 184:157–164.
226. Sintov AC, Shapiro L. New microemulsion vehicle facilitates percutaneous penetration in vitro and cutaneous drug bioavailability in vivo. J Control Release 2004; 95:173–183.
227. Chen HB, Chang XL, Weng T, Zhao XZ, Gao ZH, Yang YJ, Xu HB, Yang XL. A study of microemulsion systems for transdermal delivery of triptolide. J Control Release 2004; 98:427–436.
228. Mei ZN, Chen HB, Weng T, Yang YJ, Yang XL. Solid lipid nanoparticles and microemulsion for topical delivery of triptolide. Eur J Pharm Biopharm 2003; 56:189–196.
229. Osborne DE, Ward AJI, O'Neill KJ. Microemulsions as topical drug delivery vehicles: I. Characterization of a model system. Drug Dev Ind Pharm 1988; 14:1203–1219.
230. Osborne DW, Middleton CA, Rogers RL. Alcohol-free microemulsions. J Dispersion Sci Technol 1988; 9:415–423.
231. Sonneville-Aubrun O, Simonnet JT, L'Alloret F. Nanoemulsions: a new vehicle for skincare products. Adv Colloid Interf Sci 2004; 108/109:145–149.
232. Puranajoti P, Patil RT, Sheth PD, Bommareddy G, Dondeti P, Egbaria K. Design and development of topical microemulsion for poorly water-soluble antifungal agents. J Appl Res Clin Exp Therapeutics 2002; 2:87–95.
233. Ruth H, Attwood D, Ktistis G, Taylor CJ. Phase studies and particle size analysis of oil-in-water phospholipid microemulsions. Int J Pharm 1995; 116:253–261.
234. Benita S, Muchtar S. Ophthalmic compositions. Patent No. EP 0521 799 A1 (1992).
235. Jarvinen K, Jarvinen T, Urtti A. Ocular absorption following topical delivery. Adv Drug Deliv Rev 1995; 16:3–19.
236. Siebenbrodt I, Keipert S. Poloxamer-systems as potential ophtalmic microemulsions. Eur J Pharm Biopharm 1993; 39:25–30.
237. Muchtar S, Almog S, Torracca MT, Saettone MF, Benita S. A submicron emulsion as ocular vehicle for delta-8-tetrahydrocannabinol: effect on intraocular pressure in rabbits. Ophthalmic Res 1992; 24:142–149.

238. Garti N, Leser ME, Yaghmur A, Clement V, Watzke HJ. Improved oil solubilization in O/W food-grade microemulsions in the presence of polyols and ethanol. J Agric Food Chem 2001; 49:2552–2562.
239. Spernath A, Yaghmur A, Aserin A, Hoffman RE, Garti N. Self-diffusion nuclear magnetic resonance, microstructure transitions, and solubilization capacity of phytosterols and cholesterol in winsor IV food-grade microemulsions. J Agric Food Chem 2003; 51:2359–2364.
240. Garti N, Amar I, Yaghmur A, Spernath A, Aserin A. Interfacial modification and structural transitions induced by guest molecules solubilized in U-type nonionic microemulsions. J Dispersion Sci Technol 2003; 24:397–410.
241. Amar I, Aserin A, Garti N. Microstructure transitions derived from solubilization of lutein and lutein esters in food microemulsions. Colloids Surf B: Biointerf 2004; 33:143–150.
242. De Campo L, Yaghmur A, Garti N, Leser ME, Folmer B, Glatter O. Five-component food-grade microemulsions: structural characterization by SANS. J Colloid Int Sci 2004; 274:251–267.
243. Garti N, Yaghmur A, Aserin A, Spernath A, Elfakess R, Ezrahi S. Solubilization of active molecules in microemulsions for improved environmental protection. Colloids Surf A: Physicochem Eng Aspects 2003; 230:183–190.
244. Amar I, Aserin A, Garti N. Solubilization patterns of lutein and lutein esters in food grade nonionic microemulsions. J Agric Food Chem 2003; 51:4775–4781.

13

Self-Emulsifying Oral Lipid-Based Formulations for Improved Delivery of Lipophilic Drugs

Jean-Sébastien Garrigue and Grégory Lambert
Novagali Pharma S.A., Batiment Genavenir IV, Evry, France

Simon Benita
Department of Pharmaceutics, School of Pharmacy, Faculty of Medicine, The Hebrew University of Jerusalem, Jerusalem, Israel

DEFINITION

Self-emulsifying drug delivery systems (SEDDS) or self-emulsifying oil formulations (SEOF) are defined as isotropic mixtures of natural or synthetic oils, solid or liquid surfactants or, alternatively, one or more hydrophilic solvents, and cosolvents/surfactants (1–6). Upon mild agitation followed by dilution in aqueous media such as the gastrointestinal (GI) fluids, these systems can form fine oil-in-water (o/w) emulsions or microemulsions [self-microemulsifying drug delivery systems (SMEDDS)]. Self-emulsifying formulations spread readily in the GI tract, and the digestive motility of the stomach and the intestine provide the agitation necessary for self-emulsification (4). SEDDS typically produce emulsions with a droplet size between 100 and 300 nm, while SMEDDS form transparent microemulsions with a droplet size of less than 50 nm. When compared with emulsions, which are sensitive and metastable dispersed forms, SEDDS are physically stable formulations that are easy to manufacture. Thus, for lipophilic drug compounds that exhibit dissolution rate–limited absorption, these systems may offer an improvement in the rate and extent of absorption and results in more reproducible blood–time profiles.

It should be noted that the water-in-oil-version of SEDDS has also been investigated. These systems can be liquid but also semisolid depending on the excipient choice. They are traditionally designed for the oral route. Indeed, the motility of the gut and the stomach enables an agitation favouring the in situ process of emulsification. These preparations can be given as soft or hard gelatine capsules (HGC) for an easy administration and precise dosage. Drinkable solution is also an available option with an emulsification of the SEDDS in a glass of water prior to ingestion. Most commercial SEDDS are available as capsules or both.

An efficient emulsification was previously described as a system giving an emulsion that had the droplet size arbitrarily smaller than five microns (4). Further reductions of the emulsion size lead to the development of SMEDDS.

It should be noted that their use by other routes, such as the rectal route, is not excluded (7). Kim and Young demonstrated in rats that an indomethacin-containing SEDDS filled in hollow-type gelatin suppositories gave a 41% increase in AUC, compared with the classical form, when rectally administered. Another example is illustrated by Novagali Pharma, which has developed specifically designed subcutaneous and intramuscular SEDDS.

INTRODUCTION

The oral route has been the major route of drug delivery for the chronic treatment of many diseases. However, oral delivery of 50% of the drug compounds is hampered because of the high lipophilicity of the drug itself. Nearly 40% of new drug candidates exhibit low solubility in water, which leads to poor oral bioavailability (BA), high intra- and intersubject variability, and lack of dose proportionality. Thus, for such compounds, the absorption rate from the GI lumen is controlled by dissolution (8). Modification of the physicochemical properties of the compound, such as salt formation and particle size–reduction, may be one approach to improve the dissolution rate of the drug. However, these methods have their own limitations. For instance, salt formation of neutral compounds is not feasible and the synthesis of weak acid and weak base salts may not always be practical (9). Moreover, the salts that are formed may convert back to their original acid or base forms and lead to aggregation in the GI tract (9). Particle size–reduction may not be desirable in situations where handling difficulties and poor wettability are experienced for very fine powders (9). To overcome these drawbacks, various other formulation strategies have been adopted including the use of cyclodextrins, nanoparticles, solid dispersions, and permeation enhancers (10). Indeed, in some selected cases, these approaches have been successful. In recent years, much attention has focused on lipid-based formulations to improve the oral BA of poorly water-soluble drug compounds. In fact, the most popular approach is the incorporation of the drug compound into inert lipid vehicles such as oils, surfactant dispersions, self-emulsifying formulations, emulsions, and liposomes (1–5,11–17). Among these, an effective method is to incorporate the drug into a self-emulsifying liquid, whereby emulsification to form o/w emulsion occurs spontaneously on mixing with water. These drug delivery systems represent a possible alternative to tablets and capsules as the system may be expected to self emulsify in the contents of the stomach. Consequently, the drug may be rapidly emptied from the stomach and distributed in fine droplet form throughout the GI tract. This would theoretically confer the advantage of a large surface area from which dissolution may occur along with a reduction in the irritation caused by the prolonged direct contact between drugs used in conventional dosage forms and the gut wall.

DRUG DELIVERY ISSUES

Among the most important parameters for the successful launch of a new chemical entity (NCE) in the marketplace, the ability to develop an effective oral dosage form is often critical. For two decades, the pharmaceutical industry has massively used combinatorial chemistry for the discovery of new drug molecules. This technology, which utilizes increasingly available drug receptor structural information, often results in the synthesis of polycyclic, highly hydrophobic compounds with poor aqueous solubility. Indeed, it has been estimated that nearly half of all potential new drug molecules identified by combinatorial chemistry are classified as poorly water-soluble. Although the inherent lipophilicity of many of these compounds often allows for easy permeation of GI barrier membranes, the poor aqueous solubility combined with slow dissolution rate frequently results in low and erratic absorption. This is mostly relevant for compounds requiring high levels of systemic exposure to achieve clinical efficacy and for compounds with a narrow therapeutic index (for example, cyclosporine A), where irregular BA translates into either unexpected toxicity or treatment inefficacy.

Conventional formulations have proven cost-effective, relatively easy to produce and are well suited to drugs with an efficacy dose at which they can be readily solubilized in the gastrointestinal tract (GIT). This formulation approach would be applicable to drugs that are either highly soluble or highly potent and to drugs that are poorly soluble but with doses that are sufficiently small to permit rapid and complete solubilization in the GIT.

Approaches for Improved Oral Delivery of Lipophilic Drugs

Salt Selection

For compounds possessing ionizable functional groups, the method of choice for improving absorption is through salt formation. Salts of poorly soluble compounds typically dissolve more quickly in the GI tract, thus improving absorption, which is limited by the slow dissolution of the parent free acid or base compound. Limitations to this approach involve identifying a nontoxic counter-ion that will yield a pharmaceutically acceptable salt. Additional salt requirements include the ability to prepare, in sufficient yield, a chemically stable, non- or minimally hygroscopic crystalline product with a sufficiently high melting point to withstand milling and formulation processing.

Prodrug Synthesis

A less frequently employed method for improving NCE solubility involves attaching a solubilizing moiety, such as an amino acid, which is cleaved by an endogenous enzyme subsequent to absorption, thereby regenerating the active parent molecule. The added complexity of prodrug development, which requires additional studies for verifying bioconversion and demonstrating lack of toxicity, has limited the application of this approach.

Another infrequently used method for improving the BA of poorly soluble compounds involves stabilizing the drug in the amorphous, or "high-energy solid" state, thereby facilitating its dissolution relative to that of the stable crystalline form. Although this technology has been effectively applied in a few instances, it is not a method of choice, as the metastable nature of most amorphous solids can spontaneously revert to the less soluble, stable crystalline form, resulting in a concomitant reduction in BA. While several novel approaches to improving NCE solubility have been proposed and have in some instances formed the basis for the creation of companies, all are basically variations of the three central themes discussed below.

Particle Size–Reduction

Particle size–reduction effectively increases the surface area–to–volume ratio, thereby increasing the dissolution rate in the GI tract and promoting absorption. Various technologies used for particle size–reduction include micronization, nanonization, and, most recently, supercritical fluid–technology, which deposits nanometer-sized drug particles on a pharmaceutically acceptable support matrix. Particle size–reduction must include a method for preventing aggregation, which commonly occurs after milling, and effectively negates any improvement in dissolution performed by particle size–reduction.

Solubilization in the Excipient Matrix

Formulations that solubilize a drug in a lipid- or water-miscible cosolvent system can provide dramatic improvements in BA by eliminating the need for preabsorptive dissolution in the GIT. Because of their immiscibility with water, lipids can maintain a poorly soluble drug in solution unlike water-miscible cosolvent systems, which often precipitate the drug following the dilution in the predominantly aqueous GI fluids. Lipid systems can also be formulated to have self-emulsifying properties, which provides further enhancement to absorption by increasing the surface area of the lipid dispersion presented to the GIT.

Inclusion Complexes

Cyclodextrins are water-soluble molecules containing a hydrophobic pocket, which can form a dissociable inclusion complex with a hydrophobic drug molecule. This typically results in improved water solubility and BA for poorly water-soluble drugs. The application of this approach is limited by the ability of the cyclodextrin to form the inclusion complex, which is strongly dependent on the molecular characteristics of the drug molecule and can be very specific.

Nanoparticles and liposomes, which have been extensively investigated for such a purpose, and discussed in other chapters of this book, are not included in the scope of this chapter.

Candidate Compounds for Oral Lipid-Based Formulations

The Biopharmaceutical Classification System (BCS), which defines BA as interrelated functions of compound permeability and solubility, is useful for identifying candidate compounds for which oral–lipid-based formulations could be expected to improve BA (8). Lipid-based formulations can potentially improve BA for properly selected compounds in every BCS category (Table 1). However, BCS Category 2 compounds, or those possessing poor water solubility and high membrane permeability, tend to exhibit the most substantial enhancements in BA when formulated in a solubilizing lipid excipient. In addition to being hydrophobic, these compounds should also possess excellent solubility (50–100 mg/mL) in a dietary triglyceride (e.g., soybean oil), as the application of these excipients can be limited by poor lipid solubility of the drug. If drug solubility is sufficiently low, use of these excipients will be rendered impractical by requiring an unacceptably large amount of excipient to formulate a dosage unit. As discussed above, administering these compounds as a lipid solution enhances BA primarily by overcoming the absorptive barriers of poor aqueous solubility and slow dissolution in the GI. The exact mechanism(s) by which

Table 1 The Potential Bioavailability Improvement of Active Ingredients Categorized by the Biopharmaceutical Classification System Using Oral Lipid-Based Formulations

Aqueous solubility	Membrane permeability	Type[a]	Potential formulation type	Potential benefit(s) of the system
High	High	I	Microemulsion w/o	Stabilization + chemical and enzymatic protection against hydrolysis (+ efflux)
High	Low	III	Microemulsion w/o	Stabilization + chemical and enzymatic protection against hydrolysis, improved bioavailability (+ efflux)
Low	High	II	S(M)EDDS, o/w	Enhancement of dissolution and solubilization, improved bioavailability [b,c]
Low	Low	IV	S(M)EDDS, o/w	Enhancement of dissolution and solubilization, improved bioavailability [b,c] (+ efflux)

[a]According to the classification of Amidon et al. (8)
[b]Enhancement of the speed and/or rate of absorption
[c]Possible chemical and enzymatic protection against hydrolysis (e.g., acidic pH, peptidases)
Abbreviations: o/w, oil-in-water; S(M)EDDS, self-microemulsifying drug delivery systems.

lipids enhance absorption of hydrophobic drug molecules is not fully understood, but is believed to involve transfer into the bilesalt–mixedmicellar phase, from which absorption across the intestinal epithelium readily occurs. Other mechanisms by which lipids may improve BA include mitigation of intestinal efflux via the p-glycoprotein transporter, reduction in intestinal first-pass metabolism by membrane-bound cytochrome enzymes, and permeability enhancing–changes in intestinal membrane fluidity. Lipids can also direct lipid-soluble drugs into the intestinal lymph, from which drugs enter the systemic blood circulation directly, thereby circumventing potential hepatic first-pass metabolism.

COMPOSITION OF SEDDS

Pouton and Wakerly, pioneers of new lipid-based formulations such as SEDDS, revealed that the self-emulsification process is specific to the nature of the oil–surfactant pair (1,2,18). The process also depends on the oil nature, the surfactant concentration and the oil/surfactant ratio, and the temperature at which self-emulsification occurs. Furthermore, it has also been demonstrated that only very specific pharmaceutical excipient combinations could lead to efficient emulsification resulting in microemulsions in the best case (3,4,19–21).

Examples of drug-loaded SEDDS and SMEDDS are depicted in Tables 2 and 3 respectively.

Oils

The oil represents the most important excipient in the SEDDS formulation. Indeed, it can solubilize relevant amounts of the poorly water-soluble drug, facilitate

Table 2 Examples of SEDDS Designed for the Oral Delivery of Lipophilic Drugs

Type of delivery system	Oil	Surfactant(s)	% w/w	Solvent(s)	Drug compound	Drug content (%)	References
SEDDS	A mixture of mono- and diglycerides of oleic acid	Solid, polyglycolyzed mono-, di- and triglycerides (HLB = 14), Tween 80 (HLB = 15)	80 or 20	–	Ontazolast	7.5	20
SEDDS (Sandimmune)	Olive oil	Polyglycolyzed glycerides (HLB = $\frac{3}{4}$)	30	Ethanol	CsA	10	22
SEDDS	Medium-chain saturated fatty acids, peanut oil	Medium-chain mono- and diglycerides, Tween 80, PEG25 glyceryl trioleate, polyglycolyzed glycerides (HLB = 6–14)	5–60	–	A naphthalene derivative	5	4
SEDDS	Medium-chain saturated fatty acids	PEG25 glyceryl trioleate	25	–	5-(5-(2,6-dicholoro-4-(dihydro-2-oxazolyl)phenoxy)-pentyl)-3-methylisoxazole)	35	3
SEDDS (positively charged)	Ethyl oleate	Tween 80	25	Ethanol	CsA	10	23
SEDDS (positively charged)	Ethyl oleate	Tween 80	25	Ethanol	Progesterone	2.5	24
SEDDS	Myvacet 9–45 or Captex 200	Labrasol (HLB = 14) or Labrafac CM10 (HLB = 10); lauroglycol (HLB = 4)	5–30; 0–25	–	CoQ_{10}	5.66	25
SEDDS (Norvir)	Oleic acid	Polyoxyl 35 castor oil	NA	Ethanol	Ritonavir	8	26
SEDDS (Fortovase)	dl-alpha tocopherol	Medium-chain mono- and diglycerides,	NA	–	Saquinavir	16	26

Abbreviations: HLB, hydrophilic–lipophilic balance; SEDDS, self-emulsifying drug delivery systems; C_sA, cyclosporin A; NA, not available.
Source: From Ref. 26.

Table 3 Examples of SMEDDS Designed for the Oral Delivery of Lipophilic Drugs

Type of delivery system	Oil	Surfactant (s)	% w/w	Solvent (s)	Drug compound	Drug content (%)	References
SMEDDS	–	Polyglycolized glycerides (HLB = 1–14)	96	-	Indometha-cin	4	21
SMEDDS (Sandimmun Neoral)	Hydrolyzed corn oil	Polyglycolized glycerides, POE–castor oil derivative	NA	Glycerol	CsA	10	6
SMEDDS (Sandimmun Neoral)	Hydrolyzed corn oil	Polyglycolized glycerides, POE–castor oil derivative	NA	Ethanol	CsA	10	26
SMEDDS	Triglyceride (LLL, LML, MLM)	Maisine 35–1, Cremophor EL	58	Ethanol	Halofantrine	5	27
SMEDDS (supersaturable)	Glyceryl dioleate	Cremophor EL, PEG400	55–58	Ethanol	Paclitaxel (± CsA)	5.7–6.25	28
SMEDDS	dl-alpha tocopherol	TPGS, tyloxapol, DOC-Na	62	Ethanol	Paclitaxel	3	29

Abbreviations: HLB, hydrophilic–lipophilic balance; NA, not available; SMEDDS, self-microemulsifying drug delivery systems; TPGS, D–alpha–tocopheryl polyethylene glycol1000 succinate; C_sA, cyclosporin A.
Source: From Ref. 26.

Table 4 List of Typical Oil, Fatty, and Lipid Compounds Used in the Formulation of SEDDS and SMEDDS

Fatty-acids, salts and esters	Aluminum monostearate, calcium stearate, ethyl oleate, isopropyl myristate, isopropyl palmitate, magnesium stearate, oleic acid, polyoxyl 40 stearate, proprionic acid, sodium stearate, stearic acid, purified stearic acid, and zinc stearate
Fatty alcohols	Benzyl alcohol, butyl alcohol, cetostearyl alcohol, cetyl alcohol, cetyl esters wax, lanolin alcohols, octyldodecanol, oleyl alcohol, and stearyl alcohol
Oils and oil esters	Almond oil, castor oil, cod liver oil, corn oil, cottonseed oil, diacetylated monoglycerides, ethiodized oil injection, glyceryl behenate, glyceryl monostearate, hydrogenated castor oil, hydrogenated vegetable oil, light mineral oil, mineral oil, mono- and diglycerides, mono- and diacetylated monoglycerides, oil-soluble vitamins, olive oil, orange flower oil, peanut oil, peppermint oil, perflubron, persic oil, polyoxyl 35 castor oil, polyoxyl 40 hydrogenated castor oil, rose oil, safflower oil, sesame oil, soybean oil, squalane, tocopherol excipient, vitamin E, and vitamin E PEG succinate
Phospholipids	Lecithin and derivatives
Waxes	Caranuba wax, emulsifying wax, hard fat, hydrophilic ointment, hydrophilic petrolatum, microcrystalline wax, paraffin, petrolatum, rose water ointment, synthetic paraffin, white wax, yellow ointment, and yellow wax

Abbreviation: PEG, polyethylene glycol.

self-emulsification, facilitate absorption if digestible, and increase the fraction of lipophilic drug transported via the lymphatic system, thereby increasing absorption from the GIT, depending on the molecular nature of the triglyceride (TG) (30,31). Various compounds of oil, fatty acids and lipids have been used and are listed in Table 4, while typical chemical structures are presented in Figure 1. The role and the contribution of the lymphatic pathway are described further in the section "The Story of Oral Cyclosporin A."

Both long-chain triglyceride (LCT) and medium-chain triglyceride (MCT) oils with different degrees of saturation have been used in the design of SEDDS. Unmodified edible oils provide the most natural basis for lipid vehicles, but their poor ability to dissolve large amounts of hydrophobic drugs and their relative inefficiency of self-emulsification markedly reduced their use in SEDDS. Modified or hydrolyzed vegetable oils have contributed widely to the success of SEDDS owing to their formulation and physiological advantages (6). Nowadays, MCTs are largely replaced by novel semisynthetic MCT derivatives, which can be defined as amphiphilic compounds exhibiting surfactant properties.

MCT are medium-chain fatty acid esters of glycerol. Medium-chain fatty acids are fatty acids containing 6 to 12 carbon atoms. These fatty acids are constituents of coconut and palm kernel oils and are also found in camphor tree drupes. Coconut and palm kernel oils are also called lauric oils because of their high content of the 12-carbon fatty acid, lauric or dodecanoic acid.

MCTs used for nutritional and other commercial purposes are derived from lauric oils. In the process of producing MCTs, lauric oils are hydrolyzed to medium-chain fatty acids and glycerol. The glycerol is drawn off from the resultant mixture, and the medium-chain fatty acids are fractionally distilled. The medium-chain fatty acid fraction used commercially is mainly comprised of the eight-carbon caprylic or octanoic acid, and the 10-carbon capric or decanoic acid. There are much

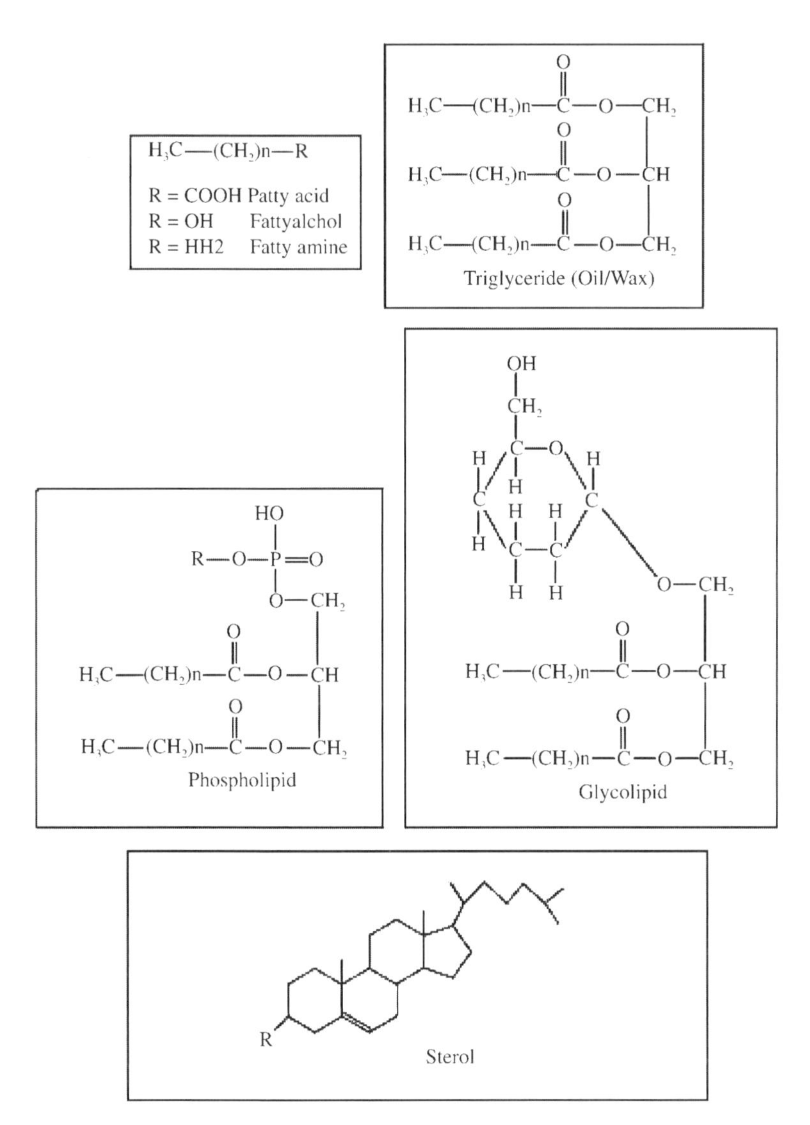

Figure 1 Chemical structure of various oil, fatty, and lipid compounds.

smaller amounts of the six-carbon caproic or hexanoic acid and the 12-carbon lauric acid in commercial products. The caprylic- and capric-rich mixture is finally re-esterified to glycerol to produce MCTs that are mainly glycerol esters of caproic (C_6), caprylic (C_8), capric (C_{10}), and lauric acid (C_{12}) in a ratio of approximately 2:55:42:1. MCTs are represented by the chemical structure in Figure 2.

MCT is also known as fractionated coconut oil. In a process called interesterification, long-chain fatty acids, such as oleic and linoleic acid, are introduced into the final product. MCT derivatives produced by the interesterification process are referred to as structural lipids or structural TGs. Unlike most natural oils of animal or vegetable origin, MCT is stable and resistant to oxidation owing mainly to the saturation of the medium-chain fatty acids.

As discussed in the section "The Story of Oral Cyclosporin A," digestibility is a relevant parameter when considering BA enhancement of SEDDS-carried drugs.

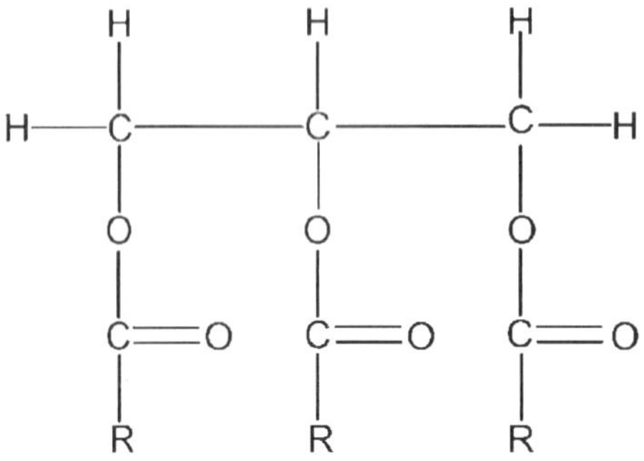

Figure 2 Chemical structure of MCT. *Abbreviation*: MCT, medium-chain triglycerides.

It should be noted that MCT is rapidly absorbed from the small intestine, intact or hydrolyzed, after ingestion and is transported to the liver via the portal circulation. MCT is therefore considered to facilitate the uptake of some lipophilic drugs when ingested. Medium-chain fatty acids are transported into hepatocytes and converted to medium-chain fatty acyl CoA esters. Medium-chain fatty acyl CoAs (mainly of caprylic and capric acids) are transported into mitochondria, where they are metabolized to acetoacetate and beta-hydroxybutyrate. The first mitochondrial enzyme in this process is medium-chain acyl CoA dehydrogenase. Acetoacetate and beta-hydroxybutyrate may be further metabolized in the liver to carbon dioxide, water, and energy, and may enter some other metabolic pathways in the liver or be transported by the systemic circulation to other tissues, where they undergo metabolism and mainly produce CO_2, H_2O, and energy. Very little ingested MCT is deposited in the body as fat.

Surfactants

Surfactant molecules may be classified based on the nature of the hydrophilic group within the molecule. The four main groups of surfactants are defined as follows:

1. *Anionic surfactants*, where the hydrophilic group carries a negative charge, such as carboxyl ($RCOO^-$), sulphonate (RSO_3^-), or sulphate ($ROSO_3^-$). Examples of pharmaceutical importance include potassium laurate, $CH_3(CH_2)_{10}COO^-\ K^+$, and sodium lauryl sulphate, $CH_3(CH_2)_{11}SO_4^-\ Na^+$.
2. *Cationic surfactants*, where the hydrophilic group carries a positive charge (e.g., quaternary ammonium halides, $R_4N^+Cl^-$). Examples of pharmaceutical importance include cetrimide, a mixture consisting mainly of tetradecyl (ca. 68%), dodecyl (ca. 22%), and hexadecyltrimethylammonium bromides (approximately 7%), as well as benzalkonium chloride, a mixture of alkylbenzyldimethylammonium chlorides of the general formula $[C_6H_5CH_2N^+(CH_3)_2R]Cl^-$, where R represents a mixture of the alkyls from C_8H_{17} to $C_{18}H_{37}$.
3. *Ampholytic surfactants (also called zwitterionic surfactants)*, where the molecule contains, or can potentially contain, both a negative and a positive charge (e.g., the sulfobetaines, $RN^+(CH_3)_2CH_2CH_2SO_3^-$). Examples of pharmaceutical importance include*N*-Dodecyl-*N*,*N*-Dimethylbetaine, $C_{12}H_{25}N^+(CH_3)_2CH_2COO^-$.

4. *Nonionic surfactants*, where the hydrophilic group carries no charge but derives its water solubility from highly polar groups such as hydroxyl or polyoxyethylene (OCH_2CH_2O) groups. Examples of pharmaceutical importance include polyoxyethylated glycol monoethers (e.g. cetomacrogol), sorbitan esters (Spans®) and polysorbates (Tweens®).

Because of their unique functional properties, surfactants find a wide range of uses in pharmaceutical preparations. These include, depending on the type of product, improving the solubility or stability of a drug in liquid preparations such as SEDDSs, and stabilizing and modifying the texture of a semisolid preparation. In addition to their use as excipients to improve the physical and chemical characteristics of the formulation, surfactants may be included to improve the efficacy or biological performance of the product. The properties of surfactants are such that they can alter the thermodynamic activity, solubility, diffusion, disintegration, and dissolution rate of a drug. Each of these parameters influences the rate and extent of drug absorption. Furthermore, surfactants can exert direct effects on biological membranes thus altering drug transport across the membrane. The overall effect of inclusion of a surfactant in a pharmaceutical formulation is complex and may be beyond those initially intended. This specific aspect is addressed later, in the section "The Story of Oral Cyclosporin A."

Emulsification is one of the most important applications of surface-active agents in pharmaceutical systems. The phenomenon has been extensively studied and many books and chapters of books have been devoted to the subject. Several compounds exhibiting surfactant properties may be employed for the design of self-emulsifying systems, the most widely recommended ones being the nonionic surfactants with a relatively high hydrophilic–lipophilic balance (HLB). The commonly used emulsifiers are various solid or liquid ethoxylated polyglycolyzed glycerides and polyoxyethylene 20 oleate (Tween 80). Safety is a major determining factor in choosing a surfactant. Emulsifiers of natural origin are preferred as they are considered to be safer than the synthetic surfactants (6,20,32,33). However, these excipients have a limited self-emulsification capacity. Nonionic surfactants are less toxic than ionic surfactants but they may lead to reversible changes in the permeability of the intestinal lumen (2,34). Usually the surfactant concentration ranges between 30% and 60% w/w to form stable SEDDS. It is very important to determine the surfactant concentration properly as large amounts of surfactants may cause GI irritation.

The surfactant involved in the formulation of SEDDS should have a relatively high HLB (Table 2) and hydrophilicity, so that immediate formation of o/w droplets and/or rapid spreading of the formulation in the aqueous media (good self-emulsifying performance) can be achieved. For an effective absorption, the precipitation of the drug compound within the GI lumen should be prevented and the drug should be kept solubilized for a prolonged period of time, at the site of absorption (4,35). Surfactants are amphiphilic in nature and they can dissolve or solubilize relatively high amounts of hydrophobic drug compounds. The lipid mixtures with higher surfactant and cosurfactant/oil ratios lead to the formation of SMEDDS (6,21,36,37). There is a relationship between the droplet size and the concentration of the surfactant being used. In some cases, increasing the surfactant concentration could lead to droplets with smaller mean droplet size, such as in the case of a mixture of saturated C_8–C_{10} polyglycolized glycerides [Labrafac CM-10 (Gattefosse, St. Priest, France)]. This could be explained by the stabilization of the oil droplets as a result of the localization of the surfactant molecules at the oil–water interface. In contrast to this, the mean droplet size may increase with increasing surfactant concentrations in some cases (1,38,39). This phenomenon could be attributed to the interfacial disruption elicited by enhanced water penetration into the oil droplets mediated by the increased surfactant concentration and leading to ejection of oil droplets into the aqueous phase (40). A very detailed in vitro study carried out utilizing the Caco-2 cell culture model evaluated the excipient effects of a novel SEDDS containing paclitaxel (Bristol-Myers Squibb, Princeton, New Jersey, U.S.A.) and devoid of Cremophor EL (BASF AG, Ludwigshafen, Germany) for IV/oral administration (41). The authors mainly focused on the influence of the nonionic and ionic surfactant ratio, i.e., Tyloxapol (Ruger Chemical Co. Inc., Irvington, New Jersey, U.S.A.) and sodium deoxycholate ratio, respectively. The drug incorporation efficiency was shown to be increased by approximately fivefold, as compared with the marketed formulation, and the excipients exhibited little or no toxicity up to a certain dilution range. In the same study, the cytotoxicity of tyloxapol, which can form lyotropic liquid crystal (LC) structures upon contact with water, appeared to be masked by sodium deoxycholate (42). It was suggested that the localization of the two surfactants at the interface between the oil droplets of the SEDDS formulations and the Caco-2 cell membrane could be the reason for this observation.

As discussed in the section "The Story of Oral Cyclosporin A," surfactants and surfactant-like compounds traditionally exhibit a permeation enhancement activity. Some of these excipients as described in Table 5 can be used for the oral formulation of SEDDS. The main advantage of these excipients is their general acceptance for use in oral drug administration (43,44).

Cosolvents

The production of an optimum SEDDS requires relatively high concentrations (generally more than 30% w/w) of surfactant (see Tables 2 and 3). Organic solvents such as ethanol, propylene glycol (PG), and polyethylene glycol (PEG) are suitable for oral delivery, and they enable the dissolution of large quantities of either the hydrophilic surfactant or the drug in the lipid base (46). These solvents can even act as cosurfactants in microemulsion systems. Alternately, alcohols and other volatile cosolvents have the disadvantage of evaporating into the shells of the soft gelatin, or hard, sealed gelatin capsules in conventional SEDDS leading to drug precipitation.

Table 5 Traditional Surfactants and Surfactant-Like Intestinal Absorption Enhancers

Intestinal absorption enhancers	
Nonionic surfactants	Polysorbates and polyoxyethylene alkyl esters and ethers
Ionic surfactants	Sodium lauryl sulphate and dioctyl sulfosuccinate
Steroid and steroid esters including bile salts	Sodium cholate, sodium deoxycholate (29,41,45), taurocholate, glycocholate, taurodeoxycholate, sodium taurodihydrofusidate, and cholesterol esters
Fatty-acids/fatty-acid esters/fatty alcohols	Oleic acid, lauric acid, capric acid, heptanoic acid, stearic acid, palmitoleic acid, palmitelaidic acid, octadecanoic acid, sucrose laurate, and isopropyl myristate
Phospholipids	Phosphatidylcholines, lysophosphatidylcholine, and monooleoyl phosphatidylethanolamine
Oils	Monoolein, cocoa butter, cardamom oil, and tricaprylin

Thus, alcohol-free formulations have been designed, but their lipophilic drug dissolution ability may be limited (6).

Others

Other components might be pH adjusters, flavors, or antioxidant agents. Indeed, a characteristic of lipid products, particularly those with unsaturated lipids, is peroxide formation with oxidation. Free radicals such as $ROO^{\bullet}$, $RO^{\bullet}$, and $OH^{\bullet}$ can damage the drug and induce toxicity. Lipid peroxides may also form due to auto-oxidation, which increases with unsaturation level of the lipid molecule. Hydrolysis of the lipid may be accelerated due to the pH of the solution or from processing energy such as ultrasonic radiation. Lipophilic antioxidants (e.g., α-tocopherol, propyl gallate, ascorbyl palmitate, or BHT) may therefore be required to stabilize the oily content of the SEDDS.

Drug Incorporation

Drugs with low aqueous solubility present a major challenge during formulation as their high hydrophobicity prevents them from being dissolved in most approved solvents. The novel synthetic hydrophilic oils and surfactants usually dissolve hydrophobic drugs to a greater extent than conventional vegetable oils. The addition of solvents, such as ethanol, PG, and PEG, may also contribute to the improvement of drug solubility in the lipid vehicle.

The efficiency of drug incorporation into a SEDDS is generally specific to each case depending on the physicochemical compatibility of the drug/system. In most cases, there is an interference of the drug with the self-emulsification process up to a certain extent leading to a change in the optimal oil/surfactant ratio. The efficiency of a SEDDS can be altered either by halting charge movement through the system by direct complexation of the drug compound with some of the components in the mixture through its interaction with the LC phase, or by penetration into the surfactant interfacial monolayer (3,5,38,47). The interference of the drug compound with the self-emulsification process may result in a change in droplet size distribution that can vary as a function of drug concentration (3). It should be pointed out that emulsions with smaller oil droplets in more complex formulations are more prone to changes caused by addition of the drug compound (6,36). Hence, the design of an optimal SEDDS requires preformulation solubility and phasediagram–studies to be conducted.

Ternary Diagrams

The use of pseudoternary diagrams is not recent. This technique was mainly used to map the microemulsion areas (composition ranges) (48). Gursoy et al. used a pseudoternary phase diagram to map the optimal composition range for three key excipients according to the resulting size following self-emulsification, in vitro cell toxicity, stability upon dilution, and viscosity (41). Therefore, several SEDDS formulations were prepared utilizing a pseudoternary phase diagram. During the phase diagram studies, ethanol (EtOH), vitamin E, and drug concentrations were kept constant at 30% (w/w), 5% (w/w), and 3% (w/w), respectively. The remaining 62% of the formulation comprised varying amounts of D–alpha–tocopheryl polyethylene glycol1000 succinate (TPGS) (5–80%), DOC–Na (5–60%), and Tyloxapol (5–80%). The distribution of the formulations on the phase diagram indicated the presence of three main regions

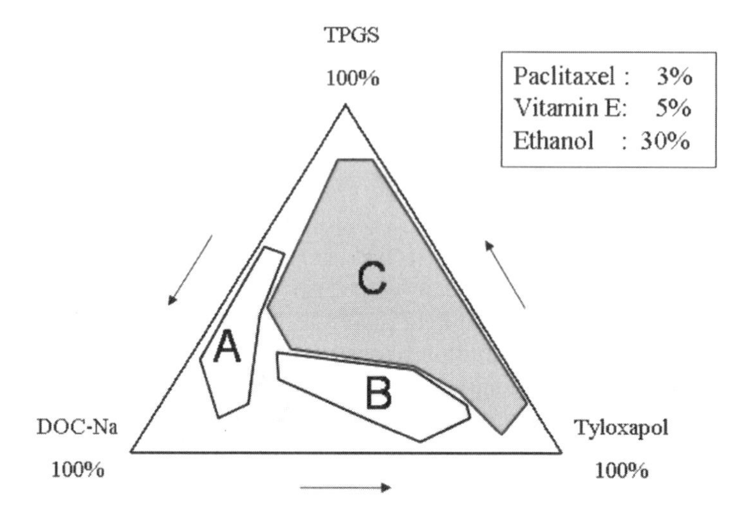

Figure 3 Pseudoternary phase diagram illustrating different dispersion types: (**A**) microemulsions (1 μm), (**B**) submicron emulsions (50–100 nm), and (**C**) microemulsions(1–10 nm). *Source*: From Ref. 49.

(Fig. 3). In region A, opaque microemulsions were formed. In region B, which covers a much narrower area between regions A and C, submicron emulsions were formed. The remaining area on the phase diagram, region C, included the transparent microemulsions. The increase in the amount of Tyloxapol and/or TPGS led to the formation of microemulsions whereas an increase in the amount of deoxycholic acid sodium salt (DOC–Na) favored the formation of macroemulsions. Physically most stable (i.e., no drug precipitation within four hours) microemulsions were formed when the amount of DOC–Na was below 50% and Tyloxapol was above 10% (w/w). The particle size of the emulsions formed in region A was around 1 μm whereas the submicron emulsions formed in region B had a particle size around 50–200 nm. In contrast to this, the microemulsion formulations in region C exhibited a particle size always below 10 nm. The use of such a pseudoternary phase diagram is a good tool for screening of optimal SEDDS composition. It enabled these authors to select the optimal composition (compromise between various parameters). In a comparable manner, paclitaxel SMEDDS formulations were optimized, in terms of droplet size and lack of drug precipitation following aqueous dilution, using a ternary phase diagram (45).

Diagrams use might be preferably associated with a plan of experience model in order to reduce the number of compositions and assessments. Nazzal and Khan proposed a response surface methodology for the optimization of Ubiquinone SMEDDS (49). They used a three-factor, three-level Box-Behnken design for the optimization procedure with the amounts of Polyoxyl 35 castor oil (X1), medium-chain mono- and diglyceride (X2), and lemon oil (X3) as the independent variables. The response variable was the cumulative percentage of ubiquinone emulsified in 10 minutes. Different ubiquinone release rates were obtained. The amount released ranged from 11% to 102.3%. Turbidity profile revealed three regions that were used to describe the progress of emulsion formation: lag phase, pseudolinear phase, and plateau turbidity. An increase in the amount of surfactant decreased turbidity values and caused a delay in lag time. Addition of cosurfactant enhanced the release rates. Increasing the amount of the eutectic agent was necessary to overcome drug precipitation, especially at higher loading of surfactants and cosurfactants. Mathematical equations and response surface plots were used to relate the dependent and independent variables. The optimization model predicted a 93.4% release with X1, X2, and X3 levels of 35, 35, and 30, respectively. The observed responses were in close agreement with the predicted values of the optimized formulation. This demonstrated the reliability of the optimization procedure in predicting the dissolution behavior of a self-emulsified drug delivery system.

Solubilization Properties

Four typical lipid-based formulations are depicted in Table 6 based on the formulation composition. They differ in their capacity to solubilize lipophilic drugs and resulting droplet size following aqueous dilution. The lipid solution exhibits very limited solubilization ability because of the lack of surfactant. Its aqueous dispersion leads to a crude o/w emulsion with droplet size greater than 1 μm and does not affect the solubilization ability of the system. SEDDS type 1 comprised of surfactants with HLB smaller than 12, shows a solvent capacity similar to that of lipid solution but its droplet size ranges from 100 to 250 nm. Because of the presence of a surfactant with relatively moderate HLB, this SEDDS type is able to dissolve the lipophilic drug without the assistance of a cosolvent, whereas in SEDDS type 2 prepared with surfactants having an HLB value superior to 12 ("hydrophilic surfactant"), the ability of the formulation to dissolve lipophilic drugs is affected and cosolvent addition is needed to restore the solubilization ability within the SEDDS (Table 6). Finally, increasing concomitantly the concentration of surfactants and cosolvent to a marked level at the expense of the oil in the formulation leads to microemulsion formation following aqueous dilution. The microemulsions exhibit a very small particle size range and their formation is attributed to the presence of the cosolvents. Their purpose is to make the interfacial film fluid by wedging themselves between the surfactant molecules and therefore resulting in a bicontinuous structure. Microemulsions can solubilize marked amounts of lipophilic compounds.

Table 6 Classification System for Lipidic Formulations and Major Characteristics

Parameters	Lipid solution	SEDDS type 1	SEDDS type 2	SMEDDS
Oil (%)	100	40—80	40–80	<20
Surfactants (%)	0	20–60 (HLB<12)	20–60 (HLB>12)	20—50 (HLB>11)
Cosolvents (%)	0	0	0–40	50–100
Droplet size (nm)	>1000	100–250	100–250	< 100
Ease of self-emulsification	Very low	Low to medium	Medium to high	High
Solvent capacity	Low	Low to medium	Medium to high	High
Loss of solvent capacity upon dilution	None	None to slight	Slight to moderate	Moderate to high
Digestibility significance on absorption	Crucial	Not crucial but likely to occur	Not crucial but may be inhibited	Not required and not likely to occur

Abbreviations: HLB, hydrophilic–lipophilic balance; SEDDS, self-emulsifying drug delivery systems; SMEDDS, self-microemulsifying drug delivery systems.
Source: From Ref. 40.

Figure 4 SEDDS-filled Licaps® hard gelatin capsules. *Abbreviation*: SEDDS, self-emulsifying drug delivery systems.

CHARACTERIZATION OF SEDDS

Aspect

It can be seen from Figure 4 that the SEDDS formulation is transparent and homogeneous. The formulation can be filled in soft gelatin capsules (SGCs) or sealed hard gelatin capsules as in Figure 4.

The primary means of self-emulsification assessment is visual evaluation (24,38,40). The efficiency of self-emulsification could be estimated by determining the rate of emulsification and droplet size distribution. Turbidity measurements can be carried out to determine the rapid equilibrium reached by the dispersion and the reproducibility of this process (40).

Mechanism of Self-Emulsification

Normally, as illustrated in Figure 5, the lipid clear solution is spontaneously emulsified upon contact with an aqueous solution phase.

The droplet size of the emulsion is a crucial factor in self-emulsification performance, because it determines the rate and extent of drug release as well as absorption

Figure 5 Self-emulsification process upon contact of a lipid clear solution with an aqueous phase.

(4,50). Photon correlation spectroscopy is a useful method for determination of emulsion droplet size, especially when the emulsion properties do not change upon infinite aqueous dilution, a necessary step in this method (1,4,24,38,51). However, microscopic techniques should be employed at relatively low dilutions for accurate droplet-size evaluation (24,51).

Emulsion droplet polarity is also a very important factor in characterizing emulsification efficiency (4). The HLB, chain length and degree of alkyl chain unsaturation of the fatty acid, molecular weight of the hydrophilic portion, and concentration of the emulsifier have an impact on the polarity of the oil droplets. Polarity represents the affinity of the drug compound for oil and/or water and the type of forces involved. Rapid release of the drug into the aqueous phase is promoted by polarity.

The mechanism by which self-emulsification occurs is not yet well understood. Nevertheless, it has been suggested that self-emulsification takes place when the entropy change favoring dispersion is greater than the energy required to increase the surface area of the dispersion (52). Spontaneous emulsification is a process that occurs when two immiscible liquids are placed in contact with each other and emulsify without the aid of any external thermal or mechanical energy source (Fig. 5). Depending on the liquids involved, the presence of appropriate surfactants, pH, or other imposed electrical potentials, completion of the spontaneous emulsification process may take from a few seconds to several days (53). This phenomenon is also achieved by inversion of the coexisting liquid phases. It has found a variety of industrial applications such as SEDDS and agricultural sprays and pesticides with potential applications in enhanced oil recovery and detergency (54–57).

The free energy of a conventional emulsion formulation is a direct function of the energy that is required to create a new surface between the oil and water phases. The two phases of the emulsion tend to separate with time to reduce the interfacial area and thus the free energy of the systems. The conventional emulsifying agents stabilize emulsions that result from aqueous dilution by forming a monolayer around the emulsion droplets, reducing the interfacial energy, and forming a barrier to coalescence. In contrast, emulsification occurs spontaneously with SEDDS because the free energy required to form the emulsion is either low and positive, or negative (6). It is necessary for the interfacial structure to show no resistance against surface shearing for emulsification to take place (25). The ease of emulsification was suggested to be related to the ease of water penetration into the various LC or gel phases formed on the surface of the droplet (2,29,58). The interface between the oil and aqueous continuous phases is formed upon addition of a binary mixture (oil/nonionic surfactant) to water (6). This is followed by the solubilization of water within the oil phase as a result of aqueous penetration through the interface. This will occur until the solubilization limit is reached close to the interphase. Further aqueous penetration will lead to the formation of the dispersed LC phase. In the end, everything that is in close proximity to the interface will be LC, the actual amount of which depends on the surfactant concentration in the binary mixture. Thus, following gentle agitation of the self-emulsifying system, water will rapidly penetrate into the aqueous cores and lead to interface disruption and droplet formation. As a consequence of the LC interface formation surrounding the oil droplets, SEDDS become very stable to coalescence. Detailed studies have been carried out to determine the involvement of the LC phase in the emulsion formation process (1,2,18,29). Also, particle size analysis and low frequency dielectric spectroscopy were utilized to examine the self-emulsifying properties of a series of Imwitor 742 (Sasol Germany, Hamburg, Germany) (a mixture of mono- and diglycerides of capric and caprylic acids)/Tween® 80 systems (5,38). The results suggested that there might be a complex relationship between LC formation and emulsion formation (38). Moreover, the presence of the drug compound may alter the emulsion characteristics, probably by interacting with the LC phase (5). Nevertheless, the correlation between the LC formation and spontaneous emulsification has still not been established (5,40).

A method to determine the spontaneity (S) of the spontaneous emulsification process quantitatively was proposed by Lopez-Montilla et al., using a laser diffraction particle size analysis technique (59). In practice, one can only observe the rapid formation of cloudy dispersions or disappearing of the SEDDS. It is very difficult to measure the kinetics of such phenomena, although some recently new advances in video imaging, laser, and light-scattering techniques for size distribution are contributing to overcoming the actual technical limitations. The technique currently used in the industry to measure the S of an emulsification process is known as the Collaborative Pesticide Analytical Committee of Europe test (CPAC test). In this technique, a 1 mL bulb pipe is vertically supported with the tip placed about 4 cm above the surface of water at the 100 mL graduation mark in a 100 mL graduated cylinder (60). The oil content in the bulb is allowed to fall freely into the water, and the ease of emulsification is visually evaluated and expressed in a qualitative fashion as good, moderate, or bad. This method presents serious disadvantages including: (i) the data obtained cannot be meaningfully compared with data obtained in other laboratories as this technique relies on visual appreciation; (ii) most oils are lighter than water thus potentially slowing down the emulsification rate; and (iii) the rate at which the oil will disperse into the water phase strongly depends on the difference in density between the oil and water. However, the CPAC test is and has been widely used despite its poor interlaboratory reproducibility, mainly because of its simplicity and ease of application. Therefore, Lopez-Montilla et al. proposed an alternative technique, referred to as the Specific Interfacial Area Test (59). Their innovative technique relies on the fact that the S of an emulsification process should account not only for the rate of emulsification but also for the volume fraction of the final internal phase as well as for the droplet size distribution of the resulting emulsion. Indeed, emulsification is known to be an energy-driven process, which is directly related to the formation of the new interface. The interfacial free energy increases as the interfacial area grows, owing to the breakage of droplets into smaller droplets, while the dispersed volume remains constant. In the case of a spontaneous process, the required interfacial free energy is provided by the excess internal energy of the system upon mixing of the two liquids. Consequently, the S is directly related to the volume of the system. The minimal energy (ΔG_{int}) required to create new interfacial area is then given by the integral of the interfacial tension (γ) with respect to the increase in interfacial area (dA) namely, $\Delta G_{\text{int}} = \int \gamma \, dA$, where both γ and A are time-dependent parameters.

Lopez-Montilla et al. studied the rate of increase of the specific interfacial area and the equilibrium specific interfacial area for different self-emulsifying systems formed by the surfactant Brij 30 (Uniqemal/ICI Surfactants, A Uniquema Business Unit, PO Box 90, Wilton Centre, Middlesbrough, Cleveland, TS90 8JE U.K.) dissolved in linear alkyl oils (C8–C16), when put in contact with ultra pure water (59). The experimental results confirmed the effectiveness of the proposed method.

Size

There are many techniques to determine the mean and particle size distribution of SEDDS and SMEDDS. Some of these techniques are described in Figure 6.

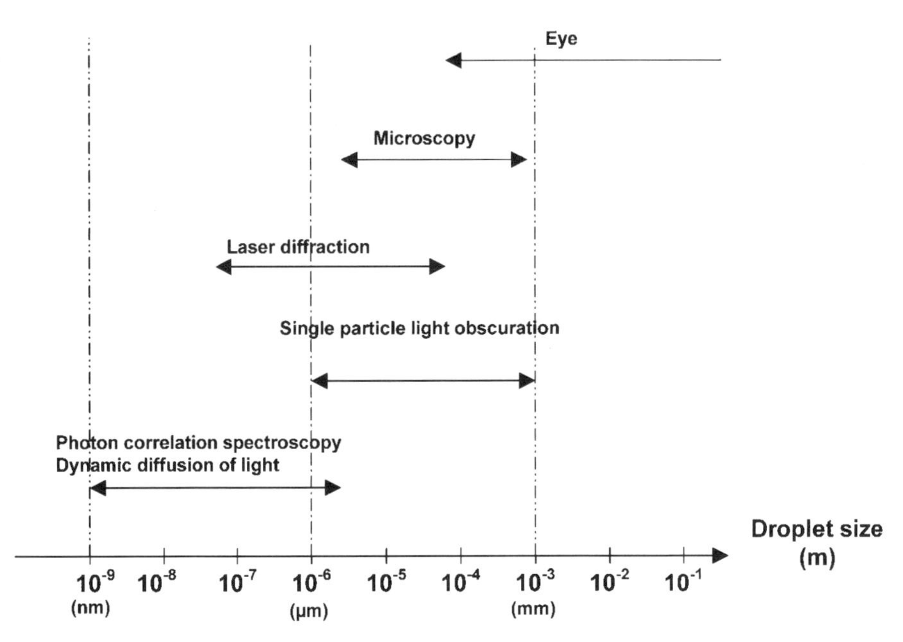

Figure 6 Particle size measurement techniques used for the evaluation of SEDDS and SMEDDS emulsification process. *Abbreviations*: SEDDS, self-emulsifying drug delivery systems; SMEDDS, self-microemulsifying drug delivery systems.

The reduction of droplet size to values below 50 nm leads to the formation of SMEDDS, which are stable, isotropic, and clear o/w dispersions (6,21,33,36,37). Pseudoternary phase diagrams in which the ratio of two or more of the components is kept constant while three other excipient concentrations are typically varied, can be constructed to describe such systems (6,39,41,61). Normally, the oil, surfactant, and cosurfactant or cosolvent ratios are changed in an attempt to identify the self-emulsifying regions and/or other types of dispersions (41,62). Finally, appropriate experimental conditions (optimum excipient concentrations) are established by means of ternary diagram studies allowing formulation of the required SEDDS and/or SMEDDS. The characterization of SMEDDS can be made utilizing dye solubilization, dilutability by the dispersed phase excess, and conductance measurements (6).

Zeta Potential

The charge of the oil droplets of SEDDS is another property that should be assessed (49,51). The oil droplets in conventional SEDDS normally exhibit a negative charge, probably because of the presence of free fatty acids; however, incorporation of a cationic lipid, such as oleylamine, at a concentration range of 1% to 3%, will yield cationic SEDDS. Thus, such systems have a positive ξ-potential value of about 35 to 45 mV (23,24,30). This positive ξ-potential value is preserved following the incorporation of the drug compounds.

Recently, it was proved that the absorptive cells, as well as the other cells in the body, are negatively charged with respect to the mucosal solution in the lumen (5). Positively charged formulations differ from negatively charged formulations with respect to their interaction with biological components in the GI environment. Positively charged droplets should be attracted to the negatively charged physiological

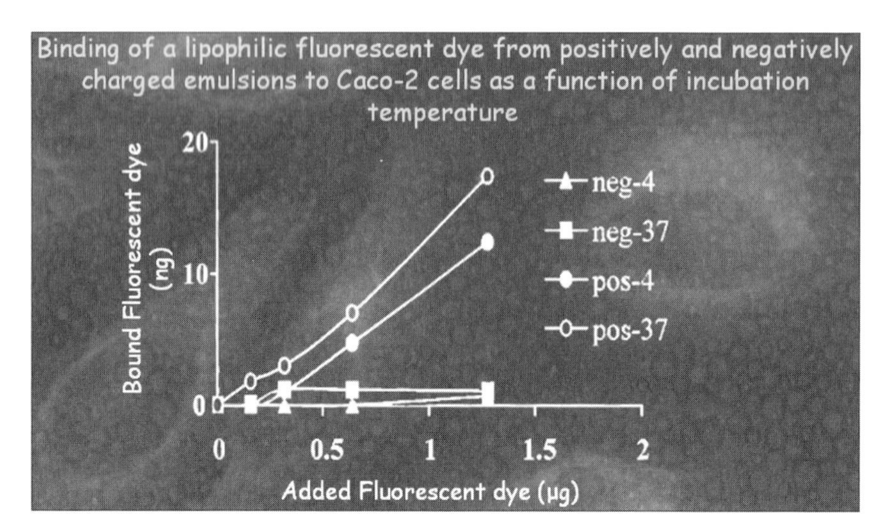

Figure 7 Binding of a lipophilic fluorescent dye from positively and negatively charged emulsions to Caco-2 cells as a function of incubation time. *Source*: From Ref. 23.

compounds that are naturally occurring in the lumen. It was shown that positively charged emulsion droplets formed by appropriate SEDDS dilution undergo electrostatic interaction with the Caco-2- monolayer (Fig. 7) and the mucosal surface of the everted rat intestine resulting in an increase in the oral BA of lipophilic drug (23,24,30).

BIOPHARMACEUTICAL ASPECTS

The release of the drug compound from SEDDS takes place upon its partitioning into the intestinal fluids during droplet transport and disintegration along the GI tract. It was proposed that two main factors, which include small particle size and polarity of the resulting oil droplets, determine the efficient release of the drug compound from SEDDS (4). In o/w microemulsions, however, the impact of the polarity of the oil droplets is not very significant because the drug compound reaches the capillaries incorporated within the oil droplets (6,33,36,44).

Many animal studies assessing the oral BA of hydrophobic drugs formulated in o/w emulsions indicated better absorption profiles, but the use of these systems is limited because of their poor physical stability and the large volumes needed (5,13–16,47,62). Thus, SEDDS may be a promising alternative to orally administered emulsions because of their relatively high physical stability and ability to be delivered in standard SGCs. A higher BA of hydrophobic drugs incorporated in SEDDS was reported earlier (4,20,63). A study carried out in nonfasted dogs, to assess the oral BA of a lipophilic naphthalene derivative formulated in SEDDS, demonstrated three-fold higher C_{max} and area under the curve (AUC), compared with other dosage forms (4). In another study using rats, the oral BA of the lipophilic anti-inflammatory drug, ontazolast, was substantially improved compared to the bioavilabity brought about by the suspension formulation, when it was administered in a lipid-based formulation, such as emulsion, glyceryl oleate (Peceol) solution, and SEDDS (20). A multiple-dosage study was conducted in humans diagnosed with HIV infection, who were given an HIV

Table 7 Mean Pharmacokinetic Parameters (n = 8) (± SD) of Vitamin E in Either an Oil Solution (Natophérol, Soybean Oil) or SEDDS in Human Healthy Volunteers (SEDDS: Tween 80, Span 80, and palm oil; (4:2:4) + 333 UI/mL of Vitamin E)

Parameters	Oily solution	SEDDS
C_{max} (μg/mL)	3.0 ± 2.1	6.6 ± 2.3
T_{max} (h)	12.0 ± 2.8	7.5 ± 2.2
$AUC_{0-\infty}$(μg.h/mL)	94.6 ± 80.0	210.7 ± 63.0
$T_{1/2}$(h)	18.0 ± 2.8	20.2 ± 3.0

Abbreviation: SEDDS, self-emulsifying drug delivery systems.
Source: From Ref. 64.

protease inhibitor (PI) orally either as a SEDDS or as an elixir (21,63). The SEDDS gave greater AUC values in addition to higher C_{max} and C_{min} values, compared with the elixir. Furthermore, a SEDDS formulation of vitamin E exhibited, as depicted in Table 7, two to three-fold higher Cmax and AUC than that exhibited by a marketed lipid solution of vitamin E, Natophérol® (Abbott Laboratories, Illinois, U.S.A.), in healthy volunteers (64). In addition, Kim et al. reported a 9.8 and five-fold increase in C_{max} and AUC in rats, respectively, of biphenyl dimethyl dicarboxylate (BDD), when incorporated in a SEDDS formulation compared to the results on using micronized powder formulation based on calcium carboxymethylcellulose, as presented in Table 8 (65).

Effect of Surfactants

In the context of oral dosage, surfactants may play a role in reducing the rate of gastric emptying and retarding the movement of drug to the absorption site by increasing the viscosity of the formulation. This is thought to be especially true of polyoxyethylene derivatives. Bile salts, which are physiological surfactants, have been shown to affect the rate of gastric emptying. The presence of bile salts in the stomach has also been shown to affect ionic movement across the gastric mucosa, thus increasing the movement of hydrogen and chloride ions out of the lumen. Surfactants may also affect the rate and extent of drug absorption by exerting an influence on the permeability of the biomembrane. Competitive binding of the surfactant

Table 8 Pharmacokinetic Parameters of Biphenyl Dimethyl Dicarboxylate BDD in a SMEDD Formulation and Dispersed in Calcium Carboxy Methylcellulose Following Oral Administration in the Rat

Parameters	BDD SMEDDS	Micronized powder mixture of BDD in Ca–CMC(2:1)
C_{max} (μ g/L)	1.550 ± 0.706	0.158 ± 0.165
T_{max} (h)	1.833 ± 1.125	1.254 ± 1.025
AUC (μ g.h/L)	9.829 ± 2.255	1.955 ± 0.712

Abbreviation: BDD, biphenyl dimethyl dicarboxylate; Ca-CMC, calcium carboxy methylcellulose; SMEDDS, self-microemulsifying drug delivery systems.
Source: From Ref. 65.

to the membrane protein is considered to be partially responsible for enhanced drug absorption, in many cases. Alternatively, the enhancement may be due to allosteric rearrangement of the membrane protein, which is triggered by the binding of one or more permeating species. The membrane effects of surfactants are explained by a combination of membrane–surfactant binding, disruption of membranes through solubilization into lipoproteins (LP), proteins, and mixed micelles, protein–protein interactions, and selective solubilization of some membrane components by the surfactant. The structure of the surfactant may play a role in determining the range and extent of the influence of a particular surfactant on drug absorption. It appears that the greatest effect is achieved by molecules having a C12–C16 hydrocarbon chain, polyoxyethylene chain lengths between 10 and 20, and molecular areas between 1.0 nm^2 and 1.6 nm^2 (66). These effects, in the case of drugs with low aqueous solubility, arise from an increase in drug solubility in addition to the higher absorption rate (67,68).

Surfactants, at high concentrations, exhibit some toxicity and have the ability in many cases to disrupt membranes. Both ionic and nonionic surfactants have been shown to assist the breakdown of the mucous layer covering the epithelium and, at high concentrations, are thought to interfere with the membrane itself, which may lead to disruption of membrane metabolism, especially the enzyme systems associated with the membrane. Adverse reactions to drug formulation agents including surfactants have been reviewed by Weiner and Bernstein (69).

Surfactants increase permeability by interfering with the lipid bilayer of the single layer of the epithelial cell membrane, which, with the unstirred aqueous layer, forms the rate-limiting barrier to drug absorption/diffusion (70). Therefore, most drugs are absorbed via the passive transcellular route. Surfactants partition into the cell membrane and disrupt the structural organization of the lipid bilayer leading to permeation enhancement. They also exert their absorption enhancing effects by increasing the dissolution rate of the drug (71).

Cyclosporin A (CsA) is a potent immunosuppresive drug used in organ transplantation. It is a cyclic undecapeptide with very poor aqueous solubility (72). Thus, extensive studies were carried out to improve the oral BA of CsA, which eventually led to the formulation of Sandimmune® (Novartis, Basel, Switzerland) and later on to an even better-performing microemulsion formulation of CsA, Sandimmune Neoral® (Novartis, Basel, Switzerland), as will be described in detail later in this chapter (36,37,40).

Coenzyme Q_{10} (CoQ_{10}) is a lipid-soluble compound that is used as an antioxidant and in the treatment of cardiovascular disorders including angina pectoris, hypertension, and congestive heart failure (73). The drug is poorly absorbed from the GI tract, possibly because of its high molecular weight and water insolubility (73). At present, CoQ10 is available on the market as oil-based and powder-filled capsule formulations, which exhibit high variations in oral BA (74). Thus, a new SEDDS approach was evaluated for improved oral BA of CoQ10. An optimized formulation determined on the basis of mean emulsion droplet diameter containing acetylated monoglycerides Myvacet® 9–45 (Quest International BV, Naarden, The Netherlands), Labrafac® CM-10, and PG monolaurate (lauroglycol) was developed. The respective compositions of the SEDDS and powder formulations are presented in Table 9. The oral BA studies carried out on dogs resulted in a two-fold higher C_{max} and BA with CoQ_{10} SEDDS when compared with the effects of powder formulation, which contained sodium lauryl sulphate as the wetting agent and lactose as the bulking agent (Table 10).

Tipranavir, a potent anti-HIV drug of the new class of nonpeptidic PIs, was incorporated into a new SEDDS formulation (62). When compared with the initial

Table 9 Composition of the Formulations of Coenzyme Q10 Used for the Pharmacokinetic Study

Composition of SEDDS formulation	Quantity (mg per capsule)	Composition of powder formulation	Quantity (mg per capsule)
CoQ 10	30.0	CoQ 10	30.0
Myvacet 9–45	188.0	Lauryl sulfate de sodium	0.3
Labrafac CM-10	235.0	Lactose	269.7
Lauroglycol	47.0		

Abbreviations: CoQ 10, coenzyme Q_{10}; SEDDS, self-emulsifying drug delivery systems.
Source: From Ref. 39.

formulation, which is a hard filled capsule, the new SEDDS formulation administered in a SGC led to approximately a two-fold higher BA (71). Saquinavir (SQV), a potent and well-tolerated anti-HIV drug, is currently used as a PI in highly active antiretroviral therapy regimens (72,75). At present, the drug is available in HGC [Invirase® (Hoffmann-La Roche Inc., New Jersey 07110, U.S.A.)] and SGC formulations [Fortovase® (Hoffmann-La Roche Inc., New Jersey 07110, U.S.A.)]. Following a single administration of 600 mg SQV, the BA of the drug in HGC is much lower than that of the SGC formulation. The significant improvement in BA (331%) of SQV with SGC is attributed to capmul, a glyceride type excipient (medium-chain mono- and diglycerides) used in the SGC formulation (75). Capmul dissolves the drug to a high extent, and the drug is rapidly released. However, this excipient has adverse effects such as diarrhea. Therefore, a new approach has been tested to keep the BA of SQV high while lowering the side effects of capmul. It has been shown in healthy subjects boosted with SQV/ritonavir (RTV) 1000 mg/10 mg BID, that SQV in a HGC could be absorbed well and tolerated better than SQV in a SGC. But, RTV may also lead to side effects as its formulation contains polyoxyl 35 castor oil (77).

It has also been shown that the amount of free drug and extent of absorption were affected by micellar solubilization of lipophilic drugs with high concentrations of surfactants in the formulation (78,79). The intestinal absorption of griseofulvin in rats was reported to decrease in the presence of 20 mM taurocholate as a result of micellar solubilization (78). Also, in vitro permeability studies conducted utilizing the Caco-2 cell line demonstrated a decrease in permeability of CsA in the presence

Table 10 Mean Pharmacokinetic Parameters (n = 4) (± SD) of Coenzyme Q10 in Dogs Following Adminitration of a SEDDS Formulation and a Solid Formulation

Parameters	SEDDS/Formulation I (mean±SD)	Powder/Formulation II (mean±SD)
AUC (g.h/mL)	61.29 ± 14.1	27.41 ± 7.6
C_{max} (g/mL)	1.39 ± 0.4	0.61 ± 0.13
T_{max} (h)	6.2 ± 1.8	5.8 ±1.2

Abbreviation: SEDDS, self-emulsifying drug delivery systems.
Source: From Ref. 39.

of surfactants, such as Cremophor EL or RH40 and TPGS, at concentrations above 0.02% w/v, which was attributed to micellar solubilization (79).

Effect of Lipids

It is important to note that lipids have an impact on the oral BA of the drug compound. They exert their effects possibly through several complex mechanisms that can lead to alteration in the biopharmaceutical properties of the drug, such as increased dissolution rate of the drug and solubility in the intestinal fluid, protection of the drug from chemical as well as enzymatic degradation in the oil droplets, and the formation of LPs promoting the lymphatic transport of highly lipophilic drugs (5,20,62,80–84). The absorption profile and the blood/lymph distribution of the drug compound are affected by the acid chain length of the TG, saturation degree, and volume of the lipid administered. Generally, compounds processed by the intestinal lymph are transported to the systemic circulation along with the lipid core of LPs, and as such require coadministered lipid to stimulate LP formation (85). Short and medium-chain fatty acids (with a carbon chain length shorter than 12 carbon atoms) are transported to the systemic circulation by the portal blood and are not incorporated to a great extent in chylomicrons (86). In contrast, long-chain fatty acids and monoglycerides are re-esterified to TGs within the intestinal cell, incorporated into chylomicrons, and secreted from the intestinal cell by exocytosis into the lymph vessels. In addition to the stimulation of the lymphatic transport, administration of lipophilic drugs with lipids may enhance drug absorption into the portal blood when compared with nonlipid formulations (87).

Bile salts, monoglycerides, cholesterol, lecithin, and lysolecithin further emulsify the large fat droplets upon entering the intestine, and smaller droplets of 0.5 to 1.0 μm mean diameter are formed. Pancreatic lipase then catalyzes the metabolism of these droplets, which later on form mixed micelles with bile salts (6,20,33,88). Following their penetration through the aqueous layer and mucin, mixed micelles and microemulsions are absorbed either by pinocytosis, diffusion, or endocytosis (33). The drug compound then reaches the systemic circulation via the portal vein or lymphatic system.

In a recent study, a statistical experimental design and multivariate optimization strategy was used to evaluate and predict the effect of different lipid combinations in SEDDS on the oral BA of CsA (89). The lipid vehicle of the SEDDS, galactolipids (GL), exhibits good self-emulsifying properties and is nonionic unlike phospholipids, which are charged. Thus, GL could be safer for long-term use (81). In formulations containing GL, MCT and monoglycerides (MCM), increasing the drug content from 12.5% to 30% led to an approximately two-fold decrease in BA. But, the droplet-size distribution, which is known to influence the rate and extent of drug release and absorption, appeared to have no effect on BA. The type of the lipid excipient and the lipid ratio within the SEDDS formulation were reported to have a significant impact on CsA oral BA. For instance, the formulation containing sphingolipids, cholesterol, and MCT was shown to result in almost no absorption while the SEDDS comprising fractionated oat oil (FOO) was reported to yield a BA comparable to the reference product. Thus, a SEDDS formulation including FOO and MCM at a 1:1 ratio, in which the FOO contained 50% neutral lipids and 50% polar lipids (mixed phospholipids and GL) was developed. The drug content was 10% in the SEDDS formulation, which was shown to be approximately bioequivalent to the reference product, Sandimmune Neoral® (89).

Digestibility

Lipids, unlike many excipients, whether present in food or as discreet pharmaceutical additives, are processed both chemically and physically within the GI tract before absorption and transport into the portal blood (or mesenteric lymph). Indeed, most of the effects mediated by formulation-based lipids or the lipid content of food are mediated by means of the products of lipid digestion–molecules that may exhibit *very* different physicochemical and physiological properties when compared with the initial excipient or food constituent. Therefore, although administered lipids have formulation properties in their own right, many of their effects are mediated by species that are produced after transformation or "activation" in the GIT. An under standing of the luminal and/or enterocyte-based processing pathways of lipids and lipid systems is therefore critical to the effective design of lipid-based delivery systems.

The general process of lipid digestion is well known and well described in a number of recent publications (90–94). Ingested TGs are digested by the action of lingual lipase in the saliva and gastric lipase and the pancreatic lipase/colipase complex in the stomach and small intestine, respectively. These sequential processes convert essentially water-insoluble, nonpolar TG into progressively more polar diglycerides, monoglycerides, and fatty acids. The end point (chemically) of digestion of one molecule of TG is the liberation of two molecules of fatty acid and one molecule of 2-monoglyceride.

In addition to the chemical breakdown of ingested lipids, the physical properties of lipid digestion products are markedly altered to facilitate absorption. Initial lipid digestion products become crudely emulsified on emptying from the stomach into the duodenum (because monoglycerides and diglycerides have some amphiphilic, emulsifying properties, and gastric emptying provides sufficient shear to provoke emulsification). The presence of partially digested emulsion in the small intestine leads to the secretion of bile salts and biliary lipids from the gallbladder that stabilize the surface of the lipid emulsion and reduce its particle size, presenting a larger lipid surface area to the pancreatic lipase/colipase digestive enzymes. In the presence of sufficient bile salt concentrations, the products of lipid digestion are finally incorporated into bile salt micelles to form a solubilized system consisting of fatty acids, monoglycerides, bile salts, and phospholipids—the so-called intestinal mixed micellar phase. The intestinal mixed micellar phase coexists with a number of physical species in the small intestine, including multilamellar and unilamellar lipid vesicles, simple lipid solutions, and fatty acid soaps (95,96).

The complexity and dynamism of the postdigestive intestinal contents (in terms of the interconversion and equilibrium-driven transfer of lipids across the various dispersed species) is a likely contributor to the uncertainty in defining the effects of lipids on drug absorption. Conversely, a more complete understanding of this preabsorptive phase and its interaction with lipophilic drugs will enhance appreciation of the effects of lipids on drug absorption and improve the ability to select appropriate lipid excipients.

Solubilization of lipid digestion products in intestinal mixed micelles enhances their dissolution and dramatically increases the GI lumen-enterocyte concentration gradient that drives absorption by means of passive diffusion. Micelles, however, are not absorbed intact, and lipids are thought to be absorbed from a monomolecular intermicellar phase in equilibrium with the intestinal micellar phase (97–99). The dissociation of monomolecular lipid from the micellar phase appears to be stimulated by the presence of an acidic microclimate associated with the enterocyte surface (100,101). In addition to passive diffusion, growing evidence suggests that active uptake processes, mediated by transport systems located in the enterocyte membrane, are also involved in the absorption of (in particular) fatty acids into the enterocytes (94).

After absorption, the biological fate of lipid digestion products is defined primarily by the chain length of the absorbed lipid. Short- and medium-chain length fatty acids are much more water-soluble than longer-chain lipids, and they diffuse relatively unhindered across the enterocyte into the portal blood (86). Long-chain lipids, however, are trafficked through the endoplasmic reticulum, re-esterified to TG, assembled into lymph LPs, and secreted into the intestinal lymph (91). Subsequently, long-chain lipids are transported through the intestinal lymph and into the central lymph, before entering the systemic circulation at the junction of the thoracic lymph duct and the left internal jugular vein in the neck. After entering the systemic circulation, the poor water solubility of lipids dictates their association with endogenous carrier systems such as plasma proteins and plasma LPs. These carrier systems facilitate the distribution of lipids to peripheral tissues, where they are either stored as fat deposits and metabolized as an energy source, or used as a structural building block in lipidic structures such as membranes (102).

The inherent physicochemical similarities between many lipophilic drugs and dietary and/or formulation-derived lipids in terms of high partition coefficients and low water solubilities suggest that the processes that control lipid digestion, absorption, and distribution may similarly affect the disposition of lipophilic drugs. Therefore, the coadministration of lipids might be expected to have an impact on the disposition of lipophilic drugs in the following ways:

1. By stimulating the release of biliary and pancreatic secretions, thereby providing an intestinal micellar phase into which a poorly water-soluble drug may become solubilized—increasing its effective solubility, dissolution rate, lumen-to-enterocyte concentration gradient, and consequently the extent of absorption. Increasing evidence suggests that coadministered lipids also have significant effects on drug absorption and metabolism at a cellular level through attenuation of enterocyte-based metabolic and anti-transport processes.
2. By enhancing the formation and turnover of lymph LPs through the enterocyte and provoking, or improving, the targeting of orally administered lipophilic drugs to the intestinal lymphatics.
3. By altering the relative proportions and constituents of plasma LPs and changing the degree of binding of lipophilic drugs to discreet LP subclasses. The presence of specific receptors for LP subclasses such as the low-density LP receptor suggests that alteration of LP-binding profiles may have a significant impact on both pharmacokinetic issues such as drug clearance and volume of distribution and on pharmacodynamic end points such as toxicity and activity.

Lipids may also have effects on gastric transit (in terms of delaying gastric emptying) and intestinal permeability (enhancing the absorption of poorly permeable compounds).

de Smidt et al. investigated whether further lipolysis of the dispersed lipidic material is required for final transfer to the enterocyte membranes (103). To assess the relative roles of lipid vehicle dispersion and vehicle digestibility, the authors studied the oral absorption of penclomedine (Pcm) from a series of Pcm-containing SEDDS (103). Three formulations were developed from MCT/tocophersolan (TPGS) mixtures, leading to emulsions having three sizes [160 nm, 720 nm, and mm-sized ("crude" oil)]; with or without the inclusion of tetrahydrolipstatin (THL), a known lipase inhibitor. Oral absorption of Pcm was studied after administration of small volumes of these formulations in the conscious rat. Kinetic evaluation was performed using population analysis. Formulations with particle size 160 nm had the highest relative BA (set at $F=1$), whereas administration in particle size 720 nm had slightly lower BA ($F=0.79$). Coinclusion of THL yielded similar BA for these two SEDDS. "Crude" oil formulations had $F=0.62$ (without THL) and 0.25 (with THL). The data in the current investigation emphasize the prominent role of increased vehicle dispersion relative to digestibility in the absorption of Pcm from MCT–TPGS in submicron emulsions. Only with Pcm administered as undispersed MCT was absorption more dependent on the action of lipase, as BA was inhibited two-fold by the coincorporation of THL.

Dispersion

Simple suspensions and solutions of drugs in lipids have been shown to enhance the oral BA of a number of poorly water-soluble compounds, including phenytoin, progesterone, and cinarrizine (15,104–108). In these examples, BA enhancement appears to have been mediated by way of improved drug dissolution from lipid solutions (compared with aqueous suspensions) and enhanced drug solubility in the lipid/bile salt-rich GI contents. Optimal BA enhancement was generally provided by lipids in which the drug was most soluble, although factors including the solubility of the lipid in the GI fluids (short-chain lipids typically dissolve in the intestinal lumen leading to drug precipitation) and the ability of long-chain lipids to stimulate lymphatic transport complicate choice of the optimal lipid.

As a consequence of the intestinal processing that lipids undergo before absorption, there has been significant interest in assessing the "digestibility" of formulation lipids as a potential indicator of in vivo BA enhancement. In this regard, digestible lipids such as dietary fats (TGs, diglycerides, fatty acids, phospholipids, cholesterol, etc) are generally more effective in terms of BA enhancement than indigestible oils such as mineral oil (14,109). However, more complex correlations of lipid chain length (medium-chain versus long-chain lipids) or lipid class (TGs versus diglycerides or monoglycerides) with digestibility and BA enhancement have met with little success.

The degree of dispersion of a lipid-based delivery system appears to have the most marked effect on the BA of a coadministered drug, and this has stimulated many of the most recent articles in the literature. Clearly, by decreasing the particle size of a dispersed formulation, the surface area available for lipid digestion and drug release or transfer is enhanced. In this regard, the BA of griseofulvin, phenytoin, Pcm, danazol, REV 5901, and, more recently, ontazolast has been shown to be enhanced after administration in an emulsion formulation compared with the administration of tablet, aqueous solution, or suspension formulation (20,104,110–112). It is not clear in these cases how much more efficient the emulsion formulation would have been compared with a simple lipid solution. In many cases the relatively complex nature of lipid-based formulations in terms of lipid class, chain length, degree of dispersion, and choice of surfactant makes explanation of the mechanistic information difficult. For example, the BA of vitamin E following administration of vitamin E acetate is greater when administered in a MCT-based emulsion than in a

LCT-based lipid solution. However, the differential roles of lipid dispersion or lipid class (MCT vs. LCT) cannot be separated (113,114).

Although emulsion formulations show great promise for the enhancement of lipophilic drug BA, the limited acceptability of oral emulsions has led to the more recent development of SEDDS. These anhydrous systems composed of an isotropic mixture of drug, lipid, and surfactant are generally filled into a soft or sealed HGC. Following administration, the capsule ruptures, and an emulsion is spontaneously formed on contact with the intestinal fluids. The optimized interfacial properties (e.g., low interfacial tension) of these systems facilitate spontaneous emulsification and also result in the formation of emulsions with particle sizes that are generally lower than those formed with conventional emulsions ($< 1\,\mu m$), providing additional benefits in terms of enhanced surface areas of interaction.

The most recent development (in terms of physicochemical/particle size approaches) in the design of lipid-based delivery systems has been the use of microemulsions, microemulsion preconcentrates, or SMEDDS, typified by the Sandimmune Neoral® formulation (see section "The Story of Oral Cyclosporin A"). Microemulsions are defined as isotropic, transparent, and thermodynamically stable (in contrast to conventional emulsions) mixtures of a hydrophobic phase (lipid), a hydrophilic phase (often water), a surfactant, and, in many cases, a cosurfactant. From a lipid formulation perspective, microemulsions are generally regarded as the ultimate extension of the "decreased particle size/increased surface area" mantra, because emulsion particle sizes are usually less than 50 nm. Microemulsions also have additional pharmaceutical advantages in terms of their solubilizing capacity, thermodynamic stability, and capacity for stable, infinite dilution.

The Lymphatic Pathway Opportunity

Following absorption, most drugs and xenobiotics traverse the enterocytes and are absorbed into the portal blood. A small number of highly lipophilic drugs, however, are transported to the systemic circulation by means of the intestinal lymphatic pathway. The GI lymphatic system is a specific transport pathway through which dietary lipids, fat-soluble vitamins, and water-insoluble peptide type molecules (e.g., CsA) can gain access to the systemic circulation (90,101,115–117). Drugs transported by way of the GI lymphatic system bypass the liver and avoid potential hepatic first-pass metabolism. Lymphatic delivery of immunomodulatory and low–therapeutic index protein and peptide drugs, used in the treatment of cancer cell metastases and HIV, presents an opportunity to maximize therapeutic benefit while minimizing general systemic drug exposure (118). Furthermore, lymphatic drug transport may promote drug incorporation into the body's lipid-handling system, thus offering the potential to manipulate drug distribution and residence time within the body.

Drug delivery to the intestinal lymphatics confers two primary advantages over conventional absorption by means of the portal blood. First, transport through the intestinal lymph avoids presystemic hepatic metabolism and therefore enhances the concentration of orally administered drugs reaching the systemic circulation. Second, from a site-specific delivery or targeting perspective, the lymphatics contain relatively high concentrations of lymphocytes, and therefore provide attractive targets for cytokines such as interferon and immunomodulators, in general (88,90,91). Furthermore, the lymphatics serve as the primary conduit for the dissemination of many tumor metastases and therefore show promise as a target for cytotoxics, and may provide an efficient route of delivery to HIV-infected T cells, because recent findings have suggested that a significant proportion of the HIV viral burden resides in the lymphoid tissue (92,119,123).

With exceptions, including halofantrine, DDT, and the lipophilic vitamins, the extent of lymphatic transport (as a proportion of the dose) is generally low as can be noted from Table 11. However, the compounds described in Table 11 are hydrophobic (as evidenced by the high log P values), and their BA is often low. Therefore, although the absolute extent of lymphatic transport may be low, the lymphatic contribution to the small fraction that is absorbed may be high, and alterations in the extent of lymphatic transport may have a significant effect on the extent of oral BA.

Intestinal Lymphatic System

The lymphatic system is an elaborate network of specialized vessels distributed throughout the vascular regions of the body. The primary and well-recognized function of the lymphatic system is to drain the capillary beds and return extracellular fluid to the systemic circulation, thus maintaining the body's water balance. However, the structure and function of the lymphatics throughout the body are not uniform and in specific areas the lymphatics perform a specialized role (125). For example, the intestinal lymphatic system is responsible for the transport of dietary fat and lipid-soluble vitamins to the systemic circulation (126,127).

Lipid Absorption from the Small Intestine

Most of the lipids are absorbed from the jejunum, with the exception of bile, which remains in the small intestine lumen in order to facilitate further digestion (93,128). The salts of the bile are finally absorbed in the distal ileum and are transported back to the liver by the portal blood in a cycle that constitutes the entero-hepatic circulation (128–130).

Lipid absorption occurs when the micellar solution of lipids comes into contact with the microvillus membrane of the enterocytes. The lipids are transported across the enterocyte membrane primarily by an energy independent process, which relies on the maintenance of an inward diffusion gradient. This gradient can partly be achieved by the attachment of the fatty acids to specific intracellular binding proteins. However, the ultimate driving force for absorption probably comes from the rapid re-esterification of the lipids, which is an ATP-dependent process, depending upon activation of fatty acids to acyl-CoA esters.

The major digestive products of TGs are monoglyceride and fatty acid while the major digestive product of biliary and dietary phosphatidylcholine is lysophosphatidylcholine. These digestive products are absorbed primarily by the enterocytes through simple diffusion. However, the absorption of cholesterol by the enterocytes is specific, as the plant sterol, sitosterol, which bears considerable resemblance to cholesterol, is poorly absorbed. Following entry into the enterocytes, the monoglycerides, fatty acids, and cholesterol are transported within the cell to the endoplasmic reticulum by fatty acid–binding protein and sterol carrier protein. Through the monoglyceride pathway, the digestive byproducts of TGs, monoglycerides, and fatty acids are resynthesized to form TG in the endoplasmic reticulum. This TG is then transported to the Golgi apparatus where it is packaged into chylomicrons and released into the lymphatics (129).

The transport and metabolism of the absorbed cholesterol is much lower than that of triacylglycerols. The estimated half-life for absorbed cholesterol in the enterocyte is about 12 hours. During absorption the cholesterol becomes incorporated into the

Table 11 Summary of Intestinal Lymphatic Transport Data

Compound	Log P	Lipid solubility	Dosing vehicle	Cumulative lymphatic transport (% dose)	Collection period	Model	References
DDT	6.19	97.5 mg/mL (peanut oil)	200 μL oleic acid	33.5%	10 hr	Anesthetized rat/mesenteric lymph duct/intraduodenal dose	85
HCB	6.53	7.5 mg/mL (peanut oil)	200 μL oleic acid	2.3%	24 hr	Anesthetized rat/mesenteric lymph duct/intraduodenal dose	124
Cyclosporin	2.92	>30 mg/mL (sesame oil)	2 mL/kg 8% HCO-60 micellar solution 2 mL/kg sesame oil	2.14%	6 hr	Anesthetized rat/thoracic lymph duct/oral dosing	116,117
Penclomedine	5.48	177 mg/mL (soybean oil)	0.5 mL 10% soybean oil emulsion	1.5%	12 hr	Anesthetized rat/ mesenteric lymph duct/intraduodenal dosing	103
Halofantrine	8.5	>50 mg/mL (peanut oil)	50 μL peanut oil	16.7%	12 hr	Conscious rat/mesenteric lymph duct/oral dosing	
Ontazolast	4.0	55 mg/mL (soybean oil)	10 mL/kg 20% soybean oil emulsion 10 mL/kg 1% Methocel/0.2% PS80 suspension	1.2%	8 hr	Conscious rat/mesenteric lymph duct/oral dosing	20

Abbreviations: DDT, dichlorodiphenyltrichloro ethane; HCB, hexachlorobenzene.

membranes of the enterocytes and diluted with endogenous cholesterol. A large proportion of the cholesterol that is transported from the enterocyte is esterified, mainly with oleic acid. The rate of esterification of cholesterol may regulate the rate of lymphatic transport of cholesterol. Two enzymes have been proposed to be involved in the esterification, cholesterol esterase and acyl-CoA cholesterol acyltransferase.

Not until the 1950s was the quantitative significance of the cholesterol lymphatic pathway known (130,131). Biggs et al. demonstrated that following an intragastric dose of [^{3}H] cholesterol, very little isotropically labeled cholesterol appeared in the plasma of rats with thoracic lymph duct cannulas (131). Chaikoff et al. recovered more than 94% of absorbed labeled cholesterol in the thoracic duct lymph of rats (132). Similar results have been reported in rabbits and in a human subject with chyluria, confirming that in mammals, the absorbed cholesterol is transported by the intestinal lymphatics and not by the portal system (133,134).

Contribution to the Enhanced Absorption of Lipophilic Drugs into the Systemic Circulation

The majority of orally administered drugs gain access to the systemic circulation by direct absorption into the portal blood. However, for some water-insoluble compounds, transport by way of the intestinal lymphatic system may provide an additional route of access to the systemic circulation. Exogenous compounds absorbed through the intestinal lymph appear to be generally transported in association with the lipid core of intestinal LPs (predominantly TG-rich chylomicrons), thereby requiring coadministered lipid to stimulate LP formation. Delivery into the bloodstream by way of the intestinal lymphatics has been suspected to contribute to the overall absorption of a number of highly lipophilic compounds including cyclosporine, naftifine, probucol, mepitiostane, halofantrine (see section 6.4.6), testosterone undecanoate, and polychlorinated biphenyls (116,117,124,135–149).

Lymph from the intestinal lymphatic system (as well as hepatic and lumbar lymph) drains through the thoracic lymph duct into the left internal jugular vein and then to the systemic circulation (94). Thus, the drug transport by way of the intestinal lymphatic system may increase the percentage of drug that can gain access to the systemic circulation. In addition, the process of intestinal lymphatic drug transport often continues over time periods longer than that typically observed for drug absorption through the portal vein. Consequently, drug transport through the lymph may be utilized to prolong the time course of drug delivery to the systemic circulation. Preliminary findings published by Hauss et al. suggest that the incorporation of a water-insoluble agent, ontazolast (a potent inhibitor of leukotriene B_4), into lipid-based formulations composed of a mixture of monoglycerides, diglycerides, and TGs increased the amount of drug that reached the systemic circulation and was transported through the lymph (20). Charman et al. have done similar work with another hydrophobic compound: halofantrine (147,150,151).

Evaluation and Assessment of Intestinal Lymphatic Transport

A number of animal models have been described for the assessment of intestinal lymphatic drug transport (20,62,151,152). Lymphatic transport studies are commonly first conducted in the laboratory rat, with subsequent investigations in larger, more complicated models (such as dog or pig). However, the utility of lymph fistulation in large animals is limited by considerable logistical and economic constraints. Ideally, sampling strategies for lymphatic transport studies should provide the capacity to estimate both the extent of lymphatic transport and the extent of portal blood absorption to estimate the overall BA of the drug/formulation. This strategy enables the unambiguous determination of the extent of lymphatic transport relative to absorption via the portal blood, and the total BA of the drug/formulation. As lymphatic transport can be affected by experimental factors such as the site of lymphatic cannulation and the period of fasting prior to dosing, it is important to standardize procedures when comparing studies (153,154).

The triple-cannulated anesthetized rat model (where the mesenteric lymph duct, jugular vein, and duodenum are accessed) has been used for the assessment of lymphatic transport (20,151,153). General anesthesia precludes oral dosing in the anesthetized model and consequently drug and lipid formulations are administered intraduodenally. This limitation thus circumvents the inherent emulsifying action of the stomach and the potential effects of lipids on gastric emptying. Thus, the conscious rat model best represents the in vivo situation in terms of both lack of anesthetic effects and the ability to orally administer drug formulations. Previously reported methods for collecting lymph from the rat required total restraint of the animal and fluid replacement, by intravenous or intraduodenal infusion, to maintain lymph output (4,93). A rat model has been developed to allow collection of mesenteric lymph for five days from conscious, minimally restrained animals with a cannula and no signs of physical distress (20,151). This model obviates the need for total restraint or general anesthesia, both of which are known to influence intestinal lymphatic transport of test compounds in unpredictable ways (151). Animals are provided free access to an electrolyte solution, which they consume in sufficient quantity to maintain adequate lymph output without the need for the previously required infusions for fluid replacement. The rat is the appropriate experimental animal to investigate oral absorption and lymphatic transport because intestinal characteristics (e.g., anatomical, metabolic, and biochemical characteristics) of these animals are similar to those found in humans (29,45–47). Specifically, the intestinal processing and absorption of dietary lipids are similar in rats and humans (24).

Proposed Mechanisms that Govern the Lymphatic Transport of Water-Insoluble Drugs

Although the mechanisms by which drugs gain access into the intestinal lymphatic system through the enterocyte are not fully elucidated, there is growing evidence that supports the hypothesis that the majority of drugs transported by the lymphatics are associated with the TG core of chylomicrons (98,155–161). In addition to this, Charman and Stella proposed that there are two important factors—the drug's diffusion/partition behavior and lipid solubility—which appear to be the prerequisites for the lymphatic transport of water-insoluble drugs (162).

Diffusion and Partition Behavior of Water-Insoluble Drugs. The extent of a drug's partitioning between the portal blood and intestinal lymph may be estimated from a comparison of the relative rates of drug mass transfer by each route. In this regard, the rate of fluid flow in the intestinal lymphatic system is approximately 500-fold less than that in the portal blood, and during peak lipid transport, the lipid content of the lymph is only in the order of 1% to 2% (w/v) (159). Thus, the effective mass ratio between lymph lipid and the portal blood is in the order of 1:50,000. Consequently, the selective lymphatic transport of small molecular weight, water-soluble drugs is unlikely if the route of absorption (portal blood vs lymph) is governed by the relative rates of fluid flow. However, this ratio suggests that for similar extents of absorption and transport by the portal blood and intestinal lymph (not taking into account metabolic conversion, chemical stability, and/or BA considerations), a candidate molecule should have a log octanol/water partition coefficients (log P) in the region of 5 (highly water-insoluble). Hauss et al. reported that when ontazolast, which has an octanol/water log P = 4.0 was incorporated into lipid-based formulations composed of a mixture of monoglycerides, diglycerides and TGs, a significantly greater amount of drug was transported by the lymph than suspension control (20). Caliph et al. studied the effects of short-, medium-, and long-chain fatty acid–based vehicles on the absolute oral BA and intestinal lymphatic transport of halofantrine (87). They reported that increases in lymphatic drug transport appeared to correlate with increases in lymphatic lipid transport.

These initial studies provide evidence that lymphatic transport contributes to the overall oral absorption of water-insoluble compounds incorporated into lipid-based formulations such as SEDDS (87). However, a more comprehensive investigation of these initial findings needs to be done (150). Finally, Figure 8 provides the overall benefits achieved by incorporating lipophilic drugs within SEDDS as a function of log P.

Lipid Solubility of Water-Insoluble Drugs. In addition to a high partition coefficient being a prerequisite for lymphatic transport, lipid solubility is a further important parameter to consider (Table 11). Charman and Stella reported the relationship between lipid solubility and lymphatic transport of two highly water-insoluble compounds [dichlorodiphenyltrichloroethane (DDT) and hexachlorobenzene (HCB)] which have similar octanol/water partition coefficients yet different solubilities (161,162). Both compounds would be regarded as highly lipophilic, as evidenced by their high octanol/water partition coefficients (DDT, 6.2 vs. HCB 6.5). However, the 13-fold higher TG solubility of DDT compared with that of HCB

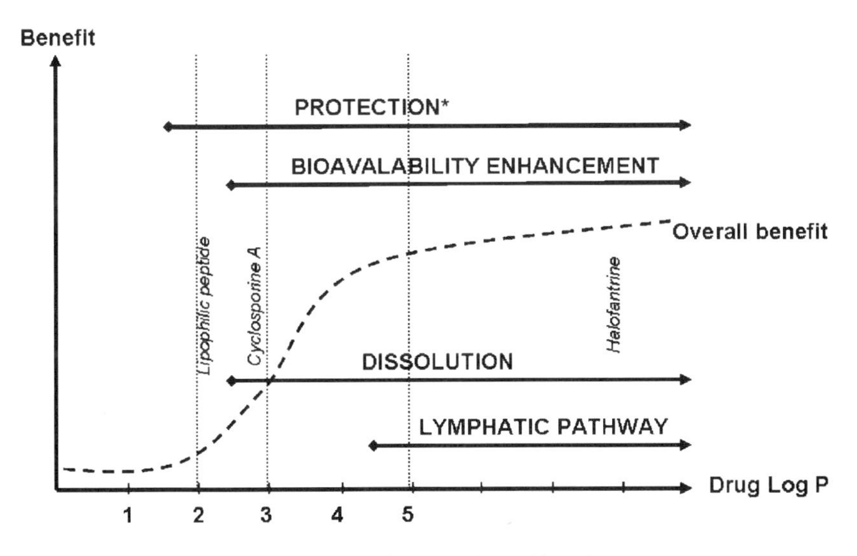

Figure 8 Lipophilic compound benefits achieved following incorporation within a SEDDS for oral use as a function of drug log P. *Abbreviations*: SEDDS, self-emulsifying drug delivery systems.

(DDT, 9.75 +/−0.15 vs. HCB 0.75 +/−0.05) solubility in peanut oil (g solute/100 mL) is reflected in the 14.6-fold increase in the extent of intestinal lymph transport reported in an anesthetized rat model (159,160). Hauss et al. previously observed that when ontazolast concentration in the lymph was correlated to chylomicron TG in the lymph, SEDDS formulations (consisting of mainly mixed TGs), which promote more rapid absorption of ontazolast, also favored lymphatic drug transport (20). These findings suggest that solubility in chylomicron of TGs may be a determining factor for promoting lymphatic transport. Taken together, these studies provide preliminary evidence that TG solubility may play a major role in promoting the lymphatic transport and increased oral absorption of lipophilic compounds.

Halofantrine-Containing SEDDS

Halofantrine (Hf), the chemical structure of which is presented in Figure 9 is a new phenanthrenemethanol antimalarial molecule which is orally active, well tolerated, and is finding increasing use in the treatment of malaria associated with multidrug resistant strains of *Plasmodium falciparum*. However, Hf is extremely lipophilic (log P $\approx$ 8) and poorly water soluble (<10 μg/mL), and the BA of Hf after oral administration of Hf HCl tablets is low and variable.

Researchers including Charman, Porter, Humberstone, Caliph, Wasan, et al. conducted a series of investigations into the lymphatic absorption of Hf (27,147,149–151,163–175). Coadministration of Hf with food increases BA up to three-fold in humans and 10-fold in beagle dogs (148,173). However, the clinical application of coadministration with food is limited because of possible uncontrolled increases in plasma levels that lead to cardiac side effects including a lengthening of the QTc interval. In an attempt to increase the oral BA of Hf, three SEDDS formulations were developed, the details of which are given in Table 12 (174). The commercial Hf formulation is formulated as a hydrochloride salt. However, to improve the lipid solubility of Hf in lipid-based formulations, the free base of Hf was used in the study. All three formulations consisted of an isotropic mix of drug, lipids, surfactant, and ethanol filled into SGCs. On capsule rupture and interaction with an aqueous environment (in vitro screens used distilled water or 0.1 N HCl), the systems designated as SMEDDS produced a clear/translucent microemulsion (particle size <50 nm) and the SEDDS produced a bright white emulsion (particle size approximately 200 nm). The three formulations were designed to probe the effects of both particle size (emulsion vs. microemulsion) and lipid class (long chain vs. medium chain) on Hf BA. Formulations were optimized by the construction of partial phase diagrams, in which the lipid phase consisted of either a 2:1 w/w ratio of Captex 355 to Capmul MCM (for the medium-chain

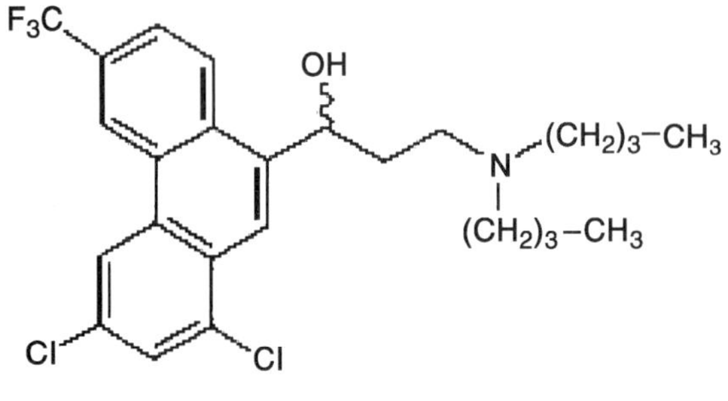

Figure 9 Halofantrine chemical structure.

Table 12 Halofantrine SEDDS Compositions

Components	MCT-SEDDS (%w/w)	MCT-SMEDDS (%w/w)	LCT-SMEDDS (%w/w)
Halofantrine	5	5	5
Captex 355	46.7	33.3	–
Capmul MCM	23.3	16.7	–
Soybean oil	–	–	29
Maisine 35–1	–	–	29
Cremophor EL	15	35	30
Ethanol (absolute)	10	10	7

Source: From Ref. 174.

system), or a 1:1 mass ratio of soybean oil to Maisine 35/1 in the case of the long-chain systems. The quantity of Hf and ethanol was kept constant.

In a previous study in male beagle dogs, the absolute BA of Hf HCl was 8.6 ± 53% (148). The mean absolute BA of the free base of Hf after administration in the lipid-based formulations ranged from 52% to 67% as presented in Table 13 (174). This compares favorably with previous data in which the absolute BA of Hf from the commercially available tablet formulation (250 mg Hf HCl) in fasted beagles was 8.6 ± 5.3%. The capacity of dispersed lipid-based formulations to enhance the BA of Hf is further supported by a recent study in fasted beagles, according to which the absolute oral BA of crystalline Hf base was found to be even lower.

No particle size–effect was seen with the medium-chain formulations, suggesting that the differences in particle size of the emulsion and microemulsion formulations were either not important or not sufficiently large to cause any difference in the intestinal processing. Although not statistically significant (possibly because of the size of the study), there was a trend towards higher BA with the long-chain formulation, indicating that for the formulations that produce very fine dispersions, lipid class effects may be more important than small variations in particle size. Significant intestinal lymphatic transport of Hf base has been described in the rat (using higher lipid doses), and this could explain the improved performance of the long-chain systems. However, it is unclear what role lymphatic transport has in these studies in which the total lipid load is much lower.

Table 13 Bioavailability Parameters Obtained in Male Beagle Dogs (Mean ± SD, n = 3) After Oral Administration of 50 mg of Hf Base Formulated as Either a Long-Chain Self-microemulsifying Formulation (SMEDDS) or a Medium-Chain SEDDS, or SMEDDS

Parameters	MCT-SEDDS	MCT-SMEDDS	LCT-SMEDDS
C_{max} (ng/mL)	363 ± 156	374 ± 198	704 ± 308
T_{max} (h)	2.8 ± 0.8	4.2 ± 1.5	2.3 ± 0.5
$AUC^{(0\text{-}inf)}$ (ng.h/mL)	5313 ± 1956	5426 ± 2841	6973 ± 2388
Fabs (%)	51.6 ± 19.2	52.7 ± 24.0	67.3 ± 21.0

Abbreviations: SEDDS, self-emulsifying drug delivery systems; SMEDDS, self-microemulsifying drug delivery system.
Source: From Ref. 174.

From the overall results of the reported studies on the BA of Hf, it can therefore be hypothesized that medium-chain fatty acid enhanced the absorption into the systemic blood circulation whereas long-chain fatty acid enhanced the lymphatic transport. Thus, the absorption profile of a drug formulated into a SMEDDS could be manipulated by varying the medium- and LCT content in the formulation to improve the oral BA of highly lipophilic drugs.

Absorption Barriers

Physiological Barriers

Aqueous Boundary Layer. The aqueous boundary layer or the unstirred water layer (UWL) is a more or less stagnant layer, about 30 μm to 100 μm in thickness, composed of water, mucus, and glycocalyx adjacent to the intestinal wall and created by incomplete mixing of the lumenal contents near the intestinal mucosal surface (175). The glycocalyx is made up of sulfated mucopolysaccharides, whereas mucus is composed of glycoproteins (mucin), enzymes, and electrolytes. Until recently, the resistance of the UWL to intestinal absorption was believed to be correlated to the effective intestinal permeability values of the solutes. However, considerable evidence suggests instead that the available surface of the apical membrane of the intestinal mucosa is the main barrier to both actively and passively absorbed solutes (175). It is also interesting to note that coadministration of food and prokinetic (motility-inducing) agents such as cisapride tend to decrease the thickness of UWL by increasing segmental and propagative contractions respectively, which may have implications for drug dissolution in the GIT (176). The reverse is true for some viscous soluble dietary fibers, such as pectin, guar gum, and sodium carboxymethylcellulose, which may increase the thickness of UWL by reducing intraluminal mixing and could possibly decrease the intestinal exsorption of lipophilic drugs like quinidine and thiopental (175,177–180).

Intestinal Epithelial Barrier. The intestinal epithelial layer that lines the GIT represents the major physical barrier to oral drug absorption. Structurally, it is made up of a single layer of columnar epithelial cells, primarily enterocytes, and intercalated goblet cells (mucus-secreting cells) joined at their apical surfaces by tight junctions or zonula occludens. These tight junctions are formed by the interaction of membrane proteins at the contact surfaces present between cells and are responsible for restricting the passage of hydrophilic molecules during the paracellular transport. In fact, electrophysiological studies have suggested that the epithelium gets tighter as it progresses distally, which has been implicated in a reduced paracellular absorption in the colon (181).

The epithelium is supported underneath by lamina propria and a layer of smooth muscle called muscularis mucosa (3–10 cells thick). These three layers, namely the epithelium, lamina propria, and muscularis mucosa, together constitute the intestinal mucosa (182). On the apical surface, the epithelium along with lamina propria projects to form villi. The cell membranes of epithelial cells that comprise the villi contain uniform microvilli, which give the cells a fuzzy border called the brush border. These structures, although they greatly increase the absorptive surface area of the small intestine, provide an additional enzymatic barrier as the intestinal digestive enzymes are contained in the brush border. In addition, on the top of the epithelial layer lies another layer, the UWL, as previously described. The metabolic and biochemical components of the epithelial barrier will be discussed elsewhere.

Efflux

There may be other possible reasons for enhanced uptake of hydrophobic and/or lipophilic drugs formulated as SEDDS from the GIT, such as a decrease in the P-gp drug efflux (183). In addition to a multidrug efflux pump, phase I metabolism by the intestinal cytochrome P450s is now becoming recognized as a significant factor in oral drug BA (184). In some cases, as shown recently, excipients incorporated in SEDDS/SMEDDS can inhibit both presystemic drug metabolism and intestinal efflux, mediated by P-glycoprotein (P-gp) resulting in an increased oral absorption of cytotoxic drugs (82,83).

Presystemic Metabolism

Orally administered drugs are subject to presystemic metabolism, which is comprised of three subtypes of mechanisms:

1. Lumenal metabolism: This may be triggered by digestive enzymes secreted from the pancreas (amylase, lipases, and peptidases including trypsin and α-chymotrypsin), and those derived from the bacterial flora of the gastrointestinal tract, especially within the lower part of the GIT.
2. First-pass intestinal metabolism. This includes brush-border metabolism and intracellular metabolism. The former occurs at the surface of the enterocytes by the enzymes present within the brush border membrane. Furthermore, the brush border activity is generally greater in the proximal small intestine (duodenum $\approx$ jejunum $>$ ileum $\gg$ colon) and involves enzymes such as alkaline phosphatase, sucrase, isomaltase, and a considerable number of peptidases (185). The intracellular metabolism occurs in the cytoplasm of enterocytes and involves the major class of phase I metabolizing enzymes (i.e., cytochrome P450s, in particular CYP3A4), several phase II conjugating enzymes, and others, such as esterases. Important examples of intestinal phase II metabolism include sulfation of isoproterenol and terbutaline, and glucuronidation of morphine and propofol (186). It is obvious that intestinal epithelium as a site of preabsorptive metabolism may significantly contribute to the low BA of therapeutic peptides and ester-type drugs like aspirin, although it could serve as a key site for targeted delivery of ester or amide prodrugs (186).
3. First-pass hepatic metabolism. As an absorbed drug reaches the liver through the portal circulation, a fraction of the administered dose is biotransformed before it reaches the systemic circulation. This is known as first-pass hepatic metabolism. In comparison with intestine, the liver dominates the process of first-pass metabolism for most drugs by virtue of its large mass, multiplicity of enzyme families present, and its unique anatomical position (187). However, this is not true for some drugs, such as terbutaline, which undergoes sulfation predominantly in the small intestine, as well as for midazolam, for which first-pass extraction by intestinal mucosa and liver appears to be comparable ($44 \pm 14\%$ vs. $43 \pm 24\%$) (186). It is also important to emphasize here that efficiency of first-pass metabolism (both hepatic and intestinal) varies considerably among different animal species and human subjects, which should be taken into account when deriving the estimates of oral BA, particularly in the case of poorly soluble drugs.

THE STORY OF ORAL CYCLOSPORIN A

Perhaps the best-known example of the BA-enhancing effects of microemulsion formulations is that of CsA. CsA is a cyclic undecapeptide with potent immunosuppressive activity, which is lipophilic ($\log P = 2.92$) and poorly water soluble. The chemical structure is presented in Figure 10.

Early cyclosporin studies showed that immunosuppressive effects were evident after administration as a lipid solution in olive oil (a TG consisting primarily of glycerides of long-chain fatty acids) but not after administration in Migliol 812 (a synthetic TG of C_{10-12} fatty acids) (188). Subsequent studies confirmed these data and showed that the absorption of cyclosporin was significantly improved after administration with a lipid vehicle that is made up of glycerides of long-chain fatty acids, therefore having a better effect when compared with that of using vehicles with MCT (188). Attempts at correlation with the relative digestabilities of the LCT and the MCT were unsuccessful, suggesting that the rate of digestion of the lipid was not a limiting factor (116). Behrens et al. subsequently proposed that the improved intestinal absorption of CsA from LCT vehicles was a function of the improved capacity of the products of LCT digestion to intercalate into intestinal mixed micelles (189).

The first commercial formulation of cyclosporin (Sandimmune)® contained drug, LCT, ethanol, and polyglycolized LCT, administered as either an oral solution/dispersion in water or milk or filled into a SGC. The soft-gel formulation produced a crude o/w emulsion on capsule rupture in situ, as illustrated in Figure 11. The biopharmaceutical performance of the Sandimmune formulation, however, was relatively poor, because absorption was variable, incomplete (≈30%), and affected by food (22,190–192).

Subsequent studies showed that absorption could be enhanced by administration of emulsions with smaller particle sizes, and this resulted in starting a number of

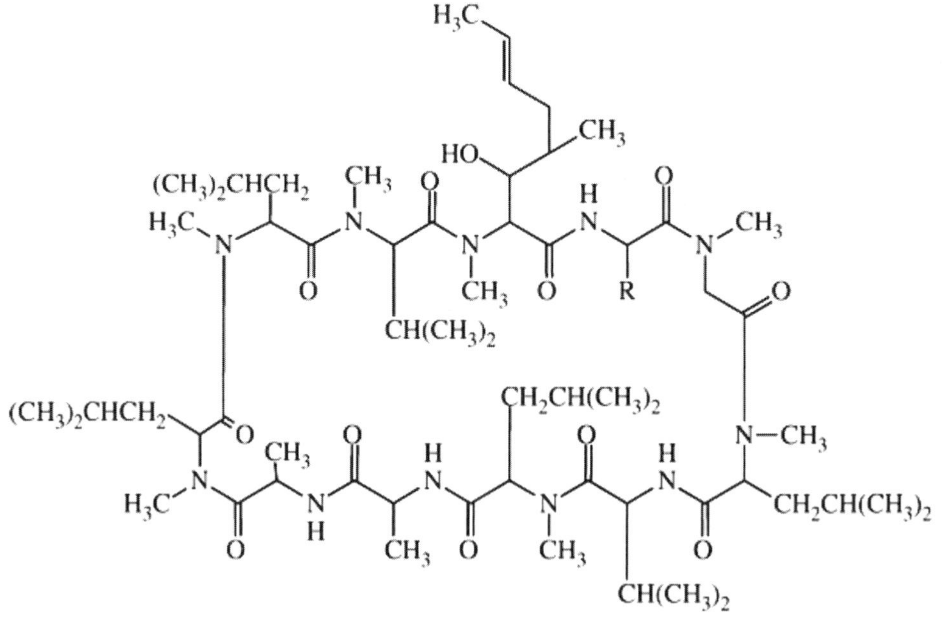

Figure 10 Cyclosporin A chemical structure.

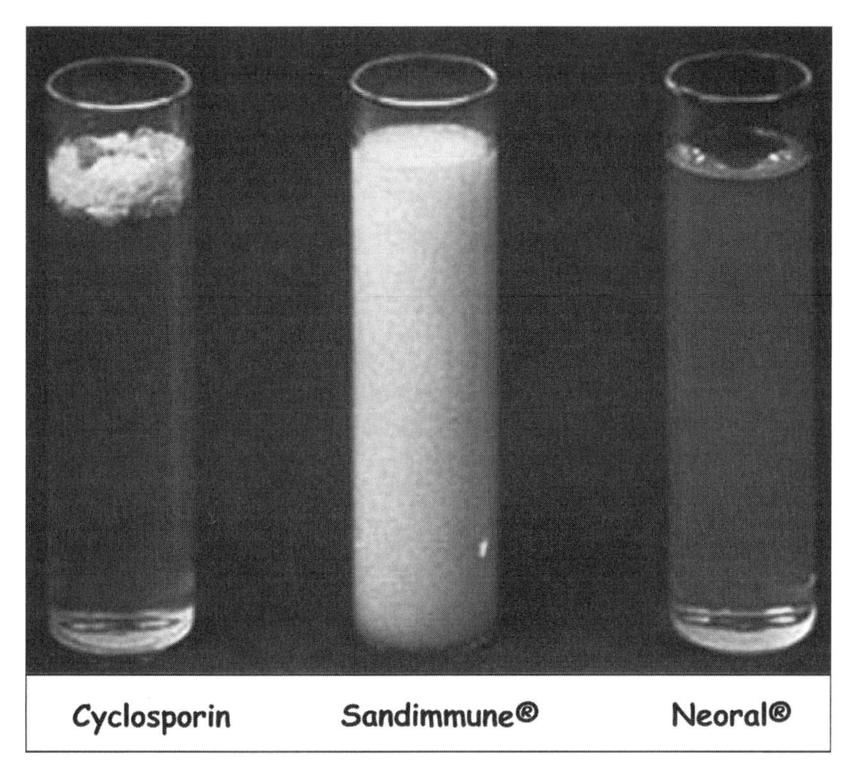

Figure 11 Visual aspect of various formulations of Cyclosporin A, Sandimmune to Neoral following aqueous dilution.

preclinical and clinical studies examining the benefit of microemulsion formulations for the enhancement of cyclosporin BA (50,193–195). With some exceptions (e.g., solid formulations with slow dissolution profiles (195)), and notwithstanding the problems of cross-study and cross-species comparisons, the data showed that the use of microemulsions with smaller particle sizes could increase the oral BA of cyclosporin compared with that of the Sandimmune formulation. As a caveat, however, a report on the BA of cyclosporin after oral administration, in either simple lipid solution combinations of long-chain monoglycerides and TGs or as a predispersed emulsion-like formulation containing the same lipids, has suggested that the particle size of an emulsion formulation or an emulsion formed on dilution/emulsification in the intestine may not be as important in dictating the BA as was previously thought (196).

The current proprietary cyclosporin formulation, Sandimmune Neoral, results in a transparent microemulsion formulation as shown in Figure 11. Although the formulation details of the Neoral formulation are not generally available, the relative bioavailabilities of the Neoral formulation and the initial Sandimmune formulation have been reported. In a dose linearity study, the relative BA of the Neoral formulation compared with the Sandimmune formulation varied from 1.74, at a 200-mg dose, to 2.39, at an 800-mg dose, illustrating the usefulness of the microemulsion formulation and suggesting an approximate two-fold increase in BA from the microemulsion formulation (197). Further studies showed that the absorption of cyclosporin from the Neoral formulation was significantly less variable and less

Table 14 Pharmacokinetic Parameters of a 200 mg Oral Administration of Cyclosporin A ($n = 48$ patients)

Parameters	Dose 200 mg	
	Sandimmune	Neoral®
T_{max} (h)	2.1	1.5
C_{max} (ng/mL)	558 ± 228	1026 ± 218
AUC (ng·h/mL)	2013 ± 614	3471 ± 1029

Source: From Ref. 200.

dependent on bile flow than oral Sandimmune and that its absorption was unaffected by food (197–199). In terms of its apparent lack of reliance on bile for absorption, it is not known whether cyclosporin is absorbed from the formulation directly or just requires much lower bile salt concentrations to facilitate absorption.

The pharmacokinetic dose proportionality and the relative BA of cyclosporin from a microemulsion formulation (Sandimmune Neoral) were compared with those of the commercial formulation (Sandimmune) over the dosage range of 200 to 800 mg (200). Single oral administrations were given as SGCs in an open, randomized study in 48 healthy volunteers. Whole blood–cyclosporin concentrations were determined by a specific monoclonal radioimmunoassay. In comparison to Sandimmune, the absorption rate (maximum concentration) and systemic availability (area under the curve) of cyclosporin were greater for Sandimmune Neoral® at all dose levels investigated. The data for the 200 mg dose are presented in Table 14. Sandimmune's AUC increased in a less than proportional manner with respect to dose, whereas that for Sandimmune Neoral was consistent with linear pharmacokinetics. Because of this difference, no global assessment of relative BA could be performed. The relative BA of cyclosporin from Sandimmune Neoral ranged from 174% to 239% compared to Sandimmune, depending on the dose level. The improvements in oral BA and dose linearity of cyclosporin exposure after administration as Sandimmune Neoral should facilitate more accurate dosage titration in the clinical setting.

THE ORAL PACLITAXEL CHALLENGE

The chemical paclitaxel structure is presented in Figure 12.

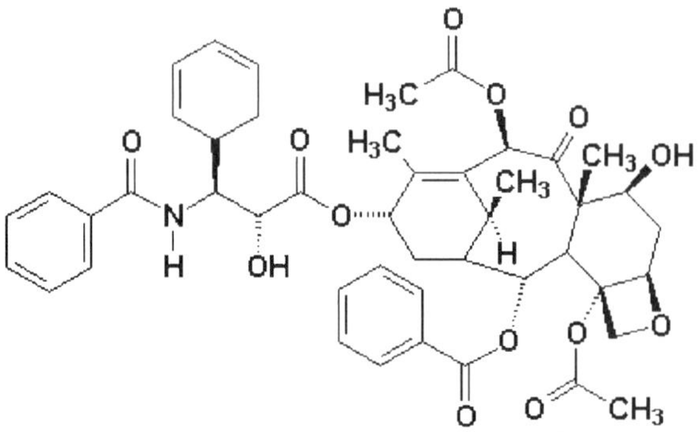

Figure 12 Paclitaxel chemical structure.

Paclitaxel is one of the most potent chemotherapeutic agents, which are currently employed for the treatment of solid tumors (201,202). However, it is a very lyophobic compound; insoluble in most pharmaceutically acceptable solvents. Therefore, it is formulated in a 1:1 combination of polyoxyethylated castor oil (Cremophor® EL) and dehydrated ethanol for IV use, currently marketed as Taxol® (Bristol-Myers Squibb, Princeton, New Jersey, U.S.A.). In addition, paclitaxel is a substrate for both P-gp and cytochrome P450 (CYP3A4) (61). Thus, extensive efforts have been devoted to the design and development of new formulations for the possible oral administration of paclitaxel that would have an improved BA. Recently, a supersaturable SEDDS (S-SEDDS) was developed for the oral delivery of paclitaxel (28). The fact that a high content of surfactants in orally administered SEDDS could lead to adverse effects at the GI tract and a decrease in intestinal absorption of the drug as a result of a decrease in free drug, led to the development of this new SEDDS in which the surfactant concentration was lowered and a cellulose polymer, hydroxypropyl methylcellulose (HPMC), was used as a viscosity enhancer in an attempt to reduce/prevent drug precipitation upon GI fluid dilution. The resultant formulations were stable and contained paclitaxel at a temporarily supersaturated state. In vivo studies conducted in Sprague–Dawley rats supported these results. Oral administration of a 10 mg/kg dose of paclitaxel as S-SEDDS containing 5% HPMC yielded almost five-fold higher AUC 0-∞ values compared with the marketed formulation. These findings demonstrated that the presence of HPMC in the paclitaxel SEDDS prevented macroscopic precipitation of the drug from the formulation, and thus provided a supersaturable formulation with improved oral BA (87). However, it should be emphasized that the plasma concentration of paclitaxel did not reach the therapeutic level when the drug was administered alone as SMEDDS at different doses, despite the presence of well-known moderate P-gp inhibitors, TPGS and Cremophor in the formulation (28,61,82,83). The low plasma concentration and poor oral BA of paclitaxel were not only because of the overexpression of P-gp by the intestinal cells but also the significant first-pass extraction by the cytochrome P450-dependent process. It can be deduced from the overall data presented that, for now, SMEDDS alone cannot overcome the efflux effect of the P-gp in the case of paclitaxel. Thus, when different doses of paclitaxel SMEDDS were coadministered with 40 mg CsA/kg, there was a substantial increase in the C_{max} and AUC values compared with those obtained with paclitaxel SMEDDS alone in rats (61). Furthermore, when coadministered with CsA, there was a significant improvement in the relative BA (by 168%) of the drug in SMEDDS as compared to after administration of Taxol. SMEDDS might have a delayed positive effect on the P-gp inhibitory effect of CsA either by increasing its oral absorption or by enhancing the interaction of CsA with cytochrome P450 at the level of the mature villus tip enterocytes of the small intestine, leading to further improvement in paclitaxel oral BA (22,202). These findings were recently confirmed in the S-SEDDS study in which the co-administration of CsA was again needed to markedly increase the BA of paclitaxel, as depicted in Figure 13 (28).

FUTURE AND PROSPECTS

SEDDS are a promising approach for the formulation of drug compounds with poor aqueous solubility. The oral delivery of hydrophobic drugs can be made possible by SEDDS, which have been shown to substantially improve oral BA. The efficiency of

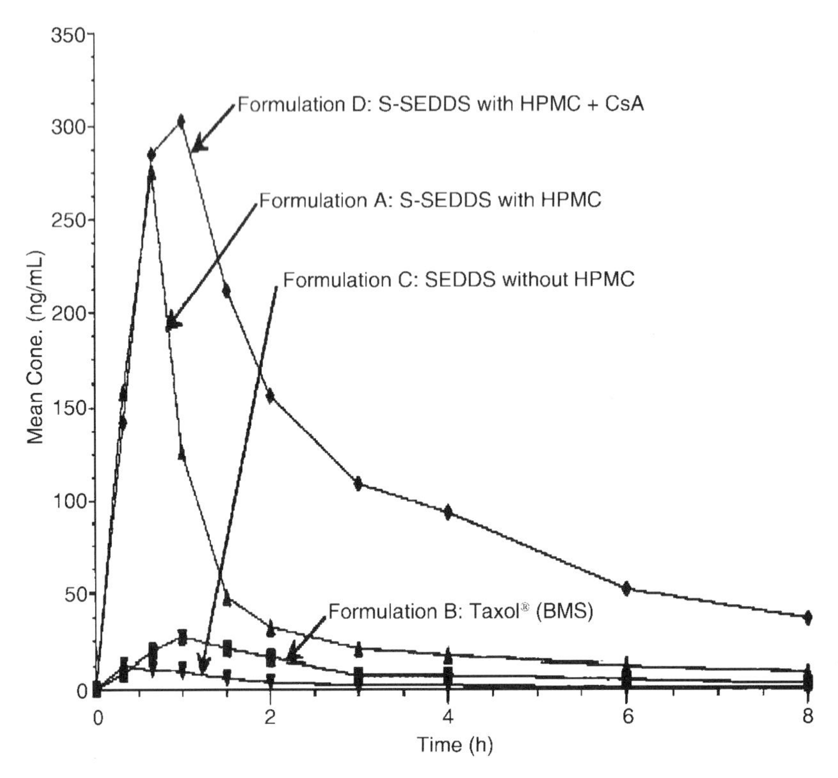

Figure 13 Mean plasma concentration–time profiles of paclitaxel in rats after oral administration of 10 mg/kg paclitaxel using the formulations indicated. *Source*: From Ref. (28).

the SEDDS formulation is case-specific in most instances; thus, composition of the SEDDS formulation should be determined very carefully. As a relatively high concentration of surfactants is generally employed in the SEDDS formulation, toxicity of the surfactant being used should be taken into account. In fact, a compromise must be reached between the toxicity and self-emulsification ability of the surfactant that is considered for use. The size and charge of the oil droplet in the emulsion formed are two other important factors that affect GI absorption efficiency. At present, drug products including CsA, RTV, and SQV, which are designed as SEDDS, are readily available on the market. As nearly 40% of new drug compounds are hydrophobic, it appears that more drug products will be formulated as SEDDS for the pharmaceutical market in the very near future.

REFERENCES

1. Wakerly MG, Pouton CW, Meakin BJ. Evaluation of the self-emulsifying performance of a non-ionic surfactant–vegetable oil mixture. J Pharm Pharmacol 1987; 39:6.
2. Wakerly MG, Pouton CW, Meakin BJ, Morton FS. Self-emulsification of vegetable oil-non-ionic surfactant mixtures. ACS Symp Ser 1986:242–255.
3. Charman SA, Charman WN, Rogge MC, Wilson TD, Dutko FJ, Pouton CW. Self-emulsifying drug delivery systems: formulation and biopharmaceutic evaluation of an investigational lipophilic compound. Pharm Res 1992; 9(1):87–93.

4. Shah NH, Carvagal MT, Patel CI, Infeld MH, Malick AW. Self-emulsifying drug delivery systems (SEDDS) with polyglycolyzed glycerides for improving in vitro dissolution and oral absorption of lipophilic drugs. Int J Pharm 1994; 106:15–23.
5. Craig DQM. The use of self emulsifying systems as a means of improving drug delivery. B.T. Gattefossé 1993; 86.
6. Constantinides PP. Lipid microemulsions for improving drug dissolution and oral absorption: physical and biopharmaceutical aspects. Pharm Res 1995; 12(11):1561–1572.
7. Kim JY, Ku YS. Enhanced absorption of indomethacin after oral or rectal administration of a self-emulsifying system containing indomethacin to rats. Int J Pharm 2000; 194(1):81–89.
8. Amidon GL, Lennernas H, Shah VP, Crison JR. A theoretical basis for a biopharmaceutic drug classification: the correlation of in vitro drug product dissolution and in vivo bioavailability. Pharm Res 1995; 12(3):413–420.
9. Serajuddin AT. Solid dispersion of poorly water-soluble drugs: early promises, subsequent problems, and recent breakthroughs. J Pharm Sci 1999; 88(10):1058–1066.
10. Aungst BJ. Novel formulation strategies for improving oral bioavailability of drugs with poor membrane permeation or presystemic metabolism. J Pharm Sci 1993; 82(10): 979–987.
11. Burcham DL, Maurin MB, Hausner EA, Huang SM. Improved oral bioavailability of the hypocholesterolemic DMP 565 in dogs following oral dosing in oil and glycol solutions. Biopharm Drug Dispos 1997; 18(8):737–742.
12. Aungst BJ, Nguyen N, Rogers NJ, et al. Improved oral bioavailability of an HIV protease inhibitor using gelucire 44/14 and Labrasol vehicles. B.T. Gattefossé 1994; 97: 49–54.
13. Toguchi H, Ogawa K, Iga K, Yashiki T, Shimamoto T. Gastro-intestinal absorption of ethyl 2-chloro-3-[4-(2-methyl-2-phenylpropyloxy)phenyl]propionate from different dosage forms in rats and dogs. Chem Pharm Bull (Tokyo) 1990; 38(10):2792–2796.
14. Palin KJ, Wilson CG. The effect of different oils on the absorption of probucol in the rat. J Pharm Pharmacol 1984; 36(9):641–643.
15. Stella V, Harlan J, Yata N, Okada H, Lindenbaum S, Higuchi T. Enhancement of bioavailability of a hydrophobic amine antimalarial by formulation with oleic acid in a soft gelatin capsule. J Pharm Sci 1978; 67(10):1375–1377.
16. Kararli TT, Needham TE, Griffin M, Schoenhard G, Ferro LJ, Alcorn L. Oral delivery of a renin inhibitor compound using emulsion formulations. Pharm Res 1992; 9(7): 888–893.
17. Schwendener RA, Schott H. Lipophilic 1-beta-D-arabinofuranosyl cytosine derivatives in liposomal formulations for oral and parenteral antileukemic therapy in the murine L1210 leukemia model. J Cancer Res Clin Oncol 1996; 122(12):723–726.
18. Pouton CW. Self-emulsifying drug delivery systems: assessment of the efficiency of emulsification. Int J Pharm 1985; 27:335–348.
19. Chanana GD, Sheth BB. Particle size reduction of emulsions by formulation design-II: effect of oil and surfactant concentration. PDA J Pharm Sci Technol 1995; 49(2):71–76.
20. Hauss DJ, Fogal SE, Ficorilli JV, et al. Lipid-based delivery systems for improving the bioavailability and lymphatic transport of a poorly water-soluble LTB4 inhibitor. J Pharm Sci 1998; 87(2):164–169.
21. Karim A, Gokhale R, Cole J. HIV protease inhibitor SC-52151: a novel method of optimizing bioavailability profile via a microemulsion drug delivery system. Pharm Res 1994:S368.
22. Ptachcinski RJ, Venkataramanan R, Burckart GJ. Clinical pharmacokinetics of cyclosporin. Clin Pharmacokinet 1986; 11(2):107–132.
23. Gershanik T, Haltner E, Lehr CM, Benita S. Charge-dependent interaction of self-emulsifying oil formulations with caco-2 cells monolayers: binding, effects on barrier function and cytotoxicity. Int J Pharm 2000; 211(1–2):29–36.

24. Gershanik T, Benita S. Positively charged self-emulsifying oil formulation for improving oral bioavailability of progesterone. Pharm Dev Technol 1996; 1(2):147–157.
25. Dabros T, Yeung A, Masliyah J, Czarnecki J. Emulsification through area contraction. J Colloid Interface Sci 1999; 210(1):222–224.
26. Gursoy RN, Benita S. Self-emulsifying drug delivery systems (SEDDS) for improved oral delivery of lipophilic drugs. Biomed Pharmacother 2004; 58:173–182.
27. Holm R, Porter CJ, Mullertz A, Kristensen HG, Charman WN. Structured triglyceride vehicles for oral delivery of halofantrine: examination of intestinal lymphatic transport and bioavailability in conscious rats. Pharm Res 2002; 19(9):1354–1361.
28. Gao P, Rush BD, Pfund WP, Huang T, Bauer JM, Morozovich W, Kuo MS, Hagemen MJ. Development of a supersaturable SEDDS (S-SEDDS) formulation of paclitaxel with improved oral bioavailability. J Pharm Sci 2003; 92(12):2386–2398.
29. Rang MJ, Miller CA. Spontaneous emulsification of oils containing hydrocarbon, nonionic surfactant, and oleyl alcohol. J Colloid Interface Sci 1999; 209(1):179–192.
30. Gershanik T, Benita S. Self-dispersing lipid formulations for improving oral absorption of lipophilic drugs. Eur J Pharm Biopharm 2000; 50(1):179–188.
31. Lindmark T, Nikkila T, Artursson P. Mechanisms of absorption enhancement by medium chain fatty acids in intestinal epithelial Caco-2 cell monolayers. J Pharmacol Exp Ther 1995; 275(2):958–964.
32. Yuasa H, Sekiya M, Ozeki S, Watanabe J. Evaluation of milk fat-globule membrane (MFGM) emulsion for oral administration: absorption of alpha-linolenic acid in rats and the effect of emulsion droplet size. Biol Pharm Bull 1994; 17(5):756–758.
33. Georgakopoulos E, Farah N, Vergnault G. Oral anhydrous non-ionic microemulsions administrated in softgel capsules. B.T. Gattefossé 1992; 85:11–20.
34. Swenson ES, Milisen WB, Curatolo W. Intestinal permeability enhancement: efficacy, acute local toxicity, and reversibility. Pharm Res 1994; 11(8):1132–1142.
35. Serajuddin AT, Sheen PC, Mufson D, Bernstein DF, Augustine MA. Effect of vehicle amphiphilicity on the dissolution and bioavailability of a poorly water-soluble drug from solid dispersions. J Pharm Sci 1988; 77(5):414–417.
36. Meinzer A, Mueller E, Vonderscher J. Microemulsion – a suitable galenical approach for the absorption enhancement of low soluble compounds? B.T. Gattefossé 1995; 88: 21–26.
37. Vonderscher J, Meinzer A. Rationale for the development of Sandimmune Neoral. Transplant Proc 1994; 26(5):2925–2927.
38. Craig DQM, Barker SA, Banning D, Booth SW. An investigation into the mechanisms of self-emulsification using particle size analysis and low frequency dielectric spectroscopy. Int J Pharm 1995; 114:103–110.
39. Kommuru TR, Gurley QB, Khan MA, Reddy JK. Self-emulsifying drug delivery systems (SEDDS) of coenzyme Q10: formulation development and bioavailability assessment. Int J Pharm 2001; 212(2):233–246.
40. Pouton CW. Lipid formulations for oral administration of drugs: non-emulsifying, self-emulsifying and "self-microemulsifying" drug delivery systems. Eur J Pharm Sci 2000; 11(Suppl 2):S93–S98.
41. Gursoy N, Garrigue JB, Razafindratsita A, Lambert G, Benita S. Excipient effects on in vitro cytotoxicity of a novel paclitaxel self-emulsifying drug delivery system. J Pharm Sci 2003; 92(12):2411–2418.
42. Staub NC Sr, Longworth KE, Serikov V, Jerome EH, Elsasser T. Detergent inhibits 70–90% of responses to intravenous endotoxin in awake sheep. J Appl Physiol 2001; 90:1788–1797.
43. Constantinides PP, Scalart JP, Lancaster C, et al. Formulation and intestinal absorption enhancement evaluation of water-in-oil microemulsions incorporating medium-chain glycerides. Pharm Res 1994; 11:1385–1390.

44. Unowsky J, Behl CR, Beskid G, Sattler J, Halpern J, Cleeland R. Effect of medium chain glycerides on enteral and rectal absorption of beta-lactam and aminoglycoside antibiotics. Chemotherapy 1988; 34:272–276.
45. Yang S, Gursoy RN, Lambert G, Benita S. Enhanced oral absorption of paclitaxel in a novel self-microemulsifying drug delivery system with or without concomitant use of P-glycoprotein inhibitors. Pharm Res 2004; 21(2):261–270.
46. Lambert G, Razafindratsita A, Gararigue JB, Yang SC, Gursoy RN, Benita S. Self-emulsifying drug delivery system for poorly soluble drugs (taxoids and oral paclitaxel formulation). Filed in March 2002: PCT 02290513.7.
47. Malcolmson C, Lawrence MJ. A comparison of the incorporation of model steroids into non-ionic micellar and microemulsion systems. J Pharm Pharmacol 1993; 45(2):141–143.
48. Denis J. How to formulate superior microemulsions. 16th SCC Congress, New York, 1988.
49. Nazzal S, Khan MA. Response surface methodology for the optimization of ubiquinone self-nanoemulsified drug delivery system. AAPS Pharm Sci Tech 2002; 3(1):E3.
50. Tarr BD, Yalkowsky SH. Enhanced intestinal absorption of cyclosporine in rats through the reduction of emulsion droplet size. Pharm Res 1989; 6(1):40–43.
51. Gershanik T, Benzeno S, Benita S. Interaction of a self-emulsifying lipid drug delivery system with the everted rat intestinal mucosa as a function of droplet size and surface charge. Pharm Res 1998; 15(6):863–869.
52. Reiss H. Entropy-induced dispersion of bulk liquids. J Colloid Interface Sci 1975; 53:61–70.
53. Davies JT. Interfacial Phenomena. New York: Academic, 1961.
54. Becher DZ. Encyclopedia of Emulsion Technology. New York: Marcel Dekker, 1983.
55. Pillai V, Kanicky JR, Shah DO. Handbook of microemulsion science and technology. New York: Marcel Dekker, 1999.
56. Raney KH, Benton WJ, Miller CA. J Colloid Interface Sci 1987; 117(1):282.
57. Rosen MJ. Surfactants and Interfacial Phenomena. New York: Wiley, 1989.
58. Groves MJ, Mustafa RM, Carless JE. Phase studies of mixed phosphated surfactants, n-hexane and water. J Pharm Pharmacol 1974; 26(8):616–623.
59. Lopez-Montilla JC, Herrera-Morales P, Shah DO. New method to quantitatively determine the spontaneity of the emulsification process. Langmuir 2002; 18(11):4258–4262.
60. Groves MT, Mustafa RM. Measurement of the "spontaneity" of self-emulsifiable oils. J Pharm Pharmacol 1974; 26(9):671–681.
61. Yang S, Benita S. Enhanced absorption and drug targetting by positively charged submicron emulsions. Drug Dev Res 2000; 50:476–486.
62. Charman WN. Lipid vehicle and formulation effects on intesinal lymphatic drug transport. In: Chasmar WN, Stella VJ, eds. Lymphatic Transport of Drugs. Boca Raton, FL: CRC Press, 1992:113–179.
63. Fischl MA, Richman DD, Flexner C, et al. Phase I/II study of the toxicity, pharmacokinetics, and activity of the HIV protease inhibitor SC-52151. J Acquir Immune Defic Syndr Hum Retrovirol 1997; 15(1):28–34.
64. Julianto T, Yuen KH, Noor AM. Improved bioavailability of vitamin E with a self emulsifying formulation. Int J Pharm 2000; 200(1):53–57.
65. Kim CK, Cho Y, Gao Z. Preparation and evaluation of biphenyl dimethyl dicarboxylate microemulsions for oral delivery. J Control Release 2001; 70:149–155.
66. Attwood D, Florence AT. Surfactant Systems. In: Their Chemistry, Pharmacy and Biology. London: Chapman and Hall, 1983.
67. O'Reilly JR, Corrigan OI, O'Driscoll CM. The effect of simple micellar systems on the solubility and intestinal-absorption of clofazimine (B663) in the anesthetized rat. Int J Pharm 1994; 105(2):137–146.
68. O'Reilly JR, Corrigan OI, O'Driscoll CM. The effect of mixed micellar systems, bile-salt fatty-acids, on the solubility and intestinal-absorption of clofazimine (B663) in the anesthetized rat. Int J Pharm 1994; 109(2):147–154.

69. Weiner M, Bernstein IL. Adverse Reactions to Drug Formulation Agents. New York: Marcel Dekker, Inc., 1989.
70. Artursson P, Karlsson J. Correlation between oral drug absorption in humans and apparent drug permeability coefficients in human intestinal epithelial (Caco-2) cells. Biochem Biophys Res Commun 1991; 175(3):880–885.
71. Kim HJ, Yoon KA, Hahn M, Park ES, Chi SC. Preparation and in vitro evaluation of self-microemulsifying drug delivery systems containing idebenone. Drug Dev Ind Pharm 2000; 26(5):523–529.
72. Kishi T, Okamoto T, Takahashi T, Goshima K, Yamagami T. Cardiostimulatory action of coenzyme Q homologues on cultured myocardial cells and their biochemical mechanisms. Clin Investig 1993; 71(8 suppl):S71–S75.
73. Ritschel WA. Microemulsion technology in the reformulation of cyclosporine: the reason behind the pharmacokinetic properties of Neoral. Clin Transplant 1996; 10(4): 364–373.
74. Greenberg S, Frishman WH. Co-enzyme Q10: a new drug for cardiovascular disease. J Clin Pharmacol 1990; 30(7):596–608.
75. Greenberg RN, Feinberg J, Goodrich J, Pilson RS, Siemon-Hryczyk P. Long-term efficacy and safety of twice-daily saquinavir soft gelatin capsules (SGC), with or without nelfinavir, and three times daily saquinavir-SGC, in triple combination therapy for HIV infection: 100-week follow-up. Antivir Ther 2003; 8(1):37–42.
76. Gill J, Feinberg J. Saquinavir soft gelatin capsule: a comparative safety review. Drug Saf 2001; 24(3):223–232.
77. Kurowski M, Sternfield T, Sawyer A, Hill A, Möcklinghoff C. Pharmacokinetic and tolerability profile of twice-daily saquinavir hard gelatin capsules and saquinavir soft gelatin capsules boosted with ritonavir in healthy volunteers. HIV Med 2003; 4(2):94–100.
78. Cardiello PG, Monhaphol T, Mahanontharit A, et al. Pharmacokinetics of once-daily saquinavir hard-gelatin capsules and saquinavir soft-gelatin capsules boosted with ritonavir in HIV-1-infected subjects. J Acquir Immune Defic Syndr 2003; 32(4):375–379.
79. Poelma FG, Breas R, Tukker JJ. Intestinal absorption of drugs. III. The influence of taurocholate on the disappearance kinetics of hydrophilic and lipophilic drugs from the small intestine of the rat. Pharm Res 1990; 7(4):392–397.
80. Chiu YY, Higaki K, Neudeck BL, Barnett JL, Welage LS, Amidon GL. Human jejunal permeability of cyclosporin A: influence of surfactants on P-glycoprotein efflux in Caco-2 cells. Pharm Res 2003; 20(5):749–756.
81. Matsuoka K, Kuranaga Y, Moroi Y. Solubilization of cholesterol and polycyclic aromatic compounds into sodium bile salt micelles (part 2). Biochem Biophys Acta 2002; 1580(2–3):200–214.
82. Chervinsky DS, Brecher ML, Hoelcle MJ. Cremophor-EL enhances taxol efficacy in a multi-drug resistant C1300 neuroblastoma cell line. Anticancer Res 1993; 13(1):93–96.
83. Dintaman JM, Silverman JA. Inhibition of P-glycoprotein by D-alpha-tocopheryl polyethylene glycol 1000 succinate (TPGS). Pharm Res 1999; 16(10):1550–1556.
84. Kawakami K, Yoshikawa T, Moroto Y, et al. Microemulsion formulation for enhanced absorption of poorly soluble drugs. I. Prescription design. J Control Release 2002; 81(1–2): 65–74.
85. Pocock DE, Vost A. DDT absorption and chylomicron transport in rat. Lipids 1974; 9(6):374–381.
86. Kiyasu JY, Bloom B, Chaikoff IL. The portal transport of absorbed fatty acids. J Biol Chem 1952; 199(1):415–419.
87. Caliph SM, Charman WN, Porter CJ. Effect of short-, medium-, and long-chain fatty acid-based vehicles on the absolute oral bioavailability and intestinal lymphatic transport of halofantrine and assessment of mass balance in lymph-cannulated and non-cannulated rats. J Pharm Sci 2000; 89(8):1073–1084.
88. Porter CJ, Charman WN. In vitro assessment of oral lipid based formulations. Adv Drug Deliv Rev 2001; 50(Suppl 1):S127–S147.

89. Odeberg JM, Kaufman P, Kroon KG, Höglund P. Lipid drug delivery and rational formulation design for lipophilic drugs with low oral bioavailability, applied to cyclosporine. Eur J Pharm Sci 2003; 20(4–5):375–382.
90. Shiau YF. Mechanisms of intestinal fat absorption. Am J Physiol 1981; 240(1): G1–G9.
91. Bisgaier CL, Glickman RM. Intestinal synthesis, secretion, and transport of lipoproteins. Annu Rev Physiol 1983; 45:625–636.
92. Thomson AB, Keelan M, Garg ML, Clandinin MT. Intestinal aspects of lipid absorption: in review. Can J Physiol Pharmacol 1989; 67(3):179–191.
93. Carey MC, Small DM, Bliss CM. Lipid digestion and absorption. Annu Rev Physiol 1983; 45:651–677.
94. Tso P, Karlstad MD, Bistrian BR, DeMichele SJ. Intestinal digestion, absorption, and transport of structured triglycerides and cholesterol in rats. Am J Physiol 1995; 268(4Pt 1): G568–G577.
95. Hernell O, Staggers JE, Carey MC. Physical–chemical behavior of dietary and biliary lipids during intestinal digestion and absorption. 2. Phase analysis and aggregation states of luminal lipids during duodenal fat digestion in healthy adult human beings. Biochemistry 1990; 29(8):2041–2056.
96. Staggers JE, Hernell O, Stafford RJ, Carey MC. Physical–chemical behavior of dietary and biliary lipids during intestinal digestion and absorption. 1. Phase behavior and aggregation states of model lipid systems patterned after aqueous duodenal contents of healthy adult human beings. Biochemistry 1990; 29(8):2028–2040.
97. Hoffman NE, Simmonds WJ, Morgan RG. The effect of micellar solubilization on mucosal metabolism of absorbed glyceryl-1-monoether. Aust J Exp Biol Med Sci 1972; 50(7):803–812.
98. Simmonds WJ. The role of micellar solubilization in lipid absorption. Aust J Exp Biol Med Sci 1972; 50(4):403–421.
99. Westergaard H, Dietschy JM. The mechanism whereby bile acid micelles increase the rate of fatty acid and cholesterol uptake into the intestinal mucosal cell. J Clin Invest 1976; 58(1):97–108.
100. Shiau YF. Mechanism of intestinal fatty acid uptake in the rat: the role of an acidic microclimate. J Physiol 1990; 421:463–474.
101. Shiau YF, Kelemen RJ, Reed MA. Acidic mucin layer facilitates micelle dissociation and fatty acid diffusion. Am J Physiol 1990; 259(4 Pt 1):G671–G675.
102. Spady DK. Lipoproteins in biological fluids and compartments: synthesis, interconversions, and catabolism. Targeted Diagn Ther 1991; 5:1–44.
103. de Smidt PC, Campanero MA, Troconiz IF. Intestinal absorption of penclomedine from lipid vehicles in the conscious rat: contribution of emulsification versus digestibility. Int J Pharm 2004; 270(1–2):109–118.
104. Chakrabarti S, Belpaire FM. Biovailability of phenytoin in lipid containing dosage forms in rats. J Pharm Pharmacol 1978; 30(5):330–331.
105. Hargrove JT, Maxson WS, Wentz AC. Absorption of oral progesterone is influenced by vehicle and particle size. Am J Obstet Gynecol 1989; 161(4):948–951.
106. Abrams LS, Weintraub HS, Patraick JE, McGuire JL. Comparative bioavailability of a lipophilic steroid. J Pharm Sci 1978; 67(9):1287–1290.
107. Yamaoka Y, Roberts RD, Stella VJ. Low-melting phenytoin prodrugs as alternative oral delivery modes for phenytoin: a model for other high-melting sparingly water-soluble drugs. J Pharm Sci 1983; 72(4):400–405.
108. Tokumura T, Tsushima Y, Tatsuishi K, Kayano M, Machida Y, Nagai T. Enhancement of the oral bioavailability of cinnarizine in oleic acid in beagle dogs. J Pharm Sci 1987; 76(4):286–268.
109. Yamahira Y, Noguchi T, Takenaka H, Maeda M. Absorption of diazepam from a lipid-containing oral dosage form. Chem Pharm Bull (Tokyo) 1979; 27(5):1190–1198.

110. Bates TR, Carrigan PJ. Apparent absorption kinetics of micronized griseofulvin after its oral administration on single- and multiple-dose regimens to rats as a corn oil-in-water emulsion and aqueous suspension. J Pharm Sci 1975; 64(9):1475–1481.
111. Charman WN, Rogge MC, Boddy AW, Berger BM. Effect of food and a monoglyceride emulsion formulation on danazol bioavailability. J Clin Pharmacol 1993; 33(4):381–386.
112. Serajuddin AT, Sheen PC, Mufson D, Bernstein DF, Augustine MA. Physicochemical basis of increased bioavailability of a poorly water-soluble drug following oral administration as organic solutions. J Pharm Sci 1988; 77(4):325–329.
113. Kimura T, Fukui E, Kageyu A, et al. Enhancement of oral bioavailability of d-alpha-tocopherol acetate by lecithin-dispersed aqueous preparation containing medium-chain triglycerides in rats. Chem Pharm Bull (Tokyo) 1989; 37(2):439–441.
114. Fukui E, Kurohara H, Kageyu A, Kurosaki Y, Nakayama T, Kimura T. Enhancing effect of medium-chain triglycerides on intestinal absorption of d-alpha-tocopherol acetate from lecithin-dispersed preparations in the rat. J Pharmacobiodyn 1989; 12(2):80–86.
115. Borgstrom B, Dahlquist A, Lundh G, Sjovall J. Studies of intestinal digestion and absorption in the human. J Clin Invest 1957; 36(10):1521–1536.
116. Ueda CT, Lemaire H, Gjell G, Nussbaumer K. Intestinal lymphatic absorption of cyclosporin A following oral administration in an olive oil solution in rats. Biopharm Drug Dispos 1983; 4(2):113–124.
117. Reymond JP, Sucker H, Vonderscher J. In vivo model for cyclosporin intestinal absorption in lipid vehicles. Pharm Res 1988; 5(10):677–679.
118. Tucker G. Drug delivery: lymphing along? Lancet 1993; 341(8856):1314–1315.
119. Pantaleo G, Graziosi C, Demarest JF, et al. HIV infection is active and progressive in lymphoid tissue during the clinically latent stage of disease. Nature 1993; 362(6418): 355–358.
120. Pantaleo G, Graziosi C, Butini L, et al. Lymphoid organs function as major reservoirs for human immunodeficiency virus. Proc Natl Acad Sci U.S.A. 1991; 88(21):9838–9842.
121. Pantaleo G, Graziosi C, Fauci AS. The role of lymphoid organs in the immunopathogenesis of HIV infection. Aids 1993; 7(suppl 1):S19–S23.
122. Pantaleo G, Graziosi C, Fauci AS. New concepts in the immunopathogenesis of human immunodeficiency virus infection. N Engl J Med 1993; 328(5):327–335.
123. Pantaleo G, Graziosi C, Fauci AS. The role of lymphoid organs in the pathogenesis of HIV infection. Semin Immunol 1993; 5(3):157–163.
124. Busbee DL, Yoo JS, Norman JO, Joe CO. Polychlorinated biphenyl uptake and transport by lymph and plasma components. Proc Soc Exp Biol Med 1985; 179(1): 116–122.
125. Zilversmit DB. The composition and structure of lymph chylomicrons in dog, rat, and man. J Clin Invest 1965; 44(10):1610–1622.
126. McDonald GB, Weidman M. Partitioning of polar fatty acids into lymph and portal vein after intestinal absorption in the rat. Q J Exp Physiol 1987; 72(2):153–159.
127. Thomson AB, Schoeller C, Keelan M, Smith L, Clandini MT. Lipid absorption: passing through the unstirred layers, brush-border membrane, and beyond. Can J Physiol Pharmacol 1993; 71(8):531–555.
128. Tso P, Balint JA. Formation and transport of chylomicrons by enterocytes to the lymphatics. Am J Physiol 1986; 250(6 Pt 1):G715–G726.
129. Tso P, Simmonds WJ. The absorption of lipid and lipoprotein synthesis. Lab Res Methods Biol Med 1984; 10:191–216.
130. Wilson MD, Rudel LL. Review of cholesterol absorption with emphasis on dietary and biliary cholesterol. J Lipid Res 1994; 35(6):943–955.
131. Biggs MW, Friedman M, Byers SO. Intestinal lymphatic transport of absorbed cholesterol. Proc Soc Exp Biol Med 1951; 78(2):641–643.
132. Chaikoff IL, Bloom B, Siperstein MD, et al. C14 cholesterol. I. Lymphatic transport of absorbed cholesterol-4-C14. J Biol Chem 1952; 194(1):407–412.

133. Rudel LL, Morris MD, Felts JM. The transport of exogenous cholesterol in the rabbit. I. Role of cholesterol ester of lymph chylomicra and lymph very low density lipoproteins in absorption. J Clin Invest 1972; 51(10):2686–2692.
134. Hellman L, Frazell EL, Rosenfeld RS. Direct measurement of cholesterol absorption via the thoracic duct in man. J Clin Invest 1960; 39:1288–1294.
135. Porter CJ, Charman WN. Intestinal lymphatic drug transport: an update. Adv Drug Deliv Rev 2001; 50(1–2):61–80.
136. Horst HJ, Holtje WJ, Dennis M, Coert H, Geelen J, Voigt KD. Lymphatic absorption and metabolism of orally administered testosterone undecanoate in man. Klin Wochenschr 1976; 54(18):875–879.
137. Gallo-Torres HE. Intestinal absorption and lymphatic transport of d,1-3,4-3H2-a-tocopheryl nicotinate in the rat. Int Z Vitaminforsch 1970; 40(4):505–514.
138. Sugihara J, Furuuchi S. Lymphatic absorption of hypolipidemic compound, 1-O-[p-(myristyloxy)-alpha-methylcinnamoyl] glycerol (LK-903). J Pharmacobiodyn 1988; 11(2):121–130.
139. Sugihara J, Furuuchi S, Ando H, Takashima K, Harigaya S. Studies on intestinal lymphatic absorption of drugs. II. Glyceride prodrugs for improving lymphatic absorption of naproxen and nicotinic acid. J Pharmacobiodyn 1988; 11(8):555–562.
140. Sugihara J, Furuuchi S, Nakano K, Harigaya S. Studies on intestinal lymphatic absorption of drugs. I. Lymphatic absorption of alkyl ester derivatives and alpha-monoglyceride derivatives of drugs. J Pharmacobiodyn 1988; 11(5):369–376.
141. Adachi I, Liu HX, Horikoshi I, Ueno M, Sato H. Possibility of lymphatic absorption of epidermal growth factor from intestine. Yakugaku Zasshi 1993; 113(3):256–263.
142. Grimus RC, Schuster I. The role of the lymphatic transport in the enteral absorption of naftifine by the rat. Xenobiotica 1984; 14(4):287–294.
143. Ichihashi T, Kinoshita H, Yamada H. Absorption and disposition of epithiosteroids in rats. 2. Avoidance of first-pass metabolism of mepitiostane by lymphatic absorption. Xenobiotica 1991; 21(7):873–880.
144. Ichihashi T, Kinoshita H, Takagishi Y, Yamada H. Intrinsic lymphatic partition rate of mepitiostane, epitiostanol, and oleic acid absorbed from rat intestine. Pharm Res 1991; 8(10):1302–1306.
145. Ichihashi T, Kinoshita H, Takagishi Y, Yamada H. Effect of bile on absorption of mepitiostane by the lymphatic system in rats. J Pharm Pharmacol 1992; 44(7):565–569.
146. Ichihashi T, Kinoshita H, Takagishi Y, Yamada H. Effect of oily vehicles on absorption of mepitiostane by the lymphatic system in rats. J Pharm Pharmacol 1992; 44(7):560–564.
147. Porter CJ, Charman SA, Charman WN. Lymphatic transport of halofantrine in the triple-cannulated anesthetized rat model: effect of lipid vehicle dispersion. J Pharm Sci 1996; 85(4):351–356.
148. Humberstone AJ, Porter CJ, Charman WN. A physicochemical basis for the effect of food on the absolute oral bioavailability of halofantrine. J Pharm Sci 1996; 85(5):525–529.
149. Wasan KM, Ranaswamy M, McIntosh MP, Porter CJ, Charman WN. Differences in the lipoprotein distribution of halofantrine are regulated by lipoprotein apolar lipid and protein concentration and lipid transfer protein I activity: in vitro studies in normolipidemic and dyslipidemic human plasmas. J Pharm Sci 1999; 88(2):185–190.
150. Porter CJ, Charman SA, Numberstone AJ, Charman WN. Lymphatic transport of halofantrine in the conscious rat when administered as either the free base or the hydrochloride salt: effect of lipid class and lipid vehicle dispersion. J Pharm Sci 1996; 85(4):357–361.
151. Porter CJ, Charman WN. Intestinal lymphatic drug transport; an update. Adv Drug Deliv Rev 2001; 50(1,2):61–80.
152. Porter CJ. Drug delivery to the lymphatic system. Crit Rev Ther Drug Carrier Syst 1997; 14(4):333–393.

153. Hauss D, Fogal S, Ficorilli J. Chronic collection of mesenteric lymph from conscious, tethered rats. Contemp Top Lab Anim Sci 1998; 37(3):56–58.
154. Kararli TT. Comparison of the gastrointestinal anatomy, physiology, and biochemistry of humans and commonly used laboratory animals. Biopharm Drug Dispos 1995; 16(5): 351–380.
155. Levet-Trafit B, Gruyer MS, Marianovic M, Chou RC. Estimation of oral drug absorption in man based on intestine permeability in rats. Life Sci 1996; 58(24):PL359–PL363.
156. McDonald GB, Sunders DR, Weidman M, Fisher I. Portal venous transport of long-chain fatty acids absorbed from rat intestine. Am J Physiol 1980; 239(3):G141–G150.
157. Liu HX, Adachi I, Horikoshi I, Ueno M. Promotion of intestinal drug absorption by milk fat globule membrane. Yakugaku Zasshi 1991; 111(9):510–514.
158. Sato H, Liu HX, Adachi I, Ueno M, Lemaire M, Horikoshi I. Enhancement of the intestinal absorption of a cyclosporine derivative by milk fat globule membrane. Biol Pharm Bull 1994; 17(11):1526–1528.
159. Reininger EJ, Sapirstein LA. Effect of digestion on distribution of blood flow in the rat. Science 1957; 126(3284):1176.
160. Wasan KM. The role of lymphatic transport in enhancing oral protein and peptide drug delivery. Drug Dev Ind Pharm 2002; 28(9):1047–1058.
161. Sieber SM. The lymphocytic absorption of p,p'-DDT and some structurally-related compounds in the rat. Pharmacology 1976; 14(5):443–454.
162. Charman WN, Stella VJ. Estimating the maximal potential for intestinal lyphatic transport of lipophilic drug molecules. Int J Pharm 1986; (34):175–178.
163. Humberstone AJ, Porter CJ, Charman WN. A physicochemical basis for the effect of food on the absolute oral bioavailability of halofanturine. J Pharm Sci 1996; 85(5): 525–529.
164. Porter CJ, Caliph SM, Charman WN. Differences in pre- and post-prandial plasma lipid profiles affect the extraction efficiency of a model highly lipophilic drug from beagle dog plasma. J Pharm Biomed Anal 1997; 16(1):175–180.
165. Humberstone AJ. Association of halofantrine with postprandially derived plasma lipoproteins decreases its clearance relative to administration in the fasted state. J Pharm Sci 1998; 87(8):936–942.
166. Khoo SM, Porter CJ, Charman WN. The formulation of Halofantrine as either non-solubilizing PEG 6000 or solubilizing lipid based solid dispersions: physical stability and absolute bioavailability assessment. Int J Pharm 2000; 205(1–2):65–78.
167. Gao P, Rush BD, Pfund WP, et al. Development of a supersaturable SEDDS (S-SEDDS) formulation of paclitaxel with improved oral bioavailability. J Pharm Sci 2003; 92(12):2386–2398.
168. Khoo SM, Edwards GA, Porter CJ, Charman WN. A conscious dog model for assessing the absorption, enterocyte-based metabolism, and intestinal lymphatic transport of halofantrine. J Pharm Sci 2001; 90(10):1599–1607.
169. Holm R, Porter CJ, Edwards GA, Mullertz A, Kristensen HG, Charman WN. Examination of oral absorption and lymphatic transport of halofantrine in a triple-cannulated canine model after administration in self-microemulsifying drug delivery systems (SMEDDS) containing structured triglycerides. Eur J Pharm Sci 2003; 20(1):91–97.
170. Khoo SM, Shackleford DM, Porter CJ, Edwards GA, Charman WN. Intestinal lymphatic transport of halofantrine occurs after oral administration of a unit-dose lipid-based formulation to fasted dogs. Pharm Res 2003; 20(9):1460–1465.
171. Taillardat-Bertschinger A, Perry CS, Galland A, Pankerd RJ, Charman WN. Partitioning of halofantrine hydrochloride between water, micellar solutions, and soybean oil: Effects on its apparent ionization constant. J Pharm Sci 2003; 92(11):2217–2228.
172. Khoo SM, Porter JH, Edwards GA, Charman WN. Metabolism of halofantrine to its equipotent metabolite, desbutylhalofantrine, is decreased when orally administered with ketoconazole. J Pharm Sci 1998; 87(12):1538–1541.

173. Milton KA, Edwards G, Ward SA, Orme ML, Breckenridge AM. Pharmacokinetics of halofantrine in man: effects of food and dose size. Br J Clin Pharmacol 1989; 28(1): 71–77.
174. Khoo S, Humberstone AJ, Porter JH, Edwards GA, Charman WN. Formulation design and bioavailability assessment of lipidic self-emulsifying formulations of Halofantrine. Int J Pharm 1998; 167:155–164.
175. Lennernäs H. Human intestinal permeability. J Pharm Sci 1998; 87(4):403–410.
176. Hörter D, Dressman JB. Influence of physicochemical properties on dissolution of drugs in the gastrointestinal tract. Adv Drug Del Rev 1997; 25(1):3–14.
177. Fuse K, Bamba T, Hosoda S. Effects of pectin on fatty acid and glucose absorption and on thickness of unstirred water layer in rat and human intestine. Dig Dis Sci 1989; 34(7):1109–1116.
178. Johnson IT, Gee JM. Effect of gel-forming gums on the intestinal unstirred layer and sugar transport in vitro. Gut 1981; 22(5):398–403.
179. Cerda JJ, Robbins FL, Byugin CW, Gerencser GA. Unstirred water layers in rabbit intestine: effects of guar gum. J Parenter Enteral Nutr 1987; 11(1):63–66.
180. Huang JD. Role of unstirred water layer in the exsorption of quinidine. J Pharm Pharmacol 1990; 42(6):435–437.
181. Davis GR, Santa Ana GA, Morawski SG, Fordtran JS. Permeability characteristics of human jejunum, ileum, proximal colon and distal colon: results of potential difference measurements and unidirectional fluxes. Gastroenterology 1982; 83(4):844–850.
182. Wang W. Oral protein delivery. J. Drug Target 1996; 4(4):195–232.
183. Yu L, Bridgers A, Polli J, et al. Vitamin E–TPGS increases absorption flux of an HIV protease inhibitor by enhancing its solubility and permeability. Pharm Res 1999; 16(12): 1812–1817.
184. Woo JS, Lee CH, Shim CR, Hwang SJ. Enhanced oral bioavailability of paclitaxel by coadministration of the P-glycoprotein inhibitor KR30031. Pharm Res 2003; 20(1): 24–30.
185. Barthe L, Woodley J, Houin G. Gastrointestinal absorption of drugs: methods and studies. Fundam Clin Pharmacol 1999; 13(2):154–168.
186. Thummel KE, Kunze KL, Shen DD. Enzyme-catalyzed processes of first-pass hepatic and intestinal drug extraction. Adv Drug Deliv Rev 1997; 27(2,3):99–127.
187. Thummel KE. Preface. Adv Drug Deliv Rev 1997; 27(2,3):97–98.
188. Reymond JP, Sucker H. In vitro model for cyclosporin intestinal absorption in lipid vehicles. Pharm Res 1988; 5(10):673–676.
189. Behrens D, Fricker R, Bodoky A, Drewe J, Harder F, Heberer M. Comparison of cyclosporin A absorption from LCT and MCT solutions following intrajejunal administration in conscious dogs. J Pharm Sci 1996; 85(6):666–668.
190. Frey FJ, Horber FF, Frey BM. Trough levels and concentration time curves of cyclosporine in patients undergoing renal transplantation. Clin Pharmacol Ther 1988; 43(1):55–62.
191. Ptachcinski RJ, Burckart GJ, Venkataramanan R. Cyclosporine concentration determinations for monitoring and pharmacokinetic studies. J Clin Pharmacol 1986; 26(5): 358–366.
192. Gupta SK, Manfro RC, Tomlanorich SJ, Gamberhoglio JG, Garovoy MR, Benet LZ. Effect of food on the pharmacokinetics of cyclosporine in healthy subjects following oral and intravenous administration. J Clin Pharmacol 1990; 30(7):643–653.
193. Ritschel WA. Microemulsions for improved peptide absorption from the gastrointestinal tract. Methods Find Exp Clin Pharmacol 1991; 13(3):205–220.
194. Ritschel WA, Adolph S, Ritschel GB, Schroeder T. Improvement of peroral absorption of cyclosporine A by microemulsions. Methods Find Exp Clin Pharmacol 1990; 12(2):127–134.
195. Drewe J, Meier R, Vonderscher J, et al. Enhancement of the oral absorption of cyclosporin in man. Br J Clin Pharmacol 1992; 34(1):60–64.

196. Bojrup M, Qi Z, Bjorkman S, et al. Bioavailability of cyclosporine in rats after intragastric administration: a comparative study of the L2-phase and two other lipid-based vehicles. Transpl Immunol 1996; 4(4):313–317.
197. Mueller EA, Kovarik JM, van Bree JB, Grevel J, Lucker PW, Kutz K. Influence of a fat-rich meal on the pharmacokinetics of a new oral formulation of cyclosporine in a crossover comparison with the market formulation. Pharm Res 1994; 11(1):151–155.
198. Kovarik JM, Mueller EA, van Bree JB, Tetzloff W, Kuktz K. Reduced inter- and intraindividual variability in cyclosporine pharmacokinetics from a microemulsion formulation. J Pharm Sci 1994; 83(3):444–446.
199. Trull AK, Tan KK, Tan L, Alexander GJ, Jamieson NV. Absorption of cyclosporin from conventional and new microemulsion oral formulations in liver transplant recipients with external biliary diversion. Br J Clin Pharmacol 1995; 39(6):627–631.
200. Mueller EA, Kovarik JM, van Bree JB, Tetzloff W, Grevel J, Kutz K. Improved dose linearity of cyclosporine pharmacokinetics from a microemulsion formulation. Pharm Res 1994; 11(2):301–304.
201. Rowinsky EK, Cazenave LA, Donehower RC. Taxol: a novel investigational antimicrotubule agent. J Natl Cancer Inst 1990; 82(15):1247–1259.
202. Rowinsky EK, Donehower RC. Taxol: twenty years later, the story unfolds. J Natl Cancer Inst 1991; 83(24):1778–1781.

14
Recent Advances in Heparin Delivery

Nathalie Ubrich and Philippe Maincent
INSERM U734–EA 3452, Laboratoire de Pharmacie Galénique, Faculté de Pharmacie, Nancy, Cedex, France

INTRODUCTION

Heparins are the standards of anticoagulants used in the prophylaxis and the treatment of deep vein thrombosis and pulmonary embolism (1–4). Currently, their most common indications are in the prevention of venous clots following orthopedic, pelvic, or abdominal surgery as well as angioplasty or heart surgery. This prophylactic therapy is recommended for at least 10 to 14 days after surgery and represents a huge pharmaceutical market worldwide.

Heparins are polyanionic glycosaminoglycans, which are heterogeneous about molecular size, anticoagulant activity, and pharmacokinetic properties (5,6). Their molecular weight ranges from 1000 to 30,000 Da, with a mean of 15,000 and 4500 Da for unfractionated heparins (UFH) and low molecular weight heparins (LMWH), respectively. Their anticoagulant feature and clearance are influenced by their chain length: the higher-molecular weight species are cleared from the circulation more rapidly than the lower-molecular weight species. This differential clearance explains the longer half-life of LMWH derived from UFH by either chemical or enzymatic depolymerization to yield fragments that are approximately one-third the size of UFH.

For many years, LMWH tended to replace UFH for many indications in most countries (7). Indeed, LMWH potentially have significant advantages over UFH and oral anticoagulants for the prevention and the treatment of thromboembolic diseases. LMWH provide higher subcutaneous bioavailabilities and longer half-lives which reduce the injections to one or two daily administrations. They do not require monitoring of the coagulation level, owing to their more predictable anticoagulant response, and hospitalization as the treatment can be initiated or completed in the outpatient setting without a nurse and the risk of bleeding complications. These are the major advantages that can reduce the cost in health care (8). In addition, LMWH have clinical and practical advantages over UFH in pregnant women in improved safety (lower incidence of osteoporosis, thrombocytopenia), and possibly, allergic skin reactions and pregnancy outcome in women with the antiphospholipid syndrome (9,10).

However, one of the drawbacks of both UFH and LMWH in their clinical use is their parenteral administration owing to their lack of oral absorption, presumably, because of their negative charge and/or molecular size. Indeed, various barriers to oral absorption of peptides, proteins, and polar macromolecular drugs include physicochemical parameters such as size, surface charge and aqueous solubility, degradation from pH effects and/or the action of digestive or metabolizing enzymes (11). Several approaches including the use of pharmaceutical adjuvants, chemical modifications, emulsions, micellar solutions, polymeric particulate systems, proteases inhibitors, or absorption enhancers have been applied to improve the oral absorption of macromolecules (12–17).

An oral heparin formulation would be an important discovery in the care of patients, as the oral administration is the physiological route that is most well accepted and tolerated by patients. For many years, several attempts to develop oral heparin formulations have been reported in the literature. After a brief description of the main mechanisms of the coagulation and the action of heparin, this chapter focuses on the current challenges, mainly absorption enhancers and drug delivery systems based on encapsulation, that are used to improve the oral absorption of heparins.

BLOOD AND MECHANISM OF ACTION OF HEPARINS

The coagulation cascade is defined as a series of linked, proteolytic reactions leading to the generation of thrombin. Briefly, under physiological conditions, the coagulation is triggered by the exposure of tissue factor (TF) to blood flow. This is followed by the activation of factor X to form factor Xa. This proteolysis results from the action of the TF-VIIa complex or the phospholipids-IXa-VIIIa-calcium. When TF is in excess, factor X activation results from the TF-VIIa complex, whereas it depends on IXa factor when the amount of TF is limited. Factor Xa triggers thereafter the conversion of prothrombin to thrombin (Fig. 1). Once thrombin is formed, two pairs of fibrinopeptides, released from fibrinogen, polymerize to form a clot. If this process occurs within the vasculature, it leads to thrombosis.

The addition of heparin exogenously results in its interaction with the plasma cofactor, antithrombin (AT), through the well-defined pentasaccharide sequence, to

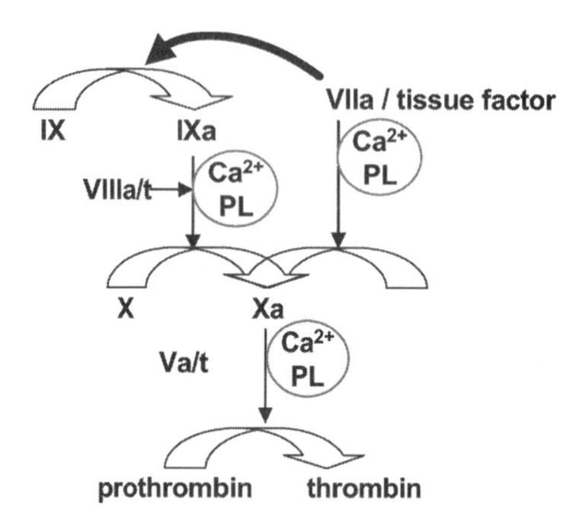

Figure 1 The main steps of blood coagulation.

exercise its anticoagulant activity. Once heparin is bound to AT, a conformational change in AT occurs, thereby converting AT from a slow to a rapid inhibitor. The heparin–AT complex inactivates a number of coagulation enzymes including thrombin (IIa) and factors IXa, Xa, XIa, and XIIa, with a higher sensitivity to thrombin and factor Xa.

EVALUATION OF HEPARIN EFFICIENCY

To evaluate the absorption of heparin after intradigestive administration, various methods are available. Most of them are based on the biological activity of heparin detected in biological fluids. Methods used reflect mainly the ability of heparin to inhibit factor IIa or Xa and are based on (i) the activity of the coagulation enzymes, expressed in anti-IIa or anti-Xa IU, causing the hydrolysis of the substrate by factor IIa or Xa followed by a change in color (chromogenic substrate) or fluorescence (fluorescent substrate) which is proportional to factor IIa or Xa concentration, (ii) the clotting time determined by the activated partial thromboplastin time (APTT). This chronometric method corresponds to the clotting time of blood plasma recalcified in the presence of cephalin (a phospholipidic platelet substitute) and an activator of contact factors, and (iii) the prothrombin time exploring the extrinsic coagulation pathway. It determines the clotting time of blood plasma recalcified in the presence of thromboplastin. All these biological methods are used both in humans and in animals.

Another biological method that is based on the variation of the size of a thrombus is tested in animals. It does not quantify the biological activity of heparin, but rather demonstrates the pharmacological action of heparin after its administration. Indeed, arterial or venous thrombosis models in animals are induced with perfusion chambers or microsurgical sutures to evaluate the efficiency of thrombolytic agents correlated with the morphological change or dissolution of the thrombus (18,19).

The results obtained from these methods are mainly related to the amount of coagulation factors presenting inter- and intraspecies variabilities as mentioned in Table 1 (20). Moreover, by using a clearly defined biological test, differences may be observed according to the substrate, activator, type of tube, or TF used. Indeed, the time values with the APTT test differ significantly according to the activator of contact factors used (Kaolin, silica). Also, the clotting time is significantly different when experiments are performed in glass or siliconized tubes. So, it is of prime importance to specify accurately the conditions of the test employed.

ORAL DELIVERY OF HEPARINS

Once drugs are orally administered, their absorption depends on various parameters, mainly their solubility in the gastrointestinal fluid, size, and charge as well as pH and enzyme environment throughout the gastrointestinal tract (GIT). Crossing the intestinal wall is possible via three main mechanisms, namely, diffusion by passive transport, by facilitated transcellular diffusion through the lipophilic absorptive cells, and by active carrier mediated transport systems or transcytosis. In addition to these three main transport processes, the absorbed moieties often undergo efflux from the cells, largely because of the action of a membrane transporter, the P-glycoprotein (21).

The small intestine is the major digestive and absorptive site of the GIT owing to its extended surface area through the circular folds of about 10 mm height and the microvilli of the absorptive cells of the mucosa.

Table 1 Coagulation Factors and Tests (APTT, Prothrombin Time with Simplastin and Clotting Time in Glass Tubes) in Some Vertebrate Animals for Normal Human Range

	Rabbit ($n=6$) (New Zealand)	Monkey ($n=9$)	Rat ($n=6$) (Wistar)	Dog ($n=6$)	Pig ($n=6$)	Human (normal range)
Factor						
I (mg/dL)	196 ± 27	305 ± 107	164 ± 16	215 ± 50	348 ± 78	150–450
II (U/mL)	2.40 ± 0.30	1.14 ± 0.10	0.45 ± 0.09	1.43 ± 0.30	0.68 ± 0.30	0.70–1.30
V (U/mL)	100 +	1.05 ± 0.10	2.82 ± 0.90	6.02 ± 0.70	2.80 ± 0.60	0.65–1.45
VII (U/mL)	3.00 ± 0.80	1.85 ± 0.30	3.91 ± 0.90	4.70 ± 1.40	1.92 ± 0.40	0.50–1.30
X (U/mL)	1.80 ± 0.50	0.80 ± 0.20	0.30 ± 0.07	2.18 ± 0.40	2.54 ± 0.90	0.75–1.25
VIII (U/mL)	3.20 ± 0.80	1.09 ± 0.20	1.74 ± 0.70	4.49 ± 1.00	6.42 ± 0.80	0.75–1.40
IX (U/mL)	1.20 ± 0.50	1.01 ± 0.20	0.77 ± 0.20	1.12 ± 0.40	3.20 ± 0.30	0.65–1.70
XI (U/mL)	2.50 ± 0.70	1.00 ± 0.40	0.63 ± 0.20	2.79 ± 0.60	1.67 ± 0.30	0.75–1.40
XII (U/mL)	0.80 ± 0.20	0.93 ± 0.40	1.28 ± 0.20	1.06 ± 0.20	5.65 ± 1.40	0.50–1.45
Fletcher (U/mL)	0.03 ± 0.01	–	1.26 ± 0.30	0.32 ± 0.10	2.00 ± 0.50	0.50–1.50
Fitzgerald (U/mL)	0.05 ± 0.03	–	0.30 ± 0.08	0.30 ± 0.05	3.05 ± 0.30	0.50–1.50
Antithrombin III (U/mL)	0.80 ± 0.10	0.79 ± 0.20	1.00 ± 0.07	0.87 ± 0.08	1.09 ± 0.20	0.80–1.20
Test						
Prothrombin time (sec)	7.5 ± 1.5	11.5 ± 1.4	10.1 ± 0.10	7.60 ± 1.00	12.6 ± 1.10	10–13
Simplastin APTT (sec)	32.8 ± 4.5	35.5 ± 2.9	18.8 ± 1.80	12.3 ± 1.10	23.0 ± 5.40	24–34
Clotting time (min) Glass	4 ± 1.70	5 ± 1.3	4 ± 0.80	4 ± 0.70	6 ± 3.80	6–12

Abbreviation: APTT, activated partial thromboplastin time.
Source: Adapted from Ref. 20.

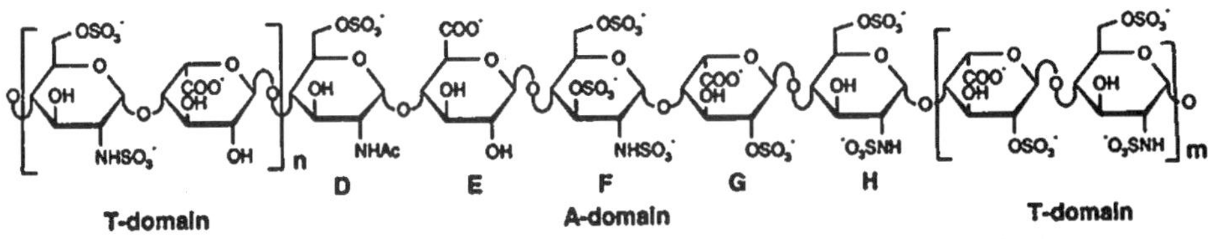

Figure 2 Structure of an anticoagulant heparin.

Owing to their large molecular size and negative charge conferred by sulfate and carboxylate groups, which hamper their absorption, several pharmaceutical approaches were evaluated for maximizing the oral absorption of heparins (Fig. 2) (22). Among them, the coadministration of absorption enhancers and the design of pharmaceutical formulations, such as liposomes, micelles, microparticles, and nanoparticles, have shown an increased permeability through the intestinal wall.

Intestinal Absorption of Heparins Associated with Absorption Enhancers

One possible way to improve the oral absorption of heparins is to use absorption enhancers that are either coadministered with heparins or linked to them. Absorption enhancers may be defined as formulation components that temporarily disrupt the intestinal wall by increasing the membrane fluidity after solubilization of membrane phospholipids, decreasing the mucus viscosity, or opening tight junctions, thus improving the permeation of the drugs (11). In contrast, the mechanism is not always very well understood.

Heparin Complexes

Early research works describing a slight gastrointestinal absorption of heparin were mostly focused on its complexation with sodium salt of ethylene diaminetetraacetic acid (EDTA), sulfonated surfactants, and amino acids (23–26). However, although these adjuvants were successful in slightly improving the oral or intestinal absorption of heparin, they have some undesirable effects like a mucosal damage of the intestine. A simultaneous administration of heparin and organic acids also led to an increase of heparin absorption after oral administration in mice, but again to a low extent (27). Indeed, organic acids may suppress the ionization of the drug by complexation, thus facilitating its permeation through the intestinal wall. However, after neutralization of heparin with spermine or lysine no oral absorption was observed in rats (28).

Synthetic Absorption Enhancers

The American company Emisphere Technologies synthesized a family of novel compounds that are effective in promoting the oral absorption of various drugs that are not available for oral administration when used alone (29–31). As part of their research program, they focused also on the oral absorption of heparin and developed new delivery agents based on non-α-amino acids that allow the absorption of heparin in rats and monkeys (32–34). Numerous compounds were synthesized and tested in vivo. The results indicated that heparin absorption depends on the structure of these compounds. Based on structure–activity relationship, the optimal absorption was achieved with an acid length of seven to nine methylene units and with 2-hydroxybenzamide that led to the subsequent discovery of sodium *N*-[8-(2-hydroxybenzoyl)amino]caprylate (SNAC), which is a novel delivery agent for unfractionated heparin. When the delivery agent was combined with heparin and administered as a single dose orally or intracolonically in an aqueous propylene glycol solution in rats and monkeys, a dramatic increase in both plasma heparin concentration (measured by the antifactor Xa activity) and clotting time (evaluated by the APTT) was observed. A 5- to 10-fold improvement in APTT was obtained compared to the baseline. An estimated relative bioavailability of 8.3% (at a heparin and delivery agent dose of 15 and 150 mg/kg, respectively) to 39% (15 mg/kg heparin and 300 mg/kg delivery agent) can be achieved following oral administration in monkeys, the calculations being based on APTT after the comparison of the areas under the curves of the oral and the subcutaneous doses (Fig. 3) (33,35). Such differences in bioavailabilities show the dramatic and critical influence of the amount of SNAC on heparin absorption. These results were also confirmed after a single oral administration of the combination of SNAC/heparin in a rat venous thrombosis model, involving a reduction of the mean thrombus weight and a significant increase in APTT values compared to animals receiving oral heparin or oral SNAC alone (36,37). It was demonstrated from the data after oral administration of SNAC combined with heparin in rabbits undergoing iliac artery balloon dilatation or endovascular stent implantation that heparin is effective to prevent restenosis by inhibiting neointimal growth and hyperplasia (38). Another delivery agent, sodium *N*-[10-(2-hydroxybenzoyl)amino]decanoate (SNAD), that facilitates the gastrointestinal absorption of LMWH was also specifically synthesized. When mixed together, SNAD allowed efficient oral absorption of LMWH from aqueous solutions and tablets in rats, dogs, and monkeys (39). Moreover, the combination of oral LMWH and SNAD prevents deep vein thrombosis: a decrease in thrombus weight, indicated it was almost completely dissolved and an increase of plasma levels of antifactor Xa were observed in a vena caval thrombosis in a porcine model and a jugular venous thrombosis in the rat (40,41). It was suggested that SNAC and SNAD interact with heparins forming noncovalent complexes, which are more lipophilic than heparin alone, allowing the passive diffusion of these complexes across the gastrointestinal mucosa via the transcellular pathway. Once in the bloodstream, the delivery agents dissociate from the heparins that prevent their ability to effect anticoagulation (42). However, the toxicological issues of these carriers must be raised. These efficient oral heparin absorption results in animals after the combination of UFH with SNAC and LMWH with SNAD led Emisphere to human trials validation (43).

Similar results demonstrated the effectiveness of SNAC as an oral heparin absorption enhancer after oral administration in healthy volunteers with escalating tolerated doses of heparin ranging from 30,000 to 150,000 IU (44). Results from the Phase I and II studies indicate that significant amounts of heparin are being orally delivered as measured by coagulation parameters such as APTT, antifactors IIa and Xa, and TFPI. Results of the Phase II trial showed that the oral formulations were comparable to injectable heparin in both safety and activity when used to prevent venous thromboembolism following elective hip arthroplasty (45). A Phase III clinical trial of liquid oral heparin started in December 1999 with the objective to demonstrate the safety and superior efficacy of oral heparin dosed for 30 days postoperatively following total hip replacement surgery, compared to Lovenox® (enoxaparin) from Aventis (46). However, owing to the lack of superiority over enoxaparin, Emisphere decided to stop the development of its oral heparin solution. However,

(A)

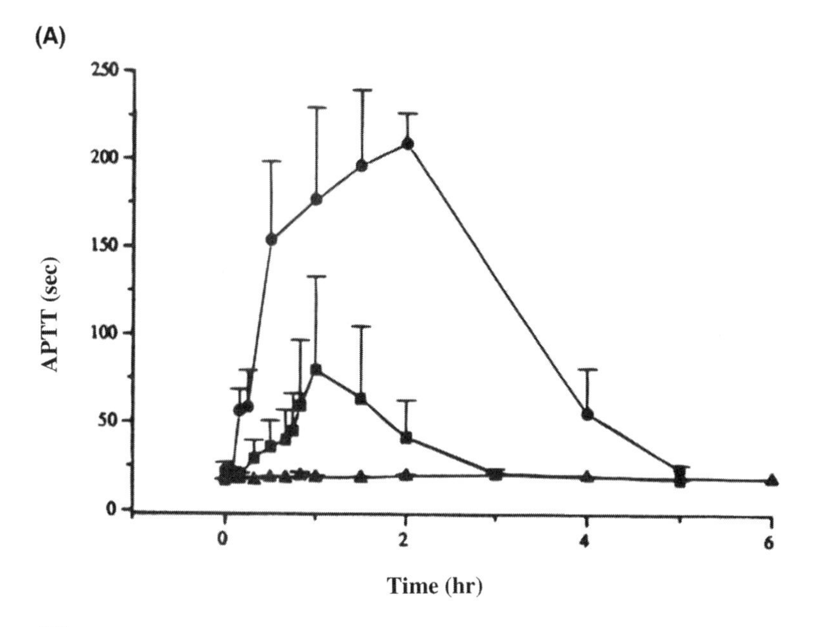

(B)

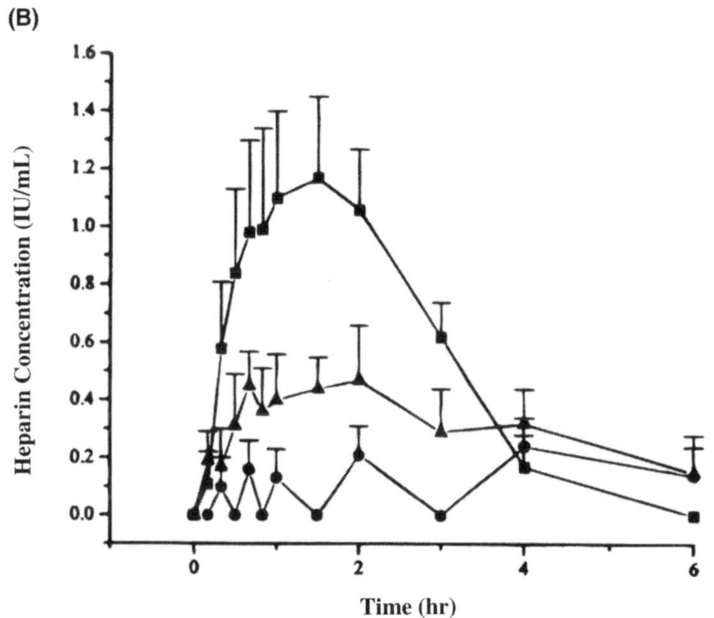

Figure 3 (**A**) Pharmacodynamic profile following the oral administration of SNAC in combination with varying heparin doses in cynomolgus monkeys. (**B**) Pharmacokinetic profile following the oral administration of SNAC in combination with varying heparin doses in cynomolgus monkeys. Squares represent 150 mg/kg SNAC in combination with 30 mg/kg (4971 IU/kg) heparin. Circles represent 100 mg/kg (16570 IU/kg) heparin. Triangles represent 150 mg/kg SNAC in combination with 15 mg/kg (2486 IU/kg) heparin. The data are plotted as mean ± SEM. *Abbreviations*: SNAC, sodium *N*-[8-(2-hydroxybenzoyl)amino] caprylate. *Source*: From Ref. 32.

further research works are still ongoing with other macromolecules including peptides and proteins.

More recently, the surfactant Labrasol, which contains saturated polyglycolyzed C_6–C_{14} glycerides consisting of a mixture of mono-, di-, and triglycerides of C_8 and C_{10} fatty acids, mono- and diesters of poly(ethylene glycol) (PEG), and free PEG 400 was found to enhance the intestinal absorption of LMWH (200 IU/kg) combined with Labrasol (50 mg/kg) in rats after in situ administration to the small intestine (47). As evidenced by the plasma antifactor Xa levels (peak concentration 0.50 ± 0.03 IU/mL), the jejunum was found to be the best site for Labrasol-enhanced absorption of LMWH, compared to duodenum and ileum. The results indicated that Labrasol induced changes in the permeability of the intestinal epithelium in a transient and reversible manner. Also, by increasing the dose of Labrasol from 50 to 200 mg/kg for a similar dose of LMWH (200 IU/kg), decreased plasma anti-Xa levels were obtained, indicating the critical importance of the Labrasol-to-LMWH ratio to improve the intestinal absorption of LMWH in order to obtain a therapeutic effect.

Bile Acids

Bile acids are natural amphipathic molecules that form micelles above a critical micelle concentration. Studies carried out with some modified bile acids led the researchers to explore the opportunity of coupling drug entities to bile acids to enhance the oral bioavailability of drugs (48). New heparin derivatives were synthesized by coupling unfractionated heparin with several kinds of hydrophobic agents to increase the hydrophobic property of UFH (49). Among these UFH derivatives, UFH–DOCA, the conjugate of UFH and deoxycholic acid (DOCA), synthesized by coupling the amino groups of heparin to the carboxylic group of DOCA, showed the highest absorption in the GIT after oral administration in rats (50). Significant increases in both clotting time and concentration of heparin–DOCA in the plasma were observed (Figs. 4 and 5) with the increase of dosage forms, without damage to the gastrointestinal wall.

Similar oral absorption of LMWH was observed after conjugation with DOCA and oral administration of the conjugate in rats (51). The pharmacokinetic results showed that the clotting time was in the range of the therapeutic window, which was 1.5 to 2.5 times the baseline, and the absolute bioavailability of this oral dosage form was 7.8%. UFH and LMWH combined with DOCA were found to show anticoagulant effect following oral administration in rats, and these conjugates could be used as new oral anticoagulant agents.

Polymeric Absorption Enhancers

Some polymers, such as chitosan and polyacrylates, show mucoadhesive properties and/or are able to favor the opening of the tight junctions, thus allowing the paracellular permeation of drugs. The efficacy of mono-*N*-carboxymethyl chitosan (MCC), a polyampholytic chitosan derivative, as an intestinal absorption enhancer of LMWH was demonstrated both in vitro on Caco-2 cell monolayers (by decreasing the transepithelial electrical resistance and by increasing the paracellular permeation of LMWH) and in vivo after intraduodenal administration in rats of the LMWH polymeric solution (7200 anti-Xa U/kg) (52). A delayed increase of the intestinal absorption of LMWH is observed with therapeutic anti-Xa levels extended for three hours after administration and subsequently sustained for three hours, indicating

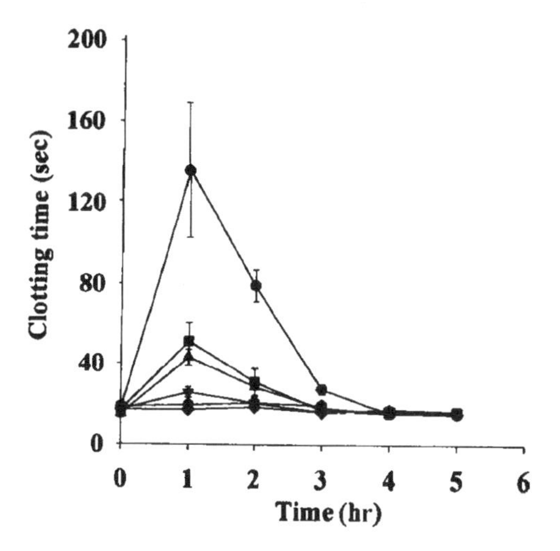

Figure 4 Clotting time profiles of heparin-DOCA after administering orally in rats. The clotting time was measured by APTT assay; raw heparin: (◆) 100 mg/kg, (⬣) the physical mixture of heparin (200 mg/kg) and DOCA (200 mg/kg); heparin-DOCA: (▼) 50 mg/kg, (▲) 80 mg/kg, (■) 100 mg/kg, (●) 200 mg/kg. The data are plotted as mean ± S.D., $n = 9$. *Abbreviations*: DOCA, deoxycholic acid; APTT, activated partial thromboplastin time. *Source*: From Ref. 50.

that carboxymethyl-modified chitosan should be the suitable polymeric carriers for effective oral delivery of anionic macromolecules. Another study was conducted to investigate the potential enhancing effect of the poly(acrylate) Carbopol® 934P (C934P) on the intestinal absorption of LMWH in rats and pigs (53). After intraduodenal administration of LMWH (10,800 and 5000 anti-Xa U/kg in rats and pigs,

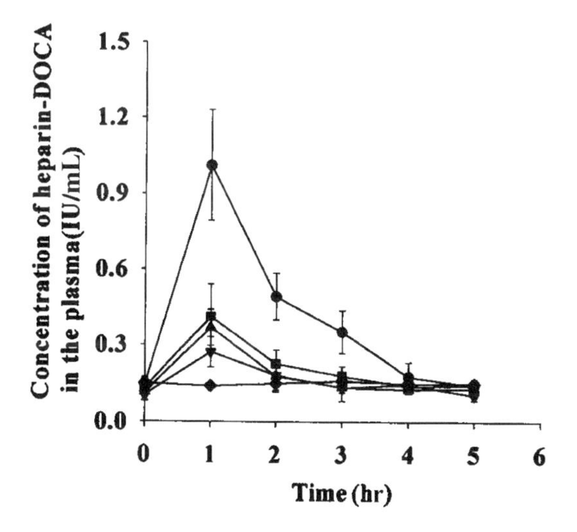

Figure 5 Change of the concentration profile of heparin-DOCA according to the dose amount after administering orally in rats. The concentration was measured by FXa chromogenic assay—raw heparin: (◆) 100 mg/kg; heparin-DOCA: (▼) 50 mg/kg, (▲) 80 mg/kg, (■) 100 mg/kg, (●) 200 mg/kg. The data are plotted as mean ± S.D., $n = 9$. *Abbreviations*: DOCA. deoxycholic acid. *Source*: From Ref. 50.

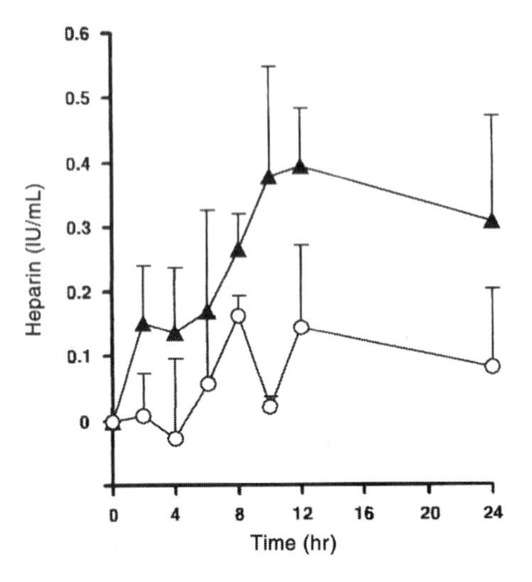

Figure 6 Comparison of the concentration profiles of heparin in blood after oral administration of heparin incorporated in tablets of poly(acrylic acid)-cysteine/GSH (▲) and in tablets comprised of unmodified poly(acrylic acid) (○). *Source*: From Ref. 54.

respectively) with C934P, significant plasma anti-Xa levels considered sufficient to prevent thrombi formation were obtained, indicating that the mucoadhesive C934P led to the increase of the intestinal absorption of LMWH. C934P has the advantage in that it can form a hydrogel with mucoadhesive properties useful for the design of oral dosage forms and is a Food and Drug Administration approved material. However, high doses of LMWH were administered to have therapeutic anticoagulant effect, which induces a significant loss of inactive heparin. More recently, thiolated polymers (thiomers) in combination with reduced glutathione were shown to improve the uptake of hydrophilic macromolecules from the GIT (54). The mechanism involving the absorption enhancement seems to be based on the thiol groups of the polymer that inhibit protein tyrosine phosphatase involved in the sealing process of tight junctions, via a GSH-mediated mechanism. An absolute bioavailability of $19.9 \pm 9.3\%$ LMWH was thereby obtained after oral administration of LMWH combined with poly(acrylic acid)-cysteine and GSH (Fig. 6).

Drug Delivery Systems

One of the most challenging and innovative fields in pharmaceutical sciences is the development of new drug delivery systems. Such new formulations would help a lot in formulating many peptides, proteins, and other macromolecules using biotechnologies. The gold key would be delivery systems allowing the oral absorption of biotech products. For many years, liposomes have been successfully developed mostly for intravenous administration of cytostatic drugs. Although promising results obtained by the oral route did not lead to the marketing of dosage forms, research is still ongoing with many drugs including heparins. Micelles are other alternative formulations, which can easily be manufactured; some interesting results have been described with heparins. Many drugs have also been encapsulated either in micro- or nanoparticles purposely increasing their bioavailability, prolonging their biological effect, or preserving their activity. Very promising results obtained by our research group will be reported below.

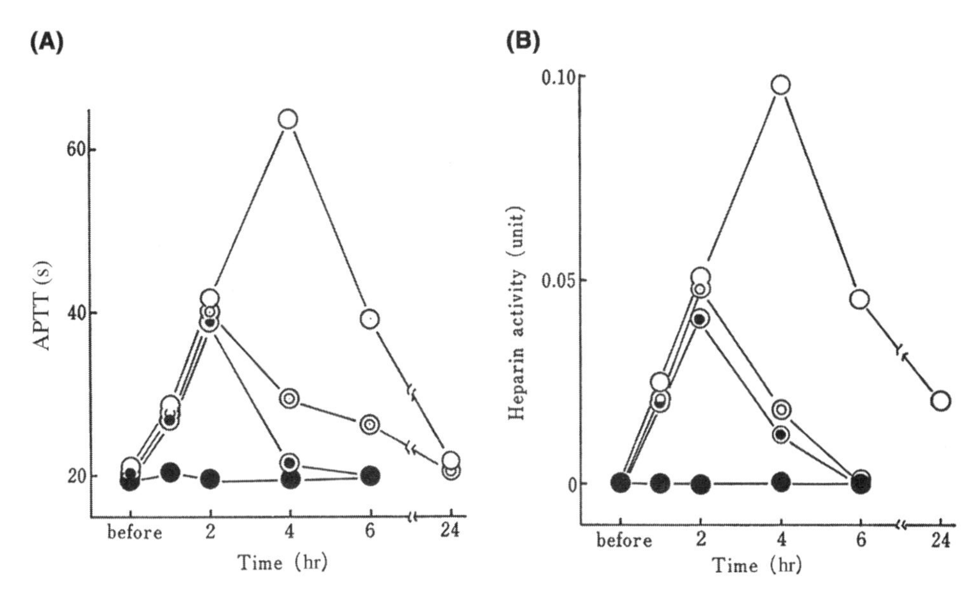

Figure 7 (**A**) Time dependences of APTT after oral administration of heparin in beagles. (**B**) Time dependences of heparin activity in the blood after oral administration of heparin in beagles. (○) Heparin (150,000 units) entrapped in liposomes, (◎) heparin (150,000 units) alone, (⊙) heparin (150,000 units) + normal saline containing liposomes, (●) normal saline containing liposomes as a control. *Abbreviation*: APTT, activated partial thromboplastin time. *Source*: From Ref. 55.

Liposomes

Liposomes, used as drug carriers, were loaded with heparin and administered orally to Beagle dogs in comparison with a control solution of heparin (55). The results were compared by measuring both the clotting time based on APTT and heparin activity in the blood: the prolongation in APTT was highest after oral administration of heparin entrapped in liposomes and reached 64 seconds after four hours, whereas a maximum in clotting time (40 seconds) was observed two hours after administration of heparin without liposomes (Fig. 7A). An increase in heparin activity showed similar time dependence with a peak concentration of 0.1 IU four hours after dosing of heparin-liposomes (Fig. 7B). Although high doses of heparin were administered orally to dogs (150,000 IU), the clotting time and the heparin activity evaluated in blood plasma did not reach peak values, and the duration of action did not last longer than when heparin was administered alone as a solution. Surprisingly, for a heparin aqueous solution, which is supposed not to be absorbed, the authors claimed a slight effect obtained after oral administration but with high doses. These results evidenced that heparin entrapped in liposomes was absorbed, although the extent of the absorption was small in spite of high doses administered.

Micelles

The use of monoolein-bile salts–mixed micelles appeared to be a suitable and safe adjuvant that was not harmful to the mucosal membrane and affected the gastrointestinal absorption of heparin to an appreciable extent after administration into loops of the small and the large intestine of rats (56). The lack of intestine permeation of heparin is mainly because of its high charge density, while its combination

Table 2 Plasma Anti-Xa Activity at Different Times After an Intraduodenal Administration of Four Heparin Salts at the Dose of 75 mg/kg of Heparinic Acid

	Anti-Xa (μg/mL)			
Product	0 min	36 min	60 min	120 min
ITF 300	0.0 ± 0.0	1.6 ± 0.4	4.1 ± 0.7	3.0 ± 0.8
ITF 331	0.0 ± 0.0	2.1 ± 0.4	3.1 ± 0.7	2.0 ± 0.8
ITF 1175	0.0 ± 0.0	1.4 ± 0.5	1.6 ± 0.4	0.3 ± 0.1
Sodium heparin	0.0 ± 0.0	0.0 ± 0.0	0.0 ± 0.0	1.2 ± 0.7

Note: Data are the mean values ± S.D. of seven administrations.
Source: From Ref. 59.

with lipophilic amines allows a reduction in the effective ionic charges by shielding the anionic charges of heparin and, consequently, leads to its neutralization (57). It has also been demonstrated that heparin, having multiple negative charges, immediately induces fusogenic behavior of cationic membranes composed of artificial lipids and natural phospholipids through electrostatic interactions between the membrane surface and heparin. The hydrophobic counterions tend to reduce such electrostatic interactions by ion-pairing effects on the anionic heparin, especially with diamine structure of the counterion, thus forming aggregates comparable to micelles. The unfractionated heparin–aminic counterion salt, ITF-300, induced plasma anti-Xa activity when administered orally to conscious dogs as enteric-coated pellets (58). In addition, when the heparin compounds containing diamine counterions, ITF-300 and ITF-331, were administered intraduodenally in rabbits, the heparin bioavailability, as demonstrated by the plasma anti-Xa activity, the APTT, and the thrombin time with peaks 60 minutes after administration, were markedly enhanced (Tables 2–4) (59). In contrast, with the heparin–monoamine salt, ITF-1175, in which the counterion is not a diamine but *N*,*N*-dibutylpropanolamine, the plasma anti-Xa activity was only slightly enhanced after intraduodenal administration. These experiments indicated that the diamine structure of the counterion seemed to be essential in the membrane permeability to favor heparin absorption and to obtain plasma anticoagulant activity. Moreover, the counterions of lipophilic diamines yield stable ion-pairs with the anionic heparin, enhancing the lipophilicity of the heparin salts and thus their permeation through the intestinal wall. Similar

Table 3 Plasma APTT in Rabbits at Different Times After Intraduodenal Administration of Four Heparin Salts at the Dose of 75 mg/kg of Heparinic Acid

	APTT ratio (time/basal time)		
Product	30 min	60 min	120 min
ITF 300	1.64 ± 0.12	> 3.70 ± 0.70	> 2.45 ± 0.49
ITF 331	1.65 ± 0.13	> 2.14 ± 0.64	1.69 ± 0.36
ITF 1175	1.66 ± 0.15	> 1.75 ± 0.14	1.18 ± 0.08
Sodium heparin	1.37 ± 0.21	1.16 ± 0.03	1.64 ± 0.42

Note: Data are the mean values ± S.D. of seven administrations.
Abbreviation: APTT, activated partial thromboplastin time.
Source: From Ref. 59.

Table 4 Plasma TT in Rabbits at Different Times After Intraduodenal Administration of Four Heparin Salts at the Dose of 75 mg/kg of Heparinic Acid

	TT ratio (time/basal time)		
Product	30 min	60 min	120 min
ITF 300	2.27 ± 0.49	>7.86 ± 1.54	>7.59 ± 1.68
ITF 331	>4.41 ± 1.66	>5.19 ± 1.81	>3.31 ± 1.53
ITF 1175	>2.99 ± 0.99	>2.93 ± 0.95	>1.52 ± 0.14
Sodium heparin	1.49 ± 0.31	1.11 ± 0.04	>4.17 ± 1.92

Note: Data are the mean values ± S.E. of seven administrations.
Source: From Ref. 59.

results were confirmed with low molecular weight heparin diamine salts after intraduodenal administration in rabbits (60). However, although the counterion technology was a good technology for enhancing heparins' bioavailability and permeability through biological membranes without affecting their activity, the acute toxicity of the counterions hampered their practical and clinical development.

Microparticles and Nanoparticles

Microencapsulation often provides an elegant way to protect, control the release, and enhance the oral bioavailability of poorly absorbable drug (61). Nanoencapsulation offers the same advantages as were demonstrated, for example, for insulin (62). Micro- and nanoencapsulation provide discrete particles, which differ mainly by their average diameter: it is considered that particles less than 1 μm diameter are called nanoparticles and greater than 1 μm diameter are called microparticles. In addition, both microparticles and nanoparticles can be of a matrix or a vesicular structure corresponding to microspheres (or nanospheres) and microcapsules (or nanocapsules), respectively. Many drugs have been encapsulated according to different techniques. The most popular ones are coacervation, heat denaturation or solvent evaporation after emulsification, ionotropic gelation, spray-drying, and many others. The selection of a particular technique depends mostly on the physicochemical characteristics of the drug of interest. The most critical parameter is the aqueous solubility, which directly influences the choice of the external continuous phase. For the encapsulation of large hydrophilic molecules, such as peptides or proteins, very few techniques, because of some very critical issues, can be used. Such critical issues include not only the yield and the encapsulation ratio, based on the high cost of some new entities, but also the high sensitivity to denaturation by solvents, heat, shear stress, etc. In the recent past, the technique of double emulsion was proposed for a successful encapsulation process of peptides and proteins. Indeed, this technique generally allows high encapsulation rates without denaturation of the peptide or protein of interest.

Another interesting technique, as it does not require organic solvent, is the thermal condensation of α-amino acids or acylated α-amino acids allowing the spontaneous formation of proteinoid microspheres under acidic conditions (63).

Proteinoid Microspheres. A slight absorption of heparin in rats and humans was demonstrated, as evidenced by an increase in APTT, following oral administration of heparin-loaded microspheres (63,64). By thermal condensation, mixtures of α-amino acids or acylated α-amino acids undergo spontaneous molecular self-assembly and form proteinoid microspheres. These proteinoids have a number of properties that make them attractive. Indeed, they are pH-dependent forming microspheres at a low pH and breaking up at high pH, allowing the delivery of the drug into the bloodstream without degradation. These microspheres are stable to enzymatic or acid conditions until the pH reaches the titration point during which the microspheres dissolve, releasing their content (65). However, in addition to a limited absorption of heparin after oral administration of these microspheres, a significant loss of heparin occurs during the manufacturing process, involving an economic loss in the event of scale-up.

Polymeric Carriers. The following section will review recent advances in the field of heparin delivery carried out by our research group in Nancy (France).

As heparin is a hydrophilic macromolecule (although less sensitive to issues such as stress or heat), it was decided to encapsulate this drug inside polymeric carriers according to the double emulsion technique, so far reserved for peptides/proteins. Thus heparin-loaded micro- and nanoparticles were prepared by the water-in-oil-in-water (w/o/w) emulsion and evaporation method.

The main characteristics of this technique when applied to the preparation of micro- and nanoparticles are presented in Figure 8.

Briefly, for heparin-loaded microparticles, 1 mL of aqueous heparin (either UFH or LMWH, 5000 IU/mL) was first emulsified by vigorous magnetic stirring for three minutes in methylene chloride containing the polymers (0.25 g). The resulting water-in-oil (w/o) emulsion was then poured into 1000 mL of a PVA aqueous

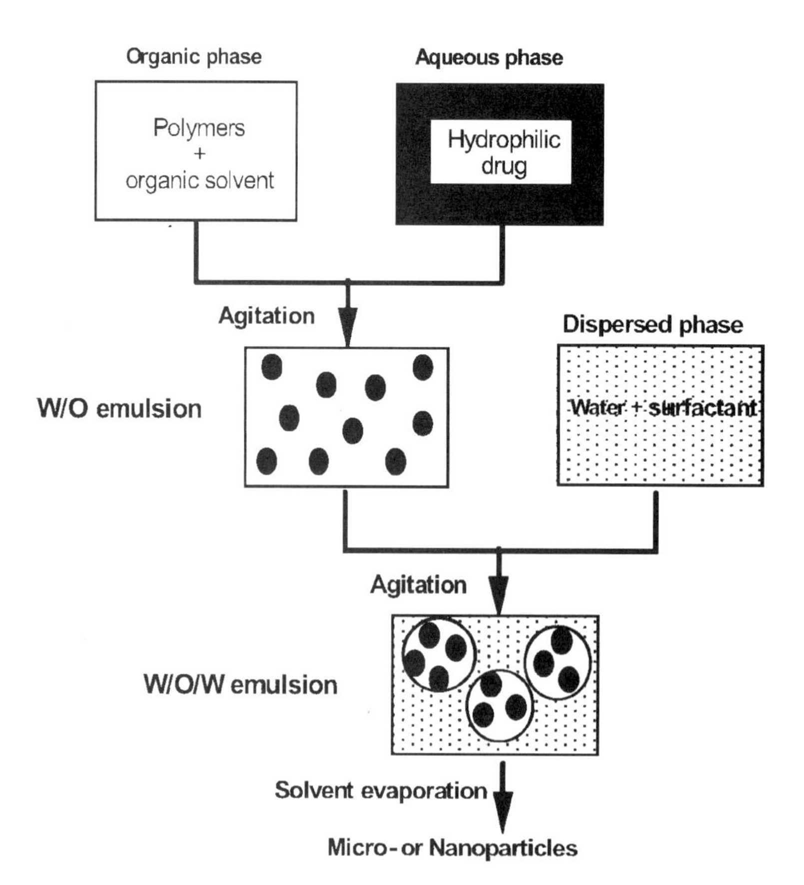

Figure 8 Flowchart of the micro- and nanoparticle preparation technique.

solution (0.1%). A w/o/w emulsion was formed by extensive stirring with a three-bladed propeller for two hours at room temperature until the organic solvent was totally removed. Upon solvent evaporation, the polymer precipitates, and the fmicroparticles core solidifies. Microparticles were then collected by filtration, washed with deionized water, and dried at room temperature. For nanoparticles, the technique was slightly modified. First, the w/o emulsion was homogenized with an ultrasound probe for 30 seconds at 60 W. Second, the first emulsion was added to 40 mL of a PVA aqueous solution (0.1%) and again homogenized by sonication for one minute at 60 W, involving the formation of the w/o/w emulsion. After evaporation of methylene chloride under reduced pressure for 15 minutes, nanoparticles were isolated by centrifugation at 45,000× *g* for 30 minutes. The supernatant was discarded, and nanoparticles were resuspended in deionized water (3 mL) and stored at 4°C (66–69).

The prepared micro- or nanoparticles were then characterized according to the state-of-the-art techniques. Colorimetric Azure II or nephelometric methods were used to measure drug entrapment efficiency indirectly from the external aqueous phase for UFH or LMWH, respectively. Correlations with the anti-Xa activity were also performed with a chromogenic substrate by using a standard kit. Micro- and nanoparticle sizes were measured by laser diffraction and photon correlation spectroscopies, respectively, and zeta-potential of nanoparticles was recorded with an electrophoretic technique. In vitro drug release was followed by suspending 50 mg of either nano- or microparticles in 20 ml of saline phosphate buffer medium (pH 7.4): at appropriate intervals, the amount of heparin (either UFH or LMWH) released in the medium was determined by either colorimetric or nephelometric method.

Besides the choice of a particular encapsulation technique, another important key feature of micro- and nanoencapsulation is the choice of a relevant polymer. Whereas for parenteral administration a biodegradable polymer is mandatory, other routes of administration such as the oral route offer a larger choice including the use of nonbiodegradable polymers. Four polymers of interest have been selected for the manufacture of both micro- and nanoparticles: poly-ε-caprolactone (PCL, M_w 42,000) and poly(D,L-lactic-co-glycolic acid 50/50) (PLGA, M_w 40,000), which are biodegradable polymers, and nonbiodegradable positively charged copolyesters of acrylic and methacrylic acids (Eudragit® RS and RL, M_w 150,000). The cationic charge of Eudragit® is because of quaternary ammonium groups: Eudragit® RL bears a higher positive charge than Eudragit® RS because of a higher density of quaternary ammonium groups. Heparin (UFH or LMWH) micro- and nanoparticles were prepared either with pure polymers or a blend of biodegradable and nonbiodegradable polymers; initial studies were carried out with a 50/50 ratio, whereas, in later studies, other ratios were evaluated (40/60 and 60/40).

We will first consider the encapsulation results obtained with (i) microparticles and (ii) nanoparticles before considering the in vivo results.

In Vitro Study. Microparticles: Regardless of the type of heparin (UFH or LMWH), it was possible to obtain spherical and discrete particles in the size range of about 100 to 150 μm (67,69). This rather small particle size range is generally obtained with the double emulsion method.

Tables 5 and 6 summarize the main physicochemical parameters obtained with the different types of microparticles.

First, it can be seen that microparticles obtained with the Eudragit® polymers display a relatively smaller size than those obtained with the biodegradable polymers. The latter ones also show a trend in the reduction of diameter when compared with microparticles manufactured with the pure biodegradable polymers. This

Table 5 Microencapsulation Efficiency, Drug Loading, and Mean Diameter of Heparin-Loaded (a) and Drug-Free (b) Microparticles Prepared by the Double Emulsion Method with a Single Polymer (250 mg) or Blends of Polymers (125/125 mg)

Polymer	Drug loading[a] (IU/g polymer)	Entrapment efficiency[a] (%)	Mean diameter (a) (μm)	Mean diameter (b) (μm)
RS	9952 ± 798	49 ± 4	96	71
RL	15960 ± 688	80 ± 3	80	–
PCL	4566 ± 700	24 ± 4	128	119
PLGA	5360 ± 749	27 ± 4	125	125
RS/PCL	7277 ± 722	36 ± 4	129	94
RL/PCL	9032 ± 466	45 ± 2	103	–
RS/PLGA	10520 ± 908	52 ± 4	87	86
RL/PLGA	12750 ± 785	64 ± 4	128	–
RS/RL	13310 ± 303	67 ± 2	88	–
PCL/PLGA	3506 ± 929	17 ± 5	82	100

[a]Data expressed as mean ± S.D. ($n \geq 3$); –, not determined.
Abbreviations: PCL, poly-ε-caprolactone; PLGA, poly (D,L-lactic-co-glycolic acid).
Source: From Ref. 67.

observation has been correlated with the surface tensioactive properties of Eudragit®, as was demonstrated earlier (70). Indeed, the presence of ammonio-methyl groups on Eudragit® allows a reduction in the interfacial tension between the organic and the aqueous phase in the second emulsion, resulting in a decrease in size of the droplets and, consequently, the microparticles. However, there is no difference in diameter between microparticles made with the two types of Eudragit®, showing that the increase in positive charges does not allow a stronger decrease in

Table 6 Mean Diameter, Encapsulation Efficiency, and Drug Loading of Unloaded and Loaded LMWH Microparticles Prepared by the w/o/w Emulsion and Solvent Evaporation Method with Biodegradable (PCL and PLGA) and Nonbiodegradable (Eudragit® RS and RL) Polymers Used Alone or in Combination (ratio 1/1)

Polymers	Drug-free microparticles	LMWH-loaded microparticles		
	Mean diameter (μm)	Mean diameter (μm)	Encapsulation efficiency (%)	Drug loading (IU/g polymer)
PCL	117 ± 35	129 ± 19	16.4 ± 1.9	3280 ± 380
PLGA	111 ± 11	142 ± 42	28.8 ± 6.4	5760 ± 1280
RS	59 ± 6	91 ± 15	39.5 ± 2.0	7900 ± 400
RL	< 1	60 ± 5	47.0 ± 1.2	9400 ± 240
RS/PCL	70 ± 3	144 ± 7	30.9 ± 4.7	6180 ± 940
RS/PLGA	59 ± 4	95 ± 8	23.8 ± 4.9	4760 ± 980
RS/RL/PLGA	63 ± 4	126 ± 41	18.3 ± 5.8	3660 ± 1160

Note: Drug-loaded microparticles were formulated with 5000 IU of LMWH. Data are expressed as mean ± S.D. ($n = 3$).
Abbreviations: LMWH, low molecular weight heparins; PCL, poly-ε-caprolactone; PLGA, poly (D,L-lactic-co-glycolic acid).
Source: From Ref. 69.

size. Another interesting observation is the difference in morphology that can be observed by scanning electronic microscopy (SEM) as shown in Figure 9.

Indeed, Eudragit® microparticles appeared porous and brittle with an irregular shape, whereas those prepared with both PCL and PLGA were smooth and spherical. On the contrary, unloaded Eudragit® RS microparticles were smaller and had a smoother and pore-free surfaces.

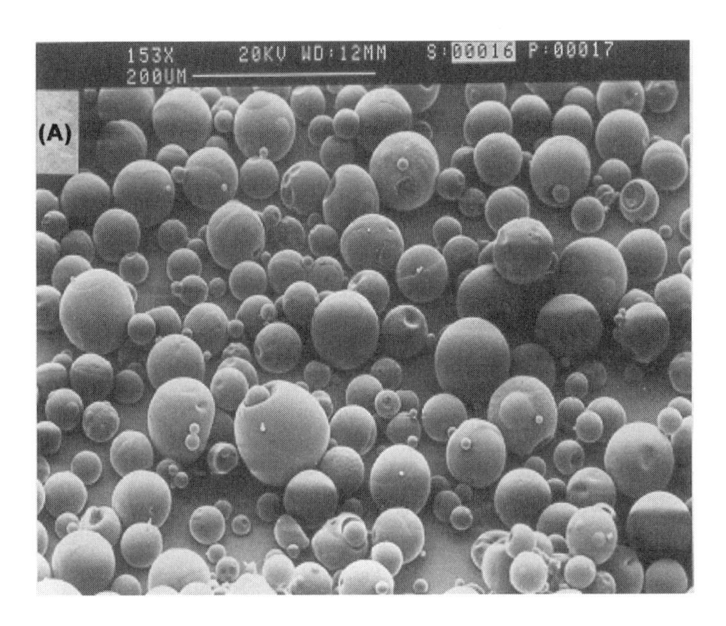

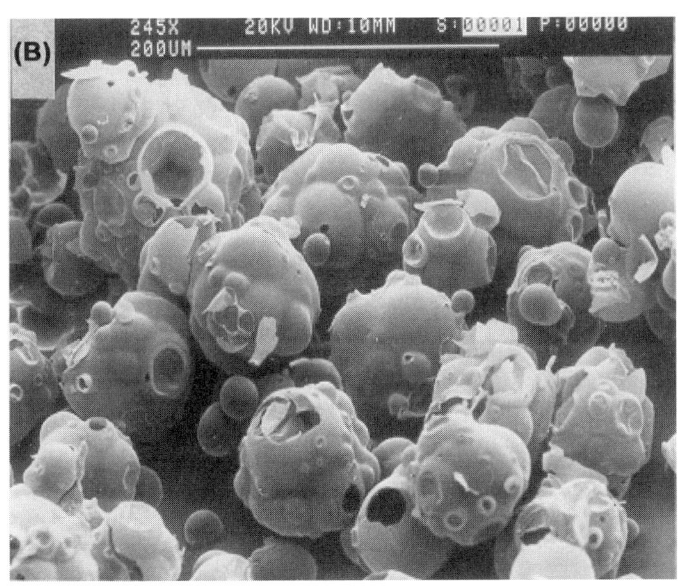

Figure 9 SEM photographs of heparin-loaded microparticles prepared with (**A**) PCL and (**B**) Eudragit® RS. *Abbreviations*: SEM, scanning electron microscopy; PCL, poly-ε-caprolactone. *Source*: From Ref. 67.

Also, for the less hydrophobic polymers, i.e., Eudragit®, the mean diameter seems also to be related to the reduced viscosity of the organic phase as shown in Table 7. Indeed, the values of the reduced viscosity of the methylene chloride solution containing the polymer were the lowest for both types of Eudragit®. These results confirm that the lower the viscosity, the smaller the microparticles. However, the significant difference in reduced viscosity between PCL and PLGA did not lead to a significant difference in size between both types of microparticles. This means that other physicochemical parameters should probably be taken into account.

The encapsulation results in drug loading (IU/g of polymer) and encapsulation efficiency for UFH and LMWH, respectively, are presented in Tables 5 and 6 (67,69). The highest encapsulation values were obtained with Eudragit® polymers used alone. This is probably because of the electrostatic interactions between the quaternary ammonium groups of the polymers and the carboxyl and sulphate groups of both types of heparins. Eudragit® RL bearing more positively charged groups (8.8–12%), compared to Eudragit® RS (4.5–6.8%), exhibited the highest entrapment efficiency. Microparticles prepared with a single biodegradable polymer (PCL or PLGA) displayed lower drug content. The lowest one was obtained with PCL owing to the more hydrophobic character of this polymer when compared to PLGA, thus not favoring the encapsulation of LMWH, a very hydrophilic drug. The lower entrapment efficiency obtained with PCL and PLGA can also be explained by the high solubility of heparin in water. We know that diffusion and drug loss across the droplet interface occur mainly during the first minutes of the second emulsification step, as the polymer precipitates, rapidly hindering any further leakage. Organic solvents with high water solubility result in rapid precipitation and hardening of polymer. However, the limited solubility of methylene chloride in water can be expected to prolong the solvent-swollen conditions, involving a slower precipitation of the polymer, and thus an increase of the drug leakage into the external aqueous phase.

The order in the encapsulation efficiency was however expected since the Eudragit® polymers were chosen, in the first place, to favor an electrostatic interaction between the anionic heparin and the cationic polymers. Another interesting feature concerns the comparison between UFH and LMWH encapsulation efficiencies. Indeed, the LMWH encapsulation efficiency is always lower than that of UFH. As LMWH is obtained from enzymatic degradation or chemical modification of UFH, involving shorter chains with consequently less negatively charged groups, the electrostatic interactions between LMWH and the positively charged polymers are much lower. And, the smaller molecular size of LMWH and its very hydrophilic

Table 7 Reduced Viscosity of Various Polymeric Solutions Obtained from Biodegradable Polymers (PCL and PLGA) and Nonbiodegradable Polymers (Eudragit® RS and RL) Dissolved in Methylene Chloride (2.5%, w/v) and Measured with a Ubbelohde Viscosimeter Equipped with a Glass Capillary (Ø: 0.56 mm)

Polymers	PCL	PLGA	Eudragit® RS	Eudragit® RL
Reduced viscosity (dL/g)	1.18	0.37	0.235	0.20

Abbreviations: PCL poly-ε-caprolactone; PLGA, poly (D,L-lactic-co-glycolic acid.
Source: From Ref. 69.

nature may also induce more diffusion toward the continuous aqueous phase during the emulsification and solidification processes.

In contrast, the question of the polymer miscibility can be raised for microparticles prepared with a blend of polymers. Indeed, as given later, these microparticles will display the highest in vivo activity. The polymer miscibility may influence the internal structure of microparticles and thus the encapsulation efficiency and the heparin release profile as well. Therefore, blends of biodegradable and nonbiodegradable polymers dissolved in methylene chloride were film casted in Petri dishes. A very heterogeneous distribution of each polymer was observed, suggesting either no or very limited miscibility of the polymers. This was confirmed by confocal laser scanning microscopy (CLSM) after labeling LMWH with fluorescein isothiocyanate (FITC) according to a previously described technique (69).

Based on the difference in fluorescence distribution for each microparticle, various internal structures may be assumed. As it is shown in Figure 10A, LMWH was mostly located at the outside surface of Eudragit® RS microparticles, definitely demonstrating that it is mainly bound onto quaternary ammonium groups of Eudragit® polymers which are preferentially directed toward the external aqueous phase during the emulsification process. Contrarily, the internal structure of RS/PLGA microparticles (Fig. 10B) displays a distribution of FITC-LMWH both at the outside surface and in the core of the particles confirming the very limited miscibility between the biodegradable and the nonbiodegradable polymers. This dual distribution of the FITC–LMWH conjugate was also observed with RS/PCL microparticles (Fig. 10C), although, it was much less marked. However, this later observation confirms the reduced affinity of LMWH for the most hydrophobic polymer, i.e., PCL, as also confirmed by the drug-loading data compared with the less hydrophobic PLGA polymer.

Consequently, it can be concluded that both types of heparin are mainly entrapped by ionic interactions with Eudragit® polymers, encapsulated only in the core of biodegradable microparticles (for both PCL and PLGA, pictures not shown), and encapsulated and bound by electrostatic interactions in microparticles prepared with blends of biodegradable and polycationic, nonbiodegradable polymers.

The drug release from a drug delivery system depends strongly on the properties of the polymeric carriers. The composition of the system, i.e., polymers, charge, and physicochemical properties (crystallinity, hydrophobicity/hydrophilicity, etc.) determines the permeability to both solvent and drug. Illustrated in Figures 11 and 12 are in vitro release profiles obtained for each formulation prepared with one or two polymers for both UFH and LMWH, respectively (67,69).

All formulations were characterized by an initial burst effect, which was expected, since for microparticles prepared by the w/o/w method, water-soluble drugs tend to migrate to the aqueous external phase, thereby concentrating at the surface of the microparticles and producing the effect (71). This initial burst release strongly depends on the formulation. It is very low for pure Eudragit® microparticles and is more important for the pure biodegradable particles or the microparticles prepared with the blends of biodegradable and nonbiodegradable polymers. Another common characteristic of all the release profiles is the incomplete heparin release after 24 hours. In the best case, around 60% to 70% of UFH and LMWH were released from PCL and PLGA microparticles, respectively. A longer period of time (>24 hours) would have probably allowed a 100% release of heparin within the biodegradable microparticles because of the slow erosion of the polymer matrix. In contrast, heparin (UFH or LMWH) released after 24 hours ranged from 2% to 15% for Eudragit® microparticles

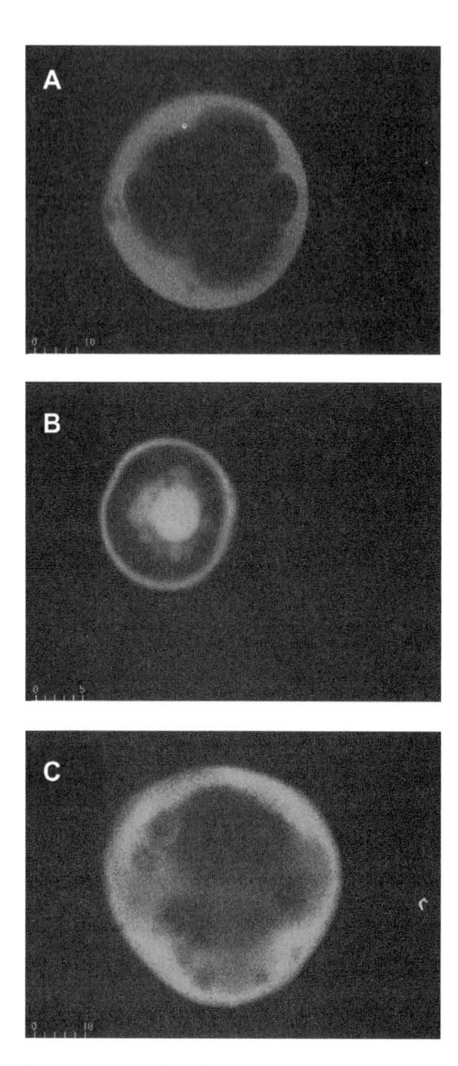

Figure 10 Confocal laser scanning microscopy images of microparticles prepared by the double emulsion method with Eudragit® RS (**A**) and blends (ratio 1/1) of RS/PLGA (**B**) and RS/PCL (**C**) after labeling of LMWH with FITC. Fluorescence excitation was performed at 488 nm. Scale bars are shown in μm. *Abbreviations*: PLGA, poly (D,L-lactic-co-glycolic acid); PCL, poly-ε-caprolactone; LMWH, low molecular weight heparins; FITC, fluorescein isothiocyanate. *Source*: From Ref. 69.

used alone or blended with biodegradable polymers. These results are not surprising if the strong interactions between the polycationic polymers and heparin are taken into account. Indeed, such strong interactions could not be disrupted at the experimental pH. Furthermore, this hypothesis is corroborated by the fact that the release of heparin is lower from microparticles prepared with Eudragit® RL compared to Eudragit® RS. Owing to a higher quaternary ammonium groups content in Eudragit® RL than in Eudragit® RS, a higher electrostatic binding was expected between Eudragit® RL and heparin. One may also assume that the association of Eudragit® with biodegradable polymers involves an additional decrease in the flexibility of PCL or PLGA chains compared with PCL and PLGA used alone.

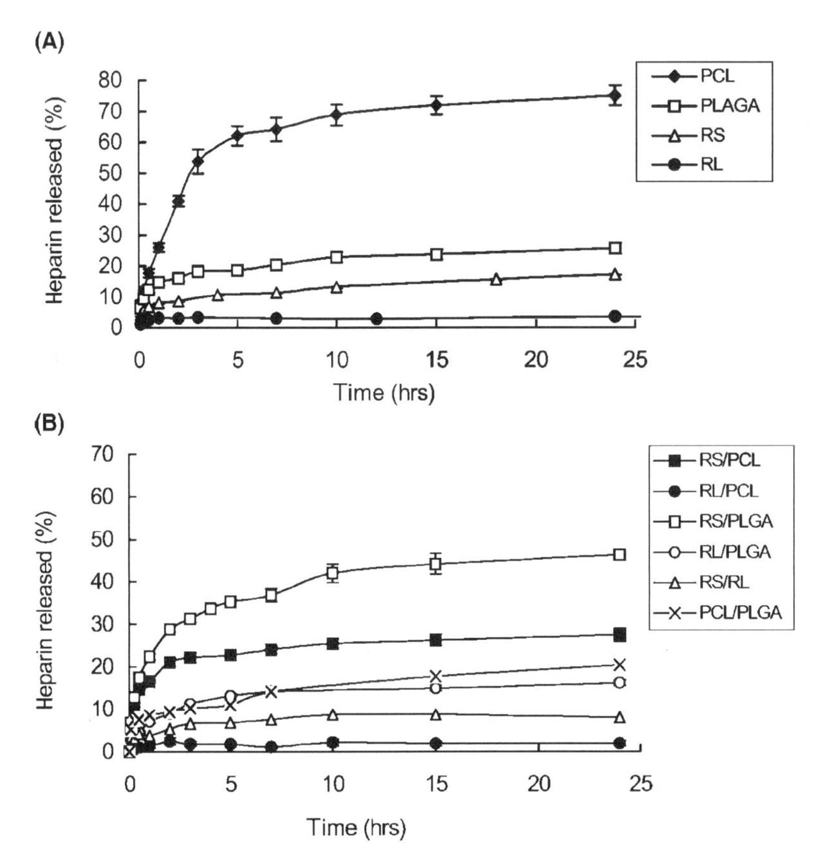

Figure 11 Release profiles of heparin from microparticles prepared either (**A**) with a single biodegradable or nonbiodegradable polymer or (**B**) with blends of two polymers. Experiments were performed in phosphate buffer at 37°C and pH 7.4. Data are shown as mean ± S.D. ($n = 3$). *Source*: From Ref. 67.

It was also important to verify whether or not heparin (both UFH and LMWH) was retaining its biological properties because of the encapsulation process, which involves both strong steps of shear and evaporation as well as a momentary interface contact with the organic solvent. Therefore, the amount of heparin release from microparticles has been determined by a biological method based on the measurement of the anti-Xa activity. The results are reported in Figures 13 and 14 for UFH and LMWH microparticles, respectively.

As reported in both figures, a good correlation was observed for the amount of heparin released after 24 hours, determined by the colorimetric Azure II technique for UFH or the nephelometry method for LMWH, and the antifactor Xa activity. As heparin retains almost 90% of its biological activity in spite of the use of organic solvent and shear stress during the encapsulation process, it can be claimed that the released heparin retains its ability to bind and inactivate factor Xa in vivo.

Nanoparticles: The double emulsion method was slightly modified to obtain nanoparticles by introducing two emulsification steps by sonication with an ultrasound probe. This modification allowed the successful preparation of heparin-loaded nanoparticles whose characteristics are summarized in Tables 8 and 9 (66,68).

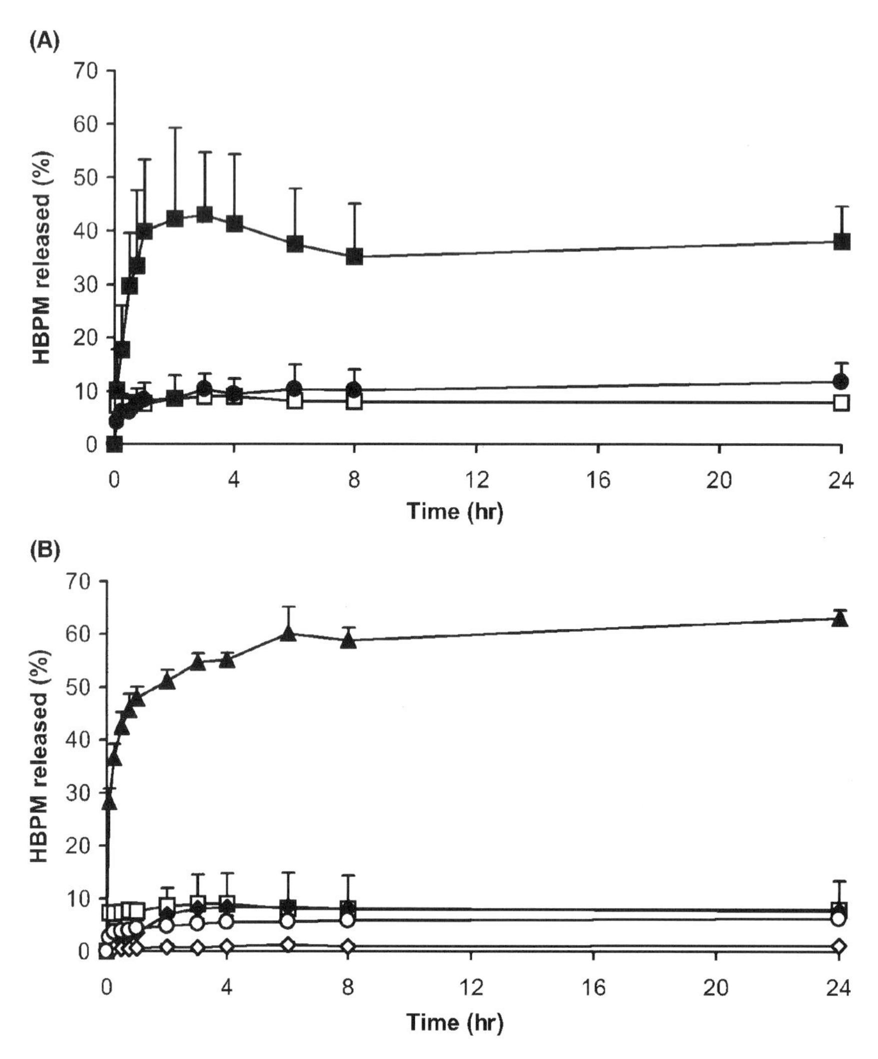

Figure 12 Release profiles of LMWH from polymeric microparticles prepared by the double emulsion with 5000 IU of LMWH into the internal aqueous phase (1 mL) and biodegradable or nonbiodegradable polymers used alone or in combination (250 mg, ratio 1/1). (**A**) PCL (■), Eudragit RS (□), RS/PCL (●). (**B**) Eudragit RS (□), Eudragit RL (♦), PLGA (▲), RS/PLGA (◇), RS/RL/PLGA (○). Experiments were performed in phosphate buffer at 37°C and pH 7.4. Data shown as mean ± S.D. ($n = 3$). *Abbrivations*: LMWH, low molecular weight heparins; PLGA, poly (D,L-lactic-co-glycolide); PCL, poly-ε-caprolactone. *Source*: From Ref. 69.

As it was already observed for microparticles, the two Eudragit® polymers lead to smaller nanoparticles than those prepared with pure biodegradable polymers or blends of polymers. The average diameter is always less than 0.5 μm. Incorporation of heparin into nanoparticles causes a slight increase in the average diameter that can be explained by the hydrophilic properties of heparin possibly partially located at the outer surface. The same trend in size (Eudragit® RS nanoparticles > Eudragit® RL nanoparticles) may be observed and is probably related to the lower overall tensio-active properties of Eudragit® RS bearing less charges. The decrease in reduced

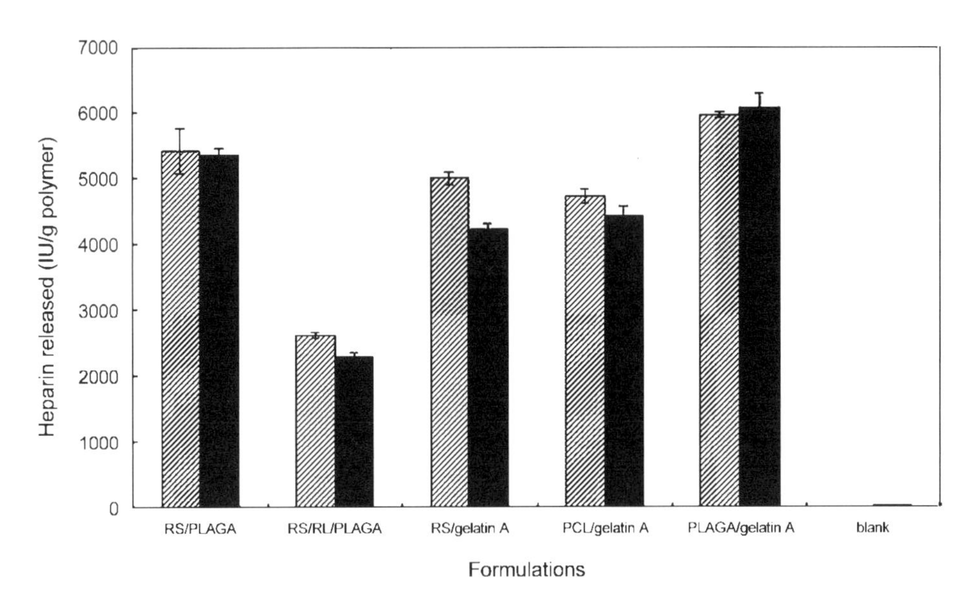

Figure 13 Comparison of the amount of UFH released after 24 hours from polymeric microparticles determined by the colorimetric method with Azure II (*black*) and by the anti-Xa activity with a chromogenic substrate (*hatched*). Blank corresponded to unloaded microparticles. Data are shown as mean ± S.D. ($n = 3$). *Abbreviation*: UFH, unfractionated heparins. *Source*: From Ref. 67.

viscosity (Table 7) may also explain the lower diameter observed when Eudragit® is part of the nanoparticles matrix because during the emulsification process, the lower the viscosity of the dispersed phase, the smaller the mean diameter.

The zeta potential values of unloaded nanoparticles reflect the charges of the raw polymers. Indeed, the polycationic Eudragit® bearing positive charges,

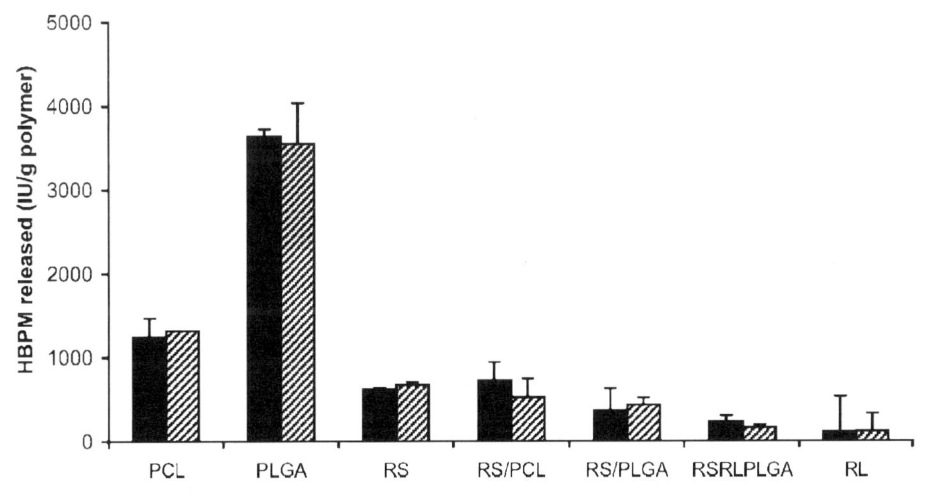

Figure 14 Comparison of the amount of LMWH released after 24 hours from various polymeric microparticles suspended in phosphate buffer and determined by nephelometry (*black*) and the biological activity based on the antifactor Xa activity with a chromogenic substrate (*hatched*). Data shown as mean ± S.D. ($n = 3$). *Abbreviation*: LMWH, low molecular weight heparins. *Source*: From Ref. 69.

Table 8 Encapsulation Efficiency, Drug Loading, Surface Potential, and Mean Diameter of Unloaded and UFH-loaded Nanoparticles Prepared by the Double Emulsion Method with a Single Polymer (250 mg) or Mixtures of Biodegradable and Nonbiodegradable Polymers (ratio 1/1)

NP	Blank nanoparticles		Loaded-nanoparticles		Entrapment efficiency (%)	Drug loading (IU/g of polymer)
	Mean size (nm)	Zeta potential (mv)	Mean size (nm)	Zeta potential (mv)		
RL	191 ± 3	55.6 ± 4.5	266 ± 8	−38.4 ± 2	97 ± 2	19480 ± 490
RS	225 ± 10	30.2 ± 5.5	269 ± 16	−22.4 ± 0.5	59 ± 1	11830 ± 140
PLGA	259 ± 6	−3.9 ± 0.4	267 ± 4	−4.5 ± 0.1	14 ± 4	2792 ± 801
PCL	278 ± 8	−2.5 ± 1.4	285 ± 10	−1.6 ± 0.2	8 ± 1	1673 ± 209
RS/PLGA	243 ± 6	32.1 ± 1.4	273 ± 7	−17.3 ± 1.4	35 ± 2	7101 ± 431
RL/PLGA	214 ± 4	45.7 ± 3.2	269 ± 8	−37.2 ± 3.3	49 ± 4	9752 ± 721
RS/PCL	272 ± 7	33.4 ± 3	286 ± 6	−20.0 ± 0.7	28 ± 2	5657 ± 324
RL/PCL	225 ± 5	48.0 ± 1.5	304 ± 3	−33.6 ± 1.9	53 ± 2	10660 ± 321
RS/RL/PLGA	224 ± 4	43.0 ± 1.6	275 ± 3	−30.7 ± 2.0	38 ± 1	7498 ± 138

Note: Drug-loaded nanoparticles were formulated with 5000 IU of heparin. Data are shown as mean ± S.D. ($n = 4$).
Abbreciations: UFH, unfractionated heparins; PCL, poly-ε-caprolactone; PLGA, poly (D,L-lactic-co-glycolie acid 50/50.
Source: From Ref. 66.

conferred by the quaternary ammonium groups, presented the highest potential surface: +55 mV for Eudragit® RL which carries 8.8% to 12% of ammonium groups versus +30 mV for Eudragit® RS characterized by 4.5% to 6.8% of positively charged groups. Blends of PCL and PLGA with Eudragit® RS and/or RL

Table 9 Mean Diameter, Surface Potential, Encapsulation Efficiency, and Drug Loading of Unloaded and Loaded LMWH Nanoparticles Prepared by the w/o/w Emulsion and Solvent Evaporation Method with Biodegradable (PCL and PLGA) and Nonbiodegradable (Eudragit RS and RL) Polymers Used Alone or in Combination (ratio 1/1)

Polymers	Blank nanoparticles		LMWH-loaded nanoparticles			
	Mean diameter (nm)	Zeta potential (mV)	Mean diameter (nm)	Zeta potential (mV)	Encapsulation efficiency (%)	Drug loading (IU/g polymer)
PCL	379 ± 68	−3.3 ± 1.9	489 ± 68	−5.8 ± 0.7	16.0 ± 2.6	3207 ± 528
PLGA	339 ± 14	−5.5 ± 0.7	390 ± 53	−9.1 ± 0.3	10.6 ± 3.3	2121 ± 651
RS	225 ± 21	35.7 ± 2.9	301 ± 38	−26.7 ± 0.8	37.9 ± 3.1	7587 ± 620
RL	187 ± 11	52.8 ± 3.8	240 ± 23	−45.7 ± 1.7	56.0 ± 2.3	11207 ± 463
RS/PCL	304 ± 42	32.7 ± 1.0	408 ± 16	−26.8 ± 1.2	31.0 ± 2.3	6203 ± 462
RS/PLGA	346 ± 23	35.5 ± 1.3	294 ± 51	−26.5 ± 0.2	23.3 ± 4.0	4653 ± 803
RS/RL/PLGA	359 ± 17	37.1 ± 0.9	294 ± 45	−44.7 ± 0.4	24.9 ± 2.5	4987 ± 496

Note: Drug-loaded nanoparticles were formulated with 5000 IU of LMWH. Data are expressed as mean ± S.D. ($n = 3$).
Abbreviations: PCL, poly-ε-caprolactone; PLGA, poly (D,L-lactic-co-glycolic acid 50/50; LMWH, low molecular weight heparins.
Source: From Ref. 68.

(ratio 50/50) also showed a positive surface potential but lower than that obtained with Eudragit® polymers used alone. On the contrary, a dramatic change occurred when LMWH was encapsulated within nanoparticles. Indeed, the zeta potential with Eudragit® used alone or in combination with PCL and PLGA was dramatically and significantly modified, ranging from strong positive to strong negative values. This is the consequence of the incorporation of heparin, a negatively charged drug bearing sulphate and carboxyle groups bound through ionic interactions, onto the positively charged groups of Eudragit®. Unloaded PCL and PLGA nanoparticles exhibited a zeta potential close to neutrality that became slightly negative after heparin encapsulation.

As reported in Tables 8 and 9, the entrapment efficiency within the various polymeric nanoparticles was significantly affected by the nature of the polymer (66,68). The encapsulation efficiency ranged from 8% to 97% and reached the highest values when Eudragit® RS and RL were used for the nanoparticle preparation. As observed for microparticles, the encapsulation efficiency is better for UFH than for LMWH. Although the charge density of LMWH is similar to that of UFH, its overall charge is much lower, probably leading to fewer electrostatic bindings between the drug and the polycationic polymers. In addition, the smaller size and molecular weight of LMWH, as well as its very hydrophilic nature, can lead to an increased diffusion of the drug into the external aqueous phase before the precipitation of the polymer(s) and, consequently, to a decrease in encapsulation. As observed by CLSM with microparticles, it can be assumed that LMWH was mainly encapsulated inside the core of PCL and PLGA nanoparticles, but it was mainly located at the outer surface of Eudragit® nanoparticles.

The release profiles of heparin (UFH and LMWH) from nanoparticles prepared from a single polymer and from mixtures of either PCL or PLGA with Eudragit® are illustrated in Figures 15 and 16 (66,68).

The highest release percentages are obtained for nanoparticles prepared with a single biodegradable polymer, i.e., PCL or PLGA. Each release profile displays a low and biphasic release pattern for each dosage form. After an initial burst stage, during which small amounts of heparin were released rapidly over one hour, the drug release profiles displayed a plateau, characterized by very slow and incomplete subsequent release, for an extended period of time, resulting from the diffusion of the drug dispersed into the polymeric matrices. Moreover, as expected and probably because of strong ionic interactions between Eudragit® RS or RL and heparin, very low heparin (both UFH and LMWH) was released from nanoparticles prepared with these two polymers, especially Eudragit® RL. Nanoparticles prepared with PLGA, alone or in combination with Eudragit®, exhibited higher drug release compared with those prepared with PCL. This could be explained by the high hydrophobicity of PCL, which reduced the wettability of nanoparticles, and the diffusion of the drug to the external aqueous phase. However, this phenomenon was more obvious for UFH than for LMWH. The combination of Eudragit® RS and/or RL with PCL or PLGA did not influence the drug release compared with the Eudragit® polymers used alone. Other experiments were performed at various pH (from three to nine) but similar results were observed. Because of the very low amount of LMWH released and the polyester nature of the polymers, esterases were also added in the dissolution medium in an attempt to enhance the drug release.

As for microparticles, it was also important to verify whether or not heparin (both UFH and LMWH) was retaining their biological properties because of both

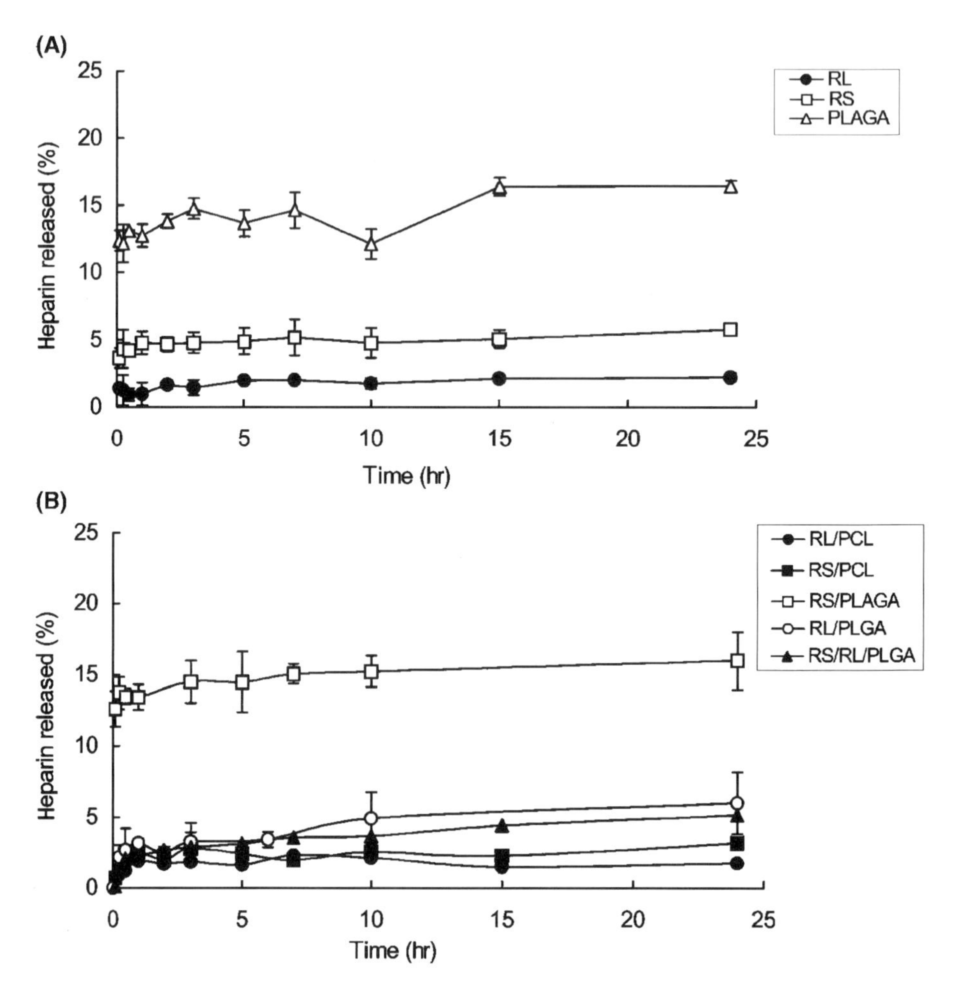

Figure 15 Release profiles of UFH from (**A**) PLGA (△), Eudragit RS (□) and Eudragit RL (●) nanoparticles, (**B**) RS/PLGA (□), RS/PCL (■), RL/PLGA (○), RL/PCL (●) and RS/RL/PLGA (▲) nanoparticles prepared with 5000 IU of heparin within the internal aqueous phase (1 mL). Experiments were performed in phosphate buffer at 37°C and pH 7.4. Data are shown as mean ± S.D. ($n = 3$). *Abbreviations*: UFH, unfractionated heparins; PCL, poly-ε-caprolactone; PLGA, poly (D,L-lactic-co-glycolic acid 50/50). *Source*: From Ref. 66.

the nature of the encapsulation process, which involves even stronger steps of shear (ultrasound energy) than for microparticles, and the presence of esterases. Therefore, the amount of heparin released from nanoparticles was determined by a biological method based on the measurement of the anti-Xa activity. The results are reported in Figures 17 and 18 for UFH and LMWH nanoparticles, respectively.

The results show that heparin (both UFH and LMWH) was unaltered by the double emulsion and solvent evaporation processes, as it preserved its antithrombotic activity (66,68). Also, as demonstrated for microparticles, there is a good correlation between the amount of heparin released from loaded nanoparticles after 24 hours as determined by the biological method based on the anti-Xa activity.

The objectives of this in vitro encapsulation work were to formulate micro- and nanoparticles able to successfully incorporate heparin (UFH or LMWH). Different

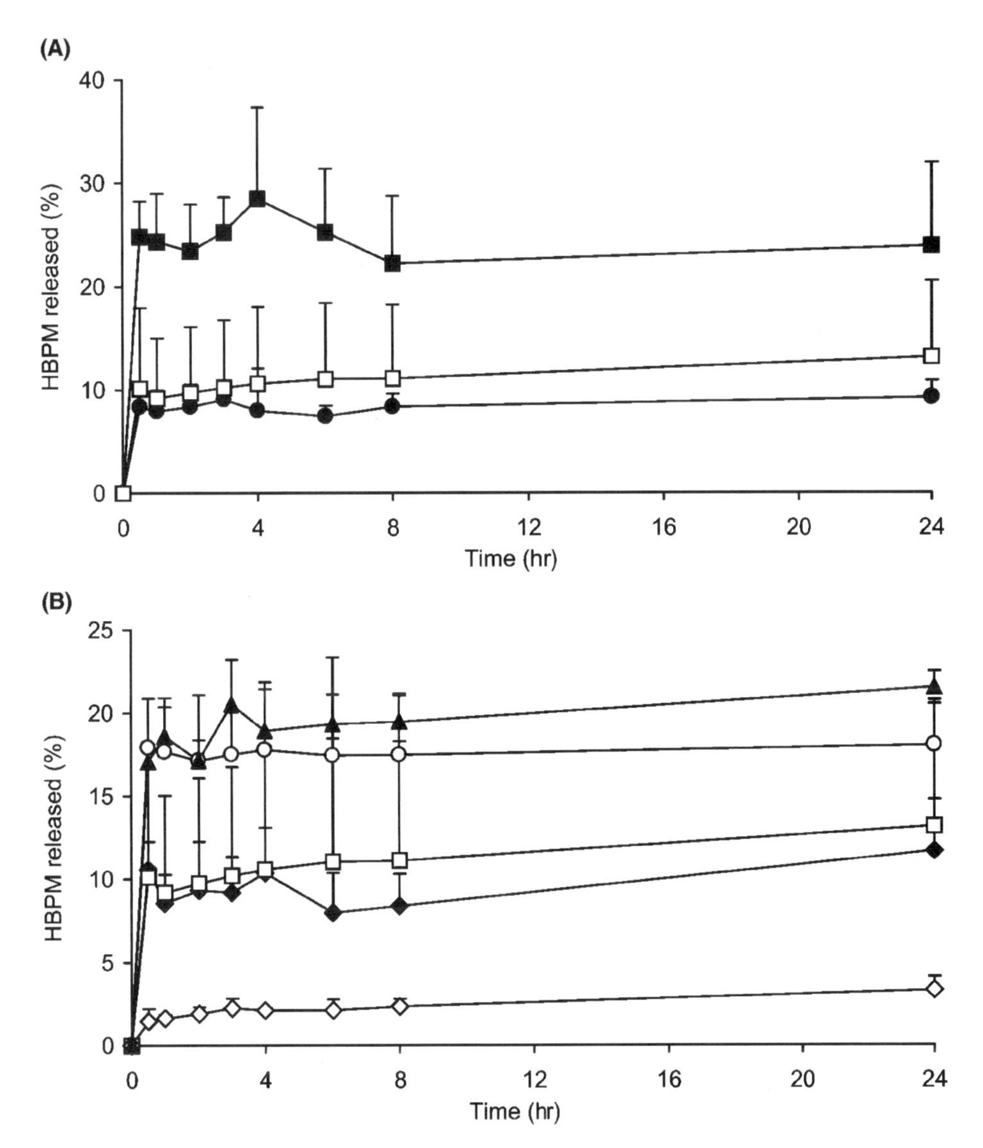

Figure 16 Release profiles of LMWH from polymeric nanoparticles prepared by the double emulsion with 5000 IU of LMWH into the internal aqueous phase (1 mL) and biodegradable or nonbiodegradable polymers used alone or in combination (250 mg, ratio 1/1). (**A**) PCL (■), Eudragit RS (□), RS/PCL (●). (**B**) Eudragit RS (□), Eudragit RL (◇), PLGA (▲), RS/PLGA (◆), RS/RL/PLGA (○). Experiments were performed in phosphate buffer at 37°C and pH 7.4. Data shown as mean ± S.D. ($n = 3$). *Abbreviations*: LMWH, low molecular weight heparins; PCL, poly-ε-caprolactone; PLGA, poly (D,L- lactic-co-glycolic acid 50/50). *Source*: From Ref. 68.

types of particles were prepared based on biodegradable and nonbiodegradable polycationic polymers. The highest encapsulation efficiencies were obtained with the polycationic polymers owing to strong electrostatic interactions with the anionic heparin. Although heparin was demonstrated to retain its biological activity after encapsulation, it must be noted that the in vitro release of heparin was always

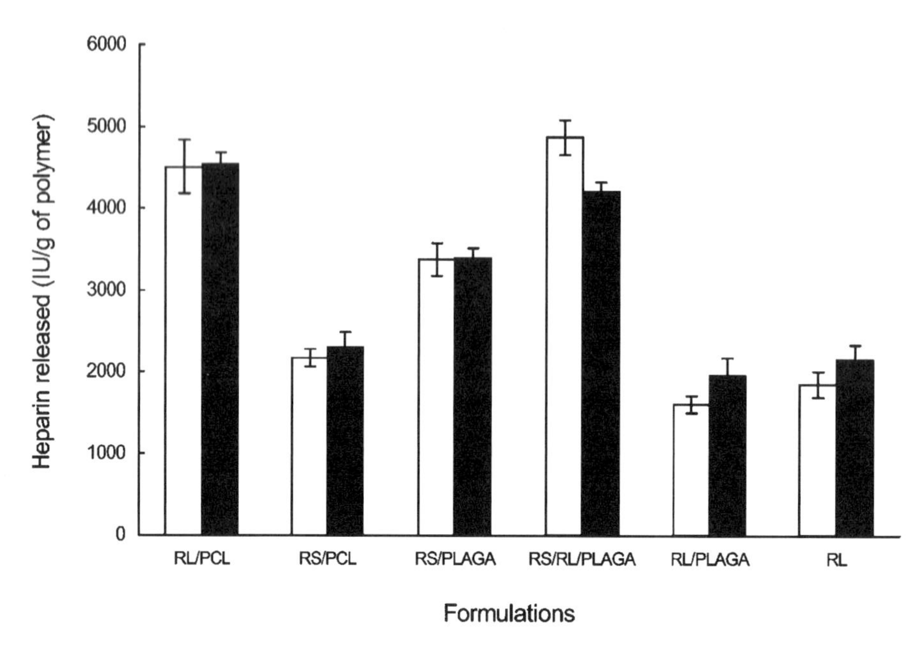

Figure 17 Comparison of the amount of UFH released from various polymeric nanoparticles after 24 hours in the dissolution medium containing esterase and determined by the colorimetric method with Azure II (*black*) and by the anti-Xa activity with a chromogenic substrate (*white*). Data are shown as mean ± S.D. ($n = 3$). *Abbreviation*: UFH, unfractionated heparins. *Source*: From Ref. 66.

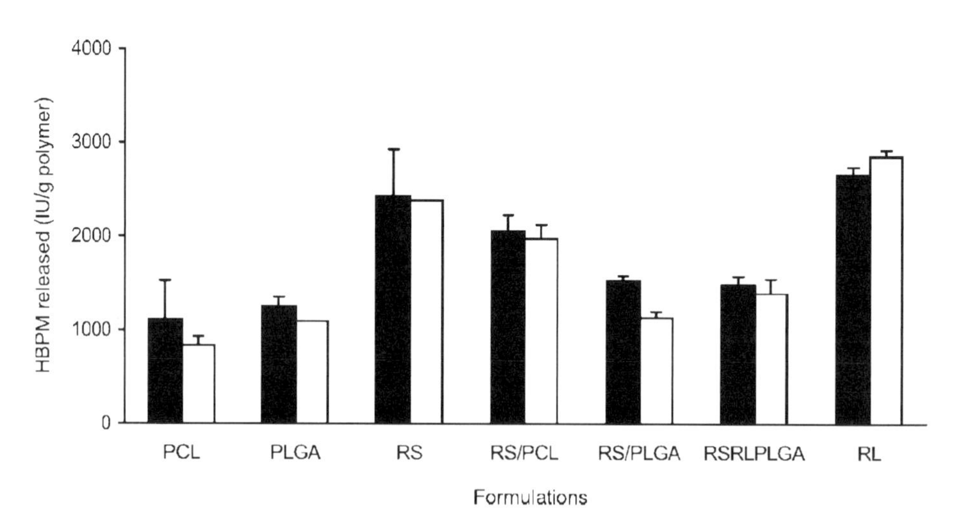

Figure 18 Comparison of the amount of LMWH released after 24 hours from various polymeric nanoparticles suspended in phosphate buffer containing esterases (50 units/mL added every six hours) and determined by nephelometry (*black*) and the biological activity based on the antifactor Xa activity with a chromogenic substrate (*hatched*). Data shown as mean ± S.D. ($n = 3$). *Abbreviation*: LMWH, low molecular weight heparins. *Source*: From Ref. 68.

incomplete after 24 hours; that could affect the in vivo performance of the different systems owing to digestive transit times. Therefore, it was very important to follow up the formulation work by bioavailability studies, which will be presented below.

In Vivo Study. The first experiments were carried out in male adult New Zealand rabbits. A second set of experiments was performed in Beagle dogs. The clotting time, measured by the APTT, and/or the plasma heparin concentration, evaluated by the anti-Xa activity, was determined in citrated blood plasma before and 1, 2, 3, 4, 5, 6, 7, 8, 10, and 24 hours after administration of each dosage form. Oral Administration of Microparticles: The first set of experiments was carried out to evaluate the potential oral absorption of UFH after administration in rabbits. Polymeric microparticles containing 2000 IU of UFH were loaded into gelatin hard capsules (size 1) and administered to overnight fasted rabbits by oral route. A solution of UFH (2000 IU), administered either intravenously (ear vein, bolus) or orally (gavage), and drug-free microparticles, administered by the oral route, were used as controls. Based on the weight of rabbits, the UFH dose was approximately 600 IU/kg. In Figure 19, the results obtained after a single oral administration in rabbits are displayed (72).

The normal clotting time in rabbits is approximately 13 to 15 seconds. From the results of Figure 19, it is obvious that the formulations can be divided into two groups

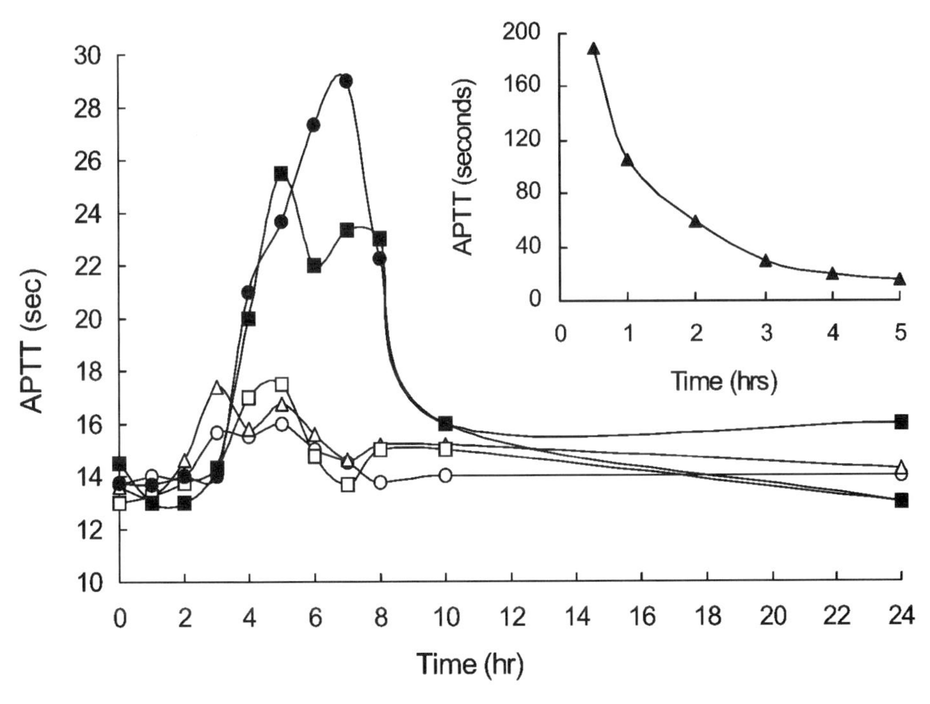

Figure 19 Activated partial thromboplastin time values (seconds) as a function of time after both oral administration in rabbits ($n = 5$) of microparticles loaded with UFH (2000 IU) and prepared with blends of biodegradable and nonbiodegradable polymers, RS/PLGA (■) and RS/RL/PLGA (●), with gelatin A within the aqueous internal phase, PCL/gelatin A (○), PLGA/gelatin A (□), RS/gelatin A (△) and intravenous administration of an aqueous heparin solution (2000 IU) (inset). Standard deviations are not represented for sake of clarity. *Abbreviations*: UFH, unfractionated heparins; PLGA, poly (D,L-lactic-co-glycolic acid 50/50). *Source*: From Ref. 72.

of in vivo profiles corresponding to the microparticles prepared with gelatin A (in the inner aqueous phase to increase the viscosity and, consequently, decrease the UFH leakage) and those prepared with blends of Eudragit® and PLGA. For gelatin-containing formulations, the APTT value increased very slightly from the baseline to peak between 15 and 17 seconds after three to five hours following oral administration. In addition, there is no statistical difference in APTT in the two-hour interval in which the clotting time was somehow higher than its initial value. The oral administration of an aqueous UFH solution did not display any biological activity at all. In the second group represented by Eudragit® RS/PLGA and Eudragit® RS/RL/PLGA (25/25/50) microparticles, five to eight hours after administration, a significant anticoagulant activity was detected with maximal APTT values between 25 and 32 seconds. The delay in clotting time versus the other type of microparticles can be explained by the slower release of heparin from microparticles prepared with Eudragit® and PLGA. In addition, the release was also prolonged, because the overall activity lasted for six hours (between 4 and 10 hours). The biological activity can also be compared with results observed after intravenous administration of an aqueous solution of heparin (Fig. 19, inset); although the clotting time was much higher initially, the anticoagulant activity lasted only four hours. Based on the area under the curve of the clotting time as a function of time between 0 and 24 hours, a tentative pharmacological bioavailability was calculated (Table 10).

The longer and higher clotting time observed with blends of Eudragit® and PLGA is confirmed by a larger bioavailability compared to gelatin microparticles. Because no absorption of heparin was observed after the oral administration of the aqueous solution, its absolute bioavailability is of course nil. The presented results definitely demonstrate the absorption of UFH from microparticles manufactured with blends of biodegradable and nonbiodegradable polymers. The absolute bioavailability is very important (around 50%) for this new drug delivery system.

Once the oral absorption of UFH was demonstrated, it was also important to verify if the results could be confirmed with LMWH. In addition, it was also decided

Table 10 Main Pharmacokinetic Parameters Obtained for UFH-Loaded Microparticles After Oral Administration in Rabbits

Formulations	RS/RL/ PLGA	RS/PLGA	RS/gelatin A[a]	PLGA/ gelatin A	PCL/ gelatin A[a]	Heparin in solution (IV route)
Maximal APTT (seconds)	25–30	22–32	17–20	16–18	15–18	ND
t_{max} (h)	6–8	5–8	3–6	3–6	3–6	ND
$AUC_{0\to24}$/kg (sec h/kg)	33.1 ± 2.2^b	37.6 ± 3.5^b	15.8 ± 2.2	6.1 ± 0.8	3.6 ± 0.1	77.9 ± 7.9
Absolute bioavailability (%)	42.5 ± 1.5^b	48.3 ± 4.4^b	13.1 ± 0.4	7.9 ± 1.0	4.7 ± 0.2	100

Note: The data are shown as mean ± SEM ($n = 5$).
[a]$n = 3$
[b]Statistically different from RS/gelatin A, PLGA/gelatin A, PCL/gelatin A at $P < 0.05$ (Student's t-test).
Abbreviations: PLGA, poly (D,L-lactic-co-glycolic acid 50/50); PCL, poly-ε-caprolactone; APTT, activated partial thromboplastin time; UFH, unfractionated heparins; IV, intravenous; ND, not determined.
Source: From Ref. 72.

to test different ratios between the biodegradable and the nonbiodegradable polymers. Therefore three ratios were studied namely 40/60, 50/50, and 60/40. The administered dose was 600 IU/kg and the microparticles were administered in gelatin hard capsules (size 1). Figure 20 shows the profile of anti-Xa activity in rabbits for the three studied ratios of polymers.

From these results, it can be seen that the in vivo results depend very much on the polymer ratios, although LMWH absorption was confirmed. First the lag time depends on the ratio of Eudragit® RS polymer. The higher the Eudragit® content, the longer the lag time which may reflect a longer adhesion on the negatively charged mucus of the GIT. This observation is also connected to the C_{max} values whose rank order is the following: 50/50 > 40/60 > 60/40. Based on the area under the curve, the results with UFH were confirmed with LMWH.

Finally, a third set of experiments was carried out with LMWH microparticles prepared with a 50/50 ratio of biodegradable (PLGA) and nonbiodegradable (Eudragit® RS) polymers but in another animal species, i.e., dog. The same dose as before was administered orally (600 IU/kg).

As it can be observed, the in vivo anti-Xa activity is extremely close in these two animal species (Fig. 21). This third experiment was important to demonstrate the absorption of LMWH in dogs before going further into clinical studies. The same key features can be found in both species: a lag time of absorption of about three hours after oral administration of microparticles and an extremely close duration of activity, of around eight hours for the two species.

Oral Administration of Nanoparticles: Lyophilized polymeric nanoparticles containing 2000 IU of UFH were resuspended in water before oral gavage through a cannula to rabbits fasted overnight (73). A solution of UFH was also administered both intravenously and orally, and unloaded nanoparticles also administered by oral gavage were used as controls. The anti-Xa activity as well as the clotting time was measured at appropriate intervals as a function of time. The Eudragit® RL/PCL

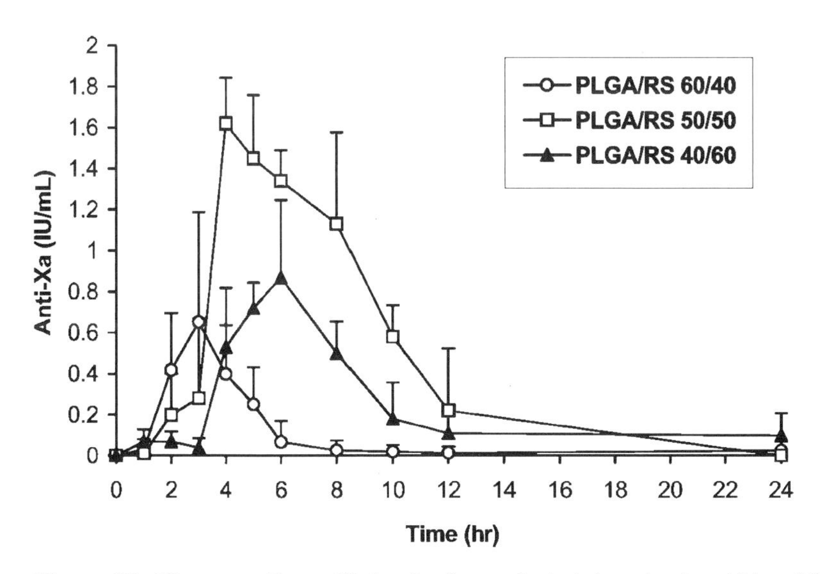

Figure 20 Plasma antifactor Xa levels after oral administration in rabbits of LMWH-loaded microparticles (600 anti-Xa U/Kg) prepared with various ratios of PLGA and Eudragit RS ($n = 6$). *Abbreviation*: LMWH, low molecular weight heparins.

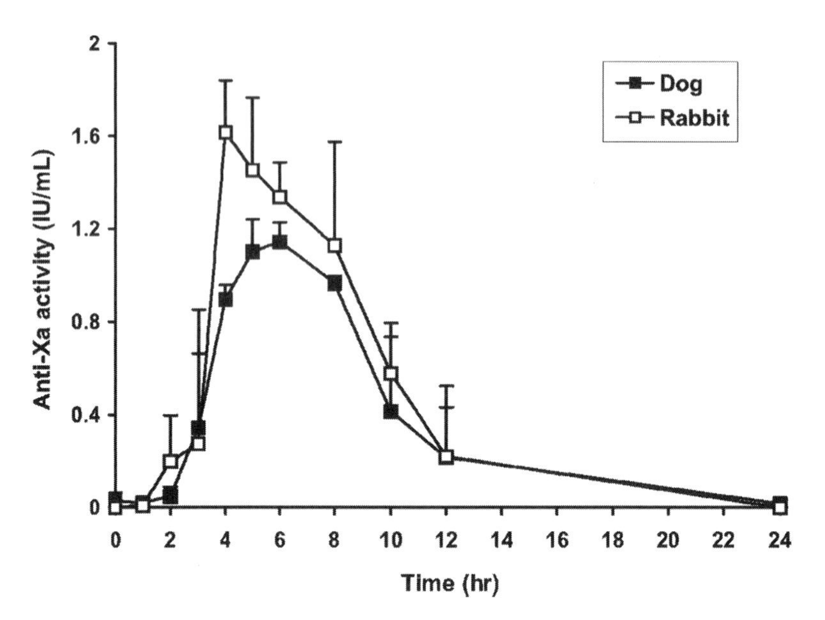

Figure 21 Comparative absorption profiles after oral administration of LMWH-loaded PLGA/Eudragit RS microparticles (ratio 50/50) in dogs and rabbits (600 anti-Xa U/Kg) ($n = 6$). *Abbreviations*: LMWH, low molecular weight heparins; PLGA, poly (D,L-lactic-co-glycolic acid 50/50).

(ratio 50/50), Eudragit® RS/PLGA (ratio 50/50), and Eudragit® RS/RL/PLGA (ratio 25/25/50) formulations, which afford a suitable drug entrapment efficiency and the highest drug releases, were selected for the in vivo study. In Figure 22 is illustrated both the amount of antifactor Xa activity and the mean clotting time determined by the APTT (which reflects mainly the anti-IIa activity).

As shown in Figure 22A, antifactor Xa activity was detected after oral administration of each formulation containing 2000 IU of heparin. The highest antifactor Xa response (peak concentration of 0.16 ± 0.01 IU/mL seven hours after dosing) was observed in rabbits receiving Eudragit® RL/PCL nanoparticles loaded with heparin. Lower peak concentrations (0.12 ± 0.05, 0.12 ± 0.04, and 0.13 ± 0.06 IU/mL) were observed three, six, and seven hours, respectively, after dosing with Eudragit® RS/RL/PLGA, Eudragit® RL, and Eudragit® RS/PLGA nanoparticles loaded with UFH. A detectable prolongation of antifactor Xa activity (0.04 IU/mL) was measured up to seven hours after oral administration of each formulation. As displayed in Figure 22B, the normal APTT in rabbits is 12 to 14 seconds. There is a good temporal relationship between Figures 22A and B. Indeed, from two hours after dosing of each formulation, an increase in anticoagulant activity was detected for all formulations (Fig. 22B), with a maximal APTT value of 24 seconds (corresponding to around a twofold increase) six and eight hours after oral administration of Eudragit® RL/PCL and Eudragit® RS/PLGA nanoparticles, respectively.

Compared with heparin administered intravenously, a decrease in both the maximal plasma concentration ($C_{\max}$) and the area under the curve was observed for all the polymer carrier formulations (Table 11).

The best absolute bioavailability (23%) (based on the areas under the anti-Xa curves) was observed with Eudragit® RL/PCL nanoparticles, whereas lower figures

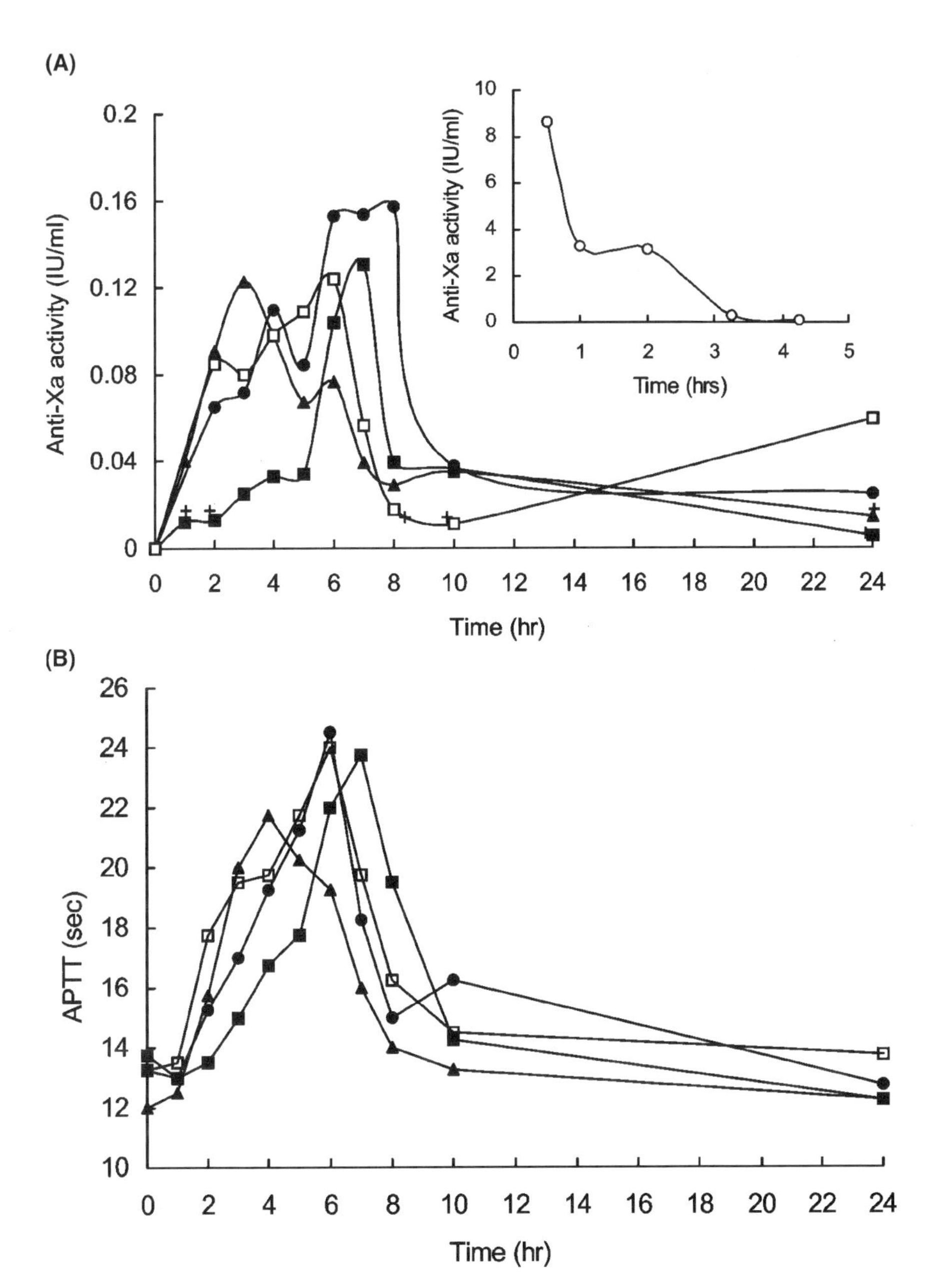

Figure 22 Mean prolongation of (**A**) antifactor Xa activity and (**B**) APTT over 24 hours following oral administration of UFH-loaded nanoparticles (2000 IU) prepared with blends of biodegradable and nonbiodegradable polymers (ratio 1/1), RL/PCL (●), RS/RL/PLGA (▲), RS/PLGA (■), and Eudragit RL (□). The inset represents the mean prolongation of antifactor Xa activity after intravenous administration of an aqueous solution of heparin (2000 IU) in rabbits. Data are shown as mean of four rabbits. +: Anti-Xa activity beyond the limit of detection. Standard deviations are not represented for sake of clarity. *Abbreviations*: APTT, activated partial thromboplastin time; UFH, unfractionated heparins; PCL, poly-ε caprolactone; PLGA, poly (D,L-lactic-co-glycolic acid 50/50). *Source*: From Ref. 73.

Table 11 Main Pharmacokinetic Parameters of UFH-Loaded Nanoparticles After Oral Administration in Rabbits (2000 IU)

Formulations	RL/PCL	RS/RL/PLGA[a]	RS/PLGA[a]	RL	Heparin in solution (IV route)
C_{max} (IU/mL)	0.16 ± 0.01	0.12 ± 0.05	0.13 ± 0.06	0.12 ± 0.04	Not determined
t_{max} (h)	6–8	3–5	6–8	5–6	Not determined
$AUC_{0\to24}$/kg (IU h/mL/kg)	0.74 ± 0.06[b]	0.39 ± 0.08	0.29 ± 0.03	0.32 ± 0.07	3.30 ± 0.59
Absolute bioavailability (%)	22.73 ± 5.46[b]	12.10 ± 2.41	9.07 ± 0.81	9.87 ± 2.06	100

Note: AUC indicates area under the curve. Data are mean ± SEM ($n = 4$).
[a]$n = 3$.
[b]Statistically different from RS/RL/PLGA, RS/PLGA or RL nanoparticles at $P < 0.05$ (Student's *t*-test).
Abbreviations: UFH, unfractionated heparins; PCL, poly-ε-caprolactone; PLGA, poly (D,L-lactic-co-glycolic acid 50/50.
Source: From Ref. 73.

were obtained for the other formulations. The bioavailability figures presented in Table 11 confirm that not all the encapsulated heparin is totally absorbed. The best bioavailability was obtained with Eudragit® RL/PCL nanoparticles. The three other types of nanoparticles presented no statistical difference in absolute bioavailability. These results, especially those obtained with Eudragit® RL/PCL nanoparticles, are very promising considering the low heparin dose administered orally. Indeed, our goal was to show the potential of heparin-loaded nanoparticles by using the same dose as that commonly administered by the intravenous route, i.e., 600 IU/kg, in treatment. This corresponds to a worst-case protocol, because it is well known that oral bioavailability is always lower than the intravenous one.

As determined for microparticles, it was also of interest to confirm that similar results using nanoparticles could be obtained with LMWH. Therefore, in a second set of in vivo trials, Eudragit® RS/PCL nanoparticles were administered orally in rabbits at two doses (200 and 600 IU/kg). As usual with these nanoparticles, a lag time of about two to three hours was observed (Fig. 23). In the studied dose range, a linear relationship was observed between each dose and its respective AUC. In terms of absolute bioavailability, high and similar figures were obtained with each dose (50%) demonstrating the absence of dose effect between 200 and 600 IU/kg.

Absorption Mechanisms: The results obtained with micro- and nanoparticles with both UFH and LMWH definitely show the absorption of heparin from multiparticular systems and open up the discussion of the absorption mechanisms. Considering the rather large mean diameter of microparticles, direct absorption either through the GIT or after uptake by Peyer's patches can be discarded. In contrast, multiparticular dosage forms increase enormously the surface area of contact with the GIT. This enhanced contact is also responsible for an increase in the concentration gradient, which could partially explain the observed increase in bioavailability. Furthermore, it is well known (and confirmed in our studies) that heparin administered

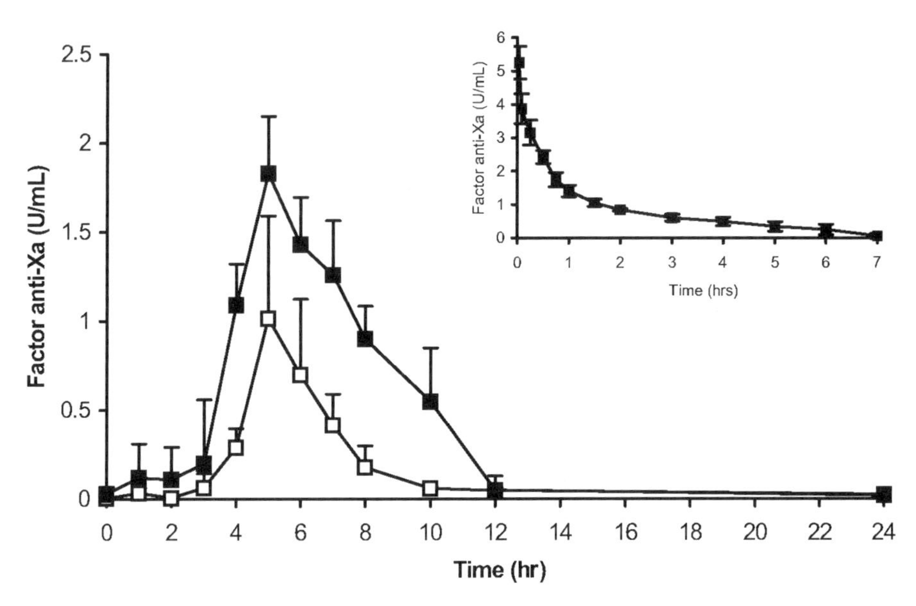

Figure 23 Time course of antifactor Xa activity after a single oral administration of LMWH-loaded NP at 600 anti-Xa U/Kg (NP600, ■) and 200 anti-Xa U/Kg (NP200, □) in fasted rabbits. Inset: mean antifactor Xa activity versus time after intravenous administration of LMWH in solution (200 anti-Xa U/Kg) in rabbits. Data are mean $\pm$ S.D. ($n \geq 6$). *Abbreviation*: LMWH, low molecular weight heparins.

orally as a solution is not absorbed because of its size and strong negative charge. The optimal anticoagulant activity of microparticles prepared with blends of PLGA and Eudragit® may be related to a low release of heparin in the acidic medium of the stomach, although they were neither gastroresistant nor presented in enterically coated hard capsules. Indeed, in the same conditions in vitro, heparin release was initially faster from PCL and PLGA microparticles compared to Eudragit® RS/PLGA and Eudragit® RS/RL/PLGA microparticles. Owing to this observation, it can be postulated that the amount of heparin still available for absorption is much higher with microparticles prepared with blends of polymers. On contact with the intestinal wall, heparin (UFH or LMWH), which was preserved from early gastric clearance, can be released by diffusion through the polymeric matrix—the release could still be increased by the partial biodegradation of PLGA. Another phenomenon could also contribute to the increase in the duration of action. During heparin release, the positively quaternary ammonium groups will be continuously unmasked. The mucus layer protecting the GIT is negatively charged—the electrostatic interactions between microparticles and mucus would increase the adhesion and, consequently, the retention time of microparticles. The dramatic increase in bioavailability of heparin in both Eudragit® RS/PLGA and Eudragit® RS/RL/PLGA microparticles results from a combination of the potential mechanisms mentioned previously.

The absorption mechanism of heparin from nanoparticles may be complicated by the absorption of the nanoparticles together with heparin. Indeed, for particles with a diameter less than 1 μm, three possible uptake mechanisms have been suggested for oral absorption of nanoparticles: uptake via a paracellular pathway, intracellular uptake and transport via the epithelial cells of the intestinal mucosa, and lymphatic uptake via the M cells and the Peyer's patches. However, like heparin alone, the heparin-loaded nanoparticles are negatively charged which does not favor the absorption of intact nanoparticles through the GIT. Conversely, it has been shown that nanoparticles are able to coat the gastrointestinal mucosa, thus increasing the surface area of the intestine in contact with the drug and, consequently, the drug gradient concentration toward the blood. Although it is difficult to predict the mechanism of heparin absorption in our study, the influence of the polycationic polymer seems important. As assumed for microparticles, some residual and non-complexed cationic charges, unmasked during heparin release, may indeed still increase the residence time of nanoparticles next to the absorption surface of the GIT and thus reinforce the heparin gradient through the GIT.

Research work is currently in progress to determine the exact absorption mechanisms of heparin (UFH or LMWH).

CONCLUSION

Research on the oral delivery of heparin has been carried out continuously for the last three decades (74). The formulations of such oral delivery systems could benefit hundreds of thousands of patients whose treatment requires daily parenteral injections. Early developments used UFH, which was the only heparin available at that time. Owing to its advantages, LMWH have mostly replaced UFH in clinical situations. This also explains the major use of LMWH in current developments in drug delivery systems. It is only recently, because of Emisphere permeation enhancers, that a major step in oral heparin delivery was achieved. Indeed, although it finally failed, Phase III studies were carried out. With the SNAC absorption enhancer, the best reported bioavailability was around 40%. Another interesting innovation concerns the use of bile acids, which also provided the absorption of heparin. Curiously, there are very few reports on heparin micro- and nanoencapsulation despite the tremendous number of drugs which have been encapsulated to enhance their bioavailability. One of the reasons could be the low encapsulation efficiency obtained with most classical polymers. The idea of blending a biodegradable polyester with polycationic polymethacrylate derivative polymers gives the possibility of achieving sufficient core loading corresponding to the amount to be administered either in animal models or in humans. It was demonstrated that PCL/Eudragit® RS and PLGA/Eudragit® RS micro- or nanoparticles have increased the absolute bioavailability of UFH and LMWH in two animal species. Studies are still ongoing to better explain the absorption mechanisms to initiate the first clinical trials in humans in the near future.

REFERENCES

1. Hirsh J. Drug therapy. Heparin. N Engl J Med 1991; 324:1565–1574.
2. Harrison L, McGinnis J, Crowther M, Ginsberg J, Hirsh J. Assessment of outpatient treatment of deep vein thrombosis with low molecular weight heparin. Arch Intern Med 1998; 158:2001–2003.
3. Hyers TM, Agnelli G, Hull R, et al. Antithrombotic therapy for venous thromboembolic disease. Chest 1998; 114:561S–578S.
4. Wells PS, Kovacs MJ, Bormanis J, et al. Expanding eligibility for outpatient treatment of deep vein thrombosis and pulmonary embolism with low molecular weight heparin. Arch Intern Med 1998; 158:1809–1812.

5. Hirsh J, Warkentin TE, Raschke R, Granger C, Ohman EM, Dalen JE. Heparin and low molecular weight heparin. Mechanisms of action, pharmacokinetics, dosing considerations, monitoring, efficacy and safety. Chest 1998; 114(5 suppl):489S–510S.
6. Hirsh J, Anand SS, Halperin JL, Fuster V. Guide to anticoagulant therapy: heparin: a statement for healthcare professionals from the American Heart Association. Circulation 2001; 103(24):2994–3018.
7. Fareed J, Hoppensteadt D, Jeske W, Clarizio R, Walenga JM. Low molecular weight heparins: a developmental perspective. Exp Opin Invest Drugs 1997; 6:705–733.
8. Hull RD, Pineo GF, Raskob GE. The economic impact of treating deep vein thrombosis with low molecular weight heparin: outcome of therapy and health economy aspects. Hemostasis 1998; 28:8–16.
9. Greer IA. Exploring the role of low-molecular-weight heparin in pregnancy. Semin Thrombosis Hemostasis 2002; 28:25–31.
10. Donohoe S, Quenby S, Mackie I, et al. Fluctuations in levels of antiphospholipid antibodies and increased coagulation activation markers in normal and heparin-treated antiphospholipid syndrome pregnancies. Lupus 2002; 11:11–20.
11. Mahato RI, Narang AS, Thoma L, Miller DD. Emerging trends in oral delivery of peptide and protein drugs. Crit Rev Therap Drug Carrier Syst 2003; 20:153–214.
12. Drewe J, Fricker G, Vonderscher J, Beglinger C. Enteral absorption of octreotide: absorption enhancement by polyoxyethylene-24-cholesterol ether. Br J Pharmacol 1998; 108:298–303.
13. Yamada K, Murakami M, Yamamoto A, Takada K, Muranishi S. Improvement of intestinal absorption of thyrotropin-releasing hormone by chemical modifications with lauric acid. J Pharm Pharmacol 1992; 44:717–721.
14. Tarr BD, Yalkowsky SH. Enhanced intestinal absorption of cyclosporine in rats through the reduction of emulsion droplet size. Pharm Res 1989; 6:40–43.
15. Watnasirichaikul S, Davies NM, Rades T, Tucker IG. Preparation of biodegradable insulin nanocapsules from biocompatible microemulsion. Pharm Res 2000; 17:684–689.
16. Bernkop-Schnurch A. The use of inhibitory agents to overcome the enzymatic barrier to perorally administered therapeutic peptides and proteins. J Contr Release 1998; 52:1–16.
17. Leone-Bay A, Paton DR, Weidner JJ. The development of delivery agents that facilitate the oral absorption of macromolecular drugs. Med Res Rev 2000; 20:169–186.
18. Sakariassen KS, Hansen SR, Cadroy Y. Methods and models to evaluate shear-dependent and surface reactivity-dependent antithrombotic efficacy. Thromb Res 2001; 104:149–174.
19. Shieh SJ, Chiu HY, Shi GY, Wu CM, Wang JC, Chen CH. A novel platelet-rich arterial thrombosis model in rabbits. Simple, reproducible, and dynamic real-time measurement by using double-opposing inverted-sutures model. Thromb Res 2001; 103:363–376.
20. Lewis JH. Comparative Hemostasis in Vertebrates. New York: Plenum Press, 1996.
21. Lin JH, Yamazaki M. Role of P-glycoprotein in pharmacokinetics: clinical implications. Clin Pharmacokinet 2003; 42:59–98.
22. Petitou M, Herault JP, Bernat A, Driguez PA, Duchaussoy P, Lormeau JC, Herbert JM. Synthesis of thrombin-inhibiting heparin mimetics without side effects. Nature 1999; 398:417–422.
23. Windsor E, Gronheim GE. Gastrointestinal absorption of heparin and synthetic heparinoids. Nature 1961; 190:263–264.
24. Tidball CS, Lipman RI. Enhancement of jejunal absorption of heparinoid sodium ethylenediaminetetraacetate in the dog. Proc Soc Exp Biol Med 1962; 111:713–715.
25. Engel RH, Riggi SJ. Intestinal absorption of heparin facilitated by sulphated or sulphonated surfactant. Proc Soc Exp Biol Med 1969; 130:706–710.
26. Koh TY, Bacharucha KR. Intestinal absorption of stable heparinic acid complex. J Lab Clin Med 1972; 80:47–55.
27. Sue TK, Jaques LK, Yven E. Effect of acidity, cations and alcoholic fractionation on absorption of heparin from gastrointestinal tract. Can J Physiol Pharmacol 1976; 54: 613–617.

28. Morton AK, Edwards HE, Allen JC, Philipps GO. An evaluation of the oral administration of commercial and fractionated samples in rats. Int J Pharm 1981; 9:321–335.
29. Leone-Bay A, Ho KK, Agarwal R, et al. 4-[4-2-hydroxybenzoyl)amino]phenylbutyric acid as novel oral delivery agent for recombinant human growth hormone. J Med Chem 1996; 39:2571–2578.
30. Leone-Bay A, Santiago N, Achan D, et al. *N*-Acylated a-amino acids as novel oral delivery agents. J Med Chem 1995; 38:4263–4269.
31. Leone-Bay A, Leipold H, Sarubbi D, Variano B, Rivera T, Baughman RA. Oral delivery of sodium cromolyn: preliminary studies in vivo and in vitro. Pharm Res 1996; 13(2):222–226.
32. Rivera TM, Leone-Bay A, Paton DR, Leipold HR, Baughman RA. Oral delivery of heparin with sodium N-[8-(2-hydroxybenzoyl)amino]caprylate: pharmacological considerations. Pharm Res 1997; 14(12):1830–1834.
33. Leone-Bay A, Paton DR, Freeman J, et al. Synthesis and evaluation of compounds that facilitate the gastrointestinal absorption of heparin. J Med Chem 1998; 41:1163–1171.
34. Leone-Bay A, Freeman J, O'Toole D, et al. Studies directed at the use of a parallel synthesis matrix to increase throughput in an in vivo assay. J Med Chem 2000; 43:3573–3576.
35. Leone-Bay A, Paton DR, Variano B, et al. Acylated non-α-amino acids as novel agents for the oral delivery of heparin sodium, USP. J Contr Release 1998; 50:41–49.
36. Gonze MD, Manord JD, Leone-Bay A, et al. Orally administered heparin for preventing deep venous thrombosis. Am J Surg 1998; 176:176–178.
37. Gonze MD, Salartash K, Sternbergh WC III, et al. Orally administered unfractionated heparin with carrier agent is therapeutic for deep venous thrombosis. Circulation 2000; 101:2658–2661.
38. Welt FGP, Woods TC, Edelman ER. Oral heparin prevents neointimal hyperplasia after arterial injury. Circulation 2001; 104:3121–3124.
39. Leone-Bay A, O'Shaughnessy C, Agarwal R, et al. Oral low molecular weight heparin absorption. Pharm Tech March 2002; 1:38–46.
40. Salartash K, Lepore M, Gonze MD, et al. Treatment of experimentally induced caval thrombosis with oral low molecular weight heparin and delivery agent in a porcine model of deep venous thrombosis. Ann Surg 1999; 231(6):789–794.
41. Salartash K, Gonze MD, Leone-Bay A, Baughman R, Sternbergh WC III, Money SR. Oral low-molecular weight heparin and delivery agent prevents jugular venous thrombosis in the rat. J Vasc Surg 1999; 30:526–532.
42. Malkov D, Wang HZ, Dinh S, Gomez-Orellana I. Pathway of oral absorption of heparin with sodium N-[8-(2-hydroxybenzoyl)amino]caprylate. Pharm Res 2002; 19(8): 1180–1184.
43. Money SR, York JW. Development of oral heparin therapy for prophylaxis and treatment of deep venous thrombosis. Cardiovasc Surg 2001; 9(3):211–218.
44. Baughman RA, Kapoor SC, Agarwal RK, Kisicki J, Catella-Lawson F, FitzGerald GA. Oral delivery of anticoagulant doses of heparin. A randomized, double-blind, controlled study in humans. Circulation 1998; 98:1610–1615.
45. Berkowitz SD, Marder VJ, Kosutic G, Baughman RA. Oral heparin administration with a novel drug delivery agent (SNAC) in healthy volunteers and patients undergoing elective total hip arthroplasty. J Thromb Haemost 2003; 1:1914–1919.
46. Pineo GF, Hull RD, Marder VJ. Orally active heparin and low molecular weight heparin. Curr Opin Pulm Med 2001; 7:344–348.
47. Rama Prasad YV, Minamimoto T, Yoshikawa Y, et al. In situ intestinal absorption studies of low molecular weight heparin in rats using Labrasol as absorption enhancer. Int J Pharm 2004; 271:225–232.
48. Swaan PW, Szoka FC, Øie S. Use of the intestinal and hepatic bile acid transporters for drug delivery. Adv Drug Del Rev 1996; 20:59–82.
49. Lee Y, Moon HT, Byun Y. Preparation of slightly hydrophobic heparin derivatives which can be used for solvent casting in polymeric formulation. Thromb Res 1998; 92:149–156.

50. Lee Y, Kim SH, Byun Y. Oral delivery of new heparin derivatives in rats. Pharm Res 2000; 17(10):1259–1264.
51. Lee Y, Nam JH, Shin HC, Byun Y. Conjugation of low-molecular-weight heparin and deoxycholic acid for the development of a new oral anticoagulant agent. Circulation 2001; 104:3116–3120.
52. Thanou M, Nihot MT, Jansen M, Verhoef JC, Junginger HE. Mono-*N*-carboxymethyl chitosan (MCC), a polyampholytic chitosan derivative, enhances the intestinal absorption of low molecular weight heparin across the intestinal epithelia in vitro and in vivo. J Pharm Sci 2001; 90(1):38–46.
53. Thanou M, Verhoef JC, Nihot MT, Verheijden JHM, Junginger HE. Enhancement of the intestinal absorption of low molecular weight heparin (LMWH) in rats and pigs using carbopol® 934P. Pharm Res 2001; 18(11):1638–1641.
54. Bernkop-Schnürch A, Kast CE, Guggi D. Permeation enhancing polymers in oral delivery of hydrophilic macromolecules: thiomer/GSH systems. J Control Rel 2003; 93:95–103.
55. Ueno M, Nakasaki T, Horikoshi I, Sakuragawa N. Oral administration of liposomally-entrapped heparin to beagle dogs. Chem Pharm Bull 1977; 25(5):1159–1161.
56. Muranishi S, Tokunaga Y, Tanigushi K, Sezaki H. Potential absorption of heparin from the small intestine and the large intestine in the presence of monoolein mixed micelles. Chem Pharm Bull 1977; 25:1159–1161.
57. Dal Pozzo A, Acquasaliente M, Geron MR, Andriuoli G. New heparin complexes active by intestinal absorption. I. Multiple ion pairs with organic compounds. Thromb Res 1989; 56:119–124.
58. Andriuoli G, Bossi M, Caramazza I, Zoppetti G. Heparin by alternative routes of administration. Haemostasis 1990; 20(suppl 1):154–158.
59. Zoppetti G, Caramazza I, Murakami Y, Ohno T. Structural requirements for duodenal permeability of heparin–diamine complexes. Biochem Biophys Acta 1992; 1156:92–98.
60. Andriuoli G, Caramazza I, Galimberti G, et al. Intraduodenal absorption in the rabbit of a novel heparin salt. Haemostasis 1992; 22:113–116.
61. Zinutti C, Barberi-Heyob M, Hoffman M, Maincent P. In-vivo evaluation of sustained release microspheres of 5-FU in rabbits. Int J Pharm 1998; 166:231–234.
62. Damgé C, Michel C, Aprahamian M, Couvreur P. New approach for oral administration of insulin with polyalkylcyanoacrylate nanocapsules as drug carriers. Diabetes 1988; 37:246–251.
63. Steiner S, Rosen R. Delivery system for pharmacological agents encapsulated with proteinoids. U.S. Patent 4,925,673 (1990).
64. Leone-Bay A, Leipold H, Agarwal R, Rivera T, Baughman RA. The evolution of an oral heparin dosing solution. Drugs Future 1997; 22:885–891.
65. Ma X, Santiago N, Chen YS, Chaudhary K, Milstein S, Baughman RA. Stability study of drug-loaded proteinoid microsphere formulations during freeze-drying. J Drug Targ 1994; 2:9–21.
66. Jiao Y, Ubrich N, Marchand-Arvier M, Vigneron C, Hoffman M, Maincent P. Preparation and in vitro evaluation of heparin-loaded polymeric nanoparticles. Drug Del 2001; 8(3):135–141.
67. Jiao Y, Ubrich N, Hoffart V, et al. Preparation and characterization of heparin-loaded polymeric microparticles. Drug Dev Ind Pharm 2002; 28(8):1033–1041.
68. Hoffart V, Ubrich N, Simonin C, et al. Low molecular weight heparin-loaded polymeric nanoparticles: formulation, characterization, and release characteristics. Drug Dev Ind Pharm 2002; 28(9):1091–1099.
69. Hoffart V, Ubrich N, Lamprecht A, et al. Microencapsulation of low molecular weight heparin into polymeric particles designed with biodegradable and nonbiodegradable polycationic polymers. Drug Del 2003; 10:1–7.
70. Chernysheva YV, Babak VG, Kildeeva NR, et al. Effect of the type of hydrophobic polymers on the size of nanoparticles obtained by emulsification-solvent evaporation. Mendeleev Commun 2003; 2:65–68.

71. Jameela SR, Suma N, Jayakrishman A. Protein release from poly(ε-caprolactone) microspheres prepared by melt encapsulation and solvent evaporation techniques: a comparative study. J Biomat Sci Polymer 1997; 8(6):457–466.
72. Jiao Y, Ubrich N, Hoffart V, et al. Anticoagulant activity of heparin following oral administration of heparin-loaded microparticles in rabbits. J Pharm Sci 2002; 91:760–768.
73. Jiao Y, Ubrich N, Marchand-Arvier M, et al. In vitro and in vivo evaluation of oral heparin-loaded polymeric nanoparticles in rabbits. Circulation 2002; 105:230–235.
74. Ubrich N, Hoffart V, Vigneron C, Hoffman M, Maincent P. Digestive absorption of heparin with alternative formulations. STP Pharm Sci 2002; 12(3):147–155.

15
Particulate Systems for Oral Drug Delivery

María José Blanco-Príeto
Centro Galénico, Farmacia y Tecnologí́a Farmacéutica, Universidad de Navarra, Pamplona, Spain

Florence Delie
School of Pharmaceutical Sciences (EPGL), University of Geneva, Quai Ernest-Ansermet, Geneva, Switzerland

INTRODUCTION

The low oral bioavailability of some drugs is clearly a limitation to their development as a medicine. Transport mechanisms across the intestinal barrier are numerous and vary greatly depending on the physicochemical properties of a given substance. Depending on the molecule, inefficient oral delivery may result from several mechanisms. Indeed, the drug may be unstable in the gastrointestinal (GI) tract, which is very rich in ubiquitous enzymes with a wide array of activity. On the other hand, the intestinal epithelium represents a rather impermeable barrier. Only small molecules can cross the intestinal barrier via the paracellular route. For other drugs not using existing receptors or receptor-like transport, it is necessary to have the perfect partition coefficient to be able to pass through the bilipidic layer of the cell membrane and still be able to stay dissolved in the hydrophilic environment of extracellular and cytosolic environments. Thus, basically, very hydrophilic or lipophilic drugs are unable to cross the intestinal barrier. Furthermore, after absorption, drugs may also suffer from a significant hepatic first pass, fast elimination, or unwanted large distribution. For these drugs, encapsulation in polymeric particles may offer several advantages by isolating the molecule from the external milieu, promoting absorption, and providing a controlled release and/or a targetable system. Moreover, encapsulation will decrease adverse effects, in particular, of irritating compounds by preventing contact with the mucosal wall.

Polymeric particles have been shown to cross the intestinal wall, although only in minute quantities. The size of the particles, as well as the nature of the polymer, are critical parameters involved in particle uptake by the GI tract. Therefore, this chapter will first introduce a brief overview of the physiology of particle absorption. Then, the studies using polymeric particles to improve oral drug delivery will be reviewed. A great deal of the work found in the literature has focused on peptide and protein delivery; however, particles may also be of interest for other drugs.

ABSORPTION OF POLYMERIC PARTICULATES FROM THE GI TRACT

Oral absorption of particulates was described as early as 1844 (1,2). Several studies conducted after either chronic or single administration defined the parameters and the mechanisms involved in intestinal absorption of particles. A better knowledge of the mechanisms implicated in particle absorption is needed to design new polymers and better-targeted systems for the oral route.

Physiologically, GI functions are to digest and to absorb nutrients, water, and vitamins from food. On the other hand, it is also designed as a barrier to restrain the entry of pathogens, toxins, and undigested macromolecules. The histological architecture of the enteric wall is schematically depicted in Figure 1. The GI tract is lined with an epithelium made of a mosaic of cells among which absorptive cells (enterocytes) and goblet cells (secreting the mucus) may be distinguished. These cells are tightly held together and form a strong barrier covered by a layer of mucus. Lymphoid follicles, part of the gut-associated lymphoid system (GALT), involved in the development of the mucosal immune response, are interspersed in the enterocyte layer. Lymphoid follicles may be diffusely distributed or clustered in so-called Peyer's patches. The number and location of Peyer's patches vary widely between species and individuals and are also age dependent (3). These follicles are overlaid by the follicle-associated epithelium (FAE), which is comprised of enterocytes, M cells differentiated from the enterocytes, and a few goblet cells. It is the site where antigens are first encountered. FAE with the M cells has been described as a privileged place for particle uptake.

Volkheimer et al. (4–7) extensively studied the oral absorption of various materials of different size and composition in different species. This series of studies aimed

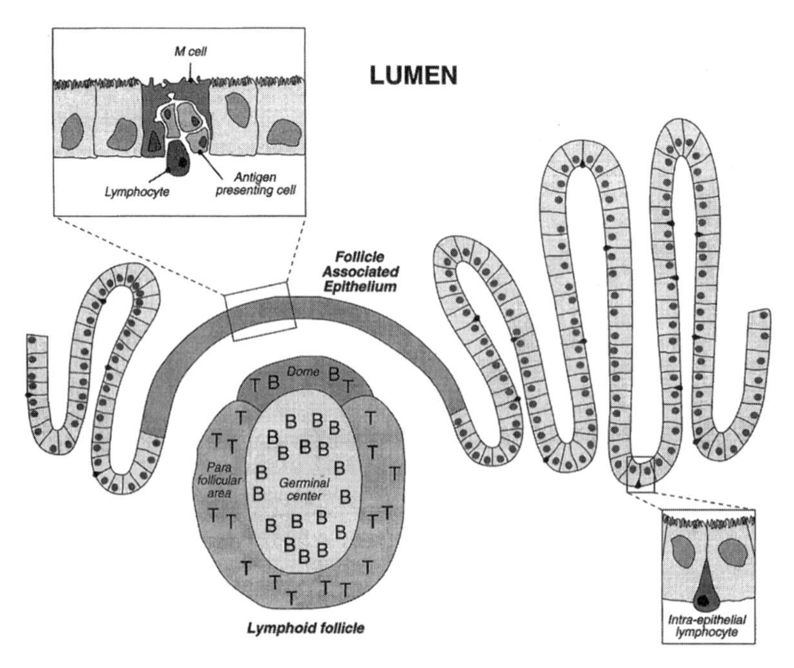

Figure 1 Schematic representation of the intestinal epithelium with a view of a Peyer's patch and detail of the follicle associated epithelium.

to provide a better knowledge of the absorption of large, solid, undissolved food particles from the intestinal lumen. These observations led to the conclusion that the "persorbability" of particles is limited by their size and "hardness"; the absorption occurred both in lymph and portal blood (2). The best results were obtained with hard particles with a diameter between 7 and 70 μm in humans. They also observed the passage in blood of particles as large as 150 μm made of polyvinyl chloride (PVC) rounded quartz particles (7).

These results attracted the interest of several nutritionists, physiologists, and toxicologists but the ingestion of such big particles remains unexplained. However, later studies were conducted neither in humans nor with the same material. Despite the interspecies specificity, the contradictory results may be related to one characteristic of the main material that Volkheimer et al. (4–7) used, namely, starch. More recent studies demonstrated that starch particles were able to open the tight junctions (TJ) of Caco-2 cells (8–10). The opening of the TJ in the presence of starch could give the large particles a way to cross the epithelial barrier. This explanation could also clarify the role of gut motility on particle absorption. Indeed, absorption rate increases with stimulants such as neostigmine or caffeine and decreases in the presence of atropine or barbituric (6).

Another group conducted numerous studies and defined more clearly the main characteristics of the "absorbability" of particles (11). They first described the uptake of latex particles (2 μm) in Peyer's patch and non-Peyer's patch areas (12). This study established that the absorption occurred in both tissues and was dose dependent. It was also observed that particles tended to accumulate in macrophages and they were transported in lymph rather than blood. In another report, the absorption of styrene divinylbenzene (SDB, 5.7 μm) and carbonized SDB (15.8 μm) particles in mice was investigated (13). In contrast to Volkheimer's work, no large particles were found in blood or tissues (liver, spleen, and lungs) neither after acute nor after chronic administration (60 days). In the 5.7-μm particle fed group, particles were found in Peyer's patches, mesenteric lymph nodes, and lungs. The authors assumed that the particles reached the lungs by the lymphatic pathway, therefore confirming data from Bertheusen et al. (14) assigning the lungs as an eliminatory organ for particulates. It was also demonstrated that particle accumulation takes place in both germ-free and conventional mice, indicating that intestinal flora is not essential for particle absorption (15). To define the role of the composition of the particles in the absorption rate, mice were fed for three months with different materials listed in Table 1 (16). Only carbon and iron dioxide were clearly and unequivocally present in intestinal segments containing Peyer's patches. These were the smallest particles used but were also the most hydrophobic. However, the observations were done at the light microscopy level and on small samples. Thus, the

Table 1 Different Materials Administered Orally

Material	Size (μm)
Chrysotile asbestos	1–5 (needles)
Quartz	<5
Carmine	2
Carbon	0.05
Iron oxide	0.1

Source: From Ref. 16.

authors did not rule out having missed some particles. The comparison of fluorescent particle absorption in aged and young mice showed that more particles were accumulated in the intestine of aged mice and more in the lungs of young mice (17). Bhagat and Dalal (18) confirmed this observation. Consistent with these results, Seifert et al. (19) compared the uptake of 1-μm fluorescein isothiocyanate–labeled polystyrene (PS) particles in thoracic lymph samples in young rats (six to eight weeks) versus five- and nine-month-old animals. The following ranking was found: five-month-old animals absorbed more particles than nine-month-old and younger animals absorbed fewer particles.

In summary, the studies by LeFevre and coworkers demonstrated that particle uptake occurs in both Peyer's patch and non-Peyer's patch areas of the intestine and the absorption is size dependent. The absorbed particles are transported into the mesenteric lymph nodes and few are found in other organs.

The uptake of particles is described as very rapid and usually the authors relate that the absorption can be detected in their first samples: 8 minutes, 10 minutes, 1 hour, and 6 hours (2,20–22).

Several authors tried to locate the preferential absorption sites of particles. Kreuter et al. (23) followed the absorption of radiolabeled poly(hexylcyanoacrylate) nanoparticles (0.2–0.3 μm). They observed a very irregular distribution of the radioactivity along the intestinal mucosa with up to a 1000-fold variation. Michel et al. (24) administered insulin loaded poly(isobutylcyanoacrylate) (PIBCA) nanoparticles (0.3 μm) at various sites of rat intestine. The subsequent hypoglycemia was more pronounced when the particles were administered in the ileum as compared to jejunum, duodenum, colon, and stomach. However, only the pharmacological activity of the drug was monitored and this study did not address physiological evidence of particulate uptake.

Some authors describe an absorption exclusively in the Peyer's patches, whereas others observed the translocation of particles in non-Peyer's patch areas (25–28). Wells et al. (29) demonstrated the absorption of PS particles in dogs, using isolated intestinal loops devoid of perceptible Peyer's patches. However, even in the absence of macroscopically visible Peyer's patches, absorption could occur in invisible lymphoid follicles. Indeed, constituted Peyer's patches are rather rare in dogs and GALT is essentially represented by isolated and discrete lymphoid follicles. On the other hand, Pappo and Ermak (20) clearly demonstrated, by immunochemistry, that the M cells were responsible for particulate uptake. Cytochemistry and scanning electron microscopy (SEM) of rabbit Peyer's patches after administration of fluorescent particles (0.67 μm) in isolated intestinal loops showed that prior to absorption, particles are adsorbed to the cell surface. The adsorption of particles occurs preferentially at the periphery of the lymphoid follicle where the M cells are located and the authors concluded that the absorption is restricted to the M cells. Jepson et al. (30) confirmed the primary role of the M cells by using antivimentin staining in rabbit tissue after instillation of fluorescent PS particles (0.46 and 0.67 μm) in intestinal loops containing Peyer's patches (30). The uptake seems to be preceded by the adhesion of the particles at the cell surface followed by the internalization (31). However, the particles are rarely described in the cytoplasm of M cells but more frequently associated with the macrophages present in the central pocket (20,32,33). The majority of the endocytosed microspheres are delivered to leukocytes located in the central hollow of the M cells.

Several authors have described the presence of particles in enterocytes (21,32, 34). Likewise, McClean et al. (28) observed little discrimination between Peyer's patch and non-Peyer's patch areas after administration of polylactic acid (PLA) micro- and nanoparticles in rats and rabbits (28). The role of enterocytes in particle uptake was further established by the observations of the absorption of starch, e.g., PS, poly(acrylic) acid, conjugates of *n*-butylcyanoacrylate, poly(methylidene malonate 2.1.2 (PMM2.1.2) and poly(lactide-co-glycolide) (PLGA) particles in the Caco-2 cell line. These cells originating from a human colorectal cancer provide an excellent model for permeability studies (8,9,35–43). Caco-2 cells are polarized and grow as monolayers similar to the normal small intestine villus cells; no analogy with M cells has been described under normal conditions of culture. The coculture of Caco-2 and intestinal lymphocytes appears to induce morphological changes of enterocytes, which then look like M cells (44). This may provide an interesting tool to study intestinal uptake of particles, however, it seems that some difficulties in reproducing this model have been encountered (45). Recently, the coculture model was adapted (46,47). The cells obtained by these methods displayed the known markers of M cells phenotype (alkaline phosphatase, villin, $\alpha5\beta1$ integrin, lectins, E-selectin, etc.) and might represent a good model to develop new vaccination particulate systems.

Several authors, using different methodologies have studied the fate of internalized particles in vivo. After uptake by M cells, the particles are phagocytosed by macrophages and transported in mesenteric lymph nodes (25,26,29). The distribution to the different organs of the body seems to involve the macrophages (13,26,48). Hematogenous translocation has been described by Matsuno et al. (34) for very small Percoll particles (0.02–0.03 µm). The particles that are able to reach the blood circulation are then phagocytosed by the mononuclear phagocyte system (MPS) as the Kupfer cells of the liver or the spleen (49). Recently, Pinto-Alphandary et al. have used PIBCA nanocapsules loaded with gold-labeled insulin to follow under transmission electron microscopy (TEM) the fate of particles (60–250 nm) after intragastric administration to rats (50). Particles were found in both Peyer's patch and non-Peyer's patch areas; however, translocation through enterocytes appeared to be limited to particles smaller than 150 nm. On the other hand, in Peyer's patch areas, nanoparticles took the appearance of "rings" and gold particles attached to insulin were scattered in the surrounding milieu, suggesting that the polymeric wall had been at least partially hydrolyzed by the esterases present in the M cell cytosol. Intact nanocapsules were seen in underlying tissues, even in blood capillaries, showing that they could cross the epithelium. Although no specific mechanism could be identified, this study emphasized the fact that nanoparticles can cross the intestinal epithelium either via Peyer's patch or via non-Peyer's patch areas.

Particle size appears to be very critical for absorption and biodistribution of particles. Volkheimer et al. (4,5) set the limit up to 150 µm, but more recent studies established that 10 µm is the real threshold, at least for PS and PLGA. Eldridge et al. (25) reported that, after administration of PLGA microspheres to mice, small particles (<5 µm) are transported into the lymph within the macrophages whereas larger particles (>5 µm) remain in the Peyer's patches. Accordingly, Jani et al. observed the absence of PS particles with a diameter larger than 3 µm in the different organs they studied (51). These results and, particularly, the absence of particles in lungs and heart were confirmed in another study (49). These observations are partly in discordance with LeFevre et al. (13) who looked for particles in mice fed with 5.7 µm latex particles for 60 days. Particles were found in Peyer's patches, lungs, mesenteric lymph nodes, and liver even 77 days after the treatment had been stopped but no particles were found in the spleen. More recently, Carr et al. (52) described the absorption and distribution of 2 and 6 μm latex particles. It appeared that 6-μm particles were more efficient in terms of volume translocated to the lymph nodes implying that, although fewer particles are found, more drug would be transported with this type of larger particle.

An important part of the work on the intestinal absorption of particulates has been done with PS latex beads, which have the advantages of being commercially available, suitable for fluorescent labeling, and homogeneous in size. Furthermore, they can be easily coated with proteins or adhesion factors. However, since they are not biodegradable, latex particles cannot be considered as suitable drug carriers. Therefore, different authors have been interested in other polymers more suitable for drug administration. The uptake of microspheres by M cells is preceded by an adhesion step (20,31). It is, therefore, likely that the interaction of the particles with the cell membrane plays a determinant role in particle transport through the epithelium. This interaction is mainly influenced by the polymer or the material, by the charge of the polymer, and by a possible coating (13,15,31,53–55). From these few studies, it appears that the mechanisms involved and the location of the uptake do not vary according to the different polymers used (except for starch particles already mentioned above). The absorption efficiency is, however, greatly influenced by the size and the nature of particle composition. LeFevre et al. (16) pointed out the importance of the composition of the particles as well as the size. Jani et al. (54) confirmed that the uptake rate is greater with nonionized particles compared to the carboxylated ones. Eldridge et al. (53) compared the absorption rate for microspheres (from 1 to 10 μm) prepared with 10 polymers with different degrees of hydrophobicity (Table 2). After oral administration of a single dose, Peyer's patches were harvested and the microspheres were counted. The results showed that 7 out of 10 polymers are absorbed and are predominantly present in the dome of the Peyer's patches. The efficiency of uptake correlates with the relative hydrophobicity of the polymer. The uptake occurred only in Peyer's patches and was restricted to particles <10 μm. Jepson et al. (56) compared the absorption of PLGA and latex microspheres of the same size and found that the PLGA

Table 2 Targeted Absorption of 1 to 10 μm Microspheres with Various Excipients by PP Following Single Oral Administration

Polymer	Biodegradability	Absorption by PP[a]
Poly(styrene)	No	Very good
Poly(methymetacrylate)	No	Very good
Poly(hydroxybutyrate)	Yes	Very good
Poly(DL-lactide)	Yes	Good
Poly(L-lactide)	Yes	Good
85:15 Poly(DL-lactide-co-glycolide)	Yes	Good
50:50 Poly(DL-lactide-co-glycolide)	Yes	Good
Cellulose acetate hydrogen phtalate	No	None
Cellulose triacetate	No	None
Ethyl cellulose	No	None

[a]The number of particles was scored as follows: 1000 to 1500, very good; 200 to 1000, good; <10, none.
Abbreviation: PP, Peyer's patches.
Source: From Ref. 53.

microspheres were absorbed with an order of magnitude lower than latex. When binding to M cells occurred, PLGA microspheres were efficiently translocated across the cells. However, these particles were randomly distributed over the intestinal epithelium and the binding to enterocytes was much greater than with PS particles. McClean et al. showed that PLA particles (nanospheres of 0.3–2 μm and microspheres of 2–30 μm) were absorbed from both Peyer's patch and non-Peyer's patch areas in rabbits (28).

To date, not all the particles' characteristics (hydrophobicity, charge, zeta potential, coating with lectins or other adhesion factors, etc.) have been investigated in a systematic manner. This is due in part to the difficulties of quantifying the absorption as well as the interindividual and interspecies differences encountered with in vivo studies (57). This may explain partly the discrepancies observed between the different reports.

In summary, many studies have been conducted on particle absorption from the GI tract, and some discrepancies exist, especially regarding the uptake mechanisms, the location of the absorption site as well as the quantity of particle taken up, however, several conclusions can be drawn. The absorption takes place primarily, but not exclusively, in the Peyer's patches at the level of the M cells. Uptake is very fast and seems to be the result of a transcellular mechanism, but a paracellular pathway, although very unlikely, cannot be excluded. The interface particulate/cell membranes appear to play an important role and all the parameters involved in the binding of particles at the cell surface have not yet been identified. The modification of particle surface properties due to the nature of the polymer or to the binding of specific proteins or lectins can modulate the absorption rate.

Although very scarce, particle absorption is still sufficient to enhance the oral bioavailability of some drugs as will be discussed in the next part of this chapter. Efforts need to be made, however, in developing new polymers and new surface characteristics to increase the particle affinity for the epithelium surface. On the other hand, one has to bear in mind that particle resorption is not a prerequisite. Indeed, the presence of a polymeric wall provides a protection from the GI environment and may favor a prolonged contact with the epithelium that may be sufficient to increase the bioavailability of certain drugs.

USE OF POLYMERIC PARTICLES FOR ORAL ADMINISTRATION

The use of particle formulations for poorly absorbed drugs has been chosen for several reasons, depending on the active substance. The presence of a polymeric wall around the drug will provide a good shield against the attack from the various and widespread enzymes existing in the GI tract. This is especially crucial for peptide and protein molecules. Because encapsulation enables a slow release profile into degradable polymeric particles, controlled release of compounds may be achieved. Furthermore, the preferential uptake of particles by the Peyer's patches, part of the GALT, makes them very attractive for inducing mucosal immunity; this issue will not be discussed in this chapter.

This part of the chapter will be devoted to the use of polymeric particles as a drug delivery system. The main contribution has been done in the field of peptides and proteins, especially with insulin; however, several studies have shown interest in particles for oral delivery with other drug models.

Peptides and Proteins

Bioavailability of peptide and protein drugs after oral administration is very low because of their instability in the GI tract and low permeability through the intestinal mucosa. Indeed, the development of a dosage form improving oral absorption of peptides drugs is one of the greatest challenges in pharmaceutical research (58,59). Therefore, studies maximizing the bioavailability of orally administered peptide drugs have been ongoing for many years. Approaches including particulate drug delivery such as nanoparticles or microparticles have raised a lot of interest.

Insulin

Insulin is the mainstay of drug therapy for patients with insulin-dependent diabetes mellitus (or Type I diabetes) (60). The disease is primarily caused by autoimmune destruction of insulin secreting β cells of the pancreas (61). Without adequate insulin secretion, patients cannot properly utilize glucose and typically have markedly elevated blood glucose (hyperglycemia) while the intracellular glucose level is generally low. The chronic complications of constantly elevated blood sugar are serious and include retinopathy, nephropathy, neuropathy, cardiovascular diseases, and peripheral vascular diseases as well as increased susceptibility to infections (61).

The control of glucose homeostasis is complex. Most patients need to self-administer twice-daily subcutaneous injections of insulin for a strict control of glycemia. Considering that this therapy imposes serious limitations on the patient's lifestyle much effort has been made to develop an oral dosage form for insulin including the use of polymeric particles. These different formulations are summarized in Table 3.

The first attempt to seek hypoglycemic effects after oral administration of insulin-loaded nanoparticles to diabetic rats was reported by Couvreur et al. (62). Insulin was adsorbed on the surface of 200 nm poly(alkylcyanoacrylate) nanoparticles prepared by polymerization of different monomers, i.e., methyl, ethyl, isobutyl, and butyl cyanoacrylates. No decrease in glucose level was observed after oral administration (52.5 IU kg^{-1}) to diabetic rats. Insulin, however, remained active after subcutaneous administration. This result suggests that the peptide was not protected from the proteolytic degradation in the GI tract.

Subsequently, 200 nm particles were directly prepared from cross-linked insulin (63). Nanoparticles were administered to rats and mice intragastrically by gastric tube. The glucose blood level in some animals was reduced to approximately 15% to 20% of the initial level three hours after administration. However, the hypoglycemic effect was quite erratic and some animals responded poorly to the treatment whereas others developed hypoglycemic shocks. The authors speculated that these particles were partially absorbed from the GI tract of mice and of normal and diabetic rats. The nanoparticles appeared to exert a slower but more pronounced response than a similarly administered commercially available Actrapid® formulation, although no dose–response relationship was established. Moreover, the high doses of nanoparticles needed would preclude the development of a commercially viable product.

The encapsulation of the peptide inside PIBCA oil-filled nanocapsules was proposed by Damgé et al. (64). Following subcutaneous injection, the hypoglycemic effect of insulin was prolonged in streptozotocin-induced diabetic rats when compared to free insulin. Surprisingly, insulin-loaded nanocapsules provided persistently reduced blood glucose levels, 50% to 60% of the initial concentration for 6 to 20 days after oral administration at doses of 12.5 to 50 IU kg^{-1}. Nanocapsules protected

Table 3 Characteristics of Orally Administered Particles Loaded with Insulin

Polymer	Size	Insulin dose	Animal model	References
PMCA, PHCA, PIBCA, PBCA	200 nm	52.5 IU kg^{-1}	Rat	62
Insulin	200 nm	66.6 mg NP/animal	Rat, mice	63
PIBCA	200 nm	12.5–50 IU kg^{-1}	Rat	64
	200 nm	100 IU kg^{-1}	Dog	68
	145 nm	100 IU kg^{-1}	Rat	69
	305 nm	2.07 mg kg^{-1}	Rat	73
	265 ± 40 nm	57 IU kg^{-1}	Rat	70
	60 – 300 nm	57 IU kg^{-1}	Rat	50
	400 ± 90	50 IU kg^{-1}	Rat	74
	< 500 nm	75 IU kg^{-1}	Rat	75
	200 nm	50 IU kg^{-1}	Rat	78
PLGA	104–169 nm	10 and 20 IU kg^{-1}	Rat	79
PLA	2–5 μm	2.5 mg of insulin	Rat	82
Eudragit®	180–500 μm	20 and 50 IU kg^{-1}	Rat	89
PVA-GS	5–25 μm	75 IU kg^{-1}	Rat	92
Chitosan	5 μm	24 IU/rat	Rat	93
Chitosan	250–400 nm	7, 14, and 21 IU kg^{-1}	Rat	96
Chitosan	120, 350, and 1000 nm	10 IU kg^{-1}	Rat	95
Poly(FA:PLGA)	0.1–10 μm	20 IU/rat	Rat	97
PLGA + FAO	<5 μm	19.2 IU/animal	Rat	81
HPMC–AS	24–36 μm	50 IU kg^{-1}	Rat	98
HPMCP + SNAC	30–500 μm	100 IU kg^{-1}	Rat	99

Abbreviations: FAO, fumaric acid oligomer; IU, international units; NP, nanoparticle; PMCA, poly(methylcyanoacrylate); PHCA, poly(hexylcyanoacrylate); PIBCA, poly (isobutyl-cyanoacrylate); PBCA, poly(butylcyanoacrylate); PLGA, poly(lactide-co-glycolide); PLA, poly(lactide acide); PVA-GS, poly(vinyl alcohol)-gel spheres; FA, fumaric acid; HPMC-AS, hydroxypropylmethylcellulose acetate succinate; HPMCP, hydroxypropylmethylcellulose phthalate; SNAC, sodium *N*-(8-(2-hydroxybenzoyl)amino) caprylate.

insulin against proteolytic degradation and induced a delayed urinary excretion of the hormone (65). In fed diabetic rats, 100 IU kg^{-1} of encapsulated insulin were able to reduce glycemia by 25% for six days after oral administration. Thus, insulin-associated nanocapsules induced a sustained reduction in glycemia when given orally to diabetic rats, but efficacy was more obvious in fasted than in fed diabetic animals (64). When insulin nanocapsules were administered to normal rats, fasted glycemia remained unaffected regardless of the dose; however, the hyperglycemia peak was reduced after a glucose overload. The authors suggested that the nanocapsules could be retained in the stomach and/or in the liver for long periods, slowly releasing insulin in the blood. A chemical interaction between the peptide and the polymer was also thought to be responsible for such a long lasting effect (64). However, recent studies have shown that insulin was located inside the oily core of the nanocapsules and there was no evidence of interaction between the polymer forming the nanocapsule envelope and the peptide (66–68). In another study performed by the same group, a decrease of the glycemia as well as a regulation of other endocrine pancreatic dysfunctions, namely hyperglucagonemia and hypersomatostatinemia, after intragastric administration of insulin-loaded PIBCA nanocapsules was observed in alloxan-induced diabetic dogs (68,69).

To determine the preferential site of absorption of insulin-loaded nanocapsules, Michel et al. (24) injected a nanocapsule suspension at different locations of the GI lumen. Absorption of insulin from isobutylcyanoacrylate nanocapsules, monitored by blood glucose, was shown to be dependent on the site of absorption. The greatest blood glucose reduction of 65% was observed when particles were administered in the ileum while a 50% reduction was recorded after administration in the duodenum and jejunum and a 30% reduction was seen in the colon. Furthermore, it was shown that insulin was protected in vitro from proteolysis from pepsin, trypsin, and α-chymotrypsin by the nanocapsules. The hypoglycemic effect, which was observed from the second day, lasted for 3 to 18 days. Aboubakar et al. (70) showed that PIBCA nanocapsules could fully protect insulin from gastric proteolysis. However, the protection of insulin in the intestinal environment was only partial, since the peptide was quite rapidly released in this medium (70). Nevertheless, in vivo, nanocapsules containing Texas Red-labeled insulin were transported throughout the small intestine with a peak detected at one hour after administration, leading to a deposition of still intact nanocapsules along the intestinal mucosa. Furthermore, fluorescence microscopy and confocal microscopy investigations strongly suggest that nanocapsules were involved in the intestinal absorption of insulin. Pinto-Alphandary et al. (50) investigated the fate of insulin, Texas Red-labeled insulin, or gold-labeled insulin encapsulated into PIBCA nanocapsules by fluorescence and TEM after administration of a single dose by intragastric force-feeding. After 90 minutes, ileum was isolated and observed by fluorescence and TEM. The fate of nanocapsules appeared different depending on the presence or the absence of M cells in the intestinal segment. Indeed, in M cell–free epithelium, apparently intact nanocapsules could be seen in the underlying tissue, suggesting they could cross the epithelium and carry the encapsulated peptide. In M cell–containing epithelium, nanocapsules appeared degraded in the vicinity of macrophages. The authors emphasized that intestinal absorption of the nanocapsules was observed without artifacts forcing the nanocapsules to stay in the stomach.

The nanocapsules prepared by Damgé et al. (64) were formed by interfacial polymerization of a cyanoacrylic monomer. An oily phase, Miglyol 812, containing the monomer and insulin was mixed with an aqueous dispersion medium in the presence of a surfactant, Poloxamer 188. The presence of these additives may play a role in the protection of insulin, its passage through the intestinal barrier, and its binding to insulin receptors eliciting the biological response. To clarify the exact role of each compound, another formulation in which insulin was attached to polymeric nanospheres was prepared by polymerization of isobutyl cyanoacrylate in an acidic medium (71). In this system, however, insulin was encapsulated as nanospheres and not as nanocapsules as presented in the first study (64). Insulin was covalently attached to the polymer, inducing, therefore, a significant modification of the molecular weight of PIBCA. In vitro, nanospheres protected insulin from degradation by proteolytic enzymes, especially when dispersed in an oily medium (Miglyol 812) containing surfactant (Poloxamer 188 and deoxycholic acid) (71). In the same medium, insulin-loaded nanospheres ($100\,\mathrm{IU\,kg^{-1}}$), administered orally in streptozotocin-induced diabetic rats, produced a 50% decrease of fasted glycemia from the second hour up to 10 to 13 days. This effect was shorter (two days) or absent when nanospheres were dispersed in water with or without surfactants. Furthermore, when administered in suspension in 1% Poloxamer and 0.01% deoxycholic acid, ^{14}C-labeled nanospheres loaded with ^{125}I-insulin increased the uptake of ^{125}I-insulin or its metabolites from the GI tract. In addition, blood and liver ^{125}I-insulin concentrations were higher with a delayed excretion when compared to free ^{125}I-insulin. Two hours after the oral administration of the double-labeled insulin-loaded nanospheres, the ^{125}I radioactivity from insulin appeared as a high rate in the blood while the ^{14}C radioactivity from the polymer was surprisingly very low. In contrast, in the gut, the level of ^{125}I was low and that of ^{14}C was high. Thus, an early release of insulin from the polymer seemed to occur, insulin then being rapidly absorbed in a systemic way. Surfactants and Miglyol could be involved in this latter step. Finally, ^{14}C- and ^{125}I radioactivities disappeared progressively as a function of time, in parallel to biological effects. The authors concluded that insulin-loaded nanospheres could be considered as a convenient oral delivery system for insulin with the prerequisite that they were dispersed in an oily phase containing surfactants. This suggests that some components of the nanocapsules can act as absorption promoters (64,72). In a complementary study, Lowe and Temple (73) investigated insulin blood levels in rats after administration of insulin encapsulated in PIBCA nanoparticles. The resulting pharmacokinetic profiles were characteristic of a sustained delivery (Fig. 2). A relatively higher plasma concentration was observed at the later time points but was balanced by lower initial concentrations; thus, there was no significant enhancement of absorption. This suggests that the nanocapsules slowly released the insulin into the intestinal lumen, with small amounts being absorbed.

In another work, Cournarie et al. (74) evaluated the biological activity of insulin given orally as PIBCA nanocapsules to streptozotocin-induced diabetic rats. Thirty minutes to one hour after oral administration, significant levels of insulin were detected in the plasma. However, peptide concentrations were very heterogeneous from one rat to another and no decrease of glycemia could be observed. In addition, parenteral injection of insulin decreased glycemia by 50% in normal rats and by only 25% in diabetics, suggesting that an insulin-resistance was developed by streptozotocin-induced diabetic rats. The authors of this study wonder if streptozotocin-induced diabetic rats constitute a suitable model to study the oral efficiency of insulin-encapsulated nanocapsules.

Radwan (75) investigated the in vivo activity after oral and subcutaneous administration of insulin-loaded PIBCA nanospheres in aqueous suspension with or without sodium cholate (0.5%) and Pluronic F68® (0.5%) as surfactants to alloxan-induced diabetic rats. Insulin absorption was evaluated by its hypoglycemic effect.

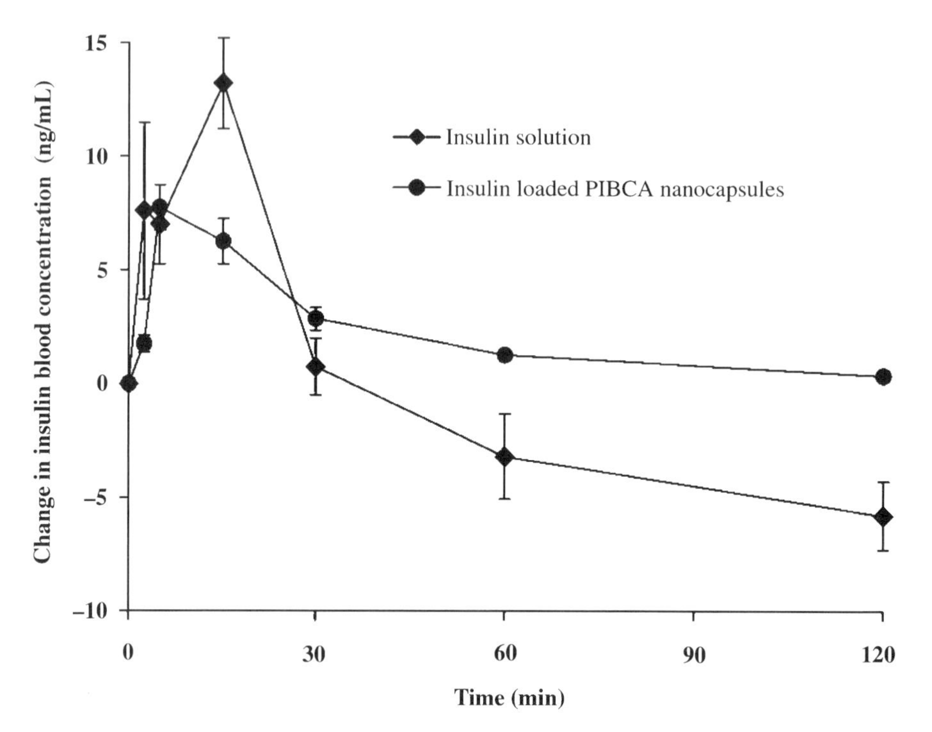

Figure 2 Plasma insulin levels following duodenal administration of insulin solutions (♦) ($n = 3$) and insulin-loaded PIBCA nanocapsules (•) ($n = 5$) corresponding to 2.07 mg kg^{-1} of insulin. *Abbreviation*: PIBCA, poly(isobutylcyanoacrylate). *Source*: From Ref. 73.

The results showed a significant pharmacological availability after oral and subcutaneous dosages of insulin-loaded nanospheres with surfactants. The presence of surfactants with PIBCA nanospheres improved the oral pharmacological availability by 49.2% compared to the administration of insulin-loaded nanocapsules without surfactants. The conclusion from this study was that PIBCA nanospheres, in the presence of surfactants, would be useful not only in improving insulin GI absorption, but also in sustaining its systemic action by lowering the blood glucose to an acceptable level.

Recently, insulin has also been encapsulated in water-containing PACA nanocapsules (76–78). These nanocapsules were prepared in situ in a biocompatible water-in-oil microemulsion by interfacial polymerization. The microemulsion consisted of a mixture of medium-chain mono-, di-, and triglycerides as the oil component, polysorbate 80 and sorbitan mono-oleate as surfactants, and an aqueous solution of insulin. These PIBCA nanocapsules, dispersed in a biocompatible microemulsion, could facilitate the intestinal absorption of the encapsulated peptide following intragastric administration, as suggested by the reduced blood glucose level observed in diabetic rats (78).

Barichello et al. (79) prepared PLGA nanoparticles by a precipitation-solvent evaporation method similar to that described by Fessi et al. (80). Although insulin was then preferentially surface-bound, thus not providing a specific protection, a strong hypoglycemic effect was observed after administration of the PLGA-nanoparticle suspension (10 or 20 IU kg^{-1}) to ileum loops in rats.

Carino et al. (81) encapsulated zinc insulin in various polyester (PLA and PLGA) and polyanhydride [poly-fumaric anhydride-co-sebacic anhydride, P(FA:SA), fumaric acid (FAO), and sebacic acid oligomers (SAO)] nanosphere formulations using phase inversion nanoencapsulation. A specific formulation, 1.6% zinc insulin in PLGA with FAO and iron oxide additives was active orally when administered to normal fasted rats (19.2 IU/animal). This formulation was shown to have 11.4% of the efficacy of intraperitoneally delivered insulin and was able to control plasma glucose levels when faced with a simultaneously administered glucose challenge. The authors hypothesized that several properties of the formulation, including size (<5 μm), release kinetics, bioadhesivity and ability to cross the GI epithelium, contributed to its oral efficacy.

Ma et al. (82) developed PLA microcapsules loaded with insulin. The particles were prepared by a water-in-oil-in-water emulsion/solvent extraction. A small dose (2.5 mg) of insulin encapsulated in PLA microcapsules (2–5 μm) were administered orally to alloxan-induced diabetic rats. The blood glucose lowering effect was quite erratic. Although in 68% of diabetic rats peripheral blood glucose level decreased by 68.5%, in 21% the blood glucose decreased by 39.7%, and in 12% of the diabetic rats no effect was found (the number of animals used in the study was 66). The authors speculated that these differences might be the results of variability in absorption of insulin in the GI tract.

Enteric coatings are widely used in the pharmaceutical industry and have been experimented with extensively in attempts to deliver insulin orally (83–88). These coatings were designed to protect insulin from the harsh environment of the stomach and small intestine before releasing the drug into the colon.

A similar approach has been studied by Morishita et al. (89–91) using Eudragit® insulin microspheres containing a protease inhibitor, aprotinin. Strong protection was observed against enzymatic degradation (89). Furthermore, when insulin (20 IU kg^{-1}) and aprotinin within Eudragit L100 microspheres were administered orally to normal and diabetic rats, significant and prolonged hypoglycemic effects were observed (91). Nevertheless, insulin microspheres without protease inhibitor did not produce any marked hypoglycemic response in either group of rats. Different polyacrylic coatings, namely, Eudragit L100, soluble at pH above 6, Eudragit S100, soluble at pH above 7 and a mixture of both polymers were compared (90). The three types of insulin microspheres were expected to release insulin at different sites throughout the small intestine. However, when microspheres were administered to nondiabetic rats by gavaging (50 IU kg^{-1}), only 10% reduction in glucose level was observed. The greatest hypoglycemic effect was seen after the administration of Eudragit L100 microspheres. Again, the biological effect of each pH-sensitive formulation was significantly enhanced by aprotinin.

Kimura et al. (92) examined the absorption of insulin administered orally (75 IU kg^{-1}) as poly(vinyl alcohol)-gel spheres (PVA-GS) with or without protease inhibitors in streptozotocin-induced diabetic rats. Further, they assessed the absorption of insulin through the intestinal mucosa of diabetic rats by an in situ loop model. Intragastric administration of PVA-GS containing insulin and a protease inhibitor (aprotinin or bacitracin), caused a significant and prolonged reduction of blood glucose level, suggesting insulin absorption. The bioavailability of insulin, estimated from the hypoglycemic effect, was about 2% in the presence of either protease inhibitors. Insulin absorption estimated by plasma insulin levels was larger in the ileum and in the large intestine than in the jejunum. The prolonged residence time of PVA-GS at the absorption site, the lower intestine, and the synchronous release of insulin and the protease inhibitors from PVA-GS are the two major hypotheses for the improved bioavailability of insulin administered as PVA-GS containing a protease inhibitor.

Takeuchi et al. (93) developed mucoadhesive liposomal dosage forms to facilitate the enteral absorption of insulin. The liposomes were coated with several polymers including chitosan for its mucoadhesive properties. An adhesion test was carried out using different parts of the small intestine. The highest mucoadhesion was found in the ileum. The chitosan-coated liposomes containing insulin were administered orally, through an intragastric tube, to normal rats (24 IU/rat). A marked reduction in basal blood glucose level was observed 30 minutes after administration. The lowered glucose level was maintained for more than 12 hours, which suggested mucoadhesion of the chitosan-coated liposomes and sustained release of insulin. The authors assumed that the insulin released from the liposomes in the mucous layer could be absorbed without being enzymatically degraded. Recently, the use of chitosan and its derivatives as excipients to improve the bioavailability of many perorally given peptide drugs [such as insulin, calcitonin (CT), and buserelin] has been reviewed (94). Pan et al. (95) developed chitosan nanoparticles with a size range of 250 to 400 nm and insulin association up to 80%. The authors investigated the ability of chitosan nanoparticles to enhance intestinal absorption of the peptide and to increase the relative pharmacological bioavailability of insulin by monitoring the plasma glucose level of alloxan-induced diabetic rats, after oral administration of insulin-loaded chitosan nanoparticles (7, 14, and 21 IU kg^{-1}). A dose of 14 IU kg^{-1} administrated as chitosan nanoparticles induced a hypoglycemic effect significantly greater than in the control group. When the doses increased this effect appeared earlier and an elongation of hypoglycemic state was observed. Furthermore, the authors found that particles of 345 nm in diameter resulted in a greater hypoglycemic effect than 123 nm nanoparticles in diabetic rats. The authors concluded that nanoparticles of different sizes had different capabilities of protecting insulin from degradation in the harsh stomach environment and enhancing the absorption of it. Nevertheless, the mechanism of this protection is still unknown. In another work, the same group investigated more deeply the size-dependent drug effect of insulin-loaded chitosan particles (120, 350, and 1000 nm) in the alloxan-induced diabetic rats after an oral administration (10 IU kg^{-1}) (96). Particles of different sizes had various degrees of hypoglycemic effect ten hours after administration ($P < 0.05$ vs. control insulin solution). The best effect was obtained after administration of the 350-nm nanoparticles: plasma glucose levels remarkably decreased with a maximal effect at 15 hours that lasted 35 hours. Moreover, the relative pharmacological availabilities relative to subcutaneous injection were 10.2% (120 nm), 14.9% (350 nm), and 7.3% (1000 nm), respectively. From this study, it was concluded that bioadhesive chitosan particles were helpful in increasing the relative pharmacological bioavailability of insulin, and the distinct advantage of proper particle size helped to increase the peptide effects.

Mathiowitz et al. (97) have developed microspheres made from a blend of poly(fumaric anhydride) and poly(lactide-co-glycolide) to encapsulate insulin. The microspheres were administered orally to normal rats (20 IU/rat). These materials adhered to the small intestine wall and released insulin based on degradation of the polymeric carrier. A 30% to 50% decrease of the basal glucose level was observed in normal rats. Nevertheless, no studies were made using a diabetic rat model. The same system also enhanced the biological activity of dicumarol and plasmid DNA absorption after oral administration.

Nagareya et al. (98) prepared hydroxypropylmethylcellulose acetate succinate (HPMC-AS, AS-HG type; high content of succinyl group) microspheres containing bovine insulin. HPMC-AS was employed as enteric polymer material. A formulation containing 9% insulin and 4% lauric acid was administered rectally to normal rats (corresponding to 50 IU kg^{-1}). Glycemia reached minimum level 0.5 hour after administration, then gradually rose to normal as shown in Figure 3. This suggests that bovine insulin was released from the microspheres and absorbed through the GI tract with the aid of lauric acid, thereby showing a pharmacological effect.

GI absorption of insulin by coadministering of hydroxypropylmethylcellulose phthalate (HPMCP) enteric microspheres with sodium *N*-(8-(2-hydroxybenzoyl)amino) caprylate (SNAC) as an absorption enhancer was recently studied (99). The W/O/W and O_1/O_2 emulsion solvent evaporation methods were used to prepare insulin–HPMCP microspheres and the size of the particles was 30 and 500 μm, respectively. The particles were orally administered to normal rats (100 IU kg^{-1}) together with SNAC and the hypoglycemic effect was monitored. Microspheres prepared by multiple emulsions had a very weak hypoglycemic effect; it seems that the porous surface observed in the microspheres could not protect the insulin from being degraded by GI enzymes. In contrast (O_1/O_2) prepared microspheres had an intense and lasting hypoglycemic effect. After the administration of the drug, the blood glucose level displayed a remarkable decrease versus the control insulin solution for two to five hours. The authors hypothesized that the insulin-loaded microspheres (O_1/O_2) coadministered with SNAC could enhance insulin absorption by taking advantage of protection from both enzyme degradation and improvement of drug permeability.

A successful oral formulation of insulin has been the goal for many investigators since the protein's initial discovery by Banting and Best in 1922. In 1972, Earle (100) tried to administer insulin orally, 10 units of insulin per tablet, together with a vasodilator incorporated, which apparently allows the insulin to be absorbed, when placed on or under the tongue. Despite its size, insulin crossed the oral mucosa in normal and diabetic subjects. Although the efficacy was lower than after injection,

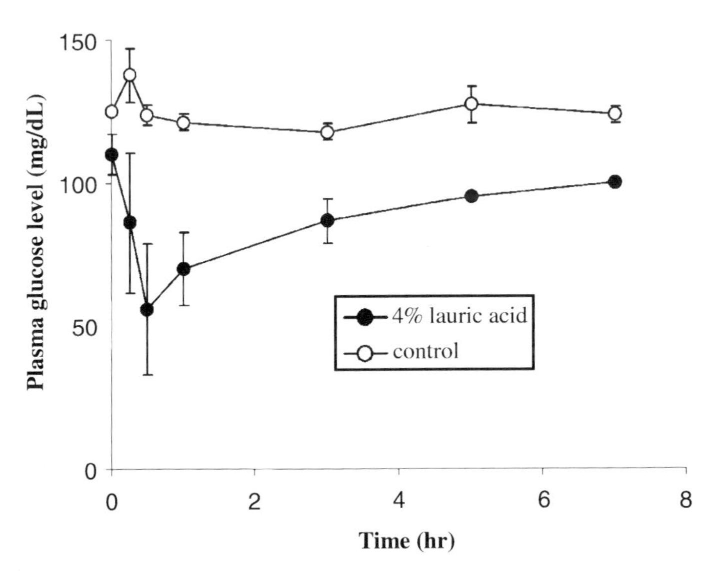

Figure 3 Plasma glucose levels following rectal administration of insulin loaded HPMC-AS microspheres in saline (○) or in the presence of 4% lauric acid (•). The dosing corresponded to 50 IU kg^{-1} insulin. Each value represents the mean of two to three experiments. *Abbreviation*: HPMC-AS, hydroxypropylmethylcellulose acetate succinate. *Source*: From Ref. 98.

the results were encouraging enough to warrant a reexamination of oral administration. Recently, several reviews reported on the oral delivery of peptides (59,101,102). Although great progress has been made using nanoparticles, microspheres, enteric coating, permeation enhancers or a combination of two or more strategies, a lot of work still needs to be done to have an oral formulation of insulin that could supplant, wholly or in part, the need for injections.

Calcitonin (CT)

CT is a single chain polypeptide hormone, discovered by Copp in 1961, that regulates numerous physiological processes (103). It decreases blood calcium levels and inhibits bone resorption by directly affecting osteoclast activity, and it is suggested to have an anabolic osteoblastic effect on bone (104). Clinically, CTs of different origin are used in the treatment of symptomatic Paget's disease, hypercalcemia, and postmenopausal osteoporosis. Therapeutically, salmon CT (sCT) is about 40 times more potent than human CT in lowering blood calcium levels because it has a higher circulating half-life and a greater affinity for the receptor (105). For Paget's disease, the peptide has to be injected subcutaneously or intramuscularly daily or three times a week. Therefore, an oral formulation yielding therapeutic plasma levels of peptide over an extended period of time is highly desirable.

The different approaches used for the encapsulation of CT and other peptides into polymeric particles are summarized in Table 4. For review of transmucosal delivery of CT or mucoadhesive systems for proteins, see Refs. 106–108.

Human CT was entrapped in polyacrylamide nanospheres (39 nm) and in PIBCA nanocapsules (88 nm) before administration to rats (73). With polyacrylamide nanospheres, no sustained release was observed in vitro and no protection was provided from proteases either. With PIBCA nanocapsules, proteolytic hydrolysis of human CT in the presence of pancreatin and porcine bile was slower than the free peptide in solution. CT nanocapsules were dosed duodenally through a small incision to rats at 0.2 mg kg^{-1} and compared with the same dose of CT dissolved in 0.1% acetic acid. The plasma pharmacokinetic profiles were consistent with increased survival time of the peptide in the intestine, and higher plasma concentrations in the later time samples as compared to the controls (Fig. 4). However, the nanocapsules gave no significant overall enhancement of peptide absorption. This led to the conclusion that the nanocapsules released the CT into the intestinal lumen, with small amounts being absorbed and the major part still being degraded.

In 1996, Vranckx et al. (109) developed poly(butyl cyanoacrylate) (PBCA) nanocapsules (50 nm) with an aqueous core wherein sCT was dissolved. Nanocapsules together with deoxycholic acid and sodium laurylsulfate as absorption enhancers were administered orally (20 IU kg^{-1}) to conscious fasted rats. Calcemia was monitored over a period of 28 hours. After oral nanocapsule administration, a hypocalcemic effect was estimated at 45% of the effect obtained with the same dose of free CT administered intravenously. Furthermore, when an emulsion containing the absorption enhancers and free sCT was administered orally, no biological effect was observed.

PS nanoparticles with different surface hydrophilic polymeric chains have been developed using free radical copolymerization between hydrophilic macromonomers and hydrophobic styrene (110–112). Sakuma et al. (113–117) carried out a series of experiments to study the potential of these nanoparticles as carriers for oral sCT delivery in rats. They reported a decrease in calcium blood concentration and an

Table 4 Characteristics of Orally Administered Particles Loaded with Calcitonin and Other Peptides

Drug	Polymer	Preparation method	Size	References
Calcitonin	Polyacrylamide	Emulsion polymerization	39 nm	73
	PIBCA	Interfacial polymerization	88 nm	
	PBCA	Interfacial polymerization	50 nm	109
	Coated PS	Copolymerization	380–500 nm	113, 116
	Derivatized amino acids	Self-aggregation	0.3–1000 nm	123–125
	PLGA-chitosan	Solvent-diffusion	300–1000 nm	122
	PLGA	Emulsion solvent-diffusion	650 nm	121
Cyclosporin A	PIHCA	Interfacial polymerization	213 nm	130
	CyA	Nanoprecipitation	250–900 nm	131
	PCL	Nanoprecipitation	100 nm	132, 134
	PCL	Nanoprecipitation	100–160 nm	135
	PCL	Nanoprecipitation	105 ± 44 nm	136
	Chitosan	Emulsification solvent diffusion	150 nm	137
	Gelatin A		140 nm	
	Na glycocholate		105 nm	
MTP-Chol	PIBCA	Interfacial polymerization	180 nm	138
Vasopressin	PHMA	Suspension polymerization	315–400 μm	139, 140
LHRH	LHRH-*n*-BCA	Copolymerization	100 nm	38, 141
Interferon-γ	PLA	Emulsion–evaporation	1 μm	142
Octreotide	PIBCA	Interfacial polymerization	260 nm	71
HIV protease inhibitor	PLA	Salting out	300–500 nm	150, 151
	Eudragit	Emulsion–diffusion	270 nm	

Abbreviations: BCA, butylcyanoacrylate; CyA, cyclosporin A; LHRH, luteinizing hormone releasing hormone; PBCA, poly(butyl cyanoacrylate); PCL, poly-ε-caprolactone; PHMA, poly(2-hydroxyethyl methacrylate); PIBCA, poly(isobutylcyanoacrylate); PLA, poly(lactic acid); PS, polystyrene; HIV, human immunodeficiency virus.

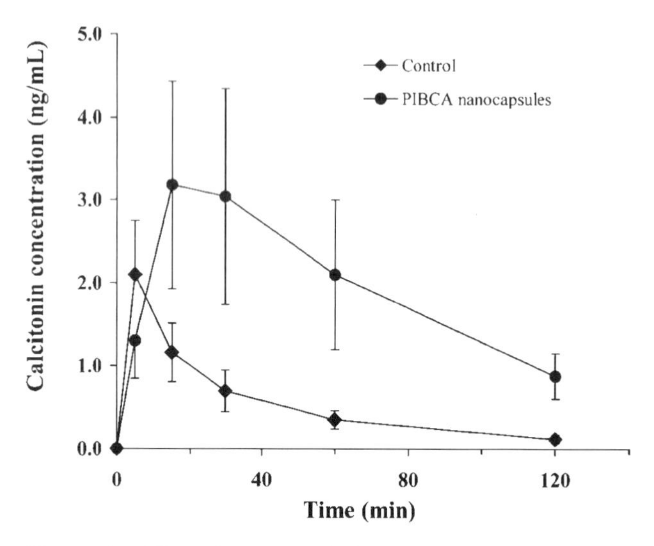

Figure 4 Effect of nanocapsule formulation on the pharmacokinetics of duodenally administered calcitonin (0.2 mg kg^{-1}): control solution (♦) ($n = 6$) and PIBCA nanocapsules (•) ($n = 11$). *Abbreviation*: PIBCA, poly(isobutylcyanoacrylate). *Source*: From Ref. 73.

increase in plasma sCT concentration after oral administration of sCT within nanoparticles, demonstrating that these nanoparticles enhanced sCT absorption from the GI tract (113,115). The authors hypothesized that the absorption enhancement of sCT by these nanoparticles resulted mainly from both bioadhesion of nanoparticles containing sCT to the GI mucosa and an increase in the stability of sCT in the GI tract (115). The stabilizing effect of nanoparticles on the enzymatic degradation of sCT was examined in vitro (114). sCT was protected against pepsin or trypsin hydrolysis by nanoparticles except those covered with poly(*N*-vinylacetamide) (PNVA) chains, which did not enhance sCT absorption via the GI tract in vivo. This stabilizing effect was affected by the structure of the polymeric chains. Nanoparticles with a poly(*N*-isopropylacrylamide) (PNIPAAm) covered surface completely inhibited sCT degradation by pepsin. However, they did not increase sCT stability in the presence of trypsin. The degradation of sCT by trypsin was inhibited totally by nanoparticles with surface poly(methacrylic acid) (PMAA) chains, even though sCT stability in the presence of pepsin was only slightly improved when encapsulated. Nanoparticles having poly(vinylamide) (PVAm) chains on their surface stabilized sCT in the presence of either enzyme. The stabilizing effects result most likely from the physicochemical interactions between the enzyme and the nanoparticles.

PS nanoparticles having surface hydrophilic polymeric chains loaded with radiolabeled diazirine were used to demonstrate the mucoadhesion of nanoparticles in the GI tract (116). Radioiodinated diazirine is a hydrophobic compound that is often used to selectively radiolabel hydrophobic lipids of membranes and was used in this study as a polymer label (118). A mixture of sCT and nanoparticles was given orally to rats (0.25 mg of sCT and 25 mg of nanoparticles kg^{-1}). As a control, an aqueous solution of sCT was administered per ounce to rats under the same conditions. The change in blood ionized calcium concentration after oral administration of sCT with nanoparticles showed that the in vivo enhancement of sCT absorption by radiolabeled nanoparticles was the same as nonlabeled nanoparticles (115). The ranking order of absorption enhancement was PNIPAAm, PVAm, and PMAA nanoparticles (Fig. 5) (116). The

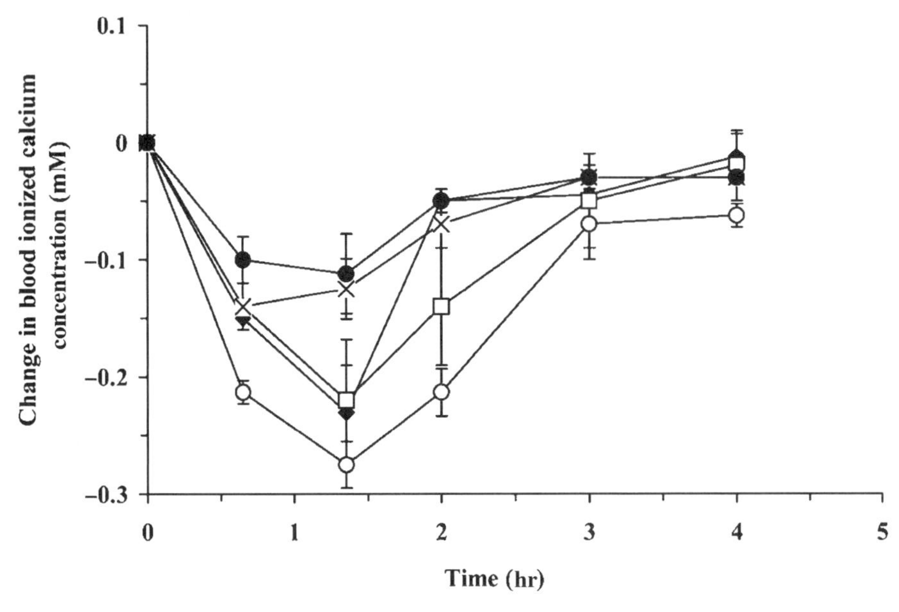

Figure 5 Concentration–time profiles of ionized calcium in blood after oral administration of sCT in aqueous solution (•), or associated with PNIPAAm nanoparticles (○), PNVA nanoparticles (×), PVAm nanoparticles (□) or PMAA nanoparticles (♦) in rats (0.25 mg sCT with 25 mg nanoparticles in 2.5 mL dosing solution kg^{-1} rat). The radioiodinated diazirine was incorporated in nanoparticles. Each value represents the mean ± S.D ($n = 3$). *Abbreviations*: PMAA, poly(methacrylic acid); PNIPAAm, poly(*N*-isoproprylacrylamide); PNVA, poly(*N*-vinyl acetamide); PVAm, poly(vinylamide); SCT, salmon calcitonin. *Source*: From Ref. 116.

GI transits of PNIPAAm, PVAm, and PMAA were slower than that of PNVA nanoparticles, which does not enhance sCT absorption at all. The slow transit rates were likely due to mucoadhesion of the nanoparticles. The strength of mucoadhesion depended on the structure of the surface polymeric chains, and PNIPAAm nanoparticles exhibited the strongest mucoadhesion effect. The authors suggested that there was no interaction between PNVA nanoparticles and the mucous layer, demonstrating that there is a good correlation between mucoadhesion and enhancement of sCT absorption. It was also speculated that the penetration enhancing effect was likely to be mediated by decreasing the adhesion of nanoparticles to the gastric mucosa. Therefore, more amount of drug is available for absorption from intestinal membrane (113). In addition, it may be necessary to examine the uptake of nanoparticles in Peyer's patches and the opening of tight junctions between intestinal epithelial cells by nanoparticles (119,120).

These studies provide very interesting data on both improving calcitonin availability and the influence of surface characteristics in particle uptake. However, as a nonbiodegradable polymer, PS is not suitable for drug delivery.

Kawashima et al. (121) evaluated the effect of mucoadhesive polymeric nanospheres on the absorption of elcatonin (calcitonin from eel) using PLGA nanospheres coated with chitosan. The reduction in blood calcium was monitored after oral administration of uncoated and chitosan-coated PLGA nanospheres in rat and compared to the administration of a solution. Three dosages were used: 125, 250, and 500 IU kg^{-1}. For all the groups, calcemia fell temporarily by about 80% to 85% one hour after intragastric administration of either the solution or suspended nanospheres containing the peptide. No differences in calcium levels were seen between the solution and uncoated particle. For the chitosan-coated particles, the hypocalcemic effect lasted longer, up to 12 hours at the doses of 125 or 250 IU kg^{-1}, and a sustained effect up to 36 hours after administration was achieved when the coated particles were administered at the dose of 500 IU kg^{-1}. This finding indicates that the chitosan-coated PLGA nanospheres are able to overcome their mucoadhesion ability due to the turnover of the mucosal surface in the small intestine. The authors suggested that this might be due to the deep penetration of chitosan-coated nanospheres into the mucus layer and their adhesion to mucus tissue, where the drug is slowly released from the system.

Yoo and Park (122) formulated sCT-oleate complexes into PLGA nanoparticles. Hydrophobic ion pairing between sCT and amphiphilic molecules gives the peptide higher solubility in an organic solvent and a higher loading efficiency can be obtained. The sCT nanoparticles (30 or 60 μg kg^{-1}) or sCT dissolved in PBS (160 μg kg^{-1}) were administered to rats through oral gavage, and plasma concentration of sCT was measured as a function of time. Two hours postfeeding of sCT nanoparticles, the plasma concentration of sCT soared to 600 pg mL^{-1} and then slowly decreased. However, when sCT was orally administered solubilized in phosphate buffer, negligible amounts of the peptide were detected. The authors stated that most likely, native sCT was degraded by digestive enzymes in the gut, and even if some had reached the absorption site, it had barely been taken up by intestinal cells. In contrast, PLGA nanoparticles could protect encapsulated sCT from the digestive environment, and they might be readily transported within intestinal epithelial cells at the absorption site by an endocytosis process, eventually resulting in the delivery of sCT into the blood stream.

A series of studies were carried out using derivatized amino acids to prepare CT-loaded microspheres (123,124). At low pH, these compounds form microspheres that dissolve readily under neutral conditions. These pH-dependent microspheres were administered by oral gavage to anesthetized rats and by nasogastric gavage to conscious rhesus monkeys (124). The delivery of the drug with this formulation led to a 25% decrease in calcemia one hour after treatment in rats (10 μg kg^{-1} calcitonin), compared to free calcitonin or empty microspheres which led to a serum calcium decrease of less than 3%. A higher dose of calcitonin (60 μg kg^{-1}) administered to primates caused a 15% decrease in serum calcium six hours after dosing. Similar experiments were carried out with microspheres prepared from modified soybean protein hydrolyzate (125). These microspheres were dosed in rats both by oral gavage or by intraduodenal injection and in male cynomolgus monkeys by nasogastric gavage. In agreement with the previous study from Leone-Bay et al. (123,124), significant levels of sCT were found in the systemic circulation. The authors pointed out the low cost of these microspheres, which makes them attractive for both medicinal as well as industrial usage.

Cyclosporin A

Cyclosporin A (CyA) is a potent immunosuppressive agent, widely used for the inhibition of graft rejection in organ transplantation. The drug is a highly specific inhibitor of T-lymphocytes, which play an important role in the induction of immune response (126). However, the formulation of CyA for oral use presents a considerable challenge as it is practically insoluble in water (127). Moreover, this compound has a poor and variable peroral bioavailability in humans and severe nephropathies have been associated with the use of CyA (128,129). For this reason, it could be of interest to incorporate this peptide into a carrier able to enhance peroral bioavailability and to reduce the distribution to the kidneys.

Bonduelle et al. (130) entrapped tritiated [^{3}H]-CyA into poly(isohexyl-cyanoacrylate) nanocapsules. The formulation was administered perorally by gavage to fasted NMRI mice and compared to a commercially available emulsion and a control emulsion (same preparation as nanocapsules without polymer). The area under blood concentration versus times (AUC) curves of radiolabeled CyA were notably higher after peroral administration of nanocapsules than after the commercial or control emulsions and resulted from both a better absorption from the GI tract and a slower elimination from the blood. Concentrations returned to baseline after approximately 24 hours for the emulsions and 18 days for the nanocapsules. Moreover, a reduced uptake by the liver, the spleen and kidneys was observed with particles compared to the emulsions, suggesting a reduction in side effects associated with the currently available formulations.

Ford et al. (131) formulated CyA nanospheres in the absence of particle-forming polymer. The drug was dissolved in acetone and mixed rapidly with an aqueous solution of surfactant stabilizers (polysorbate 80 and sodium dodecyl sulfate) to give a colloidal precipitate of spherical particles (250–900 nm). The particle matrix was made of the drug itself. The oral absorption of CyA from these nanospheres was assessed in male beagle dogs and compared to Neoral® microemulsion formulation. Results showed that CyA could be absorbed from nanospheres but the extent of absorption was markedly lower than Neoral. The relative bioavailability of cyclosporin A from nanospheres was only 3% based on comparison of the AUC values for the two formulations. In conclusion, the ability of this system to facilitate oral absorption of CyA in dogs is poor compared to an efficiently absorbed microemulsion formulation.

Molpeceres et al. (132–134) studied the pharmacokinetics of CyA incorporated into poly-ε-caprolactone (PCL) nanoparticles as an alternative formulation with lower surfactant concentration than the commercially available Sandimmun-emulsion (SIM-EM)® and Sandimmun Neoral microemulsion (SIM-Neoral)®. The formulations were administered by oral gavage to male and female Wistar rats as CyA loaded PCL nanoparticles or as an emulsion in whole milk. The average AUC values after CyA dosage as PCL nanoparticles were four and two times higher than the corresponding AUCs calculated after SIM-EM and SIM-Neoral treatment, respectively. The authors concluded that PCL nanoparticles are a good carrier for increasing the oral bioavailability of CyA.

Saez et al. evaluated the effect that a slight nanoparticle size variation induced by freeze-drying had on the oral bioavailability and pharmacokinetics of CyA loaded into PCL nanoparticles (135). The authors compared the bioavailability of the drug administered as 100 nm particles or 160 nm obtained from the same batch after lyophilization. A single CyA oral dose (5 mg kg^{-1}) was given to rats by gavage. The maximal concentration (C_{max}) was obtained two hours after administration and the value was higher with the 100 nm nanoparticles. However, nanoparticles of 160 nm induced a significantly larger mean residence time (MRT) and higher drug levels in the liver. These changes can be considered useful if a selective accumulation of CyA in the liver is sought but a potential increase in its hepatotoxicity is then expected. Consequently, the effect of particle size changes because of the freeze-drying of PCL on the pharmacokinetics of loaded drugs should be carefully evaluated. Nevertheless, if particle size has been clearly demonstrated to be a major factor influencing particle absorption by the GI tract (see first part of this chapter), it seems, however, difficult to understand why such a slight difference (100 to 160 nm) could induce such different pharmacokinetic profiles. The difference in size is subtle and the two populations must have an overlap due to the polydispersity. In the absence of SEM pictures it is difficult to check the size dispersion of the two batches. On the other hand, it is not impossible that the lyophilisation step may also induce a redistribution of the peptide in the polymeric matrix resulting thus in a different drug release pattern. This hypothesis might easily be checked with a release kinetic study in buffer.

Varela et al. (136) evaluated, from a comparative standpoint the in vivo nephrotoxicity of CyA incorporated in 100-nm PCL nanoparticles or in its free form (Sandimmum®) after oral administration to rats (10 mg/kg/day for three days). The authors measured CyA concentrations in blood, urine, liver, spleen, and kidney at 24 hours postdosing and the nephrotoxicity induced by the treatment was determined by measuring different markers in plasma and in glomeruli. When CyA was associated with PCL nanoparticles, a higher drug concentration was achieved in blood and tissues. However, despite the higher plasma level, the nephrotoxicity obtained with each formulation was comparable. In vitro, the immunosuppressive effect evaluated by a lymphocyte proliferation assay did not evidence a significant difference until the dose reached 25 μM. These authors suggested that nanoparticles may restrict the pharmacological and toxic effects of the absorbed drug fraction by unknown mechanisms. Nevertheless, an efficient drug targeting to lymphocytes cannot be neglected.

El-Shabouri developed positively charged nanoparticles loaded with CyA as an attempt to improve the GI uptake and the overall bioavailability of the drug (137). Lecithin and poloxamer 188 were used as emulsifiers and chitosan hydrochloride and gelatin-A were selected as positively charged inducing agents. The bioavailability and pharmacokinetics of CyA after administration in beagle dogs of these nanoparticles were compared to those carrying a negative charge obtained by using sodium glycocholate as negatively charged inducing agent and the microemulsion (Neoral) used as a reference. Each animal received a single oral dose equivalent to 100 mg CyA and mean plasma levels of the drug were determined. The relative bioavailability of CyA from chitosan-nanoparticles was increased by about 73%, and increased only by about 18% from gelatin nanoparticles, and in the case of sodium glycocholate nanoparticles, it was decreased by approximately 36%. Thus, from these results, it was evident that, in spite of the fact that sodium glycocholate-nanoparticles have the smallest size (104 nm) relative to that of chitosan (148 nm) or of gelatin (139 nm); they gave the lowest $C_{\max}$ and the lowest AUCs. These authors suggest that these results could be attributed to the anionic charge conferred to the particles by sodium glycocholate used in the formulation, since the interaction of the nanoparticles to the negatively charged surface of the intestinal mucosa might be hindered.

Muramyltripeptide-Cholesterol

The use of immunotherapy with macrophage activators has been suggested to treat metastatic tumors as an alternative to conventional therapy. Indeed, when activated by immunomodulators, cells of the macrophage-monocyte lineage become cytostatic towards neoplastic cells regardless of their resistance to conventional therapeutic agents. Muramyldipeptide (MDP) has interesting immunomodulating properties in vitro, but due to its hydrophilicity, it is cleared too rapidly to produce an antimetastatic effect in vivo. Therefore, the use of lipophilic derivatives and carriers has been investigated.

Yu et al. (138) encapsulated the lipophilic MTP cholesterol (MTP-Chol) in PIBCA nanocapsules and in liposomes, and administered these to mice carrying an experimental model of liver metastasis. When administered intravenously, nanocapsules and an emulsion of MTP-Chol were effective as a prophylactic treatment (started before tumor cell inoculation). In contrast, it had no effect when initiated once metastases had started to grow. Intragastric administration of nanocapsules containing MTP-Chol, starting two days before the tumor inoculation, gave a significant but limited antimetastatic effect without obvious dose-dependency. Interestingly, empty nanocapsules at a dose of 50 μg also produced a reduction in the metastases, but this observation was not reproduced at a higher dose (100 μg). A similar effect was observed after treatment with liposomes or MTP-Chol–associated emulsion, despite likely differences in the stability of these forms in the intestinal environment. It is possible that the observed results reflect the effect of MTP-Chol released from the carriers and absorbed as the free drug.

Vasopressin

Vasopressin, also called antidiuretic hormone, is an octapeptide secreted by the neuronal cells of the hypothalamic nuclei and stored in the posterior lobe of the neurohypophysis. It stimulates the contraction of the muscular tissue of the capillaries and arterioles, raising blood pressure. Furthermore, it has a specific effect on epithelial cells of the distal portion of the uriniferous tubule, inducing resorption of water independently of solutes, resulting in the concentration of urine and dilution of blood serum. At high doses, vasopressin stimulates uterine and GI smooth muscles. Vasopressin is used primarily as an antidiuretic drug in the treatment of acute or chronic diabetes insipidus but it may also be used to stimulate smooth muscle tissue, especially as a vasoconstrictor to control hemorrhages.

The absorption across rat intestinal tissue of 9-desglycinamide, 8-arginine vasopressin used as a model peptide was studied in rat intestinal loops (in vitro model), in a chronically isolated loop in rats (in situ model) and after intraduodenal administration in vivo (139,140). A controlled release bioadhesive drug delivery system consisting of microspheres of poly(2-hydroxyethyl methacrylate) with a mucoadhesive coating of polycarbophil was tested. This preparation was compared to an aqueous solution of the peptide in a suspension of polycarbophil particles, as a control for a fast-release formulation. The bioadhesive formulation led to a slight improvement of peptide absorption in vitro in comparison to the control, but no effect was found in vivo. In contrast, bioavailability was significantly increased in all three models from the polycarbophil suspension compared to the free drug in solution. A prolongation of the absorption phase in vitro and in the chronically isolated loop in situ suggested that the polymer was able to protect the peptide from proteolysis. The enhancement in bioavailability was reduced with increasing complexity of the model. This interrelationship suggests that the observed improvement in vasopressin absorption and the mucoadhesive properties of the polymer are associated. It was concluded that a fast-release oral dosage form for peptide drugs on the basis of polycarbophil appears to be possible. However, a very important limitation is the inability of these delivery systems to renew the bio(muco) adhesive surface. High mucus turnover rates impede the prolonged adhesion on the mucosal surface.

Luteinizing Hormone Releasing Hormone

To enhance oral peptide absorption, Hillery et al. (38) synthesized a novel drug polymer conjugate that formed its own nanoparticulate delivery system—the copolymerized peptide particle system (CPP). Using the peptide luteinizing hormone releasing hormone (LHRH) as a model drug to investigate the viability of this approach, a polymerizable derivative of LHRH was prepared by conjugating the peptide with vinylacetic acid. The LHRH–vinylacetate conjugate was then copolymerized with *n*-butylcyanoacrylate. The copolymerized system was formulated as particles of an average diameter of 100 nm, to which LHRH was covalently linked. The particles were stable in vitro when incubated three hours in intestinal medium, mucosal scrapings and serum. Moreover, in vitro transport studies using the Caco-2 cell model suggested that absorption in vivo might take place.

In a further study, CPPs were administered orally to male Wistar rats (141). Using double antibody radioimmunoassay (RIA), it was shown that LHRH was present in plasma over 12 hours, whereas the free peptide or a LHRH-vinylacetate derivative in a buffer did not show detectable absorption, confirming the protective effect of CPP encapsulation. The maximum plasma uptake, 1.6% of the administered dose of LHRH, was detected three hours after single dosing with the CPP. Significant levels of the peptide were detected up to 12 hours. Multiple (two or five days) dosing led to a higher LHRH uptake. The authors pointed out that the RIA may have underestimated the extent of oral uptake of LHRH, because of incomplete extraction of LHRH and also due to the shielding effects of the intact particles, CPP degradation products, and plasma proteins which might be associated with CPP metabolites. The conclusion was that at least some of the CPP particles remained intact in the GI tract to protect the peptide and, this proportion of intact particles was responsible for the detected uptake of LHRH. The chemical conjugation of LHRH within a protective particulate matrix is a viable approach to promote oral delivery.

Interferon-γ

Interferons are a complex group of proteins and glycoproteins with antiviral, immunomodulatory and differentiation activity. Because there are considerable obstacles to interferon absorption from the stomach, conventional routes for interferon administration involve injection, usually via the subcutaneous route. Eyles et al. (142) employed microsphere technology to enhance the oral bioavailability of interferon-γ. They have investigated the oral uptake and distribution of poly(L-lactide) microsphere-encapsulated radiolabeled interferon-γ in male Wistar rats. As a control, another group of rodents received free radiolabeled interferon-γ attached to empty microspheres. When 1 μg of free radiolabeled interferon-γ was administered by gavage, a quite different distribution of radioactivity was recorded after 15 and 240 minutes in comparison to the group receiving 1 μg in loaded microspheres. Whereas blood and tissue (liver, spleen, and kidneys) radioactivity, collected after 15 minutes from animals dosed with microencapsulated interferon, was higher than the control group, the thyroid activity was significantly lower. For microspheres with higher interferon loading (30 μg) the difference between the two experimental groups was not as significant. Moreover, fifteen minutes after gavage, high activity was detected in both the proximal and distal small intestine, although accumulation per mass unit was much higher in the proximal segments. It is noteworthy that the uptake of radiolabeled interferon-γ into the Peyer's patches was high relative to other intestinal regions after oral administration of trace-loaded spheres. These results led to the tentative conclusion that microencapsulation of interferon markedly affects the oral uptake, and possibly also post-absorption pharmacokinetic parameters of the drug.

Octreotide, a Somatostatin Analog

Octreotide is a somatostatin analog with greater pharmacological activity, being 70, 23, and 3 times more potent at inhibiting growth hormone, glucagons, and insulin secretion, respectively (143,144). This octapeptide has been used in man, for the treatment of acromegaly, insulin-dependent diabetes, and secreting and nonsecreting tumors among others. When administered orally, it has a low bioavailability (145–149).

To improve biological response of octreotide, Damgé et al. (71) have incorporated the peptide into PIBCA-nanocapsules of 260-nm diameter. Two biological effects of the peptide, namely, insulin response to glucose and the inhibition of prolactin secretion in response to oestrogens as well as plasma level of octreotide, were recorded after subcutaneous and peroral administration in male Wistar rats. When administered subcutaneously, the octreotide-loaded nanocapsules ($20\,\mu g\,kg^{-1}$) suppressed the insulinaemia peak induced by intravenous glucose overload and depressed insulin secretion over 48 hours, preventing the secretory rebound; however glycemia was unaffected. Moreover, the plasma octreotide concentrations increased 2.7 times in parallel with the octreotide dose. Administered by oral gavage, to oestrogen-treated rats, octreotide-loaded nanocapsules (200 and $1000\,\mu g\ kg^{-1}$) significantly improved the reduction of prolactin secretion by 72% and 88%, respectively, compared with 32% and 54% with free octreotide at the same doses. Accordingly, a slight increase in plasma octreotide level was observed.

HIV-1 Protease Inhibitors

Peptidomimetic inhibitors of the human immunodeficiency virus type 1 (HIV-1) proteases are effective antiviral agents that strongly inhibit virus replication. As a consequence of their peptidic structure and high lipophilicity, most HIV-1 protease inhibitors reported so far display poor oral bioavailability. The peptidomimetic compound CGP 57813 is a potent inhibitor of HIV-1. It is a highly lipophilic substance with a relatively short half-life. To improve its bioavailability, Leroux et al. (150) nanoencapsulated the compound into PLA and pH sensitive methacrylic acid copolymers (Eudragit® L100–55 and Eudragit S100 which dissolve above pH 5.5 and 7, respectively). CGP 57813 formulations were administered orally, by gavage, to female BALB/c mice. Only nanoparticles made of Eudragit soluble at low pH (Eudragit L100–55) provided sufficient plasma levels of CGP 57813. The good results obtained with Eudragit L100–55 nanoparticles can be explained by the increase of surface area, by the relatively rapid dissolution of the polymer provided by the nanoparticles formulation, and by the high level of dispersion of the compound within the polymeric matrix. Eudragit S100 nanoparticles gave intermediate plasma levels, probably because of the incomplete dissolution of the particles at the site of absorption. Finally, PLA nanoparticles provided very low and fluctuating drug plasma levels. It was concluded that the passage of intact PLA nanoparticles across the GI mucosa was too low to provide sufficient exposure to the antiviral drug.

pH-sensitive nanoparticles were further administered orally to beagle dogs and the influence of food on the pharmacokinetics was evaluated (151). When free CGP 57813 was administered to fasted dogs as an aqueous suspension, no drug was detected in the plasma (Fig. 6). On the other hand, incorporation of CGP 57813 in Eudragit nanoparticles substantially improved the bioavailability of the compound in dogs but in contrast with the results obtained in mice, the AUC following the administration of CGP 57813 nanoparticles was relatively low. The AUC was slightly increased in fed animals. The presence of food appears to promote the absorption, most likely by increasing the intestinal bulk, slowing the GI transit and stimulating GI secretions. The same positive effect of food was observed with Eudragit S100 nanoparticles. When these particles were administered to fasted dogs, no detectable plasma levels of the compound were observed (Fig. 6). In contrast, the same particles administered to feed dogs gave the highest plasma concentrations. In the absence of food, the nanoparticles may be eliminated too rapidly before starting to dissolve. Recently, similar results were obtained in fed beagle dogs with CGP 70726, another HIV-1 protease inhibitor exhibiting 70-fold lower water solubility than CGP 57813 (152). The results of these studies show that besides the administration conditions, the selection of an optimal pH-sensitive formulation strongly depends on the animal species.

In summary, peptides and proteins are primarily delivered by parenteral routes. Such therapies may result in poor patient compliance because of the inconvenience of repeated injections. Oral delivery, the preferred route in drug administration presents two major problems for peptides: a rapid digestion by proteolytic enzymes in the GI tract and low permeability through biological membranes. To overcome these problems, an interesting approach consists of creating oral formulations which can act as carriers protecting the drug from degradation and releasing it slowly into the systemic circulation.

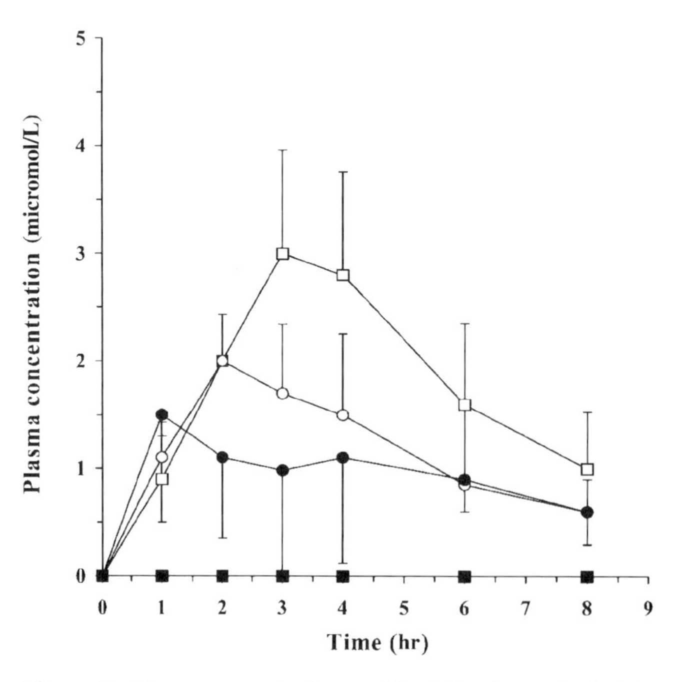

Figure 6 Plasma concentration profiles following oral administration of CGP 57813 (100 mg kg^{-1}) from Eudragit L100–55 (•, fasted; ○, fed dogs) and S100 (■, fasted; □, fed dogs) nanoparticles ($n = 4$). *Source*: From Ref. 151.

Other Active Drugs

The main body of literature about the use of particulates for oral drug delivery is devoted to peptides and proteins whether for sustained delivery or more frequently for vaccine purposes. Nonetheless, the potential of particle delivery has been investigated for other drugs. The encapsulation of active compounds into polymeric particles is a rationale to prevent gastro-intestinal irritation, reduce drug hydrolysis, favor enteric absorption or provide a controlled release system. The main characteristics of the different systems are summarized in Table 5.

Vincamine-loaded hexylcyanoacrylate nanoparticles (200 nm) were administered orally to rabbits (153). The comparison of pharmacokinetic parameters showed a significant improvement of bioavailability when vincamine was administered under the form of nanoparticles with an absolute bioavailability of 22% for the free drug and 36% when administered in the form of nanoparticles. Interestingly, the nanoparticle formulation gave a faster and higher peak of absorption. In vitro release studies showed that in the nanoparticle suspension, 75% of vincamine were in the free form, whereas the remaining 25% were still associated to the polymer. Taking this information into account, recalculated parameters showed an absolute bioavailability of 78% for the vincamine associated with nanoparticles. This increased bioavailability of vincamine when administered as nanoparticles was the result of a better stability in the gastrointestinal lumen, a prolonged contact with the stomach wall, and a lower stomach and/or liver first pass effect.

Calcium channel antagonists of the dihydropyridine family such as nifedipine, darodipine, and isradipine are characterized by an intense and very rapid hypotensive activity mediated by a vasodilation that does not last very long because of a fast elimination. Therefore, the treatment must be repeated at least two to three times a day and patients are potentially submitted to a transitory cerebral ischemia. Patients may also experience dizziness, headaches, flushing, or peripheral edema. Polymeric encapsulation would provide a sustained release profile, therefore reducing the initial hypotensive activity and most likely diminishing the frequency of dosing. Maincent's group published several reports on the feasibility of such an approach using different drugs and different polymers. In the first study, darodipine was encapsulated in PIBCA nanocapsules (154). After oral administration in renovascular hypertensive rats, the overall pharmacological activity was the same whether the animals had received the PEG 400 solution of the free drug or daropidine-loaded nanoparticles. However, in the group receiving the daropidine-loaded PIBCA particles, the hypotensive peak was not as pronounced as in the case of the solution, therefore preventing the adverse effect of a rapid vasodilation and the cerebral ischemia risk accompanying the treatment. Blood concentrations of the drug were in agreement with the observed pharmacological activity. The bioavailability was the same whether the drug was encapsulated or not, showing that the nanocapsules did not diminish the intestinal absorption of darodipine. The peak was lower than with the free drug and the onset was slightly delayed. The interest of nanoparticles for the oral delivery of calcium blockers was further demonstrated with nifedipine encapsulated in potentially less toxic polymers (155). Two biodegradable polymers, PCL and PLGA and nonbiodegradable Eudragit RL and RS were used to prepare the nanoparticles to be administered in spontaneously hypertensive rats (SHR). All the polymeric carriers gave a decreased C_{max} that may be explained by the slow release of the drug from the particles, as demonstrated by in vitro release studies, parallel to a moderate antihypertensive effect especially with Eudragit. The AUC

Table 5 Physicochemical Properties and Preparation Methods Used for Particulate Systems in the Formulation of Poorly Bioavailable Drugs

Drug	Polymer	Method	Size	References
Vincamine	PHCA	Emulsion–polymerization	200–230 nm	153
Darodipine	PIBCA	Interfacial polymerization	150 nm	154
Nifedipine	Eudragit RL and RS	Interfacial deposition	170 nm	155
	PCL		210 nm	
	PLGA		120 nm	
Heparin	PCL	Emulsion–evaporation	80 to 280 μm	157
	PLGA		270–310 nm	158
Indomethacin	PLA	Interfacial deposition	230 nm	159
	PIBCA	Interfacial polymerization	237 nm	
	PLA	Interfacial deposition	130 nm	160
Methotrexate	Chitosan	Emulsification–diffusion	>150 μm	161
Atovaquone	PLA	Dessolvation	200–270 nm	165
Ampicilin	PVAL	Emulsion–precipitation	<40 μm 45%: 1–5 μm	166
Theophylline	Chitosan			
AZT	PHCA	Emulsion–polymerization	230 nm	168
Avarol	PBCA	Emulsion–polymerization	60–320 nm 450–970 nm	170
Antitubercular, INH, RIF, PZA	PLGA	Emulsion–evaporation	1.1–2.2 μm	171
			190–290 nm	172
Dicumarol DNA	Poly FA:SA	Phase inversion nanoencapsulation	0.1–5 μm	97

Abbreviations: AZT, azidothymidine; PLA, poly(lactic) acid; PIBCA, poly(isobutyl-cyanoacrylate); PHCA, poly(hexylcyanoacrylate); PCL, poly-ε-caprolactone; PLGA, poly (lactide-co-glycolide); PVAL, poly(vinyl alcohol) FA:SA, fumaric acid:sebacic acid.

and the mean residence time were increased when the drug was encapsulated. The results obtained with the Eudragit may be explained by a slight bioadhesive effect due to the presence of quaternary ammonium groups in the polymer backbone that are prone to interact with negatively charged mucopolysaccharides. More recent investigations are underway with another agent, isradipine, encapsulated in PLA, PCL, and different types of PLGA (156).

These studies have shown that polymeric particles led to an efficient sustained release of dihydropyridine calcium channel antagonists. Nanoparticles appeared to be safer than the free drug solution by reducing the intensity of the initial vasodilatation.

Because of a very short half-life and a lack of oral absorption of the drug, heparinotherapy is characterized by multiple and daily injections. Heparin is a negatively charged high molecular weight glycosaminoglycane used as an anticoagulant for the treatment and prevention of thrombosis or embolisms. Oral treatment would increase greatly the quality of life of patients in need of this treatment. Micro- and nanoparticles made of biodegradable PLGA or PCL and nonbiodegradable Eudragit RS and RL used alone or in combination have been designed to encapsulate heparin prior to oral administration in rabbits (157,158). After a lag time of four hours, microparticles (80 to 280 μm) made of a mixture of Eudragit RS and PLGA or Eudragit RS, Eudragit RL and PLGA increased the clotting time assayed by an APTT test (157). The effect lasted for four hours and the absolute bioavailability compared to the same dose given intravenously was 42.5% and 48.3%, respectively. Such large particles are not expected to cross the intestinal epithelium intact and the activity of the heparin-loaded particles can be explained by the fact the macromolecule is protected from the GI environment, the negative charge of the particles and of the heparin itself will favor an electrostatic interaction with the mucus layer and thus increase the retention time at the mucosa level. Finally, the large surface contact generated by the particles will increase the concentration gradient thus favoring the release of the heparin at the absorption site. The preparation of nanoparticles (270 to 310 nm) led to the same drug loading as the formation of microparticles, 10% to 95% depending on the polymers used (158). After administration to the animals, the lag time was shorter as compared to the microparticles, but the overall activity was lower than with the microparticles since the maximum absolute bioavailability was observed with the mixture of Eudragit RL and PCL at 23% compared to the intravenous administration. These results might be explained by the slow diffusion of the drug from the polymeric carrier as shown by the in vitro release study. Although not commented by the authors, it is likely that the smaller size of the particles renders them more sensitive to intestinal tract degradation. These two studies, however, show clearly that encapsulation into polymeric nano- or microparticles-induced oral absorption of heparin.

Nonsteroidal anti-inflammatory drugs are well-known for the gastrointestinal ulceration they induce after oral administration. Part of this adverse effect is mediated by a direct contact of the drug with the mucosa. Ammoury et al. (159) encapsulated indomethacin into PLA and PIBCA nanocapsules of the same size (approximately 230 nm) and the same drug loading. After a one-hour jejunal infusion in rat, no significant difference in the rate of drug absorption was observed between PIBCA particles and the drug in solution while the fraction absorbed from PLA particle suspension was significantly lower. Regardless of the polymer, the encapsulation of the drug led to the protection of the mucosa against the ulceration observed with indomethacin solution. Following oral administration in rat, no significant difference was observed between the overall pharmacokinetic parameters of the solution and both polymer-encapsulated formulations. However, C_{max} and the AUC were slightly smaller for the encapsulated indomethacin leading to a relative bioavailability of 77% and 84% for PLA and PIBCA nanocapsules, respectively. T_{max} was delayed by one hour for the PLA-encapsulated indomethacin compared to PIBCA-encapsulated and free drug, suggesting that the nanocapsules followed different absorption pathways or different release mechanisms or kinetics. The thromboxane B2 levels determined at 4 and 24 hours after administration demonstrated that the unloaded nanocapsules had no effect on inflammatory markers. In indomethacin treated groups, a significant decrease was observed when compared to the control groups. However, no statistical difference was seen between free indomethacin and encapsulated drug regardless of the polymer. At 24 hours, the pharmacological activity of indomethacin-PLA nanocapsules was still maintained at a level under 10% of the initial value whereas for the two other groups, the thromboxane B2 levels started to come back to normal.

This study shows that it is possible to protect mucosae from indomethacin-induced ulceration with the nanocapsule approach; the release mechanism of the drug from both polymers greatly influenced the pharmacokinetic profiles as well as the pharmacological effect. This study, however, was conducted with nanocapsule suspensions having a short shelf life. A later study showed the instability of the PLA nanocapsules after freeze-drying (160). Lyophilization in the presence of cryoprotecting agents provided stable formulations giving the same rate of protection of the gastric mucosa as obtained by Ammoury et al. (159).

The antineoplastic agent methotrexate has a very short elimination half-life when given orally. Furthermore, it induces gastric irritation, diarrhea, ulcerative somatitis and hemorrhagic enteritis. Shielding the drug inside polymeric particles is thus expected to reduce these adverse effects by preventing contact with the mucosae. Therefore, Singh and Udupa loaded methotrexate in chitosan microspheres (161). Although the microspheres were quite large (more than 150 μm in diameter), the drug given as particles was absorbed and was eventually more effective than the free drug solution. Indeed, encapsulated and free methotrexate provided a better rate of survival in mice bearing the Ehrlich ascites carcinoma compared to an untreated group. After particle administration, the pharmacokinetic profiles showed an increased AUC without a higher peak of absorption when compared to the free drug. The sustained plateau observed with the particles supports the hypothesis of a slow release profile. The bioadhesive properties of chitosan may also explain this effect (37,162). The anti-irritating properties have not been investigated, the authors stating that chitosan particles are supposed to swell and float in gastric acidic medium. However, chitosan is soluble at low pH and it would have been interesting to study the release profile in gastric juice. The protection of methotrexate in low pH solutions was demonstrated by Madhan Kumar and Panduranga Rao with 1 to 5 μm particles made of Prot A7 proteinoid. Unfortunately, no in vivo experiment was reported for this system (163).

Atovaquone is an antiparasitic drug active against toxoplasma cysts and therefore displays interesting possibilities for the treatment of toxoplasmosis-related encephalitis, especially for AIDS patients. However, it suffers from low and erratic oral bioavailability and it presents variable sensitivity against the different strains of *Toxoplasma gondii* (164). PLA nanocapsules (200 nm in size) were prepared to improve both intestinal absorption and intracellular uptake to enhance the inhibitory effect on less sensitive strains (165). In vitro, nanocapsules were as effective as the atovaquone suspension against the development of *T. gondii* strain grown in MRC5 fibroblasts. In an acute mice model, the group receiving unloaded nanocapsules had a 100% death rate at day 7 after infection. On the other hand, in the atovaquone-loaded nanocapsule group, the survival rate was 70% at 30 days whereas in the group treated with atovaquone suspension, all the mice died within 12 days. Furthermore, in the nanocapsule group, parasites were undetectable in the blood and no cyst was found in the brain of the surviving animals. In the chronic infection model, the number of brain cysts and the titration of parasitic burden showed a marked efficacy of atovaquone. When compared to the suspension, nanoencapsulated atovaquone exhibited an increased activity and a significant reduction in parasitic burden was observed; however, the cyst count in the brain was not reduced.

Sugimoto et al. (166) tested the ability of poly(vinyl alcohol)-gel microspheres as an intestinal drug delivery system. Theophylline and ampicillin were used as drug models. The theophylline-encapsulated preparation did not show a significant difference to the solution; however, the maximal absorption peak demonstrated a delay characteristic of a controlled delivery system. In contrast, the bioavailability of poorly absorbed ampicillin was enhanced by the encapsulation in PVAL gel microspheres and the effect was further enhanced in the presence of chitosan.

After parenteral or oral administration, particles have been shown to have a tropism for the mononuclear and phagocyte system (22,167). Therefore, particle systems present an attractive approach for targeting drugs to these cells, especially in the case of AIDS for which macrophages represent the main circulatory reservoir of the HIV. Furthermore, by controlling the body distribution of the potentially toxic drug, side effects would certainly be decreased. Azidothymidine (AZT) was entrapped in poly(hexylcyanoacrylate) (PHCA) particles and the fate of the particles after oral dosing was traced in reticuloendothelial cell–containing organs by following the radioactivity attached to the drug (168). The body distribution of the AZT-loaded particles was further confirmed by radioluminography (169). AUC in the liver was 30% higher with the nanoparticles than with the free drug and higher concentrations were also reported in blood and brain indicating that such a formulation would be very interesting for this kind of application.

PBCA particles were used to improve the oral bioavailability of avarol, a cytostatic and antiviral drug, active against HIV (170). Although it is stable in the GI environment and suitable for oral administration, oral bioavailability of avarol is limited by poor water solubility. Two size range batches (around 200 and 750 nm, respectively) with the same drug load were administered to rats. Both nanoparticle preparations led to an eight- to ninefold increase in AUC when compared with avarol solubilized in Solutol. However, no difference was observed between the preparations implying that particle size did not matter, therefore, suggesting that the particles themselves were not translocated from the stomach lumen. This hypothesis was corroborated by the fact that no particle was found in the blood.

Chemotherapy of tuberculosis involves daily administration of a cocktail of three to four drugs for at least six to eight months. This very heavy treatment results most of the time in a high rate of noncompliance leading to the spreading of the disease and development of emerging resistant bacterial strains. Therefore, the development of a sustained release formulation would prevent these troubles by facilitating patients' compliance. Khuller et al. (171,172) have developed antitubercular drugs loaded in PLGA-micro- and nanoparticles. Particles were loaded either with rifampicine (RIF) or isoniazid (INH) or pyrazinamide (PZA). In a first study, drugs were administered encapsulated in microparticles (1.1 to 2.2 μm) or free either alone or as a combination of the three drugs to rats and pharmacokinetic profiles were compared (171). Encapsulation of the drugs induced a prolonged residence time of the compounds, from 24 to 72 hours for INH, from 48 to 72 hours for RIF and from 48 to 108 hours for PZA. All the plasma levels were above the minimal inhibitory concentration (MIC) and no significant differences were observed whether the drugs were administered alone or together. Based on these results, mice were inoculated with *Mycobacterium tuberculosis* $H_{37}Rv$ and treated either with the free drugs daily or with nanoparticles (190–290 nm) containing the drugs once every 10 days (172). Animals were sacrificed at day 46 after the beginning of the chemotherapy and signs of infection were sought in lung and spleen homogenates. Results showed no bacterial proliferation in both groups treated either with free drug or encapsulated. On the other hand, untreated groups developed colonies in lungs and spleen. This study shows that it is possible to reduce the number of dosings without changing the dose given to the animals. Further, it was also shown that drugs given as particulates did not induce the hepatotoxicity that may be expected with those drugs.

Although less abundant than peptide and protein data, the literature provides stimulating reports establishing the interest of using polymeric particles to increase the oral absorption of poorly absorbed molecules. This approach has also proven to be effective in modulating pharmacokinetic parameters, especially prolonging half-life, or reducing side effects because of the irritant activity of the compounds.

CONCLUSIONS

Absorption of polymeric particles after oral administration has been widely documented in the literature. The rate of uptake appears to be very low and very controversial data are found, however, drug effect is reported for several systems associating various drugs to different types of polymers.

Part of the limitation in the development of oral particles may be technological, e.g., low rate of encapsulation in a given polymer to reach a targeted size. On the other hand, one may also raise some concern that dosing drugs, especially proteins, in the form of particulates would increase the risk of developing immunity against the active molecule with severe side effects leading to an inefficient and even harmful treatment. Indeed, the main place for the absorption of particles is still thought to be the Peyer's patches, the very site of initiation of the GI mucosal immunity. However, such a concern does not exist for small peptides and other molecules and, efforts should focus on developing these systems.

The studies presented in this chapter show that polymeric particles, primarily because of their stability, may improve the oral bioavailability of poorly absorbed drugs. Much interesting and encouraging data have been generated, demonstrating the principle of the concept. However, pharmacokinetic data are generally quite erratic and the mechanisms of uptake are still not well defined. It is also clear that the ideal particulate system does not exist and because of the problems encountered, vary based on the drug; a customized solution will be needed for each compound. Part of the success of such a strategy lies in designing better absorbed particles such as systems offering better interaction with the GI wall, thus permitting higher blood concentrations to be achieved. In addition, further understanding of the physiology of particle uptake would promote the optimization of drug delivery systems based on particulate carriers. It should be possible, in the future, to use bacterial and viral mechanisms of uptake and selectivity to engineer carriers, which will have a high selectivity for mucosal sites.

REFERENCES

1. Kreuter J. Peroral administration of nanoparticles. Adv Drug Del Rev 1991; 7:71–86.
2. Volkheimer G. Persorption of particles: physiology and pharmacology. Adv Pharmacol Chemother 1977; 14:163–187.
3. Cornes JS. Number, size, and distribution of Peyer's patches in the human small intestine. Gut 1965; 6:225–233.
4. Volkheimer G, Schultz FH. The phenomenon of persorption. Digestion 1968; 1:213–218.
5. Volkheimer G, Schulz FH, Aurich I, Strauch S, Beuthin K, Wendlandt H. Persorption of particles. Digestion 1968; 1:78–82.
6. Volkheimer G, Schulz FH, Hofmann I, et al. The effect of drugs on the rate of persorption. Pharmacology 1968; 1:8–14.
7. Volkheimer G. Hematogenous dissemination of ingested polyvinyl chloride particles. Proc Ann NY Acad Sci 1975; 246:164–171.
8. Björk E, Isaksson U, Edman P, Artursson P. Starch microspheres induce pulsatile delivery of drugs and peptides across the epithelial barrier by reversible separation of the tight junctions. J Drug Target 1994; 2:501–507.
9. Edman P, Björk E, Rydén L. Microspheres as nasal delivery system for peptide drugs. J Contr Rel 1992; 21:165–172.
10. Lueßen HL, Lehr CM, Rentel CO, et al. Bioadhesive polymers for the peroral delivery of peptide drugs. J Contr Rel 1994; 29:329–338.
11. LeFevre ME, Joel DD. The Peyer's patch epithelium: an imperfect barrier. In: Schiller CM, ed. Toxicology of Intestinal Function. New York: Raven Press, 1984:45–56.
12. LeFevre ME, Vanderhoff JW, Laissue JA, Joel DD. Accumulation of 2 μm latex particles in mouse Peyer's patches during chronic latex feeding. Experientia 1978; 31(4):120–122.
13. LeFevre ME, Hancock DC, Joel DD. Intestinal barrier to large particulates in mice. J Toxicol Env Health 1980; 6:691–704.
14. Bertheusen KJ, Diemer NN, Proelstholm J, Klinken L. Pulmonary excretion of carbon black injected into the cerebral ventricles of the rat. Acta Pathol Microbiol Scand Sect A 1978; 86:90–92.
15. LeFevre ME, Joel DD, Schidlovsky G. Retention of ingested latex particles in Peyer's patches of germfree and conventional mice (42133). Proc Soc Exp Biol Med 1985; 179:522–528.
16. LeFevre ME, Warren JB, Joel DD. Particles and macrophages in murine Peyer's patches. Expl Cell Biol 1985; 53:121–129.
17. LeFevre ME, Boccio AM, Joel DD. Intestinal uptake of fluorescent microspheres in young and aged mice (42825). Proc Soc Exp Biol Med 1989; 190:23–27.
18. Bhagat HR, Dalal PS. Oral vaccination by microspheres. In: Cohen S, Bernstein H, eds. Microparticulate Systems for the Delivery of Proteins and Vaccines. New York, Basel, Hong Kong: Marcel Dekker, 1996:381–399.
19. Seifert J, Haraszti B, Sass W. The influence of age and particle number on absorption of polystyrene particles from rat gut. J Anat 1996; 189:483–486.
20. Pappo J, Ermak TH. Uptake and translocation of fluorescent latex particles by rabbit Peyer's patch follicle epithelium: a quantitative model for M cell uptake. Clin Exp Immunol 1989; 76:144–148.
21. Sanders E, Ashworth CT. A study of particulate intestinal absorption and hepatocellular uptake. Use of polystyrene latex particles. Exp Cell Res 1961; 22:137–145.
22. Jani PU, McCarty DE, Florence AT. Nanospheres and microspheres uptake via Peyer's patches: observation of the rate of uptake in the rat after a single oral dosage. Int J Pharm 1992; 86:239–246.
23. Kreuter J, Muller U, Munz K. Quantitative and microautoradiographic study on mouse intestinal distribution of polycyanoacrylate nanoparticles. Int J Pharm 1989; 55:39–45.
24. Michel C, Aprahamian M, Defontaine L, Couvreur P, Damgé C. The effect of site of administration in gastrointestinal tract on the absorption of insulin from nanocapsules in diabetic rats. J Pharm Pharmacol 1991; 43:1–5.

25. Eldridge JH, Gilley RM, Staas JK, Moldoveanu Z, Tice TR. Biodegradable microspheres: vaccine delivery system for oral immunization. Curr Top Microbiol Immunol 1989; 146:59–66.
26. Jeurissen SHM, Kraal G, Sminia T. The role of Peyer's patches in intestinal humoral immune responses is limited to memory formation. Adv Exp Med Biol 1987; 216A:257–265.
27. LeFevre ME, Joel DD. Intestinal absorption of particulate matter. Life Sci 1977; 21:1403–1408.
28. McClean S, Prosser E, Meehan E, et al. Binding and uptake of biodegradable poly-DL-lactide micro- and nanoparticles in intestinal epithelia. Eur J Pharm Sci 1998; 6:153–163.
29. Wells CL, Maddaus MA, Erlandsen SL, Simmons RL. Evidence for the phagocytic transfer of intestinal particles in dogs and rats. Infect Immun 1988; 56:278–282.
30. Jepson MA, Simmons NL, Savidge TC, James PS, Hirst BH. Selective binding and transcytosis of latex microspheres by rabbit intestinal M cells. Cell Tissue Res 1993; 271:399–405.
31. Porta C, James PS, Phillips AD, Savidge TC, Smith MW, Cremaschi D. Confocal analysis of fluorescent bead uptake by mouse Peyer's patch follicle associated M cells. Exp Physiol 1992; 77:929–932.
32. Brown WR, Le M, Reid R, Boedeker E. Histochemical and immunohistochemical characterization of rabbit Peyer's patches. Gastroenterology 1994; 106(suppl 4):A657.
33. Jepson MA, Simmons NL, Hirst GL, Hirst BH. Identification of M cells and their distribution in intestinal Peyer's patches and appendix. Cell Tissue Res 1993; 273:127–136.
34. Matsuno K, Schaffner T, Gerber HA, Ruchti C, Hess MW, Cottier H. Uptake by enterocytes and subsequent translocation to internal organs, e.g., the thymus, of Percoll microspheres administered *per os* to suckling mice. J Reticuloendot Soc 1983; 33:263–273.
35. Cruz N, Alvarez X, Berg RD, Deitch EA. Bacterial translocation across enterocytes: results of a study of bacterial–enterocyte interactions utilizing Caco-2 cells. Shock 1994; 1(1):1–7.
36. Delie F, Rubas W. Caco-2 monolayers as a tool to examine intestinal uptake of particulates. Proc Int Symp Contr Rel Bioact Mater 1996; 23:149–150.
37. Lehr CM, Bouwstra JA, Kok W, Noach ABJ, de Boer AG, Junginger HE. Bioadhesion by mean of specific binding of tomato lectin. Pharm Res 1992; 9(4):547–553.
38. Hillery AM, Toth I, Shaw AJ, Florence AT. Co-polymerised peptide particles (CPP) I: synthesis, charaterisation and in vitro studies on a novel oral nanoparticulate delivery system. J Contr Rel 1996; 41:271–281.
39. Le Visage C, Couvreur P, Mysiakine E, Breton P, Bru N, Fattal E. In vitro and in vivo evaluation of poly(methylidene malonate 2.1.2) microparticles behavior for oral administration. J Drug Target 2001; 9(2):141–153.
40. Desai MP, Labhasetva V, Walter E, Levy RL, Amidon GL. The mechanism of uptake of biodegradable microparticles in Caco-2 cells is size dependent. Pharm Res 1997; 14(11):1568–1573.
41. Pietzonka P, Rothen-Rutishauser B, Langguth P, Wunderli-Allenspach H, Walter E, Merkle HP. Transfer of lipophilic markers from PLGA and polystyrene nanoparticles to caco-2 monolayers mimics particle uptake. Pharm Res 2002; 19(5):595–601.
42. Artursson P. Cell cultures as models for drug absorption across the intestinal mucosa. Crit Rev Ther Drug Car Syst 1991; 8(4):305–330.
43. Delie F, Rubas W. A human colonic cell line sharing similarities with enterocytes as a model to examine oral absorption. Advantages and limitations of the Caco-2 model. Crit Rev Ther Drug Carr Syst 1997; 14(3):221–286.
44. Kernéis S, Bogdanova A, Kraehenbuhl JP, Pringault E. Conversion by Peyer's patch lymphocytes of human enterocytes into M cells that transport bacteria. Science 1997; 277:949–952.
45. Karlsson J, Axling S, Nyberg E, Artursson P. An improved model for studies of particle transport across intestinal epithelial cells converted to M cells. AAPS Annu Meet Pharm Sci 1998; 1(1).

46. Gullberg E, Leonard M, Karlsson J, et al. Expression of specific markers and particle transport in a new human intestinal M-cell model. Biochem Biophys Res Com 2000; 279(3):808–813.
47. Tyrer P, Ruth FA, Kyd J, Harvey M, Sizer P, Cripps A. Validation and quantitation of an in vitro M-cell model. Biochem Biophys Res Com 2002; 299(3):377–383.
48. Joel DD, Laissue JA, LeFevre ME. Distribution and fate of ingested carbon particles in mice. J Reticuloendothel Soc 1978; 24:477–487.
49. Jani PU, Florence AT, McCarthy DE. Further histological evidence of the gastrointestinal absorption of polystyrene nanospheres in the rat. Int J Pharm 1992; 84: 245–252.
50. Pinto-Alphandary H, Aboubakar M, Jaillard D, Couvreur P, Vauthier C. Visualization of insulin-loaded nanocapsules: in vitro and in vivo studies after oral administration to rats. Pharm Res 2003; 20:1071–1084.
51. Jani PU, Haklbert GW, Langridge J, Florence AT. Nanoparticles uptake by the rat gastrointestinal mucosa: quantitation and particle size dependency. J Pharm Pharmacol 1990; 42:821–826.
52. Carr KE, Hazzard RA, Reid S, Hodges GM. The effect of size on uptake of orally administered latex microparticles in the small intestine and transport to mesenteric lymph nodes. Pharm Res 1996; 13(8):1205–1209.
53. Eldridge JH, Hammond CJ, Meulbroek JA, Staas JK, Gilley RM, Tice TR. Controlled vaccine release in the gut-associated lymphoid tissues. 1. Orally administered biodegradable microspheres target the Peyer's patches. J Contr Rel 1990; 11:205–214.
54. Jani P, Haklbert GW, Langridge J, Florence AT. The uptake and translocation of latex nanospheres and microspheres after oral administration to rats. J Pharm Pharmacol 1989; 41:809–812.
55. Pappo J, Ermak TH, Steger HJ. Monoclonal antibody-directed targeting of fluorescent polystyrene microspheres to Peyer's patch M cells. Immunology 1991; 73:277–280.
56. Jepson MA, Simmons NL, O'Hagan DT, Hirst BH. Comparison of poly(DL-lactide-co-glycolide) and polystyrene microsphere targeting to intestinal M cells. J Drug Targ 1993; 1:245–249.
57. Delie F. Evaluation of nano- and microparticle uptake by the gastrointestinal tract. Adv Drug Del Rev 1998; 34(2,3):221–233.
58. Lee VHL, Yamamoto A. Penetration and enzymatic barriers to peptide and protein absorption. Adv Drug Deliv Rev 1990; 4:171–207.
59. Mahato RI, Narang AS, Thoma L, Miller DD. Emerging trends in oral delivery of peptide and protein drugs. Crit Rev Ther Drug Carrier Syst 2003; 20(2,3):153–214.
60. Chien YW. Human insulin: basic sciences to therapeutics uses. Drug Dev Ind Pharm 1996; 22:753–789.
61. Foster DW. Diabetes mellitus. In: Fauci AS, Braunwald E, Iselbacher KJ, eds. Harrison's Principles of Internal Medicine. New York: McGraw-Hill, 1998:2060–2080.
62. Couvreur P, Lenaerts V, Kante B, Roland M, Speiser PP. Oral and parenteral administration of insulin associated to hydrolysable nanoparticles. Acta Pharm Technol 1980; 26(4):220–222.
63. Oppenheim RC, Stewert NF, Gordon L, Patel HM. The production and evaluation of orally administred insulin nanoparticles. Drug Dev Ind Pharm 1982; 8(4): 531–546.
64. Damgé C, Michel C, Aprahamian M, Couvreur P. New approach for oral administration of insulin with polyalkylcyanoacrylate nanocapsules as drug carrier. Diabetes 1988; 37:246–251.
65. Damgé C, Michel C, Aprahamian M, Couvreur P, Devissaguet JP. Nanocapsules as carriers for oral peptide delivery. J Contr Rel 1990; 13:233–239.
66. Aboubakar M, Puisieux F, Couvreur P, Deyme M, Vauthier C. Study of the mechanism of insulin encapsulation in poly(isobutylcyanoacrylate) nanocapsules obtained by interfacial polymerization. J Biomed Mater Res 1999; 47:568–576.

67. Aboubakar M, Puisieux F, Couvreur P, Vauthier C. Physico-chemical characterization of insulin-loaded poly(isobutylcyanoacrylate) nanocapsules obtained by interfacial polymerization. Int J Pharm 1999; 183:63–66.
68. Damgé C, Hillaire-Buys D, Puech R, Hoetzel A, Michel C, Ribes G. Effects of orally administered insulin nanocapsules in normal and diabetic dogs. Diab Nutr Metab 1995; 8:3–9.
69. Damgé C, Vranckx H, Balschmidt P, Couvreur P. Poly(alkylcyanoacrylate) nanospheres for oral administration of insulin. J Pharm Sci 1997; 86:1403–1409.
70. Aboubakar M, Couvreur P, Pinto-Alphandary H, et al. Insulin-loaded nanocapsules for oral administration: in vitro and in vivo investigation. Drug Develop Res 2000; 49:109–117.
71. Damgé C, Vonderscher J, Marbach P, Pinget M. Poly(alkyl cyanoacrylate) nanocapsules as a delivery system in the rat for octreotide, a long-acting somatostatin analogue. J Pharm Pharmacol 1997; 49:949–954.
72. Vauthier C, Dubernet C, Fattal E, Pinto-Alphandary H, Couvreur P. Poly(alkylcyanoacrylates) as biodegradable materials for biomedical applications. Adv Drug Del Rev 2003; 55:519–548.
73. Lowe PJ, Temple CS. Calcitonin and insulin in isobutylcyanoacrylate nanocapsules: protection against proteases and effect on intestinal absorption in rats. J Pharm Pharmacol 1994; 46:547–552.
74. Cournarie F, Auchere D, Chevenne D, Lacour B, Seiller M, Vauthier C. Absorption and efficiency of insulin after oral administration of insulin-loaded nanocapsules in diabetic rats. Int J Pharm 2002; 242:325–328.
75. Radwan MA. Enhancement of absorption of insulin-loaded polyisobutylcyanoacrylate nanospheres by sodium cholate after oral and subcutaneous administration in diabetic rats. Drug Dev Ind Pharm 2003; 27(9):981–989.
76. Watnasirichaikul S, Davies NM, Rades T, Tucker IG. Preparation of biodegradable insulin nanocapsules from biocompatible microemulsion. Pharm Res 2000; 17:684–689.
77. Watnasirichaikul S, Tucker IG, Davies NM. Effect of formulation variables on characteristics of poly(ethylcyanoacrylate) nanocapsules prepared from w/o microemulsions. Int J Pharm 2002; 235:237–246.
78. Watnasirichaikul S, Rades T, Tucker IG, Davies NM. In-vitro release and oral bioactivity of insulin in diabetic rats using nanocapsules dispersed in biocompatible microemulsion. J Pharm Pharmacol 2002; 54:473–480.
79. Barichello JM, Morishita M, Takayama K, Nagai T. Encapsulation of hydrophilic and lipophilic drugs in PLGA nanoparticles by the nanoprecipitation method. Drug Dev Ind Pharm 1999; 25(4):471–476.
80. Fessi H, Puisieux F, Devissaguet JP, Ammoury N, Benita S. Nanocapsule formation by interfacial polymer deposition following solvent displacement. Int J Pharm 1989; 55:R1–R4.
81. Carino GP, Jacob JS, Mathiowitz E. Nanospheres based oral insulin delivery. J Contr Rel 2000; 65:261–269.
82. Ma XY, Pan GM, Lu Z, et al. Preliminary study of oral polylactide microcapsulated insulin in vitro and in vivo. Diab Obes Metab 2000; 2:243–250.
83. Hosny EA, Khan Ghilzai NM, Al-Dhawalie AH. Effective intestinal absorption of insulin in diabetic rats using enteric coated capsules containing sodium salicylate. Drug Dev Ind Pharm 1995; 21:1583–1589.
84. Hosny EA, Ghilzai NM, Al-Najar TA, Elmazar MM. Hypoglycemic effect of oral insulin in diabetic rabbits using pH-dependent coated capsules containing sodium salicylate without and with sodium cholate. Drug Dev Ind Pharm 1998; 24:307–311.
85. Hosny EA, Al-Shora HI, Elmazar MM. Oral delivery of insulin from enteric-coated capsules containing sodium salicylate: effect on relative hypoglycemia of diabetic beagle dogs. Int J Pharm 2002; 237:71–76.

86. Hosny EA, Ghilzai NM, Elmazar MM. Promotion of oral insulin absorption in diabetic rabbits using pH-dependent coated capsules containing sodium cholate. Pharm Acta Helv 2004; 72:203–207.
87. Saffran M, Kumar S, Savariar C, Burnham JC, Williams F, Neckers DC. A new approach to the oral administration of insulin and other peptide drugs. Science 1986; 233:1081–1084.
88. Touitou E, Rubinstein A. Targeted enteral delivery of insulin to rats. Int J Pharm 1986; 30:95–99.
89. Morishita I, Morishita M, Takayama K, Machida Y, Nagai T. Hypoglycemic effect of novel oral microspheres of insulin with protease inhibitor in normal and diabetic rats. Int J Pharm 1992; 78:9–16.
90. Morishita I, Morishita M, Takayama K, Machida Y, Nagai T. Enteral insulin delivery by microspheres in 3 different formulations using Eudragit L100 and S100. Int J Pharm 1993; 91:29–37.
91. Morishita M, Morishita I, Takayama K, Machida Y, Nagai T. Novel oral microspheres of insulin with protease inhibitor protecting from enzymatic degradation. Int J Pharm 1992; 78:1–7.
92. Kimura T, Sato K, Sugimoto K, et al. Oral administration of insulin as poly(vinyl alcohol)-gel spheres in diabetic rats. Biol Pharm Bull 1996; 19:897–900.
93. Takeuchi H, Yamamoto H, Niwa T, Hino T, Kawashima Y. Enteral absorption of insulin in rats from mucoahesive chitosan-coated liposomes. Pharm Res 1996; 13(6):896–901.
94. Bernkop-Schnürch A. Chitosan and its derivatives: potential excipients for peroral peptide delivery systems. Int J Pharm 2000; 194:1–13.
95. Pan Y, Li YJ, Zhao HY, et al. Bioadhesive polysaccharide in protein delivery system: chitosan nanoparticles improve the intestinal absorption of insulin in vivo. Int J Pharm 2002; 249(1,2):139–147.
96. Pan Y, Zheng JM, Zhao HY, Li YJ, Xu H, Wei G. Relationship between drug effects and particle size of insulin-loaded bioadhesive microspheres. Acta Pharmacol Sin 2002; 23(11):1051–1056.
97. Mathiowitz E, Jacob JS, Jong YS, et al. Biologically erodable microspheres as potential oral drug delivery systems. Nature 1997; 386:410–414.
98. Nagareya N, Uchida T, Matsuyama K. Preparation and characterization of enteric microspheres containing bovine insulin by a w/o/w emulsion solvent evaporation method. Chem Pharm Bull 1998; 46:1613–1617.
99. Qi R, Ping QN. Gastrointestinal absorption enhancement of insulin by administration of enteric microspheres and SNAC to rats. J Microencaps 2004; 21:37–45.
100. Earle MP. Experimental use of oral insulin. Isr J Med Sci 1972; 8:899–900.
101. Allémann E, Leroux JC, Gurny R. Polymeric nano- and microparticles for the oral delivery of peptides and peptidomimetics. Adv Drug Del Rev 1998; 34(2,3):171–189.
102. Carino GP, Mathiowitz E. Oral insulin delivery. Adv Drug Del Rev 1999; 35:249–257.
103. Copp H. Remembrance: calcitonin–discovery and early development. J Endoc Soc 1992; 131:1007–1008.
104. Hosking DJ. Treatment of severe hypercalcemia with calcitonin. Metab Bone Dis Rel Res 1980; 2:207–212.
105. Nuesch E, Schmidt R. Comparative pharmacokinetics of calcitonins. In: Milan Pecile A, ed. Proc Int Symp Excerpta Medica Int Congr Ser 1981; 540:352–364.
106. Sakuma S, Hayashi M, Akashi M. Design of nanoparticles composed of graft copolymers for oral peptide delivery. Adv Drug Del Rev 2001; 47:21–37.
107. Takeuchi H, Yamamoto H, Kawashima Y. Mucoadhesive nanoparticulate systems for peptide drug delivery. Adv Drug Del Rev 2001; 47:39–54.
108. Torres-Lugo M, Peppas NA. Transmucosal delivery systems for calcitonin: a review. Biomaterials 2000; 21:1191–1196.

109. Vranckx H, Demoustier M, Deleers M. A new nanocapsule formulation with hydrophilic core: application to the oral administration of salmon calcitonin in rats. Eur J Pharm Biopharm 1996; 42(5):345–347.
110. Akashi M, Chao D, Yashima E, Miyauchi N. Graft copolymers having hydrophobic backbone and hydrophilic branches, V. Microspheres obtained by the copolymerization of poly(ethylene glycol) macromonomer with methyl methacrylate. J Appl Polym Sci 1990; 39:2027–2030.
111. Chen MQ, Kishida A, Akashi M. Graft copolymers having hydrophobic backbone and hydrophilic branches, XI. Preparation and thermosensitive properties of polystyrene microspheres having poly(*N*-isopropyl-acrylamide) branches on their surface. J Polym Sci Polym Chem 1996; 34:2213–2220.
112. Riza M, Tokura S, Iwasaki M, Yashima E, Kishida A, Akashi M. Graft copolymers having hydrophobic backbone and hydrophilic branches, X. Preparation and properties of water-dispersible polyanionic microspheres having poly(methacrylic acid) branches on their surface. J Polym Sci Polym Chem 1995; 33:1219–1225.
113. Sakuma S, Suzuki N, Kikuchi H, et al. Oral peptide delivery using nanoparticles composed of novel graft copolymers having hydrophobic backbone and hydrophilic branches. Int J Pharm 1997; 149:93–106.
114. Sakuma S, Suzuki N, Kikuchi H, et al. Absorption enhancement of orally administered salmon calcitonin by polystyrene nanoparticles having poly(*N*-isopropylacrylamide) branches on their surfaces. Int J Pharm 1997; 158:69–78.
115. Sakuma S, Ishida Y, Sudo R, et al. Stabilization of salmon calcitonin by polystyrene nanoparticles having surface hydrophilic polymeric chains, against enzymatic degradation. Int J Pharm 1997; 159:181–189.
116. Sakuma S, Sudo R, Suzuki H, Akashi M, Hasashi M. Mucoadhesion of polystyrene nanoparticles having surface hydrophilic polymeric chains in the gastrointestinal tract. Int J Pharm 1999; 177:161–172.
117. Sakuma S, Suzuki N, Sudo R, Hiwatari K, Kishida A, Akashi M. Optimized chemical structure of nanoparticles as carriers for oral delivery of salmon calcitonin. Int J Pharm 2002; 239:185–195.
118. Brunner J, Semenza G. Selective labeling of hydrophobic core of membranes with 3-(trifluoromethyl)-3-(m-[1251]iodophenyl)diazirine, a carbene-generating reagent. Biochemistry 1981; 20:7174–7182.
119. Florence AT, Hillery AM, Hussain N, Jani PU. Nanoparticles as carriers for oral peptide absorption: studies on particle uptake and fate. J Contr Rel 1995; 36:39–46.
120. Borchard G, Luessen HL, de Boer AG, Verhoef JC, Lehr C-M, Junginger HE. The potential of mucoadhesive polymers in enhancing intestinal peptide drug absorption. III: effects of chitosan-glutamate and carbomer on epithelial tight junctions in vitro. J Contr Rel 1996; 39:131–138.
121. Kawashima Y, Yamamoto H, Takeuchi H, Kuno Y. Mucoahesive DL-lactide/glycolide copolymer nanospheres coated with chitosan to improve oral delivery of elcatonin. Pharm Dev Technol 2000; 5:77–85.
122. Yoo HS, Park GT. Biodegradable nanoparticles containing protein-fatty acid complexes for oral delivery of salmon calcitonin. J Pharm Sci 2004; 93:488–495.
123. Leone-Bay A, Santiago N, Achan D, et al. N-acylated α-amino acids as novel oral delivery agents for proteins. J Med Chem 1995; 38:4263–4269.
124. Leone-Bay A, McInnes C, Wang N, et al. Microsphere formation in a series of derivatized α-amino acids: properties, molecular modeling, and oral delivery of salmon calcitonin. J Med Chem 1995; 38:4257–4262.
125. Milstein SJ, Barantsevich EN, Grechanovski VA, Sarubbi DJ. pH-dependent microspheres from modified soybean protein hydrolysate. J Microencaps 1996; 13:651–665.
126. Urata T, Arimori K, Nakano M. Modification of release rates of cyclosporin A from poly(L-lactic acid) microspheres by fatty acid esters and in vivo evaluation of the microspheres. J Contr Rel 1999; 58:133–141.

127. Ismailos G, Reppas C, Dressman JB, Macheras P. Unusual solubility behavior of cyclosporin A in aqueous media. J Pharm Pharmacol 1991; 43:287–289.
128. Wood MJ, Maurer G, Niedelberger W, Beveridge T. Cyclosporin: pharmacokinetics, metabolism and drug interactions. Transplant Proc 1983; 15:2409–2412.
129. Mihatsch MJ, Thiel J, Spichtin TP, Harder F, Torhorts J. Morphological findings in kidney transplants after treatment with cyclosporin. Transplant Proc 1983; 15: 2821–2835.
130. Bonduelle S, Foucher C, Leroux JC, Chouinard F, Cadieux C, Lenaerts V. Association of cyclosporin to isohexylcyanoacrylate nanospheres and subsequent release in human plasma in vitro. J Microencaps 1992; 9:173–182.
131. Ford J, Woolfe J, Florence AT. Nanospheres of cyclosporin A: poor oral absorption in dogs. Int J Pharm 1999; 183(1):3–6.
132. Molpeceres J, Chacon M, Berges L, Guzman M, Aberturas MR. Increased oral bioavailability of cyclosporin incorporated in polycaprolactone nanoparticles. III Spanish-Portuguese Conference on Controlled Drug Delivery, Lisbon, 1998:127–128.
133. Molpeceres J, Chacon M, Berges L, Guzman M, Aberturas MR. Dose-linearity of cyclosporin pharmacokinetics in rats: polycaprolactone nanoparticles versus Sandimmun emulsion. In: Proc 2nd World meeting on Pharmaceutics, Biopharmaceutics and Pharmaceutical Technology, Paris, 1998:633–634.
134. Molpeceres J, Chacon M, Guzman M, Berges L, del Rosario AM. A polycaprolactone nanoparticle formulation of cyclosporin-A improves the prediction of area under the curve using a limited sampling strategy. Int J Pharm 1999; 187(1):101–113.
135. Saez A, Guzman M, Molpeceres J, Aberturas MR. Freeze-drying of polycaprolactone and poly(D,L-lactic-glycolic) nanoparticles induce minor particle size changes affecting the oral pharmacokinetics of loaded drugs. Eur J Pharm Biopharm 2000; 50:379–387.
136. Varela MC, Guzman M, Molpeceres J, Aberturas MR, Rodriguez-Puyol D, Rodriguez-Puyol M. Cyclosporine-loaded polycaprolactone nanoparticles: immunosupression and nephrotoxicity in rats. Eur J Pharm Sci 2001; 12:471–478.
137. El Shabouri MH. Positively charged nanoparticles for improving the oral bioavailability of cyclosporin-A. Int J Pharm 2002; 249:101–108.
138. Yu WP, Barratt GM, Devissaguet JP, Puisieux F. Anti-metastatic activity in vivo of MDP-L-alanyl-cholesterol (MTP-Chol) entrapped in nanocapsules. Int J Immunopharmacol 1991; 13(2,3):167–173.
139. Lehr CM, Bouwstra JA, Tukker JJ, Junginger HE. Intestinal transit of bioadhesive microspheres in an in situ loop in the rat—A comparative study with copolymer and blends based on poly(acrylic acid). J Contr Rel 1990; 13:51–62.
140. Lehr CM, Bouwstra JA, Kok W, et al. Effect of the mucoadhesive polymer polycarbophil on the intestinal absorption of a peptide drug in the rat. J Pharm Pharmacol 1992; 44:402–407.
141. Hillery AM, Toth I, Florence AT. Co-polymerised peptide particles (CPP) II: oral uptake of a novel co-polymeric nanoparticulate delivery system for peptides. J Contr Rel 1996; 42:65–73.
142. Eyles JE, Alpar HO, Conway BR, Keswick M. Oral delivery and fate of poly(lactic acid) microsphere-encapsulated interferon in rats. J Pharm Pharmacol 1997; 49:669–674.
143. Bauer W, Briner U, Doepfner W, et al. SMS 201–995: a very potent and selective octapeptide analogue of somatostatin with prolonged action. Life Sci 1982; 31:1133–1140.
144. Pless J, Bauer W, Briner U, Doepfner W, Marbach P, Maurer R. Chemistry and pharmacology of SMS 201–995, a long-acting octapeptide analogue of somatostatin. Scand J Gastroenterol 1996; 21:54–64.
145. Andersen M, Hansen TB, Bollersley J, Bjerre P, Schroder HD, Hagen C. Effect of 4 weeks of octreotide treatment on prolactin, thyroid stimulating hormone and thyroid hormones in acromegalic patients. A double blind controlled cross-over study. J Endocrinol Invest 1995; 18:840–846.

146. Orskov L, Moller N, Bak JF, Porksen N, Schmitz O. Effects of the somatostatin analogue, octreotide, on glucose metabolism and insulin sensitivity in insulin-dependent diabetes mellitus. Metab Clin Exp 1996; 45:211–217.
147. Arnold R, Trautmann ME, Creutzfeldt W, Benning M, Benning R, Neuhaus C. Somatostatin analogue octreotide and inhibition of tumor growth in metastatic endocrine gastroenteropancreatic tumors. Gut 2003; 38:430–438.
148. Robbins RJ. Somatostatin and cancer. Metab Clin Exp 1996; 45:98–100.
149. Williams G, Burrin JM, Ball JA, Joplin GF, Bloom SR. Effective and lasting growth-hormone suppression in active acromegaly with oral administration of somatostatin analogue SMS 201–995. Lancet 1986; 2(8510):774–778.
150. Leroux JC, Cozens R, Roesel JL, et al. Pharmacokinetics of a novel HIV-1 protease inhibitor incorporated into biodegradable or enteric nanoparticles following intravenous and oral administration to mice. J Pharm Sci 1995; 84(12):1388–1391.
151. Leroux JC, Cozens RM, Roesel JL, Galli B, Doelker E, Gurny R. pH-sensitive nanoparticles: an effective means to improve the oral delivery of HIV-1 protease inhibitor in dogs. Pharm Res 1996; 13(12):485–487.
152. De Jaeghere F, Leroux JC, Allémann E, et al. Evaluation of pH-sensitive nanoparticles for oral delivery of HIV-1 protease inhibitors. Proc Coll Drug Carriers 3rd Exp Meet Berlin, 1997.
153. Maincent P, LeVerge R, Sado P, Couvreur P, Devissaguet JP. Disposition kinetics and oral bioavailability of vincamine-loaded polyalkyl cyanoacrylate nanoparticles. J Pharm Sci 1986; 75(10):955–958.
154. Hubert B, Atkinson J, Guerret M, Hoffman M, Devissaguet JP, Maincent P. The preparation and acute antihypertensive effects of a nanocapsular form of darodipine, a dihydropyridine calcium entry blocker. Pharm Res 1991; 8(6):734–738.
155. Kim YI, Fluckiger L, Hoffman M, Lartaud-Idjouadiene I, Atkinson J, Maincent P. The antihypertensive effect of orally administered nifedipine-loaded nanoparticles in spontaneously hypertensive rats. Brit J Pharmacol 1997; 120:399–404.
156. Leroueil-Le Verger M, Fluckiger L, Kim YI, Hoffman M, Maincent P. Preparation and characterization of nanoparticles containing an antihypertensive agent. Eur J Pharm Biopharm 1998; 46:137–143.
157. Jiao Y, Ubrich N, Hoffart V, et al. Anticoagulant activity of heparin following oral administration of heparin-loaded microparticles in rabbits. J Pharm Sci 2002; 91(3):760–768.
158. Jiao Y, Marchand-Arvier M, Vigneron C, Hoffman M, Lecompte T, Maincent P. In vitro and in vivo evaluation of oral heparin–loaded polymeric nanoparticles in rabbits. Circulation 2002; 105:230–235.
159. Ammoury N, Fessi H, Devissaguet JP, Dubrasquet M, Benita S. Jejunal absorption, pharmacological activity, and pharmacokinetic evaluation of indomethacin-loaded poly(DL-lactide) and poly(isobutyl-cyanoacrylate) nanocapsules in rats. Pharm Res 1991; 8(1):101–108.
160. de Chasteigner S, Fessi H, Cavé G, Devissaguet JP, Puisieux F. Gastro-intestinal tolerance study of a freeze-dried oral dosage form of indomethacin-loaded nanocapsules. STP Pharma Sci 1995; 5:242–246.
161. Singh UV, Udupa N. Methotrexate loaded chitosan and chitin microspheres—in vitro characterization and pharmacokinetics in mice bearing Ehrlich ascite carcinoma. J Microencaps 1998; 15(5):581–594.
162. He P, Davis SS, Illum L. In vitro evaluation of the mucoadhesive properties of chitosan microspheres. Int J Pharm 1998; 166:75–68.
163. Madhan Kumar AB, Panduranga Rao K. Preparation and characterization of pH-sensitive proteinoid microspheres for the oral delivery of methotrexate. Biomaterials 1998; 19(7–9):725–732.
164. Hughes WT, Kennedy W, Shenep JL, Flynn PM, Hetheringtin SV, Fullen F. Safety and pharmacokinetics of 566C80, a hydroxynaphtoquinone with anti-pneumocystis carnii

activity: a phase I study in human immunodeficiency virus (HIV) infected men. J Inf Dis 1991; 163:843–848.

165. Sordet F, Aumjaud Y, Fessi H, Derouin F. Assessment of the activity of atovaquone-loaded nanocapsules in the treatment of acute and chronic murine toxoplasmosis. Parasite 1998; 5:223–229.
166. Sugimoto K, Yoshida M, Yata T, Higaki K, Kimura T. Evaluation of poly(vinyl alcohol)-gel spheres containing chitosan as dosage form to control gastrointestinal transit time of drugs. Biol Pharm Bull 1998; 21:1202–1206.
167. Illum L, Davis SS, Wilson CG, Thomas N, Frier M, Hardy JG. Blood clearance and organ deposition of intravenously administered colloidal particles. The effects of particles size, nature and shape. Int J Pharm 1982; 12:135–146.
168. Löbenberg R, Araujo L, Kreuter J. Body distribution of azidothymidine bound to nanoparticles after oral administration. Eur J Pharm Biopharm 1997; 44:127–132.
169. Löbenberg R, Maas J, Kreuter J. Improved body distribution of ^{14}C-labelled AZT bound to nanoparticles in rats determined by radioluminography. J Drug Targ 1997; 5(3):171–179.
170. Beck P, Kreuter J, Muller WEG, Schatton W. Improved peroral delivery of avarol with polybutylcyanoacrylate nanoparticles. Eur J Pharm Biopharm 1994; 40(3):134–137.
171. Ain Q, Sharma S, Garg SK, Khuller GK. Role of poly [D,L-lactide-co-glycolide] in development of a sustained oral delivery system for antitubercular drug(s). Int J Pharm 2002; 239(1,2):37–46.
172. Pandey R, Zahoor A, Sharma S, Khuller GK. Nanoparticle encapsulated antitubercular drugs as a potential oral drug delivery system against murine tuberculosis. Tuberculosis 2003; 83(6):373–378.

16

Vesicles as a Tool for Dermal and Transdermal Delivery

P. L. Honeywell-Nguyen and J. A. Bouwstra
Leiden/Amsterdam Center for Drug Research, Leiden University, Einsteinweg, RA Leiden, The Netherlands

DERMAL AND TRANSDERMAL DRUG DELIVERY

The skin covers a total surface area of approximately 1.8 m^2 and provides the contact between the human body and its external environment (1). This large and outermost layer of the human body is easily accessible and hence attractive as a noninvasive delivery route for selected drug compounds. Dermal and transdermal drug delivery can have many advantages as compared to other routes of drug administration. Dermal drug delivery is the topical application of drugs to the skin in the treatment of localized skin diseases. The advantage of dermal drug delivery is that high concentrations of drugs can be localized at the site of action (i.e., the skin), reducing the systemic drug levels and therefore also reducing the systemic side effects. Transdermal drug delivery, on the other hand, uses the skin as an alternative route for the delivery of systemically acting drugs. This drug delivery route for systemic therapy can have several advantages as compared to conventional oral drug administration. First of all, it circumvents the variables that could influence gastrointestinal absorption such as pH, food intake, and gastrointestinal motility. Secondly, it circumvents the first-pass hepatic metabolism and is therefore suitable for drugs with a low bioavailability. Thirdly, transdermal drug delivery can give a constant, controlled drug input. This would reduce the need for frequent drug intake, especially of drugs with a short biological half-life. Furthermore, variations in drug plasma levels can be avoided, reducing the side effects in particular of drugs with a narrow therapeutic window. Finally, transdermal drug delivery is easy and painless, which in turn will increase patient compliance (2).

Despite these advantages of the skin as a site of drug delivery, only less than 10 drugs are currently available in the market as a transdermal delivery system. These transdermal delivery systems contain drugs including fentanyl, nitroglycerine, scopolamine, clonidine, nicotine, estradiol, and testosterone (3–10). By far the most important reason for such few transdermal delivery systems is the fact that human skin is highly impermeable for most drug compounds. Previously, many attempts have been made and many methods have been employed in order to improve drug delivery across the intact human skin. This chapter will deal with vesicles, which were developed as a vehicle in order to enhance dermal and transdermal drug delivery.

THE SKIN BARRIER FUNCTION

Human skin consists of two distinctive layers: the upper layer, the epidermis, and the underlying layer of connective tissue, the dermis. The epidermis can be further distinguished into two parts: the nonviable and viable epidermis. The nonviable epidermis is often referred to as the stratum corneum (SC) (11). This outermost layer has been identified as the main barrier for the penetration of drugs across the skin (12,13).

The Nonviable Epidermis or SC

The barrier function of mammalian skin is mainly attributed to its outermost layer, the SC (12,13). The SC constitutes a barrier to water loss and entry of harmful compounds from the external environment. Hence, the SC is also the skin layer that controls the transport of drugs into and across the skin. One of the most convincing evidences for the barrier function of the SC is provided by the simple fact that stripping off this layer will result in a significant increase in the skin's permeability to water and other compounds (14,15).

The SC consists of corneocytes embedded into a matrix of intercellular lipid lamellae. The structure of the SC is often compared to a brick wall, with the corneocytes as the bricks surrounded by the mortar of the intercellular lipid lamellae (16). The corneocytes are flat, hexagonal shaped cells, which are composed of densely packed keratin surrounded by a cornified envelope. The cornified envelope has a lipoidal exterior formed by a monolayer of chemically bound lipids. This anchors them to the hydrophilic interior of the corneocytes as well as to the intercellular lipid lamellae. Although the corneocytes comprise 85% of dry weight of the SC, it is the remaining 15% that deserves more focus and attention. This 15% of the SC consists of SC lipids, which are mainly composed of ceramides, cholesterol, and fatty acids (17,18). It has been generally accepted that the SC lipids play an essential role in the barrier properties of the SC (19,20). These barrier properties are based on the composition and in particular the highly organized crystalline lamellar structure of the intercellular lipid matrix (21–24). This unique structural arrangement results in a practically impermeable barrier for many compounds, including water (13,25,26).

The SC is 15 to 20 μm thick, although there is a large variation between different sites of the body. It consists of approximately 15 cell layers, but again this is largely dependent on the site of the body. At sites of increased pressure, such as the sole of the feet, the SC thickness and cell layers could be increased manyfold. The upper three to five layers of the SC undergoes desquamation and is replaced continuously by the migration of germinal cells to the skin surface. During this process the germinal cells flatten, lose their nuclei, and undergo dehydration and intercellular keratin formation, becoming terminally differentiated corneocytes that are desquamated from the skin surface. This process of migration and desquamation illustrates that the SC is not only a structurally rigid barrier, but also a most versatile biologic barrier of the human body (1,11,27,28).

A remarkable characteristic of the SC is its low water content. The SC contains significantly lower amounts of water as compared to the underlying viable skin layers. This gives further evidence for the location of the skin barrier and also suggests that the material properties of the SC are quite unique (1). The distribution profile of water in the skin has been determined in vitro by Warner et al. (29) and Von Zglinicki et al. (30), using electron probe analysis and X-ray microanalysis, and in vivo by Caspers et al. (31) using confocal Raman microspectroscopy. The results from these studies show similar water concentration profiles for the SC. The water concentration in the SC rises from 15% to 30% at the skin surface to about 40% at the junction between the SC and stratum granulosum. This is followed by a very steep rise to a water concentration of 70% in the underlying viable epidermis and dermis. Hence, these studies have shown a steep water or osmotic gradient across the SC. However, two recent studies have shown that this increase in the water content from the superficial corneocyte layers to the deepest cell layers near the viable epidermis is not gradual. Using cryo-scanning electron microscopy Bouwstra et al. (32) visualized human skin in vitro that had been fully hydrated after equilibration over pure water for 24 hours (hydration level of 300%, by weight. The results have shown that at full hydration there was an enormous increase in the swelling of the cells. However, the corneocytes in the deepest cell layers near the viable epidermis showed no swelling at full hydration of the SC. This striking feature was also demonstrated in vivo by Warner et al. (33), who reported that the innermost two to four corneocytes contain considerably less water than the bulk of the SC. Hence, the water content in the deepest layers of the SC close to the junction with the viable epidermis is remarkably low in comparison to the water content in its superficial or central regions, both in vitro and in vivo. This suggests that the osmotic gradient in the SC is not unidirectional. This osmotic gradient and its direction could be important in the transport and mechanism of action of so-called elastic vesicles. This type of vesicles will be discussed extensively in the later section.

The Viable Epidermis

The viable epidermis is located directly underneath the SC. The main cell type in the viable epidermis is the keratinocyte. Keratinocytes are stratified epithelial cells, which are responsible for the formation of the SC. Other cell types in the viable epidermis include melanocytes, Langerhans cells, and Merkell cells. The viable epidermis can be divided into three layers, each containing keratinocytes in different stages of cell differentiation. The basal cell layer above the dermis, the stratum basale, consists of germinal cells that divide and move upwards from the dermal–epidermal junction towards the skin surface. During this migration, the cells change morphologically as well as histochemically. At the stratum spinosum, the cells appear round. However, these cells soon become more flattened as they move upwards from the stratum spinosum towards the stratum granulosum, the last living cellular layer of the viable epidermis. The cell organelles and nuclei gradually disappear and increased keratinization takes places. This differentiation process eventually results in terminally differentiated cells—the corneocytes of the SC (1,11). As the keratinocytes approach the stratum granulosum–SC junction, they produce membrane-bound granules termed as lamellar bodies (34). These organelles are spheroid in shape and contain flattened lamellar disks. At the junction of the SC, the lamellar bodies fuse with the cell membrane and subsequently extrude the lamellar disks into the intercellular spaces. These lamellar disks fuse to form the intercellular lipid lamellae that constitute the main barrier of the SC (35).

The Dermis

The dermis is located underneath the viable epidermis', and forms the bulk of the skin. It consists of an amorphous mucopolysaccharide ground substance containing a dense network of connective tissue fibers, mainly collagen, elastin, and reticulum. The dermis contains blood vessels, lymph vessels, and nerves. It also contains the pilosebaceous units and sweat glands, which penetrate the dermis and the epidermis to reach the surface of the skin (11,27). The rich blood supply in the dermis participates in many processes, including body temperature regulation, immune responses, and pain- and pressure-regulating mechanisms. It also delivers nutrients to the skin and removes waste products from the skin (1). Hence, the blood vessels in the dermis also remove locally applied drug compounds from the skin and subsequently distribute them into the systemic circulation. This way the dermis functions as a "sink," ensuring a maximal concentration gradient across the epidermis and hence facilitating percutaneous absorption (2).

Routes of Drug Penetration

Drugs applied to the skin surface can serve two purposes. Dermal delivery is aimed at treating localized skin diseases. In this case, it is required that the drug penetrates the outer skin layers to reach its site of action within the skin, with little or no systemic uptake. On the other hand, transdermal delivery systems are designed to obtain therapeutic systemic blood levels. Hence, it is required that the drug reaches the dermis where it can be taken up for systemic circulation. However, in either case of dermal or transdermal drug delivery, the drug has to cross the outer layer of the skin, the SC. Since this layer is the main barrier of the skin, transport across the SC is the rate-limiting step in both dermal and transdermal drug delivery.

There are two potential pathways for a molecule to cross the SC: (a) the transappendageal route and (b) the transepidermal route (1,27,36). These routes of drug penetration are shown in Figure 1. The transappendageal route involves transport of drugs via the sweat glands (not shown) and the pilosebaceous units. This route bypasses the intact SC and is therefore also known as a "shunt" route. The transappendageal route, however, is not considered to be very significant, as the appendages only contribute 0.1% to the total surface area of the skin. Hence, transport of most of the drug compounds occurs via the transepidermal route, which involves transport across the intact continuous SC. Two pathways through the intact SC can be distinguished: (a) the intercellular route, along the lipid domains between the corneocytes and (b) the transcellular route, crossing through the corneocytes and the intervening lipids (Fig. 1). The intercellular route has been thought to be the route of preference for most drug molecules (37). Hence, many techniques have been aimed to disrupt and weaken the highly organized intercellular lipid lamellae in an attempt to enhance drug transport across the intact skin (38–40).

VESICLES AS SKIN DELIVERY SYSTEMS

General

One of the methods to enhance drug transport across the skin is the use of vesicles. The basic structure of vesicles is illustrated in Figure 2. Vesicles are hollow colloidal particles, consisting of amphiphilic molecules. These amphiphilic molecules consist

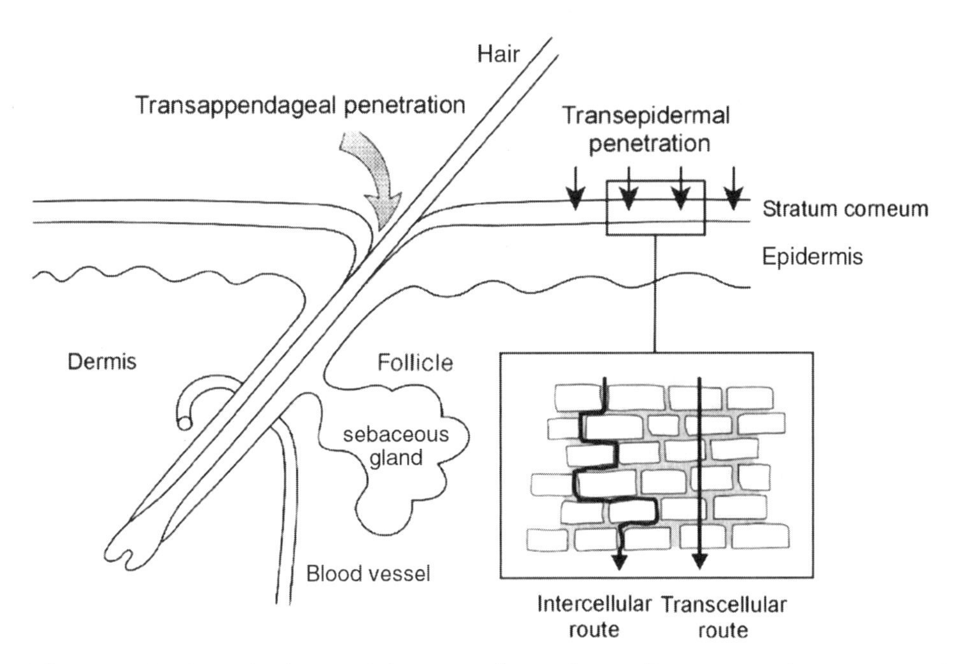

Figure 1 A schematic picture of the penetration pathways. Drug penetration can occur via the transappendageal or the transepidermal route. Transappendageal penetration occurs via the hair follicles and sweat glands, while transepidermal penetration occurs via the intact SC. The transepidermal route can be divided into two different penetration pathways. The intercellular route involves transport along the lipid lamellae between the corneocytes, while the transcellular route travels through the corneocytes and the intervening lipids. *Abbreviation*: SC, stratum corneum. *Source*: From Ref. 1.

of a polar hydrophilic head group and an apolar hydrophobic tail. Due to their amphiphilic properties, these molecules can form—in the presence of excess of water—one (unilamellar vesicles) or more (multilamellar vesicles) concentric bilayers that surround an equal number of aqueous compartments (41). Both water-soluble

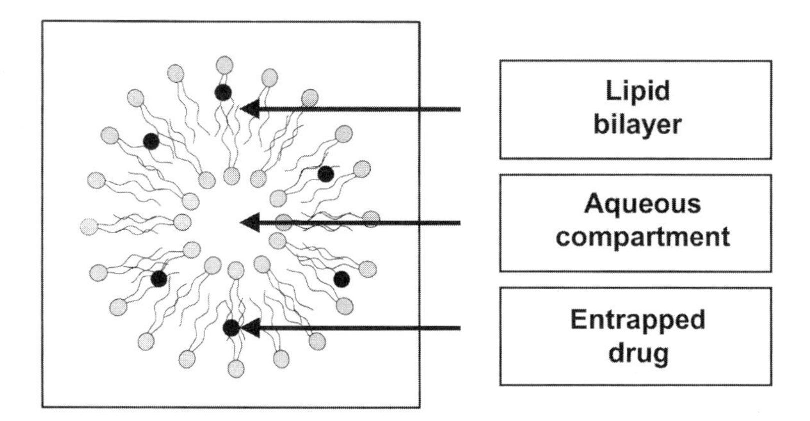

Figure 2 Schematic picture of a vesicle. Vesicles are hollow colloidal particles consisting of a lipid bilayer surrounding an aqueous internal compartment. Drug molecules can be entrapped into vesicles.

and water-insoluble drugs can be entrapped into the vesicles. Hydrophilic drugs can be entrapped into the internal aqueous compartment, while lipophilic drugs can be entrapped in the vesicle bilayer or partition between the bilayer and the aqueous phase (42).

A wide variation of lipids and surfactants can be used to prepare vesicles. Most commonly, the vesicles are composed of phospholipids or nonionic surfactants (43,44). These are referred to as liposomes and niosomes or nonionic surfactant vesicles. The composition of the vesicles influences their physicochemical characteristics such as size, charge, phase state, lamellarity, and bilayer elasticity. These physicochemical characteristics in turn have a significant effect on the behavior of the vesicles and hence also on their effectiveness as a drug delivery system.

The rationale for using vesicles in dermal and transdermal drug delivery is manifold (39,42):

a. Vesicles might act as drug carriers to deliver entrapped drug molecules into or across the skin.
b. Vesicles might act as penetration enhancers owing to the penetration of the individual lipid components into the SC and subsequently the alteration of the intercellular lipid lamellae within the skin layer.
c. Vesicles might serve as a depot for sustained release of dermal active compounds.
d. Vesicles might serve as a rate-limiting membrane barrier for the modulation of systemic absorption, hence providing a controlled transdermal delivery system.
e. The individual components of vesicles might have additional useful (cosmetic) properties.
f. Vesicles are biodegradable, minimally toxic, and relatively nonimmunogenic.

Despite extensive research and a high interest from the pharmaceutical and cosmetic industry, the use of vesicles is yet controversial. Two major questions remain subjects for discussion:

1. What is the effectiveness of vesicles?
   a. Do they enhance drug deposition in the skin (dermal delivery)?
   b. Do they enhance drug transport across the skin (transdermal delivery)?
   c. Or do they enhance both dermal and transdermal delivery?
2. What is the interaction between vesicles and the skin?

The answers to the above two questions are required in order to obtain a sound understanding of the mechanism of action of vesicles.

The Effectiveness of Vesicles as Skin Drug Delivery Vehicles

Early Studies in the 1980s

The first papers to report on the effectiveness of vesicles for skin delivery were published in the early 1980s. In two studies using white New Zealand rabbits, Mezei and Gulasekharam (45,46) demonstrated that liposomal encapsulation of triamcinolone acetonide significantly increased the in vivo drug disposition in the epidermis and dermis, the sites where drug activity was desired. On the other hand, percutaneous absorption was greatly reduced, hence reducing the risk of systemic side effects. In a second series of studies, Ganesan et al. and Ho et al. (47,48) suggested that the skin transport of highly polar compounds entrapped in liposomes will be slower in comparison to simple solutions of the drug, because of the slow release rate of the compound out of the liposomes. Ho et al. and Ganesan et al. also reported that liposome-entrapped lipophilic compounds permeated the skin with similar facility to that of the free drug solution. It was concluded that more total drug might be delivered by liposomes as compared to a simple aqueous solution, partly due to the increased solubility of lipophilic compounds in liposomes. However, two other studies by Komatsu et al. (49) did not demonstrate the beneficial effect of liposomes. Using guinea pigs, Komatsu et al. reported that liposomes were not more effective than a control ointment in the in vivo enhancement of butylparaben across the skin. In a following in vitro study, Komatsu et al. (50) suggested that entrapment of butylparaben in liposomes even reduced the percutaneous drug absorption. This was in line with the later obtained results by Knepp et al. (51,52), who also demonstrated that the drug enhancement effect of liposomes was rather questionable. Knepp et al. reported that the application of liposome-entrapped progesterone dispersed in an agarose gel reduced the in vitro drug transport across mouse skin as compared to the application of free progesterone in an agarose gel. It should be noted, however, that in the studies by Komatsu et al. and Knepp et al. the thermodynamic activities of the different test formulations were not kept equal. This would result in unequal driving forces and hence could complicate the comparison between vesicles and the control system. Interestingly, Knepp et al. also found that the transdermal delivery of progesterone was very much dependent on the lipid composition of the liposomes, while the release of progesterone from different liposome formulations into the buffer was essentially lipid independent.

It is evident that the results from the above mentioned studies were not entirely consistent with each other. This trend set by the early studies in the 1980s would continue into the next two decades. Large variations were found in the literature concerning the effectiveness of vesicles, hence enhancing the controversy of vesicles as transdermal delivery vehicles. Several transport studies have reported that vesicles only enhanced the drug deposition in the skin, suggesting that vesicles are only useful for topical dermal delivery (53–60). Others, however, have suggested that vesicles could deliver systemic drug concentrations, and hence are suitable candidates for transdermal delivery (61–67). The inconsistent results found in literature could, at least partly, be explained by the fact that vesicles with different compositions and physicochemical characteristics were used in different studies.

The Effect of the Physicochemical Properties of Vesicles on Drug Transport

Results from the early studies by Knepp et al. (51) have already shown that the vesicle composition and its physicochemical properties can have a significant effect on the drug permeation. In this aspect, several studies have demonstrated that the thermodynamic state of vesicles is an important feature, which plays a vital role in their effectiveness as a transdermal delivery vehicle. In a previous study, Hofland et al. (68) investigated the in vitro penetration of estradiol from gel-state and liquid-state vesicles across human SC. The results have demonstrated that the penetration rate of estradiol from liquid-state vesicles was manyfold higher than its penetration rate from gel-state vesicles. This was the case for estradiol encapsulated into gel-state and liquid-state vesicles as well as for a pretreatment protocol using drug-free vesicles. The results of Hofland et al. were confirmed by two later studies by Van Kuijk-Meuwissen et al. (69,70). Using confocal laser scanning microscopy, Van Kuijk-Meuwissen et al. investigated the penetration of a lipophilic fluorescent marker from gel-state and liquid-state vesicles into human skin in vitro and into rat skin in vivo. The images from both in vitro and in vivo studies revealed that the label incorporated into gel-state vesicles did not penetrate as deep into the skin in comparison to the label incorporated into liquid-state vesicles. Similar findings were also reported by Ogiso et al. (71), who incorporated the fluorescent probe nile red into liquid-state and gel-state vesicles. Histochemical examination of the skin following treatment with these vesicles suggested that liquid-state vesicles rapidly enter the SC in contrast to the gel-state vesicles. More recently, Perez-Cullell et al. (72) have investigated the penetration of fluorescein encapsulated into liquid-state and gel-state vesicles across human skin in vivo using the tape stripping technique. Perez-Cullell et al. reported that the skin penetration of fluorescein was higher when applied in liquid-state vesicles in comparison to application in gel-state vesicles.

From the above study, it is evident that the thermodynamic state of the vesicles is of special importance, where liquid-state vesicles have been shown to be more effective than gel-state vesicles. Other physicochemical properties, such as size and lamellarity, might also influence the effectiveness of vesicles as a delivery vehicle, although probably to a lesser extent than the thermodynamic state. Du Plessis et al. (73) investigated the transport of cyclosporin A across mouse, hamster, and pig skin from vesicles with three different sizes: 0.06, 0.3, and 0.6 μm. The intermediate particle size of 0.3 μm resulted in both the highest reservoir in the deeper skin strata as well as the highest drug concentration in the receiver compartment. Du Plessis et al. suggested that there might be an optimum particle size for optimal drug delivery. Fresta and Puglisi (74) have reported that multilamellar vesicles showed a significant reduction in penetration into the skin as compared to unilamellar vesicles. In the latter study it is unclear whether vesicle lamellarity plays an important role or whether the difference in transport is caused by the fact that the multilamellar vesicles were significantly larger in size as compared to the unilamellar vesicles.

The Interactions of Vesicles with Animal and Human Skin

Mechanisms of Vesicle–Skin Interactions

The exact mechanism of vesicle–skin interactions is not yet fully understood, despite numerous studies using different visualization techniques including freeze-fracture electron microscopy (FFEM) and confocal laser scanning microscopy. Several mechanisms behind the vesicle–skin interactions have been described in the literature. It has been suggested that vesicle–skin interactions can occur either at the skin surface or in the deeper layers of the SC. Abraham and Downing (75) and Hofland et al. (76,77) have demonstrated adsorption and fusion of vesicles onto the skin surface, resulting in the formation of lamellae and rough structures on top of the outermost corneocytes. Furthermore, Hofland et al. (76,77) have also demonstrated changes in the deeper layers of the SC after treatment of the skin with liposomes and niosomes (vesicles prepared from surfactants). Irregular structures as well as vesicular structures were visualized in the intercellular lipid lamellar regions of the SC after treatment with $C_{12}EO_3$ niosomes or NAT 106 liposomes. Hofland et al. and Abraham et al. suggested that vesicular lipids could penetrate molecularly dispersed into the intercellular matrix, where they mix with SC lipids modifying the lipid bilayers and inducing new structures. In another study, Zellmer et al. (78) have made a similar conclusion. Using differential scanning calorimetry of human SC treated with liposomes, Zellmer et al. suggested that individual lipid molecules can penetrate into the SC and disturb the highly organized structure of the intercellular lipid lamellae.

From the above studies there is no doubt that individual vesicular lipids could penetrate into the SC. However, it is still debatable whether vesicles themselves can enter the SC as intact entities. In the first two papers about vesicles as skin delivery vehicles, Mezei et al. (45,46) claimed that liposomes can enter the SC intactly and thereby increase the skin absorption. This was supported by Foldvari et al. (79), who reported that liposomes could penetrate intactly even through the SC. Using electron microscopy with colloidal iron as a marker, Foldvari et al. visualized the presence of intact liposomes in the dermis. However, it seems that there is a size restriction to penetration, as liposomes larger than 0.7 μm were not observed. Interestingly, the authors quoted that the liposomes are "very flexible lipid vesicles," that can be filtered through a polycarbonate filter with pore sizes much smaller than their own diameter. In recent years, it has been suggested that a novel type of flexible or elastic vesicles, also called Transfersomes®, could even penetrate intact through the skin and reach the systemic circulation under influence of a natural occurring transepidermal osmotic gradient (80,81).

The Effect of the Physicochemical Properties of Vesicles on Vesicle–Skin Interactions

The wide variation of different interactions reported in the literature could be explained by the fact that vesicle–skin interactions are strongly dependent on the composition and the physicochemical properties of the vesicular system (82,83). As with vesicle effectiveness, the ability of vesicles to interact with the skin is also strongly dependent on their thermodynamic state. Liquid-state vesicles have been shown to have superior characteristics to gel-state vesicles for interactions with the skin. Hofland et al. (76) applied liquid-state and gel-state niosomes occlusive to human skin in vitro. After treatment with liquid-state vesicles, vesicular structures and other changes in the intercellular lipid domains in the deeper layers of the SC were visualized with FFEM. However, no ultrastructural changes in the skin were found when gel-state niosomes were applied. This suggests that liquid-state vesicles or their components can enter the deeper layers of the SC where they can modify the intercellular lipid lamellae, while gel-state vesicles or their components remain on the skin surface. This superior nature of liquid-state vesicles for skin interactions is the most likely explanation for the fact that they are also more effective in enhancing drug transport into and across the skin. This is in agreement with Coderch et al. (84), who found a correlation between the skin penetration and the fluidity of the vesicle bilayers. The authors suggested that the role of vesicles in percutaneous absorption could be attributed to the incorporation of vesicle lipids into the intercellular domains, with subsequent modification of the fluidity of the skin lipids that is directly related to the fluidity behavior of the vesicles themselves.

Despite numerous contradictory and inconsistent results concerning the use of vesicles as skin delivery vehicles, one aspect has been generally accepted based on the research for the past 20 years. We can conclude that vesicles will only have suitable characteristics for skin interactions and hence will only have the optimal effectiveness as a delivery vehicle when they possess the appropriate lipid composition and physicochemical properties.

In addition to the physicochemical properties of the vesicles, the mode of application could also play a significant role in both the skin interactions and the effectiveness of vesicles as skin delivery vehicles. In this context, the difference between occlusive and nonocclusive application is of importance, especially for the elastic vesicles. Although water is a good penetration enhancer for most compounds, these elastic vesicles seem to function better under nonocclusive applications. Cevc et al. (80) have suggested that the transport of these elastic vesicles (Transfersomes) is driven by the osmotic gradient across the skin. Occlusion would eliminate this osmotic gradient and is therefore detrimental for the actions of the elastic vesicles. The fact that elastic vesicles function better under nonocclusive application was demonstrated by Van Kuijk-Meuwissen et al. (70). Using confocal laser scanning microscopy, the authors showed that the label from elastic vesicles penetrated deeper into the skin after nonocclusive application as compared to after occlusive application. The drug transport and the skin interactions of the elastic vesicles are described in detail below.

ELASTIC VESICLES

General

The interest of both the pharmaceutical and cosmetic industry in skin delivery has prompted the development and investigation of a wide variety of vesicular systems with different physicochemical characteristics. In recent years it has come to light that not only the liquid-state of the vesicles is important, but also the bilayer elasticity might have a significant effect on the behavior of vesicles. In the early 1990s a novel type of liquid-state vesicles was introduced—the "elastic vesicles." These are liquid-state vesicles that are characterized by elastic, deformable bilayers. Cevc et al. (80,85,86) introduced the first generation of elastic vesicles, also referred to as Transfersomes, consisting of phospholipids such as phosphatidylcholine and an edge activator such as sodium cholate. Later, Van den Bergh et al. (87) introduced a second series of elastic and rigid vesicles consisting of only surfactants. Vesicle elasticity was obtained using the same principle as Cevc et al. This principle is based on the combination of stabilizing and destabilizing molecules within one vesicle bilayer. It is a fact that both stabilizing and destabilizing molecules are present within one vesicle bilayer and their ability to move and redistribute within this bilayer gives the vesicles their elasticity.

Van den Bergh et al. used the bilayer-forming surfactant sucrose laurate ester (L-595) as the stabilizing molecule and micelle-forming surfactant octaoxyethylene laurate ester (PEG-8-L) as the destabilizing molecule. Incorporation of a micelle-forming component would result in destabilization of the lipid bilayer thereby increasing the vesicle elasticity. Hence, vesicle elasticity and vesicle stability are strongly dependent upon the ratio of the bilayer- and micelle-forming surfactant within the vesicle membrane (Fig. 3). By changing this ratio, a series of vesicles can be prepared, ranging from very rigid to very elastic.

The Effectiveness of Elastic Vesicles as Skin Delivery Vehicles

Several studies have shown that elastic vesicles were more effective than the conventional rigid liposomes in the enhancement of drug transport across animal and

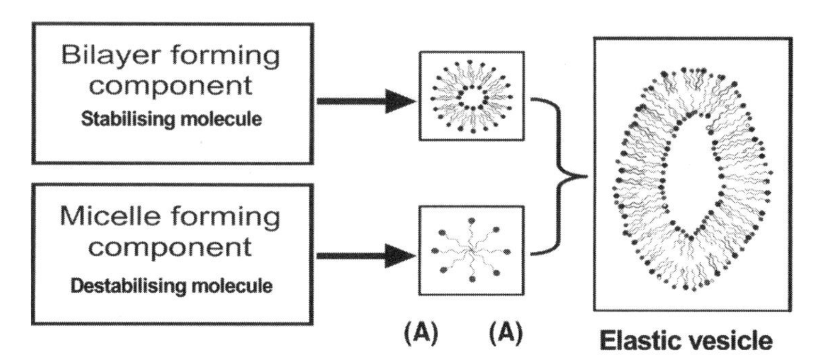

Figure 3 The composition of surfactant-based elastic vesicles. The surfactant-based elastic vesicles consist of stabilizing and destabilizing molecules. Bilayer-forming components are used as stabilizing molecules and micelle-forming components are used as destabilizing molecules. Incorporation of a micelle-forming component would result in destabilization of the lipid bilayer thereby increasing the vesicle elasticity. Vesicle elasticity is therefore strongly dependent on the ratio of the bilayer-forming and micelle-forming components.

human skin. Cevc et al. (63) performed the first efficacy studies using Transfersomes. These Transfersomes were claimed to have a very high deformability. It was suggested that Transfersomes applied under nonocclusive conditions could penetrate the skin barrier spontaneously and subsequently be distributed throughout the entire body (63). Using lidocaine as a model drug, Transfersomes were applied to the skin of rats and humans in vivo. It was reported that Transfersomes were significantly better than standard liposomes and had comparable effectiveness to that of the corresponding subcutaneous injections of similar drug quantities (88). Similar findings were also reported for Transfersomes encapsulated with corticosteroids (65). When using a higher dose of applied drug, Transfersomes were able to transport enough drug into the blood circulation to bring the systemic corticosteroid concentration close to the value achieved by a subcutaneous injection of a somewhat lower drug quantity. However, when using a lower dose of applied drug, corticosteroids were retained in the skin with little systemic absorption. Cevc et al. concluded that Transfersomes improved the site-specificity of corticosteroid delivery and were active at doses several times lower than those currently used in dermatological formulations for skin diseases. In the next few studies, Cevc et al. (64) demonstrated that Transfersomes were also able to transport high molecular weight compounds, such as peptides and proteins, across the skin. In an in vivo study using insulin, it was reported that dermally applied Transfersomes induced therapeutically significant hypoglycemia in mice and humans. Transfersomes were claimed to be nearly as efficient as insulin injection needles. Furthermore, it was suggested that Transfersomes could also be used for noninvasive transdermal immunization (89,90). Gap junction proteins incorporated into Transfersomes and applied to the intact skin induced antigen-specific antibody titers, marginally higher than subcutaneous injections of gap junction proteins in vesicle suspensions. The application of gap junction proteins in traditional rigid liposomes did not give rise to a significant biological response. In the latest paper concerning the effectiveness of Transfersomes, a formulation of diclofenac in these ultradeformable vesicles—also referred to as Transfenac—was used. Cevc and Blume (91) reported that the application of Transfenac resulted in 10 times higher concentrations of diclofenac in the tissues under the skin in comparison to the application of a commercially available diclofenac hydrogel. It was concluded that Transfenac had the potential to replace combined oral or topical diclofenac administrations in humans.

Following the astonishing encouraging results obtained by Cevc et al. other investigators have studied the potential benefit of using elastic vesicles for transdermal drug delivery. These studies have confirmed the fact that the elastic vesicles were more effective as compared to the conventional rigid vesicles; however, they have never achieved the exceptionally high transport rates that were demonstrated earlier by Cevc et al. Using elastic vesicles consisting of phosphatidylcholine in combination with an edge-activator (sodium cholate, Span 80, or Tween 80), El Maghraby et al. (92–94) showed that elastic vesicles were superior to conventional rigid liposomes as well as an aqueous control in the enhancement of estradiol transport across human cadaver skin in vitro. In another study, El Maghraby et al. (95) compared phosphatidylcholine/sodium cholate elastic vesicles with four traditional liposomes for the skin delivery of 5-fluorouracil. Elastic vesicles were found to be superior to the traditional liposomes in the skin delivery of 5-fluorouracil. This difference was most remarkable when elastic vesicles were compared to the most rigid liposome formulations. The authors concluded that elastic vesicles were promising drug delivery systems. However, as an aqueous ethanolic receptor phase was needed to liberate the drug from the skin in this study, El Maghraby et al. suggested that elastic vesicles only improved skin deposition, hence are only useful for dermal drug delivery. The latter was also concluded by Trotta et al. (96), who used elastic vesicles to enhance the transport of dipotassium glycyrrhizinate. The findings showed that skin fluxes of dipotassium glycyrrhizinate were below the detection limit, while skin deposition increased 4.5-fold in comparison to an aqueous control.

In 1998, Van den Bergh et al. introduced elastic vesicles consisting solely of surfactants, a bilayer-forming surfactant and a micelle-forming surfactant. The bilayer-forming surfactant is L-595, the micelle-forming surfactant is PEG-8-L. This was in contrast to Cevc et al. and El Maghraby et al. who used phosphatidylcholine in combination with a surfactant. The surfactant-based elastic vesicles of Van den Bergh et al. (97) have shown to be more effective than rigid vesicles in enhancing the penetration of ^3H_2O across hairless mouse skin. In addition, pergolide was successfully incorporated into a series of elastic and rigid vesicles and subsequently used in transport studies (98). The results have shown that the elastic vesicles were superior to the rigid vesicles in the enhancement of pergolide transport across the skin. In the second study using pergolide, transport studies were performed using the L-595 and PEG-8-L (50/50) elastic vesicles. Several aspects of elastic vesicle delivery were examined to elucidate the possible mechanisms of action of these vesicles (99). It was observed that occlusion reduced the permeation enhancement activity of elastic vesicles, but increased the overall drug transport since water was an excellent penetration enhancer for this particular model drug. A strong correlation was found between the saturated concentration of pergolide incorporated into the vesicle formulation and the pergolide transport, both of which were greatly influenced by the pH of the vesicle formulation. All the results with pergolide have suggested that for optimal drug delivery it is required that drug molecules be applied together with the elastic vesicles. This strongly points to the fact that a penetration enhancing effect is not the main or the only mechanism of action of the elastic vesicles, but that these vesicles act as drug carrier systems. Based on the results obtained, a mechanism of action for the elastic vesicles was proposed (Fig. 4). It was suggested that drug transport from elastic vesicles consists of four essential steps. For optimal drug delivery it is essential that drug molecules are associated with vesicle bilayers in high amounts

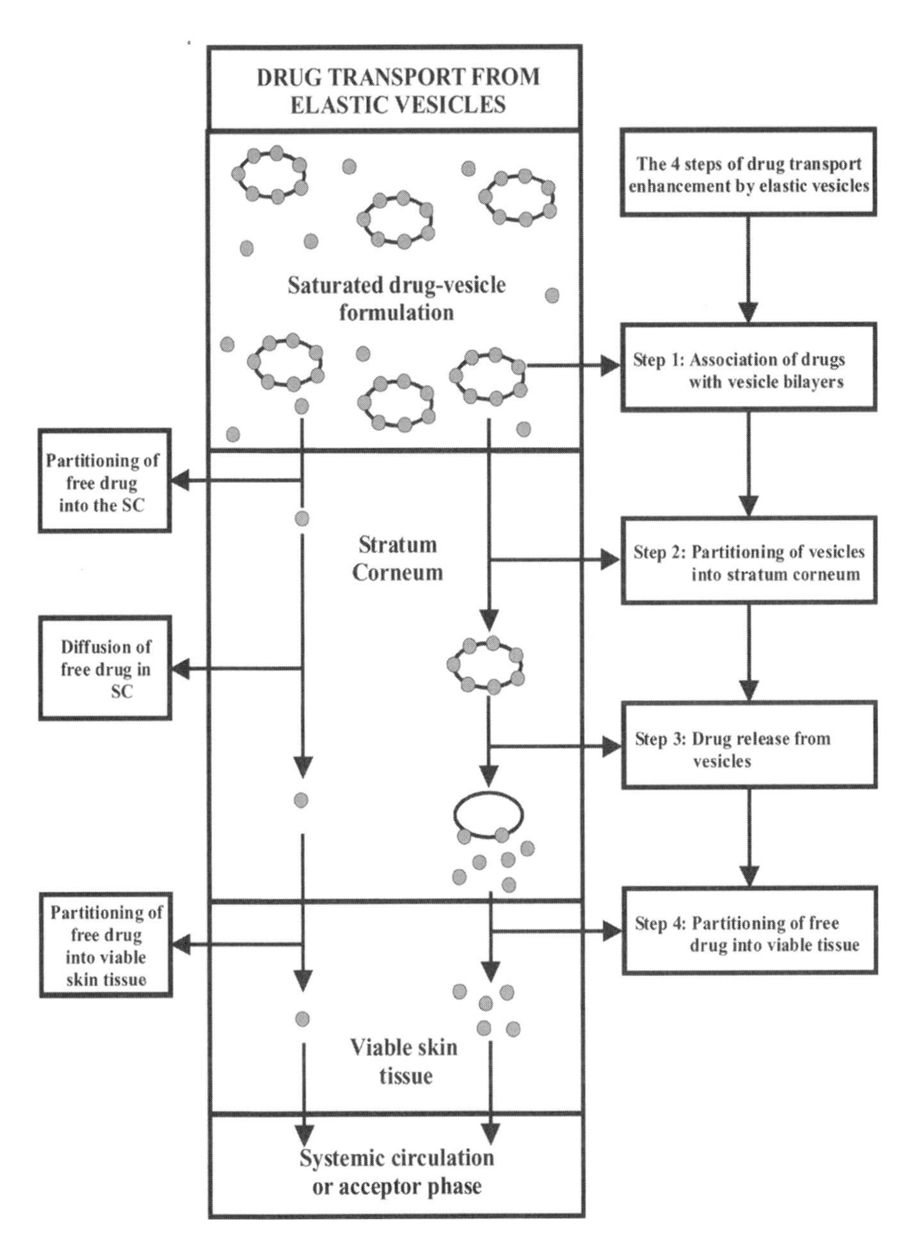

Figure 4 The proposed mechanism of action of the elastic vesicles. It can be suggested that the elastic vesicles promote drug transport by acting as drug carrier systems according to four essential steps. For optimal drug transport it is essential to associate high amounts of drugs with the vesicle bilayers (Step 1). Vesicles partition in the SC (Step 2). Hereby the vesicles will enhance drug transport into the skin, as vesicles carry vesicle-bound drug molecules into the SC. The vesicles themselves will remain in the SC and will not penetrate into the deeper skin layers in significant quantities. Vesicle-bound drug molecules are released from the vesicles (Step 3) with subsequent transport of free-drug molecules into the viable skin tissue (Step 4). *Abbreviation*: SC, stratum corneum.

(Step 1). Vesicles partition into the SC, whereby vesicles will carry drug molecules into the skin (Step 2). Most of the vesicles themselves remain in the SC and do not penetrate into the deeper layers of the skin. Vesicle-bound drug molecules are released from the vesicles (Step 3) after which the free-drug can diffuse through the deeper layers of the SC and partition into the viable skin tissue (Step 4). This mechanism of action was supported by the findings of subsequent studies, which are described below. The findings of permeation studies on lidocaine and rotigotine supported the findings on pergolide: namely, it is important to associate drugs with vesicles, and a penetration enhancement effect is not the main mechanism of action of elastic vesicles (100,101).

The Interactions of Elastic Vesicles with Animal and Human Skin

The interactions of elastic vesicles with animal and human skin are one of the most debatable, yet one of the most interesting issues in vesicular research. This debate was initiated by a paper published in 1992, reporting that ultradeformable Transfersomes could penetrate the intact skin and could reach as far as the systemic circulation (80). It was suggested that the driving force for such vesicle transport was provided by the natural occurring transepidermal osmotic gradient. Hence, it was concluded that occlusion is detrimental for the actions of elastic vesicles. In a very recent study, Cevc et al. (81) suggested that Transfersomes could even penetrate the skin as intact vesicles and do not disintegrate during this transport process. The authors labelled Transfersomes with two different fluorescent markers and applied the formulation nonocclusively onto intact murine skin in vivo. Confocal laser scanning microscopy images of the skin revealed a practically indistinguishable distribution for the labels in the SC, which was indicative of the fact that fragmentation of Transfersomes had not occurred. Furthermore, using size exclusion chromatography, Cevc et al. reported that the chromatograms obtained from sera of mice that received Transfersomes on the skin were indistinguishable from the results measured with the original vesicle suspension. The authors concluded that Transfersomes permeate across the skin barrier as intact vesicles rather than as bilayer fragments. Many other researchers, however, remain skeptical towards both the concept of vesicle permeation across the intact skin as well as the concept of vesicle transport as intact entities. Furthermore, it should be stressed that using confocal laser scanning microscopy it is very difficult to draw conclusions about intact vesicle permeation. With this technique only the model fluorescent label can be visualized, which might not necessarily correspond to the vesicles themselves.

It has been suggested that the transport of elastic vesicles into or possibly across the skin occurs via channel-like structures. Using confocal laser scanning microscopy and a lipophilic label, Schatzlein and Cevc (102) visualized an intercluster and an intercorneocyte pathway within the intercellular lipid lamellae of murine skin. It was postulated that these regions contain structural irregularities within the intercellular lipid lamellae, and that these irregularities can function as "virtual channels" through which elastic vesicles can penetrate the skin. In contrast to studies relating to the elastic vesicle transport using animal skin by Cevc et al., Schatzlein et al. and Van den Bergh et al. investigated the elastic vesicle transport using human skin. Using two-photon excitation microscopy, Van den Bergh et al. studied the in vitro penetration pathways of a lipophilic fluorescent label in surfactant-based elastic and rigid vesicles. The results demonstrated that the elastic vesicles could alter the penetration pathway of the fluorescent label. When incorporated into elastic vesicles,

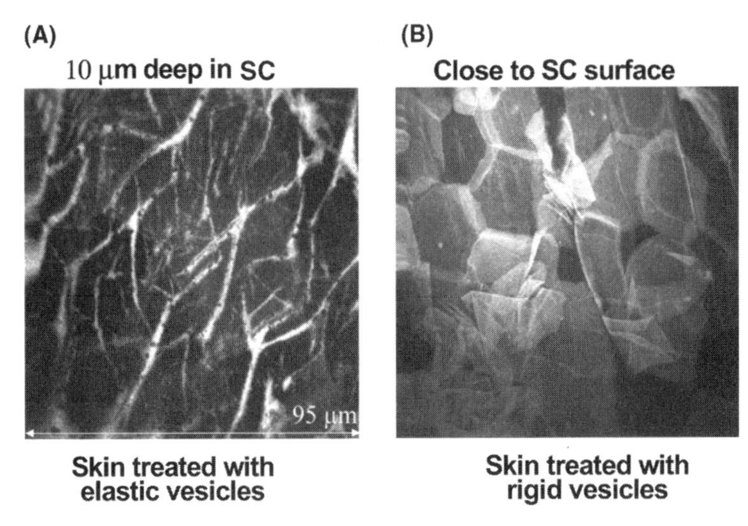

Figure 5 Two photon excitation images of skin treated with elastic vesicles and rigid vesicles. (**A**) An *xy* image (95 mm×95 mm) of skin treated with elastic vesicles recorded at a depth of 10 mm. After one hour of application, already a fine meshwork of channels was observed without any distinguishable polygonal cell contours. (**B**) An *xy* image (95 mm×95 mm) recorded at a depth of three after an hour treatment with rigid vesicles. Cell contours are clearly visible indicating a homogenous distribution of the label in the upper SC layers. *Abbreviation*: SC, stratum corneum.

the transport of the fluorescent label was via a fine meshwork of thread-like channels. This was not seen when the label was incorporated into rigid vesicles (Fig. 5) (103). Although the results from Schatzlein et al. and Van den Bergh et al. were in agreement concerning the concept of vesicle transport via channel-like structures, the channels visualized in both studies did not resemble each other. Van den Bergh et al. have visualized a much finer meshwork of channels as compared to the honeycomb-like system of intercluster and intercorneocyte pathways by Schatzlein et al. Furthermore, in the study by Schatzlein et al. the corneocyte shapes were clearly detectable, while these were not visible in the study by Van den Bergh et al. It is unclear whether these morphological differences were caused by the different experimental methods used (e.g., differences in vesicle composition, animal skin vs. human skin, in vivo vs. in vitro situation) or whether the channels found in the two studies are entirely different entities. Interestingly, the microscopic images obtained by Van den Bergh et al. illustrated that the lipophilic fluorescent label was always confined to the SC. This suggests that the elastic vesicle components do not travel beyond the SC, which is in contradiction to the claim of Cevc et al. that the elastic vesicles could even reach the systemic circulation. Van den Bergh et al. (97,103) also investigated the vesicle–skin interactions using transmission electron microscopy and FFEM. After treatment with elastic vesicles they found lamellar stacks in the intercellular lipid spaces, in vivo using murine skin and in vitro using excised human skin. Van den Bergh et al. suggested that the lamellar stacks could disorganize the intercellular lipid bilayers and thereby create possible penetration pathways for drug molecules. In contrast to the elastic vesicles, treatment with rigid vesicles did not induce any changes in the skin ultrastructure. Furthermore, Van den Bergh et al. did not visualize any abnormal structures in the viable epidermis or the dermis using electron microscopic techniques, confirming their previous observation that the elastic vesicles do not travel beyond the SC.

Very recently three in vivo studies were performed with human volunteers. In the first two studies the transport of vesicles into human SC was investigated with tape stripping in combination with FFEM. In the last study tape stripping was used in combination with attenuated total reflectance-Fourier transformed infrared spectroscopy (ATR-FTIR). The combination of FFEM and tape stripping allows the visualization of the SC structure at various depths. Hence, it provides information on the intact permeation of vesicles into the SC, but it does not give information on the quantitative aspects of vesicular transport. ATR-FTIR in combination with tape stripping provides a quantitative assessment as well as information on the penetration depth of vesicles into human SC.

Tape Stripping and FFEM

The first tape stripping and FFEM study was performed to visualize the interactions between L-595/PEG-8-L (50/50) elastic, L-595/PEG-8-L (100/0) rigid vesicles, and PEG-8-L micelles and human skin in vivo and in vitro (104). No ultrastructural changes were visualized in the skin after treatment with rigid vesicles. The rigid vesicles are fused on the surface of the skin (Fig. 6). Treatment with PEG-8-L micelles resulted in rough, irregular fracture planes. This suggests that PEG-8-L micelles could act as penetration enhancers. The results for the elastic vesicles were most remarkable and clearly indicated that they have very different modes of action as compared to the micelles. The elastic vesicles showed an extremely fast partitioning into the SC. After one hour of nonocclusive application, intact vesicles were visualized in the deeper layers of the SC. Interestingly, these vesicles were accumulated within channel-like structures that are located within the intercellular lipid lamellae (Fig. 6). No other alterations in the structure of the intercellular lipid lamellae were

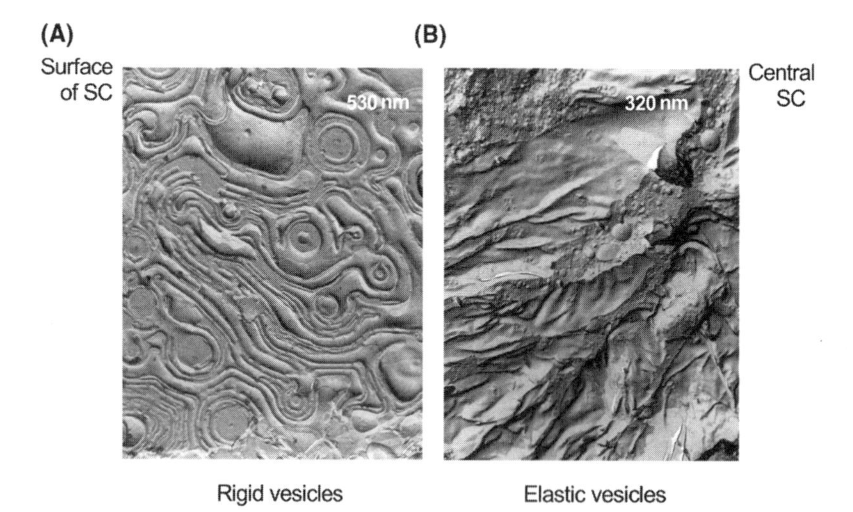

Figure 6 (**A**) In vivo interactions between rigid vesicles and human skin. A micrograph of the first tape strip of skin treated with rigid vesicles after one hour of application showing large areas with vesicle fusion. (**B**). In vivo interaction between elastic vesicles and human skin. Micrograph of the ninth tape strip showing elastic vesicles in channel-like regions. *Arrows*: intact vesicles.

observed. Previously, Van den Bergh et al. (103) had also reported channel-like penetration pathways of a model lipophilic fluorescent label incorporated into elastic vesicles of similar composition. These channels (Fig. 5) had similar size and dimension as those visualized in this study, suggesting that both model compounds and intact elastic vesicles follow the same route through human SC (Fig. 6). No intact vesicles were observed in the deepest layers of the SC, suggesting that elastic vesicles do not penetrate into the viable epidermis in large quantity. Results from the in vitro studies with dermatomed human skin showed similar features, indicating an excellent in vivo and in vitro correlation, which provides more confidence that past or future in vitro transport of elastic vesicles studies could be used to predict the in vivo outcome. The results of this tape stripping and FFEM study are in good agreement with the mode of action detailed in Figure 4. The fact that (a) elastic vesicles show a fast partitioning into the SC, (b) both model compounds and elastic vesicles follow the same route through human SC, and (c) no other abnormalities were found in the intercellular lipid lamellae, all point out that elastic vesicles do not act as penetration enhancers, but more likely act as drug carrier systems.

In the second tape stripping and FFEM study various additional parameters were investigated for the elastic vesicles to further investigate their transport into the skin (105). As with the previous tape stripping and FFEM study, intact elastic vesicles were found in the deeper layers of the SC after one hour of nonocclusive treatment. After four hours' nonocclusive treatment, vesicle material was also observed mainly in the channel-like regions. However, extensive vesicle fusion was seen and only very few intact vesicles were found in the SC, suggesting that elastic vesicles do not remain intact in the SC for a long period of time. When the application was changed from nonocclusive to occlusive, interestingly, only a few intact vesicles were visualized in the deeper layers of the SC. However, the presence of lipid plaques was frequently observed, both inside and outside the channel-like regions. Hence, it can be postulated that occlusion impairs the transport of intact elastic vesicles into the skin, but vesicle components or vesicle fragments can penetrate into the SC, where they can form lipid plaques. This is in agreement with the osmotic theory of Cevc et al. (80), who postulated that the transport of elastic vesicles into the skin is driven by an osmotic gradient across the skin.

Although using microscopy it was possible to visualize vesicle material in the skin and changes in the skin lipid organization, it was not possible to quantify the amount of vesicle material in the skin. In another study a quantitative assessment was performed to transport elastic and rigid surfactant-based vesicles and a model drug from these vesicle formulations into human skin in vivo, by Honeywell-Nguyen et al. (106). This was investigated using the tape stripping technique with subsequent analysis of the tape strips for the presence of vesicle material and the presence of the model compound. A deuterium labelled phospholipid (DLP) was added to the elastic and rigid vesicles to serve as a marker for quantification using ATR-FTIR. Because of its very high lipophilicity, it can be assumed that DLP was entirely bound to the vesicles, and hence would be an excellent marker for the distribution of vesicle material within the SC. Ketorolac was chosen, based on its physicochemical properties, as a model compound and was incorporated into elastic and rigid vesicle formulations at saturated concentrations. Distribution profiles of DLP and ketorolac obtained after elastic and rigid vesicle treatment showed a remarkable difference between the elastic and rigid vesicles. DLP added in elastic vesicle formulations could rapidly enter the deeper layers of the SC and reach almost as deep as the SC-viable epidermal junction. DLP applied in rigid vesicle formulations, however, did not penetrate very deep into the SC. Furthermore, higher amounts of ketorolac were found in the deeper layers of the SC after elastic vesicle treatment as compared to after rigid vesicle treatment. When comparing the SC distribution profiles of DLP and ketorolac after elastic vesicle treatment, it was observed that there was a good correlation between these profiles as a function of depth in the upper and middle layers of the SC. However, in the deeper layers of the SC, no correlation was observed. In the deeper layers of the SC the relative amount of DLP increased while the relative amount of ketorolac decreased. This strongly suggests that ketorolac was associated with elastic vesicle material in the upper and middle layers of the SC, but was not associated with vesicle material in the lowest layers of the SC, and hence must have been partly released from the vesicles. This is in excellent agreement with Steps 2, 3, and 4 of the proposed mechanism of action in Figure 4. Furthermore, in concordance with the tape stripping and FFEM studies, it was also observed that the relative amount of DLP found in the deeper layers of the SC was very much lower than what was found closer to the skin surface, suggesting that elastic vesicle material does not penetrate beyond the SC in significant quantities. This supports the hypothesis that elastic vesicles mainly remain in the SC and that drug molecules are released from these vesicles with subsequent transport of free-drug molecules into the viable skin layers (Step 3 in Fig. 4).

In summary, it is evident from the aforementioned studies that elastic vesicles have superior characteristics to rigid vesicles for the interaction with animal and human skin. However, the exact mode and nature of elastic vesicle transport into and possibly across the skin is not yet clear, even though in the last five years in vivo techniques have shed some light on the mode of action of elastic vesicles. This is necessary to assess the full potential applications of elastic vesicles as skin delivery vehicles, such as the delivery of large molecules or targeting to certain sites and cells within the skin. This could give rise to the development of very interesting and novel transdermal and dermal drug delivery systems.

REFERENCES

1. Schaefer H, Redelmeier TE. Skin Barrier. Principles of Percutaneous Absorption. KargerBasel1996.
2. Wiechers JW. The barrier function of the skin in relation to percutaneous absorption of drugs. Pharm Weekbl 1989; 11:185–198.
3. Muijsers RB, Wagstaff AJ. Transdermal fentanyl: an updated review of its pharmacological properties and therapeutic efficacy in chronic cancer pain control. Drugs 1902; 61:2289–2307.
4. Todd PA, Goa KL, Langtry HD. Transdermal nitroglycerin (glyceryl trinitrate). A review of its pharmacology and therapeutic use. Drugs 1990; 40:880–902.
5. Parrott AC. Transdermal scopolamine: a review of its effects upon motion sickness, psychological performance, and physiological functioning. Aviat Space Environ Med 1989; 60:1–9.
6. Noerr B. Transdermal clonidine. Neonatal Netw 2000; 19:67–69.
7. Burris JF. The USA experience with the clonidine transdermal therapeutic system. Clin Auton Res 1993; 3:391–396.
8. Gore AV, Chien YW. The nicotine transdermal system. Clin Dermatol 1998; 16:599–615.
9. Wiseman LR, McTavish D. Transdermal stradiol/norethisterone. A review of its pharmacological properties and clinical use in postmenopausal women. Drugs Aging 1994; 4:238–256.

10. McClellan KJ, Goa KL. Transdermal testosterone. Drugs 1998; 55:253–258.
11. Barry BW. In: Dermatological Formulations: Percutaneous Absorption. New York: Marcel Dekker, 1983.
12. Blank IH, Scheuplein RJ. Transport into and within the skin. Br J Dermatol 1969; 81:4–10.
13. Scheuplein RJ, Blank IH. Permeability of the skin. Physiol Rev 1971; 51:702–747.
14. Roy SD, Flynn GL. Transdermal delivery of narcotic analgesics: pH, anatomical, and subject influences on cutaneous permeability of fentanyl and sufentanil. Pharm Res 1990; 7:842–847.
15. Higo N, Hinz RS, Lau DT, Benet LZ, Guy RH. Cutaneous metabolism of nitroglycerin in vitro. II. Effects of skin condition and penetration enhancement. Pharm Res 1992; 9:303–306.
16. Elias PM. Epidermal lipids, barrier function, and desquamation. J Invest Dermatol 1983; 80(suppl):44s–49s.
17. Wertz PW. Epidermal lipids. Semin Dermatol 1992; 11:106–113.
18. Weerheim A, Ponec M. Determination of stratum corneum lipid profile by tape stripping in combination with high-performance thin-layer chromatography. Arch Dermatol Res 2001; 293:191–199.
19. Williams ML, Elias PM. The extracellular matrix of stratum corneum: role of lipids in normal and pathological function. Crit Rev Ther Drug Carrier Syst 1987; 3:95–122.
20. Wertz PW. Lipids and barrier function of the skin. Acta Derm Venereo 2000; 208(suppl):7–11.
21. Bouwstra JA, Honeywell-Nguyen PL, Gooris GS, Ponec M. Structure of the skin barrier and its modulation by vesicular formulations. Prog in Lip Res 2003; 42:1–36.
22. Bouwstra JA, Dubbelaar FE, Gooris GS, Ponec M. The lipid organisation of the skin barrier. Acta Derm Venereol 2000; 208(suppl):23–30.
23. Pilgram GS, Engelsma-van Pelt AM, Bouwstra JA, Koerten HK. Electron diffraction provides new information of human stratum corneum lipid organization studied in relation to depth and temperature. J Invest Dermatol 1999; 113:403–409.
24. Bouwstra JA, Gooris GS, Bras W, Downing DT. Lipid organization in pig stratum corneum. J Lipid Res 1995; 36:685–695.
25. Grubauer G, Feingold KR, Harris RM, Elias PM. Lipid content and lipid type as determinants of the epidermal permeability barrier. J Lipid Res 1989; 30:89–96.
26. Landmann L. The epidermal permeability barrier. Anat Embryol (Berl) 1988; 178:1–13.
27. Siddiqui O. Physicochemical, physiological, and mathematical considerations in optimizing percutaneous absorption of drugs. Crit Rev Ther Drug Carrier Syst 1989; 6:1–38.
28. Kalia YN, Merino V, Guy RH. Transdermal drug delivery. Clinical aspects. Dermatol Clin 1998; 16:289–299.
29. Warner RR, Myers MC, Taylor DA. Electron probe analysis of human skin: determination of the water concentration profile. J Invest Dermatol 1998; 90:218–224.
30. Von Zglinicki T, Lindberg M, Roomans GM, Forslind B. Water and ion distribution profiles in human skin. Acta DermVenereol 1993; 73:340–343.
31. Caspers PJ, Lucassen GW, Carter EA, Bruining HA, Puppels GJ. In vivo confocal raman microspectroscopy of the skin: noninvasive determination of molecular concentration profiles. J Invest Dermatol 2001; 116:434–442.
32. Bouwstra JA, De Graaff A, Gooris GS, Nijsse J, Wiechers JW, Van Aelst AC. Water distribution and related morphology in human stratum corneum at different hydration levels. J Invest Dermatol 2003; 120:750–758.
33. Warner RR, Stone KJ, Boissy YL. Hydration disrupts human stratum corneum ultrastructure. J Invest Dermatol 2003; 120:275–284.
34. Odland GF, Holbrook K. The lamellar granules of epidermis. Curr Probl Dermatol 1981; 9:29–49.
35. Landmann L. Epidermal permeability barrier: transformation of lamellar granule-disks into intercellular sheets by a membrane-fusion process, a freeze-fracture study. J Invest Dermatol 1986; 87:202–209.

36. Moser K, Kriwet K, Naik A, Kalia YN, Guy RH. Passive skin penetration enhancement and its quantification in vitro. Eur J Pharm Biopharm 2001; 52:103–112.
37. Potts RO, Guy RH. Predicting skin permeability. Pharm Res 1992; 9:663–669.
38. Barry BW. Novel mechanisms and devices to enable successful transdermal drug delivery. Eur J Pharm Sci 2001; 14:101–114.
39. Schreier H, Bouwstra JA. Liposomes and niosomes as topical drug carriers: dermal and transdermal drug delivery. J Contr Rel 1994; 30:1–15.
40. Suhonen TM, Bouwstra JA, Urtti A. Chemical enhancement of percutaneous absorption in relation to stratum corneum structural alterations. J Contr Rel 1999; 59:149–161.
41. Gregoriadis G, Florence AT. Liposomes in drug delivery. Clinical, diagnostics and ophthalmic potential. Drugs 1993; 45:15–28.
42. Martin GP, Lloyd AW. Basic principles of liposomes for drug use. In: Braun-Falco O, Korting HC, Maibach HI, eds. Liposome Dermatics. Berlin, Heidelberg: Springer-Verlag, 1992:21–26.
43. Crommelin DJA, Schreier H. Liposomes. In: Kreuter J, ed. Colloidal Drug Delivery Systems. New York: Marcel Dekker, 1994:73–190.
44. Bouwstra JA, Hofland HEJ. Niosomes. In: Kreuter J, ed. Colloidal Drug Delivery Systems. New York: Marcel Dekker, 1994:191–217.
45. Mezei M, Gulasekharam V. Liposomes—a selective drug delivery system for the topical route of administration. Lotion dosage form. Life Sci 1980; 26:1473–1477.
46. Mezei M, Gulasekharam V. Liposomes-a selective drug delivery system for the topical route of administration: gel dosage form. J Pharm Pharmacol 1982; 34:473–474.
47. Ganesan MG, Weiner ND, Flynn GL, Ho NFH. Influence of liposomal drug entrapment on percutaneous absorption. Int J Pharm 1984; 20:139–154.
48. Ho NFH, Ganesan MG, Weiner ND, Flynn GL. Mechanisms of topical delivery of liposomally entrapped drugs. J Contr Rel 1985; 2:61–65.
49. Komatsu H, Higaki H, Okamoto H, Miyagawa K, Hashida M, Sezaki H. Preservative activity and in vivo percutaneous penetration of butylparaben entrapped in liposomes. Chem Pharm Bull 1986; 34:3415–3422.
50. Komatsu H, Okamoto H, Miyagawa K, Hashida H, Sezaki H. Percutaneous absorption of butylparaben from liposomes in vitro. Chem Pharm Bull 1986; 34:3423–3430.
51. Knepp VM, Szoka FC, Guy RH. Controlled drug release from a novel liposomal delivery system. II. Transdermal delivery characteristics. J Contr Rel 1990; 12:25–30.
52. Knepp VM, Hinz RS, Szoka FC, Guy RH. Controlled drug release from a novel liposomal delivery system. I. Investigation of transdermal potential. J Contr Rel 1988; 5:211–221.
53. Wohlrab W, Lasch J. Penetration kinetics of liposomal hydrocortisone in human skin. Dermatologica 1987; 174:18–22.
54. Wohlrab W, Lasch J. The effect of liposomal incorporation of topically applied hydrocortisone on its serum concentration and urinary excretion. Dermatol Monatsschr 1989; 175:348–352.
55. Masini V, Bonte F, Meybeck A, Wepierre J. Cutaneous bioavailability in hairless rats of tretinoin in liposomes or gel. J Pharm Sci 1993; 82:17–21.
56. Katahira N, Murakami T, Kugai S, Yata N, Takano M. Enhancement of topical delivery of a lipophilic drug from charged multilamellar liposomes. J Drug Target 1999; 6:405–414.
57. Egbaria K, Ramachandran C, Weiner N. Topical delivery of cyclosporin: evaluation of various formulations using in vitro diffusion studies in hairless mouse skin. Skin Pharmacol 1990; 3:21–28.
58. Fresta M, Puglisi G. Corticosteroid dermal delivery with skin-lipid liposomes. J Contr Rel 1995; 44:141–151.
59. Bernard E, Dubois JL, Wepierre J. Percutaneous absorption of a new antiandrogen included in liposomes or in solution. Int J Pharm 1995; 126:235–243.

60. Kim MK, Chung SJ, Lee MH, Cho AR, Shim CK. Targeted and sustained delivery of hydrocortisone to normal and stratum corneum-removed skin without enhanced skin absorption using a liposome gel. J Contr Rel 1997; 46:243–251.
61. Touitou E, Dayan N, Bergelson L, Godin B, Eliaz M. Ethosomes-novel vesicular carriers for enhanced delivery: characterization and skin penetration properties. J Contr Rel 2000; 65:403–418.
62. Dayan N, Touitou E. Carriers for skin delivery of trihexyphenidyl HCl: ethosomes vs. liposomes. Biomaterials 2000; 21:1879–1885.
63. Cevc G, Schatzlein D, Blume G. Transdermal drug carriers: basic properties, optimization and transfer efficiency in the case of epicutaneously applied peptides. J Contr Rel 1995; 36:3–16.
64. Cevc G, Gebauer D, Stieber J, Schatzlein A, Blume G. Ultraflexible vesicles, transfersomes, have an extremely low pore penetration resistance and transport therapeutic amounts of insulin across the intact mammalian skin. Biochim Biophys Acta 1998; 1368:201–215.
65. Cevc G, Blume G, Schatzlein A. Transfersomes-mediated transepidermal delivery improves the regio-specific and biological activity of corticosteroids in vivo. J Contr Rel 1997; 45:211–236.
66. Guo J, Ping Q, Zhang L. Transdermal delivery of insulin in mice by using lecithin vesicles as a carrier. Drug Del 2000; 7:113–116.
67. Guo J, Ping Q, Sun G, Jiao C. Lecithin vesicular carriers for transdermal delivery of cyclosporin A. Int J Pharm 2000; 194:201–207.
68. Hofland HE, van der Geest R, Bodde HE, Junginger HE, Bouwstra JA. Estradiol permeation from nonionic surfactant vesicles through human stratum corneum in vitro. Pharm Res 1994; 11:659–664.
69. Van Kuijk-Meuwissen ME, Mougin L, Junginger HE, Bouwstra JA. Application of vesicles to rat skin in vivo: a confocal laser scanning microscopy study. J Contr Rel 1998; 56:189–196.
70. Van Kuijk-Meuwissen ME, Junginger HE, Bouwstra JA. Interactions between liposomes and human skin in vitro, a confocal laser scanning microscopy study. Biochim Biophys Acta 1998; 1371:31–39.
71. Ogiso T, Ninaka N, Iwaki H. Mechanism for enhancement effect of lipid disperse system on percutaneous absorption. J Pharm Sci 1996; 85:57–64.
72. Perez-Cullell N, Coderch L, de la Maza A, Parra JL, Estelrich J. Influence of the fluidity of liposome compositions on percutaneous absorption. Drug Deliv 2000; 7:7–13.
73. Du Plessis J, Ramachandran C, Weiner N, Muller DG. The influence of particle size of liposomes on the deposition of drug into skin. Int J Pharm 1994; 103:277–282.
74. Fresta M, Puglisi G. Application of liposomes as potential cutaneous drug delivery systems. In vitro and in vivo investigation with radioactively labelled vesicles. J Drug Target 1996; 4:95–101.
75. Abraham W, Downing DT. Interaction between corneocytes and stratum corneum lipid liposomes in vitro. Biochim Biophys Acta 1990; 1021:119–125.
76. Hofland HE, Bouwstra JA, Spies F et al. Interactions between non-ionic surfactant vesicles and human stratum corneum in vitro. J Liposome Res 1995; 5:241–263.
77. Hofland HE, Bouwstra JA, Bodde HE, Spies F, Junginger HE. Interactions between liposomes and human stratum corneum in vitro: freeze fracture electron microscopical visualization and small angle X-ray scattering studies. Br J Dermatol 1995; 132:853–866.
78. Zellmer S, Pfeil W, Lasch J. Interaction of phosphatidylcholine liposomes with the human stratum corneum. Biochim Biophys Acta 1995; 1237:176–182.
79. Foldvari M, Gesztes A, Mezei M. Dermal drug delivery by liposome encapsulation: clinical and electron microscopic studies. J Microencapsul 1990; 7:479–489.
80. Cevc G, Blume G. Lipid vesicles penetrate into intact skin owing to the transdermal osmotic gradients and hydration force. Biochim Biophys Acta 1992; 1104:226–232.

81. Cevc G, Schatzlein A, Richardsen H. Ultradeformable lipid vesicles can penetrate the skin and other semi-permeable barriers unfragmented. Evidence from double label CLSM experiments and direct size measurements. Biochim Biophys Acta 2002; 1564:21–30.
82. Betz G, Imboden R, Imanidis G. Interactions of liposome formulations with human skin in vitro. Int J Pharm 2002; 229:117–129.
83. Kirjavainen M, Urtti A, Jaaskelainen I, et al. Interaction of liposomes with human skin in vitro-the influence of lipid composition and structure. Biochim Biophys Acta 1996; 1304:179–189.
84. Coderch L, Fonollosa J, De Pera M, Estelrich J, de la Maza A, Parra JL. Influence of cholesterol on liposome fluidity by EPR. Relationship with percutaneous absorption. J Contr Rel 2000; 68:85–95.
85. Cevc G, Schatzlein A, Gebauer D, Blume G. Ultra-high efficiency of drug and peptide transfer through the intact skin by means of novel drug-carriers, transfersomes. In: Brain KR, James VF, Walters KA, eds. Prediction of Percutaneous Penetration. Vol. 3B. Cardiff: STS Publishing, 1993:226–236.
86. Cevc G. Transfersomes, liposomes and other lipid suspensions on the skin: permeation enhancement, vesicle penetration, and transdermal drug delivery. Crit Rev Ther Drug Carrier Syst 1996; 13:257–388.
87. Van den Bergh BA. Elastic liquid state vesicles as a tool for topical drug delivery. Ph.D thesis, University of Leiden, Netherlands, 1999.
88. Planas ME, Gonzalez P, Rodriguez L, Sanchez S, Cevc G. Noninvasive percutaneous induction of topical analgesia by a new type of drug carrier, and prolongation of local pain insensitivity by anesthetic liposomes. Anesth Analg 1992; 75:615–621.
89. Paul A, Cevc G, Bachhawat BK. Transdermal immunization with large proteins by means of ultradeformable drug carriers. Eur J Immunol 1995; 25:3521–3524.
90. Paul A, Cevc G, Bachhawat BK. Transdermal immunisation with an integral membrane component, gap junction protein, by means of ultradeformable drug carriers, transfersomes. Vaccine 1998; 16:188–195.
91. Cevc G, Blume G. New, highly efficient formulation of diclofenac for the topical, transdermal administration in ultradeformable drug carriers, transfersomes. Biochim Biophys Acta 2001; 1514:191–205.
92. El Maghraby GM, Williams AC, Barry BW. Skin delivery of oestradiol from lipid vesicles: importance of liposome structure. Int J Pharm 2000; 204:159–169.
93. El Maghraby GM, Williams AC, Barry BW. Oestradiol skin delivery from ultradeformable liposomes: refinement of surfactant concentration. Int J Pharm 2000; 196:63–74.
94. El Maghraby GM, Williams AC, Barry BW. Skin delivery of oestradiol from deformable and traditional liposomes: mechanistic studies. J Pharm Pharmacol 1999; 51:1123–1134.
95. El Maghraby GM, Williams AC, Barry BW. Skin delivery of 5-fluorouracil from ultradeformable and standard liposomes in vitro. J Pharm Pharmacol 2001; 53:1069–1077.
96. Trotta M, Peira E, Debernardi F, Gallarate M. Elastic liposomes for skin delivery of dipotassium glycyrrhizinate. Int J Pharm 2002; 241:319–327.
97. Van den Bergh BA, Bouwstra JA, Junginger HE, Wertz PW. Elasticity of vesicles affects hairless mouse skin structure and permeability. J Contr Rel 1999; 62:367–379.
98. Honeywell-Nguyen PL, Frederik PM, Bomans PHH, Junginger HE, Bouwstra JA. Transdermal delivery of pergolide from surfactant-based elastic and rigid vesicles: characterization and in vitro transport studies. Pharm Res 2002; 19:991–997.
99. Honeywell-Nguyen PL, Bouwstra JA. The in vitro transport of pergolide from surfactant-based elastic vesicles through human skin: a suggested mechanism of action. J Contr Rel 2003; 86:145–156.
100. Honeywell-Nguyen PL, van den Bussche MM, Junginger HE, Bouwstra JA. The effect of surfactant-based elastic and rigid vesicles on the penetration of lidocain across human skin, STP. Pharma Sci 2002; 12:57–262.

101. Honeywell-Nguyen PL, Arenja S, Bouwstra JA. Skin penetration and mechanisms of action in the delivery of the D_2-agonist rotigotine from surfactant-based elastic vesicle formulations. Pharm Res 2003; 20:1619–1625.
102. Schatzlein A, Cevc G. Non-uniform cellular packing of the stratum corneum and permeability barrier function of intact skin: a high-resolution confocal laser scanning microscopy study using highly deformable vesicles (Transfersomes). Br J Dermatol 1998; 138:583–592.
103. Van den Bergh BA, Vroom J, Gerritsen H, Junginger HE, Bouwstra JA. Interactions of elastic and rigid vesicles with human skin in vitro: electron microscopy and two-photon excitation microscopy. Biochim Biophys Acta 1999; 1461:155–173.
104. Honeywell-Nguyen PL, de Graaff AM, Groenink HWW, Bouwstra JA. The in vivo and in vitro interactions of elastic and rigid vesicles with human skin. Biochim Biophys Acta 2002; 1573:130–140.
105. Honeywell-Nguyen PL, de Graaff AM, Groenink HWW, Bouwstra JA. The in vivo transport of elastic vesicles into human skin: effects of occlusion, volume and duration of application. J Contr Rel 2003; 90:243–255.
106. Honeywell-Nguyen L, Gooris GS, Bouwstra JA. Quantitative assessment of the transport of elastic and rigid vesicle components and a model drug from these vesicle formulations into human skin In vivo. J Invest Dermatol 2004; 123:902–910.

17

Lipid and Polymeric Colloidal Carriers for Ocular Drug Delivery

Simon Benita
Department of Pharmaceutics, School of Pharmacy, Faculty of Medicine, The Hebrew University of Jerusalem, Jerusalem, Israel

S. Tamilvanan
Department of Pharmaceutics, School of Pharmacy, Addis Ababa University, Addis Ababa, Ethiopia and Department of Pharmaceutics, School of Pharmacy, Faculty of Medicine, The Hebrew University of Jerusalem, Jerusalem, Israel

INTRODUCTION

Drugs are commonly applied to the eye for a localized action on the surface or in the anterior segment of the eye. However, drug delivery in ocular therapeutics is a challenging problem especially if the compound is lipophilic. Due to physiological and anatomical constraints, only a small fraction of the administered drug in eye drops is ocularly absorbed (1–3). The cornea is comprised of three layers, which account for its poor permeability characteristics: (i) the outer epithelium, which is lipophilic, (ii) the stroma, which constitutes ~90% of the thickness of the cornea and is hydrophilic, and (iii) the inner endothelium consisting of a single layer of flattened epithelium-like cells. Because the cornea has both hydrophilic and lipophilic structures, it presents an effective barrier to the absorption of both hydrophilic and lipophilic compounds. The compounds that penetrate the cornea best exhibit a log *P* octanol/water of two (4). Furthermore, for the treatment of different extra- and intra-ocular etiological conditions such as glaucoma, uveitis, keratitis, dry eye syndromes, cytomegalovirus retinitis, acute retinal necrosis, proliferative vitreoretinopathy, macular degenerative disease, etc., several lipophilic and poorly water soluble drugs have become available in recent years. These drugs represent a formulation challenge for scientists who design simple eye drop aqueous solutions because of aqueous solubility limitations. In addition, daily administrations are needed and most of the traditional ophthalmic dosage forms are clearly not only uncomfortable for the patient but also not efficient in combating some of the current virulent ocular diseases especially of the posterior segment. Therefore, lipid and polymeric colloidal carriers are more and more in demand to overcome the drawbacks of aqueous medicated collyres for the treatment of different extra- and intra-ocular etiological conditions. They are being explored to improve the bioavailability and duration of therapeutic action of ocular drugs. It is now recognized that severe ocular diseases, mainly of the back of the eye, cannot be treated by topical administration of the conventional ocular aqueous dosage forms (5). The most plausible alternative to improve delivery of active ingredients, in particular the lipophilic compounds, is the use of lipid or polymeric colloidal carriers. It is estimated that 40% or more of bioactive substances being identified through combinatorial screening programs are poorly soluble in water (6,7). Consequently, the drug molecules belonging to these categories cannot be easily incorporated into aqueous-cored/based eye drop dosage forms at adequate concentrations, and thus, the clinical efficacy of highly lipophilic drugs is being impeded. Dosage forms for ocular topical application of lipophilic drugs consist of oil drops, lotions, ointments, gels, and ocular systems, most of them being still ineffective and uncomfortable to the patients (8). Thus, lipid and polymeric colloidal particles represent an attractive alternative to efficiently deliver active ingredients especially if they are lipophilic and often exhibit hydrophobic character.

Oil-in-water (o/w) lipid emulsion systems and nanoparticles, respectively, are the widely investigated lipid- and polymer-based colloidal carriers to resolve the solubility problems of poorly water soluble drugs. Nanoparticles are submicron colloidal systems usually made of polymers. According to the preparation method nanospheres (NSs) or nanocapsules (NCs) can be obtained. NCs are vesicular systems in which the drug is confined to a core (usually oil) surrounded by a polymeric membrane. NSs are matrix systems in which a drug is dispersed or dissolved in the polymeric matrix (9). Other colloidal carriers such as liposomes or niosomes are not covered in this chapter. Similarly the traditional nanosuspension type collyria (tiny drug particles dispersed in aqueous surfactant solutions) are also not discussed in this chapter.

For nearly a decade, o/w lipid emulsions containing either anionic or cationic droplets have been recognized as an interesting and promising ocular topical delivery vehicle for lipophilic drugs (5,10). The natural biodegradability, sub-micrometer droplet size range, sterilizability, substantial drug solubilization either at the innermost oil phase or at the o/w interface, minimized side effects, and improved ocular bioavailability make the lipid emulsion a promising ocular delivery vehicle or carrier. In accordance with these statements, for the first time, an anionic lipid emulsion containing cyclosporine A (CsA) (0.05%) (Restasis™, Allergan, Irvine, California, U.S.) was recently approved by the FDA and is now available at U.S. pharmacies for the treatment of chronic dry eye disease. Furthermore, an over-the-counter product that features a nonmedicated (blank) anionic emulsion formulation, Refresh Endura®, has been launched in the U.S. market for eye lubricating purposes in patients suffering from moderate to severe dry eye syndrome. Taking into consideration that the NSs (which will be interchangeably denominated in this chapter nanoparticles) are polymeric matrix colloidal particulate systems, it is more likely that following ocular instillation, they may have some advantages over the lipid colloidal carriers, mainly physical stability and potential for controlled release of encapsulated drugs. Additionally, they have prolonged residence time in the precorneal area, particularly in the case of ocular inflammation and/or infection. In contrast to the matrix type nanoparticles, the NCs which are reservoir or vesicular type carriers are unable to retain the incorporated active principle over prolonged periods of time under sink conditions (11,12).

The main purpose of this chapter is to describe the briefly nonexhaustive ocular pathologies that may be treated efficiently by the drug loaded in emulsions, nanoparticles, and NCs. This will be followed by the definition and classification of various emulsion-based systems and a description about nanoparticles and NCs. The discussion of current pharmaceutical know-how on these colloidal carriers is beyond the scope of this chapter. Finally, the ocular biofate of these colloidal carriers as ophthalmic drug delivery systems will be discussed.

TOPICALLY TREATED OCULAR PATHOLOGIES

The therapeutic treatment of ocular diseases has been evolving since the early 1990s owing to an ever increasing number of new chemical entities available in the ophthalmic market. Concomitantly, progress is being made by numerous investigators in the development of novel dosage forms for efficient ocular delivery. It is believed that this trend will further expand in the near future, and novel ocular delivery systems based on lipid and polymeric colloidal carriers may be proposed and may have a beneficial impact on the prognosis of ocular diseases described briefly below.

Glaucoma

Glaucoma is a group of eye diseases that gradually steals sight without warning and often without symptoms. Vision loss is caused by damaging the optic nerve. This nerve acts like an electric cable with over a million wires and is responsible for carrying the images that we see to the brain. It was once thought that high intraocular pressure (IOP) was the main cause of this optic nerve damage. Although IOP is clearly a risk factor, it is now known that other factors must also be involved because even people with "normal" IOP can experience vision loss due to glaucoma.

The two main types of glaucoma are open angle glaucoma, or primary open angle glaucoma (POAG), and angle closure glaucoma. POAG is the most common form of glaucoma, affecting about three million Americans. It happens when the eye's drainage canals become clogged over time. The IOP rises because the correct amount of fluid cannot drain out of the eye. With open angle glaucoma, the entrances to the drainage canals are clear and should be working correctly. The clogging problem occurs inside the drainage canals, like the clogging that can occur inside the pipe below the drain in a sink. Most people have no symptoms and no early warning signs. If open angle glaucoma is not diagnosed and treated, it can cause a gradual loss of vision. This type of glaucoma develops slowly and sometimes without noticeable sight loss for many years. It usually responds well to medication, especially if found early and treated.

Angle closure glaucoma, also known as acute glaucoma or narrow angle glaucoma, occurs much more rarely and is very different from open angle glaucoma in that the eye pressure usually goes up very quickly. This happens when the drainage canals get blocked or covered over, like the clog in a sink when something is covering the drain. With angle closure glaucoma, the iris and the cornea are not as wide and open as they should be. The outer edge of the iris bunches up over the drainage canals, when the pupil enlarges too much or too quickly. This can happen when entering a dark room. Symptoms of angle closure glaucoma may include headaches, eye pain, nausea, rainbows around lights at night, and very blurred vision.

The therapy of glaucoma aims either to reduce the rate of aqueous humor secretion or to increase the aqueous humor outflow. The first approach is achieved by using beta-adrenergic agonists (epinephrine) and antagonists (timolol) or carbonic anhydrase inhibitors (acetazolamide), whereas the second approach is based on the use of parasympathomimetic agents (pilocarpine) (13). However, when such medical therapy is unsuccessful argon laser trabeculoplasty or surgical actions are required.

Dry Eye Syndrome or Keratoconjunctivitis Sicca

Keratoconjunctivitis sicca (KCS) or dry eye describes the changes in the eye which result from inadequate tear volume either due to lack of tear production (aqueous tear-deficient KCS) or due to excessive loss of tears arising from an accelerated evaporation because of poor tear quality (evaporative KCS).

To understand dry eye it is better to know how tears help keep the cornea healthy. The cornea is the optically clear portion of the eye that allows entry of light into the eye (see diagram showing the major parts of eye as depicted in Fig. 1).

Like all living tissues, the cornea requires a supply of oxygen and energy to remain healthy. Oxygen and nutrients are supplied to most tissues by the blood that moves through the area in blood vessels. The healthy cornea has no blood vessels, if it did it would not be clear; so the oxygen and nutrients are supplied through the tri-layered tear film.

The outermost layer of the tear film is an oily layer supplied by glands in the eyelids. This layer helps prevent evaporation of the next aqueous layer. The middle layer is the liquid aqueous layer produced by the main tear gland and a gland in the third eyelid. This is the layer that is decreased in the dry eye. The innermost layer in direct contact with the cornea is a mucous layer produced by glands located in the folds of the eyelid. The mucous layer helps the aqueous layer to adhere tightly to the surface of the cornea.

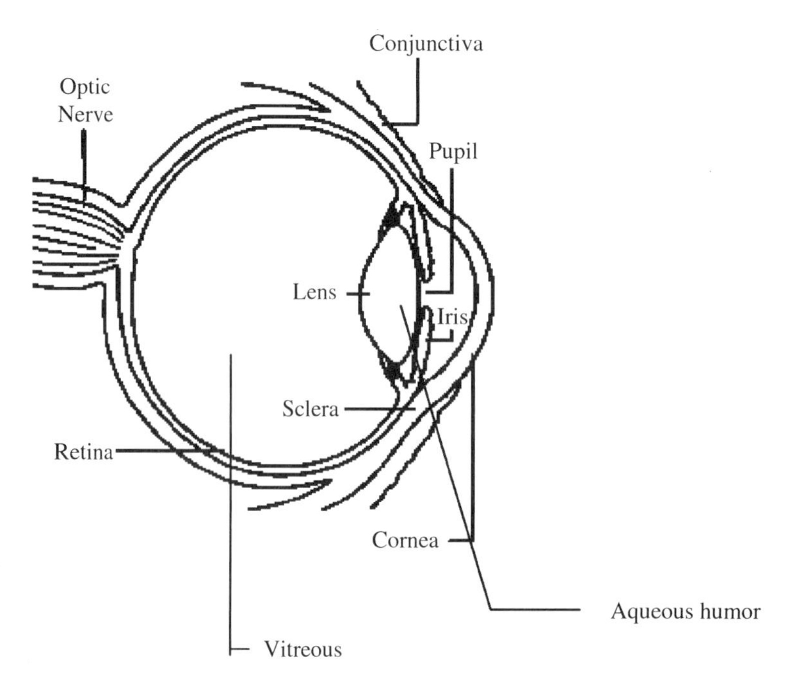

Figure 1 Diagram showing major parts of the eye.

A breakdown in the tear film and a loss of the aqueous layer cause dry eye. This loss results in dryness to areas of the corneal surface or in more advanced cases, drying to the entire corneal surface. When the cornea is deprived of oxygen and nutrients, it rapidly undergoes destructive changes. These changes result in brown pigmentation, scar tissue growth, ulcer development, and blood vessel growth across the cornea leading to partial vision loss. When a patient has "dry eye" where there is a lack of the watery layer of the tears, the oil and mucous layers are increased. This leads to a thick, mucoid, greenish discharge that sticks to the hairs around the eye. Patients complain of itching, burning, photophobia, gritty sensation, pulling sensation, pressure behind the eye, or a foreign body sensation. Some patients note a flood of tears after severe irritation. Symptoms are aggravated by prolonged visual efforts, such as reading, working on the computer, driving, or watching television.

Diagnosis is made by collecting a history about the condition, an examination, and a number of testing procedures (14). These tests include the Schirmer tear test, which measures the production of watery layer. A Schirmer test is performed by using standardized strips of filter paper placed, without topical anesthesia, at the junction between the middle and the lateral third of the lower lid. Wetting of the paper (to five millimeters or less) after five minutes on two successive occasions confirms the diagnosis of aqueous tear-deficient dry eye. Fluorescein stain is used to define possible breaks in the corneal surface and the rate of the tear breakup. In addition, rose bengal stain may be used to evaluate the health of the outer layer of the cornea called the epithelium.

The treatment of dry eye syndrome relies mainly on the use of artificial tears to fulfill the physiological role of normal tears such as lowering the surface tension of the tear film, forming a hydrophilic layer and enhancing the tear volume. Other possible treating agents include vitamin A derivatives and mucolytic (*N*-acetylcysteine) and tear stimulating products (Bromhexine and Gammalinolenic acid). These treatments are not considered effective, and newer delivery systems are being investigated to improve the pharmacological treatment. A recent introduction in this list is CsA incorporated in anionic emulsion formulation.

Ocular Inflammation

Inflammation of the ocular surface (and particularly of the cornea) may be driven by external factors (trauma, foreign body, or viral and bacteriological infections) or by the hosts internal factors (aberrant immune or inflammatory responses). Blepharitis and conjunctivitis are the common cases of inflammation. Blepharitis is a chronic inflammation of the eyelids. It is a very common condition that can be associated with a low grade bacterial infection (staphylococcal blepharitis) or a generalized skin condition (seborrheic blepharitis). The inflammation of the conjunctiva can be the consequence of bacterial (*Staphylococcus*, *Pneumococcus*, and *Haemophilus* species), viral (Adenovirus), or fungal infections.

Due to the fact that the symptomatology of ocular surface affections (inflammation) is similar despite the different etiologies, the treatments may be based on the use of a topical steroidal active compound (corticosteroids in their various forms and formulations) associated with nonsteroidal anti-inflammatory drugs (NSAIDs) such as diclofenac and flurbiprofen, an immunosuppressive agent (azathioprine), antimicrobial agents (tobramycin, ofloxacin), and mydriatic-cycloplegic drugs (pilocarpine).

Herpes Simplex Keratitis

The keratitis caused by the herpes simplex virus (HSV) typically presents as a unilateral "red eye" with a variable degree of pain or ocular irritation. Photophobia and epiphora are common; however, vision may or may not be affected, depending upon the location and extent of the corneal lesion. Herpes simplex is actually the most common virus found in humans. It is transmitted via bodily fluids, usually saliva, and may affect the skin and mucous membranes of the host. A vesicular skin rash and follicular conjunctivitis with the initial infection may be seen, but these are less common with recurrent HSV. In recurrent attacks, the virus invades and replicates within the corneal epithelium. As the cells die, an ulcerative keratitis results. Disciform stromal scarring, conjunctivitis, and uveitis are common sequelae. A more common sign is secondary uveitis. A dendritic corneal ulcer is the hallmark of HSV infection, accompanied by stromal keratitis in more severe presentations. These ulcers may begin as nondescript punctate keratopathies, but quickly coalesce to form the familiar branching patterns, which stain brightly with sodium fluorescein dye. Because the virus invades and compromises the epithelial cells surrounding the ulcer, the leading edges (the so-called "terminal end-bulbs") will stain with rose bengal or lissamine green.

Corneal epithelial disease secondary to HSV infection must be managed aggressively and quickly to prevent deeper penetration. The treatment of choice is topical trifluridine (1%) given at two-hour intervals, nine times daily. A cycloplegic (homatropine 2% three to four times daily or scopolamine 0.25% twice to four times daily) may need to be prescribed, again depending upon the severity of the uveitic response. Topical steroids in cases of active epithelial HSV keratitis should be avoided. From the studies it is shown that the virus replicates more rapidly in the presence of steroids, prolonging the course of the disease. The clinically significant use of oral acyclovir (400 mg, five times a day) or another oral antiviral for recalcitrant ulcers is yet to be proved. However, it has been shown recently that the use of oral acyclovir 400 mg q.i.d. significantly reduces the recurrence of herpes simplex keratitis in immunocompetent patients.

A new development in the management of herpes simplex keratitis has come in the form of topical acyclovir ointment (Zovirax). The ointment should be placed in the lower cul-de-sac five times per day at four-hour intervals. At this point, toxicity seems to be low.

Ocular Infections

Bacterial Conjunctivitis

The eye has a battery of defenses to prevent bacterial invasion. These include lysozyme and immunoglobulins in the tear film, constant movement of the tear film with blinking, and colonization of the surface of the eye with nonpathogenic bacteria which help to prevent pathogenic bacteria from replicating and invading ocular tissues. When any of these defense mechanisms breaks down, pathogenic bacterial infection is possible. Invading bacteria, and the exotoxins they produce, are antigenic. This induces an antigen–antibody immune reaction and triggers the inflammatory cascade. In a normal, healthy person the eye will fight to return to homeostasis, and the bacteria will eventually be eradicated. However, a large inoculum of external organisms may overcome the ocular defense mechanisms causing a bacterial conjunctivitis. The most commonly encountered organisms are *Staphylococcus aureus*, *Haemophilus influenzae*, *Streptococcus pneumoniae*, and *Pseudomonas aeruginosa*.

In cases of hyperacute bacterial conjunctivitis, the patient will present with similar signs and symptoms, albeit much more severe. The most common infectious organisms in hyperacute conjunctivitis are *Neisseria gonorrhoeae* and *Corynebacterium diphtheroids*. There is more danger in hyperacute bacterial conjunctivitis because these organisms can more readily penetrate through an intact corneal epithelium. The infection often starts in one eye and may spread to the other. There is usually a thick mucopurulent discharge. There may be mild photophobia and discomfort, but usually not severe pain. Visual function is normal in most cases.

Bacterial Keratitis

When the normal ocular defense mechanisms such as the epithelial glycocalyx or the corneal epithelium are compromised, infection may occur. Possible causes include direct corneal trauma, chronic eyelid disease, tear film abnormalities affecting the ocular surface, and hypoxic trauma from contact lens wear. Pathogenic bacteria colonize the corneal stroma and immediately become antigenic, both directly and indirectly, by releasing enzymes and toxins. This sets up an antigen–antibody immune reaction which initiates the inflammatory cascade. Polymorphonuclear leukocytes (PMNs) aggregate in the cornea creating an infiltrate. The PMNs phagocytize and digest the bacteria. The collagen stroma is damaged by bacterial and leukocytic enzymes and undergoes degradation, necrosis, and thinning. With severe thinning the cornea may perforate, creating the possibility of endophthalmitis. As the cornea heals scarring occurs which may result in decreased vision. Gram-positive organisms are predominantly responsible for bacterial keratitis in more northern climates while in southern climates, *P. aeruginosa* is most commonly isolated from ulcers. *P. aeruginosa* is also the most common organism isolated from contact lens related bacterial keratitis. The patient presents with unilateral ocular redness, pain, and photophobia. Visual acuity is usually reduced, and profuse tearing is common. There is a focal stromal infiltrate with an overlying area of epithelial excavation. There is likely to be a thick, ropy, mucopurulent discharge. Corneal edema is common and usually surrounds the infiltrate. There is usually injection of the episcleral and conjunctival vessels often out of proportion to the size of the corneal epithelial defect. In severe cases, there will be a pronounced anterior chamber reaction, often with hypopyon. The IOP may be low due to secretory hypotony of the ciliary body or may be elevated due to blockage of the trabecular meshwork by inflammatory cells. Often, the eyelids will also be edematous.

Chlamydial Conjunctivitis

Chlamydia trachomatis is an intracellular parasite that contains its own DNA and RNA. The sub-group A causes chlamydial infections, the serotypes A, B, Ba, and C cause trachoma, and serotypes D through K produce adult inclusion conjunctivitis. The modes of ocular transmission may be hand contact from a site of genital infection to the eye, laboratory accidents, a mother infecting the newborn during birth, shared cosmetics, and occasionally an improperly chlorinated hot tub. Diagnostic testing for chlamydia can be performed by culturing the organism or by using an enzyme immunoassay or direct fluorescent antibody testing. Conjunctival scrapings for Giemsa staining may show intracytoplasmic inclusion bodies in conjunctival epithelial cells. Chlamydial (inclusion) conjunctivitis typically affects sexually active teens and young adults and is the most frequent infectious cause of neonatal conjunctivitis in the United States. The Centers for Disease Control recognizes chlamydia as one of the major sexually transmitted pathogens, estimating approximately four to six million new cases per year. Women seem to be more susceptible than men. Ocular signs and symptoms include an eye infection that has persisted for over three weeks despite treatment with topical antibiotics. Conjunctival injection, superficial punctate keratitis, superior corneal pannus, peripheral subepithelial infiltrates, iritis, and follicles (most dense in the inferior cul-de-sac) may all be present. Mucopurulent, stringy, or mucous discharge is common. A palpable preauricular node is almost always present.

Viral Conjunctivitis

Viral conjunctival infections are thought to be transmitted by airborne respiratory droplets or direct transfer from one's fingers to the conjunctival surface of the eyelids. After an incubation period of five to 12 days, the disease enters the acute phase, causing watery discharge, conjunctival hyperemia, and follicle formation.

Most viral infections produce a mild, self-limiting conjunctivitis, but some have the potential to produce severe, disabling visual difficulties. The two most common self-limiting forms of viral conjunctivitis are epidemic keratoconjunctivitis and pharyngoconjunctival fever. The key clinical signs of both conditions include conjunctival injection, tearing, serous discharge, edematous eyelids, pinpoint subconjunctival hemorrhages, pseudomembrane formation, and palpable preauricular lymph nodes. In severe cases conjunctival scarring and symblepharon formation (adherence of the bulbar and palpebral conjunctivas) may occur. Both conditions are highly contagious. Patients will usually report recent contact with individuals who had either red eyes or an upper respiratory infection. Both forms tend to affect one eye and then spread to the other eye within a few days.

All of the above-described ocular infections are usually treated by combining oral and topical aqueous-based antimicrobial formulations.

LIPID AND POLYMERIC COLLOIDAL CARRIERS: DESCRIPTION AND CLASSIFICATION

Three main types of formulations, namely, o/w submicron emulsions, microemulsion, and water-in-oil-in-water (w/o/w) or oil-in-water-in-oil (o/w/o) multiple emulsions as depicted in the flow diagram (Fig. 2), are included in the lipid-based colloidal carrier category.

Interestingly, all of these emulsion-based ocular delivery systems exhibit almost similar refractive index values to that of the conventional aqueous-based/cored

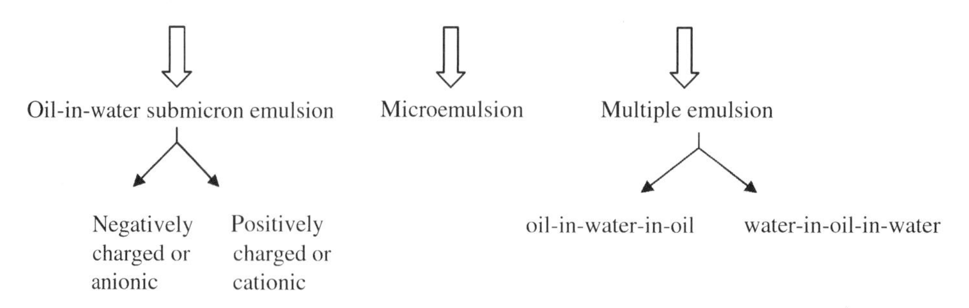

Figure 2 Flow diagram for lipid-based colloidal carrier for ocular (topical) drug delivery.

eyedrops. Thus, all these emulsions are not expected to produce visual blurring problems after topical application into the patient's eye (5). With regard to this issue, the ocular utility of semi-solid w/o emulsion-based ointments will not be discussed in this chapter although interested readers will find pertinent information on this type of emulsion/ointment dosage form in the literature (15–17).

The o/w submicron emulsions are also termed in the literature as macroemulsion, miniemulsion, and nanoemulsion. They are liquid-in-liquid heterogeneous dispersions in which one immiscible liquid (oil) is dispersed as droplets in another liquid (water) with the help of surface active agents called emulsifiers. They are milky in appearance. The oil droplet size of the final emulsion has been reduced to the range of 100 to 500 nm by means of appropriate manufacturing processes combining conventional emulsification techniques with high-speed mixers, ultrasonics, and/or two-stage homogenizers, etc. Based on the ability of the chosen emulsifier molecules alone or in specific combinations, a negative or positive surfaces charge can be formed around the emulsified droplets' surfaces during and after emulsification. Depending on these surface modifications, o/w emulsions can be classified into two types: negatively charged (anionic) and positively charged (cationic) o/w emulsions, respectively. Considering the fact that the corneal surfaces are negatively charged at physiological pH, the cationic emulsions were developed because the positive charges would cause them to be attracted electrostatically to the negatively charged proteoglycans of the cell surfaces, a process that can lead to adherence of opposite surface charges. These electrostatic interactions and adherences would appear to enhance drug absorption.

In contrast to o/w submicron emulsions, microemulsion vehicles are formed spontaneously when admixing appropriate quantities of the components, without requiring energy. They are also thermodynamically stable. The formulation of water in oil and o/w microemulsions usually involves a combination of three to five basic components, namely, oil, water, surfactant, cosurfactant, and osmotic agent. The tendency toward a w/o or an o/w microemulsion is dependent on the properties of both the oil and surfactant and the oil-to-water ratios. Furthermore, they are transparent or opalescent and single optically isotropic liquids. They are often described either as swollen micelles or as critical solutions (18). The oil droplet size of the microemulsion is typically in the range of 1 to 30 nm and therefore possibly produces only a weak scattering of visible light. The microemulsion has an o/w, a w/o, or a bicontinuous structure, in which neither oil nor water surrounds the other.

In addition to the two above-described structurally distinguished and completely different emulsion-based ocular drug delivery systems, there is one more liquid retentive-natured but more complicated system called multiple emulsions in which individual oil or water droplets of an o/w and a w/o emulsion enclose small water or oil droplets, respectively, so forming a w/o/w and an o/w/o emulsion. The particle size of the multiple emulsions normally ranges from 1 to 2 μm.

Nanoparticles exhibit particle sizes ranging from 0.1 to 0.5 μm. The polymers used to produce nanoparticles for ocular drug delivery purposes are biodegradable in nature, obtained from both natural and synthetic origins. Upon topical instillation, nanoparticles are expected to be retained in the cul-de-sac and the drug is slowly released from the particles through diffusion, chemical reaction, or polymer degradation or by an ion exchange mechanism.

Polymerization of monomers and nanoprecipitation of preformed polymers at the o/w or w/o interface in the presence of suitable stabilizers are the two known techniques to prepare NCs. Because NCs are vesicular (heterogenous) systems, the drug is confined to a cavity surrounded by a single polymeric membrane. The inner core may be aqueous or composed of a lipophilic solvent, usually an oil. Avoidance of burst release of the entrapped drug molecules, protection from drug degradation both during storage and after administration, and greater drug loading percentage if achieved can be considered clear advantages of NCs over nanoparticles.

O/W SUBMICRON EMULSIONS

The two types of o/w submicron emulsions, namely, anionic and cationic emulsions, will be discussed in this section. A nonexhaustive list with referenced publications having the publications related to ocular use of o/w submicron emulsions is shown in Table 1. The main advantage of this type of emulsion system, besides the escalated solubilization of lipophilic drugs, is the presence of relatively low concentrations of emulsifiers because high concentrations of emulsifiers usually lead to ocular irritation. Additionally, the emulsions exhibit a small enough and desired droplet size even after subjection of the final emulsion to autoclave sterilization.

Anionic Emulsion

A submicron emulsion based on soybean oil and oleic acid (7:3 ratio) as an oil vehicle and the emulsifier combinations, egg yolk phospholipids (Ovathin 160) and poloxamer

Table 1 Nonexhaustive List of O/W Submicron Emulsion for Ocular Drug Delivery

Emulsion type	Drug used	Indications	References
Negatively charged emulsion	Δ^8-THC	Glaucoma	19
Negatively charged emulsion	Pilocarpine base and indomethacin	Glaucoma and inflammation	20
Negatively charged emulsion	Adaprolol maleate	Glaucoma	21,22
Negatively charged emulsion	Indomethacin	Inflammation	23,24
Negatively charged emulsion	Dexanabinol and pilocarpine base	Glaucoma	25–27
Negatively charged emulsion	Pilocarpine base	Glaucoma	28–31
Positively charged emulsion	Piroxicam	Inflammation	32
Positively and negatively charged emulsion	Indomethacin	Inflammation	33
Negatively charged emulsion	CsA	Dry eye syndrome	34–43
Positively and negatively charged emulsion	CsA	Dry eye syndrome	44–46
Positively and negatively charged emulsion	Miconazole	Fungal Infections	47

Δ^8-THC and dexanabinol are derivatives of *cannabis sativa*.
Abbreviations: CsA, cyclosporine A; Δ^8-THC, delta-8-tetrahydrocannabinol.

180, was developed for ocular administration of delta-8-tetrahydrocannabinol (Δ^8-THC) (19). Δ^8-THC is an antiglaucoma drug naturally occurring as a constituent in *Cannabis sativa.* Due to its extremely low aqueous solubility, this drug is a suitable candidate for incorporation in the oil phase of the emulsion. The fresh emulsion exhibited a droplet size value of 130 ± 41 nm and a potential zeta value of -59 mV, while following storage at 4°C over nine months the same emulsion presented a droplet size and zeta potential value of 146 ± 47 nm and -57.1 mV, respectively. An intense and long-lasting IOP-depressant effect was observed after ocular application (50 μL) of the THC emulsion, 0.4% (weight of water), to rabbits with ocular hypertension (chymotrypsin model). Lesser effects were observed in normotensive rabbits. No irritation effect of either the emulsion vehicle or THC emulsion on the rabbit eyes was detected. Although the number of trials on animals (six rabbits for each group) was low, the reproducibility was good, and these results underline the promising properties of submicron emulsions as vehicles for lipophilic ophthalmic drugs. Nevertheless, no comparison with other dosage forms was made, and therefore this study did not provide enough evidence regarding the benefits of the emulsion carrier compared to others.

By changing the oil vehicle combination (soybean oil and oleic acid) used in the previous study (THC emulsion) to a single oil vehicle, medium chain triglycerides (MCT), dexanabinol (HU-211) and pilocarpine base were incorporated into a newly developed submicron emulsion suitable for topical ocular administration (25–27). In these particular emulsions, the emulsion stabilization was achieved by phospholipids (Lipoid E80®) and small amounts of amphoteric surfactants, Miranol MHT® (lauroamphodiacetate and sodium tridecethsulfate), and Miranol C_2M® (cocoamphodiacetate). HU-211 is a synthetic cannabinoid, which like Δ^8-THC is insoluble in water; yet, unlike THC it is nonpsychotropic. As a potential drug in glaucoma therapy, HU-211 requires a suitable ocular vehicle due to its lipophilic properties. Pilocarpine, a widely used antiglaucoma drug, is mainly formulated as a hydrochloride salt dissolved in an aqueous solution. The use of the aqueous preparation is limited by the need for frequent instillation (every six hours) due to very poor absorption and rapid drainage resulting in poor compliance and severe side effects such as miosis and synechiae formation.

Naveh et al. (26) conducted a randomized double-blind trial on three groups, each consisting of 12 healthy rabbits, and determined the effects following a single administration of the emulsion containing 1.7% pilocarpine base as opposed to an aqueous solution containing 2% pilocarpine hydrochloride (equivalent to 1.7% pilocarpine base). Pilocarpine base is known to be water soluble, and its incorporation into the emulsion system was studied to improve its ocular bioavailability. The percentage of reduction of IOP was calculated based on a baseline, which provides the values of the IOP according to the time before the treatment. The effects that were obtained using the emulsion and marketed eye drops (Pilocarpine Hydrochloride 2% eye drops) were different. The third rabbit group served as control. The eye drops modified the IOP earlier (two hours after instillation), but the duration of its pharmacological action was shorter (five hours). Eleven hours after the instillation, the control values were the same as the values observed on the treated rabbits. The instillation of the pilocarpine base-laden emulsion demonstrated a pharmacological action only five hours later. However, the IOP kept going down until the 29th hour. Furthermore, the decrease in maximum pressure was only 18% for regular eye drops and 28.5% for the emulsion eye drops. The emulsion provided both a sustained and delayed pharmacological action compared to the pharmacological action of regular eye drops. This observation led to the conclusion that the emulsion dosage forms have a distinct advantage compared to regular eye drops which must be administered four times a day due to the short duration of pharmacological action.

A partitioning study between oil and water has been carried out using the ultrafiltration technique at low pressure, indicating that only 5.9% and 14.9% of the total pilocarpine emulsion content was dissolved in the oil droplets and located at the oil–water interface, respectively. About 79.2% of the water soluble pilocarpine base was located in the water phase of the emulsion. Nevertheless, during the experiment, an ocular hypotensive effect was also noted in the contralateral eyes of pilocarpine emulsion treated rabbits while this effect was negligible in rabbits treated with aqueous pilocarpine. It appears that the retention of 20.8% of the total pilocarpine content in the internal oil phase and the oil–water interface of the emulsion are sufficient to concomitantly increase the corneal concentration of pilocarpine and enhance the ocular absorption of the drug through the cornea, probably due to the presence of tiny oily droplets and surfactant in the emulsion formulation. The authors deduced that this combined effect would account for the prolonged hypotensive effect and systemic absorption and contralateral effect.

Unlike pilocarpine emulsion, HU-211-loaded emulsion exerted its pharmacological action at 1.5 hours post–single dose instillation into normotensive rabbits (27). IOP reduction effect persisted for more than six hours. In addition, in comparison to the pilocarpine base-laden emulsion that acted for a longer period of time, HU-211 in the submicron emulsion induced a short and transitory (but significant) ocular hypotensive effect in the contralateral eye, indicating that some systemic absorption had occurred. In comparison, treatment with pilocarpine in the submicron emulsion exerted a pronounced and significant contralateral reduction in IOP, indicative of prolonged systemic absorption.

Zurowska-Pryczkowska et al. (28) have studied how submicron emulsion as a vehicle influences chemical stability of pilocarpine and concomitantly how the drug incorporation affects the physical stability of submicron emulsions. These authors prepared the negatively charged emulsions based on soybean oil (10% w/w) and Lipoid E80 (1.2% w/w) as emulsifier. To study the factors influencing the emulsion stability, the following technological factors were varied in the course of emulsion preparation: chemical form of the drug (base or as hydrochloride salt) and method of drug incorporation (de novo or ex tempore) into the emulsion, pH value (6.5 or 5.0) and presence of additional excipients, that is, poloxamer or Tween 80® as coemulgators, and methylcellulose and sodium carmellose (1% w/w) as viscosity increasing agents. The mean diameter of oil droplets in the resulting emulsions measured by a laser diffractometer was 0.6 to 0.7 μm; this was larger than in a drug-free emulsion where a 0.33 μm value was measured. In the presence of higher concentration of Lipoid E80 (2.4% w/w) or coemulgators (poloxamer 2.0% w/w or Tween 80 0.5% w/w) the mean droplet size in the emulsions was the same as in a drug-free system. However, the emulsions containing poloxamer were not stable during storage at 4°C indicating an occurrence of an unknown incompatibility between poloxamer and pilocarpine. In contrast, Tween 80 was compatible with the system, and on the basis of the droplet sizes after six months' storage, the preparations containing this cosurfactant could be selected as the most satisfying one. Although pH 5.0 may promote hydrolysis of triglycerides and therefore change the emulsion pH slightly during storage, it is the most favorable for chemical stability of the drug and therefore is the chosen pH. In comparison to the results of Naveh et al. (26), which indicate that at pH 4.8, as much as 5.9% of the total pilocarpine was distributed to the oily phase of the MCT oil-based emulsion, this study tends to show that practically all drug was present in the aqueous phase, irrespective of the pH value (5.0 vs. 6.5). Pilocarpine salt rather than base should be used for preparing emulsions. Considering the fact that when viscosity of eye drops is increased, the bioavailability of pilocarpine may be improved, the viscosity of the emulsions was increased by the addition of traditionally used agents such as methylcellulose or sodium carmellose (1% w/w). However, an intensive creaming was noted in such preparations and thus made the preparations a esthetically unacceptable. In a subsequent paper from the same group on in vivo evaluation using normotensive rabbits, it was shown that the submicron emulsion formulated with pilocarpine hydrochloride at pH 5.0 could be indicated as a preparation offering prolonged pharmacological action (miotic effect) together with satisfactory chemical stability (29). However, the bioavailability arising from such a formulation is not significantly improved when compared to an aqueous solution. Introduction of pilocarpine into the oil phase of the emulsion in the form of ion pair with oleic acid or simply a base does not lead to improvement of either physicochemical or pharmacological properties of the preparation. Emulsions at pH 8.5 could be the most promising regarding drug bioavailability but only extemporaneous preparations can be considered because of the fast degradation of pilocarpine.

To verify the extension of the residence time of the emulsion in the conjunctival sac, Beilin et al. (30) added a fluorescent marker to the formulations. After the instillation of fluorescent-marked emulsion and regular eye drops at 1, 2, 3, 5, 10, and 15 minutes, 200 μL of saline solution were used to rinse the eye. The intensity of fluorescence was then determined. One minute after the topical instillation into the eye, $39.9 \pm 10.2\%$ of the fluorescence was measured for the emulsion whereas only $6.8 \pm 1.8\%$ for regular eye drops. Furthermore, the areas under the curves of the pupillary diameters according to time were 7.5 ± 0.7 mm/hr for the emulsion dosage form and 5.7 ± 0.6 mm/hr for regular eye drops, the difference between the two values being significant ($P < 0.05$). Therefore it is reasonable to consider that the emulsion formulation provided a much more delayed pharmacological action due to a delayed residence time in the conjunctival sac following a more significant contact time with the cornea.

Anselem et al. (21) and Melamed et al. (22) prepared a submicron emulsion containing adaprolol maleate, a novel soft β-blocking agent. The size of the emulsion droplets was 100 ± 30 nm. According to these authors, no ocular irritation was noticed in the group of 40 healthy volunteers following topical application of the dosage forms. They observed a delayed IOP-depressant effect following the topical instillation of the adaprolol maleate emulsion.

Aviv et al. (20) prepared emulsions based on two different compositions and tested them in vivo both in rabbits and in human eyes. They used indomethacin-loaded emulsion for animal study (Table 2) and pilocarpine base-loaded emulsion for human testing (Table 3).

At 30 minutes post-dosing of 50 μL of the emulsion or 1% aqueous solution, the indomethacin concentration in the anterior chamber of the rabbit eye was two times higher for the regular aqueous solution than for the emulsion. The area under the curve (AUC) was 2.2 times larger for the emulsion, which means an increase of the bioavailability for the colloidal dosage form. Following a single instillation of pilocarpine base-laden emulsion or corresponding placebo, namely, emulsion without any drug, into the eyes of 20 healthy volunteers, the pupillary diameter and the variation of IOP were determined. The effect of emulsion, which contains pilocarpine base, was obvious as compared to the placebo and was noticed within one hour from instillation. The return to the initial values was noticed within

Table 2 Composition of Indomethacin-Loaded Submicron Emulsion Used for Rabbit Study

Components	Amount (% w/w)	Function
Indomethacin	0.20	Drug
MCT	17.0	Oil internal phase
Lipoid E80	3.00	Emulsifier
Tween 80	1.00	Emulsifier
Vitamin E	0.02	Antioxidant agent
Methyl parahydroxybenzoate	0.10	Preservative agent
Propyl parahydroxybenzoate	0.02	Preservative agent
Glycerol	2.25	Osmotic agent
EDTA	0.10	Chelating agent
Distilled water	Up to 100	External phase

Abbreviation: MCT, medium chain triglycerides.
Source: From Ref. 20.

12 hours. Furthermore, this human trial also showed that a pressure variation follows a systemic absorption in the contralateral eye. Another randomized human trial, conducted by Garty et al. (31), compared the activity of the pilocarpine base-laden emulsion instilled twice daily with a generic dosage form instilled four times a day to 40 hypertensive patients for seven days. No local side effects were observed. The IOP decreased by 25% in both of the formulations during this time period. No significant difference was noticed between the two treatments. These results proved that the emulsion extended the action of the drug, and two daily administrations have the same result as four instillations of regular eye drops.

To predict rates of drug transport across ocular tissue barriers such as cornea, sclera, and conjunctiva, an ex vivo penetration study of a drug-loaded submicron emulsion through an excised rabbit cornea (ex vivo permeation study) needs to be carried out. Although it is a generally recognized fact that an ex vivo model does not take into account the composition of aqueous humor, the composition of tear fluid, the mechanical stress of the eyelids, and tear flow, such a study was done by Muchtar et al. (23), who prepared indomethacin-loaded emulsion based on MCT oil and Lipoid E80 and Miranol MHT emulsifier combination and used a novel corneal diffusion assembly to

Table 3 Composition of Pilocarpine Base-Loaded Submicron Emulsion Used for Human Study

Components	Amount (% w/w)	Function
Pilocarpine base	1.70	Drug
MCT	4.25	Oil internal phase
Lipoid E-75®	0.75	Emulsifier
Tyloxapol	1.00	Emulsifier
Vitamin E	0.02	Antioxidant agent
Chlorbutanol	0.20	Preservative agent
Thiomersal	0.01	Preservative agent
Glycerol	2.25	Osmotic agent
Distilled water	Up to 100	External phase

Abbreviation: MCT, medium chain triglycerides.
Source: From Ref. 20.

conduct the diffusion kinetics of the drug from emulsion and marketed Indocollyre® eye drops. The coefficients of permeability were 6.9×10^{-5} cm/sec for the indomethacin incorporated into an emulsion and 1.83×10^{-5} cm/sec for the conventional aqueous eye drops containing the same drug. Nevertheless, as the two compositions did not have the same pH (3.8 for the emulsion and 6.8 for the aqueous eye drops), it is difficult to arrive at conclusions. The pH of the Indocollyre is high due to the large proportion of ionized indomethacin at this pH ($pK_a = 4.5$) and its hydrosolubility. Furthermore, the emulsion was not formulated at pH 6.8 due to its instability. The increase in indomethacin permeation (3.8 times compared to Indocollyre) could be attributed to the low ionization of the drug at emulsion pH of 3.8 resulting in a better permeation through the lipophilic epithelium layer. A similar increase in the transcorneal flux of indomethacin through isolated rabbit cornea has been reported by other groups (24). These authors included indomethacin in different colloidal systems including a negatively charged submicron emulsion which comprised Miglyol 840 oil and an emulsifier combination consisting of lecithin and poloxamer 188. They concluded that the main factor responsible for the increased corneal penetration of indomethacin is the colloidal nature of the carriers, and therefore, it is possible to define the colloidal carrier systems including the submicron emulsion as authentic ocular drug vehicles.

The aqueous solubility of CsA, a lipid-soluble cyclic endecapeptide, is approximately between 20 and 30 µg/mL. Thus, no adequate aqueous formulations are available for ocular administration of the drug. Moreover, if CsA is administered orally for the treatment of KCS, the accompanying side effects due to systemic circulation may cause adverse reactions such as hypertrichosis or renal dysfunction. In addition, the concentration of CsA present in the oral formulations is limited due to the drug's hydrophobic nature. A pharmaceutical anionic emulsion comprising CsA (0.05–0.2% w/w) and based on castor oil, Tween® 80, pemulen, glycerol, and water was developed and exhibited a bimodal distribution size in the submicron range with a potential zeta value of approximately −50 mV (34–36). The emulsion was demonstrated to be stable at room temperature for at least 18 months without showing any crystallization of CsA (35). After a single dose instillation of this topical CsA emulsion into the rabbit eye, the drug was found to penetrate into extraocular tissues at concentrations adequate for local immunomodulation while penetration into intraocular tissues was far less marked, and absorption into the blood was minimal. Furthermore, because castor oil was used in developing the emulsion dosage form, there may be additional benefits to patients with dry eye disease arising from the long ocular retention time of the emulsion vehicle. The castor oil droplets in the emulsion may form a lipid layer over the tear film, possibly reducing the evaporation of the limited natural tears produced by these patients while the emulsion remains in the eye. In a randomized, multicenter, double-masked, parallel group, dose response–controlled Phase 2 clinical study (12 weeks in 162 patients), conducted by Stevenson et al. (37) on human subjects, it was observed that ophthalmic emulsions containing different CsA concentrations of 0.05%, 0.1%, 0.2%, and 0.4% w/w were safe, well tolerated, and they significantly improved the ocular signs and symptoms of moderate to severe dry eye disease. In another published report (38), the same group observed in two Phase 3 pivotal trials, with 405 patients (study 002) and 472 patients (study 003), respectively, with a similar dose response–controlled clinical trial that the novel ophthalmic formulations of CsA at concentrations of 0.05% and 0.1% were safe and effective in the treatment of moderate to severe dry eye disease yielding improvements in both objective and subjective measures. A much larger Phase 3 study of CsA topical ophthalmic emulsions conclusively showed that CsA is practically undetectable in the blood of patients treated with these ophthalmic preparations, and therefore the emulsion treatment is safe and probably does not exert any systemic side effects (39). However it is generally agreed that devising valid efficacy endpoints is a major problem in clinical trials for KCS, as no single objective or subjective endpoint is specific for diagnosing the disease. Indeed, it needs to be pointed out that subjective symptoms, such as itchy or dry eyes, often do not correlate with objective signs, such as staining techniques to evaluate the cornea and conjunctiva or the Schirmer Tear Test, which measures aqueous production, with or without anesthesia. Therefore, a large European multicenter clinical trial has been conducted in a similar complementary way to evaluate efficacy and safety of topical CsA in the treatment of moderate to severe KCS (40). As a part of this clinical study, impression cytology (IC) specimens were taken in a large series of patients as a tertiary complementary test, to investigate the expression of immune-related markers by the conjunctival epithelium, confirm the presence of inflammation in KCS, and monitor these markers throughout a six-month treatment with CsA. A previously validated method of flow cytometry in image cytometry (48,49) was used to quantify the percentage of conjunctival cells expressing various inflammation- and apoptosis-related antigens, human leukocyte antigen-DR class II antigens, CD40, CD40 ligand, F, and the apoptotic marker APO2.7, and to objectively measure their levels of expression (48,49). Results at baseline, before the randomized treatment, showed the highest levels of HLA DR expression and to a lesser extent of the other markers in the conjunctival epithelium in KCS, both in Sjögren's (an autoimmune disorder characterized by destruction of the lacrimal gland) and non-Sjögren's syndrome eyes (41–43). Furthermore, the effects on these markers following the topical administration of two concentrations (0.05% and 0.1% w/w) of negatively charged CsA topical emulsions and of their vehicle were monitored using flow cytometry in IC specimens repeatedly obtained during the six-month treatment clinical trial (40). Specimens were processed and analyzed in a masked manner by flow cytometry, using monoclonal antibodies (mAb) directed to HLA DR, CD40, CD40 ligand, FAS, and the apoptotic marker APO2.7. Topical CsA strikingly reduced HLA DR and to a lesser extent other inflammatory and apoptotic markers, whereas the vehicle used as a control tear substitute had almost no effect. It is not surprising, therefore, that treatment with tear substitute (Refresh Lubricant Eye Drops, Allergan, Inc., Irvine, California, U.S.) alone is often insufficient to moderate the disease, and immunomodulatory therapy may be needed to eliminate the inflammatory cycle. It was already mentioned in the introduction of this chapter that the CsA ophthalmic emulsion and a blank emulsion were released to the markets for dry eye treatment. Furthermore, an interesting and very elaborate review recently described and summarized the main pharmaceutical systems and devices for topical and intraocular delivery of CsA to eye (50).

Cationic Emulsion

The residence time of the formulation depends both on the particle size and on the nature of the surface charge. To compare and show the influence of emulsion droplet surface charge on the indomethacin ocular tissue distribution, cationic and anionic emulsions were prepared based on MCT oil (33). The Lipoid E80–stearylamine–poloxamer 188 and Lipoid E80–deoxycholic acid–poloxamer 188 emulsifier combinations were respectively used for making the emulsion droplets positively and negatively charged. Indocollyre (0.1%) served as a commercial control eye drop formulation. An ocular pharmacokinetics study in rabbits was conducted for both the emulsions and the control eye drops. At a fixed instillation volume of 50 μL, it was found that whatever the formulations, the highest concentration of indomethacin is achieved in the cornea followed by conjunctiva, sclera-retina, and aqueous humor. However, the positively charged emulsion provided significantly higher drug levels than the control solution and negatively charged emulsion only in the aqueous humor and sclera-retina. In addition, the enhanced penetration effect cannot be attributed to the slight increase in viscosity of the emulsions (two cPs) as compared to the marketed collyre solution because there was a clear difference in behavior between the positively and negatively charged emulsion formulations, although both types of emulsions presented the same viscosity. Calvo et al. (24) observed in confocal scanning microscope experiments that the colloidal carriers including the submicron emulsion are able to penetrate into the epithelial cells without causing damage to the cell membrane. This evidence suggests that the incorporation of colloidal particles and oil droplets into the corneal epithelia is mediated probably through an endocytic mechanism. It could, therefore, be anticipated that the endocytic effect is more pronounced with the positively charged emulsion than with the negatively charged emulsion owing to the negatively charged moieties of the cell membranes. Indeed, the positively charged emulsion markedly enhanced the penetration of indomethacin through the different ocular tissues by strongly adhering to the cornea and increasing the residence time as a result of the electrostatic interaction between the oil droplets and the cornea. Therefore, again independent ex vivo studies (33) were carried out to demonstrate the spreading properties of positively charged blank emulsions onto the isolated rabbit cornea compared to saline and negatively charged emulsions (Table 4).

From Table 4, it appears that the positively charged emulsion has a contact angle ~50% lower than that of the negatively charged emulsion. In addition, the spreading coefficient of the positively charged emulsion on the cornea is four times higher than that of the negatively charged emulsion. Almost spontaneous spreading occurs when the positively charged emulsion is placed on the cornea because of the electrostatic attraction between the negatively charged cornea and the positively charged droplets of the emulsion. This difference is due to the presence of the positive charge conferred by the stearylamine, because all other ingredients of the positively and negatively charged emulsions are the same.

Table 4 Wettability Results Using Young-Dupre and Spreading Coefficient Equations [$\gamma_{cv} - \gamma_{cL} = \gamma_{LV} \cos\theta$ and $S = \gamma_{sv} - (\gamma_{LV} + \gamma_{CL})$, respectively]

Phases	Average contact angle	Average γ_{CL} (mN/m)	Spreading coefficient (mN/m)
Cornea–air	–	67.5	–
Cornea–saline	70	43	−47
Cornea–blank positively charged emulsion	21.2	34.7	−2.4
Cornea–blank negatively charged emulsion	38	35.6	−8.6

Source: From Ref. 33.

To substantiate and strengthen the influence of vehicle surface charges (positive vs. negative) on the therapeutic potential of ocular tissue disorder, additional studies were carried out.

The objective of Klang et al. (32,51) was to examine if piroxicam, an NSAID with free radical scavenger properties, could relieve rabbit corneal ulceration induced by alkali burns. Due to aqueous solubility limitations, there is neither an aqueous intravenous nor an ocular marketed preparation of piroxicam. Therefore, it is considered the most appropriate NSAID candidate to be incorporated into a positively charged submicron emulsion delivery system. Two formulations were tested: a formulation with piroxicam and another formulation without piroxicam. The oil selected to make emulsion formulation was MCT. The values of zeta potential were +41 mV for the blank emulsion and +36.19 mV for the piroxicam emulsion. Ophthalmic topical administration of 0.1% piroxicam-loaded positively charged emulsion relieved rabbit corneal ulceration induced by alkali burns more effectively than did the controls (blank emulsion, piroxicam solution, or saline). In addition, while piroxicam in aqueous solution inhibited the epithelization regeneration process of the eye, piroxicam incorporated into an emulsion did not interrupt this process. This is probably due to piroxicam localization in the internal oil phase of the emulsion, which prevented piroxicam from interfering with the epithelial healing process.

To further elucidate the influence of emulsion surface charge on ocular tissue distribution, CsA was incorporated successfully into a positively and negatively charged submicron emulsion delivery system (44,45). The positively and negatively charged emulsions containing CsA were prepared principally following the respective compositions as described in the indomethacin emulsions but instead of MCT oil, castor oil was selected because it dissolved the highest drug amount per unit volume. It is therefore possible to design o/w emulsion with low oil content while still dissolving the required therapeutic dose of CsA. In addition, the viscosity and the refractive index of the submicron emulsion formulation were measured and found to be 1.5 and 1.345 cPs, respectively, close to the viscosity and the refractive index of water indicating that the emulsion is virtually suitable for ocular use. Physicochemical characterizations such as particle size (220–240 nm), creaming rate, and drug content suggested that the submicron emulsion formulation was able to withstand thermic and mechanical shocks while showing virtually no sign of drug nanoprecipitation or crystallization within the emulsion upon storage at 4°C, 25°C, and 37°C over a period of up to six months. On the other hand, although the CsA-loaded negatively charged submicron emulsion prepared based on deoxycholic acid had a zeta potential value of −40 mV, it was not stable after a few weeks of storage at 4°C. A remarkable fact arising from the results of in vivo study in rabbits is that the positively charged emulsion provided higher CsA levels than the negatively charged emulsion formulation. In comparison to the negatively charged emulsion, the CsA-loaded positively charged emulsion enhanced the drug bioavailability in the conjunctiva and cornea target tissues (Table 5) (5). This is probably due to the interaction between the positive charge of the emulsion vehicle surface and the negative charge of the corneal and conjunctival surface at physiological pH.

In addition, the ocular surface seems to serve not only as a barrier to drug penetration but also as a reservoir for CsA because the therapeutic concentration level was nevertheless maintained in the conjunctiva and cornea even after completion of the study at 480 minutes following positively charged emulsion instillation. Furthermore, because of its high molecular weight, CsA should poorly permeate through the tight junction epithelial surface of the cornea. The fact that high levels of drug were detected in the cornea while the aqueous humor concentration was

Table 5 Ocular Pharmacokinetics Parameters Obtained Following Instillation of One Single 50 μL of CsA-Loaded Emulsions in Rabbit Eyes

	T_{max}(min)		C_{max}(μg/g)		$AUC_{0-8\ hrs}$ (μg min g^{-1})	
Tissues	Emuls (−)	Emuls (+)	Emuls (−)	Emuls (+)	Emuls (−)	Emuls (+)
Cornea	60	120	0.68 ± 0.30	1.02 ± 0.44	143.1	328.95
Conjunctiva	15	60	0.64 ± 0.67	2.29 ± 1.78	206.18	373.58
Iris	180	180	0.22 ± 0.09	0.25 ± 0.16	70.43	90.53
Sclera-retina	60	60	0.14 ± 0.16	0.25 ± 0.30	23.78	44.63
Aqueous humor	120	120	0.07 ± 0.03	0.09 ± 0.02	14.10	15.00

Emuls (−), negatively charged emulsion; Emuls (+), positively charged emulsion.
Abbreviation: CsA, cyclosporine A; AUC, area under the curve.
Source: From Ref. 45.

relatively low indicates that an endocytic mechanism of penetration is likely to occur, rendering the corneal epithelial layer a drug reservoir. This is also supported by the well-known stromal resistance to the diffusion of lipophilic drugs. The reservoir effect, as observed after instillation of the positively charged emulsion, has important clinical consequences because, following an initial loading dose, the ocular surface tissues may act to supply a therapeutic quantity of drug to other tissues over an extended period of time. Associated with poloxamer 188 and phospholipids, a cationic primary amine, stearylamine, has been used to obtain all of the above-described cationic lipid emulsion effects. Moreover, the stability and ocular tolerance following topical instillation into the eye of these cationic lipid emulsion vehicles were investigated (47,52,53). The promising results obtained with CsA-loaded cationic lipid emulsion paved the way for the formulation to obtain approval recently from regulatory authorities to undergo Phase I clinical study for the free drug cationic lipid emulsion (46).

MICROEMULSIONS

A detailed referenced listing of microemulsions and multiple emulsions investigated as ophthalmic vehicles is shown in Table 6. Vandamme (10) reviewed the recent development and future challenges of microemulsions as ocular drug delivery systems. Unfortunately, in this review, submicron emulsion and microemulsion were viewed together despite significant structural and composition differences between these two systems.

Significant contributions in the field of pharmaceutical microemulsions have been provided by the research group led by Maria Rosa Gasco at the University of Turin, Italy (54–58,66). Two β-blockers namely, timolol and levobunolol, were incorporated as an ion pair complex with octanoic acid into two different microemulsion systems (54–56). Ion pair formation was found to increase the lipophilicity of these drugs and, subsequently, therefore increase the solubility of the formed complex within the internal oil phase of the microemulsions. This was confirmed by a partition coefficient study of these drugs (in their base forms) in the internal and external phases of the microemulsions through a lipoid artificial membrane.

Table 6 Micro- and Multiple Emulsions Investigated as Ophthalmic Vehicles

Emulsion type	Drug used	Indications	References
Microemulsion	Timolol	Glaucoma	54,55
Microemulsion	Levobunolol	Glaucoma	56
Microemulsion	Prednisone	Inflammation	57
Microemulsion	Nitrofurazone Phenylbutazone Prednisone Betamethasone Menadione	Infection and inflammation	58
Microemulsion	Atropine Chloramphenicol Indomethacin Diclofenac	Infection and inflammation	59–61
Microemulsion	Pilocarpine hydrochloride	Glaucoma	62
Multiple emulsion	Prednisolone	Inflammation	63,64
Multiple emulsion	Hydrocortisone	Inflammation	65

A linear relationship between the percentage of octanoic acid used and apparent oil/water partition coefficient was also obtained indicating the dependence of lipophilicity increment on the counterion concentration (octanoic acid). Therefore, Gasco et al. (55) selected a formulation as described in Table 7 and compared the pharmacological action of the microemulsion to the effects observed after administration of an aqueous solution containing timolol (1.4% volume of water) or timolol combined with octanoic acid (timolol/octanoate molar ratio 1:10, with 0.33% w/w) in vivo on rabbits.

The AUC values for timolol in aqueous humor following 50 μL instillation at every four minutes from the microemulsion and the ion pair solution (both in total 150 μL) were 3.5 and 4.2 times higher, respectively, than the AUC value observed for timolol solution alone (total 300 μL). Although the ocular bioavailability of the microemulsion does not reach the maximum, its pharmacokinetic profile is different from the aqueous solutions, and even the aqueous humor concentration of timolol was observed beyond the two-hour post-instillation time point. The data suggest

Table 7 Composition of Timolol Base-Loaded Microemulsion

Components	Amount (% w/w)	Function
Timolol base	2.60	Drug
Egg lecithin	28.70	Surfactant
Isopropylmyristate	11.70	Internal phase
Octanoic acid	4.70	Ion pairing agent
1-Butanol	14.90	Cosurfactant
Vitamin E	0.01	Antioxidant agent
Isotonic phosphate buffer (pH = 7.4)	Up to 100	External phase

Source: From Ref. 55.

the occurrence of a reservoir effect of the drug coupled with good corneal penetration properties of ion-paired timolol particularly dissolved in the internal phase of the microemulsion. Further investigation with a microemulsion containing another β-blocking agent, levobunolol, as an ion pair was carried out to evaluate the permeation through different membranes (including one artificial lipid membrane and hairless mouse skin) and the eye irritation in rabbits (56). The emulsions contained soy phosphatidylcholine (SPC) and sodium octyl phosphate or sodium cholate as surfactants and sodium hexanoate or monobutyrylglycerol as cosurfactant. The study confirmed the validity of the ion pair approach to increase the permeability of drugs through membranes simulating the corneal epithelium. Ion pair incorporation into the microemulsion system provided a delayed release effect, potentially useful in ocular delivery. Microemulsion prepared with SPC and sodium cholate proved nonirritating in rabbit eyes.

There are some reports detailing the delay in diffusion kinetics of microemulsified drugs in comparison to the corresponding conventional aqueous solution control using a hydrophilic artificial membrane (59–62). However, precautions must be taken when extrapolating such results obtained from diffusion cell experiments to the results in vivo following instillation into rabbit eyes owing to different physiological conditions. The volume in the donor compartment of the in vitro diffusion cell apparatus remains constant whereas in vivo the drainage eliminates part of the drug present in the aqueous phase and dilutes significantly the drug in the tears.

MULTIPLE EMULSIONS

Initially one important problem with multiple emulsions was their intrinsic instability. However, the stabilization of these formulations has been studied and improved (67). Few reports are available concerning the use of multiple emulsions as a vehicle for ocular topical use (63–65,67). Multiple emulsions containing prednisolone were prepared using a two step emulsification technique. The effect of various concentrations of hydrophilic and lipophilic emulsifiers (Tween 20 and Span 60) in multiple liquid paraffin–based emulsions on the drug bioavailability and duration of action as a controlled release formulation was studied following instillation into the rabbit's eye. The parameters of ocular activity such as AUC, time for maximum response, and half value duration were assessed for the o/w/o and w/o/w multiple emulsions (63,64). Although the presence of the hydrophilic emulsifier molecule in the multiple emulsions reduced ocular bioavailability, the increased concentration of lipophilic emulsifier enhanced the bioavailability considerably. Therefore the bioavailability of the drug can be controlled by the proper choice and concentration of the emulsifiers in the preparation of emulsions. Furthermore, the hydrophile–lipophile balance (HLB) values calculated for a surfactant mixture in a multiple emulsion were found to be reversely correlated to the values of the parameters of ocular activity. Decreasing the HLB value of the surfactant mixture in a w/o/w emulsion favors the bioavailability as well as the intensity and duration of drug action. In another report from the same research group, the effect of different emulsions (o/w, w/o/w, and o/w/o) containing hydrocortisone on IOP of rabbits was studied (65). The drug-loaded o/w emulsion showed a maximum bioavailability increase followed by the drug loaded in w/o/w– and o/w/o–multiple emulsions. According to the authors, o/w/o–multiple emulsions may however represent an optimal delivery system for hydrocortisone in terms of maximizing drug ocular bioavailability while minimizing animal intravariations.

NANOPARTICLES

Native Nanoparticles

Encapsulation of ocular active agents into a polymeric colloidal delivery vehicle can be achieved through several manufacturing techniques: emulsion polymerization, quasiemulsion solvent diffusion, solvent deposition, simple sorption or nanoprecipitation, desolvation of proteins, heat denaturation, ionic interaction gelation, and micellar and interfacial polymerization (68–72). Among these techniques, emulsion polymerization is the most commonly applied one as far as synthetic polymer-based nanoparticles for ophthalmic drug delivery are concerned (73). The mechanisms involved to prepare drug-loaded nanoparticles are solubilization, entrapment, encapsulation, and ionic or covalent linkage to the carrier (74). Natural and synthetic polymers (and inorganic inert substances) of choice include albumin, chitosan, calcium phosphonates, gelatin, poly(alkyl cyanoacrylates) (PACA), polyacrylamides, poly(epsilon-caprolactone) and polymethylmethacrylates. Table 8 is a list of nanoparticles developed so far through various manufacturing techniques for different ophthalmic therapeutic agents.

Hydrocortisone was loaded onto albumin nanoparticles by sorption (75). When compared to the two controls (one comprised of a 0.03% saturated drug solution, the other of a 0.2% micellar drug solution), hydrocortisone-loaded nanoparticles elicited a sustained drug transport through porcine cornea in the ex vivo study. The distribution of both 0.2% hydrocortisone preparations (nanoparticles and solution) was then evaluated under in vivo conditions in healthy and inflamed eyes of rabbits. In all tissues the level of drug was higher in the inflamed eye than in the healthy one due to increased cell permeability as a result of inflammatory processes. The application of nanoparticles led to lower hydrocortisone tissue concentrations than the reference solution due to the strong binding of hydrocortisone onto the particle system and the resulting slow release. An exception occurred with the reference solution in the conjunctiva in which less drug was found in the inflamed tissue than in the normal one, because enhanced lacrimation led to increased drug drainage. In contrast, the corresponding nanoparticle preparation was more efficiently retained at the inflamed conjunctiva than at the normal one. Consequently, in the inflamed eye, hydrocortisone-loaded nanoparticles enabled targeting to the precorneal area away from the inner segments of the eye.

A CsA-loaded chitosan nanoparticle was prepared using a modified ionic gelation technique (78). These nanoparticles had a mean size of 293 nm, a zeta potential of +37 mV, and high CsA association efficiency and loading (73% and 9%, respectively). In vitro release studies, performed under sink conditions in pure water, revealed a fast release during the first hour followed by a more gradual drug release during a 24-hour period. In vivo experiments showed that following topical instillation of 10 μL of various formulations (CsA nanoparticles and CsA solutions either in chitosan or in water) containing 320 μg/mL of CsA into rabbit eyes, drug concentrations in the cornea and conjunctiva were higher particularly for nanoparticles than for the other two controls. The increase in drug concentration was attributed to the electrostatic interaction occurring between the chitosan positive surface and the cornea/conjunctiva negative surface and to the colloidal nature of nanoparticles. A recent study from the same research group assessed the potential of drug-free chitosan nanoparticles for ocular drug delivery by investigating their interaction with the ocular mucosa in vivo and also their toxicity in conjunctival cell cultures (79). Fluorescent tagged nanoparticles were prepared by ionotropic gelation. The stability of the particles in the presence of lysozyme was investigated by determining the size and their interaction with mucin, through viscosity

Table 8 Biodegradable and Biocompatible Polymers and an Inert Substance–Based Nanoparticle for Ophthalmic Drug Delivery

Polymer/substance	Drug	Encapsulation technique	Indications	References
Albumin	Hydrocortisone	Sorption	Inflammation	75
Albumin	Pilocarpine	Sorption	Glaucoma	76
Calcium phosphate	7-OH-DPAT	Ionic reaction	Glaucoma	77
Chitosan	CsA	Ionic-interaction gelation	Dry eye syndrome	78,79
Gelatin	Pilocarpine and hydrocortisone	Desolvation	Glaucoma and inflammation	71
Poly(butyl cyanoacrylate)	Pilocarpine	Emulsion polymerization	Glaucoma	80
Poly(butyl cyanoacrylate)	Progesterone	Emulsion polymerization	Inflammation	81
Poly(butyl cyanoacrylate)	Amikacin	Emulsion polymerization	Infection	82
Poly(butyl cyanoacrylates)	Betaxolol	Emulsion polymerization	Glaucoma	83
Poly(hexyl cyanoacrylate)	Pilocarpine	Emulsion-polymerization	Glaucoma	84
Poly(ε-caprolactone)	Betaxolol	Emulsion polymerization	Glaucoma	85
Poly(ε-caprolactone)	Carteolol	Emulsion polymerization	Glaucoma	86
Poly(ε-caprolactone)	Indomethacin	Nanoprecipitation	Inflammation	87,88
Poly(methyl methacrylate)	Pilocarpine	Emulsion polymerization	Glaucoma	89
Eudragit RS100 and RL100	Flurbiprofen	Quasiemulsion solvent diffusion	Inflammation	90
Eudragit RS100	Ibuprofen	Quasiemulsion solvent diffusion	Inflammation	91,92
Eudragit RS100	Diflunisol	Quasiemulsion solvent diffusion	Inflammation	93
Eudragit RL100	Cloricromene	Quasiemulsion solvent diffusion	Inflammation	94,95
Polyethyl-2-cyanoacrylate with PEG	Acyclovir	Micellar polymerization with simple adsorption process	Infection	96
PLA with PEG	Acyclovir	Solvent deposition	Infection	97
Poly(ε-caprolactone) with hyaluronic acid	Drug free	Solvent deposition	Vehicle	98

Abbreviations: 7-OH-DPAT, 7-hydroxy-2-dipropyl-aminotetralin; CsA, cyclosporine A; PLA, poly-D,L-lactic acid; PEG, polyethylene glycol.

measurements of the mucin dispersion. The in vivo interaction of chitosan-fluorescent nanoparticles with the rabbit cornea and conjunctiva was analyzed by spectrofluorimetry and confocal microscopy. Their potential toxicity was assessed in a human conjunctival cell line by determining cell survival and viability. The results show that they were stable upon incubation with lysozyme and did not affect the viscosity of a mucin dispersion. In vivo studies showed that the amounts of chitosan fluorescent in the cornea and conjunctiva were significantly higher for chitosan-fluorescent nanoparticles than for a control chitosan-fluorescent solution, with these amounts being fairly constant for up to 24 hours. Confocal studies suggest that nanoparticles penetrate into the corneal and conjunctival epithelia. Cell survival at 24 hours after incubation with chitosan was high, and the viability of the recovered cells was near 100%. Thus, chitosan nanoparticles are promising vehicles for ocular drug delivery (79).

A biodegradable calcium phosphate nanoparticle containing 7-hydroxy-2-dipropyl-aminotetralin, a dopamine D_2/D_3 receptor agonist, was prepared following an ionic reaction between calcium chloride and dibasic sodium phosphate in sodium citrate medium (77). The size of particles ranged from 70 to 1075 nm with a majority of the bigger particles in the 701 nm range and the smaller particles in the 196 nm range. The IOP lowering effect of the nanoparticle associated drug and the drug dissolved in cellobiose solution was monitored in pigmented and nonpigmented rabbits eyes following topical application. In nonpigmented rabbits, a more pronounced and sustained IOP reduction was observed for the nanoparticulated drug compared to the free drug solution. Furthermore, the observed IOP lowering effect of the drug in solution was markedly decreased in pigmented rabbits in comparison to nonpigmented rabbits. However, the drug in nanoparticle form elicited a dose dependent pharmacological activity indicating that calcium phosphate nanoparticles act not only to achieve controlled and targetable drug delivery but also to prevent the drug binding to the anterior segment of the pigmented rabbit's eye.

Gelatin nanoparticles encapsulating hydrophilic (pilocarpine HCl) or hydrophobic (hydrocortisone) compounds as model drugs were produced using a desolvation method (71). The influence of a number of preparation parameters on the particle properties was investigated. For the pilocarpine HCl–loaded spheres, pH influence during particle preparation on the size was observed. Slightly negative zeta potential values were measured for all samples. In the case of pilocarpine HCl–loaded spheres, no influence of the gelatin type or the pH level was observed, which could be attributed to the shielding effect of ions present in the dispersion medium. When hydrocortisone was entrapped, a difference in zeta potential value between gelatin type A and gelatin type B particles was measured. To increase its aqueous solubility, hydrocortisone was complexed with two types of cyclodextrins: neutral (hydroxypropyl-β-cyclodextrin, HP-β-CD) and cationic (2-hydroxy-3-trimethyl-ammoniopropyl cyclodextrin, Cat. CD). The drug entrapment efficiencies for both drugs were lower but still ranged between 35% and 57%. Compared to the aqueous drug solutions, a sustained release for both drugs was observed when studied using a Franz type diffusion cell having dialysis membrane (molecular weight cut off of 10–12 kDa) to separate the donor and acceptor compartment. The release kinetics of pilocarpine HCl in distilled water is close to zero order, and no significant differences were measured among the various preparations. In the case of hydrocortisone, the release data suggest a difference in release rate depending on the type of cyclodextrin employed, indicating the occurrence of probable interaction between gelatin and cyclodextrins (71).

Hui and Robinson (81) and Wood et al. (84), in their pioneering work, demonstrated that ocular bioavailability and duration of therapeutic action of the drug could be improved significantly by entrapping the drug into mucoadhesive acrylate/alkylcyanoacrylate polymers to increase precorneal retention. These earlier observations support the idea that colloidal suspensions of the mucoadhesive PACA nanoparticles could increase the residence time of the drug in the tear film thereby prolonging the penetration of drugs into the ocular tissues. Mucoadhesive ocular colloidal drug delivery systems aim to improve bioavailability, reduce frequency of administration, and promote localization of drugs in specified regions. Indeed, PACA is conveniently used for the preparation of nanoparticles loaded with many drug candidates probably because of its mucoadhesive characteristics (80,82,83,85,99–104).

PACA can be synthesized by a dispersion polymerization process carried out in an aqueous phase at a low pH and employing a polymeric stabilizing agent. Drugs can be incorporated into the matrix of the nanoparticles or adsorbed onto the surface of the colloidal carrier depending on the sequence of addition of the drug, either before or after the polymerization process. Harmia et al. (80) demonstrated however that pilocarpine adsorbed onto nanoparticles induced longer miosis compared to drugs incorporated into the particles. Furthermore, the transport pathway of poly (butyl cyanoacrylate) nanoparticles' through the rabbit cornea and conjunctiva was studied using fluorescent microscopy (105). Rhodamine 6G or propidium iodide was used as the fluorescent laser dye. Incubation of the fluorescent-tagged nanoparticles with excised rabbit cornea and conjunctiva for about 30 minutes in standard perfusion cells showed an uptake of nanoparticles possibly by an endocytosis process or lysis of the cell wall by nanoparticles metabolic degradation products. Nevertheless, despite these previous positive results, the actual potential of the PACA nanoparticles is limited because of the likely induced disruption caused to the corneal epithelium cell membrane. However, the use of poly(epsilon caprolactone) can possibly be considered because it did not elicit any adverse effect to the corneal epithelium cell membrane including disruption as it was observed in the case of PACA (105).

Poly (d, l-lactic-coglycolic acid) nanoparticles (50:50) containing either 6-coumarin or Texas red bovine serum albumin were used to investigate the characteristics and mechanisms of nanoparticle uptake in primary cultured rabbit conjunctival epithelial cells (106). The effect of size was studied using various average particle sizes (100 nm, 800 nm, and 10 μm). The effect of cytochalasin D (an actin filament inhibitor), nocodazole (a microtubule inhibitor), and metabolic inhibitors (sodium azide or dinitrophenol) on nanoparticle uptake was also investigated. The 100-nm particles had the highest uptake in cells in comparison with 800-nm and 10-μm particles. In addition to being size-dependent, PLGA nanoparticle internalization was concentration- incubation time-, and temperature-dependent in primary cultured rabbit conjunctival epithelial cells. Maximum uptake of nanoparticles at 37°C occurred at two-hours incubation time, and particles uptake was saturable over a 0.1 to 4 mg/mL concentration range. The uptake of encapsulated Texas red bovine serum albumin in primary cultured rabbit conjunctival epithelial cells at four hours was 28% higher than free bovine serum albumin application. The two-hour uptake of 0.5 μg/mL nanoparticles in cells decreased significantly when cells were preincubated with 0.1 μg/mL cytochalasin D but not in the presence of nocodazole despite a noticeable decrease in uptake pattern. These findings suggest that PLGA nanoparticle uptake in primary cultured rabbit conjunctival epithelial cells occurs most likely by adsorptive-type endocytosis (106). Internalization appeared to be energy-dependent and possible, to inhibit by vesicle formation blocker cytochalasin D. It also displayed a characteristic punctate distribution pattern under confocal microscopy following uptake. Because clathrin protein was implicated in the endocytosis of PLGA nanoparticles in vascular smooth muscle cells, the same research group sought to examine the expression of clathrin and caveolin-1 in primary cultured rabbit conjunctival epithelial cells and to determine whether they play a role in PLGA nanoparticle endocytosis (107). PLGA (50:50) nanoparticles (100 nm in diameter) containing 6-coumarin (fluorescent marker, 0.05% w/v) were used in this molecular mechanistic study. The pharmacological treatments were meant to disrupt formation of clathrin-coated vesicles (hypertonic challenge and intracellular K^+ depletion) and caveolae (nystatin and filipin) on apical uptake of nanoparticles in primary cultured rabbit conjunctival epithelial cells. Transferrin was chosen as a marker for clathrin-dependent endocytosis from the basolateral aspect, whereas cholera toxin B subunit was chosen as a marker for caveolae-mediated endocytosis. The staining pattern of nanoparticles in cells was compared with that of clathrin heavy chain (HC) and caveolin-1 under fluorescent confocal microscopy to examine possible colocalization using clathrin HC and caveolin-1 mouse mAb. Two pairs of primers were designed (based on conserved regions of clathrin and caveolin-1 gene in different species) to amplify a 744-bp and 152-bp fragment of clathrin HC and caveolin-1 gene, respectively. Reverse transcription–polymerase chain reaction (RT–PCR) to detect the message for clathrin HC and caveolin-1 was performed using total RNA prepared from freshly isolated primary cultured rabbit conjunctival epithelial cells. Human embryonic kidney (HEK) 293 cells were used as the positive control for clathrin gene expression, whereas rabbit heart muscle and HEK cells were used as the positive control for caveolin-1 gene expression. The RT–PCR products were separated using a 2% agarose gel electrophoresis. Western blot analysis was performed to detect the expression of both clathrin and caveolin-1 proteins in rat cerebral endothelial cells (RCECs) using mouse mAbs. "Henrietta Lacks" cells (HeLa) and A431 epidermoid cells were used as positive controls. The effect of transfection of primary cultured rabbit conjunctival epithelial cells (using Lipofectamine 2000™ reagent) with specific antisense oligonucleotides designed against the rabbit clathrin isoform on clathrin protein expression and PLGA nanoparticle uptake was investigated. Apical uptake of nanoparticles in primary cultured rabbit conjunctival epithelial cells was decreased by 45% and 35%, respectively, as a result of K^+ depletion and hypertonic media treatments. Likewise, the same treatments significantly decreased the basolateral uptake of fluorescein isothiocyanate (FITC-) transferrin by 50%. In contrast, nystatin and filipin had no effect on apical uptake of nanoparticles and cholera toxin B subunit in RCECs, suggesting a lack of involvement of caveolae in the internalization of these two agents. Confocal microscopy showed fluorescent staining of the cell membrane in the presence of clathrin mAb, but not in the presence of caveolin-1 mAb, with partial overlap with a nanoparticle staining pattern. RT–PCR confirmed the presence of the clathrin HC gene, but not the caveolin-1 gene, in primary cultured rabbit conjunctival epithelial cells as indicated by a 744-bp fragment of the gene. However, caveolin-1 gene was detected in other rabbit tissues such as the epithelium of the cornea, trachea, and heart muscle, as indicated by a 152-bp fragment of the gene. Western blot analysis revealed a clathrin HC band (180 kDa) in primary cultured rabbit conjunctival epithelial cells and HeLa cells. However, caveolin-1 protein (22 kDa) was not detected in primary cultured rabbit conjunctival epithelial cells, but was detected in A431 cells. Transfection of RCECs with antisense oligonucleotide directed against clathrin HC resulted in knockdown of the clathrin HC protein in a concentration dependent manner. However, clathrin HC protein knockdown had no effect on apical uptake of nanoparticles in primary cultured rabbit conjunctival epithelial cells. Overall, endocytosis of nanoparticles in primary cultured rabbit conjunctival epithelial cells occurs mostly independently of clathrin- and caveolin-1-mediated pathways. In addition, the gene and protein expression of clathrin HC, but not caveolin-1, was identified in rabbit conjunctival epithelial cells (107).

Polymethylmethacrylate and Eudragit® RS100 and RL100 in the form of nanoparticles have been widely studied for ocular applications because they were shown to be well tolerated in animals through the modified Draize test (89–92,108).

Eudragits (RS100 and RL100) are copolymers of poly(ethylacrylate, methyl-methacrylate, and chlorotrimethyl-aminoethyl methacrylate), containing an amount of quaternary ammonium groups between 4.5% to 6.8% and 8.8% to 12% for RS100 and RL100, respectively. Because both resins are insoluble at physiological pH values and capable of swelling, they represent good materials for making the dispersion of drugs through a modified quasiemulsion solvent diffusion technique (109). During such a swelling in water, these resins are able to produce positive charges (+35 mV for RS100 polymeric suspension in water) on the nanoparticle surfaces due to the protonation of quaternary ammonium groups of these polymers. In addition, it is known that acidic compounds, such as ibuprofen, interact with eudragits by means of electrostatic binding between their carboxyl moiety and the quaternary ammonium groups of the polymer (93,110). In spite of the drug:polymer electrostatic binding during nanoparticle suspension preparation by the modified quasiemulsion solvent diffusion technique, the suspended particles maintained the positive charge suggesting that the drug molecules are mainly dispersed within the polymeric matrix, besides the drug fraction that is adsorbed onto the polymers' surface which cannot neutralize completely the available positively charged moieties. Therefore, an improvement in the pharmacological action would be expected following application of eudragits-based suspension formulations into the eye possibly via an electrostatic adhesion between the negatively charged corneal epithelium cell membranes and positively charged suspended polymeric nanoparticles. Indeed, the above-described publications demonstrated the advantages of eudragits formulations in comparison to classical drug solutions.

Eudragit RL100 nanosuspensions loaded with cloricromene [a synthetic coumarin derivative, AD6, which has recently been proposed for the treatment of uveitis (111)] were prepared, characterized, and optimized (94). In aqueous solution, the native esteric form of this agent is quite unstable with time due to the slow conversion into an insoluble acid form. Thus, AD6 incorporation into the Eudragit RL100 polymer nanoparticle delivery system is expected to prevent the conversion of this agent into the insoluble acid form. Furthermore, the nanoparticle delivery system is also expected to ensure a slow and prolonged release of this agent at the intraocular level, especially at the aqueous humor, which could improve the therapeutic effects of AD6. Following single dose instillation of AD6-loaded RL100 nanoparticles or the reference AD6 aqueous solution into the rabbit eye, both of the formulations showed rapid conversion of AD6 into its active acid form, MET, in the aqueous humor (95). However, the nanoparticle formulation ensured a much higher concentration of MET compared with the aqueous solution, showing a similar T_{max} value (60 minutes) but with a 1.7-fold increase of the C_{max} values (0.68 ± 0.07 µg/mL and 1.16 ± 0.33 µg/mL for aqueous solution and RL100 nanoparticle formulation, respectively) and $AUC_{0\text{-}240}$ values (107.7 ± 10.32 µg min/mL and 190.85 ± 24.11 µg min/mL for aqueous solution and RL100 nanoparticle formulation, respectively).

Surface Modified Nanoparticles

Keeping in mind the well-known fact that polyethylene glycol (PEG) has the ability to modulate the interfacial properties of the colloidal carriers, the application potentialities of such surface modified systems both in terms of mucoadhesion and improved drug ocular permeation were investigated recently on PACA nanoparticles loaded with acyclovir (97). Through a simple adsorption process, the PEG coating onto the PACA particles was achieved. It has to be added that although the colloidal carrier provided a higher ocular bioavailability of the drug following topical instillation into the rabbit eye, no significant difference was observed between coated and uncoated NSs. The plausible reason is that the PEG moieties may have interacted with PACA particle surface very weakly.

To firmly anchor the PEG together with the participating carrier surface, poly-d,l-lactic acid (PLA) is used as the polymeric matrix for NS colloidal suspensions containing acyclovir (97). PEG-coated PLA NSs were prepared by using pegylated 1,2-distearoyl-3-phosphatidylethanolamine, which could firmly anchor PEG moieties to the surface of the colloidal carrier by inserting the lipophilic phospholipid part into the hydrophobic core of the PLA colloidal matrix. The ocular tolerability of PLA NS was evaluated by a modified Draize test. PEG-coated and uncoated PLA NSs showed a sustained acyclovir release and were highly tolerated by the eye. Both types of PLA preparations were able to increase the aqueous humor levels of acyclovir and to improve the ocular pharmacokinetic profile, but the efficacy of the PEG-coated NSs was significantly higher than that of the simple PLA particles. Thus, PEG-coated PLA NSs can be proposed as a potential ophthalmic delivery system for the treatment of ocular viral infections (herpes simplex keratitis), allowing a better compliance and an increased intraocular level of the antiviral agent.

A bioadhesive polysaccharide, hyaluronic acid, was proposed for coating poly(epsilone-caprolactone) (PECL) NSs to enhance the bioavailability of drugs in ocular delivery (98). In various studies the mucoadhesive properties of hyaluronic acid in ocular delivery were demonstrated but, contrary to chitosan, no permeability enhancing property has yet been described (76,112,113). A recent review details the coating methods using various other polysaccharides such as dextran, pectin, heparin, etc. on the polymeric- and iron oxide-cored nanoparticles (70). With the exception of hyaluronic acid, the ocular applicability of other polysaccharide-decorated nanoparticles has not yet been reported despite their potential in this area.

Presently, Piloplex® and Glaupex® are the two pilocarpine-loaded poly(methylmethacrylate-acrylic acid) and PACA nanoparticles, respectively, which have been developed for ophthalmic drug delivery (104). Nevertheless, the commercial development of nanoparticles remains limited for various reasons including sterilization-induced stability and slightly improved performance of nanoparticles produced following a long manufacturing process compared to the simple techniques involved in making classical aqueous solutions.

NANOCAPSULES

Native NCs

In this area, the research group of Professor Alonso (114) of Spain has done a significant amount of work. Table 9 lists the NCs used for ophthalmic drug delivery purposes.

NCs developed earlier utilized poly(isobutylcyanoacrylate) polymer to encapsulate betaxolol chlorohydrate (83). These NCs induced only a marginal decrease in the IOP compared with the activity obtained with the commercial form and by other carriers. More promising results have been shown with another antiglaucomatous drug, pilocarpine (122). Furthermore, when incorporated into a Pluronic® gel, the contact of the poly(isobutylcyanoacrylate) NCs containing pilocarpine with the absorbing tissue was prolonged further leading to sustained release. This approach yielded a significant increase in the bioavailability of the drug and should undergo further development in the future. However, in view of the issue of corneal

Table 9 NCs as Ocular Drug Delivery Carrier

Polymers	Drug	Encapsulation technique	Indications	References
Poly(isobutyl cyanoacrylates)	Betaxolol	Emulsion-polymerization	Glaucoma	83
Poly(ε-caprolactone)	Metipranolol	Interfacial deposition	Glaucoma	115
Poly(ε-caprolactone)	Carteolol	Interfacial deposition	Glaucoma	86
Poly(ε-caprolactone)	Indomethacin	Interfacial deposition	Inflammation	87,88
Poly(ε-caprolactone)	Fluorescence	Interfacial deposition	Vehicle	116
Poly(ε-caprolactone)	CsA	Interfacial deposition	Dry eye syndrome	117,118
Chitosan	Indomethacin	Interfacial deposition	Inflammation	119–121

Abbreviations: NCs, nanocapsules; CsA, cyclosporine A.

epithelium cell membrane disruption caused by PACA, metipranolol containing NCs were prepared from PECL polymer (105,115). It has also been reported that the corneal epithelium cells specifically have taken up PECL NCs without damaging the cell membrane (116). In an attempt to elucidate the positive behavior of PECL NCs, Marchal-Heussler et al. (85) compared particles prepared by interfacial polymerization of isobutylcyanoacrylate monomers with the ones formed from poly(lactide) and PECL by nanoprecipitation. They have examined the corneal penetration of betaxolol associated to PECL NCs and nanoparticles and observed that the penetration was favorable for the NCs with respect to the nanoparticles. A similar comparison between NCs and nanoparticles was made using PECL and carteolol combination, respectively, as a polymer and antiglaucomatous drug candidate (86). NCs displayed a better antiglaucoma effect than nanoparticles because the drug was entrapped in the oil core of the carrier, thus more readily available to the eye. Additionally, in comparison to aqueous eye drops, a decline in the cardiovascular effects was produced particularly after applying NCs indicating the marked reduction of the undesired noncorneal absorption of the drug.

Calvo et al. (87,88) showed an interesting comparative study regarding the efficacy of three colloidal carriers: nanoparticles, NCs, and nanoemulsions. PECL and indomethacin were, respectively, selected to represent polymer and drug models. Irrespective of the difference in the inner structure or chemical composition between these colloidal systems, the in vitro corneal penetration of encapsulated indomethacin was found to be very similar (more than threefold that of the commercial aqueous drops) for the three formulations investigated. Moreover, 300% increase in drug ocular bioavailability was observed after instillation of colloidal systems into the rabbit eye in comparison with the value obtained for a commercial solution. The similar behavior of the three colloidal carriers suggests that any of their specific ingredients (PECL, lecithin, and oil) must have acted as a penetration enhancer or an endocytotic stimulator. Thus, all the three systems were considered "authentic colloidal carriers" to improve the transcorneal permeation of drug.

In another study, NCs composed of an oil core (Migliol 840) surrounded by a PECL coat were prepared following an interfacial deposition technique and evaluated as potential vehicles for the topical ocular application of CsA (117). NCs had a mean size in the range of 210–270 nm, a negative zeta potential value of between −55 and −60 mV, and a maximum loading capacity of 50% (CsA/PECL ratio). However, the NCs produced at this high CsA load displayed a thick spongeous polymer coating around the oily nanodroplets. Following the application of CsA-loaded NCs and oil solution of CsA into the rabbit eye, up to five times higher drug levels at the cornea and a fourfold increase in the AUC value were observed for the nanoencapsulated form of the drug as compared to the oil solution. The efficacy of this topical formulation has also been observed on a penetrating keratoplasty rejection model in the rat (118). Le Bourlais et al. (123) also proposed an alternative preparation of cyclosporin NCs made from PACA dispersed in a poly(acrylic acid) gel. This formulation considerably reduced the toxicity of PACA on the cornea and promoted absorption of the drug. As stated already, the other possible ocular delivery systems and devices investigated for CsA ocular topical delivery were compiled comprehensively in a review (50).

Cationic Polymer-Coated NCs

Because the ocular therapy of topically applied drugs is influenced by the residence time of formulations in the negatively charged cornea, NCs coated with cationic polymers were investigated (119). Polysaccharide-like chitosan and polyaminoacid like poly-L-lysine (PLL) were intentionally selected as coating polymers to confer a cationic surface onto PECL NCs while indomethacin served as the drug candidate in this study. Following the preparation of PECL NCs by interfacial deposition technique, the PLL was directly adsorbed onto preformed NCs whereas the chitosan was included in the NCs formation medium. Uncoated NCs showed a zeta potential value of -39.9 ± 0.3 mV, the PLL-coated formulation had 27.9 ± 3.5 mV, and the chitosan-coated system possessed 37.1 ± 1.8 mV. Indomethacin in vitro release, carried out under sink conditions using a bulk-equilibrium reverse dialysis technique at 37°C in 0.1 M phosphate buffer of pH 7.4, from NCs was independent of the investigated carrier composition, and 85% of the drug diffused out from the colloidal systems within two hours. The in vivo studies in rabbits showed that in comparison to commercial eye drops, uncoated, PLL- and chitosan-coated PECL NCs significantly increased the concentration of indomethacin in the cornea and aqueous humor. However, with respect to uncoated NCs, the ability of coating polymers used to enhance the ocular bioavailability of the drug was substantially different. While the chitosan coating increased the drug ocular bioavailability twice, the PLL coating failed to increase when compared to uncoated PECL NCs. The observed advantageous effect of chitosan-coated PECL NCs was associated probably to the mucoadhesive characteristic nature of the polycationic polysaccharide polymer, which was absent in the PLL case. In another study from the same group, they found that coating the negatively charged surface of PECL NCs with cationic polymers could prevent the particle degradation caused by the adsorption of lysozyme, a positively charged enzyme found in tear fluid (120). A recent study from the same group reported the influence of different surface coatings such as chitosan and PEG onto the Rhodamine 6G-loaded PECL NCs in their interaction with the ocular mucosa ex vivo and in vivo (121). The ex vivo studies revealed that the developed NCs, especially the chitosan-coated ones, enhanced the penetration of the encapsulated dye through the cornea. Both of the coated NCs were able to enter the corneal epithelium by a transcellular pathway, but the penetration rate was dependent on the coating composition, as revealed by the confocal laser scanning microscopy examinations. The PEG coating onto the NCs accelerates the transport across the whole epithelium, whereas the chitosan coating favors the retention of the NCs in the superficial layers of the epithelium. In vivo data regenerated by the same authors corroborated these ex vivo results indicating that the surface composition of colloidal drug carriers affects their biodistribution in the eye.

FUTURE DIRECTIONS IN OCULAR DRUG DELIVERY USING LIPID AND POLYMERIC COLLOIDAL CARRIERS

Following ocular topical application, lipid and polymeric colloidal carriers were shown to enhance the absorption of a wide range of therapeutic agents demonstrating the capacity and versatility of this approach. Micro- and multiple emulsions may have some advantages in ocular topical delivery of water soluble drugs. Although o/w submicron emulsions with anionic surface show improvement in ocular bioavailability of water insoluble drugs, cationization of submicron emulsions holds promising application for drug absorption enhancement and for ferrying compounds across cell membranes. However, the importance and the full potential of the cationization strategy in emulsions are yet to be explored. Furthermore, cationic emulsions provide an interesting opportunity to be used as drug delivery ocular vehicles for numerous therapeutic compounds that can range in size from small molecules to macromolecules such as oligonucleotides. In comparison to native nanoparticles, surface modified nanoparticles improve the efficacy of treatment in ocular diseases through increased bioavailability of encapsulated drugs. Due to the presence of both oil core and polymer coat in their structure, NCs can be considered a promising carrier for ocular drug delivery. It is anticipated that more efforts will be invested in the future to develop new methods of preparation and expand the applications of NCs in the ophthalmology market (124).

REFERENCES

1. Le Bourlais CL, Acar L, Zia H, Sado PA, Needham T, Leverge R. Ophthalmic drug delivery systems—recent advances. Prog Retin Eye Res 1998; 17:35–58.
2. Kaur IP, Kanwar M. Ocular preparations: the formulation approach. Drug Dev Ind Pharm 2002; 28:473–493.
3. Kaur IP, Garg A, Singla AK, Aggarwal D. Vesicular systems in ocular drug delivery: an overview. Int J Pharm 2004; 269:1–14.
4. Prausnitz MR, Noonan JS. Permeability of cornea, sclera and conjunctiva: a literature analysis for drug delivery to the eye. J Pharm Sci 1998; 87:1479–1488.
5. Tamilvanan S, Benita S. The potential of lipid emulsion for ocular delivery of lipophilic drugs. Eur J Pharm Biopharm 2004; 58:357–368.
6. Lipinski CA. Avoiding investment in doomed drugs. Is poor solubility an industry-wide problem? Curr Drug Dis 2001; 17–19.
7. Lipinski C. Poor aqueous solubility—an industry wide problem in drug discovery. Am Pharm Rev 2002; 5:82–85.
8. Ding S. Recent development in ophthalmic drug delivery. Pharm Sci Tech Today 1998; 1:328–335.
9. Lambert G, Fattal E, Couvreur P. Nanoparticulate systems for the delivery of antisense oligonucleotides. Adv Drug Deliv Rev 2001; 47:99–112.

10. Vandamme TF. Microemulsions as ocular drug delivery systems: recent developments and future challenges. Prog Retin Eye Res 2002; 21:15–34.
11. Magenheim B, Levy MY, Benita S. A new in vitro technique for the evaluation of drug release profile for colloidal carriers—ultrafiltration technique at low pressure. Int J Pharm 1993; 94:115–223.
12. Santos Magalhaes NS, Fessi H, Puisieux F, Benita S, Seiller M. An in-vitro release kinetic examination and comparative evaluation between submicron emulsions and polylactic acid nanocapsules of clofibride. J Microencapsulation 1995; 12:195–205.
13. Leino M, Urtti A. Recent developments in anti-glaucoma drug research. In: Reddy IK, ed. Ocular Therapeutics and Drug Delivery. Lancaster, PA: Technomic, 1996: 245.
14. Kulkarni SP, Banumathi B, Thoppil SO, Bhogte CP, Vartak D, Amin PD. Dry eye syndrome. Drug Dev Ind Pharm 1997; 23:465.
15. Sieg JW, Robinson JR. Vehicle effects on ocular drug availability II: evaluation of pilocarpine. J Pharm Sci 1977; 66:1222–1228.
16. Sieg JW, Robinson JR. Vehicle effects on ocular drug availability III: shear-facilitated pilocarpine release from ointments. J Pharm Sci 1979; 68:724–728.
17. Puglisi G, Spampinato S, Giammona G, et al. Effect of different ointment bases on ocular anti-inflammatory activity of 4-biphenylacetic acid in the rabbit. Drug Des Deliv 1989; 5:341–352.
18. Danielsson I, Lindman B. The definition of microemulsion. Colloid Surf 1981; 3:391–392.
19. Muchtar S, Almog S, Torracca MT, Saettone MF, Benita S. A submicron emulsion as ocular vehicle for delta-8-tetrahydrocannabinol: effect on intraocular pressure in rabbits. Ophthal Res 1992; 24:142–149.
20. Aviv H, Friedman D, Bar-Ilan A, Vered M. Submicron emulsions as ocular drug delivery vehicles. PCT WO 94/05298, 1995.
21. Anselem S, Beilin M, Garty N. Submicron emulsion as ocular delivery system for adaprolol maleate, a soft β-blocker. Pharm Res 1993; 10:S205.
22. Melamed S, Kurtz S, Greenbaum A, Haves JF, Neumann R, Garty N. Adaprolol maleate in submicron emulsion, a novel soft β-blocking agent, is safe and effective in human studies. Invest Ophthalmol Vis Sci 1994; 35:1387–1390.
23. Muchtar S, Abdulrazik M, Frucht-Pery J, Benita S. Ex vivo permeation study of indomethacin from a submicron emulsion through albino rabbit cornea. J Control Rel 1997; 44:55–64.
24. Calvo P, Vila-Jato JL, Alonso MJ. Comparative in vitro evaluation of several colloidal systems, nanoparticles, nanocapsules, and nanoemulsions, as ocular drug carriers. J Pharm Sci 1996; 85:530–536.
25. Muchtar S, Benita S. Emulsions as drug carriers for ophthalmic use. Colloids Surf A: Physicochem Eng Aspects 1994; 91:181–190.
26. Naveh N, Muchtar S, Benita S. Pilocarpine incorporated into a submicron emulsion vehicle causes an unexpectedly prolonged ocular hypotensive effect in rabbits. J Ocular Pharmacol 1994; 10:509–520.
27. Naveh N, Weissman C, Muchtar S, Benita S, Mechoulam R. A submicron emulsion of HU-211, a synthetic cannabinoid, reduces intraocular pressure in rabbits. Graefes Arch Clin Exp Ophthalmol 2000; 238:334–338.
28. Zurowska-Pryczkowska K, Sznitowska M, Janicki S. Studies on the effect of pilocarpine incorporation into a submicron emulsion on the stability of the drug and the vehicle. Eur J Pharm Biopharm 1999; 47:255–260.
29. Sznitowska M, Janicki S, Zurowska-Pryczkowska K, Mackiewicz J. In vivo evaluation of submicron emulsions with pilocarpine: the effect of pH and chemical form of the drug. J Microencapsulation 2001; 18:173–181.
30. Beilin M, Bar-Ilan A, Amselem S. Ocular retention time of submicron emulsion (SME) and the miotic response to pilocarpine delivered in SME. Invest Ophthalmol Vis Sci 1995; 36:S166.

31. Garty N, Lusky M, Zalish M, et al. Pilocarpine in submicron emulsion formulation for treatment of ocular hypertension: a phase II clinical trial. Invest Ophthalmol Vis Sci 1994; 35:2175.
32. Klang SH, Siganos CS, Benita S, Frucht-Pery J. Evaluation of a positively-charged submicron emulsion of piroxicam in the rabbit corneum healing process following alkali burn. J Control Rel 1999; 57:19–27.
33. Klang S, Abdulrazik M, Benita S. Influence of emulsion droplet surface charge on indomethacin ocular tissue distribution. Pharm Devel Tech 2000; 5:521–532.
34. Acheampong AA, Shackleton M, Tang-Liu DD-S, Ding S, Stern ME, Decker R. Distribution of cyclosporin A in ocular tissues after topical administration to albino rabbits and beagle dogs. Curr Eye Res 1999; 18:91–103.
35. Ding S, Olejnik O. Cyclosporin ophthalmic oil/water emulsions. Formulation and characterization. Pharm Res 1997; 14:S41.
36. Ding S, Tien W, Olejnik O. Nonirritating emulsions for sensitive tissue. US Patent 5474979 (to Allergan Inc, USA), 12 December 1995.
37. Stevenson D, Tauber J, Reis BL. Efficacy and safety of cyclosporin A ophthalmic emulsion in the treatment of moderate-to severe dry eye disease: a dose-ranging randomized trial. The Cyclosporin A Phase 2 Study Group. Ophthalmology 2000; 107:967–974.
38. Sall K, Stevenson OD, Mundorf TK, Reis BL. Two multicenter, randomized studies of the efficacy and safety of cyclosporin ophthalmic emulsion in moderate-to-severe dry eye disease. CsA Phase 3 Study Group. Ophthalmology 2000; 107:631–639.
39. Small DS, Acheampong A, Reis B, et al. Blood concentrations of cyclosporin A during long-term treatment with cyclosporin A ophthalmic emulsions in patients with moderate to severe dry eye disease. J Ocular Pharmacol Ther 2002; 18:411–418.
40. Brignole F, Pisella PJ, De Saint-Jean M, Goldschild M, Goguel A, Baudouin C. Flow cytometric analysis of inflammatory markers in KCS: 6-month treatment with topical cyclosporin A. Invest Ophthalmol Vis Sci 2001; 42:90–95.
41. Brignole F, De Saint-Jean M, Goldschild M, Goguel A, Baudouin C. Flow cytometric analysis of inflammatory markers in conjunctival epithelial cells of patients with dry eyes. Invest Ophthalmol Vis Sci 2000; 41:1356–1363.
42. Kunert KS, Tisdale AS, Stern ME, Smith JA, Gipson IK. Analysis of topical cyclosporin treatment of patients with dry eye syndrome: effect on conjunctival lymphocytes. Arch Ophthalmol 2000; 118:1489–1496.
43. Turner K, Pflugfelder SC, Ji Z, Feuer WJ, Stern M, Reis BL. Interleukin-6 levels in the conjunctival epithelium of patients with dry eye disease treated with cyclosporine ophthalmic emulsion. Cornea 2000; 19:492–496.
44. Tamilvanan S, Khoury K, Gilhar D, Benita S. Ocular delivery of cyclosporin A. I. Design and characterization of cyclosporin A-loaded positively-charged submicron emulsion. STP Pharma Sci 2001; 11:421–426.
45. Abdulrazik M, Tamilvanan S, Khoury K, Benita S. Ocular delivery of cyclosporin A. II. Effect of submicron emulsion's surface charge on ocular distribution of topical cyclosporin A. STP Pharma Sci 2001; 11:427–432.
46. Etheridge J. Novagali launches phase 1 trial, negotiates second-round funds. Bioworld Int 2003; 8:4.
47. Wehrle P, Korner D, Benita S. Sequential statistical optimization of a positively-charged submicron emulsion of miconazole. Pharm Dev Technol 1996; 1:97–111.
48. Brignole F, De Saint-Jean M, Goldschild M, Goguel A, Baudouin C. Expression of Fas-Fas ligand antigens and apoptotic marker APO2.7 by the human conjunctival epithelium: positive correlation with class II HLA DR expression in inflammatory ocular surface disorders. Exp Eye Res 1998; 67:687–697.
49. Baudouin C, Brignole F, Becquet F, Pisella PJ, Goguel A. Flow cytometry in impression cytology specimens: a new method for evaluation of conjunctival inflammation. Invest Ophthalmol Vis Sci 1997; 8:1458–1464.

50. Lallemand F, Felt-Baeyens O, Besseghir K, Behar-Cohen F, Gurny R. Cyclosporine A delivery to eye: a pharmaceutical challenge. Eur J Pharm Biopharm 2003; 56:307–318.
51. Benita S. Prevention of topical and ocular oxidative stress by positively charged submicron emulsion. Biomed Pharmacother 1999; 53:193–206.
52. Klang S, Frucht-Pery J, Hoffman A, Benita S. Physicochemical characterization and acute toxicity evaluation of a positively-charged submicron emulsion vehicle. J Pharm Pharmacol 1994; 46:986–993.
53. Klang SH, Baszkin A, Benita S. The stability of piroxicam incorporated in a positively charged submicron emulsion for ocular administration. Int J Pharm 1996; 132:33–44.
54. Gallarate M, Gasco MR, Trotta M. Influence of octanoic acid on membrane permeability of timolol from solutions and from microemulsions. Acta Pharm Technol 1988; 34:102–105.
55. Gasco MR, Gallarate M, Trotta M, Bauchiero L, Gremmo E, Chiappero O. Microemulsions as topical delivery vehicles: ocular administration of timolol. J Pharm Biomed Anal 1989; 7:433–439.
56. Gallarate M, Gasco MR, Trotta M, Chetoni P, Saettone MF. Preparation and evaluation in vitro of solutions and o/w microemulsions containing levobunolol as ion pair. Int J Pharm 1993; 100:219–225.
57. Gasco MR, Gallarate M, Pattarino F. On the release kinetics of prednisone from oil in water microemulsions. Pr Ed (II Farmaco Ed) 1988; 10:325–330.
58. Trotta M, Gasco MR, Morel S. Release of drugs from oil–water microemulsions. J Control Rel 1989; 10:237–243.
59. Siebenbrodt I, Keipert S. Versuche zur entwicklung und charakterisierung ophthalmologisch verwendbarer tensidhaltiger mehrkomponentensysteme. Pharmazie 1991; 46:435–438.
60. Siebenbrodt I, Keipert S. Poloxamer-systeme als potentielle ophthalmika. Teil 1: Viskose polymertensidlosungen. PZ Wiss 1992; 137:135–141. Chem Abstr 1992; 117:178192m.
61. Siebenbrodt I, Keipert S. Poloxamer-systems as potential ophthalmics. II. Microemulsions. Eur J Pharm Biopharm 1993; 39:25–30.
62. Hasse A, Keipert S. Development and characterization of microemulsions for ocular application. Eur J Pharm Biopharm 1997; 43:179–183.
63. Safwat SM, Kassem MA, Attia MA, El-Mahdy M. The formulation–performance relationship of multiple emulsions and ocular activity. J Control Rel 1994; 32:259–268.
64. Kassem MA, Safwat SM, Attia MA, El-Mahdy M. Influence of the phase volume ratio of multiple emulsions on the ocular activity of prednisolone. STP Pharma 1995; 5: 309–315.
65. Kassem MA, Attia MA, Safwat SM, El-Mahdy M. Preparation and evaluation of hydrocortisone multiple emulsions in rabbit's eye. Pharm Ind 1994; 56:584–588.
66. Gasco MR. Microemulsions in the pharmaceutical field: perspective and applications. In: Solans C, Kunieda H, eds. Industrial Applications of Microemulsions. New York: Marcel Dekker, 1997:97–122.
67. Vaziri A, Warburton B. Improved stability of w/o/w multiple emulsions by addition of hydrophilic colloid components in the aqueous phases. J Microencapsulation 1995; 12:1–5.
68. Gref R. Surface-engineered nanoparticles as drug carriers. In: Baraton MI, ed. Synthesis, Functionalization and Surface Treatment of Nanoparticles. Stevenson Ranch, California, USA: American Scientific Publishers, 2002:233–256.
69. Vauthier C, Dubernet C, Fattal E, Pinto-Alphandary H, Couvreur P. Poly(alkylcyanoacrylates) as biodegradable materials for biomedical applications. Adv Drug Deliv Rev 2003; 55:519–548.
70. Lemarchand C, Gref R, Couvreur P. Polysaccharide-decorated nanoparticles. Eur J Pharm Biopharm 2004; 58:327–341.
71. Vandervoort J, Ludwig A. Preparation and evaluation of drug-loaded gelatin nanoparticles for topical ophthalmic use. Eur J Pharm Biopharm 2004; 57:251–261.

72. Müller-Goymann CC. Physicochemical characterization of colloidal drug delivery systems such as reverse micelles, vesicles, liquid crystals and nanoparticles for topical administration. Eur J Pharm Biopharm 2004; 58:343–356.
73. Zimmer A, Kreuter J. Microspheres and nanoparticles used in ocular delivery systems. Adv Drug Deliv Rev 1995; 16:61–73.
74. Kreuter J. Evaluation of nanoparticles as drug delivery systems. I. Preparation methods. Pharm Acta Helv 1983; 58:196–209.
75. Zimmer AK, Maincent P, Thouvenot P, Kreuter J. Hydrocortisone delivery to healthy and inflamed eyes using a micellar polysorbate 80 solution or albumin nanoparticles. Int J Pharm 1994; 110:211–222.
76. Zimmer A, Chetoni P, Saettone MF, Zerbe H, Kreuter J. Evaluation of pilocarpine-loaded albumin particles as controlled drug delivery systems for the eye. II. Co-administration with bioadhesive and viscous polymers. J Control Rel 1995; 33:31–46.
77. Chu TC, He Q, Potter DE. Biodegradable calcium phosphate nanoparticles as a new vehicle for delivery of potential ocular hypotensive agent. J Ocular Pharmacol Ther 2002; 18:507–514.
78. De Campos AM, Sánchez A, Alonso MJ. Chitosan nanoparticles: a new vehicles for the improvement of the delivery of drugs to the ocular surface. Application to cyclosporin A. Int J Pharm 2001; 224:159–168.
79. De Campos AM, Diebold Y, Carvalho EL, Sánchez, Alonso MJ. Chitosan nanoparticles as new ocular drug delivery systems: in vitro stability, in vivo fate, and cellular toxicity. Pharm Res 2004; 21:803–810.
80. Harmia T, Kreuter J, Speiser P, Boye T, Gurny R, Kubis A. Enhancement of the miotic response of rabbits with pilocarpine-loaded polybutylcyanoacrylate nanoparticles. Int J Pharm 1986; 33:187–193.
81. Hui H, Robinson J. Ocular delivery of progesterone using a bioadhesive polymer. Int J Pharm 1985; 26:203–213.
82. Losa C, Calvo P, Castro E, Vila-Jato JL, Alonso MJ. Improvement of ocular penetration of amikacin sulphate by association to poly(butylcyanoacrylate) nanoparticles. J Pharm Pharmacol 1991; 43:548–552.
83. Marchal-Heussler L, Maincent P, Hoffman M, Spittler J, Couvreur P. Antiglaucomatous activity of betaxolol chlorohydrate sorbed onto different isobutylcyanoacrylate nanoparticle preparations. Int J Pharm 1990; 58:115–122.
84. Wood RW, Li VHK, Kreuter J, Robinson JR. Ocular deposition of poly-hexyl-2-cyano[3-^{14}C] acrylate nanoparticles in the albino rabbit. Int J Pharm 1985; 23:175–183.
85. Marchal-Heussler L, Fessi H, Devissaguet JP, Hoffman M, Maincent P. Colloidal drug delivery systems for the eye. A comparison of the efficacy of three different polymers: polyisobutylcyanoacrylate, polylactic-co-glycolic acid, poly-epsilon-caprolactone. STP Pharm Sci 1992; 2:98–104.
86. Marchal-Heussler L, Sirbat D, Hoffman M, Maincent P. Poly(epsilon-caprolactone) nanocapsules in carteolol ophthalmic delivery. Pharm Res 1993; 10:386–390.
87. Calvo P, Vila Jato JL, Alonso MJ. Comparitive in vitro evaluation of several colloidal systems, nanoparticles, nanocapsules, and nanoemulsions, as ocular drug carriers. J Pharm Sci 1996; 85:530–536.
88. Calvo P, Alonso MJ, Vila-Jato JL, Robinson JR. Improved ocular bioavailability of indomethacin by novel ocular drug carriers. J Pharm Pharmacol 1996; 48:1147–1152.
89. Harmia T, Speiser P, Kreuter J. Optimization of pilocarpine loading onto nanoparticles by sorption procedures. Int J Pharm 1986; 33:45–54.
90. Pignatello R, Bucolo C, Spedalieri G, Maltese A, Puglisi G. Flurbiprofen-loaded acrylate polymer nanosuspensions for ophthalmic application. Biomaterials 2002; 23:3247–3255.
91. Pignatello R, Bucolo C, Ferrara P, Maltese A, Puleo A, Puglisi G. Eudragit RS100® nanosuspensions for the ophthalmic controlled delivery of ibuprofen. Eur J Pharm Sci 2002; 16:53–61.

92. Bucolo C, Maltese A, Puglisi G, Pignatello R. Enhanced ocular anti-inflammatory activity of ibuprofen carried by an Eudragit RS100® nanoparticle suspension. Ophthal Res 2002; 34:319–323.
93. Pignatello R, Amico D, Chiechio S, Giunchedi P, Spadaro C, Puglisi G. Preparation and analgesic activity of Eudragit RS100 microparticles containing diflunisol. Drug Deliv 2001; 8:35–45.
94. Pignatello R, Bucolo C, Maltese A, Maugeri F, Puglisi G. Eudragit® RL 100 nanosuspensions loaded with cloricromene for ophthalmic application: characterization and optimization of formulative parameters. Proceedings of the International Meeting on Pharmaceutics, Biopharmaceutics and Pharmaceutical Technology, March 15–18, 2004, Nuremberg, pp. 855–856.
95. Bucolo C, Maltese A, Maugeri F, Busà B, Puglisi G, Pignatello R. Eudragit RL 100 nanoparticle system for the ophthalmic delivery of cloricromene. J Pharm Pharmacol 2004; 56:841–846.
96. Fresta M, Fontana G, Bucolo C, Cavallaro G, Giammona G, Puglisi G. Ocular tolerability and in vivo bioavailability of poly(ethylene glycol) (PEG)-coated polyethyl-2-cyanoacrylate nanosphere encapsulated acyclovir. J Pharm Sci 2001; 90:288–297.
97. Giannavola C, Bucolo C, Maltese A, et al. Influence of preparation conditions on acyclovir-loaded poly-d,l-lactic acid nanospheres and effect of PEG coating on ocular drug bioavailability. Pharm Res 2003; 20:584–590.
98. Barbault-Foucher S, Gref R, Russo P, Guechot J, Bochot A. Design of polyepsilon-caprolactone nanospheres coated with bioadhesive hyaluronic acid for ocular delivery. J Control Rel 2002; 83:365–375.
99. Zimmer A, Mutschler E, Lambrecht G, Mayer D, Kreuter J. Pharmacokinetic and pharmacodynamic aspects of an ophthalmic pilocarpine nanoparticle-delivery system. Pharm Res 1994; 11:1435–1442.
100. Diepold R, Kreuter J, Himber J, et al. Comparison of different models for the testing of pilocarpine eyedrops using conventional eyedrops and a novel depot formulation (nanoparticles). Graefe's Arch Clin Exp Ophthalmol 1989; 227:188–193.
101. Harmia T, Speiser P, Kreuter J. A solid colloidal drug delivery system for the eye: encapsulation of pilocarpin in nanoparticles. J Microencapsulation 1986; 3:3–12.
102. Kreuter J. Nanoparticles and liposomes in ophthalmic drug delivery; biopharmaceutical, technological and clinical aspects. In: Saettone MF, Bucci G, Speiser P, eds. Ophthalmic Drug Delivery. Fidia Res. Series. Vol. 11. Padua: Liviana Press, 1987:101–106.
103. Zimmer A, Kreuter J. Microspheres and nanoparticles used in ocular drug delivery systems. Adv Drug Deliv Rev 1995; 16:61–73.
104. Kreuter J. Nanoparticles as bioadhesive ocular drug delivery systems. In: Lenaerts V, Gurny R, eds. Bioadhesive Drug Delivery Systems. Boca Raton, Florida: CRC Press, 1990:203–212.
105. Zimmer A, Kreuter J, Robinson JR. Studies on the transport pathway of PBCA nanoparticles in ocular tissues. J Microencapsulation 1991; 8:497–504.
106. Qaddoumi MG, Ueda H, Yang J, Davda J, Labhasetwar V, Lee VHL. The characteristics and mechanisms of uptake of PLGA nanoparticles in rabbit conjunctival epithelial cell layers. Pharm Res 2004; 21:641–648.
107. Qaddoumi MG, Gukasyan HJ, Davda J, Labhasetwar V, Kim KJ, Lee VHL. Clathrin and caveolin-1 expression in primary pigmented rabbit conjunctival epithelial cells: role in PLGA nanoparticle endocytosis. Mol Vis 2003; 9:559–568.
108. Pignatello R, Bucolo C, Puglisi G. Ocular tolerability of Eudragit RS100 and RL100 nanosuspensions as carriers for ophthalmic controlled drug delivery. J Pharm Sci 2002; 91:2636–2641.
109. Eudragit Technical Sheets. Weiterstadt, Germany: Röhm Pharma GmbH.
110. Jenquin MR, Liebowitz SM, Sarabia RE, McGinity JW. Physical and chemical factors influencing the release of drugs from acrylic resin films. J Pharm Sci 1990; 79:811–816.

111. Bucolo C, Cuzzocrea S, Mazzon E, Caputi AP. Effects of cloricromene, a coumarin derivative, on endotoxin-induced uveitis in Lewis rats. Invest Ophthalmol Vis Sci 2003; 44:1178–1184.
112. Saettone MF, Chetoni P, Torracca MT, Burgalassi S, Giannaccini B. Evaluation of muco-adhesive properties and in vivo activity of ophthalmic vehicles based on hyaluronic acid. Int J Pharm 1989; 51:203–212.
113. Durrani AM, Farr SJ, Kellaway IW. Influence of molecular weight and formulation pH on the precorneal clearance rate of hyaluronic acid in the rabbit eye. Int J Pharm 1995; 118:243–250.
114. Alonso MJ. Nanomedicines for overcoming biological barriers. Biomed Pharmacother 2004; 58:168–172.
115. Losa C, Marchal-Heussler L, Orallo F, Vila Jato JL, Alonso MJ. Design of new formulations for topical ocular administration: polymeric nanocapsules containing metipranolol. Pharm Res 1993; 10:80–87.
116. Calvo P, Thomas C, Alonso MJ, Vila-Jato JL, Robinson JR. Study of the mechanism of interaction of poly(-caprolactone) nanocapsules with the cornea by confocal laser scanning microscopy. Int J Pharm 1994; 103:283–291.
117. Calvo P, Sánchez A, Martinez J, et al. Polyester nanocapsules as new topical ocular delivery systems for cyclosporin A. Pharm Res 1996; 13:311–315.
118. Ramon JJ, Calonge M, Gomez S, et al. Efficacy of topical cyclosporin-loaded nanocapsules on keratoplasty rejection in the rat. Curr Eye Res 1998; 17:39–46.
119. Calvo P, Vila-Jato JL, Alonso MJ. Evaluation of cationic polymer-coated nanocapsules as ocular drug carriers. Int J Pharm 1997; 153:41–50.
120. Calvo P, Vila-Jato JL, Alonso MJ. Effect of lysozyme on the stability of polyester nanocapsules and nanoparticles: stabilization approaches. Biomaterials 1997; 18:1305–1310.
121. De Campos AM, Sánchez A, Gref R, Calvo P, Alonso MJ. The effect of a PEG versus a chitosan coating on the interaction of drug colloidal carriers with the ocular mucosa. Eur J Pharm Sci 2003; 20:73–81.
122. Desai SD, Blanchard J. Pluronic® F127-based ocular delivery systems containing biodegradable polyisobutylcyanoacrylate nanocapsules of pilocarpine. Drug Deliv 2000; 7:201–207.
123. Le Bourlais CA, Chevanne F, Turlin B, et al. Effect of cyclosporin A formulations on bovine corneal absorption: ex vivo study. J Microencapsulation 1997; 14:457–467.
124. Heurtault B, Saulnier P, Pech B, Proust JE, Benoit JP. A novel phase inversion-based process for the preparation of lipid nanocarriers. Pharm Res 2002; 19:875–880.

18

The Use of Drug-Loaded Nanoparticles in Cancer Chemotherapy

Jean-Christophe Leroux
University of Montreal, Centre ville, Montreal, Quebec, Canada

Angelica Vargas, Eric Doelker, Robert Gurny, and Florence Delie
School of Pharmaceutical Sciences (EPGL), University of Geneva, Quai Ernest-Ansermet, Geneva, Switzerland

INTRODUCTION

Currently, the limiting factor in cancer chemotherapy is the lack of selectivity of anticancer drugs toward neoplastic cells. Generally, rapidly proliferating cells such as those of the bone marrow or the gastrointestinal tract are affected by the cytotoxic action of these drugs. This results in a narrow therapeutic index of most anticancer compounds. Furthermore, the emergence of resistant cell sublines during the chemotherapeutic treatment may require the use of higher doses of anticancer drugs or the elaboration of dosing protocols combining different anticancer drugs. This, in return, may enhance the toxicity of the treatment. To decrease the toxicity and to enhance the selectivity of existing drugs, many drug delivery systems have been developed in recent years (1,2). These systems include soluble drug–polymer conjugates, polymeric micelles, nanoparticles (NP), liposomes, and microparticles (3–8). NP are attracting increased attention and growing interest for drug targeting, because they can be easily prepared with well-defined biodegradable polymers (9). In the first studies, the rationale behind using NP for cancer therapy was based on the fact that certain neoplastic cells have been found to exhibit an enhanced endocytotic activity (10). Recent studies have shown that encapsulating chemotherapeutic agents in NP can also influence pharmacokinetic parameters and the biodistribution of the drug. Therefore, it is expected to help tumor accumulation with a resulting limitation of the adverse effects. High concentration in neoplastic tissues can be obtained based on peculiarities of the vascular system surrounding tumors. Indeed, the vascular system of tumors is highly disorganized and presents an enhanced permeability due to the tumor-increased needs in oxygen and nutrients. This phenomenon called the enhanced permeability and retention (EPR) effect provides an interesting gateway for small NP to penetrate more readily into tumoral sites (11–13). This retention effect is further potentialized by an impaired lymphatic drainage reducing carrier clearance from the tumor.

The present chapter covers the developments and progresses made in the delivery of anticancer drugs associated with polymeric NP and the interactions of the latter with cancer cells and tissues. Throughout the chapter, the term NP refers to submicronic carriers (<1 μm) consisting of either nanospheres (NS), matrix-type systems, or nanocapsules (NC), vesicular systems (14). Because many submicronic carriers have been in the past named microparticles, it is possible that some studies dealing with the present topic but using the term microparticles have been unintentionally omitted.

IN VITRO UPTAKE OF NP BY TUMORAL CELLS

The interaction of NP with cultured cells may be either passive or mediated via determined moieties such as monoclonal antibodies (MAb) which can bind specifically to targeted cells.

Interaction of Uncoated NP with Cultured Cells

In vitro studies involving NP were initially performed to determine if malignant cells were able to internalize NP to a significant extent (Table 1). However, targeting drug carriers to homogeneous cell cultures is of limited value because in this case the carriers are directly added to the target cells. In these conditions, no anatomical barriers prevent the NP from acceding to target cells, and no extensive clearance of the drug is encountered (11,15). Furthermore, it still remains unclear if an increase of drug efficiency in vivo resulting from the binding of the drug to NP is related to a greater uptake of the drug by tumoral cells in vitro (16–18). Coculture (culture of different cell lines in the same dish) has recently been developed to investigate more deeply the relationship between the uptake by one cell line and the effect on a cancer cell line (19). So far, in vitro experiments have led to inconsistent results regarding any increase of efficacy of NP-bound anticancer drugs. In fact, the free drug remained often more or as effective as its bound counterpart (20–23). These results may be explained by the limited access of the bound drug to the internal cell compartments. Astier et al. (24) have reported that doxorubicin-loaded poly(alkyl methacrylate) (DXR PAMA) NS were more effective in inhibiting human U-937 leukemia cell growth than free DXR. However, U-937 cells belong to a monocyte-like cell line with marked phagocytic properties.

On the other hand, in vitro studies help in understanding the mechanisms of action of bound anticancer drugs. By coupling covalently DXR to polyglutaraldehyde NS, Tökés et al. (22) and Rogers et al. (21) have demonstrated that DNA intercalation of DXR was not essential for its pharmacological action and that the cell membrane was a major site of action of these DXR NS. Moreover, in vitro experiments have revealed the role of the NP polymer in the cytotoxic action of drug-loaded NP. In another study where it was observed that the binding of DXR to poly(alkyl cyanoacrylate) (PACA) NS could increase its efficiency it was also found that the polymers used contributed to the cytotoxicity (25). Although, as stated by the authors, the polymer itself could act as a second chemotherapeutic agent, it has to be pointed out that colloidal carriers have a propensity to concentrate in the reticuloendothelial system (RES) and thus could possibly damage the latter, especially if they are loaded with highly cytotoxic agents (9,11). These observations

Table 1 Uptake and Activity of NP In Vitro

Year	Drug	Carrier	Polymer	Size (nm)	Cells	Comments	References
1976	6-MP	NS	Albumin	500–1200	Hela and KB cells Human glioblastoma	Phagocytosis of a large number of NS by all cell types	(163)
1979	None	NS	Albumin Gelatin	500	EMT6 mouse mammary tumor WEH1-3 mouse myelomonocytic SP-1 rat squamous cells	No uptake of albumin NS Gelatin NS were taken up only by EMT6 and SP-1 cells	(164)
1982	DXR	NS	PGL	450	CCRF-CEM human leukemia L1210 Mouse leukemia 180 mouse sarcoma	DXR covalently bound to NS. Bound drug retained its full activity for sensitive cell lines. Resistant cell sublines became over 10 times more sensitive to bound drug	(22)
1983	DXR	NS	PGL	450	DXR-R-rat liver cell line RLC Carcinogen-altered DXR-R-hepatocytes. Normal rat hepatocytes	Bound drug was more effective than free drug against resistant cell sublines. Free drug equally effective in killing normal hepatocytes at low concentrations, more effective at high concentrations	(21)
1983	DXR	NS	PGL	450	Variable human cancer specimens	For lung and ovarian cancers, more than half of tumor specimens demonstrated an increased sensitivity to bound drug	(165)
1983	Dactino-mycin	NS	PBCA	81	Mouse peritoneal macrophages	Bound drug uptake was more important when 20% fetal calf serum was present in culture medium	(166)
1984	DXR	NS	PGL	450	L1210 mouse leukemia	Bound drug incubated with resistant cells was 12 times more active than free drug in reducing LD_{50}	(30)

(*Continued*)

Table 1 Uptake and Activity of NP In Vitro (*Continued*)

Year	Drug	Carrier	Polymer	Size (nm)	Cells	Comments	References
1984	5-FU DXR	NS	Albumin	<1000	B16 mouse melanoma MMTV mouse breast adenocarcinoma	Potency of bound drug lower than that of free drug	(20)
1987	None	NS	PACA	100–500	Osteogenic sarcomas T and KN Ewing's sarcoma Malignant fibrous histiocytoma Human fetal lung fibroblasts	Polymers with shorter side chains more toxic to the cells Very high amounts of polymer used No convincing evidence of phagocytosis	(18)
1988	DXR	NS	PAMA	250–370	U-937 human leukemia	Bound drug three times more effective than free drug. Decreased efflux of bound drug versus free drug from the cells	(24)
1988	None	NS	PAMA	300	Human hepatocytes Human hepatoma	Twofold increase in the uptake of NS by the hepatoma cells versus hepatocytes	(101)
1988	DXR	NS	PAMA	250–370	Rc-Pa renal carcinoma	Bound drug less effective than free drug	(167)
1989	DXR	NS	PIBCA PIHCA	190 342	DC3F Chinese hamster lung cell line	Bound drug 3.7–5.5 more cytotoxic than free drug for the sensitive cell line. In the resistant subline sensitivity was restored with bound drug and markedly improved with a mixture of free drug and unloaded NS	(25)
1989	DXR	NS	PAMA	250–311	Normal rat hepatocytes	Increase of uptake for bound drug	(168)
1989	MTP-Chol	NC	PIBCA PLA	183 336	Rat alveolar macrophages	The NS were able to activate the macrophages but high concentrations of polymer might inhibit macrophage functions	(50)

1991	Ce_6	NS	Polystyrene	1 μm	MGH-U1 human bladder carcinoma	50-fold higher uptake of Ce_6 when incorporated into NS than free Ce_6	(65)
						Ce_6–NS conferred at least a 10-fold enhancement of phototoxicity, and were retained for a longer period intracellularly than is free Ce_6	(63)
						Ce_6–NS-treated cells displayed light dose dependant morphologic alterations, which were not observed with free Ce_6	(64)
1991	DXR	NS	PIHCA	300	MCF7 human breast carcinoma	Bound drug less effective than free drug for sensitive cell lines but decreased LD_{50} by 150 for resistant sublines	(23)
1992	DXR	NS	PIHCA	300	MCF7 human breast carcinoma SKVBL1 human ovarian carcinoma K562 human erythroblastic leukemia P-388 mouse monzocytic leukemia LR73 Chinese hamster ovarian cells	In resistant cell lines the sensitivity was restored with bound drug. A mixture of unloaded NS and free drug was not more effective than free drug alone on MCF7 resistant cells	(31)
1992	MTP-Chol	NC	PLA	250	Rat alveolar macrophages	PIBCA NC more toxic than PLA NC. At low concentrations, bound drug more effective than free.	(27)
1993			PIBCA			MDP in stimulating the production of nitric oxide	(52)
1994	DXR	NS	PIBCA	230	P-388 mouse leukemia	Bound drug decreased 10-fold IC_{50} in the resistant subline. Part of the	(32)
			PIHCA			toxicity was due to the polymer itself	(33)
1994	DXR	NS	PIBCA	170	P-388 mouse leukemia	Increase in the uptake of bound drug	(34)

(Continued)

Table 1 Uptake and Activity of NP In Vitro (*Continued*)

Year	Drug	Carrier	Polymer	Size (nm)	Cells	Comments	References
1994	DXR	NS	PIHCA	200–300	C6 rat glioblastoma	In the resistant cell lines the sensitivity was partially or completely restored	(35)
1994	MTP-Chol	NC	PLA	n.m.	Rat alveolar macrophages	Increased cytostatic activity versus free MDP	(51)
1995	ZnPcS AlNc SiNc	NC	PIHCA Poloxamer-coated-PIHCA	180	V-79 Chinese hamster cells	$ZnPcS_2$ (amphiphilic derivative) activity was not significantly modified by incorporation into the NC, whereas $ZnPcS_4$ (hydrophilic derivative) was markedly more active when incorporated into the NC	(66)
1995	MTP-Chol	NC	PLA	200	RAW 264.7 mouse macrophages	At low concentrations, bound drug more effective than free MDP in stimulating the production of nitric oxide	(46)
1996	DXR	NP	Cholesterol-bearing pullulan	18	HeLa cells: uterocervical carcinoma cells	NP were less active than the free drug	(169)
1997	DXR	NS	PIHCA PIBCA	243 186	P-388: murine malignant leukemia cell line P-388/ADR: resistant cell line. Direct and indirect contact (by filter separation) between cells and NS	PIBCA associated DXR gave higher internalization rate that the free drug (15 folds in 6 hr), however, PIHCA loaded NS led to a similar accumulation as the free drug	(38)
1999	DXR	NS	PLGA–DXR conjugates	300	HepG2: human hepatoblastoma cell line	The released fraction from the nanoparticle was slightly less effective than the free DXR	(170)
1999	DXR	NS	PACA	300	P-388: murine malignant leukemia cell line P-388/ADR: resistant cell line J774.A1: macrophage-monocyte cell line	In sensitive cell, free and encapsulated DXR exhibited the same cytotoxicity. In resistant cells, free DXR had no effect whereas encapsulated DXR had the same effect versus regular cells	(37)

2000	DXR and/or cyclosporin A	NS	PACA	<300	P-388/ADR: mouse leukemia expressing the MDR gene	The best synergistic effect on resistant cells was obtained when both CyA and DXR were associated within a single particle	(36)
2000	DXR	NS	PLGA–DXR conjugates	200	HepG2: human hepatoblastoma cell line	Increased uptake with the NP, slightly lower IC_{50} with the NP	(171)
2000	DXR	NS	PACA	140	M5076 and J774.A1 in coculture separated by a membrane	Particles incubated with macrophages were still able to affect the cell growth of M5076	(19)
2000	*m*-THPC	NC	PLA PLA-PEG Poloxamer-coatedPLA	180 ± 60	Macrophage-like J774 cells and HT29 human adenocarcinoma cells	Surface-modified NC showed a reduced uptake by phagocytic cells. Different photodamage was observed with NC	(68)
2001	DXR	NS	Chitosan	259	A375 human melanoma C26 murine colorectal carcinoma cell	No difference between DXR-loaded chitosan-TPP. Nanoparticles incorporating dextran sulphate and DXR solution. DXR complexed-chitosan nanoparticles had a decreased cytostatic activity	(172)
2001	Paclitaxel	NS	Albumin	150–200	IGROV-1 ovarian carcinoma cell line. IGROV-1/Pt1 resistant to cisplatin. A341 squamous cell of the cervix	Paclitaxel NP exhibit the same activity as the free drug with the ovarian derived cells. NP were more active on A341 expressing increased sensitivity to taxane	(156)
2002	Paclitaxel	NS	PLGA RG502 RG755	<200	NCI-H69 SCLC human small cell lung cancer cell line	After 168 hrs of incubation, encapsulated paclitaxel was more efficient than Cremophore EL formulation at 2.5 μg/mL. At 25 μg/mL, the efficacy was the same	(59)

(Continued)

Table 1 Uptake and Activity of NP In Vitro (*Continued*)

Year	Drug	Carrier	Polymer	Size (nm)	Cells	Comments	References
2002	5-FU	NS	Albumin	<700	HEp-2 cancer cell line	5-FU NP were more efficient than the free drug	(173)
2003	Tamoxifen or rhodamine	NS	PCL	100–300	MCF-7 estrogen positive breast cancer cells	Particles were readily internalized. Intracellular concentrations were only higher at the lowest dose when compared to the drug solution	(174)
2003	p-THPP	NS	PLA PLGA	sub 130	EMT-6 mouse mammary tumor	Phototoxicity of p-THPP depends on polymer nature. 50:50 PLGA-NS loaded with p-THPP induced at least three times higher PDT effect than free drug and were more toxic than other formulations	(71)
						Higher internalization of NS formulations relative to free drug	(69)
2003	HPPH	NS	Silica	30	UCI-107 and HeLa	Significant cell death after PDT was induced by HPPH NP. HPPH-polysorbate 80 micelles showed similar results, but not-loaded micelles exhibited some cell toxicity	(72)
2004	Paclitaxel	NS	PLGA	<600	HT-29 human colon cancer cell line	NP were three times more efficient than the commercial formulation	(60)

Abbreviations: NS, nanospheres; G-MP, 6-mercaptopurine; PGL, polyglutaraldehyde; DXR, doxorubicin; PBCA, poly(butyl cyanoacrylate); 5-FU, 5-fluorouracil; PACA, poly (alkyl cyanoacrylate; PAMA, poly(alkyl methacrylate); PIBCA, poly(isobutyl cyanoacrylate); PIHCA, poly(isohexyl cyanoacrylate); MTP-Chol, muramyl dipeptide-L-alanyl-cholesterol; PLA, poly(lactic acid); Ce_6, chlorine e_6; NC, nanocapsules; MDP, muramyl dipeptidel; ZnPcS, sulfonated zinc phthalocyanides; NP, nanoparticles; n.m., not mentioned; TPP, sodium tripolyphosphate; PLGA, poly(lactic-co-glycolic acid); HPPH, 2-devinyl-2-(1-hexgloxyethyl) pyropheophorbide; PDT, photodynamic therapy: *m*-THPC, meta-tetra(hydroxy phenyl)chlorin.

were further confirmed with the use of DXR-NP to overcome the multidrug resistance in P-388 cells. These studies have recently been reviewed by Vauthier et al. (26) and will be discussed in this chapter. Selection of appropriate polymers remains a key issue because it was shown that even well established biodegradable polymers such as poly(lactic acid) (PLA) cannot be considered as inert vehicles (27,28). Regarding PACA NS, it was demonstrated that the most cytotoxic polymers for tumoral cells were those with shorter side chains, confirming previous results obtained with normal cells (18,29). However, while evaluating the polymer toxicity from in vitro tests the fact that the amount of NS incubated with tumoral cells is often excessively high and corresponds sometimes to what is administered to one mouse must be taken into account (18).

One of the most interesting properties of anticancer-drug–loaded NP is their capacity for overcoming pleiotropic drug resistance. In vitro studies evaluating the efficacy of NP on multidrug resistant cells have been until recently performed only with DXR (21–23,25,30–38). The mechanisms involved in pleiotropic drug resistance are complex. Among them, the overexpression of a membrane glycoprotein, namely, the P-glycoprotein (P-gp), which pumps out the anticancer drug from the inside of the tumoral cells has been extensively investigated (2,39–41). By coupling DXR to NS, DXR efflux from tumoral U-937 cells was considerably reduced (Fig. 1) (24,31). Accordingly, increased drug retention, possibly because of the inability of P-gp to reject NS-bound DXR outside the cell, may partly explain how drug resistance can be counteracted. Colin de Verdière et al. (34) demonstrated that poly(isobutyl cyanoacrylate) (PIBCA) NS were not endocytosed by P-388 leukemic

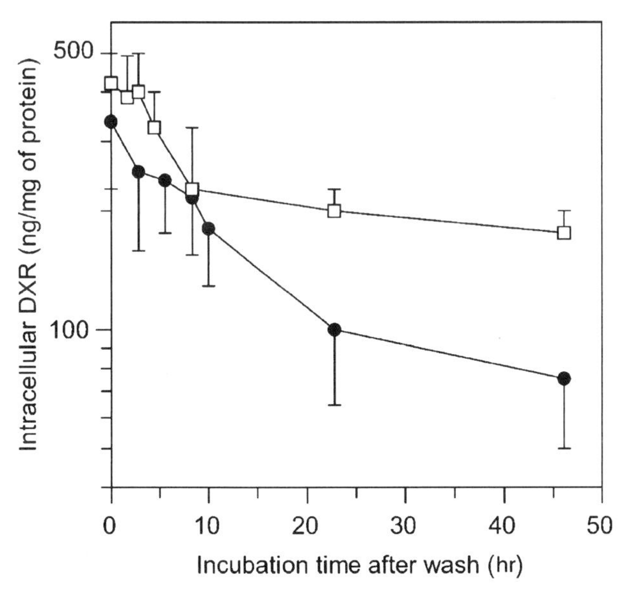

Figure 1 Efflux of DXR from U-937 cells. Exponentially growing cells were exposed for four hours at 37°C to 0.5 μg/mL of free DXR (•) or NS-bound DXR (□) (4×10^4 NS/cell). After washing in PBS, the cells were resuspended in drug-free medium for 48 hours and harvested at various times. Each point represents the mean of two separate experiments performed in quadruplicate (bars, SD). *Abbreviations*: DXR, doxorubicin; NS, nanospheres. *Source*: From Ref. 24.

murine cells and that the accumulation of DXR in the resistant cells could not be explained, in this case, by a reduced efflux. Resistant cells treated with either alginate, PLA or poly(lactic-co-glycolic acid) (PLGA) DXR-loaded NP, were equally sensitive to the free drug whereas cells treated with PACA loaded NS were more sensitive to the drug (23,31,38,42). Interestingly, the intracellular concentration of DXR in P-388 resistant cells was six times higher when PIBCA NP were administered, whereas it was not different when poly(isohexyl cyanoacrylate) (PIHCA) NS were used (38). The polymer degradation by-products were found to play a major role in the efficacy of DXR NS because the intracellular drug concentration and the cytotoxicity were both increased when DXR was incubated in the presence of PACA degradation products (38). Actually, it was demonstrated that degradation products can form ion pair complexes with DXR (38,43). The interaction of cells with DXR-loaded NP without direct contact by placing particles on a filter located on the top of a cell layer led to similar activity of DXR either in sensitive or in non-sensitive cells. This study suggests that the direct contact between cells and NS is not required for the drug to be active. Comforting this observation, similar intracellular DXR concentrations were found in the presence of cytochalasin B, an endocytosis inhibitor (44). Following these observations, a mechanism was proposed to explain how PIBCA NP were able to overcome the multidrug resistance in P-388 cells. Particles first adhere to the cell membrane; DXR is then released from the carrier. When the polymer starts to degrade, an ion pair is formed between the drug and the by-products. This complex is then able to cross the cell membrane without being recognized by the P-gp. The principle of this mechanism is depicted in Figure 2.

Because particles can be rapidly taken up by macrophages after systemic administration, macrophage-like cells, J774.A1 cells, were cultured on a filter placed over P-388 cancer cells (19,37). DXR-containing formulations were introduced in the upper compartment in contact with J774.A1 cells, and the P-388 cell growth was evaluated. In the coculture system, multidrug resistance was only partially overcome. The efficacy of free and encapsulated DXR was decreased by a factor of four- and sevenfold, respectively, in the coculture experiment as compared to the results obtained in the monoculture. This study shows that even after interaction with macrophages, DXR-NP remains cytotoxic against P-388 cells.

Another mechanism also seems to be involved in the increased efficiency of DXR NS against DXR-resistant cells. DXR bound to NS can efficiently interact with the membrane of resistant cells, inducing a perforation of the cytoplasmic membrane (21). This mechanism is dependent on the NS density per cell because for the same drug concentration, the activity of DXR increases with higher NS densities (24). This issue is potentially important for the elaboration of improved in vivo drug administration protocols. Indeed, the drug activity may be dependent not only on its dose but also on the amount of NP administered. The interest of NP for overcoming the multidrug resistance has recently been established with another active compound, in paclitaxel, that was encapsulated into lipid NC (45). These particles interacted with efflux pumps and as a result, an inhibition of multidrug resistance was observed in glioma cells in culture or implanted in rats.

Immunomodulators that activate macrophages to make them cytotoxic for tumor cells have attracted increasing interest for their incorporation into colloidal carriers. Immunomodulators such as muramyl dipeptide (MDP) have a very short half life which prevents accumulation of the drug into macrophages. Therefore, a lipid derivative of MDP, namely, muramyl dipeptide-L-alanyl-cholesterol (MTP-Chol), was incorporated into PLA and PIBCA NC (27). These NC are made from

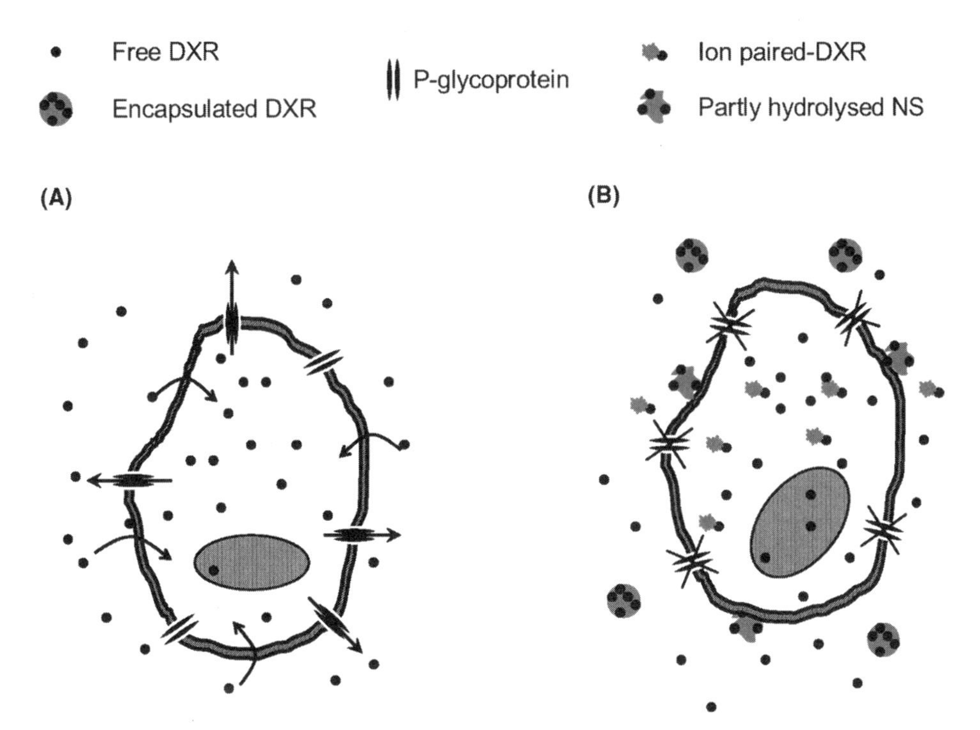

Figure 2 Proposed mechanism of action for the reversion of multidrug resistance in P-388/ADR cells. (**A**) Resistant cell in contact with free DXR, (**B**) resistant cells in contact with DXR-loaded poly(cyanoacrylate) NP. *Abbreviations*: DXR, doxorubicin; NP, nanoparticles. *Source*: From Refs. 26, 38.

an oily core surrounded by a polymeric wall and can be loaded with lipophilic drugs (46–49). MTP-Chol–loaded NC were able to activate rat alveolar macrophages for a cytostatic effect on tumoral cells (50). However, the NC formulation was not superior to the nanoemulsion, micellar solution, and liposome formulations. The possible mechanisms of action of MTP-Chol NC are depicted in Figure 3. At low immunomodulator concentrations MTP-Chol NC are generally more effective than free MDP, probably because of an improved intracellular delivery by phagocytosis (46,51). NC containing MTP-Chol were found to be slightly toxic for rat alveolar macrophages (27,50,52). This toxicity could be explained by an increased sensitivity of activated macrophages to nitric oxide that is produced consecutively to the activation of the macrophages by the immunomodulator. Because MTP-Chol NC are internalized by phagocytosis, which is an energy-dependent process, it is possible that the internalization of the immunomodulator by this mechanism could sensitize the macrophages to nitric oxide (27).

In recent years there has been a growing interest for the molecular therapies of cancer (53). For instance, the insertion of new or foreign genes into a tumoral cell may represent an alternative to conventional drug therapy. Antisense oligonucleotides also represent a promising active compound by specifically inhibiting gene expression (54–56). However, these molecules are highly sensitive to biological degradation and suffer from a poor cellular internalization. NP provide a shield against biological hydrolysis and may increase intracellular concentrations after

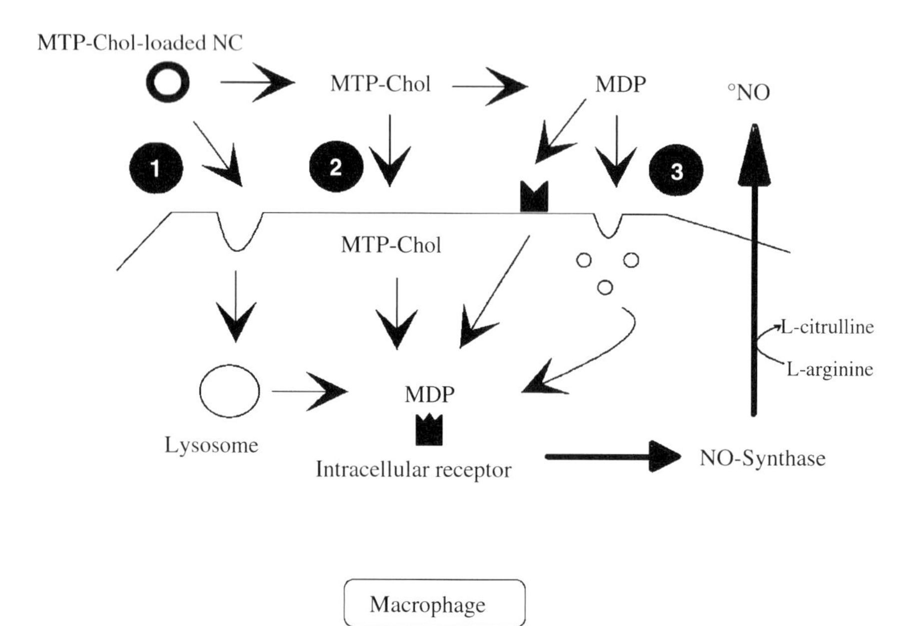

Figure 3 Different pathways by which encapsulated MTP-Chol or free hydrophilic MDP could enter rat alveolar macrophages and induce an activation of these cells mediated by a release of NO. (1) Phagocytosis of intact NC; (2) Passive diffusion of released MTP-Chol; (3) Uptake of MDP by receptor mediated endocytosis or pinocytosis. *Abbreviations*: MTP-Chol, muramyl dipeptide-L-alanyl-cholesterol; MDP, muramyl dipeptide; NP, nanoparticles; NC, nanocapsules.

endocytosis. Because a chapter of this book is devoted to gene therapy, it will not further be developed here.

Paclitaxel is a very potent antineoplasic drug isolated from the bark of Pacific yew (*Taxus brevifolia*). It has an excellent clinical efficacy against a wide array of cancers such as ovarian cancer, head and neck cancer, small cell and non small cell lung cancer, colon cancer, melanoma, or Kaposi's sarcoma [see Ref. (57)]. Its clinical use is, however, limited by its low water solubility requiring the use of a mixture of polyoxyethylated castor oil, Cremophor EL, and ethanol to have an injectable formulation, Taxol®. This excipient is well known for several adverse effects such as hypersensitivity, nephrotoxicity, neurotoxicity, and cardiotoxicity. Thus, before treatment, patients must receive a premedication consisting of corticosteroids and antihistamine drugs. Among the different approaches aimed at improving paclitaxel delivery, polymeric NP were investigated (58).

Paclitaxel was entrapped in PLGA with different lactic to glycolic ratios before incubation with National Cancer Institute (NCI)-H69 cells (small cell lung cancer cell line), and the growth inhibition was compared with Taxol (59). The unloaded NS were shown to be nontoxic. Globally, NS were less efficient than Taxol in terms of cell growth. However, the vehicle, Cremophor, proved to be toxic, participating likely to the drug effect. At 25 μg/mL, Taxol and NP were equally potent (Taxol was efficient at an earlier time of incubation, 72 hours vs. 168 hours), and at 2.5 μg/mL, NP were more active.

Vitamin E succinate with polyethylene glycol (TPGS)-coated PLGA NP was also investigated on HT-29, a human colon adenocarcinoma cell line (60). Coated NP proved to be better internalized than uncoated ones. Further, cell viability was three times lower when the drug was incubated under the form of coated NS compared to Taxol.

These works present the interest of NP as an alternative formulation for the intravenous (i.v.) administration of highly lipophilic drugs such as paclitaxel. This approach has also been investigated with lipophilic photosensitizers (PS) used in a newly developed approach for cancer treatment, photodynamic therapy (PDT).

PDT is based on the administration of tumor localizing PS, followed by the exposure of the neoplastic area to light absorbed by the PS. This therapy results in a sequence of photochemical and photobiological processes that trigger irreversible damage to the irradiated tissues. The photodynamic treatment is thought to be selective per se in that the toxicity to the tumor tissue is induced by the local activation of the PS, while normal tissues not exposed to light are spared (61). In spite of this, selectivity is not absolute, and photosensitization of the skin and some non-neoplastic tissues (e.g., liver, spleen, and kidney) can occur (62).

The interest in using polymeric NP was demonstrated by Bachor et al. (63–65) who conjugated the porphyrin derivative, chlorine e_6 (Ce_6) with polystyrene 1000 nm particles. Association of PS to the particles increased the intracellular uptake of the drug whereas the free drug stayed in the membrane of MGH-U1 cells (human bladder carcinoma cell line). After irradiation at 659 nm, nanoparticle-associated Ce_6 (0.43 μM) induced a decrease in colony formation resulting in a total growth inhibition at a light dose of $5\,J/cm^2$ whereas the treatment with unconjugated Ce_6 did not produce any effect (63). Further, Lenaerts et al. (66) studied the effect of encapsulation on the activity of different phthalocyanine derivatives into PIHCA NC. The more hydrophobic compounds were able to express some phototoxic activity once encapsulated, showing the interest of this approach for nonsoluble compounds. The authors also suggested that encapsulation would reduce the degree of aggregation of the drug that is noxious to its activity. This point is considered to be of key importance for the success of the PDT because the tendency of highly hydrophobic PS to undergo aggregation, in contact with aqueous environments, has a deleterious effect on the photosensitized oxidative process (67). Indeed, due to the lack of physiologicaly acceptable solvents for hydrophobic PS, their encapsulation allows for overcoming formulation problems encountered especially when i.v. administration is needed.

The *meta*-tetra(hydroxyphenyl)chlorin was entrapped in plain or surface modified with polyethyleneglycol (PEG) PLA NP (68). The cellular uptake evaluated by the total fluorescence intensity showed that encapsulation decreased the intracellular uptake of the drug, and a higher rate of internalization was obtained when the drug was solubilized in a solution made of ethanol, PEG, and water (20/30/50 in volume) compared to the encapsulated drug. Plain PLA NS were less internalized than either poloxamer or PEG-coated particles. Nevertheless, a quite similar phototoxic activity was observed for all treated groups. This strongly suggests that the PS do not follow the same cellular pathway whether they are encapsulated or free. This was further demonstrated in a recent publication where the degree of internalization of meso-tetra(4-hydroxyphenyl)porphyrin (p-THPP)–loaded NP sharply dropped when incubation was done at +4°C whereas it did not affect the uptake of the free drug (69). Endocytosis of NP leads to a higher intracellular concentration whereas free drug due to its hydrophobic nature tends to concentrate into the cell membrane where it is less active. Konan et al. (70,71) studied the encapsulation of p-THPP in 50:50, 75:25 PLGA, and PLA sub-200-nm NP. The phototoxic activity of the drug either in NP or dissolved in dimethyl sulfoxide (DMSO) was investigated on EMT-6 mammary cancer cells. NP proved to be more toxic for the same drug concentration and the same light dose. Further, PLGA 50:50-loaded NP were more active than other polymers. This might be related to a faster release profile due to the more hydrophilic nature of this polymer. Similar results with an increase in cellular death were also observed with 2-devinyl-2-(1-hexyloxyethyl) pyropheophorbide (HPPH) loaded into fine silica NP of 30 nm in diameter (72).

Encapsulation possibly leads to a different intracellular localization of the hydrophobic PS favoring light activated phototoxicity. One must, however, be very careful in extrapolating these data to in vivo situations. Indeed, these in vitro experiments were obtained on cell monolayers directly in contact with light in conditions that are rarely encountered in vivo.

As for any therapy against cancer, one of the essential goals in PDT is the enhancement of the ratio between the concentration of the photoactive agent in the tumor and the tissue from which the tumor originated (73). Although few in vivo studies have been carried out, advantages of the surface modification of NP to increase the accumulation of the photosensitizer in the tumor have been demonstrated. PIHCA NP coated with poloxamer 407 (66) and PLA NP coated with PEG 20,000 enhanced the tumor uptake of the encapsulated compound (74). Indeed, the PEG-coated NP improved the PDT response of the tumor (75).

These studies document the interest of the nanoparticle approach for photodynamic treatment improvement. Encapsulation would make it possible to reduce the dose administered to the patients and would enable the use of a lower light dose then reducing the potential collateral damages to the neighboring tissues.

Interaction of Surface-Modified NP with Cultured Cells

The basic principle of targeting is to bring a drug to the right cells by using a recognition signal that would be attached to the particle surface, thus increasing local concentration of the chemotherapeutic agent. To achieve tumor-specific targeting, nanoparticle formulations should be able to recognize specific cell determinants belonging only to target cells. An example depicting monoclonal antibody–driven particles is given in Figure 4. For details regarding the different approaches and preparation modes of this type of particle, the reader is referred to chapter 5.

Approaches for the specific targeting of cancer cells and tumors are usually based on three strategies. The first one involves the needs of cancer cells for more blood supply and the presence of newly developing vascularization, called neovascularization. The endothelial cells forming these new vessels express at their membranes specific markers such as cell adhesion molecules (CAM) which represent target sites. Secondly, due to their high rate of division, cancer cells have increased needs for nutrients; therefore, receptors for specific nutrients, such as folate, and vitamins are overexpressed. Coating particles with the ligand to the receptors will provide a higher rate of internalization in these cells. Finally, cancer cells are characterized by the membrane expression of specific proteins against which it is possible to raise MAb which will specifically target particles to the desired cells.

Cancer sites are the localization of an active neovascularization, due to tumor growth requiring more blood supply; therefore, vascular targeting has been developed with integrin-coated NP for chemotherapeutic agents. For instance, Blackwell et al.

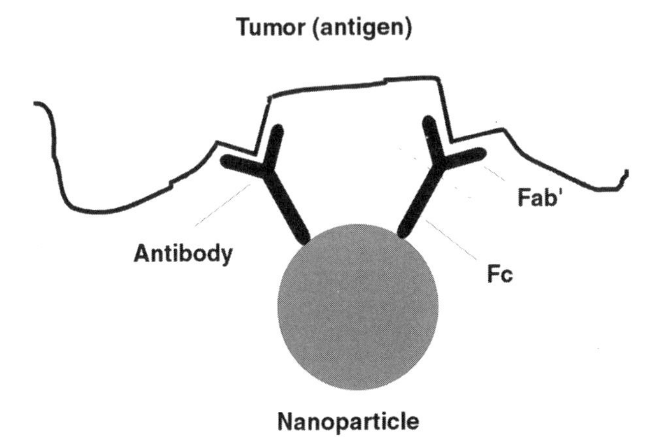

Figure 4 Schematic diagram of the attachment of monoclonal antibody to a nanoparticle. *Source*: From Ref. 183.

coated polystyrene NP with humanized HuEP5C7.g2 MAb to bind E- and P-selectin to target adhesion molecules specific for newly growing endothelial tissue such as expressed in the vessels surrounding tumors (76). In vitro, those particles specifically adhered to selectin-expressing cells suggesting that this system may be used as a drug delivery carrier for anticancer drugs.

Due to their high rate of division, cancer cells may overexpress the folate receptor at their surface. This characteristic was used to target cancer cells with folate-coated NP. Using superparamagnetic magnetite NP, it was shown that folic acid-coated NP had a better rate of uptake in human breast cancer cells (BT20) (77). This strategy was also applied to particles made of a new copolymer, poly[aminopoly(ethylene glycol) cyanoacrylate-co-hexadecyl cyanoacrylate] (78). Indeed, when in contact with cells, folate-coated-pegylated NS exhibited a 10-fold higher affinity to the folate receptors as compared to free folate molecules due to their aggregation form (79). This led to a stronger interaction with cancer cells. Folate-coated NS were preferentially taken up by cells expressing folate receptors (KB 3-1) but not by MCF-7 cells not expressing the receptor.

Thiamin can also be used to target cancer cells as was shown with gadolinium-loaded NP coated with thiamine residues (80). Particles interacted more with cells overexpressing the thiamine transporter as compared to control breast cancer cells.

As the needs for iron increase in dividing cells, transferrin receptors are present in higher number at the cancer cell membranes (81). NP were prepared based on a complex made of cyclodextrin and PEG, and their surface was modified with transferrin conjugates (82). Those particles were able to induce a higher transfection rate of K562 leukemia cells with an increase over the non-targeted NP.

DISTRIBUTION AND PHARMACOKINETICS OF ANTICANCER DRUGS COUPLED TO NP

The localization of NP in specific tissues may be dependent only on the intrinsic characteristics of the carrier (e.g., nature of the polymer, size, and surface properties) or be partly governed by an external mechanism (e.g., magnetic field). Therefore, this section makes a distinction between the natural distribution of a nonmagnetized carrier and that of the magnetically directed carrier.

Distribution of Anticancer Drugs Coupled to Nonmagnetic NP

The tissue distribution and pharmacokinetics of an anticancer drug can be altered by its incorporation into NP (Table 2). Generally, following i.v. administration, NP are rapidly and extensively taken up by RES (9). Accordingly, as soon as a few minutes after the injection of NP, the anticancer drug mainly accumulates in the liver and spleen (83–90). Thereafter, the RES drug level gradually decreases over several days depending on the biodegradability of the polymer and on the drug release kinetics. Following i.v. injection of ^{14}C-DXR PIHCA NS, Verdun et al. (88) found that radioactivity concentration in the liver remained high during the first day and decreased rapidly during day two and day three, as a consequence of fecal excretion. After eight days, only about 30% of the labeled drug could be detected in the liver. Indeed, PACA NP are biodegradable and, accordingly, are rapidly eliminated from the organism (5,91). This is not the case with other nanoparticle polymers such as PAMA. PAMA NP are well tolerated in vivo but are very slowly biodegradable (9,92). Considering that some tumoral diseases may alter the RES, the long term consequences of the accumulation of PAMA NP following multiple dosings of anticancer drugs should be seriously considered (93). Rolland suggested that the clinical use of DXR-loaded PAMA NS would still be compatible with a monthly injection protocol (92). However, because of the importance of the RES in the host defense, it seems that the use of such polymers is very limited when other well characterized biodegradable polymers are available (93,94).

Couvreur et al. (95) reported no increase of liver DXR concentration following i.v. administration of ^{3}H-DXR-loaded poly(methyl cyanoacrylate) (PMCA) NS. In fact, an increase of DXR concentration in the gut and lungs was noticed. Although this unusual distribution pattern may be due to the higher hydrophilicity of PACA made of short side chains, these results have not been further exploited, probably because of the higher toxicity of PMCA NS (29,96). Bapat and Boroujerdi reported that DXR concentrations in the liver and spleen following i.v. injection of DXR-loaded poly(butyl cyanoacrylate) (PBCA) NS were lower than in the control and did not increase over time (97). These low concentrations were attributed to the slow breakdown of the polymer in the body. Nevertheless, they did not specify if their extraction procedure was able to separate the free drug from its NS-bound counterpart. According to Kattan et al. (98) most of conventional methods of drug determination do not discriminate between nanoparticle-associated drugs and free drugs.

Because NP are mainly taken up by the RES, nonphagocytic cells such as heart and muscle cells should not theoretically accumulate NP. The incorporation of anticancer drugs into NP would thus appear suitable for avoiding the deposition of highly toxic compounds into nontarget organs. DXR is a potent antitumor agent, which is active against a wide spectrum of malignancies, but it is associated with acute and delayed cardiotoxicity (39). When administered in the nanoparticulate form, the concentration of DXR in the mouse heart was generally lower than in the control (88,97,99). Surprisingly, when using rabbits as experimental animal models, Rolland (87) did not find any significant difference in tissue concentration between free and NS-bound DXR in the spleen, lungs, and heart. He suggested that the bound drug could still produce a decrease of cardiac toxicity due to its binding to other subcellular compartments not involved in toxicity, and by a slow release of the

Table 2 Distribution and Pharmacokinetics of NP

Year	Drug	Polymer	Size (nm)	Route of adminis-tration	Tumor	Host	Comments	References
1977	5-FU	Albumin	400–1000	i.v.	None	ICR mice	Increased and prolonged distribution of the drug in the liver when associated with NS	(83)
1979	5-FU	Albumin	660	i.v.	None	JCL-ICR mice	Bound drug mainly taken up by the liver. The amount of bound drug trapped in the lungs was increased by decreasing the solidification temperature of the NS	(84)
					Ehrlich ascites s.c.	ICR mice	No preferential accumulation of NP in the tumor	
1980	5-FU	Albumin	1000	i.p.	Ehrlich ascites i.p.	ICR mice	Higher concentrations of bound drug in the ascites versus free drug	(137)
				i.t.			Elimination of drug retarted with NS	
1980	Dactinomycin	PMCA	200–300	i.v.	None	Wistar rats	Similar hepatic distribution for free and bound drug. Enhanced accumulation of bound drug in the small gut and in the lungs	(95)
		PECA					Bound drug mainly taken up by the liver	
1980	Vinblastin	PECA	200–300	i.v.	None	Wistar rats	Bound drug mainly taken up by the RES	(95)
1980	Dactinomycin	PBCA	276	i.v.	None	Wistar rats	Bound drug mainly taken up by the RES	(86)

(*Continued*)

Table 2 Distribution and Pharmacokinetics of NP (*Continued*)

Year	Drug	Polymer	Size (nm)	Route of adminis-tration	Tumor	Host	Comments	References
1981	MMC	Gelatin	280	i.v.	None	Wistar rats	Similar distribution of bound and free drug	(85)
	MMC–dextran conjugate						Bound drug mainly taken up by the liver	
1983	None	PIBCA	254	i.v.	Lewis lung carcinoma s.c.	C57 mice	NS mainly taken up by the RES. NS concentration in the tumor was maximal when NS had totally disappeared from the blood stream	(102)
1983	None	PIBCA coated with specific MAb	200	i.v.	Colorectal sarcoma s.c.	Mice	Most of the carrier was localized inside the tumor 24 hrs after injection of the NS	(175)
1983	5-FU	PBCA	50–100	i.p.	S180 sarcoma s.c.	Swiss mice	Bound drug mainly found in the RES and kidney. Concentration in tumor similar to free drug	(104)
1984	None	PHCA coated with specific MAb	130	i.v.	788T osteogenic sarcoma s.c.	CBA mice	No significant uptake of the uncoated or MAb-coated NS by the tumoral nodule	(122)
1985	Dactinomycin	PIBCA PHCA	n.m.	i.v.	None	NMRI mice	Biexponential kinetic for bound and free drug. Dactinomycin was more rapidly cleared from the blood compartment when bound to PIBCA NS versus PHCA NS	(100)
1986	None	PIBCA	n.m.	i.v.	S447 sarcoma	Rats	Tumoral tissue concentrated the carrier 4 hrs after injection of NS	(29)
1986	None	Polyacryla-	250	i.v.	None	Balb/c mice	Erythrocytes were eliminated faster by	(176)

		mide coated with specific MAb anti-TNP				bearing TNS-labeled erythrocytes	the RES when the MAb were administered bound to the NS	
1986	None	PHCA	200–300	i.v.	Osteogenic sarcoma s.c.	Nude mice	NS mainly taken up by the RES. The amount of NS detected in the tumor after 7 days was higher than in the muscle but still inferior to 1%.	(105)
1986	DXR	PIBCA	180–190	i.v.	None	NMRI mice	Prolonged plasmatic concentrations for bound drug. Reduction of cardiac concentrations	(99)
1988	Nitrosourea	PBCA	100	i.p.	B16 melanoma s.c.	C57 mice	Bound drug mainly taken up by the spleen. No control performed with free drug	(177)
1988	None	Albumin coated with specific MAb	500	i.v.	Lewis lung carcinoma s.c.	C57 mice	MAb-coated NS were only slightly localized in the carcinoma	(121)
1988	DXR	PAMA coated or not with poloxamer 188	300	i.v.	None	New Zealand rabbits	Bound drug showed prolonged elevation of plasma concentrations and reduction of total clearance. No difference in cardiac concentrations between free and bound drug. No difference between the pharmacokinetic parameters of uncoated and poloxamer-coated NS	(87)
1989	5-Fu	PGL	270	i.v.	None	Albino rats	High deposition of the drug in the liver, gut and lungs. No control performed with free drug	(178)

(Continued)

Table 2 Distribution and Pharmacokinetics of NP (*Continued*)

Year	Drug	Polymer	Size (nm)	Route of adminis-tration	Tumor	Host	Comments	References
1989	DXR	PAMA	300	i.v.	Hepatocellular carcinoma	Human	Bound drug showed prolonged elevation of plasma concentrations and reduction of total clearance. Study performed with 4 patients	(92)
1990	5-FU	MC	540	i.v.	None	Albino rats	Preferential deposition of the drug in the lungs.	(179)
		EC	472				No control performed with free drug	
1990	DXR	PIHCA	260	i.v.	None	NMRI mice	Bound drug mainly taken up by the RES. Reduction of cardiac and kidney concentrations	(88)
1990	DXR	PIHCA	300	i.v.	Liver metastasis of M 5076 reticulum cell sarcoma	C57-BL/6 mice	Rate of elimination of DXR was in the same range for liver and neoplastic tissue and was not influenced by the nanoparticulate form.	(16)
							Concentration of drug in neoplastic tissue 5-fold higher for bound drug versus free drug 24 hrs after administration	
1992	DXR	PIHCA	300	i.v.	Variable	Human	Important variability of DXR plasmatic levels between patients. Study performed with 3 patients.	(98)
1993	DXR	PIBCA	100–220	i.v.	None	CD rats	Concentration of DXR in liver, spleen, kidneys, heart, and lungs was lower for bound drug than for free drug	(97)

1995	$ZnPcI_4$	PIHCA coated with poloxamer 407	180	i.v.	EMT-6 mouse mammary tumor	Balb/c mice	Higher tumor uptake of the encapsulated compound, relatively low uptake by the RES	(66)
1995	$ZnPcF_{16}$	PLA	464	i.v.	EMT-6 mouse mammary tumor	Balb/c mice	PEG-coated NS gave higher blood levels of the compound and lower RES uptake than plain NS	(74)
		PLA coated with PEG 20,000	988					
1997	Mitoxantrone	PBCA	253	i.v.	B16-melanoma	B6D2F1/Shoe mice	NP increased drug concentration in the tumor, heart and spleen.	
		Poloxamine coated-PBCA	266				Tumor growth was significantly reduced	(90)
1999	DXR	PBCA	270	i.v.	None	Wistar rats	NP prevent heart distribution, only coated particles were able to deliver DXR to the brain	(89)
		Polysorbate 80 coated PBCA						
2002	None	PHDCA	135–161	i.v.	9L gliosarcoma, intracerebral	Fisher rat	Long-circulating particles accumulated 3.1 times more in the tumor	(109)
		PEG-coated PHDCA	146–161					
2003	Irinotecan (CPT-11)	PLA, PEG-PPG-PEG	118	s.c.	Sarcoma 180	ddY mice	CPT-11 NP had a larger AUC, longer MRT compared to the free drug	(144)

Abbreviations: 5-FU, 5-fluorouracil; AUC, area under the curve; DXR, doxorubicin; EC, ethylcellulose; i.p., intraperitoneal; i.t., intratumoral; i.v., intravenous; MAb, monoclonal antibodies; MC, methylcellulose; MMC, mitomycin C; MRT, mean residence time; n.m., not mentioned; NP, nanoparticles; NS, nano spheres; PAMA, poly(alkyl methacrylate); PBCA, poly(butyl cyanoacrylate); PECA, poly(ethyl cyanoacrylate); PEG, polyethylene glycol; PGL, polyglutaraldehyde; PHCA, poly(hexyl cyanoacrylate); PHDCA, poly(hexadecylcyanoacrylate); PIBCA, poly(isobutyl cyanoacrylate); PIHCA, poly(isohexyl cyanoacrylate); PLA, poly(lactic acid); PMCA, poly(methyl cyanoacrylate); PPG, poly(propylene glycol); RES, reticuloendothelial system; s.c., subcutaneous; TNP, trinitrophenyl; $ZnPcF_{16}$, hexadecafluorinated zinc phthalocyanine; $ZnPcI_4$, tetraiodo zinc phthalocyanine.

drug from the NS. In the same study, no differences were found between the pharmacokinetic parameters of the bound and free DXR (87). These findings are in disagreement with other studies, where the administration of DXR-loaded NS resulted in a reduction of DXR blood clearance during the first few minutes after i.v. injection (88,100).

Although it has been shown that some phagocytic cells may have a higher endocytotic activity than normal cells, it still remains unclear if nontargeted NP can concentrate in tumors (10,101). According to Grislain et al. (102), four hours after i.v. administration of PBCA NS to mice bearing a subcutaneous (s.c.) Lewis lung carcinoma, accumulation of the carrier in the tumoral tissue was higher than in the underlying tissue, and the concentration of the NS in the lungs was as much as five times higher than that of the liver (102). The authors concluded that the deposition of the carrier in the lungs was possibly the consequence of the NS uptake by lung metastases of the primary tumor. However, as underlined by Douglas et al. (96), no histologic examinations were performed in this study and because the Lewis lung carcinoma is an hemorrhagic tumor that possesses poorly formed sinusoidal channels in which an endothelial lining is lacking, it is possible that the conditions encountered in this case were optimal for the uptake of NS by tumoral cells (96,103). Other results on the distribution of NS-bound anticancer drugs to tumors are not as conclusive (84,104,105). Even when high concentrations of bound drugs were found in the tumor in comparison to the underlying tissue, the amount of NS detected in the targeted area was still relatively low (105).

Chiannilkuchai et al. (16) have administered DXR-loaded PIHCA NS to mice bearing reticulosarcoma M 5076 metastases. They found that the drug concentrated dramatically in the healthy hepatic tissue, which in turn could act as a reservoir for the anticancer drug (Fig. 5). This probably enabled a higher exposure of liver metas-

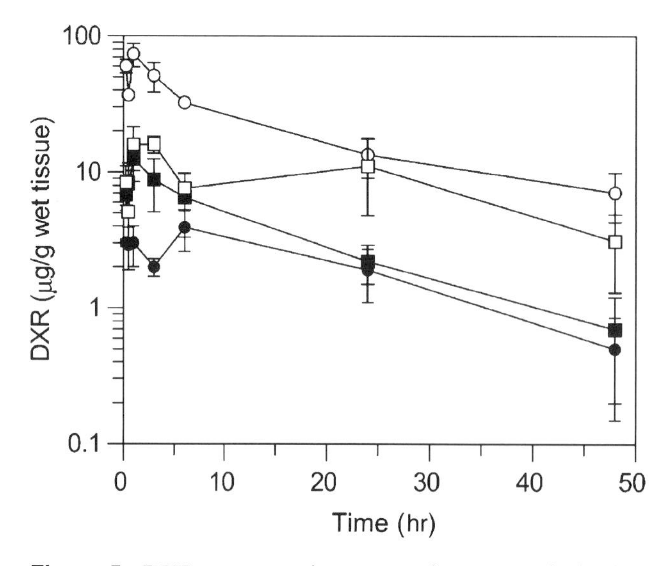

Figure 5 DXR concentration versus time curves in healthy hepatic (*circles*) and tumoral (*squares*) tissues after i.v. administration of DXR (10 mg/kg corresponding to 133 mg/kg PIHCA) in its free (solid symbols) or NS-bound form (open symbols). *Abbreviations*: i.v., intravenous; DXR, doxorubicin; PIHCA, poly(isohexyl cyanoacrylate); NS, nanospheres. *Source*: From Ref. 16.

tases to the drug. These pharmacokinetic findings were consistent with histologic examinations where a considerable number of NS were detected in the Kupffer cells whereas NS could not be found in neoplastic cells (16).

Brain targeting is one of the most challenging issues in drug delivery, and the use of NP to reach the brain after systemic administration has been the focus of several reports (106). Blood–brain barrier is made of a tight endothelial cell junction of capillaries within the brain and limits the xenobiotic penetration, thus limiting drug delivery at this site. PIBCA NS loaded with DXR were coated with polysorbate 80 by adsorption (89). A decrease in concentration of DXR in the heart was observed when administered as NS (bare or polysorbate 80-coated NS) as compared to aqueous or polysorbate 80 DXR solutions. A high concentration in the brain was reported in animals receiving the coated NS whereas the concentration when treated with other preparations did not reach the detection limit.

To study the influence of surface properties, the fate of uncoated NP was compared with ^{14}C-labeled-polymethil methacrylate (PMMA) NP coated with different hydrophilic surfactants (polysorbate 80, poloxamine 908- and poloxamer 407) administered intravenously in nude mice bearing three tumor models: a murine B16 melanoma inoculated intramuscularly, a human breast cancer, MaTu, administered subcutaneously, and a glioblastoma, U-373, implanted in the brain (107). In melanoma and breast cell models, poloxamine 908 and poloxamer 407–coated particles achieved high concentration levels whereas uncoated or polysorbate-coated NS gave low undistinguishable levels. In the glioma model, poloxamine 908- and poloxamer 407–coated NS gave higher concentrations; however, no difference in concentration was seen with the hemisphere not bearing the tumor. Polysorbate-coated particles did not induce a high concentration in the brain tumor. These results are partly in discordance with the previous report by Gulyaev et al. (89); however, the polymer was not the same, and this may explain a different behavior. This explanation is supported by the report by Olivier et al. (108) who studied the penetration of dalargine associated with NS in the presence or absence of polysorbate 80. Higher penetration was observed when dalargine was associated with polysorbate 80–coated NS, and this effect was not seen with polystyrene NS decorated with the same surfactant. It was concluded that a non-specific permeabilization of the blood–brain barrier was related to a likely toxicity of the carrier.

In another study, radiolabeled-long-circulating PEG-coated poly(hexadecylcyanoacrylate) NP were administered intravenously to rats bearing intracerebral 9L gliosarcoma, and the distribution was compared to noncoated NS (109). Both types of particles induced a significant accumulation in the tumor; however, long-circulating particle concentration was 3.1 times higher than the conventional NS. Concentration in non-tumoral tissue was also higher, and normal to cancer ratio was 11, four hours after administration for pegylated NS. The Tumor accumulation mechanism was hypothesized to be due to the leakier vascularization of the brain tumor and the difference between both types of particles related to the longer circulation time of pegylated NS that did not accumulate in the liver and other RES organs. Particles proved to be large enough to stay in the tumor whereas the hydrophilic tracer sucrose did not, confirming the role of the EPR effect for particle retention into tumor tissue.

Numerous studies have demonstrated the efficacy of coating polystyrene NS with block copolymers of poloxamer and poloxamine and PEG for reducing NS uptake by macrophages and allowing a longer in vivo circulation time of the NS or their accumulation in the bone marrow (110–116). Studies performed with "stealth liposomes" either stabilized with PEG or with ganglioside GM1 have shown a greater accumulation in tumors of the stabilized carrier over the control liposomes (117,118). Such investigations need to be performed to a greater extent with biodegradable NP (119,120). A study carried out by our group demonstrated that coating PLA NS with PEG 20,000 could produce a substantial increase of the blood circulation time of the photoactivable compound hexadecafluorinated zinc phthalocyanine ($ZnPcF_{16}$), as compared to plain spheres (Table 2) (74). After 24 and 168 hours, the cumulated uptake of the compound in the liver and spleen represented 61% and 44% for plain NS versus 50% and 29% for PEG-coated NS, respectively. The reduction of the uptake of the NS by the RES and the resulting longer blood circulation time were associated with a threefold increase of the compound concentration in the tumor after 24 hours. Those long-circulating or stealth particles are taken up by mononuclear phagocytic system organs to a lesser extent as compared with conventional particles which makes them able to extravasate more readily in sites with leaky vasculature such as tumor sites. However, better clinical efficiency could be obtained with smaller particles and covalently bound PEG.

Distribution of MAb-coated NS has not led to the results anticipated based on previous in vitro experiments (121,122). Apart from NS clearance by the RES, poor localization of the NS in the tumors could be explained by several factors including competitive displacement of adsorbed MAb, secondary coating process where the MAb-coated NS receive an opsonic layer, and most probably, an insufficient access of the NS to tumor cells (11,122). It may be this latter issue has so far been underestimated. Initial enthusiasm for therapies using MAb, drug–MAb complex, and MAb-macromolecular systems has been dampened by the fact that macromolecules cannot reach all regions of a tumor in adequate quantity (3,123,124). One method of increasing the transport of antibodies to tumoral cells could be for instance to use lower molecular weight agents, for example, antibody fragments $F(ab)_2$ and Fab (124). NP are confronted by a greater problem in that they are much bigger than antibody molecules. Accordingly, it can be anticipated that the use of MAb-coupled NS may still have potential applications but in a limited number of cases such as nonsolid tumors, intratumoral (i.t.) therapy, and for the treatment of solid tumors which are blood supplied by discontinuous capillaries or sinusoidal blood channels (11).

The first clinical trial using anticancer-drug-loaded NS was carried out by Rolland in 1989 (92). He compared the pharmacokinetics of 20 to 30 mg PAMA NS-bound DXR intravenously administered to four hepatoma patients against 20 mg of free DXR administered to one hepatoma patient. This resulted in a 15% to 48% increase in the dose-normalized area under the drug concentration–time curve for the patients receiving the NS-bound DXR (1,92). Three years later, Kattan et al. (98) conducted a clinical trial using DXR-loaded PIHCA NS. The pharmacokinetic profiles of DXR were examined on three cancerous patients with refractory solid tumors receiving 60 to 75 mg/m^2 DXR-loaded NS. These profiles were compared to those obtained with the administration of free DXR at the same dose to the same patients. Contrary to the previous study, three controls with free DXR were evaluated in this trial, and it was found that the variability of kinetic parameters of free and bound DXR was such that no convincing conclusions could be drawn from the results. Any other comparison of this study and the one from Rolland et al. (92) is difficult because the plasma concentrations were monitored between 0.5 and 24 hours in the first study and between 5 and 90 minutes in the other one.

Distribution of Anticancer Drugs Coupled to Magnetic NP

To minimize colloidal carrier uptake by the RES and to enhance their extravasation from the capillaries, magnetic NS were developed in the late seventies and have been reviewed in detail by Gupta and Hung (125). Only considerations related to their use in cancer chemotherapy will be discussed in this section. These NS (200–1000 nm) contain a magnetically active component (e.g., Fe_3O_4) and can be guided by an externally placed electromagnet. An electromagnetic field ranging from 1000 to 8000 G is generally applied at the target site for five to 60 minutes after dosing (126–134).

Very high anticancer-drug concentrations in the target area (generally a section of the tail) can be obtained following the administration of magnetic NS. For instance, Senyei et al. (127) showed that as late as 60 minutes postinjection, 0.05 mg/kg DXR administered intra-arterially (i.a.) via magnetic NS resulted in almost twice the local tissue concentration than what was achieved with a 100-fold higher i.v. dose solution. The use of magnetic NS can increase the extravasation and partly overcome the limited access of conventional NP to extravascular tumors. Widder et al. (128) demonstrated that 30 minutes after infusion of albumin NS to rats bearing s.c. Yoshida sarcoma, 1000-nm NS were found in the extravascular compartment adjacent to tumor cells and occasionally in tumor cells. By 24 and 72 hours, NS were still detectable within tumor cells. Furthermore, it was shown elsewhere that DXR-loaded NS could traverse the vascular endothelium as early as two hours after dosing and that the pharmacodynamic characteristics of the drug were not altered by its entrapment into NS (129).

To evaluate the efficiency of drug delivery via magnetic NS, Gallo et al. (130) introduced two indices, namely, the relative exposure (r_e) and the targeting efficiency (t_e):

$$r_e = \frac{(AUC_i)_{NS}}{(AUC_i)_S} \tag{1}$$

$$t_e = \frac{(AUC)_{target\ site}}{(AUC_i)_{nontarget\ site}} \tag{2}$$

where AUC_i is the dose-normalized area under the drug concentration–time curve (AUC) for the ith tissue and the subscripts NS and S refer to the NS and the solution dosage forms, respectively; $(AUC)_{target\ site}$ and $(AUC)_{nontarget\ site}$ are the AUC for the target site and nontarget site, respectively.

These two indices provide more valuable information on the drug delivery than simple drug concentrations at the target site at different time points and should also be used for non-magnetic NS. Using these parameters, it was found that DXR delivery to a selected target area can be greatly improved by its incorporation into magnetic NS (130,131). Magnetic NS can also decrease the nontarget tissue exposure to DXR. For instance, the administration of 2.0 mg/kg DXR to rats as a solution or via magnetic albumin NS provided r_e values inferior to 0.6 for the heart and kidney and an r_e of 1.63 at the target site (130). The efficiency of drug targeting is highly dependent on the strength of the magnetic field (131). Table 3 suggests that the 1000 G magnetic field was only partly able to control the distribution of DXR because the liver uptake of the drug remained high. Considering a possible human application of magnetic NS, this observation emphasizes the importance of defining the minimal magnetic field strength to apply to ensure an acceptable drug targeting.

Table 3 Drug t_e of magnetic albumin nanoparticles in the delivery of 0.4 mg/kg doxorubicin administered i.a. (tail) to Wistar rats, in the presence or absence of a 1000 G or 8000 G magnetic field applied at target site for 30 minutes

	t_e		
Tissue	Control	1000 G	8000 G
Tail			
Upstream from target site	0.62	1.38	15.54
Target site	1.00	1.00	1.00
Downstream from target site	0.43	0.88	6.12
Liver	0.51	0.99	6.73

Abbreviation: t_e, targeting efficiency.
Source: From Ref. 131.

Moreover, it was shown that the pharmacokinetics of magnetic NS could be altered by the dose administered (133,134). Indeed, a reduction of the carrier dose has proved to increase the t_e. The reduction of the dose decreased more particularly the amount of DXR delivered to the liver and heart. As underlined by Gupta et al. (133), such a dose dependence of kinetic parameters is important because the carrier dose may often need to adapt to different factors such as the weight of the patient and the volume of the target tissue.

Finally it has to be pointed out that, in most cases, the target area was easily accessible and that simple i.t. injection of drug-loaded NS could provide similar results. The magnetic carriers were also often administered via the arterial route. Although this mode of administration allows the carrier to avoid the hepatic first pass effect, it suffers from a lack of clinical convenience. Furthermore, in the case of disseminated tumoral diseases or nonsuperficial tumors, it can be anticipated that such highly specific tissue distribution of anticancer drugs obtained with animals bearing s.c. tumors may not be as easily achievable. Target areas other than the mouse or rat tail should, in the future, be further investigated. Ibrahim et al. (135) showed that the kidney levels of DXR could be enhanced by its incorporation into 220-nm-PIBCA magnetic NS injected intravenously. However, in this study, the data were collected only 10 minutes after dosing, providing no indication on the possible retention of the carrier at the target site.

IN VIVO ACTIVITY AND TOXICITY OF ANTICANCER DRUGS COUPLED TO NP

Preliminary studies regarding the in vivo activity of PACA NS against sarcoma implanted subcutaneously demonstrated the efficiency of carried anticancer drugs in reducing the tumor size (Table 4) (29,104,136).

The increased efficiency of the nanoparticle dosage form was first attributed to an improved delivery of the bound drug to malignant cells (104,136). Although this hypothesis may be relevant for some specific tumors, the modification of the drug's pharmacokinetic parameters, as discussed earlier, may be more important (11). Generally, an increase of life span was noted for the animals treated with nanoparticle-bound anticancer drugs. This prolonged survival rate was observed

Table 4 Activity and Toxicity of NP In Vivo

Year	Drug	Carrier	Polymer	Size (nm)	Dose regimen	Tumor	Host	Comments	References
1979	5-FU	NS	Albumin	n.m.	1 mg i.p. on day 1	Ehrlich ascites i.p.	ICR mice	Greater increase in life span versus free drug	(138)
1980	Dactinomycin	NS	PMCA	200	111–222 g/kg i.v. × 2 days on day 10–18	S250 sarcoma s.c.	LOU/dec rats	More effective than free drug in reducing tumor size. Increase in toxicity	(136)
1980	5-FU	NS	Albumin	1000	0.5 mg i.p. × 1–3 days on day 1, 5, 9	Ehrlich ascites i.p.	ICR mice	Greater suppression of tumor growth versus free drug. Greater increase in life span	(137)
					0.5 mg i.t.× 1–3 days on day 1, 5, 9	Ehrlich ascites s.c.	ICR mice	Higher suppression of tumor growth versus free drug	
1981	MMC MMC-dextran-conjugate	NS	Gelatin	280	5–10 mg/kg equivalent MMC i.p. on day 1	P-388 leukemia i.p.	BDF1 mice	As effective as free drug	(85)
1981	DXR	Magnetic NS	Albumin	1000	0.5 mg/kg i.a. × 1 day on day 6–8	Yoshida sarcoma s.c.	Holtzman rats	75% remission for the NS group. No death or metastases occurred in the NS group	(158)
1981	DXR	NS	Albumin	1440	300 μg i.v. on day 0	AH 7974 ascites i.v. (portal vein)	Donryu rats	Greater increase in life span versus free drug	(139)
1982	DXR	NS	PIBCA	n.m.	3–15 mg/kg i.v. × 2–5 days	None	NMRI mice	Reduction in toxicity	(148)
1983	5-FU	NS	PBCA	50–100	6.25–25 mg/kg i.p. on day 0–5	S180 sarcoma s.c.	Swiss mice	More effective than free drug in reducing tumor weight. Increase in toxicity	(104)

(*Continued*)

Table 4 Activity and Toxicity of NP In Vivo (*Continued*)

Year	Drug	Carrier	Polymer	Size (nm)	Dose regimen	Tumor	Host	Comments	References
1983	DXR	Magnetic NS	Albumin	1000	0.5 or 2.5 mg/kg i.a. on day 6–8	Yoshida sarcoma s.c.	Holtzman rats	77% of rats receiving magnetic NS had total remission. Death in 90–100 % of control group	(159)
1984	5-FU	NS	Albumin	<1000	50–400 mg/kg i.v. on day 2, 4, 7	B16 melanoma s.c.	C57/BL6 mice	Potency of cytotoxic NS lower than free drug but enhancement of therapeutic index	(20)
1984	Vindesine	Magnetic NS	Albumin	1000	0.5 or 2.5 mg/kg i.a.	Yoshida sarcoma s.c.	Holtzman rats	85% of rats receiving magnetic NS had total remission. Death for the majority of control group	(160)
1985	Methotrexate	NS	PBCA	n.m.	400 mg/kg s.c. on day 13	Human osteosarcoma	Nude mice	50% reduction of tumor growth versus free drug	(180)
1986	DXR	NS	PIBCA	n.m.	7.5 mg/kg i.v. × 3 days	None	BDF1 mice	Reduction in toxicity	(29)
1986	DXR	NS	PIBCA	n.m.	5 mg/kg i.v. on day 8 or day 12	S130 myeloma s.c.	LOU/dec rats	As effective as free drug in reducing tumor size. Prolonged survival compared to free drug	(29)
1986	Dactinomycin	NS	PHCA	n.m.	222 µg/kg i.v. on day 14, 19	S250 sarcoma s.c.	Rats	More effective than free drug in reducing tumor size. More long term survivors	(29)
1986	DXR	NS	PIBCA PIHCA	175 342	5–25 mg/kg i.v. × 1–3 days on day 1–9	L1212 leukemia i.v. P-388 leukemia i.v.	B6D2F1 mice	Greater increase in life span versus free drug. Pulmonary complications for high doses of PIBCA NS	(140)

1989	DXR	NS	PIHCA	250	2.5–12.5 mg/kg i.v. × 1–2 days on day 7–14	M5076 hystiocyto-sarcoma i.v.	C57/BL6 mice	More effective than free drug in reducing the number of hepatic metastases. Survival prolonged compared to free drug	(141, 142)
1990	DXR	Magnetic NS	Albumin	800	2 or 8 mg/kg i.t. on day 12	Sarcoma s.c.	AS inbred rats	Slightly more effective than free drug	(181)
1991	MTP-Chol	NC	PIBCA	180	3–10 μg i.v. × 3–4 days on day -2, 2, 6, 10	M5076 histiocytosar-coma i.v.	C57/BL6 mice	Effective in reducing the number of hepatic metastases only when a pretreatment at day -2 was administered	(149)
					50–100 μg p.o. × 4 days on day -2, 2, 6, 10			Significant but limited anti-metastatic activity	
1992	MTP-Chol	NC	PIBCA PLA PVCAC	180	2.5–20 μg i.v. × 1–4 days on day 2, 6, 10	M5076 histiocytosar-coma i.v.	C57/BL6 mice	Anti-metastatic activity only when given as a prophylactic treatment	(150)
1992	DXR	NS	PIHCA	300	15–90 mg/m^2 i.v. every 28 days (max. cumulative dose 180 mg/m^2)	Variable	Human	Apparent reduction of cardio- and myelotoxicity. Study performed with 21 patients	(98)
1992	DXR	NS	PIHCA	300	2 mg/kg i.p. on day 1, 2, 3, 4	P-388-DXR-R-leukemia i.p.	DBA2 mice	Greater increase in life span versus free drug	(31)

(Continued)

Table 4 Activity and Toxicity of NP In Vivo (*Continued*)

Year	Drug	Carrier	Polymer	Size (nm)	Dose regimen	Tumor	Host	Comments	References
1992	MMC	NS	PBCA	364	2 or 4 mg/kg i.p.	None	Virgin and pregnant mice	Reduction of toxicity and genotoxicity	(182)
1993	Mitoxantrone	NS	PBCA	177–280	5 mg/kg i.v. on day 1 5 mg/kg i.v. on day 1, 8, 15	P-388 leukemia i.p. B16 melanoma i.m.	B6D2F1/ Shoe mice	No convincing evidence of increased efficiency. No reduction of toxicity. Some premature deaths	(120)
1994	DXR	NS	PIBCA	235	11 m/kg i.v.	None	CD1 mice	Increased bone marrow toxicity versus free drug	(146)
			PIHCA	240					
1995	DXR	NS	PIBCA	150	5 m/kg i.v.	None	Sprague-Dawley rats	Reduction of mortality but increase of renal toxicity versus free drug	(147)
1996	$ZnPcF_{16}$	NS coated with PEG 20,000	PLA	988	1, 2, 5 μmol/kg i.v. on day 10	EMT-6 mouse mammary tumor	Balb/c mice	Higher suppression of tumor growth versus o/w emulsion formulation	(75)
					Light treatment 24, 48 or 72 hrs after i.v.			Improved PDT response and prolonged tumor sensitivity to PDT	

1996	Paclitaxel	NS	PVP	50–60	Taxol®: 8 mg/kg NP: 160 μg/animal six times on a period of 15 days i.v.	B16F10 murine melanoma	C57/bl6 mice	From the day 26, a significant decrease in tumor volume was observed whereas with Taxol®, the tumor kept growing although to a lesser extent than the controls	(145)
2000	Gadolinium	NS	Chitosan		2400 μg/animal i.t.	B16F10 melanoma s.c.	C57/BL6 mice	Tumor growth in the NP treated group was significantly reduced compared to the group treated with solution	(155)
2000	DXR	NS	PLGA–DOX conjugates	200	free DXR: 240 μg/kg/day 12 days peritumorally NP DXR: 2.4 mg/kg equiv. DXR day 10 and 11 NP DXR-conj: 1.2 mg/kg equiv. DXR day 10 and 11	EL4 Thymoma cell, s.c.	C57/BL6 mice	Single shot DXR NP give the same results as daily shot of free DXR. No difference was seen between conjugated DXR and regular DXR NP	(171)

(*Continued*)

Table 4 Activity and Toxicity of NP In Vivo (*Continued*)

Year	Drug	Carrier	Polymer	Size (nm)	Dose regimen	Tumor	Host	Comments	References
2001	Dextran-DXR	NS	Chitosan	100	i.v. 8 mg/kg/wk for 4 wks	J774A.1, n.m.	Balc/c mice	NP induced a tumor regression and increased survival time compared to the free conjugate and the free drug	(143)
2002	5-FU	NS	Albumin	<700	3 mg/kg/day for 15 days i.v.	Dalton ascite lymphoma cells	Swiss albino mice	A 2-fold decrease in tumor growth was observed with the NP compared to the free drug	(173)
2003	Irinotecan (CPT-11)	NS	Poloxamer-coated PLA	118	20 mg/kg i.v.at day 7, 8, 9 after tumor inoculation	Sarcoma 180 s.c.	ddY mice	CPT-11 showed no effect compared to the control whereas CPT-11 loaded NP reduced significantly the volume of the tumor	(144)

Abbreviations: 5-FU, 5-fluorouracil; DXR, doxorubicin; i.a., intra-arterial; i.m., intramuscular; i.p., intraperitoneal; i.t., intratumoral; i.v., intravenous; MMC, mitomycin C; MTP-Chol, muranyl dipeptide-L-alanyl-cholesterd; n.m., not mentioned; NC, nanocapsules; NS, nanospheres; NP, nanoparticles; p.o., per oral; PBCA, poly(butyl cyanoacrylate); PDT, photodynamic therapy; PEG, polyethylene glycol; PHCA, poly(hexyl cyanoacrylate); PIBCA, poly(isobutyl cyanoacrylate); PIHCA, poly(isohexyl cyanoacrylate) PLA, poly(latic acid); PLGA, poly(lactic-co-glycolic acid); PMCA, polyuinyl pyrrolidone; s.c., subcutancous; $ZnPcF_{16}$, hexadecafluorinated zinc phthalocyanine.

Table 5 Effect of Free DXR or DXR-Loaded PIHCA NS on the Survival of DBA2 Mice Bearing P-388-DXR-Resistant Ascitis (i.p. Injection of 10^6 P-388-DXR-Resistant Cells at Day 0)

	50% survival time (days)		
Groups of 10 mice	Nontreated (C)	DXR (T)	DXR-loaded NS (T)
1	11	14	21
2	14	15	21
3	11	11	17
4	11	14	18
Mean	11.7	13.5	20.2
T/C		115%	164%
P (T vs. C)		Not significant	0.003

DXR or DXR-loaded NS was given by i.p. route at days 1 to 4 at a dose of 2 mg/kg body weight/day; the activity of unloaded NS was not significant. P is the probability that T (treated) is statistically different from C (control), using the Student's *t*-test.
Abbreviations: DXR, doxorubicin; PIHCA, poly(isohexyl cyanoacrylate); i.p., intra peritonea.
Source: From Ref. 31.

for animals bearing solid tumors as well as leukemia (29,31,137–145). Interestingly, Cuvier et al. (31) demonstrated that it was possible, using DXR-loaded NS, to significantly prolong the survival time of mice inoculated with DXR-resistant leukemia cells (Table 5). This bypass of tumor cell resistance confirmed the previous in vitro observations. Allémann et al. (75) showed that the photodynamic activity of $ZnPcF_{16}$ against EMT-6 mouse mammary tumors could be greatly improved after incorporation of the compound in PEG-coated NS. One week after treatment, 63% of the mice had no macroscopic sign of tumor regrowth and at three weeks complete healing was observed for these mice. In contrast, with an oil-in-water emulsion formulation only 14% tumor regression was observed.

In vivo investigations have revealed the importance of the binding of the drug to the carrier on the activity of the nanoparticle suspension. A loss of efficiency may occur if a mixture of unloaded NP and free drug is administered instead of the drug-loaded nanoparticle formulation (140,142). This indicates that in vivo, the nanoparticle polymer is less likely to exert a cytostatic effect, as was noted in vitro.

On the other hand, anticancer drugs bound to NP have sometimes been associated with a higher acute toxicity, namely, premature deaths (104,120,136). The increase of toxicity could be attributed to the accumulation of the carrier in the organs of the RES or to the injection of agglomerated NS that may have provoked embolisms (104,120,136). This confirms the importance of developing drug carriers that can avoid the RES (when it is not the target site) and of making sure that the nanoparticle suspension is always fully dispersed before the injection. An increase of the bone marrow and renal toxicity has also been reported for DXR-loaded NS (146,147).

Conversely, a reduction of the general toxicity of anticancer drugs bound to NP has been demonstrated in a number of cases (20,29,148). From these findings, one can infer that even if the incorporation of a cytostatic agent into NP is not always followed by an increase of drug efficiency, a reduction of the toxicity may produce an increase of the therapeutic index (20).

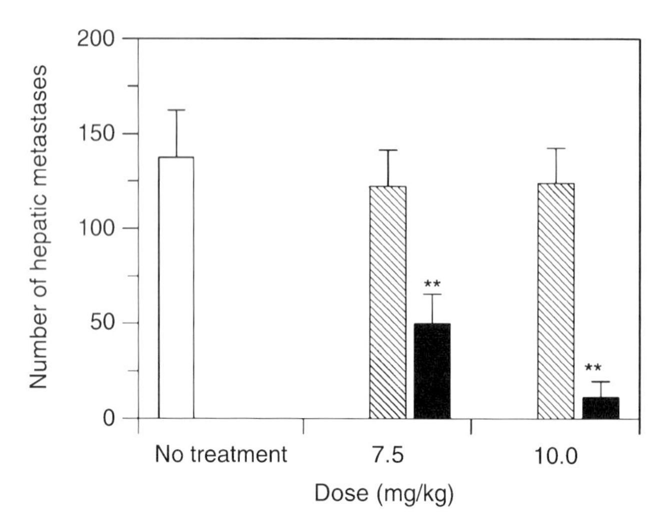

Figure 6 Number of hepatic metastases for C57/BL6 mice after treatment at days 11 and 13 with i.v. free DXR (*shaded columns*) or NS-bound DXR (*closed columns*) compared with control (*open column*). The mice were inoculated i.v. at day 0 with 7×10^3 tumoral cells (reticulum cell sarcoma M 5076). The study was performed with groups of 8 to 10 mice (mean $\pm$ SD). $**$, significantly different from control ($P < 0.001$, Student's t-test). *Abbreviations*: i.v., intravenous; DXR, doxorubicin; NS, nanospheres. *Source*: From Ref. 142.

One of the most promising applications of anticancer-drug-loaded NP may be their use in the treatment of hepatic metastases (141,142). Administration of DXR-loaded PIHCA NS to mice i.v. inoculated with tumoral cells resulted in a marked reduction of the number of hepatic metastases (Fig. 6) and in an increase of survival time (Fig. 7). These interesting results may be partly explained by the fact that the

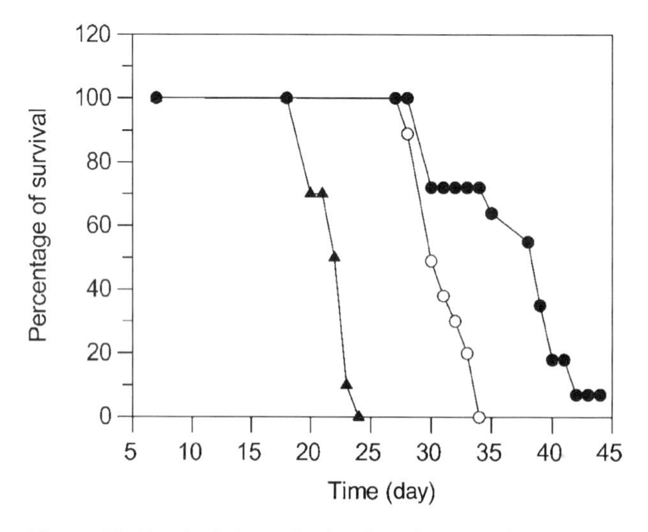

Figure 7 Survival time of mice (bearing hepatic metastases) untreated (▲) or treated with 10 mg/kg i.v. of free DXR (○) or PIHCA NS-bound DXR (•). Single administration on day 7. *Abbreviations*: DXR, doxorubicin; NS, nanosphere; PIHCA, poly(isohexyl cyanocrylate). *Source*: From Ref. 142.

tumoral cells inoculated were macrophagic in origin (142). Thus, these cells should have a propensity to take up colloidal drug carriers. The activity of the immunomodulator MTP-Chol coupled to NC against metastases has also been evaluated. Indeed, MTP-Chol coupled to NC was partially able to inhibit the metastatic proliferation that normally occurs following the i.v. injection of tumoral cells. However, this formulation was efficient only when given as a prophylactic treatment (149–151). According to Yu et al. (150), this would correspond to patients in the clinic undergoing surgery for a primary tumor and who could develop metastases. Furthermore, the efficacy of MTP-Chol–loaded NC is limited, and this formulation will certainly need to be combined with other chemotherapeutic drugs to ensure optimal results. It is well established that opsonization of nanoparticles by some specific plasma proteins enhances the uptake of the carrier by macrophages (114,152). Accordingly, one possible approach to increase the efficiency of encapsulated immunomodulators could be for instance to use nanoparticles precoated with opsonic proteins. Encouraging results have already been obtained in vitro and in vivo with protein-coated poly(L-lactic-co-glycolic acid) microspheres leading to an increase of macrophage activity towards tumoral cells (153). Peroral administration of MTP-Chol-loaded NC have led to inconsistent results and further studies are needed to evaluate the potentialities of this route of administration (149).

DXR was coupled with dextran to be encapsulated into hydrogel chitosan NP (143). Treatment with the NP shows a significant decrease in tumor volume by a factor of threefold compared to unloaded chitosan particles and DXR solution when administered to J774A.1 tumor bearing mice. Dextran–DXR conjugates induced a decrease in tumor size as well, although to a lesser extent than NS. With DXR NS, a survival rate of 50% was observed at day 100 after tumor implantation. At that time point, dextran–DXR conjugates induced a survival rate of 25% whereas the untreated controls died at day 45, and unloaded particles and DXR in solution gave a survival rate of 60 days.

Irinothecan is a potent prodrug of the family of campothecin, alkaloid isolated from *Campotheca acuminata*. Its development is however limited by side effects such as bone marrow suppression or severe diarrhea. PLA NP coated with PEG-block-poly(propylene glycol)-block-poly(ethylene glycol) copolymer NP were injected to mice bearing sarcoma 180 solid tumor and to rats for activity and pharmacokinetic studies, respectively (144). Drug-loaded particles induced a significant reduction of the tumor volume whereas free drug was not effective as compared to the control group receiving saline. Neither body weight decrease nor diarrhea was observed in any of the groups. Particles induced a sustained plasma concentration over 24 hours whereas the drug was cleared from the plasma in eight hours when administered as a solution, possibly as a result of the avoidance of the RES system.

Gadolinium-157 is an anticancer agent used in radiotherapy (154). Gadopentetic acid–chitosan particles were prepared as a gadolinium drug delivery system (155). C57BL/6 mice inoculated with B16F10 melanoma cell line (a radioresistant cell line) received two shots of either particle suspension or dimeglumine gadopentetate solution (Magnevist®, a magnetic resonance imaging diagnostic agent) at 16 hours' interval. Eight hours after the last administration, the tumor mass was irradiated by thermal neutron. When mice were irradiated, the drug solution was as efficient as the untreated control animals, whereas nanoparticle treated mice had a high survival rate with 50% of the mice still alive at the end of the study, whereas all the other groups died within 21 days. The inefficiency of the solution might be explained by the long interval between the administration and the irradiation. Pharmacokinetic studies showed a high concentration of gadolinium in the tumor when administered as NS corresponding to 74% of the dose whereas it was of 0.7% when administered as a solution. As a consequence of high drug concentration in the tumor, skin damage after irradiation was observed in the group receiving the NP. This study proved the interest of NS for radiotherapy because it may first increase the tumor concentration of the active compound, leading therefore to a decrease in the administered dose and, second, would make it possible to decrease the irradiation intensity limiting the side effects of the treatment.

A phase I study performed with DXR-loaded PIHCA NS has shown that the NS formulation seems to be relatively well tolerated (98). Allergic reactions occurring at the beginning of the study were attenuated by increasing the infusion time from 10 to 60 minutes. Moreover, dextran 70, an adjuvant contained in the drug mixture, was suspected to be the causative agent of the remaining allergic episodes. Although unexpected side effects such as fever or bone pain were recorded, there was no cardiac toxicity among the 18 patients evaluated. It was not possible to affirm that DXR-loaded NS can completely avoid cardiotoxicity because the number of patients was small and the maximal cumulative dose was low (180 mg/m^2). Normally, the range of cumulative dose required to produce cardiotoxicity is 50 to 700 mg/m^2 (39). As for free DXR, the dose limiting factor of the DXR-loaded NS was neutropenia.

To improve the efficacy of paclitaxel, very fine NP (50–60 nm) made of polyvinylpyrrolidone cross-linked with *N,N′*-methylene bisacrylamide were prepared (145). Activity was assessed on C57/BL6 mice inoculated with a murine melanoma model. The encapsulated drug induced a regression of tumor volume effective at day 26 after administration. In the same model, Taxol produced only a partial reduction in tumor volume although, in this case, tumor growth was significantly slower as compared to the controls receiving either saline or unloaded NS. At day 60, survival rate was of 50% for the group treated with NS whereas all mice died in 50 days in the group treated with Taxol. Histological investigation on normal tissue suggested that particles reduced as well the toxicity of the drug.

Recently, clinical efficacy of paclitaxel incorporated in albumin NP (ABI-007) was tested in patients with squamous cell carcinoma of the head and neck and anal canal (156). The formulation was administered intra-arterially to 43 patients every four weeks for three cycles. Myelosuppression (grade 4 neutropenia) was observed in three patients. Non-hematological toxicities included total alopecia, neurologic, skin, gastrointestinal and ocular toxicities, and flu-like syndrome. This study proved the interest of NS for paclitaxel administration, especially because no premedication was necessary. When treated with Taxol, patients received a drug to avoid allergic reactions expected with polyoxyethylated castor oil used to solubilize the drug. Similarly, the drug was better tolerated when administered intravenously to 19 patients suffering from breast cancer and melanoma (157). No hypersensitivity was observed, and the maximum tolerated dose was determined at 300 mg/m^2.

Interesting results have been obtained with magnetic NS (158–160). The administration of anticancer-drug-loaded magnetic NS to rats bearing s.c. Yoshida sarcoma (tail) was associated with a high rate of tumor remission. No deaths or widespread metastases occurred in the NS-treated groups. As discussed in section 3.1, such results could be explained partly by the facilitated accessibility of the tumor to the magnetic field and by the i.a. injection of the magnetic NS.

The interest of coating particles to target cancer site has been tested recently. Gene-loaded cationic polymerized lipid-based NP were prepared and a $\alpha_V\beta_3$ ligand was covalently attached to their surface (161). The specific targeting effect was demonstrated when the NP loaded with the luciferase gene were administered to tumor bearing mice; specific localization was achieved in the tumor whereas no luciferase activity was seen in the liver, heart, or lung. In addition, the activity in the tumor was suppressed when free soluble $\alpha_V\beta_3$ ligand was coinjected. Tumor regression was also obtained with gene-loaded NP covalently coupled to a $\alpha_V\beta_3$ ligand directed to target neovascularization surrounding tumor sites. Tumor regression and disappearance was associated in 75% of the cases with a decrease of blood vessel density.

Among the studies involving NP coupled to anticancer drugs some important issues have sometimes been neglected. In a few cases, the treatment was initiated at the same moment or a few hours following the tumor cell inoculation (104,120,137–139). According to Poste (162), the implantation of tumor cells into host tissue disrupts the local microvasculature and enhances the vascular permeability at the injection site. Vascular repair is generally achieved within two to three days. Premature injection of NP may result in unusually high amounts of cytostatic agent in the tumor. Accordingly, positive results obtained with NP injected much later after the inoculation of tumoral cells provide more reliable information (Table 4). Furthermore, transplanted tumors growing subcutaneously fail to provide a sufficiently relevant model for evaluating the effectiveness of agents in treating metastases (162).

In some studies (Table 4), failures in demonstrating any difference between the activity of a nanoparticle drug formulation over the solution dosage form may be caused by the presence of a too high amount of unbound anticancer drug in the nanoparticle suspension. In vivo, administration of a nanoparticle suspension containing some unbound drug may be of little consequence when the entrapment efficiency of the preparation procedure is high. But, when the proportion of the unbound drug in the preparation reaches, for instance, 85% to 92% of the total administered dose, it seems unlikely to anticipate important variations of the activity of the nanoparticle suspension from the solution formulation (120).

CONCLUDING REMARKS

Over the last 30 years, considerable progress has been made in the preparation of well characterized nanoparticle formulations loaded with a variety of anticancer agents (9). Despite this considerable amount of work, still little is known about the type of neoplasia which may respond beneficially to a nanoparticulate dosage form. More in vitro and in vivo experiments are needed to determine the best conditions requiring the administration of anticancer-drug-loaded NP. Increased efficiency of nanoparticle formulations against some drug-resistant leukemia and strong activity against hepatic metastases are examples which illustrate the possible applications of NP in the future. Recent studies have also pointed out that NP offer interesting delivery systems for new types of treatment under development at the present time. PDT or radiotherapy agents administered as solid particle suspensions will induce a better targeting of the diseased tissues and should definitely be a safer way to administer water insoluble drugs. NP may also be used for sustained release delivery of cytostatic agents in some specific circumstances. The extensive uptake of the NP by the RES and their limited access to extravascular tumors are major obstacles to an efficient drug targeting. Surface modifications of drug-loaded biodegradable NP may increase the half life of the carrier and/or allow its accumulation in a wider variety of target areas. This latter approach may open new and attractive perspectives in the chemotherapy of cancer.

REFERENCES

1. Gupta PK. Drug targeting in cancer chemotherapy: a clinical perspective. J Pharm Sci 1990; 79:949–962.
2. Zee-Cheng RKY, Cheng CC. Delivery of anticancer drugs. Methods Find Exp Clin Pharmacol 1989; 11:439–529.
3. Duncan R. Drug–polymer conjugates: potential for improved chemotherapy. Anti-Cancer Drugs 1992; 3:175–210.
4. Jones MC, Leroux JC. Polymeric micelles—a new generation of colloidal drug carriers. Eur J Pharm Biopharm 1999; 48:101–111.
5. Couvreur P. Polyalkylcyanoacrylates as colloidal drug carriers. Crit Rev Ther Drug Carrier Syst 1988; 5:1–20.
6. Couvreur P. Liposomes en clinique humaine: le point sur les résultats thérapeutiques. In: Delattre J, Couvreur P, Puisieux F, Philippot JR, Schubert F, eds. Les Liposomes. Paris: Inserm, 1993:214–243.
7. Flandroy PMJ, Grandfils C, Jérôme RJ. Clinical applications of microspheres in embolization and chemoembolization: a comprehensive review and perspectives. In: Rolland A, ed. Pharmaceutical Particulate Carriers. Therapeutic Applications. New York: Marcel Dekker, 1993:321–366.
8. Menei P, Benoit JP. Implantable drug-releasing biodegradable microspheres for local treatment of brain glioma. Acta Neurochir Suppl 2003; 88:51–55.
9. Allémann E, Gurny R, Doelker E. Drug loaded nanoparticles—preparation methods and drug targeting issues. Eur J Pharm Biopharm 1993; 39:173–191.
10. Kreuter J. Evaluation of nanoparticles as drug-delivery systems III: materials, stability, toxicity, possibilities of targeting, and use. Pharm Acta Helv 1983; 58:242–250.
11. Poste G. Drug targeting in cancer therapy. In: Receptor-mediated targeting of drugs. In: Gregoriadis G, Poste G, Senior J, Trouet A, eds. New York: Plenum Press, 1984: 427–475.
12. Jain RK. Transport of molecules across tumor vasculature. Cancer Metastasis Rev 1987; 6:559–593.
13. Duncan R. Polymer conjugates for tumour targeting and intracytoplasmic delivery. The EPR effect as a common gateway? Pharm Sci Technol Today 1999; 2(11):441–449.
14. De Jaeghere F, Doelker E, Gurny R. Nanoparticles. In: Mathiowitz E, ed. The Encyclopedia of Controlled Drug Delivery. New York: Wiley and Sons Inc., 1999:641–664.
15. Bellamy WT. Prediction of response to drug therapy. A review of in vitro assays. Drugs 1992; 44:690–708.
16. Chiannilkulchai N, Ammoury N, Caillou B, Devissaguet JP, Couvreur P. Hepatic tissue distribution of doxorubicin-loaded nanoparticles after i.v. administration in reticulosarcoma M 5076 metastasis-bearing mice. Cancer Chemother Pharmacol 1990; 26:122–126.
17. Maassen S, Fattal E, Müller RH, Couvreur P. Cell cultures for the assessment of toxicity and uptake of polymeric particulate drug carriers. STP Pharma 1993; 3:11–22.
18. Gipps EM, Groscurth P, Kreuter J, Speiser PP. The effects of polyalkylcyanoacrylate nanoparticles on human normal and malignant mesenchymal cells in vitro. Int J Pharm 1987; 40:23–31.
19. Soma CE, Dubernet C, Barratt G, Benita S, Couvreur P. Investigation of the role of macrophages on the cytotoxicity of doxorubicin and doxorubicin-loaded nanoparticles on M5076 cells in vitro. J Contr Rel 2000; 68(2):283–289.
20. Oppenheim RC, Gipps EM, Forbes JF, Whitehead RH. Development and testing of proteinaceous nanoparticles containing cytotoxics. In: Davis SS, Illum L, McVie JG, Tomlinson E, eds. Microspheres and Drug Therapy. Amsterdam: Elsevier, 1984: 117–128.
21. Rogers KE, Carr BI, Tökés ZA. Cell surface-mediated cytotoxicity of polymer-bound adriamycin against drug-resistant hepatocytes. Cancer Res 1983; 43:2741–2748.

22. Tökés ZA, Rogers KE, Rembaum A. Synthesis of adriamycin-coupled polyglutaraldehyde microspheres and evaluation of their cytostatic activity. Proc Natl Acad Sci USA 1982; 79:2026–2030.
23. Treupel L, Poupon MF, Couvreur P, Puisieux F. Vectorisation de la doxorubicine dans des nanosphères et réversion de la résistance pléiotropique des cellules tumorales. C R Acad Sci Paris 1991; 313:171–174.
24. Astier A, Doat B, Ferrer MJ, et al. Enhancement of adriamycin antitumor activity by its binding with an intracellular sustained-release form, polymethacrylate nanospheres, in U-937 cells. Cancer Res 1988; 48:1835–1841.
25. Kubiak C, Couvreur P, Manil L, Clausse B. Increased cytotoxicity of nanoparticle-carried adriamycin in vitro and potentiation by verapamil and amiodarone. Biomaterials 1989; 10:553–556.
26. Vauthier C, Dubernet C, Chauvierre C, Brigger I, Couvreur P. Drug delivery to resistant tumors: the potential of poly(alkyl cyanoacrylate) nanoparticles. J Contr Rel 2003; 93(2):151–160.
27. Morin C, Barratt G, Fessi H, Devissaguet JP, Puisieux F. Nitric oxide synthesis in rat alveolar macrophages is stimulated by an encapsulated immunomodulator. Proc 6th Int Conf Pharm Technol 1992; 5:50–60.
28. Smith A, Hunneyball IM. Evaluation of poly(lactic acid) as a biodegradable drug delivery system for parenteral administration. Int J Pharm 1986; 30:215–220.
29. Couvreur P, Grislain L, Lenaerts V, Brasseur F, Guiot P, Biernacki A. Biodegradable polymeric nanoparticles as drug carrier for antitumor agents. In: Guiot P, Couvreur P, eds. Polymeric Nanoparticles and Microspheres. Boca Raton: CRC Press, 1986:27–93.
30. Tökés ZA, Ross KL, Rogers KE. Use of microspheres to direct the cytotoxic action of adriamycin to the cell surface. In: Davis SS, Illum L, McVie JG, Tomlinson E, eds. Microspheres and Drug Therapy. Amsterdam: Elsevier, 1984:139–149.
31. Cuvier C, Roblot-Treupel L, Millot JM, Lizard G, Chevillard S, Manfait M, Couvreur P, Poupon MF. Doxorubicin-loaded nanospheres bypass tumor cell multidrug resistance. Biochem Pharmacol 1992; 44:509–517.
32. Nemati F, De Verdière AC, Dubernet C, et al. Cytotoxicity and uptake of free doxorubicin and doxorubicin-loaded nanoparticles in sensitive and multidrug-resistant leukaemic murine cells. Proc 6th Int Conf Pharm Technol 1992; 3:93–104.
33. Némati F, Dubernet C, De Verdière AC, et al. Some parameters influencing cytotoxicity of free doxorubicin and doxorubicin-loaded nanoparticles in sensitive and multidrug resistant leukemic murine cells: incubation time, number of nanoparticles per cell. Int J Pharm 1994; 102:55–62.
34. De Verdière AC, Dubernet C, Nemati F, Poupon MF, Puisieux F, Couvreur P. Uptake of doxorubicin from loaded nanoparticles in multidrug-resistant leukemic murine cells. Cancer Chemother Pharmacol 1994; 33:504–508.
35. Bennis S, Chapey C, Couvreur P, Robert J. Enhanced cytotoxicity of doxorubicin encapsulated in polyisohexylcyanoacrylate nanospheres against multidrug-resistant tumour cells in culture. Eur J Cancer 1994; 30A:89–93.
36. Soma CE, Dubernet C, Bentolila D, Benita S, Couvreur P. Reversion of multidrug resistance by co-encapsulation of doxorubicin and cyclosporin A in polyalkylcyanoacrylate nanoparticles. Biomaterials 2000; 21(1):1–7.
37. Soma CE, Dubernet C, Barratt G, et al. Ability of doxorubicin-loaded nanoparticles to overcome multidrug resistance of tumor cells after their capture by macrophages. Pharm Res 1999; 16(11):1710–1716.
38. De Verdière AC, Dubernet C, Némati F, et al. Reversion of multidrug resistance with polyalkylcyanoacrylate nanoparticles: towards a mechanism of action. Br J Cancer 1997; 76(2):198–205.
39. Speth PAJ, Van Hoesel QGCM, Haanen C. Clinical pharmacokinetics of doxorubicin. Clin Pharmacokinet 1988; 15:15–31.

40. Ling V. Multidrug resistance and P-glycoprotein expression. Ann N Y Acad Sci 1987; 507:7–8.
41. Ban T. Pleiotropic, multidrug-resistant phenotype and P-glycoprotein: a review. Chemotherapy 1992; 38:191–196.
42. Nemati F, Dubernet C, Fessi H, et al. Reversion of multidrug resistance using nanoparticles in vitro: influence of the nature of the polymer. Int J Pharm 1996; 138(2):237–246.
43. Pepin X, Attali L, Domrault C, et al. On the use of ion-pair chromatography to elucidate doxorubicin release mechanism from polyalkylcyanoacrylate nanoparticles at the cellular level. J Chromatogr B Biomed Sci Appl 1997; 702(1–2):181–191.
44. Colin de Verdiere A, Dubernet C, Nemati F, Poupon MF, Puisieux F, Couvreur P. Uptake of doxorubicin from loaded nanoparticles in multidrug-resistant leukemic murine cells. Cancer Chemother Pharmacol 1994; 33(6):504–508.
45. Garcion E, Heutault B, Lamprecht R, et al. A new generation of anticancer drug loaded colloidal vectors reverts multidrug resistance in glioma and reduce. European Conference on Drug Delivery and Pharmaceutical Technology, May 10–12, 2004, Sevilla, Spain, #33.
46. Seyler I, Morin C, Barratt G, Appel M, Devissaguet JP, Puisieux F. Induction of macrophage NO-synthase by an immunomodulator entrapped within polymeric nanocapsules. Eur J Pharm Biopharm 1995; 41(1):49–54.
47. Al Khouri Fallouh N, Roblot-Treupel L, Fessi H, Devissaguet JP, Puisieux F. Development of a new process for the manufacture of polyisobutylcyanoacrylate nanocapsules. Int J Pharm 1986; 28:125–132.
48. Bonduelle S, Foucher C, Leroux JC, Chouinard F, Cadieux C, Lenaerts V. Association of cyclosporin to isohexylcyanoacrylate nanospheres and subsequent release in human plasma in vitro. J Microencapsul 1992; 9:173–182.
49. Labib A, Lenaerts V, Chouinard F, Leroux JC, Ouellet R, Van Lier JE. Biodegradable nanospheres containing phthalocyanines and naphthalocyanines for targeted photodynamic tumor therapy. Pharm Res 1991; 8:1027–1031.
50. Barrat GM, Yu WP, Fessi H, et al. Delivery of MDP-L-alanyl-cholesterol to macrophages: comparison of liposomes and nanocapsules. Cancer J 1989; 2:439–443.
51. Morin C, Barratt G, Fessi H, Devissaguet JP, Puisieux F. Improved intracellular delivery of a muramyl dipeptide analog by means of nanocapsules. Int J Immunopharmacol 1994; 16:451–456.
52. Morin C, Barratt G, Fessi H, Devissaguet JP, Puisieux F. Biodegradable nanocapsules containing a lipophilic immunomodulator: drug retention and tolerance towards macrophages in vitro. J Drug Targ 1993; 1:157–164.
53. Chabner BA. Biological basis for cancer treatment. Ann Int Med 1993; 118:633–637.
54. Lebedeva I, Benimetskaya L, Stein CA, Vilenchik M. Cellular delivery of antisense oligonucleotides. Eur J Pharm Biopharm 2000; 50(1):101–120.
55. Wagner RW, Fanagan WM. Antisense technology and prospects for therapy of viral infections and cancer. Mol Med Today 1997:31–38.
56. Wagner E, Kircheis R, Walker GF. Targeted nucleic acid delivery into tumors: new avenues for cancer therapy. Biomed Pharmacother 2004; 58(3):152–161.
57. Spencer CM, Faulds D. Paclitaxel. A review of its pharmacodynamic and pharmacokinetic properties and therapeutic potential in the treatment of cancer. Drugs 1994; 48(5):794–847.
58. Singla AK, Garg A, Aggarwal D. Paclitaxel and its formulations. Int J Pharm 2002; 235(1–2):179–192.
59. Fonseca C, Simoes S, Gaspar R. Paclitaxel-loaded PLGA nanoparticles: preparation, physicochemical characterization and in vitro anti-tumoral activity. J Contr Rel 2002; 83(2):273–286.
60. Feng SS, Mu L, Win KY, Huang G. Nanoparticles of biodegradable polymers for clinical administration of paclitaxel. Curr Med Chem 2004; 11(4):413–424.

61. Reddi H. Role of delivery vehicles for photosensitizers in the photodynamic therapy of tumours. J Photochem Photobiol B 1997; 37:189–195.
62. Kessel D. Delivery of photosensitizing agents. Adv Drug Deliv Rev 2004; 56(1):7–8.
63. Bachor R, Shea CR, Gilles R, Hasan T. Photosensitized destruction of human bladder carcinoma cells treated with chlorin e_6-conjugated microspheres. Proc Natl Acad Sci USA 1991; 88:1580–1584.
64. Bachor R, Shea CR, Belmonte SJ, Hasan T. Free and conjugated chlorin e_6 in the photodynamic therapy of human bladder carcinoma cells. J Urol 1991; 146:1654–1658.
65. Bachor R, Scholz M, Shea CR, Hasan T. Mechanism of photosensitization by microsphere-bound chlorin e_6 in human bladder carcinoma cells. Cancer Res 1991; 51(15):4410–4414.
66. Lenaerts V, Labib A, Chouinard F, et al. Nanocapsules with a reduced liver uptake: targeting of phthalocyanines to EMT-6 mouse mammary tumour in vivo. Eur J Pharm Biopharm 1995; 41(1):38–43.
67. Aveline BM, Hasan T, Redmond RW. The effects of aggregation, protein binding and cellular incorporation on the photophysical properties of benzoporphyrin derivative monoacid ring A (BPDMA). J Photochem Photobiol B 1995; 30(2–3):161–169.
68. Bourdon O, Mosqueira V, Legrand P, Blais J. A comparative study of the cellular uptake, localization and phototoxicity of meta-tetra(hydroxyphenyl) chlorin encapsulated in surface-modified submicronic oil/water carriers in HT29 tumor cells. J Photochem Photobiol B 2000; 55(2–3):164–171.
69. Konan YN, Chevallier J, Gurny R, Allémann E. Encapsulation of p-THPP into nanoparticles: cellular uptake, subcellular localization and effect of serum on photodynamic activity. Photochem Photobiol 2003; 77(6):638–644.
70. Konan YN, Cerny R, Favet J, Berton M, Gurny R, Allémann E. Preparation and characterization of sterile sub-200 nm meso-tetra(4-hydroxylphenyl)porphyrin-loaded nanoparticles for photodynamic therapy. Eur J Pharm Biopharm 2003; 55(1):115–124.
71. Konan YN, Berton M, Gurny R, Allémann E. Enhanced photodynamic activity of meso-tetra(4-hydroxyphenyl)porphyrin by incorporation into sub-200 nm nanoparticles. Eur J Pharm Sci 2003; 18(3–4):241–249.
72. Roy I, Ohulchanskyy TY, Pudavar HE, et al. Ceramic-based nanoparticles entrapping water-insoluble photosensitizing anticancer drugs: a novel drug-carrier system for photodynamic therapy. J Am Chem Soc 2003; 125(26):7860–7865.
73. Lange N. Controlled drug delivery in photodynamic therapy and fluorescence-based diagnosis of cancer. In: Mycek MA, Pogue BW, eds. Handbook of Biomedical Fluorescence. New York, Basel: Marcel Dekker, 2003:563–635.
74. Allémann E, Brasseur N, Benrezzak O, et al. PEG-coated poly(lactic acid) nanoparticles for the delivery of hexadecafluoro zinc phthalocyanine to EMT-6 mouse mammary tumours. J Pharm 1995; 47:382–387.
75. Allémann E, Rousseau J, Brasseur N, Kudrevich SV, Lewis K, Van Lier JE. Photodynamic therapy of tumours with hexadecafluoro zinc phthalocyanine formulated in PEG-coated poly(lactic acid) nanoparticles. Int J Cancer 1996; 66(6):821–824.
76. Blackwell JE, Dagia NM, Dickerson JB, Berg EL, Goetz DJ. Ligand coated nanosphere adhesion to E- and P-selectin under static and flow conditions. Ann Biomed Eng 2001; 29(6):523–533.
77. Zhang Y, Kohler N, Zhang M. Surface modification of supramagnetic magnetite nanoparticles and their intracellular uptake. Biomaterials 2002; 23:1553–1561.
78. Stella B, Arpicco S, Peracchia MT, et al. Design of folic acid-conjugated nanoparticles for drug targeting. J Pharm Sci 2000; 89(11):1452–1464.
79. Stella B, Marsaud V, Arpicco S, et al. Biological characterization of folic acid-conjugated nanoparticles in cellular models. Proc Int Symp Control Rel Bioact Mater 2001:5200.

80. Oyewumi MO, Liu S, Moscow JA, Mumper RJ. Specific association of thiamine-coated gadolinium nanoparticles with human breast cancer cells expressing thiamine transporters. Bioconjug Chem 2003; 14(2):404–411.
81. Thorstensen K, Romslo I. The transferrin receptor: its diagnostic value and its potential as therapeutic target. Scand J Clin Lab Invest suppl 1993; 215:113–120.
82. Bellocq NC, Pun SH, Jensen GS, Davis ME. Transferrin-containing, cyclodextrin polymer-based particles for tumor-targeted gene delivery. Bioconjug Chem 2003; 14(6):1122–1132.
83. Sugibayashi K, Morimoto Y, Nadai T, Kato Y. Drug-carrier property of albumin microspheres in chemotherapy. I. Tissue distribution of microsphere-entrapped 5-fluorouracil in mice. Chem Pharm Bull 1977; 25:3433–3434.
84. Sugibayashi K, Morimoto Y, Nadai T, Kato Y, Hasegawa A, Arita T. Drug-carrier property of albumin microspheres in chemotherapy. II. Preparation and tissue distribution in mice of microsphere-entrapped 5-fluorouracil. Chem Pharm Bull 1979; 27(1): 204–209.
85. Yoshioka T, Hashida M, Muranishi S, Sezaki H. Specific delivery of mitomycin C to the liver, spleen and lung: nano- and microspherical carriers of gelatin. Int J Pharm 1981; 8:131–141.
86. Kante B, Couvreur P, Lenaerts V, et al. Tissue distribution of [^{3}H]actinomycin D adsorbed on polybutylcyanoacrylate nanoparticles. Int J Pharm 1980; 7:45–53.
87. Rolland A. Pharmacokinetics and tissue distribution of doxorubicin-loaded polymethacrylic nanoparticles in rabbits. Int J Pharm 1988; 42:145–154.
88. Verdun C, Brasseur F, Vranckx H, Couvreur P, Roland M. Tissue distribution of doxorubicin associated with polyisohexylcyanoacrylate nanoparticles. Cancer Chem Pharm 1990; 26:13–18.
89. Gulyaev AE, Gelperina SE, Skidan IN, Antropov AS, Kivman GY, Kreuter J. Significant transport of doxorubicin into the brain with polysorbate 80-coated nanoparticles. Pharm Res 1999; 16(10):1564–1569.
90. Reszka R, Beck P, Fichtner I, Hentschel M, Richter J, Kreuter J. Body distribution of free, liposomal and nanoparticle-associated mitoxantrone in B16-melanoma-bearing mice. J Pharm Exp Ther 1997; 280(1):232.
91. Couvreur P, Vauthier C. Polyalkylcyanoacrylate nanoparticles as drug carrier: present state and perspectives. J Contr Rel 1991; 17:187–198.
92. Rolland A. Clinical pharmacokinetics of doxorubicin in hepatoma patients after a single intravenous injection of free or nanoparticle-bound anthracycline. Int J Pharm 1989; 54:113–121.
93. O'Mullane JE, Artursson P, Tomlinson E. Biopharmaceutics of microparticulate drug carriers. Ann N Y Acad Sci 1987; 507:120–140.
94. Widder KJ, Senyei AE, Ranney DF. Magnetically responsive microspheres and other carriers for the biophysical targeting of antitumor agents. Adv Pharm Chem 1979; 16:213–271.
95. Couvreur P, Kante B, Lenaerts V, Scailteur V, Roland M, Speiser P. Tissue distribution of antitumor drugs associated with polyalkylcyanoacrylate nanoparticles. J Pharm Sci 1980; 69:199–202.
96. Douglas SJ, Davis SS, Illum L. Nanoparticles in drug delivery. Crit Rev Ther Drug Carrier Syst 1987; 3:233–261.
97. Bapat N, Boroujerdi M. Effect of colloidal carriers on the disposition and tissue uptake of doxorubicin: II. Conjugation with isobutylcyanoacrylate nanoparticles. Drug Dev Ind Pharm 1993; 19:2667–2678.
98. Kattan J, Droz JP, Couvreur P, et al. Phase I clinical trial and pharmacokinetic evaluation of doxorubicin carried by polyisohexylcyanoacrylate nanoparticles. Invest New Drugs 1992; 10:191–199.
99. Verdun C, Couvreur P, Vranckx H, Lenaerts V, Roland M. Development of a nanoparticle controlled-release formulation for human use. J Contr Rel 1986; 3:205–210.

100. Grislain L, Couvreur P, Roland M. Modification de la pharmacocinétique de molécules associées aux nanoparticules. S T P Pharma 1985; 1:1038–1042.
101. Bégué JM, Rolland A, Merdrignac G, et al. Etude de la pénétration de nanoparticules de polyméthacrylate dans des cultures primaires d'hépatocytes de rat et d'homme et des lignées cellulaires d'hépatomes: application à la vectorisation d'agents cytostatiques utilisés en chimiothérapie anticancéreuse. Arch Int Phys Biochem 1988; 96:A413.
102. Grislain L, Couvreur P, Lenaerts V, Roland M, Deprez-De Campeneere D, Speiser P. Pharmacokinetics and distribution of a biodegradable drug-carrier. Int J Pharm 1983; 15:335–345.
103. James SE, Salsbury AJ. Effect of (+/−)-1,2-bis(3,5-dioxopiperazin-1-yl)propane on tumor blood vessels and its relationship to the antimetastasic effect in the Lewis lung carcinoma. Cancer Res 1974; 34:839–842.
104. Kreuter J, Hartmann HR. Comparative study on the cytostatic effects and the tissue distribution of 5-fluorouracil in a free form and bound to polybutylcyanoacrylate nanoparticles in sarcoma 180-bearing mice. Oncology 1983; 40:363–366.
105. Gipps EM, Arshady R, Kreuter J, Groscurth P, Speiser PP. Distribution of polyhexyl cyanoacrylate nanoparticles in nude mice bearing human osteosarcoma. J Pharm Sci 1986; 75:256–258.
106. Kreuter J. Nanoparticulate systems for brain delivery of drugs. Adv Drug Deliv Rev 2001; 47(1):65–81.
107. Lode J, Fichtner I, Kreuter J, Berndt A, Diederichs JE, Reszka R. Influence of surface-modifying surfactants on the pharmacokinetic behavior of ^{14}C-poly (methylmethacrylate) nanoparticles in experimental tumor models. Pharm Res 2001; 18(11):1613–1619.
108. Olivier JC, Fenart L, Chauvet R, Pariat C, Cecchelli R, Couet W. Indirect evidence that drug brain targeting using polysorbate 80-coated polybutylcyanoacrylate nanoparticles is related to toxicity. Pharm Res 1999; 16(12):1836–1842.
109. Brigger I, Morizet J, Aubert G, et al. Poly(ethylene glycol)-coated hexadecylcyanoacrylate nanospheres display a combined effect for brain tumor targeting. J Pharm Exp Ther 2002; 303(3):928–936.
110. Dunn SE, Brindley A, Davies MC, Davis SS, Illum L. Studies on in-vitro uptake by Kupffer cells and the in-vivo biodistribution of a range of novel polymeric colloids. J Pharm Pharmacol 1992; 44(suppl):1082.
111. Rudt S, Müller RH. In vitro phagocytosis assay of nano- and microparticles by chemiluminescence. II. Effect of surface modification by coating of particles with poloxamer on the phagocytic uptake. J Contr Rel 1993; 25:51–59.
112. Rudt S, Wesemeyer H, Müller RH. In vitro phagocytosis assay of nano- and microparticles by chemiluminescence. IV. Effect of surface modification by the coating particles with poloxamine and Antarox CO on the phagocytic uptake. J Contr Rel 1993; 25: 123–132.
113. Leroux JC, De Jaeghere F, Anner BM, Doelker E, Gurny R. An investigation on the role of plasma and serum opsonins on the internalization of biodegradable poly(D,L-lactic acid) nanoparticles by human monocytes. Life Sci 1995; 57:695–703.
114. Leroux JC, Gravel P, Balant L, et al. Internalization of poly(D,L-lactic acid) nanoparticles by isolated human leukocytes and analysis of plasma proteins adsorbed onto the particles. J Biomed Mater Res 1994; 28:471–481.
115. Tröster SD, Kreuter J. Influence of the surface properties of low contact angle surfactants on the body distribution of ^{14}C-poly(methyl methacrylate) nanoparticles. J Microencapsul 1992; 9:19–28.
116. Illum L, Davis SS. Targeting of colloidal particles to the bone marrow. Life Sci 1987; 40:1553–1560.
117. Papahadjopoulos D, Allen TM, Gabizon A, et al. Sterically stabilized liposomes: improvement in pharmacokinetics and antitumor therapeutic efficacy. Proc Natl Acad Sci USA 1991; 88:11460–11464.

118. Unezaki S, Maruyama K, Ishida O, Takahashi N, Iwatsuru M. Enhanced tumor targeting of doxorubicin by ganglioside GM1-bearing long-circulating liposomes. J Drug Targ 1993; 1:287–292.
119. Douglas SJ, Davis SS, Illum L. Biodistribution of poly(butyl 2-cyanoacrylate) nanoparticles in rabbits. Int J Pharm 1986; 34:145–152.
120. Beck P, Kreuter J, Reszka R, Fichtner I. Influence of polybutylcyanoacrylate nanoparticles and liposomes on the efficacy and toxicity of the anticancer drug mitoxantrone in murine tumour models. J Microencapsul 1993; 10:101–114.
121. Akasaka Y, Ueda H, Takayama K, Machida Y, Nagai T. Preparation and evaluation of bovine serum albumin nanospheres coated with monoclonal antibodies. Drug Design Deliv 1988; 3:85–97.
122. Illum L, Jones PDE, Baldwin RW, Davis SS. Tissue distribution of poly(hexyl 2-cyanoacrylate) nanoparticles coated with monoclonal antibodies in mice bearing human tumor xenografts. J Pharm Exp Ther 1984; 230:733–736.
123. Kosmas C, Linardou H, Epenetos A. Review: advances in monoclonal antibody tumour targeting. J Drug Targ 1993; 1:81–91.
124. Jain RK. Physiological barriers to delivery of monoclonal antibodies and other macromolecules in tumors. Cancer Res 1990; 50:814s–819s.
125. Gupta PK, Hung CT. Magnetically controlled targeted micro-carrier systems. Life Sci 1989; 44:175–186.
126. Widder KJ, Senyei AE, Scarpelli DG. Magnetic microspheres: a model system for site specific drug delivery in vivo. Proc Soc Exp Biol Med 1978; 58:141–146.
127. Senyei AE, Reich SD, Gonczy C, Widder KJ. In vivo kinetics of magnetically targeted low-dose doxorubicin. J Pharm Sci 1981; 70:389–391.
128. Widder KJ, Marino PA, Morris RM, Howard DP, Poore GA, Senyei AE. Selective targeting of magnetic albumin microspheres to the Yoshida sarcoma: ultrastructural evaluation of microsphere disposition. Eur J Cancer Clin Oncol 1983; 19:141–147.
129. Gupta PK, Hung CT, Rao NS. Ultrastructural disposition of adriamycin-associated magnetic albumin microspheres in rats. J Pharm Sci 1989; 78:290–294.
130. Gallo JM, Gupta PK, Hung CT, Perrier DG. Evaluation of drug delivery following the administration of magnetic albumin microspheres containing adriamycin to the rat. J Pharm Sci 1989; 78:190–194.
131. Gupta PK, Hung CT. Albumin microspheres. V. Evaluation of parameters controlling the efficacy of magnetic microspheres in the targeted delivery of adriamycin in rats. Int J Pharm 1990; 59:57–67.
132. Gallo JM, Hung CT, Gupta PK, Perrier DG. Physiological pharmacokinetic model of adriamycin delivered via magnetic albumin microspheres in the rat. J Pharmacokinet Biopharm 1989; 17:305–326.
133. Gupta PK, Hung CT. Effect of carrier dose on the multiple tissue disposition of doxorubicin hydrochloride administered via magnetic albumin microspheres in rats. J Pharm Sci 1989; 78:745–748.
134. Gupta PK, Hung CT. Targeted delivery of low dose doxorubicin hydrochloride administered via magnetic albumin microspheres in rats. J Microencapsul 1990; 7:85–94.
135. Ibrahim A, Couvreur P, Roland M, Speiser P. New magnetic drug carrier. J Pharm Pharmacol 1983; 35:59–61.
136. Brasseur F, Couvreur P, Kante B, et al. Actinomycin D adsorbed on polymethylcyanoacrylate nanoparticles: increased efficiency against an experimental tumor. Eur J Cancer 1980; 16:1441–1445.
137. Morimoto Y, Akimoto M, Sugibayashi K, Nadai T, Kato Y. Drug-carrier property of albumin microspheres in chemotherapy. IV. Antitumor effect of single-shot or multiple-shot administration of microsphere-entrapped 5-fluorouracil on Ehrlich ascites or solid tumor in mice. Chem Pharm Bull 1980; 28:3087–3092.

138. Sugibayashi K, Akimoto M, Morimoto Y, Nadai T, Kato Y. Drug-carrier property of albumin microspheres in chemotherapy. III. Effect of microsphere-entrapped 5-fluorouracil on Ehrlich ascites carcinoma in mice. J Pharm Dyn 1979; 2:350–355.
139. Morimoto Y, Sugibayashi K, Kato Y. Drug-carrier property of albumin microspheres in chemotherapy. V. Antitumor effect of microsphere-entrapped adriamycin on liver metastasis of AH 7974 cells in rats. Chem Pharm Bull 1981; 29:1433–1438.
140. Brasseur F, Verdun C, Couvreur P, Deckers C, Roland M. Evaluation expérimentale de l'efficacité thérapeutique de la doxorubicine associée aux nanoparticules de polyalkylcyanoacrylate. Proc 4th Int Conf Pharm Technol 1986; 5:177–186.
141. Chiannilkulchai N, Driouich Z, Benoit JP, Parodi AL, Couvreur P. Nanoparticules de doxorubicine: vecteurs colloidaux dans le traitement des métastases hépatiques chez l'animal. Bull Cancer 1989; 76:845–848.
142. Chiannilkulchai N, Driouich Z, Benoit JP, Parodi AL, Couvreur P. Doxorubicin-loaded nanoparticles: increased efficiency in murine hepatic metastases. Sel Cancer Ther 1989; 5:1–11.
143. Mitra S, Gaur U, Ghosh PC, Maitra AN. Tumour targeted delivery of encapsulated dextran–doxorubicin conjugate using chitosan nanoparticles as carrier. J Contr Rel 2001; 74(1–3):317–323.
144. Onishi H, Machida Y, Machida Y. Antitumor properties of irinotecan-containing nanoparticles prepared using poly(DL-lactic acid) and poly(ethylene glycol)-block-poly(propylene glycol)-block-poly(ethylene glycol). Biol Pharm Bull 2003; 26(1):116–119.
145. Sharma D, Chelvi TP, Kaur J, et al. Novel Taxol formulation: polyvinylpyrrolidone nanoparticle-encapsulated Taxol for drug delivery in cancer therapy. Oncol Res 1996; 8:281–286.
146. Gibaud S, Andreux JP, Weingarten C, Renard M, Couvreur P. Increased bone marrow toxicity of doxorubicin bound to nanoparticles. Eur J Cancer 1994; 30A:820–826.
147. Manil L, Couvreur P, Mahieu P. Acute renal toxicity of doxorubicin (adriamycin)-loaded cyanoacrylate nanoparticles. Pharm Res 1995; 12:85–87.
148. Couvreur P, Kante B, Grislain L, Roland M, Speiser P. Toxicity of polyalkylcyanoacrylate nanoparticles. II. Doxorubicin-loaded nanoparticles. J Pharm Sci 1982; 71:790–792.
149. Yu WP, Barrat GM, Devissaguet JP, Puisieux F. Anti-metastatic activity in vivo of MDP-L-alanyl-cholesterol (MTP-CHOL) entrapped in nanocapsules. Int J Immunopharmacol 1991; 13:167–173.
150. Yu WP, Foucher C, Barratt G, Fessi H, Devissaguet JP, Puisieux F. Anti-metastatic activity of muramyl dipeptide-L-alanyl-cholesterol incorporated into various types of nanocapsules. Proc 6th Int Conf Pharm Technol 1992; 3:83–92.
151. Barratt GM, Puisieux F, Yu WP, Foucher C, Fessi H, Devissaguet JP. Anti-metastatic activity of MDP-L-alanyl-cholesterol incorporated into various types of nanocapsules. Int J Immunopharmacol 1994; 16:457–461.
152. Blunk T, Hochstrasser DF, Sanchez JC, Müller BW, Müller RH. Colloidal carriers for intravenous drug targeting: plasma protein adsorption patterns on surface modified latex particles evaluated by two-dimensional polyacrylamide gel electrophoresis. Electrophoresis 1993; 14:1382–1387.
153. Tabata Y, Ikada Y. Protein precoating of polylactide microspheres containing a lipophilic immunopotentiator for enhancement of macrophage phagocytosis and activation. Pharm Res 1989; 6:296–301.
154. Shih JL, Brugger RM. Gadolinium as a neutron capture therapy agent. Med Phys 1992; 19(3):733–744.
155. Tokumitsu H, Hiratsuka J, Sakurai Y, Kobayashi T, Ichikawa H, Fukumori Y. Gadolinium neutron-capture therapy using novel gadopentetic acid–chitosan complex nanoparticles: in vivo growth suppression of experimental melanoma solid tumor. Cancer Lett 2000; 150(2):177–182.
156. Damascelli B, Cantu G, Mattavelli F, et al. Intraarterial chemotherapy with polyoxyethylated castor oil free paclitaxel, incorporated in albumin nanoparticles (ABI-007):

Phase II study of patients with squamous cell carcinoma of the head and neck and anal canal: preliminary evidence of clinical activity. Cancer 2001; 92(10):2592–2602.
157. Ibrahim NK, Desai N, Legha S, et al. Phase I and pharmacokinetic study of ABI-007, a Cremophor-free, protein-stabilized, nanoparticle formulation of paclitaxel. Clin Cancer Res 2002; 8(5):1038–1044.
158. Widder KJ, Morris RM, Poore G, Howard DP Jr, Senyei AE. Tumor remission in Yoshida sarcoma-bearing rats by selective targeting of magnetic albumin microspheres containing doxorubicin. Proc Natl Acad Sci USA 1981; 78:579–581.
159. Widder KJ, Morris RM, Poore GA, Howard DP, Senyei AE. Selective targeting of magnetic albumin microspheres containing low-dose doxorubicin: total remission in Yoshida sarcoma-bearing rats. Eur J Cancer Clin Oncol 1983; 19:135–139.
160. Morris RM, Poore GA, Howard DP, Sefranka JA. Selective targeting of magnetic albumin microspheres containing vindesine sulfate: total remission in Yoshida sarcoma-bearing rats. In: Davis SS, Illum L, McVie JG, Tomlinson E, eds. Microspheres and Drug Therapy. Amsterdam: Elsevier, 1984:439.
161. Hood JD, Bednarski M, Frausto R, et al. Tumor regression by targeted gene delivery to the neovasculature. Sci 2002; 296:2404–2407.
162. Poste G, Kirsh R. Site-specific (targeted) drug delivery in cancer chemotherapy. Biotech 1983; 1:869–878.
163. Kramer PA, Burnstein T. Phagocytosis of microspheres containing an anticancer agent by tumor cells in vitro. Life Sci 1976; 19:515–520.
164. Oppenheim RC, Stewart NF. The manufacture and tumour cell uptake of nanoparticles labelled with fluorescein isothiocyanate. Drug Dev Ind Pharm 1979; 5:563–571.
165. Tökés ZA, Rogers KE, Daniels AM, Daniels JR. Increased cytotoxic effects by polymer-bound adriamycin are mediated through the cell surface. Proc Am Assoc Cancer Res 1983; 24:255.
166. Guiot P, Couvreur P. Quantitative study of the interaction between polybutylcyanoacrylate nanoparticles and mouse peritoneal macrophages in culture. J Pharm Belg 1983; 38:130–134.
167. Astier A, Benoit G, Doat B, Ferrer MJ, Le Verge R. Activité antitumorale de l'adriamycine liée à des nanoparticules sur 2 lignées cellulaires cancéreuses humaines. J Pharm Clin 1988; 7:43–55.
168. Rolland A, Bégué JM, Le Verge R, Guillouzo A. Increase of doxorubicin penetration in cultured rat hepatocytes by its binding to polymethacrylic nanoparticles. Int J Pharm 1989; 53:67–73.
169. Akiyoshi K, Taniguchi I, Fukui H, Sunamoto J. Hydrogel nanoparticle formed by self-assembly of hydrophobized polysaccharide. Stabilization of adriamycin by complexation. Eur J Pharm Biopharm 1996; 42(4):286–290.
170. Yoo HS, Oh JE, Lee KH, Park TG. Biodegradable nanoparticles containing doxorubicin-PLGA conjugate for sustained release. Pharm Res 1999; 16(7):1114–1118.
171. Yoo HS, Lee KH, Oh JE, Park TG. In vitro and in vivo anti-tumor activities of nanoparticles based on doxorubicin-PLGA conjugates. J Contr Rel 2000; 68(3):419–431.
172. Janes KA, Fresneau MP, Marazuela A, Fabra A, Alonso MJ. Chitosan nanoparticles as delivery systems for doxorubicin. J Contr Rel 2001; 73(2–3):255–267.
173. Santhi K, Dhanaraj SA, Joseph V, Ponnusankar S, Suresh B. A study on the preparation and anti-tumor efficacy of bovine serum albumin nanospheres containing 5-fluorouracil. Drug Dev Ind Pharm 2002; 28(9):1171–1179.
174. Chawla JS, Amiji MM. Cellular uptake and concentrations of tamoxifen upon administration in poly(epsilon-caprolactone) nanoparticles. AAPS Pharm Sci 2003; 5(1):E3.
175. Couvreur P, Aubry J. Monoclonal antibodies for the targeting of drugs: application to nanoparticles. In: Breimer DD, Speiser P, eds. Topics in Pharmaceutical Sciences. Amsterdam: Elsevier, 1983:305–316.

176. Laakso T, Andersson J, Artursson P, Edman P, Sjöholm I. Acrylic microspheres in vivo. X. Elimination of circulating cells by active targeting using specific monoclonal antibodies bound to microparticles. Life Sci 1986; 38:183–190.
177. Simeonova M, Ivanova T, Raikova E, Georgieva M, Raikov Z. Tissue distribution of polybutylcyanoacrylate nanoparticles carrying spin-labelled nitrosourea. Int J Pharm 1988; 43:267–271.
178. Mukherji G, Murthy RSR, Miglani BD. Preparation and evaluation of polyglutaraldehyde nanoparticles containing 5-fluorouracil. Int J Pharm 1989; 50:15–19.
179. Mukherji G, Murthy RSR, Miglani BD. Preparation and evaluation of cellulose nanospheres containing 5- fluorouracil. Int J Pharm 1990; 65:1–5.
180. Kreuter J. Factors influencing the body distribution of polyacrylic nanoparticles. In: Buri P, Gumma A, eds. Drug Targeting. Amsterdam: Elsevier, 1985:51–63.
181. Gupta PK, Hung CT, Lam FC. Application of regression analysis in the evaluation of tumor response following the administration of adriamycin either as a solution or via magnetic microspheres to the rat. J Pharm Sci 1990; 79:634–637.
182. Blagoeva P, Balansky RM, Mircheva TJ, Simeonova MI. Diminished genotoxicity of mitomycin C and farmorubicin included in polybutylcyanoacrylate nanoparticles. Mutat Res 1992; 268:77–82.
183. Illum L, Jones PDE, Kreuter J, Bladwin RW, Davis SS. Adsorption of monoclonal antibodies to polyhexylcyanoacrylate nanoparticles and subsequent immunospecific binding to tumour cells in vitro. Int J Pharm 1983; 17:65–83.

19

Development of 5-FU–Loaded PLGA Microparticles for the Treatment of Glioblastoma

Nathalie Faisant
INSERM U646, 'Ingénierie de la Vectorisation Particulaire', Université d'Angers, Immeuble IBT, Angers, France

Jean-Pierre Benoit
INSERM U646, 'Ingénierie de la Vectorisation Particulaire', Université d'Angers, Angers, France

Philippe Menei
INSERM U646, 'Ingénierie de la Vectorisation Particulaire', Université d'Angers, Immeuble IBT and Department of Neurosurgery, CHU Angers, Angers, France

INTRODUCTION

Malignant gliomas represent 13% to 22% of brain tumors and regardless of the method of treatment, average survival of patients is less than one year (1). Despite surgery, external beam radiation therapy, and systemic chemotherapy, these tumors tend to recur within a few centimeters of their original location (2).

Several factors such as the blood-brain barrier (BBB), the intense neovascularization of the tumor, and its aggressiveness are responsible for the poor success of glioma therapy. The BBB forms a physiological barrier that prevents the influx of molecules from the bloodstream into the brain. Therefore, many chemotherapeutic agents do not diffuse into the central nervous system (CNS) (3).

To improve drug delivery to brain tumors, several strategies have been studied and reviewed (4,5). A first approach is to take advantage of the natural permeability properties of the BBB of some species by modifying chemotherapeutic agents in order to improve, for example, their lipophilicity, or by linking them to a carrier capable of traversing the BBB. A second approach consists of first disrupting the BBB with intra-arterial hyperosmolar mannitol, and then delivering the therapeutic agent. A third approach is to encapsulate a drug in a colloidal carrier that has the ability to interact with the BBB; for example, poly(butylcyanoacrylate) nanoparticles and immunoliposomes have displayed some interesting properties (6,7). A fourth approach is local delivery of the therapeutic agent to the tumor site, either via catheters or via sustained-release polymers. This approach offers the advantage of sustained local exposure to concentrated amounts of the drug while avoiding significant systemic exposure. In addition, this approach could be particularly appropriate for gliomas, as 80% to 90% of them recur within 2 cm of the original site of resection. Catheters have been in clinical use for many years and several systems have been tested; however, these devices are all limited by mechanical pitfalls and infections, and none has been proven superior over the others in the treatment of malignant gliomas.

An alternative approach to local intratumoral delivery into the brain is offered by biocompatible sustained-release devices made of polymers. A variety of local brain delivery systems such as liposomes, gelatin–chondroitin sulfate–coated microspheres, gelatin sponges (4,5,8), etc., have been developed and tested. Biocompatible polyanhydride implants loaded with carmustine 1,3-bis (2-chloroethyl)-1-nitrosourea (BCNU) (Gliadel®, Baltimore, Maryland, U.S.) have been extensively studied and they are the most advanced new strategy in glioma therapy; clinical trials have shown their safety and efficacy in newly diagnosed gliomas and recurrences, and also a significant increase of survival time in patients compared to placebo treatment (9–11). However, the limitation of these implants is their large size of several centimeters, which allows neither a real intratumoral or intraparenchymal implantation, nor a stereotaxic administration.

Our group has developed 5-fluorouracil (5-FU)–loaded poly(lactide-co-glycolide) (PLGA) microspheres for intracranial implantation, which has been used since the mid-1990s. Administered as a suspension of 5-FU–loaded micrometric particles, this device allows implantation to take place within the tissue that is surrounding the resection cavity. The rationale of this approach was to prevent the preferential recurrence of glioblastoma from the brain parenchyma near the resection site. Another advantage of this method is the possibility of stereotaxic implantation in the vicinity, or within the tumor itself in the case of inoperable gliomas.

The choice of 5-FU as the entrapped anticancer drug for such intracranial chemotherapy stems from its properties; this hydrophilic and antimetabolic drug does not cross the BBB easily, its anticancer activity may be improved by sustained administration, and it is a powerful radiosensitizer (12). 5-FU is a pyrimidine, which during administration acts on the synthesis of nucleic acids, destroying only the cells that duplicate their DNA. In a malignant glioma, only a small percentage of the cells (14–44%) are in division at any given time (13). Moreover, as there is no cell division in a healthy brain, 5-FU has specific antitumoral action against malignant glial cells. Therefore, 5-FU is a good candidate for the development of a sustained-release device for intracerebral implantation.

PLGA is a well-known biocompatible copolymer. Its degradation products are lactic and glycolic acids, which are natural metabolic products. It has been extensively used as the material for surgical sutures. Its biocompatibility, in the case of implantation in the brain for the present application, will be discussed in this chapter.

5-FU–loaded PLGA microspheres, for intracranial implantation in the treatment of glioblastoma, have been evaluated by our group for their feasibility, biocompatibility, and efficacy in animal and human patients. This chapter presents the different developmental steps, including technological, pharmaceutical, and medical approaches.

MICROSPHERE FORMULATION AND DEVELOPMENT

Basic Formulation

The formulation of 5-FU–loaded microspheres was set up in our laboratory in the mid-1990s by Boisdron-Celle et al. (14). Basically, microspheres were produced by an emulsion-evaporation technique. 5-FU crystals were suspended within an organic phase (methylene chloride/acetone) containing PLGA 50:50. This solid-in-oil

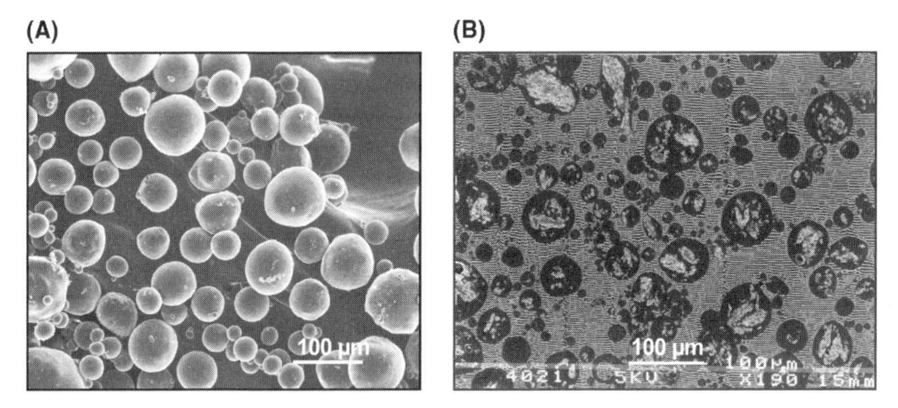

Figure 1 Scanning electron photomicrographs of 5-FU–loaded microspheres before 5-FU release. (**A**) Surfaces of sterilized microspheres, (**B**) Cross sections after inclusion into a resin. The hatched regions correspond to the resin; the dark gray regions correspond to PLGA; bright gray regions correspond to 5-FU crystals. *Abbreviations*: 5-FU, 5-fluorouracil; PLGA, poly(lactide-co-glycolide).

dispersion was emulsified into an outer aqueous polyvinyl alcohol solution by mechanical stirring at room temperature. The emulsion was transferred into a large volume of deionized water. The resulting solvent extraction allowed microspheres to form, which were collected by filtration, washed, frozen in liquid nitrogen, and freeze-dried.

Figure 1 shows the scanning electron microscopy (SEM) photographs of the microspheres. The main characteristics of the microspheres were the mean size of the population, the core loading (final 5-FU/microsphere, w/w ratio), and the 5-FU release patterns, which classically showed a rapid release (burst effect) during the first 24 hours followed by a more sustained release until 100% recovery was achieved. It was shown that the microparticle structure depended directly on the experimental conditions governing the precipitation of the core material. Thus the size of the 5-FU crystals, the organic/aqueous phase ratio, the theoretical drug loading, and the microparticle size played predominant roles on drug loading and release patterns (14).

Therefore, different process variables led to the preparation of several batches (200–250 mg) of microspheres with variable releases, which were tested in preclinical trials.

Pilot Production of Microspheres

For animal trials, the previously mentioned lab-scale process, involving small-capacity glassware and a number of manual manipulations, was used to prepare the microspheres. Consequently, a lack of reproducibility was observed. Furthermore, the characterization tests used up a great deal of the product, and hence much material was lost.

To increase reproducibility and a production capacity to envisage the fabrication of clinical batches, a scale-up by a factor of 13 was set up allowing the preparation of batches of 5 g. Its conception and running are presented in Figure 2 (15).

For the preparation of clinical batches of 5-FU–loaded microspheres, several improvements were thus made. Considering the high solubility of 5-FU in water (12 mg/mL), three points had to be taken into account to meet certain specifications. Firstly, the manufacturing process should limit the loss of 5-FU in the aqueous phase

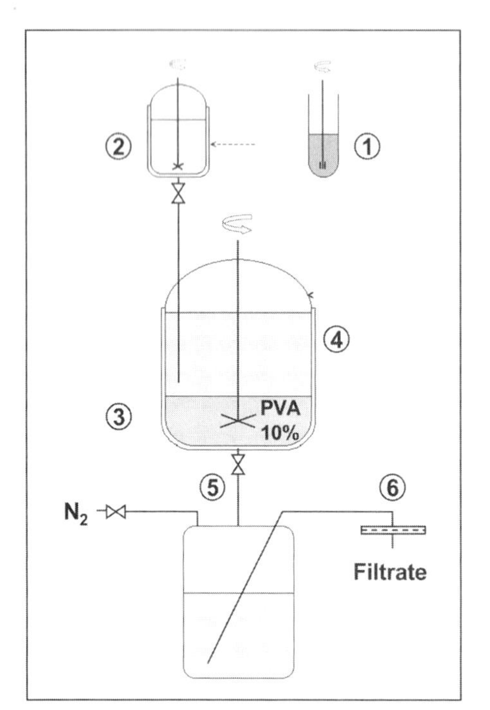

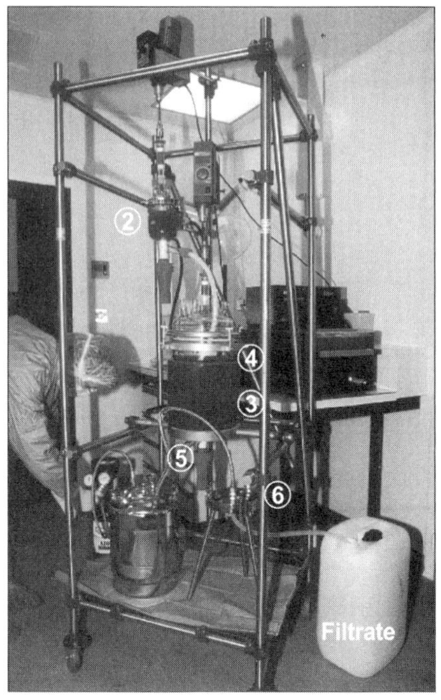

Figure 2 Equipment and procedure for the preparation of 5-FU–loaded microspheres. *Organic phase preparation*—5-FU was obtained as crystallized powder (particle mean size: 371 ± 19 μm) and subsequently milled in a planetary micromill (Pulverisette 7, Fritsch, Idar-Oberstein, Germany). *Step 1*: 4 g 5-FU powder was added to 40 mL dichloromethane and dispersed with a homogenizer (13,500 rpm, 3.5 minutes; Ultra Turrax, Ika, T25 basic/S25N-10G, Staufen, Germany). *Step 2*: The suspension was then transferred into a 150 mL thermojacketed reactor and a further 5 mL of dichloromethane was added as a rinse. Finally, 5 g of PLGA was added, the reactor hermetically closed, and the organic phase was stirred with a three-bladed paddle (four hours, room temperature) to allow total dissolution of the polymer. *Emulsion preparation— Step 3*: A tube to a larger (6 L) thermojacketed reactor containing 1.5 L of aqueous phase (polyvinyl alcohol, 10%) connected the 150 mL reactor. This system was cooled to 2°C before emulsion preparation. The organic phase was then transferred and emulsified in the aqueous phase (PVA 10%) by stirring with a four-bladed paddle (375 rpm, 4.75 minutes). *Solvent extraction— Step 4*: Adding 4.5 L of cold water into the emulsion performed solvent extraction. Stirring at the same speed for two minutes allowed the microspheres to harden. *Microsphere filtration and further treatment— Step 5*: The suspension of microspheres was transferred into a 10 L pressure tank. *Step 6*: The microspheres were filtered under nitrogen pressure (0.8 bar, filtration system supplied by Sartorius, Palaiseau, France) using a cellulose ester filter membrane (3 μm pore size, Millipore, Saint Quentin en Yvelines, France). Microspheres were then freeze-dried and sieved (125 μm). Residual solvents were removed at 37°C under vacuum for 72 hours. Finally, microspheres were placed in vials as dose fractions and sterilized under gamma radiation (19.6 kGy). *Abbreviations*: 5-FU, 5-fluorouracil; PLGA, poly(lactide-co-glycolide); PVA, polyvinyl alcohol.

during the emulsion and the extraction steps to obtain correct core loading and encapsulation efficiency. Secondly, the distribution of 5-FU crystals within the microspheres should be controlled to limit early burst; the burst effect may, in part, be attributed to the porosity of the microspheres and to the presence of some drug crystals on the particle surface (14). Thirdly, radiosterilization is known to induce important alterations of the polymer that may lead to accelerated rates of release (16). It was therefore necessary to define the conditions for sterilization.

To limit the loss of 5-FU during emulsification, a more viscous organic phase was prepared by increasing the PLGA/solvent ratio from 8.3% to 11% and by decreasing the temperature of the suspension and the emulsion. For example, in the case of the same initial 5-FU/PLGA ratio, mean core loadings were 16.3% and 21.6% for operating temperatures of 4°C and 2°C, respectively, during manufacture.

During emulsification, several parameters were fixed in order to control the distribution of crystals within the final product. Firstly, dispersing them within dichloromethane using a Turrax® (IKA® Works, Inc., Wilmington, North Carolina, U.S.) homogenizer standardized the size of the drug crystals. Secondly, using dichloromethane alone reduced the rate of solvent extraction. Earlier attempts used a dichloromethane/acetone mixture to accelerate microsphere hardening (14,17). Slower hardening limited the entrapment of crystals near the surface of the microspheres, and this could be responsible for the high burst effect. Finally, the emulsion time was adjusted to obtain a good degree of homogeneity of 5-FU crystals within the microspheres and therefore control the in vitro rate of release of 5-FU. An extended emulsion time resulted in some solubilization of 5-FU crystals located near or on the surface of the droplets. Consequently, these microspheres had smoother surfaces and exhibited a lower burst effect; 72%, 41%, and 33% for emulsion times of 2.5, 3.5, and 4.5 minutes, respectively.

In terms of reproducibility of the process, several technical improvements were achieved. All temperatures were precisely controlled by the circulation of a cooling solution in thermojacketed glass cells. A direct connection between the small and the large reactors was established. Emulsion and extraction steps were performed in the same reactor, which is not a classical procedure for the emulsion–extraction method. Finally, the use of draining valves permitted the rapid transfer of solutions/dispersions. In addition, this pilot study was set up in a "Class 10,000" clean room with all the equipment being sterilized prior to microsphere production. Possible sources of contamination were thereby limited; bacteriological controls on the microspheres were negative both before and after radiosterilization.

Thanks to these modifications, we were able to produce 5-FU–loaded microspheres in accordance with the specifications that will be precised later in this chapter. Quite a high degree of reproducibility between batches was observed (CV = 5% for drug loading, 15% for mean particle size, and 12% for burst effect at 24 hours, $n = 9$).

Understanding 5-FU Release Mechanisms

All microparticle batches exhibited a bimodal drug release pattern in vitro; a significant burst effect was followed by a zero-order release phase over three weeks (Fig. 3).

By following the physicochemical changes occurring within the system during drug release, we investigated the underlying release mechanisms (18). It was shown that drug diffusion predominantly controlled drug release, but a significant contribution of the polymer degradation process was noted. Upon water imbibition, the average molecular weight of the macromolecules decreased, leading to increased drug diffusion coefficients. Importantly, the breakdown of the polymeric network occurred only after drug exhaustion, and thus did not contribute to the control of the in vitro release process. Two mathematical models were developed, and allowed an accurate quantitative description of the resulting drug release kinetics (18,19). The impact of gamma irradiation to sterilize the microspheres was investigated. With

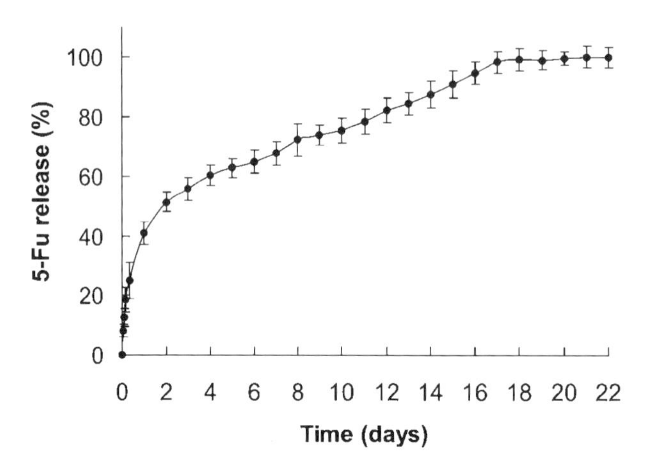

Figure 3 Typical in vitro release profile of 5-FU from clinical batches of microspheres ($n = 7$). *Abbreviation*: 5-FU, 5-fluorouracil.

increasing gamma irradiation doses, the average polymer molecular weight logically decreased monotically, leading to an accelerated initial burst effect but without changing the zero-order release phase (20). Therefore, in the case of clinical batch production, the radiation dose had to be decreased to 19.6 kGy in order to limit the difference between in vitro release kinetics before and after radiation (21). Mean 24-hour burst effects were 25.7% ± 2.7% and 40.6% ± 4.3% before and after radiosterilization, respectively.

PRECLINICAL TRIALS

We established the brain compatibility of the PLGA microspheres and studied their fate and efficacy after stereotaxic implantation in the rat brain.

Biocompatibility

Our group was the first to study the brain's reaction to PLGA microsphere implantation in the rat brain. The biodegradation and tissue reaction of radiosterilized and stereotaxically implanted poly(D,L-lactide-co-glycolide) microspheres in the rat brain were studied by routine staining, immunohistochemistry, transmission electron microscopy, and SEM. The brain tissue reaction was a nonspecific astrocytic proliferation and a macrophage–microglia cell reaction, typically found after any implantation or injection into the CNS. Some foreign-body giant cells were observed and the inflammatory and macrophagous reactions decreased dramatically after one month and almost ended after two months by which time the microspheres were totally biodegraded (22). It was concluded that PLGA microspheres did not induce any complementary reaction to mechanical trauma because of the implantation itself. These results were confirmed and completed by other groups with intracerebral PLGA microparticles or implants (23–26). More recently, we carried out a biocompatibility study of PLGA particles prepared by the double emulsion technique. It was shown that the implantation of PLGA microspheres (size greater than 30 μm) into the striatum did not disturb the three-dimensional ultrastructure of the tissues; intact nerve cells and axons were observed in contact with particles, even during their degradation process (27). Moreover, microspheres degraded identically in the brain and in gelatin, thus showing that the degradation process was mainly because of hydrolysis of the polymer's ester bonds, without any enzymatic activity or phagocytosis. However, Nicholas et al. (28) demonstrated that phagocytosis of small fluorescent dye–containing microspheres (size less than 7.5 μm) was evident in vivo at one week and one month after injection, which was primarily the result of microglial cell activity. The phagocytosis could therefore be dependent on the size of the particles. These data suggest that the efficacy of a drug delivery system in the CNS by the means of microspheres can be the result of not only the slow diffusion of the drug from the particles, but also of the controlled release by the glial cells after phagocytosis of drug-containing microspheres (29). In the case of 5-FU–loaded microspheres, the mean particle size range was 20 to 40 μm, and the previously mentioned phenomenon should be negligible.

The biocompatibility and tissue reaction of 5-FU–loaded microspheres were also controlled. No sign of clinical or histological toxicity was observed when compared to blank microspheres, indicating that 5-FU had no specific cytotoxic effect (30). On the contrary, Chen et al. (24) observed specific drug-induced necrosis in cases where carboplatin-loaded microspheres were used. Similarly, we observed a neurological toxicity after intracranial implantation of BCNU-loaded microspheres in rats (unpublished results).

Efficacy Trials

Once the feasibility of intracerebral implantation of PLGA microspheres was proved, 5-FU–loaded microspheres were tested for their efficacy on tumor-bearing rats. In all the experiments performed with PLGA-loaded microspheres, the malignant brain tumor used as the tumoral model was the C6 glioma cell line, initially produced by Benda et al. (31). For implantation in the brain, anesthetized rats were placed in a stereotaxic head frame and a burr hole was drilled (anterior 0 mm, lateral 3 mm, depth 7 mm, according to Bregma) (Fig. 4).

Tumor cells were thus inoculated into the striatum and the time assigned for tumor growth was 6 to 12 days prior to treatment depending on the protocol. For implantation purposes, microspheres were suspended in a sodium-carboxymethylcellulose (CMC) sterile solution (1.2% CMC, 0.9% polysorbate 80, 3.8% mannitol, all w/w), and 10 to 20 μL suspensions were injected at the same coordinates.

In a first set of experiments, the efficacy of 5-FU–loaded microspheres was assessed by the time the rat survived (30). Compared to an injection of 5-FU solution, implantation of blank microspheres and two types of 5-FU–loaded microspheres [Fast Release 1 (FR1) and Slow Release 1 (SR1) releasing 100% of encapsulated 5-FU within 3 days and 18 days, respectively] were tested. Only intratumoral implantation of the SR1 type 5-FU–loaded microspheres significantly decreased mortality and the average survival time was doubled. No significant effect of the injection of 5-FU solution or implantation of blank microspheres could be observed.

These first results clearly proved that the tested microspheres could be active and that a sustained release of 5-FU for a sufficient period of time was necessary to increase the survival time of glioma-bearing rats.

However, rat survival was an indirect criterion to measure the effect. How did released 5-FU act on the tumor?

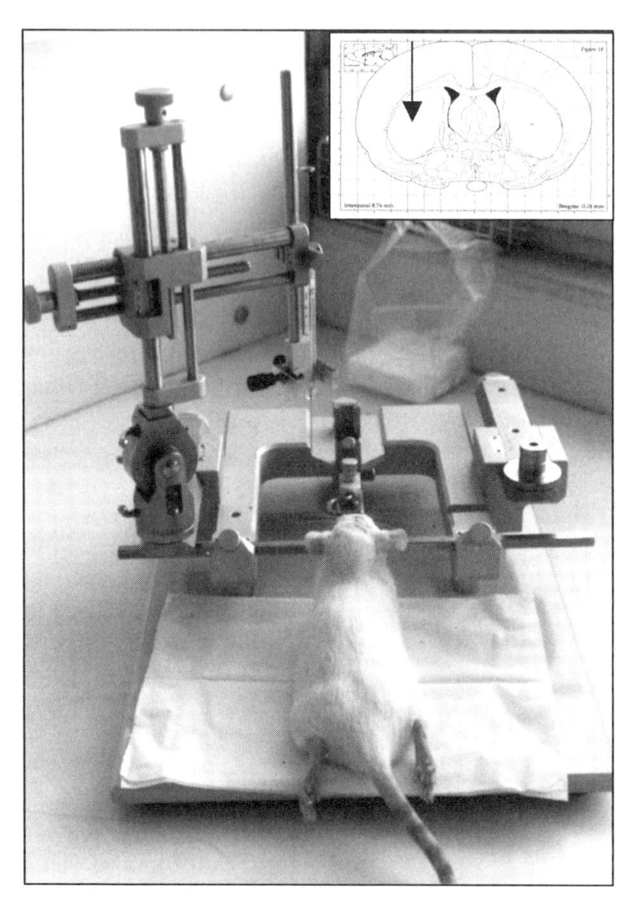

Figure 4 Stereotaxic head frame for microsphere implantation into rat brain.

This point was investigated by a second study that involved following the mean tumor volume evaluated by magnetic resonance imaging (MRI) (32). This technique had previously been optimized for rat brain tumor imaging (33). Here again, a 5-FU bolus injection was compared to a sustained delivery of the same dose of 5-FU administered via two types of microspheres (FR2 and SR2 releasing 100% of encapsulated 5-FU in 8 and 18 days, respectively). In terms of rat survival, both microsphere batches induced a significant 50% increase in the animals' life span. However, neither treatment cured any of the animals from this highly invasive tumor. These findings were not consistent with previous reports, but experimental conditions were significantly different in terms of tumor induction/therapy (therapy started 12 days after inoculation instead of seven days) and release profiles (FR2 was slower than FR1). However, MRI images showed that the local and sustained delivery of 5-FU slowed tumor development in the vicinity of the microspheres by a factor of three, compared to an intratumoral bolus injection.

In order to optimize the microsphere therapeutic system, the formulation and, more specifically, the drug delivery kinetics, were examined (34). Three formulations, prepared by modifying process parameters as previously mentioned, were tested: a "fast release" (FR3: releasing 58% of encapsulated 5-FU in 24 hours and 100% within seven days); a "slow release" (SR3: releasing 22% in 24 hours and 100% within 25 days); and an "intermediate release" (IR1: releasing 28% in 24 hours and 100% within 20 days). All the treated animals, whatever the release profile considered was, displayed a comparable 50% increase in life span versus control, but only a low proportion of them were cured (11%). However, the formulations that led to the best tumor response were the FR3 and IR1, showing stagnation in the tumor growth rate. In fact, all our experimental data confirmed that the 5-FU release pattern played a predominant role in controlling local tumor development. Particularly, a burst effect followed by a sustained release over three weeks seemed to limit tumor growth better. Although a good correlation between in vitro release and tumor response was systematically observed, the real profile of release was found to be different in vitro than in vivo. Menei et al. (30) observed the presence of 5-FU crystals within microspheres in the brain even after 12 days or 20 days whereas 100% release was obtained in vitro after 72 hours and 18 days, respectively. These observations indicated a longer sustained delivery period in vivo than in vitro. This was confirmed by a previous study, which showed constant 5-FU plasma levels during the first 23 days after intramuscular implantation of FR1 microspheres in rabbits (unpublished data). Clinical trials in patients, detailed later in this chapter, confirmed that significant concentrations of 5-FU were detectable in the cerebrospinal fluid (CSF) even one month after implantation (35). In the brain, such a difference of 5-FU release rates between in vitro and in vivo may be explained by adsorption of brain proteins to the microsphere surface as previously observed by other authors (36). As previously mentioned, the phagocytosis of small drug-containing microspheres by the glial cells could also control and limit release (29).

Once the drug is released, the area that could be impregnated by the drug within the brain is of importance. Observed tumor regrowth seemed to be controlled locally by the rate of release, but not totally stopped. After local and sustained delivery of 5-FU, tumor regrowth took place mainly in the cortical area instead of the striatal zone. These observations could be explained based on the poor diffusion of the drug within the brain; 5-FU released from the loaded microspheres implanted in the striatum remained in the vicinity of the microspheres and therefore could not target cortical cells. On the other hand, though the bolus injection was performed in the striatum, the exposure time of tumor cells was not long enough to control tumor growth.

The diffusion of 5-FU within the brain was investigated by using tritiated-5-FU microspheres (37). To monitor 5-FU diffusion from the implantation site, tissue combustion was performed on animals implanted with [H3]-drug microspheres. T2-weighted nuclear MRI was performed on animals implanted with magnetite-loaded microspheres to determine microsphere localization after deposit. This study showed a limited distribution of 5-FU in the brain; a maximum of 3 mm diffusion was noted if the microspheres were considered to remain where they were deposited. However, such studies were difficult to handle with small animals. On the contrary, in the case of monkey brains, a high concentration of carmustine, a very lipophilic drug could be observed, within 3 mm, with significant concentrations within 5 cm, as long as 30 days after the implantation of polyanhydride pellets (38). Therefore, it is obvious that the hydrophilicity/lipophilicity balance of the encapsulated drug may play an important role in its diffusion potential within the brain.

The fact that 5-FU diffusion was limited to the vicinity of the implantation site supported the choice of multipoint administration of the 5-FU during surgery to cover the whole area of the tumor location. Microspheres, with multiple, close injection sites, could guarantee a local, sustained delivery of the drug, with minimal damage to the surrounding healthy tissues.

Finally, as 5-FU is a well-known radiosensitizer, the synergistic effect with its anticancer effect was investigated by Roullin et al. (39). The implantation of two types of microspheres (FR3 or FR4) on day 12 after inoculation was followed by radiotherapy from day 13 at a total dose of 36 Gy given in nine fractions over three weeks. 5-FU microspheres associated with radiotherapy caused a 47% complete remission rate as opposed to the 8% rate with radiotherapy alone. Drug (FR4) delivery over three weeks produced slightly better survival results (57%) compared to one-week–sustained release (FR3) (41%). Magnetic resonance images showed exponentially increasing tumor volumes during the first half of the radiotherapy cycle, followed by a decrease, and tumor disappearance if survival exceeded 120 days. This study indicated that, as part of an efficient therapeutic strategy, it seems important to favor the presence of the radiosensitizer 5-FU during the whole radiotherapy cycle (mainly three weeks) with respect to a massive concentration of the antitumor agent at the beginning of the radiotherapy.

In parallel to these experiments, development of a similar 5-FU delivery system, using a new polymer, poly(methylene malonate 2.1.2), with a slow degradation rate, was achieved by our group. The formulation was similar and efficacy studies were performed on two animal models, the C6, as previously done for PLGA microspheres, and the F98 glioma, known for its aggressiveness. A significant increase in the life span was observed in both cases compared to the controls (40–42), confirming the impact of slow release of 5-FU within brain tissue, whatever the matrix and the glioma model. However, a strong inflammatory and immune reaction was observed and was attributed to polymer degradation products (29). This toxicity prohibits the use of such a polymer in the human brain and underlined the importance of biocompatibility studies.

Owing to the results obtained for 5-FU-PLGA–loaded microspheres, specifications in terms of release profile and the utilization of this new therapeutic device could be proposed. Firstly, the release profile should present a burst effect (releasing a massive amount of drug at the beginning of the treatment) followed by a constant release for three weeks. Secondly, a multisite injection should be performed as 5-FU hardly diffuses within the tissue. Thirdly, the treatment should be coupled with radiotherapy to enhance 5-FU anticancer activity.

For the following clinical trials, the desired specifications for the implantable 5-FU–loaded microspheres were chosen as follows: in all cases, the mean particle sizes were 30 to 50 μm, to avoid the needle plugging during injection into the brain tissue; the core loading range was 19% to 24% by weight. In vitro release after radiosterilization reached 100% recovery within at least 18 days and the burst effect within 24 hours was less than 45% of the total drug content after radiosterilization.

APPLICATION TO GLIOMA THERAPY AFTER TUMOR RESECTION: PHASE I–II AND IIB STUDIES

Based on preclinical results, a Phase I–II clinical trial was conducted on eight patients with newly diagnosed glioblastoma (35). The protocol is illustrated in Figure 5.

After a complete macroscopic resection of the tumor and histological confirmation, 5-FU–loaded microspheres were suspended in a sodium-CMC sterile solution (1.2% CMC, 0.9% polysorbate 80, 3.8% mannitol, all w/w) and were implanted all around the wall of the surgical resection cavity, every 1 cm^2, to a depth of 2 cm. A total volume of 1.5 to 2.5 mL (depending on the dose) of microsphere

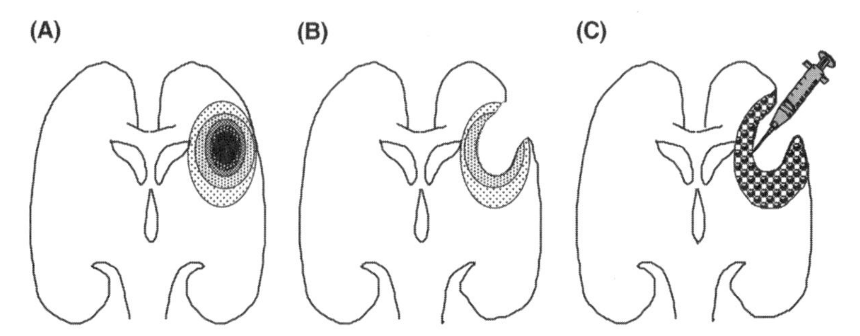

Figure 5 Principle of the in vivo implantation of 5-FU–loaded microspheres after resection of the tumor. (**A**) Detection of the tumor confirmed by histological observations, (**B**) resection of all visible tumor, some tumor cells staying in place around the cavity, (**C**) implantation of the suspension of 5-FU–loaded microspheres all around the wall of the resection cavity, every 1 cm^2, to a depth of 2 cm. A total volume of 1.5 to 2.5 mL of MS suspension was injected with 100 μL per injection site. *Abbreviations*: 5-FU, 5-fluorouracil; MS, microsphere.

suspension was injected with 100 μL per injection site. Two doses of 5-FU were studied sequentially: three patients received 70 mg 5-FU (group I) and five received 132 mg (group II). External beam radiotherapy was initiated before the seventh post-surgical day: 59.4 Gy, split in 33 fractions of 1.8 Gy each given as five fractions per week. Patients were followed up by clinical examination, MRI, and 5-FU assays in the blood and CSF.

Plasmatic 5-FU levels were very low during the course of the study, and were finally undetectable by day 30. In the CSF, sustained concentrations of 5-FU were found and remained at least one month after microsphere implantation, irrespective of the initial dose of microspheres administered. This longer in vivo release profile compared to in vitro release profile was previously observed in animal studies (22). Systemic tolerance to the treatment was good; one case of recurrent brain swelling was observed at the higher dose studied (group II). A local recurrence of the tumor occurred in all patients of group I. In group II, one patient died 31 weeks after surgery because of a wide infiltration of the right temporal lobe by neoplastic cells from the resection site. Two patients died of a recurrence distant to the original tumor site and outside the irradiated volume. The two other patients unusually survived. One died four years after the diagnosis of a cerebellar recurrence. The last patient was still in remission eight years after treatment.

This first study confirmed that PLGA 5-FU–loaded microspheres were safe systems to be implanted into the brain. Moreover, a dose of 132 mg seemed to be able to limit local recurrence of the tumor in three out of five cases.

Based on these data, a multicenter Phase IIb clinical trial was set up to compare the effect of peroperative implantation of 5-FU microspheres followed by early radiotherapy (arm A) and early radiotherapy alone (arm B) in 95 patients with high-grade glioma. The administered dose was 130 mg of 5-FU (arm A) and the protocol was as described for Phase I–II.

A total of 77 (arm A: 38; arm B: 39) of the 95 enrolled patients were treated according to the study treatment plan and all were evaluated with the intention to treat for efficacy and safety.

Median overall survival was almost two months longer in arm A (15.2 months) than in arm B (13.5 months). In the case of patients with complete resection, a 2.9 month difference in overall survival was reported in favor of arm A versus arm B. This difference, however, did not reach statistical significance. Moreover, the speed of tumor progression was not different for the two treatment arms (6.5 months in arm A and 6.3 months in arm B). In terms of recurrence, all relapses that occurred in patients with incomplete resection occurred locally, whereas 16% of relapses in patients with complete resection were because of distant recurrences. In terms of adverse effects, safety of the treatment was considered acceptable, with the prophylactic use of high doses of steroids.

This second clinical trial confirmed the feasibility of such a treatment, but without showing a significant effect of the implantation of 5-FU–loaded microspheres in the wall of the surgical cavity after resection of the tumor. However, a tendency to increase survival time and to slightly limit local recurrence was observed in comparison to early radiotherapy without implantation. Methodological issues may be mentioned to explain the absence of a statistically significant effect: the open status of the study (for example, more patients underwent chemotherapy and/or a second surgery in arm B than in arm A), or the randomization based on a diagnostic assumption by the investigator.

Taken together with an acceptable safety profile, these first efficacy results could justify a phase III study, appropriately powered to assess the impact of 5-FU microspheres on overall survival time.

STEREOTAXIC IMPLANTATION IN MALIGNANT GLIOMA: PHASE I STUDY

In the case of deep and nonoperable malignant gliomas, resection is impossible and the only way to administer drugs into the brain is the stereotaxic technique. Therefore, because of the good tolerance observed in the previous studies and because microspheres can be easily injected by stereotaxy, a Phase I study was designed including 10 patients with newly diagnosed malignant glioma. The mean size of the tumor (calculated on axial MRI images by multiplying the largest cross-sectional diameter measured in centimeters by the largest diameter perpendicular to it) was 1034 cm^2.

Microsphere implantation was scheduled to be either concomitant to the stereotaxic biopsy, after an extemporaneous histological examination (for three patients) or during second surgery (for seven patients).

After the histological diagnosis was confirmed, 2.5 mL of a microsphere suspension (132 mg 5-FU) was injected inside the tumor by stereotaxy, using a stereotaxic frame (Leksell®, Elekta, Linac House, West Sussex, U.K.) ($n = 8$) (Fig. 6), or a frameless stereotaxy procedure with computer assisted neurosurgery ($n = 1$) (Brain Lab neuronavigation).

The number of trajectories was adapted to the size and shape of the tumor. Microsphere implantation was well tolerated except for the four patients who had a single trajectory, who experienced a transitory worsening of preexisting neurological symptoms. There was no radical edema, nor hematological complication. 5-FU was detected in the CSF and in the blood of some patients at very low concentrations. These very low blood concentrations of 5-FU explain the lack of hematological complications or healing problems. The median overall survival was 40 weeks with two long survivors (71 and 89 weeks) who presented a small or cystic tumor. We can assume that these cases could represent the best indications and/or that the dose should be adapted to tumor size.

Figure 6 Stereotaxic headframe for microsphere implantation into human brain.

This study showed that stereotaxic intratumoral implantation of biodegradable drug-releasing microspheres is feasible. This treatment could be applied with only one stereotaxic procedure: immediately after the diagnosis of a brain tumor is confirmed by histological observation, microsphere implantation could be performed using the same trajectory. In this case, the only limitation is some bleeding following the biopsy, as observed in one patient. The treatment can also be applied during a second surgery if the diagnosis cannot be done on fresh sections.

Therefore, stereotaxic, intratumoral implantation of biodegradable drug-releasing microspheres is a promising approach for the treatment of nonoperable brain tumors.

CONCLUSION

The targeting of drugs in the CNS by the implantation of biodegradable microspheres remains a promising procedure and offers numerous theoretical advantages. The potential applications of these biodegradable microspheres for neurological diseases are legion. In the field of cancer therapy, the present developments of 5-FU–loaded microspheres have shown their biocompatibility and ability to limit tumor growth in rats. This material is easily implanted by stereotaxy in precise and functional areas of the brain without causing damage to the surrounding tissue. However, their efficacy on survival time needs to be confirmed by extensive clinical trials. Using the same strategy, several chemotherapeutic agents, reviewed by Wang et al. (5), have been successfully incorporated into polymeric microparticulate devices. Immunotherapy, which mainly consists of the local delivery of cytokines to generate and maintain an immune response to cancer cells, can take advantage of the sustained release provided by loaded microparticles (43,44). This approach is very promising and could overcome the limits of diffusion and activity of classical chemical agents.

In the field of neurodegenerative diseases, neurotransmitters, neuromodulators, neurohormones, or trophic factors, all of which play a substantial role in the activity and maintenance of the CNS, can be delivered specifically to restricted regions of the brain, thanks to such devices. Our group has already obtained very promising results in animal models of neurodegenerative diseases (Alzheimer's, Parkinson's, Huntington's diseases) using nerve growth factor-(NGF) or glial cell derived neurotrophic factor-(GDNF)-releasing microspheres (45–48).

Moreover, another field of application concerns the insertion of such polymeric microspheres in other areas of the CNS such as the subarachnoid or the epidural space. Thus, to provide long-lasting anesthesia, and to decrease systemic drug toxicity, bupivacaine-loaded polylactide (PLA) microspheres were developed (49). Our group has developed and proved the feasibility of intrathecal injection of baclofen-PLGA–loaded microspheres for the treatment of spasticity (50).

Although the CNS area target is identified and the neuroactive drug selected, work still remains before microparticle devices can be used therapeutically. For precise clinical purposes, the pharmaceutical development and preclinical and clinical evaluations of microparticulate devices will require many years of work.

REFERENCES

1. Fine HA, Dear KB, Loeffler JS, Black PM, Canellos GP. Meta-analysis of radiation therapy with and without adjuvant chemotherapy for malignant gliomas in adults. Cancer 1993; 71:2585–2597.
2. Hochberg F, Pruitt A. Assumptions in the radiotherapy of glioblastoma. Neurology 1980; 30:907–911.
3. Abbott NJ, Romero IA. Transporting therapeutics across the blood-brain barrier. Mol Med Today 1996; 2:106–113.
4. Pardridge WM. Drug delivery to the brain. J Cereb Blood Flow Metab 1997; 17:713–731.
5. Wang PP, Frazier J, Brem H. Local drug delivery to the brain. Adv Drug Deliv Rev 2002; 54:987–1013.
6. Kreuter J. Nanoparticulate systems for brain delivery of drugs. Adv Drug Deliv Rev 2001; 47:65–81.
7. Zhang Y, Zhu C, Pardridge WM. Antisense gene therapy of brain cancer with an artificial virus gene delivery system. Mol Ther 2002; 6:67–72.
8. Gutman RL, Peacock G, Lu DR. Targeted drug delivery for brain cancer treatment. J Contr Rel 2000; 65:31–41.
9. Brem H, Piantadosi S, Burger PC, et al. Placebo-controlled trial of safety and efficacy of intraoperative controlled delivery by biodegradable polymers of chemotherapy for recurrent gliomas*1. The Lancet 1995; 345:1008–1012.
10. Brem H, Gabikian P. Biodegradable polymer implants to treat brain tumors. J Contr Rel 2001; 74:63–67.
11. Westphal M, Hilt DC, Bortey E, et al. A phase 3 trial of local chemotherapy with biodegradable carmustine (BCNU) wafers (Gliadel wafers) in patients with primary malignant glioma. Neuro-oncology 2003; 5:79–88.
12. Lawrence TS, Maybaum J. Fluoropyrimidines as radiation sensitizers. Semin Radiat Oncol 1993; 3:20–28.
13. Hoshino T, Wilson CB. Cell kinetic analyses of human malignant brain tumors (gliomas). Cancer 1979; 44:956–962.
14. Boisdron-Celle M, Menei P, Benoit JP. Preparation and characterization of 5-fluorouracil-loaded microparticles as biodegradable anticancer drug carriers. J Pharm Pharmacol 1995; 47:108–114.
15. Faisant N, Benoit JP, Menei P. Utilisation de microsphères biodégradables libérant un agent anticancéreux pour le traitement du glioblastome. France FR 9906207, Wo 00/69413, 2000.

16. Spenlehauer G, Vert M, Benoit JP, Chabot F, Veillard M. Biodegradable cisplatin microspheres prepared by the solvent evaporation method: morphology and release characteristics. J Contr Rel 1988; 7:217–229.
17. Bodmeier R, McGinity JW. Solvent selection in the preparation of poly(-lactide) microspheres prepared by the solvent evaporation method. Int J Pharm 1988; 43:179–186.
18. Faisant N, Siepmann J, Benoit JP. PLGA-based microparticles: elucidation of mechanisms and a new, simple mathematical model quantifying drug release. Eur J Pharm Sci 2002; 15:355–366.
19. Siepmann J, Faisant N, Benoit JP. A new mathematical model quantifying drug release from bioerodible microparticles using Monte Carlo simulations. Pharm Res 2002; 19:1885–1893.
20. Faisant N, Siepmann J, Oury P, et al. The effect of gamma-irradiation on drug release from bioerodible microparticles: a quantitative treatment. Int J Pharm 2002; 242: 281–284.
21. Geze A, Venier-Julienne MC, Cottin J, Faisant N, Benoit JP. PLGA microsphere bioburden evaluation for radiosterilization dose selection. J Microencapsul 2001; 18:627–636.
22. Menei P, Daniel V, Montero-Menei C, Brouillard M, Pouplard-Barthelaix A, Benoit JP. Biodegradation and brain tissue reaction to poly(-lactide-co-glycolide) microspheres. Biomaterials 1993; 14:470–478.
23. Chen W, He J, Olson J, Lu DR. Carboplatin-loaded PLGA microspheres for intracerebral implantation: in vivo characterization. Drug Delivery 1997; 4:301–311.
24. Chen W, He J, Olson J, Lu D. Direct intracerebral delivery of carboplatin from PLGA microspheres against experimental malignant glioma in rats. Drug Delivery 1998; 5: 101–110.
25. Emerich DF, Tracy MA, Ward KL, Figueiredo M, Qian R, Henschel C, Bartus RT. Biocompatibility of poly(D,L-lactide-co-glycolide) microspheres implanted into the brain. Cell Transplant 1999; 8:47–58.
26. Kou JH, Emmett C, Shen P, et al. Bioerosion and biocompatibility of poly(D,L-lactic-co-glycolic acid) implants in brain. J Contr Rel 1997; 43:123–130.
27. Veziers J, Lesourd M, Jollivet C, Montero-Menei C, Benoit JP, Menei P. Analysis of brain biocompatibility of drug-releasing biodegradable microspheres by scanning and transmission electron microscopy. J Neurosurg 2001; 95:489–494.
28. Nicholas AP, McInnis C, Gupta KB, et al. The fate of biodegradable microspheres injected into rat brain. Neurosci Lett 2002; 323:85–88.
29. Fournier E, Passirani C, Montero-Menei CN, Benoit JP. Biocompatibility of implantable synthetic polymeric drug carriers: focus on brain biocompatibility. Biomaterials 2003; 24:3311–3331.
30. Menei P, Boisdron-Celle M, Croue A, Guy G, Benoit JP. Effect of stereotactic implantation of biodegradable 5-fluorouracil-loaded microspheres in healthy and C6 glioma-bearing rats. Neurosurgery 1996; 39:117–123; discussion 123–114.
31. Benda P, Lightbody J, Sato G, Levine L, Sweet W. Differentiated rat glial cell strain in tissue culture. Science 1968; 161:370–371.
32. Lemaire L, Roullin VG, Franconi F, et al. Therapeutic efficacy of 5-fluorouracil-loaded microspheres on rat glioma: a magnetic resonance imaging study. NMR Biomed 2001; 14:360–366.
33. Lemaire L, Franconi F, Saint-Andre JP, Roullin VG, Jallet P, Le Jeune JJ. High-field quantitative transverse relaxation time, magnetization transfer and apparent water diffusion in experimental rat brain tumour. NMR Biomed 2000; 13:116–123.
34. Roullin VG, Lemaire L, Venier-Julienne MC, Faisant N, Franconi F, Benoit JP. Release kinetics of 5-fluorouracil-loaded microspheres on an experimental rat glioma. Anticancer Res 2003; 23:21–25.
35. Menei P, Venier MC, Gamelin E, et al. Local and sustained delivery of 5-fluorouracil from biodegradable microspheres for the radiosensitization of glioblastoma: a pilot study. Cancer 1999; 86:325–330.

36. Saini P, Greenspan P, Lu DR. Adsorption of brain proteins on the surface of poly(D,L-lactide-co-glycolide) (PLGA) microspheres. Drug Delivery 1997; 4:129–134.
37. Roullin VG, Deverre JR, Lemaire L, et al. Anti-cancer drug diffusion within living rat brain tissue: an experimental study using [3H](6)-5-fluorouracil-loaded PLGA microspheres. Eur J Pharm Biopharm 2002; 53:293–299.
38. Fung LK, Ewend MG, Sills A, et al. Pharmacokinetics of interstitial delivery of carmustine, 4-hydroperoxycyclophosphamide, and paclitaxel from a biodegradable polymer implant in the monkey brain. Cancer Res 1998; 58:672–684.
39. Roullin VG, Mege M, Lemaire L, et al. Influence of 5-fluorouracil-loaded microsphere formulation on efficient rat glioma radiosensitization. Pharm Res 2004; 21:1558–1563.
40. Fournier E, Passirani C, Vonarbourg A, et al. Therapeutic efficacy study of novel 5-FU-loaded PMM 2.1.2-based microspheres on C6 glioma. Intl J Pharm 2003; 268:31–35.
41. Fournier E, Passirani C, Montero-Menei C, et al. Therapeutic effectiveness of novel 5-fluorouracil-loaded poly(methylidene malonate 2.1.2)-based microspheres on F98 glioma-bearing rats. Cancer 2003; 97:2822–2829.
42. Fournier E, Passirani C, Colin N, et al. Development of novel 5-FU-loaded poly(methylidene malonate 2.1.2)-based microspheres for the treatment of brain cancers. Eur J Pharm Biopharm 2004; 57:189–197.
43. Pages F, Vives V, Sautes-Fridman C, et al. Control of tumor development by intratumoral cytokines. Immunol Lett 1999; 68:135–139.
44. Rhines LD, Sampath P, DiMeco F, et al. Local immunotherapy with interleukin-2 delivered from biodegradable polymer microspheres combined with interstitial chemotherapy: a novel treatment for experimental malignant glioma. Neurosurgery 2003; 52:872–879; discussion 879–880.
45. Pean JM, Menei P, Morel O, Montero-Menei CN, Benoit JP. Intraseptal implantation of NGF-releasing microspheres promote the survival of axotomized cholinergic neurons. Biomaterials 2000; 21:2097–2101.
46. Menei P, Pean JM, Nerriere-Daguin V, Jollivet C, Brachet P, Benoit JP. Intracerebral implantation of NGF-releasing biodegradable microspheres protects striatum against excitotoxic damage. Exp Neurol 2000; 161:259–272.
47. Jollivet C, Aubert-Pouessel A, Clavreul A, et al. Long-term effect of intra-striatal glial cell line-derived neurotrophic factor-releasing microspheres in a partial rat model of Parkinson's disease. Neurosci Lett 2004; 356:207–210.
48. Jollivet C, Aubert-Pouessel A, Clavreul A, et al. Striatal implantation of GDNF releasing biodegradable microspheres promotes recovery of motor function in a partial model of Parkinson's disease. Biomaterials 2004; 25:933–942.
49. Estebe JP, Le Corre P, Malledant Y, Chevanne F, Leverge R. Prolongation of spinal anesthesia with bupivacaine-loaded (D,L-lactide) microspheres. Anesth Analg 1995; 81:99–103.
50. Cruaud O, Benita S, Benoit JP. The characterization and release kinetics evaluation of baclofen microspheres designed for intrathecal injection. Int J Pharm 1999; 177:247–257.

20

Nanoparticles as Drug Delivery Systems for the Brain

Jörg Kreuter
Institut für Pharmazeutische Technologie, Johann Wolfgang Goethe-Universität Frankfurt, Frankfurt/Main, Germany

INTRODUCTION

The blood-brain barrier (BBB) represents an insurmountable obstacle for most drugs including many essential drugs such as antineoplastic agents, antibiotics, and central nervous system active drugs (1,2). This barrier is formed by the endothelial cells surrounding the brain capillaries, which are linked by tight continuous circumferential junctions that abolish any aqueous paracellular pathways between these cells and greatly restrict the movement of polar solutes across the cerebral endothelium. A barrier function also occurs at the arachnoid membrane and in the ependymal cells surrounding the circumventricular organs of the brain (3–5). Lipophilic drugs may passively diffuse across the brain endothelial cells. However, this mechanism is restricted by molecular weight and especially by extremely effective efflux pumps that include P-glycoprotein (Pgp), sometimes referred to as multidrug resistance protein (MDR), and multiple organic anion transporter (MOAT) (5). Consequently, the BBB essentially comprises the major interface between the blood and the brain and plays a major role in the regulation of the constancy of the internal environment of the brain. However, since the brain is dependent on the blood for the delivery of substrates as well as for the removal of metabolic waste, and owing to the fact that water-soluble components and especially larger molecular weight components including larger lipophilic molecules and particles, cannot enter the brain, special receptors and transporters, as for instance the insulin, transferrin, and low-density lipoprotein (LDL) receptor, exist for the transport of essential substances across the BBB (6). Glucose and amino acids can also access the brain by similar mechanisms (5).

A number of attempts have been made to overcome the above-mentioned barrier. These include strategies like the osmotic opening of the tight junctions, and the use of prodrugs or carrier systems such as antibodies, liposomes, and nanoparticles (1,2,7–11). The opening of the tight junctions by osmotic pressure, however, is a very invasive procedure that also allows the entry of toxins and other unwanted substances into the brain, resulting in potentially significant damage. Prodrugs are either more lipophilic and thus can permeate across the lipophilic endothelial barrier or are able to take advantage of the above-mentioned transporter systems in the brain. Very often, however, a suitable chemical modification is not possible because the prodrugs cannot interact with these transporters or cannot be cleaved in the brain into suitable chemical components or they may become substrates for the efflux transporter systems. Like prodrugs, colloidal carriers such as liposomes and nanoparticles may take advantage of the biochemical transport systems (1,2). Among these systems, for instance, the LDL-receptor and the transferrin transcytosis systems may be employed in the delivery of drugs by the above particulate colloidal drug delivery systems (12,13).

DRUG DELIVERY TO THE BRAIN WITH NANOPARTICLES

One of the colloidal drug carrier systems that can be employed for the delivery of drugs to the brain is nanoparticles. Nanoparticles are defined as polymeric particles made of natural or artificial polymers ranging between about 10 and 1000 nm (1 μm) in size. Drugs can be bound in the form of a solid solution or dispersion, or can be adsorbed to the surface or be chemically attached (14).

Dalargin

The first successful drug delivery across the BBB to the brain was achieved with nanoparticles of a size of about 250 nm made of the very rapidly biodegradable polymer, poly(butyl cyanoacrylate) (1,2,15,16). The hexapeptide dalargin (Tyr-D-Ala-Gly-Phe-Leu-Arg), a Leu-enkephalin analogue with opioid activity, was adsorbed to their surface by incubation for four hours, resulting in the sorptive binding of 40% of the initial amount of dalargin. These particles were then overcoated with the surfactant polysorbate 80 (Tween® 80) by incubation for another 30 minutes, thereby achieving an equilibrium between surface-bound polysorbate and polysorbate in solution. After intravenous (IV) administration to mice, these particles yielded a dose-dependent antinociceptive (analgesic) effect in the tail-flick as well as the hot plate test (15–18). This antinociception was accompanied by a pronounced Straub effect and was totally inhibited by injection of the μ-opiate receptor antagonist naloxone, 10 minutes before injection of the nanoparticle preparation. This demonstrates that the observed effects were not caused by peripheral action and that dalargin indeed was transported across the BBB, exhibiting a direct effect on the central nervous system (CNS). In contrast to the polysorbate 80–coated dalargin-loaded nanoparticles, all of the controls including a solution of dalargin, a solution of polysorbate 80, a suspension of poly(butyl cyanoacrylate) nanoparticles, a mixture of dalargin with polysorbate 80, dalargin with nanoparticles or a mixture of all three components: dalargin, polysorbate 80, and nanoparticles, mixed immediately before injection, and dalargin bound to nanoparticles without polysorbate 80 coating did not exhibit any antinociceptive effects (Table 1).

The antinociceptive effects were not only dose-dependent but also circadian phase (day-time)–dependent (18). Significant daily rhythms in the pain reaction induced in the hot plate or in the tail-flick tests were observed with untreated mice as well as with mice that were subjected to IV injection of dalargin solution or of dalargin bound to nanoparticles coated with polysorbate 80 (Fig. 1). No difference in these pain reactions, either quantitatively or in the time of the minima and maxima, was visible between untreated control and dalargin solution. After injection of

Table 1 Analgesia in Male ICR Mice (20–22 g) Determined by Percentage (Mean ± Standard Deviation) of MPE in the Tail-Flick Test 45 Minutes After Intravenous Injection of Dalargin or of Excipients in Free Form or in Combination with Nanoparticles, to Mice ($n = 5$)

Group		% MPE
1	Suspension of empty nanoparticles (200 mg/kg)	0.75 ± 3.0
2	Polysorbate 80 solution (1%, 200 mg/kg)	12.0 ± 3.1
3	Dalargin (solution 10 mg/mL, 10 mg/kg)	9.3 ± 8.7
4	Dalargin (10 mg/kg) + polysorbate 80 (1%, 200 mg/kg)	7.8 ± 2.3
5	Dalargin (10 mg/kg) + empty nanoparticles (200 mg/kg)	1.5 ± 5.4
6	Dalargin (10 mg/kg) + empty nanoparticles (200 mg/kg) + polysorbate 80 (200 mg/kg)	12.5 ± 2.0
7	Dalargin-loaded nanoparticles (10 mg/kg)	3.7 ± 1.1
8	Polysorbate 80–coated and dalargin-loaded nanoparticles (2.5 mg/kg)	11.6 ± 9.7
9	Polysorbate 80–coated and dalargin-loaded nanoparticles (5 mg/kg)	36.8 ± 21.5[a]
10	Polysorbate 80–coated and dalargin-loaded nanoparticles 7.5 mg/kg)	51.8 ± 20.2[a]

[a]$P < 0.05$.
Abbreviation: MPE, maximally possible effect.
Source: From Ref. 16.

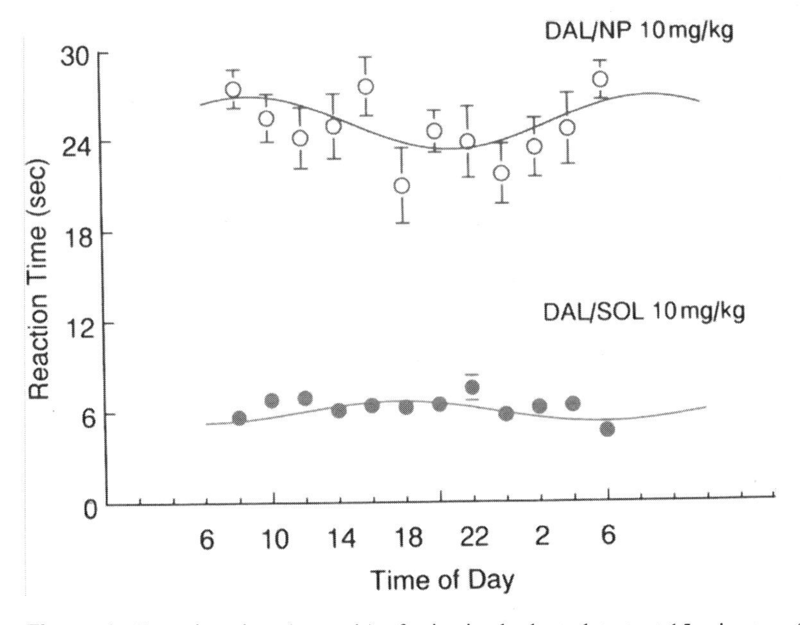

Figure 1 Reaction time (seconds) of mice in the hot plate test 15 minutes after intravenous injection of a dalargin solution (DAL/SOL) or of polysorbate 80–coated dalargin-loaded poly(butylcyanoacrylate) nanoparticles (mean ± SEM). The solid line represents the cosine fit to the data. Lights from 07:00 to 19:00. *Source*: From Ref. 18.

dalargin bound to nanoparticles coated with polysorbate 80, a strong antinociceptive effect combined with a shift of the minima and maxima of 10 to 12 hours compared to that of the controls and the dalargin solution was obtained. The clear dose-dependence of the antinociceptive effects was maintained in the maxima as well as in the minima (Fig. 2). Interestingly, the antinociceptive effect with the polysorbate-coated nanoparticles was more pronounced during the early rest period at around 10:00 a.m., whereas under baseline conditions (controls and dalargin solution), the acrophase (longest reaction times) in the response was in the evening. This could indicate that the antinociceptive effects produced by the nanoparticle-bound endorphin are more pronounced when the level of pain sensitivity is high (i.e., when the reaction time to a painful stimulus is shortest) (18). There is evidence for a circadian-time–dependent fluctuation in the permeabilities of small cerebral vesicles (19). A time-dependence in the activity of the transporter systems is also likely. Thus a circadian phase–dependency could be involved in the kinetics of the crossing of the BBB, resulting in the observed phase shift of the dalargin nanoparticle effects (18).

Besides the overcoating of the poly(butyl cyanoacrylate) nanoparticles with polysorbate 80, overcoating with polysorbate 20, 40, and 60 also enabled a transport of the nanoparticle-bound dalargin across the BBB, whereas coating with other surfactants such as poloxamers 184, 188, 338, 407, poloxamine 908, Cremophor® EZ, Cremophor® RH 40 (BASF, Ludwigshafen, Germany) and polyoxyethylene-(23)-laurylether (Brij® 35) led to no effects (Table 2) (20).

Polysorbate 85 may even enable a brain transport after oral administration of poly(butyl cyanoacrylate) nanoparticles as reported by Schroeder et al. (22). The antinociceptive effect with dalargin, obtained by this delivery route, although rather prolonged, was not as pronounced as that after an IV injection.

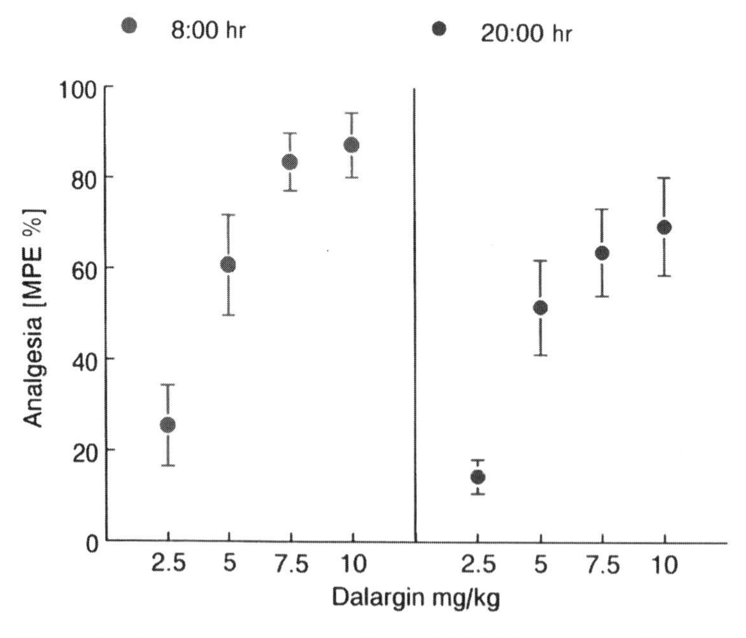

Figure 2 Dose–response curves of mice (% MPE) in the tail-flick test 15 minutes after intravenous injection of polysorbate 80–coated dalargin-loaded poly(butyl cyanoacrylate) nanoparticles (mean ± SEM) at two times of the day. Lights from 07:00 to 19:00. *Abbreviation*: MPE, maximal possible effect. *Source*: From Ref. 18.

Table 2 Maximal Possible Antinociceptive Effect [MPE (%)] Obtained After Intravenous Injection of Dalargin-Loaded Surfactant-Coated Poly(butyl cyanoacrylate) Nanoparticles and Amount of Apo E Adsorbed on the Surface of these Particles in Percentage of the Total Amount of Adsorbed Plasma Proteins

Surfactant	MPE (%)	Apo E adsorbed (%)
Uncoated	4.1 ± 1.0	0
Polysorbate 20	51 ± 19	21.6
Polysorbate 40	61 ± 41	29.7
Polysorbate 60	30 ± 36	13.9
Polysorbate 80	89 ± 22	14.6
Poloxamer 338	1.4 ± 2.4	0
Poloxamer 407	8.1 ± 2.9	0
Cremophor EL	11.7 ± 15.1	0
Cremophor RH 40	23 ± 17	0

Abbreviations: Apo E, apolipoprotein E; MPE, maximal possible effect.
Source: From Refs. 20, 21.

To determine the dalargin–brain pharmacokinetics, the brain concentrations of ^{3}H-dalargin in form of [Leucyl-4,5-^{3}H]-dalargin and of [^{3}H-Tyr]-dalargin were determined after IV injection in mice. The concentrations in brain homogenates were up to threefold higher with dalargin bound to the polysorbate 80–coated nanoparticles than with dalargin solution, and statistically, were significantly different at most time points. There were also concentration differences between brain homogenate fractions, with the highest difference, compared to the dalargin solution, in the fraction representing the synaptosomes (23,24). Nevertheless, these differences were smaller than expected in comparison with the huge difference in the pharmacological responses. However, it has to be considered that determination of the ^{3}H-radioactivity cannot discriminate between a drug that has and a drug that has not actually crossed the BBB. The observed differences between brain fractions may be indicative of the lack of efficient BBB crossing of dalargin in solution form.

Other Drugs

Other antinociceptive drugs, transported across the BBB by the polysorbate 80–overcoated poly(butyl cyanoacrylate) nanoparticles, were loperamide and kytorphin that showed effects similar to those seen with the use of dalargin (25,26). A significant antinociceptive effect with dalargin and kytorphin was also observed after polysorbate 85 coating. It is noteworthy that loperamide is not a peptide and, in contrast to dalargin and kytorphin, is very lipophilic. However, it should be mentioned that after the binding of loperamide to poly(lactic acid) nanoparticles, no antinociceptive response was obtainable after coating with polysorbate 80 or after preparation in the presence of this surfactant (unpublished results), although nanoparticles with different release characteristics were manufactured and tested (27,28). This shows that the ability of nanoparticles to enable a delivery of drugs across the BBB does not only depend on the coating material but also on the core polymer.

Besides the antinociceptive drugs, a successful brain transport with nanoparticles was shown with tubocurarine, the novel *N*-methyl-D-aspartic acid (NMDA) receptor antagonists MRZ 2/576 (8-chloro-4-hydroxy-1-oxo-1,2-dihydropyridazino[4,5-*b*]quinoline-5-oxide choline salt) and MRZ 2/596 (8-chloro-1,4-dioxo-1,2,3,4-tetrahydropyridazino[4,5-b]quinoline choline salt), amitriptiline, and, most importantly, with doxorubicin and the β-galactosidase reporter DNA (26,29–34).

Tubocurarine was chosen as a model drug that normally cannot cross the BBB but induces epileptic spikes after gaining access to the brain, for instance, by intraventricular injection. In order to test the ability of nanoparticles to transport tubocurarine from the blood across the BBB into the brain, brain perfusion experiments were performed in rats, and the development of epileptic spikes in the electroencephalo-graph (EEC) was recorded (29). A normal EEC was observed after addition of tubocurarine to the perfusate. Polysorbate 80 or uncoated tubocurarine-loaded nanoparticles in the perfusate did not change the EEC. However, the addition of polysorbate 80–overcoated tubocurarine-loaded nanoparticles to the perfusate led to the development of frequent severe spikes in the EEC comparable to those that occur following the direct intraventricular injection of the drug into the brain.

The novel NMDA receptor antagonist MRZ 2/576 is a potent but rather short-acting (5–15 minutes) anticonvulsant, following IV administration. This short duration, most probably, is due to the rapid elimination of the drug from the central nervous system by efflux pump–mediated transport processes. Accordingly, these efflux processes can be inhibited by pretreatment with probenecid, leading to a prolongation of the anticonvulsive action of MRZ 2/576 from about 15 to 150 minutes. An even more prolonged duration of the anticonvulsive activity of up to 210 minutes in mice was achieved by IV administration of the drug bound to poly(butyl cyanoacrylate) nanoparticles coated with polysorbate 80 (30,31). The action of the nanoparticles was further prolonged by up to 270 minutes by combination with probenecid pretreatment.

Another similar NMDA receptor antagonist, MRZ 2/596 is not able to cross the BBB at all, but comparable significant anticonvulsive effects were obtained after binding to polysorbate 80–coated nanoparticles (31).

Binding of amitriptyline, a tricyclic antidepressant, to polysorbate 80–coated nanoparticles led to an improvement in brain area under the curve (AUC) following IV injection of polysorbate 80–coated nanoparticles (26). This increase in brain AUC was accompanied by a reduction in serum AUC. However, the binding of valproic acid to these nanoparticles did not increase the brain concentrations of this drug (32).

Lockman et al. (33,34) produced nanoparticles made of emulsifying wax/Brij 78 or Brij 72/polysorbate 80 with a size of about 100 nm. They bound [^{3}H]-thiamine via a polyethylene glycol (PEG)-spacer introduced into the nanoparticles using distearoylphosphatidylethanolamine to target the thiamine transporters in the brain, and showed an association with the thiamine transporter. Biodistribution studies in BALB/c mice, however, showed only a very low brain uptake of emulsifying wax/Brij 78 nanoparticles labelled with [^{111}In]-indium chloride, and no significant difference was observed between emulsifying wax/Brij 78 nanoparticles with protruding PEG chains on the outside and nanoparticles with thiamine bound to the PEG chains.

Employment of Nanoparticles for the Therapy of Brain Tumors

Brain tumors, especially malignant gliomas, belong to the most aggressive human cancers. Despite numerous advances in neurosurgical operative techniques, adjuvant chemotherapy, and radiotherapy, the prognosis for patients remains very unfavorable

(35). Features responsible for the aggressive character of gliomas include rapid proliferation, diffuse growth, and invasion into distant brain areas in addition to extensive cerebral edema and high levels of angiogenesis. Glioblastomas develop a distinct neovasculature that is permeable to macromolecules and small particles. However, disruption of the BBB remains a local event, which is evident in the tumor core but absent at its growing margins. Accordingly, therapeutic drug levels have been found in necrotic tumor areas, while in peritumoral regions, the drug levels were markedly lower or nondetectable (36).

Doxorubicin is one of the most efficient anticancer drugs, but it cannot cross the BBB and, therefore, is not used against brain tumors. For this reason, the drug was bound to poly(butyl cyanoacrylate) nanoparticles and the particles were coated with polysorbate 80 (37,38). These nanoparticles achieved very high brain doxorubicin concentrations of 6 μg/g following IV injection of 5 mg/kg doxorubicin into the tail vein of rats, whereas with all three control preparations, that is, doxorubicin solution in saline, doxorubicin solution in saline plus polysorbate 80, and doxorubicin bound to nanoparticles without polysorbate-coating, the concentrations were below the detection limit of 0.1 μg/g (37). In contrast, the differences in plasma concentrations between the four preparations were not large, although sometime, statistically significant. In addition, the heart concentrations of both nanoparticle preparations were very low confirming earlier data of Couvreur et al. (40) with uncoated nanoparticles. Because the use of doxorubicin is limited by its cumulative high heart toxicity, this observation is of major importance.

The polysorbate 80–coated doxorubicin nanoparticles were then tested in rats with intracranially transplanted glioblastoma 101/8 (38). The employed animal system, the 101/8 glioblastoma in rats, is morphologically similar to human glioblastomas. A statistically significant improvement in survival time was obtained with the polysorbate 80–coated nanoparticles after IV injection of doses of 1.5 mg/kg doxorubicin on day two, five, and eight after tumor transplantation (Fig. 3). Moreover, 20% to 40% of the rats survived for half a year and were then sacrificed, and no signs of tumor could be observed by histological examination. Histology of rats sacrificed at earlier time points confirmed smaller tumor sizes and lower values for proliferation and apoptosis in this group. Untreated control animals, as well as animals receiving nanoparticles coated with polysorbate 80 without drugs, died between nine and 20 days. The other controls, doxorubicin solution in saline, doxorubicin solution in saline plus polysorbate 80, and doxorubicin bound to nanoparticles without polysorbate-coating, also prolonged the survival times although this prolongation was much less pronounced than that observed with the polysorbate-coated doxorubicin nanoparticles (Fig. 3) (38).

In the same study, only a limited dose-dependent systemic toxicity was found in the group treated with doxorubicin in saline that was not enhanced but rather alleviated in animals treated with doxorubicin bound to the polysorbate-coated nanoparticles. Signs of short-term neurotoxicity, such as increased apoptosis in areas distant from the tumor or degenerative morphological changes of neurons, were entirely absent in the group that had received doxorubicin-loaded polysorbate 80–coated nanoparticles and had been sacrificed on day 12 as well as in long-term survivors (38).

Another study focused on the assessment of the acute toxicity of empty as well as polysorbate 80–coated doxorubicin poly(butyl cyanoacrylate) nanoparticles (41). IV administration of empty nanoparticles in doses up to 400 mg/kg did not cause mortality within the period of observation (30 days) and did not affect body weight

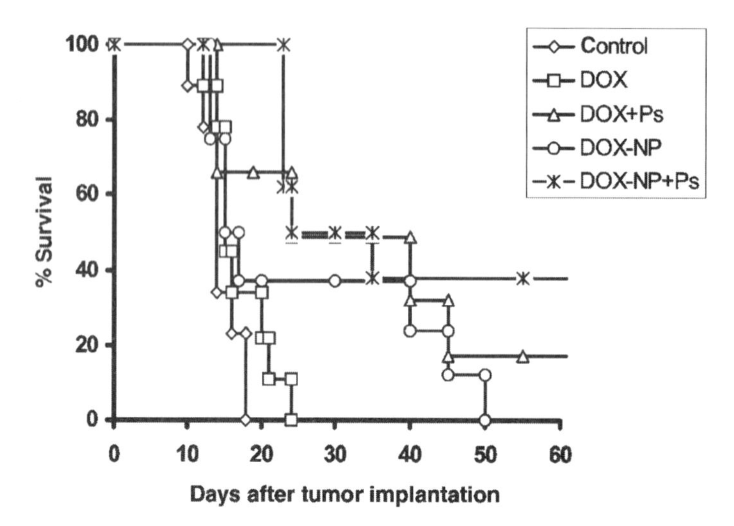

Figure 3 Antitumor efficacy of the following doxorubicin preparations (3×1.5 mg/kg) after intravenous injection in rats with intracranially implanted glioblastoma 101/8: DOX, DOX + Ps, DOX-NP, and DOX-NP + Ps. The preparation DOX-NP + Ps yielded 38% (3/8) and the preparation DOX + Ps 13% (1/8) long-time survivors that were sacrificed after 180 days. The long-term surviving animals of the DOX-NP + Ps group showed no residual tumor growth but some reactive changes, indicating a sufficient application of tumor cells during implantation. *Abbreviations*: DOX, doxorubicin in saline; DOX + Ps, doxorubicin in 1% polysorbate 80 in saline; DOX-NP, doxorubicin bound to nanoparticles; DOX-NP + Ps, doxorubicin bound to nanoparticles coated with polysorbate 80. *Source*: From Ref. 2.

or weight of internal organs. The doxorubicin nanoparticles were also tested in healthy and intracranially implanted 101/8 glioblastoma–bearing rats. Again, doxorubicin solution in saline, doxorubicin solution in saline plus polysorbate 80, and doxorubicin bound to nanoparticles without polysorbate-coating served as controls. No significant difference in toxicity was observable among these groups. The results in tumor-bearing rats were similar to those in healthy rats. Weight loss was highest after administration of doxorubicin in saline. The results of this study confirmed the above results in that the toxicity of doxorubicin bound to nanoparticles appeared to be similar or even lower than that of free doxorubicin (38).

Doxorubicin-loaded solid lipid nanoparticles (SLN) consisting of stearic acid, Epikuron® 200 (Lucas Meyer, Hamburg, Germany), and taurocholate sodium were administered intravenously to rabbits (42). Four types of SLN were used—nonstealth particles and so-called stealth nanoparticles containing 0.15%, 0.30%, and 0.45% stearic acid–PEG 2000 where the PEG chain acted as the stealth agent. The doxorubicin plasma concentrations were significantly increased and prolonged with increasing stearic acid–PEG contents, but all nanoparticles achieved much higher plasma concentrations than doxorubicin solution. The same increase with increasing stearic acid–PEG content was observed in the brain, reaching a doxorubicin concentration of 240 ng/g after administration of 1 mg/kg doxorubicin. No doxorubicin was found in the brain after administration of doxorubicin solution. The heart concentration and the liver concentration of doxorubicin were once again decreased by the nanoparticles.

SLN were also used as carriers for another anticancer drug, camptothecin (43). Camptothecin was incorporated into SLN consisting of stearic acid and soybean lecithin. The lipid phase was dispersed in water using 0.5% poloxamer 188 and 2.25% glycerol as dispersing agents, and the body distribution of camptothecin was determined in mice after IV injection at two dosages—1.3 and 3.3 mg camptothecin/kg body weight. A solution of 1.3 mg camptothecin/kg in 40% polyethylene glycol 400, 58% propylene glycol, and 2% polysorbate 80 served as a control. The use of the solution led to high plasma levels after short times and the solution was rapidly eliminated, whereas the peak plasma levels were much lower with the SLN, and the elimination was considerably prolonged, resulting in a fivefold increase in plasma camptothecin AUC with the 1.3 mg/kg dose. The brain–camptothecin AUC, after injection of the SLN at this dose, was even 10-fold higher than that of the solution. These results indicate that this preparation may be of interest for the treatment of brain tumors. However, it has to be considered that owing to the poor solubility of the drug in aqueous environments, a rather unphysiological vehicle, which may skew the results, was employed for the preparation of the solution.

Malignant brain tumors do not express wild-type tumor suppressor genes like p53 (39). This loss of function renders the tumors rather resistant to the induction of apoptosis by drugs such as doxorubicin. The delivery of wild-type suppressor genes across the BBB together with highly active chemotherapeutic drugs, consequently, is of enormous importance. Therefore, the possibility of suppressor gene delivery into the brain with nanoparticles was evaluated using β-galactosidase reporter DNA bound to poly(butyl cyanoacrylate) nanoparticles that were coated with polysorbate 80. The particles were injected intravenously to rats with an intracranially implanted F98 rat glioblastoma, five days after tumor implantation. The animals were sacrificed 24, 48, and 72 hours after nanoparticle injection. The results showed a time-dependent transport of the gene across the endothelial and glial cells. The strongest gene expression was observed in the experimental tumors in the rat brains. The injection of naked control DNA did not induce any expression at all (39).

LONG CIRCULATING NANOPARTICLES FOR BRAIN DRUG DELIVERY

A major disadvantage of colloidal drug carriers such as nanoparticles and liposomes is their rapid capture by cells of the reticuloendothelial system (RES), especially the liver and spleen, leading to a short blood circulation time and low blood concentrations of the carriers. This rapid capture and elimination is caused by the adsorption of certain plasma components on the particle surface—the so-called opsonization, (44). The circulation times can be significantly prolonged by coating with PEG chain–containing surfactants or by the covalent coupling of PEG to the nanoparticle surface (42,45–51). This influence of PEG is also reflected in the above-mentioned studies by Zara et al. (42) which showed that the plasma and the corresponding brain concentrations of doxorubicin bound to SLN increased with increasing stearic acid–PEG 2000 content of the particles.

Calvo et al. (49) measured the body distribution of ^{14}C as well as the ^{14}C concentration in different brain tissues after IV administration of [^{14}C]-poly[methoxy poly (ethylene glycol) cyanoacrylate-co-hexadecyl cyanoacrylate] nanoparticles ([^{14}C]-PEG-PHDCA nanoparticles) or [^{14}C]-poly (hexadecyl cyanoacrylate) nanoparticles ([^{14}C]-PHDCA nanoparticles) in mice and rats. Uncoated [^{14}C]-PHDCA nanoparticles and [^{14}C]-PHDCA nanoparticles coated with polysorbate 80 or with poloxamine 908 were used in these studies. The highest blood concentrations were obtained in mice and rats with the poloxamine 908–coated particles followed by the [^{14}C]-PEG-PHDCA particles. The poloxamine 908–coated particles also yielded the lowest liver concentrations in mice, but not in rats. In the brain, the PEG-PHDCA particles achieved the highest ^{14}C concentrations, followed by uncoated PHDCA nanoparticles and those coated with polysorbate 80. However, when the nanoparticle dose was reduced using the same total polysorbate concentration, the highest brain [^{14}C] levels were achieved with the polysorbate 80–coated particles, in rats and mice. The authors suggested that these higher–brain concentrations were caused by a greater BBB permeability as a result of elevated free blood polysorbate concentrations at the lower nanoparticle dose. They supported their assumption by another experiment that involved the IV injection of 5% [^{14}C]-sucrose in a 1% polysorbate solution in saline, which also led to higher [^{14}C]-sucrose levels in the brain. However, the assumption that free polysorbate 80 concentrations of up to 1% lead to an increased BBB permeability and drug transport is contradicted by all the above-reported studies with other drugs in which no increased drug flux into the brain was seen at this polysorbate concentration (15–18,20,22–26,29–31,37,38).

Another interesting finding in the above study is the observation that although poloxamine 908–coating led to higher nanoparticle plasma concentrations than PEG-modification, the brain concentrations of PEG-modified nanoparticles in both animal species were significantly higher (approximately threefold) than that with the poloxamine-coated particles (49). This may be an indication that increased plasma concentrations caused by a shielding stealth effect are not sufficient to achieve a higher drug transport into the brain and that specific interactions of the carriers with the brain capillary endothelium are required.

The [^{14}C]-PEG-PHDCA as well as the [^{14}C]-PHDCA nanoparticles showed an accumulation with a retention effect in Fischer rats bearing an intracerebrally transplanted 9L glioblastoma (49). The tumor concentrations of the pegylated nanoparticles were about three times higher than that with the normal PHDCA particles, and their tumor-to-brain ratio was 11:1. No retention effect was seen with the hydrophilic marker [^{3}H]-sucrose after coinjection with the [^{14}C]-nanoparticles, although increased ^{3}H levels were found in the tumor between three and 30 minutes. The ^{3}H concentrations in the residual brain of the tumor-bearing rats were again about three times lower than those in the tumors, and in the brain of normal control rats, the ^{3}H concentrations were about six to 10 times lower due to the intact nature of BBB. The authors attribute this tumor retention of the nanoparticles to the so-called enhanced permeability and retention (EPR) effect, which is typical for large molecules and particulates (50,51). Conversely, because sucrose possesses a rather low molecular weight, it would leave the tumor and brain area rapidly after it has gained entry through the leaky blood vessels in the tumor area (50).

The [^{14}C]-PEG-PHDCA nanoparticles also yielded much higher brain and spinal cord ^{14}C concentrations in DA/Rj rats with experimental allergic encephalitis (EAE) than normal PHDCA nanoparticles (52). In contrast, the coating of the latter particles with poloxamine 908 resulted in very low and insignificant brain and spinal cord concentrations, although this surfactant again achieved very high nanoparticle plasma levels. The concentration of the PEG-PHDCA nanoparticles was especially pronounced in the white matter and was significantly higher in the pathological situation where the BBB permeability is increased. An enhanced infiltration with macrophages containing nanoparticles was observed at the EAE lesions. This transport within macrophages could augment the overall nanoparticle transport across the BBB (52).

STABILITY OF NANOPARTICLES FOR BRAIN DELIVERY UPON STORAGE

Poly(butyl cyanoacrylate) nanoparticles are normally stored in lyophilized form after addition of a cryoprotector such as mannitol. Such particles loaded with doxorubicin were stable for at least two years without indication of any degradation and showed no loss of activity in animal experiments (unpublished observation). Sommerfeld et al. (53,54) also found that empty poly(butyl cyanoacrylate) nanoparticles stabilized during polymerization with dextran 70,000 were stable in suspension form at pH 2, for one year. After some time, the particles tended to sediment. However, this sediment was easily resuspendable by gentle shaking. Particles stabilized with poloxamer 188 or polysorbate 85 tended to form agglomerates during storage.

Nanoparticles for parenteral use are normally produced under aseptic conditions. Sterilization with 15 kGy gamma irradiation is also possible (Gelperina, personal communication). Autoclaving or treatment with formaldehyde led to the formation of agglomerates (55).

MECHANISM OF NANOPARTICLE-MEDIATED DRUG TRANSPORT TO THE BRAIN

Until today the mechanism of the transport of drugs across the BBB using nanoparticles has not been fully elucidated. A number of possibilities have been suggested for this mechanism (1,2,56):

1. An increased retention of the nanoparticles in the blood capillaries of the brain, combined with an adsorption to the capillary walls could create a higher concentration gradient that would increase the transport across the endothelial cell layer and as a result enhance the delivery to the brain.
2. The polysorbate 80, used as the coating agent, could inhibit the efflux system, especially Pgp.
3. A general toxic effect on the brain vasculature could lead to the permeabilization of the brain blood vessel endothelial cells.
4. A general surfactant effect characterized by a solubilization of the endothelial cell membrane lipids would lead to membrane fluidization and an enhanced drug permeability across the BBB.
5. The nanoparticles could lead to an opening of the tight junctions between the endothelial cells. The drug could then permeate the tight junctions in free form or together with the nanoparticles in bound form.
6. The nanoparticles may be endocytosed by the endothelial cells followed by the release of the drugs within these cells and delivery to the brain.
7. The nanoparticles with bound drugs could be transcytosed through the endothelial cell layer.

All these mechanisms could also work in combinations.

Although an increased concentration of the nanoparticles on the walls of the blood capillaries of the brain would enhance the transport across the endothelial cell layer and delivery to the brain (mechanism 1), the drugs would still be subjected to very effective efflux pumps located in the luminal membrane of the endothelial cells such as Pgp. Because loperamide, doxorubicin, and tubocurarine are known substrates for Pgp, this mechanism appears to be largely insignificant (5). Nevertheless, de Verdière et al. (57) showed in multidrug resistant P388/ADR cells, that doxorubicin-loaded poly(isobutyl cyanoacralate) nanoparticles delivered the drug efficiently to these cells after incubation, but were not endocytosed. Owing to the fact that the multidrug resistance expressed by these cells is also caused by the presence of Pgp and because the drug was not transported into the cells in nanoparticle-bound form via endocytosis, the drug must have crossed the membrane in a different way. The authors suggest that the resistance was overcome as a result of both the adsorption of nanoparticles to the cell surface and an increased doxorubicin diffusion following the formation of an ion pair with the nanoparticle degradation products (58). The accumulation of the ion pair at the plasma membrane would result in a high drug gradient close to the cells and an increased drug diffusion into the cells (57,58). At present, it is totally unclear as to whether a similar mechanism may also be taking place in the delivery of drugs across the BBB using poly(alkyl cyanoacrylate) nanoparticles.

Inhibition of these efflux systems, especially the Pgp, by polysorbate 80 (mechanism 2) is also possible because polysorbate 80 was shown to be able to inhibit Pgp (59–61). In addition, Pgp is also responsible for the multidrug resistance which represents a major obstacle in cancer chemotherapy. However, as shown in the above control in vivo experiments in which polysorbate 80 added to the drug solutions in concentrations of 1% as used with the nanoparticles, no drug transport was visible. Nevertheless, as shown by Calvo et al. (49), it may augment some of the above mechanisms and may, in part, explain the enhanced sucrose flux with this surfactant. However, at present, the significance of the inhibition of the efflux pumps by polysorbate 80 during nanoparticle-mediated drug delivery to the brain is not known (1,2).

A general toxic effect of the nanoparticles on the brain vasculature could lead to the permeabilization and disruption of the BBB (mechanism 3) was suggested by Olivier et al. (62) despite their own contradicting results. These authors confirmed the occurrence of an antinociceptive effect after IV injection of polysorbate 80–coated dalargin-loaded nanoparticles but not after the use of uncoated dalargin-loaded nanoparticles. An antinociceptive effect with uncoated nanoparticles, however, would have to occur if the nanoparticles would have caused the suggested general toxic effect. Therefore, Olivier et al. based their assumption merely on further experiments in the same study where they observed a decrease in locomotive activity in the mice after IV injection of polysorbate 80–coated poly(butyl cyanoacrylate) nanoparticles, but in this case also with uncoated nanoparticles. Again with both preparations, they also found an equal and very considerably increased flux of [^{14}C]-sucrose and [^{3}H]-inulin across the endothelial cells in an in vitro model of the BBB—the Cecchelli model. However, other experiments demonstrated that this in vitro permeabilization was most likely due to an in vitro model artifact: Olivier et al. (62) did not add any serum to the cell coculture, thus enabling a disruption of the integrity of their cell layers, but in the later experiments using the same model, serum was added, making the experimental conditions equivalent to the which is more equivalent to the in vivo situation, and no paracellular transport of sucrose and inulin across the endothelial cells was observed (62,63). These experiments demonstrate that slight changes in the in vitro models of the blood-brain barrier can lead to totally different results. In addition, according to assumptions of Olivier et al. (62), an IV administration of polysorbate 80–coated poly(butyl cyanoacrylate) nanoparticles five or 30 minutes prior to IV injection of dalargin solution would also have induced an antinociceptive effect. This, however, was not the case (Table 3) (63). Moreover, in contrast to

Table 3 Analgesia in Male ICR Mice (20–22 g) Determined by Percentage (Mean ± Standard Deviation) of MPE in the Tail-Flick Test 45 Minutes After Intravenous Injection of Dalargin (7.5 mg/kg) as a Solution or in Combination with Nanoparticles, to Mice ($n = 6$)

	% MPE
Suspension of empty nanoparticles	0.75 ± 3.2
Dalargin solution	9.3 ± 2.8
Polysorbate 80–coated dalargin-loaded PBCA nanoparticles	52 ± 20[a]
Polysorbate 80–coated nanoparticles mixed with dalargin immediately before injection	4.5 ± 5.1
Polysorbate 80–coated nanoparticles followed by injection of dalargin 5 min later	3.9 ± 4.6
Polysorbate 80–coated nanoparticles followed by injection of dalargin 30 min later	0.7 ± 4.8

[a]$P < 0.05$).
Abbreviation: MPE, maximally possible effects.
Source: From Ref. 63.

suggestions made by Olivier et al. (62), the toxicity of the nanoparticles was very low, with a lethal dose $(LD)_{50}$ significantly above 400 mg/kg (41).

The observations of Kreuter et al. (63) also do not support mechanisms 4 and 5, which suggest enhanced drug permeability due to membrane fluidization by surfactants or due to opening of the tight junctions between the endothelial cells. Electron microscope investigations showed no changes in the cell-to-cell contacts (tight junctions), and no large-scale increase in the inulin spaces was found after jugular vein injection of the polysorbate 80–coated nanoparticles, as would have been observed after significant opening of the tight junctions (24,63). If membrane fluidization had been the ruling mechanism, other surfactants would also have to be effective; this was not the case (20).

The most likely mechanism is mechanism 6—endocytotic uptake of the nanoparticles carrying the drug (56). The drug may then be released by desorption or biodegradation of the nanoparticles. Grislain et al. (21) demonstrated that poly(butyl cyanoacrylate) nanoparticles are very rapidly biodegradable in vivo and that the degradation products are rapidly excreted. Alternatively, the nanoparticles with bound drugs may be transported into the brain by transcytosis (mechanism 7). In any case, a very significant endocytotic uptake of the polysorbate 80–coated nanoparticles can be observed in vitro in brain endothelial cell tissue cultures (24,64,65). Accordingly, a significant and rapid uptake was observed at an incubation temperature of 37°C only with polysorbate-coated particles, whereas without coating, this uptake was minimal and was inhibited at 4°C, a temperature at which phagocytosis does not occur, or by a treatment with cytochalasin B, a potent phagocytic uptake inhibitor (65).

Uptake via endocytosis by the brain capillary endothelial cells is supported by the observation that apolipoprotein E (Apo E) was adsorbed onto polysorbate-coated nanoparticles by incubation in blood plasma (66). This adsorption occurred only with those surfactants that enabled the development of an antinociceptive effect in mice with the dalargin-loaded nanoparticles, whereas those that did not adsorb Apo E also yielded no significant antinociceptive effect (Table 2) (20). An antinociceptive effect also occurred after direct adsorption of Apo E without any surfactants onto the dalargin-loaded nanoparticles, although this effect was slightly lower than that with polysorbate 80 alone (56). The same was observable with Apo B but not with other apolipoproteins. Moreover, the antinociceptive effect with the polysorbate 80–coated dalargin nanoparticles was reduced by about 50% in Apo E–deficient mice (56). These findings indicate that the polysorbates act as anchors for Apo E and possibly also for Apo B. Correspondingly, Lück (66) observed that Apo E becomes displaced by other plasma components during incubation in plasma when it was adsorbed directly on to the nanoparticles without polysorbate 80–precoating.

Both Apo E and B bind to the lipoprotein receptors of cells (56,67–71). These receptors are also present in brain capillary endothelial cells (12,72–74). According to the above-described observations, the polysorbate-coated nanoparticles adsorb apolipoproteins B and/or E from the blood, after injection. Thus they then seem to mimic lipoprotein particles and are taken up by the brain capillary endothelial cells via receptor-mediated endocytosis. In this scenario, nanoparticles would act as Trojan horses for bound drugs (56,66).

Bound drugs may then be further transported into the brain by diffusion following release within the endothelial cells or, alternatively, by transcytosis (mechanism 7). In vitro transcytosis of LDL across the blood-brain barrier was observed in the Cecchelli model by Dehouck et al. (12). Furthermore, cholesterol depletion upregulated the expression of the LDL receptor in this model. It may be possible that the polysorbate-coated poly(butyl cyanoacrylate) nanoparticles can also be transcytosed. However, at present, little is known about this mechanism—transcytosis of the nanoparticles.

CONCLUSIONS

Nanoparticles represent a tool for the transport, across the BBB, of essential drugs that normally are unable to cross this barrier (1,2). The nanoparticles may be especially helpful for the delivery of larger and complex molecules like proteins or even nucleic acids and genes, for the treatment of infections and inflammations, and especially for the treatment of brain tumors. The first promising results in this direction have already been obtained in rats. Brain tumors such as glioblastomas are extremely aggressive and often impossible to treat and therefore in most cases, lead to the death of the patients six to nine months after diagnosis. One of the reasons for this is their diffuse growth and invasion into distant brain areas. By binding to nanoparticles coated with polysorbates, antitumor drugs may also be transported across the intact BBB (37). A delivery across the intact BBB would also enable a cytostatic effect at these sites that normally are not accessible by most anticancer drugs. Local drug injections and drug-loaded implants have only a limited efficacy because they are able to deliver drugs only over very short distances whereas drugs bound to polysorbate-coated nanoparticles can reach the whole brain (74).

The mechanism for the transport of nanoparticle-bound drugs across the blood-brain barrier is at present not fully elucidated. This transport to the brain depends on the coating of the particles with polysorbates, especially polysorbate 80. This material seems to act as an anchor for Apo E or B after injection into the blood stream. The particles thus seem to mimic LDL particles and could interact with the lipoprotein receptors located on the brain capillary endothelial cells, followed by their endocytosis. The drug may then be released in these cells and may diffuse into the interior of the brain, or the particles may be transcytosed.

ACKNOWLEDGMENT

The author wishes to thank Dr. Svetlana Gelperina for helpful reading of the manuscript.

REFERENCES

1. Kreuter J. Nanoparticulate systems for brain delivery of drugs. Adv Drug Deliv Rev 2001; 47:65–81.
2. Kreuter J. Transport of drugs across the blood-brain barrier by nanoparticles. Curr Med Chem: Central Nervous System Agents 2002; 2:241–249.
3. Davson H, Segal M. Physiology of the CSF and Blood-Brain Barrier. Boca Raton: CRC Press, 1996:1–192.
4. Begley D, Kreuter J. Do ultra-low frequency (ULF) magnetic fields affect the blood-brain barrier? In: Holick MF, Jung EG, eds. Biologic Effects of Light 1998. Boston: Kluwer Academic Publishers, 1999:297–301.
5. Begley D. The blood-brain barrier: principles for targeting peptides and drugs to the central nervous system. J Pharm Pharmacol 1996; 48:136–146.
6. Rapoport SI. Modulation of the blood-brain barrier permeability. J Drug Target 1996; 3:417–425.
7. Gummerloch MK, Neuwelt EA. Physiology and pharmacology of the blood-brain barrier. In: Bradbury MWB, ed. Handbook of Experimental Pharmacology. Vol. 103. Berlin, Heidelberg: Springer, 1992:525–542.
8. Pardridge WM, Buciak JL, Friden PM. Selective transport of an anti-transferrin receptor antibody through the blood-brain barrier in vivo. J Pharmacol Exp Ther 1991; 259:66–70.
9. Zhou X, Huang L. Targeted delivery of DANN by liposomes and polymers. J Contr Rel 1992; 19:269–274.
10. Chen D, Lee KH. Biodistribution of calcitonin encapsulated in liposomes in mice with particular reference to the central nervous system. Biochim Biophys Acta 1993; 1158:244–250.
11. Huwyler J, Wu D, Pardridge WM. Brain drug delivery of small molecules using immunoliposomes. Proc Nat Acad Sci USA 1996; 93:14164–14169.
12. Dehouck B, Fenart L, Dehouck M-P, Pierce A, Torpier G, Cecchelli R. A new function for the LDL receptor: transcytosis of LDL across the blood-brain barrier. J Cell Biol 1997; 138:877–889.
13. Fenart L, Casanova A, Dehouk B, et al. Evaluation of effect of charge and lipid coating on ability of 60-nm nanoparticles to cross an in vitro model of the blood-brain barrier. J Pharmacol Exp Ther 1999; 291:1017–1022.
14. Kreuter, J. Nanoparticles. In: Swarbrick J, Boylan JC, eds. Encyclopedia of Pharmaceutical Technology. Vol. 10. New York: Marcel Dekker, 1994:165–190.
15. Alyautdin R, Gothier D, Petrov V, Kharkevich D, Kreuter J. Analgesic activity of the hexapeptide dalargin adsorbed on the surface of polysorbate 80–coated poly(butyl cyanoacrylate) nanoparticles. Eur J Pharm Biopharm 1995; 41:44–48.
16. Kreuter J, Alyautdin RN, Kharkevich DA, Ivanov AA. Passage of peptides through the blood-brain barrier with colloidal polymer particles (nanoparticles). Brain Res 1995; 674:171–174.
17. Schroeder U, Sabel BA. Nanoparticles, a drug carrier system to pass the blood-brain barrier, permit central analgesic effects of i.v. dalargin injections. Brain Res 1996; 710:121–124.
18. Ramge P, Kreuter J, Lemmer B. Circadian phase-dependent antinociceptive reaction in mice after i.v. injection of dalargin-loaded nanoparticles determined by the hot-plate test and the tail-flick test. Chronobiol Int 1999; 17:767–777.

19. Mato M, Ookawara S, Tooyama K. Chronobiological studies of the blood-brain barrier. Experientia 1981; 37:1013–1015.
20. Kreuter J, Petrov VE, Kharkevich DA, Alyautdin RN. Influence of the type of surfactant on the analgesic effects induced by the peptide dalargin after its delivery across the blood-brain barrier using surfactant-coated nanoparticles. J Contr Rel 1997; 49:81–87.
21. Grislain L, Couvreur P, Lenaerts V, Roland M, Deprez-De Campenere D, Speiser P. Pharmacokinetics and distribution of a biodegradable drug-carrier. Int J Pharm 1983; 15:335–345.
22. Schroeder U, Sommerfeld P, Sabel BA. Efficacy of oral dalargin-loaded nanoparticle delivery across the blood-brain barrier. Peptides 1998; 19:777–780.
23. Schroeder U, Schroeder H, Sabel BA. Body distribution of ^{3}H-labelled dalargin bound to poly(butyl cyanoacrylate) nanoparticles after i.v. injections to mice. Life Sci 2000; 66:495–502.
24. Alyautdin RN, Reichel A, Löbenberg R, Ramge P, Kreuter J, Begley DJ. Interaction of poly(butylcyanoacrylate) nanoparticles with the blood-brain-barrier in vivo and in vitro. J Drug Target 2001; 9:209–221.
25. Alyautdin RN, Petrov VE, Langer K, Berthold A, Kharkevich DA, Kreuter J. Delivery of loperamide across the blood-brain barrier with polysorbate 80–coated polybutylcyanoacrylate nanoparticles. Pharm Res 1997; 14:325–328.
26. Schroeder U, Sommerfeld P, Ulrich S, Sabel BA. Nanoparticle technology for delivery of drugs across the blood-brain barrier. J Pharm Sci 1998; 87:1305–1307.
27. Ueda M, Kreuter J. Optimization of the preparation of loperamide-loaded poly(L-lactide) nanoparticles by high pressure emulsification-solvent evaporation. J Microencapsul 1997; 5:593–605.
28. Ueda M, Iwata A, Kreuter J. Influence of the preparation methods on the drug release behaviour of loperamide-loaded nanoparticels. J Microencapsul 1998; 15:361–372.
29. Alyautdin RN, Tezikov EB, Ramge P, Kharkevich DA, Begley DJ, Kreuter J. Significant entry of tubocurarine into the brain of rats by absorption to polysorbate 80–coated polybutylcyanoacrylate nanoparticles: an in situ brain perfusion study. J Microencapsul 1998; 15:67–74.
30. Friese A, Seiler E, Quack G, Lorenz B, Kreuter J. Enhancement of the duration of the anticonvulsive activity of a novel NMDA receptor antagonist using poly(butylcyanoacrylate) nanoparticles as a parenteral controlled release delivery system. Eur J Pharm Biopharm 2000; 49:103–109.
31. Friese A. Kleinpartikuläre Trägersysteme (Nanopartikel) als ein parenterales Arzneistofftransportsystem zur Verbesserung der Bioverfügbarkeit ZNS-aktiver Substanzen dargestellt am Beispiel der NMDA-Rezeptor-Antagonisten MRZ 2/576 und MRZ 2/596. Ph.D. thesis, J. W. Goethe-Universität Frankfurt, Frankfurt, 2000.
32. Darius J, Meyer FP, Sabel BA, Schroeder U. Influence of nanoparticles on the brain-to-serum distribution and the metabolism of valproic acid in mice. J Pharm Pharmacol 2000; 52:1043–1047.
33. Lockman PR, Koziara J, Roder KE, et al. In vitro and in vivo assessment of baseline blood-brain barrier parameters in the presence of novel nanoparticles. Pharm Res 2003; 20:705–713.
34. Lockman PR, Oyewumi MO, Koziara J, et al. Brain uptake of thiamine-coated nanoparticles. J Contr Rel 2003; 93:271–282.
35. DeAngelis LM. Brain Tumors. New Engl J Med 2001; 344:114–123.
36. Donelli MG, Zucchetti M, D'Incalci M. Do anticancer agents reach the tumor target in the human brain? Cancer Chemother Pharmacol 1992; 30:251–260.
37. Gulyaev AE, Gelperina SE, Skidan IN, Antropov AS, Kivman Gya, Kreuter J. Significant transport of doxorubicin into the brain with polysorbate 80–coated nanoparticles. Pharm Res 1999; 16:1564–1569.

38. Gelperina SE, Smirnova ZS, Khalanskiy AS, Skidan IN, Bobruskin AI, Kreuter J. Chemotherapy of brain tumours using doxorubicin bound to polysorbate 80–coated nanoparticles. Proceedings of 3rd World Meeting APV/APGI, Berlin, 3–6 April 2000, 2000: 441–442.
39. Walz CM, Ringe K, Sabel BA. Nanoparticles in brain tumor therapy. Chapter II. Controlled Release Society 30th Annual Meeting Proceedings. Glasgow 2002: # 630.
40. Couvreur P, Kante B, Grislain L, Roland M, Speiser P. Toxicity of polyalkylcyanoacrylate nanoparticles II: doxorubicin-loaded nanoparticles. J Pharm Sci 1982; 71:790–792.
41. Gelperina SE, Khalansky AS, Skidan IN, et al. Toxicological studies of doxorubicin bound to polysorbate 80-coated poly(butyl cyanoacrylate) nanoparticles in healthy rats and rats with intracranial glioblastoma. Toxicol Lett 2002; 126;131–141.
42. Zara GP, Cavalli R, Bargoni A, Fundarò A, Vighetto D, Gasco MR. Intravenous administration of non-stealth and stealth doxorubicin-loaded solid lipid nanoparticles at increasing concentration of stealth agent: pharmacokinetics and distribution in brain and other tissue. J Drug Target 2002; 10:327–335.
43. Yang SC, Lu LF, Zhu JB, Liang BW, Yang CZ. Body distribution in mice of injected camptothecin solid lipid nanoparticles and targeting effect on brain. J Contr Rel 1999; 59:299–307.
44. Kreuter J. Nanoparticles. In: Kreuter J, ed. Colloidal Drug Delivery Systems. New York: Marcel Dekker, 1994:219–342.
45. Tröster SD, Müller U, Kreuter J. Modification of the body distribution of poly (methyl methacrylate) nanoparticles by coating with surfactants. Int J Pharm 1990; 61:85–100.
46. Gref R, Minamitake Y, Peracchia MT, Trubetskoy V, Torchilin V, Langer R. Biodegradable long-circulating polymeric particles. Science 1994; 263:1600–1603.
47. Bazile D, Prud'Homme C, Bassoullet MT, et al. PEG-PLA nanoparticles avoid uptake by the mononuclear phagocytes system. J Pharm Sci 1995; 84:493–498.
48. Peracchia MT, Fattal E, Deasmaele D, et al. Stealth® PEGylated polycyanoacrylate nanoparticles for intravenous administration and splenic targeting. J Contr Rel 1999; 60:121–128.
49. Calvo P, Gouritin B, Chacun H, et al. Long-circulating PEGylated polycyanoacrylate nanoparticles as new drug carriers for brain delivery. Pharm Res 2001; 18:1157–1166.
50. Brigger I, Morizet J, Aubert G, et al. Poly(ethylene glycol)-coated hexadecylcyanoacrylate nanospheres display a combined effect for brain tumor targeting. J Pharmacol Exp Ther 2002; 303:928–936.
51. Maeda H. The enhanced permeability and retention (EPR) effect in tumour vasculature: the key role of tumour-selective macromolecular drug targeting. Adv Enzyme Regul 2001; 41:189–207.
52. Calvo P, Gouritin B, Villarroya H, et al. Quantification and localization of PEGylated polycyanoacylate nanoparticles in brain and spinal cord during experimental allergic encephalomyelitis in the rat. Eur J Neurosci 2002; 15:1317–1326.
53. Sommerfeld P, Schroeder U, Sabel BA. Long-term stability of PBCA nanoparticle suspensions suggest clinical usefulness. Int J Pharm 1997; 155:201–207.
54. Sommerfeld P, Sabel BA, Schroeder U. Long-term stability of PBCA nanoparticle suspensions. J Microencapsul 2000; 17:69–79.
55. Sommerfeld P, Schroeder U, Sabel BA. Sterilization of unloaded polybutylcyanoacrylate nanoparticles. Int J Pharm 1998; 164:113–118.
56. Kreuter J, Shamenkov D, Petrov V, et al. Apolipoprotein-mediated transport of nanoparticle-bound drugs across the blood-brain barrier. J Drug Target 2002; 10:317–325.
57. de Verdière AC, Dubernet C, Némati F, et al. Reversion of multidrug resistance with polyalkylcyanoacrylate nanoparticles: towards mechanism of action. Br J Cancer 1997; 76:198–205.

58. Poupaert JH, Couvreur P. A computationally derived structural model of doxorubicin interacting with oligomeric polyalkylcyanoacrylate in nanoparticles. J Contr Rel 2003; 92:19–26.
59. Woodcock DM, Linsenmeyer ME, Chrojnowski G, et al. Reversal of multidrug resistance by surfactants. Br J Cancer 1992; 66:62–68.
60. Nerurkar MM, Burto PS, Borchardt RT. The use of surfactants to enhance the permeability of peptides through Caco-2 cells by inhibition of an apically polarized efflux system. Pharm Res 1996; 13:528–534.
61. Cordon-Cardo C, O'Brien JP, Casals D, et al. Multidrug resistance gene (P-glycoprotein) is expressed by endothelial cells at blood-brain barrier sites. Proc Natl Acad Sci USA 1989; 86:695–698.
62. Olivier J-C, Fenart L, Chauvet R, Pariat C, Cecchelli R, Couet W. Indirect evidence that drug brain targeting using polysorbate 80–coated polybutylcyanoacrylate nanoparticles is related to toxicity. Pharm Res 1999; 16:1836–1842.
63. Kreuter J, Ramge P, Petrov V, et al. Direct evidence that polysorbate 80-coated poly(butyl cyanoacrylate) nanoparticles deliver drugs to the CNS via specific mechanisms requiring prior binding of drugs to the nanoparticles. Pharm Res 2003; 20:409–416.
64. Borchard G, Audus KL, Shi F, Kreuter J. Uptake of surfactant-coated poly(methyl methacrylate)-nanoparticles by bovine brain microvessel endothelial cell monolayers. Int J Pharm 1994; 110:29–35.
65. Ramge P, Unger RE, Oltrogge JB, et al. Polysorbate 80-coating enhances uptake of polybutylcyanoacrylate (PBCA)–nanoparticles by human, bovine and murine primary brain capillary endothelial cells. Eur J Neurosci 2000; 12:1931–1940.
66. Lück M. Plasmaproteinadsorption als möglicher Schlüsselfaktor für eine kontrollierte Arzneistoffapplikation mit partikulären Trägern. Ph.D. thesis, Freie Universität, Berlin, 1997:14–24, 137–154.
67. Brown MS, Goldstein JL. A receptor-mediated pathway for cholestrol homeostasis. Science 1986; 232:37–47.
68. Knott TJ, Pease RJ, Powell LM, et al. Complete protein sequence and identification of structural domains of human apolipoprotein B. Nature 1986; 323:734–738.
69. Yang C-Y, Chen S-H, Gianturco SH, et al. Sequence, structure receptor-binding and internal repeats of human apolipoprotein B-100. Nature 1986; 323:738–742.
70. Mahley RW. Apolipoprotein E. Cholesterol transport protein with expanding role in cell biology. Science 1988; 240:622–630.
71. Wilson C, Wardell MR, Weisgraber KH, Mahley RW, Agard DA. Three-dimensional structure of the LDL rector-binding domain of apolipoprotein E. Science 1991; 252:1817–1822.
72. Méresse S, Dehouk MP, Delorme P, et al. Bovine endothelial cells express tight junctions and monoamino oxidase activity in long-term culture. J Neurochem 1989; 53:1363–1371.
73. Dehouk B, Dehouk MP, Fruchard JC, Cecchelli R. Upregulation of the low density lipoprotein receptor at the blood-brain barrier: intercommunications between brain capillary endothelial cells and astrocytes. J Cell Biol 1994; 126:465–473.
74. Roullin V-G, Deverre J-R, Lemaire L, et al. Anti-cancer drug diffusion within living rat brain tissue: an experimental study using [^{3}H](6)-5-fluorouracil-loaded PLGA microspheres. Eur J Pharm Biopharm 2002; 53:293–299.

21

Cosmetic Applications of Colloidal Delivery Systems

Simon Benita
Department of Pharmaceutics, School of Pharmacy, Faculty of Medicine, The Hebrew University of Jerusalem, Jerusalem, Israel

Marie-Claude Martini
Institut des Sciences Pharmaceutiques et Biologiques, Lyon, France

Anne-Marie Orecchioni
Universite de Rouen, Rouen, France

Monique Seiller
Universite de Caen, Caen, France

INTRODUCTION

Cosmetic technology is constantly evolving in the production of raw materials and excipients and in the formulation of active ingredients. The new surfactant molecules, the search for original active substances and efficient combinations, and the design of novel vehicles or carriers led to the implementation of new cosmetic systems in contrast to the classic forms such as creams or gels.

The achievements of the extensive research conducted over the last two decades have resulted in the development of well-controlled innovative delivery systems. Some of these systems have been extensively investigated for their therapeutic potential and examined quite successfully for their possible cosmetic uses.

The main objective of this chapter is to concentrate and fully describe the preparation, characterization, and fate of the various delivery systems following topical application.

The recent sophisticated cosmetic preparations based on the innovative carriers are more attractive than the regular cosmetic preparations, because the sensitive and active cosmetic substances are more fully protected when incorporated in the innovative carrier systems. Also, the carriers can promote and enhance the cutaneous permeation of specific substances that normally exhibit low skin permeability. This can occur because of the stratum corneum, the outermost skin layer of dense, overlapping laminates of dead cells with each cell packed with keratin filaments in an amorphous matrix of proteins with lipids and water-soluble substances. The novel carrier therefore favors the penetration of the active substance through the stratum corneum, which is recognized as the rate-limiting barrier to the ingress of materials. This is an important feature, as the novel carrier can retain or even sustain the release of these active substances in the epidermis, thus targeting the skin.

Among the various delivery systems or carriers, the liposomes have been the most widely studied and successfully marketed by the cosmetic industry. Furthermore, other vesicular systems (such as nanocapsules, nanospheres, and multiple emulsions) also have been developed and are currently under investigation (1). If their claimed efficiency and potential are proven, then these vesicular and colloidal systems will be rapidly developed and will be present in the cosmetic market. It is believed that effective and pioneering cosmetic applications of these vesicular and colloidal systems still remain to be discovered. When discovered and developed, such systems will probably open up a new era of effective and sophisticated cosmetic preparations.

TYPES OF VESICULAR DELIVERY SYSTEMS

Liposomes

Liposomes were first studied in 1965 as models of biological membranes (2–7). By 1970, their structure and physical–chemical characteristics had led researchers in a number of fields to investigate the potential of liposomes as carriers of therapeutical active ingredients. Research on the use of liposomes in the cosmetics sector started more recently. This research has included attempts to achieve an optimal modulation in the release of active substances introduced within this type of vesicle by phospholipids. Although research on liposomes in the cosmetics field started fairly late, the well-established ability of liposomes to protect encapsulated active substances and, above all, the similarity between most of their components and cutaneous lipids have made liposomes the prime carriers of dermatological ingredients. Also, they are compatible with many active, biodegradable substances of limited toxicity. Thus, for some years now, the use of liposomes in cosmetology has continued to grow.

Liposomes are spherical vesicles with an aqueous cavity at their center (Fig. 1). The encasing envelope is made up of a varying number of bimolecular sheets (lamellae) composed of phospholipids. They are either unilamellar, and thus have a single bilayer, oligolamellar, with multiple bilayers, or multilamellar, with a large number of bilayers. Liposomes are either homogeneous, with a narrow distribution, or heterogeneous, with a broad distribution. Their size varies from approximately 15 to 3500 nm.

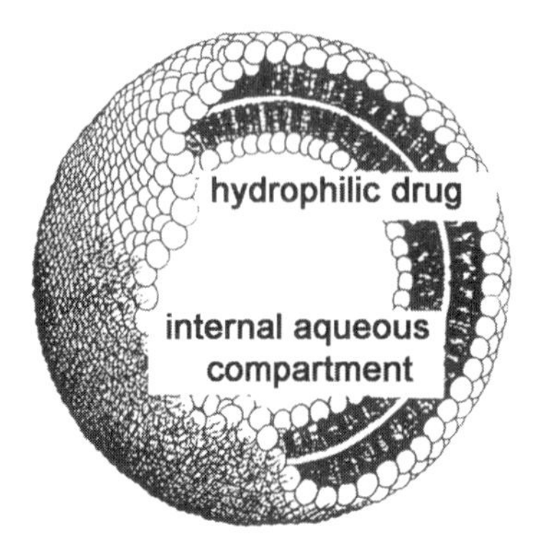

Figure 1 Schematic description of unilamellar liposomes.

The type of liposomes—small unilamellar vesicles (SUV), large multilamellar vesicles (MLV), and so forth—depends mainly on:

- the nature of the amphiphilic lipids selected;
- the composition of the iso-osmotic dispersing solution; and
- the method of preparation (ultradispersion).

Since their discovery in 1971 as carriers of active substances, a whole series of names other than liposomes can be found in the literature and have been given to commercial products depending on the nature of the components which form their envelope. Thus, if the envelope is formed of sphingolipids, the name sphingosome is given; if some stabilizing agent or active substances are introduced into the vesicles, some imaginative trade name is given, for example, Dermosome, Glycosomes, Brookosome, Phytosomes, or Marinosomes, which refers to liposome formulation with lipids obtained from marine organisms containing high concentrations of unsaturated fatty acids (8,9). More recently, ethosomes, which are liposomes comprising a high ethanol content (20–45%, v/v), are being investigated not only as noninvasive delivery carriers that enable drugs to reach the deep skin layers and/or the systemic circulation, but also for their potential cosmetic applications (10,11). The trade name Ufasome, for example, was coined to describe vesicles made of long-chain unsaturated fatty acids. Vesicles whose envelopes are made up of nonionic surfactants are called Niosomes. Their discovery resulted from the physicochemical problems exhibited by liposomes. In fact, the large extent of leakage and the mediocre stability of liposomes led a number of researchers to design new systems containing amphiphilic lipids different from those used for liposomes; thus, synthetic nonionic lipids were synthesized, allowing the formation of vesicles which present the same properties exhibited by liposomes without the disadvantages previously mentioned. These systems have also found their place in all sorts of cosmetic preparations.

These synthetic nonionic lipid systems vary in size, shape, and structure but remain in the same ranges as liposomes. They appeared as commercial products in the market shortly after liposomes. The first publication concerning original amphiphilic lipids capable of forming such vesicles and the first patents dealing with these lipids came out in 1972 and 1975, respectively.

Two other systems may be likened to liposomes, particularly as they are composed primarily of phospholipids.

1. Supramolecular biocarriers (Fig. 2) are very small vesicles (20 nm) formed by a gelified polysaccharide hydrophilic core capable of capturing the active substances in the links of a network. This central core is attached to a crown of surrounding fatty acids by covalent bonds. This whole structure is covered by an external sheet of phospholipids attached to the periphery. The hydrophilic active substances are attached with some stability to the heart of the core, and the active lipophilic substances penetrate through the double-lipid membrane. These supramolecular carriers can be either of natural or synthetic origin (Fig. 3).
2. A second liposome-like system is totally lipid in nature and acts as a carrier for lipophilic substances (Lipomicrons). The phospholipid molecules, fatty acids, and cholesterol are arranged to form spherical globules of 400 to 500 nm. Their periphery is hydrophilic, but their interior is completely lipophilic and may be loaded with liposoluble vitamins or sunscreen agents.

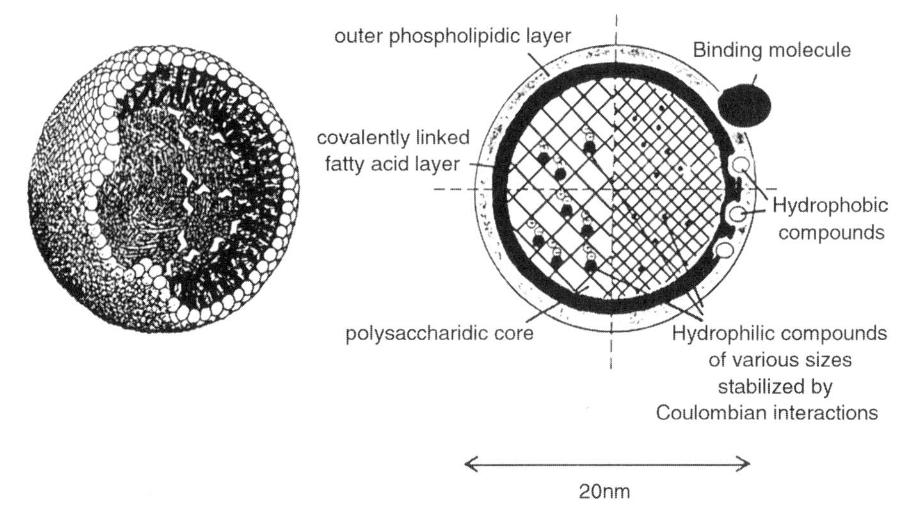

Figure 2 Schematic description of supramolecular biocarriers (biovectors).

Nanoparticulate Systems

The entities known as particulate systems are becoming increasingly popular, just as the delivery systems discussed earlier, both in the fields of cosmetology and pharmacy. Although, in principle, particulate systems are generally well known, their proliferation and sometimes their definitions cause some confusion as to their capacity as carriers and their biofate in the stratum corneum.

Depending on the size of particulate systems, which ranges from a few nanometers to a millimeter, two groups are formed, which offer completely different carrier possibilities, as the boundary is around 1 μm.

For particulate sizes less than 1 μm (nanoparticles), the insertion of vesicles between corneal cells is possible. Thus, these are true carriers of active agents, the release kinetics of which are a function of the stability and structure of the system.

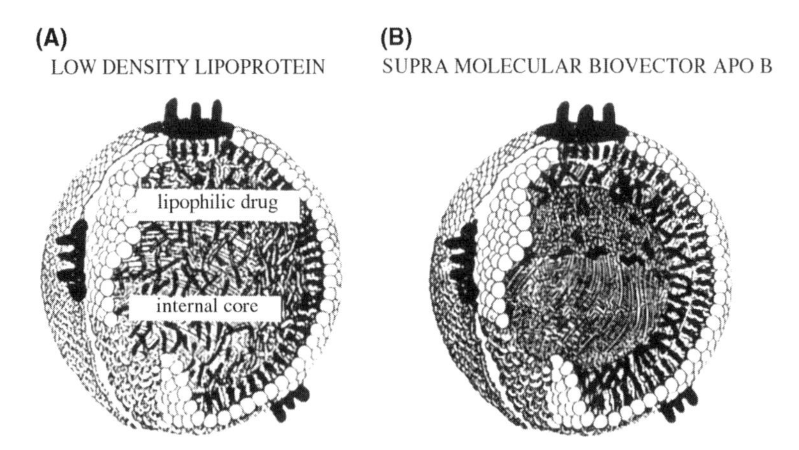

Figure 3 Comparative description between (**A**) natural and (**B**) synthetic supramolecular biocarriers.

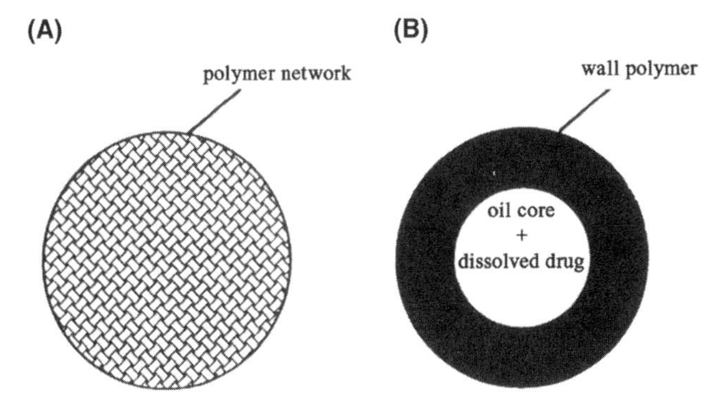

Figure 4 Schematic description of a nanosphere (**A**) and a nanocapsule (**B**).

For sizes greater than 1 μm, the fate of the particulate forms remains superficial, with the possible release of agents or surface activity likely because of the components of the matrix itself. These latter systems are not addressed in this chapter because of their lack of uniqueness, as compared with that of nanoparticles, and the tremendous lack of precision in their names and definitions. Microspheres, microcapsules, millispheres, thallaspheres, microbeads, or pearls are included in this category. All these different systems share a common spherical form, which is either a matrix or reservoir type. They vary in size ranging from 5 μm (microsphere) to several millimeters (microbeads) and in the material from which they are composed (e.g., polyamides, collagens, polysaccharides).

Among nanoparticulate systems, nanospheres are generally distinguished from nanocapsules (12). Nanospheres are matrix systems of polymers composed of solid core with a porous structure and a discontinuous envelope (Fig. 4A). Substances are absorbed upon the polymeric materials and, generally, dissolve in the polymerization environmental medium, which is most often oriented toward hydrophilic ingredients.

Nanocapsules are reservoir systems made up of a continuous polymerized envelope surrounding a liquid or gelified core (Fig. 4B). The substances are most often of a lipophilic nature, and they may be composed of a dispersion or oily mixture.

Certain nanoparticles are hybrid nanocapsules–nanospheres insofar as the active substances are intimately integrated into the gelified core of a nanocapsule.

Microemulsions

The study of microemulsions is often coupled with the study of micellar solutions because of their structural similarity. The distinctions between these two systems remain to be defined (Fig. 5).

Microemulsions are stable dispersions of a liquid in the form of spherical droplets whose diameter is less than 100 nm. They are composed of oil, water, and one or more surface-active agents (SAAs). The dispersion is micellar in nature, formed by the aggregation of amphiphilic molecules around either an aqueous core (normal micelles) or a lipid core (inverse micelles), resulting in oil-in-water (o/w) or water-in-oil (w/o) microemulsions, respectively. They form spontaneously without the use of energy, subject to the appropriate choice of SAAs that must sufficiently lower the interfacial tension to make it either negligible or a negative value.

The main characteristics of microemulsions are low viscosity associated with a Newtonian-type flow, a transparent or translucid appearance associated with the

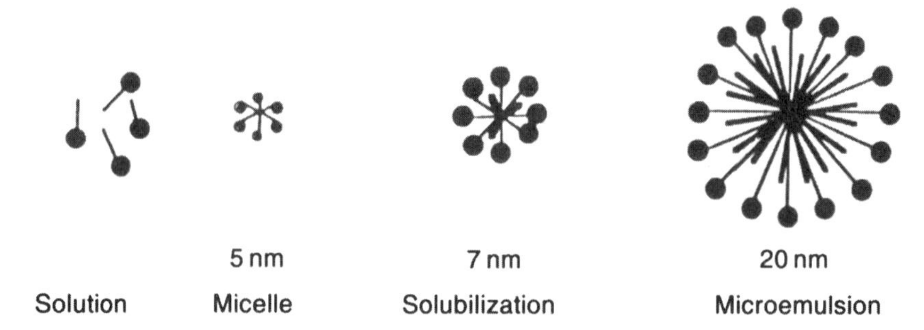

Figure 5 Formation and schematic description of a microemulsion.

isotropic character of the system, and thermodynamic stability within a specific temperature setting. Certain microemulsions may thus be obtained without heating simply by mixing the components as long as they are in a liquid state. The dilution leads to microemulsions by Ostwald ripening (13).

Multiple Emulsions

Multiple emulsions are emulsions of emulsions (14). They can comprise, in each of their three constituent phases, active hydrosoluble and/or liposoluble substances. Although interesting as cosmetological forms, multiple emulsions are not widely used in cosmetic preparation. Nevertheless, it is anticipated that multiple emulsions will be very much in demand in the coming years, particularly if their extended-release abilities are confirmed.

The use of multiple emulsions could be almost as broad as that of the simple emulsions from which they are derived and at least equal to other vesicular systems to which they are similar. One of the considerations in their favor is that they are capable of exhibiting the same properties as those of simple emulsions and, like these, are applicable directly to the skin, as they can be constituted in the form of white, oily creams with a consistency well suited to proper spreading. Multiple emulsions are emulsions in which the dispersion phase contains another dispersion phase. Thus, a water-in-oil-in-water (w/o/w) emulsion is a system in which the globules of water are dispersed in globules of oil, and the oil globules are themselves dispersed in an aqueous environment. A parallel arrangement exists in oil-in-water-in-oil (o/w/o) types of multiple emulsions in which an internal oily phase is dispersed in aqueous globules, which are themselves dispersed within an external oily phase (Fig. 6).

These emulsions are composed of at least two nonmiscible liquids. Emulsifiers, most often selected from among synthetic SAAs, are thus required to formulate them. In orienting themselves along the two interfaces, these synthetic SAAs—called primary SAA (SAA 1) and secondary SAA (SAA 2), respectively—each forms a film which, during a longer or shorter period of time, gives the system a certain degree of stability. For w/o/w-type of emulsions, the only one which will be described in this chapter, the SAA 1 molecules, which tend to be lipophilic, are arranged on the internal w/o interface, and the SAA 2 molecules, which have a hydrophilic tendency, are arranged along the external o/w surface. Thus, if the nature of the SAAs is well adapted to that of the oily phase, two monomolecular films will be formed: the apolar part of each emulsifier is localized in the oil, whereas the polar part is situated in the internal or external aqueous phase. It follows that the emulsifiers are organized into a bimolecular layer along the interfaces, forming with the oil the envelope of the

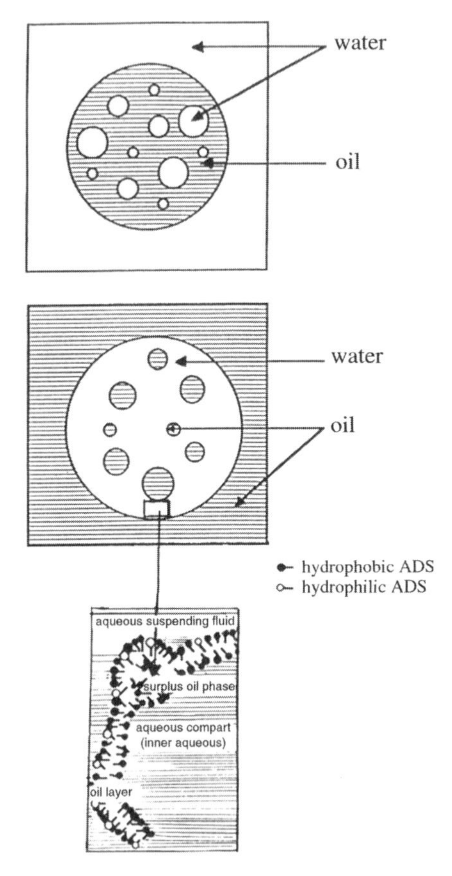

Figure 6 Schematic representation of o/w/o and w/o/w multiple emulsions. *Abbreviations*: o/w/o, oil-in-water-in-oil; w/o/w, water-in-oil-in-water.

vesicle itself. The idea that these systems could serve as useful vehicles for the transport of active ingredients was elucidated only during the 1990s, although multiple emulsions have been known for a number of decades.

Certain products now on the market are called triphase emulsions or triple-phase emulsions. Some of these are not true multiple emulsions but rather complex preparations composed, for example, of three main components, a simple gelified emulsion, a simple emulsion containing microparticles in suspension, or a simple emulsion in which three active substances are incorporated and three different activities are presented. One must remember that true multiple emulsions are also sometimes called triple emulsions. Although, in theory, quadruple and even quintuple emulsions exist, a more precise name would be double emulsion, as in these systems two emulsions coexist: one emulsion with a continuous oily phase and the other with a continuous aqueous phase.

Liquid Crystals

Discovered a century ago, liquid crystals have been used in many different fields for a number of decades, but their value in cosmetology has been confirmed only in the last 15 years (15,16). Liquid crystals are intermediate anisotropic fluids that are between the conventional solid and liquid isotropic phases. This structure is attributable to a very specific morphology of molecules that must be elongated and present

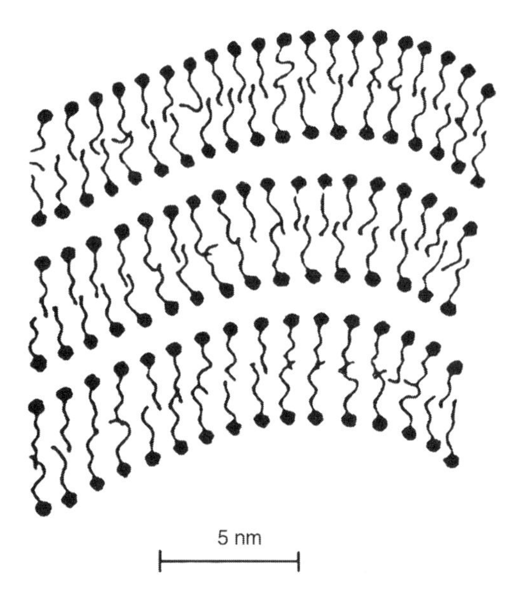

Figure 7 Schematic representation of lamellar phases.

an irregular distribution of electrical charges. These molecules must be arranged in a specific order (Fig. 7). This is true for thermotropic cholesteric liquid crystals and for lamellar phases rich in solvent (17).

Solid Lipid Nanoparticles

In the last decade, solid lipid nanoparticles (SLN), the core of which consists of a lipid compound which remains solid not only at room temperature but also at body temperature, have been investigated for different interesting applications (18). They constitute an alternative to nanoparticles formed from synthetic polymers or natural nanomolecules for biocompatibility and safety. These SLN can be referred to as o/w submicron emulsions, the internal oil phase of which is solid lipid instead of being liquid lipid. They also represent a promising carrier system for cosmetic active ingredients because of their numerous advantages over existing conventional formulations (19). SLN may offer protection to labile compounds, modulate the release of the active ingredients, and can act as occlusives to increase the water content of the skin. There is no doubt that SLN can successfully compete against classical emulsions, liposomes, microparticles, and even conventional nanoparticles. These SLN have a diameter of 5 nm to 1000 nm. As for injectable lipid emulsions, the lipid phase comprises a fatty compound of vegetal origin but in contrast to lipid emulsions, they can be stabilized not only by phospholipids but also by synthetic surfactants or polymers. The active ingredients should be miscible or soluble in the lipid core. Schematically, three types of SLN structures can be formed, depending on the formulation constituents and manufacturing process. Type I SLN corresponds to a homogenous matrix model where the lipid and the active ingredient are crystallized (or solidified) simultaneously and uniformly.

Type II SLN corresponds to a solid nanoparticle in which an outer shell enriched with the active compound is-formed, while the core is mainly composed of the solid lipid.

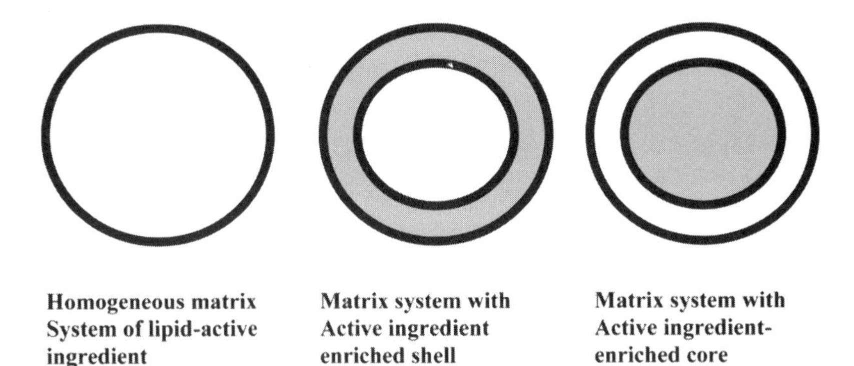

Figure 8 Different types of SLN. *Abbreviation*: SLN, solid lipid nanoparticles. *Source*: From Ref. 139.

Type III SLN corresponds to a solid nanoparticle in which the core is composed of a mixture of lipid core and active ingredient, while the other outer shell is made almost exclusively of the lipid (Fig. 8).

Each one of the three models represents an ideal type, but there can also be mixed types that can be considered as a fourth model of SLN.

There is an additional type of SLN with a solid matrix termed nanostructured lipid carrier (NLC). This carrier differs from the above-mentioned SLN by the presence of a special structure of the lipid blend which leads to a less ordered crystalline structure of the matrix allowing an enhanced uptake (or capture) of the active ingredient and avoiding expulsion of the compound during storage.

In contrast, for SLN, during storage the crystalline transition of the lipids from the α form to the β form induces a modification of the crystallinity resulting in a more ordered crystal lattice which leads to compound expulsion.

COMPOSITION

Liposomes

Phospholipids

Phospholipids and specific additives are considered the most important primary materials for vesicular delivery systems (20–26).

Some phospholipids, like phosphatidylcholine, are of natural origin, and some, like dimyristoyl or dipalmitoyl phosphatidylcholine, are of semisynthetic or synthetic origin. The phospholipids that are most commonly used are the glycerophospholipids. In these substances, glycerol, which may be considered the skeleton of the molecules, is esterified in positions 1 and 2 by long-chain fatty acids and in position 3 by phosphoric acid. A number of different hydrophilic molecules can be attached to this acid, for example, choline and ethanolamine. Although the length of the fatty acid chains and/or the degree of saturation may vary from C14 to C18 and from one to two double bonds, respectively, the lipophilic nature of the alkyl chain is relatively constant when compared with the hydrophilic part of the chain. In fact, this hydrophilic part of the chain presents different properties and has considerable influence on the characteristics of liposomes.

A few other phospholipids should be mentioned in addition to glycerophospholipids. These are less commonly used, but because of their great similarity

Table 1 Typical Formula of a Liposomal Suspension

Soya lecithin (mg)	200
Cholesterol (mg)	25
Phosphatidic acid (mg)	30
Hyaluronic acid (mg)	10
Preservative (mg)	5–10
Water (g)	to 10

to the lipids in the skin, they are good candidates for forming liposomes for dermatological use. Sphingolipids serve as a good example: they have a sphingosine skeleton upon which a number of derivatives are grafted.

All these phospholipids have the fundamental property of forming flat lamellar sheets, in which aqueous compartments alternate with lipid sheets, in the presence of water. Under certain temperatures and agitation conditions these flat bilamellar structures may fragment, fold onto themselves, and become fastened at their ends. They thus become the external wall of water-trapping vesicles, which are themselves suspended in the water. By choosing certain specific phospholipids fractions, even dry mixtures can be achieved. In the presence of water, these allow the immediate formation of unilamellar liposomes of the large unilamellar vesicle (LUV) type (proliposomes).

Additives

In addition to the phospholipids, which constitute the main envelope, two types of additives may be used (Table 1). The first is a sterol (including phytosterol, dihydrocholesterol, and cholesterol). By localizing themselves in the phospholipid bilayer, these sterols permit the phospholipids to modulate the physical and chemical characteristics of the envelope, which becomes more rigid. As a result of the modified compactness, the permeability will change according to the proportion and the location of these substances.

The second type of additive, in addition to, for example, buffers, electrolytes, pH modifiers, and preservatives, comprises ionic substances. These are anionic derivatives (phosphatidic acid, dicetylphosphate) or cationic derivatives (stearylamine). The function of these additives is to confer a negative or positive charge on vesicles, thus giving them a greater stability for their aggregation and fusion. They may also cause an increase in the interlamellar space resulting in a greater capacity for the encapsulation of certain active substances. Finally, it has been reported that the relative high concentration of ethanol in the ethosomes renders them soft and malleable, and thus they can incorporate highly lipophilic molecules (10). Thus, it is said that, except for liposomes, the essential components of the nonionic surfactant agent vesicle envelopes are, in this case, not phospholipids but rather nonionic synthetic surfactants. They are essentially SAAs containing ester or ether bonds. Their hydrophilic part consists of condensation products of polyoxyethylene, polyoses, and most of all polyglycerols (the most effective). The lipophilic part is usually composed of one or two hydrocarbon chains, between C12 and C18, either saturated or unsaturated. The additives used are the same as for liposomes and have the same function.

Nanoparticulate Systems

The primary constituents of nanospheres and nanocapsules are identical for both types of these systems. In essence, they are acrylic derivatives, most frequently of

Table 2 Typical Formula of a Nanocapsule Suspension

Polylactic acid (mg)	125
Mink oil (mL)	0.5
Poloxamer 188 (Pluronic F 68) (mg)	75
Phosphatidylcholine (mg)	75
Purified Water (mL)	to 20

the polyalkylcyanoacrylate type (PACA), or derivatives of copolymers of styrene, lactic and glycolic acids, cross-linked polysiloxanes, or biological macromolecules such as albumin, gelatin, or dextran.

To ensure and to control the particle size of nanoparticles of the SAA type, dodecylsulfate or sodium oleate and polysorbates are added. Sometimes, cross-linking agents (glutaraldehyde) or stabilizing substances (polyvinyl alcohol, cellulose derivatives), salt buffers, and protective colloids are incorporated to facilitate the formation of the individual nanoparticles in either an aqueous or an oily environment.

Recently, gliadins, extracted from wheat gluten, have been used to elaborate nanoparticles. As biopolymers, gliadins do not present the common drawbacks of synthetic materials related to the presence of monomer or initiator residues. Their biocompatibility is assumed, and, as plant proteins, they are recognized as prion-free unlike animal proteins. In addition, gliadins are hydrophobic and slightly polar. So, they are able to interact with skin keratin (27).

The active substances thus incorporated may be hydrophilic or lipophilic in the case of nanospheres or lipophilic, respectively (Table 2). It should be added that both types of active ingredients may be incorporated invariably into these two types of nanoparticles depending on the methods of fixation or incorporation used as depicted in Figure 9.

Microemulsions

Microemulsions are generally systems with four components: water, oil, surfactant, and cosurfactants (28–36).

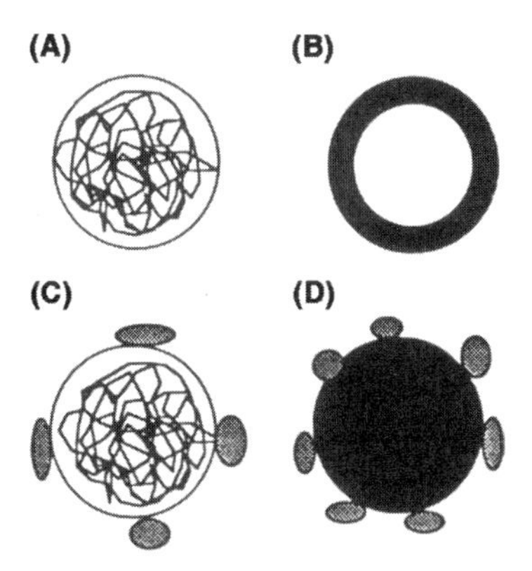

Figure 9 Description of the different incorporation patterns of active ingredients in nanospheres and nanocapsules. Active substance association: (**A**) dissolved in the nanosphere matrix, (**B**) dissolved in the liquid phase nanocapsule, (**C**) adsorbed at the nanosphere surface, and (**D**) adsorbed at the nanocapsule surface.

The aqueous phase may contain hydrophilic active ingredients and preservatives, and the fatty phase may be composed of mineral oil, silicone oil, vegetable oil, or esters of fatty acids, all of which are classic ingredients for cosmetic products. The lipid phase may also contain lyophilic active ingredients.

The surfactants chosen are generally among the nonionic group because of their good cutaneous tolerance. Certain derivatives of sugar are currently being studied (Table 3). To a lesser degree, and only for specific cases, amphoters are being investigated. The amphiphilic monolayers become more flexible with increasing temperature and more rigid with increasing pressure. The head–head repulsion of the amphiphilic molecules is enhanced at high temperature, whereas an attractive interaction between the hydrophobic tails increases at higher pressure (34,35). The cosurfactants (CoS) were originally short-chain fatty alcohols (pentanol, hexanol, benzyl alcohol). These most often include polyols, esters of polyol, derivatives of glycerol, and organic acids. Their purpose is to make the interfacial film fluid by wedging themselves between the surfactant molecules. Thus, they create a bicontinuous structure allowing a continuous phase inversion; the undetectable transition from a w/o microemulsion into a o/w microemulsion when the water-to-oil proportions are altered (36).

Multiple Emulsions

The nature of the components has no significant effect on either the multiple character or the stability of multiple emulsions provided that a specific ratio is maintained between the SAA 1 and SAA 2 rates, the hydrophile lipophile balance (HLB) of their mixture, and an optimal concentration of additives (37–40).

The primary materials capable of producing this sort of system are quite numerous and practically identical to those generally used for simple emulsions. The oily phases most frequently utilized to form multiple w/o/w emulsions are hydrocarbons, esters, and triglycerides. The emulsifiers are usually nonionic surfactant agents. Some examples are: for SAA 1 emulsions, the long hydrocarbon chain ester sorbitan, perfluorocarbon derivatives, and most importantly polymeric surfactants; for SAA 2 emulsions, esters of polyoxyethylene sorbitan, copolymers of ethylene propylene oxide, strongly ethoxylated fatty alcohols, and condensation products of polyglycerol (Table 4).

As in vesicular systems, additives are often introduced to control and increase the stability of the system. Along with electrolytes, sugars or glycols are used, most frequently hydrophilic polymers (xanthan gum, cellulose derivatives, and carboxyvinylic compounds) introduced in one of the aqueous phases, most commonly in the external phase. Lipophilic substances (waxes, acids or fatty alcohols, silicone derivatives) are introduced into the oily phase.

Table 3 Example of a Formulation of O/W Microemulsion

Sucroester	30
Glycerol	5
Jojoba oil	2
Elastin	2
Preservative	0.1
Purified water q.s.p.	100

Abbreviation: o/w, oil-in-water.

Table 4 Example of a Formulation of O/W/O Multiple Emulsion

Almond oil	25
Urea	2
Sodium lactate	3
Sorbitan oleate (Span 80)	8
Magnesium sulfate	0.7
Polyoxyethylated sorbitan stearate (Tween 60)	1
Preservative	0.1
Distilled water q.s.p.	100

Abbreviation: o/w/o, oil-in-water-in-oil.

Liquid Crystals

There are two types of liquid crystals: thermotropics and lyotropics (41). Thermotropic liquid crystals are made up of identical molecules or of a combination of molecules of the same geometrical form. They are "cholesteric" liquid crystals—so named because of their helical structure even though they are not always derived from cholesterol.

In fact, liquid crystals dispersed in cosmetic preparations are generally based on derivatives of methyl butyl phenol (Licritherm^{TM}) or equivalent molecules. There are also combinations of cholesteric and synthetic chiral nematic esters on the market.

Liquid crystals constitute more or less thick mesomorphic structures in which the molecules are arranged in a certain order, leading to characteristics somewhere between those of a liquid state and a crystalline state.

Lyotropics, one kind of liquid crystals, are systems with two or three constituents: water, oil, and surfactant (Table 5). The soluble molecules are amphiphilic. One example is compounds of glycerol stearate and polyethylene glycol, which provide a lamellar structure beyond a given concentration, generally higher than 40%. They may be formed from concentrated micellar solutions or during the production of microemulsions wherein the proportions of the various components lead to a gelified phase. These lamellar structures may also form at the interface of an emulsion.

Solid Lipid Nanoparticles (SLNs)

The SLN are constituted from a variety of lipids at concentrations close to those used in classic cosmetic formulations comprising surfactants, polymers, and water (19). The lipids most commonly used are short- and mid-chain mono-, di-, and triglycerides with relatively low melting points (close to room temperature). These lipids can be used alone or as blends. The creation of a less ordered solid lipid matrix is the prerequisite for a sufficiently high compound load. A crystallization process towards a perfect dense crystal should be avoided. Thus attempts should be made to obtain amorphous rather than crystalline SLN.

Table 5 Example of a Gel Formulation Containing Lyotropic Liquid Crystals

Oxethylated (2 OE) oleyl alcohol 30% + oxyethylated (10 OE) oleyl alcohol 70% (g)	50
Hydroxyethyl cellulose	2
Preservative	0.1
Distilled water q.s.p.	100

This can be achieved by mixing special lipids like hydroxyoctacasanyl–hydroxystearate with isopropylmyristate. In addition to lecithins, nonionic synthetic surfactants can be used such as glyceryl stearate, sorbitan stearate, and glyceryl behenate. Irrespective of the type of surfactant, it should be emphasized that it is not necessary to use high concentrations to obtain stable formulations. The polymers that can be used include tristearin, polyethylene glycol600 (PEG600), poloxamers, and polyvinylpyrolidone (PVP).

PRODUCTION

Liposomes

There are a number of processes available for the production of liposomes, and appropriate selection would depend on the type of liposome desired: multilamellar or unilamellar and large or small (42–50). Only two of these are described briefly below.

The reference method (Fig. 10) was initially developed by Bangham, allowing the formation of multilamellar vesicles. The process is carried out in four main stages:

1. Dissolution of the components of the envelope in a volatile solvent,
2. Evaporation of the solvent under reduced pressure in a rotating evaporator to form a lamellar layer of phospholipids along the wall of the evaporator,
3. Addition of a generally buffered aqueous solution at a temperature higher than the phase-transition temperature of the phospholipids, and
4. Stirring the suspension until it has cooled completely.

Dissolution of phospholipids and other constituents of the wall coating in a volatile solvent

Evaporation in a rotorevaporator

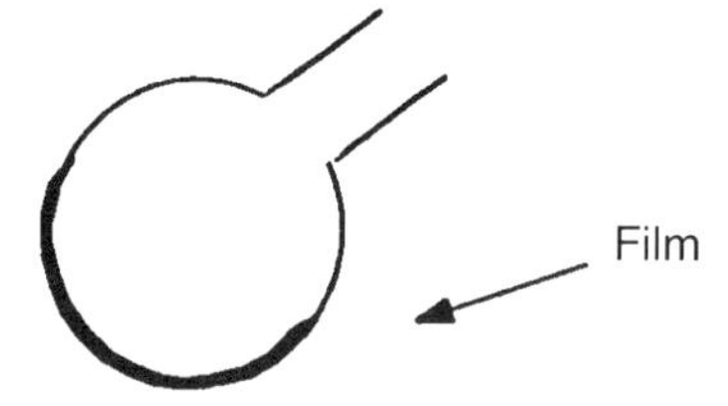

Addition of an aqueous solution

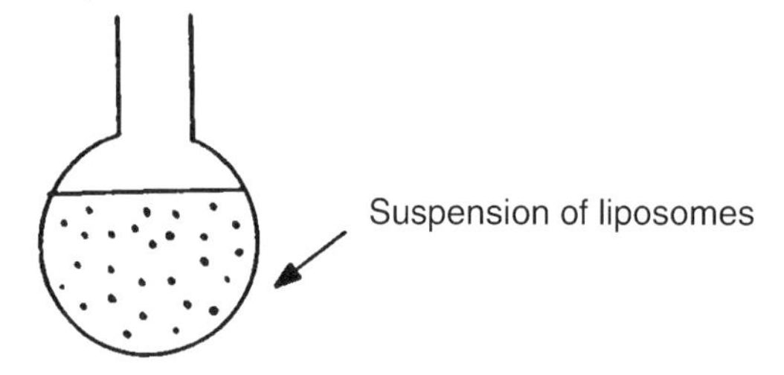

Figure 10 Description of a typical process for the preparation of multilamellar liposomes.

The active substance to be encapsulated is added to the organic solvent, if it is lipophilic, or to the aqueous solution, if it is hydrophilic. The method generally produces multilamellar liposomes and is carried out under nitrogen atmosphere. This is one of the oldest methods in use; for comparison, one of the most recent methods is mentioned below.

The stages are as follows:

1. Preparation of a first liquid phase, primarily made up of a solution of phospholipids in a single solvent or a combination of solvents,
2. Preparation of a second liquid phase, miscible in the solvent or solvents as described earlier, insoluble for phospholipids, and primarily made up of water,
3. Addition of the first phase to the second with moderate stirring followed by partial or total elimination of the pure solvent in a way which will permit the production of a colloidal suspension of liposomes adequately loaded, and
4. Just as with the preceding procedure, the active ingredients are added to the solvent most suited to them according to their solubility properties.

This procedure yields multilamellar and especially oligolamellar liposomes of a very uniform size. Small unilamellar liposomes from which multilamellar liposomes were obtained through some of the methods described earlier, for example, are subjected to high-frequency ultrasound waves during a fairly long period of time or to a high-pressure homogenization process with a two-stage homogenizing valve assembly.

It is also possible to obtain small unilamellar liposomes by other methods, which circumvent the need for ultrasound, like the method which involves injection of an ethanolic phospholipid solution into the aqueous phase or the method known as "detergent removal," which consists of preparing mixed micelles of phospholipids detergent and then eliminating the latter.

Just as unilamellar liposomes may be obtained from multilamellar ones, it is also possible to obtain large unilamellar liposomes from small unilamellar liposomes. Small vesicles fuse and provoke the formation of large lamellar structures by the addition of calcium ions, which function as complexing agents promoting the formation of large vesicles.

The above method, which requires phosphatidic acid, is less popular than two methods which make use of the injection of ether and evaporation in the inverse phase. The first of these two methods consists of dissolving the phospholipids in ether and then injecting this solution slowly into the aqueous solution of the active hydrophilic substance to be encapsulated at a specified temperature. The evaporation of the ether causes the spontaneous formation of vesicles in the aqueous phase.

The second method makes use of an emulsion produced with a continuous oily phase after dissolving phospholipids in a volatile lipophilic solvent and the active hydrophilic ingredient to be encapsulated in the aqueous phase. Large liposomes form spontaneously after evaporation of the external phase of the emulsion.

All of the methods cited have both advantages and disadvantages, which will not be detailed in this chapter, as they have been described thoroughly in the literature. Each process is characterized by its possible induced degradation, the extent of its encapsulation efficiency a which can be attained, its simplicity, its reproducibility, and its control of the particle size distribution of the liposomes obtained. As for this last point, when the liposome populations are too heterogeneous, but include the type and size of liposomes desired, common separation methods such as gel filtration, ultrafiltration, ultracentrifugation, dialysis, or membrane diffusion under pressure are employed.

None of the methods used for liposomes are satisfactory for the production of large quantities of nonionic surfactant vesicles. Indeed, large quantities of solvents are necessary; moreover, the encapsulation rate is low, as the dispersions are not rich enough in vesicles. A number of patents have been issued on specific methods that do not present any of the disadvantages mentioned earlier. The most common (and simplest) of these makes use of the formation of a lamellar phase. In summary, it involves the following stages: fusion of the amphiphilic lipid phase at a very high temperature; formation of a lamellar phase by mixing of the amphiphilic lipid phase with the aqueous phase containing the hydrophilic substances; gradual introduction of an iso-osmotic aqueous solution to the aqueous phase previously described; and homogenization, with the help of an efficient homogenizer, during a predetermined period of time, during which the preparation is allowed to reach the ambient temperature and the vesicles are formed.

Nanoparticulate Systems

As for liposomes, a large number of processes exist for obtaining nanoparticulate systems, and the process selected depends on the type of nanoparticles desired (either nanospheres or nanocapsules), their particle size, and the material used to form the nanoparticles (51–53).

Nanospheres are most commonly obtained using polymerization reactions, purified natural macromolecules, or preformed polymers. Nanocapsules are most commonly obtained through the use of interfacial polymerization reactions of monomers resulting in the formation of an insoluble polymer at the interface. A typical method for the preparation of nanospheres is described in Figure 11.

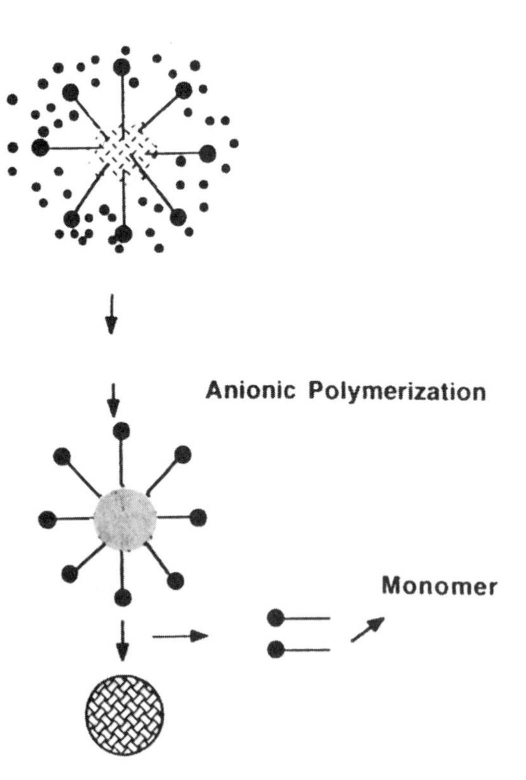

Figure 11 Schematic representation of the nanosphere preparation technique using the anionic polymerization approach.

Most of the methods used for obtaining nanospheres are based on the polymerization of monomers introduced in the dispersed phase of an emulsion or w/o microemulsion or dissolved in a nonsolvent polymer environment. Two phases are distinguished, a nucleation phase and a growing phase.

Methods of Producing Nanospheres

There are four main methods for obtaining nanospheres.

a. *Preparation in emulsion.* The continuous phase is composed of a mixture of monomer + SAA + water. The monomer + SAA combination produces micelles from 1 to 10 nm. Polymerization occurs in the interior of the micelle, and the particles expand owing to the continuous insertion of monomer molecules into the interior of the micelle until they reach a size of 200 nm. The monomer is soluble in the continuous phase. It is in this phase that free radicals (FR) are formed, which promotes the initiation of polymerization with the formation of an insoluble polymer. This process involves three stages: (i) the monomer is in the micelle; (ii) a gradual incorporation of monomer molecules into the micelle, giving rise to a polymerization initiated by FRs, and, subsequently, a reorganization of oligomers surrounded by SAA, which causes further growth of the micelle; and (iii) the fusion of two micelles, leading to a certain increase in size.
b. *Dispersion–polymerization.* The monomer is dissolved in the continuous phase, and nucleation is directly induced in the monomer solution without any diffusion. The oligomers, which have formed, then turn into aggregates stabilized by molecules of SAA. The polymer is obtained by the growth of the aggregates.
c. *Polymerization in inverse microemulsion.* Dilated nonaqueous micelles are stabilized by a layer of SAA and dispersed in the oil. The process of polymerization is continuous. The number of polymer particles, which have formed, increases over time, but their size remains constant.
   The formation of polymer particles results from the fusion of many micelles by collision. Thus, each polymer particle contains relatively few chains, which is in contrast to particles formed through other processes.
d. *Evaporation of the solvent.* An organic polymer solution is emulsified in an aqueous phase and followed by evaporation of the solvent. If the solvent and nonsolvent are not chosen carefully, the organic solvent in which the polymer is dissolved can diffuse rapidly into the aqueous solution resulting in the precipitation of the polymer. Therefore, the choice of solvent and nonsolvent is very critical. They must be conducive to the formation of nanoparticles. Once the nanospheres are formed, the polymer solvent is evaporated under reduced pressure.

Methods of Producing Nanocapsules

There are two principal methods of obtaining nanocapsules.

a. *Emulsification.* For a lipophilic substance to be encapsulated, two phases, A and B, are mixed: monomer + alcohol + lipophilic substance represents the dispersed phase, whereas water + SAA represents the continuous phase. The monomer, which is insoluble in water, is polymerized at the interface of the o/w emulsion.

For a hydrophilic substance, two phases, A and B, are mixed. The hydrophilic substance is dissolved in water which is emulsified in the lipophilic phase containing the monomer using an appropriate SAA, resulting in a w/o emulsion. The monomer can be methyl, ethyl, or butyl cyanoacrylate. The monomer, which is insoluble in water, is polymerized at the interface of the w/o emulsion.

For an inverse microemulsion (w/o), the individual size is small. Rinsing is necessary for the solvent to be eliminated.

b. *Evaporation of the solvent.* Two different environments are used: S1 = polymer in solution in an organic solvent, and S2 = polymer nonsolvent solution.

Oil to be capsulated must be miscible with S1 but not with S1 + S2. Oil must be minutely dispersed in S1 + S2. The polymer capsulates the oil in which lipophilic substances are dissolved. The encapsulation of lipophilic substances can thus be performed easily, whereas the encapsulation of hydrophilic substances is difficult. Whatever method is used, the critical point is the thickness of the membrane, which ranges between 3 nm and 10 nm, for a total size reaching between 150 nm and 800 nm.

Microemulsions

Owing to the stable thermodynamic character of microemulsions, it is straightforward and easy to obtain them (54–59).

A stable w/o emulsion obtained with a lipophilic SAA may be used as a base for preparing an o/w microemulsion. To this emulsion, a hydrophilic aqueous SAA solution is added followed by stirring. A gelified phase appears because of the cubic structure of the product. If hydrophilic SAA is again added, an o/w microemulsion is obtained.

To prepare a w/o microemulsion, an o/w emulsion stabilized either by an ionic or a nonionic SAA should be used. By titration, a CoS is added. As in the abovementioned example, a gelified phase appears which becomes fluid and results in the formation of a w/o microemulsion. However, these methods are empirical and relatively crude. Likewise, a microemulsion is usually created by the establishment of a pseudoternary diagram for which a ratio of SAA/CoS is fixed, representing a sole constituent.

The establishment of a ternary diagram is generally accomplished for the purpose of locating the microemulsion or the microemulsion zones by titration. Using a specific ratio of SAA/CoS, various combinations of oil and SAA/CoS are produced. The water is added drop by drop.

After the addition of each drop, the mixture is stirred and examined through a crossed polarized filter. The appearance (transparence, opalescence, and isotropy) is recorded, along with the number of phases. In this way, an approximate delineation of the boundaries can be obtained in which it is possible to refine through the production of compositions point-by-point beginning with the four basic components (60).

Multiple Emulsions

The operational technique plays an even more important role in the production of multiple emulsions than in the production of simple emulsions (61–66).

Four main types of protocols are recommended:

1. The two-stage procedure as depicted in Figure 12 (the most frequently utilized method). The name is not precisely indicative, as other methods involve two stages.
2. Dispersion of a lamellar phase. This method is still rarely used. It is similar to the procedure described above for obtaining nonionic surfactant vesicles.
3. Dispersion of an isotropic oil solution in water. This procedure is recommended more for the formation of w/o/w-type microemulsions.
4. Phase inversion. This procedure is becoming more commonly used. Here again, the name is inaccurate, as it does not actually involve a phase inversion.

Each of the four protocols is executed in almost identical fashion.

First, either a continuous oily phase simple emulsion, a lamellar phase, or an isotropic oil solution is produced at 70°C to 80°C ± 1. Whichever system is used, the water, the oil, and the emulsifiers (the proportions of the three components vary depending on which type of system is desired) are combined with the help of a classic turbine mixer during a period of approximately 30 minutes.

Later, for the so-called double-stage protocol, the continuous oil phase emulsion is poured slowly into the aqueous phase. For the other three procedures, it is the water which is gradually introduced either into the lamellar phase or into the isotropic oil emulsion or the continuous oily phase emulsion. The second dispersion is likewise produced by a turbine stirrer, most commonly at room temperature, during a period of approximately 30 minutes. Of course, each procedure offers both advantages and disadvantages.

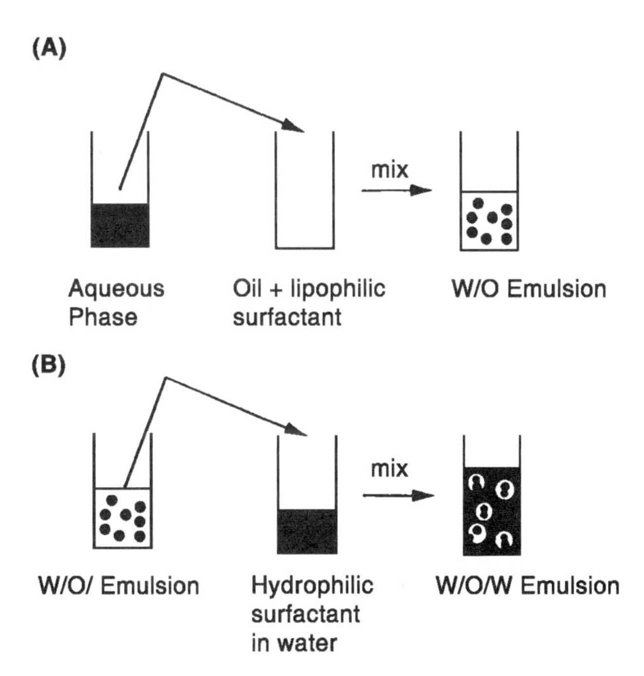

Figure 12 Preparation of a multiple emulsion by a two-step procedure. (**A**) Step 1: formulation of w/o emulsion. (**B**) Formulation of w/o/w emulsion. *Abbreviations*: w/o, water-in-oil; w/o/w, water-in-oil-in-water.

The two-stage procedure has the major advantage of having very well-regulated steps; in theory, it is possible to fix the amount of internal water. Its disadvantage is that it is not very reproducible, although this is likely because of the second emulsification, which is a critical stage in the procedure. In fact, the primary w/o emulsion is quite viscous and is difficult to disperse. This is the stage when a shearing should be executed and certain globules of oil, which have just formed, are at risk of breaking down. If this happens, some of the internal water may become mixed with some of the external water.

The process which involves dispersion in the water of a lamellar phase has the advantage of necessitating only a single emulsification stage. The initial phase is an anisotropic phase; it is thermodynamically stable and easy to achieve.

One of the limitations of this procedure stems from the fact that not all of the surfactants form a lamellar phase. Where such a phase does exist, the hydrophile lipophile balance hydrophile lipophile balance is often elevated, which is undesirable for the stability of a multiple emulsion. Furthermore, only a small quantity of oil (rarely exceeding 10%) is incorporated into the lamellar phase. This procedure is best adapted to obtain nonionic surfactant vesicles, as, in this kind of vesicle, the amount of apolar substances is smaller.

The procedure that involves the dispersion in water of an isotropic oil solution presents more or less the same advantage as the procedure detailed in the preceding paragraph, in which only one emulsification process is involved. The initial phase is a pure phase, stable, and easy to obtain.

The main disadvantage is the incorporation of a low concentration of water-soluble compounds in inverse micelles (seldom exceeding 10%).

The disadvantages, which are common to these two procedures, have to do with the necessarily elevated quantities of SAAs needed. Furthermore, because of the dispersion in water of the lamellar isotropic oil phases, in which the initial water is not truly emulsified in the form of globules, it is difficult to know whether the water transported by this dispersion still remains, soluble, and if so, then to determine the total water quantity in the internal aqueous phase.

The advantages of the phase inversion procedure are that it is easy to perform and it provides a precisely known rate of internal aqueous phase similar to the double-stage procedure. Moreover, unlike the double-stage procedure, it is reproducible. However, it entails the disadvantage of having two emulsification stages, and of being a very delicate procedure to perform. Indeed, even a very slight excess of water might be sufficient to transform the multiple emulsion into an aqueous simple emulsion type.

Liquid Crystals

Special laboratory skills are necessary to obtain thermotropic, liquid crystals (67,68). These crystals are, like any chemical molecules, added at the same time as the other constituents of the vehicle in which they must, following a change in temperature, present a certain appearance.

Obtaining lyotropic liquid crystals is quite easy, and any formulator can produce them. The process consists of mixing, usually by a simple turbine mixer, the solvent and the ingredients (e.g., fatty ethoxyl alcohol, phospholipid stearate of glycerol or PEG) which will produce the desired structure above a certain concentration level. A lamellar structure could easily be obtained. Most often, this structure is not prepared by itself, but rather is obtained during the preparation of the cosmetic formulations. Thus, a cream or gel may be formed on the spot with ingredients, which will produce liquid crystals equally well.

Solid Lipid Nanoparticle (SLN)

The main production method of SLN is by high-pressure homogenization which can be performed using either the hot or the cold homogenization procedure (18).

Hot Homogenization Method

The active compound is dissolved, solubilized, or dispersed in the lipid melted at 5°C to 10°C above its melting point. The lipid-melted phase is dispersed in a hot surfactant solution at the same temperature by high speed stirring. The obtained pre-emulsion is passed through a high-pressure homogenizer at 500 Pa for two or three homogenization cycles and then cooled to room temperature. The lipid solidifies and induces the formation of SLN. This technique can be used even with temperature-sensitive active compounds, because the exposure time to elevated temperatures is relatively short.

Cold Homogenization Method

The active compound is dissolved, solubilized or dispersed in the melted lipid. The mixture is cooled and after solidification, the lipid mass is ground to yield microparticles in the range of 50 to 100 μm. The lipid microparticles are dispersed in the cold surfactant solution by stirring. The presuspension is passed through a high-pressure homogenizer at room temperature or even lower to break down the microparticles into SLN. This technique is recommended for thermolabile and hydrophilic compounds which might partition from the liquid lipid phase to the water phase during the hot homogenization process.

The methods based on microemulsion processes or lipid nanopellets or lipospheres are not in use anymore.

CHARACTERIZATION

As all the particulate systems described earlier are dispersions, the approach used for their characterization is almost identical.

The following features pertain to all of them: particle size distribution, morphology analysis, electrical charge nature, creaming or sedimentation rate, etc. rheological behavior; and rate and extent of the encapsulated active substances. The methods of characterization utilized for all the vesicular systems are applicable to SLN. However, there are some specific important properties that need to be examined such as the crystalline nature of the lipids and the coexistence of additional colloidal structures. Differential scanning calorimetry (DSC) and X-ray diffraction are the most relevant techniques to evaluate the crystallinity of the lipids, while nuclear magnetic resonance (NMR) techniques are able to detect additional colloidal structures in the presence of SLN.

Two approaches are used for the characterization of these systems. The first one is simple and of a systematic nature. Its objective is to determine rapidly and without difficulty the type of system obtained.

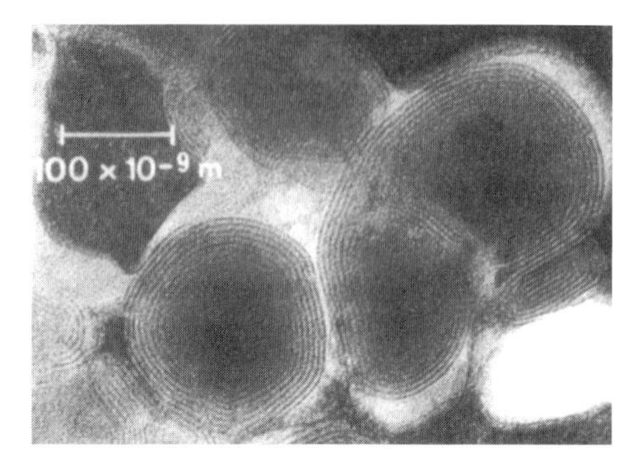

Figure 13 TEM photomicrograph of multilamellar liposomes following negative staining with phosphotungstic acid. *Abbreviation*: TEM, transmission electron microscopy.

Microscopic Analyses

Microscopic examination is the first test performed to identify the type of particulate systems obtained (69–77). Moreover, it constitutes an excellent means of following up on the physical stability of these systems as a function of prolonged storage times.

The most studied characteristic of liposomes is their dimensions. If the vesicles are medium sized (in the order of 1 μm), they can be examined under any ordinary optical microscope, a polarized light microscope, or a fluorescent microscope using a marker like carboxyfluoresceine. However, examination under an electron microscope is necessary for nanoparticles and liposomes smaller than 1 μm, as depicted in Figure 13. In any event, to measure the size and to ascertain the particle distribution of these two types of systems, other methods which are faster and more accurate than microscopic techniques must be used. As for any other dispersed system, the technique used involves the electronic counting of the particles of vesicles that have (for the most part) a diameter greater than 600 nm. Photon-correlation spectroscopy is used for size determination of vesicles of smaller size.

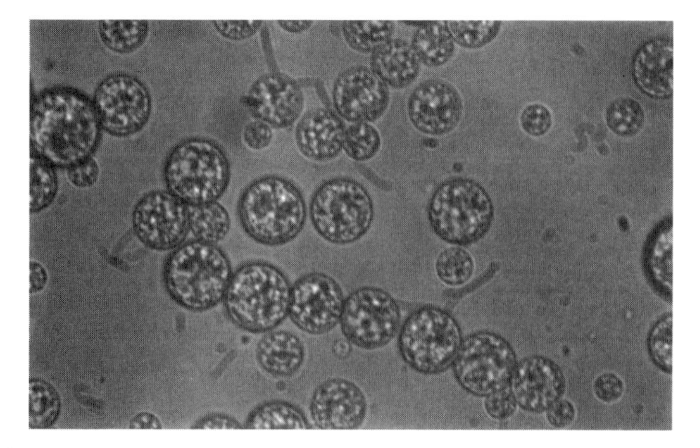

Figure 14 Photograph of a w/o/w multiple emulsion obtained with a normal light microscope (at a magnification of 1000×). *Abbreviation*: w/o/w, water-in-oil-in-water.

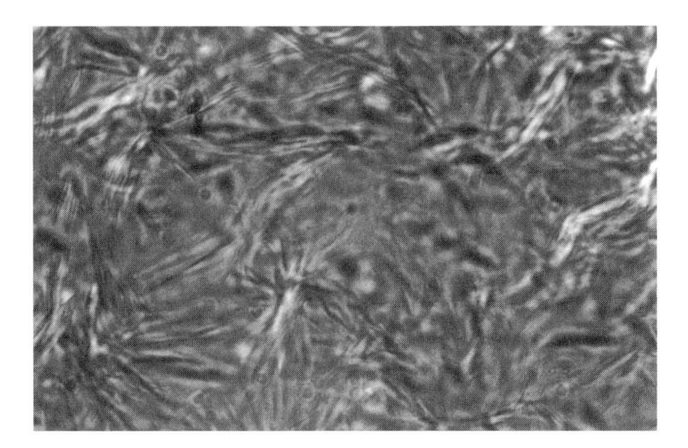

Figure 15 Photograph of lyotropic liquid crystal forms with polarizing microscope.

For multiple emulsions, optical microscopy is used as a common method of analysis, as shown in Figure 14, to keep track of all these systems and to determine the particle size distribution. This method allows the direct measurement of the size of multiple globules with a diameter greater than 0.5 μm as well as a rough estimation of the percentage of multiple globules compared with the simple globules. Furthermore, the size of the internal aqueous globules (usually in the order of a micrometer) can be determined by this method. Prior to observation, and to measure the actual size of the globules, the multiple emulsions must be well diluted with a solution whose osmotic pressure should be similar to that of the internal aqueous phase.

Indeed, dilution with a solution of lower molar concentration would, after the ingress of external water into the internal water, cause swelling and then rupture of the internal aqueous globules. Inversely, dilution with a more highly concentrated solution would precipitate the escape of the internal water and hence a shrinking of the internal aqueous globules.

Furthermore, examination under polarized light could sometimes detect a texture corresponding to a lamellar phase indicating the directional orientation which the two emulsifiers take in the oily phase.

As for the vesicular systems described earlier, microscopy following cryofracture of the multiple emulsion is used to enable a precise analysis of the internal aqueous globules. The objective here is to generate accurate information about their exact morphology.

Microemulsions are examined by DSC, small angle X-ray scattering, and generalized Fourier transformation (78,79).

Some researchers instead of obtaining a direct estimate following observation prefer to use photographs for measurement; in this way, they follow up on the size and stability of the emulsions.

Finally, examination under a polarizing microscope is also a valid tool for detecting and observing liquid crystals, as shown in Figure 15.

Rheological Analysis

To gain a deeper knowledge of particulate systems, we often turn to rheology (80–84). Because of the diversity offered by this technique, it is possible to characterize the structure of such systems and to follow their evolution over time. In rheological analyses, it means:

Viscoelastic Oscillator Analyses

The oscillatory experiment which consists of applying a sinusoidal stress and recording the consecutive strain is defined as follows:

$$\tau(\mathrm{t}) = \tau_{\mathrm{o}}\cos\ \omega t$$

$$\varepsilon(t) = \varepsilon_{\mathrm{o}}\ \cos(\omega \mathrm{t} + \delta)$$

where τ_{o} and ε_{o} are the maximal amplitudes of stress and strain, respectively, $\omega + 2\pi N$, with N the frequency of strain and δ the phase angle of stress/strain.

The basic viscoelastic parameters which describe the rheological behavior are:

$$G^{*} = \tau_{\mathrm{o}}/\epsilon_{\mathrm{o}}$$

the rigidity modulus and δ the phase angle of the stress for the strain (for a viscoelastic material, it is included between 0° and 90°; for a purely elastic material, it equals 0° and for a perfect Newtonian liquid it equals 90°); $(\tau_{\mathrm{o}})_{\mathrm{c}}$ the critical stress at which the sample becomes more viscous than elastic. The comparison of (τ_{o}) and $G^{*}/2\pi$ values gives information about the form and the homogeneity of the droplets.

Analyses of the Permanent Flow System

During the process of sweeping, the system is sheared in a cycle of increasing constraint, then constant constraint, and then diminishing constraint. This kind of test provides information about the function of shearing in the destructuration of the dispersion. It also provides information about the more or less reversible character of the dispersion.

More simply stated, in this kind of analysis, the Newtonian character of these systems, taking microemulsions, for example, can be exposed. However, microemulsions are often thickened with gelling agents, like carrageenan, and this additional physical property may alter skin permeation (85).

Incorporation Efficiency Encapsulation of the Active Substances

The assay of the active substance incorporated into one of the dispersion phases provides useful information on the encapsulation ability of the systems (86–88).

The location of the active substance in the dispersion phase depends on its affinity for the various constituents of the formulation. Hydrophilic substances are dissolved in the aqueous phase, whereas lipophilic substances are dissolved in the oily phases. The rate of encapsulation depends on a number of factors, such as the concentration of the active substance, the nature of the constituents, and the type and size of the vesicles. Three drugs of three different polarities in gliadin nanoparticles were studied: the hydrophobic vitamin E (VE), the slightly polar mixture of linalool and of linalyl acetate (LLA), and the cationic amphiphilic benzalkonium chloride (BZC). This comparative work shows that the amount of the entrapped VE and LLA is higher than that of the cationic BZC, confirming a strong interaction between gliadins and apolar compounds because of the apolarity of the proteins (89).

Prior to the determination of the amount of active substance encapsulated, it is nevertheless preferable, once the dispersed systems have been formed, to eliminate the nonencapsulated active substance. Separation is accomplished by various procedures, for example, filtration on gel, ultracentrifugation, or dialysis.

STABILITY

Liposomes

To date, the production of stable liposomes is still delicate and chancy (90–93). It should nevertheless be kept in mind that, as for all dispersion, these vesicles have a tendency toward flocculation and fusion and later sedimentation.

Likewise, instability is also associated with the increase in permeability of the envelope and the resulting leakage of the encapsulated active substance, as, for example, it is likely to occur when the rate of cholesterol in the envelope is insufficient.

Degradation of phospholipids is largely related to autoxidation, which can be markedly decreased if antioxidants are added. Furthermore, oxidation of liposomes can be avoided by using hydrogenated phospholipids.

Another cause of degradation is hydrolysis of the constituents of the membrane which results in the release of fatty acids, an increase in the fluidity of the membrane, and an increase in membrane permeability. All of these phenomena can lead to fusion of the vesicles and either partial or total leakage of the encapsulated active ingredient. Protection against hydrolysis and autoxidation can be ensured by coating the liposomal vesicles with a membrane comprising biological macromolecules, atelocollagen, and glycosaminoglycans. The second coating ensures a diminution in the liposomal membrane fluidity as a result of interaction between the phospholipids and the macromolecules. This leads to a reduction in the permeability of the membranes.

Various physical procedures, such as irradiation with gamma rays or cryodesiccation, promise to provide answers to increasing the stability of these systems in the future.

Generally speaking, the stability of liposomes in gelified aqueous or hydroalcoholic environments ranges between two and three years at temperatures between 4°C and 25°C. However, liposomes remain stable for only a few months if they are dispersed (and this is quite frequently the case in cosmetic products) in a lipid-rich environment or in a solution containing surfactants in which the phospholipids envelope dissolves or gradually becomes soluble.

Nanoparticles

As long as organic solvents, residual monomers, and polymerization inducers are eliminated, then problems in the stability of nanoparticles are practically nonexistent. Nevertheless, to avoid any eventual degradation of the polymeric materials in aqueous suspension or hydrolysis of the active substance after release, cryodesiccation can be performed without causing any alteration in the size.

Microemulsions

As microemulsions are stable thermodynamic systems within a defined temperature range, they present no problems for storage under normal conditions.

Multiple Emulsions

Multiple emulsions are unstable thermodynamic systems; over time they progress inevitably toward their rupture (94,95). In addition to the traditional factors of instability inherent in simple emulsions (separation, creaming, coalescence, phase inversion), specific unstable conditions occur within multiple emulsions such as the diffusion of internal water into the external aqueous continuous phase. These instabilities are all irreversible. They may appear separately or concurrently. Today, using very effective polymeric SAAs and thickeners that increase the viscosity of the various phases (particularly cross-linking of the intermediate oil phase through the use of chemical or physical processes like gamma radiation), we are able significantly to slow down the destabilization process of these systems (96). In fact, stable multiple emulsions can be prepared for a period of two years without any alteration of their characteristics or release of the encapsulated substance.

Liquid Crystals

Liquid crystals do not exhibit any problem of stability provided they are stored within their specific temperature ranges, as above such temperatures they lose their properties (including even lyotropic liquid crystals). It should be noted that the presence of lamellar structures at the emulsion interface greatly increases their stability.

Solid Lipid Nanoparticles

In view of their solid nature, SLN do not exhibit any stability problem. This is confirmed by the lack of degradation of SLN following autoclave sterilization. Nevertheless, some precautions should be taken to deter the lipids, following fusion, to crystallize in another form which might modify the release rate of the encapsulated active ingredient.

COSMETIC USES

Liposomes

Liposomes are the principal vehicles for the transport of active ingredients, which, depending on their molecular size and solubility properties, are localized at different sites of the liposomes (97–116).

If the active ingredients are small hydrophilic molecules, such as sugars, amino acids, peptides, or normal moisturizing factors (NMF), they are localized at the center of the vesicle or between the bilayers.

If they are lipophilic molecules, such as liposoluble vitamins and their esters, they are located in the lipid bilayer.

The transformation of these lipid vesicular carriers when they come into contact with the stratum corneum is still the subject of a strong controversy. The presence of globular structures in the first layers of corneal cells was first assumed and later confirmed through the various transmission electron microscopic (TEM) techniques. Furthermore, the presence of multilamellar structures in the deep layers of the stratum corneum has also been observed after cryofracture. Given the fact that intercellular distances are either of 6.4 or 13.4 nm, it is surprising to find entirely intact vesicular structures with an interior bilayer diameter of between 50 nm and 0.5 μm. It is now known that the molecules constituting liposomes become dispersed during the course of their intercellular migration, and their interactions with the cutaneous lipids are more or less a function of their chemical structure. Phospholipids accumulation at certain lipid sites leads to the formation of new vesicular structures at these sites.

It has been demonstrated that phosphatidylcholine, having a relatively modest hydrophilic moiety, is more likely to react than phosphatidyl inositol choline, as the latter has a heavier hydrophilic moiety group. It is thus a dynamic structure whose initial globular form is indispensable in penetrating the cutaneous barrier. Extensive and thorough knowledge about interactions between liposomes and living cells is currently being gained. These interactions studied with routes of administration other than topical may involve four different phenomena:

1. *Absorption*, facilitated by the charge on the vesicles because of the ionization of the primary or secondary component.
2. *Lipidic transfer*, through a mediation of surface proteins (here an analogy between lipids, liposomes, and membrane lipids can be seen). It is important to emphasize the role played by molecules, such as lipoproteins, in the constitution of the membrane.
3. *Endocytosis*, also influenced by the charge on the liposomes. This permits the digestion of liposomes by lysosomes inside the cell which in turn promotes the release of the active ingredient at the same location.
4. *Fusion*, which results from the insertion of constituents of the liposome membrane into the cell membrane.

This simple and attractive hypothesis has, however, not been proven to date. Because of their biomimetism with the cellular membranes, all of these lipid systems are designed as vehicles for carrying active ingredients, particularly hydrophilic ones, sometimes lipophilic, and to ensure prolonged release of such ingredients.

Although the nature of interactions between liposomes and the skin remains unexplained to date, these systems have been used continuously since the initial creation of a "liposomed" cosmetic product called "Capture" manufactured by Dior more than two decades ago.

The first cosmetic applications of liposomes were in the field of hydration of the skin. It was demonstrated that products containing liposomes, into which propylene glycol was added, possessed excellent moisturizing properties. This hydrating capacity was prolonged and extended with a product containing liposomes constituted from glucosylceramides, an ester of glucose and fatty acid and an ester of sucrose or trehalose and fatty acid. The following products have been tested:

1. Creams or lotions made up of lysophospholipids, an aqueous phase containing monovalent and polyvalent alcohols, and an oily phase that contains hydrocarbons, esters, triglycerides, fatty acids, and fatty alcohol as well as silicone.
2. Dispersions constituted from a mixture of an emulsion, containing elastin, collagen, casein, and fibroin, and a liposomal suspension, containing vitamins E, F, and A.
3. Liposomes made of cutaneous lipids dispersed in gel.

Currently, a large number of active substances used in cosmetology have been encapsulated, presented, and dispersed in hydroalcoholic liquids, aqueous gels, or creams with a continuous aqueous phase. Most cosmetic compounds, whether intended for use on the face, the hands, the body, or the hair, are appreciated by those who use them.

Because of their bilamellar structure, nonionic surfactant vesicles have the same properties as liposomes. However, as the nature of lipids found in the envelope is different, their properties are not identical. Of course, the same controversies exist for nonionic surfactant vesicles as for liposomes insofar as their passage through the skin is concerned. Even if we accept that only vesicles in the order of 20 nm in size can penetrate the corneal layer intact, it is still undeniable that vesicles with nonionic SAAs have some effect after topical application.

The first results were reported by Handjani-Vila et al. (113,114), who showed that sodium carboxylate pyrolidone has the greatest hydrating power when it is encapsulated in nonionic surfactant vesicles, rather than in emulsions or even in liposomes.

Recently, nonionic surfactant vesicles without any active ingredients, composed of fatty ethoxyl alcohols and cholesterol, were tested on human skin. Microscopic examination following cryofracture of a cross-section of skin reveals that the vesicles are located in intercellular lipid sites in the stratum corneum at a depth of a few micrometers. These results confirm, with the aid of images obtained by electron paramagnetic resonance (EPR) and plates obtained through microscopy following cryofracture of a biopsy of human skin, that intercellular restructuring of the epidermis takes place after contact with such systems (115). Like liposomes, nonionic surfactant vesicles will then restore to a delipidified corneal layer the lamellar structures normally contained therein. It was shown that incorporation of estradiol in nonionic surfactant vesicles (particularly multilamellar structures) substantially enhances penetration of this ingredient into the stratum corneum of a skin section (116).

Although the mechanism by which these carriers operate remains to be explained, they nevertheless constitute very useful systems for cosmetic applications.

Without any active substance, like liposomes, the nonionic surfactant vesicles enhance the supply of lipids and water to the stratum corneum. Even when they do not penetrate the stratum corneum intact, they facilitate the accumulation of hydro- or lipophilic active substances in the upper layers of the epidermis. Finally, like liposomes, nonionic surfactant vesicles exhibit a high cutaneous tolerance.

Preparations containing nonionic surfactant vesicles were first placed on the market at approximately the same time as the liposomes described in the preceding sections. They too provide satisfactory results for people who use them knowledgeably.

For these systems as well, the first cosmetic applications were intended as hydrating agents, and they were followed by self-tanning ingredients of the skin. Prevention of dry skin has been studied with empty vesicles. With their component ingredients, these systems work against water loss by forming an occlusive film on the surface of the epidermis.

As mentioned earlier, their hydrating capacity has been studied by encapsulating them with sodium carboxylate pyrolidone (PCNa), a component of the skin's natural hydrating system normal moisturizing factors (NMF). The hygroscopic substances which make up the NMF are often used as hydrating agents in cosmetic compositions. However, these composites, which are soluble in water, develop a weak affinity for the corneal layers once they are incorporated into aqueous solutions or in oily emulsions within water. They penetrate the stratum corneum with only moderate success and are, aside from that, easily removed by washing with water. If such substances are administered in a lipid environment, their diffusion and the resistance to washing are increased. The application of nonionic surfactant vesicles containing 10% PCNa will, even after rinsing, markedly improve the hydration of the skin by 40% to 90% as compared with the initial moisturizing level before treatment.

One of the first tests based on the study of coloration of the skin was performed with vesicles containing 0.6% of tartaric aldehyde and, as control, aqueous solutions of various concentrations of the same component. Four hours after application, both before and after washing, the intensity of coloration produced by these vesicles was as high as the intensity produced by the aqueous solution containing 10 times the quantity of tartaric aldehyde.

A second, parallel test was conducted with vesicles containing 1.5% of tartaric aldehyde and 3% dihydroxyacetone and an o/w emulsion. An intense tan was obtained with these vesicles, which was resistant to washing with soap and water, whereas with the o/w emulsion, the tan, which was already weak before washing, disappeared almost completely after washing. These nonionic surfactant vesicles have been in continuous use since their first commercial exploitation as Niosomes by L'Oreal. They are often commonly used, and a number of active substances have been encapsulated. Even if, as for liposomes, the precise mechanism by which they operate on application to the skin is yet to be determined, their biomimetic approach, like that of liposomes, has inspired a growing number of researchers, as their vesicular carriers constitute an undeniable valuable asset to cosmetology.

Nanoparticles

Although cosmetic applications of nanoparticles proliferate (numerous patents have been granted), publications, studies, or reports on their biofate in the skin following topical application have been rare. The incorporation of active cosmetic hydrophilic substances (e.g., amino acids, vegetable extracts, organic and mineral elements) in the nanospheres attempts to modulate the release of the substances in the skin. Where nanocapsules are concerned, the active substances (AS) are usually of lipophilic natures, such as α-tocopherol, and they can be composed of an oily compound or a dispersion (117). Here again, the objective is to control the release of the AS, as the molecule is protected during a shorter or longer period from biodegradation in the organism. The release profile of the AS (by, e.g., erosion, diffusion, elution) depends on the nature of the constituents. Lancome launched a cosmetic product containing nanocapsules of VE (Primordiale). As for the vesicular systems described in the preceding sections, except for the very small sizes which can be detected in the pilosebaceous apparatus, it is unlikely that they penetrate the stratum corneum intact, as their size varies from 100 to 800 nm.

Microemulsions

Over the last years, many studies have dealt with the percutaneous absorption of various active ingredients carried by microemulsions both from pharmacological and cosmetological perspectives (118–130).

Overall, hydrosoluble active ingredients have been the most sought after and (rather curiously) are the ingredients most often added to the external phase of o/w microemulsions. Various investigators have shown that microemulsions acted as absorption enhancers of both liposoluble and hydrosoluble active substances. This action was not just attributed to the high proportion of SAAs.

Still, when the hydrosoluble active element is in the internal phase of a w/o microemulsion, its physicochemical characteristics are altered. Because of this, and owing to the components of the fatty phase, it can accumulate in the cutaneous lipophilic structures or in the cutaneous fats (orotic acid). Liposoluble active ingredients have been introduced both in the internal and in the external microemulsion phases compared with w/o or o/w emulsions or with petroleum jelly (vaseline). Nevertheless, they are more frequently incorporated into the internal phase, where they are solubilized. This is the case with α-tocopherol, azelaic acid, and octyl dimethyl *para*-aminobutyric acid, for which the absorption rate is significantly increased, and which no longer behave like lipophilic compounds but like hydrophilic ones. When estradiol or prednisone is introduced into the internal o/w microemulsion phase, it is stored in the intercellular lipids of the stratum corneum. Estradiol placed in the external phase is better retained. Thus, it can be observed that depending on the formulation and the type of the microemulsion, the absorption of an active ingredient can be modulated on request (131–133).

There are numerous cosmetic products in the form of microemulsions; the use of these products ranges from body care to facial and hair treatments. They include bath oils, body thinning products, fixatives for hair, hardeners for nails, hydrating products, antiwrinkle products, products to prevent seborrhea, and antiaging serums marketed principally in France, Italy, Belgium, and the United States.

Multiple Emulsions

The behavior of the multiple emulsions on the skin following topical application has not yet been addressed (134–136). Do small-sized globule vesicles in the range of 20 nm penetrate intact? In what manner does the rupture of larger-sized globule vesicles take place? Are they reformed on contact with the intercellular lipids? For multiple emulsions, these questions do not present a problem. Although the release of the encapsulated active substance is complicated, as a number of different mechanisms exist, the multiple emulsion's behavior after application to the skin appears to be relatively simple, as it is quite similar to the behavior observed for simple emulsions. If evaporation of the water after application to the skin leading to other structures is not taken into consideration, two principal hypotheses may be proposed.

In the first hypothesis, the encapsulated active substance has a near zero diffusion rate through the oily membrane. The active substance is thus released in the internal phase only by virtue of the rupture of the multiple oily globules. This rupture takes place either by shearing of the preparation or after swelling of the internal aqueous globules. Shearing may be induced by massaging or rubbing when the cream is applied to the skin. Swelling of the internal aqueous globules may be caused by dilution of the multiple emulsion in water. It is possible that the instructions for use could require mixing immediately prior to application of a given quantity of the multiple emulsion with a specific volume of water. The strong osmotic gradient thus created causes the external water to enter the internal phase, causing swelling of the internal water globules. When these globules reach their maximum size, they burst.

In both of these cases, rupture by shearing or rupture after swelling, the active substance, having reached the external aqueous phase, becomes immediately available. The multiple emulsion behaves like a simple emulsion with a continuous aqueous phase. In this case, multiple emulsions serve the purpose of protecting (by encapsulation) an active substance or of permitting the incorporation into the same preparation of incompatible active substances which will not come into contact until the cream is used.

In the second hypothesis, the encapsulated active substances diffuse through the oily membrane, and the multiple emulsions keep their multiple globules intact when applied to the skin.

The release profile of the active substance from a multiple emulsion would, in this case, not be similar to that from a simple emulsion. It would be slower and gradual, and it would depend on the following factors: the localization and phase partition distribution of the active substance and its permeability or diffusion rate through the oily membrane; the rigidity, the viscosity, and the thickness of the interfacial film as well as the presence or absence of liquid crystals at this level; the particle size and distribution of the dispersion; and other factors.

Because they do not concentrate or accumulate the active substance at the level of the stratum corneum as the vesicular systems described above do, they at least extend the release rate of the active substance. Their fate after application to the skin has been the subject of few publications to date; only the works of Ferreira (134), Kundu (136) and, most recently, Biju et al. (1) on the subject are well known.

Multiple emulsions used as cosmetic preparations generally come in the form of lotions or creams of varying density. As soon as they are constituted, they can be used directly; unlike the vesicular systems described earlier, it is not necessary to disperse them in a gel environment or in a cream to obtain an acceptable cosmetic product.

Cosmetic applications of multiple emulsions have been patented for their composition. One example of an application is perfume encapsulated in the internal phase: very small amounts of it are released over a long period of time. However, it could be released instantaneously by intentionally breaking the primary emulsion, for example, by rubbing during application to the skin. The patents show that multiple emulsions are highly recommended for all kinds of cosmetic applications, for example, sunscreens, make-up removers, cleansers, and nutritive, hydrating, and refreshing and cooling products. The mixture of a continuous oily phase emulsion with an aqueous solution containing, in addition to hydrophilic SAAs, saccharides (such as xylose, lactose, or sorbitol), polysaccharides (such as xanthan gum, pectin, carraghenate, or dextran) or synthetic polymers (such as acrylic derivatives) allows the obtention of multiple emulsions presenting cosmetic qualities of good spreading and unctuousness, which can be modified on request. In these systems, it is possible to incorporate a number of active substances both in the internal and external aqueous phases and even in the intermediate oily phase without significantly altering their stability. One example is formula which produces a fairly viscous cream that remains stable for a period of 36 months at 25°C.

All other things being equal, w/o/w multiple emulsions are more effective than simple emulsions with a continuous aqueous phase and are more pleasant to use than simple emulsions with a continuous oily phase because of a less greasy feel. Multiple emulsions will certainly have a future in cosmetics at least as bright as that of the vesicular systems described earlier.

Simple emulsions are the building blocks of cosmetology. Multiple emulsions, like their simpler form, give skin both water and oil; they can comprise a large number of ingredients, both hydrophilic and lipophilic; they are easy to administer, as they can be applied directly to the skin; they present good cosmetic qualities.

Like any vesicular system, multiple emulsions constitute a new form which could prove extremely fruitful, even if only for the possibilities it presents for the protection of encapsulated substances and perhaps for prolonged release of substances

(137). Recently, silica microspheres comprising a high content of entrapped retinol were incorporated within a multiple emulsion of o/w/o which slowly released the active ingredient allowing an enhanced antiaging efficiency (138).

The first commercial use of a w/o/w-type multiple emulsion is Unique Moisturizing by Lancaster, was introduced to the market in 1991. Since then, other multiple emulsion formulations have been marketed showing similar effects (Table 6).

Liquid Crystals

For reasons of both cosmetic appeal owing to the colored appearance they give to preparations into which they are introduced and active substance solubilization, or simply because they increase the stability of dispersed systems, liquid crystals are enjoying a growing popularity (16).

To date, liquid crystals have been selected only rarely as absorption enhancers, because they contain such a high proportion of SAAs. In fact, poor cutaneous tolerance was observed when liquid crystals were used as the principal vehicle. After dispersion in a gel, their tolerance is no longer a problem.

Thermotropic liquid crystals were introduced in a cosmetic product for the first time by the Estee Lauder Company, with the launching of Eyxon, a translucid gel. Today, liquid crystals incorporated into microcapsules made of gelatin, which rupture on topical application, are available (Merck). Lyotropic liquid crystals are incorporated into special dermatological formulations that exhibit hydrating properties (Liphaderm).

Most of all, liquid crystals are used as excipients to protect sensitive, active substances (vitamins, antioxidants, and oils). Liquid crystals may enhance the stability of emulsions while creating a rheological barrier resulting in an increase in the viscosity and a decrease in coalescence by modification of Van der Waals forces. It is also claimed that they enhance cutaneous penetration.

Solid Lipid Nanoparticle

Like the other vesicular systems, the SLNs are actually popular for the same properties as previously described in Section 1. SLNs possess some features which make them promising carriers for cosmetic applications. It has been shown that SLNs protect labile compounds such as retinol, α-tocopherol, and coenzyme Q10 from chemical degradation (139). Depending on the produced SLN–type, controlled release of the active ingredients is possible. SLNs, with an active ingredient–enriched shell show burst release characteristics, whereas those with an active ingredient–enriched core lead to sustained release (18,19,140). In addition, SLNS exhibit specific properties since they can act as occlusives and can be used to increase the water content of the skin (141). Furthermore, SLNs show a UV-blocking potential and can act as physical sunscreens on their own (142).

The hydrating effect was measured indirectly as a result of the occlusive action of the SLN evaluated by an in vitro test, which consists of the determination of the water evaporation through a cellulose acetate membrane, resulting in the calculation of the occlusion factor F (140). It was observed that F value depended on the particle size of the SLN. A decrease in particle size leads to an increase in F reflecting the formation of a better occlusive film following application of the tiny SLN. Indeed, it has been shown, assuming a hexagonal packaging in a monolayer, that microparticles

Table 6 Multiple Emulsions Already on the Market

Company	Product name	Texture claim	Date	Active ingredients	Effects claimed
Lancaster Group (Benckiser Cosmetics)	Unique moisturizer	Triphasic w/o/w	1991	Glycerol polymethacrylate, ceramides, cholesterol, vitamin E, borage seed oil, UV filter	Moisturizer, long-lasting
Helena Rubinstein	Nutritional response	Triple-phase emulsion	1991	Amino acids, vitamins, glycoceramides, oligoelements	Moisturizer, nutritional system for dry skin
Nivea Beiersdorf	Nivea eye optimale	Triple-phase w/o/w	1995	Multivitamin complex, provitamin B5, UVA and UVB filters	Smoothing effect, immediate and long-term moisturizer, nourishing
La Prairie (Beiersdorf)	Cellular time release moisturizer intensive	Triple-phase w/o/w	1997	Vitamins E and C, glycoproteins amino acids, biotin, panthenol, sodium hyaluronate, glycerol	Smoothing effect, immediate long-term moisturizer (16 hrs), firming, nourishing
C.S. Dermatologie	Triffadiane vergetures	Triple emulsion w/o/w	1997	Centella asiatica extract	Long-lasting moisturizer, slow release of extract helps prevent stretch marks
Vichy	RetiC	Triple emulsion w/o/w	1998	VitC Retinol	Moisturizer, nutritional system for dry skin
Laboratoires Bio-Derma	Hydrabis legere	Triple emulsion w/o/w	2001	Glycerol polymethacrylate, ceramides, cholesterol, vitamin E, borage seed oil, UV filter	Moisturizer, long-lasting
Laboratoires Dermatologiques d'Uriage	Ekysed	Triple emulsion w/o/w	2003	Sulfate of dextran, vitamins K1, A, C, and E	Decongestant and regenerating cream

Abbreviation: w/o/w, water-in-oil-in-water.

Table 7 Comparison Between the Various Cosmetic Forms

	Liposomes	Nanoparticles	Microemulsions	Multiple emulsions	Liquid crystals	Solid lipidic particles
Appearance	Fluid aqueous suspensions with a bluish color	Fluid aqueous suspensions with a bluish color or oily suspension	Highly fluid transparent solution	White creams more or less viscous	Transparent formulations more or less viscous	Aqueous suspension more or less viscous with a bluish appearance
Main constituents	Phospholipids, cholesterol	Polymers, acrylic derivatives, polylactic acid, polyglycolic acid	Nonionic surfactant (sucroesters), cosurfactants: short alcohol (pentanol)	Constituents of classic emulsions: Water, oil/lipid phase, surfactants	Nonionic surfactants, cholesterol derivatives, esters of PEG	Constituents of classic emulsions: water, solid lipid phase, surfactants
Method of preparation	Easy, several methods	Relatively easy, several methods	Very easy, spontaneous	Relatively sensitive, difficult	Difficult	Relatively easy, numerous methods
Particle size distribution	20–1000 nm	100–500 nm	10–100 nm	2–10 μm	Nonspherical particles	50 nm to 1000 μm
Presentation	H	H or L	L/H or H/L	H/L/H or L/H/L	L/H or H/L	H
Stability	Poor	Very good	Excellent	Moderate	Relatively good	Good
Main activity	AI protection, targeting of AI	AI protection, targeting of AI, enhanced release of AI	AI protection, modulated permeation of AI	AI protection, enhanced release of AI, water/oil supply, pleasant touch	AI protection, stability aesthetic appeal	AI protection, AI release modulation, water and oil supply, Stability
Limitations	Need to be incorporated into a vehicle, progressive dissolution	Need to be incorporated into a vehicle	Marked concentration of surfactants	Collapse risk of droplets and immediate release of AI	Marked concentration of surfactants	Need to be incorporated into a vehicle

Abbreviations: AI, active ingredient; H, hydrophile; L, lipophile.

of 2000 nm in size yield larger pores between the particles favoring the evaporation of water hydrodynamically while nanoparticles of 200 nm in size yield smaller pores markedly reducing the evaporation of water (139). This hydrating action was investigated in vivo in a comparative study. It was observed that the established commercial formulation with SLN could increase skin hydration by 32%, while the pure commercial formulation increased the skin hydration by 24% (139). Little or no increase in elasticity was observed. This was attributed by the authors to the young age of the volunteers (25 years on average). Furthermore, the effect on wrinkle depth was studied comparing a well-known formulation effective in wrinkle treatment with the same formulation containing additionally SLN. It should be noted that no significant difference was noted between the two formulations. There is a tendency to use particulate compounds such as titanium dioxide rather than molecular UV blockers for UV protection to avoid the potential side effects induced by UV molecular blockers such as photoallergy and phototoxicity. The mechanism of protection of the particulate compounds is by simply scattering of UV rays. However, there are indications that the very small titanium dioxide particles ranging in size from 5 nm to 20 nm penetrate into the skin and can interact with the immune system (143). Surprisingly, it was discovered that highly crystalline solid lipid particles based on cetyl palmitate can also act as particulate UV blockers by scattering the light efficiently. Furthermore, it was found that the incorporation of a molecular sunscreen into the SLN matrix led to a synergistic protective effect. This means that the total amount of molecular sunscreen in the formulation can be reduced, thus further minimizing the side effects in addition to the already achieved reduction by firm incorporation of the sunscreen into the particle matrix.

CONCLUSION

Since the last decade, numerous efforts have been invested to enhance the dermatological performances of cosmetic products while keeping them safe. The main approach used for this purpose is the improvement of the encapsulation processes of active ingredients to protect them from the potential detrimental effects of the environment while improving their bioavailability and targeting to the appropriate topical sites (Table 7). It is anticipated that in the near future, sophisticated and "clever" cosmetic products will be available in the market (144).

REFERENCES

1. Biju SS, Ahuja A, Khar RK. Tea tree oil concentration in follicular casts after topical delivery: determination by high–performance thin layer chromatography using a perfused bovine udder model. J Pharm Sci 2005; 94:240–245.
2. Gregoriadis G, Wiley J, Chichester S. Liposomes as Drug Carriers. Chichester, UK: Wiley, 1988.
3. Redziniak G. Liposomes and skin: past, present, future. Pathol Biol 2003; 51:279–281.
4. Lasic DD. Liposomes and niosomes. In: Reiger MM, Rehin LD, eds. Surfactants in Cosmetics. 2nd ed. New York: Marcel Dekker, 1997:263–283.
5. Wendel A. Lecithins, Phospholipids, Liposomes in Cosmetics, Dermatology and in Washing and Cleansing Preparations. Augsburg: Verlag fuer chemische Industrie, 1994.
6. Wendel A. Lecithins, Phospholipids, Liposomes in Cosmetics, Dermatology and in Washing and Cleansing Preparations. Part II. Augsburg: Verlag fuer chemische Industrie, 1997.

7. Posner R. Liposomes. J Drugs Dermatol 2002; 1:161–164.
8. Bombardelli E. Phytosome: new cosmetic delivery system. Boll Chim Fram 1991; 130:141–148.
9. Moussaoui N, Cansell M, Denizot A. Marinosomes, marine lipid-based liposomes physical characterization and potential applications in cosmetics. Int J Pharm 2002; 242:361–365.
10. Godin B, Touitou E. Ethosomes: new prospects in transdermal delivery. Crit Rev Ther Drug Carrier Syst 2003; 20:63–102.
11. Esposito E, Menegatti E, Cortesi R. Ethosomes and liposomes as topical vehicles for azelaic acid: a preformulation study. J Cosmet Sci 2004; 55:253–264.
12. Magenheim B, Benita S. Nanoparticles characterization. A comprehensive physicochemical approach. STP Pharma Sci 1991; 4:221–241.
13. Pons R, Carrera I, Caelles J, Rouch J, Panizza P. Formation and properties of miniemulsions formed by microemulsions dilution. Adv Colloid Interf Sci 2003; 106: 129–146.
14. De Luca M, Vaution C, Rabaron A, Seiller M. Classification et obtention des emulsions multiples. STP Pharma Sci 1998; 4:679–687.
15. Caquet JP, Bernoud T. Les cristaux liquides dans les cosmetiques. Parf Cosmet Aromes 1990; 91:77–84.
16. Cioca G, Calvo L. Liquid crystals and cosmetic applications. Cosmet Toiletries 1990; 105:57–62.
17. Dorfler HD. Chirality, twist and structures of micellar lyotropic cholesteric liquid crystals in comparison to the properties of chiralic thermotropic phases. Adv Colloid Interf Sci 2002; 98:285–340.
18. Müller RH, Mehnert W, Lucks J-S, et al. Solid lipid nanoparticles (SLN)– an alternative colloidal carrier system for controlled drug delivery. Eur J Pharm Biopharm 1995; 41:62–69.
19. Müller RH, Mäder K, Gohla S. Solid lipid nanoparticles (SLN) for controlled drug delivery–a review of the state of the art. Eur J Pharm Biopharm 2000; 50:161–177.
20. Bangham AD, Hill MW, Miller NGA. Preparation and use of liposomes as a model of biological membranes. In: Korn ED, ed. Methods in Membrane Biology. New York: Plenum Press, 1974:1–68.
21. Gebicki JM, Hicks M. Preparation and properties of vesicles enclosed by fatty acid membranes. Chem Phys Lipids 1976; 16:142–160.
22. Kirby C, Clarke J, Gregoriadis G. Effect of cholesterol content of small unilamellar liposomes on their stability in vivo and in vitro. Biochem J 1980; 186:591–598.
23. Konings AWT. Lipid peroxidation in liposomes. In: Gregoriadis G, ed. Liposome Technology. Boca Raton, Florida: CRC Press, 1984:139–161.
24. Stainmess S, Fessi H, Devissaguet J, Puisieux P. Procede de preparation de systems colloidaux dispersibles de lipides amphiphiles sous forme de liposomes submicroniques. Brevet No. 88.08.874, France, June, 1988.
25. Marsh D. Handbook of Lipic Bilayers. Boca Raton, Florida: CRC Press, 1990.
26. Crommelin DJA. Influence of lipid composition and ionic strength on the physical stability of liposomes. J Pharm Sci 1990; 73:1559–1563.
27. Duclairoir C, Irach JM, Nakache E, Orecchion AM, Chabenat C, Popineau Y. Gliadin nanoparticles formation, all-trans-retinoic acid entrapment and release, size optimization. Polym Int 1999; 79:327–333.
28. Prince LM. Microemulsions: Theory and Practice. New York: Academic Press, 1977:177–177.
29. Prince LM. Micellization, solubilization and microemulsions. In: Mittal KL, ed. Micellization, Solubilization and Microemulsions. New York: Plenum Press, 1977:45–54.
30. Friberg SE, Ganzuo L. Microemulsions with esters. J Soc Cosmet Chem 1983; 34:73–81.
31. Ceglie A, Das KP, Lindman B. Microemulsion structure in four-component system for different surfactants. Colloid Surf 1987; 28:29–40.

32. Ceglie A, Das KP, Lindman B. Effect of oil on the microscopic structure in four component cosurfactant microemulsion. J Colloid Interf Sci B 1987; 115:115–120.
33. Chew CH, Gan LM. Monohexylether of ethylene glycol and diethylene glycol as microemulsion cosurfactants. J Dispersion Sci Technol 1990; 11:49–68.
34. Kawabata Y, Nagao M, Seto H, et al. Temperature and pressure effects on the bending modulus of monolayers in a ternary microemulsion. Phys Rev Lett 2004; 92:056103.
35. Nandi I, Bari M, Joshi H. Study of isopropyl myristate microemulsion systems containing cyclodextrins to improve the solubility of 2 model hydrophobic drugs. AAPS Pharm Sci Tech 2003; 4:E10.
36. Raman IA, Suhaimi H, Tiddy GJ. Liquid crystals and microemulsions formed by mixtures of a non ionic surfactant with palm oil and its derivatives. Adv Colloid Interf Sci 2003; 106:109–127.
37. Florence AT, Whitehill D. The formulation and stability of multiple emulsions. Int J Pharm 1982; 11:277–308.
38. Magdassi S, Frenkel M, Garti N. Correlation between nature of emulsifiers and multiple emulsion stability. Drug Dev Ind Pharm 1985; 11:791–798.
39. Fox C. An introduction to multiple emulsions. Cosmet Toiletries 1986; 62:101–112.
40. Prybilski C, de Luca M, Grossiord JL, Seiller M. W/o/w multiple emulsions manufacturing and formulation consideration. Cosmet Toiletries 1990; 106:97–100.
41. Suzuki T, Nakamura M, Sumida H, Shigeta I. Liquid crystal make-up remover: conditions of formation and its cleansing mechanisms. J Soc Cosmet Chem 1992; 43:21–36.
42. Saunders L, Perrin J, Gammack DB. Aqueous dispersion of phospholipids by ultrasonic radiations. J Pharm Pharmcol 1962; 14:567–572.
43. Bangham AD, Standish MM, Watkins JC. Diffusion of univalent ions across the lamellae of swollen phospholipids. J Mol Biol 1965; 13:238–252.
44. Huang CH. Studies on phosphatidylcholine vesicles. Formation and physical characteristics. Biochemistry 1969; 8:344–352.
45. Batzri S, Korn ED. Single bilayer liposomes prepared without sonication. Biochem Biophys Acta 1973; 298:1015–1019.
46. Deamer DW, Bangham AD. Large volume liposomes by an ether evaporation method. Biochem Biophys Acta 1974; 443:629–634.
47. Kremer JNH, Esker MWJ, Pathmamanohara NC, Wiersema PH. Vesicles of variable diameter prepared by a modified injection method. Biochemistry 1977; 16:3932–3935.
48. Szoka FC, Papahadjopoulos D. Procedure for preparation of liposomes with large internal aqueous surface and high capture by reverse phase evaporation. Proc Natl Acad Sci USA 1978; 75:4194–4198.
49. Barenholz Y, Amselem S, Lichtenberg D. A new method for preparation of phospholipids vesicles (liposomes). FEBS Lett 1979; 99:210–215.
50. Oshawa T, Mirua H, Harada K. A novel method for preparing liposomes with a high capacity to encapsulate protein and drugs. Chem Pharm Bull 1984; 32:2442–2445.
51. Soppimath KS, Aminabhavi TM, Kulkarni AR, Rudzinski WE. Biodegradable polymeric nanoparticles as drug delivery devices. J Control Release 2001; 70:1–20.
52. Patravale VB, Date AA, Kulkarni RM. Nanosuspensions: a promising drug delivery strategy. J Pharm Pharmacol 2004; 56:827–840.
53. Bornschein M, Melegari P, Bismarck C, Keipert S. Mikro- und Nanopartikeln als arzneistofftragersysteme unter besonderer Berucksichtigung der herstellungsmethoden. Pharmazie 1989; 44:585–593.
54. Kamenka N, Haouche G, Brun B, LIndman B. Microemulsions in zwitterionic surfactant systems. J Colloid Interf Sci 1982; 86:369–369.
55. Franz J (Sandoz SA). Compositions pharmaceutiques topiques sous forme d'une microemulsion. Fr Patent Application 2-502-951 (1982).
56. Muller BW, Kleinebudde P. Untersuchengen an sogenannten Mikroemulsions systemen 1 Teil: Unter suchungen an arzneistoffhaltign systemen. Pharm Ind 1988; 50:370–378.

57. Muller BW, Kleinebudde P. Untersuchengen an sogenannten Mikroemulsions systemen 2 Teil: Unter suchungen an arzneistoffhaltign systemen. Pharm Ind 1988; 50:1301–1306.
58. Fevrier F. Formulation de microemulsions cosmetiques. Nouv Dermatol 1991; 10: 84–87.
59. Bobin MF, Dejour N, Revillon A, Martini MC. Formulation of ultrafine emulsions with surcroesters for pharmaceutic and cosmetic applications. Congres Mondial de l'emulsion, Paris, France, 1993.
60. Djordjevic L, Primorac M, Stupar M, Krajisnik D. Characterization of caprylocaproyl macrogolglycerides based microemulsions drug delivery vehicles for an amphiphilic drug. Int J Pharm 2004; 271:11–19.
61. Matsumoto S, Kita Y, Yonesawa D. An attempt at preparing water in oil in water multiple phase emulsions. J Colloid Interf Sci 1976; 57:353–361.
62. Kavaliunas DR, Franck SG. Liquid crystal stabilization of multiple emulsions. J Colloid Interf Sci 1978; 66:586–588.
63. Matsumoto S, Sherman P. A preliminary study of w/o/w emulsions with a view to possible food application. J Texture Stud 1981; 12:243–257.
64. Matsumoto S. Development of w/o/w type dispersion during phase inversion of concentrated w/o emulsions. J Colloid Interf Sci 1983; 94:362–368.
65. Magdassi S, Frenkel M, Garti N. On the factors affecting the yield of preparation and stability of multiple emulsions. J Dispersion Sci Technol 1984; 5:49–59.
66. de Luca M. Les emulsions multiples H/L/H. Obtention, validation et liberation. These de l'Universite Paris XI, No. 191, 1991.
67. Colinet C. Contribution a l'etude de systemes mesomorphes a partir de surfactifs non ioniques. These de Pharmacie, Lyon I, 1991.
68. Coupier E. Contribution a l'etude de la stabilite des systemes mesomorphes. Influence de la temperature et des divers adjuvants. These de Pharmacie, Lyon I, 1991.
69. Schurtenberger P, Hauser H. Characterization of the size distribution of unilamellar vesicles by gel filtration quasi-elastic light scattering and electron microscopy. Biochim Biohpys Acta 1984; 778:470–480.
70. Ruf H, Georgalis Y, Grell E. Dynamic laser light scattering to determine size distributions of vesicles. Meth Enzymol 1989; 172:364–372.
71. Kojro Z, Lin SQ, Grell E, Ruf H. Determination of internal volume and volume distribution of lipid vesicles from dynamic light scattering data. Biochim Biophys Acta 1989; 985:1–8.
72. New RRC. Characterization of liposomes. In: New RRC, ed. Liposomes, A Practical Approach. McLean, Virginia: IRL Press, 1990.
73. Atwood D, Ktistis G. A light scattering study on o/w microemulsion. Int J Pharm 1989; 52:332–349.
74. Baker RC, Florence AT, Ottewill RH, Tadros TH. Investigations into the formation and characterization of microemulsions. II. Light scattering conductivity and viscosity studies of microemulsion. J Colloid Interf Sci 1984; 100:332–349.
75. Keipert S, Siebenbrodt I, Luders F, Bornschein M. Mikroemulsionen und ihre potentielle pharmazentischenutzung. Die Pharmazie 1989; 44:433–444.
76. Selser JC, Yeh Y, Baskin RJ. A light scattering measurement of membrane vesicle permeability. Biophys J 1976; 16:135–139.
77. Davis SS, Burbage AS. Electron micrography of w/o/w emulsions. J Colloid Interf 1977; 62:87–94.
78. Podlogar F, Gasperlin M, Tomsic M, Jamnik A, Rogac MB. Structural characterization of water-Tween 40/Imwitor 308-isopropyl myristate microemulsions using different experimental methods. Int J Pharm 2004; 276:115–128.
79. De Campo L, Yaghmur A, Garti N, Leser ME, Folmer B, Galtter O. Five-component food-grade microemulsions: structural and characterization by SANS. J Colloid Interf Sci 2004; 274:251–267.

80. Kita Y, Matsumoto S, Yonezawa D. Viscosimetric method for estimating the stability of w/o/w type multiphase emulsions. J Colloid Interf Sci 1977; 62:87–94.
81. Matsumoto S, Inoue T, Koda M, Ota T. An attempt to estimate stability of the oil layer in w/o/w emulsions by means of viscosimetry. J Colloid Interf Sci 1980; 77:564–574.
82. Zatz JI, Cueman GH. Assessment of stability in water in oil in water multiple emulsions. J Soc Cosmet Chem 1988; 39:211–222.
83. Elbary AA, Nour SA, Ibrahim I. Physical stability and rheological properties of w/o/w emulsions as a function of electrolytes. Pharm Ind 1990; 52:357–363.
84. Grossiord JL, Seiller M, Puisieux F. Apport des analyses rheologiques dans l'etude des emulsions multiples H/L/H. Rheologica Acta 1993; 32:168–180.
85. Valenta C, Schltz K. Influence of Carrageeenan on the rheology and skin permation of microemulsion formulations. J Control Release 2004; 95:257–265.
86. Ralston EH, Hjelmeland LM, Klauser RD, Weinstein JN, Blumenthal R. Caroxyfluoresceine as a proble for liposome cell interactions. Effect of impurities and purification of the dye. Biochim Biophys Acta 1981; 649:133–137.
87. Davis SS, Walker I. Measurement of the yield of multiple emulsion droplets by a fluorescent technique. Int J Pharm 1983; 17:203–213.
88. Omotosho JA, Whateley TL, Law TK, Florence AT. The nature of the oil phase and release of solutes from multiple w/o/w emulsions. J Pharm Pharmacol 1986; 58: 865–870.
89. Duclairoir C, Orecchioni A-M, Depraetere P, Osterstock F, Nakache E. Evaluation of gliadins nanoparticles as drug delivery systems: a study of three different drugs. Int J Pharm 2003; 253:133–144.
90. Poly PA. Liposomes d'un produit de contraste iode. Influence de la formulation sur l'encapsulation Etude de la conservation par lyophillisation. Doctoral thesis No. 65, 3eme cycle, Université Paris-Sud, 1983.
91. Micheland SH, Poly PA, Puisieux F. Etude du comportement des liposomes lors d'experiences de congelation/decongelation et de lyophillisation. Influence des cryoprotecteurs. 3rd International Conference on Pharmaceutical Technology APGI, Paris, 1983, pp. 223–233.
92. Crowe LM, Womersley C, Crowe JH, Reid D, Allep L, Rudolph A. Prevention of fusion and leakage in freeze-dried liposomes. Biochim Biophys Acta 1986; 861:131–140.
93. Law TK, Whateley TL, Florence AT. Stabilization of w/o/w multiple emulsions by interfacial complexation of macromolecules and non-ionic surfactants. J Control Release 1986; 3:279–280.
94. Oz A, Kamles P, Franck P, Sylvan G. Multiple emulsions stabilized by colloidal microcrystalline cellulose. J Dispersion Sci Tech 1989; 10:163–185.
95. de Luca M, Grossiord JL, Medard JM, Seiller M. A stable w/o/w multiple emulsion. Cosmet Toiletries 1990; 105:65–69.
96. Özer Ö, Baloglu E, Ertan G, Muguet V, Yazan Y. The effect of the type and the concentration of the lipophilic surfactant on the stability and release kinetics of the w/o/w multiple emulsions. Int J Cosmet Sci 2000; 22:459–470.
97. Kulkarni VS, Ross M, Brockway B, Wilmott J, Hawyard J. Novel method of formulating skin care products with liposomes. J Cosmet Sci 2002; 53:297–298.
98. Lampen P, Pittermann W, Heise HM, Schmitt J, Jungmann H, Kietzmann M. Penetration studies of vitamin E acetate applied from cosmetic formulations to the stratum corneum of an in vitro model using quantification by tape stripping, UV spectroscopy, and HPLC. J Cosmet Sci 2003; 54:119–131.
99. Lasch J, Zellmer S, Pfeil W, Schubert R. Interaction of liposomes with the human skin lipid barrier:hSCLLD as model system-DSC of intact stratum corneum and in situ CLSM of human skin. J Liposome Res 1995; 5:99–108.
100. Redziniak G, Meybeck A, Perrier P. New vehicle to enhance the biological performances of active cosmetic ingredient. Vol II, XIVth International Federation Societies of Cosmetic Chemists Congress, Barcelona, 1986, pp. 299–307.

101. Tholon L, Neliat G, Chesne C, Saboureau D, Perrier E, Branka JE. An in vitro, ex vivo, and in vivo demonstration of the lipolytic effect of slimming liposomes: an unexpected alpha(2)-adrenergic antagonism. J Cosmet Sci 2002; 53:209–218.
102. Wolf P, Cox P, Yarosh DB, Kripke ML. Sunscreens and T4N5 liposomes differ in their ability to protect against ultraviolet-induced sunburn cell formation, alternations of dendritic epidermal cells, and local suppression of contact hypersensitivity. J Invest Dermatol 1995; 104:287–292.
103. Topical compositions containing liposomes and acrylic gels. Estee Lauder Inc., Eur. Patent 3 196 38 (1989).
104. Trommer H, Wartewig S, Bottcher R, et al. The effects of hyaluronan and its fragments on lipid models exposed to UV irradiation. Int J Pharm 2003; 254:223–234.
105. Hanjani-Vila RM, Guesnet J. Les liposomes. Un avenir prometteur en dermatologie. Ann Dermatol Venerol 1989; 116:423–430.
106. Abraham W, Downing DT. Interaction between corneocytes and stratum corneum lipid liposomes in vitro. Biochim Biophys Acta 1990; 1021:119–125.
107. Lautenschlarger H. Liposomes in dermatological preparations. Part I. Cosmet Toiletries 1990; 105:89–96.
108. Clar E, Ribiert A. Lipid vesicles in cosmetology. Cosmet Toilet Mfg 1991; 2:138–147.
109. Ghyczy M, Roding J, Hoff E. Control of skin humidity by liposomes. Cosmet Toilet Mfg 1991; 2:148–152.
110. Hofland HEJ, Bouwstra JA, Spies F, Bodde HE, Nagelkerke JF, Cullander C, Junginger HE. Interactions between non-ionic surfactant vesicles and human stratum corneum in vitro. J Liposome Res 1995; 5:241–263.
111. Bonte F, Chevalier JM, Meybeck A. Determination of retinoic acid liposomal association level in a topical formulation. Drug Dev Indus Pharm 1994; 20:25–34.
112. Handjani-Vila RM, Vanlerberghe G. Les Niosomes. In: Buri P, Puisieux F, Benoit JB, eds. Formes Pharmaceutiques nouvelles. Aspects technologique, biopharmaceutique, et medical. Paris: Lavoisier Tec et Doc, 1985.
113. Vanlerberche G, Handjani-Vila RM. Les Niosomes des structures restructurantes pour les soins de la peau. Nouv Dermatol 1986; 5:259–262.
114. Vanlerberche G, Handjani-Vila RM, Ribier A. Les niosomes. Une nouvelle famille de vesicules a base d'amphiphiles non ioniques. Colloques Nationaux du CNRS No 938, Masson, Paris, 1989, pp. 303–311.
115. Gabrijedeie V, Sentjure M, Kristl J. Evaluation of liposomes as drug carriers into the skin by one dimensional EPR imaging. Int J Pharm 1990; 62:75–79.
116. Schreier H, Bouwstra J. Liposomes and niosomes as topical drug carriers: dermal and transdermal drug delivery. J Control Release 1994; 30:1–15.
117. Duclairoir C, Orecchioni AM, Depraetere P, Nakache E. α-tocopherol encapsulation and in vitro release from wheat gliadin nanoparticles. J Microencapsul 2002; 19:53–60.
118. Martini MC, Bobin MF, Flandin H, Cotte J. Absorption and fate of orotic acid after topical application in different vehicles. J Appl Cosmetol 1984; 2:19–26.
119. Martini MC, Bobin MF, Flandin H, Caillaud F. Role des microemulsions dans l'absorption percutanee de l'alpha-tocopherol. J Pharm Belg 1984; 39:348–354.
120. Gallarate M, Gasco MR, Rua G. In vitro release of azelaic acid from oil in water microemulsions. Acta Pharm Jugosl 1990; 40:533–538.
121. Linn EE, Pohlan RC, Byrd TK. Microemulsion for intradermal delivery of cetyl alcohol and octyl dimethyl PABA. Drug Dev Indus Pharm 1990; 16:899–920.
122. Fevrier F, Bobin MF, Lafforgue C, Martini MC. Advances in microemulsion and transepidermal penetration of tyrosine. STP Pharma Sci 1991; 1:60–63.
123. Burigana V. Mise au point de dosage de la N-acetyl hydroxyproline. Etude de l'abosrption cutanee de cette molecule. These de pharmacie, Lyon I, 1993.
124. Fevrier F, Bobin MF, Martini MC. In vitro cutaneous penetration of praline with microemulsion formulation compared to emulsion. Symposium International Federation Societies of Cosmetic Chemists, Barcelona, 1993.

125. Fevrier F. Microemulsion in cosmetology. LCR International Federation Societies of Cosmetic Chemists Congress, Barcelona, 16–19.9/1993, pp. 105–118.
126. Gloor M, Hauth A, Gehring W. O/W emulsions compromise the stratum corneum barrier and improve drug penetration. Pharmazie 2003; 58:709–715.
127. Carlotti ME, Gallarate M, Rossatto V. O/W microemulsion as a vehicle for sunscreens. J Cosmet Sci 2003; 54:451–462.
128. Rangarajan M, Zatz JL. Effect of formulation on the topical delivery of alpha-tocopherol. J Cosmet Sci 2003; 54:161–174.
129. Sintov AC, Shapiro L. New microemulsion vehicle facilitates percutaneous penetration in vitro and cutaneous drug bioavailability in vivo. J Control Release 2004; 5:173–183.
130. Kweon JH, Chi SC, Park ES. Transdermal delivery of diclofenac using microemulsions. Arch Pharm Res 2004; 27:351–356.
131. Jurcovic P, Sentjure M, Gasperlin M, Kristl J, Pecar S. Skin protection against UV induced free radicals with ascorbyl palmitate in microemulsion. Eur J Pharm Biopharm 2003; 56:59–66.
132. Kristl J, Volk B, Gasperlin M, Sentjure M, Jurcovic P. Effect of colloidal carriers on ascorbyl palmitate stability. Eur J Pharm Sci 2003; 19:181–189.
133. Gallarate M, Carlotttei ME, Trotta M, Grande AE, Talarico C. Photostability of naturally occurring whitening agents in cosmetic microemulsions. J Cosmet Sci 2004; 55:139–148.
134. Ferreira LAM. Emulsions multiples H/L/H et simples L/H et H/L. Etude comparative pur l'approche de la voie topique. These Universite de Paris-Sud, No. 370, 1994.
135. Seiller M, Orecchioni AM, Vautions C. Vesicular systems in cosmetology. In: Baran R, Maibach HI, eds. Cosmetic Dermatology. London: Martin Dunitz, 1994:27–35.
136. Kundu SC. Preparation and evaluation of multiple emulsions as controlled release topical drug delivery systems. Avail Univ. Microfilms, Dep. Abstra. Int. Order no. DA 9224664 from Dirs Abstr Int B 1990 51 (1990) No. 4 S1763–4.
137. Gallarate M, Carlotti ME, Trotta M, Bovo S. On the stability of ascorbic acid in emulsified systems for topical and cosmetic use. Int J Pharm 1999; 188:233–241.
138. Lee M-H, Oh S-G, Moon S-K, Bae S-Y. Preparation of silica particles encapsulating retinaol using O/W/O multiple emulsions. J Colloid Interf Sci 2001; 240:83–89.
139. Müller RH, Radtke M, Wissing SA. Solid lipid nanoparticles (SLN) and nanostructured lipid carriers (NLC) in cosmetic and dermatological preparations. Adv Drug Deliv Rev 2002; 54(suppl 1):S131–S155.
140. Wissing SA, Müller RH. The influence of solid lipid nanoparticles on skin hydration and viscoelasticity–in vivo study. Eur J Pharm Biopharm 2003; 56:67–72.
141. Wissing SA, Müller RH. Cosmetic applications for solid lipid nanoparticles (SLN). Int J Pharm 2003; 254:65–68.
142. Wissing SA, Müller RH. Solid nanoparticles as carrier for sunscreens: in vitro release and in vivo skin penetration. J Control Release 2002; 81:225–233.
143. Hagedorn-Leweke U, Lippold BC. Accumulation of sunscreens and other compounds in keratinous substrates. Eur J Pharm Biopharm 1998; 46:215–221.
144. Raschke T. Short review of topical encapsulation technologies. IFSCC Mag 2003; 6:207–211.

Index